Reveal ALGEBRA 2®

McGraw Hill

mheducation.com/prek-12

Cover: (t to b, l to r) Kelley Miller/National Geographic/ Getty Images; Westend61/Getty Images; skodonnell/E+/ Getty Images; Daniel Viñé Garcia/Moment/Getty Images; Aksonov/E+/Getty Images

Send all inquiries to:
McGraw-Hill Education
8787 Orion Place
Columbus, OH 43240

ISBN: 978-0-07-695912-9
MHID: 0-07-695912-0

Printed in the United States of America.

10 11 12 13 14 15 16 LWI 29 28 27 26 25 24 23

Contents in Brief

Reveal AGA® Makes Math Meaningful...

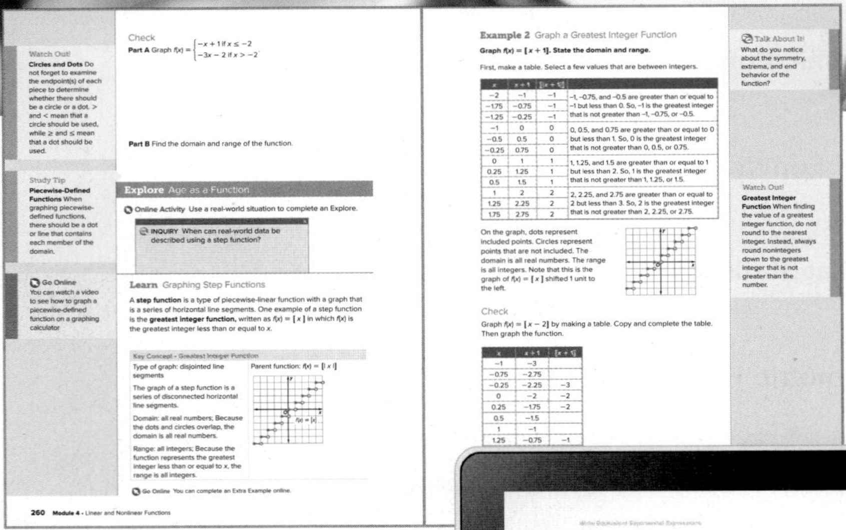

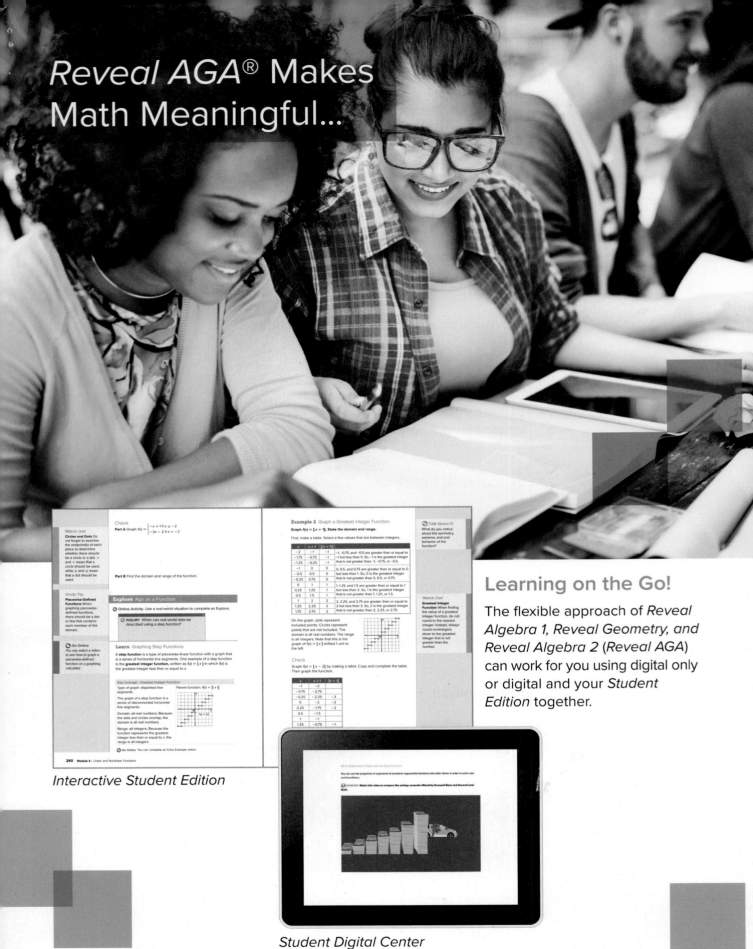

Interactive Student Edition

Learning on the Go!

The flexible approach of *Reveal Algebra 1, Reveal Geometry, and Reveal Algebra 2* (*Reveal AGA*) can work for you using digital only or digital and your *Student Edition* together.

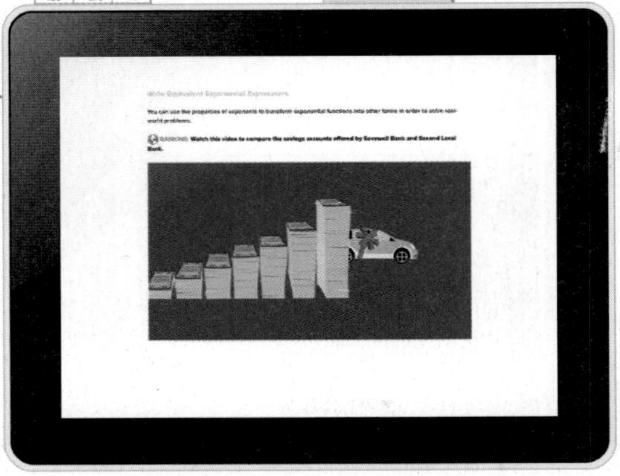

Student Digital Center

Rawpixel.com/Shutterstock, McGraw-Hill Education

...to Reveal YOUR Full Potential!

Reveal AGA® Brings Math to Life in Every Lesson

Reveal AGA is a blended print and digital program that supports access on the go. Use your student edition as a reference as you work through assignments in the Student Digital Center, access interactive content, animations, videos, eTools, and technology-enhanced practice questions.

Go Online! my.mheducation.com

WebSketchpad® Powered by The Geometer's Sketchpad®- Dynamic, exploratory, visual activities embedded at point of use within the lesson.

Animations and Videos – Learn by seeing mathematics in action.

Interactive Tools – Get involved in the content by dragging and dropping, selecting, highlighting, and completing tables.

Personal Tutors – See and hear a teacher explain how to solve problems.

eTools – Math tools are available to help you solve problems and develop concepts.

dolgachov/123RF

TABLE OF CONTENTS

Module 1
Relations and Functions

Module 2

Linear Equations, Inequalities, and Systems

TABLE OF CONTENTS

Module 3

Quadratic Functions

Module 4
Polynomials and Polynomial Functions

Module 5
Polynomial Equations

Module 6

Inverse and Radical Functions

TABLE OF CONTENTS

Module 7

Exponential Functions

Module 8

Logarithmic Functions

TABLE OF CONTENTS

Module 9

Rational Functions

Module 10

Inferential Statistics

TABLE OF CONTENTS

Module 11

Trigonometric Functions

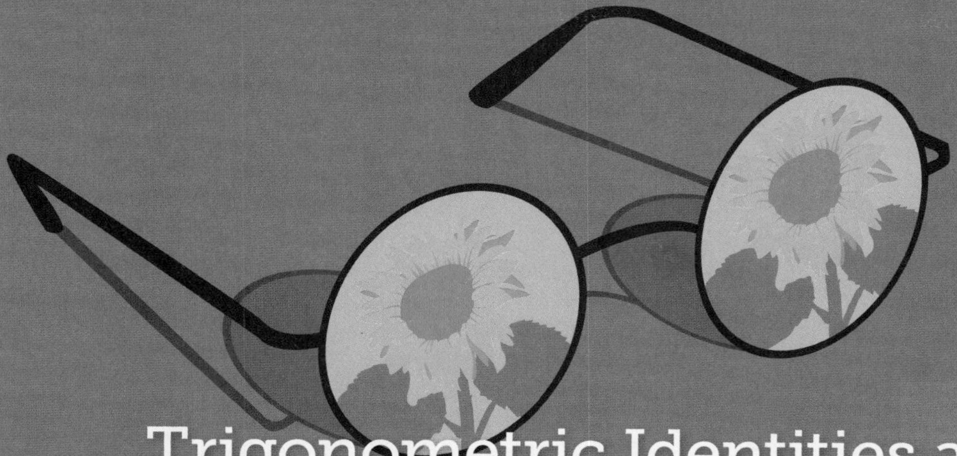

Module 12
Trigonometric Identities and Equations

Module 1

Relations and Functions

Essential Question

How can analyzing a function help you understand the situation it models?

What Will You Learn?

How much do you already know about each topic **before** starting this module?

KEY

— I don't know. — I've heard of it. — I know it!

	Before			After		
identify one-to-one and onto functions						
identify discrete and continuous functions						
identify intercepts of graphs of functions						
identify linear and nonlinear functions						
identify extrema of graphs and functions						
identify end behavior of graphs of functions						
identify graphs that display line or point symmetry						
sketch and compare graphs of functions						
graph linear functions						
graph linear inequalities in two variables						
graph piecewise, step, and absolute value functions						
translate, dilate & reflect the graphs of functions						

Foldables Make this Foldable to help you organize your notes about relations and functions. Begin with four sheets of notebook paper.

1. **Fold** each sheet of paper in half from top to bottom.
2. **Cut** along the fold. Staple the eight half-sheets together to form a booklet.
3. **Cut** tabs into the margin. The top tab is 2 lines deep, the next tab is 6 lines deep, and so on.
4. **Label** each of the tabs with a lesson number, and the final tab *vocabulary*.

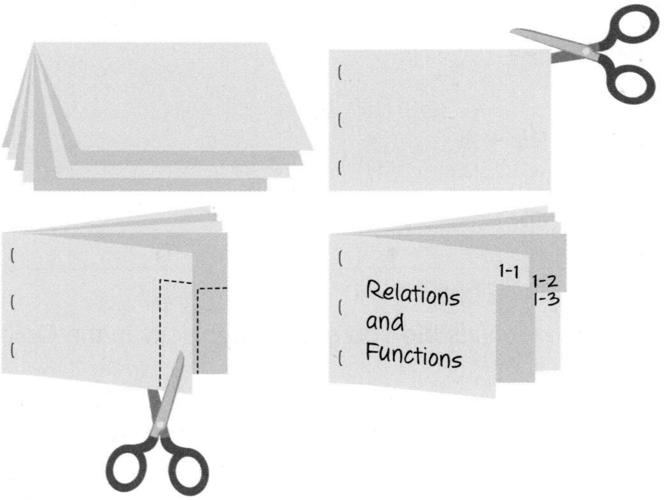

What Vocabulary Will You Learn?

- absolute value function
- algebraic notation
- boundary
- closed half-plane
- codomain
- constant function
- constraint
- continuous function
- dilation
- discontinuous function
- discrete function
- domain
- end behavior
- even functions
- extrema
- family of graphs
- greatest integer function
- identity function
- intercept
- interval notation
- line of reflection
- line of symmetry
- line symmetry
- linear equation
- linear function
- linear inequality
- maximum
- minimum
- nonlinear function
- odd functions
- one-to-one function
- onto function
- open half-plane
- parabola
- parent function
- piecewise-defined function
- point of symmetry
- point symmetry
- range
- reflection
- relative maximum
- relative minimum
- set-builder notation
- step function
- symmetry
- translation
- transformation
- *x*-intercept
- *y*-intercept

Are You Ready?

Complete the Quick Review to see if you are ready to start this module. Then complete the Quick Check.

Quick Review

Example 1

Evaluate $3a^2 - 2ab + b^2$ if $a = 4$ and $b = -3$.

$$\begin{aligned} 3a^2 - 2ab + b^2 &= 3(4^2) - 2(4)(-3) + (-3)^2 \\ &= 3(16) - 2(4)(-3) + 9 \\ &= 48 - (-24) + 9 \\ &= 48 + 24 + 9 \\ &= 81 \end{aligned}$$

Example 2

Solve $3x + 6y = 24$ for y.

$3x + 6y = 24$	Original equation
$3x + 6y - 3x = 24 - 3x$	Subtract $3x$ from each side.
$6y = 24 - 3x$	Simplify.
$\frac{6y}{6} = \frac{24}{6} - \frac{3x}{6}$	Divide each side by 6.
$y = 4 - \frac{1}{2}x$	Simplify.

Quick Check

Evaluate each expression if $a = -3$, $b = 4$, and $c = -2$.

1. $4a - 3$

2. $2b - 5c$

3. $b^2 - 3b + 6$

4. $\frac{2a + 4b}{c}$

Solve each equation for the given variable.

5. $a = 3b + 9$ for b

6. $15w - 10 = 5v$ for v

7. $3x - 4y = 8$ for x

8. $\frac{d}{6} + \frac{f}{3} = 4$ for d

How Did You Do?

Which exercises did you answer correctly in the Quick Check?

Functions and Continuity

Explore Analyzing Functions Graphically

Online Activity Use graphing technology to complete the Explore.

> **INQUIRY** How can you use a graph to analyze the relationship between the domain and range of a function?

Explore Defining and Analyzing Variables

Online Activity Use a real-world situation to complete the Explore.

> **INQUIRY** How can you define variables to effectively model a situation?

Learn Functions

A function describes a relationship between input and output values. The **domain** is the set of *x*-values to be evaluated by a function. The **codomain** is the set of all the *y*-values that could possibly result from the evaluation of the function. The codomain of a function is assumed to be all real numbers unless otherwise stated. The **range** is the set of *y*-values that actually result from the evaluation of the function. The range is contained within the codomain.

If each element of a function's range is paired with exactly one element of the domain, then the function is a **one-to-one function**. If a function's codomain is the same as its range, then the function is an **onto function**.

Example 1 Domains, Codomains, and Ranges

Part A Identify the domain, range, and codomain of the graph.

Domain	Range	Codomain
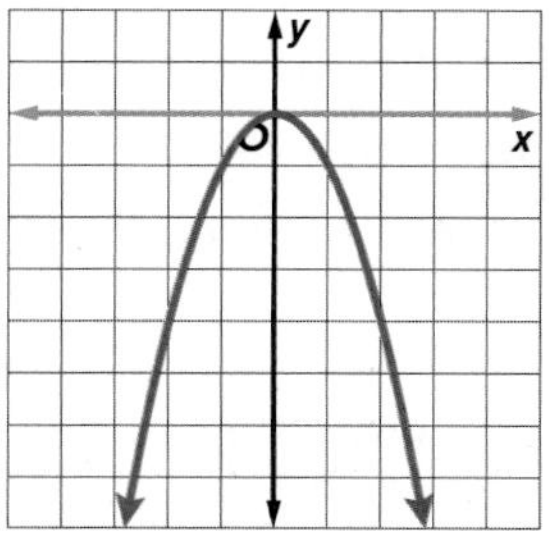	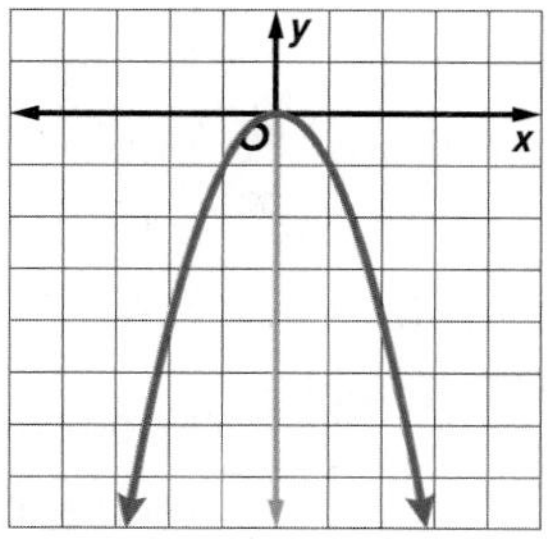	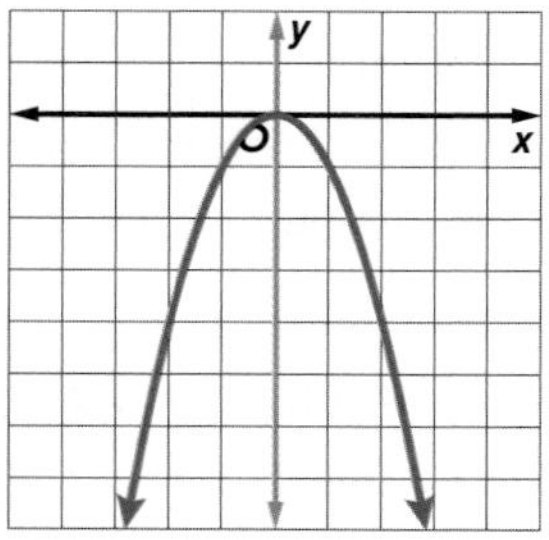
Because there are no restrictions on the *x*-values, the domain is all real numbers.	Because the maximum *y*-value is 0, the range is $y \leq 0$.	Because it is not stated otherwise, the codomain is all real numbers.

(*continued on the next page*)

Today's Goals
- Determine whether functions are one-to-one and/or onto.
- Determine the continuity, domain, and range of functions.
- Write the domain and range of functions by using set-builder and interval notations.

Today's Vocabulary
domain
codomain
range
one-to-one function
onto function
continuous function
discontinuous function
discrete function
algebraic notation
set-builder notation
interval notation

Study Tip
Horizontal Line Test Place a pencil flat at the top of the graph and move it down to represent a horizontal line.
- If there are places where the pencil intersects the graph at more than one point, then more than one element of the domain is paired with an element of the range. The function is not one-to-one.
- If there are places where the pencil does not intersect the graph at all, then there are real numbers that are not paired with an element of the domain. The function is not onto.

Part B Use these values to determine whether the function is onto.

The range is not the same as the codomain because it does not include the positive real numbers. Therefore, the function is not onto.

Check

For what codomain is $f(x)$ an onto function?

A. $y \leq 3$ B. $y \geq 3$

C. all real numbers D. $x \leq 3$

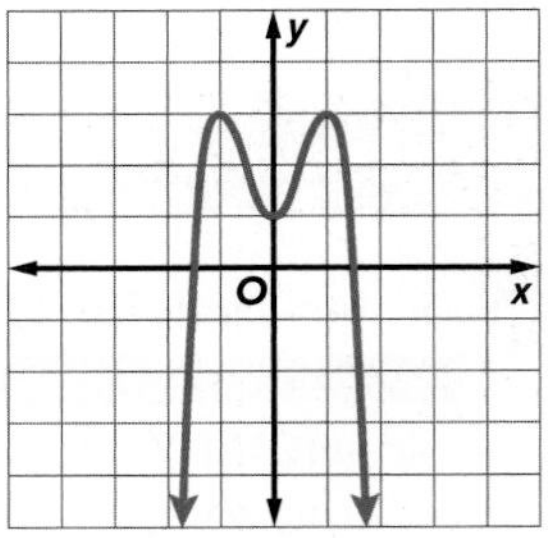

Example 2 Identify One-to-One and Onto Functions from Tables

OLYMPICS **The table shows the number of medals the United States won at five Summer Olympic Games.**

Year	Number of Gold Medals	Number of Silver Medals	Number of Bronze Medals
2016	46	37	38
2012	46	29	29
2008	36	38	36
2004	36	39	26
2000	37	24	32

Analyze the functions that give the number of gold and silver medals won in a particular year. Define the domain and range of each function and state whether it is *one-to-one, onto, both* or *neither*.

Gold Medals	Silver Medals
Let $f(x)$ be the function that gives the number of gold medals won in a particular year. The domain is in the column Year, and the range is in the column Number of Gold Medals. The function is not one-to-one because 46 in the range is paired with two values in the domain, 2016 and 2012, and 36 in the range is paired with two values in the domain, 2008 and 2004. The function is not onto because the range does not include every whole number.	Let $g(x)$ be the function that gives the number of silver medals won in a particular year. The domain is the column Year, and the range is the column Number of Silver Medals. The function is one-to-one because no two values in the domain share a value in the range. The function is not onto because the range does not include every whole number.

Go Online You can complete an Extra Example online.

Use a Source

Choose another country and research the number of medals they won in the Summer Olympic Games from 2000-2016. Are the functions that give the number of each type of medal won in a particular year *one-to-one, onto, both,* or *neither*?

Example 3 Identify One-to-One and Onto Functions from Graphs

Determine whether each function is *one-to-one, onto, both,* or *neither* for the given codomain.

***f*(*x*), where the codomain is all real numbers**

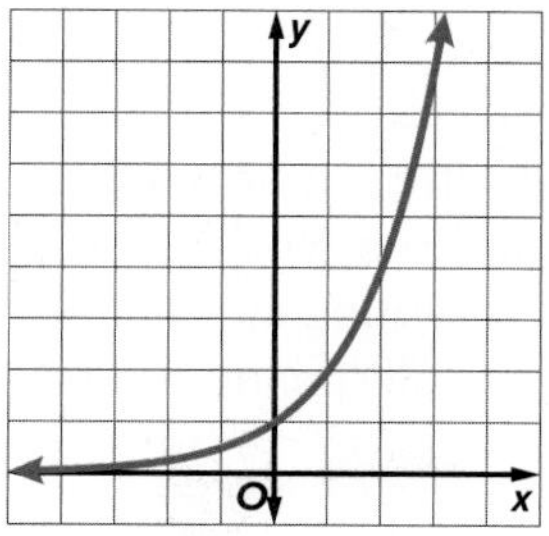

The graph indicates that the domain is all real numbers, and the range is all positive real numbers.

Every y-value is paired with exactly one unique x-value, so the function is one-to-one.

If the codomain is all real numbers, then the range is not equal to the codomain. So, the function is not onto.

***g*(*x*), where the codomain is $\{y \mid y \leq 4\}$**

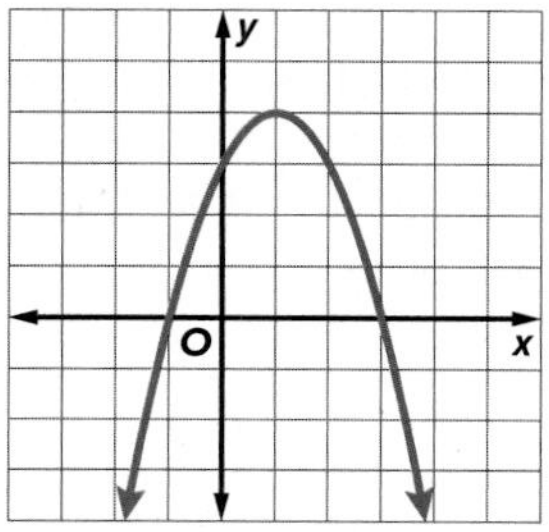

The graph indicates that the domain is all real numbers, and the range is $y \leq 4$.

Each y-value is not paired with a unique x-value; for example, $y = 3$ is paired with both $x = 0$ and $x = 2$. So the function is not one-to one.

The codomain and range are equal, so the function is onto.

***h*(*x*), where the codomain is all real numbers**

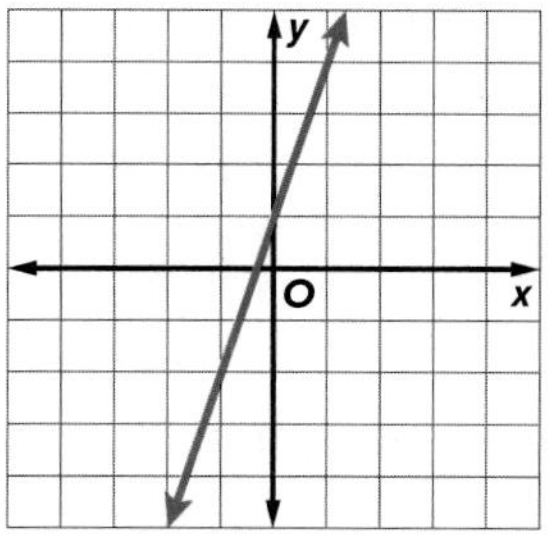

The graph indicates that the domain and range are both all real numbers.

Every y-value is paired with exactly one unique x-value, so the function is one-to-one.

The codomain and range are equal, so the function is onto.

Learn Discrete and Continuous Functions

Functions can be discrete, continuous, or neither. Real-world situations where only some numbers are reasonable are modeled by discrete functions. Situations where all real numbers are reasonable are modeled by continuous functions.

A **continuous function** is graphed with a line or an unbroken curve. A function that is not continuous is a **discontinuous function**. A **discrete function** is a discontinuous function in which the points are not connected. A function that is neither discrete nor continuous may have a graph in which some points are connected, but it is not continuous everywhere.

Study Tip

Intervals An interval is the set of all real numbers between two given numbers. For example, the interval $-2 < x < 5$ includes all values of x greater than -2 but less than 5. Intervals can also continue on infinitely in a direction. For example, the interval $y \geq 1$ includes all values of y greater than or equal to 1. You can use intervals to describe the values of x or y for which a function exists.

Go Online You can complete an Extra Example online.

Talk About It!

Does the range of the function need to be all real numbers for a function to be continuous? Justify your argument.

Example 4 Determine Continuity from Graphs

Examine the functions. Determine whether each function is *discrete, continuous,* or *neither* discrete nor continuous. Then state the domain and range of each function.

a. $f(x)$

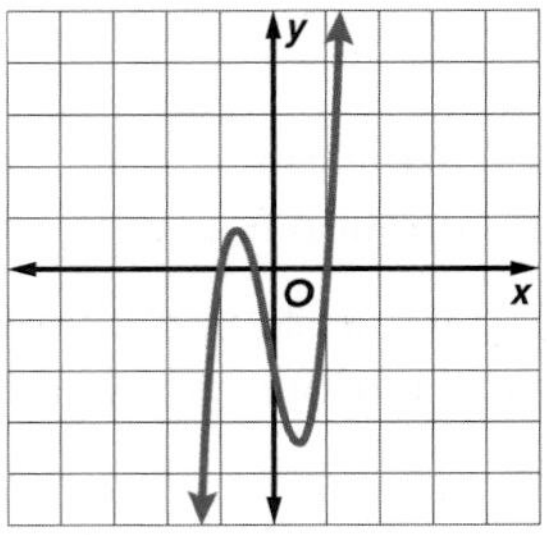

The function is continuous because the graph of the function is a curve with no breaks or discontinuities.

Because the graph of the function continues indefinitely to the left and right, its domain is all real numbers. Because the graph of the function continues indefinitely up and down, its range is all real numbers.

b. $g(x)$

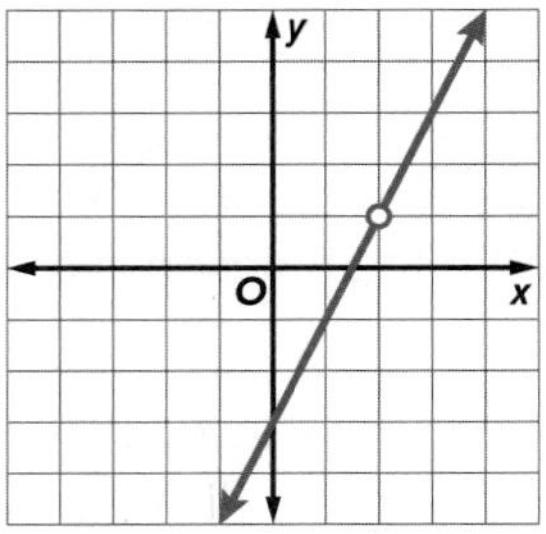

The function is neither because there are continuous sections, but there is a break at (2, 1).

Because the function is not defined for $x = 2$, the domain is all values of x except $x = 2$. The function is not defined for $y = 1$, so the range is all values of y except $y = 1$.

c. $h(x)$

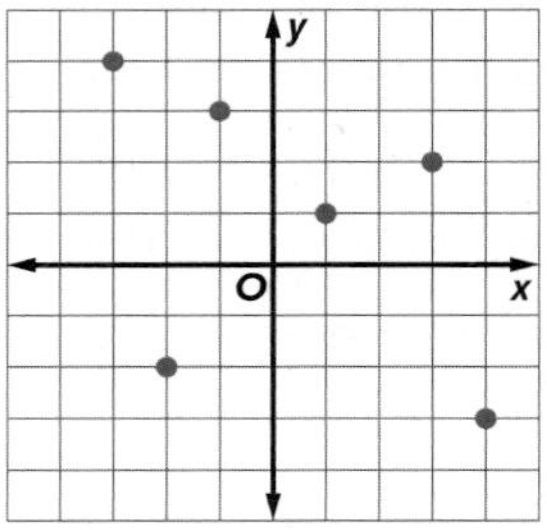

The function is discrete because it is made up of distinct points that are not connected.

The domain is $\{-3, -2, -1, 1, 3, 4\}$ and the range is $\{-3, -2, 1, 2, 3, 4\}$.

Problem-Solving Tip

Use a Graph If you are having trouble determining the continuity given the equation of a function, you can graph the function to help visualize the situation.

Example 5 Determine Continuity

BUSINESS **Determine whether the function that models the cost of coffee beans is *discrete, continuous,* or *neither* discrete nor continuous. Then state the domain and range of the function.**

COFFEE

Weight	Price
Up to 2 lbs	\$8/lb
2.5 lbs	\$20
3 lbs	\$22
5 lbs	\$35

Because customers can purchase any amount of coffee up to 2 pounds, the function is continuous over the interval $0 \le x \le 2$.

Go Online You can complete an Extra Example online.

Study Tip

Accuracy When calculating cost, the result can be any fraction of a dollar or cent, and is therefore continuous. However, because the smallest unit of currency is \$0.01, the price you actually pay is rounded to the nearest cent. Therefore, the price you pay is discrete.

For larger quantities, the coffee is sold by distinct amounts. This part of the function is discrete.

Since the domain and range are made up of neither a single interval nor individual points, the function is neither discrete nor continuous.

The domain of the function is $0 \leq x \leq 2$ or $x = 2.5, 3, 5$. This represents the possible weights of coffee beans that customers could purchase. The range of the function is $0 \leq y \leq 16$ or $y = 20, 22, 35$. This represents the possible costs of coffee beans.

Think About It!

Why does the range include values from 0 to 16 instead of 0 to 8?

Learn Set-Builder and Interval Notation

Sets of numbers like the domain and range of a function can be described by using various notations. Set-builder notation, interval notation, and algebraic notation are all concise ways of writing a set of values. Consider the set of values represented by the graph.

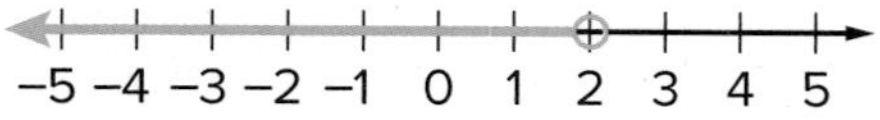

- In **algebraic notation**, sets are described using algebraic expressions. Example: $x < 2$
- **Set-builder notation** is similar to algebraic notation. Braces indicate the set. The symbol | is read as *such that*. The symbol ∈ is read *is an element of*. Example: $\{x|x < 2\}$
- In **interval notation** sets are described using endpoints with parentheses or brackets. A parenthesis, (or), indicates that an endpoint *is not* included in the interval. A bracket, [or], indicates that an endpoint is included in the interval. Example: $(-\infty, 2)$

Study Tip

Using Symbols You can use the symbol $\mathbb{R}$ to represent all real numbers in set-builder notation. In interval notation, the symbol ∪ indicates the union of two sets. Parentheses are always used with ∞ and $-\infty$ because they do not include endpoints.

Example 6 Set-Builder and Interval Notation for Continuous Intervals

Write the domain and range of the graph in set-builder and interval notation.

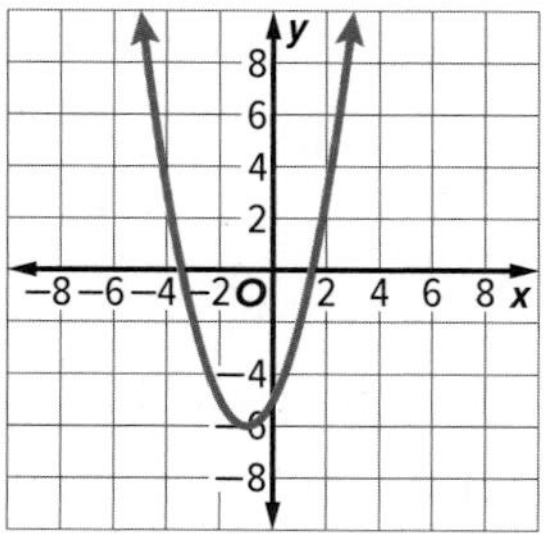

Domain

The graph will extend to include all x-values.

The domain is all real numbers.

$\{x \mid x \in \mathbb{R}\}$

$(-\infty, \infty)$

Range

The least y-value for this function is -6.

The range is all real numbers greater than or equal to -6.

$\{y \mid y \geq -6\}$

$[-6, \infty)$

Go Online You can complete an Extra Example online.

Check

State the domain and range of each graph in set-builder and interval notation.

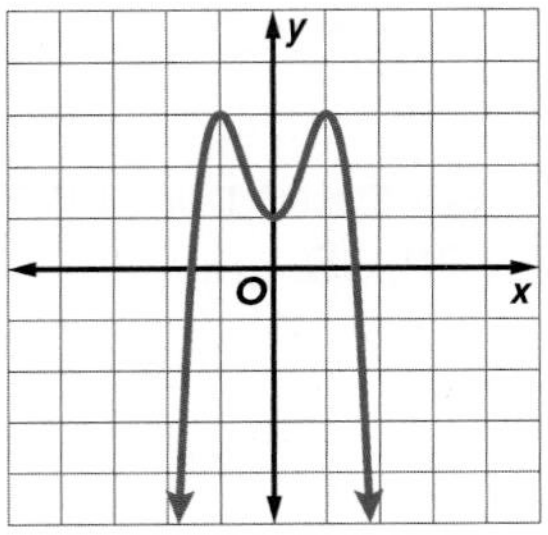

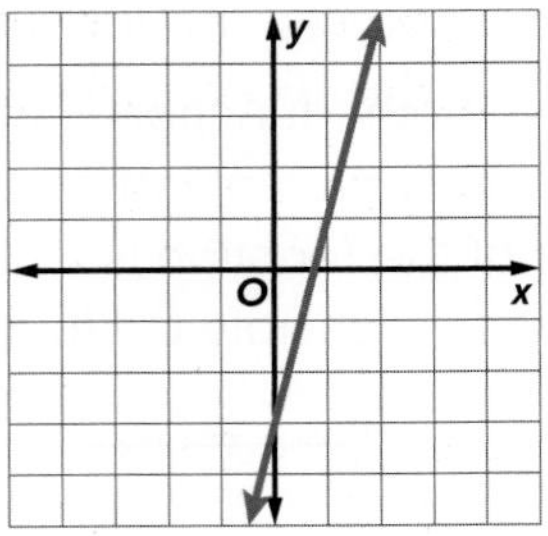

Example 7 Set-Builder and Interval Notation for Discontinuous Intervals

Write the domain and range of the graph in set-builder and interval notation.

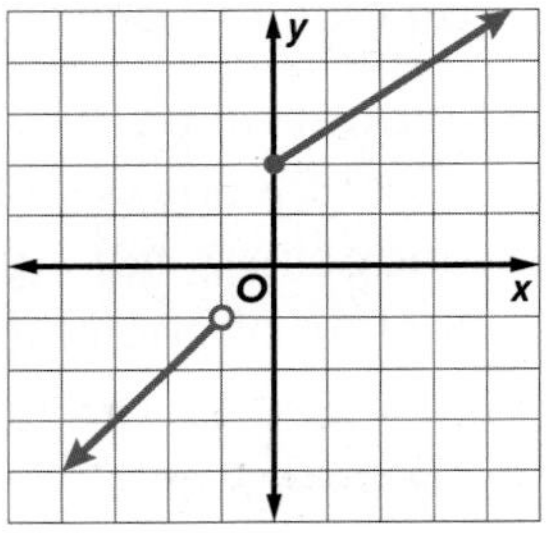

Domain

The domain is all real numbers less than −1 or greater than or equal to 0.

$\{x \mid x < -1 \text{ or } x \geq 0\}$

$(-\infty, -1) \cup [0, \infty)$

Range

The range is all real numbers less than −1 or greater than or equal to 2.

$\{y \mid y < -1 \text{ or } y \geq 2\}$

$(-\infty, -1) \cup [2, \infty)$

Check

State the domain and range of the graph in set-builder and interval notation.

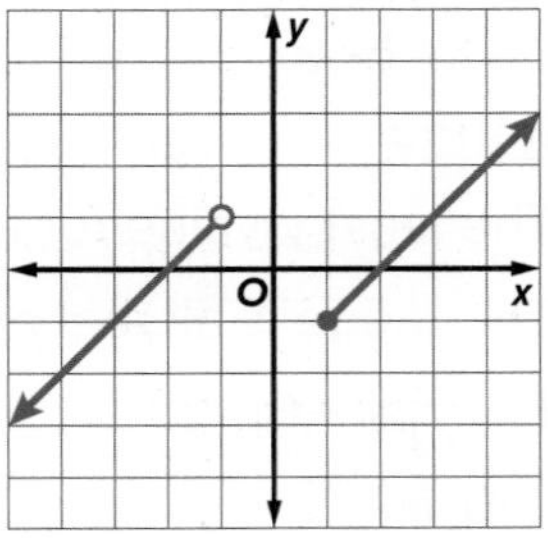

Go Online You can complete an Extra Example online.

Practice

Go Online You can complete your homework online.

Example 1

Identify the domain, range, and codomain in each graph. Then use the codomain and range to determine whether the function is onto.

1. 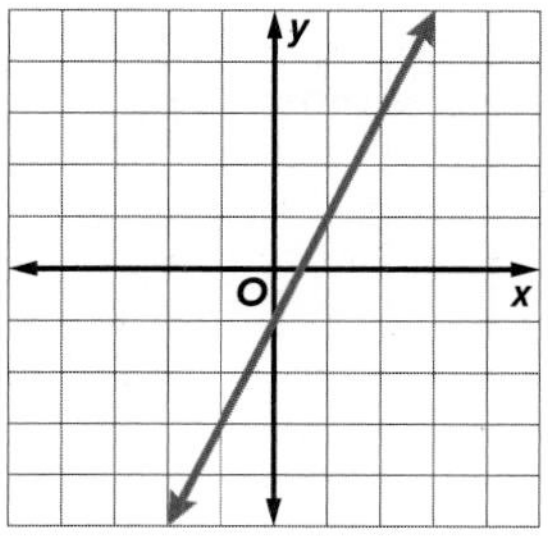

2. 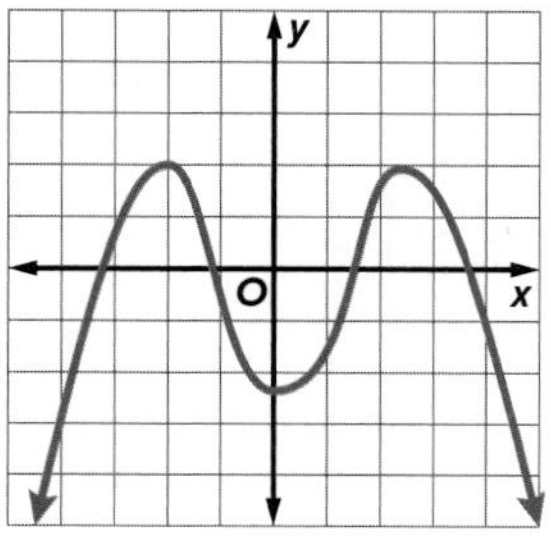

3. 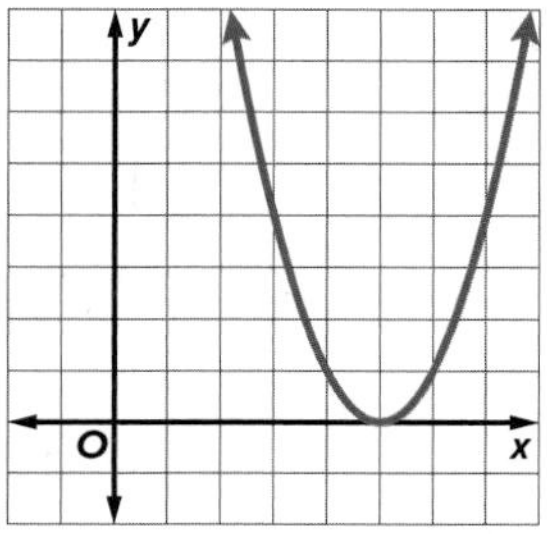

Example 2

4. SALES Cool Athletics introduced the new Power Sneaker in one of their stores. The table shows the sales for the first 6 weeks. Define the domain and range of the function and state whether it is *one-to-one, onto, both* or *neither.*

Week	1	2	3	4	5	6
Pairs Sold	8	10	15	22	15	44

5. TEMPERATURES The table shows the low temperatures in degrees Fahrenheit for the past week in Sioux Falls, Idaho. Define the domain and range of the function and state whether it is *one-to-one, onto, both,* or *neither.*

Day	1	2	3	4	5	6	7
Low Temp.	56	52	44	41	43	46	53

6. PLANETS The table shows the orbital period of the eight major planets in our Solar System given their mean distance from the Sun. Define the domain and range of the function and state whether it is *one-to-one, onto, both* or *neither.*

Planet	Mean Distance from Sun (AU)	Orbital Period (years)
Mercury	0.4	0.241
Venus	0.7	0.615
Earth	1.0	1.0
Mars	1.5	1.881
Jupiter	5.2	11.75
Saturn	9.5	29.5
Uranus	19.2	84
Neptune	30	164.8

Example 3

Determine whether each function is *one-to-one, onto, both,* or *neither.*

7. 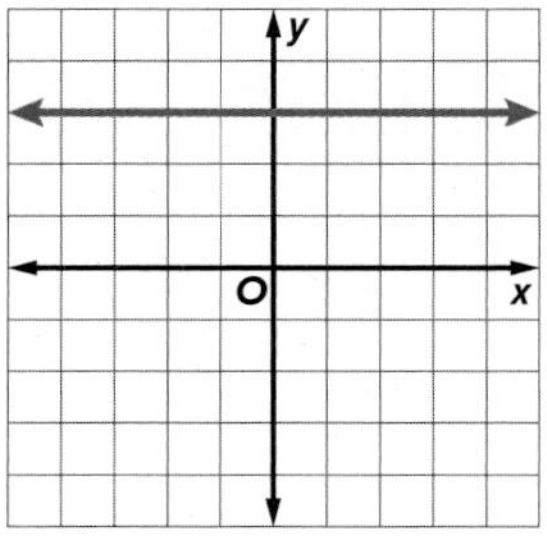

8. 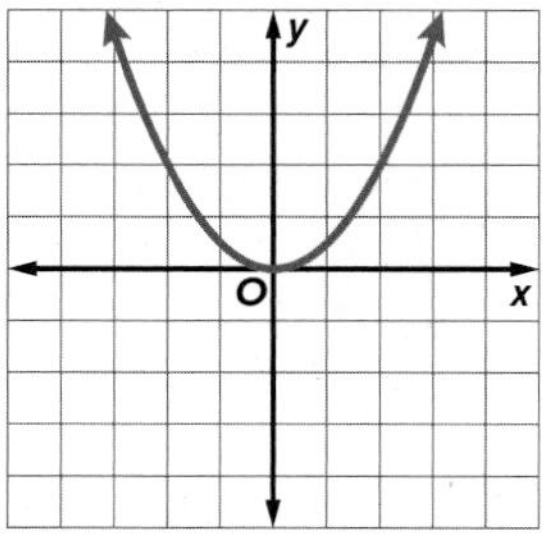

9. 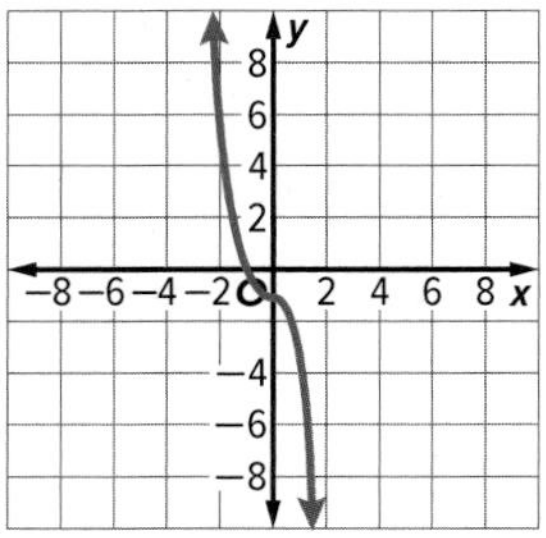

Example 4

Examine the graphs. Determine whether each relation is *discrete, continuous*, or *neither* discrete nor continuous. Then state the domain and range of each relation.

10.

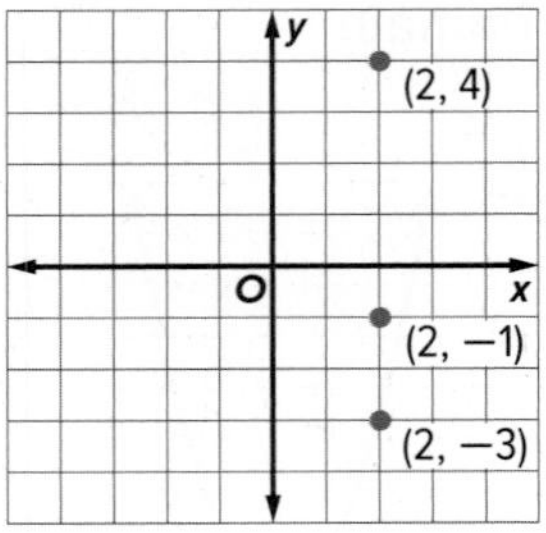

11.

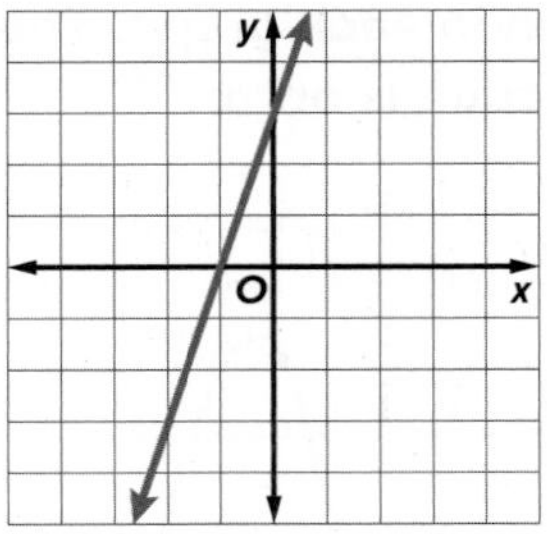

12.

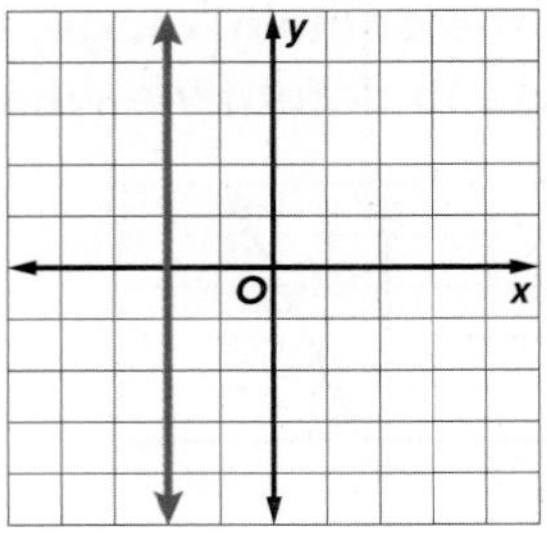

Example 5

13. PROBABILITY The table shows the outcome of rolling a number cube. Determine whether the function that models the outcome of each roll is *discrete, continuous,* or *neither* discrete nor continuous. Then state the domain and range of the function.

Roll	Outcome
1	4
2	3
3	6
4	3
5	5
6	4

14. AMUSEMENT PARK The table shows the price of tickets to an amusement park based on the number of people in the group. Determine whether the function that models the price of tickets is *discrete, continuous,* or *neither* discrete nor continuous. Then state the domain and range of the function.

Group Size	Price
up to 15 people	$45
16–50 people	$38
51–100 people	$30
101 or more people	$26

15. GROCERIES A local grocery store sells grapes for $1.99 per pound. Determine whether the function that models the cost of grapes is *discrete, continuous,* or *neither* discrete nor continuous. Then state the domain and range of the function.

Examples 6 and 7

Write the domain and range of the graph in set-builder and interval notation.

16.

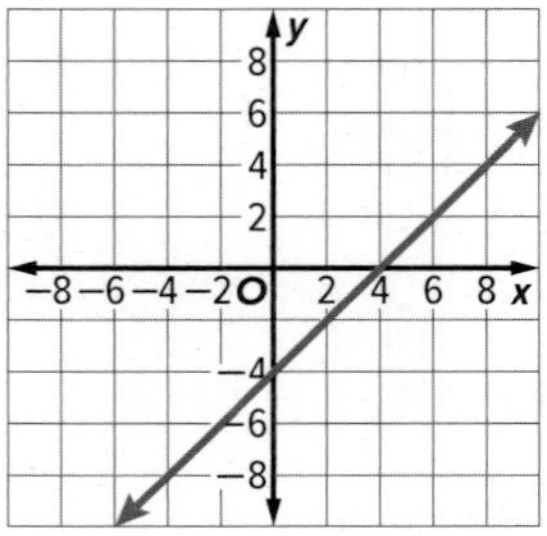

17.

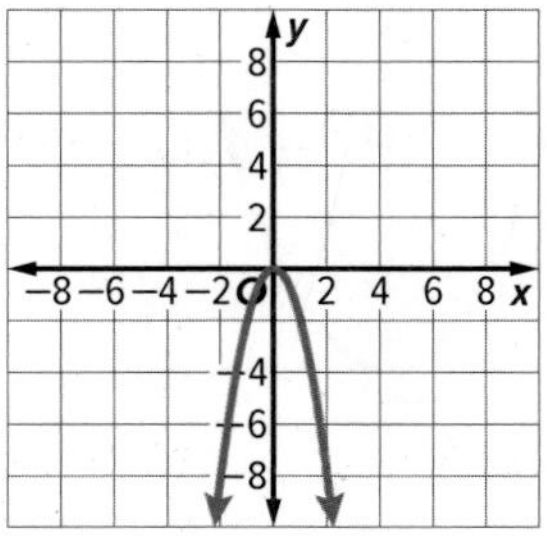

18.

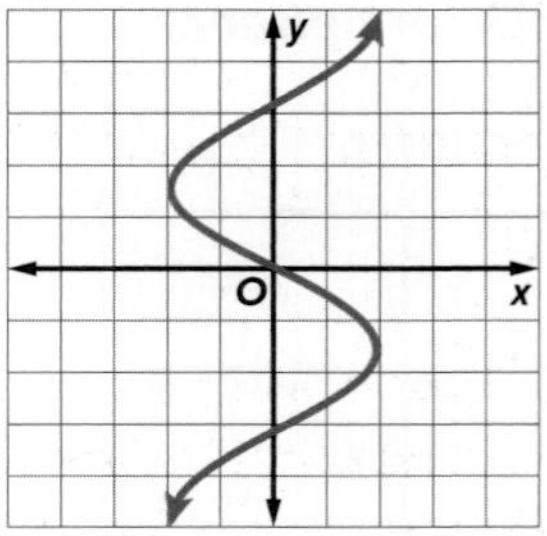

Write the domain and range of the graph in set-builder and interval notation.

19.
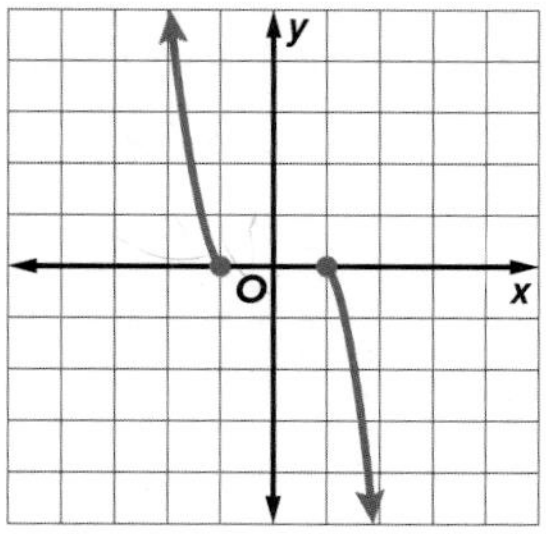

20.
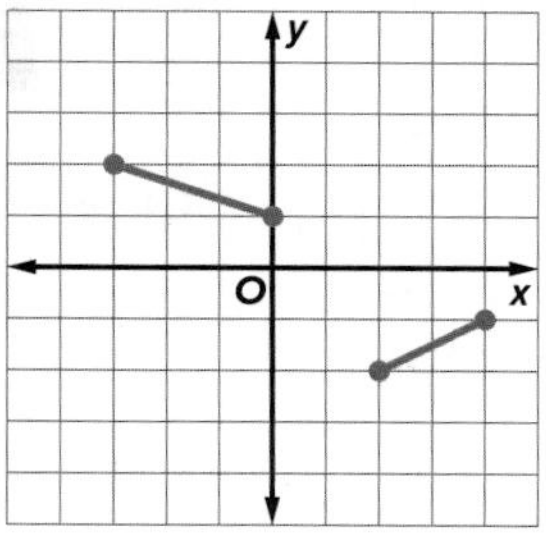

21.
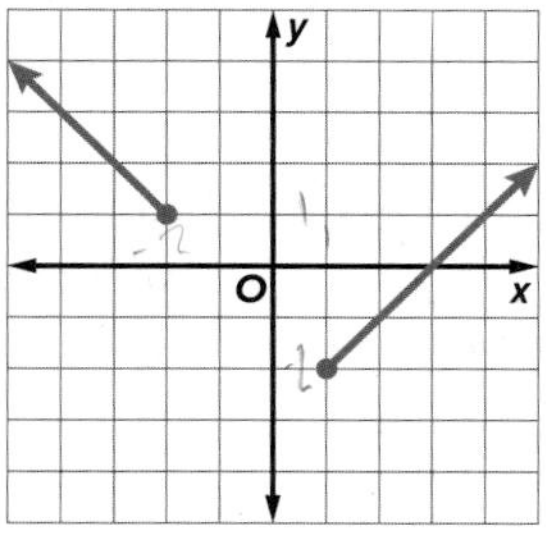

Mixed Exercises

STRUCTURE **Write the domain and range of each function in set-builder and interval notation. Determine whether each function is *one-to-one, onto, both,* or *neither.* Then state whether it is *discrete, continuous,* or *neither* discrete nor continuous.**

22.
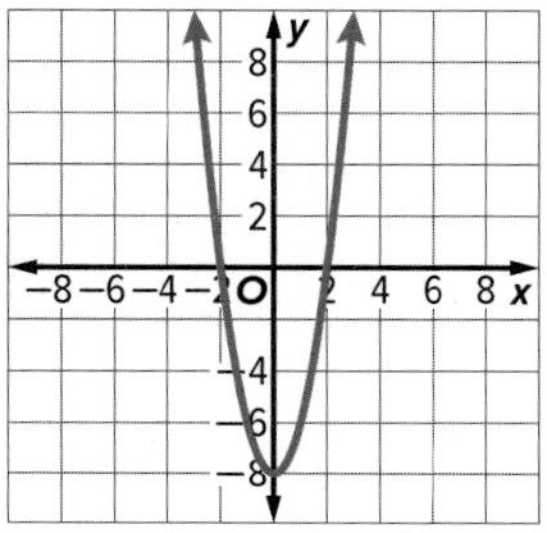

23.
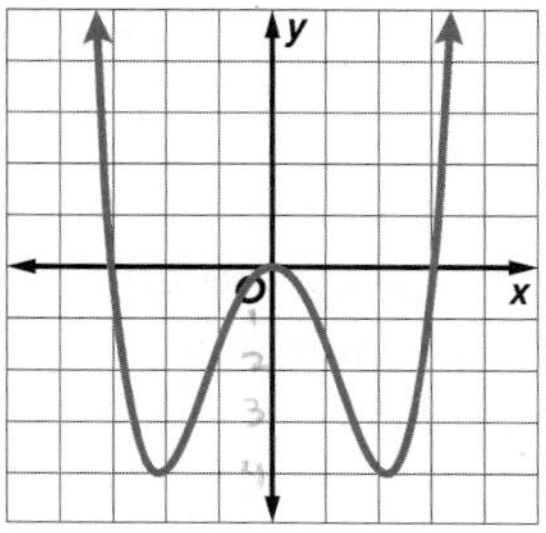

24.
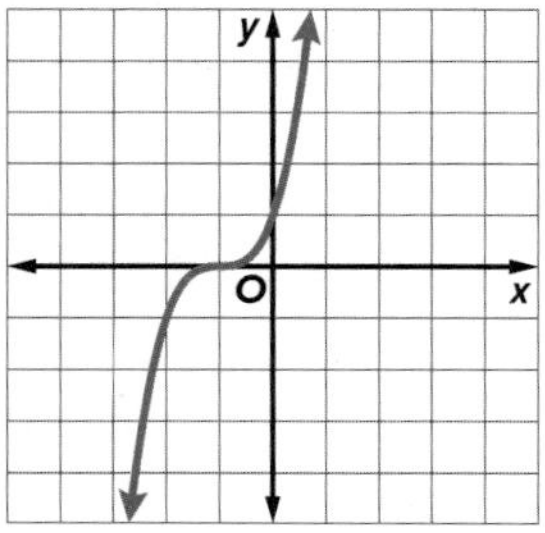

25.
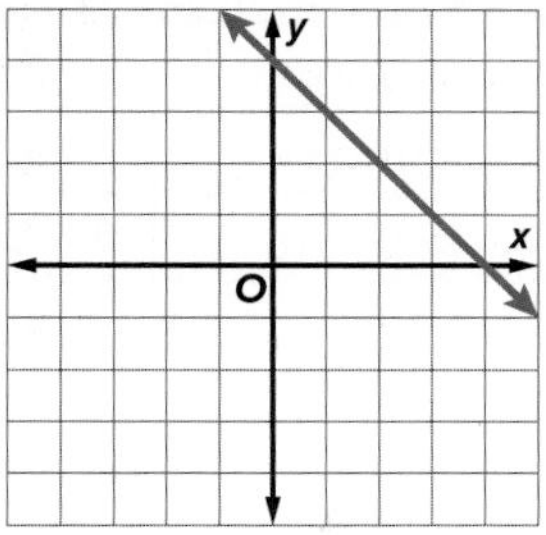

26.
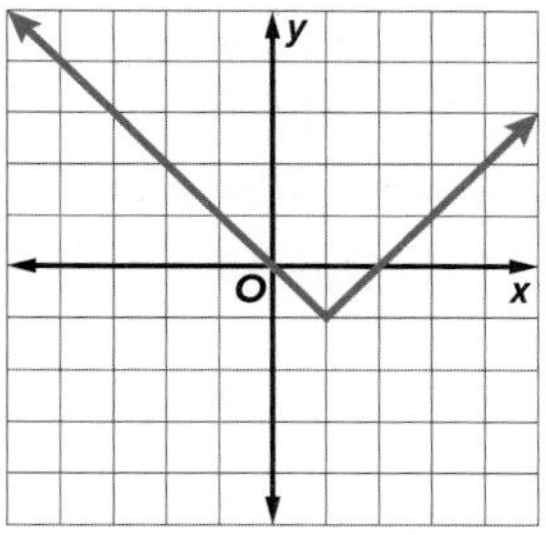

27.
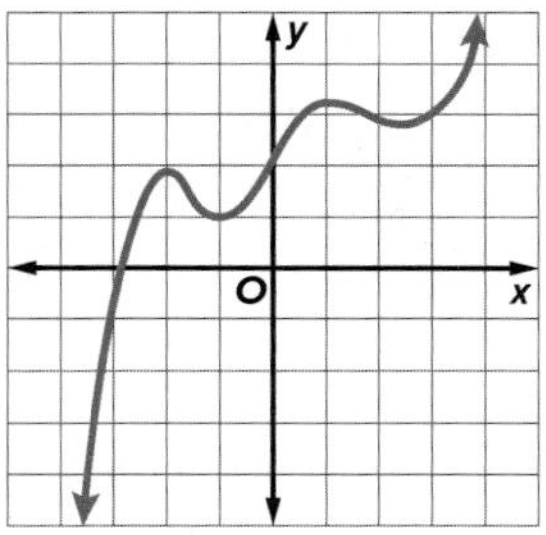

28. USE A SOURCE Research the total number of games won by a professional baseball team each season for five consecutive years. Determine the domain, range, and continuity of the function that models the number of wins.

29. SPRINGS When a weight up to 15 pounds is attached to a 4-inch spring, the length L, in inches, that the spring stretches is represented by the function $L(w) = \frac{1}{2}w + 4$, where w is the weight, in pounds, of the object. State the domain and range of the function. Then determine whether it is *one-to-one, onto, both,* or *neither* and whether it is *discrete, continuous,* or *neither* discrete nor continuous.

30. CASHEWS An airport snack stand sells whole cashews for $12.79 per pound. Determine whether the function that models the cost of cashews is *discrete, continuous,* or *neither* discrete nor continuous. Then state the domain and range of the function in set-builder and interval notation.

31. PRICES The Consumer Price Index (CPI) gives the relative price for a fixed set of goods and services. The CPI from September, 2017 to July, 2018 is shown in the graph. Determine whether the function that models the CPI is *one-to-one, onto, both,* or *neither.* Then state whether it is *discrete, continuous,* or *neither* discrete nor continuous.

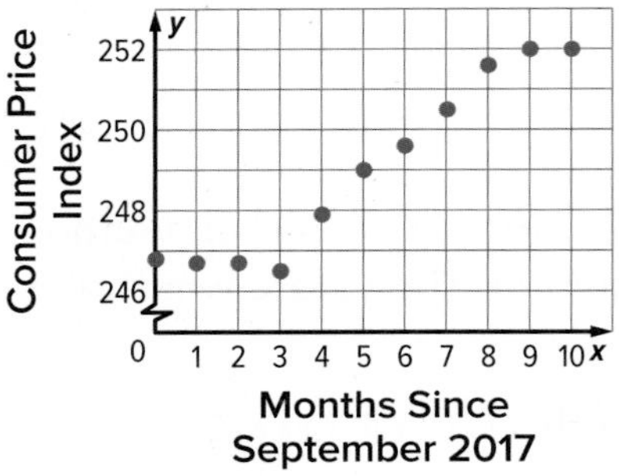

32. LABOR A town's annual jobless rate is shown in the graph. Determine whether the function that models the jobless rate is *one-to-one, onto, both,* or *neither.* Then state whether it is *discrete, continuous,* or *neither* discrete nor continuous.

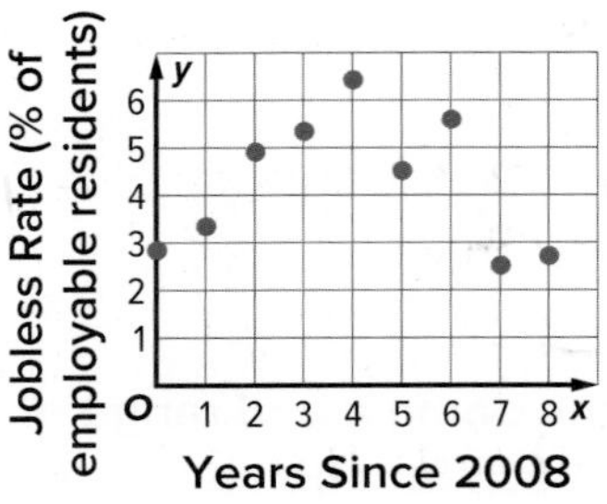

33. COMPUTERS If a computer can do one calculation in 0.0000000015 second, then the function $T(n) = 0.0000000015n$ gives the time required for the computer to do n calculations. State the domain and range of the function. Then determine whether it is *one-to-one, onto, both,* or *neither* and whether it is *discrete, continuous,* or *neither* discrete nor continuous.

34. SHIPPING The table shows the cost to ship a package based on the weight of the package. Determine whether the function that models the shipping cost is *discrete, continuous,* or *neither* discrete nor continuous. Then state the domain and range of the function in set-builder notation.

Package Weight (lbs)	Cost
up to 5 pounds	\$4
5-10 pounds	\$6
exceeds 10 pounds	\$0.65/lb

Higher-Order Thinking Skills

35. CREATE Sketch the graph of a function that is onto, but not one-to-one, if the codomain is restricted to values greater than or equal to −3.

36. ANALYZE Determine whether the following statement is *true* or *false.* Explain your reasoning.

If a function is onto, then it must be one-to-one as well.

37. PERSEVERE Consider $f(x) = \frac{1}{x}$. State the domain and the range of the function. Determine whether the function is *one-to-one, onto, both,* or *neither.* Determine whether the function is *discrete, continuous,* or *neither* discrete nor continuous.

38. PERSEVERE Use the domain {−4, −2, 0, 2, 4}, the codomain {−4, −2, 0, 2, 4}, and the range {0, 2, 4} to create a function that is neither one-to-one nor onto.

39. WRITE Compare and contrast the vertical and horizontal line tests.

Linearity, Intercepts, and Symmetry

Explore Symmetry and Functions

Online Activity Use graphing technology to complete the Explore.

INQUIRY How can you tell whether the graph of a function is symmetric?

Learn Linear and Nonlinear Functions

In a **linear function**, no variable is raised to a power other than 1. Any linear function can be written in the form $f(x) = mx + b$, where m and b are real numbers. Linear functions can be represented by **linear equations**, which can be written in the form $Ax + By = C$. The graph of a linear equation is a straight line.

A function that is not linear is called a **nonlinear function**. The graph of a nonlinear function includes a set of points that cannot all lie on the same line. A nonlinear function cannot be written in the form $f(x) = mx + b$. A **parabola** is the graph of a quadratic function, which is a type of nonlinear function.

Example 1 Identify Linear Functions from Equations

Determine whether each function is a linear function. Justify your answer.

a. $\mathbf{f(x) = \frac{6x - 5}{3}}$

$f(x) = \frac{6x - 5}{3}$	Original equation
$f(x) = \frac{6}{3}x - \frac{5}{3}$	Distribute the denominator of 3.
$f(x) = 2x - \frac{5}{3}$	Simplify.

The function can be written in the form $f(x) = mx + b$, so it is a linear function.

b. $\mathbf{5y = 4 + 3x^3}$

$5y = 4 + 3x^3$	Original equation
$5y = 3x^3 + 4$	Commutative Property

The function cannot be written in the form $f(x) = mx + b$ because the independent variable x is raised to a whole number power greater than 1. So, it is a nonlinear function.

Go Online You can complete an Extra Example online.

Today's Goals

- Identify linear and nonlinear functions.
- Identify and interpret the intercepts of functions.
- Identify whether graphs of functions possess line or point symmetry and determine whether functions are even, odd, or neither.

Today's Vocabulary

linear function
linear equation
nonlinear function
parabola
intercept
x-intercept
y-intercept
symmetry
line symmetry
line of symmetry
point symmetry
point of symmetry
even functions
odd functions

Study Tip

Linear Functions To write any linear equation in function form, solve the equation for y and replace the variable y with $f(x)$.

Example 2 Identify Linear Functions from Graphs

Determine whether each graph represents a *linear* or *nonlinear* function.

a.

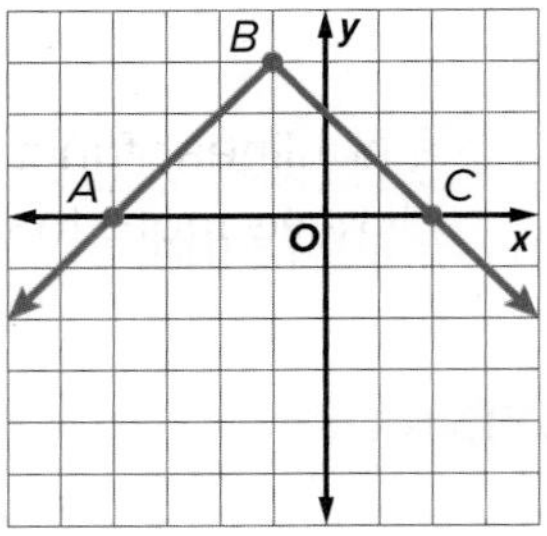

There is no straight line that will contain the chosen points *A*, *B*, and *C*, so this graph represents a nonlinear function.

b.

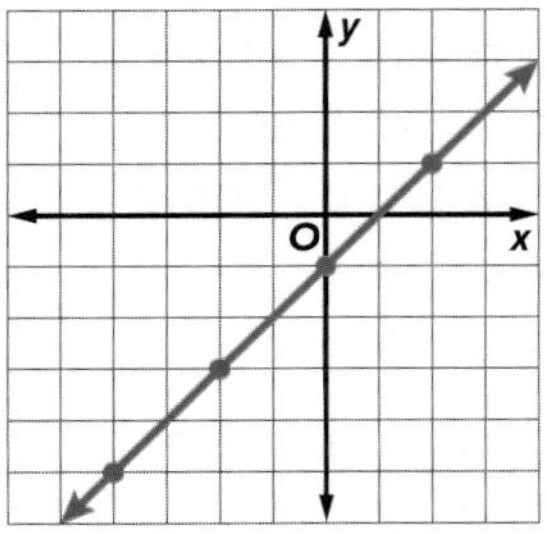

The points on this graph all lie on the same line, so this graph represents a linear function.

Example 3 Identify Linear Functions from Tables

EARNINGS **Makayla has started working part-time at the local hardware store. Her time at work steadily increases for the first five weeks. The table shows her total earnings each of those weeks. Are her weekly earnings modeled by a *linear* or *nonlinear* function?**

Week	1	2	3	4	5
Earnings ($)	85	119	153	187	221

Graph the points that represent the week and total earnings and try to draw a line that contains all the points.

Since there is a line that contains all the points, Makayla's earning can be modeled by a linear function.

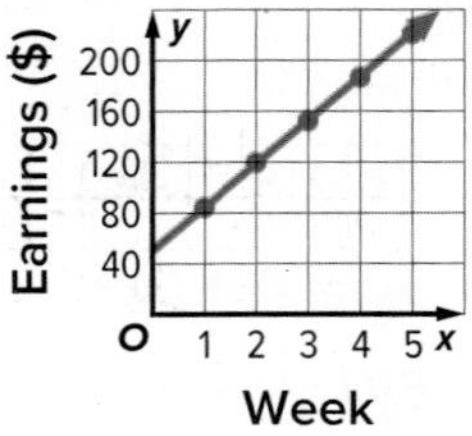

Think About It!
Why is $f(x) = \sqrt{2x} + 3$ not a linear function?

Think About It!
Are negative *x*- or *y*-values possible in the context of the situation?

Go Online You can complete an Extra Example online.

Learn Intercepts of Graphs of Functions

A point at which the graph of a function intersects an axis is called an **intercept**. An ***x*-intercept** is the *x*-coordinate of a point where the graph crosses the *x*-axis, and a ***y*-intercept** is the *y*-coordinate of a point where the graph crosses the *y*-axis.

A linear function has at most one *x*-intercept while a nonlinear function may have more than one *x*-intercept. No function will have more than one *y*-intercept.

Study Tip

Point or Coordinate *Intercept* may refer to the point or one of its coordinates. The context of the situation will often dictate which form to use.

Example 4 Find Intercepts of a Linear Function

Use the graph to estimate the *x*- and *y*-intercepts.

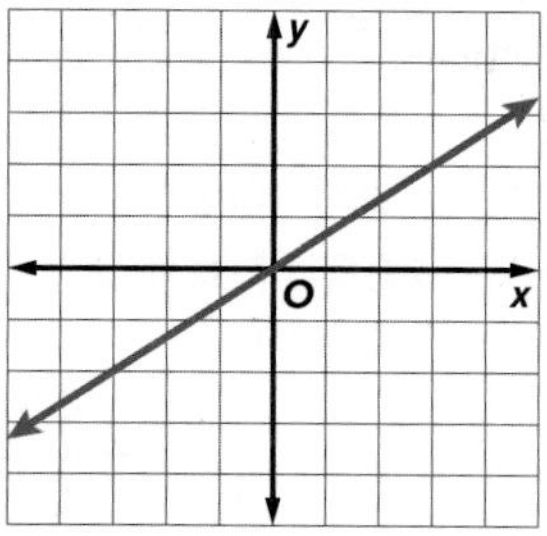

The graph intersects the *x*-axis at (0, 0), so the *x*-intercept is 0.

The graph intersects the *y*-axis at (0, 0), so the *y*-intercept is 0.

Think About It!

Describe a line that does not have two distinct intercepts.

Example 5 Find Intercepts of a Nonlinear Function

Use the graph to estimate the *x*- and *y*-intercepts.

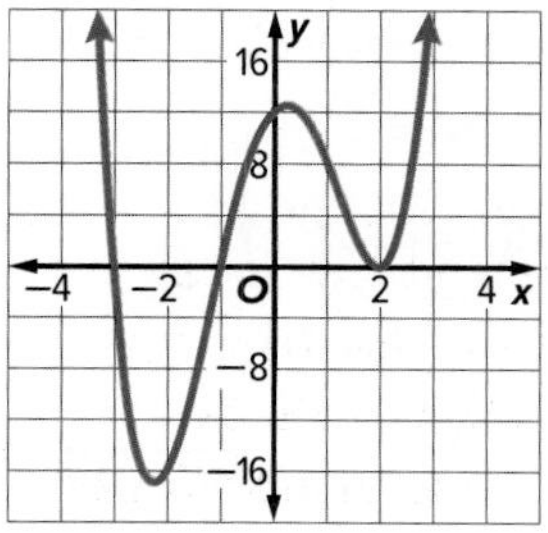

The graph appears to intersect the *x*-axis at (−3, 0), (−1, 0), and (2, 0), so the function has *x*-intercepts of −3, −1, and 2.

The graph appears to intersect the *y*-axis at (0, 12), so the function has a *y*-intercept of 12.

Think About It!

The graph of the nonlinear function has three *x*-intercepts. Can the graph have more than one *y*-intercept? Explain your reasoning.

Check

Estimate the *x*- and *y*-intercepts of each graph.

a.

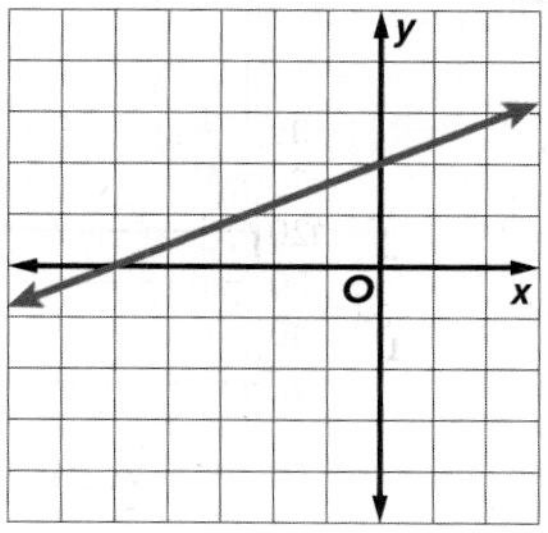

b.

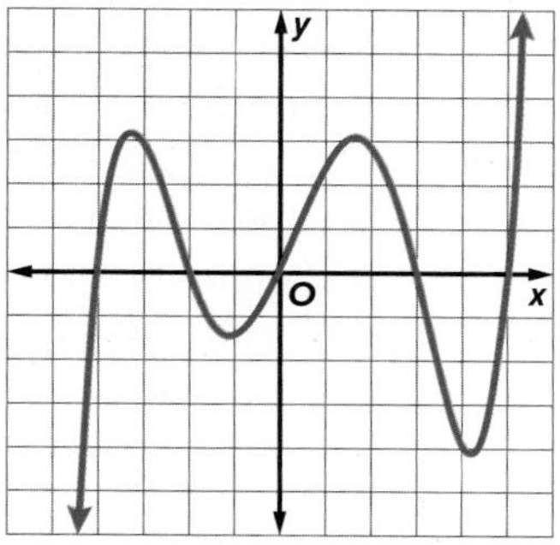

Go Online You can complete an Extra Example online.

Think About It!

Describe the domain of the function that models the rocket's height over time.

Watch Out!

Switching Coordinates A common mistake is to switch the coordinates for the intercepts. Remember that for the x-intercept, the y-coordinate is 0, and for the y-intercept, the x-coordinate is 0.

Talk About It

Can the graph of a function be symmetric in a horizontal line? Justify your answer.

Example 6 Interpret the Meaning of Intercepts

MODEL ROCKETS **Ricardo launches a rocket from a balcony. The table shows the height of the rocket after each second of its flight.**

Time (s)	Height (ft)
0	15
1	60
2	130
3	180
4	210
5	170
6	110
7	55
8	0

Part A Identify the x- and y-intercepts of the function that models the flight of the rocket.

In the table, the x-coordinate when $y = 0$ is 8. Thus, the x-intercept is 8.

In the table, the y-coordinate when $x = 0$ is 15. Thus, the y-intercept is 15.

Part B What is the meaning of the intercepts in the context of the rocket's flight?

The x-intercept is the number of seconds after the rocket is launched that it returns to the ground. The y-intercept is the height of the balcony from which the rocket is launched.

Learn Symmetry of Graphs of Functions

A figure has **symmetry** if there exists a rigid motion—reflection, translation, rotation, or glide reflection—that maps the figure onto itself.

Key Concept • Symmetry

Type of Symmetry	Description	Example
A graph has **line symmetry** if it can be reflected in a vertical line so that each half of the graph maps exactly to the other half.	The line dividing the graph into matching halves is called the **line of symmetry**. Each point on one side is reflected in the line to a point equidistant from the line on the opposite side.	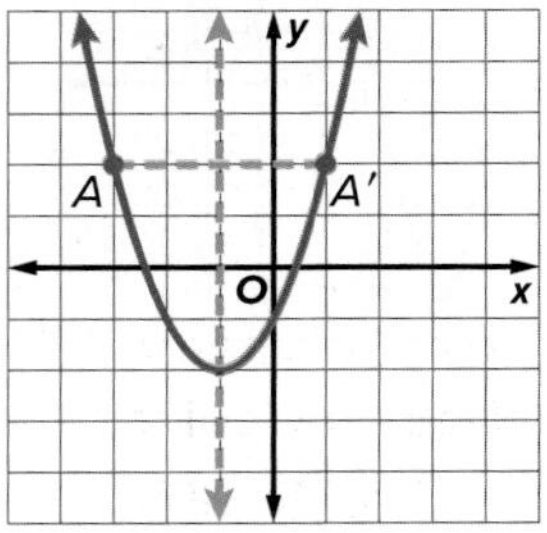
A graph has **point symmetry** when a figure is rotated 180° about a point and maps onto itself.	The point about which the graph is rotated is called the **point of symmetry**. The image of each point on one side of the point of symmetry can be found on the line through the point and the point of symmetry, equidistant from the point of symmetry.	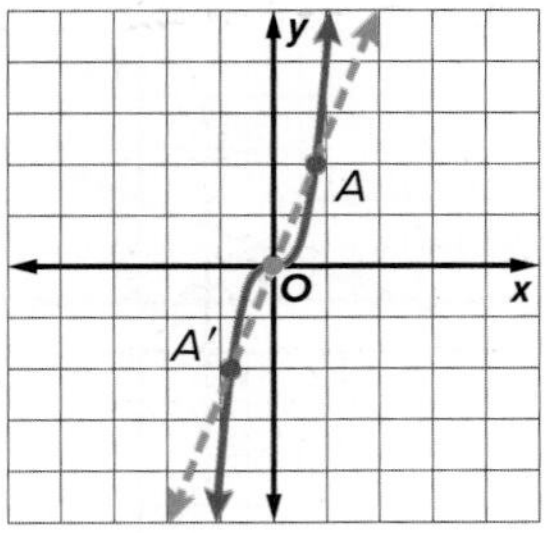

Key Concept • Even and Odd Functions		
Type of Function	**Algebraic Test**	**Example**
Functions that have line symmetry with respect to the *y*-axis are called **even functions**.	For every x in the domain of f, $f(-x) = f(x)$.	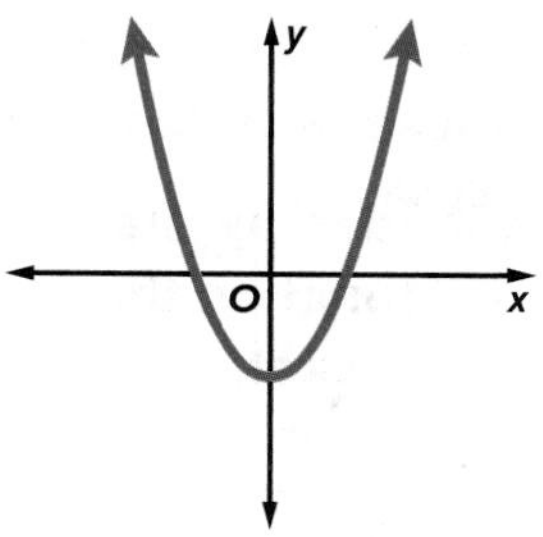
Functions that have point symmetry with respect to the origin are called **odd functions**.	For every x in the domain of f, $f(-x) = -f(x)$.	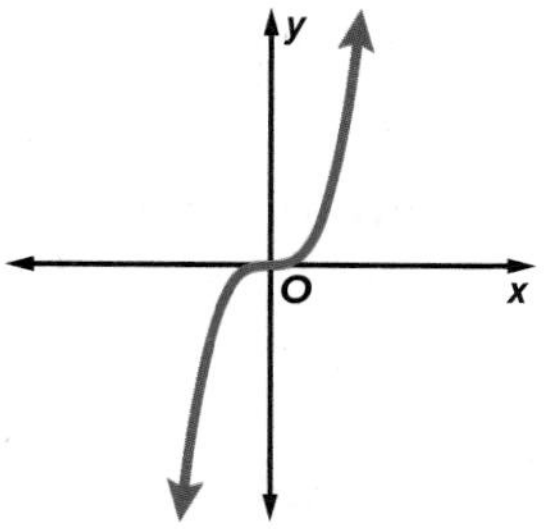

Example 7 Identify Types of Symmetry

Identify the type of symmetry in the graph of each function. Explain.

a. $f(x) = 3x + 1$

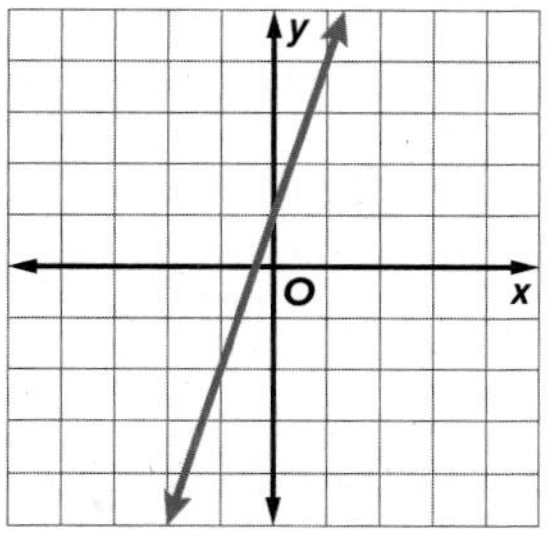

point symmetry: A 180° rotation about any point on the graph is the original graph.

b. $g(x) = -x^2 - 4x - 2$

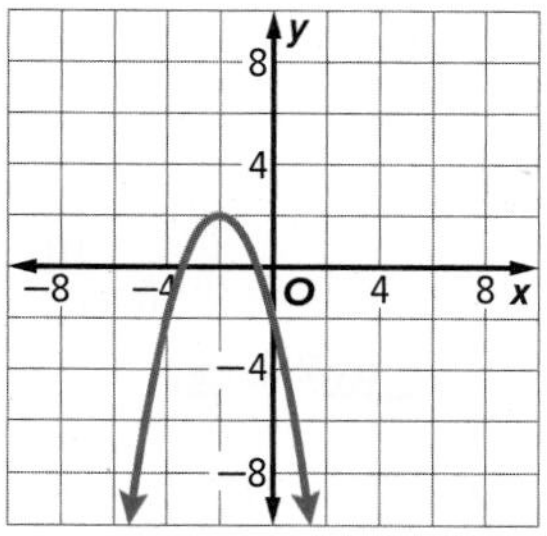

line symmetry: The reflection in the line $x = -2$ coincides with the original graph.

c. $h(x) = 3x^4 + 4x^3 - 12x^2 + 13$

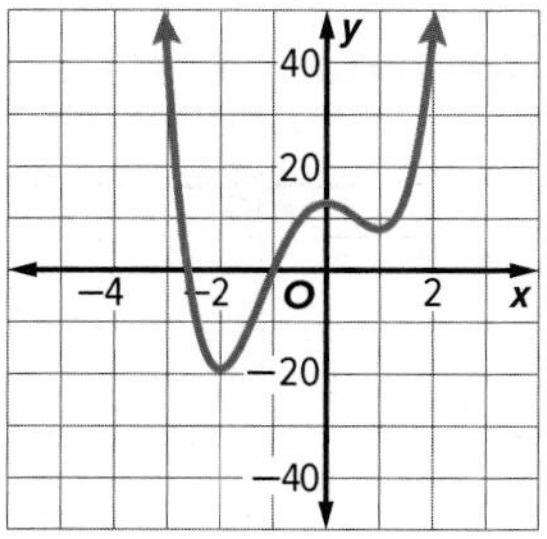

no symmetry: There is no line or point of symmetry.

d. $j(x) = x^3 - 2$

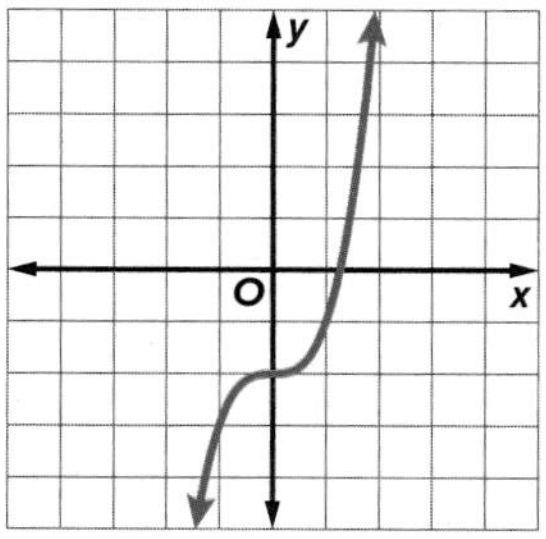

point symmetry: A 180° rotation about the point (0, −2) is the original graph.

Go Online You can complete an Extra Example online.

Think About It!
How would knowing the type of symmetry help you graph a function?

Example 8 Identify Even and Odd Functions

Determine whether each function is *even*, *odd*, or *neither*. Confirm algebraically. If the function is odd or even, describe the symmetry.

a. $f(x) = x^3 - 4x$

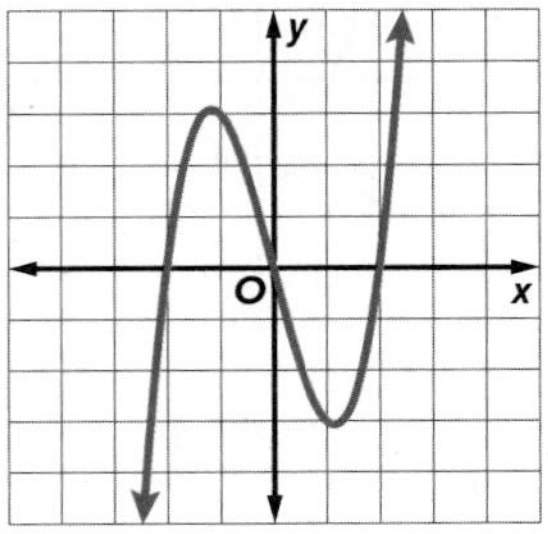

It appears that the graph of $f(x)$ is symmetric about the origin. Substitute $-x$ for x to test this algebraically.

$$
\begin{aligned}
f(-x) &= (-x)^3 - 4(-x) \\
&= -x^3 + 4x && \text{Simplify.} \\
&= -(x^3 - 4x) && \text{Distribute.} \\
&= -f(x) && f(x) = x^3 - 4x
\end{aligned}
$$

Because $f(-x) = -f(x)$ the function is odd and is symmetric about the origin.

b. $g(x) = 2x^4 - 6x^2$

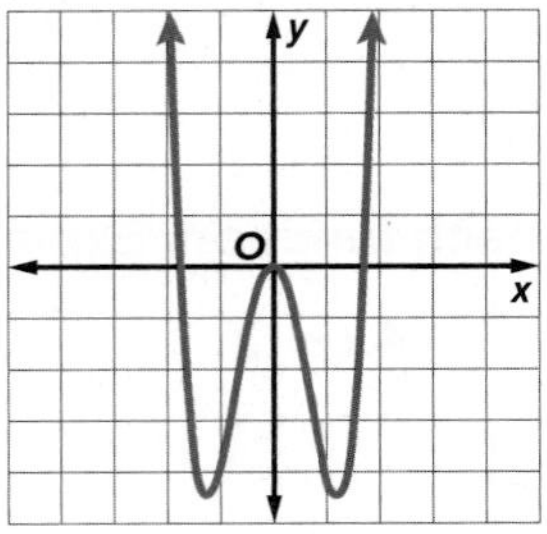

It appears that the graph of $g(x)$ is symmetric about the y-axis. Substitute $-x$ for x to test this algebraically.

$$
\begin{aligned}
g(-x) &= 2(-x)^4 - 6(-x)^2 \\
&= 2x^4 - 6x^2 && \text{Simplify.} \\
&= g(x) && g(x) = 2x^4 - 6x^2
\end{aligned}
$$

Because $g(-x) = g(x)$ the function is even and is symmetric in the y-axis.

c. $h(x) = x^3 + 0.25x^2 - 3x$

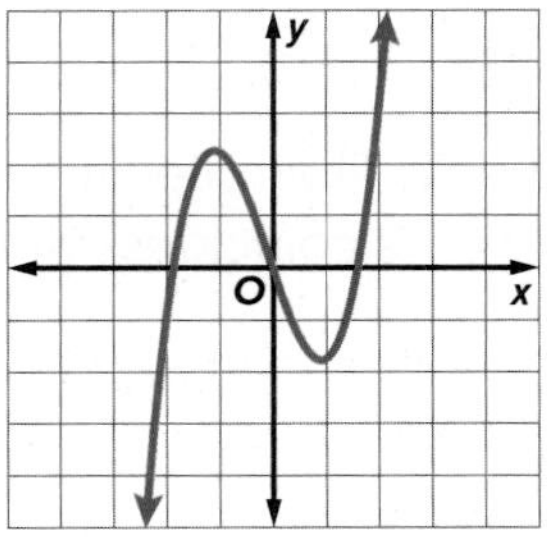

It appears that the graph of $h(x)$ may be symmetric about the origin. Substitute $-x$ for x to test this algebraically.

$$
\begin{aligned}
h(-x) &= (-x)^3 + 0.25(-x)^2 - 3(-x) \\
&= -x^3 + 0.25x^2 + 3x
\end{aligned}
$$

Because $-h(x) = -x^3 - 0.25x^2 + 3x$, the function is neither even nor odd because $h(-x) \neq h(x)$ and $h(-x) \neq -h(x)$.

Watch Out!

Even and Odd Functions Always confirm symmetry algebraically. Graphs that appear to be symmetric may not actually be.

Check

Assume that f is a function that contains the point $(2, -5)$. Which of the given points must be included in the function if f is:

even? ___?___ odd? ___?___

$(-2, -5)$ $(-2, 5)$ $(2, 5)$ $(-5, -2)$ $(-5, 2)$

Go Online You can complete an Extra Example online.

Practice

Go Online You can complete your homework online.

Example 1

Determine whether each function is a linear function. Justify your answer.

1. $y = 3x$

2. $y = -2 + 5x$

3. $2x + y = 10$

4. $y = 4x^2$

Example 2

Determine whether each graph represents a *linear* or *nonlinear* function.

5. 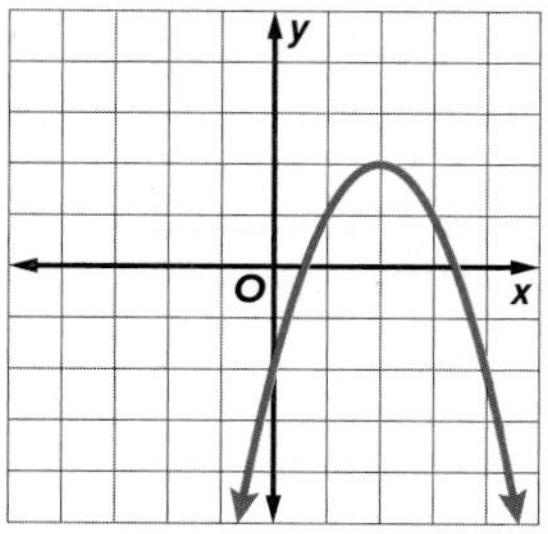

6. 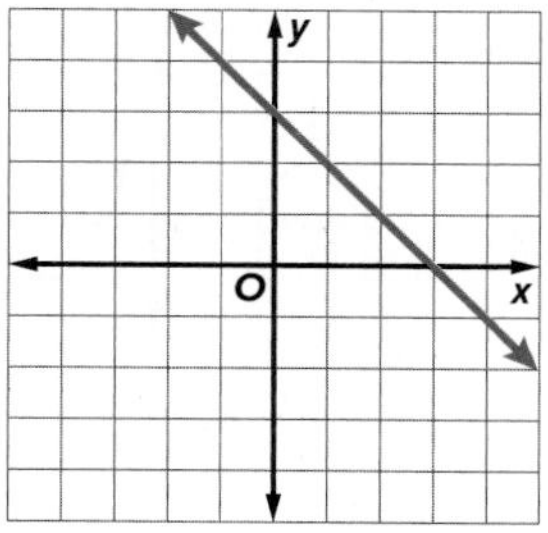

7. 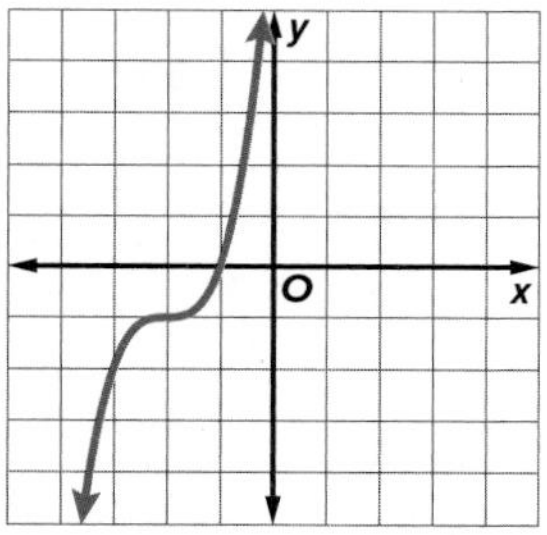

8. 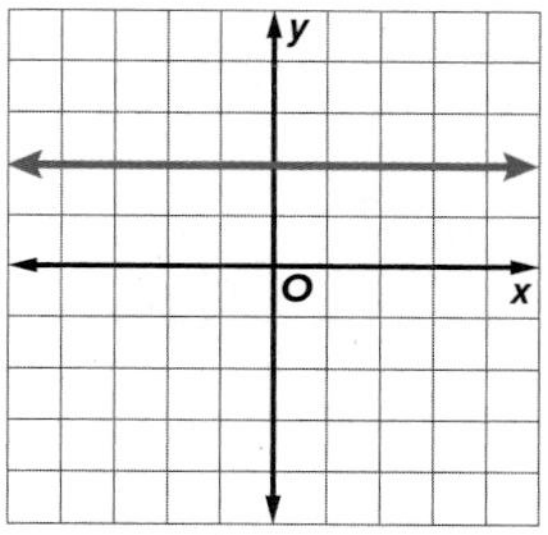

Example 3

9. MEASUREMENT The table shows a function modeling the number of inches and feet. Can the relationship be modeled by a *linear* or *nonlinear* function? Explain.

Inches	0	1	2	3	4
Feet	0	12	24	36	48

10. ASTRONOMY The table shows the velocity of *Cassini 2* space probe as it passes Saturn. Is the velocity modeled by a *linear* or *nonlinear* function? Explain.

Cassini 2 Velocity					
Time (s)	5	10	15	20	25
Velocity (mph)	50,000	60,000	70,000	60,000	50,000

Examples 4 and 5

Use the graph to estimate the *x*- and *y*-intercepts.

11.

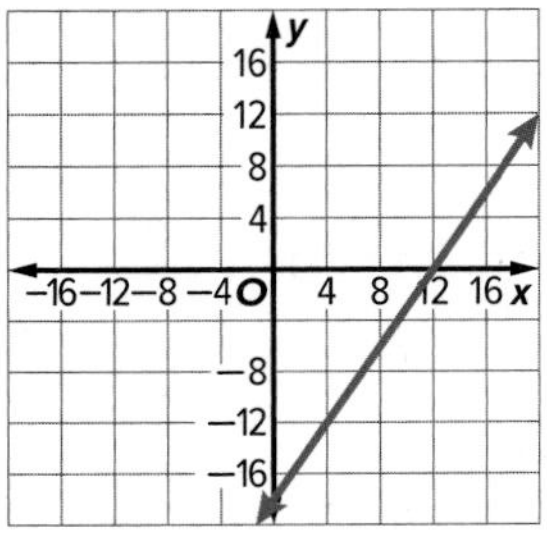

12.

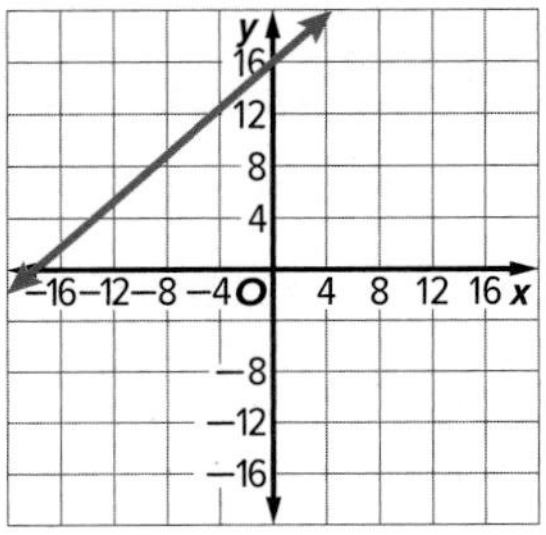

13.

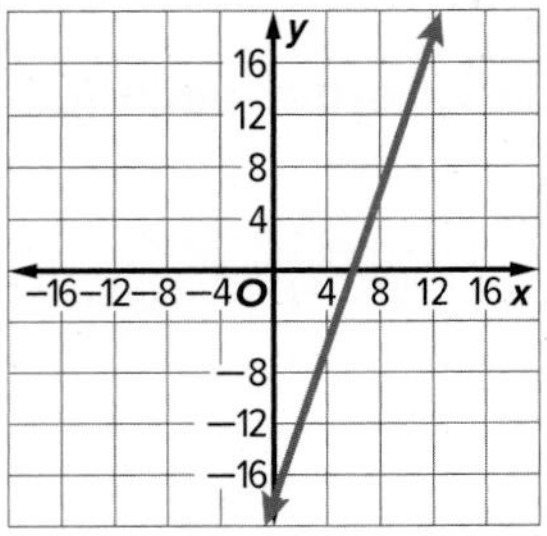

14.

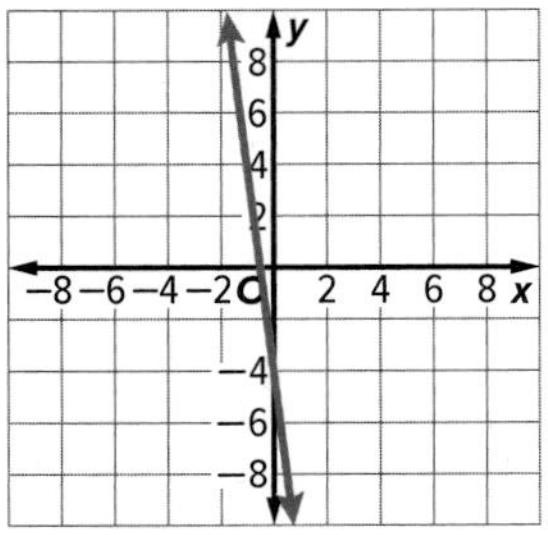

15.

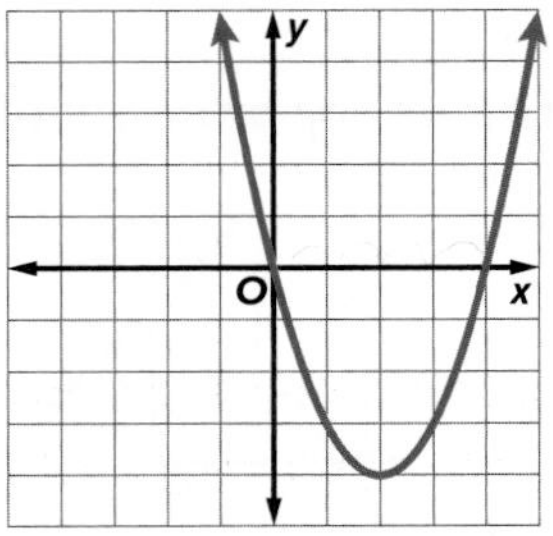

16.

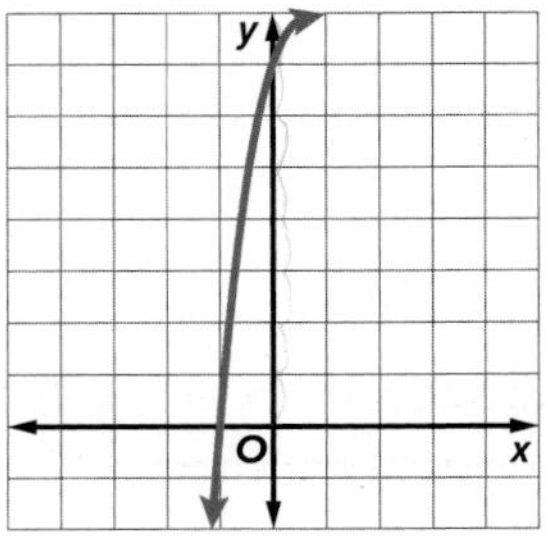

Example 6

17. MONEY At the beginning of the week, Aksa's parents deposited $20 into Aksa's lunch account. The amount of money Aksa had left after each day is shown in the table, where *x* is the number of days and *y* is the remaining balance.

Days	Account Balance
0	$20
1	$16
2	$12
3	$8
4	$4
5	$0

a. What are the *x*- and *y*-intercepts?

b. What do the *x*- and *y*-intercepts represent?

18. GOLF In golf, the first shot on every hole can be hit off a tee. The table shows the height *y* of the golf ball *x* seconds after it has been hit off the tee.

Time (s)	0	1	3	5	7
Height (in.)	3	20	36	28	0

a. What are the *x*- and *y*-intercepts?

b. What do the *x*- and *y*-intercepts represent in the context of the situation?

Example 7

Identify the type of symmetry for the graph of each function.

19.

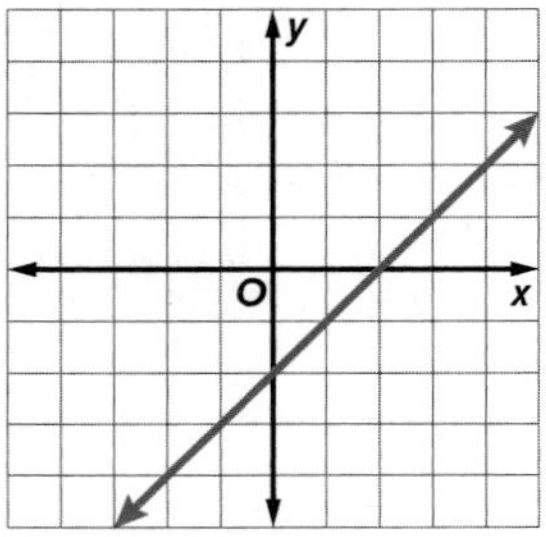

20.

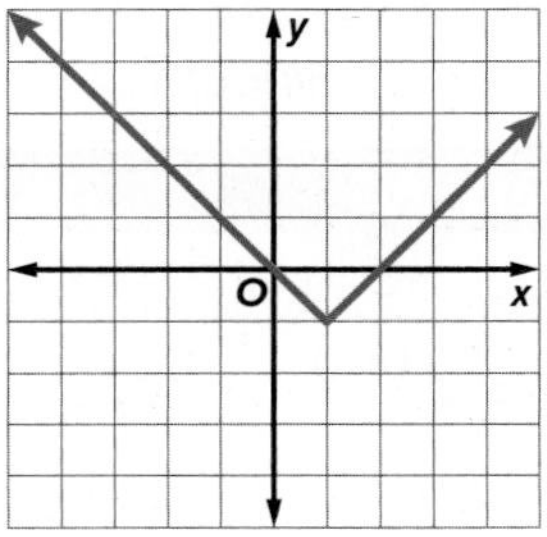

21.

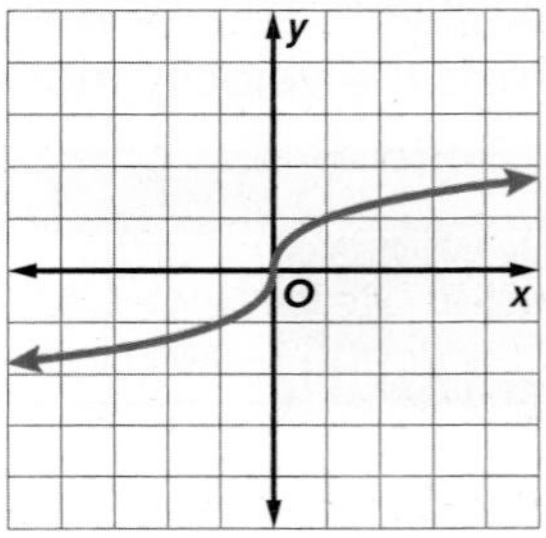

Example 8

Determine whether each function is *even*, *odd*, or *neither*. Confirm algebraically. If the function is odd or even, describe the symmetry.

22. $f(x) = 2x^3 - 8x$

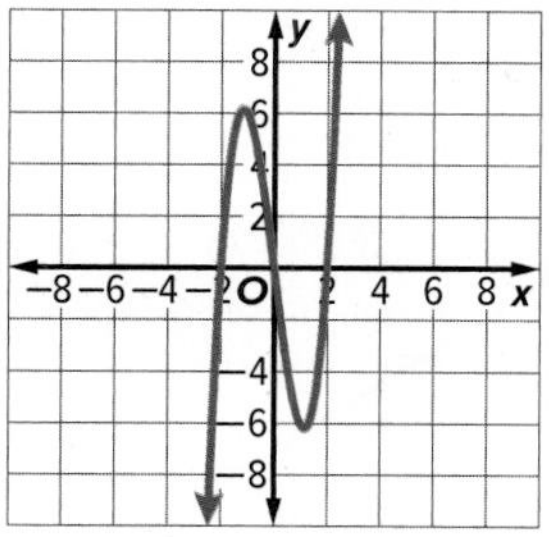

23. $f(x) = x^3 + x^2$

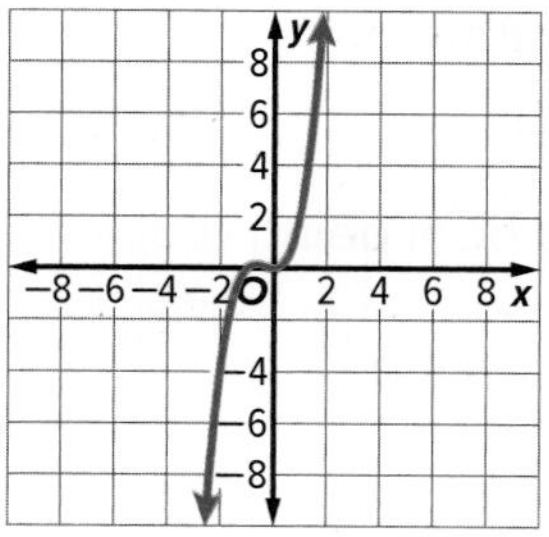

24. $f(x) = x^2 + 2$

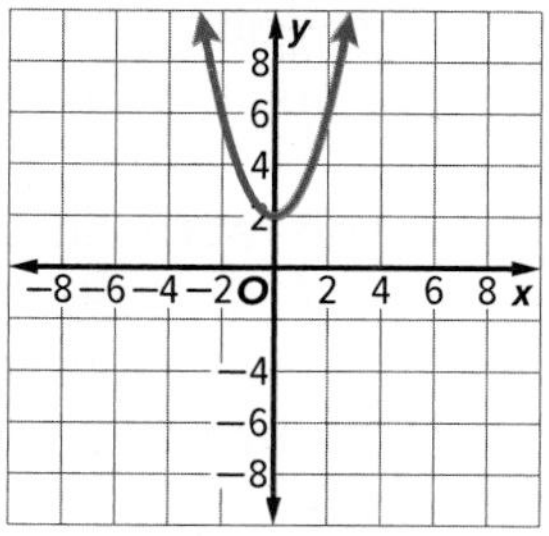

Mixed Exercises

Determine whether each equation represents a linear function. Justify your answer. Algebraically determine whether each equation is *even*, *odd*, or *neither*.

25. $-\frac{3}{x} + y = 15$

26. $x = y + 8$

27. $y = 8$

28. $y = \sqrt{x} + 3$

29. $y = 3x^2 - 1$

30. $y = 2x^3 + x + 1$

Determine whether each graph represents a *linear* or *nonlinear* function. Use the graph to estimate the *x*- and *y*-intercepts. Identify the type of symmetry in each graph.

31.

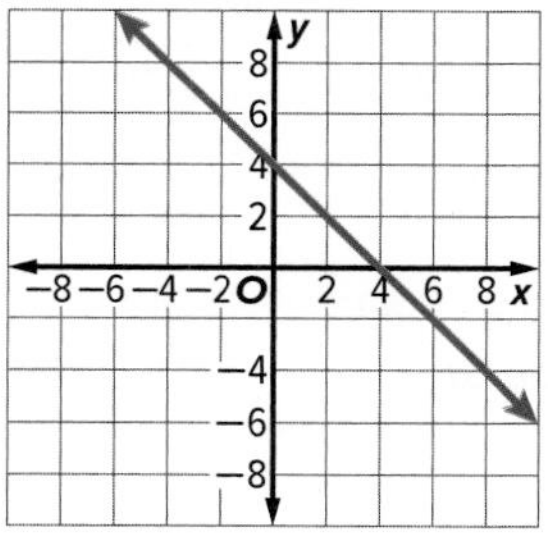

32.

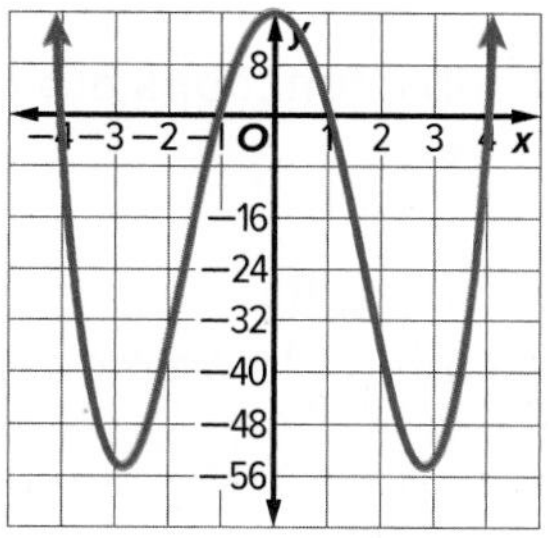

33.

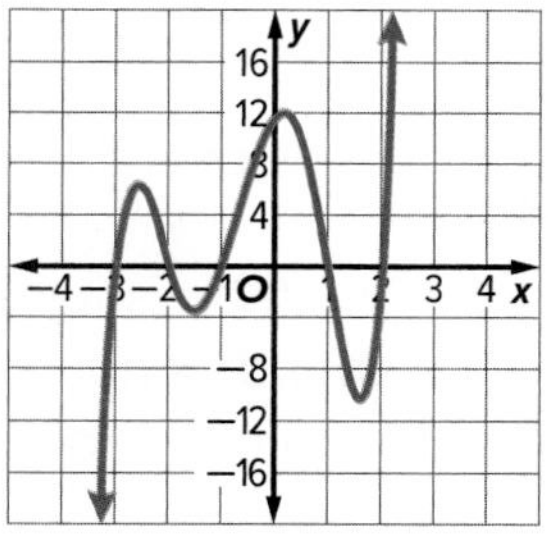

34. GAMES Pedro is creating an online racquetball game. In one play, the motion of the ball across the screen is partially modeled by the graph shown. State whether the graph has line symmetry or point symmetry, and identify any lines or points of symmetry.

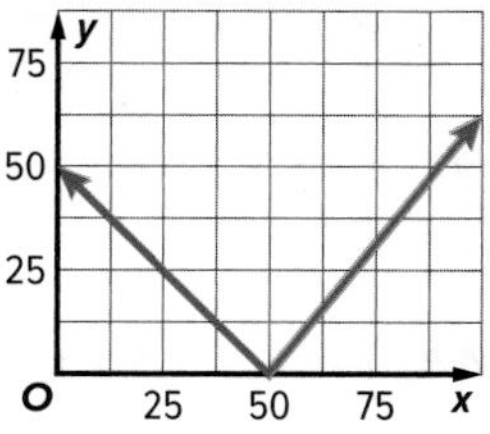

35. BASKETBALL Tiana tossed a basketball. The graph shows the height of the basketball as a function of time. If the graph of the function is extended to include the domain of all real numbers, would it have line or point symmetry? If so, identify the line or point of symmetry.

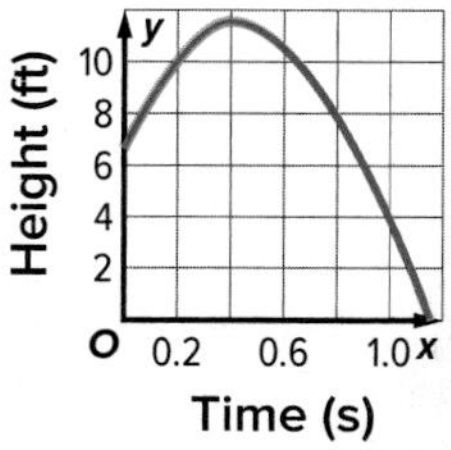

36. PROFIT Stefon charges people $25 to test the air quality in their homes. The device he uses to test air quality cost him $500. The function $y = 25x - 500$ describes Stefon's net profit, y, as a function of the number of clients he gets, x.

a. State whether the function is a linear function. Write *yes* or *no*. Explain.

b. What do the x- and y-intercepts of the function represent in terms of the situation?

37. **PLAYGROUNDS** A playground is shaped as shown. The total perimeter is 500 feet.

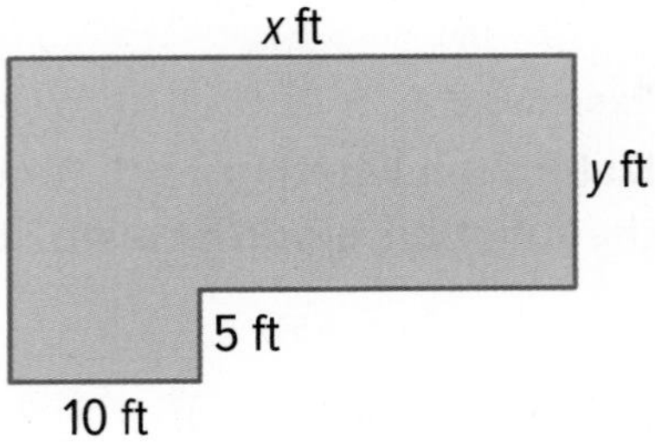

a. Write an equation that relates x and y.

b. Is the equation that relates x and y linear? Explain.

c. Graph the equation. State whether the graph has line symmetry or point symmetry.

38. **POOLS** The graph represents a 720-gallon pool being drained.

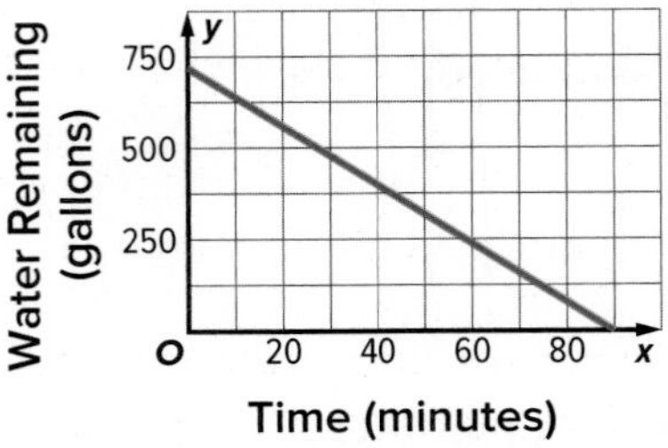

a. What are the x- and y-intercepts? What do the x- and y-intercepts represent?

b. Does the graph display line symmetry? Explain why or why not in terms of the situation.

39. **VOLUME** The function, $f(r) = \frac{4}{3}\pi r^3$ describes the relationship between the volume $f(r)$ and radius r of a sphere. Determine whether the function is *odd*, *even*, or *neither*. Explain your reasoning.

40. **USE A SOURCE** Research online to find an equation that models a car's braking distance in relation to its speed. Then identify and interpret the y-intercept of the equation.

41. **FIND THE ERROR** Javier claimed that all cubic functions are odd. Is he correct? If not, provide a counterexample.

42. **ANALYZE** The table shows a function modeling the number of gifts y Cornell can wrap if he spends x hours wrapping. Can the table be modeled by a *linear* or *nonlinear* function? Explain.

Hours	0	1	2	3	4
Gifts	0	14	28	42	56

43. **PERSEVERE** Determine whether an equation of the form $x = a$, where a is a constant, is *sometimes*, *always*, or *never* a linear function. Explain your reasoning.

44. **WHICH ONE DOESN'T BELONG?** Of the four functions shown, identify the one that does not belong. Explain your reasoning.

$y = 2x + 3$

x	y
0	4
1	2
2	0
3	−2

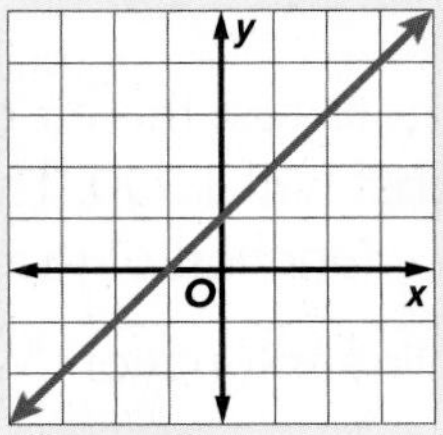

$y = 2xy$

Extrema and End Behavior

Learn Extrema of Functions

Graphs of functions can have high and low points where they reach a maximum or minimum value. The maximum and minimum values of a function are called **extrema.**

The **maximum** is at the highest point on the graph of a function. The **minimum** is at the lowest point on the graph of a function.

A **relative maximum** is located at a point on the graph of a function where no other nearby points have a greater *y*-coordinate. A **relative minimum** is located at a point on the graph of a function where no other nearby points have a lesser *y*-coordinate.

Today's Goals

- Identify extrema of functions.
- Identify end behavior of graphs.

Today's Vocabulary

extrema
maximum
minimum
relative maximum
relative minimum
end behavior

Example 1 Find Extrema from Graphs

Identify and estimate the *x*- and *y*-values of the extrema. Round to the nearest tenth if necessary.

***f*(*x*)**

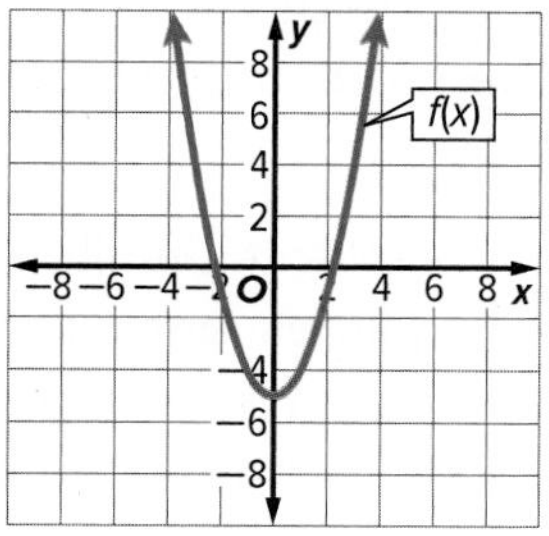

***g*(*x*)**

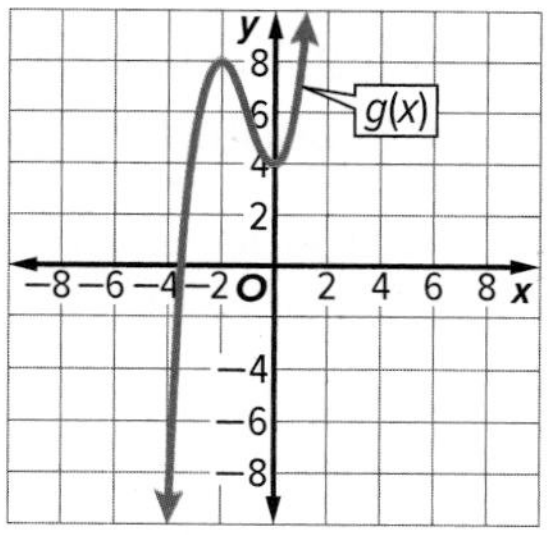

***f*(*x*):** The function $f(x)$ is decreasing as it approaches $x = 0$ from the left and increasing as it moves away from $x = 0$. Further, $(0, -5)$ is the lowest point on the graph, so $(0, -5)$ is a minimum.

***g*(*x*):** The function $g(x)$ is increasing as it approaches $x = -2$ from the left and decreasing as it moves away from $x = -2$. Further, there are no greater *y*-coordinates surrounding $(-2, 8)$. However, $(-2, 8)$ is not the highest point on the graph, so $(-2, 8)$ is a relative maximum.

The function $g(x)$ is decreasing as it approaches $x = 0$ from the left and increasing as it moves away from $x = 0$. Further, there are no lesser *y*-coordinates surrounding $(0, 4)$. However, $(0, 4)$ is not the lowest point on the graph, so $(0, 4)$ is a relative minimum.

Go Online You can complete an Extra Example online.

Watch Out!

No Extrema Some functions, like $f(x) = x^3$, have no extrema.

Study Tip

Reading in Math In this context, *extrema* is the plural form of *extreme point*. The plural of *maximum* and *minimum* are *maxima* and *minima*, respectively.

Think About It!

Why are the extrema identified on the graph of $g(x)$ relative maxima and minima instead of maxima and minima?

Example 2 Find and Interpret Extrema

SOCIAL MEDIA **Use the table and graph to estimate the extrema of the function that relates the number hours since midnight *x* to the number of posts being uploaded *y*. Describe the meaning of the extrema in the context of the situation.**

x	y
0	2.8
4	1.8
8	3.1
12	11.5
14	9.1
16	10.2
20	5.8
24	2.8

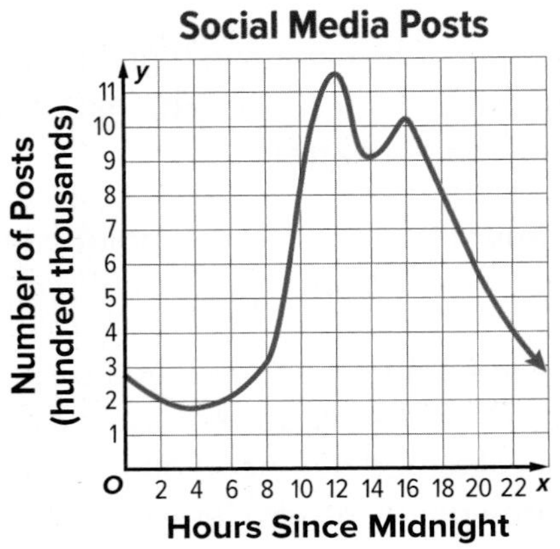

maxima The number of posts sent 12 hours after midnight is greater than the number of posts made at any other time during the day. The highest point on the graph occurs when $x = 12$. Therefore, the maximum number of posts sent is about 1,150,000 at noon.

minima The number of posts sent 4 hours after midnight is less than the number of posts made at any other time during the day. The lowest point on the graph occurs when $x = 4$. Therefore, the minimum number of posts sent is about 180,000 at 4:00 A.M.

relative maxima The number of posts sent 16 hours after midnight is greater than the number of posts during surrounding times, but is not the greatest number sent during the day. The graph has a relative peak when $x = 16$. Therefore, there is a relative peak in the number of posts sent, or relative maximum, at 4:00 P.M. of about 1,020,000 posts.

relative minima The number of posts sent 14 hours after midnight is less than the number of posts during surrounding times, but is not the least number sent during the day. The graph dips when $x = 14$. Therefore, there is a relative low in the number of posts sent, or relative minimum, at 2:00 P.M. of about 910,000 posts.

Explore End Behavior of Linear and Quadratic Functions

Online Activity Use graphing technology to complete the Explore.

INQUIRY Given the behavior of a linear or quadratic function as *x* increases towards infinity, how can you find the behavior as *x* decreases toward negative infinity or vice versa?

Go Online
You can complete an Extra Example online.

Learn End Behavior of Graphs of Functions

End behavior is the behavior of a graph as x approaches positive or negative infinity. As you move right along the graph, the values of x are increasing toward infinity. This is denoted as $x \to \infty$. At the left end, the values of x are decreasing toward negative infinity, denoted as $x \to -\infty$. When a function $f(x)$ increases without bound, it is denoted as $f(x) \to \infty$. When a function $f(x)$ decreases without bound, it is denoted as $f(x) \to -\infty$.

Example 3 End Behavior of Linear Functions

Describe the end behavior of each linear function.

a. $\boldsymbol{f(x)}$

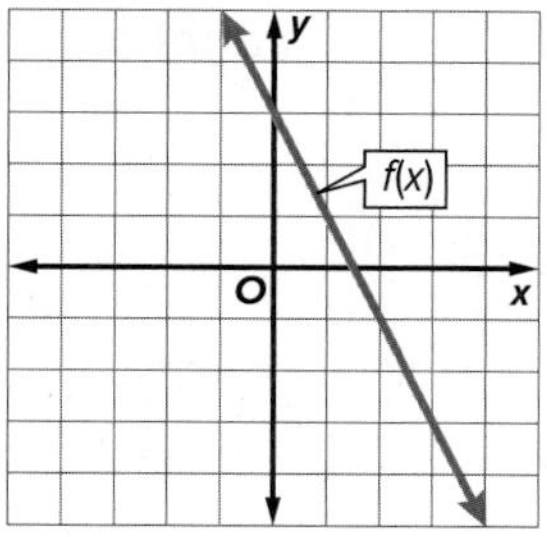

As x decreases, $f(x)$ increases, and as x increases $f(x)$ decreases. Thus, as $x \to -\infty$, $f(x) \to \infty$ and as $x \to \infty$, $f(x) \to -\infty$.

b. $\boldsymbol{g(x)}$

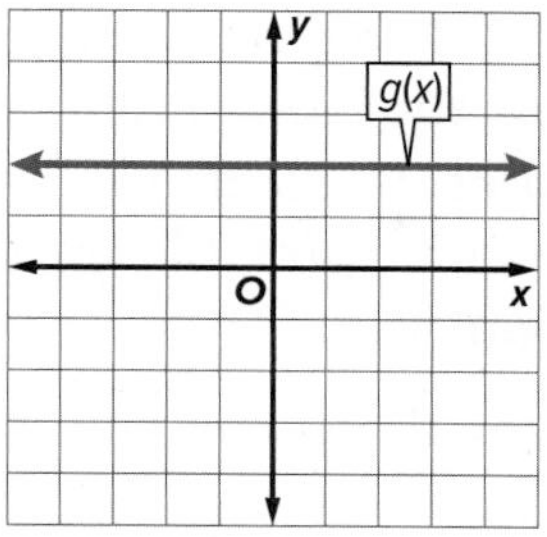

As x decreases or increases, $g(x) = 2$. Thus, as $x \to -\infty$, $g(x) = 2$, and as $x \to \infty$, $g(x) = 2$.

Check

Use the graph to describe the end behavior of the function.

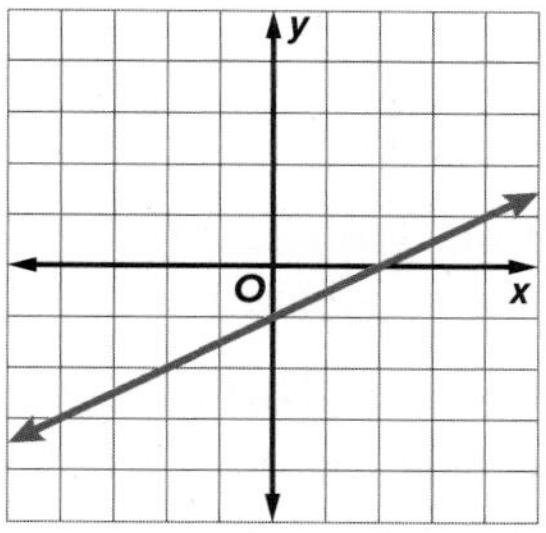

Example 4 End Behavior of Nonlinear Functions

Describe the end behavior of each nonlinear function.

a. $\boldsymbol{f(x)}$

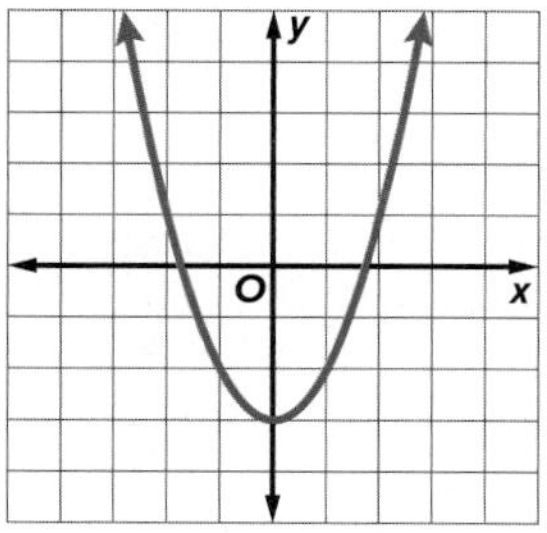

As you move left or right on the graph, $f(x)$ increases. Thus as $x \to -\infty$, $f(x) \to \infty$, and as $x \to \infty$, $f(x) \to \infty$.

b. $\boldsymbol{g(x)}$

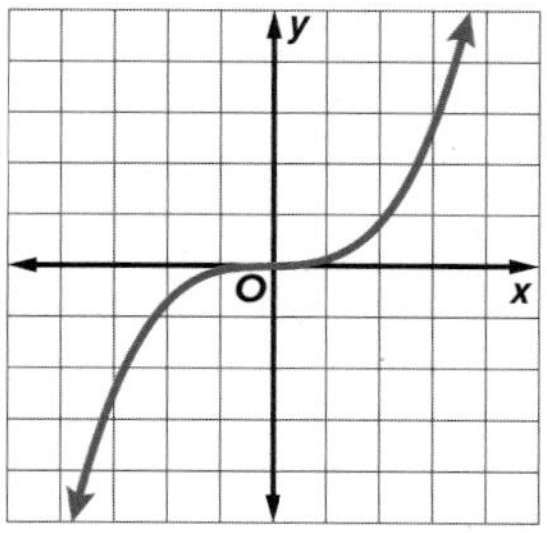

As $x \to -\infty$, $g(x) \to -\infty$, and as $x \to \infty$, $g(x) \to \infty$.

Go Online You can complete an Extra Example online.

Think About It!

For $f(x) = a$, where a is a real number, describe the end behavior of $f(x)$ as $x \to \infty$ and as $x \to -\infty$.

Talk About It!

In part **a**, the function's end behavior as $x \to -\infty$ is the opposite of the end behavior as $x \to \infty$. Do you think this is true for all linear functions where $m \neq 0$? Explain your reasoning.

Math History Minute

Júlio César de Mello e Souza (1895–1974) was a Brazilian mathematician who is known for his books on recreational mathematics. His most famous book, *The Man Who Counted*, includes problems, puzzles, and curiosities about math. The State Legislature of Rio de Janeiro declared that his birthday, May 6, be Mathematician's Day.

> **Think About It!**
> If the graph of a function is symmetric about a vertical line, what do you think is true about the end behavior of $f(x)$ as $x \to -\infty$ and as $x \to \infty$?

Check

Use the graph to describe the end behavior of the function.

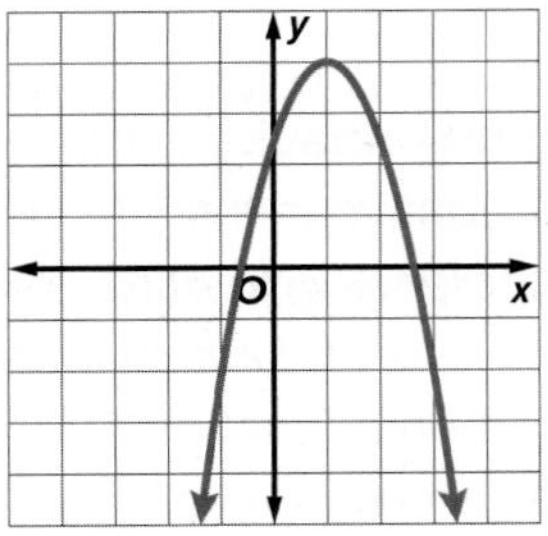

Example 5 Determine and Interpret End Behavior

DRONES **The graph shows the altitude of a drone above the ground $f(x)$ after x minutes. Describe the end behavior of $f(x)$ and interpret it in the context of the situation.**

Flight of a Drone

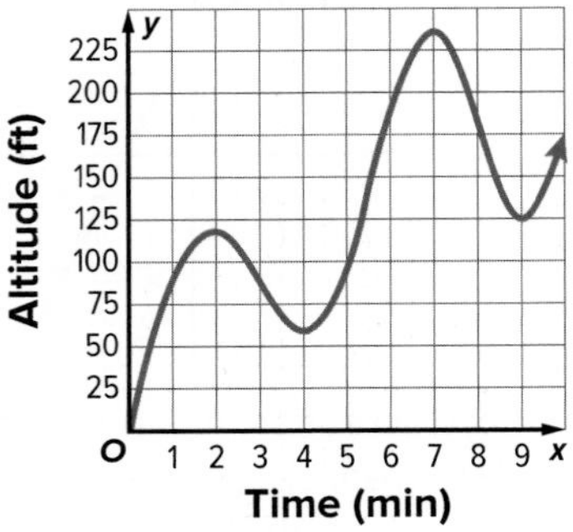

Since the drone cannot travel for a negative amount of time, the function is not defined for $x < 0$. So, there is no end behavior as $x \to -\infty$.

As $x \to \infty$, $f(x) \to \infty$. The drone is expected to continue to fly higher.

> **Study Tip**
> **Assumptions** Assuming that the drone can continue to fly for an infinite amount of time and to an infinite altitude lets us analyze the end behavior as $x \to \infty$. While there are maximum legal altitudes that a drone can fly as well as limited battery life, assuming that the time and altitude will continue to increase allows us to describe the end behavior.

Check

RIDESHARING Mika and her friends are using a ride-sharing service to take them to a concert. The function models the cost of the ride $f(x)$ after x miles. Describe the end behavior of $f(x)$ and interpret it in the context of the situation.

Ridesharing

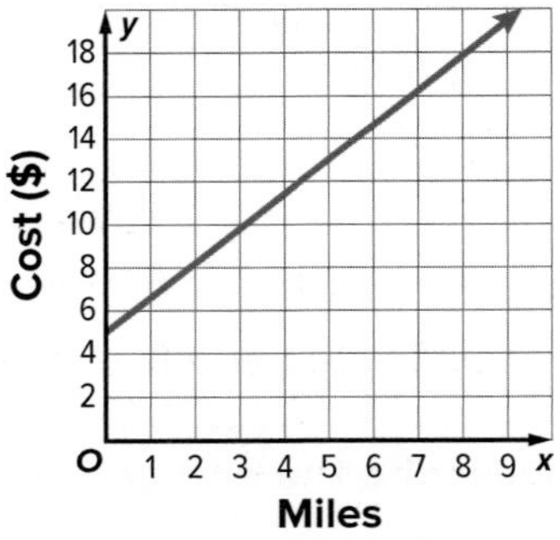

Part A

What is the end behavior of the function?

A. as $x \to -\infty$, $f(x) \to -\infty$; as $x \to \infty$, $f(x) \to -\infty$

B. as $x \to -\infty$, $f(x) \to \infty$; as $x \to \infty$, $f(x) \to \infty$

C. as $x \to \infty$, $f(x) \to -\infty$; $f(x)$ is not defined for $x < 0$

D. as $x \to \infty$, $f(x) \to \infty$; $f(x)$ is not defined for $x < 0$

Part B

What does the end behavior represent in the context of the situation?

Go Online You can complete an Extra Example online.

> **Go Online** to practice what you've learned about analyzing graphs in the Put It All Together over Lessons 1-1 through 1-3.

Practice

Go Online You can complete your homework online.

Examples 1 and 2

Identify and estimate the *x*- and *y*-values of the extrema. Round to the nearest tenth if necessary.

1.

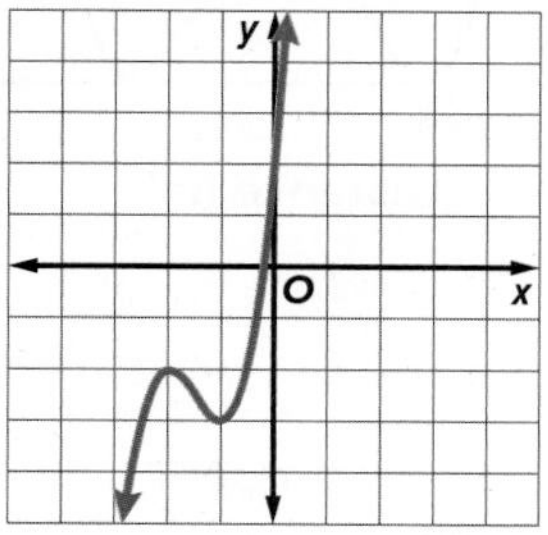

2.

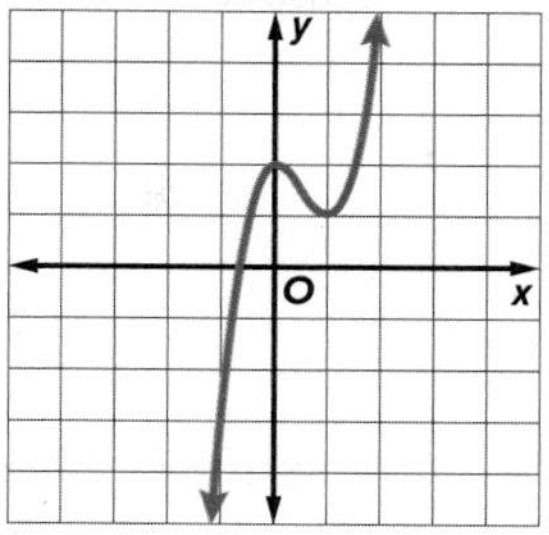

3.

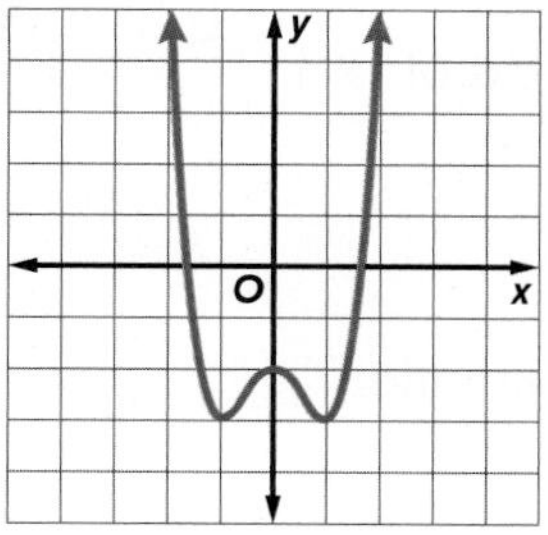

4.

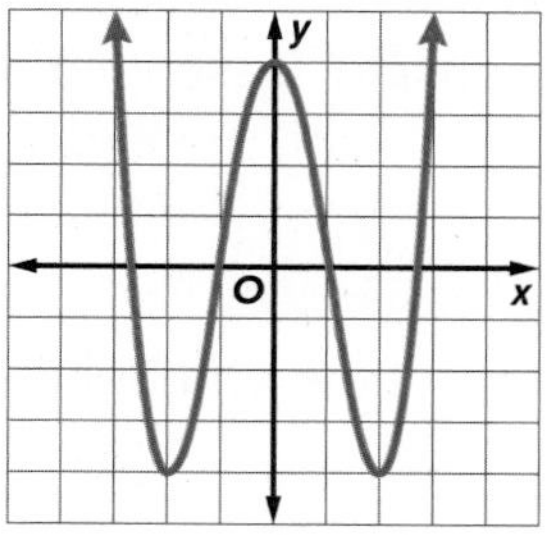

5. LANDSCAPES Jalen uses a graph of a function to model the shape of two hills in the background of a videogame that he is writing. Estimate the *x*-coordinates at which the relative maxima and relative minima occur. Describe the meaning of the extrema in the context of the situation.

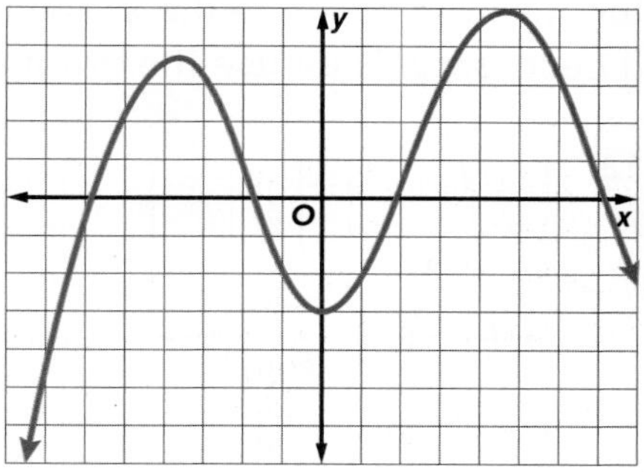

Examples 3-5

Describe the end behavior of each function.

6.

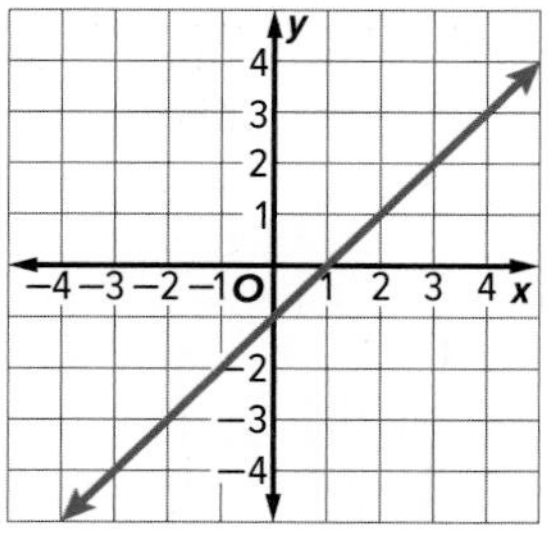

7.

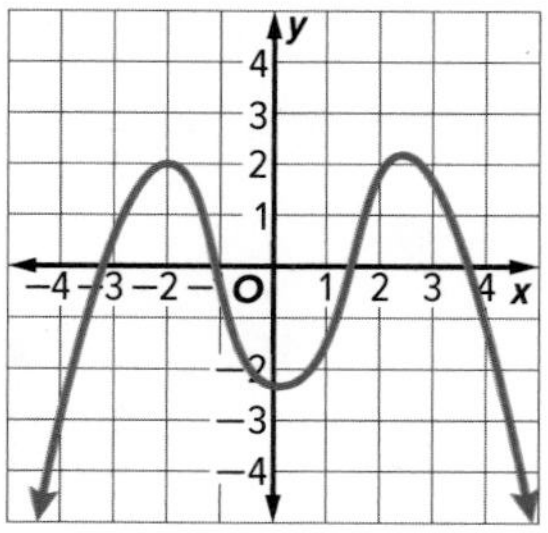

8.

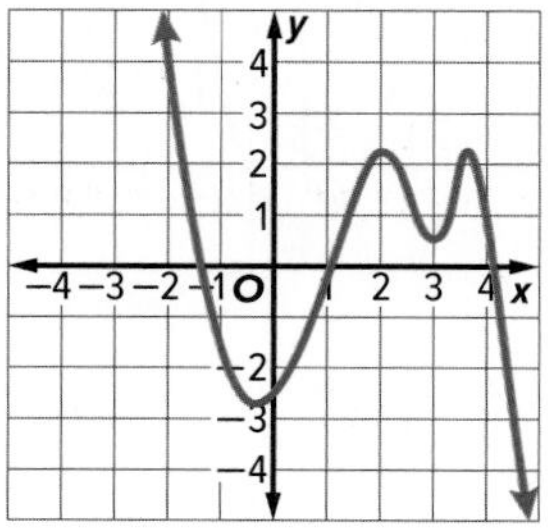

9.

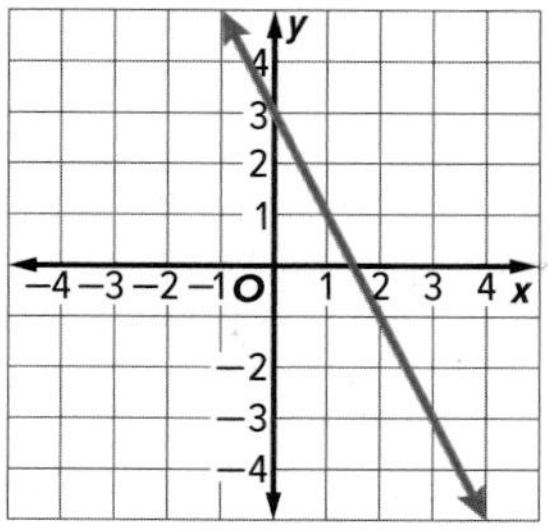

10. ROLLER COASTER The graph shows the height of a roller coaster in terms of its distance away from the starting point. Describe and interpret the end behavior in the context of the situation.

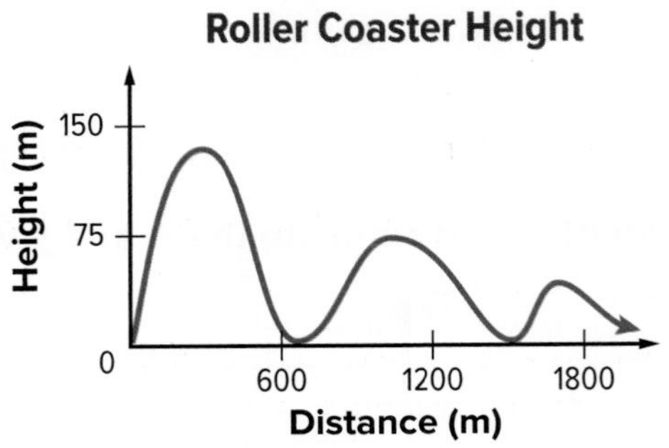

Mixed Exercises

11. MODEL The height of a fish t seconds after it is thrown to a dolphin from a 64-foot-tall platform can be modeled by the equation $h(t) = -16t^2 + 48t + 64$, where $h(t)$ is the height of the fish in feet. The graph of the function is shown.

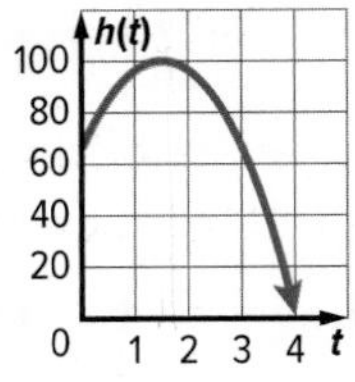

a. Estimate the t-coordinate at which the height of the fish changes from increasing to decreasing. Describe the meaning in terms of the context of the situation.

b. Describe and interpret the end behavior of $h(t)$ in the context of the situation.

Identify and estimate the x- and y-values of the extrema. Round to the nearest tenth if necessary. Then use the graphs to describe the end behavior of each function.

12.

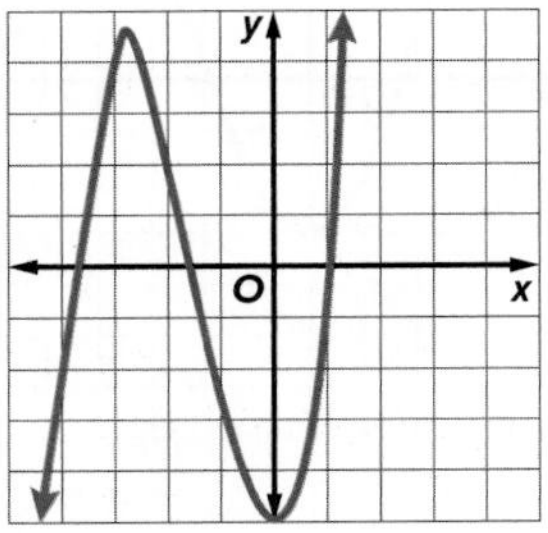

13.

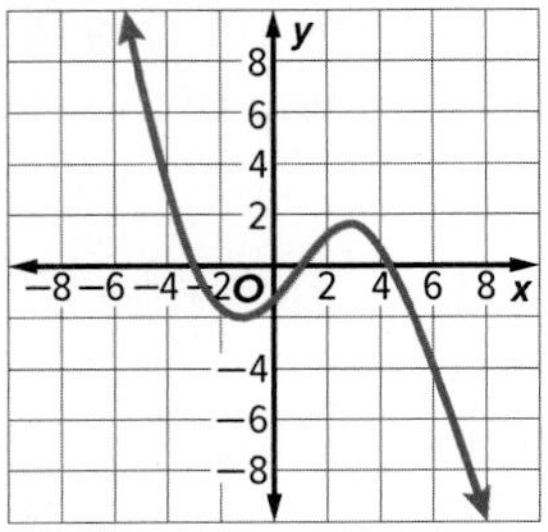

14.

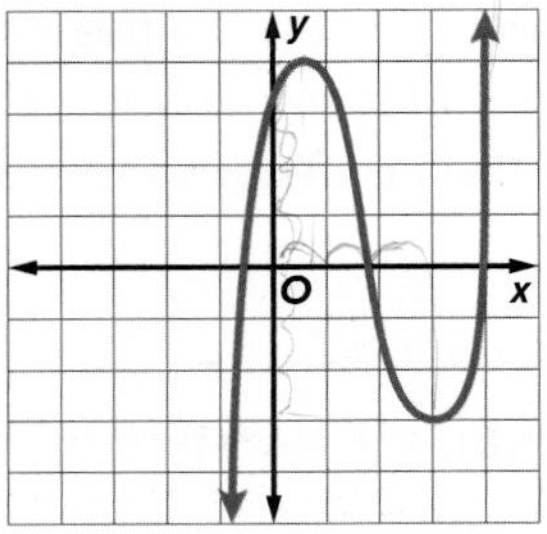

15.

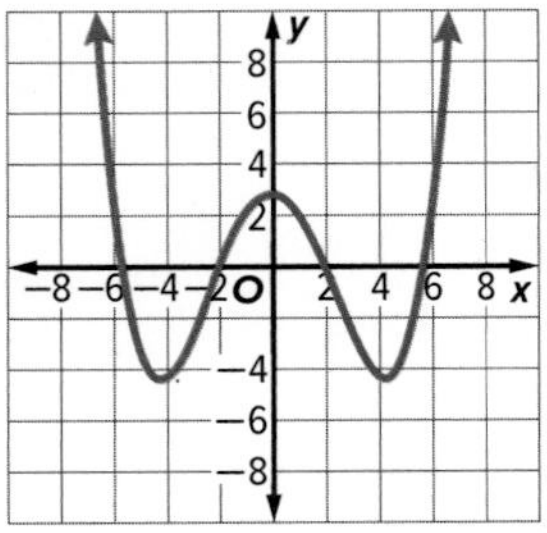

16.

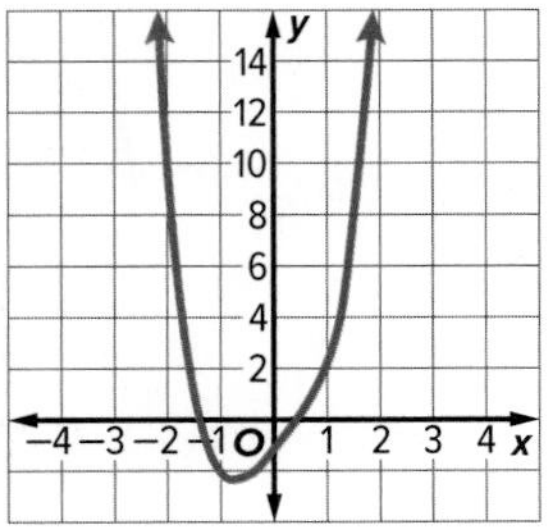

17.

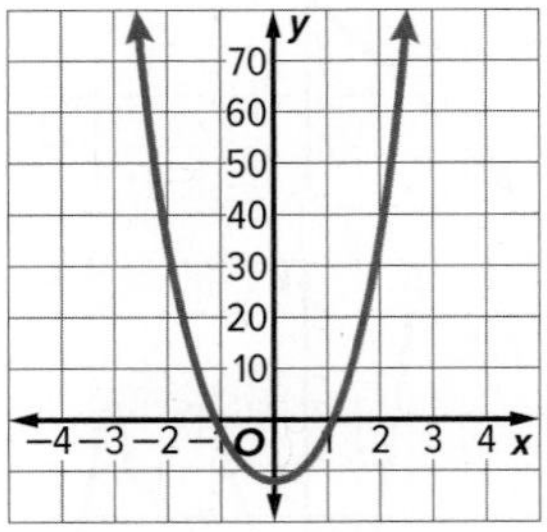

18. BUBBLES The volume of a soap bubble can be estimated by the formula $V = 4\pi r^2$, where r is its radius. The graph shows the function of the bubble's volume. Describe the end behavior of the graph.

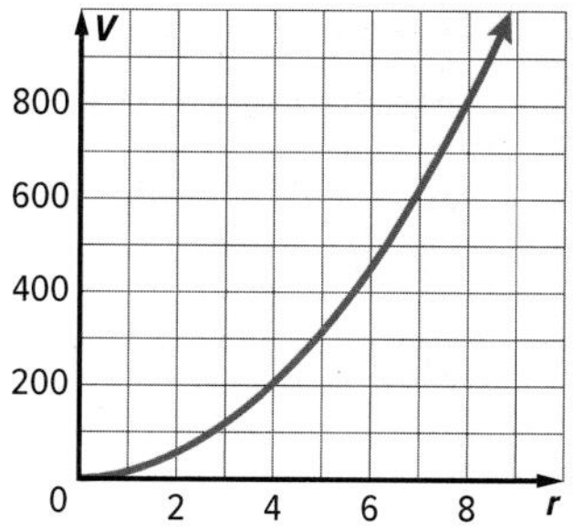

19. SCIENCE The table shows the density of water at its saturation pressure for various temperatures. Interpret the end behavior of the graph of the function as temperature increases.

Temperature (°C)	0	50	100	150	200	250	300	350
Density (g/cm³)	1.000	0.988	0.958	0.917	0.865	0.799	0.713	0.573

Identify and estimate the *x*- and *y*-values of the extrema. Round to the nearest tenth if necessary. Then describe the end behavior of each function.

20.

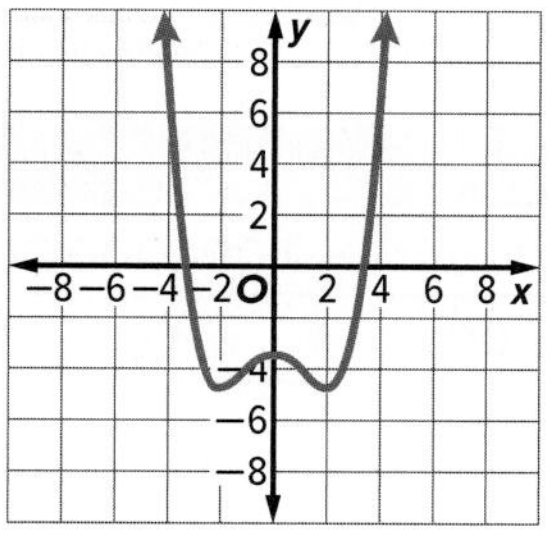

21.

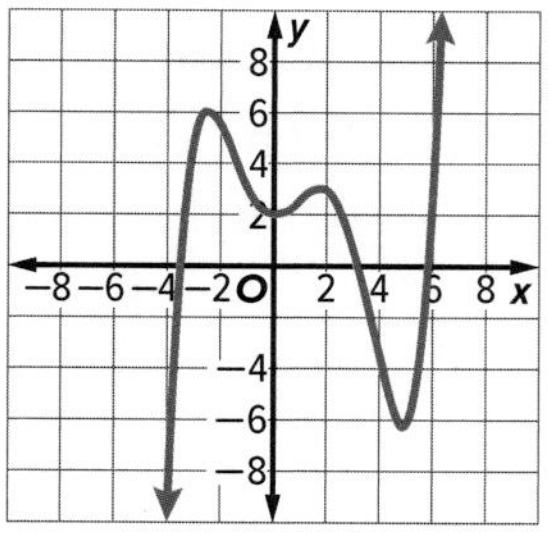

USE ESTIMATION **Use a graphing calculator to estimate the *x*-coordinates at which any extrema occur for each function. Round to the nearest hundredth.**

22. $f(x) = x^3 + 3x^2 - 6x - 6$

23. $f(x) = -2x^3 + 8$

24. $f(x) = -2x^4 + 5x^3 - 4x^2 + 3x - 7$

25. $f(x) = x^5 - 4x^3 + 3x^2 - 8x - 6$

26. CONSTRUCT ARGUMENTS Sheena says that in the graph of $f(x)$ shown below, the graph has relative maxima at B and G, and a relative minimum at A. Is she correct? Explain.

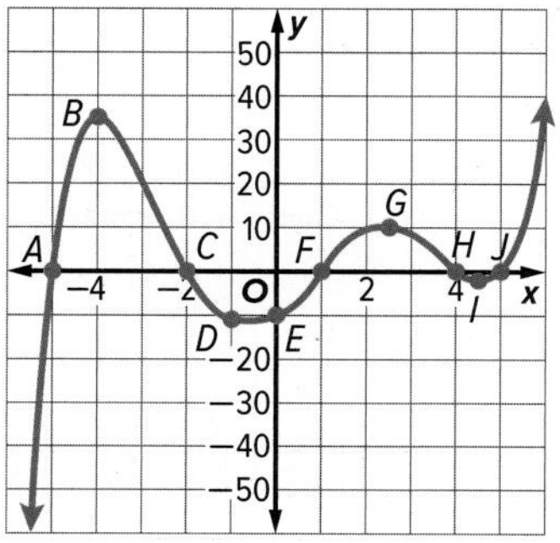

27. CHEMISTRY Dynamic pressure is generated by the velocity of a moving fluid and is given by $q(v) = \frac{1}{2}pv^2$, where p is the density of the fluid and v is its velocity. Water has a density of 1 g/cm³. What happens to the dynamic pressure of water when the velocity continuously increases?

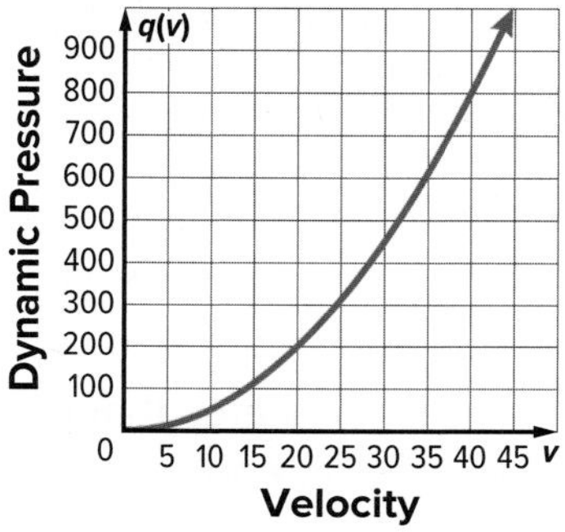

28. ENGINEERING Several engineering students built a catapult for a class project. They tested the catapult by launching a watermelon and modeled the height h of the watermelon in feet over time t in seconds.

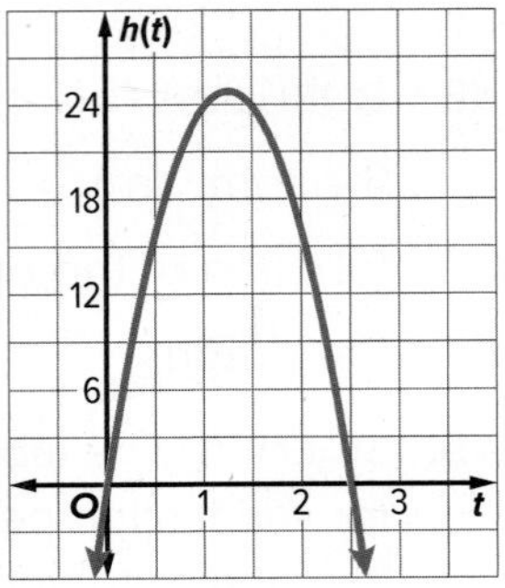

a. Considering the context of the problem, what is an appropriate domain for $h(t)$? Explain your reasoning.

b. Use the graph of $h(t)$ to find the maximum height of the watermelon. When does the watermelon reach the maximum height? Explain your reasoning.

29. DRILLING The volume of a drill bit can be estimated by the formula for a cone, $V = \frac{1}{3}\pi hr^2$, where h is the height of the bit and r is its radius. Substituting $\frac{\sqrt{3}}{3}r$ for h, the volume of the drill bit is estimated as $\frac{\sqrt{3}}{9}\pi r^3$. The graph shows the function of drill bit volume. Describe the end behavior.

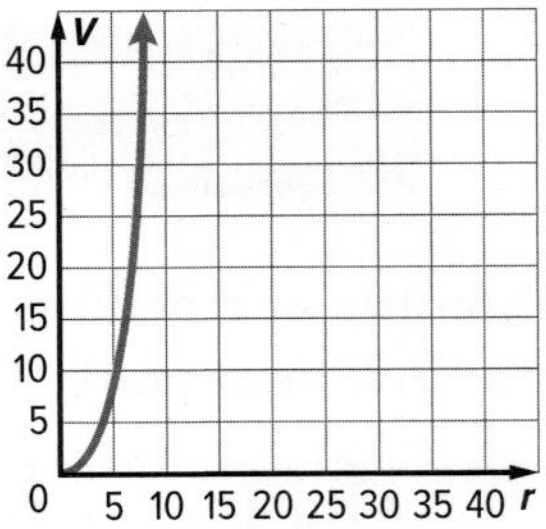

30. The table shows the values of a function. Use the table to describe the end behavior of the function.

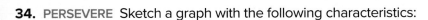

x	y
−1000	−1,001,000,000
−100	−1,010,000
−10	−1100
−1	−2
1	0
10	900
100	990,000
1000	999,000,000

31. WRITE Describe what the end behavior of a graph is and how it is determined.

32. CREATE Sketch a graph of a linear function and a nonlinear function with the following end behavior: as $x \to -\infty$, $f(x) \to \infty$ and as $x \to \infty$, $f(x) \to -\infty$.

33. ANALYZE A catalyst is used to increase the rate of a chemical reaction. The reaction rate, or the speed at which the reaction is occurring, is given by $R(x) = \frac{0.5x}{x + 10}$, where x is the concentration of the catalyst solution in milligrams of solute per liter. What does the end behavior of the graph mean in the context of this experiment?

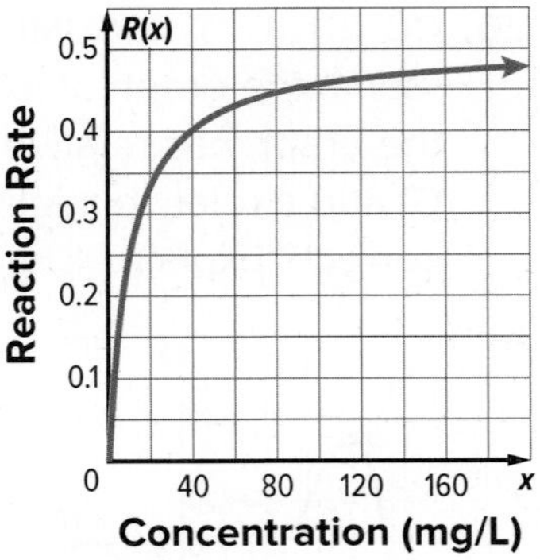

34. PERSEVERE Sketch a graph with the following characteristics:

- 2 relative maxima
- 2 relative minima
- end behavior: $x \to \infty$, $f(x) \to \infty$ and as $x \to -\infty$, $f(x) \to -\infty$

35. FIND THE ERROR Joshua states that the end behavior of the graph is: as $x \to -\infty$, $f(x) \to -\infty$ and as $x \to \infty$, $f(x) \to \infty$. What error did he make?

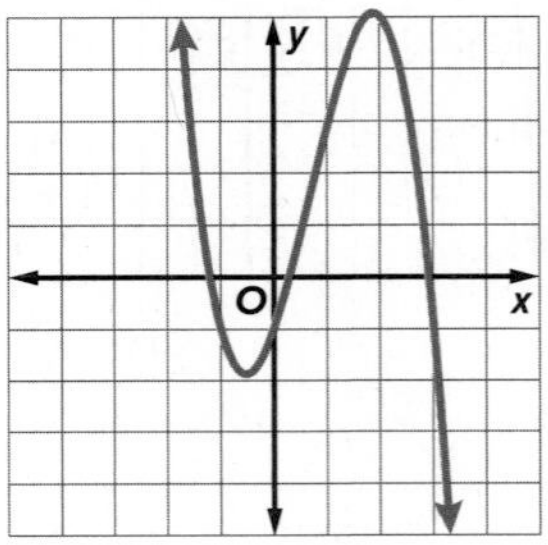

Sketching Graphs and Comparing Functions

Explore Using Technology to Examine Key Features of Graphs

Online Activity Use graphing technology to complete the Explore.

INQUIRY What can key features of a function tell you about its graph?

Today's Goal
- Sketch graphs of functions by using key features.

Learn Sketching Graphs of Functions

You can use key features of a function to sketch its graph.

Key Concept • Using Key Features

Key Feature	What it tells you about the graph of $f(x)$
Domain	the values of x for which $f(x)$ is defined
Range	the values that $f(x)$ take as x varies
Intercepts	where the graph crosses the x- or y-axis
Symmetry	where one side of the graph is a reflection or rotation of the other side
End Behavior	what the graph is doing at the right and left sides as x approaches infinity or negative infinity
Extrema	high or low points where the graph changes from increasing to decreasing or vice versa
Increasing/ Decreasing	where the graph is going up or down as x increases
Positive/Negative	where the graph is above or below the x-axis

Study Tip

Scales and Axes Before you sketch a function, consider the scales or axes that best fit the situation. You want to capture as much information as possible, so you want the scales to be big enough to easily see the extrema and x- and y intercepts, but not so big that you cannot determine the values.

Talk About It!

Given the y-intercept and for what values of x the function is positive, what other information do you need to sketch a linear function? Explain your reasoning.

Example 1 Sketch a Linear Function

Use the key features of the function to sketch its graph.

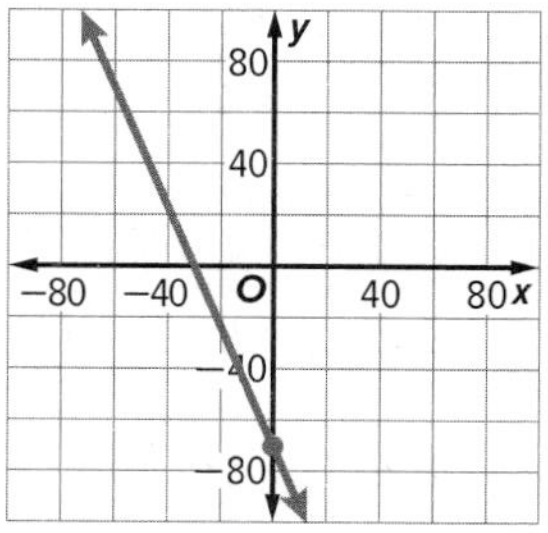

y-intercept: (0, −70)

Linearity: linear

Positive: for values of x such that $x < -30$

Decreasing: for all values of x

End Behavior: As $x \to \infty$, $f(x) \to -\infty$.
As $x \to -\infty$, $f(x) \to \infty$.

Go Online You can complete an Extra Example online.

Example 2 Sketch a Nonlinear Function

Use the key features of the function to sketch its graph.

***y*-intercept:** (0, 3)

Linearity: nonlinear

Continuity: continuous

Positive: for all values of x

Decreasing: for all values of x such that $x < 0$

Extrema: minimum at (0, 3)

End Behavior: As $x \to \infty$, $f(x) \to \infty$.
As $x \to -\infty$, $f(x) \to \infty$.

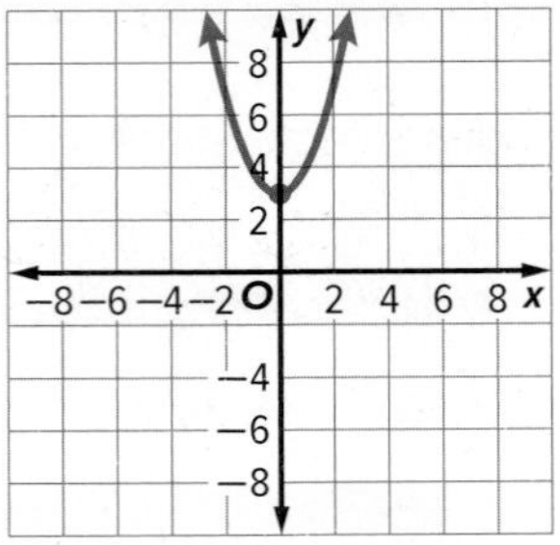

Example 3 Sketch a Real-World Function

TEST DRIVE **Hae is test driving a car she is thinking of buying. She decides to accelerate to 60 miles per hour and then decelerate to a stop to test its acceleration and brakes. It takes her 15 seconds to reach her maximum speed and 15 additional seconds to come to a stop. Use the key features to sketch a graph that shows the speed *y* as a function of time *x*.**

***y*-intercept:** Hae starts her test drive at a speed of 0 miles per hour.

Linear or Nonlinear: The function that models the situation is nonlinear.

Extrema: Hae's maximum speed is 60 miles per hour, which she reaches 15 seconds into her test drive.

Increasing: Hae increases the speed at a uniform rate for the first 15 seconds.

Decreasing: Hae decreases the speed at a uniform rate for the next 15 seconds until she reaches a stop.

End Behavior: Because Hae starts at 0 miles per hour and ends at 0 miles per hour, there is no end behavior.

Before sketching, consider the constraints of the situation. Hae cannot drive a negative speed or for a negative amount of time. Therefore, the graph only exists for positive x- and y-values.

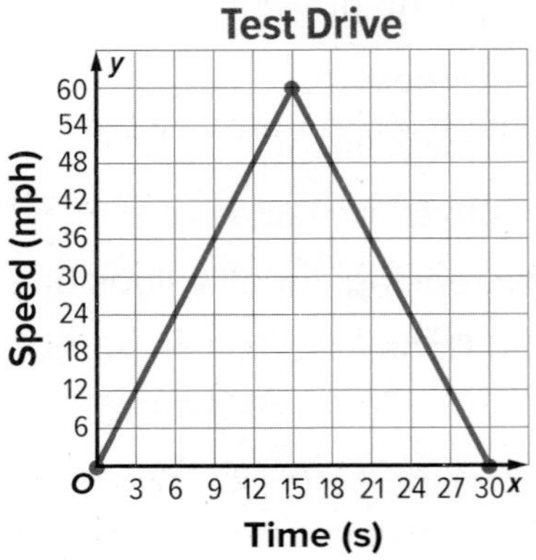

Go Online You can complete an Extra Example online.

Study Tip

Assumptions When sketching the function using the given key features, assumptions must be made. As in this example, the same key features could describe many different graphs. The key features could also be represented by a parabola, a curve that is narrower or wider, or an absolute value function.

Think About It!

Explain why the end behavior is not defined in the context of this situation.

Think About It!

Based on the graph, the speed of the car at 10 seconds is 40 miles per hour. Is it appropriate to assume that the car is traveling that exact speed at a specific time? Explain.

Example 4 Compare Properties of Linear Functions

Use the table and graph to compare the two functions.

x	$f(x)$
−6	−3
−3	−2
0	−1
3	0
6	1

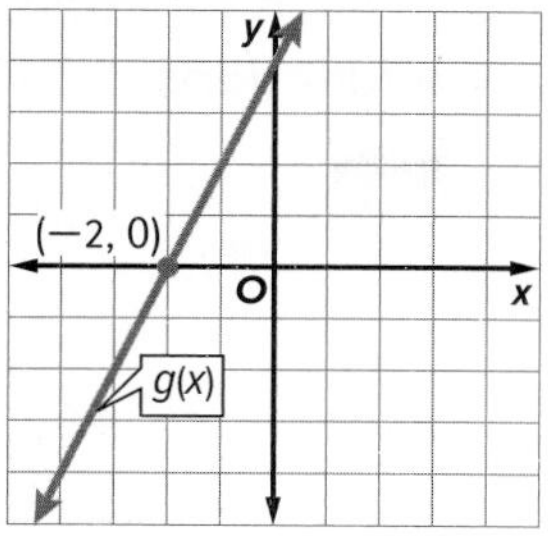

x-intercept of $f(x)$: 3
x-intercept of $g(x)$: −2.

So, $f(x)$ intersects the x-axis at a point farther to the right than $g(x)$.

x	$f(x)$
−6	−3
−3	−2
0	−1
3	0
6	1

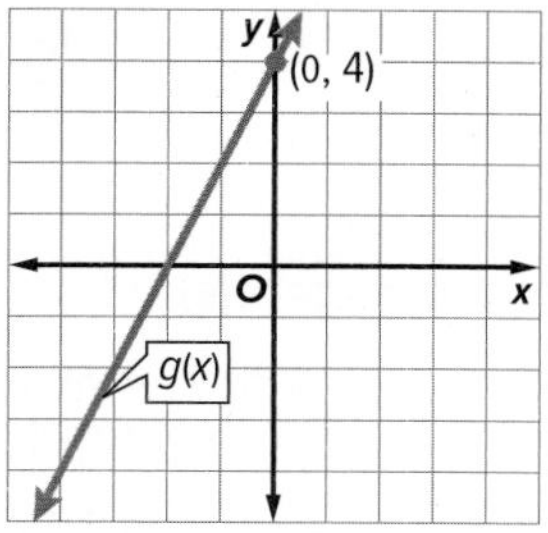

y-intercept of $f(x)$: −1,
y-intercept of $g(x)$: 4.

So, $g(x)$ intersects the y-axis at a higher point than $f(x)$.

x	$f(x)$
−6	−3
−3	−2
0	−1
3	0
6	1

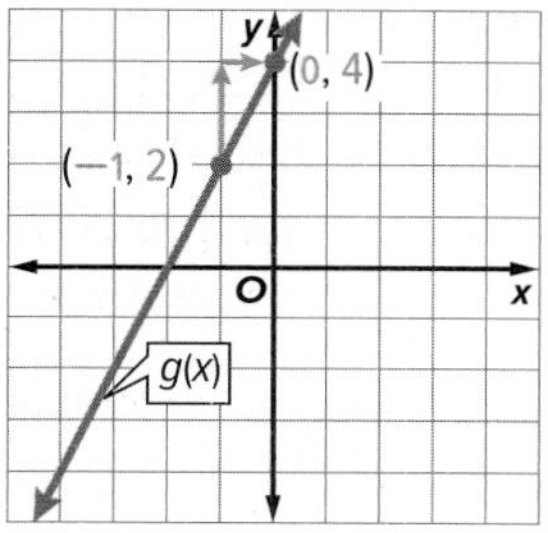

slope of $f(x)$: $\frac{1}{3}$
slope of $g(x)$: 2
Each function is increasing, but the slope of $g(x)$ is greater than the slope of $f(x)$.

So, $g(x)$ increases faster than $f(x)$.

Check

Use the table and graph to compare the two functions.

x	$f(x)$
−2	−6
−1	−4
0	−2
1	0
2	2

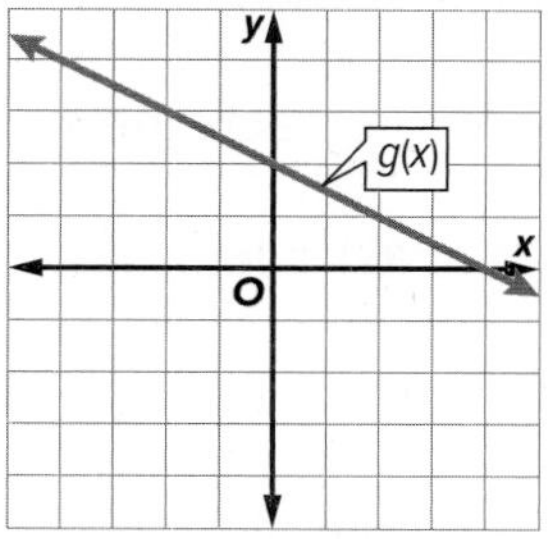

Which statements about $f(x)$ and $g(x)$ are true? Select all that apply.

A. $g(x)$ has a faster rate of change than $f(x)$.

B. The x-intercept of $g(x)$ is greater than the x-intercept of $f(x)$.

C. The y-intercept of $f(x)$ is greater than the x-intercept of $g(x)$.

D. Both functions are decreasing.

E. $f(x)$ increases while $g(x)$ decreases.

F. $f(x)$ has a faster rate of change than $g(x)$.

Go Online You can complete an Extra Example online.

Think About It!

How would a function that passes through (1, 0) with a slope of −4 compare to $f(x)$ and $g(x)$?

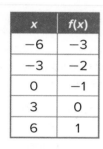

Example 5 Compare Properties of Nonlinear Functions

Examine the categories to see how to use the description and the graph to identify key features of each function. Then complete the statements to compare the two functions.

$f(x)$	$g(x)$
x-intercept: (−3.4, 0) y-intercept: (0, 1.5) relative maximum: (−2.3, 4.7) relative minimum: (−0.4, 1.1) end behavior: as $x \to -\infty$, $f(x) \to -\infty$ and as $x \to \infty$, $f(x) \to \infty$	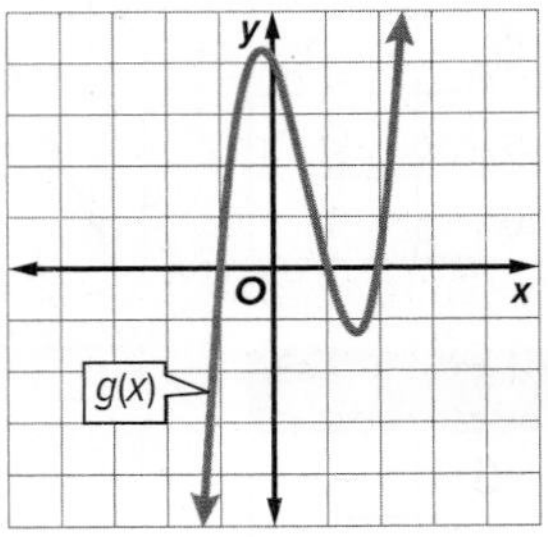
x-intercept(s)	
$f(x)$ intersects the x-axis once at (−3.4, 0).	$g(x)$ intersects the x-axis three times at (−1, 0), (1, 0), and (2, 0).
y-intercept	
$f(x)$ intersects the y-axis at (0, 1.5).	$g(x)$ intersects the y-axis at (0, 4).
Extrema	
$f(x)$ has a relative maximum of 4.7 and a relative minimum of 1.1.	$g(x)$ has a relative maximum of about 4.2 and a relative minimum of about −1.2.
End Behavior	
As $x \to -\infty$, $f(x) \to -\infty$, and as $x \to \infty$, $f(x) \to \infty$.	As $x \to -\infty$, $g(x) \to -\infty$, and as $x \to \infty$, $g(x) \to \infty$.

- The x-intercept of $f(x)$ is less than any of the x-intercepts of $g(x)$.
- The graph of $g(x)$ intersects the x-axis more times than $f(x)$.
- The y-intercept of $f(x)$ is less than the y-intercept of $g(x)$.
- So, $g(x)$ intersects the y-axis at a higher point than $f(x)$.
- The relative maximum of $f(x)$ is greater than the relative maximum of $g(x)$. The relative minimum of $f(x)$ is greater than the relative minimum of $g(x)$.
- The two functions have the same end behavior.

Go Online You can complete an Extra Example online.

Practice

Go Online You can complete your homework online.

Examples 1 and 2

Use the key features of each function to sketch its graph.

1. ***x*-intercept:** (2, 0)
   ***y*-intercept:** (0, −6)
   Linearity: linear
   Continuity: continuous
   Positive: for values $x > 2$
   Increasing: for all values of x
   End Behavior: As $x \to \infty$, $f(x) \to \infty$ and as $x \to -\infty$, $f(x) \to -\infty$.

2. ***x*-intercept:** (0, 0)
   ***y*-intercept:** (0, 0)
   Linearity: linear
   Continuity: continuous
   Positive: for values $x < 0$
   Negative: for values of $x > 0$
   Decreasing: for all values of x
   End Behavior: As $x \to \infty$, $f(x) \to -\infty$ and as $x \to -\infty$, $f(x) \to \infty$.

3. ***x*-intercept:** (5, 0)
   ***y*-intercept:** (0, 5)
   Linearity: linear
   Continuity: continuous
   Positive: for values $x < 5$
   Decreasing: for all values of x
   End Behavior: As $x \to \infty$, $f(x) \to -\infty$ and as $x \to -\infty$, $f(x) \to \infty$.

4. ***x*-intercept:** (5, 0)
   ***y*-intercept:** (0, 2)
   Linearity: linear
   Continuity: continuous
   Positive: for values $x < 5$
   Decreasing: for all values of x
   End Behavior: As $x \to \infty$, $f(x) \to -\infty$ and as $x \to -\infty$, $f(x) \to \infty$.

5. ***x*-intercept:** (−1, 0) and (1, 0)
   ***y*-intercept:** (0, 1)
   Linearity: nonlinear
   Continuity: continuous
   Symmetry: symmetric about the line $x = 0$
   Positive: for values $-1 < x < 1$
   Negative: for values of $x < -1$ and $x > 1$
   Increasing: for all values of $x < 0$
   Decreasing: for all values of $x > 0$
   Extrema: maximum at (0, 1)
   End Behavior: As $x \to \infty$, $f(x) \to -\infty$ and as $x \to -\infty$, $f(x) \to -\infty$.

6. ***x*-intercept:** (−3, 0) and (2, 0)
   ***y*-intercept:** (0, −4)
   Linearity: nonlinear
   Continuity: continuous
   Positive: for values $x < -3$ and $x > 2$
   Negative: for values of $-3 < x < 2$
   Increasing: for all values of $x > 0$
   Decreasing: for all values of $x < 0$
   Extrema: minimum at (0, −4)
   End Behavior: As $x \to \infty$, $f(x) \to \infty$ and as $x \to -\infty$, $f(x) \to \infty$.

7. ***x*-intercept:** (1, 0)
   ***y*-intercept:** (0, −1)
   Linearity: linear
   Continuity: continuous
   Positive: for values $x > 1$
   Increasing: for all values of x
   End Behavior: As $x \to -\infty$, $f(x) \to -\infty$ and as $x \to \infty$, $f(x) \to \infty$.

8. ***x*-intercept:** (−2, 0) and (2, 0)
   ***y*-intercept:** (0, −1)
   Linearity: nonlinear
   Continuity: continuous
   Symmetry: symmetric about the line $x = 0$
   Positive: for values $x < -2$ and $x > 2$
   Negative: for values of $-2 < x < 2$
   Increasing: for all values of $x > 0$
   Decreasing: for all values of $x < 0$
   Extrema: minimum at (0, −1)
   End Behavior: As $x \to -\infty$, $f(x) \to \infty$ and as $x \to \infty$, $f(x) \to \infty$.

Example 3

9. PELICANS A pelican descends to the ground. The pelican starts at a height of 6 feet. The pelican reaches the ground, at a height of 0 feet, after 3 seconds. The function that models the situation is linear. Use the key features to sketch a graph.

10. SCOOTERS Greg rides his motorized scooter for 20 minutes. Greg starts riding at 0 mph. Greg's maximum speed is 35 mph, which he reaches 5 minutes after he starts riding. Greg's speed increases steadily for 5 minutes. At the 10-minute mark, Greg decreases his speed for 2.5 minutes, then he stays at 20 mph for 5 minutes. At the 17.5-minute mark, he again decreases his speed for 2.5 minutes until he stops. Use the key features to sketch a graph.

Examples 4 and 5

11. Compare the key features of the functions represented with a graph and a table.

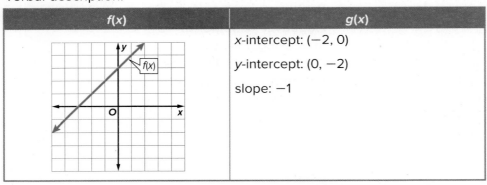

$f(x)$

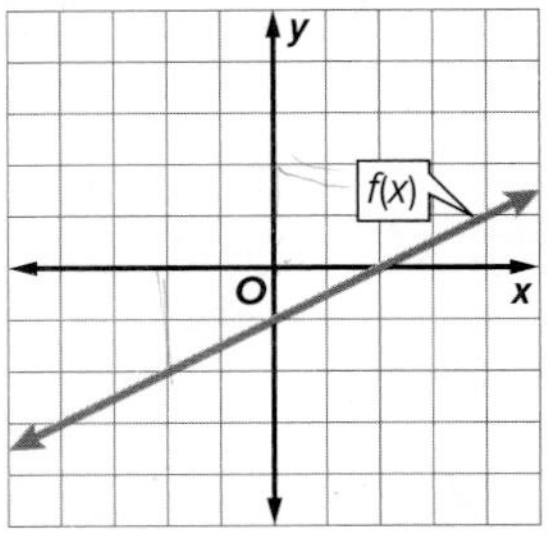

$g(x)$

x	$g(x)$
−2	−4
−1	−1
0	2
1	5
2	8

12. Compare the key features of the functions represented with a graph and a verbal description.

$f(x)$

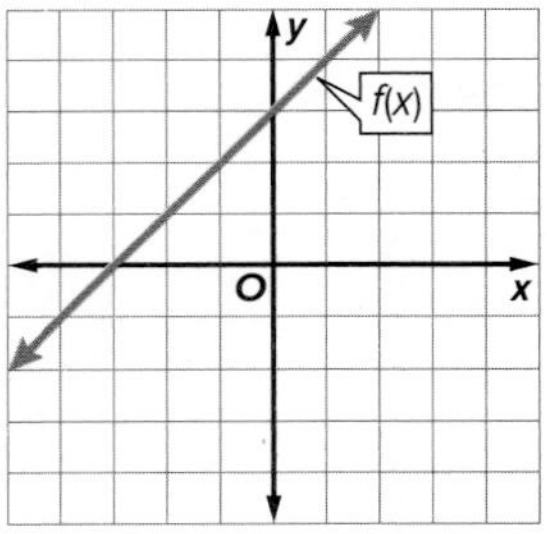

$g(x)$

x-intercept: (−2, 0)

y-intercept: (0, −2)

slope: −1

13. Compare the key features of the functions represented with a table and a verbal description.

$f(x)$

x	$f(x)$
−4	0
−3	−3
−2	−4
−1	−3
0	0

$g(x)$

x-intercept: (1, 0)

y-intercept: (0, −7)

relative maximum: none

relative minimum: none

end behavior:
as $x \to -\infty$, $g(x) \to -\infty$ and
as $x \to \infty$, $g(x) \to \infty$

14. Compare the key features of the functions represented with a graph and a verbal description.

$f(x)$	$g(x)$
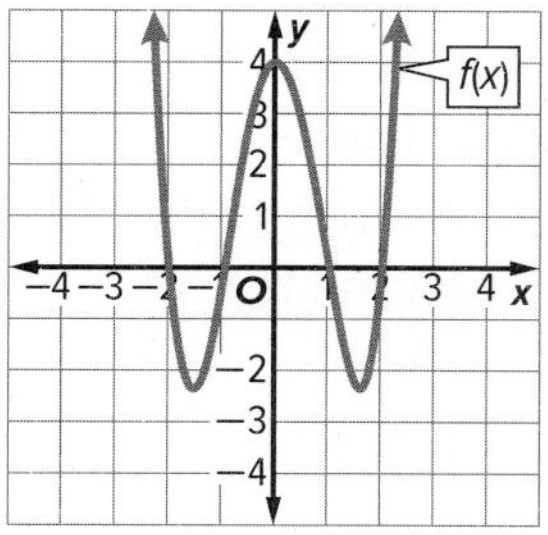	x-intercept: $(-1, 0)$, $(1, 0)$, $(2, 0)$ y-intercept: $(0, -4)$ relative maximum: $(1.37, 0.35)$ relative minimum: $(-0.37, -4.85)$, $(2, 0)$ end behavior: as $x \to -\infty$, $g(x) \to \infty$ and as $x \to \infty$, $g(x) \to \infty$

15. Compare the key features of the functions represented with a table and a verbal description.

$f(x)$:

x	$f(x)$
$-\frac{2}{3}$	1
$-\frac{1}{3}$	$\frac{3}{4}$
0	$\frac{1}{2}$
$\frac{1}{3}$	$\frac{1}{4}$
$\frac{2}{3}$	0

$g(x)$:

x-intercept: $\left(\frac{3}{8}, 0\right)$

y-intercept: $\left(0, \frac{1}{2}\right)$

slope: $-\frac{4}{3}$

16. Compare the key features of the functions represented with a graph and a table.

$f(x)$:

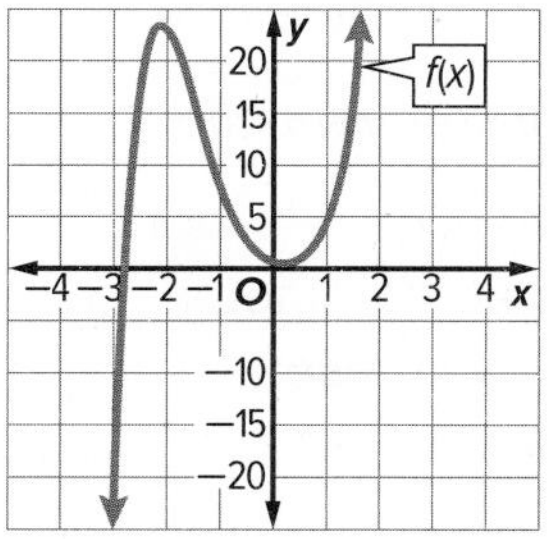

$g(x)$:

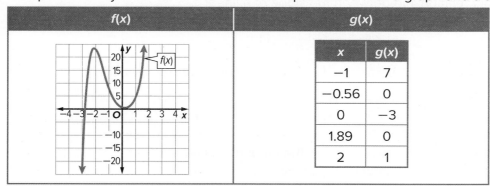

x	$g(x)$
−1	7
−0.56	0
0	−3
1.89	0
2	1

Mixed Exercises

17. USE A MODEL Sketch the graph of a linear function with the following key features. The x-intercept is 2. The y-intercept is 2. The function is decreasing for all values of x. The function is positive for $x < 2$. As $x \to -\infty$, $f(x) \to \infty$ and as $x \to \infty$, $f(x) \to -\infty$.

18. WATER Sia filled a pitcher with water. The pitcher started with 0 ounces of water. After 8 seconds the pitcher contains 64 ounces of water. The function that models the situation is linear.

a. Use the key features to sketch a graph.

b. What is the end behavior of the graph? Explain.

19. USE TOOLS Monica walks for 60 minutes. She starts walking from her house. The maximum distance Monica is from her house is 2 miles, which she reaches 30 minutes after she starts walking. At the 30-minute mark, Monica starts walking back to her house for 30 minutes until she reaches her house. Use the key features to sketch a graph.

20. USE A SOURCE Research the value of a new car after it is purchased. Use the information you collect to describe key features of a graph that represents the value of a new car x years after it is purchased. Then use the key features to sketch a graph.

21. SKI LIFTS A ski lift descends at a steady pace down a mountainside from a height of 1800 feet to ground level. If it makes no stops along the way to load or unload passengers, then the time it takes to complete its descension is 4 minutes.

a. Is the graph that relates the lift's height as a function of time linear or nonlinear? Explain.

b. Use the key features to sketch a graph.

22. CONSTRUCT ARGUMENTS Keisha babysits for her aunt for an hourly rate of \$9. The graph shows Keisha's earnings y as a function of hours spent babysitting x. Explain why the graph only exists for positive x- and y-values.

Keisha's Earnings

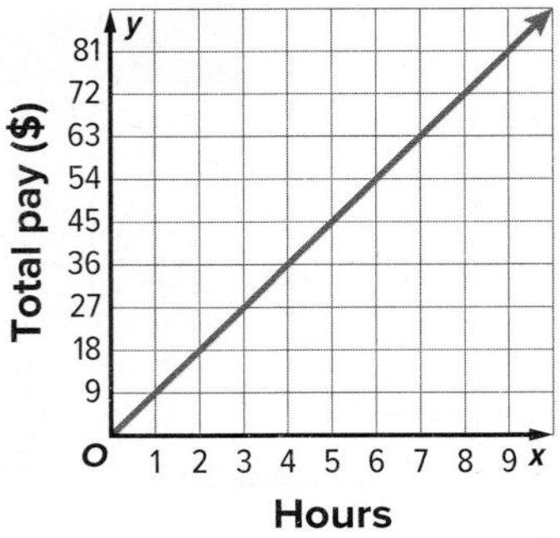

Higher-Order Thinking Skills

23. CREATE Choose a function and create a list of key features to describe the function. Then sketch the function.

24. WRITE Describe the relationship between the slope of a linear function and when the function is increasing or decreasing.

25. ANALYZE Determine whether the statement is *always*, *sometimes*, or *never* true.

A graph that has more than one x-intercept is represented by a nonlinear function.

26. PERSEVERE Deborah filled an empty tub with water for 30 minutes. The maximum amount of water in the tub is 50 gallons, which is reached 10 minutes after Deborah starts filling the tub. The amount of water in the tub increases steadily for 10 minutes. At the 10-minute mark, the amount of water in the tub starts decreasing for 20 minutes until there is no water left in the tub.

a. Use the key features to sketch a graph.

b. Describe an event that could have occurred at the 10-minute mark if Deborah continues filling the tub at the same rate from the 10-minute mark to the 30-minute mark as the rate from the 0-minute mark to the 10-minute mark.

27. FIND THE ERROR Linda and Rubio sketched a graph with the following key features. The x-intercept is 2. The y-intercept is -9. The function is positive for $x > 2$. As $x \to -\infty$, $f(x) \to -\infty$ and as $x \to \infty$, $f(x) \to \infty$. Is either graph correct based on the key features? Explain your reasoning.

Linda's Graph

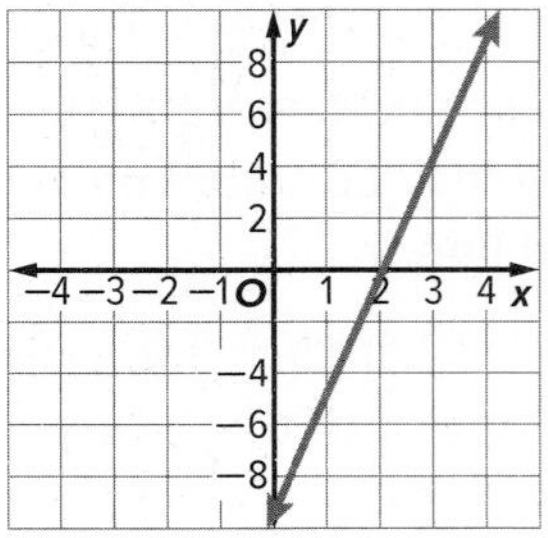

Rubio's Graph

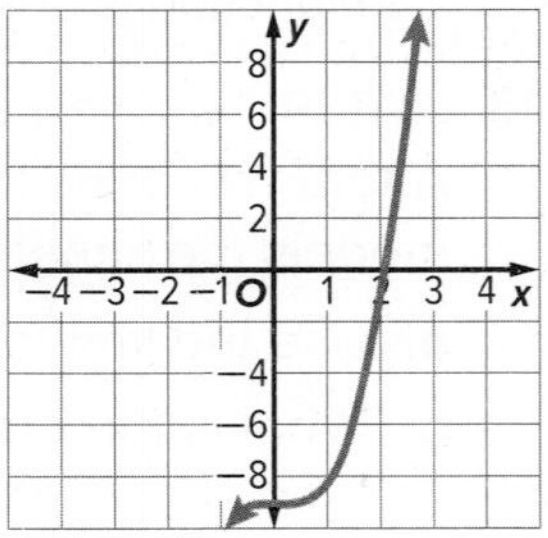

Graphing Linear Functions and Inequalities

Learn Graphing Linear Functions

The graph of a function represents all ordered pairs that are true for the function. You can use various methods to graph a linear function.

Example 1 Graph by Using a Table

Graph $x + 3y - 6 = 0$ by using a table.

Solve the equation for y. Then make a table and complete the graph.

$x + 3y - 6 = 0$ Original function

$y = -\frac{1}{3}x + 2$ Solve.

x	$-\frac{1}{3}x + 2$	y
−6	$-\frac{1}{3}(-6) + 2$	4
−3	$-\frac{1}{3}(-3) + 2$	3
0	$-\frac{1}{3}(0) + 2$	2
3	$-\frac{1}{3}(3) + 2$	1
6	$-\frac{1}{3}(6) + 2$	0

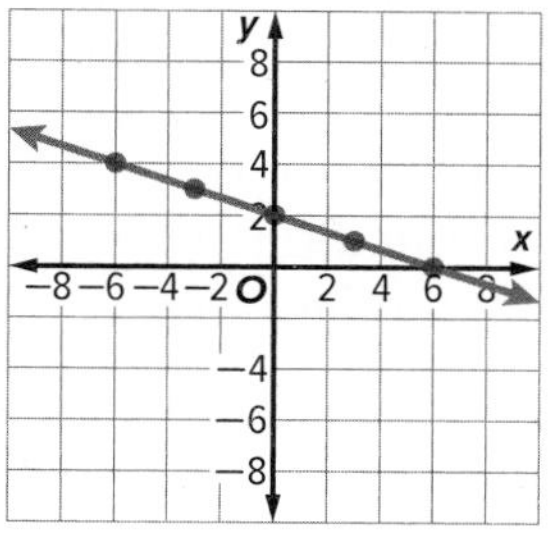

Check

Graph $6x + 2y = 10$ by using a table.

x	y
−1	?
0	?
1	?
3	?

Example 2 Graph by Using Intercepts

Graph $3x - 2y = -12$ by using the x- and y-intercepts.

To find the x-intercept, let $y = 0$.

$3x - 2y = -12$ Original function

$3x - 2(0) = -12$ Replace with 0.

$x = -4$ Simplify.

To find the y-intercept, let $x = 0$.

$3x - 2y = -12$

$3(0) - 2y = -12$

$y = 6$

(continued on the next page)

Go Online You can complete an Extra Example online.

Today's Goals
- Graph linear functions.
- Graph linear inequalities in two variables.

Today's Vocabulary

linear inequality

boundary

closed half-plane

open half-plane

constraint

Go Online

You can watch a video to see how to graph linear functions.

Study Tip

Slope Recall that slope is the ratio of the change in the y-coordinates (rise) to the corresponding change in the x-coordinates (run) as you move from one point to another along a line.

Think About It!

Explain why −6, −3, 0, 3, and 6 were selected for the x-values in the table.

Think About It!
How can you check that the graph is correct?

The x-intercept is -4, and the y-intercept is 6. This means that the graph passes through $(-4, 0)$ and $(0, 6)$.

Plot the two intercepts.

Draw a line through the points.

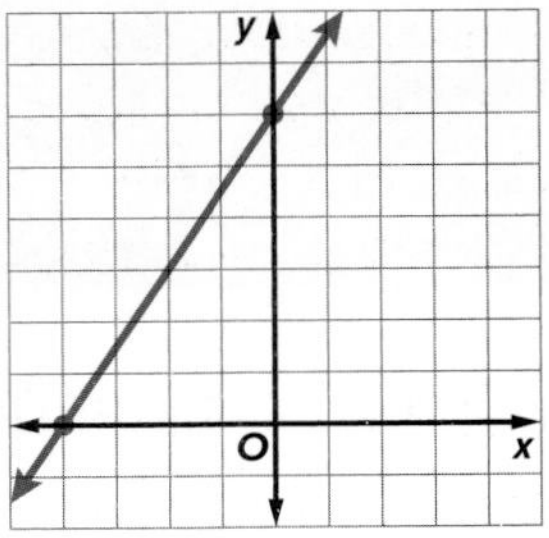

Example 3 Graph by Using the Slope and y-intercept

Graph $y = \frac{3}{2}x - 4$ by using m and b.

Follow these steps.

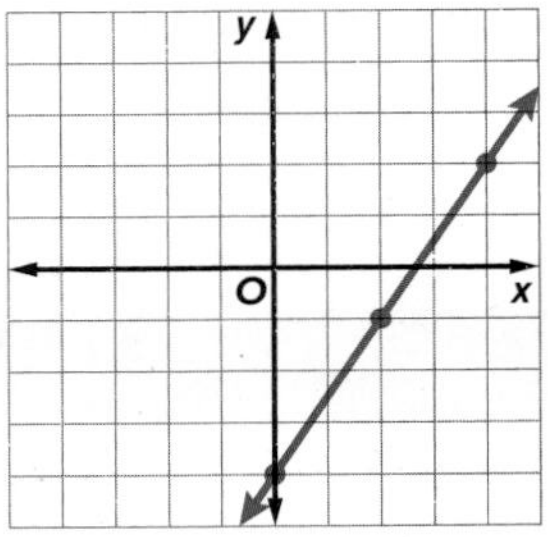

- Begin by identifying the slope m and y-intercept b of the function.

  $m = \frac{3}{2}$ $\qquad$ $b = -4$
- Use the value of b to plot the y-intercept $(0, -4)$.
- Use the slope of the line $m = \frac{3}{2}$ to plot more points. From the y-intercept, move up 3 units and right 2 units. Plot a point at $(2, -1)$.
- From the point $(2, -1)$, move up 3 units and right 2 units. Plot a point at $(4, 2)$.
- Draw a line through the points.

Explore Shading Graphs of Linear Inequalities

Online Activity Use graphing technology to complete the Explore.

INQUIRY How can you use a point to test the graph of an inequality?

Learn Graphing Linear Inequalities in Two Variables

Go Online
You can watch a video to see how to graph a linear inequality in two variables.

The graph of a **linear inequality** is a half-plane with a boundary that is a straight line. The half-plane is shaded to indicate that all points contained in the region are solutions of the inequality. A **boundary** is a line or curve that separates the coordinate plane into two half-planes.

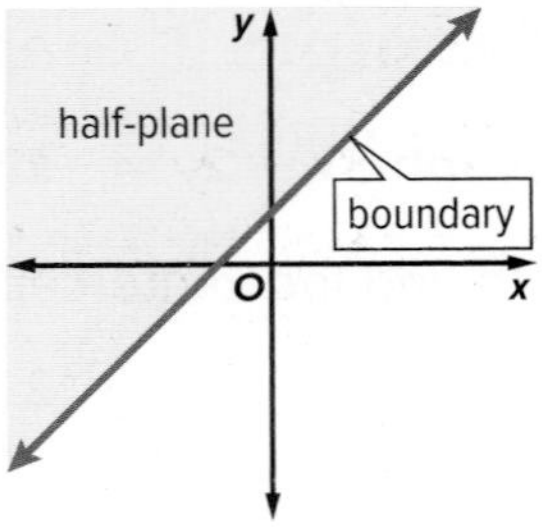

Go Online You can complete an Extra Example online.

The boundary is solid when the inequality contains $\leq$ or $\geq$ to indicate that the points on the boundary are included in the solution, creating a **closed half-plane**. The boundary is dashed when the inequality contains $<$ or $>$ because the points on the boundary do not satisfy the inequality. This results in an **open half-plane**.

A **constraint** is a condition that a solution must satisfy. Each solution of the inequality represents a viable, or possible, option that satisfies the constraint.

Example 4 Graph an Inequality with an Open Half-Plane

Graph $12 - 4y > x$.

Step 1 Graph the boundary.

$12 - 4y > x$	Original inequality
$-4y > x - 12$	Subtract 12 from each side.
$y < -\frac{1}{4}x + 3$	Divide each side by −4, and reverse the inequality symbol.

The boundary of the graph is $y = -\frac{1}{4}x + 3$. Because the inequality symbol is $>$, the boundary is dashed.

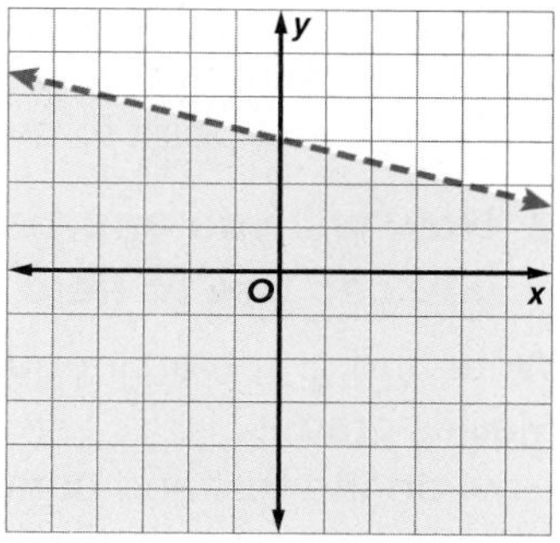

Step 2 Use a test point and shade.

Test (0, 0).

$12 - 4y > x$	Original inequality
$12 - 4(0) \overset{?}{>} 0$	Substitute.
$12 > 0$	True

Because (0, 0) is a solution of the inequality, shade the half-plane that contains the test point.

Check: Check by selecting another point in the shaded region to test.

Example 5 Graph an Inequality with a Closed Half-Plane

Graph $9 + 3y \leq 8x$.

Step 1 Graph the boundary.

Solve for y in terms of x and graph the related function.

$9 + 3y \leq 8x$	Original inequality
$3y \leq 8x - 9$	Subtract 9 from each side.
$y \leq \frac{8}{3}x - 3$	Divide each side by 3.

The related equation of $y \leq \frac{8}{3}x - 3$ is $y = \frac{8}{3}x - 3$, and the boundary is solid.

(*continued on the next page*)

Study Tip

Above or Below
Usually the shaded half-plane of a linear inequality is said to be *above* or *below* the line of the related equation. However, if the equation of the boundary is $x = c$ for some constant c, then the function is a vertical line. In this case, the shading is considered to be *to the left* or *to the right* of the boundary.

Talk About It!

Can a linear inequality ever be a function? Explain your reasoning.

Think About It!

Why should you not test a point that is on the boundary?

Think About It!
Is (3, 5) a solution of the inequality? Explain.

Step 2 Use a test point and shade.

Select a test point, such as (0, 0).

$9 + 3y \leq 8x$	Original inequality
$9 + 3(0) \overset{?}{\leq} 8(0)$	$(x, y) = (0, 0)$.
$9 \nleq 0$	False

Shade the side of the graph that does not contain the test point.

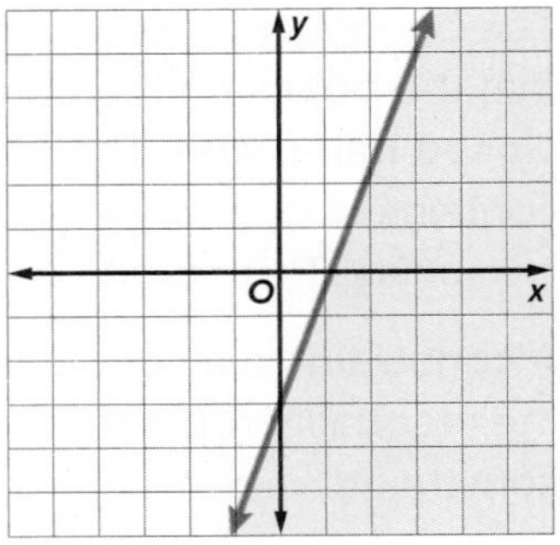

Go Online
You can watch a video to see how to graph an inequality using a graphing calculator.

Apply Example 6 Linear Inequalities

GRADES **Malik's algebra teacher determines semester grades by finding the sum of 70% of a student's test grade average and 30% of a student's homework grade average. If Malik wants a semester grade of 90% or better, write and graph the inequality that represents the constraints for Malik's test grade *x* and homework grade *y*.**

1 What is the task?

Describe the task in your own words. Then list any questions that you may have. How can you find answers to your questions?

Use the description to write the inequality. Find points on the boundary and use a test point to create the graph.

2 How will you approach the task? What have you learned that you can use to help you complete the task?

Write and graph an inequality to represent the constraints on Malik's grades.
How do the test and homework grades relate to the semester grade?

3 What is your solution?

Use your strategy to solve the problem. What inequality represents the constraints for Malik's test and homework grades? Graph the inequality.

$0.7x + 0.3y \geq 0.9$

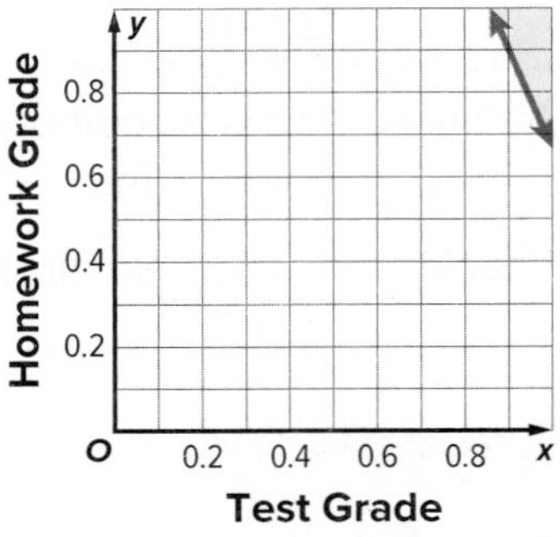

Which of these are viable solutions for Malik's test and homework grades?

- ☑ 88% test, 100% homework
- ☑ 90% test, 90% homework
- ☐ 90% test, 80% homework
- ☐ 95% test, 70% homework
- ☑ 95% test, 80% homework
- ☑ 100% test, 70% homework

Go Online
You can complete an Extra Example online.

4 How can you know that your solution is reasonable?

Write About It! Write an argument that can be used to defend your solution.
I can select a point in the shaded region, such as (0.95, 0.8) and test it in the inequality.

Practice

Go Online You can complete your homework online.

Example 1

Graph each equation by using a table.

1. $4x - 1 = y$
2. $-3 = 5x - y$
3. $y - 4 = -2x$
4. $y + x = 1$
5. $y + 3x = 1$
6. $y + 4x - 1 = 4x + 2$

Example 2

Graph each equation by using the *x*- and *y*-intercepts.

7. $3y - x = 6$
8. $2x - 3y = 6$
9. $y - x = -3$
10. $-2x + y = 4$
11. $y - 2x = -3$
12. $\frac{1}{2}x + y = 2$

Example 3

Graph each equation by using *m* and *b*.

13. $y = -\frac{5}{3}x + 12$
14. $y = \frac{2}{3}x + 6$
15. $y = 4x - 15$
16. $y - 2x = -1$
17. $y - x = -4$
18. $4 = 3x - y$

Examples 4 and 5

Graph each inequality.

19. $y > 1$
20. $y \leq x + 2$
21. $x + y \leq 4$
22. $x + 3 < y$
23. $2 - y < x$
24. $y \geq -x$
25. $x - y > -2$
26. $9x + 3y - 6 \leq 0$
27. $y + 1 \geq 2x$
28. $y - 7 \leq -9$
29. $x > -5$
30. $y + x > 1$

Example 6

31. **CRAFT FAIR** Kylie is going to try to sell two of her oil paintings at the local craft fair. She is hoping to earn at least \$400.

    a. Write the inequality that represents the constraint of the situation, where x is the price of the first oil painting, and y is the price of the second.

    b. Graph the inequality that represents the constraint on the sale.

32. BUILDING CODES A city has a building code that limits the height of buildings around the central park. The code says that the height of a building must be less than $0.1x$, where x is the distance in hundreds of feet of the building from the center of the park. Assume that the park center is located at $x = 0$. Graph the inequality that represents the building code.

33. WEIGHT A delivery crew is going to load a truck with tables and chairs. The trucks weight limitations are represented by the inequality $200t + 60c < 1200$, where t is the number of tables and c is the number of chairs. Graph this inequality.

34. ART An artist can sell each drawing for \$100 and each watercolor for \$400. He hopes to make at least \$2000 every month.

a. Write an inequality that expresses how many drawings and/or watercolors the artist needs to sell each month to reach his goal.

b. Graph the inequality.

c. If the artist sells three watercolors one month, how many drawings would he have to sell in the same month to reach \$2000?

Mixed Exercises

Graph each equation or inequality.

35. $x + y = 1$

36. $y \geq -3x - 2$

37. $x + 2y > 6$

38. $y + 2 = 3x + 3$

39. $y + 3 = 0$

40. $4x - 3y > 12$

41. $x + y = 3$

42. $2y - x = 2$

43. $4x + 3y = 12$

44. $\frac{1}{2}x + \frac{1}{4}y = 8$

45. $-\frac{1}{2}x + y = -2$

46. $y \geq \frac{3}{4}x + 6$

47. $y = -2x + 3$

48. $2x - y = 1$

49. $2y + 3 \leq 11$

50. $y + 2 = -x + 1$

51. $6x + 4y \leq -24$

52. $-2x + 5y = 2$

53. REASONING Name the x- and y-intercept for the linear equation given by $6x - 2y = 12$. Use the intercepts to graph the equation and describe the graph as *increasing*, *decreasing*, or *constant*.

54. ANIMALS During the winter, a horse requires about 36 liters of water per day and a sheep requires about 3.6 liters per day. A farmer is able to supply his horses and sheep with a total of 300 liters of water each day.

a. Write an equation or inequality that represents the possible number of each type of animal that the farmer can keep.

b. Graph the equation or inequality.

55. **COMPUTERS** A school system is buying new computers. They will buy desktop computers costing \$1000 per unit, and notebook computers costing \$1200 per unit. The total cost of the computers cannot exceed \$80,000.

a. Write an inequality that describes this situation.

b. Graph the inequality.

c. If the school wants to buy 50 desktop computers and 25 notebook computers, will they have enough money? Explain.

56. **BAKED GOODS** Mary sells giant chocolate chip and peanut butter cookies for \$1.25 and \$1.00, respectively, at a local bake shop. She wants to make at least \$25 a day.

a. Write and graph an inequality that represents the number of cookies Mary needs to sell each day.

b. If Mary decides to charge \$1.50 for chocolate chip cookies rather than \$1.25, what impact will this have on the graph of the solution set? Give an (x, y) pair that is not in the original solution set, but is in the solution set of the new revised scenario.

c. How does the graph of the inequality change if Mary wants to make at least \$50 a day? How does the graph of the inequality change if Mary wants to make no more than \$25 a day?

57. **FUNDRAISING** The school drama club is putting on a play to raise money. Suppose it will cost \$400 to put on the play and that 300 students and 150 adults will attend.

a. Write an equation to represent revenue from ticket sales if the club wants to raise \$1400 after expenses.

b. Graph your equation. Then determine four possible prices that could be charged for student and adult tickets to earn \$1400 in profit.

58. **CONSTRUCTION** You want to make a rectangular sandbox area in your backyard. You plan to use no more than 20 linear feet of lumber to make the sides of the sandbox.

a. Write and graph a linear inequality to describe this situation.

b. What are two possible sizes for the sandbox?

c. Can you make a sandbox that is 7 feet by 6 feet? Justify your answer.

d. What can you conclude about the intercepts of your graph?

59. **SPIRITWEAR** A company makes long-sleeved and short-sleeved shirts. The profit on a long-sleeved shirt is \$7 and the profit on a short-sleeved shirt is \$4. How many shirts must the company sell to make a profit of at least \$280?

a. Write and graph a linear inequality to describe this situation.

b. Write two possible solutions to the problem.

c. Which values are reasonable for the domain and for the range? Explain.

d. The point (−10, 90) is in the shaded region. Is it a solution of the problem? Explain your reasoning.

60. **MONEY** Gemma buys candles and soaps online. The scented candles cost \$9, and the hand soaps cost \$4. To qualify for free shipping, Gemma needs to spend at least \$50.

a. Write an inequality that represents the constraints on the number of candles x and the number of soaps y that Gemma must buy in order to qualify for free shipping.

b. Graph the inequality.

c. Suppose Gemma decides not to buy any soaps. Determine the number of candles she needs to buy in order to qualify for free shipping. Explain.

d. If Gemma decides not to buy any candles how many soaps will she need to buy in order to qualify for free shipping? Explain.

e. Will Gemma qualify for free shipping if she buys 2 candles and 8 soaps? Explain how you can be sure.

61. **FIND THE ERROR** Paulo and Janette are graphing $x - y \geq 2$. Is either of them correct? Explain your reasoning.

Paulo

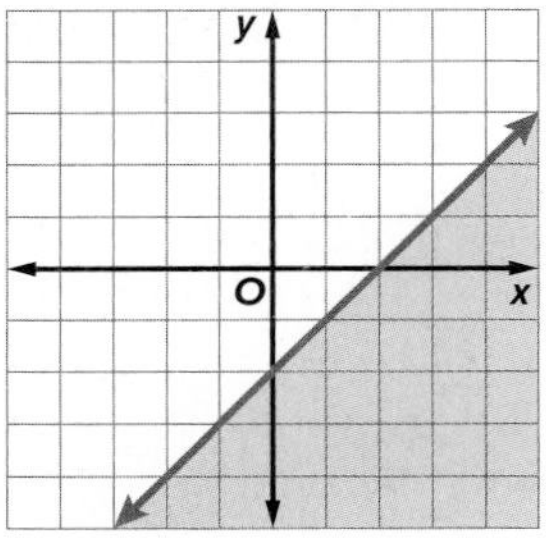

Janette

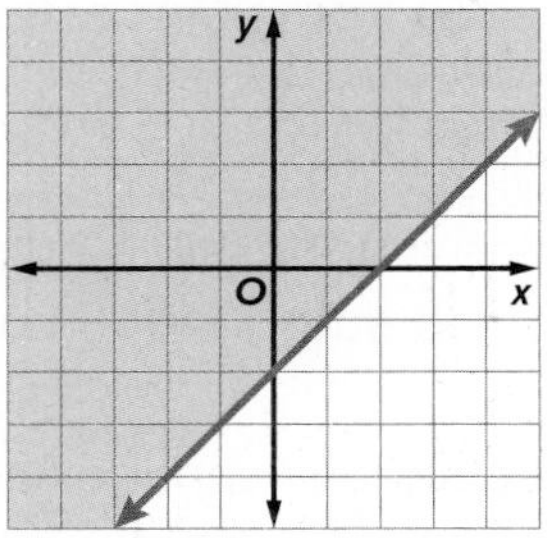

62. **CREATE** Write an inequality that has a graph with a dashed boundary line. Then graph the inequality.

63. **WRITE** You can graph a line by making a table, using the x- and y-intercepts, or by using m and b. Which method do you prefer? Explain your reasoning.

64. **ANALYZE** Write a counterexample to show that the following statement is false. *Every point in the first quadrant is a solution for* $3y > -x + 6$.

65. **PERSEVERE** Write an equation of the line that has the same slope as $2x - 8y = 7$ and the same y-intercept as $4x + 3y = 15$.

Special Functions

Explore Using Tables to Graph Piecewise Functions

Online Activity Use a table and graphing technology to complete the Explore.

INQUIRY How can you write a piecewise function when given a table of values?

Learn Graphing Piecewise-Defined Functions

A function that is written using two or more expressions is called a **piecewise-defined function**. Each of the expressions is defined for a distinct interval of the domain.

A dot is used if a point is included in the graph. A circle is used for a point that is not included in the graph.

Example 1 Graph a Piecewise-Defined Function

Graph $f(x) = \begin{cases} x - 3 \text{ if } x \leq 1 \\ 2x \text{ if } x > 1 \end{cases}$. **Then, analyze the key features.**

Step 1 Graph $f(x) = x - 3$ for $x \leq 1$.

Find $f(x)$ for the endpoint of the domain interval, $x = 1$.

$f(x) = x - 3$

$= 1 - 3$ or -2

Since 1 satisfies the inequality, place a dot at $(1, -2)$. Because $x - 3$ is defined for values of x less than or equal to 1, graph the linear function with a slope of 1 to the left of the dot.

Step 2 Graph $f(x) = 2x$ for $x > 1$.

Find $f(x)$ for the endpoint of the domain interval, $x = 1$.

$f(x) = 2x$

$= 2(1)$ or 2

Since 1 is not included in the inequality, place a circle at $(1, 2)$. Because $2x$ is defined for values greater than 1, graph the linear function with a slope of 2 to the right of the circle.

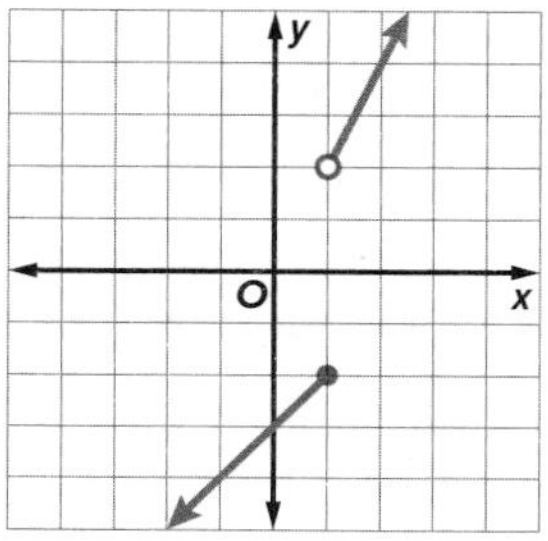

(continued on the next page)

Go Online You can complete an Extra Example online.

Today's Goals
- Write and graph piecewise-defined functions.
- Write and graph step functions.
- Graph and analyze absolute value functions.

Today's Vocabulary
- piecewise-defined function
- step function
- greatest integer function
- absolute value function
- parent function

Talk About It!
Can the piecewise-defined function in the example have $x \geq 1$ instead of $x > 1$ after the second expression? Explain.

Go Online
You can watch a video to see how to graph a piecewise-defined function using a graphing calculator.

Step 3 Analyze key features.

The function is defined for all values of x, so the domain is all real numbers.

The range is all real numbers less than or equal to -2 and all real numbers greater than 2. This can be represented symbolically as

$\{f(x) \mid f(x) \leq -2 \text{ or } f(x) > 2\}$.

The y-intercept is -3, and there is no x-intercept.

The function is increasing for all values of x.

Check

Graph $f(x) = \begin{cases} \frac{2}{3}x \text{ if } x \leq 0 \\ 3 \text{ if } 1 \leq x \leq 3 \\ -2x + 5 \text{ if } x \geq 4 \end{cases}$. Then, analyze the key features.

The domain is ___?___.

The range is ___?___.

The x- intercept is ___?___.

The y-intercept is ___?___.

For $\{x \mid x \leq 0\}$, the function is ___?___.

For $\{x \mid 1 \leq x \leq 3\}$, the function is ___?___.

For $\{x \mid x \geq 4\}$, the function is ___?___.

Think About It!

What do the domain and range represent in the context of this situation?

Example 2 Model by Using a Piecewise-Defined Function

UNIFORMS **The football coach is ordering new jerseys for the new season. The manufacturer charges $88 for each jersey when five or fewer are ordered, $75 each for an order of six to 11 jerseys, $65 each for an order of 12 to 29 jerseys, and $56 each when thirty or more jerseys are ordered.**

Part A Write a piecewise-defined function describing the cost of the jerseys.

$$f(x) = \begin{cases} 88x & \text{if } 0 < x \leq 5 \\ 75x & \text{if } 5 < x \leq 11 \\ 65x & \text{if } 11 < x \leq 29 \\ 56x & \text{if } x > 29 \end{cases}$$

Part B Evaluate the function.

What would it cost to purchase 11 jerseys? $825

What would it cost to purchase 25 jerseys? $1625

Evaluate $f(29)$. $f(29) = \$1885$

Watch Out!

Evaluating Endpoints of Intervals When evaluating a piecewise-defined function for a value of x that is an endpoint for two consecutive intervals, be careful to evaluate the function that contains that point.

Go Online You can complete an Extra Example online.

Learn Graphing Step Functions

A common type of piecewise function is a step function. A **step function** has a graph that is a series of horizontal line segments that may resemble a staircase.

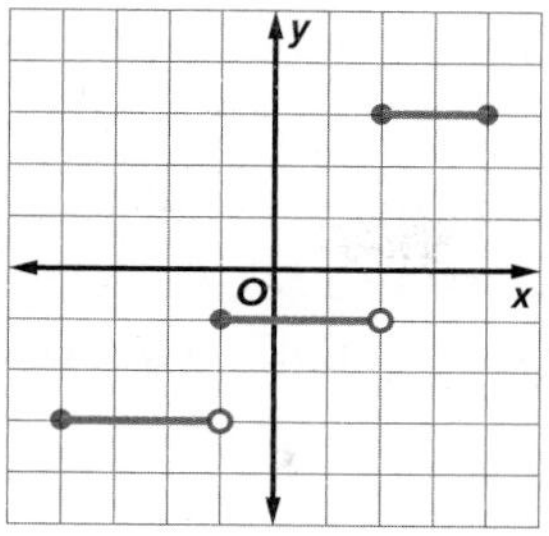

A step function is defined by a set of constant functions. The domain of a step function is an interval of real numbers. The range of a step function is a discrete set of real numbers. The graph of a step function is discontinuous because it cannot be drawn without lifting your pencil.

The **greatest integer function**, written $f(x) = [\![x]\!]$, is one kind of step function in which $f(x)$ is the greatest integer less than or equal to x. For example, $[\![10.7]\!] = 10$, $[\![-6.35]\!] = -7$, and $[\![5]\!] = 5$.

Think About It!

Why is the range of the function not expressed as $\{y \mid -3 \le y \le 3\}$?

Think About It!

Explain why the value of $[\![4.3]\!]$ is 4, but the value of $[\![-4.3]\!]$ is -5.

Example 3 Graph a Step Function

POSTAL RATES **The cost of mailing a first-class letter is determined by rates adopted by the U.S. Postal Service. The rates adopted in 2016 charge \$0.47 for letters not over 1 ounce, \$0.68 if not over 2 ounces, \$0.89 if not over 3 ounces, and \$1.10 if not over 3.54 ounces. Complete the table and draw a graph that represents the charges. State the domain and range.**

Step 1 Make a table.

Let x be the weight of a first-class letter and $C(x)$ represent the cost for mailing it. Use the given rates to make a table.

x	$C(x)$
$0 < x \le 1$	\$0.47
$1 < x \le 2$	\$0.68
$2 < x \le 3$	\$0.89
$3 < x \le 3.54$	\$1.10

Step 2 Make a graph.

Graph the first step of the function. Place a circle at (0, 0.47) since there is no charge for not mailing a letter. Place a dot at (1, 0.47) since a letter weighing one ounce will cost \$0.47 to mail. Draw a segment that connects the points.

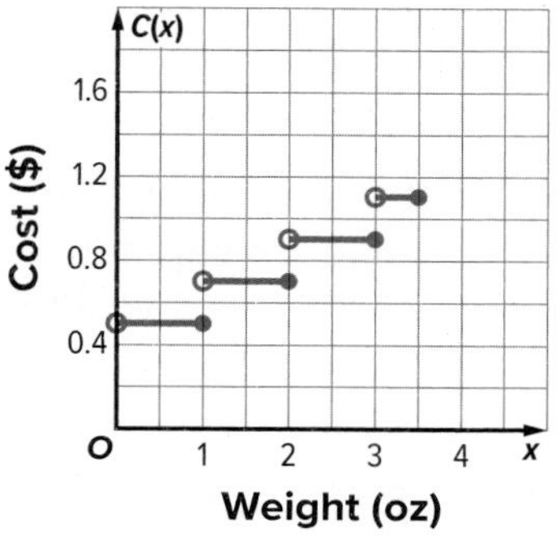

Graph the remaining steps.

Place a circle on the left end of each segment as that domain value is included with the segment below it.

(*continued on the next page*)

Step 3 State the domain and range.

The constraints for the weight of a first-class letter are more than 0 ounces up to and including 3.54 ounces. Therefore, the domain is $\{x \mid 0 < x \leq 3.54\}$.

Because the only viable solutions for the cost of mailing a first-class letter are \$0.47, \$0.68, \$0.89, and \$1.10, the range is $\{C(x) = 0.47, 0.68, 0.89, 1.10\}$.

Check

FIGURINES Chris and Joaquin design figurines for board game and toy companies. The rate they charge $R(x)$ depends on the number of hours x they spend creating the figurines. Draw a graph that represents the charges. State the domain and range.

x	$R(x)$
$0 < x \leq 5$	500
$5 < x \leq 15$	1400
$15 < x \leq 30$	2500
$30 < x \leq 50$	4000

Think About It!
Would $C(x)$ still be a function if the open points at (0, 0.47), (1, 0.68), (2, 0.89), and (3, 1.10) were closed points? Justify your argument.

Example 4 Graph a Greatest Integer Function

Complete the table and graph $f(x) = [\![2x - 1]\!]$. State the domain and range.

Step 1 Make a table.

Make a table of the intervals of x and associated values of $f(x)$.

x	$f(x)$
$-1 \leq x < -0.5$	-3
$-0.5 \leq x < 0$	-2
$0 \leq x < 0.5$	-1
$0.5 \leq x < 1$	0
$1 \leq x < 1.5$	1
$1.5 \leq x < 2$	2

Step 2 Make a graph.

Graph the first step. Place a dot at $(-1, -3)$, because $f(-1) = [\![2(-1) - 1]\!] = [\![-3]\!] = -3$. Place a circle at $(-0.5, -3)$, since every decimal value greater than -1 and up to but not including -0.5 produces an $f(x)$ value of -3.

Graph the remaining steps. Place a dot on the left end of each segment as that point is included with the segment, and place a circle on the right end because that domain value is included with the segment above it.

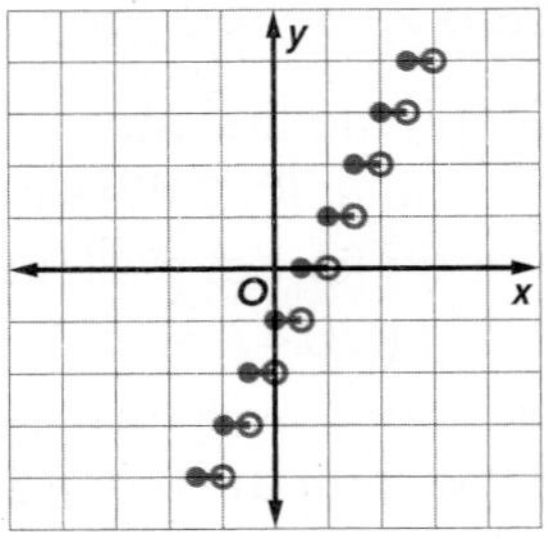

Step 3 State the domain and range.

The domain of $f(x) = [\![2x - 1]\!]$ is all real numbers. The range is all integers.

Go Online You can complete an Extra Example online.

Think About It!
Will the range of a greatest integer function always be all integers? If not, provide a counterexample.

Go Online
You can watch a video to see how to graph step functions.

Learn Graphing Absolute Value Functions

An **absolute value function** is a function that contains an algebraic expression within absolute value symbols. It can be defined and graphed as a piecewise function.

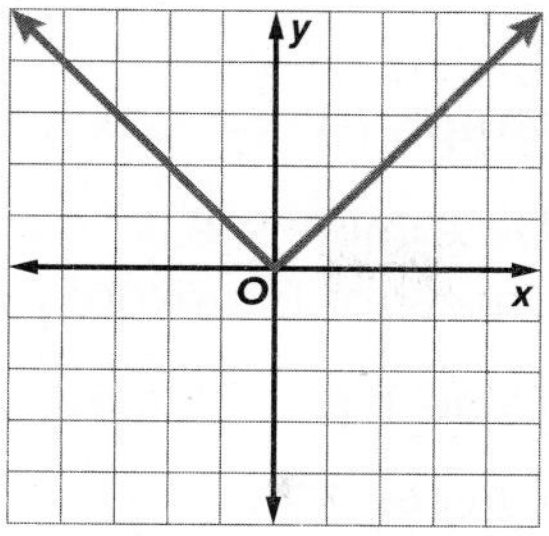

For an absolute value function, $f(x) = |x|$ is the **parent function**, which is the simplest of the functions in a family.

Key Concept • Parent Function of Absolute Value Functions	
parent function	$f(x) = \|x\|$ or $f(x) = \begin{cases} x \text{ if } x \geq 0 \\ -x \text{ if } x < 0 \end{cases}$
domain	all real numbers
range	all nonnegative real numbers
intercepts	x-intercept: $x = 0$, y-intercept: $y = 0$

> **Think About It!**
> Describe the line of symmetry of any absolute value function and compare it with the line of symmetry of the parent function.

Example 5 Graph an Absolute Value Function, Positive Coefficient

Graph $f(x) = \left|\frac{3}{4}x\right| + 3$. State the domain and range.

Create a table of values. Plot the points and connect them with two rays.

x	$f(x) = \left\|\frac{3}{4}x\right\| + 3$
−4	$\left\|\frac{3}{4}(-4)\right\| + 3 = 6$
−2	$\left\|\frac{3}{4}(-2)\right\| + 3 = 4\frac{1}{2}$
0	$\left\|\frac{3}{4}(0)\right\| + 3 = 3$
2	$\left\|\frac{3}{4}(2)\right\| + 3 = 4\frac{1}{2}$
4	$\left\|\frac{3}{4}(4)\right\| + 3 = 6$

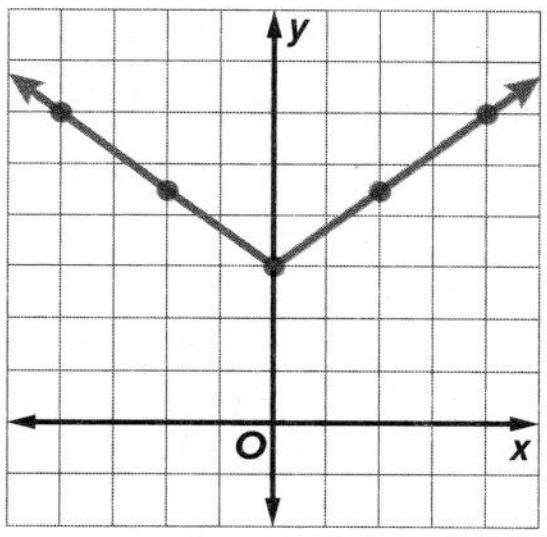

The function is defined for all values of x, so the domain is all real numbers. The function is defined only for values of $f(x)$ such that $f(x) \geq 3$, so the range is $\{f(x) \mid f(x) \geq 3\}$.

Check

Graph $f(x) = |x - 1| + 3$. State the domain and range.

> **Go Online**
> An alternate method is available for this example.

Go Online You can complete an Extra Example online.

Example 6 Graph an Absolute Value Function, Negative Coefficient

Graph $f(x) = -2|x + 1|$. State the domain and range.

Determine the two related linear equations using the two possible cases for the expression inside of the absolute value.

Case 1: $(x + 1)$ is positive.

$f(x) = -2(x + 1)$	$x + 1$ is positive, so $\|x + 1\| = x + 1$.
$= -2x - 2$	Simplify.

Case 2: $(x + 1)$ is negative.

$f(x) = -2[-(x + 1)]$	$x + 1$ is negative, so $\|x + 1\| = -(x + 1)$.
$= -2(-x - 1)$	Distributive Property
$= 2x + 2$	Simplify.

The x-coordinate of the vertex is the value of x where the two cases of the absolute value are equal.

$-2x - 2 = 2x + 2$	Set Case 1 equal to Case 2.
$-2x = 2x + 4$	Add 2 to each side.
$-4x = 4$	Subtract $2x$ from each side.
$x = -1$	Divide each side by -4.

Think About It!

How does multiplying the absolute value by a negative number affect the shape of the graph? the range?

The x-coordinate of the vertex represents the constraint of the piecewise-defined function. Write the piecewise-defined function that describes the function and use it to graph the absolute value function.

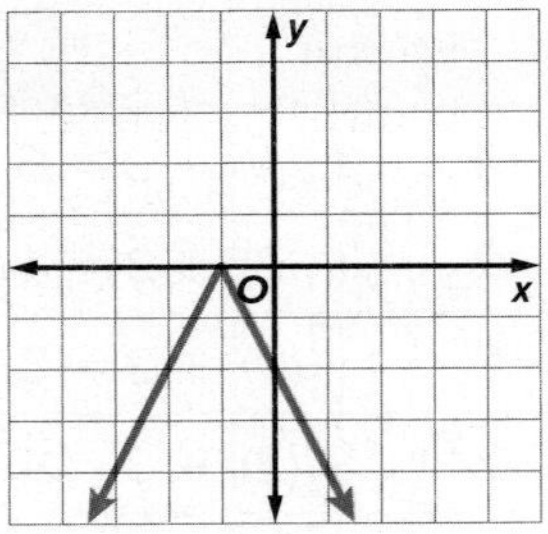

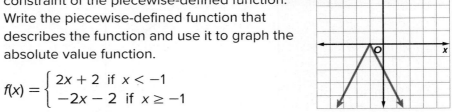

$$f(x) = \begin{cases} 2x + 2 & \text{if } x < -1 \\ -2x - 2 & \text{if } x \geq -1 \end{cases}$$

The function is defined for all values of x, so the domain is all real numbers. The function is defined only for values of $f(x)$ such that $f(x) \leq 0$, so the range is $\{f(x) \mid f(x) \leq 0\}$.

Check

Graph $f(x) = 0.25|8x| - 3$. State the domain and range.

Go Online You can complete an Extra Example online.

Practice

Go Online You can complete your homework online.

Examples 1 and 2

Graph each function. Then, analyze the key features.

1. $f(x) = \begin{cases} -1 & \text{if } x \leq 0 \\ 2x & \text{if } 0 < x \leq 3 \\ 6 & \text{if } x > 3 \end{cases}$

2. $f(x) = \begin{cases} -x & \text{if } x < -1 \\ 0 & \text{if } -1 \leq x \leq 1 \\ x & \text{if } x > 1 \end{cases}$

3. $f(x) = \begin{cases} x & \text{if } x < 0 \\ 2 & \text{if } x \geq 0 \end{cases}$

4. $h(x) = \begin{cases} 3 & \text{if } x < -1 \\ x + 1 & \text{if } x > 1 \end{cases}$

5. TILE Mark is purchasing new tile for his bathrooms. The home improvement store charges \$48 for each box of tiles when three or fewer boxes are purchased, \$45 for each box when 4 to 8 boxes are purchased, \$42 for each box when 9 to 19 boxes are purchased, and \$38 for each box when more than nineteen boxes are purchased.

a. Write a piecewise-defined function describing the cost of the boxes of tile.

b. What is the cost of purchasing 5 boxes of tile? What is the cost of purchasing 19 boxes of tile?

6. BOOKLETS A digital media company is ordering booklets to promote their business. The manufacturer charges \$0.50 for each booklet when 50 or fewer are ordered, \$0.45 for each booklet when 51 to 100 booklets are ordered, \$0.40 for each booklet when 101 to 250 booklets are ordered, and \$0.35 for each booklet when 251 or more booklets are ordered. Each order consists of a \$10 shipping charge, no matter the size of the order.

a. Write a piecewise-defined function describing the cost of ordering booklets.

b. What is the cost of purchasing 132 booklets? What is the cost of purchasing 518 booklets?

Examples 3 and 4

Graph each function. State the domain and range.

7. $f(x) = [\![x]\!] - 6$

8. $h(x) = [\![3x]\!] - 8$

9. $f(x) = [\![x + 1]\!]$

10. $f(x) = [\![x - 3]\!]$

11. PARKING The rates at a short-term parking garage are \$5.00 for 2 hours or less, \$10.00 for 4 hours or less, \$15.00 for 6 hours or less, and \$20.00 for 8 hours or less. Draw a graph that represents the charges. State the domain and range.

12. BOWLING The bowling alley offers special team rates. They charge \$30 for one hour or less of team bowling, \$45 for 2 hours or less, and \$60 for unlimited bowling after 2 hours of play. Draw a graph that represents the charges. State the domain and range.

Examples 5 and 6

Graph each function. State the domain and range.

13. $f(x) = |x - 5|$

14. $g(x) = |x + 2|$

15. $h(x) = |2x| - 8$

16. $k(x) = |-3x| + 3$

17. $f(x) = 2|x - 4| + 6$

18. $h(x) = -3|0.5x + 1| - 2$

19. $g(x) = 2|x|$

20. $f(x) = |x| + 1$

Mixed Exercises

Graph each function. State the domain and range.

21. $f(x) = \begin{cases} -3x & \text{if } x \le -4 \\ x & \text{if } 0 < x \le 3 \\ 8 & \text{if } x > 3 \end{cases}$

22. $f(x) = \begin{cases} 2x & \text{if } x \le -6 \\ 5 & \text{if } -6 < x \le 2 \\ -2x + 1 & \text{if } x > 4 \end{cases}$

23. $g(x) = \begin{cases} 2x + 2 & \text{if } x < -6 \\ x & \text{if } -6 \le x \le 2 \\ -3 & \text{if } x > 2 \end{cases}$

24. $g(x) = \begin{cases} -2 & \text{if } x < -4 \\ x - 3 & \text{if } -1 \le x \le 5 \\ 2x - 15 & \text{if } x > 7 \end{cases}$

25. $f(x) = \begin{cases} -0.5x + 1.5 & \text{if } x \le 1 \\ x - 4 & \text{if } x > 1 \end{cases}$

26. $f(x) = |x - 2|$

27. $f(x) = [\![x + 2]\!]$

28. $g(x) = 2[\![0.5x + 4]\!]$

29. $f(x) = [\![|0.5x|]\!]$

30. $g(x) = |[\![2x]\!]|$

31. $g(x) = \begin{cases} [\![x]\!] & \text{if } x < -4 \\ x + 1 & \text{if } -4 \le x \le 3 \\ -x & \text{if } x > 3 \end{cases}$

32. $h(x) = \begin{cases} -|x| & \text{if } x < -6 \\ |x| & \text{if } -6 \le x \le 2 \\ |-x| & \text{if } x > 2 \end{cases}$

33. Identify the domain and range of $h(x) = |x + 4| + 2$.

34. FINANCE For every transaction, a certain financial advisor gets a 5% commission, regardless of whether the transaction is a deposit or withdrawal. Write a formula using the absolute value function for the advisor's commission C. Let D represent the value of one transaction.

35. GAMING The graph shows the monthly fee that an an online gaming site charges based on the average number of hours spent online per day. Write the function represented by the graph.

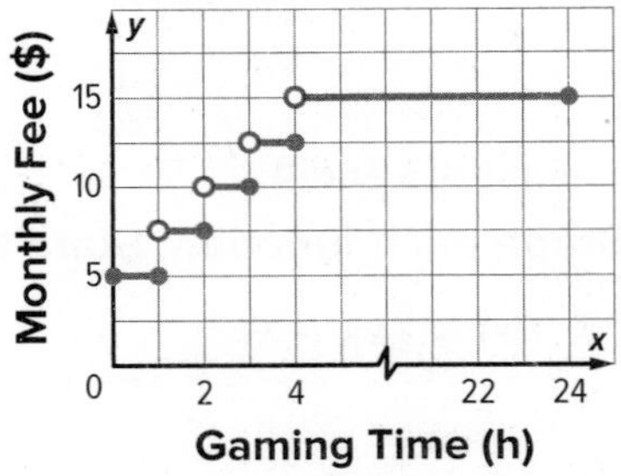

36. ROUNDING A science teacher instructs students to round their measurements as follows: If the decimal portion of a measurement is less than 0.5 mm, round down to the nearest whole millimeter. If the decimal portion of a measurement is exactly 0.5 or greater, round up to the next whole millimeter. Write a formula to represent the rounded measurements.

37. REUNIONS The cost to reserve a banquet hall is \$500. The catering cost per guest is \$17.50 for the first 40 guests and \$14.75 for each additional guest.

a. Write a piecewise-defined function describing the cost C of the reunion.

b. Use a graphing calculator to graph the function.

c. If the Cramers can spend up to \$900 on an event, what is the greatest number of guests that can attend? Explain.

38. **SAVINGS** Nathan puts $200 into a checking account when he gets his paycheck each month. The value of his checking account is modeled by $v = 200[\![m]\!]$, where m is the number of months that Nathan has been working. After 105 days, how much money is in the account?

39. **POLITICS** The approval rating $R(t)$, measured as a percent, of a class officer during her 9-month term starting in September is described by the graph, where t is her time in office.

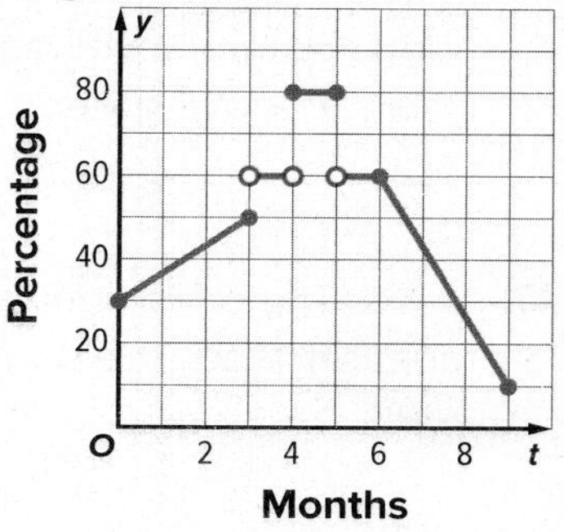

a. Formulate a piecewise-defined function $R(t)$ describing the approval rating of this class officer. Then, identify the range.

b. During which months is the approval rating increasing?

40. **STRUCTURE** Consider the functions $f(x) = 3[\![x]\!]$ and $g(x) = [\![3x]\!]$ for $0 \leq x \leq 2$.

a. Graph each function.

b. What effect does this 3 appear to have on the graphs?

c. Consider the functions $f(x) = 4[\![x]\!]$ and $g(x) = [\![4x]\!]$ for $0 \leq x \leq 2$. Graph each function.

d. What effect does this 4 appear to have on the graphs?

e. Generalize your findings from **parts a through d** to explain the differences between $f(x) = n[\![x]\!]$ and $g(x) = [\![nx]\!]$ for $0 \leq x \leq 2$, where n is any positive integer greater than or equal to 2.

41. **USE TOOLS** Use a graphing calculator to graph the absolute value of the greatest integer of x, or $f(x) = |[\![x]\!]|$. Is the graph what you expected? Explain.

Write a piecewise-defined function for each graph.

42.

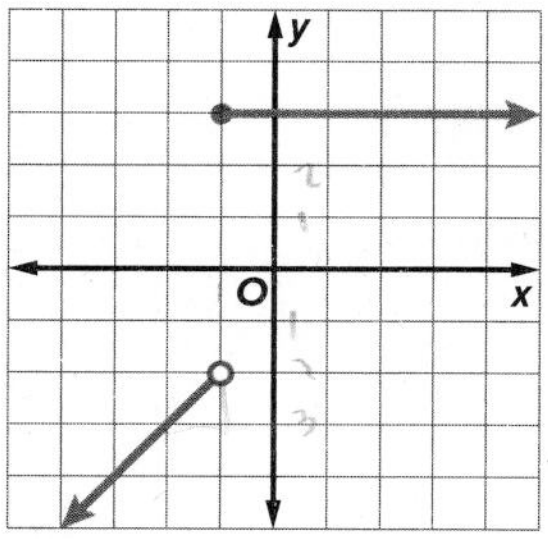

43.

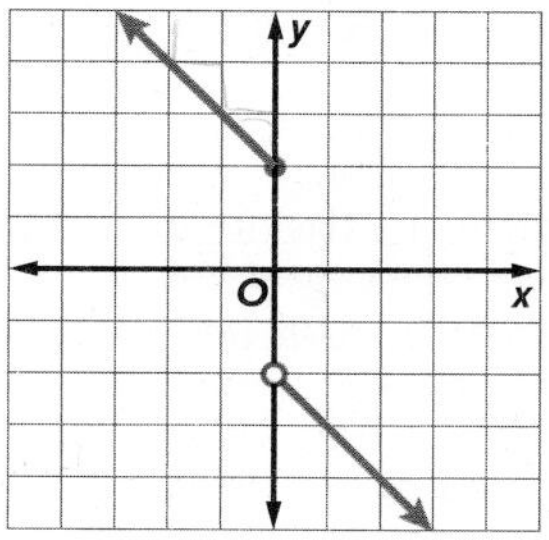

44.

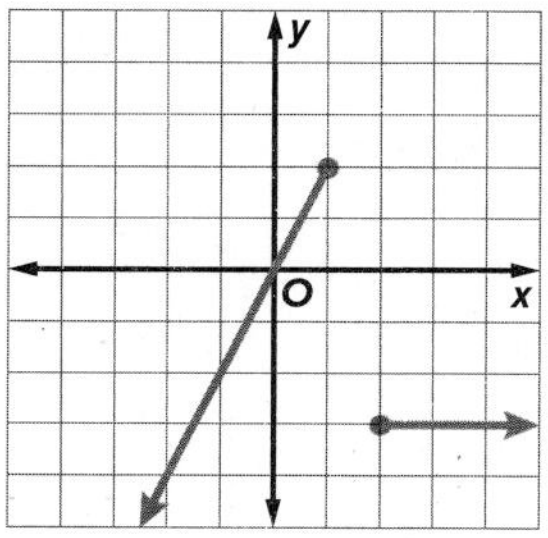

45.

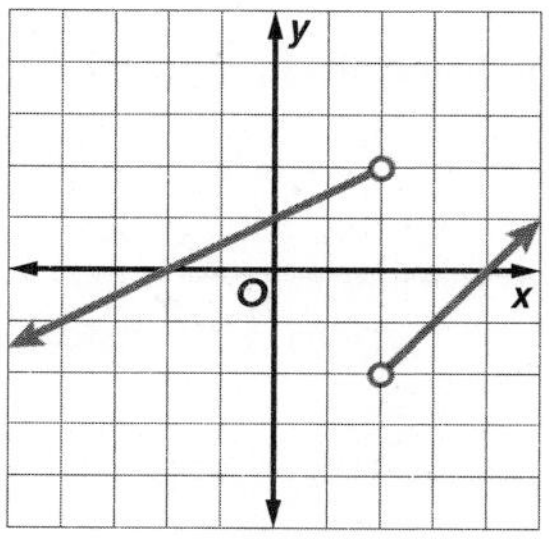

46. SKYSCRAPERS To clean windows of skyscrapers, some companies use a carriage. A carriage is mounted on a railing on the roof of a skyscraper and moves up and down using cables. A crew plans to start at the 12th floor and move the carriage down as they clean the windows on the west side of a building. If the crew members clean windows at a constant rate of 0.75 floor per hour, the absolute value function $f(t) = |0.75t - 12|$ represents the the number of floors above ground level that the carriage is after t hours. Graph the function. How far above ground level is the carriage after 4 hours?

47. TAXIS The table shows the cost C of a taxi ride of m miles. Graph the function. State the domain and range.

m	C
$0 < m \leq 1$	\$2.00
$1 < m \leq 2$	\$4.00
$2 < m \leq 3$	\$6.00
$3 < m \leq 4$	\$8.00
$4 < m \leq 5$	\$10.00
$5 < m \leq 6$	\$12.00

48. WALKING Jackson left his house and walked at a constant rate. After 20 minutes, he was 2 miles from his house. Jackson then walked back towards his house at a constant rate. After another 30 minutes he arrived at his house.

a. Jackson wants to write a function to model the distance from his house d after t minutes. Should Jackson write an absolute value function or a piecewise-defined function? Explain your reasoning.

b. Write an appropriate function to model the distance from his house d after t minutes.

c. Graph the function.

d. State the domain and range.

49. CREATE Write an absolute value relation in which the domain is all nonnegative numbers and the range is all real numbers.

50. PERSEVERE Graph $|y| = 2|x + 3| - 5$.

51. ANALYZE Find a counterexample to the statement and explain your reasoning.

In order to find the greatest integer function of x when x is not an integer, round x to the nearest integer.

52. CREATE Write an absolute value function in which $f(5) = -3$.

53. WRITE Explain how piecewise functions can be used to accurately represent real-world problems.

54. WRITE Explain the difference between a piecewise function and step function.

Transformations of Functions

Learn Translations of Functions

A **family of graphs** includes graphs and equations of graphs that have at least one characteristic in common. The parent graph is transformed to create other members in a family of graphs.

Key Concept • Parent Functions

Constant Function

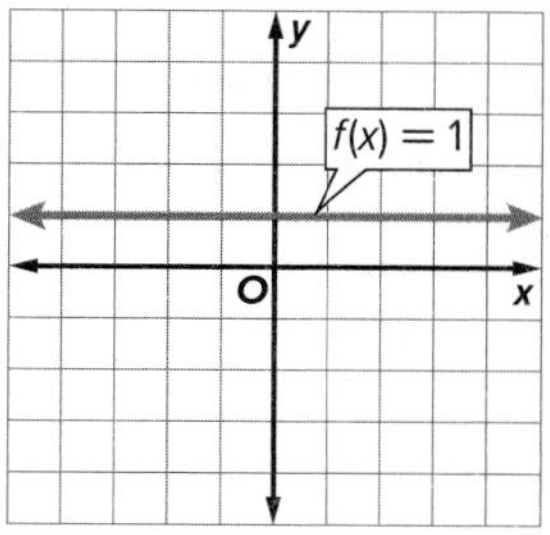

The general equation of a **constant function** is $f(x) = a$, where a is any number.

Domain: all real numbers

Range: $\{f(x) \mid f(x) = a\}$

Identity Function

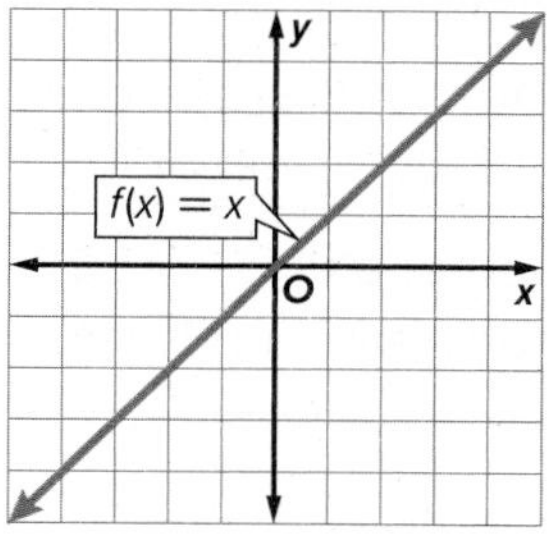

The **identity function** $f(x) = x$ includes all points with coordinates (a, a). It is the parent function of most linear functions.

Domain: all real numbers

Range: all real numbers

Absolute Value Function

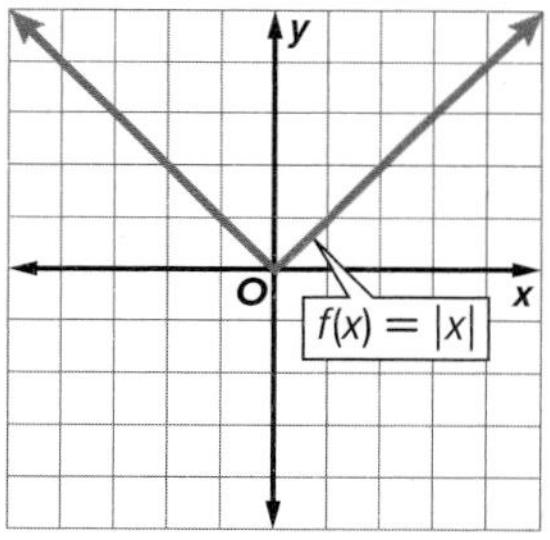

The parent function of absolute value functions is $f(x) = |x|$.

Domain: all real numbers

Range: $\{f(x) \mid f(x) \geq 0\}$

Quadratic Function

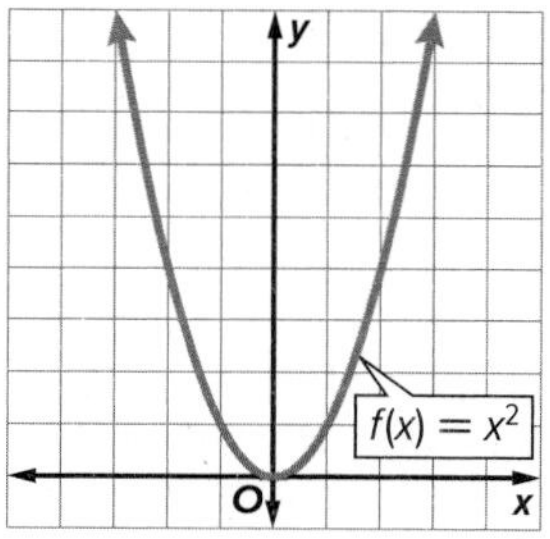

The parent function of quadratic functions is $f(x) = x^2$.

Domain: all real numbers

Range: $\{f(x) \mid f(x) \geq 0\}$

A **translation** is a transformation in which a figure is slid from one position to another without being turned.

Key Concept • Translations

Translation	Change to Parent Graph
$f(x) + k;\ k > 0$	The graph is translated k units up.
$f(x) + k;\ k < 0$	The graph is translated $\lvert k\rvert$ units down.
$f(x - h);\ h > 0$	The graph is translated h units right.
$f(x - h);\ h < 0$	The graph is translated $\lvert h\rvert$ units left.

Today's Goals

- Apply translations to the graphs of functions.
- Apply dilations to the graphs of functions.
- Apply compositions of transformations to the graphs of functions and use transformations to write equations from graphs.

Today's Vocabulary

family of graphs
constant function
identity function
transformations
translation
dilation
reflection
line of reflection

Go Online
You may want to complete the Concept Check to check your understanding.

Go Online
You can watch a video to see how to describe translations of functions.

Think About It!
Describe the vertex and axis of symmetry of a translated quadratic function, $f(x) = (x - h)^2 + k$, in terms of h and k.

Study Tip

Signs Although the translated function uses subtraction, the translation is in the form $f(x) + k$ because a constant is being added to the parent function. The value of k is negative because adding a negative number is the same as subtracting the positive opposite of that number. The translation $p(x) = (x - 3)^2$ is in the form $f(x - h)$, because the constant is being subtracted from the variable before evaluating.

Problem-Solving Tip

Use a Graph When writing the equation of a graph, use the key features of the graph to determine transformations. Notice how the maximum, minimum, intercepts, and axis of symmetry have changed from the parent function in order to determine which transformations have been applied.

Go Online

You may want to complete the Concept Check to check your understanding.

Example 1 Translations

Describe the translation in $g(x) = (x + 2)^2 - 4$ as it relates to the graph of the parent function.

Since $g(x)$ is quadratic, the parent function is $f(x) = x^2$.

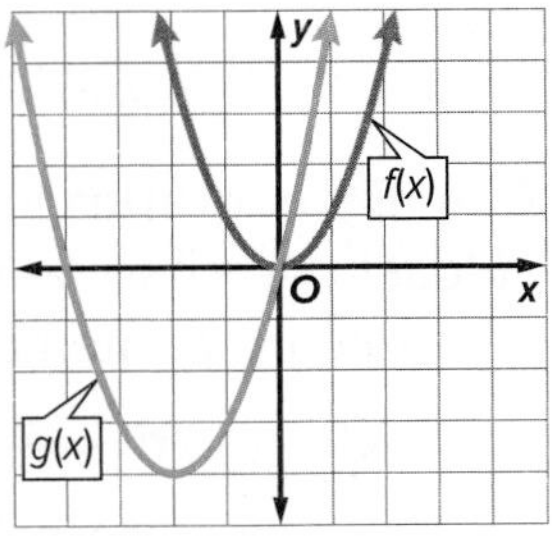

Since $f(x) = x^2$, $g(x) = f(x - h) + k$, where $h = -2$ and $k = -4$.

The constant k is added to the function after it has been evaluated, so k affects the output, or y-values. The value of k is less than 0, so the graph of $f(x) = x^2$ is translated down 4 units.

The value of h is subtracted from x before it is evaluated and is less than 0, so the graph of $f(x) = x^2$ is also translated 2 units left.

The graph of $g(x) = (x + 2)^2 - 4$ is the translation of the graph of the parent function 2 units left and 4 units down.

Example 2 Identify Translated Functions from Graphs

Use the graph of the function to write its equation.

The graph is an absolute value function with a parent function of $f(x) = |x|$. Notice that the vertex of the function has been shifted both vertically and horizontally from the parent function.

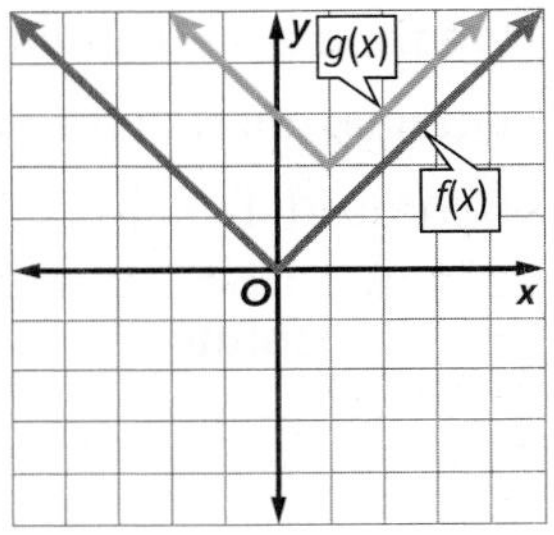

To write the equation of the graph, determine the values of h and k in $g(x) = |x - h| + k$.

The translated graph has been shifted 2 units up and 1 unit right. So, $h = 1$ and $k = 2$. Thus, $g(x) = |x - 1| + 2$.

Learn Dilations and Reflections of Functions

A **dilation** is a transformation that stretches or compresses the graph of a function. Multiplying a function by a constant dilates the graph with respect to the y-axis. Multiplying the independent variable by a constant dilates the graph with respect to the x-axis.

Key Concept • Dilations

Dilation	Change to Parent Graph
$af(x)$, $\lvert a\rvert > 1$	vertical stretch; Each y-value of $f(x)$ is multiplied by $\lvert a\rvert$.
$af(x)$, $0 < \lvert a\rvert < 1$	vertical compression; Each y-value of $f(x)$ is multiplied by $\lvert a\rvert$.
$f(ax)$, $\lvert a\rvert > 1$	horizontal compression; Each x-value of $f(x)$ is multiplied by $\frac{1}{\lvert a\rvert}$.
$f(ax)$, $0 < \lvert a\rvert < 1$	horizontal stretch; Each x-value of $f(x)$ is multiplied by $\frac{1}{\lvert a\rvert}$.

Go Online You can complete an Extra Example online.

A **reflection** is a transformation where a figure, line, or curve, is flipped in a **line of reflection**. Often the reflection is in the *x*- or *y*-axis.

When a parent function $f(x)$ is multiplied by -1, the result $-f(x)$ is a reflection of the graph in the *x*-axis. When only the variable is multiplied by -1, the result $f(-x)$ is a reflection of the graph in the *y*-axis.

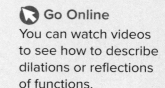

Go Online You can watch videos to see how to describe dilations or reflections of functions.

Key Concept • Reflections	
Reflection	**Change to Parent Graph**
$-f(x)$	reflection in the *x*-axis
$f(-x)$	reflection in the *y*-axis

Example 3 Vertical Dilations

Describe the dilation and reflection in $g(x) = -\frac{2}{5}x$ as it relates to the parent function.

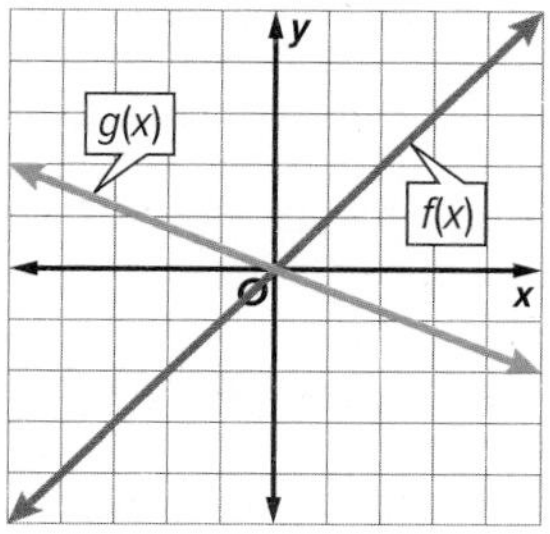

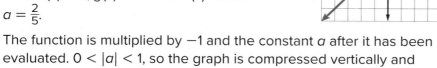

Since $g(x)$ is a linear function, the parent function is $f(x) = x$.

Since $f(x) = x$, $g(x) = -1 \cdot a \cdot f(x)$ where $a = \frac{2}{5}$.

The function is multiplied by -1 and the constant a after it has been evaluated. $0 < |a| < 1$, so the graph is compressed vertically and reflected in the *x*-axis.

The graph of $g(x) = -\frac{2}{5}x$ is the graph of the parent function compressed vertically and reflected in the *x*-axis.

Think About It!
Describe the effect of multiplying the same value of a, $-\frac{2}{5}$, by a different parent function such as $f(x) = |x|$.

Example 4 Horizontal Dilations

Describe the dilation and reflection in $g(x) = (-2.5x)^2$ as it relates to the parent function.

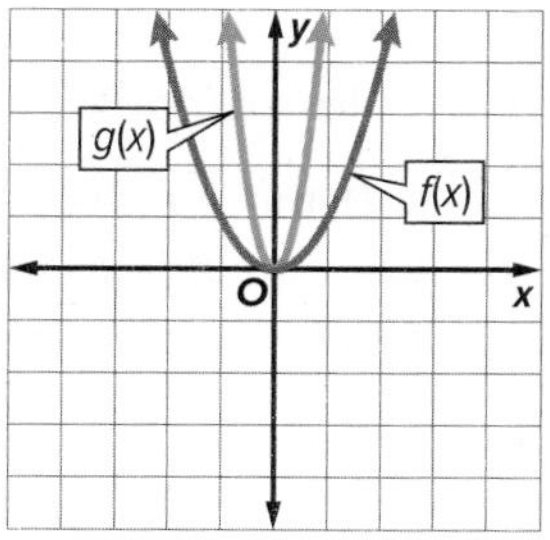

Since $g(x)$ is a quadratic function, the parent function is $f(x) = x^2$. Since $f(x) = x^2$, $g(x) = f(-1 \cdot a \cdot x)$, where $a = 2.5$.

x is multiplied by -1 and the constant a before the function is performed and $|a|$ is greater than 1, so the graph of $f(x) = x^2$ is compressed horizontally and reflected in the *y*-axis.

Think About It!
Why does the graph of $g(x) = (-2.5x)^2$ appear the same as $j(x) = (2.5x)^2$?

Go Online You can complete an Extra Example online.

Explore Using Technology to Transform Functions

Online Activity Use graphing technology to complete the Explore.

INQUIRY How does performing an operation on a function change its graph?

Think About It!
Do the values of a, h, and k affect various parent functions in different ways? Explain.

Learn Transformations of Functions

The general form of a function is $g(x) = a \cdot f(x - h) + k$, where $f(x)$ is the parent function. Each constant in the equation affects the parent graph.

- The value of $|a|$ stretches or compresses (dilates) the parent graph.
- When the value of a is negative, the graph is reflected across the x-axis.
- The value of h shifts (translates) the parent graph left or right.
- The value of k shifts (translates) the parent graph up or down.

In $g(x) = a \cdot f(x - h) + k$, each constant affects the graph of $f(x) = x^2$.

Go Online
You can watch a video to see how to graph transformations of functions using a graphing calculator.

Key Concept • Transformations of Functions

Dilation, a

If $|a| > 1$, the graph of $f(x)$ is stretched vertically. If $0 < |a| < 1$, the graph of $f(x)$ is compressed vertically.

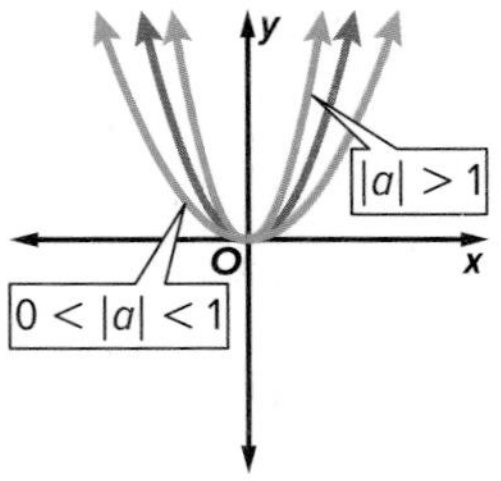

Reflection, a

If $a > 0$, the graph of $f(x)$ opens up. If $a < 0$, the graph of $f(x)$ opens down.

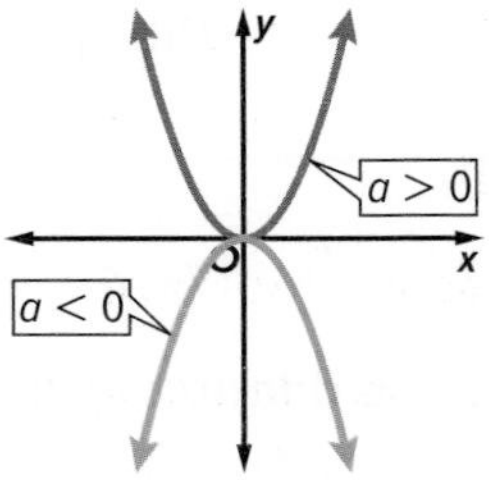

Horizontal Translation, h

If $h > 0$, the graph of $f(x)$ is translated h units right. If $h < 0$, the graph of $f(x)$ is translated $|h|$ units left.

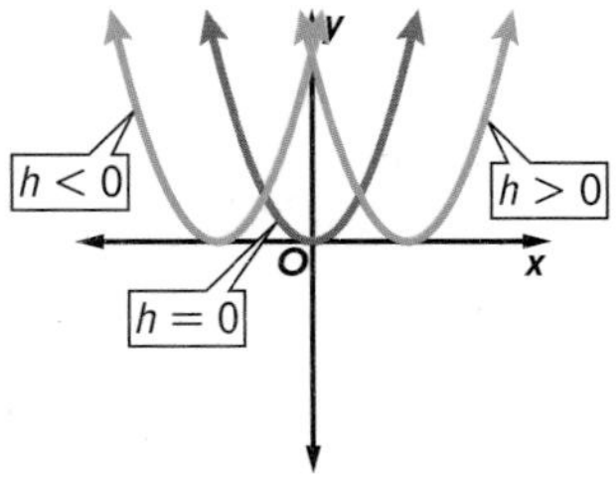

Vertical Translation, k

If $k > 0$, the graph of $f(x)$ is translated k units up. If $k < 0$, the graph of $f(x)$ is translated $|k|$ units down.

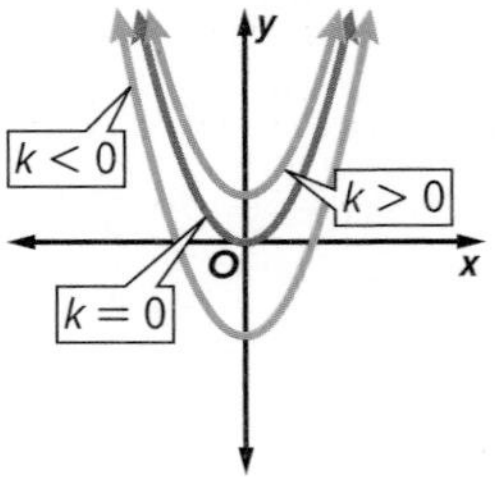

Example 5 Multiple Transformations of Functions

Describe how the graph of $g(x) = -\frac{2}{3}|x + 3| + 1$ is related to the graph of the parent function.

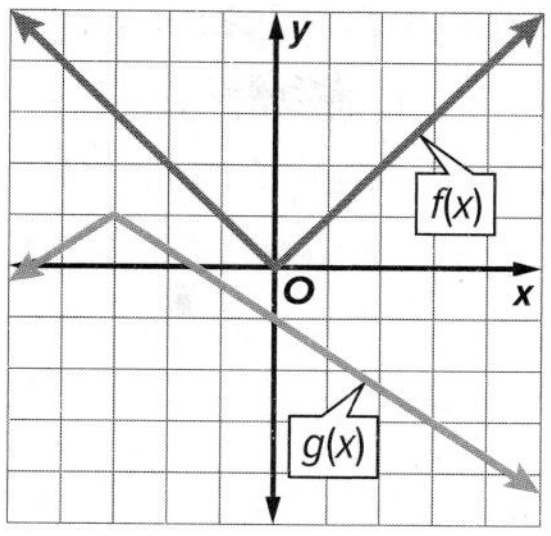

The parent function is $f(x) = |x|$.

Since $f(x) = |x|$, $g(x) = af(x - h) + k$ where $a = -\frac{2}{3}$, $h = -3$ and $k = 1$.

The graph of $g(x) = -\frac{2}{3}|x + 3| + 1$ is the graph of the parent function compressed vertically, reflected in the x-axis, and translated 3 units left and 1 unit up.

Check

Describe how $g(x) = -(0.4x + 2)$ is related to the graph of the parent function.

Go Online You can watch a video to see how to use transformations to graph an absolute value function.

Think About It!

Write an equation for a quadratic function that opens down, has been stretched vertically by a factor of 4, and is translated 2 units right and 5 units down.

Example 6 Apply Transformations of Functions

DOLPHINS **Suppose the path of a dolphin during a jump is modeled by $g(x) = -0.125(x - 12)^2 + 18$, where x is the horizontal distance traveled by the dolphin and $g(x)$ is its height above the surface of the water. Describe how $g(x)$ is related to its parent function and interpret the function in the context of the situation.**

Because $f(x) = x^2$ is the parent function, $g(x) = af(x - h) + k$, where $a = -0.125$, $h = 12$, and $k = 18$.

Translations

$12 > 0$, so the graph of $f(x) = x^2$ is translated 12 units right.

$18 > 0$, so the graph of $f(x) = x^2$ is translated 18 units up.

Dilation and Reflection

$0 < |-0.125| < 1$, so the graph of $f(x) = x^2$ is compressed vertically.

$a < 0$, so the graph of $f(x) = x^2$ is a reflection in the x-axis.

Interpret the Function

Because a is negative, the path of the dolphin is modeled by a parabola that opens down. This means that the vertex of the parabola (h, k) represents the maximum height of the dolphin, 18 feet, at 12 feet from the starting point of the jump.

Study Tip

Interpretations When interpreting transformations, analyze how each value influences the function and alters the graph. Then determine what you think each value might mean in the context of the situation.

Go Online You can complete an Extra Example online.

Example 7 Identify an Equation from a Graph

Write an equation for the function.

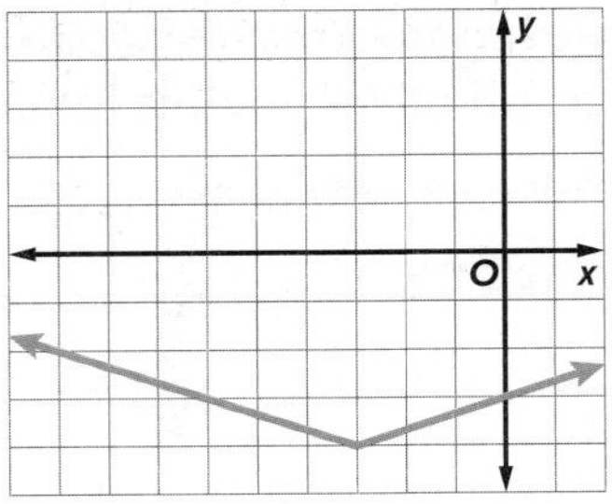

Step 1 Analyze the graph.

The graph is an absolute value function with a parent function of $f(x) = |x|$. Analyze the graph to make a prediction about the values of a, h, and k in the equation $y = a|x - h| + k$.

The graph appears to be wider than the parent function, implying a vertical compression, and is not reflected. So, a is positive and $0 < |a| < 1$.

The graph has also been shifted left and down from the parent graph. So, $h < 0$ and $k < 0$.

Step 2 Identify the translation(s).

Identify the horizontal and vertical translations to find the values of h and k.

The vertex is shifted 3 units left, so $h = -3$.

It is also shifted 4 units down, so $k = -4$.

$y = a\|x - h\| + k$	General form of the equation
$y = a\|x - (-3)\| + (-4)$	$h = -3$ and $k = -4$
$y = a\|x + 3\| - 4$	Simplify.

Step 3 Identify the dilation and/or reflection.

Use the equation from Step 2 and a point on the graph to find the value of a.

The point $(0, -3)$ lies on the graph. Substitute the coordinates in for x and y to solve for a.

$y = a\|x + 3\| - 4$	Original equation
$-3 = a\|0 + 3\| - 4$	$(0, -3) = (x, y)$
$-3 = a\|3\| - 4$	Add.
$-3 = 3a - 4$	Evaluate the absolute value.
$\frac{1}{3} = a$	Solve.

Step 4 Write an equation for the function.

Since $a = \frac{1}{3}$, $h = -3$ and $k = -4$, the equation is $g(x) = \frac{1}{3}|x + 3| - 4$.

Check

Use the graph to write an equation for $g(x)$.

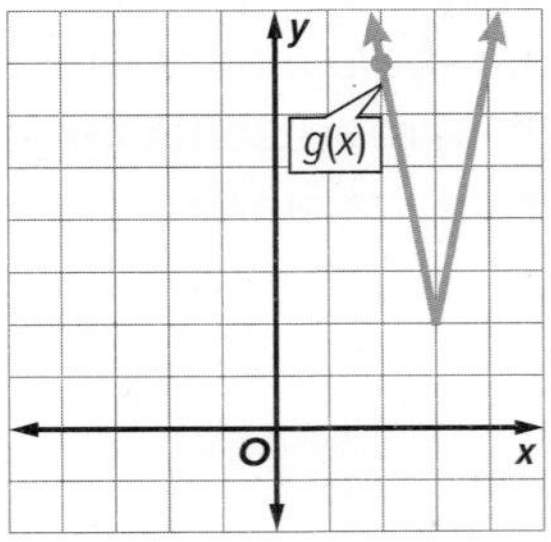

Go Online You can complete an Extra Example online.

Watch Out!

Choosing a Point When substituting for x and y in the equation, use a point other than the vertex.

Think About It!

How does the equation you found compare to the prediction you made in step 1?

Practice

Go Online You can complete your homework online.

Example 1

Describe each translation as it relates to the graph of the parent function.

1. $y = x^2 + 4$

2. $y = |x| - 3$

3. $y = x - 1$

4. $y = x + 2$

5. $y = (x - 5)^2$

6. $y = |x + 6|$

Example 2

Use the graph of each translated parent function to write its equation.

7.

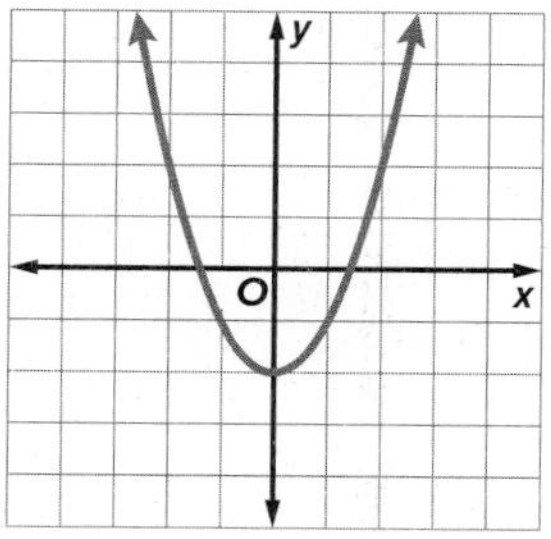

8.

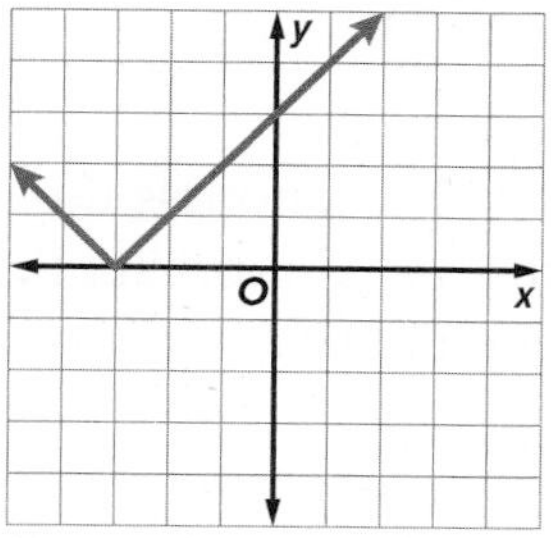

9.

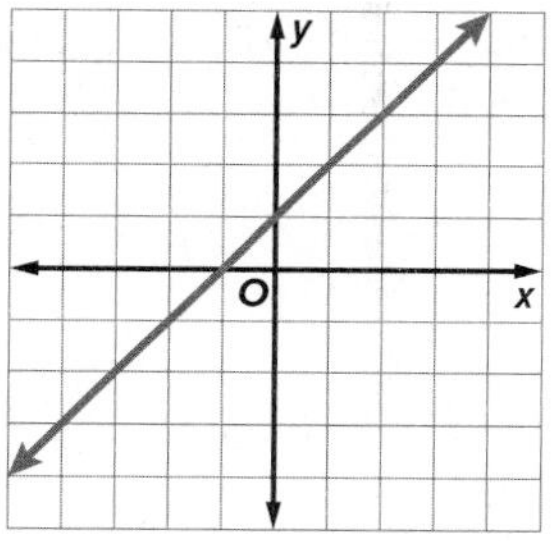

10.

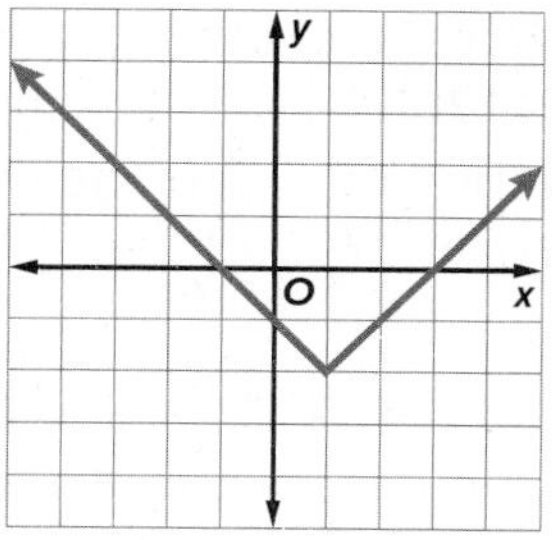

11.

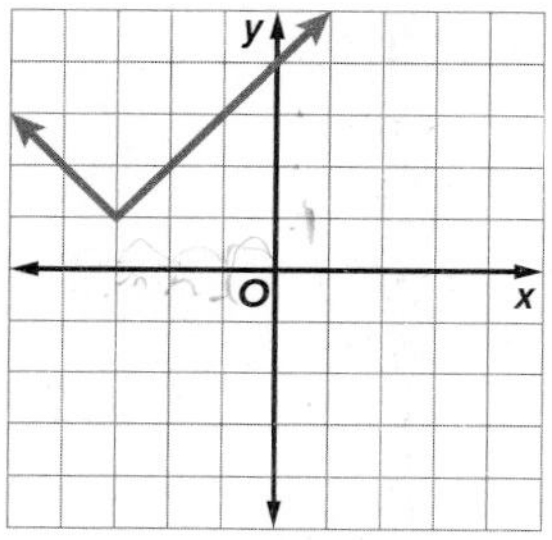

12.

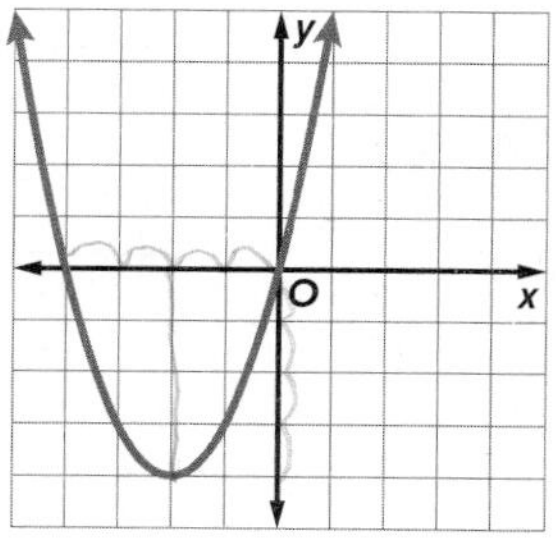

Examples 3 and 4

Describe each dilation and reflection as it relates to the parent function.

13. $y = (-3x)^2$

14. $y = -6x$

15. $y = -4|x|$

16. $y = |-2x|$

17. $y = -\frac{2}{3}x$

18. $y = -\frac{1}{2}x^2$

19. $y = |-\frac{1}{3}x|$

20. $y = \left(-\frac{3}{4}x\right)^2$

Example 5

Describe each transformation as it relates to the graph of the parent function.

21. $y = -6|x| - 4$

22. $y = 3x + 11$

23. $y = \frac{1}{3}x^2 - 2$

24. $y = \frac{1}{2}|x - 1| + 14$

25. $y = -0.8(x + 3)$

26. $y = (1.5x)^2 + 22$

Example 6

27. BILLIARDS The function $g(x) = |0.5x|$ models the path of a cue ball in a certain shot on a pool table, where the x-axis represents the edge of the table. Describe how $g(x)$ is related to its parent function and interpret the function in the context of the situation.

28. SALAD The cost for a salad depends on its weight, x, in ounces, and is described by $c(x) = 4.5 + 0.32x$. Describe how $c(x)$ is related to its parent function and interpret the function in the context of the situation.

29. TRAVEL The cost to travel x miles east or west on a train is the same. The function for the cost is $c(x) = 0.75|x| + 25$. Describe how $c(x)$ is related to its parent function and interpret the function in the context of the situation.

30. ARCHERY The path of an arrow can be modeled by $h(x) = -0.03x^2 + 6$, where x is distance and $h(x)$ is height, both in feet. Describe how $h(x)$ is related to its parent function and interpret the function in the context of the situation.

Example 7

Write an equation for each function.

31.

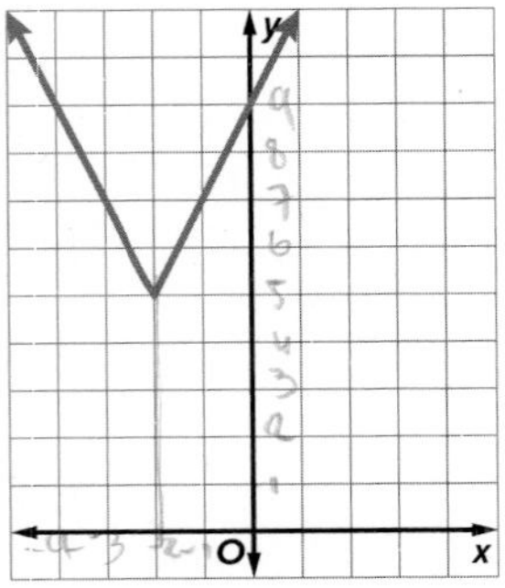

32.

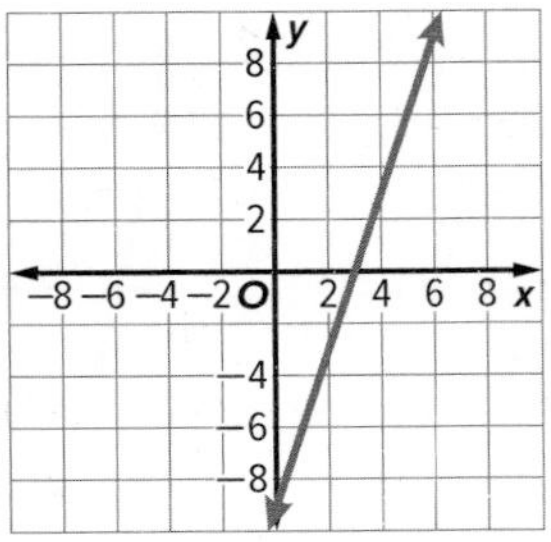

Write an equation for each function.

33.

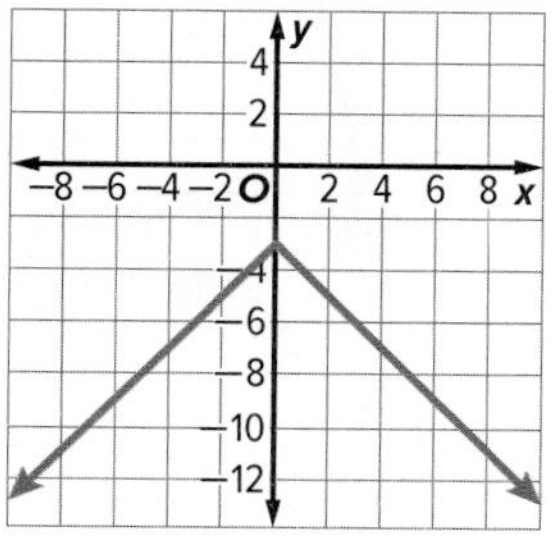

34.

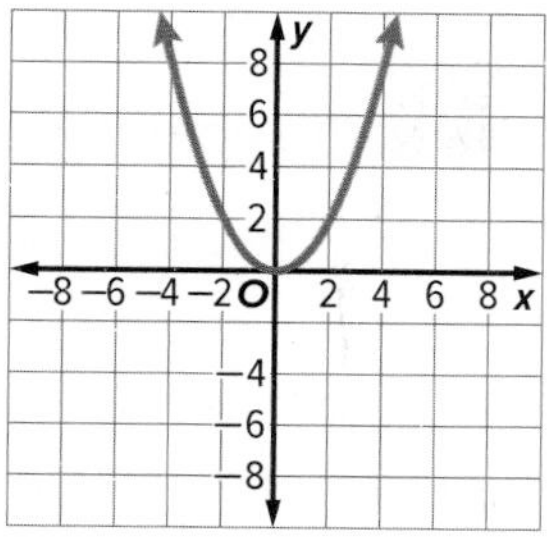

35.

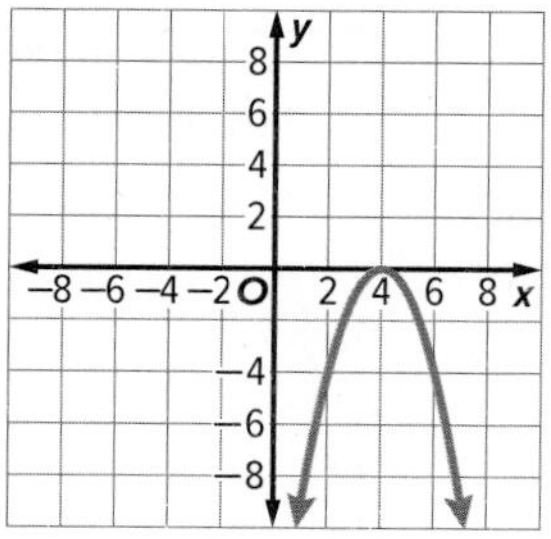

36.

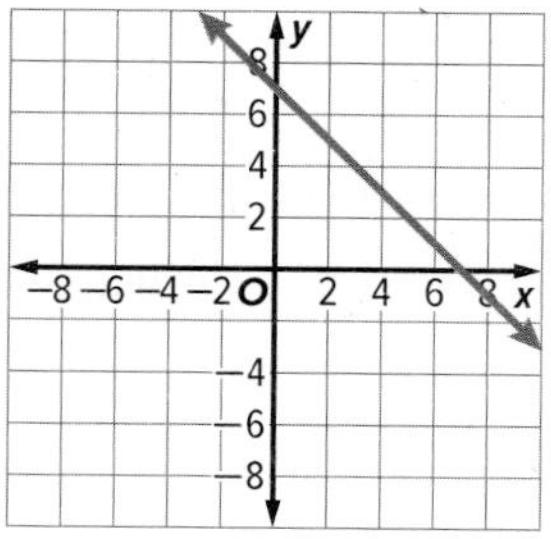

Mixed Exercises

Describe each transformation as it relates to the graph of the parent function. Then graph the function.

37. $y = |x| - 2$

38. $y = (x + 1)^2$

39. $y = -x$

40. $y = -|x|$

41. $y = 3x$

42. $y = 2x^2$

43. Describe the translation in $y = x^2 - 4$ as it relates to the parent function.

44. Describe the reflection in $y = -x^3$ as it relates to the parent function.

45. Describe the type of transformation in the function $f(x) = (5x)^2$.

46. ARCHITECTURE The cross-section of a roof is shown in the figure. Write an absolute value function that models the shape of the roof.

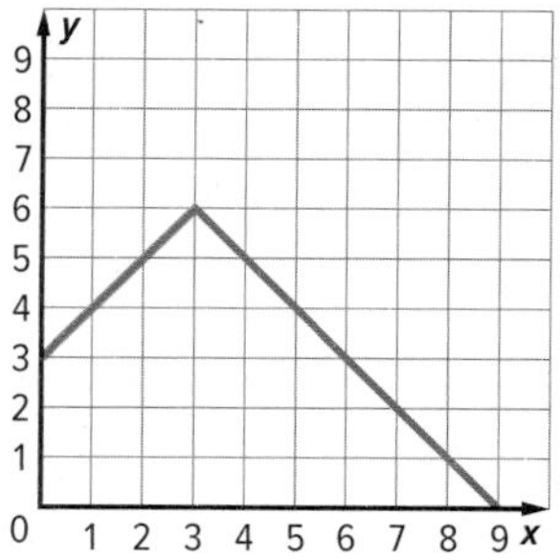

47. SPEED The speedometer in Henry's car is broken. The function $y = |x - 8|$ represents the difference y between the car's actual speed x and the displayed speed.

a. Describe the translation. Then graph the function.

b. Interpret the function and the translation in terms of the context of the situation.

48. GRAPHIC DESIGN Kassie sketches the function $f(x) = -1.25(x - 1)^2 + 18.75$ as part of a new logo design. Describe the transformations she applied to the parent function in creating her function.

49. GEOMETRY Chen made a graph to show how the perimeter of a square changes as the length of sides increase. How is this graph related to the parent function $y = x$?

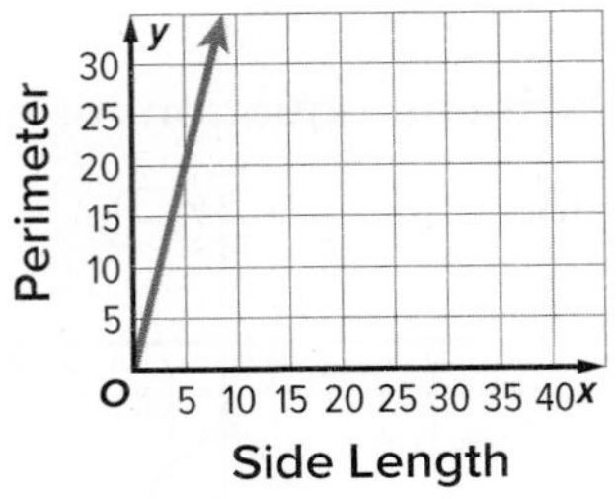

50. REASONING Compare the graph of the parent function $f(x) = |x|$ with the graphs of $g(x) = |x + 2|$ and $h(x) = |x - 3|$. How are the graphs similar? How are they different?

51. BUSINESS Maria earns $10 an hour working as a lifeguard. She drew the graph to show the relation of her income as a function of the hours she works. How did she modify the function $y = x$ to create her graph?

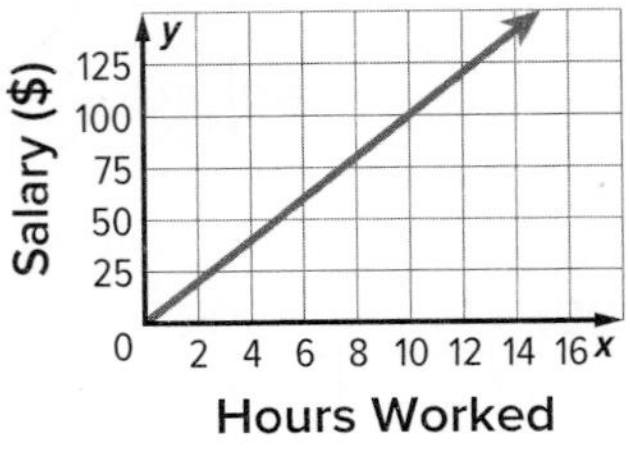

52. ANALYZE What determines whether a transformation will affect the graph vertically or horizontally? Use the family of quadratic functions as an example.

53. PERSEVERE Laura sketches the path of a model rocket that she launches.

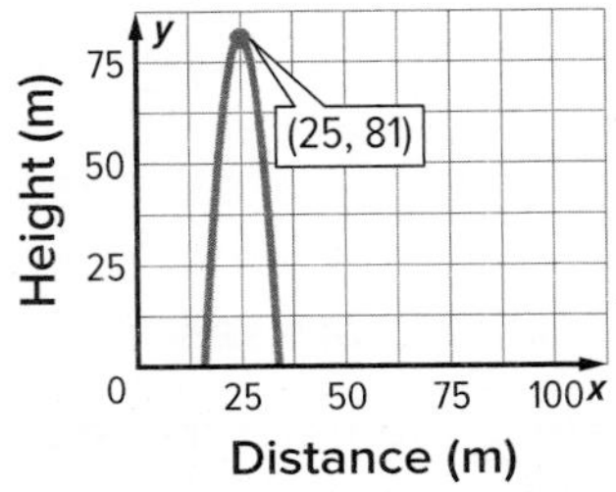

a. What type of function does the graph show?

b. In which axis has the parent function been reflected?

c. How has the graph been translated? Assume that the function has not been dilated.

d. What is the equation for the curve shown on the graph?

54. ANALYZE Graph $g(x) = -3|x + 5| - 1$. Describe the transformations of the parent function $f(x) = |x|$ that produce the graph of $g(x)$. What are the domain and range?

55. ANALYZE Consider $f(x) = |2x|$, $g(x) = x + 2$, $h(x) = 2x^2$, and $k(x) = 2x^3$.

a. Graph each function and its reflection in the y-axis.

b. Analyze the functions and the graphs. Determine whether each function is *odd*, *even*, or *neither*.

c. Recall that if for all values of x, $f(-x) = f(x)$ the function $f(x)$ is an even function. If for all values of x, $f(-x) = -f(x)$ the function $f(x)$ is an odd function. Explain why this is true.

56. PERSEVERE Explain why performing a horizontal translation followed by a vertical translation has the same results as performing a vertical translation followed by a horizontal translation.

57. CREATE Draw a graph in Quadrant II. Use any of the transformations you learned in this lesson to move your figure to Quadrant IV. Describe your transformation.

58. ANALYZE Study the parent graphs at the beginning of this lesson. Select a parent graph with positive y-values when $x \to -\infty$ and positive y-values when $x \to \infty$.

59. WRITE Explain why the graph of $g(x) = (-x)^2$ appears the same as the graph of $f(x) = x^2$. Is this true for all reflections of quadratic functions? If not, describe a case when it is false.

Review

Essential Question

How can analyzing a function help you understand the situation it models?

Module Summary

Lesson 1-1 through 1-3

Function Behavior

- The graph of a continuous function is a line or curve. The domain of a continuous function is a single interval of all real numbers.
- A linear function is a function in which no independent variable is raised to a whole number power greater than 1.
- If a vertical line intersects the graph of a relation more than once, then the relation is not a function.
- An *x*-intercept occurs when the graph intersects the *x*-axis, and a *y*-intercept occurs when the graph intersects the *y*-axis.
- A graph has line symmetry if each half of the graph on either side of a line matches the other side exactly.
- A graph has point symmetry when a figure is rotated 180° about a point and maps exactly onto the other part.
- A point is a relative maximum if there are no other nearby points with a greater *y*-coordinate. A point is a relative minimum if there are no other nearby points with a lesser *y*-coordinate.
- End behavior is the behavior of the graph at its ends. At the right end, the values of *x* are increasing toward infinity. This is denoted as $x \to \infty$. At the left end, the values of *x* are decreasing toward negative infinity, denoted as $x \to -\infty$.

Lessons 1-4 through 1-7

Graphs of Functions

- You can use key features of a function to sketch its graph. Features such as intercepts, symmetry, end behavior, extrema, and intervals where the function is increasing, decreasing, positive, or negative provide information for sketching the graph.
- A function that is written using two or more expressions is a piecewise-defined function.
- A step function has a graph that is a series of horizontal line segments.
- An absolute value function is a function that contains an algebraic expression within absolute value symbols.
- A translation moves a figure up, down, left, or right.
- A dilation shrinks or enlarges a figure proportionally. Multiplying a function by a constant dilates the graph with respect to the *x*- or *y*-axis.
- A reflection is a transformation that flips a figure in a line of reflection.

Study Organizer

Foldables

Use your Foldable to review this module. Working with a partner can be helpful. Ask for clarification of concepts as needed.

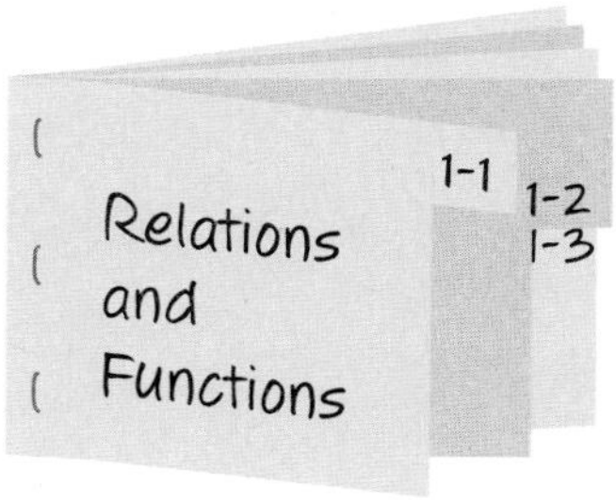

Test Practice

1. MULTIPLE CHOICE What is the domain of the function shown? (Lesson 1-1)

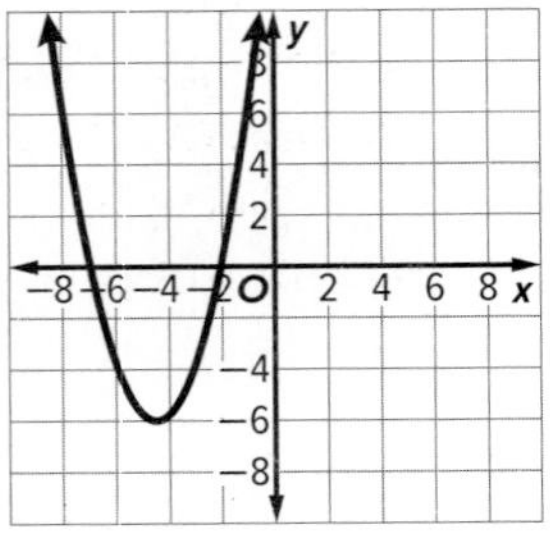

A. $(-8, 0)$

B. $[-6, \infty)$

C. $(-\infty, \infty)$

D. $(-\infty, 0)$

2. MULTIPLE CHOICE Salvatore is a plumber. He charges \$100 for all work that is completed in less than 2 hours. He charges \$250 for work that requires 2 to 5 hours, and he charges \$400 for work that takes between 5 and 8 hours.

Which best describes the domain of the function that models Salvatore's price scale? (Lesson 1-1)

A. The domain is $\{100, 150, 400\}$.

B. The domain is $\{x \mid 0 < x \leq 8\}$.

C. The domain is $\{x \mid 100 \leq x \leq 400\}$.

D. The domain is $\{x \mid x = 2, 5, 8\}$.

3. OPEN RESPONSE The table shows the amount of money Tia owed her friend over time after borrowing the money to go to a theme park. (Lesson 1-2)

Week	0	1	2	3	4	5	6
Amount	\$80	\$68	\$52	\$39	\$21	\$10	\$0

What are the coordinates of the *x*-intercept and the coordinates of the *y*-intercept? What is the meaning of the intercepts in context of the situation?

4. MULTIPLE CHOICE What type of symmetry is shown? (Lesson 1-2)

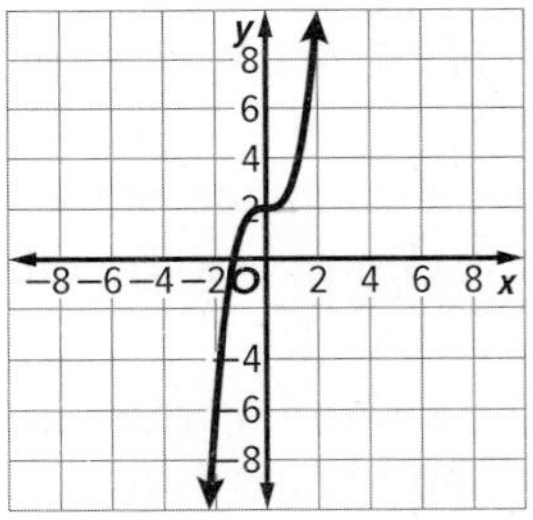

A. line symmetry

B. point symmetry

C. both line and point symmetry

D. no symmetry

5. **OPEN RESPONSE** Determine whether each of the *x*-values (–5, –4, –2, 0, 1, 5) is a *relative maximum*, *relative minimum* or *neither*. (Lesson 1-3)

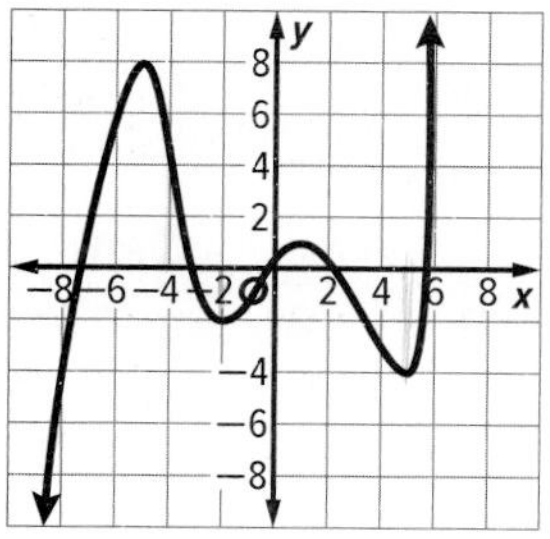

6. **OPEN RESPONSE** The graph shows the height of a ball after being thrown from a height of 26 feet.

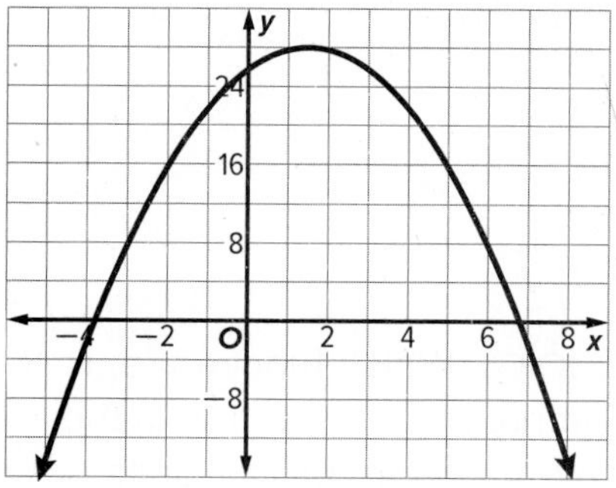

Explain why the end behavior does or does not make sense in this context. (Lesson 1-3)

7. **OPEN RESPONSE** Compare the key features of $f(x)$ and $g(x)$. (Lesson 1-4)

f(x)
$f(x) = -2x + 4$

g(x)
x-intercept: (1, 0)
y-intercept: (0, −5)
slope: 5

8. **MULTIPLE CHOICE** Sofia is sketching the graph of a function. She knows that as $x \to \infty, y \to -\infty$ and that the function has a *y*-intercept at (0, 8). Which other feature fits the sketch of the graph? (Lesson 1-4)

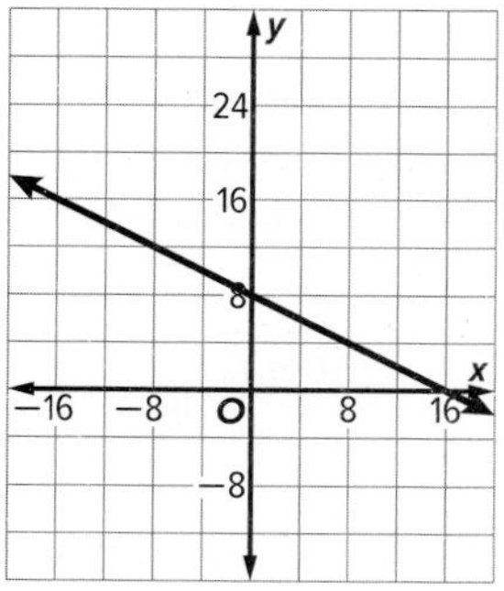

A. as $x \to -\infty, y \to -\infty$

B. *x*-intercept at (−14, 0)

C. increases for *y* in the interval $8 < x < 16$

D. decreases for *y* in the interval $-16 < x < 16$

9. MULTIPLE CHOICE A guidance counselor tells Rebekah that she needs a combined score of at least 1210 on the two portions of her college entrance exam to be eligible for the college of her choice. Suppose x represents Rebekah's score on the verbal portion and y represents her score on the math portion. Choose the equation or inequality that represents the scores Rebekah needs for eligibility. (Lesson 1-5)

A. $x + y \geq 1210$

B. $x + y = 1210$

C. $x + y > 1210$

D. $x + y \leq 1210$

10. OPEN RESPONSE The value of the Kim's portfolio is found by finding the sum of 60% of Stock A's value and 40% of Stock B's value. Kim wants her portfolio to be more than \$500,000. Write the inequality that represents the constraints for Stock A's value x and Stock B's value y. (Lesson 1-5)

11. GRAPH Graph $f(x) = [\![\frac{1}{3}x]\!] + 2$. (Lesson 1-6)

12. OPEN RESPONSE Determine if each transformation is a *translation*, *reflection*, or *dilation* from the parent function $f(x) = x^2$. (Lesson 1-7)

$$g(x) = -x^2$$
$$h(x) = (3x)^2$$
$$j(x) = (x - 4)^2$$

13. OPEN RESPONSE Describe the transformation(s) from the parent function to $f(x) = 3(x - 2)^2 + 9$. (Lesson 1-7)

14. OPEN RESPONSE If $g(x) = f(0.75x)$ then how is the graph of $g(x)$ related to the graph of $f(x)$? (Lesson 1-7)

15. MULTIPLE CHOICE Which function represents a vertical compression, reflection in the x-axis, and a translation down 3 units in relation to the parent function? (Lesson 1-7)

A. $y = -\frac{2}{3}(x - 3)^2$

B. $y = -\left(\frac{2}{3}x\right)^2 - 3$

C. $y = -\frac{2}{3}x^2 - 3$

D. $y = \frac{2}{3}(-x + 3)^2$

Module 2
Linear Equations, Inequalities, and Systems

Essential Question

How are equations, inequalities, and systems of equations or inequalities best used to model to real-world situations?

What Will You Learn?

How much do you already know about each topic **before** starting this module?

KEY

— I don't know. — I've heard of it. — I know it!

	Before			After		
	I don't know.	I've heard of it.	I know it!	I don't know.	I've heard of it.	I know it!
solve linear equations						
solve linear inequalities						
solve absolute value equations and inequalities						
write equations of linear functions in standard, slope-intercept, and point-slope form						
solve systems of equations by graphing, by substitution, and by elimination						
solve systems of inequalities in two variables						
use linear programming to find maximum and minimum values of a function						
solve systems of equations in three variables						

Foldables Make this Foldable to help you organize your notes about equations and inequalities. Begin with one sheet of paper.

1. **Fold** 2-inch tabs on each of the short sides.
2. **Fold** in half in both directions.
3. **Open** and cut as shown.
4. **Refold** along the width. Staple each pocket. Label pockets as *Solving Equations and Inequalities*, *Writing Equations for Functions*, *Systems of Equations and Inequalities*, and *Solve and Graph Inequalities*. Place index cards for notes in each pocket.

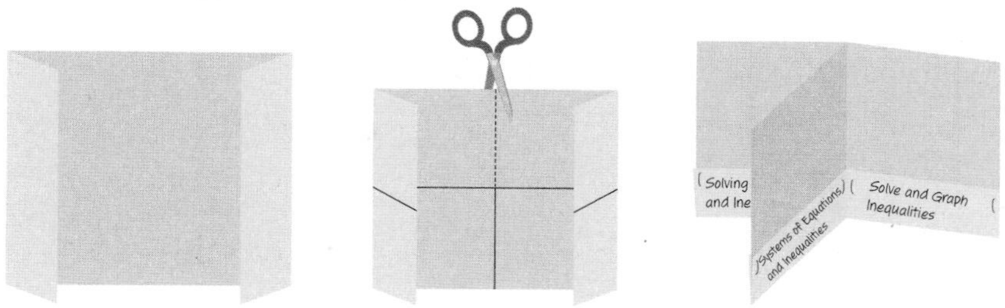

What Vocabulary Will You Learn?

- absolute value
- bounded
- consistent
- dependent
- elimination
- empty set
- equation
- extraneous solution
- feasible region
- inconsistent
- independent
- inequality
- linear programming
- optimization
- ordered triple
- root
- solution
- substitution
- system of equations
- system of inequalities
- unbounded
- zero

Are You Ready?

Complete the Quick Review to see if you are ready to start this module.
Then complete the Quick Check.

Quick Review

Example 1

Graph $2y + 5x = -10$.

Find the *x*- and *y*-intercepts.

$2(0) + 5x = -10$ $\quad$ $2y + 5(0) = -10$

$5x = -10$ $\quad$ $2y = -10$

$x = -2$ $\quad$ $y = -5$

The graph crosses the *x*-axis at $(-2, 0)$ and the *y*-axis at $(0, -5)$. Use these ordered pairs to graph the equation.

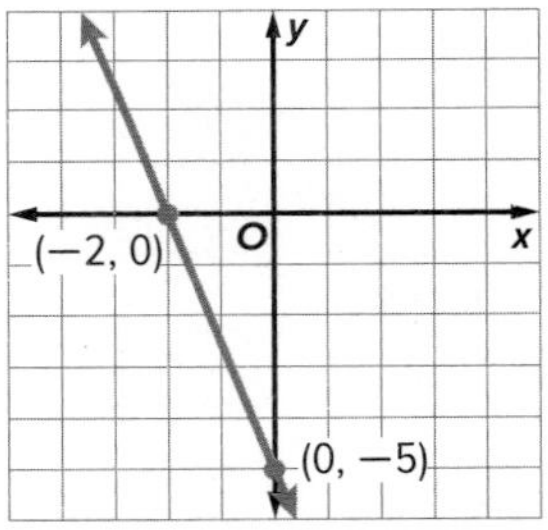

Example 2

Graph $y \geq 3x - 2$.

The boundary is the graph of $y = 3x - 2$. Since the inequality symbol is $\geq$, the boundary will be solid.

Test the point (0, 0).

$0 \geq 3(0) - 2$ $\quad (x, y) = (0, 0)$

$0 \geq -2$

Shade the region that includes (0, 0).

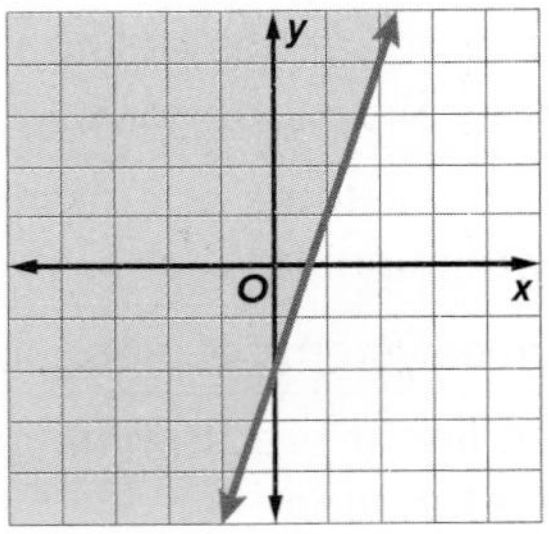

Quick Check

Graph each equation.

1. $x + 2y = 4$

2. $y = -x + 6$

3. $3x + 5y = 15$

4. $3y - 2x = -12$

Graph each inequality.

5. $y < 3$

6. $x + y \geq 1$

7. $3x - y > 6$

8. $x + 2y \leq 5$

How Did You Do?

Which exercises did you answer correctly in the Quick Check?

Solving Linear Equations and Inequalities

Learn Solving Linear Equations

An **equation** is a mathematical sentence stating that two mathematical expressions are equal. The **solution** of an equation is a value that makes the equation true. To solve equations, use the properties of equality to isolate the variable on one side.

Key Concept • Properties of Equality

Property of Equality	Symbols
Addition Property of Equality	For any real numbers a, b, and c, if $a = b$, then $a + c = b + c$.
Subtraction Property of Equality	For any real numbers a, b, and c, if $a = b$, then $a - c = b - c$.
Multiplication Property of Equality	For any real numbers a, b, and c, if $a = b$, then $ac = bc$.
Division Property of Equality	For any real numbers a, b, and c, $c \neq 0$, if $a = b$, then $\frac{a}{c} = \frac{b}{c}$.

Example 1 Solve a Linear Equation

Solve $\frac{1}{3}(2x - 57) + \frac{1}{3}(6 - x) = -4$.

$\frac{1}{3}(2x - 57) + \frac{1}{3}(6 - x) = -4$	Original equation
$\frac{2}{3}x - 19 + 2 - \frac{1}{3}x = -4$	Distributive Property
$\frac{1}{3}x - 17 = -4$	Combine like terms.
$\frac{1}{3}x = 13$	Add 17 to each side and simplify.
$x = 39$	Multiply each side by 3 and simplify.

Example 2 Write and Solve an Equation

SPACE **The diameter of Earth is 828 kilometers less than twice the diameter of Mars. If Earth has a diameter of 12,756 kilometers, what is the diameter of Mars?**

Part A Write an equation that represents the situation.

Words	The diameter of Earth	is	828 less than twice the diameter of Mars.
Variable	Let m = the diameter of Mars.		
Equation	12,756	=	$2m - 828$

Go Online You can complete an Extra Example online.

(continued on the next page)

Today's Goals
- Solve linear equations.
- Solve linear equations by examining graphs of the related functions.
- Solve linear inequalities.

Today's Vocabulary
equation
solution
root
zero
inequality

Study Tip
Justifications The properties of equality can be used as justifications. However, in most future solutions, the justifications for steps will read as "Subtract c from each side," or "Divide each side by c."

Go Online
You can watch a video to see how to solve equations in one variable using a graphing calculator.

Part B Solve the equation.

$12{,}756 = 2m - 828$	Original equation
$12{,}756 + 828 = 2m - 828 + 828$	Add 828 to each side.
$13{,}584 = 2m$	Simplify.
$\frac{13{,}584}{2} = \frac{2m}{2}$	Divide each side by 2.
$6792 = m$	Simplify.

The diameter of Mars is 6792 kilometers. This is a reasonable solution because 12,756 is a little less than $6792 \cdot 2 = 13{,}584$, as indicated in the problem.

Check

BASKETBALL In 1962, Wilt Chamberlain set the record for the most points scored in a single NBA game. He scored 28 points from free throws and made x field goals, worth two points each. If Wilt Chamberlain scored 100 points, how many field goals did he make?

Which equation represents the number of field goals that Chamberlain scored?

A. $100 = 28 + 2x$ B. $100 = 28x + 2$ C. $28 = 2x$ D. $100 = 2x$

How many field goals did Chamberlain score?

Think About It!

What does it mean to solve for a variable?

Example 3 Solve for a Variable

GEOMETRY **The formula for the perimeter of a parallelogram is $P = 2a + 2b$ where a and b represent the measures of the bases. Solve the equation for b.**

$P = 2a + 2b$	Original equation
$P - 2a = 2a + 2b - 2a$	Subtract $2a$ from each side.
$P - 2a = 2b$	Simplify.
$\frac{P}{2} - \frac{2a}{2} = \frac{2b}{2}$	Divide each side by 2.
$\frac{P}{2} - a = b$	Simplify.

Check

GEOMETRY The formula for the area A of a trapezoid is solved for h. Identify the missing justification.

$A = \frac{1}{2}h(a + b)$	Original equation
$2A = 2 \cdot \frac{1}{2}h(a + b)$	Multiplication Property of Equality
$2A = h(a + b)$	Simplify.
$\frac{2A}{(a + b)} = \frac{h(a + b)}{(a + b)}$	?
$\frac{2A}{(a + b)} = h$	Simplify.

Go Online You can complete an Extra Example online.

Learn Solving Linear Equations by Graphing

The solution of an equation is called a **root**. You can find the root of an equation by examining the graph of its related function $f(x)$. A related function is found by solving the equation for 0 and then replacing 0 with $f(x)$. A related function for $2x + 13 = 9$ is $f(x) = 2x + 4$ or $y = 2x + 4$.

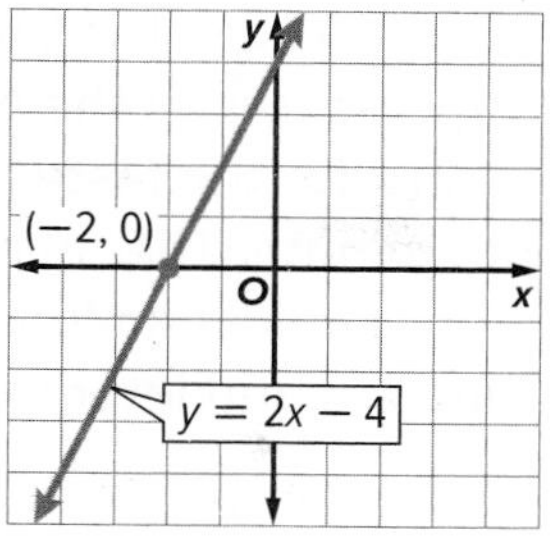

Values of x for which $f(x) = 0$ are called **zeros** of the function f. The zero of a function is the x-intercept of its graph. The solution and root of a linear equation are the same as the zero and x-intercept of its related function.

Talk About It!

Because there is typically more than one way to solve an equation for 0, there may be more than one related function for an equation. What is another possible related function of $2x + 13 = 9$? How does the zero of this function compare to the zero of $f(x) = 2x + 4$?

Example 4 Solve a Linear Equation by Graphing

Solve $\frac{1}{2}x - 11 = -8$ by graphing.

Step 1 Find a related function.

Rewrite the equation with 0 on the right side.

$\frac{1}{2}x - 11 = -8$	Original equation.
$\frac{1}{2}x - 11 + 8 = -8 + 8$	Add 8 to each side.
$\frac{1}{2}x - 3 = 0$	Simplify.

Replacing 0 with $f(x)$ gives the related function, $f(x) = \frac{1}{2}x - 3$.

Step 2 Graph the related function.

Since the graph of $f(x) = \frac{1}{2}x - 3$ intersects the x-axis at 6, the solution of the equation is 6.

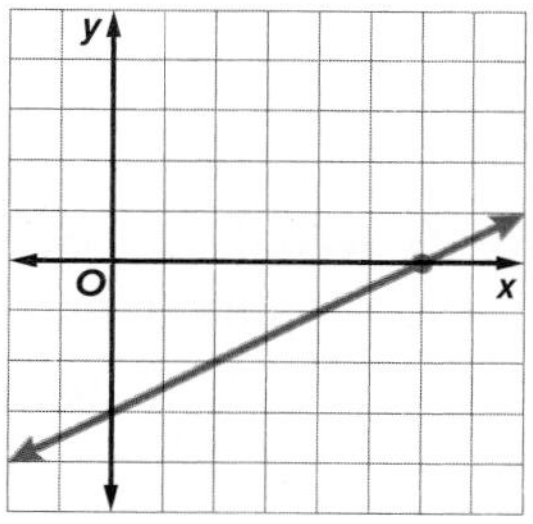

Think About It!

Explain why −3 is *not* a zero of the function.

Check

Graph the related function of $2x - 5 = -11$. Use the graph to solve the equation.

The related function is ___?___.

The solution is ___?___.

Watch Out!

Intercepts Be careful not to mistake y-intercepts for zeros of functions. The y-intercept on a graph occurs when $x = 0$. The x-intercepts are the zeros of a function because they are where $f(x) = 0$.

Go Online You can complete an Extra Example online.

Use a Source

Use available resources to find Paul Crake's actual time. How does this compare to your solution?

Study Tip

Assumptions
Assuming that the rate at which Paul Crake climbed the stairs was constant allows us to represent the situation with a linear equation. While the rate at which he climbed likely varied throughout the race, using constant rates allows for reasonable graphs and solutions.

Example 5 Estimate Solutions by Graphing

TOWER RACE **The Empire State Building Run-Up is a race in which athletes run up the building's 1576 stairs. In 2003, Paul Crake set the record for the fastest time, running up an average of about 165 stairs per minute. The function $c = 1576 - 165m$ represents the number of steps Crake had left to climb c after m minutes. Find the zero of the function and interpret its meaning in the context of the situation.**

Step 1 Graph the function.

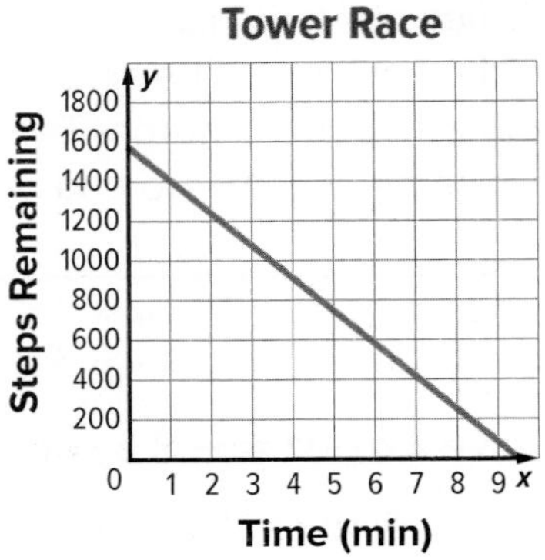

Step 2 Estimate the zero.

The graph appears to intersect the x-axis at about 9.5. This means that Paul Crake finished the race in about 9.5 minutes, or 9 minutes and 30 seconds.

Step 3 Solve algebraically.

Use the equation. Substitute 0 for c, and solve algebraically to check your solution.

$c = 1576 - 165m$	Original equation
$0 = 1576 - 165m$	Replace c with 0.
$165m = 1576$	Add $165m$ to each side.
$m \approx 9.55$	Divide each side by 165.

The solution is about 9.55. So, Paul Crake completed the Empire State Building Run-Up in about 9.55 minutes, or 9 minutes and 33 seconds. This is close to the estimated time of 9.5 minutes.

Check

DOG WALKING Bethany spends \$480 on supplies to start a dog walking service for which she plans to charge \$23 per hour. The function $y = 23x - 480$ represents Bethany's profit after x hours of dog walking.

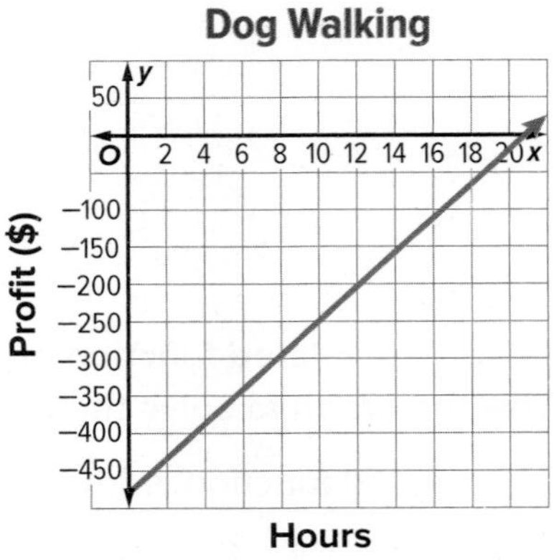

Part A The graph appears to intersect the x-axis at about __?__.

Part B Solve algebraically to verify your estimate. Round to the nearest hundredth.

Go Online You can complete an Extra Example online.

Explore Comparing Linear Equations and Inequalities

Online Activity Use a comparison to complete the Explore.

> **INQUIRY** How do the solution methods and the solutions of linear equations and inequalities in one variable compare?

Learn Solving Linear Inequalities

An **inequality** is a mathematical sentence that contains the symbol $<, \leq, >, \geq$, or $\neq$. Properties of inequalities allow you to perform operations on each side of an inequality without changing the truth of the inequality.

Key Concept • Addition and Subtraction Properties of Inequality

Symbols	For any real numbers, a, b, and c: If $a > b$, then $a + c > b + c$. If $a > b$, then $a - c > b - c$. If $a < b$, then $a + c < b + c$. If $a < b$, then $a - c < b - c$.	
Examples	$2 > 0$ $2 + 1 > 0 + 1$ $3 > 1$	$9 > 6$ $9 - 4 > 6 - 4$ $5 > 2$

Go Online You may want to complete the Concept Check to check your understanding.

Key Concept • Multiplication and Division Properties of Inequality

Symbols	For any real numbers, a, b, and c: where c is positive: If $a > b$, then $ac > bc$. If $a > b$, then $\frac{a}{c} > \frac{b}{c}$. If $a < b$, then $ac < bc$. If $a < b$, then $\frac{a}{c} < \frac{b}{c}$. where c is negative: If $a > b$, then $ac < bc$. If $a > b$, then $\frac{a}{c} < \frac{b}{c}$. If $a < b$, then $ac > bc$. If $a < b$, then $\frac{a}{c} > \frac{b}{c}$.	
Examples	$8 > -2$ $8(3) > -2(3)$ $24 > -6$	$-4 < -2$ $-4(-8) > -2(-8)$ $32 > 16$

Think About It! What does the arrow on the graph of a solution set represent?

The solution sets of inequalities can be expressed by using set-builder notation. For example, $\{x \mid x > 1\}$ represents the set of all numbers x such that x is greater than 1. The solution sets can also be graphed on number lines. Circles and dots are used to indicate whether an endpoint is included in the solution. The circle at 1 means that this point is *not* included in the solution set.

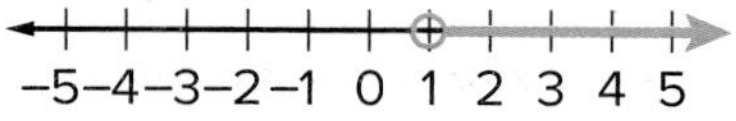

Think About It!

What does a dot on the graph of a solution set indicate?

Study Tip

Reversing the Inequality Symbol Adding the same number to, or subtracting the same number from, each side of an inequality does not change the truth of the inequality. Multiplying or dividing each side of an inequality by a positive number does not change the truth of the inequality. However, multiplying or dividing each side of an inequality by a negative number requires that the order of the inequality be reversed. In Example 6, $\geq$ was replaced with $\leq$.

Watch Out!

Reading Math Be sure to always read problems carefully. The term *at least* is used here, and can be confusing since it actually means *greater than or equal to*, and is represented by $\geq$. In this instance, Jake should intake at least 1300 mg, which means he must intake an amount greater than or equal to 1300 mg.

Example 6 Solve a Linear Inequality

Solve $-5.6n + 12.9 \geq -71.1$. Graph the solution set on a number line.

$-5.6n + 12.9 \geq -71.1$	Original inequality.
$-5.6n + 12.9 - 12.9 \geq -71.1 - 12.9$	Subtract 12.9 from each side.
$-5.6n \geq -84$	Simplify.
$\frac{-5.6n}{-5.6} \leq \frac{-84}{-5.6}$	Divide each side by -5.6, reversing the inequality symbol.
$n \leq 15$	Simplify.

The solution set is $\{n \mid n \leq 15\}$. Graph the solution set.

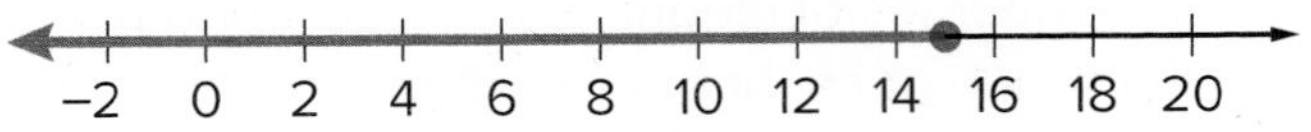

Check

What is the solution of $-p - 3 \geq -4(p + 6)$?
Graph the solution set.

Example 7 Write and Solve an Inequality

NUTRITION **The recommended daily intake of calcium for teens is 1300 mg. Jake gets 237 mg of calcium from a multivitamin he takes each morning and 302 mg from each glass of skim milk that he drinks. How many glasses of milk would Jake need to drink to meet the recommendation?**

Step 1 Write an inequality to represent the situation.

Let g = the number of glasses of milk Jake needs.

$237 + 302g \geq 1300$

Step 2 Solve the inequality.

$237 + 302g \geq 1300$	Original inequality
$302g \geq 1063$	Subtract 237 from each side.
$g \geq 3.52$	Divide each side by 302.

Step 3 Interpret the solution in the context of the situation.

Jake will need to drink slightly more than 3.5 glasses of milk to intake at least the recommended daily amount of calcium. This is a viable solution because Jake can pour part of a full glass of milk.

Go Online You can complete an Extra Example online.

Practice

Go Online You can complete your homework online.

Example 1

Solve each equation. Check your solution.

1. $6x - 5 = 7 - 9x$

2. $-1.6r + 5 = -7.8$

3. $\frac{3}{4} - \frac{1}{2}n = \frac{5}{8}$

4. $\frac{5}{6}c + \frac{3}{4} = \frac{11}{12}$

5. $2.2n + 0.8n + 5 = 4n$

6. $6y - 5 = -3(2y + 1)$

7. $-6(2x + 4) + \frac{1}{2}(8 + 3x) = -20$

8. $7(-1 + 4x) - 12x = 5$

9. $-4(10 + 3x) - (x + 8) = -9$

Example 2

Solve each problem.

10. REASONING The length of a rectangle is twice the width. Find the width if the perimeter is 60 centimeters. Define a variable, write an equation, and solve the problem.

11. GOLF Sergio and three friends went golfing. Two of the friends rented clubs for \$6 each. The total cost of the rented clubs and the green fees was \$76. What was the cost of the green fees for each person? Define a variable, write an equation, and solve the problem.

Example 3

Solve each equation or formula for the specified variable.

12. BANKING The formula for simple interest I is $I = Prt$, where P is the principal, r is the interest rate, and t is time. Solve for P.

13. MEAN The mean A of two numbers, x and y, is given by $A = \frac{x + y}{2}$. Solve for y.

14. SLOPE The slope m between two points (x_1, y_1) and (x_2, y_2) is $m = \frac{y_2 - y_1}{x_2 - x_1}$. Solve for y_2.

15. CYLINDERS The surface area of a cylinder A is given by $A = 2\pi r^2 + 2\pi rh$, where r is radius and h is height. Solve for h.

16. PHYSICS The height h of a falling object is given by $h = vt - gt^2$, where v is the initial velocity of the object, t is time, and g is the gravitational constant. Solve for v.

Example 4

Find the related function for each equation. Then graph the related function. Use the graph to solve the equation.

17. $2x + 5 = -7$

18. $-x + 3 = 6$

19. $\frac{1}{2}x + 4 = 10$

20. $\frac{1}{3}x + 1 = -5$

21. $-3x - 1 = 1$

22. $-\frac{1}{4}x + 1 = -2$

Example 5

Solve each problem.

23. LUNCH ACCOUNT At the beginning of the quarter, Nahla deposited \$100 into her lunch account. If Nahla spends an average of \$18 per week on lunches, then $m = 100 - 18w$ represents the amount of money Nahla has left in her lunch account m after w weeks of school.

a. Use the graph to estimate to the nearest tenth the number of weeks that Nahla can purchase school lunches with the money she deposited at the beginning of the quarter.

b. Solve algebraically to verify your estimate. Round to the nearest hundredth.

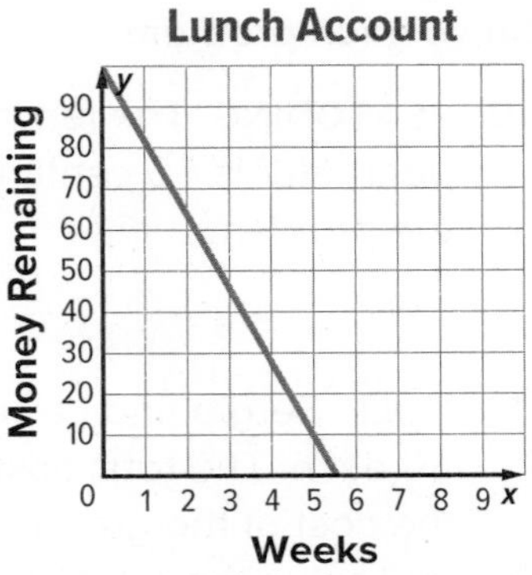

24. READING Mario has a 500-page novel which he is required to read over summer break for his upcoming language arts class.

a. If Mario reads 24 pages each day, write and graph a function to represent the number of pages that Mario has left to read p after d days.

b. Estimate the zero of the function by using the graph. Justify your response.

c. Find the zero algebraically. Interpret its meaning in the context of the situation.

Example 6

Solve each inequality. Graph the solution set on a number line.

25 $\frac{z}{-4} \geq 2$

26. $3a + 7 \leq 16$

27. $20 - 3n > 7n$

28. $7f - 9 > 3f - 1$

29. $0.7m + 0.3m \geq 2m - 4$

30. $4(5x + 7) \leq 13$

Example 7

Solve each problem.

31. INCOME Manuel takes a job translating English instruction manuals to Spanish. He will receive \$15 per page plus \$100 per month. Manuel plans to work for 3 months during the summer and wants to make at least \$1500. Write and solve an inequality to find the minimum number of pages Manuel must translate in order to reach his goal. Then, interpret the solution in the context of the situation.

32. STRUCTURE On a conveyor belt, there can only be two boxes moving at a time. The total weight of the boxes cannot be more than 300 pounds. Let x and y represent the weights of two boxes on the conveyor belt.

a. Write an inequality that describes the weight limitation in terms of x and y.

b. Write an inequality that describes the limit on the average weight a of the two boxes.

c. Two boxes are to be placed on the conveyor belt. The first box weighs 175 pounds. What is the maximum weight of the second box?

Mixed Exercises

Solve each equation. Check your solution by graphing the related function.

33. $-3b + 7 = -15 + 2b$

34. $a - \frac{2a}{5} = 3$

35. $2.2n + 0.8n + 5 = 4n$

Solve each inequality. Graph the solution set on a number line.

36. $\frac{4x - 3}{2} \geq -3.5$

37. $1 + 5(x - 8) \leq 2 - (x + 5)$

38. $-36 - 2(w + 77) > -4(2w + 52)$

39. REASONING An ice rink offers open skating several times a week. An annual membership to the skating rink costs \$60. The table shows the cost of one session for members and non-members.

Open Ice Skating Sessions	
members	\$6
non-members	\$10

Kaliska plans to spend no more than \$90 on skating this year. Define a variable then write and solve inequalities to find the number of sessions she can attend with and without buying a membership. Should Kaliska buy a membership?

40. **PRECISION** The formula to convert temperature in degrees Fahrenheit to degrees Celsius is $\frac{5}{9}(F - 32) = C$.

a. Solve the equation for F.

b. Use your result from part *a* to determine the temperature in degrees Fahrenheit when the Celsius temperature is 30.

c. At a certain temperature, a Fahrenheit thermometer and a Celsius thermometer will read the same temperature. Write and solve an equation to find the temperature.

41. **FIND THE ERROR** Steven and Jade are solving $A = \frac{1}{2}h(b_1 + b_2)$ for b_2. Is either of them correct? Explain your reasoning.

Steven

$$A = \frac{1}{2}h(b_1 + b_2)$$

$$\frac{2A}{h} = (b_1 + b_2)$$

$$\frac{2A - b_1}{h} = b_2$$

Jade

$$A = \frac{1}{2}h(b_1 + b_2)$$

$$\frac{2A}{h} = (b_1 + b_2)$$

$$\frac{2A}{h} - b_1 = b_2$$

42. **CREATE** Write an equation involving the Distributive Property that has no solution and another example that has infinitely many solutions.

43. **PERSEVERE** Solve $d = \sqrt{(x_2 - x_1)^2 + (y_2 - y_1)^2}$ for y_1.

44. **ANALYZE** Vivek's teacher made the statement, "Four times a number is less than three times a number." Vivek quickly responded that the answer is *no solution*. Do you agree with Vivek? Write and solve an inequality to justify your argument.

45. **WRITE** Why does the inequality symbol need to be reversed when multiplying or dividing by a negative number?

46. **PERSEVERE** Given $\triangle ABC$ with sides $AB = 3x + 4$, $BC = 2x + 5$, and $AC = 4x$, determine the values of x such that $\triangle ABC$ exists.

Lesson 2-2

Solving Absolute Value Equations and Inequalities

Learn Solving Absolute Value Equations Algebraically

The **absolute value** of a number is its distance from zero on the number line. The definition of absolute value can be used to solve equations that contain absolute value expressions by constructing two cases. For any real numbers a and b, if $|a| = b$ and $b \geq 0$, then $a = b$ or $a = -b$.

Step 1 Isolate the absolute value expression on one side of the equation.

Step 2 Write the two cases.

Step 3 Use the properties of equality to solve each case.

Step 4 Check your solutions.

Absolute value equations may have one, two, or no solutions.

- An absolute value equation has one solution if one of the answers does not meet the constraints of the problem. Such an answer is called an **extraneous solution**.
- An absolute value equation has no solution if there is no answer that meets the constraints of the problem. The solution set of this type of equation is called the **empty set**, symbolized by {} or Ø.

Example 1 Solve an Absolute Value Equation

Solve $2|5x + 1| - 9 = 4x + 17$. Check your solutions. Then graph the solution set.

$2|5x + 1| - 9 = 4x + 17$ Original equation

$2|5x + 1| = 4x + 26$ Add 9 to each side.

$|5x + 1| = 2x + 13$ Divide each side by 2.

Case 1

$5x + 1 = 2x + 13$

$3x + 1 = 13$

$x = 4$

Case 2

$5x + 1 = -(2x + 13)$

$7x + 1 = -13$

$x = -2$

CHECK Substitute each value in the original equation.

$2|5(4) + 1| - 9 \stackrel{?}{=} 4(4) + 17$

$33 = 33$ True

$2|5(-2) + 1| - 9 \stackrel{?}{=} 4(-2) + 17$

$9 = 9$ True

Both solutions make the equation true. Thus, the solution set is {4, −2}. The solution set can be graphed by graphing each solution on a number line.

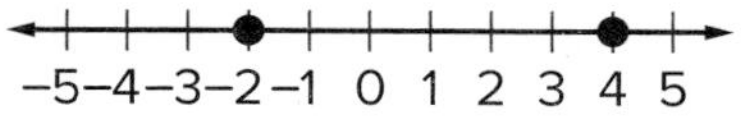

Go Online You can complete an Extra Example online.

Today's Goals
- Write and solve absolute value equations, and graph the solutions on a number line.
- Write and solve absolute value inequalities, and graph the solutions on a number line.

Today's Vocabulary
absolute value
extraneous solution
empty set

Watch Out!
Distribute the Negative For Case 2, remember to use the Distributive Property to multiply the entire expression on the right side of the equation by −1.

Check

Graph the solution set of $|5x - 3| - 6 = -2x + 12$.

Think About It!

What would the graph of the solution set of an absolute value equation with only one solution look like?

Example 2 Extraneous Solution

Solve $2|x + 1| - x = 3x - 4$. Check your solutions.

$2|x + 1| - x = 3x - 4$ Original equation

$2|x + 1| = 4x - 4$ Add x to each side.

$|x + 1| = 2x - 2$ Divide each side by 2.

Case 1

$x + 1 = 2x - 2$

$1 = x - 2$

$3 = x$

Case 2

$x + 1 = -(2x - 2)$

$x + 1 = -2x + 2$

$3x + 1 = 2$

$x = \frac{1}{3}$

There appear to be two solutions, 3 and $\frac{1}{3}$.

CHECK Substitute each value in the original equation.

$2|3 + 1| - 3 \stackrel{?}{=} 3(3) - 4$

$5 = 5$ True

$2\left|\frac{1}{3} + 1\right| - \left(\frac{1}{3}\right) \stackrel{?}{=} 3\left(\frac{1}{3}\right) - 4$

$\frac{7}{3} \neq -3$ False

Because $\frac{7}{3} \neq -3$, the only solution is 3. Thus, the solution set is $\{3\}$.

Example 3 The Empty Set

Solve $|4x - 7| + 10 = 2$.

$|4x - 7| + 10 = 2$ Original equation

$|4x - 7| = -8$ Subtract 10 from each side.

Because the absolute value of a number is always positive or zero, this sentence is never true. The solution is $\varnothing$.

Talk About It!

Is the following statement *always*, *sometimes*, or *never* true? Justify your argument. For real numbers a, b, and c, $|ax + b| = -c$ has no solution.

Check

Solve each absolute value equation.

a. $|x + 10| = 4x - 8$

b. $3|4x - 11| + 1 = 9x + 13$

c. $|2x + 5| - 18 = -3$

d. $-5|7x - 2| + 3x = 3x + 10$

Go Online You can complete an Extra Example online.

Apply Example 4 Write and Solve an Absolute Value Equation

FOOTBALL **The NFL regulates the inflation, or air pressure, of footballs used during games. It requires that footballs have an air pressure of 13 pounds per square inch (PSI), plus or minus 0.5 PSI. What is the greatest and least acceptable air pressure of a regulation NFL football?**

1 What is the task?

Describe the task in your own words. Then list any questions that you may have. How can you find answers to your questions?

I need to find the greatest and least acceptable air pressure for an NFL football. How can I write an absolute value equation to find the solution? Will there be one solution or two that make sense in this problem? I can find the answers to my questions by referencing other examples in the lesson and by checking my solutions.

2 How will you approach the task? What have you learned that you can use to help you complete the task?

I will write an equation to represent the situation. I have learned how to write and solve an equation involving absolute value.

3 What is your solution?

Use your strategy to solve the problem.

What absolute value equation represents the greatest and least acceptable air pressure?

$|x - 13| = 0.5$

How are the solutions of the equation represented on a graph?

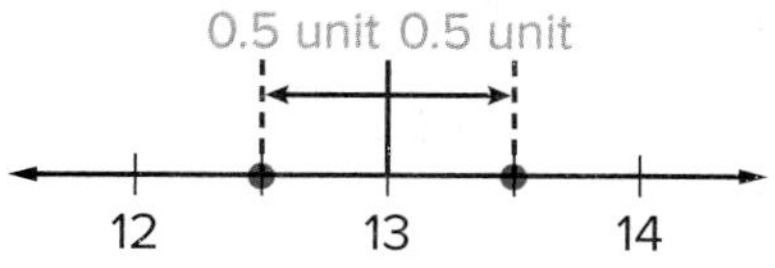

Interpret your solution. What are the greatest and least acceptable air pressures for an NFL football?

The greatest air pressure an NFL football can have is 13.5 PSI and the least is 12.5 PSI.

4 How can you know that your solution is reasonable?

Write About It! Write an argument that can be used to defend your solution. Because the distance between 13 and each solution is 0.5, both solutions satisfy the constraints of the equation. I can substitute each solution back into the original equation and check that the value makes the equation true. The pressures of 12.5 PSI and 13.5 PSI are within 0.5 PSI of 13 PSI.

Go Online You can complete an Extra Example online.

Learn Solving Absolute Value Inequalities Algebraically

When solving absolute value inequalities, there are two cases to consider. These two cases can be rewritten as a compound inequality.

Key Concept • Absolute Value Inequalities

For all real numbers a, b, c and x, $c > 0$, the following statements are true.

Absolute Value Inequality	Case 1	Case 2	Compound Inequality
$\lvert ax + b\rvert < c$	$ax + b < c$	$-(ax + b) < c$ $\frac{-(ax+b)}{-1} > \frac{c}{-1}$ $ax + b > -c$	$ax + b < c$ and $ax + b > -c$ which is $-c < ax + b < c$
$\lvert ax + b\rvert > c$	$ax + b > c$	$-(ax + b) > c$ $\frac{-(ax+b)}{-1} < \frac{c}{-1}$ $ax + b < -c$	$ax + b > c$ or $ax + b < -c$

These statements are also true for $\leq$ and $\geq$, respectively.

Think About It!
Describe a shortcut you could use to write case 2.

Example 5 Solve an Absolute Value Inequality (< or ≤)

Solve $\lvert 4x - 8\rvert - 5 < 11$. Then graph the solution set.

$\lvert 4x - 8\rvert - 5 < 11$ Original inequality

$\lvert 4x - 8\rvert < 16$ Add 5 to each side.

Because the inequality uses <, rewrite it as a compound inequality joined by the word *and*. For the case where the expression inside the absolute value symbols is negative, reverse the inequality symbol.

$4x - 8 < 16$	and	$4x - 8 > -16$
$4x < 24$		$4x > -8$
$x < 6$		$x > -2$

So, $x < 6$ and $x > -2$. The solution set is $\{x \mid -2 < x < 6\}$. All values of x between -2 and 6 satisfy the original inequality.

The solution set represents the interval between two numbers. Because the < symbols indicate that -2 and 6 are not solutions, graph the endpoints of the interval on a number line using circles. Then, shade the interval from -2 to 6.

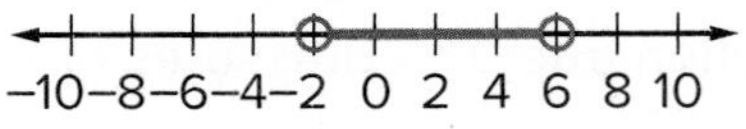

Go Online You can complete an Extra Example online.

Go Online
An alternate method is available for this example.

Study Tip
Check Your Solutions Remember to check your solutions by substituting values in each interval in the original inequality.

Example 6 Solve an Absolute Value Inequality (> or ≥)

Solve $\frac{|6x + 3|}{2} + 5 \geq 14$. Then graph the solution set.

$\frac{\|6x + 3\|}{2} + 5 \geq 14$	Original inequality
$\frac{\|6x + 3\|}{2} \geq 9$	Subtract 5 from each side.
$\|6x + 3\| \geq 18$	Multiply each side by 2.

Since the inequality uses ≥, rewrite it as a compound inequality joined by the word *or*. For the case where the expression inside the absolute value symbols is negative, reverse the inequality symbol.

$6x + 3 \geq 18$	or	$6x + 3 \leq -18$
$6x \geq 15$		$6x \leq -21$
$x \geq \frac{5}{2}$		$x \leq -\frac{7}{2}$

So, $x \geq \frac{5}{2}$ or $x \leq -\frac{7}{2}$. The solution set is $\left\{x \,\middle|\, x \leq -\frac{7}{2} \text{ or } x \geq \frac{5}{2}\right\}$.

All values of x less than or equal to $-\frac{7}{2}$ as well as values of x greater than $\frac{5}{2}$ satisfy the constraints of the original inequality.

The solution set represents the union of two intervals. Since the ≤ and ≥ symbols indicate that $-\frac{7}{2}$ and $\frac{5}{2}$ are solutions, graph the endpoints of the interval on a number line using dots. Then, shade all points less than $-\frac{7}{2}$ and all points greater than $\frac{5}{2}$.

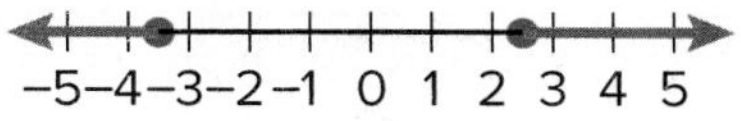

Check

Match each solution set with the appropriate absolute value inequality.

$-8|x + 14| + 7 \geq -17$

$\frac{|2x - 8|}{3} - 10 < 6$

$5|2x + 28| + 6 \geq -24$

$\frac{|3x - 12|}{4} - 13 > 5$

A. $\{x \mid x \text{ is a real number.}\}$ B. $\{x \mid 28 < x < -20\}$

C. $\{x \mid x < -20 \text{ or } x > 28\}$ D. $\{x \mid x \leq -11 \text{ or } x \geq -17\}$

E. $\{x \mid -20 < x < 28\}$ F. $\{x \mid -17 \leq x \leq -11\}$

Go Online You can complete an Extra Example online.

Watch Out!

Isolate the Expression Remember to isolate the absolute value expression on one side of the inequality symbol before determining whether to rewrite an absolute value inequality using *and* or *or*. When transforming the inequality, you might divide or multiply by a negative number, causing the inequality symbol to be reversed.

Example 7 Write and Solve an Absolute Value Inequality

SLEEP **You can find how much sleep you need by going to sleep without turning on an alarm. Once your sleep pattern has stabilized, record the amount of time you spend sleeping each night. The amount of time you sleep plus or minus 15 minutes is your sleep need. Suppose you sleep 8.5 hours per night. Write and solve an inequality to represent your sleep need, and graph the solution on a number line.**

Part A Write an absolute value inequality to represent the situation.

The difference between your actual sleep need and the amount of time you sleep is less than or equal to 15 minutes. So, 8.5 hours is the central value and 15 minutes, or 0.25 hour, is the acceptable range.

The difference between your actual sleep need and 8.5 hours is 0.25 hour. Let n = your actual sleep need.

$$|n - 8.5| \leq 0.25$$

Part B Solve the inequality and graph the solution set.

Rewrite $|n - 8.5| \leq 0.25$ as a compound inequality.

$n - 8.5 \leq 0.25$ and $n - 8.5 \geq -0.25$

$n \leq 8.75$ $n \geq 8.25$

The solution set represents the interval between two numbers. Since the $\leq$ and $\geq$ symbols indicate that 8.25 and 8.75 are solutions, graph the endpoints of the interval on a number line using dots. Then, shade the interval from 8.25 to 8.75.

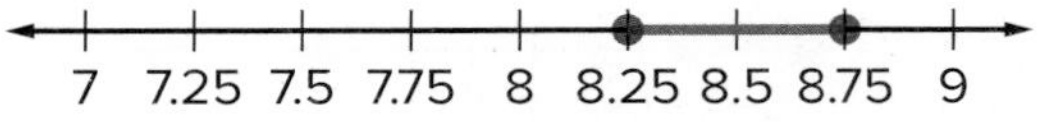

This means that you need between 8.25 and 8.75 hours of sleep per night, inclusive.

Check

FOOD A survey found that 58% of American adults eat at a restaurant at least once a week. The margin of error was within 3 percentage points.

Part A Write an absolute value inequality to represent the range of the percent of American adults who eat at a restaurant once a week, where x is the actual percent.

Part B Use your inequality from Part **A** to find the range of the percent of American adults who eat at a restaurant once a week.

Go Online You can complete an Extra Example online.

Practice

Examples 1–3

Solve each equation. Check your solutions.

1. $|8 + p| = 2p - 3$

2. $|4w - 1| = 5w + 37$

3. $4|2y - 7| + 13 = 9$

4. $-2|7 - 3y| - 6 = -14$

5. $2|4 - n| = -3n$

6. $5 - 3|2 + 2w| = -7$

7. $5|2r + 3| - 5 = 0$

8. $3 - 5|2d - 3| = 4$

Example 4

Solve each problem.

9. WEATHER The packaging of a thermometer claims that the thermometer is accurate within 1.5 degrees of the actual temperature in degrees Fahrenheit. Write and solve an absolute value equation to find the least and greatest possible temperature if the thermometer reads 87.4° F.

10. OPINION POLLS Public opinion polls reported in newspapers are usually given with a margin of error. A poll for a local election determined that Candidate Morrison will receive 51% of the votes. The stated margin of error is ±3%. Write and solve an absolute value equation to find the minimum and maximum percent of the vote that Candidate Morrison can expect to receive.

Examples 5 and 6

Solve each inequality. Graph the solution set on a number line.

11. $|2x + 2| - 7 \leq -5$

12. $\left|\frac{x}{2} - 5\right| + 2 > 10$

13. $|3b + 5| \leq -2$

14. $|x| > x - 1$

15. $|4 - 5x| < 13$

16. $|3n - 2| - 2 < 1$

17. $|3x + 1| > 2$

18. $|2x - 1| < 5 + 0.5x$

Example 7

Solve each problem.

19. RAINFALL For 90% of the last 30 years, the rainfall at Shell Beach has varied no more than 6.5 inches from its mean value of 24 inches. Write and solve an absolute value inequality to describe the rainfall in the other 10% of the last 30 years, and graph the solution on a number line.

20. MANUFACTURING A food manufacturer's guidelines state that each can of soup produced cannot vary from its stated volume of 14.5 fluid ounces by more than 0.08 fluid ounce. Write and solve an absolute value inequality to describe acceptable volumes, and graph the solution on a number line.

Mixed Exercises

Solve. Check your solutions.

21. $8x = 2|6x - 2|$

22. $-6y + 4 = |4y + 12|$

23. $8z + 20 > -|2z + 4|$

24. $-3y - 2 \leq |6y + 25|$

REASONING **Write an absolute value equation to represent each situation. Then solve the equation and discuss the reasonableness of your solution given the constraints of the absolute value equation.**

25. The absolute value of the sum of 4 times a number and 7 is the sum of 2 times a number and 3.

26. The sum of 7 and the absolute value of the difference of a number and 8 is -2 times a number plus 4.

27. MODELING A carpenter cuts lumber to the length of 36 inches. For her project, the lumber must be accurate within 0.125 in.

a. Write an inequality to represent the acceptable length of the lumber. Explain your reasoning.

b. Solve the inequality. Then state the maximum and minimum length for the lumber.

28. SAND A home improvement store sells bags of sand, which are labeled as weighing 35 pounds. The equipment used to package the sand produces bags with a weight that is within 8 ounces of the labeled weight.

a. Write an absolute value equation to represent the maximum and minimum weight for the bags of sand.

b. Solve the equation and interpret the result.

29. CONSTRUCT ARGUMENTS Megan and Yuki are solving the equation $|x - 9| = |5x + 6|$. Megan says that there are 4 cases to consider because there are two possible values for each absolute value expression. Yuki says only 2 cases need to be considered. With which person, do you agree? Will they both get the same solution(s)?

Solve each inequality. Graph the solution set on a number line.

30. $3|2z - 4| - 6 > 12$

31. $6|4p + 2| - 8 < 34$

32. $\frac{|5f - 2|}{6} > 4$

33. $\frac{|2w + 8|}{5} \geq 3$

34. $-\frac{3x|6x + 1|}{5} < 12x$

35. $-\frac{7}{8}|2x + 5| > 14$

36. TIRES The recommended inflation of a car tire is no more than 35 pounds per square inch. Depending on weather conditions, the actual reading of the tire pressure could fluctuate up to 3.4 psi. Write and solve an absolute value equation to find the maximum and minimum tire pressure.

37. PROJECTILE An object is launched into the air and then falls to the ground. Its velocity is modeled by the equation $v = 200 - 32t$, where the velocity v is measured in feet per second and time t is measured in seconds. The object's speed is the absolute value of its velocity. Write and solve a compound inequality to determine the time intervals in which the speed of the object will be between 40 and 88 feet per second. Interpret your solution in the context of the situation.

38. USE A SOURCE Research to find a poll with a margin of error. Describe the poll then write an absolute value inequality to represent the actual results.

39. CONSTRUCT ARGUMENTS Roberto claims that the solution to $|3c - 4| > -4.5$ is the same as the solution to $|3c - 4| \geq 0$, because an absolute value is always greater than or equal to zero. Is he correct? Explain your reasoning.

40. WRITE Summarize the difference between *and* compound inequalities and *or* compound inequalities.

41. WHICH ONE DOESN'T BELONG? Identify the compound inequality that does not share the same characteristics as the other three. Justify your conclusion.

$-3 < x < 5$	$x > 2$ and $x < 3$	$x > 5$ and $x < 1$	$x > -4$ and $x > -2$

42. FIND THE ERROR Ana and Ling are solving $|3x + 14| = -6x$. Is either of them correct? Explain your reasoning.

Ana	Ling
$\|3x + 14\| = -6x$	$\|3x + 14\| = -6x$
$3x + 14 = -6x$ or $3x + 14 = 6x$	$3x + 14 = -6x$ or $3x + 14 = 6x$
$9x = -14$ $\quad$ $14 = 3x$	$9x = -14$ $\quad$ $14 = 3x$
$x = -\frac{14}{9}$ $\quad$ $x = \frac{14}{3}$	$x = -\frac{14}{9}$ $\quad$ $x = \frac{14}{3}$

43. PERSEVERE Solve $|2x - 1| + 3 = |5 - x|$. List all cases and resulting equations.

ANALYZE **If *a*, *x*, and *y* are real numbers, determine whether each statement is *sometimes, always,* or *never* true. Justify your argument.**

44. If $|a| > 7$, then $|a + 3| > 10$.

45. If $|x| < 3$, then $x + 3 > 0$.

46. If y is between 1 and 5, then $|y - 3| \leq 2$.

47. CREATE Write an absolute value inequality with a solution of $a \leq x \leq b$.

Equations of Linear Functions

Today's Goals

- Write linear equations in standard form and identify values of *A*, *B*, and *C*.
- Create linear equations in slope-intercept form and by using the coordinates of two points.
- Create linear equations in point-slope form by using two points on the line or the slope and a point on the line.

Explore Arithmetic Sequences

Online Activity Use a real-world situation to complete the Explore.

INQUIRY How can you write formulas that relate the numbers in an arithmetic sequence?

Learn Linear Equations in Standard Form

Any linear equation can be written in standard form, $Ax + By = C$, where $A \geq 0$, A and B are not both 0, and A, B, and C are integers with a greatest common factor of 1.

Example 1 Write Linear Equations in Standard Form

Write $y = \frac{2}{5}x + 14$ in standard form. Identify *A*, *B*, and *C*.

$y = \frac{2}{5}x + 14$	Original equation
$-\frac{2}{5}x + y = 14$	Subtract $\frac{2}{5}x$ from each side.
$2x - 5y = -70$	Multiply each side by -5.

$A = 2$, $B = -5$, and $C = -70$.

Think About It!
Is $-2x + 2y = 2$ written in standard form? Why or why not?

Check

Write $2y = 10x - 16$ in standard form. Identify *A*, *B*, and *C*.

equation in standard form: ___?___

$A =$ ___?___ $B =$ ___?___ $C =$ ___?___

Learn Linear Equations in Slope-Intercept Form

The equation of a linear function can be written in slope-intercept form, $y = mx + b$, where m is the slope and b is the y-intercept.

The slope is $\frac{\text{rise}}{\text{run}} = \frac{2}{3}$. This value can be substituted for m in the slope-intercept form.

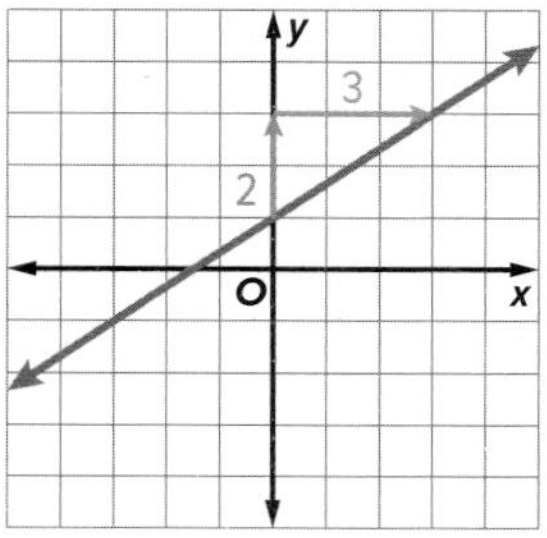

The line intersects the y-axis at 1. This value can be substituted for b in the slope-intercept form.

Go Online You can complete an Extra Example online.

Think About It!
Is the b in slope-intercept form equivalent to the B in standard form, $Ax + By = C$? If yes, explain your reasoning. If no, provide a counterexample.

Example 2 Write Linear Equations in Slope-Intercept Form

Write $12x - 4y = 24$ in slope-intercept form. Identify the slope m and y-intercept b.

$12x - 4y = 24$	Original equation
$-4y = -12x + 24$	Subtract $12x$ from each side.
$y = 3x - 6$	Divide each side by -4.

$m = 3$ $b = -6$

Check

Write $4x = -2y + 22$ in slope-intercept form.

Example 3 Interpret an Equation in Slope-Intercept Form

SHOES **The equation $3246x - 2y = -152{,}722$ can be used to estimate shoe sales in Europe from 2010 to 2015, where x is the number of years after 2010 and y is the revenue in millions of dollars.**

Part A Write the equation in slope-intercept form.

$3246x - 2y = -152{,}722$	Original equation
$-2y = -3246x - 152{,}722$	Subtract $3246x$ from each side.
$y = 1623x + 76{,}361$	Divide each side by -2.

Part B Interpret the parameters in the context of the situation.

1623 represents that sales increased by $1623 million each year.

76,361 represents that in year 0, or in 2010, sales were $76,361 million.

Example 4 Use a Linear Equation in Slope-Intercept Form

SMARTPHONES **In 2013, there were 1.31 billion smartphone users worldwide. By 2017, there were 2.38 billion smartphone users. Write and use an equation to estimate the number of users in 2025.**

Step 1 Define the variables. Because you want to estimate the number of users in 2025, write an equation that represents the number of smartphone users y after x years. Let x be the number of years after 2013 and let y be the number of billions of smartphone users.

Step 2 Find the slope. Since x is the years after 2013, (0, 1.31) and (4, 2.38) represent the number of smartphone users in 2013 and 2017, respectively. Round to the nearest hundredth.

$$m = \frac{2.38 - 1.31}{4 - 0} = 0.27$$

So, the number of users is increasing at a rate of 0.27 billion per year.

Go Online You can complete an Extra Example online.

Study Tip

Assumptions Assuming that the rate at which the number of smartphone users increases is constant allows us to represent the situation using a linear equation. While the rate at which the number of smartphone users increases may vary each year, using a constant rate allows for a reasonable equation that can be used to estimate future data.

Think About It!

When using the equation to estimate the number of smartphone users in the future, what constraint does the world's population place on the possible number of users?

Step 3 Find the *y*-intercept. The *y*-intercept represents the number of smartphone users when $x = 0$, or in 2013. So, $b = 1.31$.

Step 4 Write an equation. Use $m = 0.27$ and $b = 1.31$ to write the equation.

$y = 0.27x + 1.31$ $\quad$ $m = 0.27, b = 1.31$

Step 5 Estimate. Since 2025 is 12 years after 2013, substitute 12 for *x*.

$y = 0.27(12) + 1.31; y = 4.55$

If the trend continues, there will be about 4.55 billion users in 2025.

Think About It!

Suppose the data spanned 2 years instead of 4 years. That is, there were 1.31 billion smartphone users in 2013 and 2.38 billions users in 2015. How would this affect the rate of change abnd your estimate in **Step 5**?

Learn Linear Equations in Point-Slope Form

The equation of a linear function can be written in point-slope form, $y - y_1 = m(x - x_1)$, where *m* is the slope and (x_1, y_1) are the coordinates of a point on the line.

Example 5 Point-Slope Form Given Slope and One Point

Write the equation of a line that passes through (3, −5) and has a slope of 11 in point-slope form.

$y - y_1 = m(x - x_1)$	Point-slope form
$y - (-5) = 11(x - 3)$	$m = 11; (x_1, y_1) = (3, -5)$
$y + 5 = 11(x - 3)$	Simplify.

Check

Write the equation of a line that passes through (13, −5) and has a slope of 4.5 in point-slope form.

Example 6 Point-Slope Form Given Two Points

Write an equation of a line that passes through (1, 1) and (7, 13) in point-slope form.

Step 1 Find the slope.

$m = \frac{y_2 - y_1}{x_2 - x_1}$	Slope formula
$= \frac{13 - 1}{7 - 1}$	$(x_1, y_1) = (1, 1); (x_2, y_2) = (7, 13)$
$= \frac{12}{6}$	Simplify.
$= 2$	Simplify.

(continued on the next page)

Go Online You can complete an Extra Example online.

Talk About It!
What other values would you need to write the equation of this line in slope-intercept form? Could you determine those values from the given information?

Step 2 Write an equation.

Substitute the slope for m and the coordinates of either of the given points for (x_1, y_1) in the point-slope form.

$y - y_1 = m(x - x_1)$ Point-slope form

$y - 1 = 2(x - 1)$ $m = 2$; $(x_1, y_1) = (1, 1)$

Check

Select all the equations for the line that passes through (–1, 1) and (–2, 13).

A. $x - 1 = -12(y + 1)$ B. $y - 1 = -12(x + 1)$ C. $x + 1 = -12(y - 1)$

D. $y + 1 = -12(x - 1)$ E. $y - 2 = -12(x + 13)$ F. $x - 2 = -12(y + 13)$

G. $x + 2 = -12(y - 13)$ H. $y - 13 = -12(x + 2)$

Example 7 Write and Interpret a Linear Equation in Point-Slope Form

ARCHITECTURE **The Tower of Pisa began tilting during its construction in 1178 and continued to move until a restoration effort reduced the lean and stabilized the structure. The Tower of Pisa leaned 5.4 meters in 1993 compared to a lean of just 1.4 meters in 1350. Write an equation in point-slope form that represents the lean *y* in meters of the Tower of Pisa *x* years after its construction in 1178.**

Step 1 Find the slope. Round to the nearest hundredth.

The tower was leaning 1.4 meters in 1350, 172 years after 1178.

The tower was leaning 5.4 meters in 1993, 815 years after 1178.

$m = \frac{5.4 - 1.4}{815 - 172} = 0.006$ The lean of the Tower of Pisa increased at a rate of 0.006 meter per year.

Think About It!
Could this equation be used to estimate the lean of the Tower of Pisa for any year? Explain your reasoning.

Step 2 Write an equation.

Substitute the slope for m and the coordinates of either of the given points for (x_1, y_1) in the point-slope form.

$y - y_1 = m(x - x_1)$ Point-slope form

$y - 1.4 = 0.006(x - 172)$ $m = 0.006$; $(x_1, y_1) = (172, 1.4)$

Check

SOCIAL MEDIA In 2011, the Miami Marlins had about 11,000 followers on a social media site. In 2016, they had about 240,000 followers. Which equation represents the number of followers *y* the Miami Marlin's had *x* years after they joined the site in 2009?

A. $y - 11{,}000 = 45{,}800(x - 2)$ B. $y - 45{,}800 = 11{,}000(x - 2)$

C. $y - 11{,}000 = 45{,}800(x - 2011)$ D. $y - 2 = 45{,}800(x - 11{,}000)$

Go Online You can complete an Extra Example online.

Practice

Go Online You can complete your homework online.

Example 1

Write each equation in standard form. Identify A, B, and C.

1. $-7x - 5y = 35$
2. $8x + 3y + 6 = 0$
3. $10y - 3x + 6 = 11$
4. $\frac{2}{3}y - \frac{3}{4}x + \frac{1}{6} = 0$
5. $\frac{4}{5}y + \frac{1}{8}x = 4$
6. $-0.08x = 1.24y - 3.12$

Example 2

Write each equation in slope-intercept form. Identify the slope m and the y-intercept b.

7. $6x + 3y = 12$
8. $14x - 7y = 21$
9. $\frac{2}{3}x + \frac{1}{6}y = 2$
10. $5x + 10y = 20$
11. $6x + 9y = 15$
12. $\frac{1}{5}x + \frac{1}{2}y = 4$

Example 3

13. CHARITY The linear equation $y - 20x = 83$ relates the number of shirts collected during a charity clothing drive, where x is the number of hours since noon and y is the total number of shirts collected. Write the equation in slope-intercept form and interpret the parameters of the equation in the context of the situation.

14. GROWTH Suppose the body length y in inches of a baby snake is given by $4x - 2y = -3$, where x is the age of the snake in months until it becomes 1 year old. Write the equation in slope-intercept form and interpret the parameters of the equation in the context of the situation.

Example 4

15. PLUMBER Two neighbors, Camila and Conner, hire the same plumber for household repairs. The plumber worked at Camila's house for 3 hours and charged her \$191. The plumber worked at Conner's house for 1 hour and charged him \$107.

    a. Define variables to represent the situation.

    b. Find the slope and y-intercept. Then, write an equation.

    c. How much would it cost to hire the plumber for 5 hours of work?

16. HIKING Tim began a mountain hike near Big Bear Lake, California at 9:00 A.M. By 10:30 A.M., his elevation is 7200 feet above sea level. At 11:15 A.M., he is at an elevation of 7425 feet above sea level.

a. Define variables to represent the situation.

b. Write an equation in slope-intercept form that represent Tim's elevation since he began hiking.

c. If Tim's altitude continues to increase at the same rate, estimate his altitude at 12:30 P.M.

Example 5

Write an equation in point-slope form for the line that satisfies each set of conditions.

17. slope of -5, passes through $(-3, -8)$

18. slope of $\frac{4}{5}$, passes through $(10, -3)$

19. slope of $-\frac{2}{3}$, passes through $(6, -8)$

20. slope of 0, passes through $(0, -10)$

Example 6

Write an equation in point-slope form for a line that passes through each set of points.

21. $(2, -3)$ and $(1, 5)$

22. $(3, 5)$ and $(-6, -4)$

23. $(-1, -2)$ and $(-3, 1)$

24. $(-2, -4)$ and $(1, 8)$

Example 7

Solve each problem.

25. SALES Light truck is a vehicle classification for trucks weighing up to 8500 pounds. In 2011, 5.919 million light trucks were sold in the U.S. In 2017, 11.055 million light trucks were sold. Write an equation in point-slope form that represents the number of light trucks y sold x years after 2010.

26. RESTAURANTS In 2012, a popular pizza franchise had 2483 restaurants. In 2017, there were 2606 franchised restaurants. Write an equation in point-slope form that represents the number of restaurants y that are franchised x years after 2010.

Mixed Exercises

REGULARITY **Write linear equations in standard form, slope-intercept form, and point-slope form that satisfy each set of conditions.**

27. slope of −2, passes through (6, −7)

28. x-intercept: 3, y-intercept: 5

29. passes through (4, −1) and (8, 3)

30. slope of 0.6, passes through (1, 1)

31. STATE YOUR ASSUMPTION The surface of Grand Lake is at an elevation of 648 feet. During a drought, the water level drops at a rate of 3 inches per day. Write an equation in slope-intercept form that gives the elevation in feet y of the surface of Grand Lake after x days. Explain any assumptions you made to write the equation.

32. USE A SOURCE Go online to find the population of your city in 2010 and 2015. Write an equation in slope-intercept form to represent the population y of the city x years after 2010. State assumptions you made to write the equation.

33. PRECISION Use the graph shown to write the equation of the line in standard form that passes through the two points.

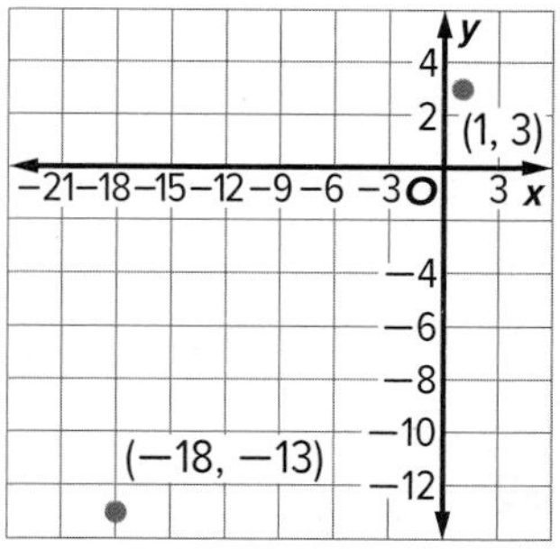

34. USE ESTIMATION In May, Jacalyn opens a savings account with an initial balance of \$200 and deposits about the same amount each month. The table shows her account balance at the beginning of each month.

Month	June	July	Aug.	Sept.	Oct.
Balance (\$)	533.95	871.86	1204.59	1541.55	1882.74

a. Write a linear equation in slope-intercept form that relates the balance of her savings account y in months x since she opened her account.

b. Estimate the balance of Jacalyn's account at the beginning of December.

35. CONSTRUCT ARGUMENTS Consider the line that passes through (3, 1) and (0, 7).

a. Given the two points, find a third point on the line. Justify your argument.

b. Explain why both $y - 1 = -2(x - 3)$ and $y - 7 = -2(x - 0)$ can be used to represent the line.

36. USE A MODEL Joe and Alisha are reading novels for book reports. Joe records the number of pages he has remaining to read after each day in the table below. Alisha records the number of pages she has remaining each day on the graph at the right.

Joe

Days	0	1	2	3	4	5
Pages Remaining	585	520	455	390	325	260

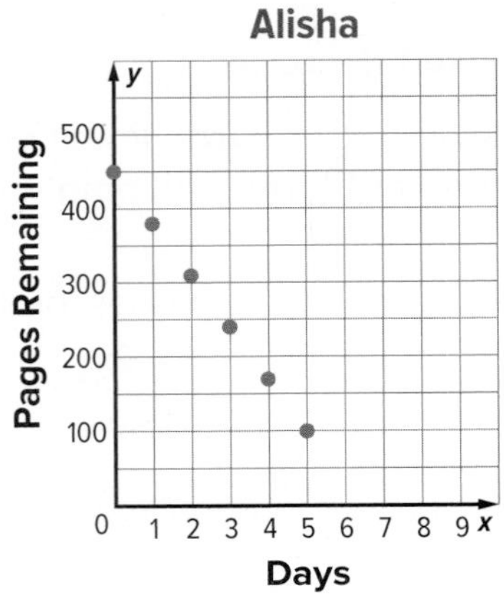

a. Describe the function that models the number of pages remaining for each student.

b. What is the y-intercept for each function? Interpret its meaning in the context of the problem.

c. Write a linear equation in slope-intercept form for the function that can be used to model the pages remaining for each student. Then write each equation in standard form.

d. After how many days will each finish with their reading? What feature of the function represents this event? Explain your answer.

e. Who is the faster reader and by how many pages per day? Support your answer.

Higher-Order Thinking Skills

37. CONSTRUCT ARGUMENTS Determine whether an equation in the form $x = a$, where a is a constant is *sometimes, always,* or *never* a function. Explain your reasoning.

38. PERSEVERE Write an equation in point-slope form of a line that passes through $(a, 0)$ and $(0, b)$.

39. CREATE Write an equation in point-slope form of a line with an x-intercept of 3.

40. WRITE Consider the relationship between hours worked and earnings. When would this situation represent a linear relationship? When would this situation represent a nonlinear relationship? Explain your reasoning.

41. FIND THE ERROR Dan claims that since $y = x + 1$ and $y = 3x + 2$ are both linear functions, the function $y = (x + 1)(3x + 2)$ must also be linear. Is he correct? Explain your reasoning.

42. PERSEVERE Write $y = ax + b$ in point-slope form.

43. WRITE Why is it important to be able to represent linear equations in more than one form?

Solving Systems of Equations Graphically

Explore Solutions of Systems of Equations

Online Activity Use graphing technology to complete the Explore.

> **INQUIRY** How is the solution of a system of equations represented on a graph?

Today's Goal
- Solve systems of linear equations by graphing.

Today's Vocabulary
system of equations
consistent
inconsistent
independent
dependent

Learn Solving Systems of Equations in Two Variables by Graphing

A **system of equations** is a set of two or more equations with the same variables. One method for solving a system of equations is to graph the related function for each equation on the same coordinate plane. The point of intersection of the two graphs represents the solution.

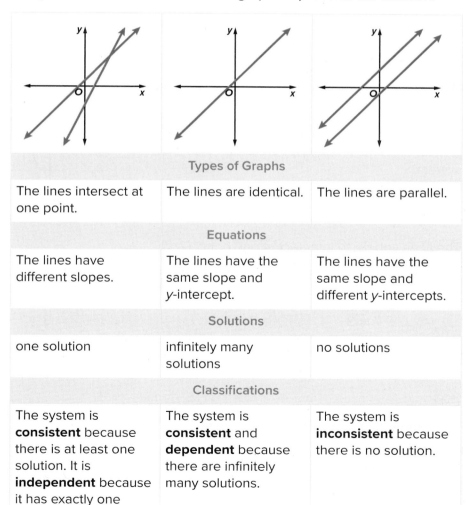

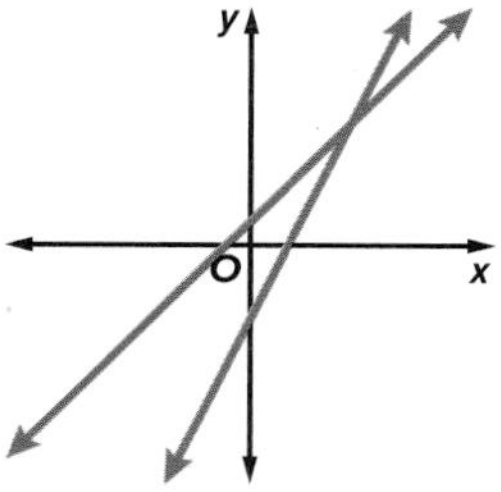	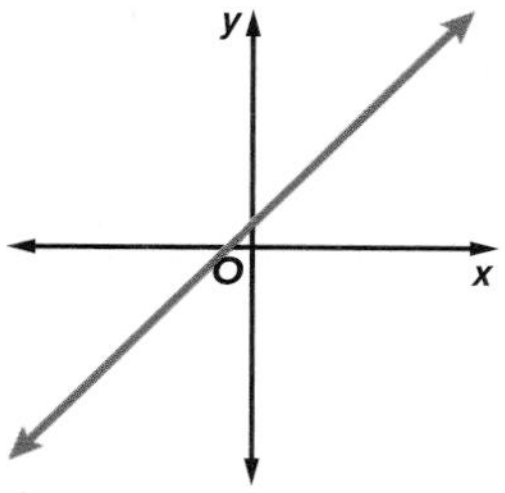	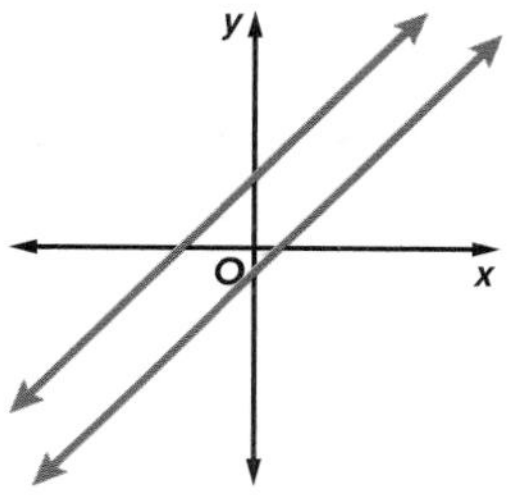
Types of Graphs		
The lines intersect at one point.	The lines are identical.	The lines are parallel.
Equations		
The lines have different slopes.	The lines have the same slope and y-intercept.	The lines have the same slope and different y-intercepts.
Solutions		
one solution	infinitely many solutions	no solutions
Classifications		
The system is **consistent** because there is at least one solution. It is **independent** because it has exactly one solution.	The system is **consistent** and **dependent** because there are infinitely many solutions.	The system is **inconsistent** because there is no solution.

Talk About It!
Explain why the intersection of the two graphs is the solution of the system of equations.

Study Tip

Number of Solutions
By first determining the number of solutions a system has, you can make decisions about whether further steps need to be taken to solve the system. If a system has one solution, you can graph to find it. If a system has infinitely many solutions or no solution, no further steps are necessary. However, you can graph the system to confirm.

Example 1 Classify Systems of Equations

Determine the number of solutions the system has. Then state whether the system of equations is *consistent* or *inconsistent* and whether it is *independent* or *dependent*.

$2y = 6x - 14$

$3x = y = 7$

Solve each equation for y.

$2y = 6x - 14 \quad \rightarrow \quad y = 3x - 7$

$3x - y = 7 \quad \rightarrow \quad y = 3x - 7$

The equations have the same slope and y-intercept. Thus, both equations represent the same line and the system has infinitely many solutions. The system is consistent and dependent.

Check

Determine the number of solutions and classify the system of equations.

$3x - 2y = -7$

$4y = 9 - 6x$

Example 2 Solve a System of Equations by Graphing

Solve the system of equations.

$5x - y = 3$

$-x + y = 5$

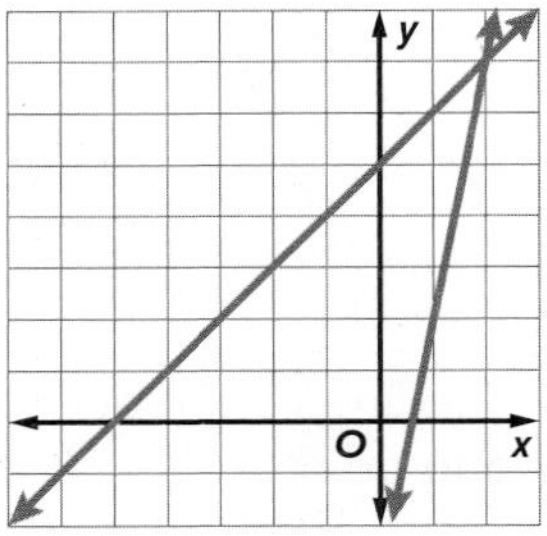

Solve each equation for y. They have different slopes, so there is one solution. Graph the system.

$5x - y = 3 \quad \rightarrow \quad y = 5x - 3$

$-x + y = 5 \quad \rightarrow \quad y = x + 5$

The lines appear to intersect at one point, (2, 7).

CHECK Substitute the coordinates into each original equation.

$5x - y = 3$	Original equation	$-x + y = 5$
$5(2) - 7 \stackrel{?}{=} 3$	$x = 2$ and $y = 7$	$-(2) + 7 \stackrel{?}{=} 5$
$3 = 3$	True	$5 = 5$

The solution is (2, 7).

Check

Solve the system of equations by graphing.

$2y + 14x = -6$

$8x - 4y = -24$

__?__

Example 3 Solve a System of Equations

Solve the system of equations.

$7x + 2y = 16$

$-21x - 6y = 24$

Solve each equation for y to determine the number of solutions the system has.

$7x + 2y = 16 \quad \rightarrow \quad y = -3.5x + 8$
$-21x - 6y = 24 \quad \rightarrow \quad y = -3.5x + -4$

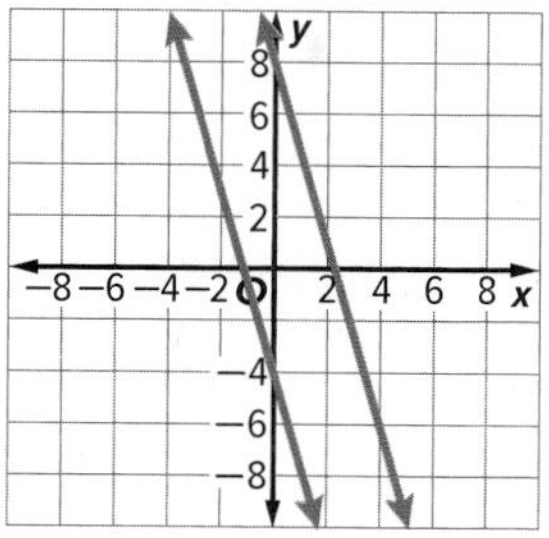

The equations have the same slope and different y-intercepts. So, these equations represent parallel lines, and there is no solution. You can graph each equation on the same grid to confirm that they do not intersect.

> **Study Tip**
>
> **Parallel Lines** Graphs of lines with the same slope and different intercepts are, by definition, parallel.

> **Think About It!**
>
> What would the graph of a system of linear equations with infinitely many solutions look like? Explain your reasoning.

Example 4 Write and Solve a System of Equations by Graphing

CARS **Suppose an electric car costs \$29,000 to purchase and \$0.036 per mile to drive, and a gasoline-powered car costs \$19,000 to purchase and \$0.08 per mile to drive. Estimate after how many miles of driving the total cost of each car will be the same.**

Part A Write equations for the total cost of owning each type of car.

Let y = the total cost of owning the car and x = the number of miles driven.

So, the equation is $y = 0.036x + 29{,}000$ for the electric car and $y = 0.08x + 19{,}000$ for the gasoline car.

Part B Examine the graph to estimate the number of miles you would have to drive before the cost of owning each type of car would be same.

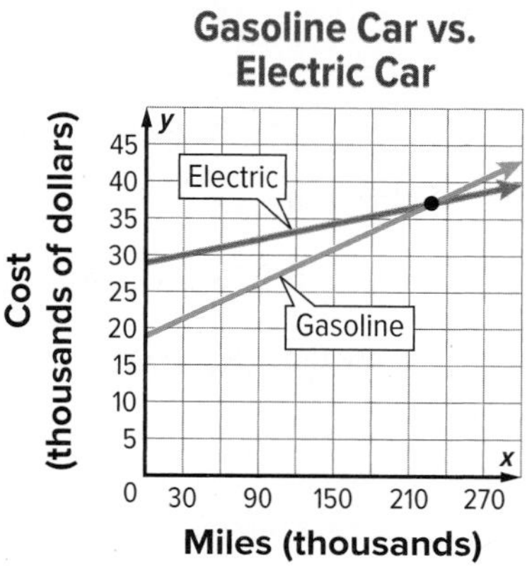

The graphs appear to intersect at approximately (225,000, 37,500). This means that after driving about 225,000 miles, the cost of owning each car will be the same.

> **Think About It!**
>
> Explain what the two equations represent in the context of the situation.

(continued on the next page)

CHECK Substitute the coordinates into each original equation.

$0.036x + 29{,}000 = y$	$0.08x + 19{,}000 = y$
$0.036(225{,}000) + 29{,}000 \stackrel{?}{=} 37{,}500$	$0.08(225{,}000) + 19{,}000 \stackrel{?}{=} 37{,}500$
$37{,}100 \approx 37{,}500$	$37{,}000 \approx 37{,}500$

The estimated number of miles makes both equations approximately true. So, our estimate is reasonable.

Example 5 Solve a System by Using Technology

Use a graphing calculator to solve the system of equations.

Step 1 Solve for *y*.

$$3.5y - 5.6x = 18.2 \rightarrow y = 1.6x + 5.2$$
$$-0.7x - y = -2.4 \rightarrow y = -0.7x + 2.4$$

Step 2 Graph the system.

Enter the equations in the **Y=** list and graph in the standard viewing window.

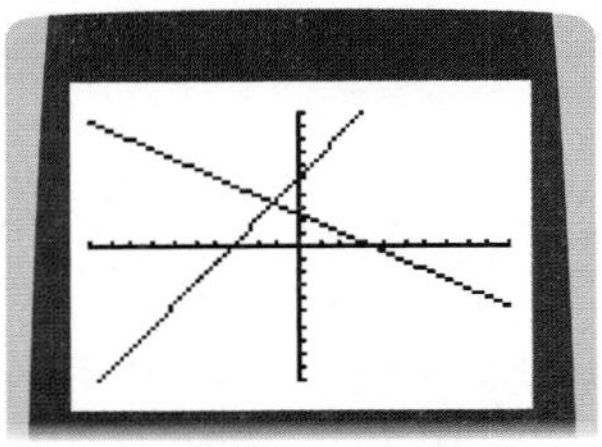

[−10, 10] scl: 1 by [−10, 10] scl: 1

Step 3 Find the intersection.

Use the **intersect** feature from the **CALC** menu to find the coordinates of the point of intersection. When prompted, select each line. Press **enter** to see the intersection. The solution is approximately (−1.22, 3.25).

Example 6 Solve a Linear Equation by Using a System

Use a graphing calculator to solve $4.5x - 3.9 = 6.5 - 2x$ by using a system of equations.

Step 1 Write a system.

Set each side of $4.5x - 3.9 = 6.5 - 2x$ equal to y to create a system of equations.

$y = 4.5x - 3.9$

$y = 6.5 - 2x$

Step 2 Graph the system.

Enter the equations in the **Y=** list and graph in the standard viewing window.

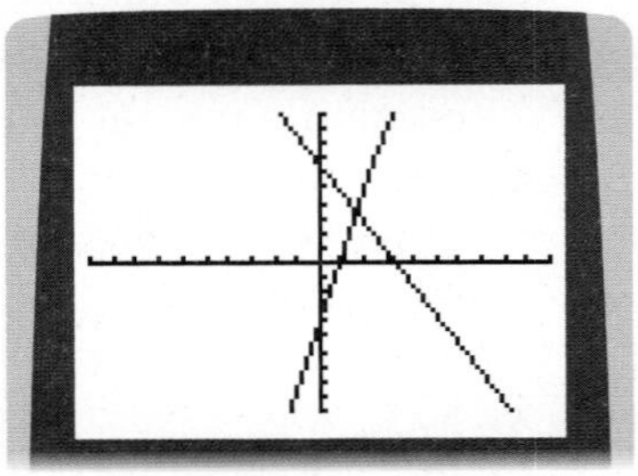

[−10, 10] scl: 1 by [−10, 10] scl: 1

Step 3 Find the intersection.

The solution is the x-coordinate of the intersection, which is 1.6.

Go Online You can complete an Extra Example online.

Watch Out!

Solving by Graphing Solving a system of equations by graphing does not usually give an exact solution. Remember to substitute the solution into both of the original equations to verify the solution or use an algebraic method to find the exact solution.

Go Online to see how to use a graphing calculator with Examples 5 and 6.

Study Tip

Window Dimensions If the point of intersection is not visible in the standard viewing window, zoom out or adjust the window settings manually until it is visible. If the lines appear to be parallel, zoom out to verify that they do not intersect.

Practice

Go Online You can complete your homework online.

Example 1

Determine the number of solutions for each system. Then state whether the system of equations is *consistent* or *inconsistent* and whether it is *independent* or *dependent*.

1. $y = 3x$
$y = -3x + 2$

2. $y = x - 5$
$-2x + 2y = -10$

3. $2x - 5y = 10$
$3x + y = 15$

4. $3x + y = -2$
$6x + 2y = 10$

5. $x + 2y = 5$
$3x - 15 = -6y$

6. $3x - y = 2$
$x + y = 6$

Examples 2 and 3

Solve the system of equations by graphing.

7. $x - 2y = 0$
$y = 2x - 3$

8. $-4x + 6y = -2$
$2x - 3y = 1$

9. $2x + y = 3$
$y = \frac{1}{2}x - \frac{9}{2}$

10. $y - x = 3$
$y = 1$

11. $2x - 3y = 0$
$4x - 6y = 3$

12. $5x - y = 4$
$-2x + 6y = 4$

Example 4

Solve each problem.

13. USE ESTIMATION Mr. Lycan is considering buying clay from two art supply companies. Company A sells 50-pound containers of clay for \$24, plus \$42 to ship the total order. Company B sells the same clay for \$28, plus \$25 to ship the total order.

a. Write equations for the total cost of ordering clay from each company.

b. Graph the equations on the same coordinate plane. Examine the graph to estimate how much Mr. Lycan would have to order for the cost of ordering clay from each company to be the same.

c. Check your estimate by substituting into each original equation. How reasonable is your estimation? Justify your reasoning.

14. USE ESTIMATION Two moving truck companies offer the same vehicle at different rates. At Haul-n-Save, the truck can be rented for \$30, plus \$0.79 per mile. At Rent It Trucks, the truck can be rented for \$75, plus \$0.55 per mile.

a. Write equations for the total cost of renting a truck from each company.

b. Graph the equations on the same coordinate plane. Examine the graph to estimate after how many miles of driving the total rental cost will be the same from each company.

c. Check your estimate. How reasonable is your estimation? Justify your reasoning.

Example 5

USE TOOLS **Use a graphing calculator to solve each system of equations. Round the coordinates to the nearest hundredth, if necessary.**

15. $12y = 5x - 15$
$4.2y + 6.1x = 11$

16. $-3.8x + 2.9y = 19$
$6.6x - 5.4y = -23$

17. $5.8x - 6.3y = 18$
$-4.3x + 8.8y = 32$

Example 6

USE TOOLS **Use a graphing calculator to solve each equation by using a system of equations. Round to the nearest hundredth, if necessary.**

18. $-4.7x + 16 = 16.79x - 80.2$

19. $0.0019x + 3.55 = 0.27x + 2.81$

20. $471 - 63x = -50.5x + 509$

21. $-47.83x - 9 = 33x + 71.019$

Mixed Exercises

Solve each system of equations by graphing.

22. $x - 3y = 6$
$2x - y = -3$

23. $2x - y = 3$
$x + 2y = 4$

24. $4x + y = -2$
$2x + \frac{y}{2} = -1$

25. LASERS A machinist programs a laser cutting machine to focus two laser beams at the same point. One beam is programmed to follow the path $y = 0.5x - 3.15$ and the other is programmed to follow $10x + 5y = 63$. Graph both equations and find the point at which the lasers are focused.

26. REASONING A high school band was selling ride tickets for the school fair. On the first day, 250 children's tickets and 150 adult tickets were sold for a total of \$550. On the second day, 180 children's tickets and 120 adult tickets were sold for a total of \$420. What is the price for each child ticket and each adult ticket?

a. Write a system of equations to represent this situation.

b. Graph the system of equations.

c. Find the intersection of the graphs. What does the point of intersection represent?

Higher-Order Thinking Skills

27. ANALYZE For linear functions a, b, and c, if a is consistent and dependent with b, b is inconsistent with c, and c is consistent and independent with d, then a will *sometimes*, *always*, or *never* be consistent and independent with d. Explain your reasoning.

28. WRITE Explain how to find the solution to a system of linear equations by graphing.

29. ANALYZE Determine if the following statement is *sometimes*, *always*, or *never* true. Explain your reasoning.

A system of linear equations in two variables can have exactly two solutions.

30. CREATE Write a system of equations that has no solution.

Solving Systems of Equations Algebraically

Learn Solving Systems of Equations in Two Variables by Substitution

One algebraic method to solve a system of equations is a process called **substitution,** in which one equation is solved for one variable in terms of the other.

Key Concept • Substitution Method

Step 1 When necessary, solve at least one equation for one of the variables.

Step 2 Substitute the resulting expression from Step 1 into the other equation to replace the variable. Then solve the equation.

Step 3 Substitute the value from Step 2 into either equation, and solve for the other variable. Write the solution as an ordered pair.

Example 1 Substitution When There Is One Solution

Use substitution to solve the system of equations.

$8x - 3y = -1$ Equation 1
$x + 2y = -12$ Equation 2

Step 1 Solve one equation for one of the variables.

Because the coefficient of x in Equation 2 is 1, solve for x in that equation.

$x + 2y = -12$ Equation 2

$x = -2y - 12$ Subtract $2y$ from each side.

Step 2 Substitute the expression. Substitute for x. Then solve for y.

$8x - 3y = -1$ Equation 1

$8(-2y - 12) - 3y = -1$ $x = -2y - 12$

$-16y - 96 - 3y = -1$ Distributive Property

$-19y - 96 = -1$ Simplify.

$-19y = 95$ Add 96 to each side.

$y = -5$ Divide each side by -19.

Step 3 Substitute to solve. Use one of the original equations to solve for x.

$x + 2y = -12$ Equation 2

$x + 2(-5) = -12$ $y = -5$

$x = -2$ Simplify.

The solution is $(-2, -5)$. Substitute into the original equations to check.

Go Online You can complete an Extra Example online.

Today's Goals
- Solve systems of equations by using the substitution method.
- Solve systems of equations by using the elimination method.

Today's Vocabulary
substitution
elimination

Go Online
You can watch a video to see how to use algebra tiles to solve a system of equations by using substitution.

Talk About It!
Describe the benefit of solving a system of equations by substitution instead of graphing when the coefficients are not integers.

Check

Use substitution to solve the system of equations.

$-5x + y = -3$

$3x - 8y = 24$

> **Think About It!**
> What can you conclude about the slopes and y-intercepts of the equations when a system of equations has no solution? when a system of equations has infinitely many solutions?

Example 2 Substitution When There Is Not Exactly One Solution

Use substitution to solve the system of equations.

$\mathbf{-5x + 2.5y = -15}$	Equation 1
$\mathbf{y = 2x - 11}$	Equation 2

Equation 2 is already solved for y, so substitute $2x - 11$ for y in Equation 1.

$-5x + 2.5y = -15$	Equation 1
$-5x + 2.5(2x - 11) = -15$	$y = 2x - 11$
$-5x + 5x - 27.5 = -15$	Distributive Property
$-27.5 = -15$	False

This system has no solution because $-27.5 = -15$ is not true.

Example 3 Apply the Substitution Method

CHEMISTRY **Ms. Washington is preparing a hydrochloric acid (HCl) solution. She will need 300 milliliters of a 5% HCl solution for her class to use during a lab. If she has a 3.5% HCl solution and a 7% HCl solution, how much of each solution should she use in order to make the solution needed?**

> **Think About It!**
> Explain what approximations were made while solving this problem and how they affect the solution.

Step 1 Write two equations in two variables.

Let x be the amount of 3.5% solution and y be the amount of 7% solution.

$x + y = 300$	Equation 1
$0.035x + 0.07y = 0.05(300)$	Equation 2

Step 2 Solve one equation for one of the variables.

$x + y = 300$	Equation 1
$x = -y + 300$	Subtract y from each side.

Step 3 Substitute the resulting expression and solve.

$0.035x + 0.07y = 15$	Equation 2
$0.035(-y + 300) + 0.07y = 15$	$x = -y + 300$
$-0.035y + 10.5 + 0.07y = 15$	Distributive Property
$0.035y = 4.5$	Simplify.
$y \approx 128.57$	Divide each side by 0.035.

Go Online You can complete an Extra Example online.

(*continued on the next page*)

Step 4 Substitute to solve for the other variable.

$x + y = 300$	Equation 1
$x + 128.57 \approx 300$	$y \approx 128.57$
$x \approx 171.43$	Simplify.

The solution of the system is (171.43, 128.57). Ms. Washington should use 171.43 mL of the 3.5% solution and 128.57 mL of the 7% solution.

Learn Solving Systems of Equations in Two Variables by Elimination

Systems of equations may be solved algebraically using **elimination,** which is the process of using addition or subtraction to eliminate one variable.

Key Concept • Elimination Method

Step 1 Multiply one or both of the equations by a number to result in two equations that contain opposite or equal terms.

Step 2 Add or subtract the equations, eliminating one variable. Then solve the equation.

Step 3 Substitute the value from Step 2 into either equation, and solve for the other variable. Write the solution as an ordered pair.

Think About It!
When using elimination, when should you add the equations, and when should you subtract the equations?

Example 4 Elimination When There Is One Solution

Use elimination to solve the system of equations.

$\mathbf{-2x - 9y = -25}$	Equation 1
$\mathbf{-4x - 9y = -23}$	Equation 2

Step 1 Multiply the equation.

Multiply Equation 2 by −1 to get opposite terms $-9y$ and $9y$.

$-4x - 9y = -23$ **Multiply by −1.** → $4x + 9y = 23$

Step 2 Add the equations.

Add the equations to eliminate the y-term and solve for x.

$-2x - 9y = -25$	Equation 1
$(+)\ 4x + 9y = 23$	Equation 2 × (−1)
$2x = -2$	Add the equations.
$x = -1$	Divide each side by 2.

Step 3 Substitute and solve.

$-4x - 9y = -23$	Substitute −1 for x in Equation 2.
$-4(-1) - 9y = -23$	$x = -1$
$-9y = -27$	Simplify.
$y = 3$	Divide each side by −9.

The solution of the system is (−1, 3).

Think About It!
Describe the benefit of using elimination instead of substitution to solve this problem.

Go Online
You can complete an Extra Example online.

Example 5 Multiply Both Equations Before Using Elimination

Use elimination to solve the system of equations.

$\mathbf{2x + 5y = 1}$	Equation 1
$\mathbf{3x - 4y = -10}$	Equation 2

Step 1 Multiply one or both equations.

Multiply Equation 1 by 3 and Equation 2 by 2.

$2x + 5y = 1$	Original equations	$3x - 4y = -10$
$3(2x + 5y) = 3(1)$	Multiply.	$2(3x - 4y) = 2(-10)$
$6x + 15y = 3$	Distribute.	$6x - 8y = -20$

Step 2 Eliminate one variable and solve.

In order to eliminate the *x*-terms, subtract the equations. Then, solve for *y*.

$6x + 15y = 3$	Equation 1 × 3
$(-)\ 6x - 8y = -20$	Equation 2 × 2
$23y = 23$	Subtract the equations.
$y = 1$	Divide each side by 23.

Step 3 Substitute and solve.

Substitute $y = 1$ in either of the original equations and solve for *x*.

$2x + 5y = 1$	Equation 1
$2x + 5(1) = 1$	$y = 1$
$2x + 5 = 1$	Multiply.
$x = -2$	Solve for *x*.

The solution of the system is $(-2, 1)$.

Think About It!
Describe the graph of this system of equations.

Example 6 Elimination Where There is Not Exactly One Solution

Use elimination to solve the system of equations.

$\mathbf{18x + 21y = 14}$	Equation 1
$\mathbf{6x + 7y = 2}$	Equation 2

Steps 1 and 2 Multiply one or both equations and add them.

Multiply Equation 2 by −3. Then add the equations.

$18x + 21y = 14$ → $18x + 21y = 14$

$6x + 7y = 2$ — Multiply by −3. → $(+)\ -18x - 21y = -6$

$$0 \neq 8$$

Because $0 \neq 8$, this system has no solution.

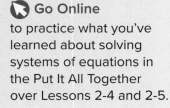

Go Online You can complete an Extra Example online.

Go Online
to practice what you've learned about solving systems of equations in the Put It All Together over Lessons 2-4 and 2-5.

Practice

Go Online You can complete your homework online.

Examples 1 and 2

Use substitution to solve each system of equations.

1. $2x - y = 9$
$x + 3y = -6$

2. $2x - y = 7$
$6x - 3y = 14$

3. $2x + y = 5$
$3x - 3y = 3$

4. $3x + y = 7$
$4x + 2y = 16$

5. $4x - y = 6$
$2x - \frac{y}{2} = 4$

6. $2x + y = 8$
$3x + \frac{3}{2}y = 12$

Example 3

Solve each problem.

7. BAKE SALE Cassandra and Alberto are selling pies for a fundraiser. Cassandra sold 3 small pies and 14 large pies for a total of $203. Alberto sold 11 small pies and 11 large pies for a total of $220. Determine the cost of each pie.

a. Write a system of equations and solve by using substitution.

b. What does the solution represent in terms of this situation?

c. How can you verify that the solution is correct?

8. STOCKS Ms. Patel invested a total of $825 in two stocks. At the time of her investment, one share of Stock A was valued at $12.41 and a share of Stock B was valued at $8.62. She purchased a total of 79 shares.

a. Write a system of equations and solve by substitution.

b. How many shares of each stock did Ms. Patel buy? How much did she invest in each of the two stocks?

Examples 4-6

Use elimination to solve each system of equations.

9. $3x - 2y = 4$
$5x + 3y = -25$

10. $5x + 2y = 12$
$-6x - 2y = -14$

11. $7x + 2y = -1$
$21x + 6y = -9$

12. $3x - 5y = -9$
$-7x + 3y = 8$

13. $x - 3y = -12$
$2x + y = 11$

14. $6w - 8z = 16$
$3w - 4z = 8$

Mixed Exercises

Use substitution or elimination to solve each system of equations.

15. $0.5x + 2y = 5$
$x - 2y = -8$

16. $h - z = 3$
$-3h + 3z = 6$

17. $-r + t = 5$
$-2r + t = 4$

18. $3r - 2t = 1$
$2r - 3t = 9$

19. $5g + 4k = 10$
$-3g - 5k = 7$

20. $4m - 2p = 0$
$-3m + 9p = 5$

21. The sum of two numbers is 12. The difference of the same two numbers is −4. Find the two numbers.

22. Twice a number minus a second number is −1. Twice the second number added to three times the first number is 9. Find the two numbers.

23. REASONING Mr. Janson paid for admission to the high school football game for his family. He purchased 3 adult tickets and 2 student tickets for a total of $22. Ms. Pham purchased 5 adult tickets and 3 student tickets for a total of $35.75. What is the cost of each adult ticket and each student ticket?

24. USE A MODEL The Newton City Park has 11 basketball courts, which are all in use. There are 54 people playing basketball. Some are playing one-on-one, and some are playing in teams. A one-on-one game requires 2 players, and a team game requires 10 players.

a. Write a system of equations that represents the number of one-on-one and team games being played.

b. Solve the system of equations and interpret your results.

25. FIND THE ERROR Gloria and Syreeta are solving the system $6x - 4y = 26$ and $-3x + 4y = -17$. Is either of them correct? Explain your reasoning.

Gloria

$6x - 4y = 26$	$-3(3) + 4y = -17$
$-3x + 4y = -17$	$-9 + 4y = -17$
$3x = 9$	$4y = -8$
$x = 3$	$y = -2$

The solution is (3, −2).

Syreeta

$6x - 4y = 26$	$6(-3) - 4y = 26$
$-3x + 4y = -17$	$-18 - 4y = 26$
$3x = -9$	$-4y = 44$
$x = -3$	$y = -11$

The solution is (−3, −11).

26. CREATE Write a system of equations in which one equation should be multiplied by 3 and the other should be multiplied by 4 in order to solve the system with elimination. Then solve your system.

27. WRITE Why is substitution sometimes more helpful than elimination?

Solving Systems of Inequalities

Explore Solutions of Systems of Inequalities

Online Activity Use a graph to complete the Explore.

INQUIRY How is a graph used to determine viable solutions of a system of inequalities?

Today's Goal
- Solve systems of linear inequalities in two variables.

Today's Vocabulary
system of inequalities
feasible region
bounded
unbounded

Learn Solving Systems of Inequalities in Two Variables

A **system of inequalities** is a set of two or more inequalities with the same variables. The **feasible region** is the intersection of the graphs. Ordered pairs within the feasible region are viable solutions. The feasible region may be **bounded**, if the graph of the system is a polygonal region, or **unbounded** if it forms a region that is open.

Key Concept • Solving Systems of Inequalities

Step 1 Graph each inequality by graphing the related equation and shading the correct region.

Step 2 Identify the feasible region that is shaded for all of the inequalities. This represents the solution set of the system.

Study Tip

Related Equation A related equation of the inequality $y \leq mx + b$ is $y = mx + b$. The inequalities $y < mx + b$, $y \geq mx + b$, and $y > mx + b$ all share this same related equation.

Example 1 Unbounded Region

Solve the system of inequalities.

$y \leq 4x - 3$ Inequality 1

$-2y > x$ Inequality 2

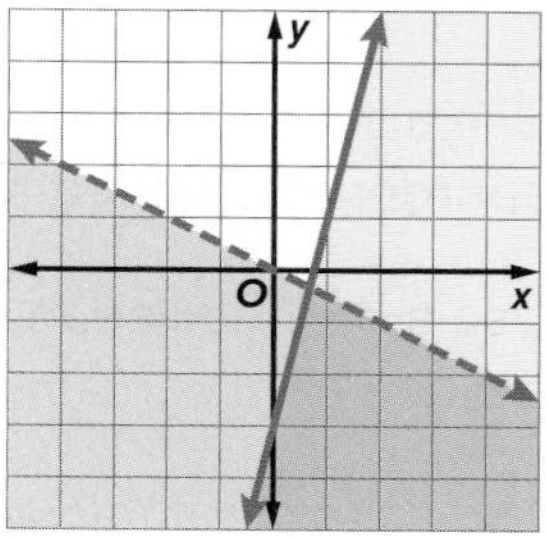

Use a solid line to graph the first boundary $y = 4x - 3$. The appropriate half-plane is shaded yellow. Use a dashed line to graph the second boundary $y = -0.5x$. The appropriate half-plane is shaded blue.

The solution of the system is the set of ordered pairs in the intersection of the graphs shaded in green. The feasible region is unbounded.

(continued on the next page)

Go Online
You can watch a video to see how to graph a system of linear inequalities.

Study Tip

Boundaries The boundaries of inequalities with symbols < and > are graphed using dashed lines, indicating that the ordered pairs on the boundary are not included in the feasible region.

CHECK

Test the solution by substituting the coordinate of a point in the unbounded region, such as (2, −3), into the system of inequalities. If the point is viable for both inequalities, it is a solution of the system.

$y \leq 4x - 3$	Original inequality	$-2(y) > x$
$-3 \stackrel{?}{\leq} 4(2) - 3$	$x = 2$ and $y = -3$	$-2(-3) \stackrel{?}{>} 2$
$-3 \leq 5$	True	$6 > 2$

Check

Graph the solution of the system of inequalities.

$y \leq \frac{1}{3}x + 2$

$y > x$

$y \leq 1$

Think About It!

How can you find the coordinates of the vertices of a polygon formed by the system of inequalities?

Example 2 Bounded Region

Solve the system of inequalities.

$\boldsymbol{y < -\frac{4}{3}x + 5}$	Inequality 1
$\boldsymbol{y \geq x - 2}$	Inequality 2
$\boldsymbol{x > 1}$	Inequality 3

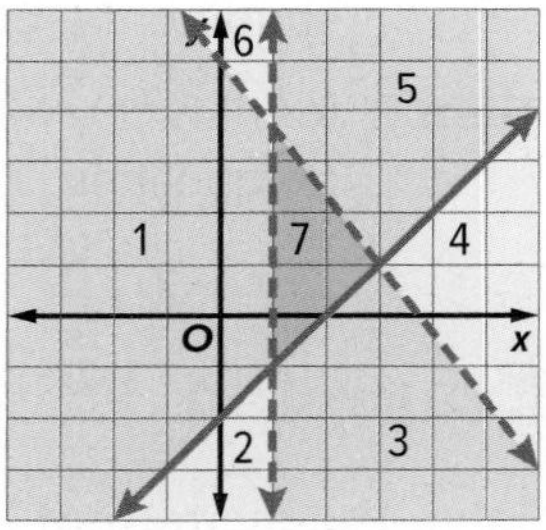

Use a dashed line to graph the first boundary $y = -\frac{4}{3}x + 5$.

The appropriate shaded area contains regions 1, 2, 3, and 7.

Use a solid line to graph the second boundary $y = x - 2$. The appropriate shaded area contains regions 1, 5, 6, and 7.

Use a dashed line to graph the third boundary $x = 1$. The appropriate shaded area contains regions 3, 4, 5 and 7.

The solution of the system is the set of ordered pairs in the intersection of the graphs, represented by region 7. The feasible region is bounded.

Go Online You can complete an Extra Example online.

Check

Graph the solution of the system of inequalities.

$y \geq \frac{4}{5}x - 3$

$y < -\frac{2}{3}x + 2$

$x \geq 0$

Example 3 Use Systems of Inequalities

TOURS **A Niagara Falls boat tour company charges $19.50 for adult tickets and $11 for children's tickets. Each boat has a capacity of 600 passengers, including 8 crew members. Suppose the company's operating cost for one boat tour is $2750. Write and graph a system of inequalities to represent the situation so the company will make a profit on each tour. Then, identify some viable solutions.**

Part A Write the system of inequalities.

Let a represent the number of adult tickets and c represent the number of children's tickets.

Inequality 1: $a + c + 8 \leq 600$

Inequality 2: $19.5a + 11c > 2750$

Inequality 3: $a \geq 0$

Inequality 4: $c \geq 0$

Part B Graph the system of inequalities.

Graph the inequalities.

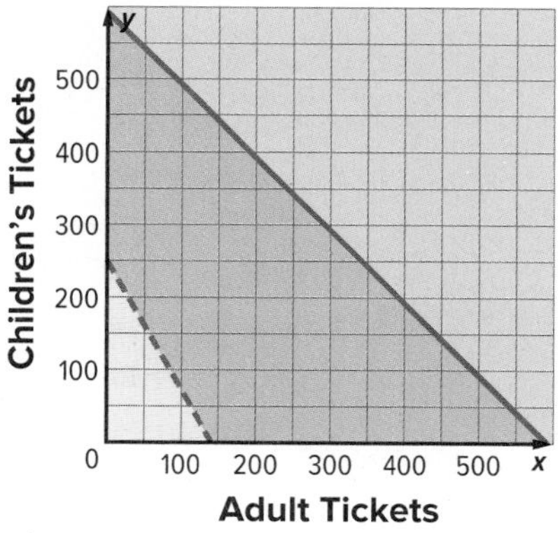

Identify feasible region.

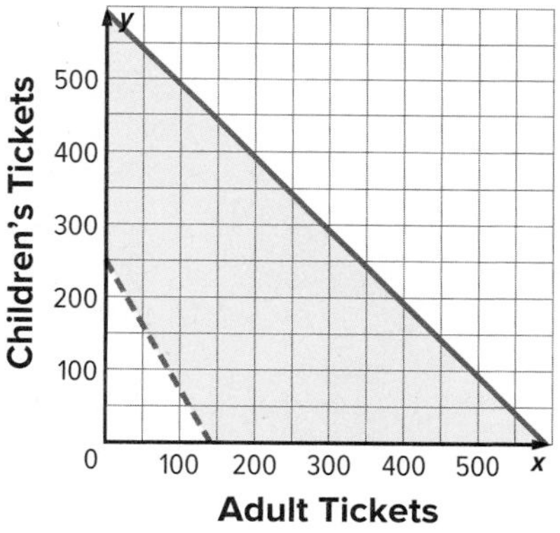

Part C Identify some viable solutions.

Passengers	Viable	Nonviable
60 adults, 100 children	☐	☒
210 adults, 350 children	☒	☐
415 adults, 200 children	☐	☒
390 adults, 240 children	☐	☒
550 adults, 0 children	☒	☐

Go Online You can complete an Extra Example online.

Talk About It!

Why is it important to label the axes given the context of this problem? Explain.

Study Tip

Consider the Context While the feasible region represents the viable solutions, the solution may be limited to only integers or only positive numbers. In this case, the touring company cannot sell a fraction of a ticket. So the solution must be given as whole numbers.

Check

FUNDRAISER The international club raised $1200 to buy livestock for a community in a different part of the world. The club can buy an alpaca for $160 and a sheep for $120. If the club wants to donate at least 8 animals, determine the system of inequalities to represent the situation.

Part A

Graph of the system of inequalities that represents the possible combinations of animals the club can donate.

Part B

Select all of the viable solutions given the constraints of the club's funds.

A. 0 alpacas, 10 sheep

B. 1 alpaca, 8 sheep

C. 3 alpacas, 6 sheep

D. 6 alpacas, 3 sheep

E. 8 alpacas, 0 sheep

Go Online You can complete an Extra Example online.

Practice

Go Online You can complete your homework online.

Example 1

Solve each system of inequalities.

1. $x - y \leq 2$
   $x + 2y \geq 1$

2. $3x - 2y \leq -1$
   $x + 4y \geq -12$

3. $y \geq \frac{x}{2} - 3$
   $y < 2x$

4. $y < \frac{x}{3} + 2$
   $y < -2x + 1$

5. $x + y \geq 4$
   $2x - y > 2$

6. $x + 3y < 3$
   $x - 2y \geq 4$

Example 2

Solve each system of inequalities.

7. $y \geq -3x + 7$
   $y > \frac{1}{2}x$
   $y < 2$

8. $x > -3$
   $y < -\frac{1}{3}x + 3$
   $y > x - 1$

9. $y < -\frac{1}{2}x + 3$
   $y > \frac{1}{2}x + 1$
   $y < 3x + 10$

10. $y \leq 0$
    $x \leq 0$
    $y \geq -x - 1$

11. $y \leq 3 - x$
    $y \geq 3$
    $x \geq -5$

12. $x \geq -2$
    $y \geq x - 2$
    $x + y \leq 2$

Example 3

13. TICKETS The high school auditorium has 800 seats. Suppose that the drama club has a goal of making at least $3400 each night of their spring play. Student tickets are $4 and adult tickets are $7.

    a. Write a system of inequalities to represent the situation.

    b. Graph the system of inequalities. In which quadrant(s) is the solution?

    c. Could the club meet its goal by selling 200 adult and 475 student tickets? Explain.

14. CONSTRUCT ARGUMENTS Anthony charges $15 an hour for tutoring and $10 an hour for babysitting. He can work no more than 14 hours a week. How many hours should Anthony spend on each job if he wants to earn at least $125 each week?

    a. Write a system of inequalities to represent this situation.

    b. Graph the system of inequalities and highlight the solution.

    c. Determine whether (4, 5), (7, 6), and (5, 10) are viable solutions given the constraints of the situation. Explain.

Mixed Exercises

Solve each system of inequalities.

15. $y \geq |2x + 4| - 2$
$3y + x \leq 15$

16. $y \geq |6 - x|$
$y \leq 4$

17. $y > -3x + 1$
$4y \leq x - 8$
$3x - 5y < 20$

18. $|x| > y$
$y \leq 6$
$y \geq -2$

19. FINANCE Sheila plans to invest $2000 or less in two different accounts. The low risk account pays 3% annual simple interest, and the high risk account pays 12% annual simple interest. Sheila wants to make at least $150 in interest this year.

a. Define the variables, then write and graph a system of inequalities to show how Sheila can split her investment between the accounts.

b. Explain why your graph for this situation is restricted to Quadrant I.

c. Give three viable solutions to meet the constraints of Sheila's investments.

20. STRUCTURE Write a system of inequalities for the graph shown.

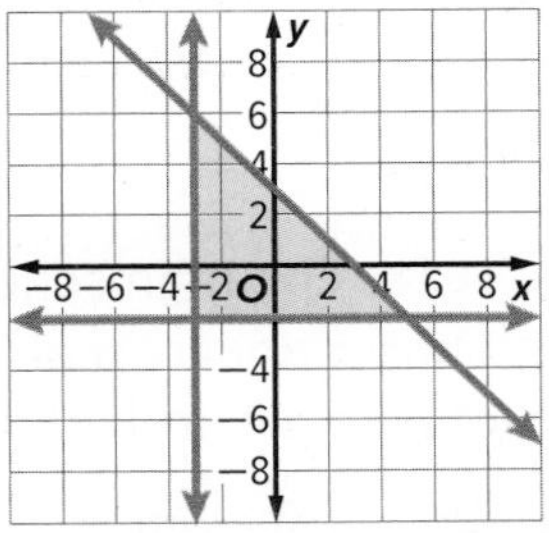

Higher Order Thinking Skills

21. PERSEVERE Find the area of the region defined by the following inequalities.

$y \geq -4x - 16$
$4y \leq 26 - x$
$3y + 6x \leq 30$
$4y - 2x \geq -10$

22. CREATE Write systems of two inequalities in which the solution:

a. lies only in the third quadrant.

b. does not exist.

c. lies only on a line.

23. ANALYZE Determine whether the statement is *true* or *false*. Justify your argument.

A system of two linear inequalities has either no points or infinitely many points in its solution.

24. WRITE Explain how you would determine whether (−4, 6) is a solution of a system of inequalities.

Optimization with Linear Programming

Learn Finding Maximum and Minimum Values

Linear programming is the process of finding the maximum or minimum values of a function for a region defined by a system of linear inequalities.

Key Concept • Linear Programming

Step 1 Graph the inequalities.

Step 2 Determine the coordinates of the vertices.

Step 3 Evaluate the function at each vertex.

Step 4 For a bounded region, determine the maximum and minimum. For an unbounded region, test other points within the feasible region to determine which vertex represents the maximum or minimum.

Example 1 Maximum and Minimum Values for a Bounded Region

Graph the system of inequalities. Name the coordinates of the vertices of the feasible region. Find the maximum and minimum values of the function for this region.

$-2 \leq x \leq 4$

$y \leq x + 2$

$y \geq -0.5x - 3$

$f(x, y) = -2x + 6y$

Steps 1 and 2 Graph the inequalities and determine the vertices.

The vertices of the feasible region are (−2, −2), (−2, 0), (4, −5), and (4, 6).

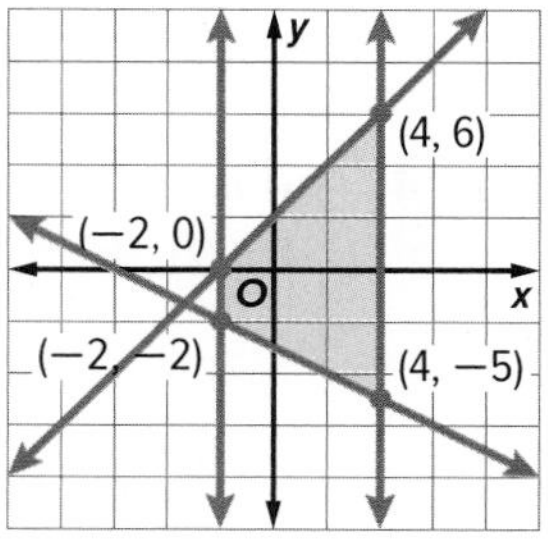

Step 3 Evaluate the function at each vertex.

(x, y)	−2x + 6y	f(x, y)
(−2, −2)	−2(−2) + 6(−2)	−8
(−2, 0)	−2(−2) + 6(0)	4
(4, −5)	−2(4) + 6(−5)	−38
(4, 6)	−2(4) + 6(6)	28

Step 4 Determine the maximum and minimum.

The maximum value is 28 at (4, 6). The minimum value is −38 at (4, −5).

Today's Goals

- Find maximum and minimum values of a function over a region.
- Solve real-world optimization problems by graphing systems of inequalities maximizing or minimizing constraints.

Today's Vocabulary

linear programming

optimization

Study Tip

Unbounded Regions An unbounded feasible region does not necessarily contain a maximum or minimum.

Study Tip

Feasible Region To determine the feasible region, you can shade the solution set of each inequality individually, and then find where they all overlap. Shading each inequality using a different color or shading style can help you easily determine the feasible region.

Example 2 Maximum and Minimum Values for an Unbounded Region

Graph the system of inequalities. Name the coordinates of the vertices of the feasible region. Find the maximum and minimum values of the function for this region.

$1 \leq y \leq 3$

$y \leq -x$

$y \geq 0.5x + 3$

$f(x, y) = -x + y$

Steps 1 and 2 Graph the inequalities and determine the vertices.

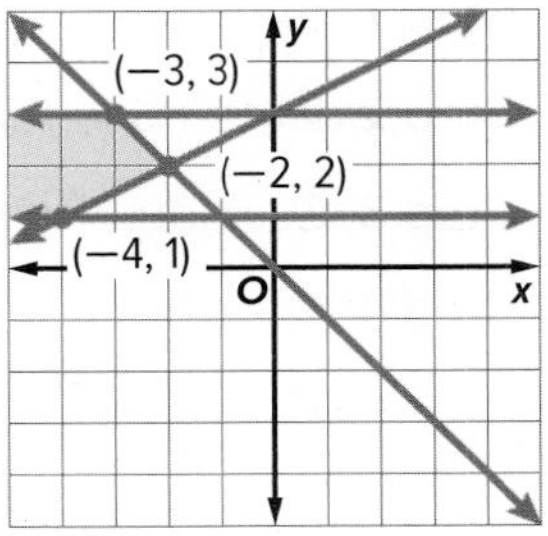

The vertices of the feasible region are (−3, 3), (−2, 2), and (−4, 1).

Notice that the region is unbounded. This may indicate that there is no minimum or maximum value.

Step 3 Evaluate the function.

Evaluate at each vertex and a point in the feasible region.

(x, y)	$-x + y$	$f(x, y)$
(−3, 3)	−(−3) + 3	6
(−2, 2)	−(−2) + 2	4
(−4, 1)	−(−4) + 1	5
(−10, 2)	−(−10) + 2	12

Step 4 The minimum value is 4 at (−2, 2). As shown by the test point (−10, 2), there is no maximum value.

Study Tip

Feasible Region To determine whether an unbounded region has a maximum or minimum for the function $f(x, y)$, you need to test several points in the feasible region to see if any values of $f(x, y)$ are greater than or less than the values of $f(x, y)$ for the vertices.

Talk About It!

The function in the Example has a minimum but no maximum on the unbounded region. Is there a function that has a maximum, but no minimum on the region? Does the function $f(x) = y$ have a maximum and/or a minimum on the region? Justify your reasoning.

Explore Using Technology with Linear Programming

Online Activity Use graphing technology to complete the Explore.

INQUIRY How can you use technology to find the maximum or minimum values of a function over a given region?

Go Online You can complete an Extra Example online.

Learn Linear Programming

Optimization is the process of seeking the optimal value of a function subject to given constraints.

Key Concept • Optimization with Linear Programming

Step 1 Define the variables.

Step 2 Write a system of inequalities.

Step 3 Graph the system of inequalities.

Step 4 Find the coordinates of the vertices of the feasible region.

Step 5 Write a linear function to be maximized or minimized.

Step 6 Evaluate the function at each vertex by substituting the coordinates into the function.

Step 7 Interpret the results.

Example 3 Optimizing with Linear Programming

GARDENING **Avoree has a 30-square-foot plot in the school greenhouse and wants to plant lettuce and cucumbers while minimizing the amount of water she uses for them. Each cucumber requires 2.25 square feet of space and uses 25 gallons of water over the lifetime of the plant. Each lettuce plant requires 1.5 square feet of space and uses 17 gallons of water. She wants to grow at least 4 of each type of plant and at least 16 plants in total. Determine how many of each plant Avoree should plot in order to minimize her water usage.**

Step 1 Define the variables.

Because the number of plants of different types determine the water usage, the independent variables should be the numbers of plants. The dependent variable in the function to be minimized should be total water used. Let c represent the number of cucumber plants and t represent the number of lettuce plants. Let $f(c, t)$ represent the water used for c cucumber plants and t lettuce plants.

Step 2 Write a system of inequalities.

Avoree wants to have at least 4 of each type of plant, so 4 must be included as minimums for both c and t in the inequalities. The total number of plants must be at least 16. Each cucumber requires 2.25 square feet of space and each lettuce plant requires 1.5 square feet of space. The total planting area of the plants must be less than or equal to 30 ft^2.

$c \geq 4$

$t \geq 4$

$c + t \geq 16$

$2.25c + 1.5t \leq 30$

Go Online You can complete an Extra Example online.

Math History Minute

In the 1960s, **Christine Darden (1942–)** became one of the "human computers" who crunched numbers for engineers at NASA's Langley Research Center. After earning a doctorate degree in mechanical engineering, Darden became one of few female aerospace engineers at NASA Langley. For most of her career, her focus was sonic boom minimization.

Step 3 Graph the system of inequalities.

Step 4 Find the coordinates of the vertices of the feasible region.

The vertices of the feasible region are (4, 12), (4, 14), and (8, 8).

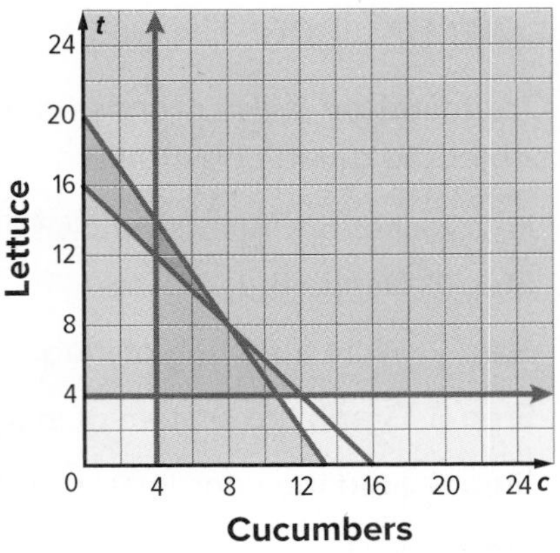

Step 5 Write a linear function to be minimized.

Because Avoree wants to minimize her water usage, the linear function will be the sum of the water usage for each plant.

$f(c, t) = 25c + 17t$

Step 6 Evaluate the function at each vertex.

(c, t)	$25c + 17t$	$f(c, t)$
(4, 12)	25(4) + 17(12)	304
(4, 14)	25(4) + 17(14)	338
(8, 8)	25(8) + 17(8)	336

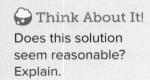

Think About It!

Does this solution seem reasonable? Explain.

Step 7 Interpret the results.

Avoree should plant 4 cucumber plants and 12 lettuce plants to minimize her water usage.

Check

SOCCER A new soccer team is being created for a professional soccer league, and they need to hire at least ten new players. They need five to eight defenders and seven to ten forwards, and they want to minimize the amount they spend on these players' salaries so they have enough money remaining to hire goalkeepers and midfielders.

Position	Minimum	Maximum	Salary per Player ($)
forward *f*	5	8	120,000
defender *d*	7	10	100,000

The least amount of money that the team can spend is $ ___?___ by hiring ___?___ forwards and ___?___ defenders.

Go Online You can complete an Extra Example online.

Practice

Go Online You can complete your homewo

Example 1

Graph the system of inequalities. Name the coordinates of the vertices of the feasible region. Find the maximum and minimum values of the function for this region.

1. $y \geq 2$
$1 \leq x \leq 5$
$y \leq x + 3$
$f(x, y) = 3x - 2y$

2. $y \geq -2$
$y \geq 2x - 4$
$x - 2y \geq -1$
$f(x, y) = 4x - y$

3. $x + y \geq 2$
$4y \leq x + 8$
$y \geq 2x - 5$
$f(x, y) = 4x + 3y$

4. $x \geq 2$
$x \leq 5$
$y \geq 1$
$y \leq 4$
$f(x, y) = x + y$

5. $x \geq 1$
$y \leq 6$
$y \geq x - 2$
$f(x, y) = x - y$

6. $x \geq 0$
$y \geq 0$
$y \leq 7 - x$
$f(x, y) = 3x + y$

Example 2

Graph the system of inequalities. Name the coordinates of the vertices of the feasible region. Find the maximum and minimum values of the function for this region.

7. $x \geq -1$
$x + y \leq 6$
$f(x, y) = x + 2y$

8. $y \leq 2x$
$y \geq 6 - x$
$y \leq 6$
$f(x, y) = 4x + 3y$

9. $y \leq 3x + 6$
$4y + 3x \leq 3$
$x \geq -2$
$f(x, y) = -x + 3y$

Example 3

10. PAINTING A painter has exactly 32 units of yellow dye and 54 units of green dye. He plans to mix as many gallons as possible of color A and color B. Each gallon of color A requires 4 units of yellow dye and 1 unit of green dye. Each gallon of color B requires 1 unit of yellow dye and 6 units of green dye. Find the maximum number of gallons he can mix.

a. Define the variables and write a system of inequalities.

b. Graph the system of inequalities and find the coordinates of the vertices of the feasible region.

c. Find the maximum number of gallons the painter can make.

11. REASONING A jewelry company makes and sells necklaces. For one type of necklace, the company uses clay beads and glass beads. Each necklace has no more than 10 clay beads and at least 4 glass beads. For every necklace, four times the number of glass beads is less than or equal to 8 more than twice the number of clay beads. Each clay bead costs $0.20 and each glass bead costs $0.40. The company wants to find the minimum cost to make a necklace with clay and glass beads and find the combination of clay and glass beads in a necklace that costs the least to make.

a. Define the variables and write a system of inequalities. Then write an equation for the cost *C*.

b. Graph the system of inequalities and find the coordinates of the vertices of the feasible region.

c. Find the number of clay beads and glass beads in a necklace that costs the least to make.

Mixed Exercises

12. REASONING Juan has 8 days to make pots and plates to sell at a local craft fair. Each pot weighs 2 pounds and each plate weighs 1 pound. Juan cannot carry more than 50 pounds to the fair. Each day, he can make at most 5 plates and 3 pots. He will make \$12 profit for every plate and \$25 profit for every pot that he sells.

a. Write linear inequalities to represent the number of pots p and plates a Juan can bring to the fair.

b. List the coordinates of the vertices of the feasible region.

c. How many pots and plates should Juan make to maximize his potential profit?

13. USE A MODEL A trapezoidal park is built on a slight incline. The ground elevation above sea level is given by $f(x, y) = x - 3y + 20$ feet. What are the coordinates of the highest point in the park and what is the elevation at that point?

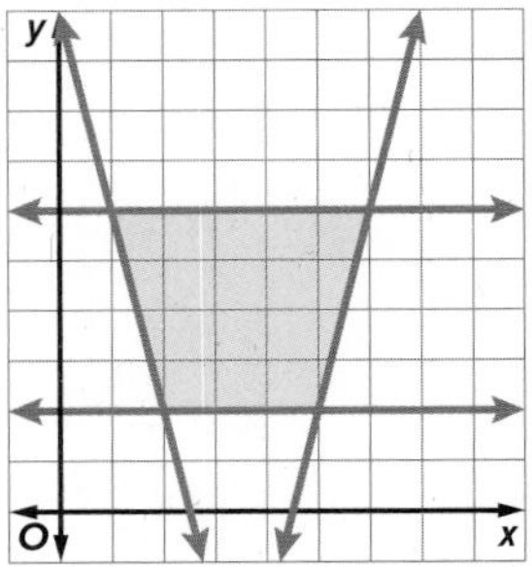

14. FOOD A zoo is mixing two types of food for the animals. Each serving is required to have at least 60 grams of protein and 30 grams of fat. Custom Foods has 15 grams of protein and 10 grams of fat and costs 80 cents per unit. Zookeeper's Best contains 20 grams of protein and 5 grams of fat and costs 50 cents per unit.

a. The zoo wants to minimize their costs. Define the variables and write the inequalities that represent the constraints of the situation.

b. Graph the inequalities. What does the unbound region represent? Determine how much of each type of food should be used to minimize costs.

15. ANALYZE Determine whether the following statement is *sometimes*, *always*, or *never* true. Explain your reasoning.

An unbounded region will not have both a maximum and minimum value.

16. WHICH ONE DOESN'T BELONG? Identify the system of inequalities that is not the same as the other three. Explain your reasoning.

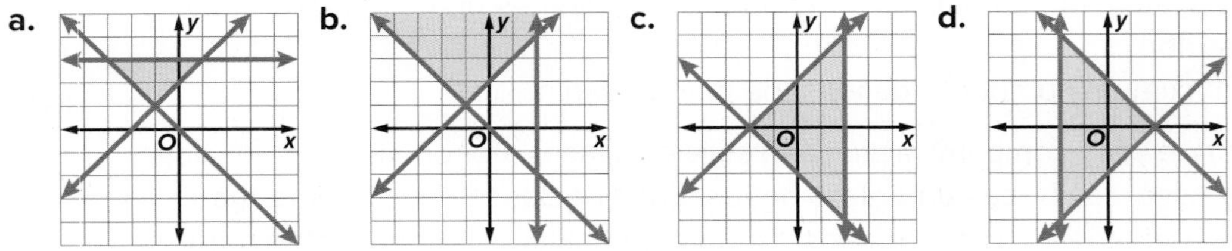

17. WRITE Upon determining a bounded feasible region, Kelvin noticed that vertices $A(-3, 4)$ and $B(5, 2)$ yielded the same maximum value for $f(x, y) = 16y + 4x$. Kelvin confirmed that his constraints were graphed correctly and his vertices were correct. Then he said that those two points were not the only maximum values in the feasible region. Explain how this could have happened.

18. CREATE Create a set of inequalities that forms a bounded region with an area of 20 units^2 and lies only in the fourth quadrant.

Systems of Equations in Three Variables

Today's Goal
- Solve systems of linear equations in three variables.

Today's Vocabulary
ordered triple

Explore Systems of Equations Represented as Lines and Planes

Online Activity Use concrete models to complete the Explore.

INQUIRY How does the way that lines or planes intersect affect the solution of a system of equations?

Learn Solving Systems of Equations in Three Variables

The graph of an equation in three variables is a plane. The graph of a system of equations in three variables is an intersection of planes.

- If the three individual planes intersect at a specific point, then there is one solution given as an ordered triple (x, y, z). An **ordered triple** is three numbers given in a specific order to locate a point in space.
- If the planes intersect in a line, every point on the line represents a solution of the system. If they intersect in the same plane, every equation is equivalent and every coordinate in the plane represents a solution of the system.
- If there are no points in common with all three planes, then there is no solution.

Talk About It!
Is it possible for a system of equations in three variables to have exactly three solutions? Justify your argument.

Example 1 Solve a System with One Solution

Solve the system of equations.

$\mathbf{4x + y + 6z = 12}$ Equation 1

$\mathbf{2x - 10y - 2z = 12}$ Equation 2

$\mathbf{3x + 8y + 19z = 38}$ Equation 3

Step 1 Eliminate one variable.

Select two of the equations and eliminate one of the variables.

$4x + y + 6z = 12$ Equation 1 $2x - 10y - 2z = 12$ Equation 2

Multiply Equation 1 by 10 to eliminate y.

$4x + y + 6z = 12$ → Multiply by 10. → $40x + 10y + 60z = 120$

Add the equations to eliminate y.

$40x + 10y + 60z = 120$ Equation 1 × 10

$(+)\quad 2x - 10y - 2z = 12$ Equation 2

$42x + 58z = 132$ Add the equations.

(continued on the next page)

Problem-Solving Tip

Identify Subgoals Before you begin solving a system of equations in 3 variables, try to identify which variable would be easiest to eliminate in Step 1. Be sure that it makes sense to select that variable to eliminate from the set of equations. As you work through the problem, check your work after each step. Check for reasonableness before moving on to the next subgoal.

Think About It!

Suppose you initially eliminated x in Step 1. How would that affect the solution? Explain your reasoning.

Use a different combination of the original equations to create another equation in two variables. Eliminate y again.

$4x + y + 6z = 12$ Equation 1

$3x + 8y + 19z = 38$ Equation 3

Multiply Equation 1 by −8 and add the equations to eliminate y.

$4x + y + 6z = 12$ **Multiply by −8.** → $-32x - 8y - 48z = -96$

Add the equations to eliminate y.

$-32x - 8y - 48z = -96$	Equation 1 × (−8).
(+) $3x + 8y + 19z = 38$	Equation 3
$-29x \quad - 29z = -58$	Add the equations.

Step 2 Solve the system of two equations.

Multiply the second equation by 2 and add the equations to eliminate z.

$42x + 58z = 132$

$-29x - 29z = -58$ **Multiply by 2.** →

$$\begin{aligned} 42x + 58z &= 132 \\ (+)\quad -58x - 58z &= -116 \\ \hline -16x \quad &= 16 \\ x &= -1 \end{aligned}$$

Use substitution to solve for z.

$42x + 58z = 132$	Equation in two variables
$42(-1) + 58z = 132$	$x = -1$
$-42 + 58z = 132$	Multiply.
$58z = 174$	Add 42 to each side.
$z = 3$	Divide each side by 58.

The result is $x = -1$ and $z = 3$.

Step 3 Solve for *y*.

Substitute the two values into one of the original equations to find y.

$4x + y + 6z = 12$	Equation 1
$4(-1) + y + 6(3) = 12$	$x = -1, z = 3$
$-4 + y + 18 = 12$	Multiply.
$y = -2$	Subtract 14 from each side.

The ordered triple is (−1, −2, 3).

Go Online You can complete an Extra Example online.

Example 2 Solve a System with Infinitely Many Solutions

Solve the system of equations.

$-x + 5y - 4z = -2$	Equation 1
$4x - 20y + 16z = 8$	Equation 2
$-x - y - z = -1$	Equation 3

Step 1 Eliminate one variable.

Select two of the equations and eliminate one of the variables.

Multiply Equation 1 by 4 and add the equations to eliminate x.

$-x + 5y - 4z = -2$ **Multiply by 4.** → $-4x + 20y - 16z = -8$

Add the equations to eliminate x.

$-4x + 20y - 16z = -8$	Equation 1 × 4.
(+) $4x - 20y + 16z = 8$	Equation 2
$0 = 0$	Add the equations.

The equation $0 = 0$ is always true. This indicates that the first two equations represent the same plane.

Step 2 Check the third plane.

Multiply Equation 1 by −1 and add the equations to eliminate x.

$-x + 5y - 4z = -2$ **Multiply by −1.** → $x - 5y + 4z = 2$

Add the equations to eliminate x.

$x - 5y + 4z = 2$	Equation 1 × (−1).
(+) $-x - y - z = -1$	Equation 3
$-6y + 3z = 1$	Add the equations.

The planes intersect in a line, because the resultant equation is in two variables. So, there are an infinite number of solutions.

Check

Solve each system of equations.

$3x - 18y + 6z = 7$
$7x - 2y + z = 1$
$-x + 6y - 2z = -2$

$3x - y - z = 2$
$-4x - 2y + 3z = 19$
$5x + 3y + z = 8$

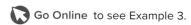

Go Online to see Example 3.

Think About It!
Is it possible to have two planes that coincide and yet the system of three equations has no solution? Explain.

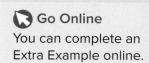

Go Online
You can complete an Extra Example online.

Example 4 Write and Solve a System of Equations

MUSEUM MEMBERSHIPS In 2016, Dali Museum in St. Petersburg, Florida offered individual, dual, and family memberships, which cost \$60, \$80, and \$100, respectively. Suppose in one month the museum sells a total of 81 new memberships, for a total of \$6420. The number of dual memberships purchased is twice that of individual memberships. Write and solve a system of equations to determine the number of new individual memberships *x*, dual memberships *y*, and family memberships *z*.

Step 1 Write the system of equations.

a total of 81 new memberships:	$x + y + z = 81$
The number of dual memberships purchased is twice that of individual memberships:	$y = 2x$
Fees for individual, dual, and family memberships, which cost \$60, \$80, and \$100, respectively, for a total of \$6420:	$60x + 80y + 100z = 6420$

Step 2 Eliminate one variable.

Substitute $y = 2x$ into Equation 1 and Equation 3 to eliminate y.

$x + y + z = 81$	Equation 1
$x + 2x + z = 81$	$y = 2x$
$3x + z = 81$	Add.
$60x + 80y + 100z = 6420$	Equation 3
$60x + 80(2x) + 100z = 6420$	$y = 2x$
$220x + 100z = 6420$	Simplify.

Step 3 Solve the resulting system of two equations.

$-300x - 100z = -8100$	Multiply new Equation 1 by -100.
$(+)\ 220x + 100z = 6420$	
$-80x = -1680$	Add to eliminate z.
$x = 21$	Solve for x.

Step 4 Substitute to find *z*.

$3x + z = 81$	Remaining equation in two variables
$3(21) + z = 81$	$x = 21$
$z = 18$	Simplify.

Step 5 Substitute to find *y*.

$y = 2x$	Equation 2
$y = 2(21)$	$x = 21$
$y = 42$	Multiply.

The solution is (21, 42, 18). So, the museum sold 21 individual memberships, 42 dual memberships, and 18 family memberships.

Go Online You can complete an Extra Example online.

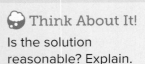

Think About It!
Is the solution reasonable? Explain.

Practice

Go Online You can complete your homework online.

Examples 1–3

Solve each system of equations.

1. $2x + 3y - z = 0$

$x - 2y - 4z = 14$

$3x + y - 8z = 17$

2. $2p - q + 4r = 11$

$p + 2q - 6r = -11$

$3p - 2q - 10r = 11$

3. $a - 2b + c = 8$

$2a + b - c = 0$

$3a - 6b + 3c = 24$

4. $3s - t - u = 5$

$3s + 2t - u = 11$

$6s - 3t + 2u = -12$

5. $2x - 4y - z = 10$

$4x - 8y - 2z = 16$

$3x + y + z = 12$

6. $p - 6q + 4r = 2$

$2p + 4q - 8r = 16$

$p - 2q = 5$

7. $2a + c = -10$

$b - c = 15$

$a - 2b + c = -5$

8. $x + y + z = 3$

$13x + 2z = 2$

$-x - 5z = -5$

9. $2m + 5n + 2p = 6$

$5m - 7n = -29$

$p = 1$

10. $f + 4g - h = 1$

$3f - g + 8h = 0$

$f + 4g - h = 10$

11. $-2c = -6$

$2a + 3b - c = -2$

$a + 2b + 3c = 9$

12. $3x - 2y + 2z = -2$

$x + 6y - 2z = -2$

$x + 2y = 0$

Example 4

13. ANIMAL NUTRITION A veterinarian wants to make a food mix for guinea pigs that contains 23 grams of protein, 6.2 grams of fat, and 16 grams of moisture. The composition of three available mixtures are shown in the table. How many grams of each mix should be used to make the desired new mix?

	Protein (g)	Fat (g)	Moisture (g)
Mix A	0.2	0.02	0.15
Mix B	0.1	0.06	0.10
Mix C	0.15	0.05	0.05

14. ENTERTAINMENT At the arcade, Marcos, Sara, and Darius played video racing games, pinball, and air hockey. Marcos spent $6 for 6 racing games, 2 pinball games, and 1 game of air hockey. Sara spent $12 for 3 racing games, 4 pinball games, and 5 games of air hockey. Darius spent $12.25 for 2 racing games, 7 pinball games, and 4 games of air hockey. How much did each of the games cost?

15. FOOD A natural food store makes its own brand of trail mix from dried apples, raisins, and peanuts. A one-pound bag of the trail mix costs $3.18. It contains twice as much peanuts by weight as apples. If a pound of dried apples costs $4.48, a pound of raisins is $2.40, and a pound of peanuts is $3.44, how many ounces of each ingredient are contained in 1 pound of the trail mix?

Mixed Exercises

Solve each system of equations.

16. $-x - 5z = -5$

$y - 3x = 0$

$13x + 2z = 2$

17. $-3x + 2z = 1$

$4x + y - 2z = -6$

$x + y + 4z = 3$

18. $x - y + 3z = 3$

$-2x + 2y - 6z = 6$

$y - 5z = -3$

19. REASONING A newspaper company has three printing presses that together can produce 3500 newspapers each hour. The fastest printer can print 100 more than twice the number of papers as the slowest press. The two slower presses combined produce 100 more papers than the fastest press. How many newspapers can each printing press produce in 1 hour?

20. USE A SOURCE A shop is having a sale on pool accessories. The table shows the orders of three customers and their total price before tax. Research the sales tax in your area to determine whether a customer who has $200 could buy 1 chlorine filter, 1 raft, and 1 large lounge chair after sales tax is applied. Justify your response.

Combo	Price Before Tax
1 Raft and 2 Chlorine Filters	$220
1 Chlorine Filter and 2 Large Lounge Chairs	$245
1 Raft and 4 Large Lounge Chairs	$315

21. TICKETS Three kinds of tickets are available for a concert: orchestra seating, mezzanine seating, and balcony seating. The orchestra tickets cost $2 more than the mezzanine tickets, while the mezzanine tickets cost $1 more than the balcony tickets. Twice the cost of an orchestra ticket is $1 less than 3 times the cost of a balcony ticket. Determine the price of each kind of ticket.

22. CONSTRUCT ARGUMENTS Consider the following system. Prove that if $b = c = -a$, then $ty = a$.

$$rx + ty + vz = a$$
$$rx - ty + vz = b$$
$$rx + ty - vz = c$$

23. PERSEVERE The general form of an equation for a parabola is $y = ax^2 + bx + c$, where (x, y) is a point on the parabola. If three points on a parabola are (2, −10), (−5, −101), and (6, −90), determine the values of a, b, and c and write the equation of the parabola in general form.

24. WRITE Use your knowledge of solving a system of three linear equations with three variables to explain how to solve a system of four equations with four variables.

25. CREATE Write a system of three linear equations that has a solution of (−5, −2, 6). Show that the ordered triple satisfies all three equations.

Lesson 2-9

Solving Absolute Value Equations and Inequalities by Graphing

Learn Solving Absolute Value Equations by Graphing

The graph of a related function can be used to solve an absolute value equation. The graph of an absolute value function may intersect the x-axis once or twice, or it may not intersect it at all. The number of times the graph intersects the x-axis corresponds to the number of solutions of the equation.

two solutions	one solution	no solution
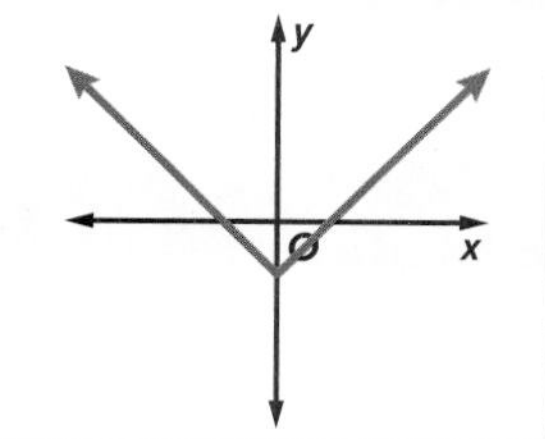	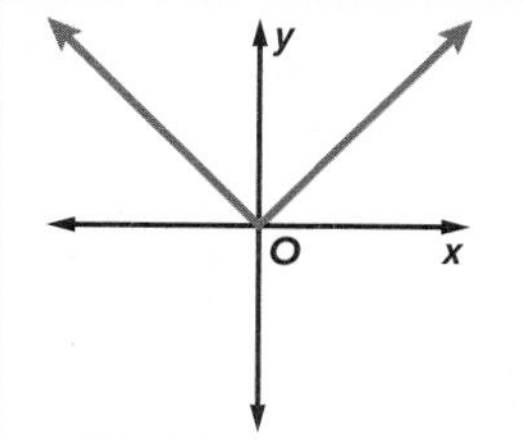	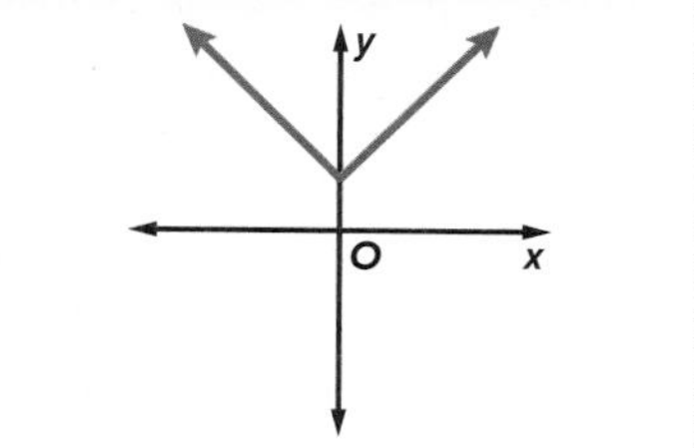

Today's Goals

- Solve absolute value equations.
- Solve absolute value inequalities.

Example 1 Solve an Absolute Value Equation by Graphing

Solve $5 + |2x + 6| = 5$ by graphing.

Step 1 Find the related function.

Rewrite the equation with 0 on the right side.

$5 + |2x + 6| = 5$ Original equation

$|2x + 6| = 0$ Subtract 5 from each side.

Replacing 0 with $f(x)$ gives the related function $f(x) = |2x + 6|$.

Step 2 Graph the related function.

Make a table of values for $f(x) = |2x + 6|$. Then graph the ordered pairs and connect them.

x	$f(x)$
−6	6
−5	4
−4	2
−3	0
−2	2
−1	4
0	6

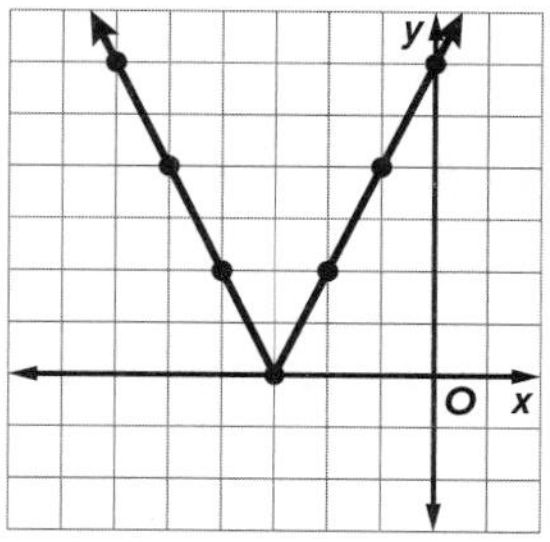

Since the graph of $f(x) = |2x + 6|$ only intersects the x-axis at −3, the equation has one solution. The solution set of the equation is {−3}.

Go Online You can complete an Extra Example online.

Think About It!

How could you use the table of values to solve the equation?

Check

Solve $|x + 1| + 9 = 13$ by graphing.

Part A Graph the related function.

Part B What is the solution set of the equation?

Go Online to see how to use a graphing calculator with Examples 2 and 3.

Example 2 Solve an Absolute Value Equation by Using Technology

Use a graphing calculator to solve $\frac{4}{5}|x - 1| + 8 = 11$.

Rewrite the equation as the system $f(x) = \frac{4}{5}|x - 1| + 8$ and $g(x) = 11$.

Enter the functions in the **Y** = list and graph. To enter the absolute value symbols, press **math** and select **abs(** from the **NUM** menu.

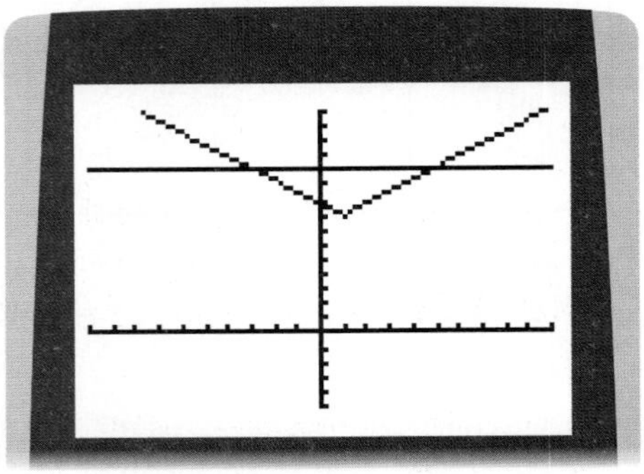

[−10, 10] scl: 1 by [−5, 15] scl: 1

Use the **intersect** feature from the **CALC** menu to find the *x*-coordinates of the points of intersection. When prompted, use the arrow keys to move the cursor to the left and right of each close to each point of intersection, and as the system $f(x) = \frac{4}{5}|x - 1| + 8$ and $g(x) = 11$. press **enter** three times.

The graphs intersect where $x = -2.75$ and $x = 4.75$. So, the solution set of the equation is $\{-2.75, 4.75\}$.

Go Online An alternate method is available for this example.

Check

Use a graphing calculator to solve $-\left|\frac{2}{3}x + 5\right| - 16 = -10$.

Example 3 Confirm Solutions by Using Technology

Solve $-3|x + 7| + 9 = 14$. Check your solutions graphically.

$-3\|x + 7\| + 9 = 14$	Original equation
$-3\|x + 7\| = 5$	Subtract 9 from each side.
$\|x + 7\| = -\frac{5}{3}$	Divide each side by −3.

Because the absolute value of a number is always positive or zero, this sentence is *never* true. The solution is ∅.

(continued on the next page)

Go Online You can complete an Extra Example online.

Use a graphing calculator to confirm that there is no solution.

Step 1 Find and graph the related function.

Rewriting the equation results in the related function $f(x) = -3|x + 7| - 5$. Enter the related function in the **Y=** list and graph.

Step 2 Find the zeros.

The graph does not appear to intersect the x-axis. Use the **ZOOM** feature or adjust the window manually to see this more clearly. Since the related function never intersects the x-axis, there are no real zeros. This confirms that the equation has no solution.

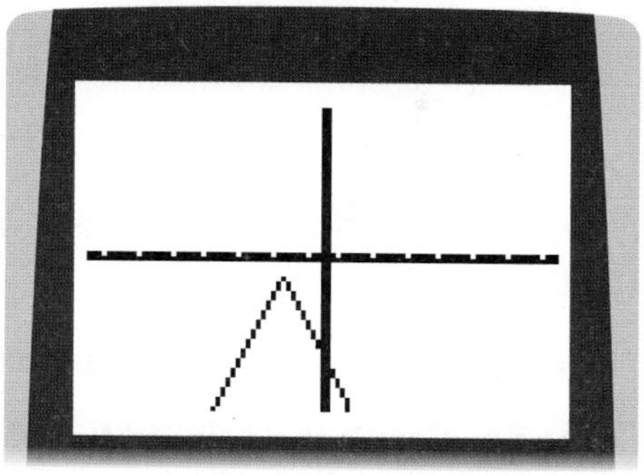

[−40, 40] scl: 1 by [−40, 40] scl: 1

> **Think About It!**
> How could you use a calculator to confirm the solutions of an equation with one or two real solutions?

Learn Solving Absolute Value Inequalities by Graphing

The related functions of absolute value inequalities are found by solving the inequality for 0, replacing the inequality symbol with an equals sign, and replacing 0 with $f(x)$.

For $<$ and $\leq$, identify the x-values for which the graph of the related function lies below the x-axis. For $\leq$, include the x-intercepts in the solution. For $>$ and $\geq$, identify the x-values for which the graph of the related function lies *above* the x-axis. For $\geq$, include the x-intercepts in the solution.

Example 4 Solve an Absolute Value Inequality by Graphing

Solve $|3x - 9| - 6 \leq 0$ by graphing.

The solution set consists of x-values for which the graph of the related function lies *below* the x-axis, including the x-intercepts. The related function is $f(x) = |3x - 9| - 6$. Graph $f(x)$ by making a table.

x	$f(x)$
0	3
1	0
2	−3
3	−6
4	−3
5	0
6	3

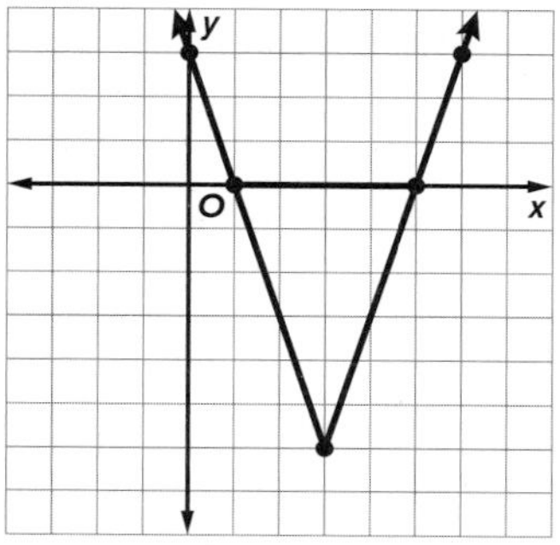

> **Talk About It!**
> Would the solution change if the inequality symbol was changed from $\leq$ to $\geq$? Explain your reasoning.

The graph lies below the x-axis between $x = 1$ and $x = 5$. Thus, the solution set is $\{x \mid 1 \leq x \leq 5\}$ or [1, 5]. All values of x between 1 and 5 satisfy the constraints of the original inequality.

Go Online You can complete an Extra Example online.

Check

Solve $|x - 2| - 3 \leq 0$ by graphing.

Part A Graph a related function.

Part B What is the solution set of $|x - 2| - 3 \leq 0$?

Example 5 Solve an Absolute Value Equation by Using Technology

Use a graphing calculator to solve $\left|\frac{5}{7}x + 2\right| - 3 > 0$.

Step 1 Graph the related function.

Rewriting the inequality results in the related function $f(x) = \left|\frac{5}{7}x + 2\right| - 3$.

Step 2 Find the zeros.

The $>$ symbol indicates that the solution set consists of x-values for which the graph of the related function lies *above* the x-axis, not including the x-intercepts.

Use the **zero** feature from the **CALC** menu to find the zeros, or x-intercepts.

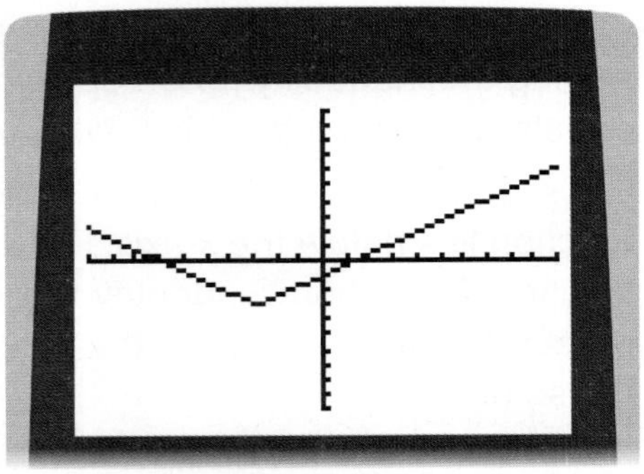

[−10, 10] scl: 1 by [−10, 10] scl: 1

The zeros are located at $x = -7$ and $x = 1.4$. The graph lies above the x-axis when $x < -7$ and $x > 1.4$. So the solution set is $\{x \mid x < -7 \text{ or } x > 1.4\}$.

Check

Use a graphing calculator to solve $\frac{1}{2}|4x + 1| - 5 > 0$.

Think About It!
The inequality in the example is solved for 0. What additional step(s) would you need to take if the given inequality had a nonzero term on the right side of the inequality symbol?

Go Online to see how to use a graphing calculator with this example.

Go Online You can complete an Extra Example online.

Practice

Go Online You can complete your homework online.

Example 1

Solve each equation by graphing.

1. $|x - 4| = 5$

2. $|2x - 3| = 17$

3. $3 + |2x + 1| = 3$

4. $|x - 1| + 6 = 4$

5. $7 + |3x - 1| = 7$

6. $|x + 2| + 5 = 13$

Example 2

USE TOOLS **Use a graphing calculator to solve each equation.**

7. $\frac{1}{2}|x - 1| + 5 = 9$

8. $\frac{3}{4}|x + 1| + 1 = 7$

9. $\frac{2}{3}|x - 2| - 4 = 4$

10. $2|x + 2| = 10$

11. $\frac{1}{5}|x + 6| - 1 = 9$

12. $3|x + 5| - 1 = 11$

Example 3

USE TOOLS **Solve each equation algebraically. Use a graphing calculator to check your solutions.**

13. $|3x - 6| = 42$

14. $7|x + 3| = 42$

15. $-3|4x - 9| = 24$

16. $-6|5 - 2x| = -9$

17. $5|2x + 3| - 5 = 0$

18. $|15 - 2x| = 45$

Example 4

Solve each inequality by graphing.

19. $|2x - 6| - 4 \leq 0$

20. $|x - 1| - 3 \leq 0$

21. $|2x - 1| \geq 4$

22. $|3x + 2| \geq 6$

23. $2|x + 2| < 8$

24. $3|x - 1| < 12$

Example 5

USE TOOLS **Use a graphing calculator to solve each inequality.**

25. $\left|\frac{1}{4}x + 4\right| - 1 > 0$

26. $\frac{2}{5}|x - 5| + 1 > 0$

27. $|3x - 1| < 2$

28. $|4x + 1| \leq 1$

29. $\frac{1}{6}|x - 1| + 1 \leq 0$

30. $\frac{1}{4}|x + 5| - 1 \leq 1$

Mixed Exercises

Solve by graphing.

31. $0.4|x - 1| = 0.2$

32. $0.16|x + 1| = 4.8$

33. $0.78|2x + 0.1| + 2.3 = 0$

34. $\left|\frac{1}{3}x + 3\right| + 1 = 0$

35. $\frac{1}{2}|6 - 2x| \leq 1$

36. $|3x - 2| < \frac{1}{2}$

USE TOOLS **Solve each equation or inequality algebraically. Use a graphing calculator to check your solutions.**

37. $\left|\frac{5}{9}x + 1\right| - 5 > 0$

38. $\frac{2}{7}\left|\frac{1}{2}x - 1\right| < 1$

39. $0.28|0.4x - 2| = 10.08$

40. REASONING A pet store sells bags of dog food that are labeled as weighing 50 pounds. The equipment used to package the dog food produces bags with a weight that is within 0.75 lbs of the advertised weight. Write an absolute value equation to determine the acceptable maximum and minimum weight for the bags of dog food. Then, use a graph to find the minimum and maximum weights.

41. SPACE The mean distance from Mars to the Sun is 1.524 astronomical units (au). The distance varies during the orbit of Mars by 0.147 au. Write an absolute value inequality to represent the distance of Mars from the Sun as it completes an orbit. Then, use a graph to solve the inequality.

42. CONSTRUCT ARGUMENTS Ms. Uba asked her students to write an absolute value equation with the solutions and related function shown in the graph. Sawyer wrote $|x + 2| = 6$ and Kaleigh wrote $2|x + 2| = 12$. Is either student correct? Justify your argument.

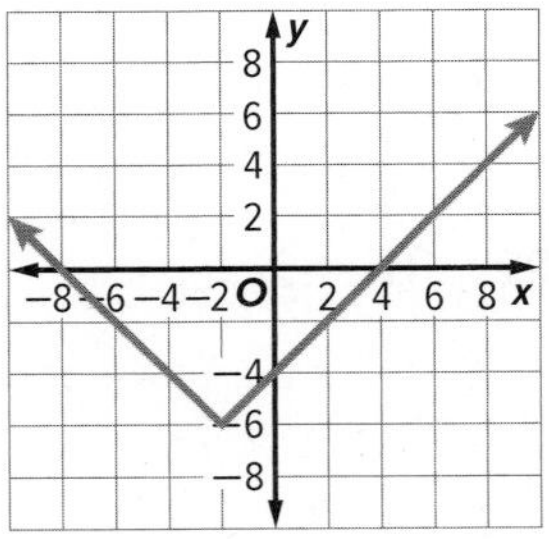

43. WRITE Compare and contrast the solution sets of absolute value equations and absolute value inequalities.

44. ANALYZE How can you tell that an absolute value equation has no solutions without graphing or completely solving it algebraically?

45. CREATE Create an absolute value equation for which the solution set is {9, 11}.

Module 2 • Linear Equations, Inequalities, and Systems

Review

Essential Question

How are equations, inequalities, and systems of equations or inequalities best used to model to real-world situations?

Module Summary

Lessons 2-1 and 2-2

Linear and Absolute Value Equations and Inequalities

- The solution of an equation or an inequality is any value that, when substituted into the equation, results in a true statement.
- An absolute value equation or inequality is solved by writing it as two cases. An absolute value equation may have 0, 1, or 2 solutions.

Lesson 2-3

Equations of Linear Functions

- Standard form: $Ax + By = C$, where A, B, and C are integers with a greatest common factor of 1, $A \geq 0$, and A and B are not both 0
- Slope-intercept form: $y = mx + b$, where m is the slope and b is the y-intercept
- Point-slope form: $y - y_1 = m(x - x_1)$, where m is the slope and (x_1, y_1) are the coordinates of a point on the line

Lessons 2-4 through 2-8

Systems of Equations and Inequalities

- The point of intersection of the two graphs a system of equations represents the solution.
- In the substitution method, one equation is solved for a variable and substituted to find the value of another variable.
- In the elimination method, one variable is eliminated by adding or subtracting the equations.
- To solve a system of inequalities, graph each inequality. Viable solutions to the system of inequalities are in the intersection of all the graphs.
- Linear programming is a method for finding maximum or minimum values of a function over a given system of inequalities with each inequality representing a constraint.
- Systems of equations in three variables can have infinitely many solutions, no solution, or one solution which is written as an ordered triple (x, y, z).
- Systems of equations in three variables can be solved by using elimination and substitution.

Lesson 2-9

Solving Absolute Value Equations and Inequalities by Graphing

- The graph of the related absolute value function can be used to solve an equation or inequality. The x-intercept(s) of the function give the solution(s) of the equation.

Study Organizer

Foldables

Use your Foldable to review this module. Working with a partner can be helpful. Ask for clarification of concepts as needed.

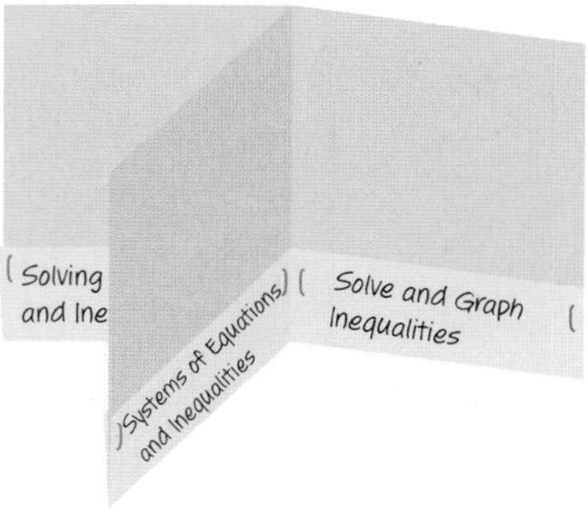

Test Practice

1. **OPEN RESPONSE** Explain each step in solving the equation $3(2x + 9) + \frac{1}{2}(4x - 8) = 55$. (Lesson 2-1)

2. **MULTIPLE CHOICE** In a single football game, a team scored 14 points from touchdowns and made x field goals, which are worth three points each. The team scored 23 points. How many field goals did the team make? (Lesson 2-1)

A. 2

B. 3

C. 4

D. 5

3. **OPEN RESPONSE** A temperature in degrees Celsius, C, is equal to five-ninths times the difference of the temperature in degrees Fahrenheit, F, and 32. Write an equation to relate the temperature in degrees Celsius to the temperature in degrees Fahrenheit. Then determine the Fahrenheit temperature that is equivalent to 25°C. (Lesson 2-1)

4. **OPEN RESPONSE** The temperature of an oven varies by as much as 7.5 degrees from the temperature shown on its display. If the oven is set to 425°, write an equation to find the minimum and maximum actual temperature of the oven. (Lesson 2-2)

5. **OPEN RESPONSE** The equation $2x + y = 10$ can be used to find the number of miles Allie has left to jog this week to reach her goal, where x is the number of days and y is the number of miles left to reach the goal. Write the equation in slope-intercept form. Then interpret the parameters in the context of the situation. (Lesson 2-3)

6. **MULTIPLE CHOICE** Thomas is driving his truck at a constant speed. The table below gives the distance remaining to his destination y in miles x minutes after he starts driving.

x	15	30	45
y	186.5	173	159.5

Which equation models the distance remaining after any number of minutes? (Lesson 2-3)

A. $y = \frac{9}{10}x - 200$

B. $y = -\frac{10}{9}x + 200$

C. $y = \frac{10}{9}x - 200$

D. $y = -\frac{9}{10}x + 200$

7. MULTIPLE CHOICE Describe the system of equations shown in the graph. (Lesson 2-4)

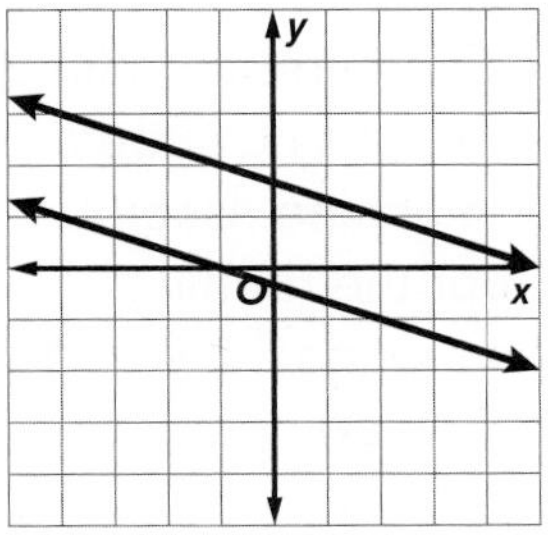

A. It has no solution, and is consistent and dependent.

B. It has no solution, and is inconsistent.

C. It has infinitely many solutions, and is consistent and dependent.

D. It has infinitely many solutions, and is inconsistent.

8. MULTIPLE CHOICE Last season, the volleyball team paid \$5 for each pair of socks and \$17 for each pair of shorts. They spent a total of \$315. This season, the price of socks is \$6 per pair and shorts are \$18. The team spent \$342 for the same number of socks and shorts as the previous season. How many pairs of socks and shorts did the team buy each season? (Lesson 2-5)

A. 12 pairs of socks; 15 shorts

B. 15 pairs of socks; 12 shorts

C. 24 pairs of socks; 11 shorts

D. 29 pairs of socks; 10 shorts

9. OPEN RESPONSE Solve this system of equations.

$5x - 2y = 1$
$2x + 8y = 7$

Round to the nearest hundredth, if necessary. (Lesson 2-5)

10. MULTIPLE CHOICE A choir concert will be held in a venue with 500 seats. The goal of the choir director is to make at least \$3250. Adult tickets cost \$8 and student tickets cost \$5. Select the system of inequalities that represents the situation, where x is the number of adult tickets sold and y is the number of student tickets sold. (Lesson 2-6)

A. $x + y \geq 3250$
$8x + 5y \leq 500$

B. $x + y \leq 3250$
$8x + 5y \geq 500$

C. $x + y \geq 500$
$8x + 5y \leq 3250$

D. $x + y \leq 500$
$8x + 5y \geq 3250$

11. OPEN RESPONSE Jazmine earns \$20 per hour for landscaping and \$15 per hour for painting. She can work no more than 12 hours per week. If Jazmine wants to earn at least \$225 in a week, could she landscape for 6 hours and paint for 6 hours to meet her goal? Justify your response. (Lesson 2-6)

12. **MULTIPLE CHOICE** Which system of inequalities represents the graph shown? (Lesson 2-6)

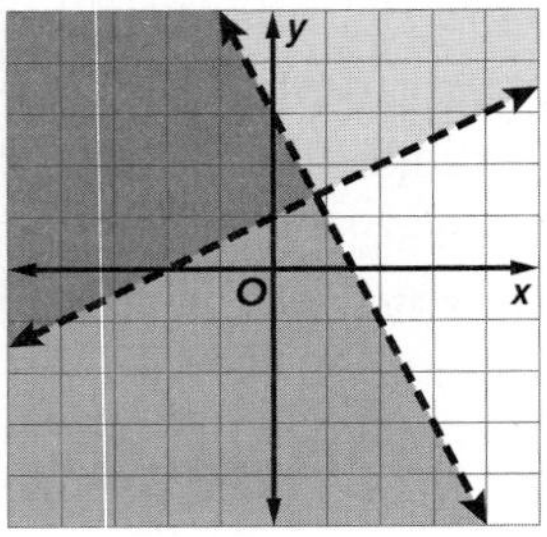

A. $x - 2y < -2$
$2x + y < 3$

B. $x - 2y < -2$
$2x + y > 3$

C. $x - 2y > -2$
$2x + y < 3$

D. $x - 2y > -2$
$2x + y > 3$

13. **MULTI-SELECT** The shaded region represents the feasible region for a linear programming problem.

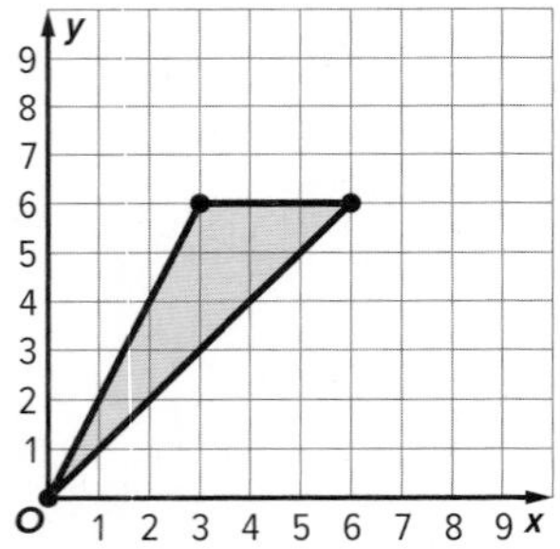

Select all the points at which a minimum or maximum value may occur. (Lesson 2-7)

A. (0, 0)

B. (1, 5)

C. (2, 6)

D. (3, 6)

E. (4, 5)

F. (5, 3)

G. (6, 6)

14. **MULTI-SELECT** A shoe manufacturer makes two types of athletic shoes—a cross-training shoe and a running shoe. Each pair of shoes is assembled by machine and then finished by hand. For the cross-training shoes, it takes 15 minutes for machine assembly and 6 minutes by hand. For the running shoes, it takes 9 minutes on the machine and 12 minutes by hand. The company can allocate no more than 900 machine hours and 500 hand hours each day. The profit is \$10 for each pair of cross-training shoes and \$15 for each pair of running shoes. Let x represent the number of cross-training shoes and y represent the number of running shoes manufactured each day.

Select the inequalities that form the boundary of the feasible region. (Lesson 2-7)

A. $x \geq 0$

B. $y \geq 0$

C. $0.1x + 0.2y \leq 500$

D. $0.25x + 0.15y \leq 900$

E. $6x + 12y \leq 500$

F. $15x + 9y \leq 900$

15. **OPEN RESPONSE** Last month, Jeremy took 2 piano lessons, 3 guitar lessons, and 1 drum lesson and spent a total of \$285. D'Asia took 1 piano lesson, 5 guitar lessons, and 2 drum lessons and spent a total of \$400. Raj took 3 piano lessons, 2 guitar lessons, and 4 drum lessons and spent a total of \$440. Write a system of equations that could be used to find the prices of each piano lesson, x, guitar lesson, y, and drum lesson, z. (Lesson 2-8)

16. **OPEN RESPONSE** Use a graphing calculator to solve the equation $2|x - 3| - 1 = 5$. (Lesson 2-9)

Module 3

Quadratic Functions

Essential Question

What are important characteristics of a quadratic, and what real-world situations can be modeled by quadratic functions and equations?

What Will You Learn?

How much do you already know about each topic **before** starting this module?

KEY

— I don't know. — I've heard of it. — I know it!

	Before			After		
find and interpret average rate of change of a quadratic function						
estimate solutions of quadratic equations by graphing						
perform operations with complex numbers						
solve quadratic equations by factoring						
solve quadratic equations by completing the square						
use the discriminant to determine the number and type of roots of a quadratic equation						
solve quadratic inequalities in two variables by graphing						
solve systems of two quadratic equations						
solve systems of nonlinear relations						

Foldables Make this Foldable to help you organize your notes about quadratic functions. Begin with one sheet of 11" by 17" paper.

1. **Fold** in half lengthwise.
2. **Fold** in fourths crosswise.
3. **Cut** along the middle fold from the edge to the last crease as shown.
4. **Refold** along the lengthwise fold and tape the uncut section at the top. Label each section with a lesson number. Close to form a booklet.

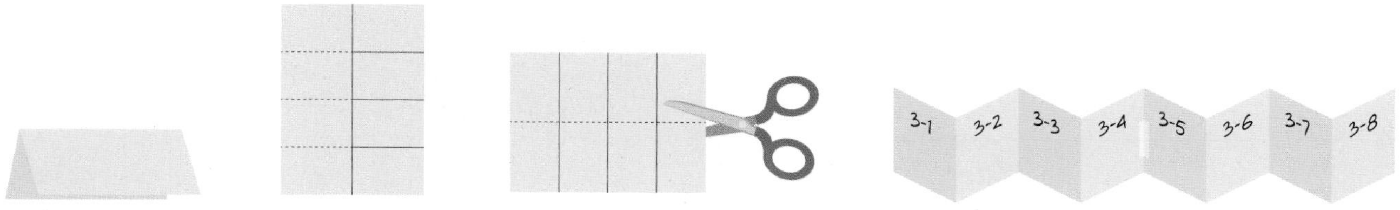

What Vocabulary Will You Learn?

- average rate of change
- axis of symmetry
- completing the square
- complex conjugates
- complex number
- difference of squares
- discriminant
- factored form
- imaginary unit *i*
- maximum
- minimum
- perfect square trinomials
- projectile motion problems
- pure imaginary number
- quadratic equation
- quadratic function
- quadratic inequality
- quadratic relations
- rate of change
- rationalizing the denominator
- standard form of a quadratic equation
- vertex
- vertex form

Are You Ready?

Complete the Quick Review to see if you are ready to start this module. Then complete the Quick Check.

Quick Review

Example 1

Given $f(x) = -2x^2 + 3x - 1$ and $g(x) = 3x^2 - 5$, find each value.

a. $f(2)$

$f(x) = -2x^2 + 3x - 1$		Original function
$f(2) = -2(2)^2 + 3(2) - 1$		Substitute 2 for x.
$= -8 + 6 - 1$ or -3		Simplify.

b. $g(-2)$

$g(x) = 3x^2 - 5$	Original function
$g(-2) = 3(-2)^2 - 5$	Substitute -2 for x.
$= 12 - 5$ or 7	Simplify.

Example 2

Factor $2x^2 - x - 3$ completely. If the polynomial is not factorable, write *prime*.

To find the coefficients of the x-terms, you must find two numbers whose product is $2(-3)$ or -6, and whose sum is -1. The two coefficients must be 2 and -3 since $2(-3) = -6$ and $2 + (-3) = -1$. Rewrite the expression and factor by grouping.

$2x^2 - x - 3$	
$= 2x^2 + 2x - 3x - 3$	Substitute $2x - 3x$ for $-x$.
$= (2x^2 + 2x) + (-3x - 3)$	Associative Property
$= 2x(x + 1) + -3(x + 1)$	Factor out the GCF.
$= (2x - 3)(x + 1)$	Distributive Property

Quick Check

Given $f(x) = 2x^2 + 4$ and $g(x) = -x^2 - 2x + 3$, find each value.

1. $f(3)$

2. $f(0)$

3. $g(4)$

4. $g(-3)$

Factor completely. If the polynomial is not factorable, write *prime*.

5. $x^2 - 10x + 21$

6. $2x^2 + 7x - 4$

7. $2x^2 - 7x - 15$

8. $x^2 - 11x + 15$

How Did You Do?

Which exercises did you answer correctly in the Quick Check?

Graphing Quadratic Functions

Explore Transforming Quadratic Functions

Online Activity Use graphing technology to complete the Explore.

INQUIRY How can you use the values of a, b, and c in the equation of a quadratic function $y = ax^2 + bx + c$ to visualize its graph?

Learn Graphing Quadratic Functions

A **quadratic function** has an equation of the form $y = ax^2 + bx + c$, where $a \neq 0$. A quadratic function has a graph that is a parabola.

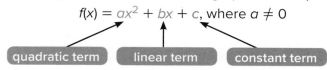

$$f(x) = ax^2 + bx + c, \text{ where } a \neq 0$$

quadratic term (ax^2) | linear term (bx) | constant term (c)

Key Concept • Graph of a Quadratic Function

$y = ax^2 + bx + c$, where $a \neq 0$

The y-intercept is c.

The **axis of symmetry** is the line about which a parabola is symmetric. Its equation is $x = -\frac{b}{2a}$.

The axis of symmetry intersects the parabola at its **vertex**. Its x-coordinate is $-\frac{b}{2a}$.

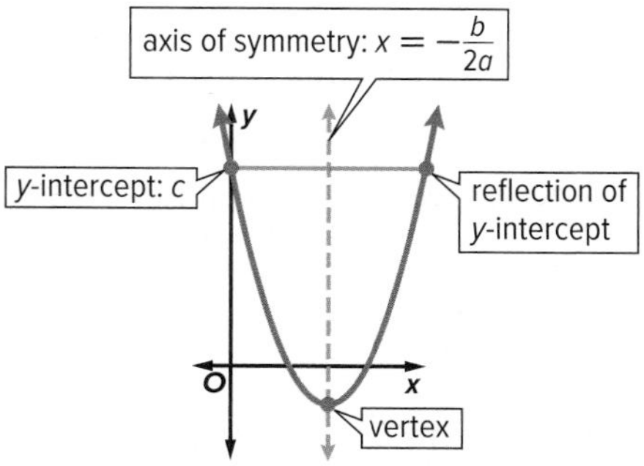

You will derive the equation for the axis of symmetry in Lesson 3-5.

Key Concept • Maximum or Minimum Values of Quadratic Graphs

The vertex of a quadratic function is either a **maximum** or **minimum**. The graph of $y = ax^2 + bx + c$, where $a \neq 0$, opens up and has a minimum when $a > 0$, and opens down and has a maximum when $a < 0$.

a is positive.

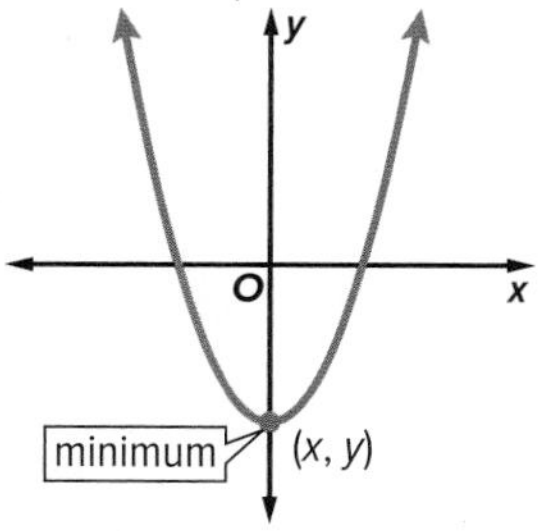

The y-coordinate is the minimum value.

a is negative.

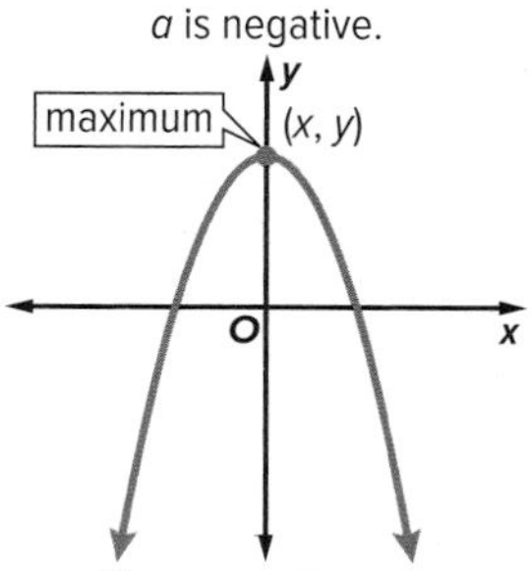

The y-coordinate is the maximum value.

Today's Goals

- Graph quadratic functions.
- Find and interpret the average rate of change of quadratic functions given symbolically, in tables, and in graphs.

Today's Vocabulary

quadratic function
axis of symmetry
vertex
maximum
minimum
rate of change
average rate of change

Think About It!

Why is it important to know whether the vertex is a maximum or minimum when graphing a quadratic function?

Watch Out!

Maxima and Minima
Just because a vertex is a maximum does not mean it will be located above the x-axis, and just because a vertex is a minimum does not mean it will be located below the x-axis.

Go Online
You can watch videos to see how to graph quadratic functions by using a table or its key features.

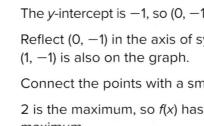

Think About It!
If you know that $f(-4) = 5$, find $f(2)$ without substituting 2 for x in the function. Justify your argument.

Example 1 Graph a Quadratic Function by Using a Table

Graph $f(x) = x^2 + 2x - 3$. State the domain and range.

Step 1 Analyze the function.

For $f(x) = x^2 + 2x - 3$, $a = 1$, $b = 2$, and $c = -3$.

c is the y-intercept, so the y-intercept is -3.

Find the axis of symmetry.

$x = -\frac{b}{2a}$ Equation of the axis of symmetry

$= -\frac{2}{2(1)}$ $a = 1, b = 2$

$= -1$ Simplify.

The equation of the axis of symmetry is $x = -1$, so the x-coordinate of the vertex is -1. Because $a > 0$, the vertex is a minimum.

Step 2 Graph the function.

x	$x^2 + 2x - 3$	$(x, f(x))$
-3	$(-3)^2 + 2(-3) - 3$	$(-3, 0)$
-2	$(-2)^2 + 2(-2) - 3$	$(-2, -3)$
-1	$(-1)^2 + 2(-1) - 3$	$(-1, -4)$
0	$(0)^2 + 2(0) - 3$	$(0, -3)$
1	$(1)^2 + 2(1) - 3$	$(-3, 0)$

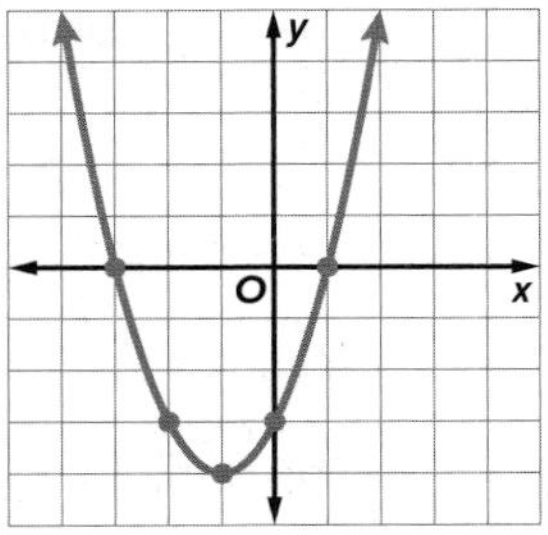

Step 3 Analyze the graph.

The parabola extends to positive and negative infinity, so the domain is all real numbers. The range is $\{y \mid y \geq -4\}$.

Think About It!
Compare the end behavior of $f(x)$ and $g(x)$.

Example 2 Compare Quadratic Functions

Compare the graph of $f(x)$ to a quadratic function $g(x)$ with a y-intercept of -1 and a vertex at $(1, 2)$. Which function has a greater maximum?

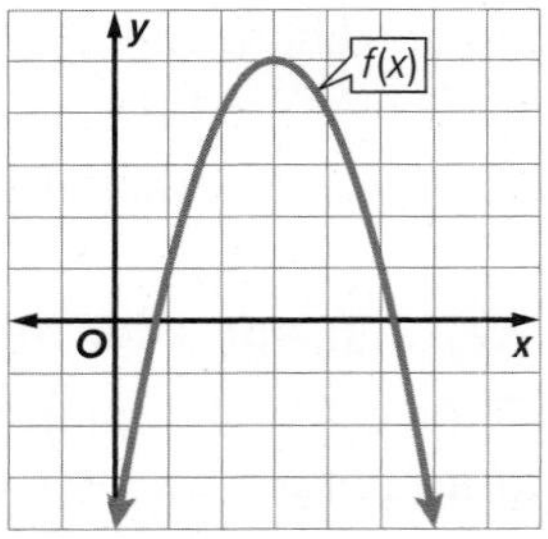

From the graph, $f(x)$ appears to have a maximum of 5. Graph $g(x)$ using the given information.

The vertex is at $(1, 2)$, so the axis of symmetry is $x = 1$.

The y-intercept is -1, so $(0, -1)$ is on the graph.

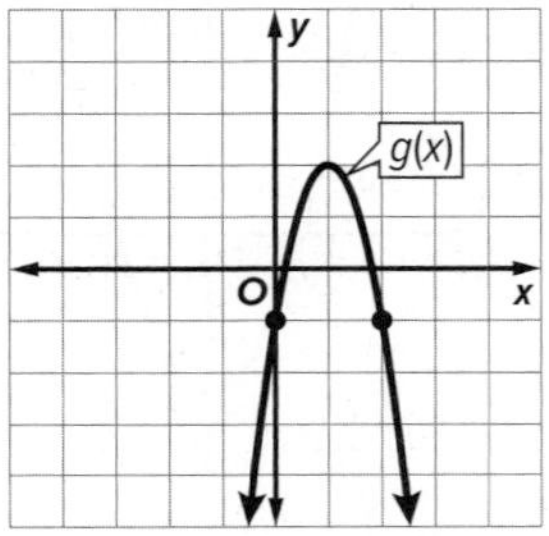

Reflect $(0, -1)$ in the axis of symmetry. So, $(1, -1)$ is also on the graph.

Connect the points with a smooth curve.

2 is the maximum, so $f(x)$ has the greater maximum.

Example 3 Use Quadratic Functions

SKIING **A ski resort has extended hours on one holiday weekend per year. Last year, the resort sold 680 ski passes at $120 per holiday weekend pass. This year, the resort is considering a price increase. They estimate that for each $5 increase, they will sell 20 fewer holiday weekend passes.**

Part A How much should they charge in order to maximize profit?

Step 1 Define the variables.

Let x represent the number of $5 price increases, and let $P(x)$ represent the total amount of money generated. $P(x)$ is equal to the price of each pass $(120 + 5x)$ times the total number of passes sold $(680 - 20x)$.

Step 2 Write an equation.

$P(x) = (120 + 5x)(680 - 20x)$	Original equation
$= 81{,}600 - 2400x + 3400x - 100x^2$	Multiply.
$= -100x^2 + 1000x + 81{,}600$	Simplify.

Step 3 Find the axis of symmetry.

Because a is negative, the vertex is a maximum.

$x = -\frac{b}{2a}$	Formula for the axis of symmetry
$= -\frac{1000}{2(-100)}$	$a = -100$, $b = 1000$
$= 5$	Simplify.

Step 4 Interpret the results.

The ski resort will make the most money with 5 price increases, so they should charge $120 + 5(5)$ or $145 for each holiday weekend pass.

Part B Find the domain and range in the context of the situation.

The domain is $\{x \mid 0 \leq x \leq 34\}$ because the number of price increases, and the number of passes sold cannot be negative. The range is $\{y \mid 0 \leq y \leq 84{,}100\}$ because the amount of money generated cannot be negative, and the maximum amount of money generated is $P(5) = -100(5)^2 + 1000(5) + 81{,}600$ or 84,100.

Check

CONCERTS Last year, a ticket provider sold 1350 lawn seats for a concert at $70 per ticket. This year, the provider is considering increasing the price. They estimate that for each $2 increase, they will sell 30 fewer tickets.

Part A How much should the ticket provider charge in order to maximize profit?

Part B Find the domain and range in the context of the situation.

Go Online You can complete an Extra Example online.

Study Tip

Assumptions You assumed that the ski resort has the ability to increase the price indefinitely and that every price increase will be $5 and will cause the resort to lose sales from exactly 20 holiday weekend passes.

Think About It!

Why is the maximum amount of money generated from holiday weekend passes $P(5)$?

Learn Finding and Interpreting Average Rate of Change

A function's **rate of change** is how a quantity is changing with respect to a change in another quantity. For nonlinear functions, the rate of change is not the same over the entire function. You can calculate the **average rate of change** of a nonlinear function over an interval.

Key Concept • Average Rate of Change

The average rate of change of a function $f(x)$ is equal to the change in the value of the dependent variable $f(b) - f(a)$ divided by the change in the value of the independent variable $b - a$ over the interval $[a, b]$.

$$\frac{f(b) - f(a)}{b - a}$$

Think About It!

Find the average rate of change of the function over the interval $[-3, 1]$. Compare it to your results from the interval $[-4, 4]$.

Example 4 Find Average Rate of Change from an Equation

Determine the average rate of change of $f(x) = -x^2 + 2x - 1$ over the interval $[-4, 4]$.

The average rate of change is equal to $\frac{f(4) - f(-4)}{4 - (-4)}$.

First find $f(-4)$ and $f(4)$.

$f(4) = -(4)^2 + 2(4) - 1$ or -9 $\qquad$ $f(-4) = -(-4)^2 + 2(-4) - 1$ or -25

Then use $f(-4)$ and $f(4)$ to find the average rate of change.

$$\frac{f(4) - f(-4)}{4 - (-4)} = \frac{-9 - (-25)}{4 - (-4)} \text{ or } 2$$

The average rate of change of the function over the interval $[-4, 4]$ is 2.

Check

Find the average rate of change of $f(x) = -2x^2 - 5x + 7$ over the interval $[-5, 5]$.

average rate of change = __?__

Talk About It!

Without graphing, how can you tell that this function is nonlinear? Justify your argument.

Example 5 Find Average Rate of Change from a Table

Determine the average rate of change of $f(x)$ over the interval $[-3, 3]$.

x	$f(x)$
−3	48
−2	21
−1	0
0	−15
1	−24
2	−27
3	−24

The average rate of change is equal to $\frac{f(3) - f(-3)}{3 - (-3)}$.

First find $f(3)$ and $f(-3)$ from the table.

$f(3) = -24$ $\qquad$ $f(-3) = 48$

$$\frac{f(3) - f(-3)}{3 - (-3)} = \frac{-24 - 48}{3 - (-3)} \text{ or } -12$$

The average rate of change of the function over the interval $[-3, 3]$ is -12.

Go Online You can complete an Extra Example online.

Check

TESTING The table shows the number of students who took the ACT between 2011 and 2015.

Year	Number of Students
2011	1,623,112
2012	1,666,017
2013	1,799,243
2014	1,845,787
2015	1,924,436

Part A

Find the average rate of change in the number of students taking the ACT from 2011 to 2015.

average rate of change = ___?___

Part B

Interpret your results in the context of the situation.

From 2011 to ___?___, the number of students taking the ACT ___?___ by an average of ___?___ students per year.

Example 6 Find Average Rate of Change from a Graph

FOOTWEAR **The graph shows the amount of money the United States has spent on sports footwear since 2005.**

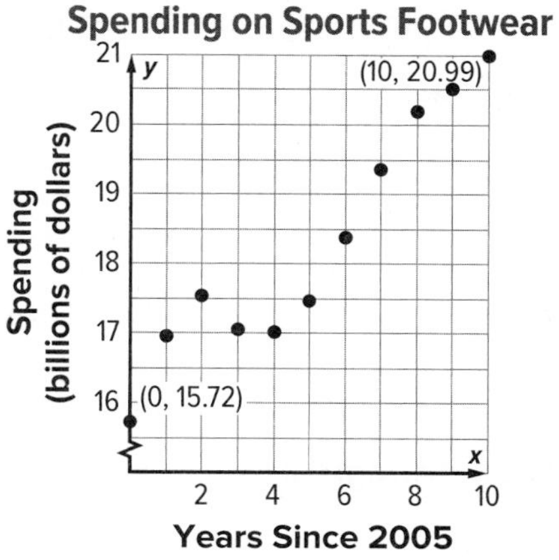

Part A

Use the graph to estimate the average rate of change of spending on sports footwear from 2005 to 2015. Then check your results algebraically.

Estimate

From the graph, the change in the y-values is approximately 5.5, and the change in the x-values is 10.

So, the rate of change is approximately $\frac{5.5}{10}$ or 0.55.

(*continued on the next page*)

Use a Source

Research the sales of another industry over a ten-year period. Then find the average rate of change during that time.

Algebraically

The average rate of change is equal to $\frac{f(10) - f(0)}{10 - 0}$.

$\frac{f(10) - f(0)}{10 - 0} = \frac{20.99 - 15.72}{10 - 0}$ or 0.527.

Part B

Interpret your results in the context of the situation.

From 2005 to 2015, the amount of money spent on sports footwear in the United States increased by an average of $527 million per year.

Check

SUPER BOWL The graph shows the number of television viewers of the Super Bowl since 2006.

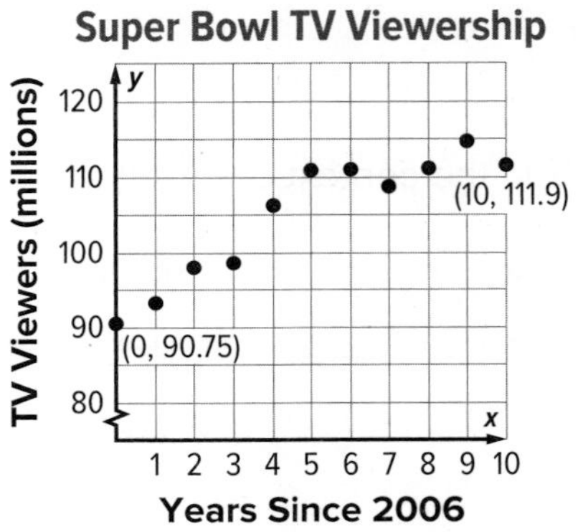

Part A

Use the graph to estimate the average rate of change in Super Bowl viewers from 2006 to 2016 to the nearest hundredth of a million. Then check your results algebraically.

estimate = ___?___ million

average rate of change = ___?___ million per year

Part B

Interpret your results in the context of the situation.

From ___?___ to ___?___, the number of Super Bowl viewers ___?___ by an average of ___?___ million viewers per year.

Go Online You can complete an Extra Example online.

Practice

Go Online You can complete your homework online.

Example 1

Graph each function. Then state the domain and range.

1. $f(x) = x^2 + 6x + 8$

2. $f(x) = -x^2 - 2x + 2$

3. $f(x) = 2x^2 - 4x + 3$

4. $f(x) = -2x^2$

5. $f(x) = x^2 - 4x + 4$

6. $f(x) = x^2 - 6x + 8$

Example 2

7. Compare the graph of $f(x)$ to a quadratic function $g(x)$ with a y-intercept of 1 and a vertex at (1, 3). Which function has a greater maximum? Explain.

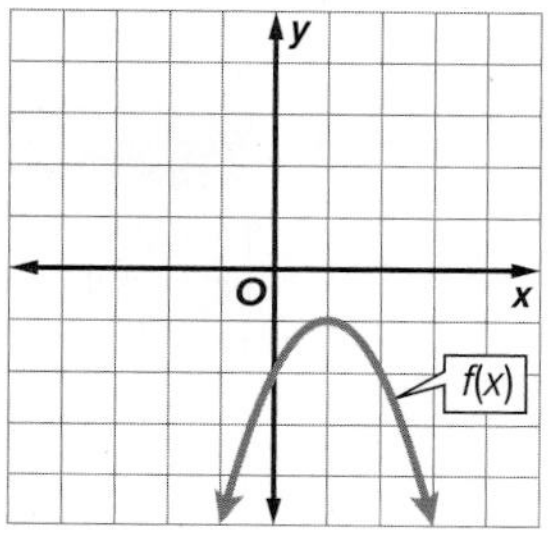

8. Compare the graph of $f(x)$ to a quadratic function $g(x)$ with a y-intercept of 0.5 and a vertex at (−1, −5). Which function has a lesser minimum? Explain.

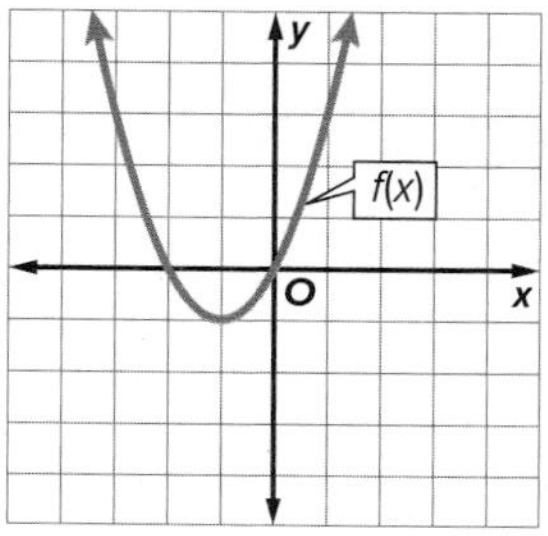

9. Compare $f(x) = x^2 - 10x + 5$ to the quadratic function $g(x)$ shown in the table. Which function has the lesser minimum? Explain.

x	$g(x)$
−10	170
−5	70
0	10
5	−10
10	10

10. Compare $f(x) = -x^2 + 6x - 15$ to the quadratic function $g(x)$ shown in the table. Which function has the greater maximum? Explain.

x	$g(x)$
−6	−26
−3	−11
0	−2
3	1
6	−2

Example 3

11. FISHING A county park sells annual permits to its fishing lake. Last year, the county sold 480 fishing permits for \$80 each. This year, the park is considering a price increase. They estimate that for each \$4 increase, they will sell 16 fewer annual fishing permits.

a. Define variables and write a function to represent the situation.

b. How much would the park have to charge for each permit in order to maximize its revenue from fishing permits?

c. Find the domain and range in the context of this situation.

12. SALES Last month, a retailer sold 120 jar candles at \$30 per candle. This month the retailer is considering putting the candles on sale. They estimate that for each \$2 decrease in price, they will sell 10 additional candles.

a. How much should the retailer charge for each candle in order to maximize its profit?

b. Find the domain and the range in the context of this situation.

Example 4

Determine the average rate of change of $f(x)$ over the specified interval.

13. $f(x) = x^2 - 10x + 5$; interval $[-4, 4]$

14. $f(x) = 2x^2 + 4x - 6$; interval $[-3, 3]$

15. $f(x) = 3x^2 - 3x + 1$; interval $[-5, 5]$

16. $f(x) = 4x^2 + x + 3$; interval $[-2, 2]$

17. $f(x) = 2x^2 - 11$; interval $[-3, 3]$

18. $f(x) = -2x^2 + 8x + 7$; interval $[-4, 4]$

Example 5

Determine the average rate of change of $f(x)$ over the specified interval.

19. interval $[-3, 3]$

x	$f(x)$
−3	0
−2	3
−1	−4
0	−3
1	0
2	5
3	12

20. interval $[-4, 4]$

x	$f(x)$
−4	−27
−2	−3
0	5
2	−3
4	−27

21. interval $[-2, 2]$

x	$f(x)$
−2	−3
−1	−3
0	−1
1	3
2	9

22. interval $[-5, 5]$

x	$f(x)$
−5	−39
−3	−15
−1	1
0	6
1	9
3	9
5	1

23. interval $[-3, 3]$

x	$f(x)$
−3	27
−2	12
−1	3
0	0
1	3
2	12
3	27

24. interval $[-2, 2]$

x	$f(x)$
−2	12
−1	5
0	0
1	−3
2	−4

Example 6

25. FOOD The graph shows the number of people in the U.S. who consumed between 8 and 11 bags of potato chips in a year since 2011.

a. Use the graph to estimate the average rate of change of consumption from 2011 to 2016. Then check your results algebraically.

b. Interpret your results in the context of the situation.

Potato Chips Consumption

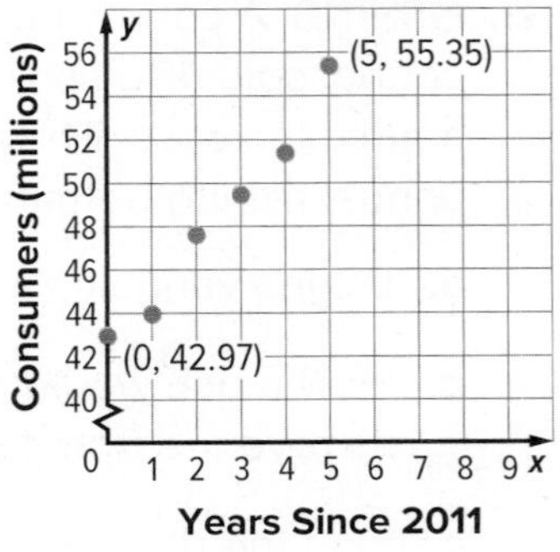

26. EARNINGS The graph shows the amount of money Sheila earned each year since 2008.

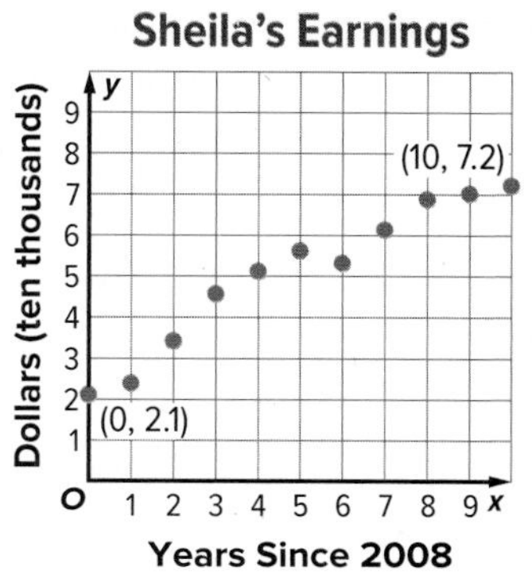

a. Use the graph to estimate the average rate of change of Sheila's earnings from 2008 to 2018. Then check your results algebraically.

b. Interpret your results in the context of the situation.

Mixed Exercises

Complete parts a-c for each quadratic function.

a. Find the y-intercept, the equation of the axis of symmetry, and the x-coordinate of the vertex.

b. Make a table of values that includes the vertex.

c. Use this information to graph the function.

27. $f(x) = 2x^2 - 6x - 9$ **28.** $f(x) = -3x^2 - 9x + 2$ **29.** $f(x) = -4x^2 + 5x$

30. $f(x) = 2x^2 + 11x$ **31.** $f(x) = 0.25x^2 + 3x + 4$ **32.** $f(x) = -0.75x^2 + 4x + 6$

Determine whether each function has a maximum or a minimum value. Then find and use the x-coordinate of the vertex to determine the maximum or minimum.

33. $f(x) = -9x^2 - 12x + 19$ **34.** $f(x) = 7x^2 - 21x + 8$

35. $f(x) = -5x^2 + 14x - 6$ **36.** $f(x) = 2x^2 - 13x - 9$

37. $f(x) = 9x - 1 - 18x^2$ **38.** $f(x) = -16 - 18x - 12x^2$

39. HEALTH CLUBS Last year, the Sports Time Athletic Club charged $20 per month to participate in an aerobics class. Seventy people attended the classes. The club wants to increase the class price. They expect to lose one customer for each $1 increase in the price.

a. Define variables and write an equation for a function that represents the situation. Make a table and graph the function.

b. Find the vertex of the function and interpret it in the context of the situation. Does it seem reasonable? Explain.

40. TICKETS The manager of a community symphony estimates that the symphony will earn $-40P^2 + 1100P$ dollars per concert if they charge P dollars for tickets. What ticket price should the symphony charge in order to maximize its profits?

41. REASONING On Friday nights, the local cinema typically sells 200 tickets at $6.00 each. The manager estimates that for each $0.50 increase in the ticket price, 10 fewer people will come to the cinema.

a. Define the variables x and y. Then write and graph a function to represent the expected revenue from ticket sales, and determine the domain of the function for the situation.

b. What price should the manager set for a ticket in order to maximize the revenue? Justify your reasoning.

c. Explain why the graph decreases from $x = 4$ to $x = 20$, and interpret the meaning of the x-intercept of the graph.

42. **USE A MODEL** From 4 feet above a swimming pool, Tomas throws a ball upward with a velocity of 32 feet per second. The height $h(t)$ of the ball t seconds after Tomas throws it is given by $h(t) = -16t^2 + 32t + 4$. For $t \geq 0$, find the maximum height reached by the ball and the time that this height is reached.

43. **TRAJECTORIES** At a special ceremonial reenactment, a cannonball is launched from a cannon on the wall of Fort Chambly, Quebec. If the path of the cannonball is traced on a graph so that the cannon is situated on the y-axis, the equation that describes the path is $y = -\frac{1}{1600}x^2 + \frac{1}{2}x + 20$, where x is the horizontal distance from the cliff and y is the vertical distance above the ground in feet. How high above the ground is the cannon?

44. **CONSTRUCT ARGUMENTS** Which function has a greater maximum: $f(x) = -2x^2 + 6x - 7$ or the function shown in the graph at the right? Explain your reasoning by copying the graph of $g(x)$ and graphing $f(x)$.

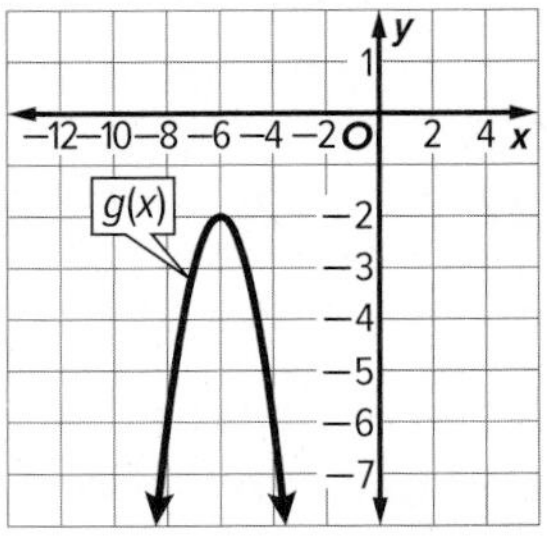

45. **FIND THE ERROR** Lucas thinks that the functions $f(x)$ and $g(x)$ have the same maximum. Madison thinks that $g(x)$ has a greater maximum. Is either of them correct? Explain your reasoning.

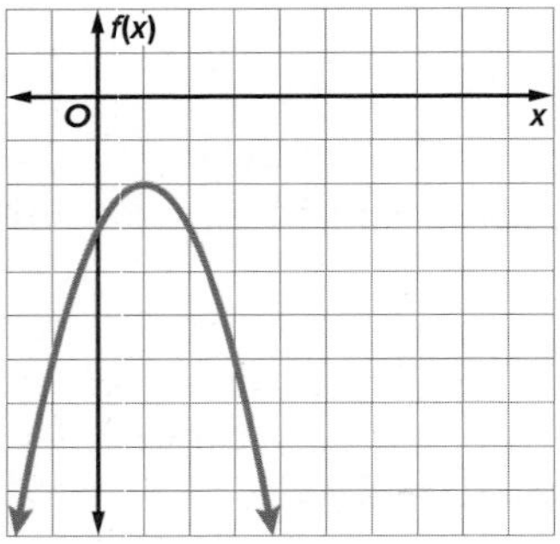

$g(x)$ is a quadratic function with x-intercepts of 4 and 2 and a y-intercept of -8.

46. **PERSEVERE** The table at the right represents some points on the graph of a quadratic function.

a. Find the values of a, b, c, and d.

b. What is the x-coordinate of the vertex of the function?

c. Does the function have a maximum or a minimum?

x	y
-20	-377
c	-13
-5	-2
-1	22
$d - 1$	a
5	$a - 24$
7	$-b$
15	-202
$14 - c$	-377

47. **WRITE** Describe how you determine whether a function is quadratic and if it has a maximum or minimum value.

48. **CREATE** Give an example of a quadratic function with each characteristic.

a. maximum of 8

b. minimum of -4

c. vertex of $(-2, 6)$

Solving Quadratic Equations by Graphing

Explore Roots of Quadratic Equations

Online Activity Use graphing technology to complete the Explore.

INQUIRY How can you use the graph of a quadratic function to find the solutions of its related equation?

Learn Solving Quadratic Equations by Graphing

A **quadratic equation** is an equation that includes a quadratic expression.

Key Concept • Standard Form of a Quadratic Equation

The **standard form of a quadratic equation** is $ax^2 + bx + c = 0$, where $a \neq 0$.

One method for finding the roots of a quadratic equation is to find the zeros of a related quadratic function. You can identify the solutions or roots of an equation by finding the *x*-intercepts of the graph of a related function. Often, exact roots cannot be found by graphing. You can estimate the solutions by finding the integers between which the zeros are located on the graph of the related function.

Example 1 One Real Solution

Solve $10 - x^2 = 4x + 14$ by graphing.

Rearrange terms so that one side of the equation is 0.

$0 = x^2 + 4x + 4$

Find the axis of symmetry.

$x = -\frac{b}{2a} = -\frac{4}{2(1)}$ or -2

Make a table of values, plot the points, and connect them with a curve.

x	y
−4	4
−3	1
−2	0
−1	1
0	4

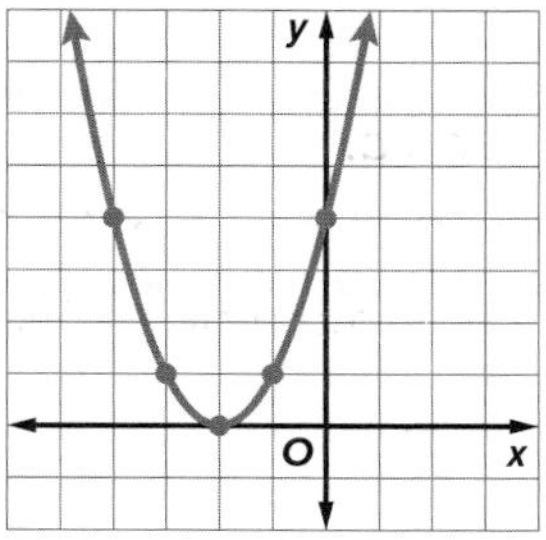

The zero of the function is −2. Therefore, the solution of the equation is −2 or $\{x \mid x = -2\}$.

Check

Solve $x^2 + 7x = 31x - 144$ by graphing. $x =$ __?__

Today's Goals

- Solve quadratic equations by graphing.

Today's Vocabulary

quadratic equation

standard form of a quadratic equation

Think About It!

How can you determine the number of solutions of a quadratic equation?

Study Tip

Solutions of Quadratic Equations A quadratic equation can have one real solution, two real solutions, or no real solutions.

Think About It!

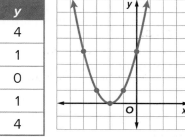

How can you find the solution of the equation from the table?

Example 2 Two Real Solutions

Use a quadratic equation to find two real numbers with a sum of 24 and a product of 143.

Let x represent one of the numbers. Then $24 - x$ will represent the other number. So $x(24 - x) = 143$.

What do you need to find?
x and $24 - x$

Step 1 Solve the equation for 0.

$x(24 - x) = 143$	Original equation
$24x - x^2 = 143$	Distributive Property
$0 = x^2 - 24x + 143$	Subtract $24x - x^2$ from each side.

Step 2 Find the axis of symmetry.

$x = -\frac{b}{2a}$	Equation of the axis of symmetry
$x = -\frac{-24}{2(1)}$	$a = 1, b = -24$
$x = 12$	Simplify.

Step 3 Make a table of values and graph the function.

x	y
14	3
13	0
12	−1
11	0
10	3

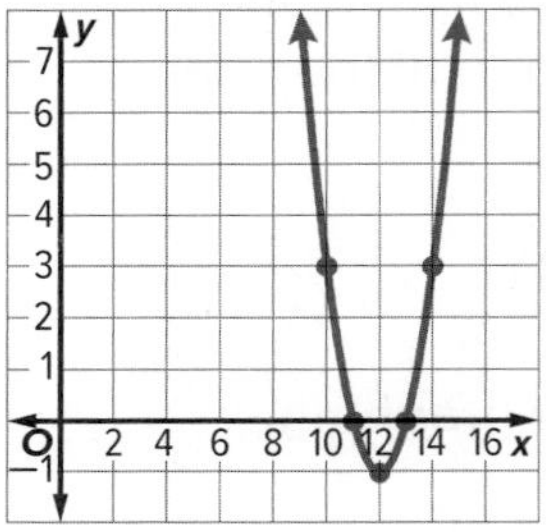

Steps 4 and 5 Find the zero(s) and determine the solution.

The zeros of the function are 11 and 13.

$x = 11$ or $x = 13$, so $24 - x = 13$ or $24 - x = 11$. Thus, the two numbers with a sum of 24 and a product of 143 are 11 and 13.

Check

Use a quadratic equation to find two real numbers with a sum of −43 and a product of 306. __?__ and __?__

Go Online You can complete an Extra Example online.

Go Online You can watch a video to see how to solve quadratic equations by graphing on a graphing calculator.

Think About It!
Explain why 9 and 15 cannot be solutions, even though their sum is 24.

Example 3 Estimate Roots

Solve $-x^2 + 3x + 7 = 0$ by graphing. If the exact roots cannot be found, state the consecutive integers between which the roots are located.

Find the axis of symmetry. $x = -\frac{b}{2a} = -\frac{3}{2(-1)}$ or $= \frac{3}{2}$

Make a table of values, plot the points, and connect them with a curve.

The x-intercepts of the graph indicate that one solution is between -2 and -1, and the other solution is between 4 and 5.

x	y
−2	−3
−1	3
0	7
1	9
2	9
3	7
4	3
5	−3

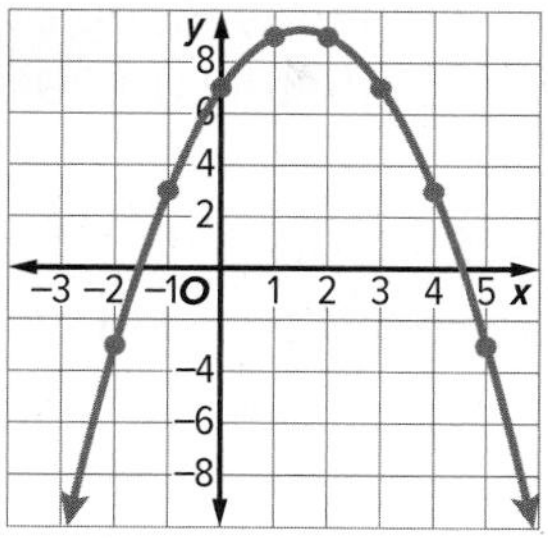

Check

Use a graph to find all of the solutions of $x^2 + 9x - 5 = 0$. State all of the pairs of consecutive integers between which the roots are located.

You can use a table to estimate the solutions of a quadratic equation. When the value of the function is positive for one value and negative for the second value, then there is at least one zero between the two values

Example 4 Solve by Using a Table

Use a table to solve $-x^2 + 5x - 1 = 0$.

Steps 1 and 2 Enter the function and view the table.

Enter $-x^2 + 5x - 1$ in the **Y=** list. Use the **TABLE** window to find where the sign of **Y1** changes. The sign changes between $x = 0$ and $x = 1$.

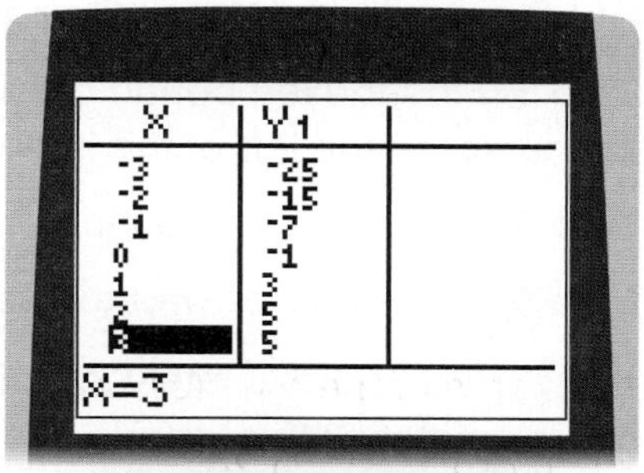

Steps 3 and 4 Edit the table settings and find a more accurate location.

Use **TBLSET** to change **ΔTbl** to 0.1 and look again for the sign change. Repeat this for 0.01 and 0.001 to get a more accurate location of one zero.

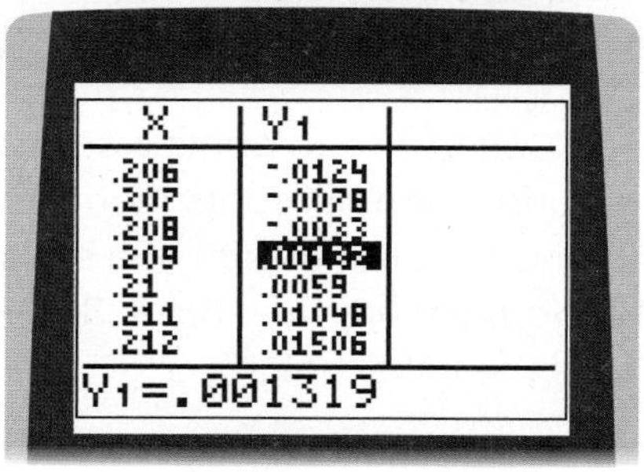

One zero is located at approximately $x = 0.209$.

(*continued on the next page*)

Go Online You can complete an Extra Example online.

Think About It!
How can you check your solutions?

Steps 5 and 6 Find the other zero and determine the solutions of the equation.

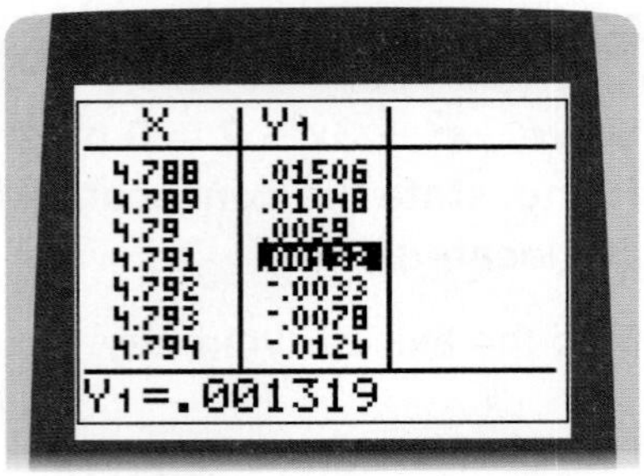

Repeat the process to find the second zero of the function.

The zeros of the function are at approximately 0.209 and 4.791, so the solutions to the equation are approximately 0.209 and 4.791.

Check

Use a table to find all of the solutions of $-x^2 - 3x + 8 = 0$.

Watch Out!
Graphing Calculator If you cannot see the graph of the function on your graphing calculator, you may need to adjust the viewing window. Having the proper viewing window will also make it easier to see the zeros.

Example 5 Solve by Using a Calculator

FOOTBALL **A kicker punts a football. The height of the ball after t seconds is given by $h(t) = -16t^2 + 50t + h_0$, where h_0 is the initial height. If the ball is 1.5 feet above the ground when the punter's foot meets the ball, how long will it take the ball to hit the ground?**

We know that h_0 is the initial height, so $h_0 = 1.5$. We need to find t when $h(t)$ is 0. Use a graphing calculator to graph the related function $h(t) = -16t^2 + 50t + 1.5$.

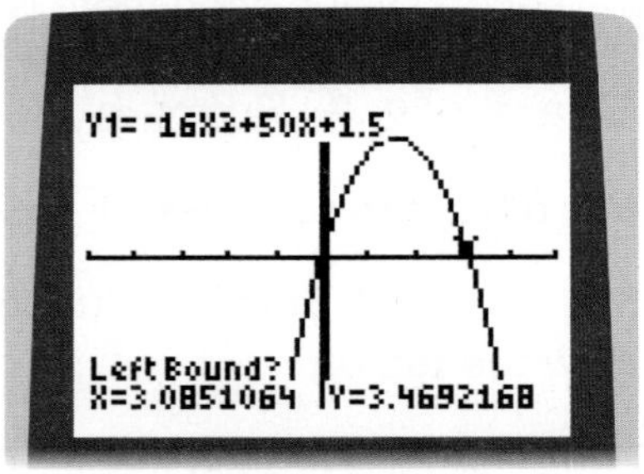

Step 1 Enter the function in the **Y=** list, and press graph.

Step 2 Use the zero feature from the **CALC** menu to find the positive zero.

Step 3 Find the left bound by placing the cursor to the left of the intercept.

Step 4 Find the right bound.

Step 5 Find and interpret the solution.

The zero is approximately 3.15. Thus, the ball hit the ground approximately 3.15 seconds after it was punted.

Think About It!
Why did you only find the positive zero?

Check

SOCCER A goalie punts a soccer ball. If the ball is 1 foot above the ground when her foots meets the ball, find how long it will take, to the nearest hundredth of a second, for the ball to hit the ground. Use the formula $h(t) = -16t^2 + 35t + h_0$, where t is the time in seconds and h_0 is the initial height.

Go Online to see how to use a TI-Nspire graphing calculator with this example.

Go Online You can complete an Extra Example online.

Practice

Go Online You can complete your homework online.

Example 1

Use the related graph of each equation to determine its solutions.

1. $x^2 + 2x + 3 = 0$

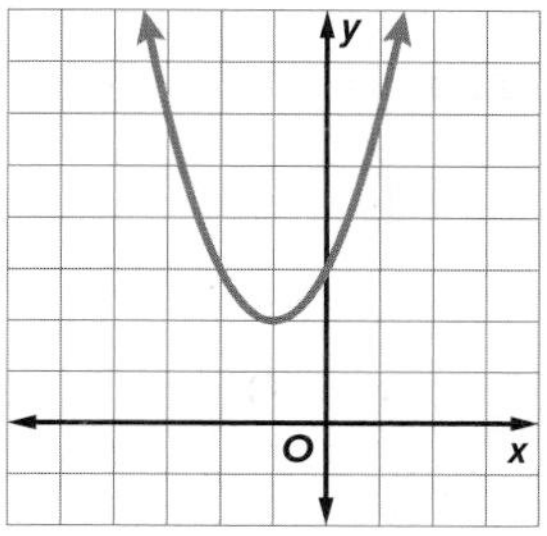

2. $x^2 - 3x - 10 = 0$

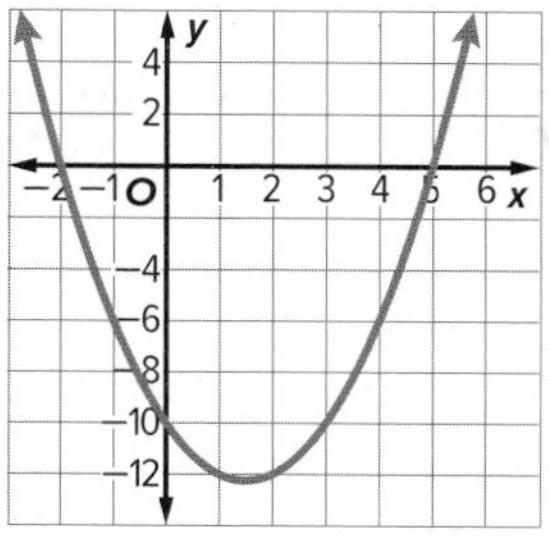

3. $-x^2 - 8x - 16 = 0$

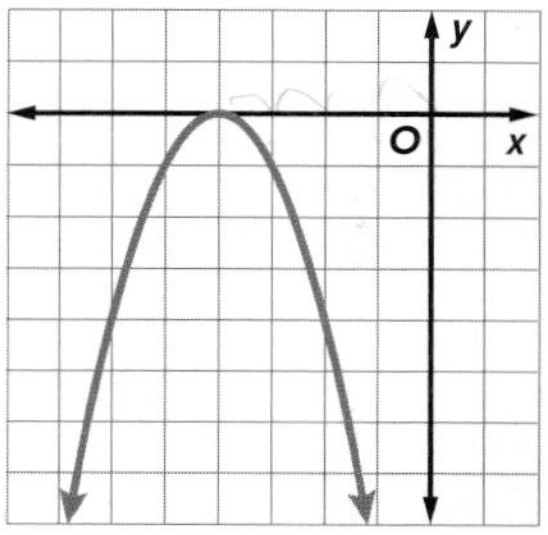

Solve each equation by graphing.

4. $x^2 - 10x + 21 = 0$

5. $4x^2 + 4x + 1 = 0$

6. $x^2 + x - 6 = 0$

7. $x^2 + 2x - 3 = 0$

8. $-x^2 - 6x - 9 = 0$

9. $x^2 - 6x + 5 = 0$

10. $x^2 + 2x + 3 = 0$

11. $x^2 - 3x - 10 = 0$

12. $-x^2 - 8x - 16 = 0$

Example 2

13. Use a quadratic equation to find two real numbers with a sum of 2 and a product of −24.

14. Use a quadratic equation to find two real numbers with a sum of −15 and a product of −54.

Example 3

Solve each equation by graphing. If the exact roots cannot be found, state the consecutive integers between which the roots are located.

15. $x^2 - 4x + 2 = 0$

16. $x^2 + 6x + 6 = 0$

17. $x^2 + 4x + 2 = 0$

18. $-x^2 - 4x = 0$

19. $-x^2 + 36 = 0$

20. $x^2 - 6x + 4 = 0$

21. $x^2 + 5x + 3 = 0$

22. $x^2 - 7 = 0$

23. $-x^2 - 4x - 6 = 0$

Example 4

Use the tables to determine the location of the zeros of each quadratic function. State the consecutive integers between which the roots are located.

24.

x	−7	−6	−5	−4	−3	−2	−1	0
$f(x)$	−8	−1	4	4	−1	−8	− 22	−48

25.

x	−2	−1	0	1	2	3	4	5
$f(x)$	32	14	2	−3	−3	2	14	32

26.

x	−6	−3	0	3	6	9	12	15
$f(x)$	−6	−1	3	5	3	−1	−6	−14

Use a table to solve each equation. If the exact roots cannot be found, approximate the roots to the nearest hundredth.

27. $-3x^2 + 3 = 0$

28. $x^2 - 3x + 2 = 0$

29. $-\frac{1}{2}x^2 + x + \frac{5}{2} = 0$

30. $x^2 - 2x - 2 = 0$

31. $-x^2 + 2x + 4 = 0$

32. $2x^2 - 12x + 17 = 0$

Example 5

PHYSICS **Use the formula $h(t) = -16t^2 + v_0t + h_0$, where $h(t)$ is the height of an object in feet, v_0 is the object's initial velocity in feet per second, t is the time in seconds, and h_0 is the initial height in feet from which the object is launched. Round to the nearest tenth, if necessary.**

33. Melah throws a baseball with an initial upward velocity of 32 feet per second. The baseball is released from Melah's hand at a height of 4 feet. Use a graphing calculator to determine how long it will take the ball to hit the ground.

34. A punter kicks a football with an initial upward velocity of 60 feet per second. The ball is 2 feet above the ground when his foot meets the ball. Use a graphing calculator to determine how long will it take the ball to hit the ground.

Mixed Exercises

Solve each equation by graphing. If the exact roots cannot be found, state the consecutive integers between which the roots are located.

35. $4x^2 - 15 = -4x$

36. $-35 = -3x - 2x^2$

37. $-3x^2 + 11x + 9 = 1$

38. $-4x^2 = 12x + 8$

39. $-0.5x^2 + 18 = -6x + 33$

40. $0.5x^2 + 0.75 = 0.25x$

41. $3x^2 + 8x = 0$

42. $2x^2 + x = 11$

43. $-0.1x^2 + 0.5x + 10 = 0$

Use a graph or table to solve each equation. If exact roots cannot be found, state the consecutive integers between which the roots are located.

44. $x^2 + 4x = 0$

45. $-2x^2 - 4x - 5 = 0$

46. $0.5x^2 - 2x + 2 = 0$

47. $-0.25x^2 - x - 1 = 0$

48. $x^2 - 6x + 11 = 0$

49. $-0.5x^2 + x + 6 = 0$

REGULARITY **Use a quadratic equation to find two real numbers that satisfy each situation, or show that no such numbers exist.**

50. Their sum is 4, and their product is −117.

51. Their sum is 12, and their product is −85.

52. Their sum is −13, and their product is 42.

53. Their sum is −8, and their product is −209.

54. BRIDGES In 1895, a brick arch railway bridge was built on North Avenue in Baltimore, Maryland. The arch is described by the equation $h = 9 - \frac{1}{50}x^2$, where h is the height in yards and x is the distance in yards from the center of the bridge. Graph this equation and describe, to the nearest yard, where the bridge touches the ground.

55. RADIO TELESCOPES The cross section of a large radio telescope is a parabola. The equation that describes the cross section is $y = \frac{2}{75}x^2 - \frac{4}{3}x - \frac{32}{3}$, where y is the depth of the dish in meters at a point x meters from the center of the dish. If $y = 0$ represents the top of the dish, what is the width of the dish? Solve by graphing.

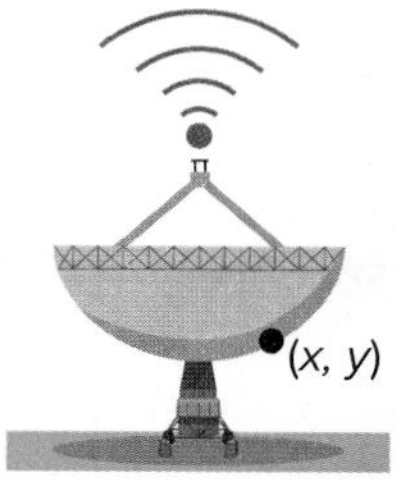

56. VOLCANOES A volcanic eruption blasts a boulder upward with an initial velocity of 240 feet per second. The height $h(t)$ of the boulder in feet, t seconds after the eruption can be modeled by the function $h(t) = -16t^2 + v_0t$. How long will it take the boulder to hit the ground if it lands at the same elevation from which it was ejected?

57. **TRAJECTORIES** Daniela hit a golf ball from ground level. The function $h = 80t - 16t^2$ represents the height of the ball in feet, where t is the time in seconds after Daniela hit it. Use the graph of the function to determine how long it took for the ball to reach the ground.

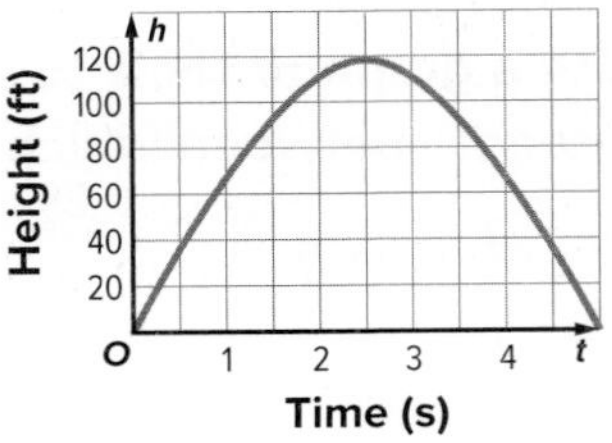

58. **HIKING** Antonia is hiking and reaches a steep part of the trail that runs along the edge of a cliff. In order to descend more safely, she drops her heavy backpack over the edge of the cliff so that it will land on a lower part of the trail, 38.75 feet below. The height $h(t)$ of an object t seconds after it is dropped straight down can also be modeled by the function $h(t) = -16t^2 + v_0t + h_0$, where v_0 is the initial velocity of the object, and h_0 is the initial height.

a. Write a quadratic function that can be used to determine the amount of time t that it takes for the backpack to land on the trail below the cliff after Antonia drops it.

b. Use a graphing calculator to determine how long until the backpack hits the ground. Round to the nearest tenth.

59. **FIND THE ERROR** Hakeem and Nandi were asked to find the location of the roots of the quadratic function represented by the table. Is either of them correct? Explain.

x	−4	−2	0	2	4	6	8	10
$f(x)$	52	26	8	−2	−4	2	16	38

Hakeem

The roots are between 4 and 6 because f(x) stops decreasing and begins to increase between x – 4 and x – 6.

Nandi

The roots are between −2 and 0 because x changes signs at that location.

60. **PERSEVERE** Find the value of a positive integer k such that $f(x) = x^2 - 2kx + 55$ has roots at $k + 3$ and $k - 3$.

61. **ANALYZE** If a quadratic function has a minimum at $(-6, -14)$ and a root at $x = -17$, what is the other root? Explain your reasoning.

62. **CREATE** Write a quadratic function with a maximum at $(3, 125)$ and roots at -2 and 8.

63. **WRITE** Explain how to solve a quadratic equation by graphing its related quadratic function.

Complex Numbers

Learn Pure Imaginary Numbers

In your math studies so far, you have worked with real numbers. However, some equations such as $x^2 + x + 1 = 0$ do not have real solutions. This led mathematicians to define imaginary numbers. The **imaginary unit *i*** is the principal square root of −1. Thus, $i = \sqrt{-1}$ and $i^2 = -1$.

Numbers of the form $6i$, $-2i$, and $i\sqrt{3}$ are called pure imaginary numbers. A **pure imaginary number** is a number of the form bi, where b is a real number and $i = \sqrt{-1}$. For any positive real number $\sqrt{-b^2} = \sqrt{b^2} \cdot \sqrt{-1}$ or bi.

The Commutative and Associative Properties of Multiplication hold true for pure imaginary numbers. The first few powers of i are shown.

$i^1 = i$ $\qquad$ $i^2 = -1$ $\qquad$ $i^3 = i^2 \cdot i$ or $-i$

Today's Goals

- Perform operations with pure imaginary numbers.
- Perform operations with complex numbers.

Today's Vocabulary

imaginary unit *i*

pure imaginary number

complex number

complex conjugates

rationalizing the denominator

Go Online You may want to complete the Concept Check to check your understanding.

Example 1 Square Roots of Negative Numbers

Simplify $\sqrt{-294}$.

$$\sqrt{-294} = \sqrt{-1 \cdot 7^2 \cdot 6} \qquad \text{Factor the radicand.}$$

$$= \sqrt{-1} \cdot \sqrt{7^2} \cdot \sqrt{6} \qquad \text{Factor out the imaginary unit.}$$

$$= i \cdot 7 \cdot \sqrt{6} \text{ or } 7i\sqrt{6} \qquad \text{Simplify.}$$

Study Tip

Square Factors When factoring an expression under a radical, look for perfect square factors.

Check

Simplify $\sqrt{-75}$.

A. $i\sqrt{75}$

B. $3i\sqrt{5}$

C. $5i\sqrt{3}$

D. $-3\sqrt{5}$

Go Online You can complete an Extra Example online.

Example 2 Products of Pure Imaginary Numbers

Simplify $\sqrt{-10} \cdot \sqrt{-15}$.

$\sqrt{-10} \cdot \sqrt{-15}$	$= i\sqrt{10} \cdot i\sqrt{15}$	$i = \sqrt{-1}$
	$= i^2 \cdot \sqrt{150}$	Multiply.
	$= -1 \cdot \sqrt{25} \cdot \sqrt{6}$	Simplify.
	$= -5\sqrt{6}$	Multiply.

Check

Simplify $\sqrt{-16} \cdot \sqrt{-25}$.

Talk About It

How can an expression with two imaginary expressions, $\sqrt{-10}$ and $\sqrt{-15}$, have a product that is real?

Example 3 Equation with Pure Imaginary Solutions

Solve $x^2 + 81 = 0$.

$x^2 + 81 = 0$	Original equation
$x^2 = -81$	Subtract 81 from each side.
$x = \pm\sqrt{-81}$	Take the square root of each side.
$x = \pm 9i$	Simplify.

Check

Solve $3x^2 + 27 = 0$.

Explore Factoring the Sum of Two Squares

Online Activity Use guiding exercises to complete the Explore.

INQUIRY Can you factor a polynomial of the form $a^2 + b^2$?

Learn Complex Numbers

Key Concept • Complex Numbers

A **complex number** is any number that can be written in the form $a + bi$, where a and b are real numbers and i is the imaginary unit. a is called the real part, and bi is called the imaginary part.

Go Online You can complete an Extra Example online.

The Venn diagram shows the set of complex numbers. Notice that all of the real numbers are part of the set of complex numbers.

Complex Numbers ($a + bi$)

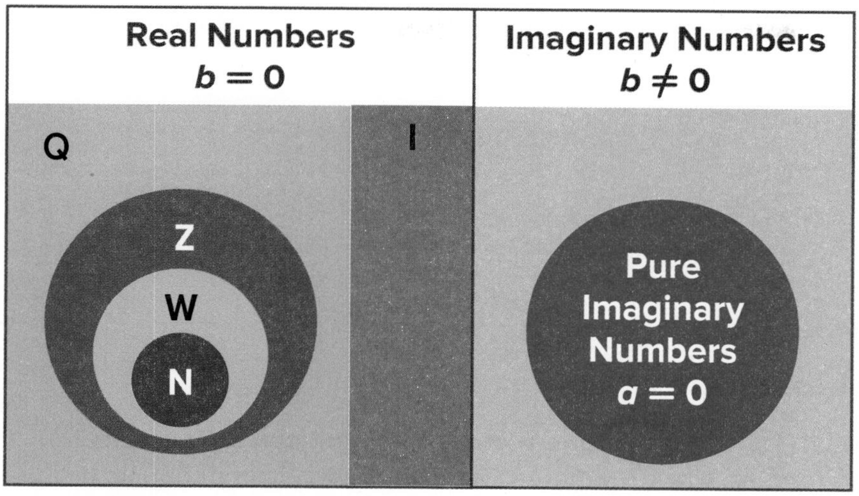

Two complex numbers are equal if and only if their real parts are equal and their imaginary parts are equal. The Commutative and Associative Properties of Multiplication and Addition and the Distributive Property hold true for complex numbers. To add or subtract complex numbers, combine like terms. That is, combine the real parts, and combine the imaginary parts.

Two complex numbers of the form $a + bi$ and $a - bi$ are called **complex conjugates**. The product of complex conjugates is always a real number.

A radical expression is in simplest form if no radicands contain fractions and no radicals appear in the denominator of a fraction. Similarly, a complex number is in simplest form if no imaginary numbers appear in the denominator of a fraction. You can use complex conjugates to simplify a fraction with a complex number in the denominator. This process is called **rationalizing the denominator**.

Example 4 Equate Complex Numbers

Find the values of x and y that make $5x - 7 + (y + 4)i = 13 + 11i$ true.

Use equations relating the real and imaginary parts to solve for x and y.

$5x - 7 = 13$	Real parts
$5x = 20$	Add 7 to each side.
$x = 4$	Divide each side by 5.
$y + 4 = 11$	Imaginary parts
$y = 7$	Subtract 4 from each side.

Check

Find the values of x and y that make $5x + 13 + (2y - 7)i = -2 + i$ true.

Go Online You can complete an Extra Example online.

Study Tip

These abbreviations represent the sets of real numbers.

Letter	Set
Q	rationals
I	irrationals
Z	integers
W	wholes
N	naturals

Think About It!

Compare and contrast the subsets of the complex number system using the Venn diagram.

Go Online
You can watch a video to see how to add or subtract complex numbers.

Example 5 Add or Subtract Complex Numbers

Simplify $(8 + 3i) - (4 - 10i)$.

$(8 + 3i) - (4 - 10i) = (8 - 4) + [3 - (-10)]i$ Commutative and Associative Properties

$= 4 + 13i$ Simplify.

Check

Simplify $(-5 + 5i) - (-3 + 8i)$.

Study Tip
Imaginary Unit
Complex numbers are often used with electricity. In these problems, j is usually used in place of i.

Example 6 Multiply Complex Numbers

ELECTRICITY **The voltage V of an AC circuit can be found using the formula $V = CI$, where C is current and I is impedance. If $C = 3 + 2j$ amps and $I = 7 - 5j$ ohms, determine the voltage.**

$V = CI$	Voltage Formula
$= (3 + 2j)(7 - 5j)$	$C = 3 + 2j$ and $I = 7 - 5j$
$= 3(7) + 3(-5j) + 2j(7) + 2j(-5j)$	FOIL Method
$= 21 - 15j + 14j - 10j^2$	Multiply.
$= 21 - j - 10(-1)$	$j^2 = -1$
$= 31 - j$	Add.

The voltage is $31 - j$ volts.

Example 7 Divide Complex Numbers

Simplify $\frac{5i}{3 + 2i}$.

Rationalize the denominator to simplify the fraction.

$\frac{5i}{3 + 2i} = \frac{5i}{3 + 2i} \cdot \frac{3 - 2i}{3 - 2i}$	$3 + 2i$ and $3 - 2i$ are complex conjugates.
$= \frac{15i - 10i^2}{9 - 4i^2}$	Multiply the numerator and denominator.
$= \frac{15i - 10(-1)}{9 - 4(-1)}$	$i^2 = -1$
$= \frac{15i + 10}{13}$	Simplify.
$= \frac{10}{13} + \frac{15}{13}i$	$a + bi$ form

Check

Simplify $\frac{2i}{-4 + 3i}$.

Go Online You can complete an Extra Example online.

Practice

Go Online You can complete your homework online.

Examples 1 and 2

Simplify.

1. $\sqrt{-48}$

2. $\sqrt{-63}$

3. $\sqrt{-72}$

4. $\sqrt{-24}$

5. $\sqrt{-84}$

6. $\sqrt{-99}$

7. $\sqrt{-23} \cdot \sqrt{-46}$

8. $\sqrt{-6} \cdot \sqrt{-3}$

9. $\sqrt{-5} \cdot \sqrt{-10}$

10. $(3i)(-2i)(5i)$

11. i^{11}

12. $4i(-6i)^2$

Example 3

Solve each equation.

13. $5x^2 + 45 = 0$

14. $4x^2 + 24 = 0$

15. $-9x^2 = 9$

16. $7x^2 + 84 = 0$

17. $5x^2 + 125 = 0$

18. $8x^2 + 96 = 0$

Example 4

Find the values of *x* and *y* that make each equation true.

19. $9 + 12i = 3x + 4yi$

20. $x + 1 + 2yi = 3 - 6i$

21. $2x + 7 + (3 - y)i = -4 + 6i$

22. $5 + y + (3x - 7)i = 9 - 3i$

23. $20 - 12i = 5x + (4y)i$

24. $x - 16i = 3 - (2y)i$

Examples 5 and 6

Simplify.

25. $(6 + i) + (4 - 5i)$

26. $(8 + 3i) - (6 - 2i)$

27. $(5 - i) - (3 - 2i)$

28. $(-4 + 2i) + (6 - 3i)$

29. $(6 - 3i) + (4 - 2i)$

30. $(-11 + 4i) - (1 - 5i)$

31. $(2 + i)(3 - i)$

32. $(5 - 2i)(4 - i)$

33. $(4 - 2i)(1 - 2i)$

34. **ELECTRICITY** Using the formula $V = CI$, find the voltage V in a circuit when the current $C = 3 - j$ amps and the impedance $I = 3 + 2j$ ohms.

Example 7

Simplify.

35. $\frac{5}{3 + i}$

36. $\frac{7 - 13i}{2i}$

37. $\frac{6 - 5i}{3i}$

Mixed Exercises

STRUCTURE **Simplify.**

38. $(1 + i)(2 + 3i)(4 - 3i)$

39. $\frac{4 - i\sqrt{2}}{4 + i\sqrt{2}}$

40. $\frac{2 - i\sqrt{3}}{2 + i\sqrt{3}}$

41. Find the sum of $ix^2 - (4 + 5i)x + 7$ and $3x^2 + (2 + 6i)x - 8i$.

42. Simplify $[(2 + i)x^2 - ix + 5 + i] - [(-3 + 4i)x^2 + (5 - 5i)x - 6]$.

ELECTRICITY **Use the formula $V = CI$, where V is the voltage, C is the current, and I is the impedance.**

43. The current in a circuit is $2 + 4j$ amps, and the impedance is $3 - j$ ohms. What is the voltage?

44. The voltage in a circuit is $24 - 8j$ volts, and the impedance is $4 - 2j$ ohms. What is the current?

45. CIRCUITS The impedance in one part of a series circuit is $1 + 3j$ ohms and the impedance in another part of the circuit is $7 - 5j$ ohms. Add these complex numbers to find the total impedance in the circuit.

46. ELECTRICAL ENGINEERING The standard electrical voltage in Europe is 220 volts.

a. Find the impedance in a standard European circuit if the current is $22 - 11i$ amps.

b. Find the current in a standard European circuit if the impedance is $10 - 5i$ ohms.

c. Find the impedance in a standard European circuit if the current is $20i$ amps.

47. FIND THE ERROR Jose and Zoe are simplifying $(2i)(3i)(4i)$. Is either of them correct? Explain your reasoning.

Jose	Zoe
$24i^3 = -24$	$24i^3 = -24i$

48. PERSEVERE Simplify $(1 + 2i)^3$.

49. ANALYZE Determine whether the following statement is *always*, *sometimes*, or *never* true. Explain your reasoning.

Every complex number has both a real part and an imaginary part.

50. CREATE Write two complex numbers with a product of 20.

51. WRITE Explain how complex numbers are related to quadratic equations.

Solving Quadratic Equations by Factoring

Explore Finding the Solutions of Quadratic Equations by Factoring

Online Activity Use graphing technology to complete the Explore.

INQUIRY How can you use factoring to solve a quadratic equation?

Learn Solving Quadratic Equations by Factoring

The **factored form** of a quadratic equation is $0 = a(x - p)(x - q)$, where $a \neq 0$. In this equation, p and q represent the x-intercepts of the graph of the related function. For example, $0 = x^2 - 2x - 3$ can be written in the factored form $0 = (x - 3)(x + 1)$ and its related graph has x-intercepts of -1 and 3.

Key Concepts • Factoring	
Using the Distributive Property	$ax + bx = x(a + b)$
Factoring Trinomials	$x^2 + bx + c = (x + m)(x + p)$ when $m + p = b$ and $mp = c$

Key Concept • Zero Product Property

Words: For any real numbers a and b, if $ab = 0$, then either $a = 0$, $b = 0$, or both a and $b = 0$.

Example: If $(x - 2)(x + 4) = 0$, then $x - 2 = 0$, $x + 4 = 0$, or both $x - 2 = 0$ and $x + 4 = 0$.

To solve a quadratic equation by factoring, first make sure that one side of the equation is 0, and factor the trinomial. Use the Zero Product Property to write separate equations. Then use the properties of equality to isolate the variable.

Example 1 Factor by Using the Distributive Property

Solve $12x^2 - 2x = x$ by factoring. Check your solution.

$12x^2 - 2x = x$	Original equation
$12x^2 - 3x = 0$	Subtract x from each side.
$3x(4x) - 3x(1) = 0$	Factor the GCF.
$3x(4x - 1) = 0$	Distributive Property
$3x = 0$ or $4x - 1 = 0$	Zero Product Property
$x = 0 \qquad x = \frac{1}{4}$	Solve.

Go Online You can complete an Extra Example online.

Today's Goals
- Solve quadratic equations by factoring.
- Solve quadratic equations by factoring special products.

Today's Vocabulary
factored form
difference of squares
perfect square trinomials

Think About It!
The equation $x^2 - 2x - 3 = 0$ could be solved by factoring, where $x^2 - 2x - 3 = (x - 3)(x + 1)$. How are the factors of the equation related to the roots, or zeros, of the related function $f(x) = x^2 - 2x - 3$?

Go Online You can watch a video to see how to use algebra tiles to factor a polynomial using the Distributive Property.

Example 2 Factor a Trinomial

Solve $x^2 - 6x - 9 = 18$ by factoring. Check your solution.

$x^2 - 6x - 9 = 18$	Original equation
$x^2 - 6x - 27 = 0$	Subtract 18 from each side.
$(x + 3)(x - 9) = 0$	Factor the trinomial.
$x + 3 = 0$ or $x - 9 = 0$	Zero Product Property
$x = -3$ $\quad x = 9$	Solve.

Think About It!

Choose two integers and write an equation in standard form with these roots. How would the equation change if the signs of the two roots were switched?

Apply Example 3 Solve an Equation by Factoring

ACCELERATION **The equation $d = vt + \frac{1}{2}at^2$ represents the displacement d of a car traveling at an initial velocity v where the acceleration a is constant over a given time t. Find how long it takes a car to accelerate from 30 mph to 45 mph if the car moved 605 feet and accelerated slowly at a rate of 2 feet per second squared.**

1 What is the task?

Describe the task in your own words. Then list any questions that you may have. How can you find answers to your questions?

Solve the equation to find the time for the car to accelerate. The acceleration is given in feet per second squared and the velocity is given in miles per hour. How do I address the difference in units?

2 How will you approach the task? What have you learned that you can use to help you complete the task?

Convert the velocity to feet per second. Then substitute the distance, velocity, and acceleration into the formula
and solve for time.

3 What is your solution?

Use your strategy to solve the problem.

What is the velocity in feet per second? 44 fps

How long does it take takes the car to accelerate from 30 mph to 45 mph? 11 s

Think About It!

Why did you not use 45 mph to solve this problem?

4 How can you know that your solution is reasonable?

Write About It! Write an argument that can be used to defend your solution.

The solutions of the equation are −55 and 11. Because time cannot be negative, $t = 11$ is the only viable solution in the context of the situation.

Go Online You can complete an Extra Example online.

Check

SALES A clothing store is analyzing their market to determine the profitability of their new dress design. If $P(x) = -16x^2 + 1712x - 44{,}640$ represents the store's profit when x is the price of each dress, find the prices at which the store makes no profit on the design.

A. \$11.25 and \$15.50

B. \$45 and \$62

C. \$50 and \$54

D. \$180 and \$248

Example 4 Factor a Trinomial Where a is Not 1

Solve $3x^2 + 5x + 15 = 17$ by factoring. Check your solution.

$3x^2 + 5x + 15 = 17$	Original equation
$3x^2 + 5x - 2 = 0$	Subtract 17 from each side.
$(3x - 1)(x + 2) = 0$	Factor the trinomial.
$3x - 1 = 0$ or $x + 2 = 0$	Zero Product Property
$x = \frac{1}{3}$ $\quad$ $x = -2$	Solve.

Check

Solve $4x^2 + 12x - 27 = 13$ by factoring. Check your solution.

Learn Solving Quadratic Equations by Factoring Special Products

Key Concept • Factoring Differences of Squares

Words: To factor $a^2 - b^2$, find the square roots of a^2 and b^2. Then apply the pattern.

Symbols: $a^2 - b^2 = (a + b)(a - b)$

Key Concept • Factoring Perfect Square Trinomials

Words: To factor $a^2 + 2ab + b^2$, find the square roots of a^2 and b^2. Then apply the pattern.

Symbols: $a^2 + 2ab + b^2 = (a + b)^2$

Not all quadratic equations have solutions that are real numbers. In some cases, the solutions are complex numbers of the form $a + bi$, where $b \neq 0$. For example, you know that the solution of $x^2 = -4$ must be complex because there is no real number for which its square is -4. If you take the square root of each side, $x = 2i$ or $-2i$.

Talk About It!

Explain how to determine which values should be chosen for m and p when factoring a polynomial of the form $ax^2 + bx + c$.

Math History Minute

English mathematician and astronomer **Thomas Harriot (1560–1621)** was one of the first, if not the first, to consider the imaginary roots of equations. Harriot advanced the notation system for algebra and studied negative and imaginary numbers.

Example 5 Factor a Difference of Squares

Solve $81 = x^2$ by factoring. Check your solution.

$81 = x^2$	Original equation
$81 - x^2 = 0$	Subtract x^2 from each side.
$9^2 - x^2 = 0$	Write in the form $a^2 - b^2$.
$(9 + x)(9 - x) = 0$	Factor the difference of squares.
$9 + x = 0$ or $9 - x = 0$	Zero Product Property
$x = -9 \qquad x = 9$	Solve.

Go Online You can watch a video to see how to use algebra tiles to factor a difference of squares.

Check

Solve $x^2 = 529$ by factoring. Check your solution.

$x =$ _?_, _?_

Example 6 Factor a Perfect Square Trinomial

Solve $16y^2 - 22y + 23 = 26y - 13$ by factoring. Check your solution.

$16y^2 - 22y + 23 = 26y - 13$	Original equation
$16y^2 - 48y + 23 = -13$	Subtract $26y$ from each side.
$16y^2 - 48y + 36 = 0$	Add 13 to each side.
$(4y)^2 + 2(4y)(-6) + (-6)^2 = 0$	Factor the perfect square trinomial.
$(4y - 6)^2 = 0$	Simplify.
$y = \frac{3}{2}$	Take the square root of each side and solve.

Think About It!
Why does this equation have one solution instead of two?

Check

Solve $16x^2 - 22x + 15 = 10x - 1$ by factoring. Check your solution.

$x =$ _?_

Example 7 Complex Solutions

Solve $x^2 = -144$ by factoring. Check your solution.

$x^2 = -144$	Original equation
$x^2 + 144 = 0$	Add 144 to each side.
$x^2 - (-144) = 0$	Write as a difference of squares.
$x^2 - (12i)^2 = 0$	$\sqrt{-144} = 12i$
$(x + 12i)(x - 12i) = 0$	Factor the difference of squares.
$x + 12i = 0$ or $x - 12i = 0$	Zero Product Property
$x = -12i \qquad x = 12i$	Solve.

Think About It!
Explain why both $(-12i)^2$ and $(12i)^2$ equal -144.

Watch Out!
Complex Numbers Remember i^2 equals -1, not 1.

Go Online You can complete an Extra Example online.

Practice

Go Online You can complete your homework online.

Examples 1 and 2

Solve each equation by factoring. Check your solution.

1. $6x^2 - 2x = 0$

2. $x^2 = 7x$

3. $20x^2 = -25x$

4. $x^2 + x - 30 = 0$

5. $x^2 + 14x + 33 = 0$

6. $x^2 - 3x = 10$

Example 3

7. GEOMETRY The length of a rectangle is 2 feet more than its width. Find the dimensions of the rectangle if its area is 63 square feet.

8. PHOTOGRAPHY The length and width of a 6-inch by 8-inch photograph are reduced by the same amount to make a new photograph with area that is half that of the original. By how many inches will the dimensions of the photograph have to be reduced?

Example 4

Solve each equation by factoring. Check your solution.

9. $2x^2 - x - 3 = 0$

10. $6x^2 - 5x - 4 = 0$

11. $5x^2 + 28x - 12 = 0$

12. $12x^2 - 8x + 1 = 0$

13. $2x^2 - 11x - 40 = 0$

14. $3x^2 + 2x = 21$

Examples 5-7

Solve each equation by factoring. Check your solution.

15. $x^2 = 64$

16. $x^2 - 100 = 0$

17. $289 = x^2$

18. $x^2 + 14 = 50$

19. $x^2 - 169 = 0$

20. $124 = x^2 + 3$

21. $4x^2 - 28x + 49 = 0$

22. $9x^2 + 6x = -1$

23. $16x^2 - 24x + 13 = 4$

24. $81x^2 + 36x = -4$

25. $25x^2 + 80x + 64 = 0$

26. $9x^2 + 60x + 95 = -5$

27. $x^2 + 12 = -13$

28. $x^2 + 100 = 0$

29. $x^2 = -225$

30. $x^2 + 4 = 0$

31. $36x^2 = -25$

32. $64x^2 = -49$

Mixed Exercises

STRUCTURE **Solve each equation by factoring. Check your solution.**

33. $10x^2 + 25x = 15$

34. $27x^2 + 5 = 48x$

35. $x^2 + 81 = 0$

36. $45x^2 - 3x = 2x$

37. $80x^2 = -16$

38. $16x^2 + 8x = -1$

39. USE A MODEL The drawing *Maisons prés de la mer* by Claude Monet is approximately 10 inches by 13 inches. Jennifer wants to make an art piece inspired by the drawing, and with an area 60% greater. If she chooses the size of her artwork by increasing both the width and height of Monet's work by same amount, what will be the dimensions of Jennifer's artwork?

40. ANIMATION A computer graphics animator would like to make a realistic simulation of a tossed ball. The animator wants the ball to follow the parabolic trajectory represented by $f(x) = -0.2(x + 5)(x - 5)$.

a. What are the solutions of $f(x) = 0$?

b. If the animator changes the equation to $f(x) = -0.2x^2 + 20$, what are the solutions of $f(x) = 0$?

41. Find two consecutive even positive integers that have a product of 624.

42. Find two consecutive odd positive integers that have a product of 323.

43. FIND THE ERROR Jade and Mateo are solving $-12x^2 + 5x + 2 = 0$. Is either of them correct? Explain your reasoning.

Jade	Mateo
$-12x^2 + 5x + 2 = 0$	$-12x^2 + 5x + 2 = 0$
$-12x^2 + 8x - 3x + 2 = 0$	$-12x^2 + 8x - 3x + 2 = 0$
$4x(-3x + 2) - (3x + 2) = 0$	$4x(-3x + 2) - (3x + 2) = 0$
$(4x - 1)(3x + 2) = 0$	$(4x + 1)(-3x + 2) = 0$
$x = \frac{1}{4}$ or $-\frac{2}{3}$	$x = -\frac{1}{4}$ or $\frac{2}{3}$

44. PERSEVERE The rule for factoring a difference of cubes is shown below. Use this rule to factor $40x^5 - 135x^2y^3$.

$$a^3 - b^3 = (a - b)(a^2 + ab + b^2)$$

45. CREATE Choose two integers. Then write an equation in standard form with those roots. How would the equation change if the signs of the two roots were switched?

46. ANALYZE Determine whether the following statement is *sometimes*, *always*, or *never* true. Explain your reasoning.

In a quadratic equation in standard form where a, b, and c are integers, if b is odd, then the quadratic cannot be a perfect square trinomial.

47. WRITE Explain how to factor a trinomial in standard form with $a > 1$.

Lesson 3-5

Solving Quadratic Equations by Completing the Square

Learn Solving Quadratic Equations by Using the Square Root Property

You can use square roots to solve equations like $x^2 - 49 = 0$. Remember that 7 and −7 are both square roots of 49 because $7^2 = 49$ and $(-7)^2 = 49$. Therefore, the solution set of $x^2 - 49 = 0$ is $\{-7, 7\}$. This can be written as $\{\pm 7\}$.

Key Concept • Square Root Property	
Words:	To solve a quadratic equation in the form $x^2 = n$, take the square root of each side.
Symbols:	For any number $n \geq 0$, if $x^2 = n$, then $x = \pm\sqrt{n}$.
Example:	$x^2 = 121$, $x = \pm\sqrt{121}$ or ± 11

Not all quadratic equations have solutions that are whole numbers. Roots that are irrational numbers may be written as exact answers in radical form or as approximate answers in decimal form when a calculator is used. Sometimes solutions of quadratic equations are not real numbers. Solutions that are complex numbers can be written in the form $a + bi$, where $b \neq 0$.

Example 1 Solve a Quadratic Equation with Rational Roots

Solve $x^2 - 4x + 4 = 25$ by using the Square Root Property.

$x^2 - 4x + 4 = 25$	Original equation
$(x - 2)^2 = 25$	Factor.
$x - 2 = \pm\sqrt{25}$	Square Root Property
$x - 2 = \pm 5$	$25 = 5(5)$ or $-5(-5)$
$x = 2 \pm 5$	Add 2 to each side.
$x = 2 + 5$ or $x = 2 - 5$	Write as two equations.
$x = 7 \quad x = -3$	Simplify.

The solution set is $\{x \mid x = -3, 7\}$.

Check

Solve $x^2 - 38x + 361 = 576$ by using the Square Root Property.

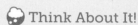

Go Online You can complete an Extra Example online.

Today's Goal

- Solve quadratic equations by using the Square Root Property.
- Complete the square in quadratic expressions to solve quadratic equations.
- Complete the square in a quadratic function to interpret key features of its graph.

Today's Vocabulary

completing the square
vertex form
projectile motion problems

Think About It!

How can you determine whether an equation of the form $x^2 = n$ will have an answer that is a whole number?

Study Tip

Square Roots When using the Square Root Property, remember to include the ± before the radical.

Go Online An alternate method is available for this example.

Example 2 Solve a Quadratic Equation with Irrational Roots

Solve $x^2 + 24x + 144 = 192$ by using the Square Root Property.

$x^2 + 24x + 144 = 192$	Original equation
$(x + 12)^2 = 192$	Factor.
$x + 12 = \pm\sqrt{192}$	Square Root Property
$x + 12 = \pm 8\sqrt{3}$	$\sqrt{192} = 8\sqrt{3}$
$x = -12 \pm 8\sqrt{3}$	Subtract 12 from each side.
$x = -12 + 8\sqrt{3}$ or $-12 - 8\sqrt{3}$	Write as two equations.
$x \approx 1.86, -25.86$	Use a calculator.

The exact solutions are $-12 - 8\sqrt{3}$ and $-12 + 8\sqrt{3}$. The approximate solutions are -25.86 and 1.86.

Watch Out!

Perfect Squares The constant, 192, on the right side of the equation is not a perfect square. This means that the roots will be irrational numbers.

Example 3 Solve a Quadratic Equation with Complex Solutions

Solve $2x^2 - 92x + 1058 = -72$ by using the Square Root Property.

$2x^2 - 92x + 1058 = -72$	Original equation
$x^2 - 46x + 529 = -36$	Divide each side by 2.
$(x - 23)^2 = -36$	Factor.
$x - 23 = \pm\sqrt{-36}$	Square Root Property
$x - 23 = \pm 6i$	$\sqrt{-36} = 6i$
$x = 23 \pm 6i$	Add 23 to each side.
$x = 23 + 6i$ or $23 - 6i$	Write as two equations.

The solutions are $23 + 6i$ and $23 - 6i$.

Think About It!

Can you solve a quadratic equation by completing the square if the coefficient of the x^2-term is not 1? Justify your argument.

Explore Using Algebra Tiles to Complete the Square

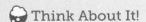

Online Activity Use algebra tiles to complete the Explore.

INQUIRY How does forming a square to create a perfect square trinomial help you solve quadratic equations?

Go Online You can complete an Extra Example online.

Learn Solving Quadratic Equations by Completing the Square

All quadratic equations can be solved by using the properties of equality to manipulate the equation until one side is a perfect square. This process is called **completing the square**.

Key Concept • Completing the Square

Words: To complete the square for any quadratic expression of the form $x^2 + bx$, follow the steps below.

Step 1 Find one half of b, the coefficient of x.
Step 2 Square the result in Step 1.
Step 3 Add the result of Step 2 to $x^2 + bx$.

Symbols: $x^2 + bx + \left(\frac{b}{2}\right)^2 = \left(x + \frac{b}{2}\right)^2$

To solve an equation of the form $x^2 + bx + c = 0$ by completing the square, first subtract c from each side of the equation. Then add $\left(\frac{b}{2}\right)^2$ to each side of the equation and solve for x.

Talk About It

For an equation in the form $ax^2 + bx + c$, if a and b are real numbers, can the value of c ever be negative? Explain your reasoning.

Example 4 Complete the Square

Find the value of c that makes $x^2 - 7x + c$ a perfect square. Then write the expression as the square of a binomial.

Step 1 Find one half of -7. $\frac{-7}{2} = -3.5$

Step 2 Square the result from Step 1. $(-3.5)^2 = 12.25$

Step 3 Add the result from Step 2 to $x^2 - 7x$. $x^2 - 7x + 12.25$

The expression $x^2 - 7x + 12.25$ can be written as $(x - 3.5)^2$.

Example 5 Solve by Completing the Square

Solve $x^2 + 18x - 4 = 0$ by completing the square.

$x^2 + 18x - 4 = 0$	Original equation
$x^2 + 18x = 4$	Add 4 to each side.
$x^2 + 18x + 81 = 4 + 81$	Add $\left(\frac{b}{2}\right)^2$ to each side.
$(x + 9)^2 = 85$	Factor.
$x + 9 = \pm\sqrt{85}$	Square Root Property
$x = -9 \pm \sqrt{85}$	Subtract 9 from each side.
$x = -9 + \sqrt{85}$ or	Write as two equations.
$x = -9 - \sqrt{85}$	
$x \approx 0.22$, or -18.22	Simplify.

The solution set is $\left\{x \mid x = -9 - \sqrt{85}, -9 + \sqrt{85}\right\}$.

Think About It!

Why do we first add 4 to each side?

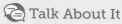

Go Online to see Example 6.

Go Online You can complete an Extra Example online.

Go Online for Example 6.

Example 7 Solve When a Is Not 1

Solve $4x^2 - 12x - 27 = 0$ by completing the square.

$4x^2 - 12x - 27 = 0$	Original equation
$x^2 - 3x - \frac{27}{4} = 0$	Divide each side by 4.
$x^2 - 3x = \frac{27}{4}$	Add $\frac{27}{4}$ to each side.
$x^2 - 3x + \frac{9}{4} = \frac{27}{4} + \frac{9}{4}$	Add $\left(\frac{b}{2}\right)^2$ or $\frac{9}{4}$ to each side.
$\left(x - \frac{3}{2}\right)^2 = 9$	Factor.
$x - \frac{3}{2} = \pm 3$	Square Root Property
$x = \frac{3}{2} \pm 3$	Add $\frac{3}{2}$ to each side.
$x = \frac{3}{2} + 3$ or $x = \frac{3}{2} - 3$	Write as two equations.
$x = \frac{9}{2}$ $\quad x = -\frac{3}{2}$	Simplify.

The solution set is $\left\{x \mid x = -\frac{3}{2}, \frac{9}{2}\right\}$.

Check

Solve $6x^2 - 21x + 9 = 0$ by completing the square.

Think About It!
Compare and contrast the solutions of this equation and the ones in the previous example. Explain.

Example 8 Solve Equations with Imaginary Solutions

Solve $3x^2 - 72x + 465 = 0$ by completing the square.

$3x^2 - 72x + 465 = 0$	Original equation
$x^2 - 24x + 155 = 0$	Divide each side by 3.
$x^2 - 24x = -155$	Subtract 155 from each side.
$x^2 - 24x + 144 = -155 + 144$	Add $\left(\frac{b}{2}\right)^2$ to each side.
$(x - 12)^2 = -11$	Factor.
$x - 12 = \pm\sqrt{-11}$	Square Root Property
$x - 12 = \pm i\sqrt{11}$	$\sqrt{-1} = i$
$x = 12 \pm i\sqrt{11}$	Add 12 to each side.
$x = 12 + i\sqrt{11}$ or $12 - i\sqrt{11}$	Write as two equations.

The solution set is $\left\{x \mid x = 12 + i\sqrt{11}, 12 - i\sqrt{11}\right\}$.

Go Online You can complete an Extra Example online.

Learn Quadratic Functions in Vertex Form

When a function is given in standard form, $y = ax^2 + bx + c$, you can complete the square to write it in vertex form.

Key Concept • Vertex Form of a Quadratic Function	
Words: The vertex form of a quadratic function is $y = a(x - h)^2 + k$.	
Symbols: Standard Form $y = ax^2 + bx + c$	Vertex Form $y = a(x - h)^2 + k$ The vertex is (h, k).
Example: Standard Form $y = 2x^2 + 12x + 16$	Vertex Form $y = 2(x + 3)^2 - 2$ The vertex is $(-3, -2)$.

After completing the square and writing a quadratic function in vertex form, you can analyze key features of the function. The vertex is (h, k) and $x = h$ is the equation of the axis of symmetry. The shape of the parabola and the direction that it opens are determined by a. The value of k is a minimum value if $a > 0$ or a maximum value if $a < 0$.

The path that an object travels when influenced by gravity is called a *trajectory*, and trajectories can be modeled by quadratic functions. The formula below relates the height of the object $h(t)$ and time t, where g is acceleration due to gravity, v is the initial velocity of the object, and h_0 is the initial height of the object.

$$h(t) = -\frac{1}{2}gt^2 + vt + h_0$$

The acceleration due to gravity g is 9.8 meters per second squared or 32 feet per second squared. Problems that involve objects being thrown or dropped are called **projectile motion problems**.

Think About It!

What is the minimum value of $y = 2(x - 3)^2 - 1$? How do you know that this value is a minimum?

Example 9 Write Functions in Vertex Form

Write $y = -x^2 - 12x - 9$ in vertex form.

$y = -x^2 - 12x - 9$	Original function
$y = (-x^2 - 12x) - 9$	Group $ax^2 + bx$.
$y = -(x^2 + 12x) - 9$	Factor out –1.
$y = -(x^2 + 12x + 36) - 9 - (-1)(36)$	Complete the square.
$y = -(x + 6)^2 + 27$	Simplify.

Watch Out!

The coefficient of the x^2-term must be 1 before you can complete the square.

Check

Write each function in vertex form.

a. $y = x^2 + 8x - 3$

$y = (x$ __?__ $)^2$ __?__

b. $y = -3x^2 - 6x - 5$

$y =$ __?__ $(x$ __?__ $)^2$ __?__

Go Online You can complete an Extra Example online.

Think About It!
How would your equation for the axis of symmetry change if the vertex form of the equation was $y = 3(x + 2)^2 - 7$? Justify your argument.

Example 10 Determine the Vertex and Axis of Symmetry

Consider $y = 3x^2 - 12x + 5$.

Part A Write the function in vertex form.

$y = 3x^2 - 12x + 5$	Original equation
$y = (3x^2 - 12x) + 5$	Group $ax^2 + bx$.
$y = 3(x^2 - 4x) + 5$	Factor.
$y = 3(x^2 - 4x + 4) + 5 - 3(4)$	Complete the square.
$y = 3(x - 2)^2 - 7$	Simplify.

Part B Find the axis of symmetry.

The axis of symmetry is $x = h$ or $x = 2$.

Part C Find the vertex, and determine if it is a maximum or minimum.

The vertex is (h, k) or $(2, -7)$. Because $a > 0$, this is a minimum.

Example 11 Model with a Quadratic Function

Think About It!
If the firework reaches a height of 241 feet after 3 seconds, what is the height of the firework after 5 seconds? Justify your answer.

FIREWORKS **If a firework is launched 1 foot off the ground at a velocity of 128 feet per second, write a function for the situation. Then find and interpret the axis of symmetry and vertex.**

Step 1 Write the function.

$h(t) = -\frac{1}{2}gt^2 + vt + h_0$	Function for projectile motion
$h(t) = -\frac{1}{2}(32)t^2 + 128t + 1$	$g = 32\,\frac{\text{ft}}{\text{s}^2}$, $v = 128\,\frac{\text{ft}}{\text{s}}$, $h_0 = 1$ ft
$h(t) = -16t^2 + 128t + 1$	Simplify.

Step 2 Rewrite the function in vertex form.

$h(t) = (-16t^2 + 128t) + 1$	Group $ax^2 + bx$.
$h(t) = -16(t^2 - 8t) + 1$	Factor.
$h(t) = -16(t^2 - 8t + 16) + 1 - 16(-16)$	Complete the square.
$h(t) = -16(t - 4)^2 + 257$	Simplify.

Step 3 Find and interpret the axis of symmetry.

Because the axis of symmetry divides the function into two equal halves, the firework will be at the same height after 2 seconds as it is after 6 seconds.

Step 4 Find and interpret the vertex.

The vertex is the maximum of the function because $a < 1$. So the firework reached a maximum height of 257 feet after 4 seconds.

Study Tip
Vertex When you interpret the vertex of a function, it is important to also consider the value of a when the function is in vertex or standard form. The value of a will tell you whether the vertex is a maximum or minimum.

Go Online You can complete an Extra Example online.

Practice

Go Online You can complete your homework online.

Examples 1–3

Solve each equation by using the Square Root Property.

1. $x^2 - 18x + 81 = 49$ **2.** $x^2 + 20x + 100 = 64$ **3.** $9x^2 - 12x + 4 = 4$

4. $4x^2 + 4x + 1 = 16$ **5.** $4x^2 - 28x + 49 = 64$ **6.** $16x^2 + 24x + 9 = 81$

7. $36x^2 + 12x + 1 = 18$ **8.** $25x^2 + 40x + 16 = 28$ **9.** $25x^2 + 20x + 4 = 75$

10. $36x^2 + 48x + 16 = 12$ **11.** $25x^2 - 30x + 9 = 96$ **12.** $4x^2 - 20x + 25 = 32$

13. $2x^2 - 20x + 50 = -128$ **14.** $2x^2 - 24x + 72 = -162$ **15.** $2x^2 + 28x + 98 = -200$

16. $x^2 - 8x + 16 = -36$ **17.** $3x^2 + 24x + 48 = -108$ **18.** $3x^2 - 24x + 48 = -363$

Example 4

Find the value of c that makes each trinomial a perfect square. Then write the trinomial as the square of a binomial.

19. $x^2 + 10x + c$ **20.** $x^2 - 14x + c$ **21.** $x^2 + 24x + c$

22. $x^2 + 5x + c$ **23.** $x^2 - 9x + c$ **24.** $x^2 - x + c$

Examples 5 and 6

Solve each equation by completing the square.

25. $x^2 - 13x + 36 = 0$ **26.** $x^2 + x - 6 = 0$ **27.** $x^2 - 4x - 13 = 0$

28. $x^2 + 3x - 6 = 0$ **29.** $x^2 - x - 3 = 0$ **30.** $x^2 - 8x - 65 = 0$

31. When the dimensions of a cube are reduced by 4 inches on each side, the surface area of the new cube is 864 square inches. What were the dimensions of the original cube?

Examples 7 and 8

Solve each equation by completing the square.

32. $2x^2 - 8x - 24 = 0$

33. $2x^2 - 3x + 1 = 0$

34. $2x^2 - 13x - 7 = 0$

35. $25x^2 + 40x - 9 = 0$

36. $2x^2 + 7x - 4 = 0$

37. $3x^2 + 2x - 1 = 0$

38. $x^2 - 4x + 12 = 0$

39. $2x^2 - 3x + 5 = 0$

40. $2x^2 + 5x + 7 = 0$

41. $x^2 - 2x + 3 = 0$

42. $x^2 = -24$

43. $x^2 - 2x + 4 = 0$

Examples 9 and 10

Write each function in vertex form. Find the axis of symmetry. Then find the vertex, and determine if it is a *maximum* or *minimum*.

44. $y = x^2 + 2x - 5$

45. $y = x^2 + 6x + 1$

46. $y = -x^2 + 4x + 2$

47. $y = -x^2 - 8x - 5$

48. $y = 2x^2 + 4x + 3$

49. $y = 3x^2 + 6x - 1$

Example 11

50. FIREWORKS The height of a firework at an amusement park celebration can be modeled by a quadratic function. Suppose the firework is launched from a platform 2 feet off the ground at a velocity of 96 feet per second. Use $h(t) = -\frac{1}{2}gt^2 + vt + h_0$, where $g = 32\ \frac{\text{ft}}{\text{s}^2}$.

a. Write a function to represent this situation.

b. Rewrite the function in vertex form.

c. Find the axis of symmetry and the vertex and interpret their meaning in the context of the situation.

51. DIVING Malik is participating in a diving championship. For each of his dives, his height above the water can be modeled by a quadratic function. The diving board is 7.5 meters above the water and Malik jumps with a velocity of 4.18 meters per second. Use $h(t) = -\frac{1}{2}gt^2 + vt + h_0$, where $g = 9.8\ \frac{\text{m}}{\text{s}^2}$.

a. Write a function in vertex form to represent this situation.

b. Find the axis of symmetry and the vertex and interpret their meaning in the context of the situation.

Mixed Exercises

PRECISION **Solve each equation. Round to the nearest hundredth, if necessary.**

52. $4x^2 - 28x + 49 = 5$

53. $9x^2 + 30x + 25 = 11$

54. $x^2 + x + \frac{1}{3} = \frac{2}{3}$

55. $x^2 + 1.2x + 0.56 = 0.91$

56. $x^2 + 0.7x + 4.1225 = 0$

57. $x^2 - 3.2x = -3.46$

58. $x^2 - 1.8x + 11.24 = 2.43$

59. $-0.3x^2 - 0.78x - 5.514 = 0$

60. $1.1x^2 - 8.8x + 22 = 2.2$

61. FREE FALL A rock falls from the top of a cliff that is 25.8 meters high. Use the formula $h(t) = -\frac{1}{2}gt^2 + vt + h_0$, where $g = 9.8\ \frac{\text{m}}{\text{s}^2}$, to write a quadratic function that models the situation. Determine to the nearest tenth of a second the amount of time it takes the rock to strike the ground. Explain your reasoning.

62. REACTION TIME Tela was eating lunch when she saw her friend Jori approach. The room was crowded and Jori had to lift his tray to avoid obstacles. Suddenly, a glass on Jori's lunch tray tipped and fell off the tray. Tela lunged forward and managed to catch the glass just before it hit the ground. The height h, in feet, of the glass t seconds after it was dropped is given by $h = -16t^2 + 4.5$. If Tela caught the glass when it was six inches off the ground, how long was the glass in the air before she caught it?

63. INVESTMENTS The amount of money A in an account in which P dollars are invested for 2 years is given by the formula $A = P(1 + r)^2$, where r is the interest rate compounded annually. If an investment of \$800 in the account grows to \$882 in two years, at what interest rate was it invested, to the nearest percent?

64. INVESTMENTS Niyati invested \$1000 in a savings account with interest compounded annually. After two years the balance in the account is \$1210. Use the compound interest formula $A = P(1 + r)^t$ to find the annual interest rate, to the nearest percent.

Write each function in vertex form. Then find the vertex.

65. $y = x^2 - 10x + 28$

66. $y = x^2 + 16x + 65$

67. $y = x^2 - 20x + 104$

68. $y = x^2 - 8x + 17$

69. AUDITORIUM SEATING The seats in an auditorium are arranged in a square grid pattern. There are 45 rows and 45 columns of chairs. For a special concert, organizers decide to increase seating by adding n rows and n columns to make a square pattern of seating $45 + n$ seats on a side.

a. How many seats are added in the expansion?

b. What is n if organizers wish to add 1000 seats?

70. DECK DESIGN The Rayburns current deck is 12 feet by 12 feet. They decide they would like to expand their deck and maintain its square shape. How much larger will each side need to be for the deck to have an area of 200 square feet?

71. VOLUME A piece of sheet metal has a length that is three times its width. It is used to make a box with an open top by cutting out 2-inch by 2-inch squares from each corner, then folding up the sides.

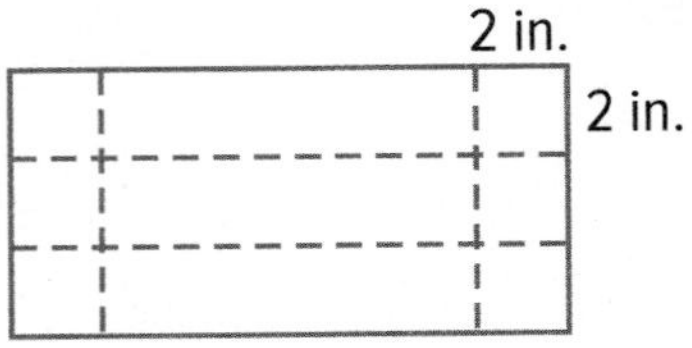

a. Define variables and write a quadratic function that represents the volume of the box in cubic inches.

b. What are the dimensions of the metal sheet that results in a box with a volume of 1.125 cubic feet?

72. CONSTRUCT ARGUMENTS Explain why the equation for the axis of symmetry for a quadratic function $y = ax^2 + bx + c$ is $x = -\frac{b}{2a}$.

73. FIND THE ERROR Alonso and Aika are solving $x^2 + 8x - 20 = 0$ by completing the square. Is either of them correct? Explain your reasoning.

Alonso	Aika
$x^2 + 8x - 20 = 0$	$x^2 + 8x - 20 = 0$
$x^2 + 8x = 20$	$x^2 + 8x = 20$
$x^2 + 8x + 16 = 20 + 16$	$x^2 + 8x + 16 = 20$
$(x + 4)^2 = 36$	$(x + 4)^2 = 20$
$x + 4 = \pm 6$	$x + 4 = \pm\sqrt{20}$
$x = 2$ or -10	$x = -4 \pm \sqrt{20}$

74. PERSEVERE Solve $x^2 + bx + c = 0$ by completing the square. Your answer will be an expression for x in terms of b and c.

75. ANALYZE Without solving, determine how many unique solutions there are for each equation. Are they *rational*, *real*, or *complex*? Justify your reasoning.

a. $(x + 2)^2 = 16$ **b.** $(x - 2)^2 = 16$ **c.** $-(x - 2)^2 = 16$

d. $36 - (x - 2)^2 = 16$ **e.** $16(x + 2)^2 = 0$ **f.** $(x + 4)^2 = (x + 6)^2$

76. CREATE Write a perfect square trinomial equation in which the linear coefficient is negative and the constant term is a fraction. Then solve the equation.

77. WRITE Explain what is means to complete the square. Describe each step.

Lesson 3-6

Using the Quadratic Formula and the Discriminant

Learn Using the Quadratic Formula

To solve any quadratic equation, you can use the Quadratic Formula.

Key Concept • Quadratic Formula

The solutions of a quadratic equation of the form $ax^2 + bx + c = 0$, where $a \neq 0$, are given by the following formula.

$$x = \frac{-b \pm \sqrt{b^2 - 4ac}}{2a}$$

Go Online You can see how the Quadratic Formula is derived.

Example 1 Real Roots, c Is Positive

CONTEST **At the World Championship Punkin Chunkin contest, pumpkins are launched hundreds of yards. The path of a pumpkin can be modeled by $h = -16t^2 + 124.4t + 42$, where h is the height in feet and t is the number of seconds after launch.**

Part A Use the Quadratic Formula to solve $0 = -16t^2 + 124.4t + 42$.

$t = \frac{-b \pm \sqrt{b^2 - 4ac}}{2a}$ — Quadratic Formula

$= \frac{-124.4 \pm \sqrt{(124.4)^2 - 4(-16)(42)}}{2(-16)}$ — $a = -16$, $b = 124.4$, $c = 42$

$= \frac{-124.4 \pm \sqrt{15{,}475.36 + 2688}}{-32}$ — Square and multiply.

$= \frac{-124.4 \pm \sqrt{18{,}163.36}}{-32}$ — Add.

$= \frac{124.4 + \sqrt{18{,}163.36}}{32}$ or $\frac{124.4 - \sqrt{18{,}163.36}}{32}$ — Multiply by $\frac{-1}{-1}$.

The approximate solutions are 8.1 seconds and −0.32 second.

Part B Interpret the roots.

The negative root does not make sense in this context because the pumpkin launches at 0 seconds. The pumpkin lands after 8.1 seconds.

Check

DIVING A diver jumps from a diving platform that is 10 feet high. Her arc can be modeled by $y = -4.9x^2 + 2.5x + 10$, where y is her height in meters and x is time in seconds.

Part A Solve $0 = -4.9x^2 + 2.5x + 10$.

Part B Interpret the roots.

The diver enters the water after approximately ___?___ seconds.

Today's Goals
- Solve quadratic equations by using the Quadratic Formula.
- Determine the number and type of roots of a quadratic equation.

Today's Vocabulary
discriminant

Think About It!
Why are the roots not at 0 and 8.1 seconds, when the pumpkin is launched and when it lands?

Go Online
You can complete an Extra Example online.

Example 2 Real Roots, c Is Negative

Solve $x^2 + 4x - 17 = 0$ by using the Quadratic Formula.

$$x = \frac{-b \pm \sqrt{b^2 - 4ac}}{2a}$$ Quadratic Formula

$$= \frac{-4 \pm \sqrt{(4)^2 - 4(1)(-17)}}{2(1)}$$ $a = 1, b = 4, c = -17$

$$= \frac{-4 \pm \sqrt{84}}{2}$$ Simplify.

$$= \frac{-4 \pm \sqrt{4} \cdot \sqrt{21}}{2}$$ Product Property of Square Roots

$$= \frac{-4 \pm 2\sqrt{21}}{2}$$ $\sqrt{4} = 2$

$$= -2 \pm \sqrt{21}$$ Divide the numerator and denominator by 2.

Check

Solve $3x^2 - 5x - 1 = 0$ by using the Quadratic Formula.

Example 3 Complex Roots

Solve $5x^2 + 8x + 11 = 0$ by using the Quadratic Formula.

$$x = \frac{-b \pm \sqrt{b^2 - 4ac}}{2a}$$ Quadratic Formula

$$= \frac{-8 \pm \sqrt{8^2 - 4(5)(11)}}{2(5)}$$ $a = 5, b = 8, c = 11$

$$= \frac{-8 \pm \sqrt{-156}}{10}$$ Simplify.

$$= \frac{-8 \pm \sqrt{-1} \cdot \sqrt{4} \cdot \sqrt{39}}{10}$$ Product Property of Square Roots

$$= \frac{-8 \pm 2i\sqrt{39}}{10}$$ Write as a complex number.

$$= \frac{-4 \pm i\sqrt{39}}{5}$$ Divide the numerator and denominator by 2.

Check

Solve $9x^2 - 3x + 18 = 0$ by using the Quadratic Formula.

Go Online You can complete an Extra Example online.

Explore The Discriminant

Online Activity Use graphing technology to complete the Explore.

> **INQUIRY** How does the discriminant of a quadratic equation relate to its roots?

Learn Using the Discriminant

In the Quadratic Formula, the **discriminant** is the expression under the radical sign, $b^2 - 4ac$. The value of the discriminant can be used to determine the number and type of roots of a quadratic equation.

Key Concept • Discriminant

Consider $ax^2 + bx + c = 0$, where a, b, and c are rational numbers and $a \neq 0$.

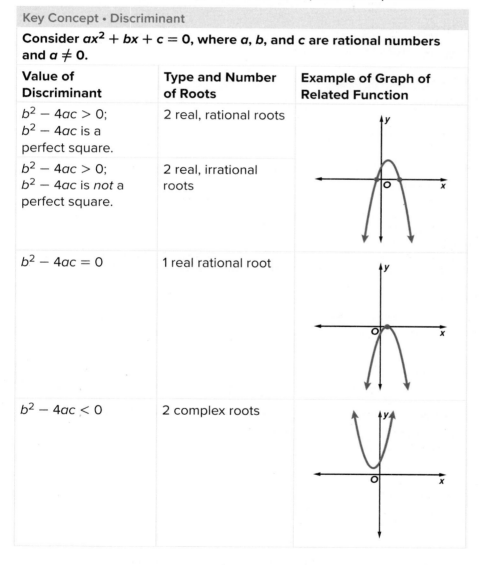

Value of Discriminant	Type and Number of Roots	Example of Graph of Related Function
$b^2 - 4ac > 0$; $b^2 - 4ac$ is a perfect square.	2 real, rational roots	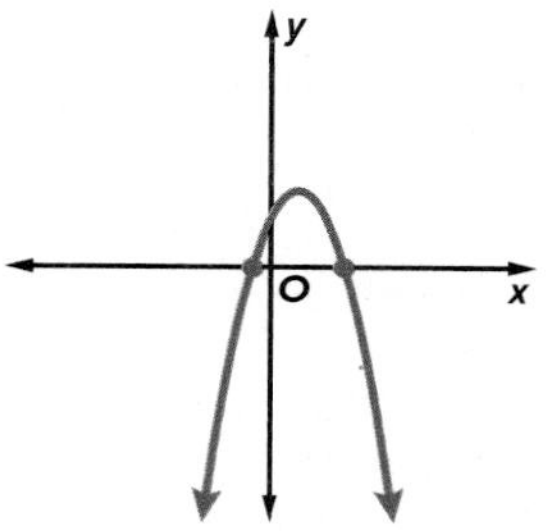
$b^2 - 4ac > 0$; $b^2 - 4ac$ is *not* a perfect square.	2 real, irrational roots	
$b^2 - 4ac = 0$	1 real rational root	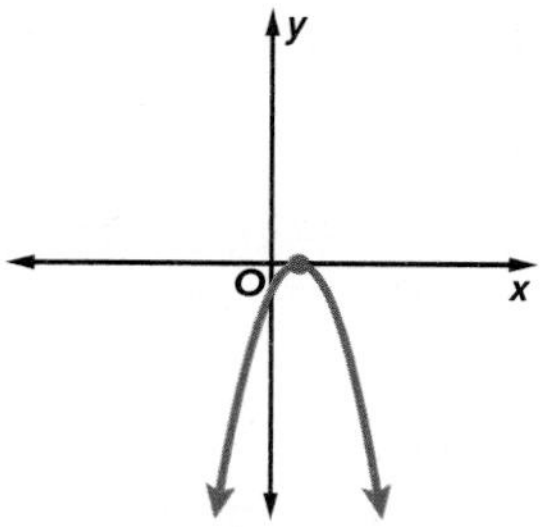
$b^2 - 4ac < 0$	2 complex roots	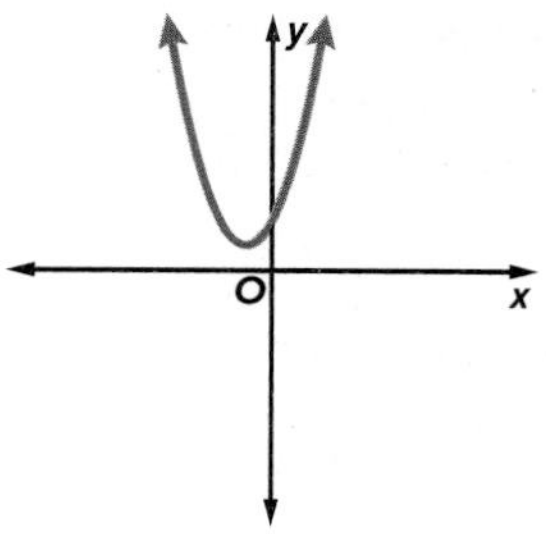

Talk About It

Why are the roots of a quadratic equation complex if the discriminant is negative?

Go Online You can complete an Extra Example online.

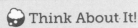

Think About It!

Is it possible for a quadratic equation to have zero real or complex roots?

Example 4 The Discriminant, Real Roots

Examine $2x^2 - 10x + 7 = 0$.

Part A Find the value of the discriminant for $2x^2 - 10x + 7 = 0$.

$a = 2 \qquad b = -10 \qquad c = 7$

$$\begin{aligned} b^2 - 4ac &= (-10)^2 - 4(2)(7) \\ &= 100 - 56 \\ &= 44 \end{aligned}$$

Part B Describe the number and type of roots for the equation.

The discriminant is nonzero, so there are two roots. The discriminant is positive and not a perfect square, so the roots are irrational.

Check

Examine $2x^2 + 8x + 8 = 0$.

Part A Find the value of the discriminant for $2x^2 + 8x + 8 = 0$.

$b^2 - 4ac =$ ___?___

Part B Describe the number and type of roots for the equation.

There is/are ___?___ root(s).

Example 5 The Discriminant, Complex Roots

Examine $-5x^2 + 10x - 15 = 0$.

Part A Find the value of the discriminant for $-5x^2 + 10x - 15 = 0$.

$a = -5 \qquad b = 10 \qquad c = -15$

$$\begin{aligned} b^2 - 4ac &= (10)^2 - 4(-5)(-15) \\ &= 100 - 300 \\ &= -200 \end{aligned}$$

Part B Describe the number and type of roots for the equation.

The discriminant is nonzero, so there are two roots. The discriminant is negative, so the roots are complex.

Check

Examine $10x^2 - 4x + 7 = 0$.

Part A Find the value of the discriminant for $10x^2 - 4x + 7 = 0$.

$b^2 - 4ac =$ ___?___

Part B Describe the number and type of roots for the equation.

There is/are ___?___ root(s).

Go Online You can complete an Extra Example online.

Go Online
to practice what you've learned in Lessons 3-2 and 3-4 through 3-6.

Practice

Go Online You can complete your homework online.

Example 1

Solve each equation by using the Quadratic Formula.

1. $x^2 + 8x + 15 = 0$

2. $x^2 - 18x + 72 = 0$

3. $12x^2 - 22x + 6 = 0$

4. $4x^2 - 6x = -2$

5. $x^2 + 8x + 5 = 0$

6. $-8x^2 + 4x = -5$

7. FOOTBALL A quarterback throws a football to a receiver. The path of a football can be modeled by the quadratic function $h = -16t^2 + 45t + 4$, where h is the height in feet and t is the number of seconds after the football is thrown. If the ball is overthrown and the receiver does not touch the ball, how long will it take the football to hit the ground?

Examples 2 and 3

Solve each equation by using the Quadratic Formula.

8. $x^2 + 2x - 35 = 0$

9. $4x^2 + 19x - 5 = 0$

10. $2x^2 - x - 15 = 0$

11. $3x^2 + 5x = 2$

12. $x^2 + x - 8 = 0$

13. $8x^2 + 5x - 1 = 0$

14. $x^2 - x - 5 = 0$

15. $16x^2 - 24x - 25 = 0$

16. $x^2 - 6x + 21 = 0$

17. $x^2 + 25 = 0$

18. $3x^2 + 36 = 0$

19. $8x^2 - 4x + 1 = 0$

20. $2x^2 + 2x + 3 = 0$

21. $x^2 - 14x + 53 = 0$

22. $4x^2 + 2x + 9 = 0$

23. $3x^2 - 6x + 11 = 0$

Examples 4 and 5

Find the value of the discriminant for each quadratic equation. Then describe the number and type of roots for the equation.

24. $x^2 - 8x + 16 = 0$

25. $x^2 - 11x - 26 = 0$

26. $3x^2 - 2x = 0$

27. $20x^2 + 7x - 3 = 0$

28. $5x^2 - 6 = 0$

29. $x^2 - 6 = 0$

30. $x^2 + 8x + 13 = 0$

31. $5x^2 - x - 1 = 0$

32. $x^2 - 2x - 17 = 0$

33. $x^2 + 49 = 0$

34. $x^2 - x + 1 = 0$

35. $2x^2 - 3x = -2$

Mixed Exercises

REGULARITY **Describe the discriminant of the related equation of each graph. Then determine the type and number of roots.**

36.

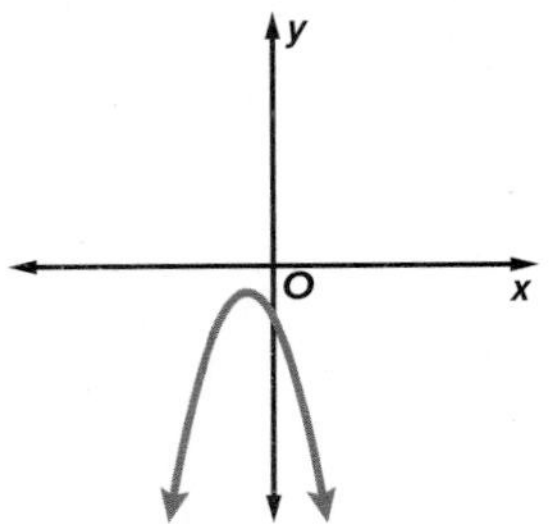

37.

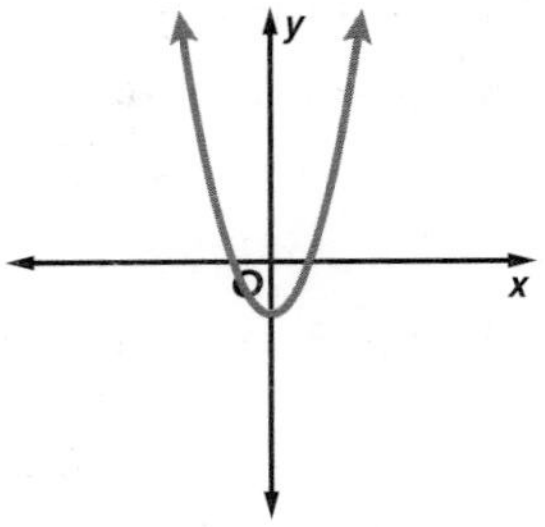

38.

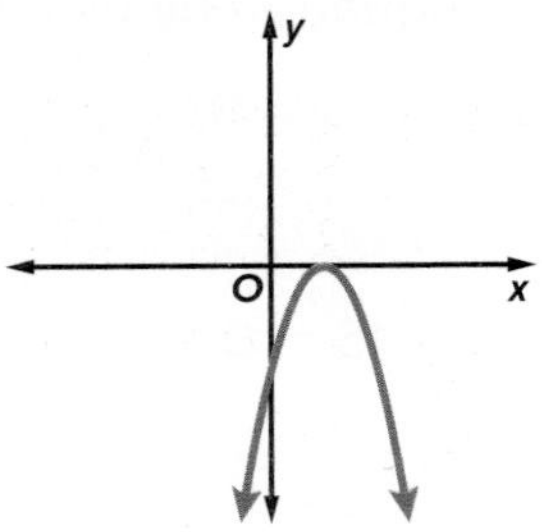

Use the discriminant to describe the number and type of roots for each equation. Then solve each equation by using the Quadratic Formula.

39. $4x^2 - 4x + 17 = 0$

40. $8x - 1 = 4x^2$

41. $7x^2 - 5x = 0$

42. $x^2 + 10x + 24 = 0$

43. $x^2 - 11x + 24 = 0$

44. $12x^2 + 9x + 1 = 0$

45. $3x^2 - 16x + 16 = 0$

46. $r^2 - \frac{3r}{5} + \frac{2}{5} = 0$

47. $2x^2 + 10x + 11 = 0$

48. USE A MODEL The height $h(t)$ in feet of an object t seconds after it is propelled up from the ground with an initial velocity of 60 feet per second is modeled by the equation $h(t) = -16t^2 + 60t$. When will the object be at a height of 56 feet?

49. SPORTS Natalya Lisovskaya set the women's shot put world record of 22.63 meters. Her throw can be modeled by $h = -4.9t^2 + 13.7t + 1.6$, where t is time in seconds and h is the height in meters. About how long was the shot in the air?

50. STOPPING DISTANCE A car's stopping distance d is the sum of the distance traveled during the time it takes the driver to react and the distance traveled while braking. This is represented as $d = vt + \frac{v^2}{2\mu g}$, where v is the initial velocity in feet per second, t is the driver's reaction time in seconds, μ is the coefficient of friction, and g is acceleration due to gravity. Use $g = 32$ ft/s^2.

a. Assume $\mu = 0.8$ for rubber tires on dry pavement and the average reaction time of 1.5 seconds. Copy and complete the table. Round to the nearest tenth.

Velocity, v (ft/s)	15		55		
Stopping Distance, d (ft)		91.25		200.7	245

b. Make different assumptions. Copy and complete the table. Round to the nearest tenth. coefficient of friction μ = ___?___
reaction time t = ___?___

Velocity, v (ft/s)	15		55		
Stopping Distance, d (ft)		91.25		200.7	245

c. How did your different assumptions affect the data you found? Interpret the information in the context of the situation.

51. GEOMETRY A rectangular box has a square base and a height that is one more than 3 times the length of a side of the base. If the sides of the base are each increased by 2 inches and the height is increased by 3 inches, the volume of the box increases by 531 cubic inches. Define a variable and write an equation to represent the situation. Then find the dimensions of the original box.

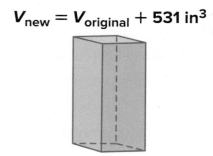

$$V_{new} = V_{original} + 531 \text{ in}^3$$

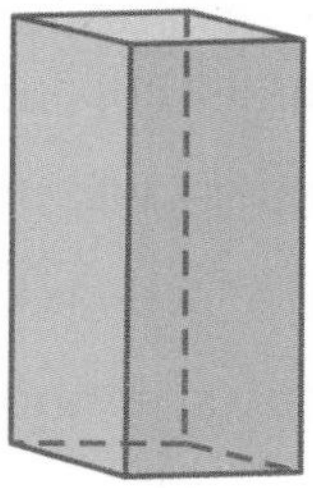

52. GAMES A carnival game has players hit a pad with a large rubber mallet. This fires a ball up a 20-foot vertical chute toward a target at the top. A prize is awarded if the ball hits the target. Explain how to find the initial velocity in feet per second for which the ball will fail to hit the target. Assume the height of the ball can be modeled by the function $h(t) = -16t^2 + vt$, where v is the initial velocity.

53. WHICH ONE DOESN'T BELONG? Use the discriminant to determine which of these equations is different from the others. Explain your reasoning.

$x^2 - 3x - 40 = 0$	$12x^2 - x - 6 = 0$	$12x^2 + 2x - 4 = 0$	$7x^2 + 6x + 2 = 0$

54. FIND THE ERROR Tama and Jonathan are determining the number of solutions of $3x^2 - 5x = 7$. Is either of them correct? Explain your reasoning.

Tama	Jonathan
$3x^2 - 5x = 7$ $b^2 - 4ac = (-5)^2 - 4(3)(7)$ $= -59$ Since the discriminant is negative, there are no real solutions.	$3x^2 - 5x = 7$ $3x^2 - 5x - 7 = 0$ $b^2 - 4ac = (-5)^2 - 4(3)(-7)$ $= 109$ Since the discriminant is positive, there are two real roots.

55. ANALYZE Determine whether each statement is *sometimes, always*, or *never* true. Explain your reasoning.

a. In a quadratic equation written in standard form, if a and c have different signs, then the solutions will be real.

b. If the discriminant of a quadratic equation is greater than 1, the two roots are real irrational numbers.

56. CREATE Sketch the corresponding graph and state the number and type of roots for each of the following.

a. $b^2 - 4ac = 0$

b. A quadratic function in which $f(x)$ never equals zero.

c. A quadratic function in which $f(a) = 0$ and $f(b) = 0$; $a \neq b$.

d. The discriminant is less than zero.

e. a and b are both solutions and can be represented as fractions.

57. PERSEVERE Find the value(s) of m in the quadratic equation $x^2 + x + m + 1 = 0$ such that it has one solution.

58. WRITE Describe three different ways to solve $x^2 - 2x - 15 = 0$. Which method do you prefer, and why?

Quadratic Inequalities

Explore Graphing Quadratic Inequalities

Online Activity Use graphing technology to complete the Explore.

INQUIRY How can you represent a quadratic inequality graphically?

Learn Graphing Quadratic Inequalities

You can graph quadratic inequalities in two variables by using the same techniques used to graph linear inequalities in two variables. A **quadratic inequality** is an inequality of the form $y > ax^2 + bx + c$, $y \geq ax^2 + bx + c$, $y < ax^2 + bx + c$, or $y \leq ax^2 + bx + c$.

Key Concept • Graphing Quadratic Inequalities

Step 1 Graph the related function.
Step 2 Test a point not on the parabola.
Step 3 Shade accordingly.

Example 1 Graph a Quadratic Inequality (< or ≤)

Graph $y \leq x^2 - 2x + 8$.

Step 1 Graph the related function.

Because the inequality is less than or equal to, the parabola should be solid.

Step 2 Test a point not on the parabola.

$y \leq x^2 - 2x + 8$	Original inequality
$0 \overset{?}{\leq} (0)^2 - 2(0) + 8$	$(x, y) = (0, 0)$
$0 \leq 8$	True

Shade the region that contains the point.

Step 3 Shade accordingly.

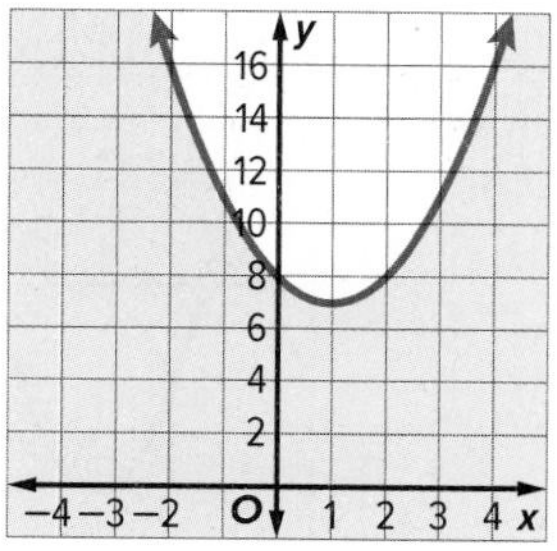

Go Online You can complete an Extra Example online.

Today's Goals

- Graph quadratic inequalities in two variables.
- Solve quadratic inequalities in two variables by graphing.

Today's Vocabulary
quadratic inequality

Think About It!
How do you know whether to make the parabola solid or dashed?

Study Tip
(0, 0) If (0, 0) is not a point on the parabola, then it is often the easiest point to test when determining which part of the graph to shade.

Example 2 Graph a Quadratic Inequality (> or ≥)

Graph $y > -5x^2 + 10x$.

Step 1 Graph the related function.

Because the inequality is greater than, the parabola should be dashed.

Step 2 Test a point not on the parabola.

Because (0, 0) is on the parabola, use (1, 0) as a test point.

$y > -5x^2 + 10x$	Original inequality
$0 \overset{?}{>} -5(1)^2 + 10(1)$	$(x, y) = (1, 0)$
$0 > 5$	False

So, (1, 0) is not a solution of the inequality.

Step 3 Shade accordingly.

Because (1, 0) is not a solution of the inequality, shade the region that does not contain the point.

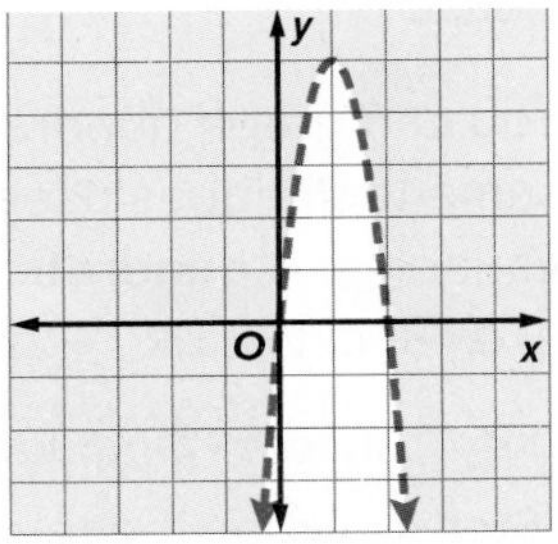

Learn Solving Quadratic Inequalities

Key Concept • Solving Quadratic Inequalities

$ax^2 + bx + c < 0$ Graph $y = ax^2 + bx + c$ and identify the x-values for which the graph lies *below* the x-axis. For ≤, include the x-intercepts in the solution.	$a > 0$ 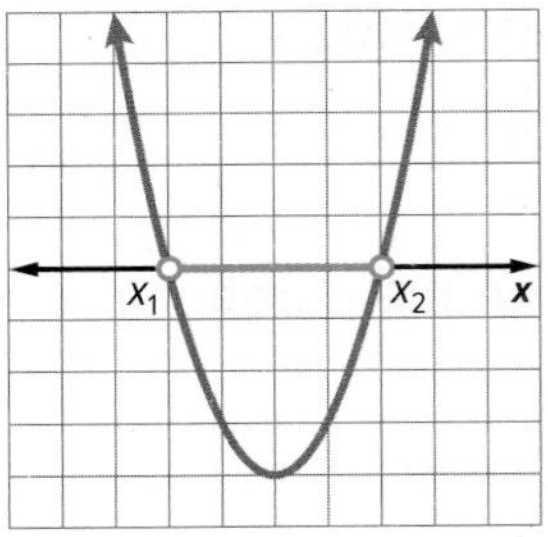 $\{x \mid x_1 < x < x_2\}$	$a < 0$ 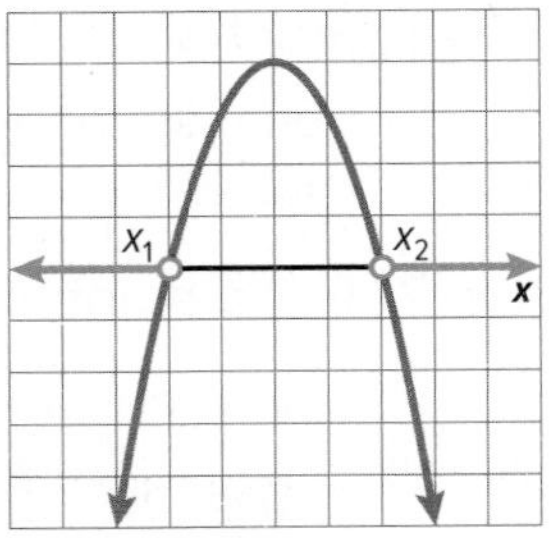 $\{x \mid x < x_1 \text{ or } x > x_2\}$
$ax^2 + bx + c > 0$ Graph $y = ax^2 + bx + c$ and identify the x-values for which the graph lies *above* the x-axis. For ≥, include the x-intercepts in the solution.	$a > 0$ 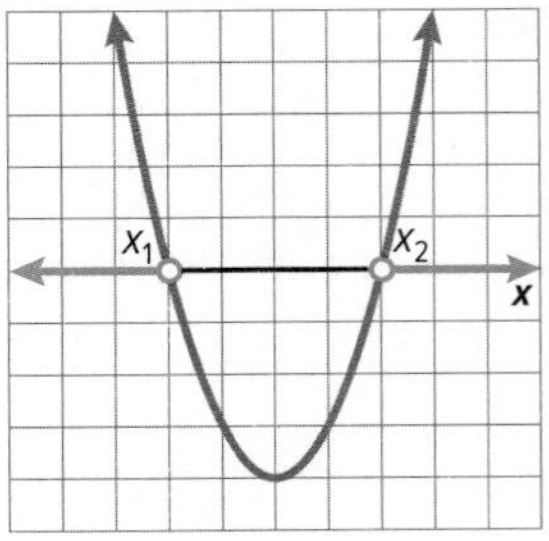 $\{x \mid x < x_1 \text{ or } x > x_2\}$	$a < 0$ 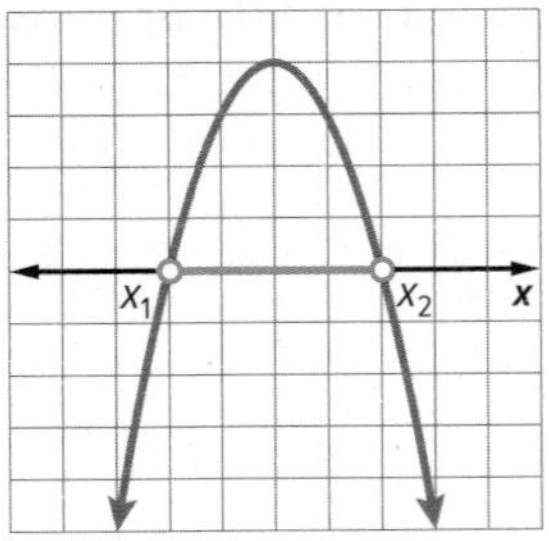 $\{x \mid x_1 < x < x_2\}$

Go Online You can complete an Extra Example online.

Example 3 Solve a Quadratic Inequality (< or ≤) by Graphing

Solve $x^2 + x - 6 < 0$ by graphing.

Because the quadratic expression is less than 0, the solution consists of x-values for which the graph of the related function lies *below* the x-axis. Begin by finding the zeros of the related function.

$x^2 + x - 6 = 0$	Related equation
$(x - 2)(x + 3) = 0$	Factor.
$x = 2$ or $x = -3$	Zero Product Property

Sketch the graph of a parabola that has x-intercepts at 2 and -3. The graph should open up because $a > 0$.

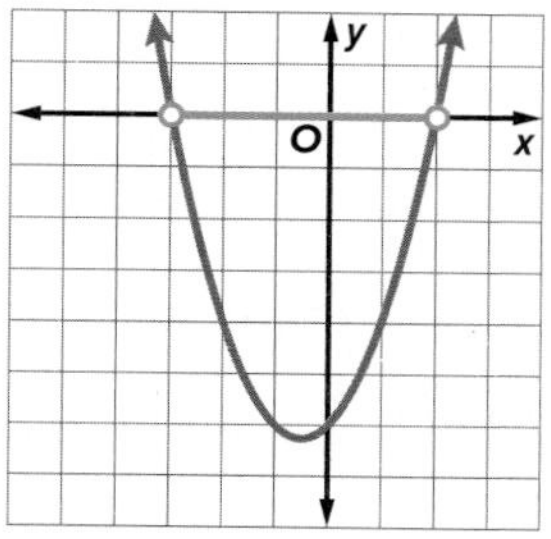

The graph lies below the x-axis between -3 and 2. Thus, the solution set is $\{x \mid -3 < x < 2\}$ or in interval notation $(-3, 2)$.

Think About It!

How could you check your solution?

Example 4 Solve a Quadratic Inequality (> or ≥) by Graphing

Solve $x^2 - 3x - 4 \geq 0$ by graphing.

Because the quadratic expression is greater than or equal to 0, the solution consists of x-values for which the graph of the related function lies *on* and *above* the x-axis. Begin by finding the zeros of the related function.

$x^2 - 3x - 4 = 0$	Related equation
$(x - 4)(x + 1) = 0$	Factor.
$x = 4$ or $x = -1$	Zero Product Property

Sketch the graph of a parabola that has x-intercepts at -1 and 4. The graph should open up because $a > 0$.

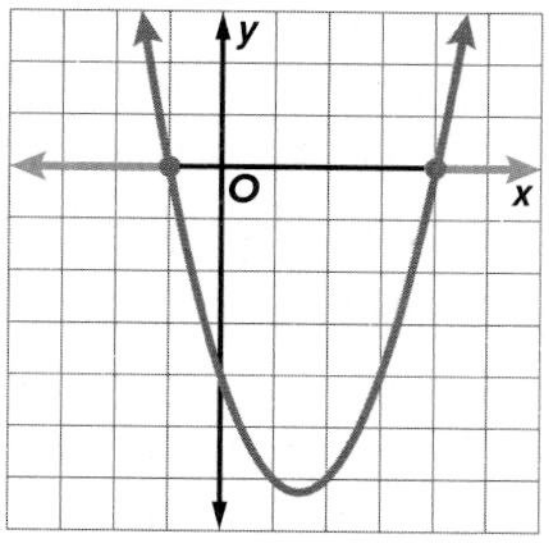

The graph lies above and on the x-axis when $x \leq -1$ or $x \geq 4$. Thus, the solution set is $\{x \mid x \leq -1 \text{ or } x \geq 4\}$ or $(-\infty, -1] \cup [4, \infty)$.

Talk About It

For a quadratic inequality of the form $ax^2 + bx + c > 0$ where $a < 0$, if the related equation has no real roots, what is the solution set? Explain your reasoning.

Check

Solve $-\frac{1}{4}x^2 + x + 1 > 0$ by graphing and write the solution set.

$\{x \mid$ ___?___ $< x <$ ___?___$\}$

Go Online You can complete an Extra Example online.

Example 5 Solve a Quadratic Inequality Algebraically

GARDENING **Marcus is planning a garden. He has enough soil to cover 104 square feet, and wants the dimensions of the garden to be at least 5 feet by 10 feet. If he wants to increase the length and width by the same number of feet, by what value can he increase the dimensions of the garden without needing to buy more soil? Create a quadratic inequality and solve it algebraically.**

Step 1 Determine the quadratic inequality.

$A = \ell w$	Area formula
$= (x + 10)(x + 5)$	$\ell = x + 10; w + 5$
$= x^2 + 15x + 50$	FOIL and simplify.

The area must be less than or equal to 104 square feet, so $x^2 + 15x + 50 \leq 104$.

Step 2 Solve the related equation.

$x^2 + 15x + 50 = 104$	Related equation
$x^2 + 15x - 54 = 0$	Subtract 104 from each side.
$(x + 18)(x - 3) = 0$	Factor.
$x = -18$ or $x = 3$	Zero Product Property

Steps 3 and 4 Plot the solutions on a number line and test a value from each interval.

Use dots because −18 and 3 are solutions of the original inequality.

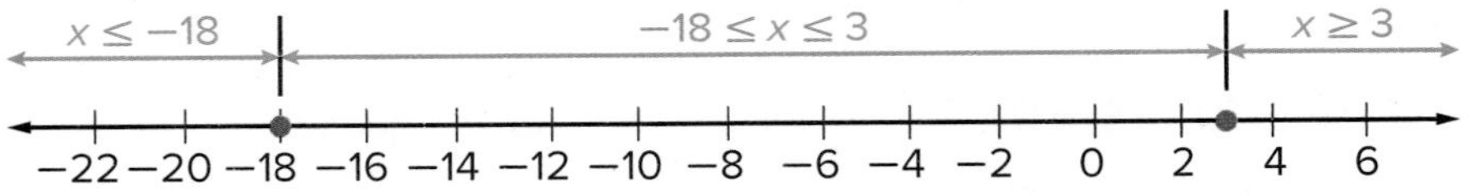

Test a value from each interval to see if it satisfies the original inequality.

Test $x = -20$, $x = 0$, and $x = 5$. The only value that satisfies the original inequality is $x = 0$, so the solution set is [−18, 3]. So, Marcus can increase the length and width up to 3 feet without needing to buy more soil. The interval $-18 \leq x \leq 0$ is not relevant because Marcus does not want to decrease the length and width or leave it as is.

Check

MANUFACTURING An electronics manufacturer can model their profits in dollars P when they sell x video players by using the function $P(x) = -0.1x^2 + 75x - 1000$. How many video players can they sell so they make $7500 or less?

The company will make $7500 or less if they make ___?___ video players or fewer and/or ___?___ video players or more.

Go Online You can complete an Extra Example online.

Practice

Go Online You can complete your homework online.

Examples 1 and 2

Graph each inequality.

1. $y \leq x^2 + 6x + 4$

2. $y < -x^2 + 4x - 6$

3. $y \leq x^2 - 4$

4. $y \leq x^2 + 4$

5. $y < 2x^2 - 4x - 2$

6. $-x^2 + 12x - 36 > y$

7. $y > x^2 + 6x + 7$

8. $y > x^2 - 8x + 17$

9. $y \geq x^2 + 2x + 2$

10. $y \geq 2x^2 + 4x$

11. $y > -2x^2 - 4x + 2$

12. $y \geq x^2 - 4x + 4$

Examples 3 and 4

Solve each inequality by graphing.

13. $x^2 - 6x + 9 \leq 0$

14. $-x^2 - 4x + 32 \geq 0$

15. $\frac{3}{4}x^2 + \frac{3}{4}x - 10 > 5$

16. $x^2 - x - 6 \leq 0$

17. $x^2 + 8x + 16 \geq 0$

18. $x^2 - 2x - 24 \leq 0$

Example 5

19. FENCING Vanessa has 180 feet of fencing that she intends to use to build a rectangular play area for her dog. She wants the play area to enclose at least 1800 square feet. What are the possible widths of the play area?

20. BUSINESS A bicycle maker sold 300 bicycles last year at a price of \$300 each. The maker wants to increase the profit margin this year, but predicts that each \$20 increase in price will reduce the number of bicycles sold by 10. How many \$20 price increases can the maker add and still expect to make a total profit of at least \$100,000?

Mixed Exercises

Solve each quadratic inequality by using a graph, a table, or algebraically.

21. $-2x^2 + 12x < -15$

22. $5x^2 + x + 3 \geq 0$

23. $11 \leq 4x^2 + 7x$

24. $x^2 - 4x \leq -7$

25. $-3x^2 + 10x < 5$

26. $-1 \geq -x^2 - 5x$

27. $x^2 + 2x + 1 \geq 0$

28. $x^2 - 3x + 2 \leq 0$

29. $x^2 + 10x + 7 \geq 0$

30. $x^2 - 5x > 14$

31. $-x^2 - 15 \leq 8x$

32. $9x \leq 12x^2$

33. $4x^2 + 4x + 1 > 0$

34. $5x^2 + 10 \geq 27x$

35. $9x^2 + 31x + 12 \leq 0$

36. REASONING Consider the equation $ax^2 + bx + c = 0$. Assume that the discriminant is zero and that a is positive. What are the solutions of the inequality $ax^2 + bx + c \leq 0$?

Write a quadratic inequality for each graph.

37. 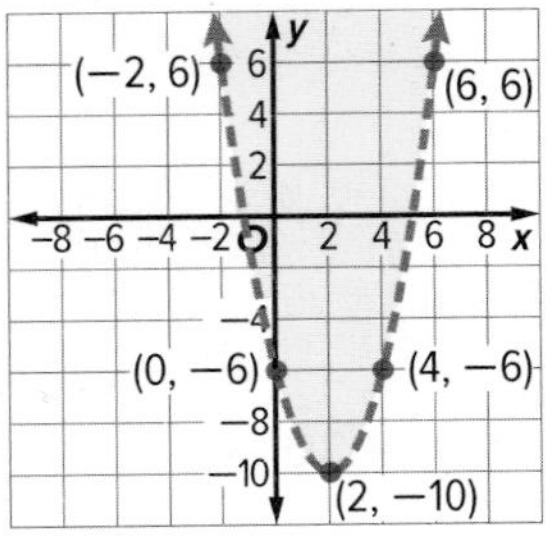

38. 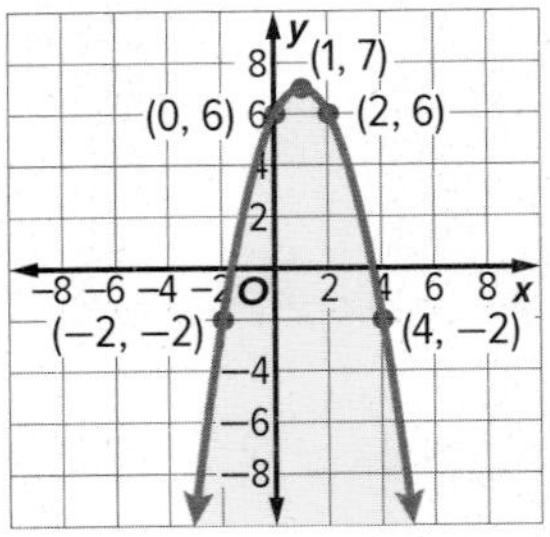

39. 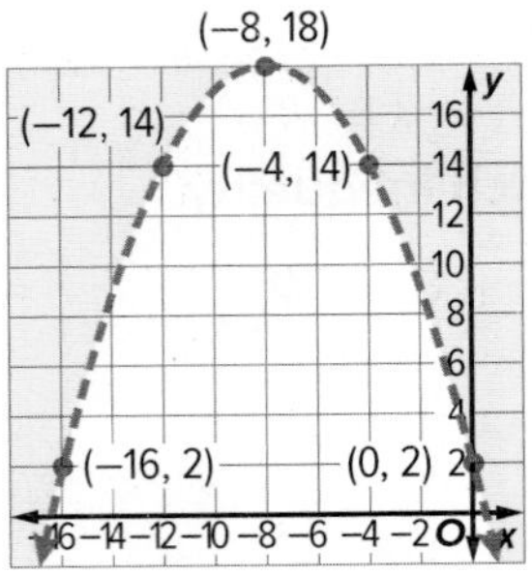

40. BASEBALL A baseball player hits a high pop-up with an initial velocity of 32 meters per second, 1.3 meters above the ground. The height $h(t)$ of the ball in meters t seconds after being hit is modeled by $h(t) = -4.9t^2 + 32t + 1.3$.

a. During what time interval is the ball higher than the camera located in the press box 43.4 meters above the ground?

b. When is the ball within 2.4 meters of the ground where the catcher can attempt to catch it?

41. CONSTRUCT ARGUMENTS Are the boundaries of the solution set of $x^2 + 4x - 12 \leq 0$ twice the value of the boundaries of $\frac{1}{2}x^2 + 2x - 6 \leq 0$? Explain.

Higher-Order Thinking Skills

42. FIND THE ERROR Don and Diego used a graph to solve the quadratic inequality $x^2 - 2x - 8 > 0$. Is either of them correct? Explain your reasoning.

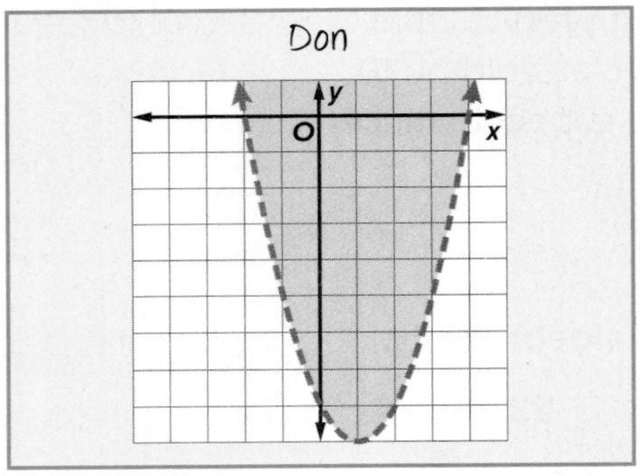

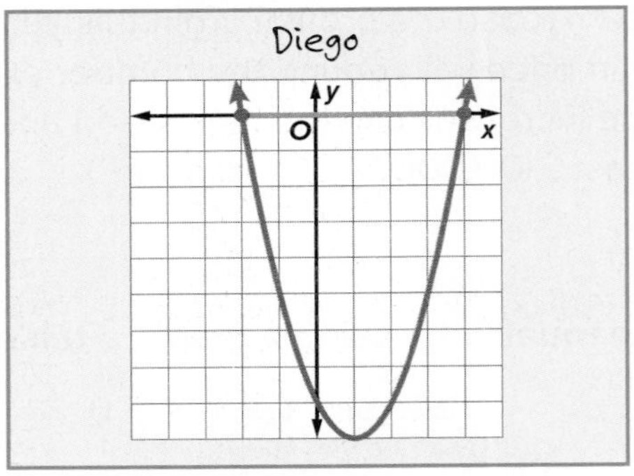

43. CREATE Write a quadratic inequality for each condition.

a. The solution set is all real numbers.

b. The solution set is the empty set.

44. ANALYZE Determine if the following statement is *sometimes*, *always*, or *never* true. Justify your reasoning.

The intersection of $y \leq -ax^2 + c$ *and* $y \geq ax^2 - c$ *is the empty set.*

45. PERSEVERE Graph the intersection of the graphs of $y \leq -x^2 + 4$ and $y \geq x^2 - 4$.

46. WRITE How are the techniques used when solving quadratic inequalities and linear inequalities similar? How are they different?

Solving Linear-Nonlinear Systems

Explore Linear-Quadratic Systems

Online Activity Use graphing technology to complete the Explore.

INQUIRY How many solutions can a linear-quadratic system of equations have?

Learn Solving Linear-Quadratic Systems

Like solving systems of linear equations, you can solve linear-quadratic systems by using graphical or algebraic methods. You can also use a system of equations to solve a quadratic equation by writing each side of the equation as a related function.

Example 1 Solve a Linear-Quadratic System by Using Substitution

Solve the system of equations.

$x = 2y^2 + 3y + 1$ (1)

$-x + y = -1$ (2)

Step 1 Solve Equation (2) for *x*.

$-x + y = -1$	Equation (2)
$-x = -y - 1$	Subtract y from each side.
$x = y + 1$	Divide each side by -1.

Step 2 Substitute for *x* in Equation (1). Then solve for *y*.

$x = 2y^2 + 3y + 1$	Equation (1)
$y + 1 = 2y^2 + 3y + 1$	$x = y + 1$
$0 = 2y^2 + 2y$	Simplify.
$0 = 2y(y + 1)$	Factor out $2y$.
$y = 0$ or $y = -1$	Zero Product Property

Step 3 Substitute the *y*-values and solve for *x*.

Case 1		Case 2
$x = y + 1$	Equation (2)	$x = y + 1$
$= 0 + 1$ or 1	Substitute for y and simplify.	$= -1 + 1$ or 0

The two solutions of the system are (1, 0) and (0, −1).

Go Online You can complete an Extra Example online.

Today's Goals
- Solve systems of linear and quadratic equations.
- Solve systems of two quadratic equations.

Today's Vocabulary
quadratic relations

Talk About It!
A system of linear equations can have infinitely many solutions. Can a linear-quadratic system have infinitely many solutions? Explain.

Study Tip
Algebra and Graphing Even when solving a system algebraically, it can be useful to graph the equations to ensure that you have the correct number of solutions.

Example 2 Solve a Linear-Quadratic System by Using Elimination

Solve the system of equations.

$x^2 = y + 5$ (1)

$-x + y = 7$ (2)

Step 1 Solve so that the ys are on the same side of each equation.

$-x + y = 7$ Equation (2)

$-x = -y + 7$ Subtract y from each side.

Step 2 Add the equations.

$$\begin{array}{r} x^2 = y + 5 \\ (+)\ -x = -y + 7 \\ \hline x^2 - x = 12 \end{array}$$

Step 3 Solve for the remaining variable.

$x^2 - x = 12$ Sum of Equations (1) and (2)

$x^2 - x - 12 = 0$ Subtract 12 from each side.

$(x - 4)(x + 3) = 0$ Factor.

$x = 4$ or $x = -3$ Zero Product Property

Step 4 Solve for the other variable.

$-x + y = 7$	Equation (2)	$-x + y = 7$
$-4 + y = 7$	Substitute x.	$-(-3) + y = 7$
$y = 11$	Simplify.	$y = 4$

The two solutions of the system are (4, 11) and (−3, 4).

Example 3 Use a System to Solve a Quadratic Equation

Use a system of equations to solve $x^2 - 2x + 6 = 4x + 1$.

Step 1 Create a system of equations.

$y = x^2 - 2x + 6$ (1)

$y = 4x + 1$ (2)

Step 2 Graph the system.

The functions appear to intersect at (1, 5) and (5, 21), so the solutions of $x^2 - 2x + 6 = 4x + 1$ are $x = 1$ and $x = 5$.

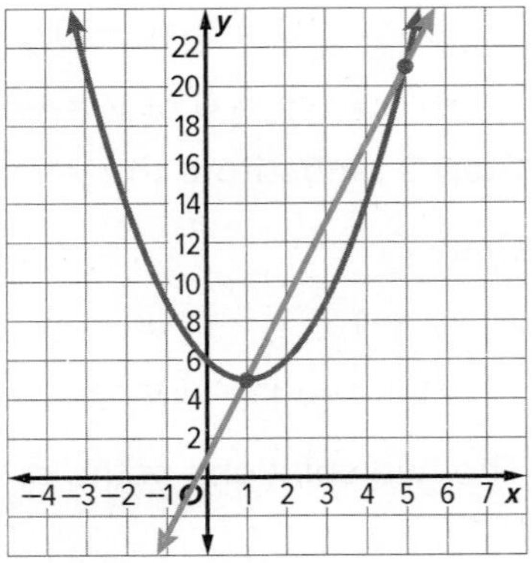

Go Online You can complete an Extra Example online.

Check

Use a system of equations to solve $2x + 11 = x^2 + x - 1$.

$x =$ __?__, __?__

Example 4 Solve a Linear-Quadratic System by Graphing

PRODUCTION **A software developer determines that her company can model their revenue R from a given product in hundreds of thousands of dollars given the unit price of the product x in dollars with the function $R = -0.1x^2 + 4x$. Create a linear-quadratic system and solve it graphically to determine the price for which the company will earn $4.2 million.**

Step 1 Create a linear-quadratic system.

The first equation is the given revenue model $R = -0.1x^2 + 4x$.

The line that represents revenue of $4.2 million is $R = 42$.

So, the linear-quadratic system is:

$R = -0.1x^2 + 4x$ (1)

$R = 42$ (2)

Step 2 Graph the system.

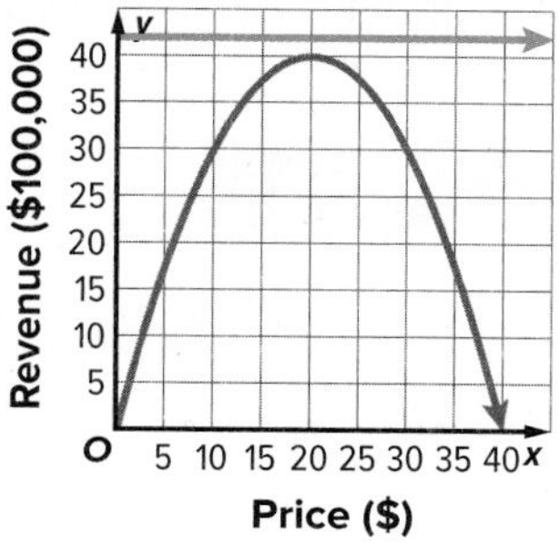

Step 3 Determine the solutions.

The graphs of the functions do not intersect at any point, so the system has 0 solutions.

Check

MOUNTAINS Engineers want to build a footbridge with steps across a steep valley. They can model the valley with the equation $y = 0.05x^2 - 4x + 80$, where y is the height above the lowest point in the valley in feet and x is the distance from where they plan to start the bridge. If they start the bridge 80 feet above ground and want it to go down an inch for every foot to the right, then at what points will the bridge start and end?

Go Online You can complete an Extra Example online.

Study Tip

R is measured in hundreds of thousands of dollars, so the value of 42 on the y-axis indicates revenue $4.2 million.

Think About It!

What does the solution set of the system mean in the context of the situation?

Go Online

You can watch videos to see how to solve systems of linear and nonlinear equations using a graphing calculator.

Learn Solving Quadratic-Quadratic Systems

Equations of parabolas with vertical axes of symmetry have the parent function $y = x^2$ and can be written in the form $y = a(x - h)^2 + k$. Equations of parabolas with horizontal axes of symmetry are of the form $x = a(y - k)^2 + h$ and are not functions. These are often referred to as **quadratic relations**. The graph of $x = y^2$ is the parent graph for the quadratic relations that have a horizontal axis of symmetry.

If a system contains two quadratic relations, it may have zero to four solutions. Just as with linear-quadratic systems, you can solve quadratic-quadratic systems by using graphical or algebraic methods.

Example 5 Solve a Quadratic-Quadratic System Graphically

Solve the system of equations by graphing.

$y = x^2$ (1)

$x = \frac{1}{5}(y - 6)^2 + \frac{6}{5}$ (2)

Think About It!
Is $x = \frac{1}{5}(y - 6)^2 + \frac{6}{5}$ a function? Explain your reasoning.

Step 1 Graph Equation (1).

Equation (1) has a vertex at (0, 0) and goes through the points (−2, 4), (−1, 1), (1, 1) and (2, 4).

Step 2 Graph Equation (2).

You can use a table of values to graph Equation (2). Because the expression on the right is set equal to x, find the value of x for several values of y.

y	$x = \frac{1}{5}(y - 6)^2 + \frac{6}{5}$	x
3	$x = \frac{1}{5}(3 - 6)^2 + \frac{6}{5}$	3
4	$x = \frac{1}{5}(4 - 6)^2 + \frac{6}{5}$	2
5	$x = \frac{1}{5}(5 - 6)^2 + \frac{6}{5}$	$\frac{7}{5}$
6	$x = \frac{1}{5}(6 - 6)^2 + \frac{6}{5}$	$\frac{6}{5}$
7	$x = \frac{1}{5}(7 - 6)^2 + \frac{6}{5}$	$\frac{7}{5}$
8	$x = \frac{1}{5}(8 - 6)^2 + \frac{6}{5}$	2
9	$x = \frac{1}{5}(9 - 6)^2 + \frac{6}{5}$	3

Think About It!
How could you check your solutions?

Step 3 Graph and solve the system.

To solve the system, graph both relations on the same coordinate plane and see where they intersect.

The relations intersect at (2, 4) and (3, 9), so those are the solutions of the system.

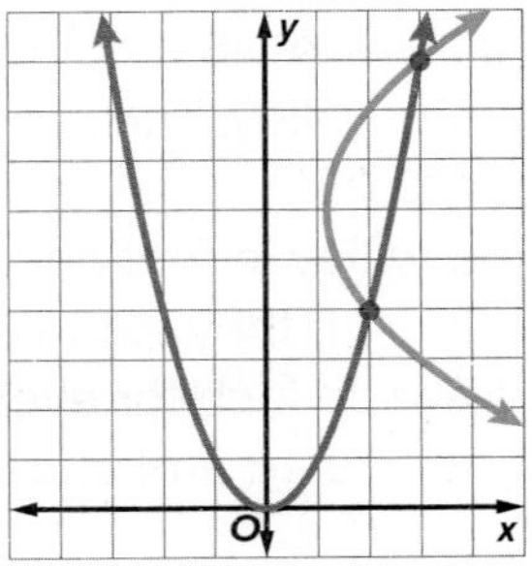

Check

Solve the system of equations.

$y = -x^2 + 5x - 6$

$3y = x^2 - x - 6$

(__?__, __?__) (__?__, __?__)

Example 6 Solve a Quadratic-Quadratic System Algebraically

Solve the system of equations.

$\mathbf{2x^2 - y = 4}$ **(1)**

$\mathbf{y = -\frac{1}{2}x^2 + 6}$ **(2)**

$2x^2 - y = 4$	Equation (1)
$2x^2 - (-\frac{1}{2}x^2 + 6) = 4$	Substitution
$2x^2 + \frac{1}{2}x^2 - 6 = 4$	Distributive Property
$\frac{5}{2}x^2 - 10 = 0$	Simplify.
$5x^2 - 20 = 0$	Multiply each side by 2.
$x^2 - 4 = 0$	Divide each side by 5.
$(x + 2)(x - 2) = 0$	Difference of Two Squares
$x = \pm 2$	Zero Product Property

Substitute −2 and 2 into one of the original equations and solve for y.

Case 1		Case 2
$y = -\frac{1}{2}x^2 + 6$	Equation (2)	$y = -\frac{1}{2}x^2 + 6$
$y = -\frac{1}{2}(2)^2 + 6$	Substitute for x.	$y = -\frac{1}{2}(-2)^2 + 6$
$= 4$	Simplify.	$= 4$

The solutions are (2, 4) and (−2,4).

Check

Solve the system of equations.

$x = \frac{1}{18}y^2$

$y^2 = -18x + 72$

(__?__, __?__) (__?__, __?__)

Go Online You can complete an Extra Example online.

Example 7 Use a System to Solve a Quadratic Equation

Use a system of equations to solve $2x^2 - 3x + 8 = 11 - 4x^2$.

Step 1 Create a system of equations.

The related equations of each side of $2x^2 - 3x + 8 = 11 - 4x^2$ are:

$y = 2x^2 - 3x + 8$ (1)

$y = 11 - 4x^2$ (2)

Step 2 Graph and solve the system.

Graph the first equation.

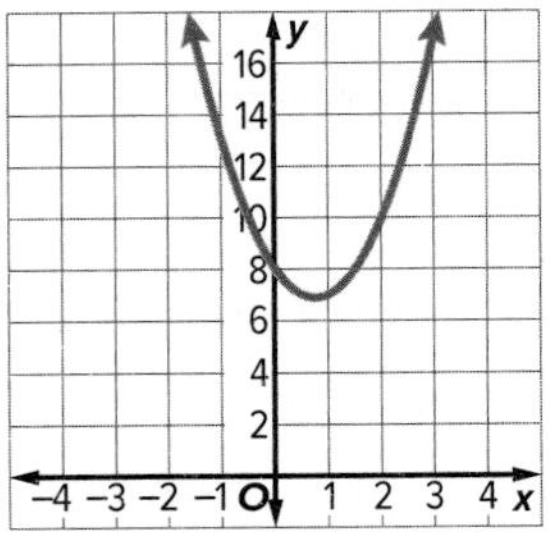

Then graph the second equation on the same coordinate plane.

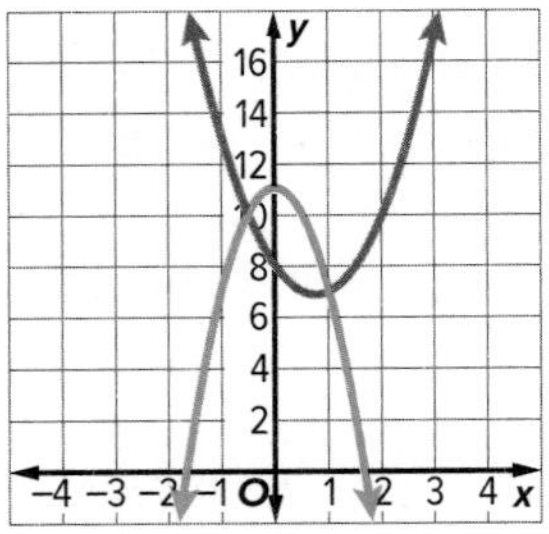

The functions appear to intersect at $\left(-\frac{1}{2}, 10\right)$ and (1, 7), so the solutions of $2x^2 - 3x + 8 = 11 - 4x^2$ are $-\frac{1}{2}$ and 1.

Check

Use a system of equations to solve $-x^2 + 3x + 14 = -\frac{1}{4}x^2 + 5$.

System Solutions: (? , ?) (? , ?)

Equation Solutions: ? , ?

Go Online You can complete an Extra Example online.

Practice

Go Online You can complete your homework online.

Examples 1 and 2

Solve each system of equations by using substitution or elimination.

1. $y = x^2 - 5$
   $y = x - 3$

2. $y = x - 2$
   $y = x^2 - 2$

3. $y = x + 3$
   $y = 2x^2$

4. $y = 3x$
   $x = y^2$

5. $y = -2x + 2$
   $y^2 = 2x$

6. $y = 2 - x$
   $y = x^2 - 4x + 2$

7. $x - y + 1 = 0$
   $y^2 = 4x$

8. $y = x - 1$
   $y = x^2$

9. $y = x$
   $y = -2x^2 + 4$

Example 3

Use a system of equations to solve each quadratic equation.

10. $x^2 + 3x + 3 = -2x - 3$

11. $x^2 + 3x + 5 = 2x + 7$

12. $2x^2 + 4x - 3 = 9x$

13. $4x^2 - 6x - 3 = -3x - 2$

14. $3x^2 - 4x + 2 = 2 - 2x$

15. $x^2 - 4x + 5 = 2x + 12$

Example 4

16. **CIVIL ENGINEERING** For safety, roads are designed to ensure that rainwater runs off to the sides. Many roads are highest in the center so that a cross section of the road is a parabola. Suppose the surface of a road can be modeled by $y = -\frac{1}{300}x^2 + \frac{3}{4}$ where x is the distance in feet from the center of the road and y is the height in feet of the road above the shoulder.

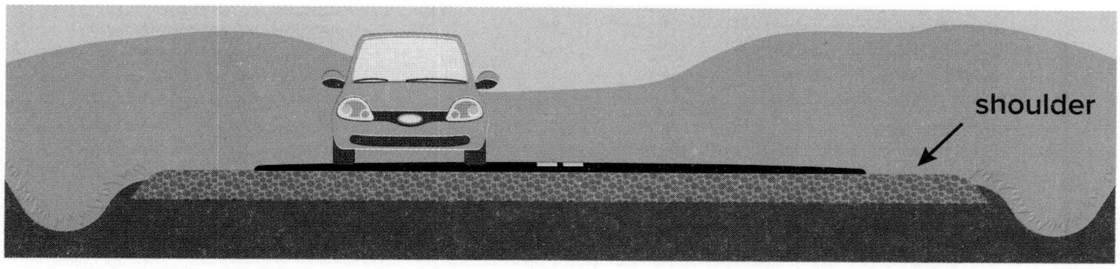

   a. How wide is the road?

   b. If the water level during a flash flood is 6 inches above the shoulder, about how wide is the portion of the road that is not under water?

17. **SMALL BUSINESS** A small business owner determines that the profit P in dollars from sales of a specific item can be modeled by $P = 2x^2 + 30x$, where x is the selling price of the item in dollars.

   a. Create a linear-quadratic system to determine the price for which the business will earn \$50,000.

   b. Solve the system in **part a** to determine the price for which the business will earn \$50,000. Round to the nearest hundredth, if necessary.

   c. What does the solution set of the system mean in the context of the situation?

Example 5

Solve each system of equations by graphing.

18. $y = x^2$
$x = y^2$

19. $y = \frac{1}{2}x^2$
$x = y^2 + 2y + 1$

20. $y = -2x^2$
$y = 6x^2$

21. $x = (y - 2)^2$
$y = x^2 + 2$

22. $x = \frac{1}{2}y^2 + 1$
$y = \frac{1}{2}(x - 1)^2$

23. $x = \frac{1}{4}(y - 1)^2 + 1$
$y = \frac{1}{4}(x - 1)^2 + 1$

Example 6

Solve each system of equations algebraically.

24. $2x^2 - y = 8$
$y = x^2 + 8$

25. $x^2 - y = 4$
$y = 2x^2$

26. $x = \frac{1}{4}y^2$
$y^2 = -4x + 18$

27. $x = \frac{1}{2}y^2$
$y^2 = -4x + 12$

28. $2y = x^2$
$y = x^2 - 8$

29. $y = x^2 + 3x - 5$
$2y = x^2 + 9x - 6$

Example 7

Use a system of equations to solve each quadratic equation.

30. $2x^2 + 5x + 3 = 2 - 2x^2$

31. $3x^2 - x + 2 = x^2 + 8$

32. $-3x^2 + 3x + 5 = 9 - 4x^2$

33. $2x^2 + 4x + 10 = x^2 + 6$

34. $4x^2 + 2x + 7 = 3x^2 + 6$

35. $5x^2 - 5x - 7 = 3x^2 - 4$

Mixed Exercises

Use the related graphs of each system of equations to determine the solutions.

36. $y = \frac{1}{4}x^2 + \frac{1}{2}x - 2$
$y = -\frac{1}{4}x^2 - \frac{1}{2}x + 2$

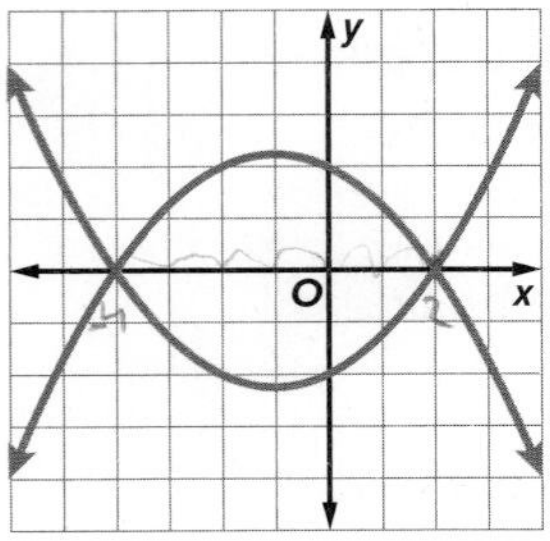

37. $y = x^2 + 4x - 1$
$y = -2x^2 - 8x - 13$

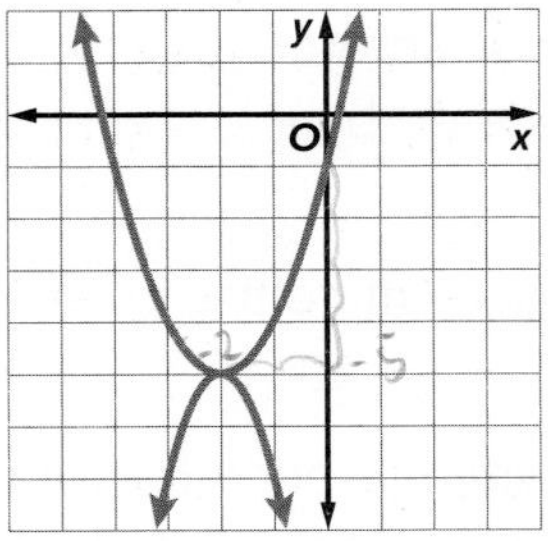

38. $y = 2x^2 - 4x + 6$
$y = -\frac{1}{2}x^2 - 2x + 1$

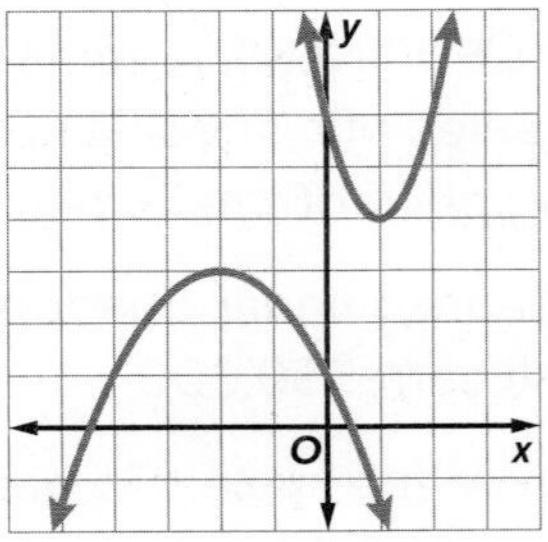

Solve each system of equations by graphing, substitution, or elimination.

39. $y = x^2 + 9x + 8$
$y = 7x + 8$

40. $y = x^2 + 7x + 12$
$y = x + 7$

41. $y = x + 3$
$y = 2x^2 - x - 1$

42. $y = \left(\frac{1}{2}x\right)^2$
$4x = y^2$

43. $y = 2x^2$
$x = y^2 - 2y + 1$

44. $y = -8x^2$
$y = 4x^2 + 1$

45. $2x^2 - y = 2$
$y = -x^2 + 4$

46. $x^2 - y = -4$
$y = 4x^2 - 2$

47. $y = x^2$
$y^2 = -3x + 5$

48. $y = -x - 3$
$y = x^2 - 5$

49. $y = -x$
$y^2 = 2x$

50. $y = 3x - 4$
$y = x^2 + 9x + 20$

51. $x - 2 = (y - 2)^2$
$y = x^2 + 5$

52. $3x - 1 = \frac{1}{8}y^2$
$y = x^2 + 5x - 4$

53. $x = (y + 2)^2$
$y = (x + 2)^2$

54. $x = \frac{1}{4}y^2$
$y^2 = -8x + 6$

55. $4y + 3 = x^2$
$y = x^2 - 24$

56. $y = x^2 + x + 5$
$y = 2x^2 + 3x + 2$

Use a system of equations to solve each quadratic equation.

57. $2x^2 + x - 1 = -10x + 12$

58. $3x^2 - 5x - 2 = -16 - 8x$

59. $x^2 + 3x + 2 = x + 5$

60. $x^2 - 2x - 3 = -\frac{1}{3}x^2 + \frac{2}{3}x + \frac{23}{3}$

61. $3x^2 + 4x + 1 = -x^2 - 2x - 1$

62. $2x^2 - 6x + 4 = -x^2 + 3x - 2$

63. $2x^2 + 6x + 5 = x^2 - 4$

64. $4x^2 + 2x + 7 = 3x^2 + 6$

65. $5x^2 - 5x - 7 = 3x^2 - 4$

66. $x^2 + 4x + 4 = x + 4$

67. $x^2 + 5x + 12 = -4x - 6$

68. $2x^2 - x - 6 = -3x - 2$

69. ROCKETS Two model rockets are launched at the same time from different heights. The height in meters for one rocket after t seconds is modeled by $y = -4.9t^2 + 48.8t$. The height for the other rocket is modeled by $y = -4.9t^2 + 46.7t + 1.5$.

a. After how many seconds are the rockets at the same height? Round to the nearest tenth.

b. From what height would the slower rocket have to be launched so that the rockets land at the same time? Justify your reasoning.

70. PACKAGING A manufacturer is making two different packages, measured in inches, as shown in the figure. The manufacturer wants the surface area of the packages to be the same.

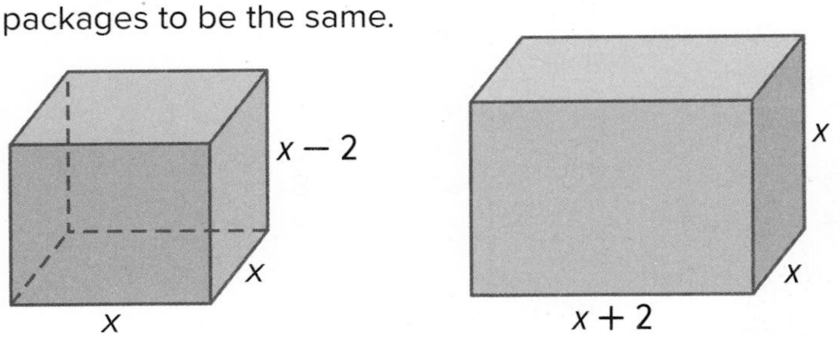

a. Create a quadratic equation that can be used to find the value of x, when the surface area of the packages is the same.

b. Solve the quadratic equation using any method.

c. Find and interpret the solution in the context of the situation.

71. VOLLEYBALL The function $h(t) = -16t^2 + vt + h_0$ models the height in feet of a volleyball, where v represents the initial velocity, h_0 represents the initial height, and t represents the time in seconds since the ball was hit. Suppose a player bumps a volleyball when it is 3 feet from the ground with an initial velocity of 25 feet per second.

a. If the net is approximately 7 feet 4 inches, could the volleyball clear the net? Justify your reasoning.

b. The player on the other side of the net jumps 1 second after the ball is hit so that the path of her hands can be described by $h(t) = -16(t - 1)^2 + 12(t - 1) + 7$. If the ball clears the net, is it possible that she blocks the ball? Justify your reasoning.

c. What assumptions did you make?

72. PERSEVERE Describe three linear-quadratic systems of equations—one with no solution, one with 1 solution, and one with 2 solutions.

73. CREATE Write a system of two quadratic equations in which there is one solution.

74. WRITE Describe a real-life situation that can be modeled by a system with a quadratic function and a linear function.

75. FIND THE ERROR Danny and Carol are solving the system $y = x^2 + 3x - 9$ and $-4x + y = 3$. Is either of them correct? Explain your reasoning.

Danny

$-4x + (x^2 + 3x - 9) = 3$

$x^2 - x - 9 = 3$

$x^2 - x - 12 = 0$

$(x + 3)(x - 4) = 0$

$x = -3$ or $x = 4$

Carol

$x^2 + 3x - 9 = -4x + 3$

$x^2 + 7x - 12 = 0$

$x \approx -8.42$ or $x \approx 1.42$

Module 3 • Quadratic Functions

Review

Essential Question

What characteristics of quadratic functions are important when analyzing real-world situations that are modeled by quadratic functions?

Module Summary

Lessons 3-1 and 3-2

Graphs of Quadratic Functions

- When the coefficient on the x^2-term is positive, the parabola opens up. When it is negative, the parabola opens down.
- The average rate of change for a parabola over the interval $[a, b]$ is $\frac{f(b) - f(a)}{b - a}$.
- The solutions of an equation in one variable are the x-intercepts of the graph of a related function.

Lesson 3-3

Complex Numbers

- The imaginary unit i is the principal square root of -1. Thus, $i = \sqrt{-1}$ and $i^2 = -1$.
- Two complex numbers of the form $a + bi$ and $a - bi$ are called complex conjugates.

Lesson 3-4 through 3-6

Solving Quadratic Equations

- For any real numbers a and b, if $ab = 0$, then either $a = 0$, $b = 0$, or both a and $b = 0$.
- You can solve a quadratic equation by graphing, by factoring, by completing the square, or by using the Quadratic Formula.
- To solve a quadratic equation of the form $x^2 = n$, take the square root of each side.
- The solutions of a quadratic equation of the form $ax^2 + bx + c = 0$, where $a \neq 0$, are given by the formula $x = \frac{-b \pm \sqrt{b^2 - 4ac}}{2a}$.

Lesson 3-7

Quadratic Inequalities

- To graph a quadratic inequality, graph the related function, test a point not on the parabola and shade accordingly.
- For $ax^2 + bx + c < 0$, graph $y = ax^2 + bx + c$ and identify the x-values for which the graph lies below the x-axis. For $\leq$, include the x-intercepts in the solution.
- For $ax^2 + bx + c > 0$, graph $y = ax^2 + bx + c$ and identify the x-values for which the graph lies *above* the x-axis. For $\geq$, include the x-intercepts in the solution.

Lesson 3-8

Systems Involving Quadratic Equations

- You can use the substitution method or the elimination method to solve a system that includes a quadratic equation.
- If a system contains two quadratic relations, it may have anywhere from zero to four solutions.

Study Organizer

Foldables

Use your Foldable to review this module. Working with a partner can be helpful. Ask for clarification of concepts as needed.

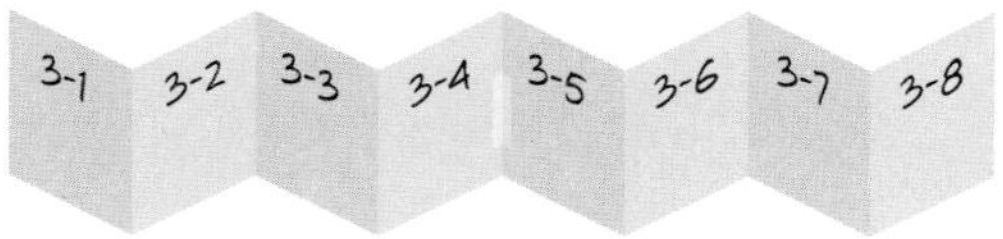

Test Practice

1. MULTIPLE CHOICE Which is the graph of $f(x) = -x^2 - 2x + 3$? (Lesson 3-1)

A.
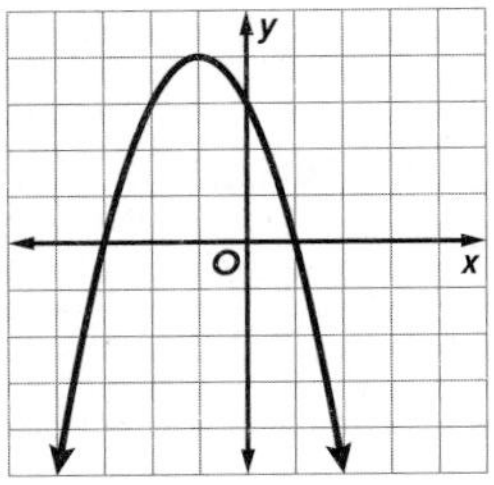

B.
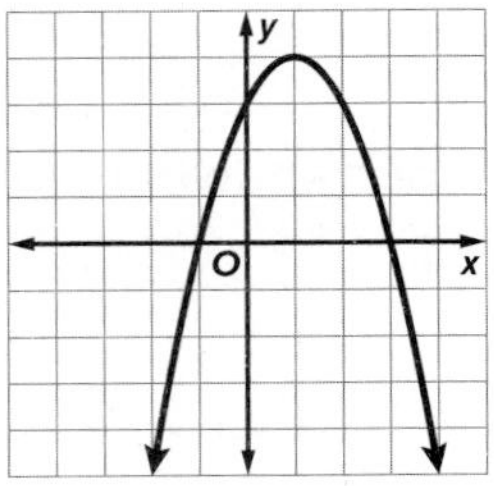

C.
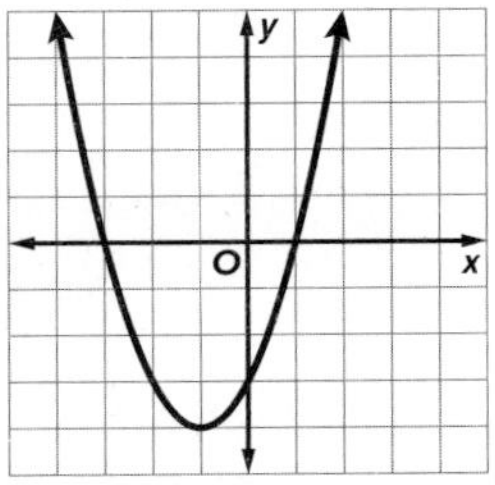

D.
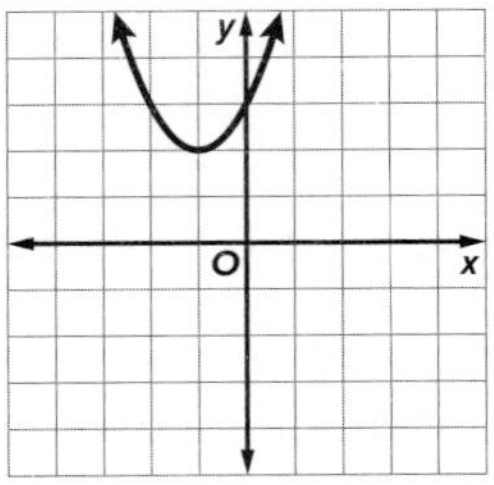

2. OPEN RESPONSE At a concert, a T-shirt cannon launches a T-shirt upward. The height of the T-shirt in feet a given number of seconds after the launch is shown in the table.

Time (s)	Height (ft)
1	74
2	116
3	126
4	104
5	50

Find and interpret the average rate of change in the height between 1 and 3 seconds after launch. (Lesson 3-1)

3. OPEN RESPONSE Use a quadratic equation to find two real numbers with a sum of 31 and a product of 210. (Lesson 3-2)

4. MULTI-SELECT The graph of $f(x) = x^2 - x - 6$ is shown.

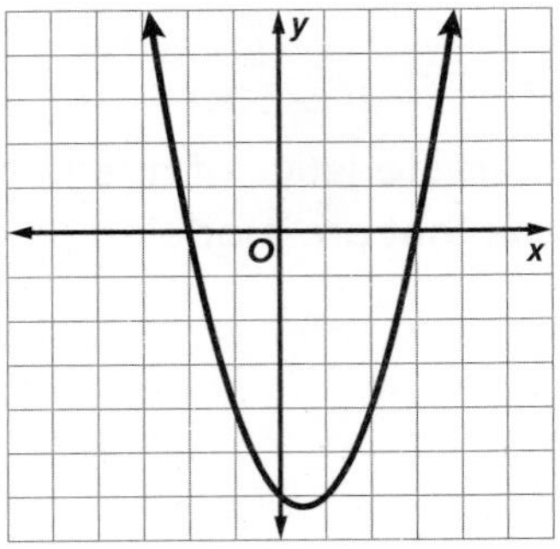

Find the solutions of $x^2 - x - 6 = 0$. Select all that apply. (Lesson 3-2)

A. -6

B. -3

C. -2

D. 2

E. 3

F 6

5. MULTIPLE CHOICE Simplify $\sqrt{-9} \cdot \sqrt{-49}$. (Lesson 3-3)

A. $-21i$

B. $21i$

C. -21

D. 21

6. OPEN RESPONSE Every complex number can be written in the form $a + bi$. For the complex number $8 - 4i$, identify the values of a and b. (Lesson 3-3)

7. **OPEN RESPONSE** Solve $4x^2 - 64 = 8x - 4$ by factoring. (Lesson 3-4)

8. **MULTIPLE CHOICE** A rectangular lawn has a width of 30 feet and a length of 45 feet. A diagram of the lawn is shown.

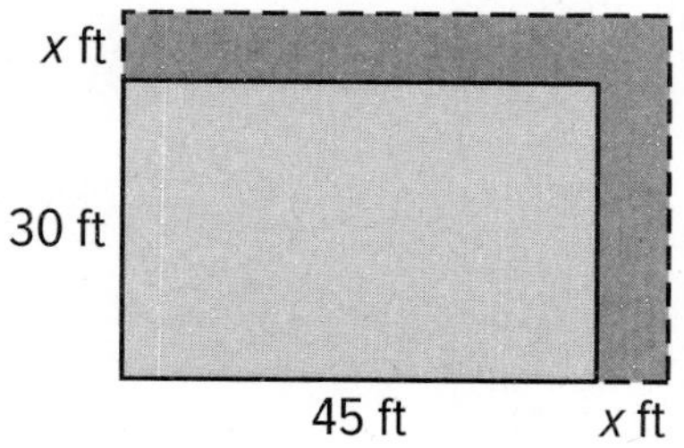

A landscape designer wants to increase the length and width of the lawn by the same amount so that the total area will be 2200 square feet. By how many feet should the designer increase the length and width of the lawn? (Lesson 3-4)

A. 10 feet

B. 19 feet

C. 28 feet

D. 29 feet

9. **MULTIPLE CHOICE** Solve $x^2 + 24x + 150 = 0$ by completing the square. (Lesson 3-5)

A. $12 \pm \sqrt{6}$

B. $12 \pm i\sqrt{6}$

C. $-12 \pm \sqrt{6}$

D. $-12 \pm i\sqrt{6}$

10. **OPEN RESPONSE** Use the square root property to solve $x^2 - 16 = 0$. (Lesson 3-5)

11. **MULTI-SELECT** Use the Square Root Property to find the solutions of $x^2 + 10x + 25 = 81$. Select all that apply. (Lesson 3-5)

A. -14

B. -9

C. -4

D. 4

E. 9

F. 14

12. **OPEN RESPONSE** The height of a firework shell, in meters, t seconds after launch can be modeled by the function $f(t) = -4.9t^2 + 80t$. To the nearest tenth of a second, how many seconds does it take to first reach a height of 300 meters? Explain your reasoning. (Lesson 3-6)

13. MULTIPLE CHOICE Solve $4x^2 - 6x - 5 = 0$ by using the Quadratic Formula. (Lesson 3-6)

A. $\frac{3 \pm \sqrt{11}}{4}$

B. $\frac{-3 \pm \sqrt{29}}{4}$

C. $\frac{3 \pm \sqrt{29}}{4}$

D. $\frac{-3 \pm \sqrt{11}}{4}$

14. MULTIPLE CHOICE The graph of $y = x^2 - 4x + 3$ is shown. Select the values for which $x^2 - 4x + 3 < 0$. (Lesson 3-7)

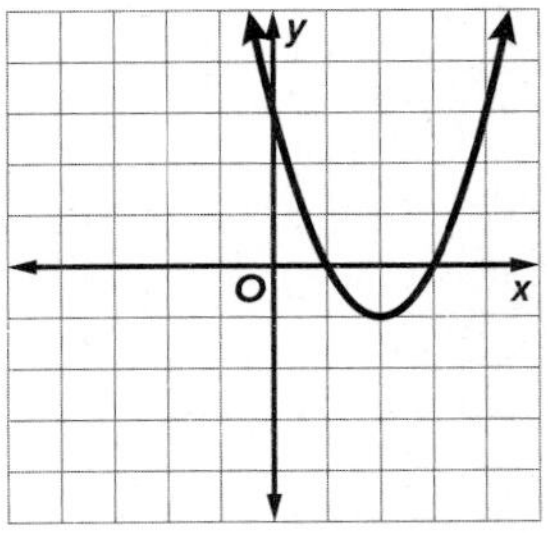

A. $x < 1, x < 3$

B. $x < 1, x > 3$

C. $x > 1, x < 3$

D. $x > 1, x > 3$

15. OPEN ENDED Caleb wants to add an L-shaped deck along two sides of his garden. The garden is a rectangle 15 feet wide and 20 feet long. The deck width will be the same on both sides and the total area of the garden and deck cannot exceed 500 square feet. How wide can the deck be? (Lesson 3-7)

16. MULTI-SELECT Solve the system of equations. Which ordered pair is part of the solution set? Select all that apply. (Lesson 3-8)

$$\begin{cases} y + 3 = x \\ y = x^2 - 5 \end{cases}$$

A. $(-2, -1)$

B. $(-2, 1)$

C. $(-1, -4)$

D. $(-1, 1)$

E. $(1, -4)$

F. $(1, 4)$

G. $(2, -1)$

H. $(2, -4)$

17. OPEN RESPONSE What are the solutions of this system of equations? (Lesson 3-8)

$$\begin{cases} y = (x - 4)^2 - 3 \\ y = x - 1 \end{cases}$$

18. MULTI-SELECT What is the solution to the system of equations? Select all that apply. (Lesson 3-8)

(1) $y = x^2 - 2x - 1$

(2) $y = x + 3$

A. $(-4, 1)$

B. $(-3, 0)$

C. $(-1, 2)$

D. $(1, 0)$

E. $(1, 4)$

F. $(4, 7)$

Module 4

Polynomials and Polynomial Functions

Essential Question

How does an understanding of polynomials and polynomial functions help us understand and interpret real-world events?

What Will You Learn?

How much do you already know about each topic **before** starting this module?

KEY

— I don't know. — I've heard of it. — I know it!

	Before			After		
graph power functions						
graph polynomial functions						
use the location principle to find zeros of a function						
identify extrema of graphs of functions						
add and subtract polynomials						
multiply polynomials						
divide polynomials using long division						
divide polynomials using synthetic division						
expand powers of binomials						

Foldables Make this Foldable to help you organize your notes about polynomials and polynomial functions. Begin with one sheet of $8\frac{1}{2}$-by-14-inch paper.

1. **Fold** a 2-inch tab along the bottom of a long side.
2. **Fold** along the width into thirds.
3. **Staple** the outer edges of the tab.
4. **Label** the tabs *Polynomials, Polynomial Functions,* and *Polynomial Graphs*.

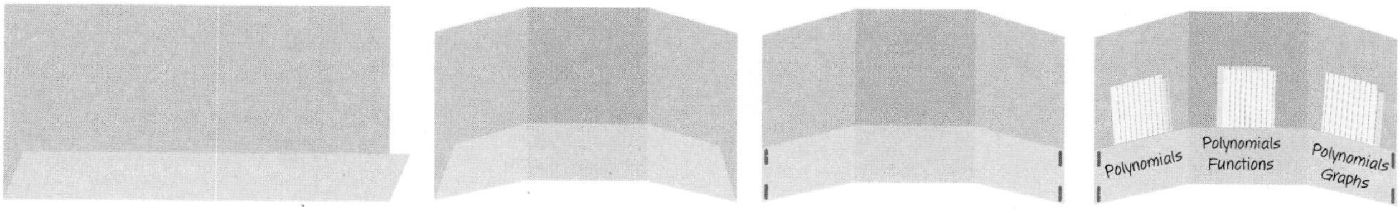

What Vocabulary Will You Learn?

- binomial
- closed
- degree
- degree of a polynomial
- FOIL method
- leading coefficient
- monomial function
- Pascal's triangle
- polynomial in one variable
- polynomial function
- power function
- quartic function
- quintic function
- standard form of a polynomial
- synthetic division
- trinomial

Are You Ready?

Complete the Quick Review to see if you are ready to start this module.
Then complete the Quick Check.

Quick Review

Example 1

Rewrite $2xy - 3 - z$ as a sum.

$2xy - 3 - z$	Original expression
$= 2xy + (-3) + (-z)$	Rewrite using addition.

Example 2

Use the Distributive Property to rewrite $-3(a + b - c)$.

$-3(a + b - c)$	Original expression
$= -3(a) + (-3)(b) + (-3)(-c)$	Distributive Property
$= -3a - 3b + 3c$	Simplify.

Quick Check

Rewrite each difference as a sum.

1. $-5 - 13$

2. $5 - 3y$

3. $5mr - 7mp$

4. $3x^2y - 14xy^2$

Use the Distributive Property to rewrite each expression without parentheses.

5. $-4(a + 5)$

6. $-1(3b^2 + 2b - 1)$

7. $-\frac{1}{2}(2m - 5)$

8. $-\frac{3}{4}(3z + 5)$

How Did You Do?

Which exercises did you answer correctly in the Quick Check?

Polynomial Functions

Explore Power Functions

Online Activity Use graphing technology to complete the Explore.

INQUIRY How do the coefficient and degree of a function of the form $f(x) = ax^n$ affect its end behavior?

Learn Graphing Power Functions

A **power function** is any function of the form $f(x) = ax^n$ where a and n are nonzero real numbers. For a power function, a is the **leading coefficient** and n is the **degree**, which is the value of the exponent. A power function with positive integer n is called a **monomial function**.

Key Concept • End Behavior of a Monomial Function

Degree: even
Leading Coefficient: positive
End Behavior:

As $x \to -\infty$, $f(x) \to \infty$.
As $x \to \infty$, $f(x) \to \infty$.

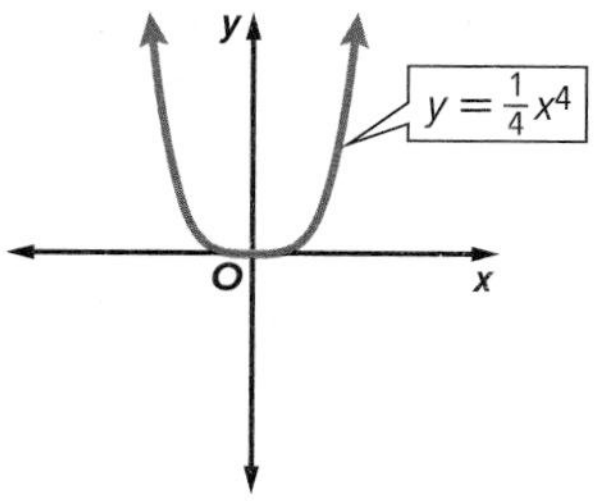

Domain: all real numbers
Range: all real numbers ≥ 0

Degree: odd
Leading Coefficient: positive
End Behavior:

As $x \to -\infty$, $f(x) \to -\infty$.
As $x \to \infty$, $f(x) \to \infty$.

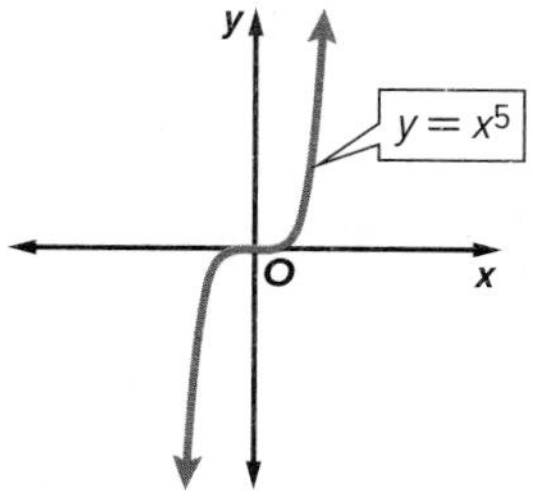

Domain: all real numbers
Range: all real numbers

Degree: even
Leading Coefficient: negative
End Behavior:

As $x \to -\infty$, $f(x) \to -\infty$.
As $x \to \infty$, $f(x) \to -\infty$.

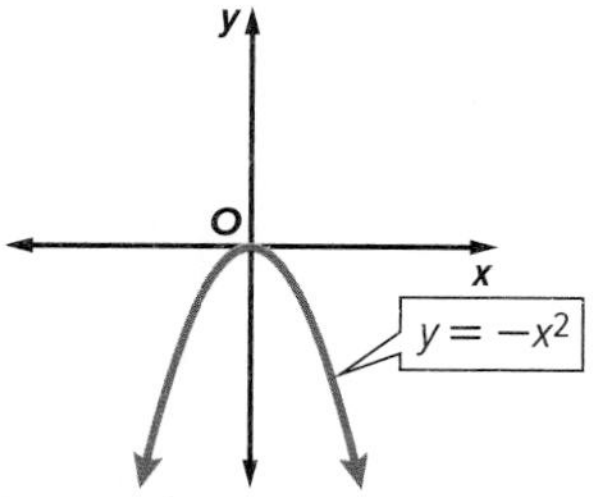

Domain: all real numbers
Range: all real numbers ≤ 0

Degree: odd
Leading Coefficient: negative
End Behavior:

As $x \to -\infty$, $f(x) \to \infty$.
As $x \to \infty$, $f(x) \to -\infty$.

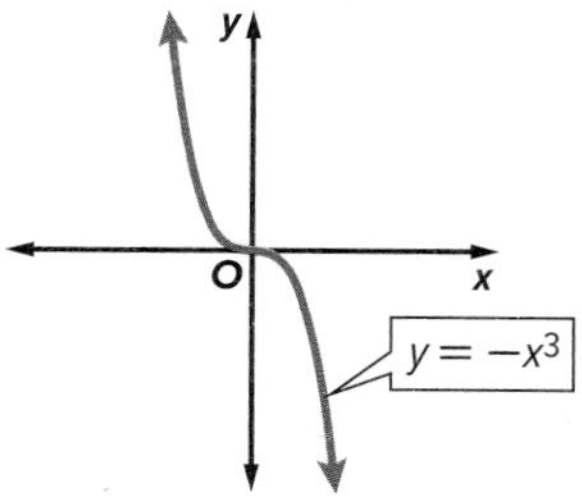

Domain: all real numbers
Range: all real numbers

Today's Goals

- Graph and analyze power functions.
- Graph and analyze polynomial functions.

Today's Vocabulary

power function
leading coefficient
degree
monomial function
polynomial in one variable
standard form of a polynomial
degree of a polynomial
polynomial function
quartic function
quintic function

Talk About It!

Is $f(x) = \sqrt{x}$ a power function? a monomial function? Explain your reasoning.

Key Concept • Zeros of Even and Odd Degree Monomial Functions

Odd-degree monomial functions always have at least one real zero. Even-degree monomial functions may have any number of real zeros or no real zeros.

Example 1 End Behavior and Degree of a Monomial Function

Describe the end behavior of $f(x) = -2x^3$ using the leading coefficient and degree, and state the domain and range.

The leading coefficient of $f(x)$ is -2, which is negative.

The degree is 3, which is odd.

Because the leading coefficient is negative and the degree is odd, as $x \to -\infty$, $f(x) \to \infty$ and as $x \to \infty$, $f(x) \to -\infty$.

Because this is a monomial function, the domain is all real numbers. Because the leading coefficient is negative and the degree is odd, the range is all real numbers.

Check

Describe the end behavior, domain, and range of $f(x) = -10x^6$.

end behavior: As $x \to -\infty$, $f(x) \to$ __?__ and as $x \to \infty$, $f(x) \to$ __?__.

domain: ____?____ range: ____?____

Go Online You can watch a video to see how to graph power functions on a TI-84.

Think About It! Interpret the domain and range given the context of the situation.

Example 2 Graph a Power Function by Using a Table

PRESSURE **For water to flow through a garden hose at a certain rate in gallons per minute (gpm), it needs to have a specific pressure in pounds per square inch (psi). Through testing and measurement, a company that produces garden hoses determines that the pressure P given the flow rate F is defined by $P(F) = \frac{3}{2}F^2$. Graph the function $P(F)$, and state the domain and range.**

Steps 1 and 2 Find a and n. Then state the domain and range.

For $P(F) = \frac{3}{2}F^2$, $a = \frac{3}{2}$, and $n = 2$.

The domain is all real numbers. Because a is positive and n is even, the range is all real numbers ≥ 0.

Steps 3–5 Create a table of values and graph the ordered pairs.

F	$\frac{3}{2}F^2$	$P(F)$
-2	$\frac{3}{2}(-2)^2$	6
-1	$\frac{3}{2}(-1)^2$	1.5
0	$\frac{3}{2}(0)^2$	0
1	$\frac{3}{2}(1)^2$	1.5
2	$\frac{3}{2}(2)^2$	6

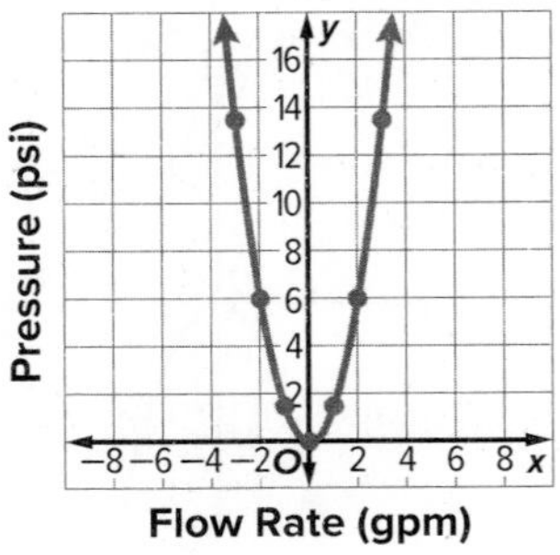

Go Online You can complete an Extra Example online.

Explore Polynomial Functions

Online Activity Use graphing technology to complete the Explore.

> **INQUIRY** How is the degree of a function related to the number of times its graph intersects the x-axis?

Go Online
You can learn how to graph a polynomial function by watching the video online.

Learn Graphing Polynomial Functions

A **polynomial in one variable** is an expression of the form $a_nx^n + a_{n-1}x^{n-1} + \ldots a_2x^2 + a_1x + a_0$, where $a_n \neq 0$, a_{n-1}, a_1, and a_0 are real numbers, and n is a nonnegative integer. Because the terms are written in order from greatest to least degree, this polynomial is written in **standard form**. The **degree of a polynomial** is n and the leading coefficient is a_n.

A **polynomial function** is a continuous function that can be described by a polynomial equation in one variable. You have learned about constant, linear, quadratic, and cubic functions. A **quartic function** is a fourth-degree function. A **quintic function** is a fifth-degree function. The degree tells you the maximum number of times that the graph of a polynomial function intersects the x-axis.

Think About It!
If a polynomial function has a leading coefficient of 4, can you determine its end behavior? Explain your reasoning.

Example 3 Degrees and Leading Coefficients

State the degree and leading coefficient of each polynomial in one variable. If it is not a polynomial in one variable, explain why.

a. $2x^4 - 3x^3 - 4x^2 - 5x + 6$ degree: 4 leading coefficient: 2

b. $7x^3 - 2$ degree: 3 leading coefficient: 7

c. $4x^2 - 2xy + 8y^2$ This is not a polynomial in one variable. There are two variables, x and y.

d. $x^5 + 12x^4 - 3x^3 + 2x^2 + 8x + 4$ degree: 5 leading coefficient: 1

Think About It!
Jamison says the leading coefficient of $4x^2 - 3 + 2x^3 - x$ is 4. Do you agree or disagree? Justify your reasoning.

Check

Select the degree and leading coefficient of $11x^3 + 5x^2 - 7x - \frac{6}{x}$.

A. degree: 3, leading coefficient: 11

B. degree: 11, leading coefficient: 3

C. This is not a polynomial in one variable. There are two variables, x and y.

D. This is not a polynomial in one variable. The term $\frac{6}{x}$ has the variable with an exponent less than 0.

Go Online You can complete an Extra Example online.

Watch Out!
Leading Coefficients
If the term with the greatest degree has no coefficient shown, as in part **d**, the leading coefficient is 1.

Think About It!

What values of x make sense in the context of the situation? Justify your reasoning.

Example 4 Evaluate and Graph a Polynomial Function

SUN **The density of the Sun, in grams per centimeter cubed, expressed as a percent of the distance from the core of the Sun to its surface can be modeled by the function $f(x) = 519x^4 - 1630x^3 + 1844x^2 - 889x + 155$, where x represents the percent as a decimal. At the core $x = 0$, and at the surface $x = 1$.**

Part A Evaluate the function.

Find the core density of the Sun at a radius 60% of the way to the surface.

Because we need to find the core density at a radius 60% of the way to the surface, $x = 0.6$. So, replace x with 0.6 and simplify.

$$f(x) = 519x^4 - 1630x^3 + 1844x^2 - 889x + 155$$
$$= 519(0.6)^4 - 1630(0.6)^3 + 1844(0.6)^2 - 889(0.6) + 155$$
$$= 67.2624 - 352.08 + 663.84 - 533.4 + 155$$
$$= 0.6224 \frac{g}{cm^3}$$

Study Tip

Axes Labels Notice that the x-axis is measuring the percent of the radius, not the actual length of the radius.

Part B Graph the function.

Sketch a graph of the function.

Substitute values of x to create a table of values. Then plot the points, and connect them with a smooth curve.

x	$f(x)$
0.1	82.9619
0.2	38.7504
0.3	14.4539
0.4	3.4064
0.5	0.1875
0.7	1.7819
0.8	1.9824
0.9	0.7859

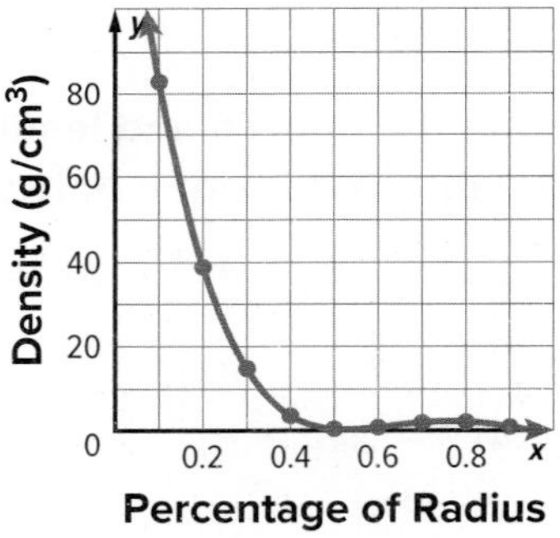

Check

CARDIOLOGY To help predict heart attacks, doctors can inject a concentration of dye in a vein near the heart to measure the cardiac output in patients. In a normal heart, the change in the concentration of dye can be modeled by $f(x) = -0.006x^4 + 0.140x^3 - 0.053x^2 + 1.79x$, where x is the time in seconds.

Part A Find the concentration of dye after 5 seconds.

$f(5) =$ ___?___

Go Online You can complete an Extra Example online.

Part B Select the graph of the concentration of dye over 10 seconds.

A.

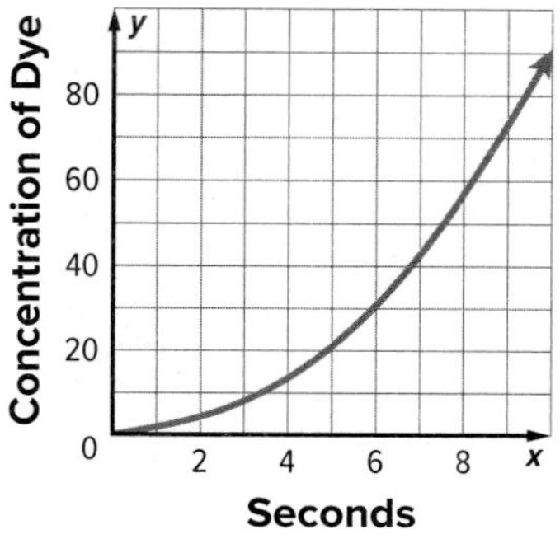

B.

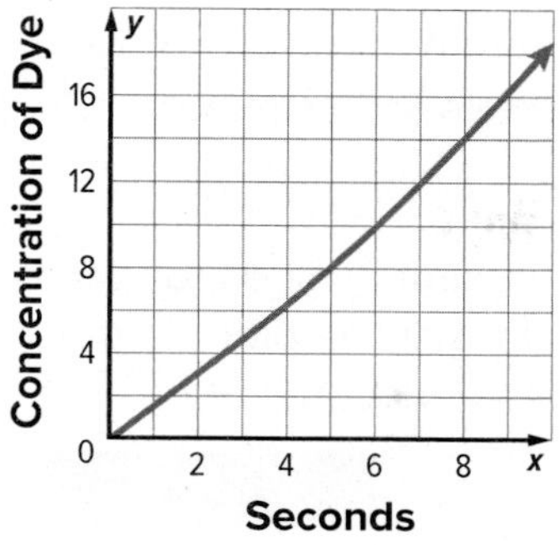

C.

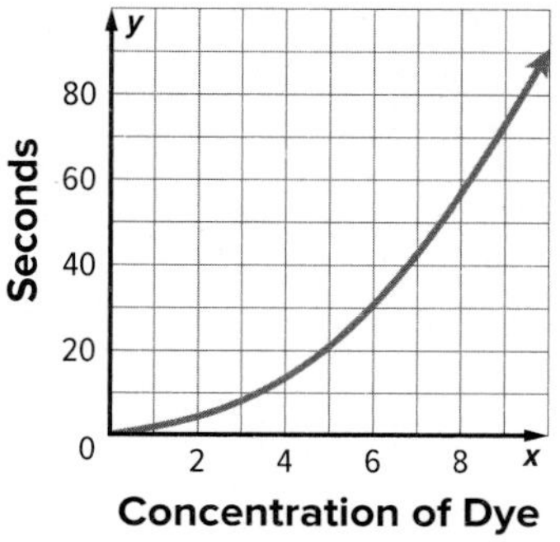

D.

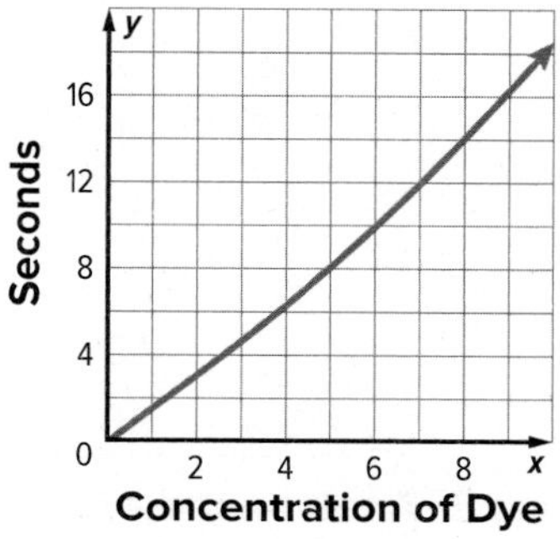

Example 5 Zeros of a Polynomial Function

Use the graph to state the number of real zeros of the function.

The real zeros occur at $x = -2$, 1, and 4, so there are three real zeros.

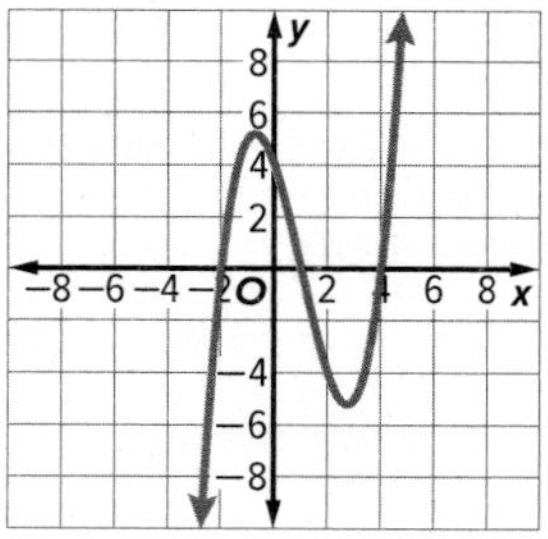

Check

Use the graph to state the number of real zeros of the function.

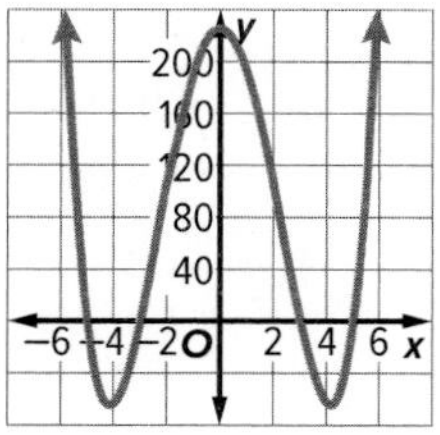

The function has __?__ real zero(s).

Go Online You can complete an Extra Example online.

Study Tip

Zeros The real zeros occur at values of x where $f(x) = 0$, or where the polynomial intersects the x-axis. Recall that odd-degree polynomial functions have at least one real zero and even-degree polynomial functions have any number of real zeros. So, the minimum number of times that an odd-degree polynomial intersects the x-axis is 1, and the minimum number of times that an even-degree polynomial intersects the x-axis is 0.

Think About It!

Find the domain and range of $f(x)$. Does $g(x)$ have the same domain and range? Explain.

Study Tip

Zeros The zeros of a polynomial function are the x-coordinates of the points at which the graph intersects the x-axis.

Example 6 Compare Polynomial Functions

Examine $f(x) = x^3 + 2x^2 - 3x$ and $g(x)$ shown in the graph.

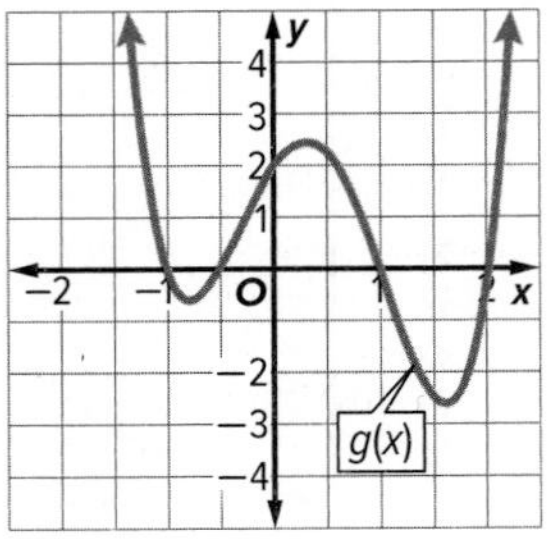

Part A Graph $f(x)$.

Substitute values for x to create a table of values. Then plot the points, and connect them with a smooth curve.

x	$f(x)$
−3	0
−2	6
−1	4
0	0
1	0
2	10
3	36

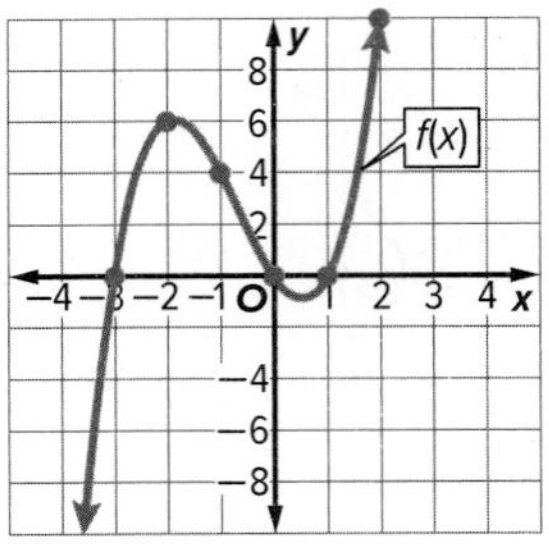

Part B Analyze the extrema.

Which function has the greater relative maximum?

$f(x)$ has a relative maximum at approximately $y = 6$, and $g(x)$ has a relative maximum between $y = 2$ and $y = 3$. So, $f(x)$ has the greater relative maximum.

Part C Analyze the key features.

Compare the zeros, x- and y-intercepts, and end behavior of $f(x)$ and $g(x)$.

zeros:

$f(x)$: −3, 0, 1
$g(x)$: The graph appears to intersect the x-axis at −1, −0.5, 1, 2

intercepts:

$f(x)$: x-intercepts: −3, 0, 1; y-intercept: 0

$g(x)$: x-intercepts: −1, −0.5, 1, 2; y-intercept: 2

end behavior:

$f(x)$: As $x \to -\infty$, $f(x) \to -\infty$, and as $x \to \infty$, $f(x) \to \infty$.

$g(x)$: As $x \to -\infty$, $g(x) \to \infty$, and as $x \to \infty$, $g(x) \to \infty$.

Go Online You can complete an Extra Example online.

Practice

Go Online You can complete your homework online.

Example 1

Describe the end behavior of each function using the leading coefficient and degree, and state the domain and range.

1. $f(x) = 3x^4$

2. $f(x) = -2x^3$

3. $f(x) = -\frac{1}{2}x^5$

4. $f(x) = \frac{3}{4}x^6$

Example 2

5. USE A MODEL The shape of a parabolic reflector inside a flashlight can be modeled by the function $f(x) = \frac{4}{3}x^2$. Graph the function $f(x)$, and state the domain and range.

6. MACHINE EFFICIENCY A company uses the function $f(x) = x^3 + 3x^2 - 18x - 40$ to model the change in efficiency of a machine based on its position x. Graph the function and state the domain and range.

Example 3

State the degree and leading coefficient of each polynomial in one variable. If it is not a polynomial in one variable, explain why.

7. $n + 8$

8. $(2x - 1)(4x^2 + 3)$

9. $-5x^5 + 3x^3 - 8$

10. $18 - 3y + 5y^2 - y^5 + 7y^6$

11. $u^3 + 4u^2t^2 + t^4$

12. $2r - r^2 + \frac{1}{r^2}$

Example 4

13. TRIANGLES Dylan drew n dots on a piece of paper making sure that no set of 3 points were collinear. The number of triangles that can be made using the dots as vertices is equal to $f(n) = \frac{1}{6}(n^3 - 3n^2 + 2n)$, when $n \geq 0$.

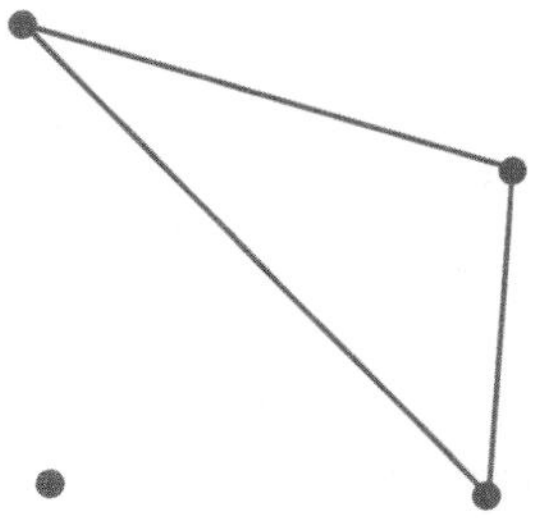

a. If Dylan drew 15 dots, how many triangles can be made?

b. Sketch a graph of the function.

14. DRILLING The volume of a drill bit can be estimated by the formula for a cone, $V = \frac{1}{3}\pi hr^2$, where h is the height of the bit and r is its radius. Substituting $\frac{\sqrt{3}}{3}r$ for h, the volume of the drill bit can be estimated by $V = \frac{\sqrt{3}}{9}\pi r^3$.

a. What is the volume of a drill bit with a radius of 3 centimeters?

b. Sketch a graph of the function in the context of the situation.

Example 5

Use the graph to state the number of real zeros of the function.

15.

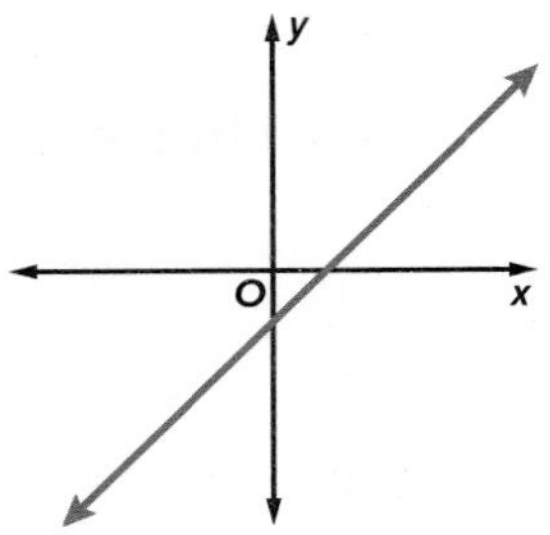

16.

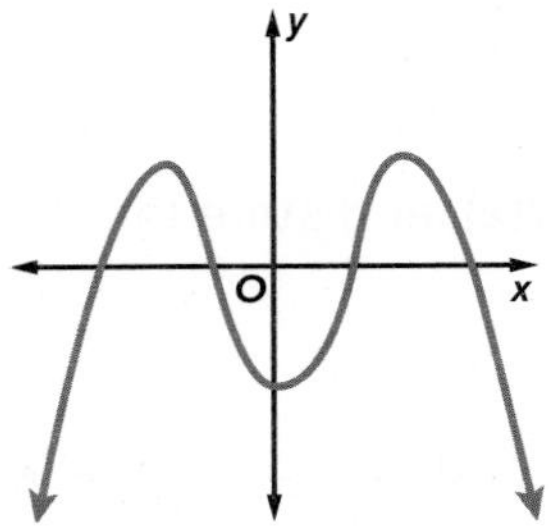

17.

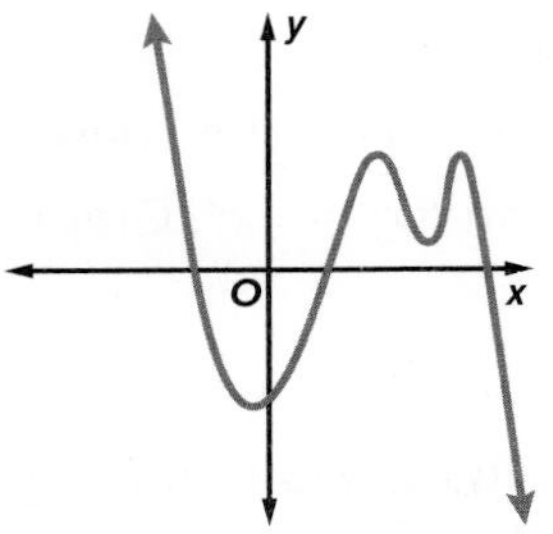

18.

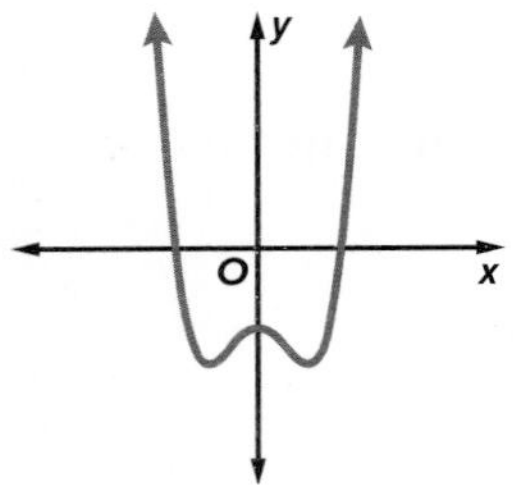

19.

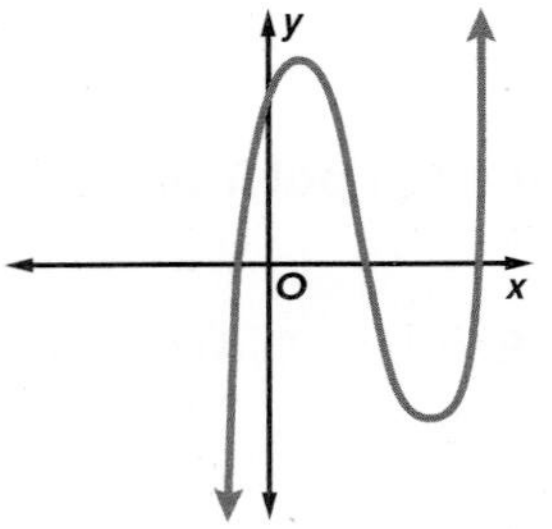

20.

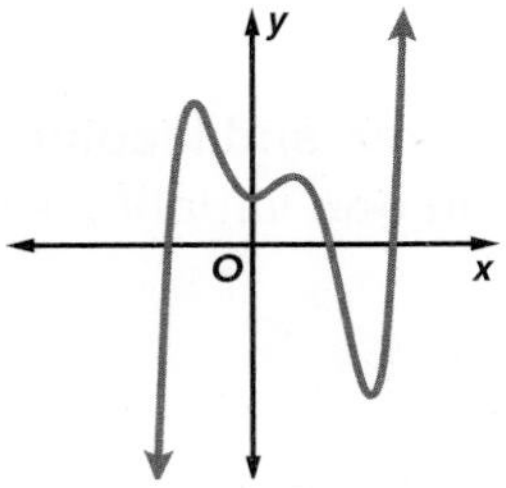

Example 6

21. Examine $f(x) = x^3 - 2x^2 - 4x + 1$ and $g(x)$ shown in the graph.

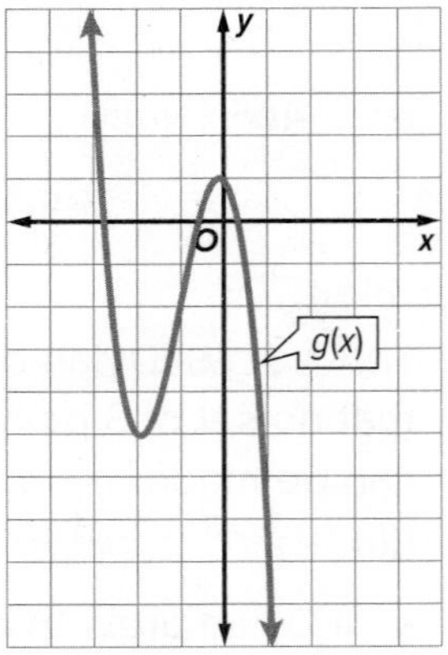

a. Which function has the greater relative maximum?

b. Compare the zeros, x- and y-intercepts, and end behavior of $f(x)$ and $g(x)$.

22. Examine the graph of $f(x)$ and $g(x)$ shown in the table.

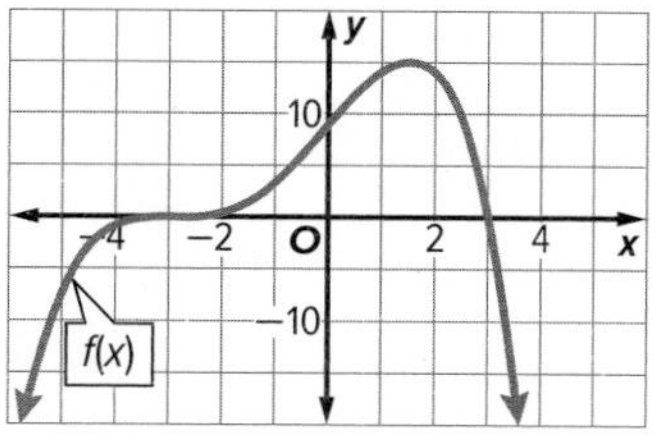

x	−5	−3	0	1.5	3
$g(x)$	7.5	0	−9	−15	0

a. Which function has the greater relative maximum?

b. Compare the zeros, x- and y-intercepts, and end behavior of $f(x)$ and $g(x)$.

Mixed Exercises

Describe the end behavior, state the degree and leading coefficient of each polynomial. If the function is not a polynomial, explain why.

23. $f(x) = -5x^4 + 3x^2$

24. $g(x) = 2x^5 + 6x^4$

25. $g(x) = 8x^4 + 5x^5$

26. $h(x) = 9x^6 - 5x^7 + 3x^2$

27. $f(x) = -6x^6 - 4x^5 + 13x^{-2}$

28. $f(x) = (5 - 2x)(4 + 3x)$

29. $h(x) = (x + 5)(3x - 4)$

30. $g(x) = 3x^7 - 4x^4 + \frac{3}{x}$

31. REASONING Describe the end behavior, and the possible degree and sign of the leading coefficient of the graph shown.

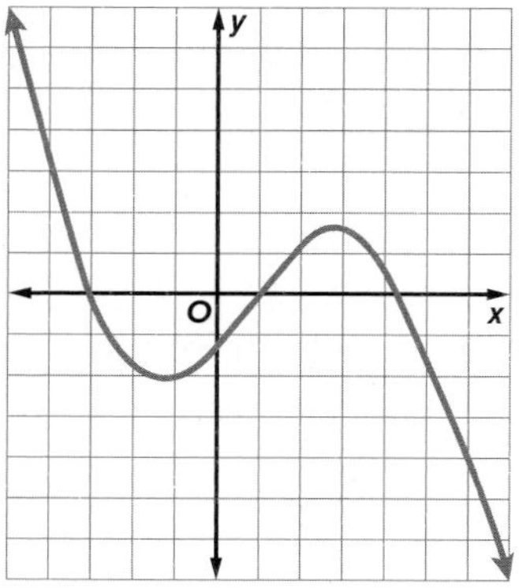

32. CONSTRUCT ARGUMENTS Explain why a polynomial function with an odd degree must have at least one real zero.

33. STRUCTURE If $f(x) = ax^3 - bx^2 + x$, determine $f(1 - x)$. Express the result in standard form. How does the end behavior of $f(1 - x)$ compare to $f(x)$?

34. COMPARING Compare the end behavior of the functions $g(x) = -3x^4 + 15x^3 - 12x^2 + 3x + 20$ and $h(x) = -3x^4 - 16x - 1$. Explain your reasoning.

35. USE A MODEL A box has a square base with sides of 10 centimeters and a height of 4 centimeters. For a new box, the height is increased by twice a number x and the lengths of the sides of the base are decreased by x. Write and graph a function to represent the volume of the new box. What new dimensions will produce a box with the greatest volume? Describe your solution process.

36. FIND THE ERROR Shenequa and Virginia are determining the number of real zeros of the graph. Is either of them correct? Explain your reasoning.

Shenequa

There are 7 real zeros because the graph intersects the x-axis 7 times.

Virginia

There are 8 real zeros because the graph intersects the x-axis 7 times, and there is a double zero.

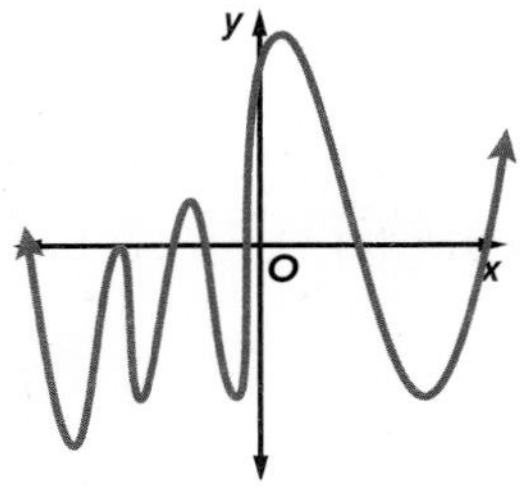

37. ANALYZE Compare the functions $g(x)$ and $f(x)$. Determine which function has the potential for more zeros and the degree of each function.

$g(x) = x^4 + x^3 - 13x^2 + x + 4$

x	−24	−18	−12	−6	0	6	12	18	24
f(x)	−8	−1	3	−2	4	7	−1	−8	5

38. PERSEVERE If $f(x)$ has a degree of 5 and a positive leading coefficient and $g(x)$ has a degree of 3 and a positive leading coefficient, determine the end behavior of $\frac{f(x)}{g(x)}$. Explain your reasoning.

39. CREATE Sketch the graph of an even-degree polynomial with 7 real zeros, one of which is a double zero, and the leading coefficient is negative.

Analyzing Graphs of Polynomial Functions

Learn The Location Principle

If the value of a polynomial function $f(x)$ changes signs from one value of x to the next, then there is a zero between those two x-values. This is called the Location Principle.

Key Concept • Location Principle

Suppose $y = f(x)$ represents a polynomial function, and a and b are two real numbers such that $f(a) < 0$ and $f(b) > 0$. Then the function has at least one real zero between a and b.

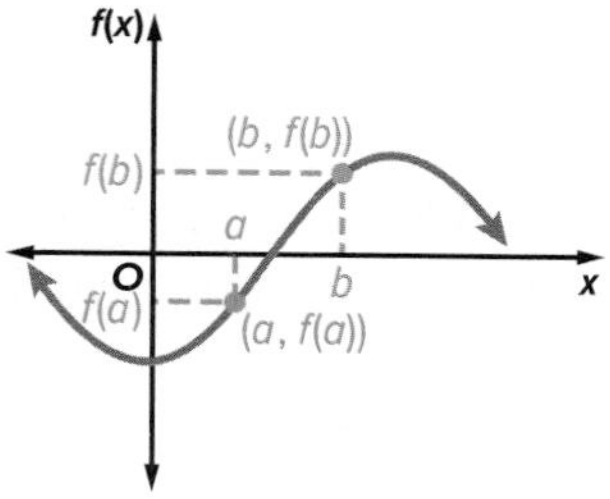

Today's Goals

- Approximate zeros by graphing polynomial functions.
- Find extrema of polynomial functions.

Think About It!

Not all real zeros can be found by using the Location Principle. Provide an example where $f(a) > 0$ and $f(b) > 0$, but there is a zero between $x = a$ and $x = b$.

Example 1 Locate Zeros of a Function

Determine the consecutive integer values of x between which each real zero of $f(x) = x^4 - 2x^3 - x^2 + 1$ is located. Then draw the graph.

Step 1 Make a table.

Because $f(x)$ is a fourth-degree polynomial, it will have as many as 4 real zeros or none at all.

x	−2	−1	0	1	2	3	4
$f(x)$	29	3	1	−1	−3	19	113

Using the Location Principle, there are zeros between $x = 0$ and $x = 1$ and between $x = 2$ and $x = 3$.

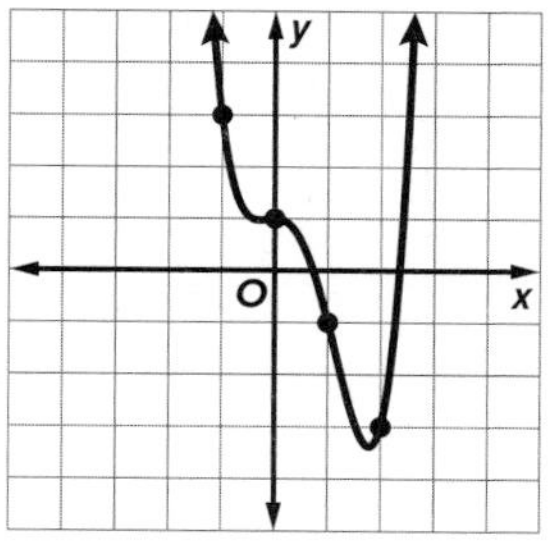

Step 2 Sketch the graph.

Use the table to sketch the graph and find the locations of the zeros.

Think About It!

How can you adjust the table on your graphing calculator to give a more precise interval for the value of each zero?

Check

Use technology to check the location of the zeros.

Input the function into a graphing calculator to confirm that the function crosses the x-axis between $x = 0$ and $x = 1$ and between $x = 2$ and $x = 3$.

You can find more accurate values of the zeros by using the **zero** feature in the CALC menu to find $x \approx 0.7213$ and $x \approx 2.3486$, which confirms the estimates.

Go Online You can complete an Extra Example online.

Check

Determine the consecutive integer values of x between which each real zero of $f(x) = 2x^4 + x^3 - 3x^2 - 2$ is located. Then draw the graph.

$x =$ __?__ and $x =$ __?__

$x =$ __?__ and $x =$ __?__

Study Tip

Turning Points Relative maxima and relative minima of a function are sometimes called turning points.

Learn Extrema of Polynomials

Extrema occur at relative maxima or minima of the function.

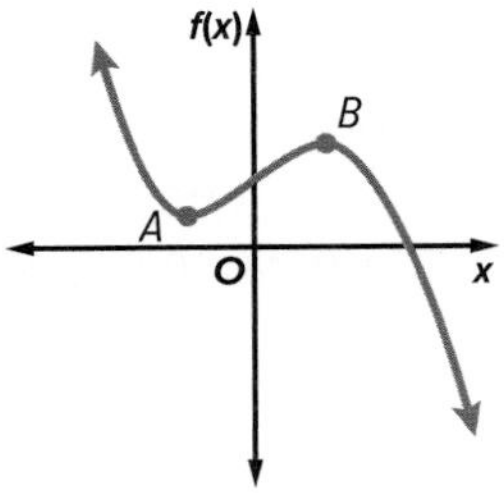

Point A is a relative minimum, and point B is a relative maximum. Both points A and B are extrema. The graph of a polynomial of degree n has at most $n - 1$ extrema.

Example 2 Identify Extrema

Use a table to graph $f(x) = x^3 + x^2 - 5x - 2$. Estimate the x-coordinates at which the relative maxima and relative minima occur.

Step 1 Make a table of values and graph the function.

x	$f(x)$
−4	−30
−3	−5
−2	4
−1	3
0	−2
1	−5
2	0
3	19

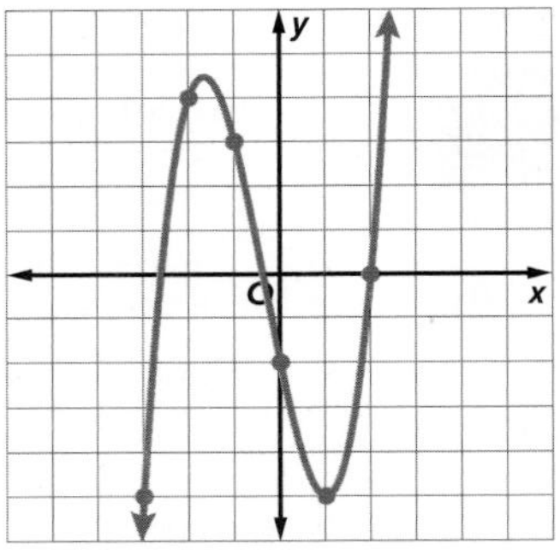

Step 2 Estimate the locations of the extrema.

The value of $f(x)$ at $x = -2$ is greater than the surrounding points indicating a maximum *near* $x = -2$.

The value of $f(x)$ at $x = 1$ is less than the surrounding points indicating a minimum *near* $x = 1$.

You can use a graphing calculator to find the extrema of a function and confirm your estimates.

Study Tip

Extrema When graphing with a calculator, keep in mind that a polynomial of degree n has at most $n - 1$ extrema. This will help you to determine whether your viewing window is allowing you to see all of the extrema of the graph.

Go Online You can complete an Extra Example online.

Check

Copy and complete the table for $f(x) = -x^4 - x^3 + 5x^2 + x - 3$ to estimate the x-coordinates at which the relative maxima and relative minima occur.

x	$f(x)$
−3	
−2	
−1	
0	
1	
2	
3	

The relative maxima occur near $x =$ __?__ and $x =$ __?__.

The relative minimum occurs near $x =$ __?__.

Example 3 Analyze a Polynomial Function

PILOTS **The total number of certified pilots in the United States is approximated by $f(x) = 0.0000903x^4 - 0.0166x^3 + 0.762x^2 + 6.317x + 7.708$, where x is the number of years after 1930 and $f(x)$ is the number of pilots in thousands. Graph the function and describe its key features over the relevant domain.**

Step 1 Graph the function.

Make a table of values. Plot the points and connect them with a smooth curve.

x	$f(x)$
0	7.708
10	131.381
20	320.496
30	507.961
40	648.356
50	717.933
60	714.616
70	658.001
80	589.356
90	571.621

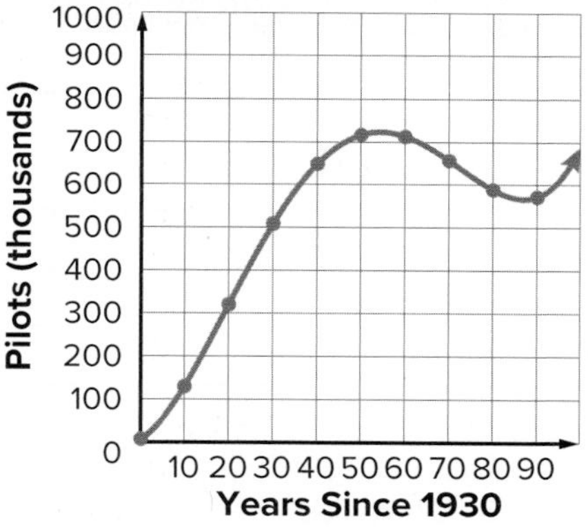

Step 2 Describe the key features.
Domain and Range

The domain and range of the function is all real numbers. Because the function models years after 1930, the relevant domain and range are $\{x \mid x \geq 0\}$ and $\{f(x) \mid f(x) \geq 7.708\}$.

(continued on the next page)

Go Online
You can learn how to graph and analyze a polynomial function on a graphing calculator by watching the video online.

Think About It!
What trends in the number of pilots does the graph suggest?

Talk About It!

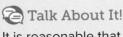

It is reasonable that the trend will continue indefinitely? Explain.

Extrema

There is a relative maximum between 1980 and 1990 and a relative minimum between 2010 and 2020 in the relevant domain.

End Behavior

As $x \to \infty$, $f(x) \to \infty$.

Intercepts

In the relevant domain, the y-intercept is at (0, 7.708). There is no x-intercept, or zero, because the function begins at a value greater than 0 and as $x \to \infty$, $f(x) \to \infty$.

Symmetry The graph of the function does not have symmetry.

Study Tip

Assumptions Determining the end behavior for the graph of a polynomial that models data assumes that the trend continues and there are no other relative maxima or minima.

Check

COINS The number of quarters produced by the United States Mint can be approximated by the function $f(x) = 16.4x^3 - 149.5x^2 - 148.9x + 3215.4$, where x is the number of years since 2005 and $f(x)$ is the total number of quarters produced in millions. Use a graph of the function to complete the table and describe its key features.

Part A Copy and complete the table.

x, Years	$f(x)$, Quarters (millions)
0	
2	
4	
6	
8	
10	

Part B Describe the key features.

The relevant domain is ___?___.

The relevant range is ___?___.

There is a relative minimum between ___?___ and ___?___.

The y-intercept is ___?___.

The graph of the function ___?___ have symmetry.

It is ___?___ to assume that the trend will continue indefinitely.

Go Online You can complete an Extra Example online.

Example 4 Use a Polynomial Function and Technology to Model

BACKPACKS **The table shows U.S. backpack sales in millions of dollars, according to the Travel Goods Association. Make a scatter plot and a curve of best fit to show the trend over time. Then determine the backpack sales in 2015.**

Year	Sales (million $)	Year	Sales (million $)
2000	1140	2008	1246
2001	1144	2009	1235
2002	1113	2010	1419
2003	1134	2011	1773
2004	1164	2012	1930
2005	1180	2013	2255
2006	1364	2014	2779
2007	1436		

Step 1 Enter the data.

Let the year 2000 be represented by 0. Enter the years since 2000 in List 1. Enter the backpack sales in List 2.

Step 2 Graph the scatter plot.

Choose the scatter plot feature in the **STAT PLOT** menu. Use List 1 for the **Xlist** and List 2 for the **Ylist**. Change the viewing window so that all the data are visible.

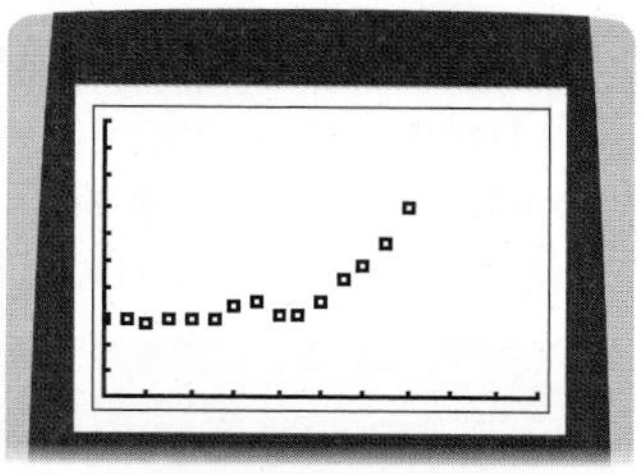

[0, 20] scl: 2, [0, 4000] scl: 400

Step 3 Determine the polynomial function of best fit.

To determine the model that best fits the data, perform linear, quadratic, cubic, and quartic regressions, and compare the coefficients of determination, r^2. The polynomial with a coefficient of determination closest to 1 will fit the data best.

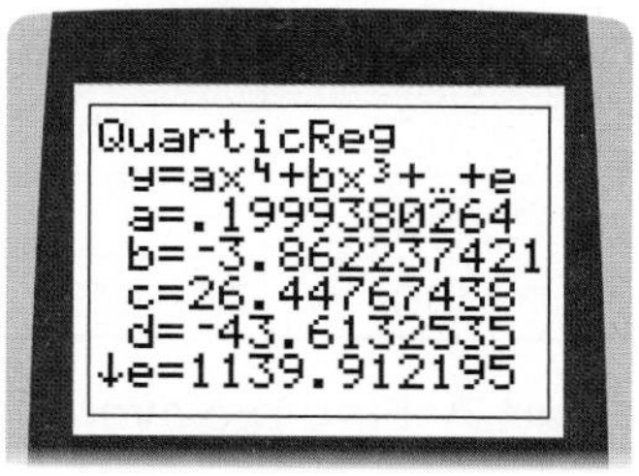

A quartic function fits the data best.

The regression equation with coefficients rounded to the nearest tenths is:

$$y \approx 0.2x^4 - 3.9x^3 + 26.4x^2 - 43.6x + 1139.9.$$

Step 4 Graph and evaluate the regression function.

Assuming that the trend continues, the graph of the function can be used to predict backpack sales for a specific year. To determine the total sales in 2015, find the value of the function for $x = 15$.

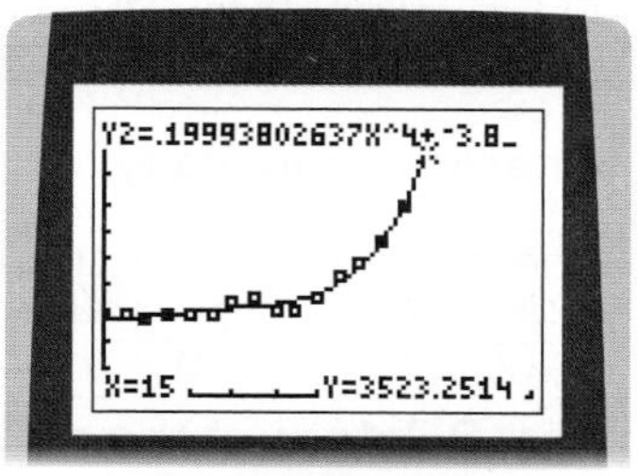

[0, 20] scl: 2, [0, 4000] scl: 400

In 2015, there were about $3.523 billion in backpack sales.

Math History Minute

By the age of 20, Italian mathematician **Maria Gaetana Agnesi (1718–1799)** had started working on her book *Analytical Institutions,* which was published in 1748. Early chapters included problems on maxima, minima, and turning points. Also described was a cubic curve called the "witch of Agnesi," which was translated incorrectly from the original Italian.

Think About It!

Explain the approximation that is made when using the model to determine the backpack sales in a specific year.

Check

TREES To estimate the amount of lumber that can be harvested from a tree, foresters measure the diameter of each tree. Determine the polynomial function of best fit, where x represents the diameter of a tree in inches and y is the estimated volume measured in board feet. Then estimate the volume of a tree with a diameter of 35 inches.

Diam (in.)	17	19	20	23	25	28	32	38	39	41
Vol (100s of board ft)	19	25	32	57	71	113	123	252	259	294

Polynomial function of best fit:
$y =$ ___?___ $x^4 +$ ___?___ $x^3 -$ ___?___ $x^2 +$ ___?___ $x -$ ___?___

The estimated volume of a 35-inch diameter tree to the nearest board foot is ___?___ of 100s board ft.

Study Tip

g-force One G is the acceleration due to gravity at the Earth's surface. Defined as 9.80665 meters per second squared, this is the g-force you experience when you stand still on Earth. On a roller coaster, you experience 0 Gs and feel weightless at the top of the hills, and you can experience a g-force of 6 Gs or more as you are pushed into your seat at the bottom of the hills.

Example 5 Find Average Rate of Change

ROCKETS **The Ares-V rocket was designed to carry as much as 75 tons of supplies and 4 astronauts to the Moon and possibly even to Mars. The table shows the expected g-force on the rocket over the course of its 200-second launch.**

Time (s)	Acceleration (Gs)	Time (s)	Acceleration (Gs)
0	1.34	120	1.46
20	1.26	140	1.93
40	1.12	160	2.47
60	1.01	180	2.84
80	1	200	2.2
100	1.15		

Think About It!

Does the average rate of change from 0 to 200 seconds accurately describe the acceleration of the launch? Justify your reasoning.

Part A Find the average rate of change.

Sketch the graph, and estimate the average rate of change of the acceleration. Then check your results algebraically.

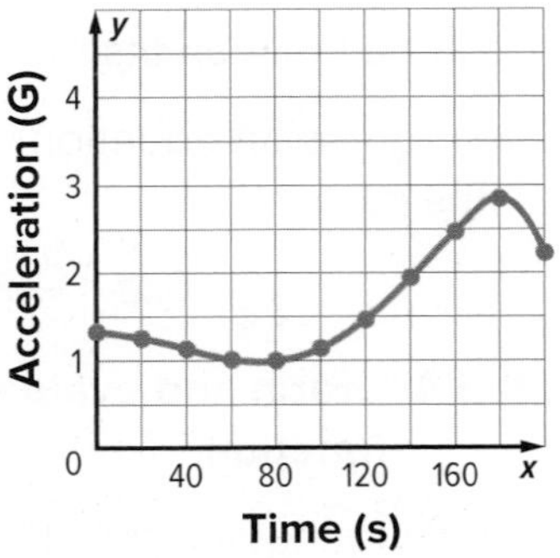

Estimate: From the graph, the change in the y-values is about 0.9, and the change in the x-values is 200. So, the rate of change is about $\frac{0.9}{200}$ or 0.0045.

Check algebraically:

The average rate of change is
$\frac{f(200) - f(0)}{200 - 0} = \frac{2.2 - 1.34}{200 - 0}$ or 0.0043.

Part B Interpret the results.

From 0 to 200 seconds, the average rate of change in acceleration was an increase of 0.0043 Gs per second.

Go Online You can complete an Extra Example online.

Practice

Go Online You can complete your homework online.

Example 1

Determine the consecutive integer values of *x* between which each real zero of each function is located by using a table. Then sketch the graph.

1. $f(x) = x^2 + 3x - 1$

2. $f(x) = -x^3 + 2x^2 - 4$

3. $f(x) = x^3 + 4x^2 - 5x + 5$

4. $f(x) = -x^4 - x^3 + 4$

Example 2

Use a table to graph each function. Then estimate the *x*-coordinates at which relative maxima and relative minima occur.

5. $f(x) = -2x^3 + 12x^2 - 8x$

6. $f(x) = 2x^3 - 4x^2 - 3x + 4$

7. $f(x) = x^4 + 2x - 1$

8. $f(x) = x^4 + 8x^2 - 12$

Example 3

9. BUSINESS A banker models the expected value v of a company in millions of dollars by using the formula $v = n^3 - 3n^2$, where n is the number of years in business. Graph the function and describe its key features over the relevant domain.

10. HEIGHT A plant's height is modeled by the function $f(x) = 1.5x^3 - 20x^2 + 85x - 84$, where x is the number of weeks since the seed was planted and $f(x)$ is the height of the plant. Graph the function and describe its key features over its relevant domain.

Example 4

11. USE ESTIMATION The table shows U.S. car sales in millions of cars. Use a graphing calculator to make a scatter plot and a curve of best fit to show the trend over time. Then use the equation to estimate the car sales in 2017. Let 2008 be represented by year 0. Round the coefficients of the regression equation to the thousandths place.

Year	Cars (millions)	Year	Cars (millions)	Year	Cars (millions)
2008	7.659	2011	6.769	2014	6.089
2009	7.761	2012	5.400	2015	7.243
2010	7.562	2013	5.635	2016	7.780

12. POPULATION The table shows the population in Cincinnati, Ohio, since 1960. Make a scatter plot and a curve of best fit to show the trend over the given time period. Then use the equation to estimate the population of Cincinnati in 2020. Let 1960 be represented by year 0.

Year	Population	Year	Population
1960	502,550	1990	364,553
1970	452,524	2000	331,258
1980	385,457	2010	296,943

13. VOLUNTEERS The table shows average volunteer hours per month for a local non-profit. Make a scatter plot and a curve of best fit to show the trend over the given time period. Then use the equation to estimate the volunteer hours for September. Let January be represented by month 1.

Month	Volunteer Hours	Month	Volunteer Hours
Jan.	48	May	100
Feb.	60	June	110
Mar.	72	July	105
Apr.	75	Aug.	93

Example 5

14. FARMS The table shows the number of farms in the U.S. at various years, according to the USDA Census of Agriculture. Find the average rate of change from 1982 to 2012. Interpret the results in the context of the situation.

Year	Farms	Year	Farms
1982	2,480,000	2002	2,130,000
1987	2,340,000	2007	2,200,000
1992	2,180,000	2012	2,110,000
1997	2,220,000		

15. SALARY The table shows the annual salary of a salesperson over time. Find the average rate of change over the given time interval. Interpret the results in context of this situation.

Year	Salary	Year	Salary
2012	$45,000	2015	$55,500
2013	$49,000	2016	$73,000
2014	$47,500	2017	$67,500

Mixed Exercises

Graph each function by using a table of values. Then, estimate the *x*-coordinates at which each zero and relative extrema occur, and state the domain and range.

16. $f(x) = x^3 - 3x + 1$

17. $f(x) = 2x^3 + 9x^2 + 12x + 2$

18. $f(x) = 2x^3 - 3x^2 + 2$

19. $f(x) = x^4 - 2x^2 - 2$

20. Determine the key features for $y = \begin{cases} x^2 & \text{if } x \leq -4 \\ 5 & \text{if } -4 < x \leq 0. \\ x^3 & \text{if } x > 0 \end{cases}$

USE TOOLS Use a graphing calculator to estimate the *x*-coordinates at which the maxima and minima of each function occur. Round to the nearest hundredth.

21. $f(x) = x^3 + 3x^2 - 6x - 6$

22. $f(x) = -2x^3 + 4x^2 - 5x + 8$

23. $f(x) = -2x^4 + 5x^3 - 4x^2 + 3x - 7$

24. $f(x) = x^5 - 4x^3 + 3x^2 - 8x - 6$

25. PRECISION Sketch the graph of a third-degree polynomial function that has a relative minimum at $x = -3$, passes through the origin, and has a relative maximum at $x = 2$. Describe the end behavior of the graph. Based on the sketch, determine whether the leading coefficient is negative or positive.

26. USE TOOLS A canister has the shape of a cylinder with spherical caps on either end. The volume of the canister in cubic millimeters is modeled by the function $V(x) = \pi(x^3 + 3x^2) + \frac{4}{3}\pi x^3$ where x represents the radius in millimeters of a canister that is $x + 3$ millimeters wide.

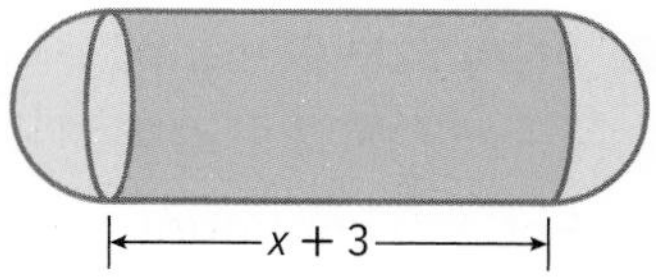

a. Use a graphing calculator to sketch the model that represents the volume of the canister. Include axes labels.

b. What is the domain of the model? Explain any restrictions that apply.

27. **CONSTRUCT ARGUMENTS** What type of polynomial function best models the data in the graph? Explain your reasoning.

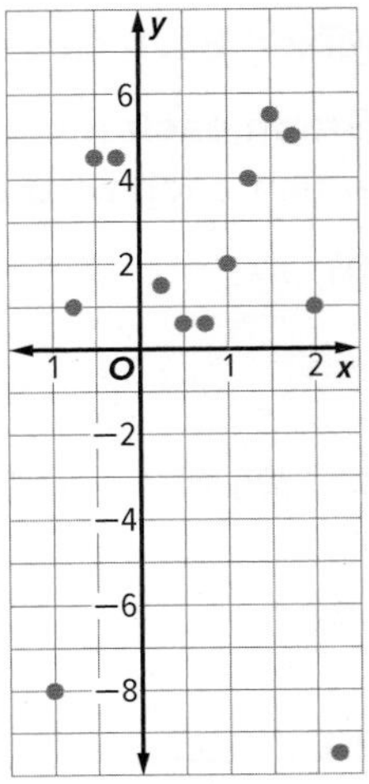

28. **FORECASTING** The table shows the number of deliveries a grocery store has made since they began to offer the service. Use a scatter plot and a curve of best fit to determine the number of deliveries the grocery store can expect to deliver 10 years after they begin the service.

Year	Deliveries	Year	Deliveries
1	60	5	175
2	193	6	156
3	235	7	195
4	210	8	328

29. **ANALYZE** Explain why the leading coefficient and the degree are the only determining factors in the end behavior of a polynomial function.

30. **ANALYZE** The table below shows the values of a cubic function. Could there be a zero between $x = 2$ and $x = 3$? Justify your argument.

x	−2	−1	0	1	2	3
$g(x)$	3	−2	−1	1	−2	−2

31. **CREATE** Sketch the graph of an odd degree polynomial function with 6 extrema and a relative extreme at $y = 0$.

32. **ANALYZE** Determine whether the following statement is *sometimes, always,* or *never* true. Justify your argument.

For any continuous polynomial function, the y-coordinate of a point where a function changes from increasing to decreasing or from decreasing to increasing is either a relative maximum or relative minimum.

33. **PERSEVERE** A function is even if for every x in the domain of f, $f(x) = f(-x)$. Is every even-degree polynomial function also an even function? Explain.

34. **PERSEVERE** A function is odd if for every x in the domain, $-f(x) = f(-x)$. Is every odd-degree polynomial function also an odd function? Explain.

Operations with Polynomials

Learn Adding and Subtracting Polynomials

A polynomial is a monomial or the sum of two or more monomials. A **binomial** is the sum of two monomials, and a **trinomial** is the sum of three monomials. The degree of a polynomial is the greatest degree of any term in the polynomial.

Polynomials can be added or subtracted by performing the operations indicated and combining like terms. You can subtract a polynomial by adding its additive inverse.

The sum or difference of polynomials will have the same variables and exponents as the original polynomials, but possibly different coefficients. Thus, the sum or difference of two polynomials is also a polynomial.

A set is **closed** if and only if an operation on any two elements of the set produces another element of the same set. Because adding or subtracting polynomials results in a polynomial, the set of polynomials is closed under the operations of addition and subtraction.

Example 1 Identify Polynomials

Determine whether each expression is a polynomial. If it is a polynomial, state the degree of the polynomial.

a. $x^6 + \sqrt[3]{x} - 4$

This expression is not a polynomial because $\sqrt[3]{x}$ is not a monomial.

b. $5a^4b + 3a^2b^7 - 9$

This expression is a polynomial because each term is a monomial. The degree of the first term is $4 + 1$ or 5, the degree of the second term is $2 + 7$ or 9, and the degree of the third term is 0. So, the degree of the polynomial is 9.

c. $\frac{2}{3}x^{-5} - 6x^{-3} - x$

The expression is not a polynomial because x^{-5} and x^{-3} are not monomials.

Check

State the degree of each polynomial.

A. $x^7 + 6x^5 - \frac{1}{3}$

B. $3c^7d^2 + 5cd - 9$

C. p^{10}

D. 25

Go Online You can complete an Extra Example online.

Today's Goals
- Add and subtract polynomials.
- Multiply polynomials.

Today's Vocabulary
binomial

trinomial

closed

FOIL method

Think About It!
Identify the like terms in $8x^5 + 3x^2 - 10$ and $4x^3 - 7x^2 + 5$.

Study Tip
Degree 0 and 1 Remember that constant terms have a degree of 0 and variable terms with no exponent indicated have a degree of 1.

Go Online
You can learn how to add and subtract polynomials by watching the video online.

Example 2 Add Polynomials

Find $(6x^3 + 7x^2 - 2x + 5) + (x^3 - 4x^2 - 8x + 1)$.

Method 1 Add horizontally.

Group and combine like terms.

$(6x^3 + 7x^2 - 2x + 5) + (x^3 - 4x^2 - 8x + 1)$ Original expression

$= (6x^3 + x^3) + (7x^2 - 4x^2) + (-2x - 8x) + (5 + 1)$ Group like terms.

$= 7x^3 + 3x^2 - 10x + 6$ Combine like terms.

Method 2 Add vertically.

Align like terms vertically and add.

$$\begin{array}{r} 6x^3 + 7x^2 - 2x + 5 \\ \underline{(+)\ x^3 - 4x^2 - 8x + 1} \\ 7x^3 + 3x^2 - 10x + 6 \end{array}$$

Align like terms.

Combine like terms.

Check

Find $(2x^3 + 9x^2 + 6x - 3) + (4x^3 - 7x^2 + 5x)$.

Talk About It!

Notice that the terms in the sum have the same variables and exponents as the terms of the original expressions: x^3, x^2, x, and a constant. Will this always be the case? Explain.

Example 3 Subtract Polynomials

Find $(2x^5 + 11x^4 + 7x - 8) - (5x^4 + 9x^3 - 3x + 4)$.

Method 1 Subtract horizontally.

Group and combine like terms.

$(2x^5 + 11x^4 + 7x - 8) - (5x^4 + 9x^3 - 3x + 4)$ Original expression

$= 2x^5 + 11x^4 + 7x - 8 - 5x^4 - 9x^3 + 3x - 4$ Distribute −1.

$= 2x^5 + (11x^4 - 5x^4) + (-9x^3) + (7x + 3x) + (-8 - 4)$ Group like terms.

$= 2x^5 + 6x^4 - 9x^3 + 10x - 12$ Combine like terms.

Method 2 Subtract vertically.

Align like terms vertically and subtract.

$$\begin{array}{r} 2x^5 + 11x^4 + 0x^3 + 7x - 8 \\ \underline{(-)\ 0x^5 + 5x^4 + 9x^3 - 3x + 4} \\ 2x^5 + 6x^4 - 9x^3 + 10x - 12 \end{array}$$

Check

Find $(8x^2 - 3x + 1) - (5x^3 + 2x^2 - 6x - 9)$.

Study Tip

Additive Inverse Distributing −1 to each term in the polynomial being subtracted is the same as finding and adding its additive inverse.

Think About It!

Why is it helpful to insert placeholders for the $0x^3$ and $0x^5$ terms? Why were placeholders not included for the x^2-term in either polynomial?

Explore Multiplying Polynomials

Online Activity Use a table to complete the Explore.

INQUIRY How is using a table to multiply polynomials related to the Distributive Property?

Learn Multiplying Polynomials

Polynomials can be multiplied by using the Distributive Property to multiply each term in one polynomial by each term in the other. When polynomials are multiplied, the product is also a polynomial. Therefore, the set of polynomials is closed under the multiplication. This is similar to the system of integers, which is also closed under multiplication. To multiply two binomials, you can use a shortcut called the **FOIL method**.

Key Concept • FOIL Method

Words: Find the sum of the products of ***F*** the *First* terms, ***O*** the *Outer* terms, ***I*** the *Inner* terms, and ***L*** the *Last* terms.

Symbols:

F L I O

Product of First Terms | Product of Outer Terms | Product of Inner Terms | Product of Last Terms

$$(2x + 4)(x - 3) = (2x)(x) + (2x)(-3) + (4)(x) + (4)(-3)$$
$$= 2x^2 - 6x + 4x - 12$$
$$= 2x^2 - 2x - 12$$

Example 4 Simplify by Using the Distributive Property

Find $2x(4x^3 + 5x^2 - x - 7)$.

$$2x(4x^3 + 5x^2 - x - 7) = 2x(4x^3) + 2x(5x^2) + 2x(-x) + 2x(-7)$$
$$= 8x^4 + 10x^3 - 2x^2 - 14x$$

Example 5 Multiply Binomials

Find $(3a + 5)(a - 7)(4a + 1)$.

Step 1 Multiply any two binomials.

$(3a + 5)(a - 7) = 3a(a) + 3a(-7) + 5(a) + 5(-7)$ FOIL Method

$= 3a^2 - 21a + 5a - 35$ Multiply.

$= 3a^2 - 16a - 35$ Combine like terms.

Step 2 Multiply the result by the remaining binomial.

$(3a^2 - 16a - 35)(4a + 1)$

$$= 3a^2(4a + 1) + (-16a)(4a + 1) + (-35)(4a + 1)$$
$$= 12a^3 + 3a^2 - 64a^2 - 16a - 140a - 35$$
$$= 12a^3 - 61a^2 - 156a - 35$$

Check

Find $(-2r - 3)(5r - 1)(r + 4)$.

Go Online You can complete an Extra Example online.

Think About It!

Why are the exponents added when you multiply the monomials?

Go Online

for an example of how to multiply two trinomials.

Go Online to see Example 6.

Apply Example 7 Write and Simplify a Polynomial Expression

BAKING **Byron is baking a three-tier cake for a birthday party. Each tier will have $\frac{1}{2}$ the volume of the previous tier. The dimensions of the first tier are shown. Find the total volume of the cake.**

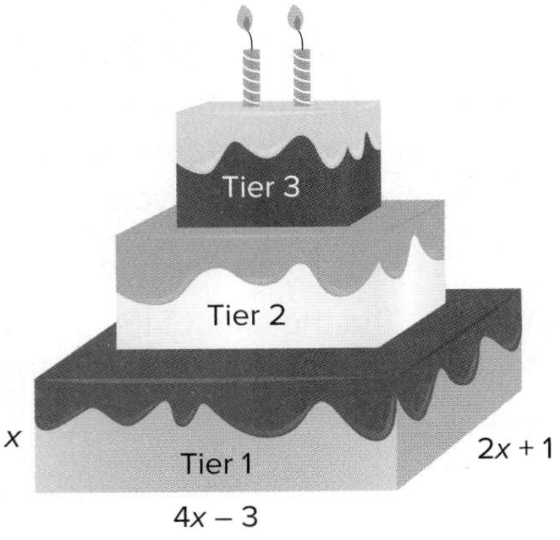

1 What is the task?

Describe the task in your own words. Then list any questions that you may have. How can you find answers to your questions?

I need to find the total volume of the cake, which is the sum of all 3 tiers. How can I represent the volume of each tier as a polynomial? Which properties will I need to know? I can find the answers to my questions by referencing other examples in the lesson.

2 How will you approach the task? What have you learned that you can use to help you complete the task?

I will find and simplify the volume of each tier and then add them together. I will use the Distributive Property and FOIL method to complete the task.

3 What is your solution?

Use your strategy to solve the problem.

What is the volume of each tier?

Tier 1: $8x^3 - 2x^2 - 3x$, Tier 2: $4x^3 - x^2 - 1.5x$,
Tier 3: $2x^3 - 0.5x^2 - 0.75x$

What is the total volume of the cake?

$14x^3 - 3.5x^2 - 5.25x$

4 How can you know that your solution is reasonable?

Write About It! Write an argument that can be used to defend your solution.

Because all the expressions are based on the expression for the volume of Tier 1, I can check that the expression for Tier 1 is correct. I can factor the expression for volume of Tier 1 to ensure that the factors are the same as the given dimensions.

$8x^3 - 2x^2 - 3x$	Expression for Tier 1
$= (8x^2 - 2x - 3)x$	Factor x from each term.
$= (4x - 3)(2x + 1)x$ ✓	Factor $8x^2 - 2x - 3$.

Think About It!
What does x represent in the polynomial expression for the volume of the cake?

Problem-Solving Tip
Solve a Simpler Problem Some complicated problems can be more easily solved by breaking them into several simpler problems. In this case, finding the volume of each tier individually simplifies the situation and makes finding the total volume easier.

Practice

Go Online You can complete your homework online.

Example 1

Determine whether each expression is a polynomial. If it is a polynomial, state the degree of the polynomial.

1. $2x^2 - 3x + 5$
2. $a^3 - 11$
3. $\frac{5np}{n^2} - \frac{2g}{h}$
4. $\sqrt{m - 7}$

Examples 2 and 3

Add or subtract.

5. $(6a^2 + 5a + 10) - (4a^2 + 6a + 12)$
6. $(7b^2 + 6b - 7) - (4b^2 - 2)$
7. $(g + 5) + (2g + 7)$
8. $(5d + 5) - (d + 1)$
9. $(x^2 - 3x - 3) + (2x^2 + 7x - 2)$
10. $(-2f^2 - 3f - 5) + (-2f^2 - 3f + 8)$
11. $(2x - 3) - (5x - 6)$
12. $(x^2 + 2x - 5) - (3x^2 - 4x + 7)$

Examples 4, 5 and 6

Multiply.

13. $3p(np - z)$
14. $4x(2x^2 + y)$
15. $-5(2c^2 - d^2)$
16. $x^2(2x + 9)$
17. $(a - 5)^2$
18. $(2x - 3)(3x - 5)$
19. $(x - y)(x^2 + 2xy + y^2)$
20. $(a + b)(a^3 - 3ab - b^2)$
21. $(x - y)(x + y)(2x + y)$
22. $(a + b)(2a + 3b)(2x - y)$
23. $(r - 2t)(r + 2t)$
24. $(3y + 4)(2y - 3)$
25. $(x^3 - 3x^2 + 1)(2x^2 - x + 2)$
26. $(4x^5 + x^3 - 7x^2 + 2)(3x - 1)$

Example 7

27. CONSTRUCTION A rectangular deck is built around a square pool. The pool has side length s. The length of the deck is 5 units longer than twice the side length of the pool. The width of the deck is 3 units longer than the side length of the pool. What is the area of the deck in terms of s?

28. VOLUME The volume of a rectangular prism is given by the product of its length, width, and height. A rectangular prism has a length of b^2 units, a width of a units, and a height of $ab + c$ units. What is the volume of the rectangular prism? Express your answer in simplified form.

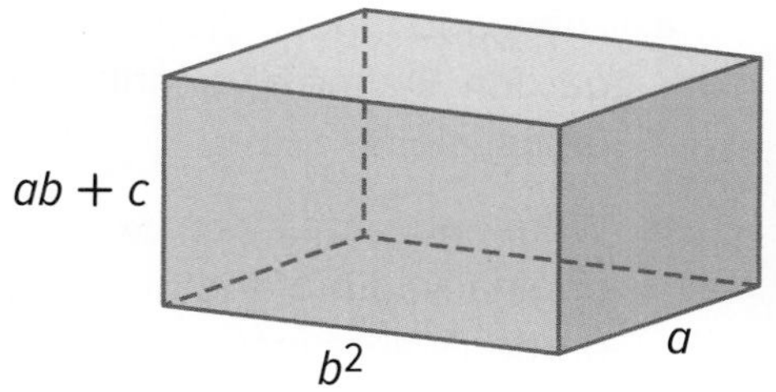

29. SAIL BOATS Tamara is making a sail for her sailboat.

a. Refer to the diagram to find the area of the sail.

b. If Tamara wants fabric on each side of her sail, write a polynomial to represent the total amount of fabric she will need to make the sail.

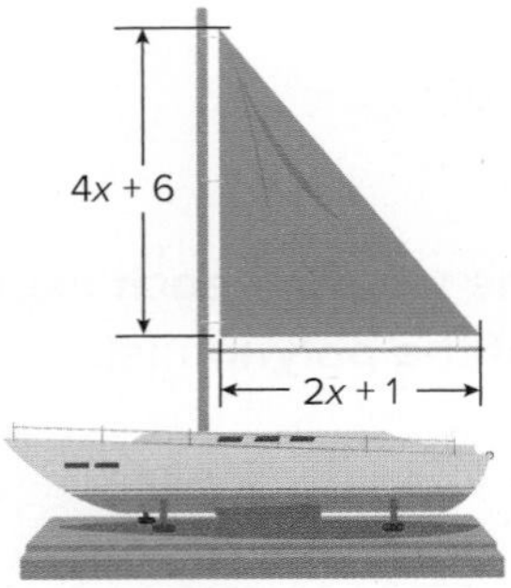

Mixed Exercises

Simplify.

30. $5xy(2x - y) + 6y^2(x^2 + 6)$

31. $3ab(4a - 5b) + 4b^2(2a^2 + 1)$

32. $\frac{1}{4}g^2(8g + 12h - 16gh^2)$

33. $\frac{1}{3}n^3(6n - 9p + 18np^4)$

34. $(g^3 - h)(g^3 + h)$

35. $(n^2 - 7)(2n^3 + 4)$

36. $(2x - 2y)^3$

37. $(4n - 5)^3$

38. $(3z - 2)^3$

39. $\frac{1}{4}(16x - 12y) + \frac{1}{3}(9x + 3y)$

40. STRUCTURE Use the polynomials $f(x) = -6x^3 + 2x^2 + 4$ and $g(x) = x^4 - 6x^3 - 2x$ to evaluate and simplify the given expression. Determine the degree of the resulting polynomial. Show your work.

a. $f(x) + g(x)$

b. $g(x) - f(x)$

41. STRUCTURE Use the polynomials $f(x) = 3x^2 - 1$, $g(x) = x + 2$, and $h(x) = -x^2 - x$ to evaluate and simplify the given expressions. Determine the degree of the resulting polynomial.

a. $f(x)g(x)$

b. $h(x)f(x)$

c. $[f(x)]^2$

42. USE A MODEL Inez wants to increase the size of her rectangular garden. The original garden is 8 feet longer than it is wide. For the new garden, she will increase the length by 25% and increase the width by 5 feet.

a. Draw and label a diagram that represents the original garden and the new garden. Define a variable and label each dimension with appropriate expressions.

b. Write and simplify an expression for the increase in area of the garden. If the original width of the garden was 10 feet, find how many square feet the garden's area increased.

43. CONSTRUCT ARGUMENTS Complete the table to show which sets are closed under the operations. Write *yes* if the set is closed under the operation. Write *no* and provide a counterexample if the set is not closed under the operation. Assume that since division by zero is undefined, it does not affect closure.

	Addition and Subtraction	Multiplication	Division
Integers			
Rational Numbers			
Polynomials			

44. STRUCTURE The polynomial $2x^2 + 3x + 1$ can be represented by the tiles shown in the figure at the right. These tiles can be arranged to form the rectangle shown. Notice that the area of the rectangle is $2x^2 + 3x + 1$ units^2.

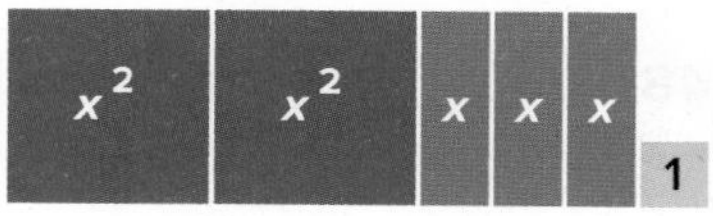

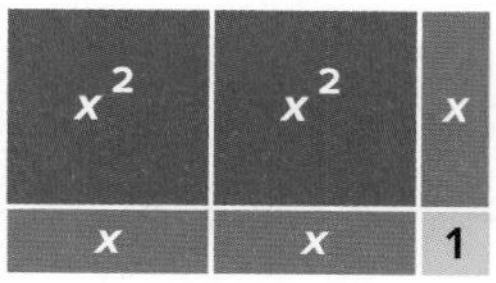

a. Find the length and width of the rectangle.

b. Find the perimeter of the rectangle. Then explain your process for finding the perimeter.

c. Select a value for x and substitute that value into each of the expressions above. For your value of x, state the length, width, perimeter, and area of the rectangle. Discuss any restrictions on the value of x.

45. BANKING Terryl invests \$1500 in two mutual funds. In the first year, one fund grows 3.8% and the other grows 6%. Write a polynomial to represent the value of Terryl's investment after the first year if he invested x dollars in the fund with the lesser growth rate.

46. GEOMETRY Consider a trapezoid that has one base that measures five feet greater than its height. The other base is one foot less than twice its height. Let x represent the height.

a. Write an expression for the area of the trapezoid.

b. Write an expression for the area of the trapezoid if its height is increased by 4 feet.

47. URBAN DEVELOPMENT The diagram represents an aerial view of a memorial in a town center. A sidewalk that is 12 feet wide with an area of 384π square feet surrounds a statue with a circular base of radius r.

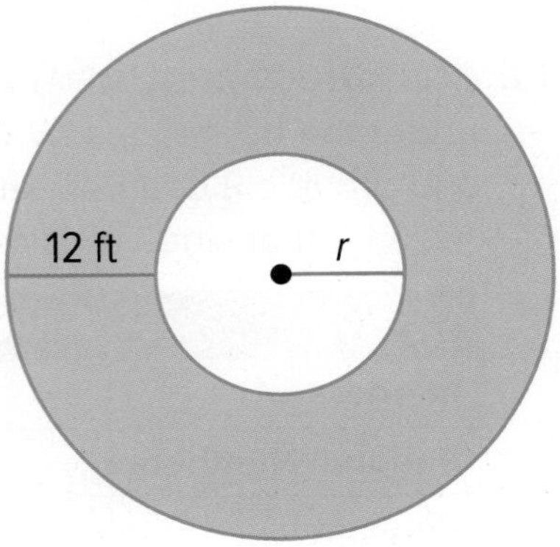

a. Find the radius of the smaller and larger circles. Show your work.

b. A nearby town wants to use the same design concept, but use two squares rather than two circles. Draw and label a diagram with two squares to represent a sidewalk with the same uniform width and area as the circular sidewalk.

c. If s represents the side length of the smaller square, write a polynomial expression for the area of the sidewalk that surrounds the smaller square.

48. ANALYZE Given $f(x)$ and $g(x)$ are polynomials, is the product always a polynomial? Justify your argument.

49. PERSEVERE Use your result from Exercise 48 to make conjectures about the product of a polynomial with m terms and a polynomial with n terms. Justify your conjecture.

a. How many times are two terms multiplied?

b. What is the least number of terms in the simplified product?

50. FIND THE ERROR Isabella found the product of $3x^2 - 4x + 1$ and $x^2 + 5x + 6$ using vertical alignment. Is her answer correct? Explain your reasoning.

Isabella's Work

$$
\begin{array}{rrrrr}
 & & 3x^2 & -\ 4x & +\ 1 \\
 & (\times) & x^2 & +\ 5x & +\ 6 \\
\hline
 & & 18x^2 & -\ 24x & +\ 6 \\
 & 15x^3 & -\ 20x^2 & +\ 5x & \\
3x^4 & -\ 4x^3 & +\ x^2 & & \\
\hline
3x^4 & +\ 11x^3 & -\ x^2 & -\ 19x & +\ 6
\end{array}
$$

51. CREATE Write an expression where two binomials are multiplied and have a product of $9 - 4b^2$.

Dividing Polynomials

Explore Using Algebra Tiles to Divide Polynomials

Online Activity Use algebra tiles to complete the Explore.

INQUIRY How can you use a model to divide polynomials?

Learn Dividing Polynomials by Using Long Division

To divide a polynomial by a monomial, find the quotient of each term of the polynomial and the monomial.

$$\frac{6x^2 - 15x}{3x} = \frac{6x^2}{3x} - \frac{15x}{3x} = 2x - 5$$

You can divide a polynomial by a polynomial with more than one term by using a process similar to long division of real numbers. This process is known as the Division Algorithm. The resulting quotient may be a polynomial or a polynomial with a remainder.

Key Concept • Division Algorithm

Words: If $f(x)$ and $g(x)$ are two polynomials in which $g(x) \neq 0$ and the degree of $g(x)$ is less than the degree of $f(x)$, then there exists a unique quotient $q(x)$ and a unique remainder $r(x)$ such that $f(x) = q(x)g(x) + r(x)$, where the remainder is zero or the degree of $r(x)$ is less than the degree of $g(x)$.

Symbols: $\frac{f(x)}{g(x)} = q(x) \rightarrow f(x) = q(x)g(x)$

$$\frac{f(x)}{g(x)} = q(x) + \frac{r(x)}{g(x)} \rightarrow f(x) = q(x)g(x) + r(x)$$

Example: $\frac{2x^2 - 5x - 3}{x - 3} = 2x + 1$

$$2x^2 - 5x - 3 = (x - 3)(2x + 1)$$
$$= 2x^2 - 5x - 3$$

$$\frac{x^2 - 4x - 1}{x + 1} = x - 5 + \frac{4}{x + 1}$$

$$x^2 - 4x - 1 = (x + 1)(x - 5) + 4$$
$$= x^2 - 4x - 5 + 4$$
$$= x^2 - 4x - 1$$

Go Online You can complete an Extra Example online.

Today's Goals

- Divide polynomials by using long division.
- Divide polynomials by using synthetic division.

Today's Vocabulary

synthetic division

Talk About It!

Is the following statement *always*, *sometimes*, or *never* true? Justify your argument.

If a quadratic polynomial is divided by a binomial with a remainder of 0, the binomial is a factor of the polynomial.

Study Tip

In algebra, *unique* means *only one*. So, there is only one quotient and remainder that will satisfy the Division Algorithm for each polynomial.

Example 1 Divide a Polynomial by a Monomial

Find $(24a^4b^3 + 18a^2b^2 - 30ab^3)(6ab)^{-1}$.

$(24a^4b^3 + 18a^2b^2 - 30ab^3)(6ab)^{-1}$

$= \frac{24a^4b^3 + 18a^2b^2 - 30ab^3}{6ab}$ Write a fraction.

$= \frac{24a^4b^3}{6ab} + \frac{18a^2b^2}{6ab} - \frac{30ab^3}{6ab}$ Sum of quotients

$= \frac{24}{6}a^{4-1}b^{3-1} + \frac{18}{6}a^{2-1}b^{2-1} - \frac{30}{6}a^{1-1}b^{3-1}$ Divide.

$= 4a^3b^2 + 3ab - 5b^2$ Simplify.

Check

Find $(9x^9y^5 + 21x^4y^4 - 12x^3y^2) \div (3x^2y^2)$.

Think About It!
How can you check the solution?

Example 2 Divide a Polynomial by a Binomial

Find $(x^2 - 5x - 36) \div (x + 4)$.

$$\begin{array}{r} x - 9 \\ x + 4\overline{)\,x^2 - 5x - 36} \\ \underline{(-)\,x^2 + 4x} \\ -9x - 36 \\ \underline{(-)-9x - 36} \\ 0 \end{array}$$

The quotient is $x - 9$ and the remainder is 0.

Check

Find $\frac{x^2 + 6x - 112}{x - 8}$.

Watch Out!

Signs Remember that you are subtracting throughout the process of long division. Carefully label the signs of the coefficients to avoid a sign error.

Example 3 Find a Quotient with a Remainder

Find $\frac{3z^3 - 14z^2 - 7z + 3}{z - 5}$.

$$\begin{array}{r} 3z^2 + z - 2 \\ z - 5\overline{)\,3z^3 - 14z^2 - 7z + 3} \\ \underline{(-)3z^3 - 15z^2} \\ z^2 - 7z \\ \underline{(-)z^2 - 5z} \\ -2z + 3 \\ \underline{(-)\,-2z + 10} \\ -7 \end{array}$$

Go Online
You can learn how to divide polynomials by using long division by watching the video online.

Go Online You can complete an Extra Example online.

The quotient is $3z^2 + z - 2$ and the remainder is -7.

Therefore, $\frac{3z^3 - 14z^2 - 7z + 3}{z - 5} = 3z^2 + z - 2 - \frac{7}{z - 5}$.

Check

Find the quotient of $(-4x^3 + 5x^2 - 2x - 9)(x - 2)^{-1}$.

Learn Dividing Polynomials by Using Synthetic Division

Synthetic division is an alternate method used to divide a polynomial by a binomial of degree 1. You may find this to be a quicker, simpler method.

Key Concept • Synthetic Division

Step 1 After writing a polynomial in standard form, write the coefficients of the dividend. If the dividend is missing a term, use 0 as a placeholder. Write the constant a of the divisor $x - a$ in the box. Bring the first coefficient down.

Step 2 Multiply the number just written in the bottom row by a, and write the product under the next coefficient.

Step 3 Add the product and the coefficient above it.

Step 4 Repeat **Steps 2** and **3** until you reach a sum in the last column.

Step 5 Write the quotient. The numbers along the bottom row are the coefficients of the quotient. The power of the first term is one less than the degree of the dividend. The final number is the remainder.

Example 4 Use Synthetic Division

Find $(3x^3 - 2x^2 - 53x - 60) \div (x + 3)$.

Step 1 Write the coefficients of the dividend and write the constant a in the box. Because $x + 3 = x - (-3)$, $a = -3$. Then bring the first coefficient down.

$$\begin{array}{r|rrrr} -3 & 3 & -2 & -53 & -60 \\ & \downarrow & & & \\ \hline & 3 & & & \end{array}$$

Step 2 Multiply by a and write the product. The product of the coefficient and a is $3(-3) = -9$.

$$\begin{array}{r|rrrr} -3 & 3 & -2 & -53 & -60 \\ & & -9 & & \\ \hline & 3 & & & \end{array}$$

Step 3 Add the product and the coefficient.

$$\begin{array}{r|rrrr} -3 & 3 & -2 & -53 & -60 \\ & & -9 & & \\ \hline & 3 & -11 & & \end{array}$$

Step 4 Repeat **Steps 2** and **3** until you reach a sum in the last column.

$$\begin{array}{r|rrrr} -3 & 3 & -2 & -53 & -60 \\ & & -9 & 33 & 60 \\ \hline & 3 & -11 & -20 & 0 \end{array}$$

(*continued on the next page*)

Think About It!

Describe a method you could use to check your answer.

Think About It!

Describe a method you could use to check your answer.

Step 5 Write the quotient. Because the degree of the dividend is 3 and the degree of the divisor is 1, the degree of the quotient is 2. The final sum in the synthetic division is 0, so the remainder is 0.

The quotient is $3x^2 - 11x - 20$.

Example 5 Divisor with a Coefficient Other Than 1

Find $\dfrac{4x^4 - 37x^2 + 4x + 9}{2x - 1}$.

To use synthetic division, the lead coefficient of the divisor must be 1.

$$\frac{(4x^4 - 37x^2 + 4x + 9) \div 2}{(2x - 1) \div 2}$$ Divide the numerator and denominator by 2.

$$= \frac{2x^4 - \frac{37}{2}x^2 + 2x + \frac{9}{2}}{x - \frac{1}{2}}$$ Simplify the numerator and denominator.

$x - a = x - \frac{1}{2}$, so $a = \frac{1}{2}$.

Watch Out!

Missing terms Add placeholders for terms that are missing from the polynomial. In this case, there are 0 x^3-terms.

Complete the synthetic division.

$\frac{1}{2}$	2	0	$-\frac{37}{2}$	2	$\frac{9}{2}$
		1	$\frac{1}{2}$	−9	$-\frac{7}{2}$
	2	1	−18	−7	1

The resulting expression is $2x^3 + x^2 - 18x - 7 + \dfrac{1}{x - \frac{1}{2}}$. Now simplify the fraction.

$$\frac{1}{x - \frac{1}{2}} = \frac{(1)^2}{\left(x - \frac{1}{2}\right) \cdot 2}$$ Multiply the numerator and denominator by 2.

$$= \frac{2}{2x - 1}$$ Simplify.

The solution is $2x^3 + x^2 - 18x - 7 + \dfrac{2}{2x - 1}$.

You can check your answer by using long division.

Check

Find $(4x^4 + 3x^3 - 12x^2 - x + 6)(4x + 3)^{-1}$.

Go Online You can complete an Extra Example online.

Practice

Go Online You can complete your homework online.

Example 1

Simplify each expression.

1. $\frac{15y^3 + 6y^2 + 3y}{3y}$

2. $(4f^5 - 6f^4 + 12f^3 - 8f^2)(4f^2)^{-1}$

3. $(6j^2k - 9jk^2) \div (3jk)$

4. $(4a^2h^2 - 8a^3h + 3a^4) \div (2a^2)$

Examples 2 and 3

Simplify by using long division.

5. $(n^2 + 7n + 10) \div (n + 5)$

6. $(d^2 + 4d + 3)(d + 1)^{-1}$

7. $(2t^2 + 13t + 15) \div (t + 6)$

8. $(6y^2 + y - 2)(2y - 1)^{-1}$

9. $(4g^2 - 9) \div (2g + 3)$

10. $(2x^2 - 5x - 4) \div (x - 3)$

Examples 4 and 5

Simplify using synthetic division.

11. $(3v^2 - 7v - 10)(v - 4)^{-1}$

12. $(3t^4 + 4t^3 - 32t^2 - 5t - 20)(t + 4)^{-1}$

13. $\frac{y^3 + 6}{y + 2}$

14. $\frac{2x^3 - x^2 - 18x + 32}{2x - 6}$

15. $(4p^3 - p^2 + 2p) \div (3p - 1)$

16. $(3c^4 + 6c^3 - 2c + 4)(c + 2)^{-1}$

Mixed Exercises

Simplify.

17. $(m^2 + m - 6) \div (m + 4)$

18. $(a^3 - 6a^2 + 10a - 3) \div (a - 3)$

19. $(2x^3 - 7x^2 + 7x - 2) \div (x - 2)$

20. $(x^3 + 2x^2 - 34x + 9) \div (x + 7)$

21. $(x^3 + 8) \div (x + 2)$

22. $(6x^3 + x^2 + x) \div (2x + 1)$

23. $(28c^3d^2 - 21cd^2) \div (14cd)$

24. $(x^4 - y^4) \div (x - y)$

25. $\dfrac{n^3 + 3n^2 - 5n - 4}{n + 4}$

26. $(a^3b^2 - a^2b + 2b)(-ab)^{-1}$

27. $\dfrac{3z^5 + 5z^4 + z + 5}{z + 2}$

28. $\dfrac{p^3 + 2p^2 - 7p - 21}{p + 3}$

29. STRUCTURE Jada used long division to divide $x^4 + x^3 + x^2 + x + 1$ by $x + 2$. Her work is shown below with three terms missing. What are A, B, and C?

$$
\begin{array}{r}
x^3 - x^2 + 3x - 5 \\
x + 2\overline{)\,x^4 + x^3 + x^2 + x + 1} \\
\underline{(-)\ x^4 + 2x^3} \\
-x^3 + A \\
\underline{(-)\ -x^3 - 2x^2} \\
3x^2 + x \\
\underline{(-)\ 3x^2 + B} \\
-5x + 1 \\
\underline{(-)\ -5x - 10} \\
C
\end{array}
$$

30. AVERAGES Bena has a list of $n + 1$ numbers and she needs to find their average. Two of the numbers are n^3 and 2. Each of the other $n - 1$ numbers are all equal to 1. Find the average of these numbers.

31. VOLUME The volume of a cylinder is $\pi(x^3 + 32x^2 - 304x + 640)$. If the height of the cylinder is $x + 40$ feet, find the area of its base in terms of x and π.

32. REASONING Rewrite $\frac{6x^4 + 2x^3 - 16x^2 + 24x + 32}{2x + 4}$ as $q(x) + \frac{r(x)}{d(x)}$ using long division. What does the remainder indicate in this problem?

33. CONSTRUCT ARGUMENTS Determine whether you have enough information to fill in the missing pieces of the long division exercise shown. If so, copy and complete the long division. Justify your response.

$$\begin{array}{r} 3x - \square \\ \square \,\big)\, \overline{9x^2 + \square} \\ \underline{9x^2 + 3x} \\ -3x + 5 \end{array}$$

34. REGULARITY Rewrite $\frac{2x^5 - 7x^4 - 15x^3 + 2x^2 + 3x + 6}{2x + 3}$ as $q(x) + \frac{r(x)}{g(x)}$ using long division.

a. Identify $q(x)$, $r(x)$, and $g(x)$.

b. How can you check your work using the expressions of $q(x)$, $g(x)$, and $r(x)$?

35. STRUCTURE When a polynomial is divided by $4x - 6$, the quotient is $2x^2 + x + 1$ and the remainder is -4. What is the dividend, $f(x)$? Explain.

36. USE A MODEL Luciano has a square garden. A new garden will have the same width and a length that is 3 feet more than twice the width of the original garden.

a. Define a variable. Copy the diagrams. Label each side of the diagrams with an expression for its length.

b. Write a ratio to represent the percent increase in the area of the garden. Use polynomial division to simplify the expression.

c. Use your expression from **part b** to determine the percent of increase in area if the original garden was a 12-foot square. Check your answer.

37. **REGULARITY** Mariella makes the following claims about the degrees of the polynomials in $\frac{f(x)}{d(x)} = q(x) + \frac{r(x)}{d(x)}$. Do you agree with each claim? Justify your answers and provide examples.

a. The degree of $d(x)$ must be less than the degree of $f(x)$.

b. The degree of $r(x)$ must be at least 1 less than the degree of $d(x)$.

c. The degree of $q(x)$ must be the degree of $f(x)$ minus the degree of $d(x)$.

38. **FIND THE ERROR** Tomo and Jamal are dividing $2x^3 - 4x^2 + 3x - 1$ by $x - 3$. Tomo claims that the remainder is −100. Jamal claims that the remainder is 26. Is either of them correct? Explain your reasoning.

39. **PERSEVERE** If a polynomial is divided by a binomial and the remainder is 0, what does this tell you about the relationship between the binomial and the polynomial?

40. **ANALYZE** What is the relationship between the degrees of the dividend, the divisor, and the quotient in any polynomial division exercise?

41. **CREATE** Write a quotient of two polynomials for which the remainder is 3.

42. **WRITE** Compare and contrast dividing polynomials using long division and using synthetic division.

43. **PERSEVERE** Mr. Collins has his class working with bases and polynomials. He wrote on the board that the number 1111 in base B has the value $B^3 + B^2 + B + 1$. The class was then given the following questions to answer.

a. The number 11 in base B has the value $B + 1$. What is 1111 (in base B) divided by 11 (in base B)?

b. The number 111 in base B has the value $B^2 + B + 1$. What is 1111 (in base B) divided by 111 (in base B)?

Powers of Binomials

Explore Expanding Binomials

Online Activity Use interactive tool to complete the Explore.

INQUIRY How can you use Pascal's triangle to write expansions of binomials?

Learn Powers of Binomials

You can expand binomials by following a set of rules and using patterns.

Key Concept • Binomial Expansion

In the binomial expansion of $(a + b)^n$,

- there are $n + 1$ terms.
- n is the exponent of a in the first term and b in the last term.
- in successive terms, the exponent of a decreases by 1, and the exponent of b increases by 1.
- the sum of the exponents in each term is n.
- the coefficients are symmetric.

Pascal's triangle is a triangle of numbers in which a row represents the coefficients of an expanded binomial $(a + b)^n$. Each row begins and ends with 1. Each coefficient can be found by adding the two coefficients above it in the previous row.

Instead of writing out the rows of Pascal's triangle, you can use the Binomial Theorem to expand a binomial. The Binomial Theorem uses combinations to calculate the coefficients of the binomial expansion.

Key Concept • Binomial Theorem

If n is a natural number, then $(a + b)^n =$

$$_nC_0a^nb^0 + {_nC_1}a^{n-1}b^1 + {_nC_2}a^{n-2}b^2 + {_nC_3}a^{n-3}b^3 + \ldots + {_nC_n}a^0b^n$$

or

$$1a^nb^0 + \frac{n!}{1!(n-1)!}a^{n-1}b^1 + \frac{n!}{2!(n-2)!}a^{n-2}b^2 + \frac{n!}{3!(n-3)!}a^{n-3}b^3 + \ldots + 1a^0b^n$$

Example 1 Use Pascal's Triangle

Use Pascal's triangle to expand $(x + y)^7$.

1	$(x + y)^0$
1 1	$(x + y)^1$
1 2 1	$(x + y)^2$
1 3 3 1	$(x + y)^3$
1 4 6 4 1	$(x + y)^4$
1 5 10 10 5 1	$(x + y)^5$
1 6 15 20 15 6 1	
1 7 21 35 35 21 7 1	

$$(x + y)^7 = x^7 + 7x^6y + 21x^5y^2 + 35x^4y^3 + 35x^3y^4 + 21x^2y^5 + 7xy^6 + y^7$$

Today's Goal

- Expand powers of binomials by using Pascal's Triangle and the Binomial Theorem.

Today's Vocabulary

Pascal's triangle

Think About It!

Both $_nC_0$ and $_nC_n$ equal 1. What does this mean for the terms of a binomial expansion? How does this relate to Pascal's triangle?

Study Tip

Combinations Recall that $_nC_r$ refers to the number of ways to choose r objects from n distinct objects. In the Binomial Theorem, n is the exponent of $(a + b)^n$, and r is the exponent of b in each term. To calculate the coefficients, remember that $n!$ represents n factorial. This is the product of all counting numbers beginning with n and counting backward to 1. For example, $3! = 3 \cdot 2 \cdot 1$. The value of 0! is defined as 1. The value of 0! is defined as 1.

Check

Write the expansion of $(c + d)^4$.

Talk About It!

Describe a shortcut you could use to write out rows of Pascal's triangle instead of adding to find every number in a row. Explain your reasoning.

Example 2 Use the Binomial Theorem

BASEBALL **In 2016, the Chicago Cubs won the World Series for the first time in 108 years. During the regular season, the Cubs played the Atlanta Braves 6 times, winning 3 games and losing 3 games. If the Cubs were as likely to win as to lose, find the probability of this outcome by expanding $(w + \ell)^6$.**

$(w + \ell)^6$

$$= {}_6C_0w^6 + {}_6C_1w^5\ell + {}_6C_2w^4\ell^2 + {}_6C_3w^3\ell^3 + {}_6C_4w^2\ell^4 + {}_6C_5w\ell^5 + {}_6C_6\ell^6$$

$$= w^6 + \frac{6!}{5!}w^5\ell + \frac{6!}{2!4!}w^4\ell^2 + \frac{6!}{3!3!}w^3\ell^3 + \frac{6!}{4!2!}w^2\ell^4 + \frac{6!}{5!}w\ell^5 + \ell^6$$

$$= w^6 + 6w^5\ell + 15w^4\ell^2 + 20w^3\ell^3 + 15w^2\ell^4 + 6w\ell^5 + \ell^6$$

By adding the coefficients, you can determine that there were 64 combinations of wins and losses that could have occurred.

$20w^3\ell^3$ represents the number of combinations of 3 wins and 3 losses. Therefore, there was a $\frac{20}{64}$ or about a 31% chance of the Cubs winning 3 games and losing 3 games against the Braves.

Study Tip

Assumptions To use the Binomial Theorem, we assumed that the teams had an equal chance of winning and losing. Assuming that allows us to reasonably estimate the probability of an outcome with only the coefficient. To find probabilities of events that are not equally likely, substitute the probability of each event for a and b in the expansion of $(a + b)_n$.

Check

GAME SHOW A group of 8 contestants are selected from the audience of a television game show. If there are an equal number of men and women in the audience, find the probability of the contestants being 5 women and 3 men by expanding $(w + m)^8$. Round to the nearest percent if necessary. __?__%

Example 3 Coefficients Other Than 1

Expand $(2c - 6d)^4$.

$(2c - 6d)^4$

$$= {}_4C_0(2c)^4 + {}_4C_1(2c)^3(-6d) + {}_4C_2(2c)^2(-6d)^2 + {}_4C_3(2c)(-6d)^3 + {}_4C_4(-6d)^4$$

$$= 16c^4 + \frac{4!}{3!}(8c^3)(-6d) + \frac{4!}{2!2!}(4c^2)(36d^2) + \frac{4!}{3!}(2c)(-216d^3) + 1296d^4$$

$$= 16c^4 - 192c^3d + 864c^2d^2 - 1728cd^3 + 1296d^4$$

Study Tip

Coefficients When the binomial to be expanded has coefficients other than 1, the coefficients will no longer be symmetric. In these cases, it may be easier to use the Binomial Theorem.

Go Online You can complete an Extra Example online.

Practice

Go Online You can complete your homework online.

Example 1

Use Pascal's triangle to expand each binomial.

1. $(x - y)^3$
2. $(a + b)^4$
3. $(g - h)^4$
4. $(m + 1)^4$
5. $(y - z)^6$
6. $(d + 2)^8$

Example 2

7. BAND A school band went to 4 competitions during the year and received a superior rating 2 times. If the band is as likely to receive a superior rating as to not receive a superior rating, find the probability of this outcome by expanding $(s + n)^4$. Round to the nearest percent if necessary.

8. BASKETBALL Oliver shot 8 free throws at practice, making 6 free throws and missing 2 free throws. If Oliver is equally likely to make a free throw as he is to miss a free throw, find the probability of this outcome by expanding $(m + n)^8$. Round to the nearest percent if necessary.

Example 3

Expand each binomial.

9. $(3x + 4y)^5$
10. $(2c - 2d)^7$
11. $(8h - 3j)^4$
12. $(4a + 3b)^6$

Mixed Exercises

Expand each binomial.

13. $\left(x + \frac{1}{2}\right)^5$
14. $\left(x - \frac{1}{3}\right)^4$
15. $\left(2b + \frac{1}{4}\right)^5$
16. $\left(3c + \frac{1}{3}d\right)^3$

17. STRUCTURE Out of 12 frames, Vince bowled 6 strikes. If Vince is as likely bowl a strike as to not bowl a strike in one frame, find the probability of this outcome. Round to the nearest percent if necessary.

18. REGULARITY A group of 10 choir members are selected at random to perform solos. If there are an equal number of boys and girls in the choir, find the probability of the choir members selected being 7 boys and 3 girls. Round to the nearest percent if necessary.

19. USE A MODEL A company is developing a robotic welder that produces circuit boards. At this stage in its development, the robotic welder only produces 50% of the circuit boards correctly. Use the Binomial Theorem to find the probability that 5 of 7 circuit boards chosen at random are correct.

20. USE A MODEL Diego flips a fair coin 12 times. What is the probability that the coin lands on tails 3 times? 5 times? 9 times?

21. REASONING A test consists of 10 true-false questions. Matthew forgets to study and must guess on every question. What is the probability that he gets 8 or more correct answers on the test? Show your work using Pascal's Triangle.

22. REGULARITY Use Pascal's Triangle to find the fourth term in the expansion of $(2x + 7)^6$. Why is it the same as the fourth term in the expansion of $(7 + 2x)^6$?

23. USE A SOURCE Research the number of judges on the Supreme Court. For most rulings, a majority is needed. How many combinations of votes are possible for a majority to be reached?

24. STRUCTURE Find the term in $(a + b)^{12}$ where the exponent of a is 5.

25. PRECISION Use the first four terms of the binomial expansion of $(1 + 0.02)^{10}$ to approximate $(1.02)^{10}$. Evaluate $(1.02)^{10}$ using a calculator and compare the value to your approximation.

Higher-Order Thinking Skills

26. PERSEVERE Find the sixth term of the expansion of $(\sqrt{a} + \sqrt{b})^{12}$.

27. ANALYZE Explain how the terms of $(x + y)^n$ and $(x - y)^n$ are the same and how they are different.

28. REGULARITY Each row of Pascal's triangle is like a palindrome. That is, the numbers read the same left to right as they do right to left. Explain why this is the case.

29. CREATE Write a power of a binomial for which the second term of the expansion is $6x^4y$.

30. WRITE Explain how to write out the terms of Pascal's triangle.

Review

Essential Question

How does an understanding of polynomials and polynomial functions help us understand and interpret real-world events?

Module Summary

Lessons 4-1 and 4-2

Polynomial Functions and Graphs

- A power function is any function of the form $f(x) = ax^n$, where a and n are nonzero real numbers. The leading coefficient is a and the degree is n.
- Odd-degree functions will always have at least one real zero.
- Even-degree functions may have any number of real zeros or no real zeros at all.
- A polynomial function is a continuous function that can be described by a polynomial equation in one variable.
- The degree of a polynomial function tells the maximum number of times that the graph of a polynomial function intersects the x-axis.
- If the value of $f(x)$ changes signs from one value of x to the next, then there is a zero between those two x-values.
- Extrema occur at relative maxima or minima of the function.

Lessons 4-3 and 4-4

Operations with Polynomials

- Polynomials can be added or subtracted by performing the operations indicated and combining like terms.
- To subtract a polynomial, add its additive inverse.
- Polynomials can be multiplied by using the Distributive Property to multiply each term in one polynomial by each term in the other.
- The set of polynomials is closed under the operations of addition, subtraction, and multiplication.
- To multiply two binomials, you can use a shortcut called the FOIL method.
- You can divide a polynomial by a polynomial with more than one term by using a process similar to long division of real numbers.
- Synthetic division is an alternate method used to divide a polynomial by a binomial of degree 1.

Lesson 4-5

Powers of Binomials

- Pascal's triangle is a triangle of numbers in which a row represents the coefficients of an expanded binomial $(a + b)^n$. Each row begins and ends with 1. Each coefficient can be found by adding the two coefficients above it in the previous row.
- You can also use the Binomial Theorem to expand a binomial. If n is a natural number, then $(a + b)^n = {}_nC_0a^nb^0 + {}_nC_1a^{n-1}b^1 + {}_nC_2a^{n-2}b^2 + {}_nC_3a^{n-3}b^3 + \ldots + {}_nC_na^0b^n$.

Study Organizer

Foldables

Use your Foldable to review this module. Working with a partner can be helpful. Ask for clarification of concepts as needed.

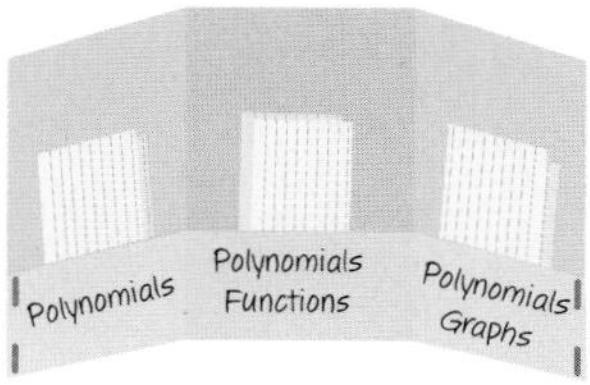

Test Practice

1. MULTIPLE CHOICE The weight of an ideal cut round diamond can be modeled by $f(d) = 0.0071d^3 - 0.090d^2 + 0.48d$, where d is the diameter of the diamond. Find the domain of the function in the context of the situation. (Lesson 4-1)

A. The domain is all real numbers.

B. The domain is $\{d \mid d > 0\}$.

C. The domain is $\{d \mid d < 0\}$.

D. The domain is $\{d \mid d > 0.48\}$.

2. OPEN RESPONSE Use the function $f(x) = 13 - 2x^2 + 6x - 9x^3$ to answer the following questions. (Lesson 4-1)

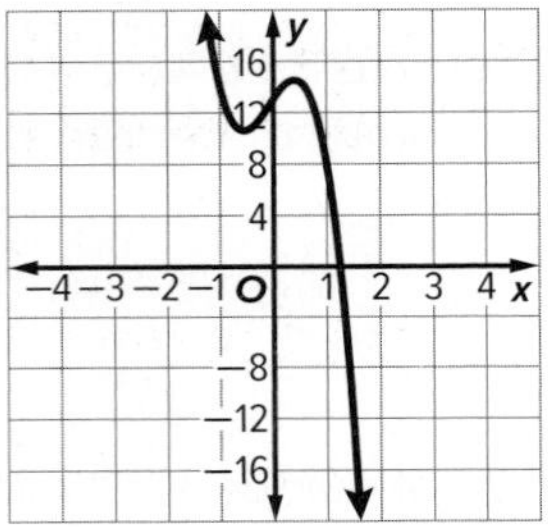

a) What is the degree?

b) What is the leading coefficient?

c) Describe the end behavior using the leading coefficient and degree.

3. MULTIPLE CHOICE The revenue of a certain business can be modeled using $f(x) = -0.01(x^4 - 11x^3 + 4x^2 - 5x + 7)$, where x is the number of years since the business was started and $f(x)$ is the revenue in hundred-thousands of dollars. Which graph represents the function? (Lesson 4-1)

A.
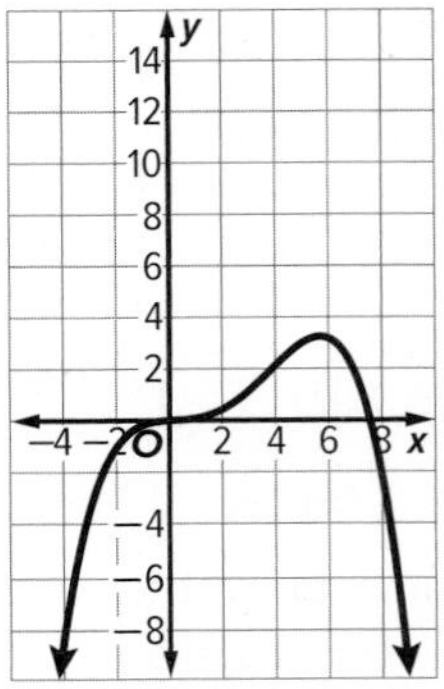

B.
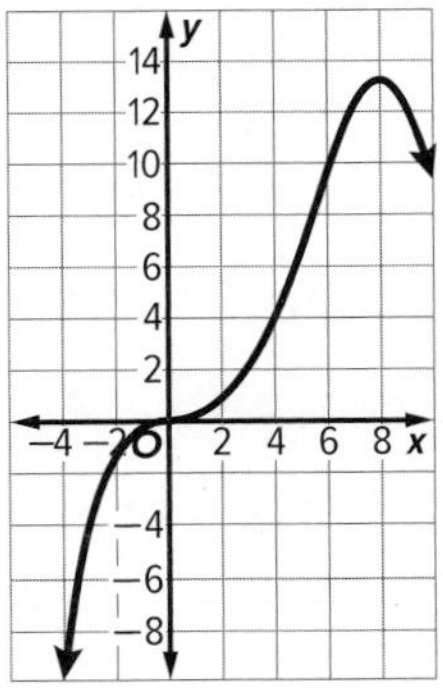

C.
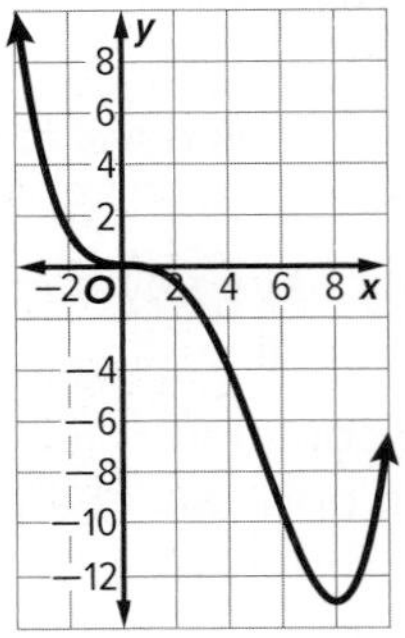

D.
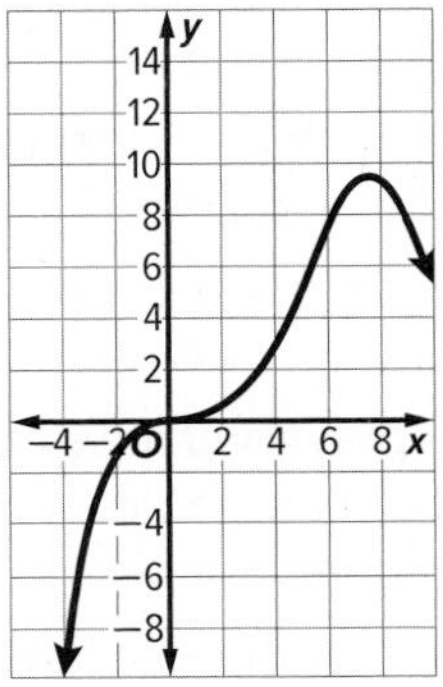

4. MULTI-SELECT Select all intervals in which a real zero is located for the function $f(x) = x^4 - 2x^3 + 3x^2 - 5$. (Lesson 4-2)

A. $x = -2$ and $x = -1$

B. $x = -1$ and $x = 0$

C. $x = 0$ and $x = 1$

D. $x = 1$ and $x = 2$

E. $x = 2$ and $x = 3$

F. $x = 3$ and $x = 4$

5. OPEN RESPONSE Describe the end behavior for $g(x) = -2x^4 - 6x^3 + 11x - 18$ as $x \to \infty$. (Lesson 4-2)

6. OPEN RESPONSE Marshall claims that there is only one real zero in the function $f(x) = 4x^3 + 7x^2 - 5x + 3$. Based on the table provided, determine whether you agree with Marshall. Then name the interval(s) in which the zero(s) is/are located. (Lesson 4-2)

x	f(x)
−3	−27
−2	9
−1	11
0	3
1	9
2	53

7. MULTIPLE CHOICE Helen started a business several years ago. The table shows her profits, in millions of dollars, for the first 7 years. Select the polynomial function of best fit that could be used to model Helen's profits. (Lesson 4-2)

x	f(x)
1	1.425
2	1.46
3	1.5
4	1.53
5	1.56
6	1.58
7	1.58

A. $f(x) = 0.001(-3.27x^2 + 53.51x + 1371)$

B. $f(x) = 0.0001(-6.944x^3 + 50.6x^2 + 250.4x + 13{,}957)$

C. $f(x) = 0.00001(9.47x^4 + 82.07x^3 - 312.5x^2 + 4203x + 138{,}500)$

D. $f(x) = 0.0001(-x^4 + 12x^3 - 77x^2 + 600x + 13{,}650)$

8. OPEN RESPONSE What is the difference?
$(7x^4 - 3x^3 + 5x^2 + 8x - 11) - (3x^4 - 9x^3 - 4x^2 + 12x + 4)$ (Lesson 4-3)

9. OPEN RESPONSE What is the sum?
$(3x^6 - 5x^3 + 8x^2 - 4x) + (-2x^6 + 7x^5 - x^3 + 6x)$
(Lesson 4-3)

10. MULTIPLE CHOICE Enrique is designing a flag for a new school club. A smaller striped square is placed as part of the design and the rest of the flag will have chevrons. (Lesson 4-3)

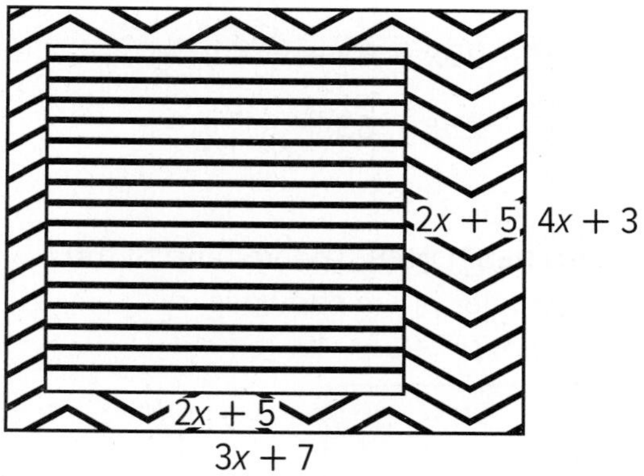

Which expression can be used to represent the area of the flag that is not striped?

A. $16x^2 + 57x + 46$

B. $8x^2 + 35x + 16$

C. $8x^2 + 17x + 4$

D. $8x^2 + 17x - 4$

11. **MULTIPLE CHOICE** Determine the quotient.
$(5x^4 + 12x^3 - 64x^2 - 95x + 132) \div (x - 3)$
(Lesson 4-4)

A. $3x^3 + 21x^2 + x - 31$

B. $3x^3 - 21x^2 - x + 31$

C. $5x^3 + 27x^2 + 17x - 44$

D. $5x^3 - 27x^2 - 17x + 44$

12. **MULTIPLE CHOICE** Determine the quotient.
(Lesson 4-4)

$$\frac{6x^3 - 71x^2 + 139x + 130}{3x + 2}$$

A. $2x^2 - 25x + 63 + \frac{8}{3}$

B. $2x^2 - 25x + 63 + \frac{8}{3x + 2}$

C. $2x^2 - 25x + 63 + \frac{4}{3}$

D. $2x^2 - 25x + 63 + \frac{4}{3x + 2}$

13. **OPEN RESPONSE** The volume of the rectangular prism shown is $45x^3 + 83x^2 + x - 12$.

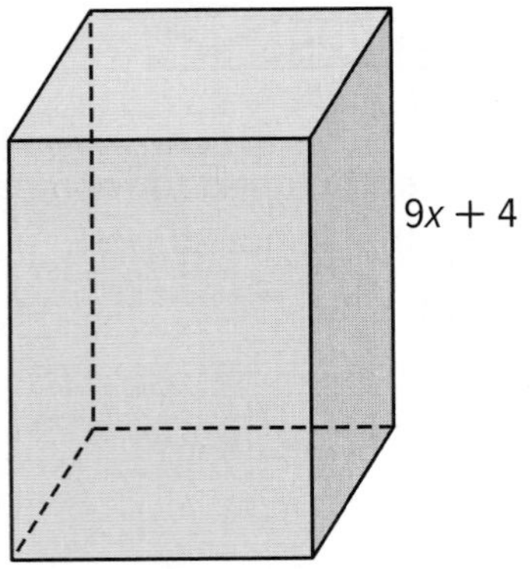

What is the area of the base? (Lesson 4-4)

14. **MULTIPLE CHOICE** Which of the following is the expansion of $(2h + f)^4$? (Lesson 4-5)

A. $2h^4 + 4h^3f + 6h^2f^2 + 4hf^3 + f^4$

B. $16h^4 + 32h^3f + 24h^2f^2 + 32hf^3 + 16f^4$

C. $16h^4 + 32h^3f + 32h^2f^2 + 8hf^3 + f^4$

D. $16h^4 + 32h^3f + 24h^2f^2 + 8hf^3 + f^4$

15. **MULTIPLE CHOICE** The first shelf on Hannah's bookshelf holds an equal number of fiction and nonfiction books. If Hannah selects 5 books randomly, what is the probability that 4 of the books will be fiction and 1 will be nonfiction?

Round your answer to the nearest tenth of a percent. (Lesson 4-5)

A. 31.3%

B. 15.6%

C. 12.5%

D. 3.1%

16. **MULTI-SELECT** Select all of the following that would be a coefficient of a term in the binomial expansion of $(x + y)^7$. (Lesson 4-5)

A. 1

B. 3

C. 7

D. 14

E. 21

F. 28

G. 30

H. 35

Module 5

Polynomial Equations

Essential Question

What methods are useful for solving polynomial equations and finding zeros of polynomial functions?

What Will You Learn?

How much do you already know about each topic **before** starting this module?

KEY

— I don't know. — I've heard of it. — I know it!

	Before			After		
solve polynomial equations by graphing						
solve polynomial equations by factoring						
solve polynomial equations in quadratic form						
prove polynomial identities						
apply the Remainder Theorem						
use the Factor Theorem to determine whether a binomial is a factor of a polynomial						
use the Fundamental Theorem of Algebra						
find zeros of polynomial functions						

Foldable Make this Foldable to help you organize your notes about polynomial equations. Begin with three sheets of notebook paper.

1. **Fold** each sheet of paper in half from top to bottom.
2. **Cut** along the fold. Staple the six half-sheets together to form a booklet.
3. **Cut** tabs into the margin. The top tab is 2 lines deep, the next tab is 6 lines deep, and so on.
4. **Label** each tab except the first with a lesson number. Use the first tab as a cover page decorating it with a graph from lesson 1.

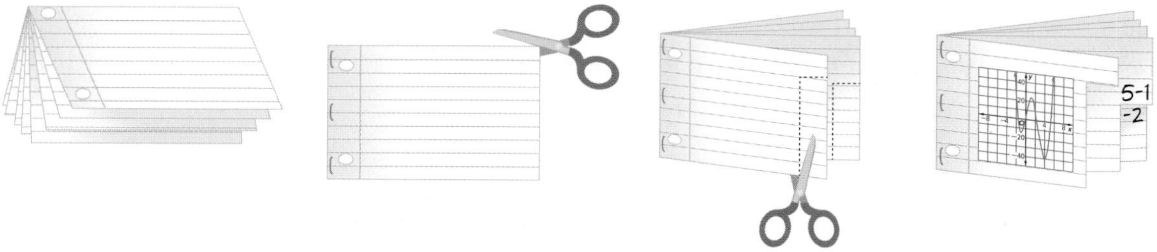

What Vocabulary Will You Learn?

- depressed polynomial
- identity
- multiplicity
- polynomial identity
- prime polynomial
- quadratic form
- synthetic substitution

Are You Ready?

Complete the Quick Review to see if you are ready to start this module. Then complete the Quick Check.

Quick Review

Example 1

Use the Distributive Property to multiply $(x^2 - 2x - 4)(x + 5)$.

$(x^2 - 2x - 4)(x + 5)$

$= x^2(x + 5) - 2x(x + 5) - 4(x + 5)$ — Distributive Property

$= x^2(x) + x^2(5) + (-2x)(x) + (-2x)(5) + (-4)(x) + (-4)(5)$ — Distributive Property

$= x^3 + 5x^2 - 2x^2 - 10x - 4x - 20$ — Multiply.

$= x^3 + 3x^2 - 14x - 20$ — Combine like terms.

Example 2

Solve $2x^2 + 8x + 1 = 0$.

$x = \frac{-b \pm \sqrt{b^2 - 4ac}}{2a}$ — Quadratic Formula

$= \frac{-8 \pm \sqrt{8^2 - 4(2)(1)}}{2(2)}$ — $a = 2, b = 8, c = 1$

$= \frac{-8 \pm \sqrt{56}}{4}$ — Simplify.

$= -2 \pm \frac{\sqrt{14}}{2}$ — $\sqrt{56} = \sqrt{4 \cdot 14}$ or $2\sqrt{14}$

The exact solutions are $-2 + \frac{\sqrt{14}}{2}$ and $-2 - \frac{\sqrt{14}}{2}$.

The approximate solutions are −0.13 and −3.87.

Quick Check

Use the Distributive Property to multiply each set of polynomials.

1. $(6x^2 - x + 2)(4x + 2)$
2. $(x^2 - 2x + 7)(7x - 3)$
3. $(7x^2 - 6x - 6)(2x - 4)$
4. $(x^2 + 6x - 4)(2x - 4)$

Solve each equation.

5. $x^2 + 2x - 8 = 0$
6. $2x^2 + 7x + 3 = 0$
7. $6x^2 + 5x - 4 = 0$
8. $4x^2 - 2x - 1 = 0$

How Did You Do?

Which exercises did you answer correctly in the Quick Check?

Solving Polynomial Equations by Graphing

Explore Solutions of Polynomial Equations

Online Activity Use graphing technology to complete the Explore.

INQUIRY How can you solve a polynomial equation by using the graph of a related polynomial function?

Learn Solving Polynomial Equations by Graphing

A related function is found by rewriting the equation with 0 on one side, and then replacing 0 with $f(x)$. The values of x for which $f(x) = 0$ are the real zeros of the function and the x-intercepts of its graph.

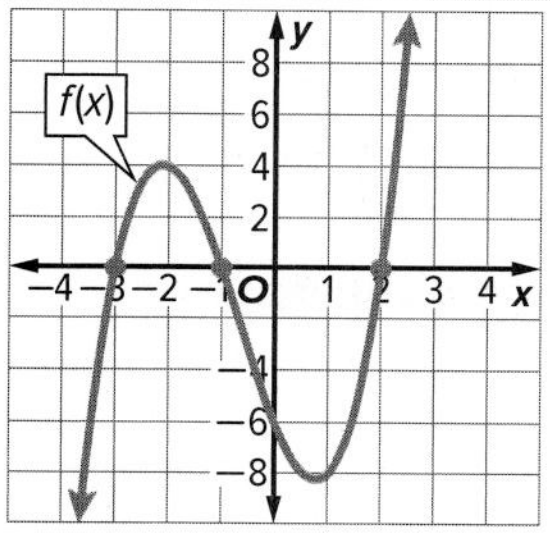

$x^3 + 2x^2 - 4x = x + 6$
- -3, -1, and 2 are solutions.
- -3, -1, and 2 are roots.

$f(x) = x^3 + 2x^2 - 5x - 6$
- -3, -1, and 2 are zeros.
- -3, -1, and 2 are x-intercepts.

Example 1 Solve a Polynomial Equation by Graphing

Use a graphing calculator to solve $x^4 + 3x^2 - 5 = -4x^3$ by graphing.

Step 1 Find a related function. Write the equation with 0 on the right.

$x^4 + 3x^2 - 5 = -4x^3$	Original equation
$x^4 + 3x^2 - 5 + 4x^3 = -4x^3 + 4x^3$	Add $4x^3$ to each side.
$x^4 + 4x^3 + 3x^2 - 5 = 0$	Simplify.

A related function is $f(x) = x^4 + 4x^3 + 3x^2 - 5$.

Step 2 Graph the related function.

Enter the equation in the **Y** = list and graph the function.

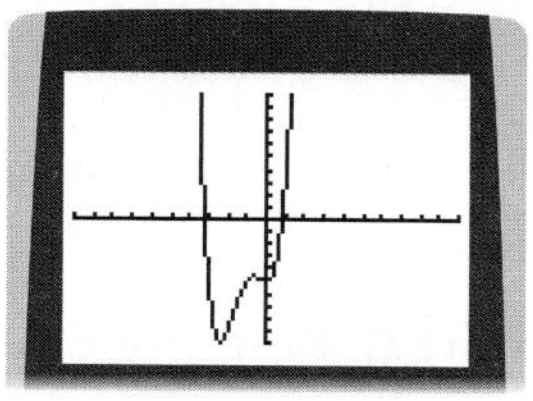

Step 3 Find the zeros.

Use the **zero** feature from the **CALC** menu. The real zeros are about -3.22 and 0.84.

Check

Use a graphing calculator to solve $4x^2 + x = \frac{1}{2}x^4 + 1$ by graphing. Round to the nearest hundredth, if necessary.

$x =$ ___?___

Today's Goals
- Solve polynomial equations by graphing.

Think About It!
How can you use the structure of the related function to determine the number of real solutions of the equation?

Talk About It!
Explain how you could use the table feature to more accurately estimate the zeros of the related function. What are the limitations of the table feature?

Example 2 Solve a Polynomial Equation by Using a System

ANIMALS **For an exhibit with six or fewer Emperor penguins, the pool must have a depth of at least 4 feet and a volume of at least 1620 gallons, or about 217 ft^3, per bird. If a zoo has five Emperor penguins, what should the dimensions of the pool shown at the right be to meet the minimum requirements?**

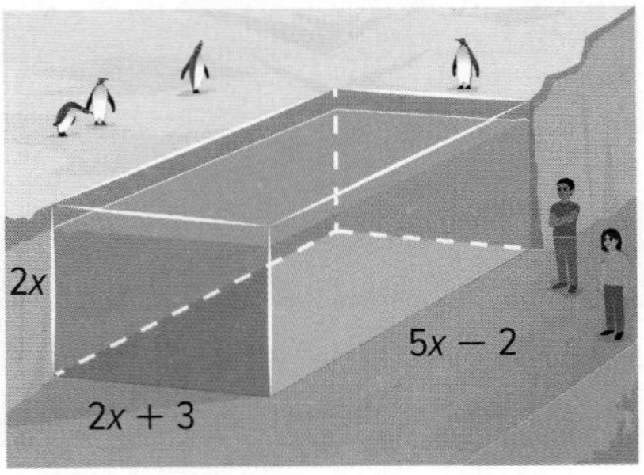

Part A Write a polynomial equation.

Use the formula for the volume of a rectangular prism, $V = \ell wh$, to write a polynomial equation that represents the volume of the pool. Let h represent the depth of the pool.

Since the minimum required volume for the pool is 217 ft^3 per penguin, or $217 \cdot 5 = 1085$ ft^3, the equation that represents the volume of the pool is $(2x + 3)(5x - 2)2x = 1085$. Simplify the equation.

$(2x + 3)(5x - 2)2x = 1085$	Volume of pool
$[2x(5x) + 2x(-2) + 3(5x) + 3(-2)]2x = 1085$	FOIL
$(10x^2 - 4x + 15x - 6)2x = 1085$	Simplify.
$(10x^2 + 11x - 6)2x = 1085$	Combine like terms.
$20x^3 + 22x^2 - 12x = 1085$	Distributive Property

So, the volume of the pool is $20x^3 + 22x^2 - 12x = 1085$.

Part B Write and solve a system of equations.

Set each side equal to y to create a system of equations.

$y = 20x^3 + 22x^2 - 12x$	First equation
$y = 1085$	Second equation

Enter the equations in the **Y** = list and graph.

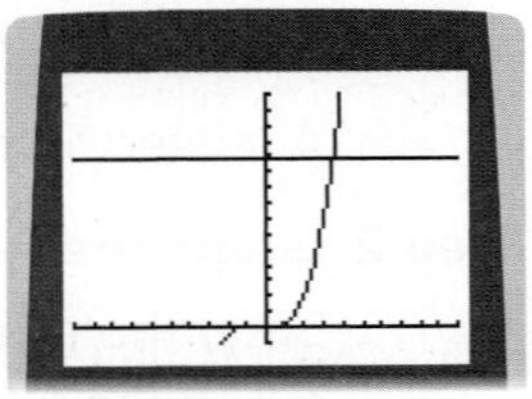

Use the **intersect** feature on the **CALC** menu to find the coordinates of the point of intersection.

The real solution is the x-coordinate of the intersection, which is 3.5.

Part C Find the dimensions.

Substitute 3.5 feet for x in the length, width, and depth of the pool.

Length: $2x + 3 = 10$ ft   Width: $5x - 2 = 15.5$ ft

Depth: $2x = 7$ ft

Go Online You can complete an Extra Example online.

Think About It!

Is your solution reasonable? Justify your conclusion.

Practice

Go Online You can complete your homework online.

Example 1

Use a graphing calculator to solve each equation by graphing. If necessary, round to the nearest hundredth.

1. $\frac{2}{3}x^3 + x^2 - 5x = -9$

2. $x^3 - 9x^2 + 27x = 20$

3. $x^3 + 1 = 4x^2$

4. $x^6 - 15 = 5x^4 - x^2$

5. $\frac{1}{2}x^5 = \frac{1}{5}x^2 - 2$

6. $x^8 = -x^7 + 3$

Example 2

7. SHIPPING A shipping company will ship a package for $7.50 when the volume is no more than 15,000 cm^3. Grace needs to ship a package that is $3x - 5$ cm long, $2x$ cm wide, and $x + 20$ cm tall.

a. Write a polynomial equation to represent the situation if Grace plans to spend a maximum of $7.50.

b. Write and solve a system of equations.

c. What should the dimensions of the package be to have the maximum volume?

8. GARDEN A rectangular garden is 12 feet across and 16 feet long. It is surrounded by a border of mulch that is a uniform width, x. The maximum area for the garden, plus border, is 285 ft^2.

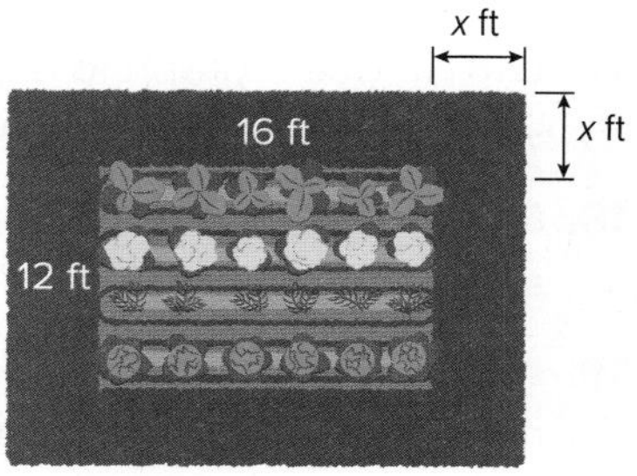

a. Write a polynomial equation to represent the situation.

b. Write and solve a system of equations.

c. What are the dimensions of the garden plus border?

9. PACKAGING A juice manufacturer is creating new cylindrical packaging. The height of the cylinder is to be 3 inches longer than the radius of the can. The cylinder is to have a volume of 628 cubic inches. Use 3.14 for π.

a. Write a polynomial equation to support the model.

b. Write and solve a system of equations.

c. What are the radius and height of the new packaging?

Mixed Exercises

Solve each equation. If necessary, round to the nearest hundredth.

10. $x^4 + 2x^3 = 7$

11. $x^4 - 15x^2 = -24$

12. $x^3 - 6x^2 + 4x = -6$

13. $x^4 - 15x^2 + x + 65 = 0$

14. BALLOON Treyvon is standing 9 yards from the base of a hill that has a slope of $\frac{3}{4}$. He throws a water balloon from a height of 2 yards. Its path is modeled by $h(x) = -0.1x^2 + 0.8x + 2$, where h is the height of the balloon in yards and x is the distance the balloon travels in yards.

a. Write a polynomial equation to represent the situation.

b. How far from Treyvon will the balloon hit the hill?

15. USE TOOLS A company models its revenue in dollars using the function $P(x) = 70{,}000\,(x - x^4)$ on the domain (0, 1) where x is the price at which they sell their product in dollars. Use a graphing calculator to sketch a graph and find the price at which their product should be sold to make revenue of \$20,000. Describe your solution process.

16. ROLLER COASTERS On a racing roller coaster, two trains start at the same time and race to see which returns to the station first. On one coaster, the height of a train on the blue track can be modeled by $f(x) = \frac{1}{20}(x^3 - 60x^2 + 900x)$ and the height of a train on the green track can be modeled by $g(x) = \frac{1}{12{,}000}(x^5 - 144x^4 + 7384x^3 - 158{,}400x^2 + 1{,}210{,}000x)$ where x is time in seconds for the first 35 seconds of the ride.

   a. What equation would determine the times when the blue and green trains are at the same height?

   b. Use a graphing calculator to sketch a graph of $f(x)$ and $g(x)$ and solve the equation from **part a**. Interpret the solution in the context of the situation.

   c. Write an equation to determine the times for which the blue train modeled by $f(x)$ is at a height of 150 feet. Use a graphing calculator to solve the equation. Interpret the solution in the context of the situation.

17. WRITE Use a graph to explain why a function with an even degree can have zero real solutions, but a function with an odd degree must have at least one real solution.

18. CREATE Write a polynomial equation and solve it by graphing a related function and finding its zeros.

19. ANALYZE Determine whether the following statement is *sometimes*, *always*, or *never* true. Justify your argument.

   If a system of equations has more than one solution, then the positive solution is the only viable solution.

20. PERSEVERE During practice, a player kicks a ball from the ground with an initial velocity of 32 feet per second. The polynomial $f(x) = -16x^2 + 32x$ models the height of the ball, where x represents time in seconds. At the same time, another player heads a ball at some distance c feet off the ground with an initial velocity of 27 feet per second. The polynomial $g(x) = -16x^2 + 27x + c$ models the height of the ball.

   a. If the balls are at the same height after 1.2 seconds, from what height did the second player head the ball?

   b. If $c > 0$, is it possible that the soccer balls are never at the same height? Is it reasonable in the context of the situation? Explain your reasoning.

21. WHICH ONE DOESN'T BELONG? Which polynomial doesn't belong? Justify your conclusion.

$x - 17 = 18x^3 + 3x^2$	$x^2 = 4x^4 + 3x^2 - 8$	$5x^2 = -2x - 11$	$-4 = 2x^5 - x^2$

Lesson 5-2

Solving Polynomial Equations Algebraically

Learn Solving Polynomial Equations by Factoring

Like quadratics, some polynomials of higher degrees can be factored. A polynomial that cannot be written as a product of two polynomials with integral coefficients is called a **prime polynomial**. Like a prime real number, the only factors of a prime polynomial are 1 and itself.

Similar to quadratics, some cubic polynomials can be factored by using polynomial identities.

Key Concept • Sum and Difference of Cubes

Factoring Technique	General Case
Sum of Two Cubes	$a^3 + b^3 = (a + b)(a^2 - ab + b^2)$
Difference of Two Cubes	$a^3 - b^3 = (a - b)(a^2 + ab + b^2)$

Polynomials can be factored by using a variety of methods, the most common of which are summarized in the table below. When factoring a polynomial, always look for a common factor first to simplify the expression. Then, determine whether the resulting polynomial factors can be factored using one or more methods.

Concept Summary • Factoring Techniques

Number of Terms	Factoring Technique	General Case
any number	Greatest Common Factor (GCF)	$2a^4b^3 + 6ab = 2ab(a^3b^2 + 3)$
two	Difference of Two Squares	$a^2 - b^2 = (a + b)(a - b)$
	Sum of Two Cubes	$a^3 + b^3 = (a + b)(a^2 - ab + b^2)$
	Difference of Two Cubes	$a^3 - b^3 = (a - b)(a^2 + ab + b^2)$
three	Perfect Square Trinomials	$a^2 + 2ab + b^2 = (a + b)^2$ $a^2 - 2ab + b^2 = (a - b)^2$
	General Trinomials	$acx^2 + (ad + bc)x + bd$ $= (ax + b)(cx + d)$
four or more	Grouping	$ax + bx + ay + by$ $= x(a + b) + y(a + b)$ $= (a + b)(x + y)$

Today's Goals

- Solve polynomial equations by factoring.
- Solve polynomial equations by writing them in quadratic form and factoring.

Today's Vocabulary

prime polynomial

quadratic form

Think About It!

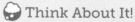

Mateo says that you could use the sum of two cubes to factor $x^{15} + y^{15}$? Is he correct? Why or why not?

Think About It!

How can you check that an expression has been factored correctly?

Example 1 Factor Sums and Differences of Cubes

Factor each polynomial. If the polynomial cannot be factored, write *prime*.

a. $8x^3 + 125y^{12}$

The GCF of the terms is 1, but $8x^3$ and $125y^{12}$ are both perfect cubes. Factor the sum of two cubes.

$8x^3 + 125y^{12}$	Original expression
$= (2x)^3 + (5y^4)^3$	$(2x)^3 = 8x^3$; $(5y^4)^3 = 125y^{12}$
$= (2x + 5y^4)[(2x)^2 - (2x)(5y^4) + (5y^4)^2]$	Sum of two cubes
$= (2x + 5y^4)(4x^2 - 10xy^4 + 25y^8)$	Simplify.

b. $54x^5 - 128x^2y^3$

$54x^5 - 128x^2y^3$	Original expression
$= 2x^2(27x^3 - 64y^3)$	Factor out the GCF.
$= 2x^2[(3x)^3 - (4y)^3]$	$(3x)^3 = 27x^3$; $(4y)^3 = 64y^3$
$= 2x^2(3x - 4y)[(3x)^2 + 3x(4y) + (4y)^2]$	Difference of two cubes
$= 2x^2(3x - 4y)(9x^2 + 12xy + 16y^2)$	Simplify.

Study Tip

Grouping When grouping 6 or more terms, group the terms that have the *most* common values.

Example 2 Factor by Grouping

Factor $14ax^2 - 16by + 20cy + 28bx^2 - 35cx^2 - 8ay$. If the polynomial cannot be factored, write *prime*.

$14ax^2 - 16by + 20cy + 28bx^2 - 35cx^2 - 8ay$	Original expression
$= (14ax^2 + 28bx^2 - 35cx^2) + (-8ay - 16by + 20cy)$	Group to find a GCF.
$= 7x^2(2a + 4b - 5c) - 4y(2a + 4b - 5c)$	Factor out the GCF.
$= (7x^2 - 4y)(2a + 4b - 5c)$	Distributive Property

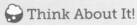

Think About It!

When factoring by grouping, what must be true about the expressions inside parentheses after factoring out a GCF from each group?

Example 3 Combine Cubes and Squares

Factor $64x^6 - y^6$. If the polynomial cannot be factored, write *prime*.

This polynomial could be considered the difference of two squares or the difference of two cubes. The difference of two squares should always be done before the difference of two cubes for easy factoring.

$64x^6 - y^6$	Original expression
$= (8x^3)^2 - (y^3)^2$	$(8x^3)^2 = 64x^6$; $(y^3)^2 = y^6$
$= (8x^3 + y^3)(8x^3 - y^3)$	Difference of squares
$= [(2x)^3 + y^3)][(2x)^3 - y^3)]$	$(2x)^3 = 8x^3$
$= (2x + y)(4x^2 - 2xy + y^2)(2x - y)(4x^2 + 2xy + y^2)$	Sum and difference of cubes

Go Online You can complete an Extra Example online.

Example 4 Solve a Polynomial Equation by Factoring

Solve $4x^3 + 12x^2 - 9x - 27 = 0$.

$4x^3 + 12x^2 - 9x - 27 = 0$	Original equation
$(4x^3 + 12x^2) + (-9x - 27) = 0$	Group to find a GCF.
$4x^2(x + 3) - 9(x + 3) = 0$	Factor out the GCFs.
$(4x^2 - 9)(x + 3) = 0$	Distributive Property
$(2x + 3)(2x - 3)(x + 3) = 0$	Difference of squares
$2x + 3 = 0$ or $2x - 3 = 0$ or $x + 3 = 0$	Zero Product Property
$x = -\frac{3}{2}, \quad x = \frac{3}{2}, \quad x = -3$	

The solutions of the equation are -3, $-\frac{3}{2}$, and $\frac{3}{2}$.

Check

Solve $x^3 + 4x^2 - 25x - 100 = 0$.

$x =$ __?__, __?__, and __?__

Example 5 Write and Solve a Polynomial Equation by Factoring

GEOMETRY **In the figure, the small cube is one fourth the length of the larger cube. If the volume of the figure is 1701 cubic centimeters, what are the dimensions of the cubes?**

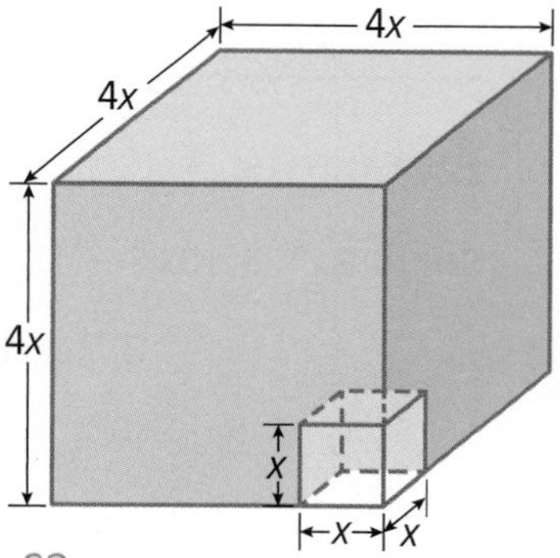

$(4x)^3 - x^3 = 1701$	Volume of figure
$64x^3 - x^3 = 1701$	$(4x)^3 = 64x^3$
$63x^3 = 1701$	Subtract.
$x^3 = 27$	Divide each side by 63.
$x^3 - 27 = 0$	Subtract 27 from each side.
$(x - 3)(x^2 + 3x + 9) = 0$	Difference of cubes
$x = 3$ or $x = \frac{-3 \pm 3i\sqrt{3}}{2}$	Solve.

Since 3 is the only real solution, the lengths of the cubes are 3 cm and 12 cm.

Learn Solving Polynomial Equations in Quadratic Form

Some polynomials in x can be rewritten in **quadratic form**.

Key Concept • Quadratic Form

An expression in quadratic form can be written as $au^2 + bu + c$ for any numbers a, b, and c, $a \neq 0$, where u is some expression in x. The expression $au^2 + bu + c$ is called the quadratic form of the original expression.

Go Online You can complete an Extra Example online.

Think About It!

The following expressions can be written in quadratic form. What do you notice about the terms with variables in the original expressions?

$2x^{10} + x^5 + 9$

$12x^6 - 20x^3 + 6$

$15x^2 + 9x^4 - 1$

Example 6 Write Expressions in Quadratic Form

Write each expression in quadratic form, if possible.

a. $4x^{20} + 6x^{10} + 15$

Examine the terms with variables to choose the expression equal to u.

$4x^{20} + 6x^{10} + 15 = (2x^{10})^2 + 3(2x^{10}) + 15$ — $(2x^{10})^2 = 4x^{20}$

b. $18x^4 + 180x^8 - 28$

If the polynomial is not already in standard form, rewrite it. Then examine the terms with variables to choose the expression equal to u.

$18x^4 + 180x^8 - 28 = 180x^8 + 18x^4 - 28$ — Standard form of a polynomial

$= 5(6x^4)^2 + 3(6x^4) - 28$ — $(6x^4)^2 = 36x^8$

c. $9x^6 - 4x^2 - 12$

Because $x^6 \neq (x^2)^2$, the expression cannot be written in quadratic form.

Check

What is the quadratic form of $10x^4 + 100x^8 - 9$?

Talk About It!

Describe how the exponent of the expression equal to u relates to the exponents of the terms with variables.

Example 7 Solve Equations in Quadratic Form

Solve $8x^4 + 10x^2 - 12 = 0$.

$8x^4 + 10x^2 - 12 = 0$	Original equation
$2(2x^2)^2 + 5(2x^2) - 12 = 0$	$2(2x^2)^2 = 8x^4$
$2u^2 + 5u - 12 = 0$	Let $u = 2x^2$.
$(2u - 3)(u + 4) = 0$	Factor.
$u = \frac{3}{2}$ or $u = -4$	Zero Product Property
$2x^2 = \frac{3}{2}$ $2x^2 = -4$	Replace u with $2x^2$.
$x^2 = \frac{3}{4}$ $x^2 = -2$	Divide each side by 2.
$x = \pm\frac{\sqrt{3}}{2}$ $x = \pm i\sqrt{2}$	Take the square root of each side.

The solutions are $\frac{\sqrt{3}}{2}$, $-\frac{\sqrt{3}}{2}$, $i\sqrt{2}$, and $-i\sqrt{2}$.

Check

What are the solutions of $16x^4 + 24x^2 - 40 = 0$?

$x =$ ___?___

Go Online You can complete an Extra Example online.

Practice

Go Online You can complete your homework online.

Examples 1-3

Factor completely. If the polynomial is not factorable, write *prime*.

1. $8c^3 - 27d^3$

2. $64x^4 + xy^3$

3. $a^8 - a^2b^6$

4. $x^6y^3 + y^9$

5. $18x^6 + 5y^6$

6. $w^3 - 2y^3$

7. $gx^2 - 3hx^2 - 6fy^2 - gy^2 + 6fx^2 + 3hy^2$

8. $12ax^2 - 20cy^2 - 18bx^2 - 10ay^2 + 15by^2 + 24cx^2$

9. $a^3x^2 - 16a^3x + 64a^3 - b^3x^2 + 16b^3x - 64b^3$

10. $8x^5 - 25y^3 + 80x^4 - x^2y^3 + 200x^3 - 10xy^3$

Example 4

Solve each equation.

11. $a^3 - 9a^2 + 14a = 0$

12. $x^3 = 3x^2$

13. $t^4 - 3t^3 - 40t^2 = 0$

14. $b^3 - 8b^2 + 16b = 0$

Example 5

15. FURNITURE A modern table is constructed with four legs made of concrete cubes with cube-shaped notches.

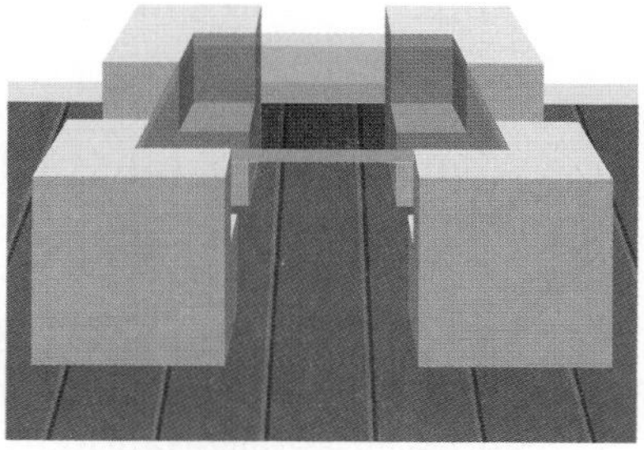

a. Define one or more variables and write an expression that represents the volume of one table leg.

b. Marisol determined that she will use 12,636 cubic inches of concrete to construct the four table legs. If the sides of the notches are 40% of the sides of the legs, how long are the sides of the legs?

Example 6

Write each expression in quadratic form, if possible.

16. $x^4 + 12x^2 - 8$

17. $-15x^4 + 18x^2 - 4$

18. $8x^6 + 6x^3 + 7$

19. $5x^6 - 2x^2 + 8$

20. $9x^8 - 21x^4 + 12$

21. $16x^{10} + 2x^5 + 6$

Example 7

Solve each equation.

22. $x^4 + 6x^2 + 5 = 0$

23. $x^4 - 3x^2 - 10 = 0$

24. $4x^4 - 14x^2 + 12 = 0$

25. $9x^4 - 27x^2 + 20 = 0$

26. $4x^4 - 5x^2 - 6 = 0$

27. $24x^4 + 14x^2 - 3 = 0$

Mixed Exercises

Factor completely. If the polynomial is not factorable, write *prime*.

28. $x^4 - 625$

29. $x^6 - 64$

30. $x^5 - 16x$

31. $8x^5y^2 - 27x^2y^5$

32. $x^6 - 4x^4 - 8x^4 + 32x^2 + 16x^2 - 64$

33. $y^9 - y^6 - 2y^6 + 2y^3 + y^3 - 1$

34. DÉCOR Each box in a set of decorative storage boxes is a cube.

a. The sides of the smallest box are 30% of the length of the sides of the largest box. The volume of the largest box is 7784 cubic inches more than the volume of the smallest box. Define one or more variables, and write an equation that represents the situation.

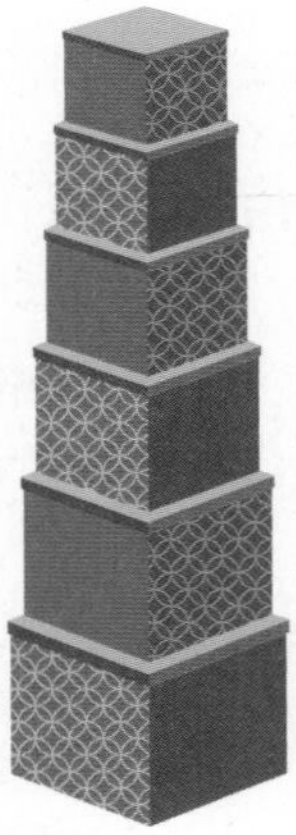

b. What are the lengths of the sides of the largest and smallest boxes?

c. Two of the other boxes have a total volume of 2457 cubic inches. The length of the sides of the smaller of these boxes is 75% of those of the larger box. Write and solve an equation to find the lengths of the sides of the boxes.

Solve each equation.

35. $x^4 + x^2 - 90 = 0$

36. $x^4 - 16x^2 - 720 = 0$

37. $x^4 - 7x^2 - 44 = 0$

38. $x^4 + 6x^2 - 91 = 0$

39. $x^3 + 216 = 0$

40. $64x^3 + 1 = 0$

41. $8x^4 + 10x^2 - 3 = 0$

42. $6x^4 - 5x^2 - 4 = 0$

43. $20x^4 - 53x^2 + 18 = 0$

44. $18x^4 + 43x^2 - 5 = 0$

45. $8x^4 - 18x^2 + 4 = 0$

46. $3x^4 - 22x^2 - 45 = 0$

47. $x^6 + 7x^3 - 8 = 0$

48. $x^6 - 26x^3 - 27 = 0$

49. $8x^6 + 999x^3 = 125$

50. $4x^4 - 4x^2 - x^2 + 1 = 0$

51. $x^6 - 9x^4 - x^2 + 9 = 0$

52. $x^4 + 8x^2 + 15 = 0$

53. **STRUCTURE** Consider the equation $x^{\frac{1}{2}} - 8x^{\frac{1}{4}} + 15 = 0$.

a. How are the exponents in the equation related?

b. How could you define *u* so that you could rewrite the equation as a quadratic equation in terms of *u*? Write the quadratic equation.

c. Solve the original equation.

Factor completely. If the polynomial is not factorable, write *prime*.

54. $21x^3 - 18x^2y + 24xy^2$

55. $8j^3k - 4jk^3 - 7$

56. $a^2 + 7a - 18$

57. $2ak - 6a + k - 3$

58. $b^2 + 8b + 7$

59. $z^2 - 8z - 10$

60. $4f^2 - 64$

61. $d^2 - 12d + 36$

62. $9x^2 + 25$

63. $y^2 + 18y + 81$

64. $7x^2 - 14x$

65. $19x^3 - 38x^2$

66. $n^3 - 125$

67. $m^4 - 1$

68. REASONING A rectangular box has dimensions of x inches, $(x + 5)$ inches, and $(x - 2)$ inches. The volume of the box is $30x$ cubic inches. Find the dimensions of the box. Explain your reasoning.

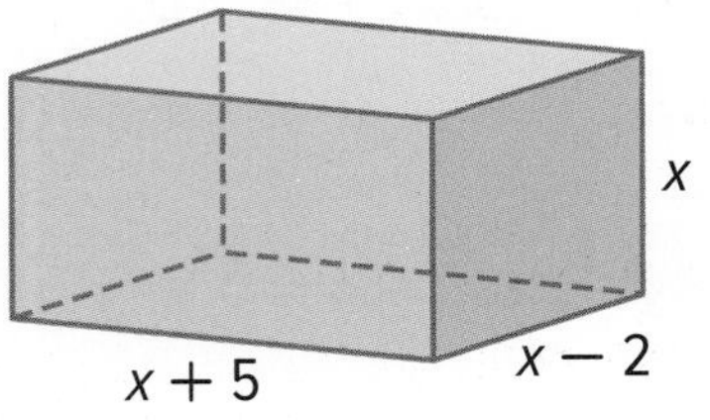

69. GEOMETRY The combined volume of a cube and a cylinder is 1000 cubic inches. If the height of the cylinder is twice the radius and the side of the cube is four times the radius, find the radius of the cylinder to the nearest tenth of an inch.

70. ANALYZE Find the solutions of $(a + 3)^4 - 2(a + 3)^2 - 8 = 0$. Show your work.

71. WRITE If the equation $ax^2 + bx + c = 0$ has solutions $x = m$ and $x = n$, what are the solutions to $ax^4 + bx^2 + c = 0$? Explain your reasoning.

72. PERSEVERE Factor $36x^{2n} + 12x^n + 1$.

73. PERSEVERE Solve $6x - 11\sqrt{3x} + 12 = 0$.

74. ANALYZE Find a counterexample to the statement $a^2 + b^2 = (a + b)^2$.

75. CREATE The cubic form of an equation is $ax^3 + bx^2 + cx + d = 0$. Write an equation with degree 6 that can be written in cubic form.

76. WRITE Explain how the graph of a polynomial function can help you factor the polynomial.

Proving Polynomial Identities

Explore Polynomial Identities

Online Activity Use graphing technology to complete the Explore.

INQUIRY How can you prove that two polynomial expressions form a polynomial identity?

Today's Goal
- Prove polynomial identities and use them to describe numerical relationships.

Today's Vocabulary
identity

polynomial identity

Learn Proving Polynomial Identities

An **identity** is an equation that is satisfied by any numbers that replace the variables. Thus, a **polynomial identity** is a polynomial equation that is true for any values that are substituted for the variables.

Unlike solving an equation, do not begin by assuming that an identity is true. You cannot perform the same operation to both sides and assume that equality is maintained.

Key Concept • Verifying Identities by Transforming One Side
- Simplify one side of an equation until the two sides of the equation are the same. It is often easier to transform the more complicated expression into the form of the simpler side.
- Factor or multiply expressions as necessary. Simplify by combining like terms.

Study Tip

Transforming One Side It is often easier to work with the more complicated side of an equation. Look at each side and determine which requires more steps to be simplified. For example, it is often easier to work on the side that involves the square or cube of an algebraic expression.

Example 1 Transform One Side

Prove that $x^3 - y^3 = (x - y)(x^2 + xy + y^2)$.

$x^3 - y^3 \stackrel{?}{=} (x - y)(x^2 + xy + y^2)$	Original equation
$\stackrel{?}{=} x(x^2) + x(xy) + x(y^2) - y(x^2) - y(xy) - y(y^2)$	Distributive Property
$\stackrel{?}{=} x^3 + x^2y + xy^2 - x^2y - xy^2 - y^3$	Simplify.
$\stackrel{?}{=} x^3 + x^2y - x^2y + xy^2 - xy^2 - y^3$	Commutative Property
$= x^3 - y^3$	True

Because the expression on the right can be simplified to be the same as the expression on the left, this proves the polynomial identity.

Go Online You can complete an Extra Example online.

Talk About It!

If you multiplied each side of the equation by a variable z, would the result still be a polynomial identity? Explain your reasoning.

Example 2 Use Polynomial Identities

TRIANGLES **Pedro claims that you can always create three lengths that form a right triangle by using the following method: take two positive integers x and y where $x > y$. Two legs of a right triangle are defined as $x^2 - y^2$ and $2xy$. The hypotenuse is defined as $x^2 + y^2$. Is Pedro correct? Explain your reasoning in the context of polynomial identities.**

To determine whether Pedro is correct, we can use information about right triangles and the expressions involving x and y to try to construct a polynomial identity. If $x^2 - y^2$ and $2xy$ are the legs of the triangle, and $x^2 + y^2$ is the hypotenuse, then it should be true that $(x^2 - y^2)^2 + (2xy)^2 = (x^2 + y^2)^2$.

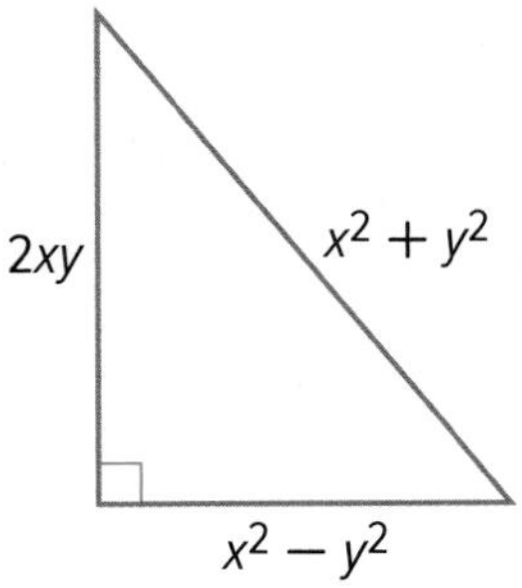

If this is an identity, you can simplify the expressions for the sides to be the same expression.

$(x^2 - y^2)^2 + (2xy)^2 \stackrel{?}{=} (x^2 + y^2)^2$	Original equation
$x^4 - 2x^2y^2 + y^4 + 4x^2y^2 \stackrel{?}{=} x^4 + 2x^2y^2 + y^4$	Square each term.
$x^4 + 2x^2y^2 + y^4 = x^4 + 2x^2y^2 + y^4$	True

Because the identity is true, this proves that Pedro is correct. His process for creating the sides of a right triangle will always work.

Check

Write in the missing explanations to prove that $x^4 - y^4 = (x - y)(x + y)(x^2 + y^2)$.

$x^4 - y^4 \stackrel{?}{=} (x - y)(x + y)(x^2 + y^2)$	Original equation
$x^4 - y^4 \stackrel{?}{=} (x^2 - y^2)(x^2 + y^2)$	?
$x^4 - y^4 \stackrel{?}{=} x^4 + x^2y^2 - x^2y^2 - y^4$	?
$x^4 - y^4 = x^4 - y^4$	?

Go Online You can complete an Extra Example online.

Math History Minute:

Former quarterback **Frank Ryan (1936–)** earned his Ph.D. in mathematics about six months after he led the Cleveland Browns to the NFL championship game of 1964, where they won 27-0. During part of his academic career, Ryan studied prime numbers, including Opperman's Conjecture that there is a prime number between n^2 and $n^2 + n$, where n is an integer. This work could eventually lead to a polynomial identity that could be used to identify prime numbers.

Use a Source

Research an application of prime numbers. How could a polynomial identity for identifying prime numbers impact the application?

Practice

Go Online You can complete your homework online.

Example 1

Prove each polynomial identity.

1. $(x - y)^2 = x^2 - 2xy + y^2$
2. $(x + 5)^2 = x^2 + 10x + 25$
3. $4(x - 7)^2 = 4x^2 - 56x + 196$
4. $(2x^2 + y^2)^2 = (2x^2 - y^2)^2 + (2xy\sqrt{2})$
5. $a^2 - b^2 = (a + b)(a - b)$
6. $x^3 + y^3 = (x + y)(x^2 - xy + y^2)$
7. $p^4 - q^4 = (p - q)(p + q)(p^2 + q^2)$
8. $a^5 - b^5 = (a - b)(a^4 + a^3b + a^2b^2 + ab^3 + b^4)$

Example 2

9. SQUARES Aponi claims that you can find the area of a square using the following method: take two positive integers x and y. The side length of the square is defined by the expression $3x + y$. The area of the square is defined by the expression $9x^2 + 6xy + y^2$. Is Aponi correct? Explain your reasoning in the context of polynomial identities.

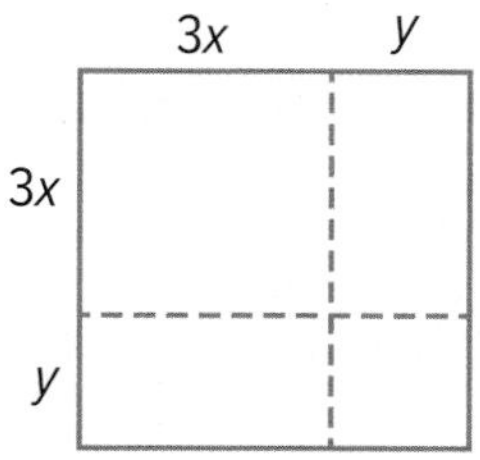

10. USE A MODEL Julio claims that you can find the area of a rectangle using the following method: take two positive integers x and y, where $x > y$. The side lengths of the rectangle are defined by the expressions $2x + y$ and $2x - y$. The area of the rectangle is defined by the expression $4x^2 - y^2$. Is Julio correct? Explain your reasoning in the context of polynomial identities.

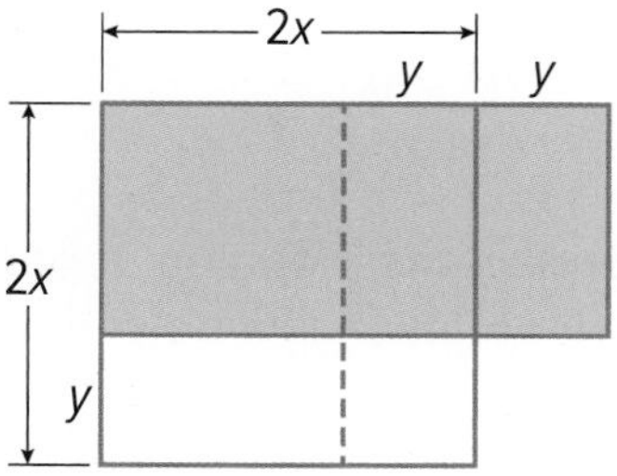

Mixed Exercises

Determine whether each equation is an identity.

11. $(x + 3)^2(x^3 + 3x^2 + 3x + 1) = (x^2 + 6x + 9)(x + 1)^3$
12. $(x + 2)(x + 1)^2 = (x^2 + 3x + 2)(x + 1)$
13. $(x + 3)(x - 1)^2 = (x^2 - 2x - 3)(x - 1)$
14. $(x + 2)^2(x^3 - 3x^2 + 3x - 1) = (x^2 + 4x + 4)(x - 1)^3$
15. $(a + b)^2 = a^2 - 2ab + b^2$

16. USE TOOLS Consider the following equation.

$(x - 2)^2(x^3 + 9x^2 + 27x + 27) = (x^2 - 4x + 4)(x + 3)^3 < 3$

a. Evaluate the expressions for each value. Copy and complete the table.

x	$(x - 2)^2(x^3 + 9x^2 + 27x + 27)$	$(x^2 - 4x + 4)(x + 3)^3$
0		
1		
2		
3		
4		

b. What conclusion can you make about the equation, based on the results in your table? Explain.

c. How can you prove your conclusion from **part b**?

USE TOOLS **Use a computer algebra system (CAS) to prove each identity.**

17. $g^6 + h^6 = (g^2 + h^2)(g^4 - g^2h^2 + h^4)$

18. $a^5 + b^5 = (a + b)(a^4 - a^3b + a^2b^2 - ab^3 + b^4)$

19. $u^6 - w^6 = (u + w)(u - w)(u^2 + uw + w^2)(u^2 - uw + w^2)$

20. $(x + 1)^2(x - 4)^3 = (x^2 - 3x - 4)(x^3 - 7x^2 + 8x + 16)$

21. WRITE Explain the meaning of polynomial identity and summarize the method for proving an equation is a polynomial identity.

22. CREATE Write and solve a system of equations using the identity $(x^2 - y^2)^2 + (2xy)^2 = (x^2 + y^2)^2$ to find the values of x and y that make a 3, 4, 5 Pythagorean triple.

23. ANALYZE Refer to Example 2. Notice that Pedro says x and y must be positive integers and x must be greater than y. Explain why these restrictions are necessary.

24. PERSEVERE Rebecca has a square garden with side length a that she wants to transform into a rectangle. Rebecca speculates that if she subtracts the same length b from one dimension of the garden and adds it to the other dimension the new rectangle's area will be smaller than the original garden in the amount of b^2. Draw a diagram and show algebraically that Rebecca is correct.

25. FIND THE ERROR George is proving the identity $a^3 + b^3 = (a + b)(a^2 - ab + b^2)$ by simplifying the right side. His work is shown. Is George correct? If not, identify and correct his error.

$(a + b)(a^2 - ab + b^2)$

$= a^3 - a^2b + ab^2 - a^2b - ab^2 + b^3$

$= a^3 - 2a^2b + b^3$

The Remainder and Factor Theorems

Today's Goals

- Evaluate functions by using synthetic substitution.
- Use the Factor Theorem to determine factors of polynomials.

Today's Vocabulary

synthetic substitution

depressed polynomial

Explore Remainders

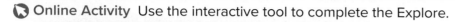

Online Activity Use the interactive tool to complete the Explore.

INQUIRY How are the divisor and quotient of a polynomial related to its factors when the remainder is zero?

Learn The Remainder Theorem

Polynomial division can be used to find the value of a function. From the Division Algorithm, we know that $\frac{f(x)}{g(x)} = q(x) + \frac{r(x)}{g(x)}$ and that $f(x) = q(x) \cdot g(x) + r(x)$, where q and r are unique and the degree of r is less than the degree of g. Suppose we were to call the dividend $p(x)$ and the divisor $x - a$. Then the Division Algorithm would be $\frac{p(x)}{x - a} = q(x) + \frac{r}{x - a}$ and $p(x) = q(x) \cdot (x - a) + r$, where a is a constant and r is the remainder. Since any polynomial can be written in this form, evaluating $p(x)$ at a gives the following.

$p(x) = q(x) \cdot (x - a) + r$	Polynomial function $p(x)$
$p(a) = q(a) \cdot (a - a) + r$	Substitute a for x.
$p(a) = q(a) \cdot (0) + r$	$a - a = 0$
$p(a) = r$	$q(a) \cdot (0) = 0$

This shows how the Remainder Theorem can be used to evaluate a polynomial at $p(a)$.

Key Concept • Remainder Theorem

Words: For a polynomial $p(x)$ and a number a, the remainder upon division by $x - a$ is $p(a)$.

Example: Evaluate $p(x) = x^2 - 4x + 7$ when $x = 5$.

Synthetic division

$$\begin{array}{r|rrr} 5 & 1 & -4 & 7 \\ & & 5 & 5 \\ \hline & 1 & 1 & 12 \end{array}$$

$p(5) = 12$

Direct substitution

$p(x) = x^2 - 4x + 7$

$p(5) = 5^2 - 4(5) + 7$

$p(5) = 12$

Applying the Remainder Theorem to evaluate a function is called **synthetic substitution**. You may find that synthetic substitution is a more convenient way to evaluate a polynomial function, especially when the degree of the function is greater than 2.

> **Study Tip**
>
> **Missing terms** Remember to include zeros as placeholders for any missing terms in the polynomial.

Example 1 Synthetic Substitution

Use synthetic substitution to find $f(-3)$ if $f(x) = -2x^4 + 3x^2 - 15x + 9$.

By the Remainder Theorem, $f(-3)$ is the remainder of $\frac{f(x)}{x - (-3)}$.

−3	−2	0	3	−15	9
		6	−18	45	−90
	−2	6	−15	30	−81

The remainder is −81. Therefore, $f(-3) = -81$.

Use direct substitution to check.

$f(x) = -2x^4 + 3x^2 - 15x + 9$	Original function
$f(-3) = -2(-3)^4 + 3(-3)^2 - 15(-3) + 9$	Substitute −3 for x.
$= -162 + 27 + 45 + 9$ or -81	True

Check

Use synthetic substitution to evaluate $f(x) = -6x^3 + 52x^2 - 27x - 31$.

$f(8) =$ ___?___

> **Think About It!**
>
> How could you use the function and synthetic substitution to estimate the number of eggs produced in 1990? What assumption would you have to make to solve this problem?

Example 2 Apply the Remainder Theorem

EGG PRODUCTION **The total production of eggs in billions in the United States can be modeled by the function $f(x) = 0.007x^3 - 0.149x^2 + 1.534x + 84.755$, where x is the number of years since 2000. Predict the total production of eggs in 2025.**

Since 2025 − 2000 = 25, use synthetic substitution to determine $f(25)$.

25	0.007	−0.149	1.534	84.755
		0.175	0.65	54.6
	0.007	0.026	2.184	139.355

In 2025, approximately 139.355 billion eggs will be produced in the United States.

Go Online You can complete an Extra Example online.

Check

KITTENS The ideal weight of a kitten in pounds is modeled by the function $f(x) = 0.009x^2 + 0.127x + 0.377$, where x is the age of the kitten in weeks. Determine the ideal weight of a 9-week-old kitten. Round to the nearest tenth.

__?__ pounds

Learn The Factor Theorem

When a binomial evenly divides a polynomial, the binomial is a factor of the polynomial. The quotient of this division is called a depressed polynomial. The **depressed polynomial** has a degree that is one less than the original polynomial.

A special case of the Remainder Theorem is called the Factor Theorem.

Key Concept • Factor Theorem

Words: The binomial $x - a$ is a factor of the polynomial $p(x)$ if and only if $p(a) = 0$.

Example:

$$\underbrace{x^3 - x^2 - 30x + 72}_{\text{dividend}} = \underbrace{(x^2 - 7x + 12)}_{\text{quotient}} \cdot \underbrace{(x + 6)}_{\text{divisor}} + \underbrace{0}_{\text{remainder}}$$

$x + 6$ is a factor of $x^3 - x^2 - 30x + 72$.

Go Online You may want to complete the Concept Check to check your understanding.

Example 3 Use the Factor Theorem

Show that $x + 8$ is a factor of $2x^3 + 15x^2 - 11x - 24$. Then find the remaining factors of the polynomial.

−8	2	15	−11	−24
		−16	8	24
	2	−1	−3	0

Because the remainder is 0, $x + 8$ is a factor of the polynomial by the Factor Theorem. So $2x^3 + 15x^2 - 11x - 24$ can be factored as $(x + 8)(2x^2 - x - 3)$. The depressed polynomial is $2x^2 - x - 3$. Check to see if this polynomial can be factored.

$2x^2 - x - 3 = (2x - 3)(x + 1)$ Factor the trinomial.

Therefore, $2x^3 + 15x^2 - 11x - 24 = (x + 8)(2x - 3)(x + 1)$.

Study Tip

Factoring Some depressed polynomials may not be factorable. In that case, the only factors are the divisor and the depressed polynomial.

Talk About It!

Suppose you were asked to determine whether $3x + 4$ is a factor of $3x^3 - 2x^2 - 8x$. Describe the steps necessary to find a solution.

Check

Select all of the factors of $3x^3 + 10x^2 - 27x - 10$.

A. $x - 2$

B. $x + 5$

C. $x + 9$

D. $x - 10$

E. $3x + 1$

F. $3x - 10$

Practice

Example 1

Use synthetic substitution to find $f(-5)$ and $f(2)$ for each function.

1. $f(x) = x^3 + 2x^2 - 3x + 1$

2. $f(x) = x^2 - 8x + 6$

3. $f(x) = 3x^4 + x^3 - 2x^2 + x + 12$

4. $f(x) = 2x^3 - 8x^2 - 2x + 5$

5. $f(x) = x^3 - 5x + 2$

6. $f(x) = x^5 + 8x^3 + 2x - 15$

7. $f(x) = x^6 - 4x^4 + 3x^2 - 10$

8. $f(x) = x^4 - 6x - 8$

Use synthetic substitution to find $f(2)$ and $f(-1)$ for each function.

9. $f(x) = x^2 + 6x + 5$

10. $f(x) = x^2 - x + 1$

11. $f(x) = x^2 - 2x - 2$

12. $f(x) = x^3 + 2x^2 + 5$

13. $f(x) = x^3 - x^2 - 2x + 3$

14. $f(x) = x^3 + 6x^2 + x - 4$

15. $f(x) = x^3 - 3x^2 + x - 2$

16. $f(x) = x^3 - 5x^2 - x + 6$

17. $f(x) = x^4 + 2x^2 - 9$

18. $f(x) = x^4 - 3x^3 + 2x^2 - 2x + 6$

19. $f(x) = x^5 - 7x^3 - 4x + 10$

20. $f(x) = x^6 - 2x^5 + x^4 + x^3 - 9x^2 - 20$

Go Online You can complete an Extra Example online.

Go Online You can complete your homework online.

Example 2

21. BUSINESS Advertising online generates billions of dollars for global businesses each year. The revenue from online advertising in the United States from 2000 to 2015 can be modeled by $y = 0.01x^3 + 0.02x^2 + x + 6$, where x is the number of years since 2000 and y is the revenue in billions of U.S. dollars.

a. Estimate the revenue from online advertising in 2008.

b. Predict the revenue from online advertising in 2022.

22. PROFIT The profit, in thousands, of Clyde's Corporation can be modeled by $P(y) = y^4 - 4y^3 + 2y^2 + 10y - 200$, where y is the number of years after the business was started. Predict the profit of Clyde's Corporation after 10 years.

Example 3

Given a polynomial and one of its factors, find the remaining factors of the polynomial.

23. $x^3 - 3x + 2$; $x + 2$

24. $x^4 + 2x^3 - 8x - 16$; $x + 2$

25. $x^3 + 5x^2 + 2x - 8$; $x + 2$

26. $x^3 - x^2 - 5x - 3$; $x - 3$

27. $2x^3 + 17x^2 + 23x - 42$; $x - 1$

28. $2x^3 + 7x^2 - 53x - 28$; $x - 4$

29. $x^4 + 2x^3 + 2x^2 - 2x - 3$; $x - 1$

30. $3x^3 - 19x^2 - 15x + 7$; $x - 7$

31. $x^3 + 2x^2 - x - 2$; $x + 1$

32. $x^3 + x^2 - 5x + 3$; $x - 1$

33. $x^3 + 3x^2 - 4x - 12$; $x + 3$

34. $x^3 - 6x^2 + 11x - 6$; $x - 3$

35. $x^3 + 2x^2 - 33x - 90$; $x + 5$

36. $x^3 - 6x^2 + 32$; $x - 4$

37. $x^3 - x^2 - 10x - 8$; $x + 2$

38. $x^3 - 19x + 30$; $x - 2$

39. $2x^3 + x^2 - 2x - 1$; $x + 1$

40. $2x^3 + x^2 - 5x + 2$; $x + 2$

41. $3x^3 + 4x^2 - 5x - 2$; $3x + 1$

42. $3x^3 + x^2 + x - 2$; $3x - 2$

43. $6x^3 - 25x^2 + 2x + 8$; $2x + 1$

44. $16x^5 - 32x^4 - 81x + 162$; $2x - 3$

Mixed Exercises

45. **REASONING** Jessica evaluates the polynomial $p(x) = x^3 - 5x^2 + 3x + 5$ for a factor using synthetic substitution. Some of her work is shown below. Find the values of a and b.

a	1	−5	3	5
		11	66	759
	1	6	69	b

46. **STATE YOUR ASSUMPTION** The revenue from streaming music services in the United States from 2005 to 2016 can be modeled by $y = 0.26x^5 - 7.48x^4 + 79.20x^3 - 333.33x^2 + 481.68x + 99.13$, where x is the number of years since 2005 and y is the revenue in millions of U.S. dollars.

a. Estimate the revenue from streaming music services in 2010.

b. What might the revenue from streaming music services be in 2020? What assumption did you make to make your prediction?

47. **NATURAL EXPONENTIAL FUNCTION** The natural exponential function $y = e^x$ is a special function that is applied in many fields such as physics, biology, and economics. It is not a polynomial function, however for small values of x, the value of e^x is very closely approximated by the polynomial function $f(x) = \frac{1}{6}x^3 + \frac{1}{2}x^2 + x + 1$. Use synthetic substitution to determine $f(0.1)$.

Find values of k so that each remainder is 3.

48. $(x^2 - x + k) \div (x - 1)$

49. $(x^2 + kx - 17) \div (x - 2)$

50. $(x^2 + 5x + 7) \div (x + k)$

51. $(x^3 + 4x^2 + x + k) \div (x + 2)$

52. If $f(-8) = 0$ and $f(x) = x^3 - x^2 - 58x + 112$, find all the zeros of $f(x)$ and use them to graph the function.

53. **REASONING** If $P(1) = 0$ and $P(x) = 10x^3 + kx^2 - 16x + 3$, find all the factors of $P(x)$ and use them to graph the function. Explain your reasoning.

54. **GEOMETRY** The volume of a box with a square base is $V(x) = 2x^3 + 15x^2 + 36x + 27$. If the height of the box is $(2x + 3)$ units, what are the measures of the sides of the base in terms of x?

55. **SPORTS** The average value of a franchise in the National Football League from 2000 to 2018 can be modeled by $y = -0.037x^5 + 1.658x^4 - 24.804x^3 + 145.100x^2 - 207.594x + 482.008$, where x is the number of years since 2000 and y is the value in millions of U.S. dollars.

    a. Copy and complete the table of estimated values. Round to the nearest million.

Year	2003	2012	2021	2025
Estimated Average Franchise Value (millions $)				

    b. What assumption did you make to make your predictions? Do you think the assumption is valid? Explain.

56. **CONSTRUCT ARGUMENTS** Divide the polynomial function $f(x) = 4x^3 - 10x + 8$ by the factor $(x + 5)$. Then state and confirm the Remainder Theorem for this particular polynomial function and factor.

57. **REGULARITY** The polynomial function $P(x)$ is symmetric in the y-axis and contains the point $(2, -5)$. What is the remainder when $P(x)$ is divided by $(x + 2)$? Explain your reasoning.

58. **STRUCTURE** Verify the Remainder Theorem for the polynomial $x^2 + 3x + 5$ and the factor $(x - \sqrt{3})$ by first using synthetic division and then evaluating for $x = \sqrt{3}$.

59. **STRUCTURE** If $(x + 6)$ is a factor of $kx^3 + 15x^2 + 13x - 30$, determine the value of k, factor the polynomial and confirm the result graphically.

60. **REGULARITY** Polynomial $f(x)$ is divided by $x - c$. What can you conclude if:

    a. the remainder is 0?
    b. the remainder is 1?
    c. the quotient is 1, and the remainder is 0?

61. **CREATE** Write a polynomial function that has a double zero of 1 and a double zero of −5. Graph the function.

62. **PERSEVERE** For a cubic function $P(x)$, $P(2) = -90$, $P(-8) = 0$, and $P(5) = 0$.

   a. Write two possible equations for $P(x)$. Explain your answer.

   b. Graph your equations from **part a**. What three points do these graphs have in common?

   c. If $P(4) = 60$, write the equation for $P(x)$.

63. **ANALYZE** Review the definition for the Factor Theorem. Provide a proof of the theorem.

64. **CREATE** Write a cubic function that has a remainder of 8 for $f(2)$ and a remainder of −5 for $f(3)$.

65. **PERSEVERE** Show that the quartic function $f(x) = ax^4 + bx^3 + cx^2 + dx + e$ will always have a rational zero when the numbers 1, −2, 3, 4, and −6 are randomly assigned to replace a through e, and all of the numbers are used.

66. **WRITE** Explain how the zeros of a function can be located by using the Remainder Theorem and making a table of values for different input values and then comparing the remainders.

67. **FIND THE ERROR** The table shows x-values and their corresponding $P(x)$ values for a polynomial function. Tyrone and Nia used the Factor Theorem to find factors of $P(x)$. Is either of them correct? Explain your reasoning.

x	−3	−1	0	1	2	4
$P(x)$	−18	0	6	2	0	122

Tyrone	Nia
$(x + 1)$ and $(x - 2)$	$(x - 6)$

Roots and Zeros

Explore Roots of Quadratic Polynomials

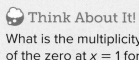

Online Activity Use graphing technology to complete the Explore.

INQUIRY Is the Fundamental Theorem of Algebra true for quadratic polynomials?

Learn Fundamental Theorem of Algebra

The zero of a function $f(x)$ is any value c such that $f(c) = 0$.

Key Concept • Zeros, Factors, Roots, and Intercepts

Words: Let $P(x) = a_n x^n + \ldots + a_1 x + a_0$ be a polynomial function. Then the following statements are equivalent.

- c is a zero of $P(x)$.
- c is a root or solution of $P(x) = 0$.
- $x - c$ is a factor of $a_n x^n + \ldots + a_1 x + a_0$.
- If c is a real number, then $(c, 0)$ is an x-intercept of the graph of $P(x)$.

Example: Consider the polynomial function $P(x) = x^2 + 3x - 18$.

The zeros of $P(x) = x^2 + 3x - 18$ are -6 and 3.

The roots of $x^2 + 3x - 18 = 0$ are -6 and 3.

The factors of $x^2 + 3x - 18$ are $(x + 6)$ and $(x - 3)$.

The x-intercepts of $P(x) = x^2 + 3x - 18$ are $(-6, 0)$ and $(3, 0)$.

Key Concept • Fundamental Theorem of Algebra

Every polynomial equation with degree greater than zero has at least one root in the set of complex numbers.

Key Concept • Corollary to the Fundamental Theorem of Algebra

Words: A polynomial equation of degree n has exactly n roots in the set of complex numbers, including repeated roots.

Examples:

$2x^3 - 5x + 2$	$-x^4 + 2x^3 - 2x$	$x^5 - 6x^3 + x^2 - 1$
3 roots	4 roots	5 roots

Repeated roots can also be called roots of multiplicity m where m is an integer greater than 1. **Multiplicity** is the number of times a number is a zero for a given polynomial. For example, $f(x) = x^3 = x \cdot x \cdot x$ has a zero at $x = 0$ with multiplicity 3, because x is a factor three times. However, the graph of the function still only intersects the x-axis once at the origin.

Today's Goals

- Use the Fundamental Theorem of Algebra to determine the numbers and types of roots of polynomial equations.
- Determine the numbers and types of roots of polynomial equations, find zeros, and use zeros to graph polynomial functions.

Today's Vocabulary

multiplicity

Think About It!

What is the multiplicity of the zero at $x = 1$ for $p(x) = (x - 1)^5$? Explain your reasoning.

Key Concept • Descartes' Rule of Signs

Let $P(x) = a_n x^n + \ldots + a_1 x + a_0$ be a polynomial function with real coefficients and $a_0 \neq 0$. Then the number of positive real zeros of $P(x)$ is the same as the number of changes in sign of the coefficients of the terms, or is less than this by an even number, and the number of negative real zeros of $P(x)$ is the same as the number of changes in sign of the coefficients of the terms of $P(-x)$, or is less than this by an even number.

Example 1 Determine the Number and Type of Roots

Solve $x^4 + 49x^2 = 0$. State the number and type of roots.

$x^4 + 49x^2 = 0$		Original equation
$x^2(x^2 + 49) = 0$		Factor.
$x^2 = 0$ or	$x^2 + 49 = 0$	Zero Product Property
$x = 0$	$x^2 = -49$	Subtract 49 from each side.
	$x = \pm\sqrt{-49}$	Square Root Property
	$x = \pm 7i$	Simplify.

The polynomial has degree 4, so there are four roots in the set of complex numbers. Because x^2 is a factor, $x = 0$ is a root with multiplicity 2, also called a double root. The equation has one real repeated root, 0, and two imaginary roots, $7i$ and $-7i$.

Study Tip

Repeated Roots If you factor a polynomial and a factor is raised to a power greater than 1, then there is a repeated root. The power to which the factor is raised indicates the multiplicity of the root. To be sure that you do not miss a repeated root, it can help to write out each factor. For example, you would write $x^2(x^2 + 49)$ as $x \cdot x(x^2 + 49)$ as a reminder that x^2 indicates a root of multiplicity 2.

Example 2 Find the Number of Positive and Negative Zeros

State the possible number of positive real zeros, negative real zeros, and imaginary zeros of $f(x) = x^5 - 2x^4 - x^3 + 6x^2 - 5x + 10$.

Because $f(x)$ has degree 5, it has five zeros, either real or imaginary. Use Descartes' Rule of Signs to determine the possible number and types of *real* zeros.

Part A Find the possible number of positive real zeros.

Count the number of changes in sign for the coefficients of $f(x)$.

$$f(x) = x^5 - 2x^4 - x^3 + 6x^2 - 5x + 10$$

yes	no	yes	yes	yes
+ to −	− to −	− to +	+ to −	− to +

There are 4 sign changes, so there are 4, 2, or 0 positive real zeros.

Part B Find the possible number of negative real zeros.

Count the number of changes in sign for the coefficients of $f(-x)$.

$$f(-x) = (-x)^5 - 2(-x)^4 - (-x)^3 + 6(-x)^2 - 5(-x) + 10$$
$$= -x^5 - 2x^4 + x^3 + 6x^2 + 5x + 10$$

no	yes	no	no	no
− to −	− to +	+ to +	+ to +	+ to +

There is 1 sign change, so there is 1 negative real zero.

Go Online You can complete an Extra Example online.

Part C Find the possible number of imaginary zeros.

Positive Real Zeros	Negative Real Zeros	Imaginary Zeros	Total Zeros
4	1	0	$4 + 1 + 0 = 5$
2	1	2	$2 + 1 + 2 = 5$
0	1	4	$0 + 1 + 4 = 5$

Talk About It!

If a polynomial has degree n and no real zeros, then how many imaginary zeros does it have? Explain your reasoning.

Check

Copy and complete the table. State the possible number of positive real zeros, negative real zeros, and imaginary zeros of $f(x) = 3x^6 - x^5 + 2x^4 + x^3 - 3x^2 + 13x + 1$. Write the rows in ascending order of positive real zeros.

Number of Positive Real Zeros	Number of Negative Real Zeros	Number of Imaginary Zeros

Learn Finding Zeros of Polynomial Functions

Key Concept • Complex Conjugates Theorem

Words: Let a and b be real numbers, and $b \neq 0$. If $a + bi$ is a zero of a polynomial function with real coefficients, then $a - bi$ is also a zero of the function.

Example: If $1 + 2i$ is a zero of $f(x) = x^3 - x^2 + 3x + 5$, then $1 - 2i$ is also a zero of the function.

When you are given all of the zeros of a polynomial function and asked to determine the function, use the zeros to write the factors and multiply them together. The result will be the polynomial function.

Example 3 Use Synthetic Substitution to Find Zeros

Find all of the zeros of $f(x) = x^3 + x^2 - 7x - 15$ and use them to sketch a rough graph.

Part A Find all of the zeros.

Step 1 Determine the total number of zeros.

Since $f(x)$ has degree 3, the function has 3 zeros.

(continued on the next page)

Step 2 Determine the type of zeros.

Examine the number of sign changes for $f(x)$ and $f(-x)$.

$f(x) = x^3 + x^2 - 7x - 15$ (no, yes, no) $f(-x) = -x^3 + x^2 + 7x - 15$ (yes, no, yes)

Because there is 1 sign change for the coefficients of $f(x)$, the function has 1 positive real zero. Because there are 2 sign changes for the coefficients of $f(-x)$, $f(x)$ has 2 or 0 negative real zeros. Thus, $f(x)$ has 3 real zeros, or 1 real zero and 2 imaginary zeros.

Step 3 Determine the real zeros.

List some possible values, and then use synthetic substitution to evaluate $f(x)$ for real values of x.

x	1	1	−7	−15
−3	1	−2	−1	−12
−2	1	−1	−5	−5
−1	1	0	−7	−8
0	1	1	−7	−15
1	1	2	−5	−20
2	1	3	−1	−17
3	1	4	5	0
4	1	5	13	37

3 is a zero of the function, and the depressed polynomial is $x^2 + 4x + 5$. Since it is quadratic, use the Quadratic Formula. The zeros of $f(x) = x^2 + 4x + 5$ are $-2 - i$ and $-2 + i$.

The function has zeros at 3, $-2 - i$ and $-2 + i$.

Part B Sketch a rough graph.

The function has one real zero at $x = 3$, so the function goes through (3, 0) and does not cross the x-axis at any other place.

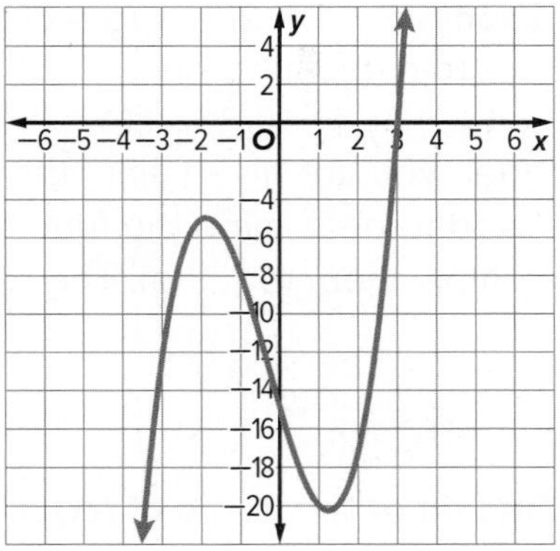

Because the degree is odd and the leading coefficient is positive, the end behavior is that as $x \to -\infty$, $f(x) \to -\infty$ and as $x \to \infty$, $f(x) \to \infty$.

Use this information and points with coordinates found in the table above to sketch the graph.

Go Online You can complete an Extra Example online.

Check

Determine all of the zeros of $f(x) = x^4 - x^3 - 16x^2 - 4x - 80$, and use them to sketch a rough graph.

Real Zeros: ___?___, ___?___

Imaginary Zeros: ___?___, ___?___

Example 4 Use a Graph to Write a Polynomial Function

Write a polynomial function that could be represented by the graph.

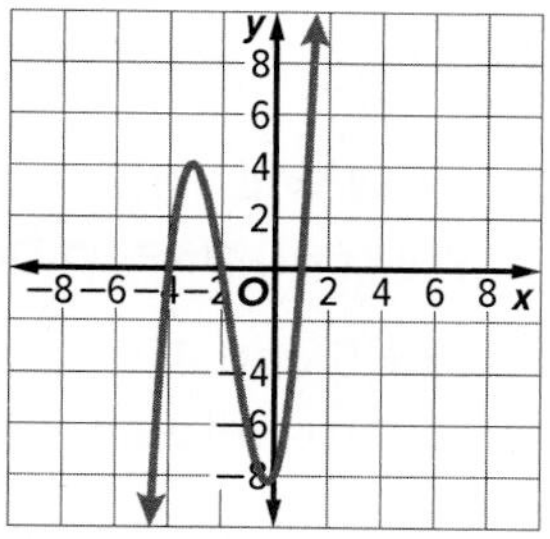

The graph crosses the x-axis 3 times, so the function is at least of degree 3. It crosses the x-axis at $x = -4$, $x = -2$, and $x = 1$, so its factors are $x + 4$, $x + 2$, and $x - 1$.

To determine a polynomial, find the product of the factors.

$y = (x + 4)(x + 2)(x - 1)$ Set the product of the factors equal to y.

$= (x^2 + 6x + 8)(x - 1)$ FOIL

$= x^3 + 5x^2 + 2x - 8$ Multiply.

A polynomial function that could be represented by the graph is $y = x^3 + 5x^2 + 2x - 8$.

Check

Write a polynomial function that could be represented by the graph.

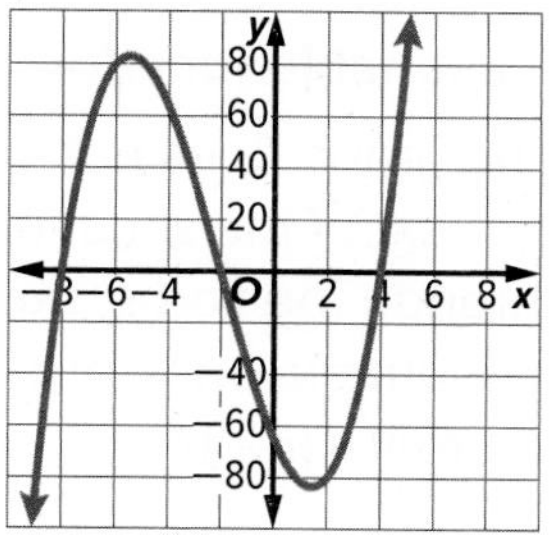

A. $y = x^3 - 6x^2 - 24x + 64$

B. $y = x^2 + 4x - 32$

C. $y = x^3 + 6x^2 - 24x - 64$

D. $y = x^3 - 64$

Go Online You can complete an Extra Example online.

Apply Example 5 Use Zeros to Graph a Polynomial Function

PROFIT MARGIN **A book publisher wants to release a special hardcover version of several Charles Dickens books. They know that if they charge \$5 or \$40, their profit will be \$0. Graph a polynomial function that could represent the company's profit in thousands of dollars given the price they charge for the book.**

1 What is the task?

Describe the task in your own words. Then list any questions that you may have. How can you find answers to your questions?

Let x represent the price that the publisher charges and let y represent the profit. I need to write and graph a polynomial function that relates x and y.

2 How will you approach the task? What have you learned that you can use to help you complete the task?

I know 5 and 40 are zeros of the function. I can use them to write factors to write an equation of the function.

3 What is your solution?

Use your strategy to solve the problem.

What is a function that represents the given information?

$y = x^2 - 45x + 200$

Graph the function.

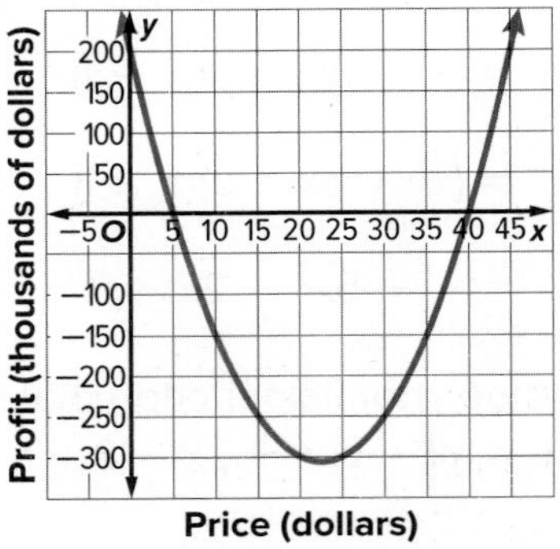

Does this function make sense in the context of the situation? If not, explain why not and write and graph a more reasonable function.

The graph does not make sense in the context of the situation. The graph passed through the zeros, but did not show reasonable book prices that would result in profit. Graph $y = -x^2 + 45x - 200$.

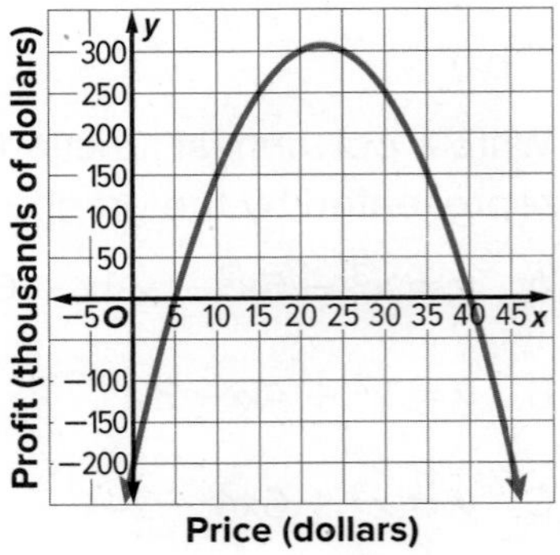

4 How can you know that your solution is reasonable?

Write About It! Write an argument that can be used to defend your solution.

With multiplying the function by −1 , the new function shows that the profit is negative when charging less than \$5 per book. This makes sense in the context of the situation.

Problem-Solving Tip

Logical Reasoning When solving a problem it is important to use logical reasoning skills to analyze the problem.

Go Online

You can complete an Extra Example online.

Practice

Go Online You can complete your homework online.

Example 1

Solve each equation. State the number and type of roots.

1. $5x + 12 = 0$
2. $x^2 - 4x + 40 = 0$
3. $x^5 + 4x^3 = 0$
4. $x^4 - 625 = 0$
5. $4x^2 - 4x - 1 = 0$
6. $x^5 - 81x = 0$
7. $2x^2 + x - 6 = 0$
8. $4x^2 + 1 = 0$
9. $x^3 + 1 = 0$
10. $2x^2 - 5x + 14 = 0$
11. $-3x^2 - 5x + 8 = 0$
12. $8x^3 - 27 = 0$
13. $16x^4 - 625 = 0$
14. $x^3 - 6x^2 + 7x = 0$
15. $x^5 - 8x^3 + 16x = 0$
16. $x^5 + 2x^3 + x = 0$

Example 2

State the possible number of positive real zeros, negative real zeros, and imaginary zeros of each function.

17. $g(x) = 3x^3 - 4x^2 - 17x + 6$
18. $h(x) = 4x^3 - 12x^2 - x + 3$
19. $f(x) = x^3 - 8x^2 + 2x - 4$
20. $p(x) = x^3 - x^2 + 4x - 6$
21. $q(x) = x^4 + 7x^2 + 3x - 9$
22. $f(x) = x^4 - x^3 - 5x^2 + 6x + 1$
23. $f(x) = x^4 - 5x^3 + 2x^2 + 5x + 7$
24. $f(x) = 2x^3 - 7x^2 - 2x + 12$
25. $f(x) = -3x^5 + 5x^4 + 4x^2 - 8$
26. $f(x) = x^4 - 2x^2 - 5x + 19$
27. $f(x) = 4x^6 - 5x^4 - x^2 + 24$
28. $f(x) = -x^5 + 14x^3 + 18x - 36$

Example 3

Find all of the zeros of each function and use them to sketch a rough graph.

29. $h(x) = x^3 - 5x^2 + 5x + 3$

30. $g(x) = x^3 - 6x^2 + 13x - 10$

31. $h(x) = x^3 + 4x^2 + x - 6$

32. $q(x) = x^3 + 3x^2 - 6x - 8$

33. $g(x) = x^4 - 3x^3 - 5x^2 + 3x + 4$

34. $f(x) = x^4 - 21x^2 + 80$

35. $f(x) = x^3 + 7x^2 + 4x - 12$

36. $f(x) = x^3 + x^2 - 17x + 15$

37. $f(x) = x^4 - 3x^3 - 3x^2 - 75x - 700$

38. $f(x) = x^4 + 6x^3 + 73x^2 + 384x + 576$

39. $f(x) = x^4 - 8x^3 + 20x^2 - 32x + 64$

40. $f(x) = x^5 - 8x^3 - 9x$

Example 4

Write a polynomial that could be represented by each graph.

41.

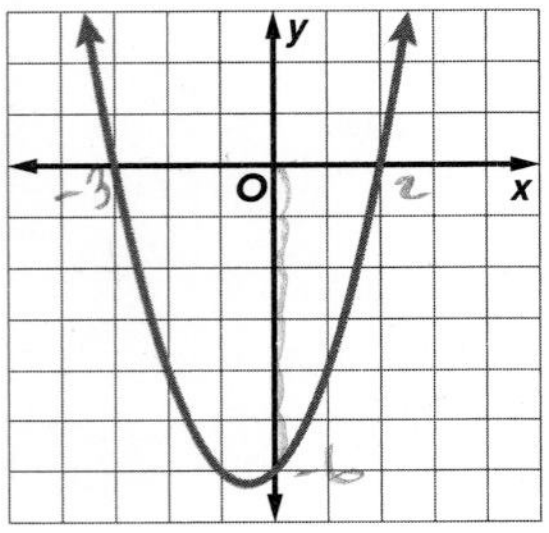

42.

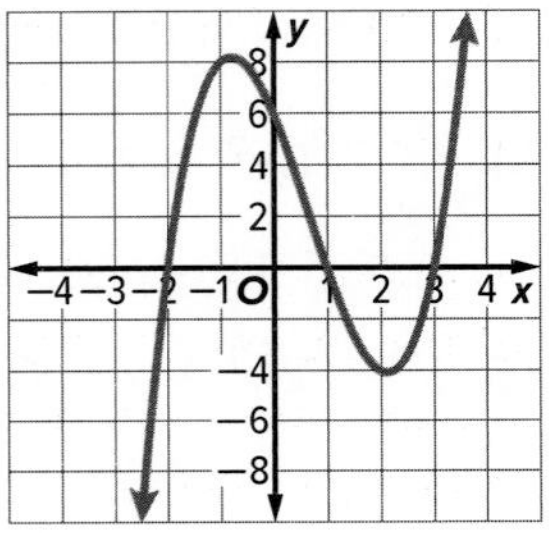

43.

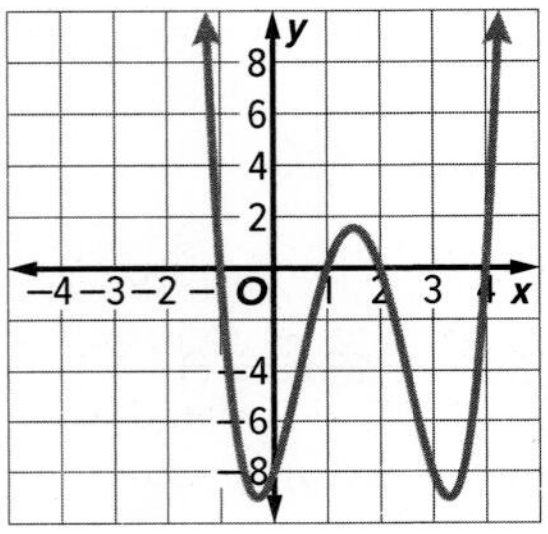

Example 5

44. FISH Some fish jump out of the water. When a fish is out of the water, its location is above sea level. When a fish dives back into the water, its location is below sea level. A biologist can use polynomial functions to model the location of fish compared to sea level. A biologist noticed that a fish is at sea level at −3, −2, −1, 1, 2, and 3 seconds from noon. Graph a polynomial function that could represent the location of the fish compared to sea level y, in centimeters, x seconds from noon.

45. BUSINESS After introducing a new product, a company's profit is modeled by a polynomial function. In 2012 and 2017, the company's profit on the product was \$0. Graph a polynomial function that could represent the amount of profit $p(x)$, in thousands of dollars, x years since 2010.

Mixed Exercises

Write a polynomial function of least degree with integral coefficients that has the given zeros.

46. 5, −2, −1

47. −4, −3, 5

48. −1, −1, $2i$

49. −3, 1, $-3i$

50. 0, −5, $3 + i$

51. −2, −3, $4 - 3i$

Sketch the graph of each function using its zeros.

52. $f(x) = x^3 - 5x^2 - 2x + 24$

53. $f(x) = 4x^3 + 2x^2 - 4x - 2$

54. $f(x) = x^4 - 6x^3 + 7x^2 + 6x - 8$

55. $f(x) = x^4 - 6x^3 + 9x^2 + 4x - 12$

56. USE A SOURCE Linear algebra is the study of linear equations. In linear algebra, the coefficients of linear equations are often organized into rectangular arrays called *matrices*. Research the eigenvalues of a matrix and how they relate to the roots of a polynomial function. What fields use linear algebra, matrices, and eigenvalues?

57. SPACE The technology for a rocket that will safely return to Earth for refueling and reuse is currently being developed. The three sections of the booster that will power the flight of the payload are cylindrical with a total volume of about 234π cubic meters. If the second stage section of the booster is x meters tall, then the interstage section $x + 3$ meters tall, and the first stage section is $5x + 6.5$ meters tall. The radius of the booster is $x - 5$ meters.

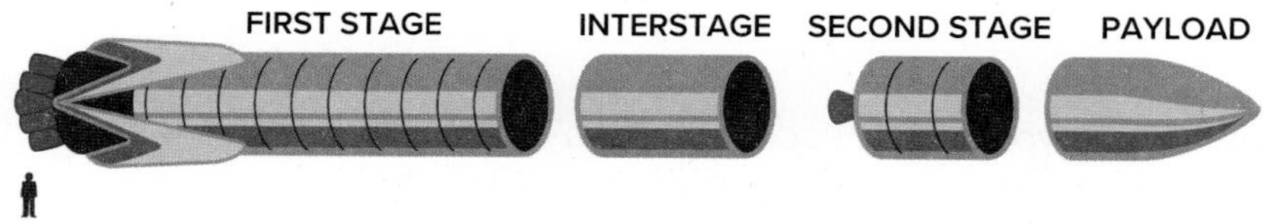

a. Write and solve an equation to represent the total volume of the booster.

b. What are the dimensions of the first stage section of the booster? Explain your reasoning.

58. CREATE Consider two polynomial functions, $f(x)$ and $g(x)$.

a. Write a polynomial function $f(x)$ of least degree with integral coefficients and zeros that include $-1 - 4i$ and $\frac{2}{3} + \frac{1}{3}i$. Explain how you found the function.

b. Write another polynomial function $g(x)$ with integral coefficients that has the same degree and zeros. How did you find this function?

c. Are you able to sketch the graphs of $f(x)$ and $g(x)$ based on the zeros? Explain your reasoning. Then sketch the graphs of $f(x)$ and $g(x)$.

59. ANALYZE Use the zeros to draw the graph of $P(x) = x^3 - 7x^2 + 7x + 15$ by hand. Discuss the accuracy of your graph, and what could be done to improve the accuracy.

60. PERSEVERE Let the polynomial function $f(x)$ have real coefficients, be of degree 5, and have zeros $4 + 3i$, $2 - 7i$, and $6 + bi$, where b is a real number.

a. What can be determined about b? Explain your reasoning.

b. Write a possible equation for $f(x)$.

61. CREATE Sketch the graph of a polynomial function with:

a. 3 real, 2 imaginary zeros **b.** 4 real zeros **c.** 2 imaginary zeros

62. PERSEVERE Write an equation in factored form of a polynomial function of degree 5 with 2 imaginary zeros, 1 real nonintegral zero, and 2 irrational zeros. Explain.

63. WHICH ONE DOESN'T BELONG? Determine which equation is not like the others. Justify your conclusion.

$r^4 + 1 = 0$	$r^3 + 1 = 0$	$r^2 - 1 = 0$	$r^3 - 8 = 0$

64. ANALYZE Provide a counterexample for each statement.

a. All polynomial functions of degree greater than 2 have at least 1 negative real root.

b. All polynomial functions of degree greater than 2 have at least 1 positive real root.

65. WRITE Explain to a friend how you would use Descartes' Rule of Signs to determine the number of possible positive real roots and the number of possible negative real roots of the polynomial function $f(x) = x^4 - 2x^3 + 6x^2 + 5x - 12$.

66. FIND THE ERROR The graph shows a polynomial function. Brianne says the function is a 4th degree polynomial. Amrita says the function is a 2nd degree polynomial. Is either of them correct? Explain your reasoning.

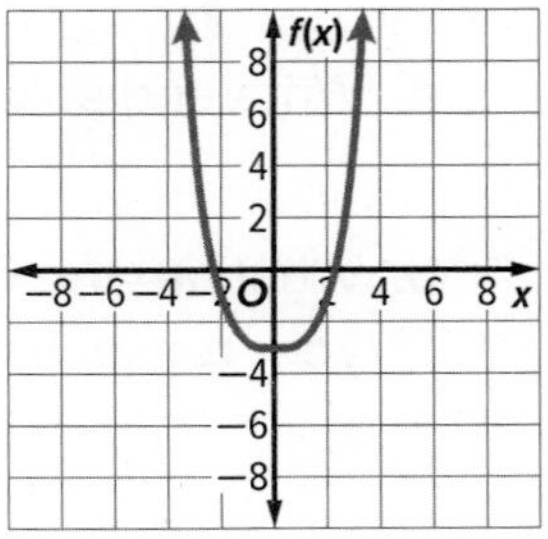

Review

Essential Question

What methods are useful for solving polynomial equations and finding zeros of polynomial functions?

Module Summary

Lessons 5-1 and 5-2

Solving Polynomial Equations

- Polynomial equations can be solved by graphing or can be solved algebraically.
- Use patterns such as the sum or difference of two cubes to factor.
  - $a^3 + b^3 = (a + b)(a^2 - ab + b^2)$
  - $a^3 - b^3 = (a - b)(a^2 + ab + b^2)$
- When factoring a polynomial, look for a common factor to simplify the expression.
- An expression in quadratic form can be written as $au^2 + bu + c$ for any numbers a, b, and c, $a \neq 0$, where u is some expression in x.

Lesson 5-3

Polynomial Identities

- An identity is an equation that is satisfied by any numbers that replace the variables.
- A polynomial identity is a polynomial equation that is true for any values that are substituted for the variables.
- To verify an identify, simplify one side of an equation until the two sides of the equation are the same.

Lesson 5-4

The Remainder and Factor Theorems

- Remainder Theorem: For a polynomial $p(x)$ and a number a, the remainder upon division by $x - a$ is $p(a)$.
- Factor Theorem: The binomial $x - a$ is a factor of the polynomial $p(x)$ if and only if $p(a) = 0$.

Lesson 5-5

Roots and Zeros

- Let $P(x) = a_nx^n + ... + a_1x + a_0$ be a polynomial function. Then the following statements are equivalent.
  - c is a zero of $P(x)$.
  - c is a root or solution of $P(x) = 0$.
  - $x - c$ is a factor of $a_nx^n + ... + a_1x + a_0$.
  - If c is a real number, then $(c, 0)$ is an x-intercept of the graph of $P(x)$.
- A polynomial equation of degree n has exactly n roots in the set of complex numbers, including repeated roots.
- The number of positive real zeros of $P(x)$ is the same as the number of changes in sign of the coefficients of the terms, or is less than this by an even number, and the number of negative real zeros of $P(x)$ is the same as the number of changes in sign of the coefficients of the terms of $P(-x)$, or is less than this by an even number.
- If $a + bi$ is a zero of a polynomial function with real coefficients, then $a - bi$ is also a zero of the function.

Study Organizer

Foldables

Use your Foldable to review this module. Working with a partner can be helpful. Ask for clarification of concepts as needed.

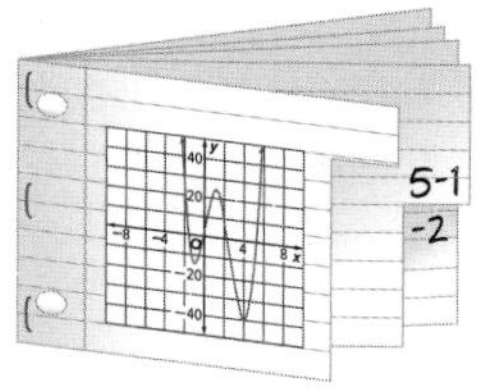

Test Practice

1. **MULTIPLE CHOICE** Which function can be used to solve $x^3 - x = 2x^2 - 2$ by graphing? (Lesson 5-1)

   A. $f(x) = x^3 - 2x^2 - x + 2$

   B. $f(x) = x - 2$

   C. $f(x) = x^3 + 2x^2 - x - 2$

   D. $f(x) = 2x^5 - 4x^3 + 2x$

2. **OPEN RESPONSE** The graph of $f(x) = x^4 - 4x^2 + x + 1$ is shown.

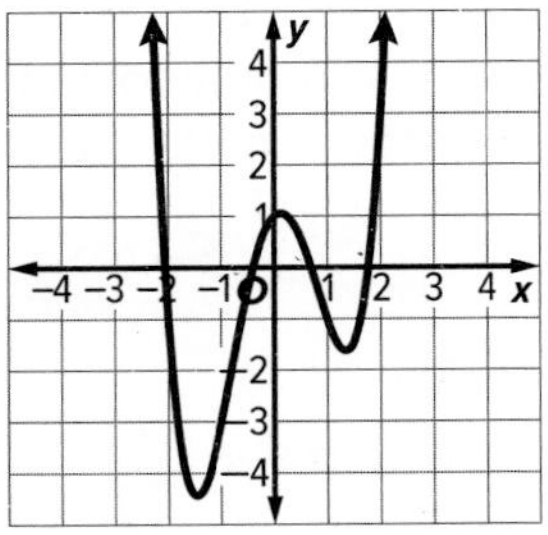

   How many real solutions does the function have? (Lesson 5-1)

3. **MULTI-SELECT** Use a graphing calculator to solve $x^3 - 10x + 4 = 4 - x$. Select all of the solutions. (Lesson 5-1)

   A. -3

   B. 0

   C. 1

   D. 3

   E. 4

   F. 7

4. **MULTIPLE CHOICE** A jewelry box is 3 inches by 4 inches by 2 inches. If increasing the length of each edge by x inches doubles the volume of the jewelry box, what is the value of x? Round your answer to the nearest hundredth if necessary. (Lesson 5-1)

   A. 0.40

   B. 0.69

   C. 0.73

   D. 1.24

5. **OPEN RESPONSE** The volume of a figure is $x^3 - 9x$. The surface area of another figure is $8x^2$. Disregarding the units, the volume of the first figure equals the surface area of the second figure. What are the possible values of x? Explain your reasoning. (Lesson 5-2)

6. **MULTI-SELECT** Find the solutions of $x^4 - x^2 - 2 = 0$. Select all that apply. (Lesson 5-2)

   A. $\pm\sqrt{2}$

   B. $\pm i$

   C. ± 1

   D. ± 2

   E. $\pm i\sqrt{2}$

$(2x+3)(x-2)$
$2x^2-4x+3x-6$
$2x^2-x-6$

7. OPEN RESPONSE Evelyn is making two rectangular table tops. The dimensions of both are shown. If both designs have the same area, what is the value of x? (Lesson 5-2)

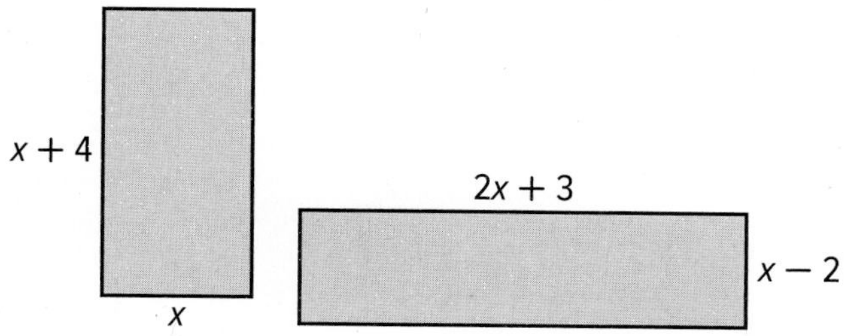

8. MULTI-SELECT Select all of the choices that are steps in the proof that $x^3 + y^3 = (x + y)(x^2 - xy + y^2)$. (Lesson 5-3)

A. $x^3 + y^3 = x(x^2) - x(xy) + x(y^2) + y(x^2) - y(xy) + y(y^2)$

B. $x^3 + y^3 = x^3 - x^2y + xy^2 + xy^2 - x^2y + y^3$

C. $x^3 + y^3 = x^3 - (x^2y + xy^2) + (x^2y - xy^2) + y^3$

D. $x^3 + y^3 = x^3 - x^2y + xy^2 + x^2y - xy^2 + y^3$

E. $x^3 + y^3 = x^3 - x^3y^3 + x^3y^3 + y^3$

F. $x^3 + y^3 = x^3 + y^3$

9. MULTIPLE CHOICE Which of the following is an equivalent expression to $(x + y)^3$? (Lesson 5-3)

A. $x^3 + y^3 + 3xy(x - y)$

B. $x^3 + y^3 + 3xy(x + y)$

C. $x^3 + y^3 + xy(x + y)$

D. $x^3 + y^3 + xy(x - y)$

10. MULTIPLE CHOICE Which of the following is an equivalent expression to $4xy$? (Lesson 5-3)

A. $(x + y)^2 - (x + y)^2$

B. $(x + y)^2 - (x - y)^2$

C. $2(x + y)^2 - (x + y)^2$

D. $(x + y)^2 - 2(x - y)^2$

11. MULTIPLE CHOICE What is the remainder when $f(x) = x^4 + x^3 - 2x^2 + 5x - 4$ is divided by $x + 3$? (Lesson 5-4)

A. -25

B. -17

C. 17

D. 25

12. MULTI-SELECT If $x - 1$ is a factor of $x^3 - 6x^2 + 11x - 6$, find the remaining factors of the polynomial. Select all that apply. (Lesson 5-4)

A. $x^2 - 5x + 6$

B. $x - 3$

C. $x + 2$

D. $x - 2$

13. **MULTIPLE CHOICE** The average price of gasoline, in dollars, from 2010 to 2016 can be modeled by the function $f(x) = 0.03x^3 - 0.4x^2 + 1.18x + 2.75$, where x represents the years since 2010. What is the estimated price of gasoline in 2011 in dollars? (Lesson 5-4)

A. \$4.70

B. \$3.56

C. \$2.42

D. \$1.14

14. **OPEN RESPONSE** Use synthetic substitution to find $f(-6)$ if $f(x) = -3x^4 - 20x^3 + 80x + 12$. (Lesson 5-4)

15. **MULTI-SELECT** Which value of x is a real root in the equation $x^4 + 3x^2 - 4 = 0$? Select all that apply. (Lesson 5-5)

A. -2

B. -1

C. 1

D. 2

E. $-2i$

F. $-i$

G. i

H. $2i$

16. **MULTI-SELECT** What are the possible numbers of positive real roots and negative real roots of $p(x) = x^5 - 2x^4 + 3x^3 - 4x^2 + 1$? Select all that apply. (Lesson 5-5)

A. 0

B. 1

C. 2

D. 3

E. 4

F. 5

17. **MULTIPLE CHOICE** A template for a shipping box is made by cutting a square with side length x inches from each corner of a rectangular piece of cardboard that is 12 inches wide and 14 inches long. Which graph could represent the relationship between the volume of a shipping box y and its height? (Lesson 5-5)

A.
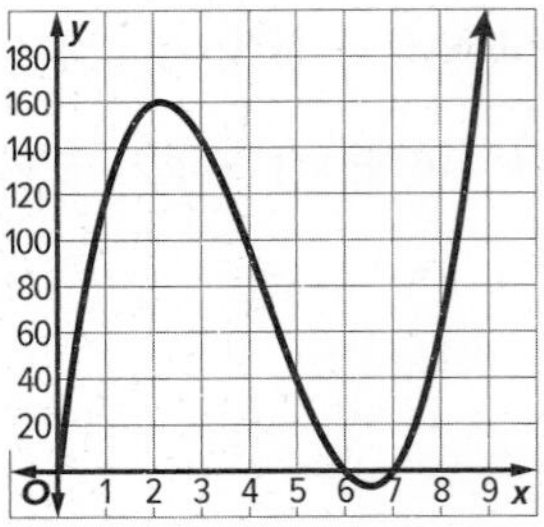

B.
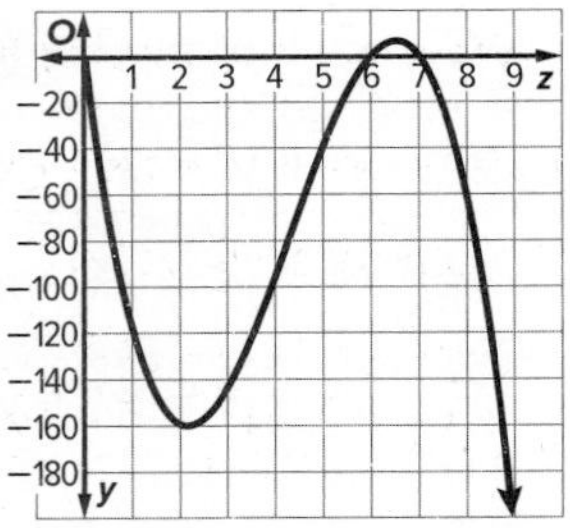

C.
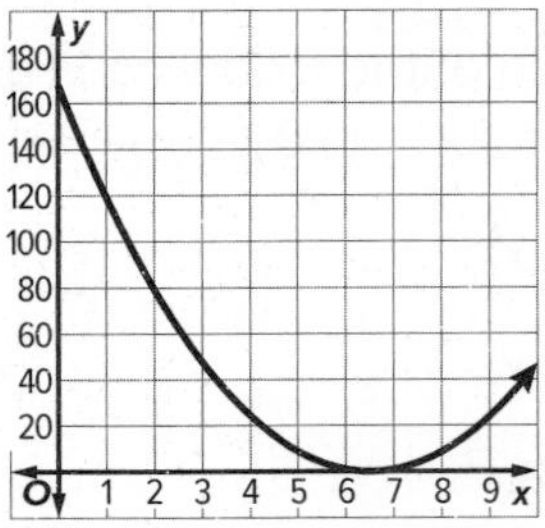

D.
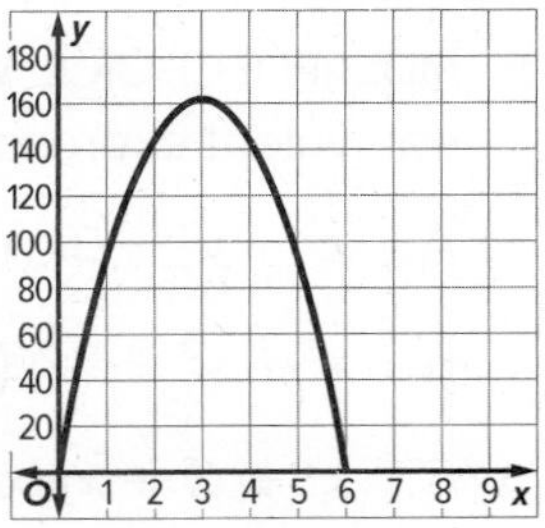

18. **MULTIPLE CHOICE** Consider the function $p(x) = x^5 - 3x^4 + 3x^3 - x^2$. Which of the following is a root of the function with multiplicity 3? (Lesson 5-5)

A. -1

B. 0

C. 1

D. 2

Module 6

Inverse and Radical Functions

Essential Question

How can the inverse of a function be used to help interpret a real-world event or solve a problem?

What Will You Learn?

How much do you already know about each topic **before** starting this module?

KEY

— I don't know. — I've heard of it. — I know it!

	Before			After		
perform operations on functions						
combine functions using a composition of functions						
verify mathematically two relations or functions are inverses						
write expressions with rational exponents						
graph square root functions						
graph cube root functions						
perform operations with radical expressions						
solve radical equations by graphing						
solve radical equations algebraically						

Foldables Make this Foldable to help you organize your notes about inverse and radical functions. Begin with three sheets of notebook paper.

1. **Stack** three sheets of notebook paper so that each is one inch higher than the previous.
2. **Align** the bottom of all the sheets.
3. **Fold** the papers and crease well. Open the papers and staple them. Label the pages with lesson titles.

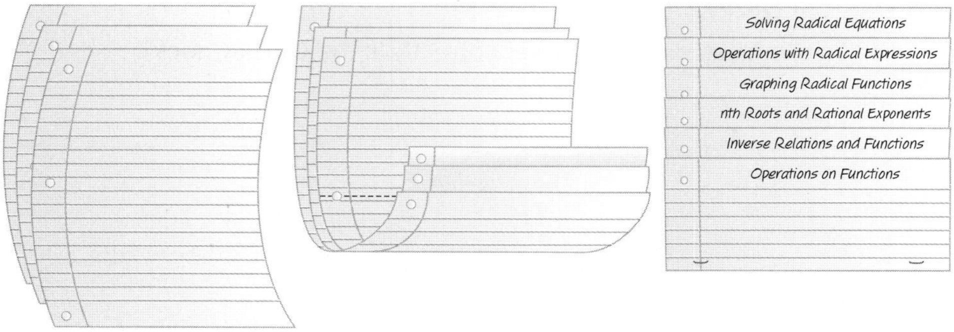

What Vocabulary Will You Learn?

- composition of functions
- conjugates
- cube root function
- index
- inverse functions
- inverse relations
- like radical expressions
- *n*th root
- principal root
- radical equation
- radical function
- radicand
- rational exponent
- square root function

Are You Ready?

Complete the Quick Review to see if you are ready to start this module. Then complete the Quick Check.

Quick Review

Example 1

Use the related graph of $0 = 3x^2 - 4x + 1$ to determine its roots. If exact roots cannot be found, state the consecutive integers between which the roots are located.

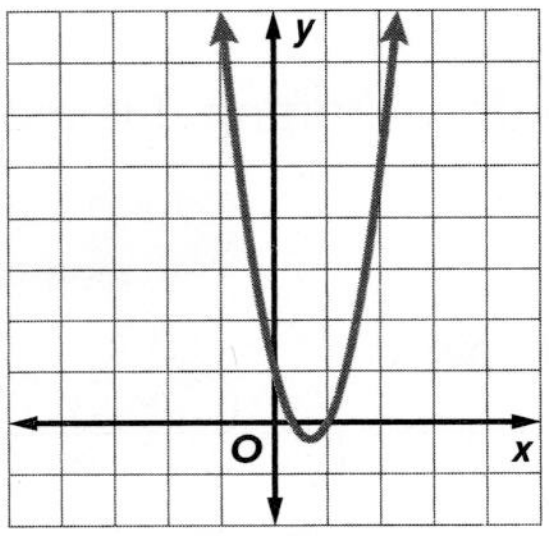

The roots are the *x*-coordinates where the graph crosses the *x*-axis.

The graph crosses the *x*-axis between 0 and 1 and at 1.

Example 2

Simplify $(3x^4 + 4x^3 + x^2 + 9x - 6) \div (x + 2)$ by using synthetic division.

$x - r = x + 2$, so $r = -2$

−2	3	4	1	9	−6
	↓	−6	4	−10	2
	3	−2	5	−1	−4

The result is $3x^3 - 2x^2 + 5x - 1 - \frac{4}{x+2}$.

Quick Check

Use the related graph of each equation to determine its roots. If exact roots cannot be found, state the consecutive integers between which the roots are located.

1. $x^2 - 4x + 1 = 0$

2. $2x^2 + x - 6 = 0$

3. $3x^2 - 3x - 1 = 0$

4. $2x^2 - 9x + 4 = 0$

Simplify each expression by using synthetic division.

5. $(5x^2 - 22x - 15) \div (x - 5)$

6. $(3x^2 + 14x - 12) \div (x + 4)$

7. $(2x^3 - 7x^2 - 36x + 36) \div (x - 6)$

8. $(3x^4 - 13x^3 + 17x^2 - 18x + 15) \div (x - 3)$

How Did You Do?

Which exercises did you answer correctly in the Quick Check?

Operations on Functions

Explore Adding Functions

Online Activity Use graphing technology to complete the Explore.

> **INQUIRY** Do you think that the graph of $f(x) + g(x)$ will be more or less steep than the graphs of $f(x)$ and $g(x)$?

Today's Goals

- Find sums, differences, products, and quotients of functions.
- Find compositions of functions.

Today's Vocabulary

composition of functions

Learn Operations on Functions

Key Concept • Operations on Functions

Operation	Definition	Example: Let $f(x) = 3x$ and $g(x) = 2x - 4$.
Addition	$(f + g)(x) = f(x) + g(x)$	$(f + g)(x) = 3x + (2x - 4)$ $= 5x - 4$
Subtraction	$(f - g)(x) = f(x) - g(x)$	$(f - g)(x) = 3x - (2x - 4)$ $= x + 4$
Multiplication	$(f \cdot g)(x) = f(x) \cdot g(x)$	$(f \cdot g)(x) = 3x(2x - 4)$ $= 6x^2 - 12x$
Division	$\left(\frac{f}{g}\right)(x) = \frac{f(x)}{g(x)}, g(x) \neq 0$	$\left(\frac{f}{g}\right)(x) = \frac{3x}{2x - 4}, x \neq 2$

To graph the sum or difference of functions, graph each function separately. Then add or subtract the corresponding functional values.

Example 1 Add and Subtract Functions

Given $f(x) = -x^2 + 3x + 1$ and $g(x) = 2x^2 - 5$, find each function.

a. $(f + g)(x)$

$(f + g)(x) = f(x) + g(x)$	Addition of functions
$= (-x^2 + 3x + 1) + (2x^2 - 5)$	$f(x) = -x^2 + 3x + 1$ and $g(x) = 2x^2 - 5$
$= -x^2 + 3x + 1 + 2x^2 - 5$	Add.
$= x^2 + 3x - 4$	Simplify.

(continued on the next page)

Go Online You can complete an Extra Example online.

Study Tip

Degree If the degree of $f(x)$ is m and the degree of $g(x)$ is n, then the degrees of $(f + g)(x)$ and $(f - g)(x)$ can be at most m or n, whichever is greater.

b. $(f - g)(x)$

$(f - g)(x) = f(x) - g(x)$	Subtraction of functions
$= (-x^2 + 3x + 1) - (2x^2 - 5)$	$f(x) = -x^2 + 3x + 1$ and $g(x) = 2x^2 - 5$
$= -x^2 + 3x + 1 - 2x^2 + 5$	Subtract.
$= -3x^2 + 3x + 6$	Simplify.

Example 2 Multiply and Divide Functions

Given $f(x) = 4x^2 - 2x + 3$ and $g(x) = -x + 5$, find each function.

a. $(f \cdot g)(x)$

$(f \cdot g)(x) = f(x) \cdot g(x)$	Multiplication of functions
$= (4x^2 - 2x + 3)(-x + 5)$	$f(x) = 4x^2 - 2x + 3$ and $g(x) = -x + 5$
$= -4x^3 + 20x^2 + 2x^2 - 10x - 3x + 15$	Distributive Property
$= -4x^3 + 22x^2 - 13x + 15$	Simplify.

b. $\left(\frac{f}{g}\right)(x)$

$\left(\frac{f}{g}\right)(x) = \frac{f(x)}{g(x)}$	Division of functions
$= \frac{4x^2 - 2x + 3}{-x + 5}, x \neq 5$	$f(x) = 4x^2 - 2x + 3$ and $g(x) = -x + 5$

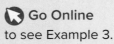

Go Online to see Example 3.

Check

Given $f(x) = -x^2 + 1$ and $g(x) = x + 1$, find each function.

$(f \cdot g)(x) =$ ____?____ $\left(\frac{f}{g}\right)(x) =$ ____?____

Learn Compositions of Functions

Key Concept • Composition of Functions

Suppose f and g are functions such that the range of g is a subset of the domain of f. Then the composition function $f \circ g$ can be described by $[f \circ g](x) = f[g(x)]$.

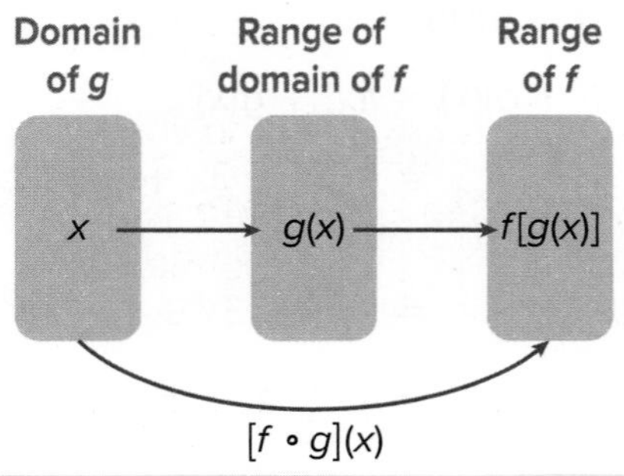

Go Online You can complete an Extra Example online.

Example 4 Compose Functions by Using Ordered Pairs

Given f and g, find $[f \circ g](x)$ and $[g \circ f](x)$. State the domain and range for each.

$f = \{(1, 12), (10, 11), (0, 13), (9, 7)\}$ $g = \{(4, 1), (5, 0), (13, 9), (12, 10)\}$

Part A Find $[f \circ g](x)$ and $[g \circ f](x)$.

To find $f \circ g$, evaluate $g(x)$ first then use the range to evaluate $f(x)$.		To find $g \circ f$, evaluate $f(x)$ first then use the range to evaluate $g(x)$.	
$f[g(4)] = f(1)$ or 12	$g(4) = 1$	$g[f(1)] = g(12)$ or 10	$f(1) = 12$
$f[g(5)] = f(0)$ or 13	$g(5) = 0$	$g[f(10)] = g(11)$	$f(10) = 11$
$f[g(13)] = f(9)$ or 7	$g(13) = 9$	$g[f(0)] = g(13)$ or 9	$f(0) = 13$
$f[g(12)] = f(10)$ or 11	$g(12) = 10$	$g[f(9)] = g(7)$	$f(9) = 7$

Because 11 and 7 are not in the domain of g, $g \circ f$ is undefined for $x = 11$ and $x = 7$. So, $g \circ f = \{(1, 10), (0, 9)\}$.

Part B State the domain and range.

$[f \circ g](x)$: The domain is the x-coordinates of the composed function, so D = {4, 5, 12, 13}. The range is the y-coordinates of the composed function, so R = {7, 11, 12, 13}.

$[g \circ f](x)$: The domain is the x-coordinates of the composed function, so D = {0, 1}. The range is the y-coordinates of the composed function, so R = {9, 10}.

Example 5 Compose Functions

Given $f(x) = 2x - 5$ and $g(x) = 3x$, find $[f \circ g](x)$ and $[g \circ f](x)$. State the domain and range for each.

Part A Find $[f \circ g](x)$ and $[g \circ f](x)$.

$[f \circ g](x) = f[g(x)]$	Composition of functions	$[g \circ f](x) = g[f(x)]$
$= f(3x)$	Substitute.	$= g(2x - 5)$
$= 2(3x) - 5$	Substitute again.	$= 3(2x - 5)$
$= 6x - 5$	Simplify.	$= 6x - 15$

Part B State the domain and range.

Because $[f \circ g](x)$ and $[g \circ f](x)$ are both linear functions with nonzero slopes, D = {all real numbers} and R = {all real numbers} for both functions.

Go Online You can complete an Extra Example online.

Study Tip

Domain and Range To ensure you have the right domain and range, it can help to graph $[f \circ g](x)$ and $[g \circ f](x)$.

Check

Given $f(x) = -x + 1$ and $g(x) = 2x^3 - x$, find $[f \circ g](x)$ and $[g \circ f](x)$. State the domain and range for each.

$[f \circ g](x) =$ ___?___ $[g \circ f](x) =$ ___?___

Domain of $[f \circ g](x)$: ___?___ Domain of $[g \circ f](x)$: ___?___

Range of $[f \circ g](x)$: ___?___ Range of $[g \circ f](x)$: ___?___

Apply Example 6 Use Composition of Functions

BOX OFFICE **A movie theater charges $8.50 for each of the *x* tickets sold. The manager wants to determine how much the movie theater gets to keep of the ticket sales if they have to give the studios 75% of the money earned on ticket sales *t*(*x*). If the amount they keep of each ticket sale is *k*(*x*), which composition represents the total amount of money the theater gets to keep?**

1 What is the task?

Describe the task in your own words. Then list any questions that you may have. How can you find answers to your questions?

I need to write functions for $t(x)$ and $k(x)$ and use them to create a composition that represents the money that the theater keeps. If the studios get 75%, what does the theater get to keep? Should the composition be $[k \circ t](x)$ or $[t \circ k](x)$?

> **Watch Out!**
>
> **Order** Remember that, for two functions $f(x)$ and $g(x)$, $[f \circ g](x)$ is not always equal to $[g \circ f](x)$. Given that the studios take their cut after the tickets have been sold, consider how that affects the order of $t(x)$ and $k(x)$.

2 How will you approach the task? What have you learned that you can use to help you complete the task?

First, I will determine functions for $t(x)$ and $k(x)$. Then, I will determine the order of the composition and simplify it. I will apply what I have learned in previous examples to complete the task.

3 What is your solution?

Use your strategy to solve the problem.

What function represents the money earned on ticket sales, *t*(*x*)? $t(x) = 8.50x$

What function represents the amount of money the theater keeps from each ticket sale, *k*(*x*)? $k(x) = 0.25x$

Because the theater uses the total earnings to determine the amount they keep from the ticket sales, what composition should be used to represent the situation? $[k \circ t](x) = 2.125x$

4 How can you know that your solution is reasonable?

Write About It! Write an argument that can be used to defend your solution.

If the theater sells 1000 tickets in a weekend, they will earn $t(1000)$, or $8500. The theater will keep 25% of $8500, which is $k(8500) = \$2125$. This is the same value as $[k \circ t](1000)$.

> **Go Online**
>
> You can complete an Extra Example online.

Practice

Go Online You can complete your homework online.

Examples 1 and 2

Find $(f + g)(x)$, $(f - g)(x)$, $(f \cdot g)(x)$, and $\left(\frac{f}{g}\right)(x)$ for each $f(x)$ and $g(x)$.

1. $f(x) = 2x$

$g(x) = -4x + 5$

2. $f(x) = x - 1$

$g(x) = 5x - 2$

3. $f(x) = x - 2$

$g(x) = 2x - 7$

4. $f(x) = x^2$

$g(x) = x - 5$

5. $f(x) = -x^2 + 6$

$g(x) = 2x^2 + 3x - 5$

6. $f(x) = 3x^2 - 4$

$g(x) = x^2 - 8x + 4$

Example 3

7. FINANCE Trevon opens a checking account that he only uses to pay fixed bills, which are expenses that are the same each month, such as car loans or rent. The checking account has an initial balance of $1750 and Trevon deposits $925 each month. The balance of the account can be modeled by $a(x) = 1750 + 925x$, where x is the number of months since the account was opened. The total of Trevon's fixed bills is modeled by $b(x) = 840x$. Define and graph the function that represents the account balance after he pays his bills.

a. Identify and write a new function to represent the account balance.

b. Graph the combined function.

8. BASEBALL A coach is ordering custom practice T-shirts and game jerseys for each of the team members. The coach orders T-shirts from a local shop that charges $7.50 for each, plus a $35 initial printer fee. The cost of the T-shirts is modeled by $t(x) = 7.5x + 35$, where x is the number of team members. He orders jerseys online, which cost $18 each with $20 shipping. The cost of the jerseys is modeled by $j(x) = 18x + 20$. Define and graph the function that represents the total cost of the T-shirts and jerseys.

a. Identify and write a new function to represent total cost.

b. Graph the combined function.

Example 4

For each pair of functions, find $f \circ g$ and $g \circ f$, if they exist. State the domain and range for each.

9. $f = \{(-8, -4), (0, 4), (2, 6), (-6, -2)\}$
$g = \{(4, -4), (-2, -1), (-4, 0), (6, -5)\}$

10. $f = \{(-7, 0), (4, 5), (8, 12), (-3, 6)\}$
$g = \{(6, 8), (-12, -5), (0, 5), (5, 1)\}$

11. $f = \{(5, 13), (-4, -2), (-8, -11), (3, 1)\}$
$g = \{(-8, 2), (-4, 1), (3, -3), (5, 7)\}$

12. $f = \{(-4, -14), (0, -6), (-6, -18), (2, -2)\}$
$g = \{(-6, 1), (-18, 13), (-14, 9), (-2, -3)\}$

Example 5

Find $[f \circ g](x)$ and $[g \circ f](x)$. State the domain and range for each.

13. $f(x) = 2x$
$g(x) = x + 5$

14. $f(x) = -3x$
$g(x) = -x + 8$

15. $f(x) = x^2 + 6x - 2$
$g(x) = x - 6$

16. $f(x) = 2x^2 - x + 1$
$g(x) = 4x + 3$

Example 6

17. USE A MODEL Mr. Rivera wants to purchase a riding lawn mower, which is on sale for 15% off the original price. The sales tax in his area is 6.5%. Let x represent the original cost of the lawn mower. Write two functions representing the price of the lawn mower $p(x)$ after the discount and the price of the lawn mower $t(x)$ after sales tax. Write a composition of functions that represents the price of the riding lawn mower. How much will Mr. Rivera pay for a riding lawn mower that originally cost $1350?

18. REASONING A sporting goods store is offering a 20% discount on shoes. Mariana also has a $5 off coupon that can be applied to her purchase. She is planning to buy a pair of shoes that originally costs $89. Will the final price be lower if the discount is applied before the coupon or if the coupon is applied before the discount? Justify your response.

Mixed Exercises

19. REASONING A bookstore that offers a 12% membership discount is currently offering 20% off each customer's total purchase when they spend more than $50. If Keshawn has $78 of books, should he request that the membership discount or the 20% off discount be applied first? Justify your response.

20. CONSTRUCT ARGUMENTS Is $[f \circ g](x)$ always equal to $[g \circ f](x)$ for two functions $f(x)$ and $g(x)$? Justify your conclusions. Provide a counterexample if needed.

If $f(x) = 3x$, $g(x) = x + 4$, and $h(x) = x^2 - 1$, find each value.

21. $f[g(1)]$

22. $g[h(0)]$

23. $g[f(-1)]$

24. $h[f(5)]$

25. $g[h(-3)]$

26. $h[f(10)]$

27. $f[h(8)]$

28. $[f \circ (h \circ g)](1)$

29. $[f \circ (g \circ h)](-2)$

30. $h[f(-6)]$

31. $f[h(0)]$

32. $f[g(7)]$

33. $f[h(-2)]$

34. $[g \circ (f \circ h)](-1)$

35. $[h \circ (f \circ g)](3)$

36. AREA Valeria wants to know the area of a figure made by joining an equilateral triangle and square along an edge. The function $f(s) = \frac{\sqrt{3}}{4}s^2$ gives the area of an equilateral triangle with side s. The function $g(s) = s^2$ gives the area of a square with side s. Write a function $h(s)$ that gives the area of the figure as a function of its side length s.

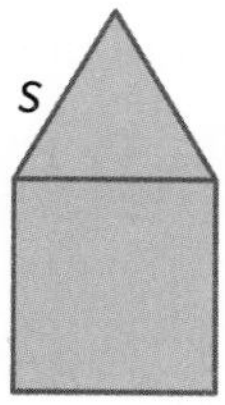

37. USE A MODEL The volume V of a weather balloon with radius r is given by $V(r) = \frac{4}{3}\pi r^3$. The balloon is being inflated so that the radius increases at a constant rate $r(t) = \frac{1}{2}t + 2$, where r is in meters and t is the number of seconds since inflation began.

a. Determine the function that represents the volume of the weather balloon in terms of time.

b. Find the volume of the balloon 12 seconds after inflation begins. Round your answer to the nearest cubic meter.

38. REASONING The National Center for Education Statistics reports data showing that since 2006, college enrollment for men in thousands can be modeled by $f(x) = 389x + 7500$, where x represents the number of years since 2006. Similarly, enrollment for women can be modeled by $g(x) = 480x + 10{,}075$. Write a function for $(f + g)(x)$ and interpret what it represents.

39. STRUCTURE The table shows various values of functions $f(x)$, $g(x)$, and $h(x)$.

x	−1	0	1	2	3	4
$f(x)$	7	−2	0	2	4	1
$g(x)$	−3	−4	−5	0	1	1
$h(x)$	0	4	1	1	5	5

Use the table to find the following values:

a. $(f + g)(-1)$ **b.** $(h - g)(0)$ **c.** $(f \cdot h)(4)$

d. $\left(\frac{f}{g}\right)(3)$ **e.** $\left(\frac{g}{h}\right)(2)$ **f.** $\left(\frac{g}{f}\right)(1)$

40. PERSEVERE If $(f + g)(3) = 5$ and $(f \cdot g)(3) = 6$, find $f(3)$ and $g(3)$. Explain.

41. CREATE Write two functions $f(x)$ and $g(x)$ such that $(f \circ g)(4) = 0$.

42. FIND THE ERROR Chris and Tobias are finding $(f \circ g)(x)$, where $f(x) = x^2 + 2x - 8$ and $g(x) = x^2 + 8$. Is either of them correct? Explain your reasoning.

Chris

$$(f \circ g)(x) = f[g(x)]$$
$$= (x^2 + 8)^2 + 2x - 8$$
$$= x^4 + 16x^2 + 64 + 2x - 8$$
$$= x^4 + 16x^2 + 2x + 58$$

Tobias

$$(f \circ g)(x) = f[g(x)]$$
$$= (x^2 + 8)^2 + 2(x^2 + 8) - 8$$
$$= x^4 + 16x^2 + 64 + 2x^2 + 16 - 8$$
$$= x^4 + 18x^2 + 72$$

43. PERSEVERE Given $f(x) = \sqrt{x^3}$ and $g(x) = \sqrt{x^6}$, determine each domain.

a. $g(x) \cdot g(x)$ **b.** $f(x) \cdot f(x)$

44. ANALYZE State whether the following statement is *sometimes*, *always*, or *never* true. Justify your argument.

The domain of two functions $f(x)$ and $g(x)$ that are composed $g[f(x)]$ is restricted by the domain of $g(x)$.

Inverse Relations and Functions

Explore Inverse Functions

Online Activity Use graphing technology to complete the Explore.

INQUIRY For what values of n will $f(x) = x^n$ have an inverse that is also a function?

Today's Goals
- Find inverses of relations.
- Verify that two relations are inverses by using compositions.

Today's Vocabulary
inverse relations

inverse functions

Learn Inverse Relations and Functions

Two relation are **inverse relation** if one relation contains elements of the form (a, b) when the other relation contains the elements of the form (b, a).

Two functions f and g are **inverse functions** if and only if both of their compositions are the identity function. The inverse of the function $f(x)$ can be written as $f^{-1}(x)$.

Key Concepts • Inverse Functions	
Words: If f and f^{-1} are inverses, then $f(a) = b$ if and only if $f^{-1}(b) = a$.	
Example: Let $f(x) = x - 5$ and represent its inverse as $f^{-1}(x) = x + 5$.	
Evaluate $f(7)$.	Evaluate $f^{-1}(2)$.
$f(x) = x - 5$	$f^{-1}(x) = x + 5$
$f(7) = 7 - 5$ or 2	$f^{-1}(2) = 2 + 5$ or 7

Not all functions have an inverse function. If a function fails the horizontal line test, you can restrict the domain of the function to make the inverse a function. Choose a portion of the domain on which the function is one-to-one. There may be more than one possible domain.

Think About It!
Write a function that does not pass the horizontal line test.

Example 1 Find an Inverse Relation

GEOMETRY **The vertices of $\triangle ABC$ can be represented by the relation $\{(2, 4), (-3, 2), (4, 1)\}$. Find the inverse of the relation. Graph both the original relation and its inverse.**

Step 1 Graph the relation.
Graph the ordered pairs and connect the points to form a triangle.

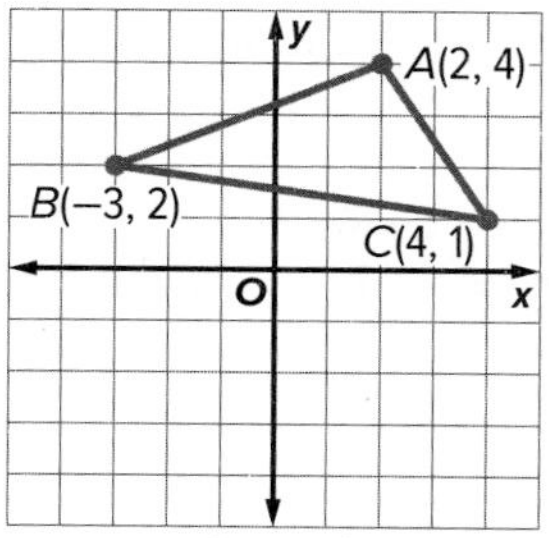

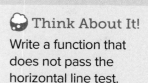

Go Online You can complete an Extra Example online.

(continued on the next page)

Step 2 Find the inverse.

To find the inverse, exchange the coordinates of the ordered pairs.
{(4, 2), (2, −3), (1, 4)}

Step 3 Graph the inverse.

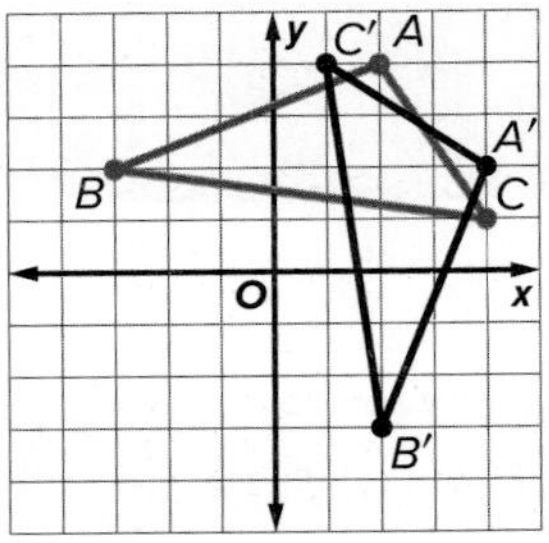

Think About It!

Describe the graph of the inverse relation.

Example 2 Inverse Functions

Find the inverse of $f(x) = 3x + 2$. Then graph the function and its inverse.

Step 1 Rewrite the function.

Rewrite the function as an equation relating x and y.

$f(x) = 3x + 2 \rightarrow y = 3x + 2$

Step 2 Exchange x and y.

Exchange x and y in the equation.

$x = 3y + 2$

Step 3 Solve for y.

$$x = 3y + 2$$
$$x - 2 = 3y$$
$$\frac{x-2}{3} = y$$

Step 4 Replace y with $f^{-1}(x)$.

Replace y with $f^{-1}(x)$ in the equation.

$y = \frac{x-2}{3} \rightarrow f^{-1}(x) = \frac{x-2}{3}$

The inverse of $f(x) = 3x + 2$ is $f^{-1}(x) = \frac{x-2}{3}$.

Step 5 Graph $f(x)$ and $f^{-1}(x)$.

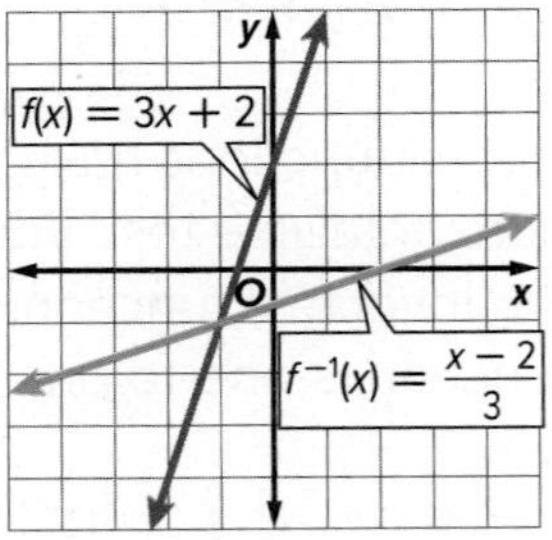

Study Tip

Inverses If $f^{-1}(x)$ is the inverse of $f(x)$, the graph of $f^{-1}(x)$ will be a reflection of the graph of $f(x)$ in the line $y = x$.

Go Online

You can learn how to graph a relation and its inverse on a graphing calculator by watching the video online.

Check

Examine $f(x) = -\frac{1}{2}x + 1$.

Part A Find the inverse of $f(x) = -\frac{1}{2}x + 1$.

$f^{-1}(x) =$ ___?___

Part B Graph $f(x) = -\frac{1}{2}x + 1$ and its inverse.

Go Online You can complete an Extra Example online.

Example 3 Inverses with Restricted Domains

Examine $f(x) = x^2 + 2x + 4$.

Part A Find the inverse of $f(x)$.

$f(x) = x^2 + 2x + 4$	Original function
$y = x^2 + 2x + 4$	Replace $f(x)$ with y.
$x = y^2 + 2y + 4$	Exchange x and y.
$x - 4 = y^2 + 2y$	Subtract 4 from each side.
$x - 4 + 1 = y^2 + 2y + 1$	Complete the square.
$x - 3 = (y + 1)^2$	Simplify.
$\pm\sqrt{x - 3} = y + 1$	Take the square root of each side.
$-1 \pm \sqrt{x - 3} = y$	Subtract 1 from each side.
$f^{-1}(x) = -1 \pm \sqrt{x - 3}$	Replace y with $f^{-1}(x)$.

Part B If necessary, restrict the domain of the inverse so that it is a function.

Because $f(x)$ fails the horizontal line test, $f^{-1}(x)$ is not a function. Find the restricted domain of $f(x)$ so that $f^{-1}(x)$ will be a function. Look for a portion of the graph that is one-to-one. If the domain of $f(x)$ is restricted to $(-\infty, -1]$, then the inverse is $f^{-1}(x) = -1 + \sqrt{x - 3}$.

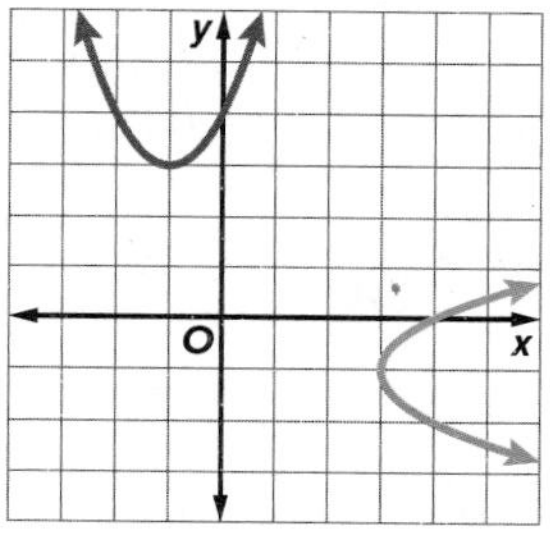

If the domain of $f(x)$ is restricted to $[-1, \infty)$, then the inverse is $f^{-1}(x) = -1 - \sqrt{x - 3}$.

> **Watch Out!**
> f^{-1} is read *f inverse* or *the inverse of f*. Note that -1 is not an exponent.

Example 4 Interpret Inverse Functions

TEMPERATURE **A formula for converting a temperature in degrees Fahrenheit to degrees Celsius is $T(x) = \frac{5}{9}(x - 32)$.**

Find the inverse of $T(x)$, and describe its meaning.

$T(x) = \frac{5}{9}(x - 32)$	Original function
$y = \frac{5}{9}(x - 32)$	Replace $T(x)$ with y.
$x = \frac{5}{9}(y - 32)$	Exchange x and y.
$\frac{9x}{5} = y - 32$	Multiply each side by $\frac{9}{5}$.
$\frac{9x}{5} + 32 = y$	Add 32 to each side.
$T^{-1}(x) = \frac{9x}{5} + 32$	Replace y with $T^{-1}(x)$.

$T^{-1}(x) =$ can be used to convert a temperature in degrees Celsius to degrees Fahrenheit.

Go Online You can complete an Extra Example online.

> **Go Online** to see Part B of the example on using the graph of $T(x)$ and $T^{-1}(x)$.

> **Think About It!**
> Find the domain of $T(x)$ and its inverse. Explain your reasoning.

Think About It!

If $j(x)$ and $k(x)$ are inverses, find $[k \circ j](x)$.

Watch Out!

Compositions of Functions Be sure to check both $[f \circ g](x)$ and $[g \circ f](x)$ to verify that functions are inverses. By definition, both compositions must result in the identity function.

Go Online to see another example on verifying inverse functions.

Talk About It!

Find the domain of the inverse, and describe its meaning in the context of the situation.

Learn Verifying Inverses

Key Concept • Verifying Inverse Functions

Words: Two functions f and g are inverse functions if and only if both of their compositions are the identity function.

Symbols: $f(x)$ and $g(x)$ are inverses if and only if $[f \circ g](x) = x$ and $[g \circ f](x) = x$.

Example 5 Use Compositions to Verify Inverses

Determine whether $h(x) = \sqrt{x + 13}$ and $k(x) = (x - 13)^2$ are inverse functions.

Find $[h \circ k](x)$.

$[h \circ k](x) = h[k(x)]$	Composition of functions
$= h[(x - 13)^2]$	Substitute.
$= \sqrt{(x - 13)^2 + 13}$	Substitute again.
$= \sqrt{x^2 - 26x + 169 + 13}$	Distribute.
$= \sqrt{x^2 - 26x + 182}$	Simplify.

Because $[h \circ k](x)$ is not the identity function, $h(x)$ and $k(x)$ are not inverses.

Check

Determine whether $f(x) = \frac{x}{9} + \frac{4}{3}$ and $g(x) = 9x + 12$ are inverses.

Explain your reasoning.

Example 6 Verify Inverse Functions

GEOMETRY **The formula for the volume of a cylinder with a height of 5 inches is $V = 5\pi r^2$. Determine whether $r = \sqrt{\frac{V}{5\pi}}$ is the inverse of the original function.**

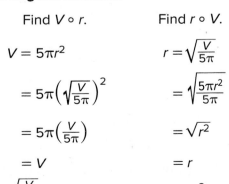

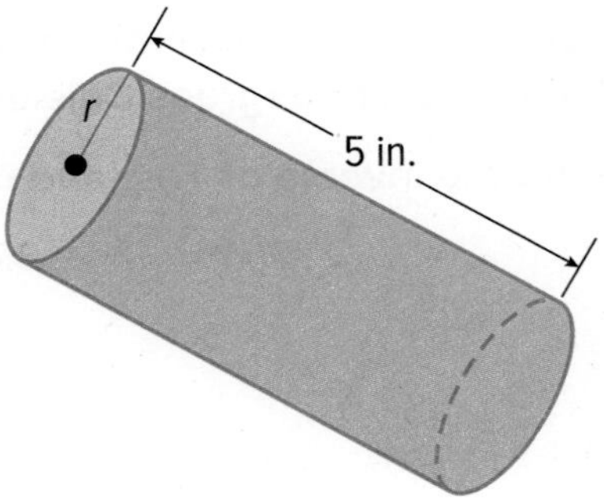

Find $V \circ r$.	Find $r \circ V$.
$V = 5\pi r^2$	$r = \sqrt{\frac{V}{5\pi}}$
$= 5\pi\left(\sqrt{\frac{V}{5\pi}}\right)^2$	$= \sqrt{\frac{5\pi r^2}{5\pi}}$
$= 5\pi\left(\frac{V}{5\pi}\right)$	$= \sqrt{r^2}$
$= V$	$= r$

$r = \sqrt{\frac{V}{5\pi}}$ is the inverse of $V = 5\pi r^2$.

Go Online You can complete an Extra Example online.

Practice

Go Online You can complete your homework online.

Example 1

For each polygon, find the inverse of the relation. Then, graph both the original relation and its inverse.

1. $\triangle MNP$ with vertices at $\{(-8, 6), (6, -2), (4, -6)\}$

2. $\triangle XYZ$ with vertices at $\{(7, 7), (4, 9), (3, -7)\}$

3. trapezoid *QRST* with vertices at $\{(8, -1), (-8, -1), (-2, -8), (2, -8)\}$

4. quadrilateral *FGHJ* with vertices at $\{(4, 3), (-4, -4), (-3, -5), (5, 2)\}$

Examples 2 and 3

Find the inverse of each function. Then graph the function and its inverse. If necessary, restrict the domain of the inverse so that it is a function.

5. $f(x) = x + 2$

6. $g(x) = 5x$

7. $f(x) = -2x + 1$

8. $h(x) = \frac{x - 4}{3}$

9. $f(x) = -\frac{5}{3}x - 8$

10. $g(x) = x + 4$

11. $f(x) = 4x$

12. $f(x) = -8x + 9$

13. $f(x) = 5x^2$

14. $h(x) = x^2 + 4$

Example 4

15. **WEIGHT** The formula to convert weight in pounds to stones is $p(x) = \frac{x}{14}$, where x is the weight in pounds.

a. Find the inverse of $p(x)$, and describe its meaning.

b. Graph $p(x)$ and $p^{-1}(x)$. Use the graph to find the weight in pounds of a dog that weighs about 2.5 stones.

16. CRYPTOGRAPHY DeAndre is designing a code to send secret messages. He assigns each letter of the alphabet to a number, where A = 1, B = 2, C = 3, and so on. Then he uses $c(x) = 4x - 9$ to create the secret code.

a. Find the inverse of $c(x)$, and describe its meaning.

b. Make tables of $c(x)$ and $c^{-1}(x)$. Use the table to decipher the message: 15, 75, 47, 3, 71, 27, 51, 47, 67.

Example 5

Determine whether each pair of functions are inverse functions. Write *yes* or *no*.

17. $f(x) = x - 1$

$g(x) = 1 - x$

18. $f(x) = 2x + 3$

$g(x) = \frac{1}{2}(x - 3)$

19. $f(x) = 5x - 5$

$g(x) = \frac{1}{5}x + 1$

20. $f(x) = 2x$

$g(x) = \frac{1}{2}x$

21. $h(x) = 6x - 2$

$g(x) = \frac{1}{6}x + 3$

22. $f(x) = 8x - 10$

$g(x) = \frac{1}{8}x + \frac{5}{4}$

Example 6

23. GEOMETRY The formula for the volume of a right circular cone with a height of 2 feet is $V = \frac{2}{3}\pi r^2$. Determine whether $r = \sqrt{\frac{3V}{2\pi}}$ is the inverse of the original function.

24. GEOMETRY The formula for the area of a trapezoid is $A = \frac{h}{2}(a + b)$. Determine whether $h = 2A - (a + b)$ is the inverse of the original function.

Mixed Exercises

Find the inverse of each function. Then graph the function and its inverse. If necessary, restrict the domain of the inverse so that it is a function.

25. $y = 4$

26. $f(x) = 3x$

27. $f(x) = x + 2$

28. $g(x) = 2x - 1$

29. $f(x) = \frac{1}{2}x^2 - 1$

30. $f(x) = (x + 1)^2 + 3$

Determine whether each pair of functions are inverse functions. Write *yes* or *no*.

31. $f(x) = 4x^2$
$g(x) = \frac{1}{2}\sqrt{x}$

32. $f(x) = \frac{1}{3}x^2 + 1$
$g(x) = \sqrt{3x - 3}$

33. $f(x) = x^2 - 9$
$g(x) = x + 3$

34. $f(x) = \frac{2}{3}x^3$
$g(x) = \sqrt[3]{\frac{2}{3}x}$

35. $f(x) = (x + 6)^2$
$g(x) = \sqrt{x} - 6$

36. $f(x) = 2\sqrt{x - 5}$
$g(x) = \frac{1}{4}x^2 - 5$

Restrict the domain of *f*(*x*) so that its inverse is also a function. State the restricted domain of *f*(*x*) and the domain of $f^{-1}(x)$.

37. $f(x) = x^2 + 5$

38. $f(x) = 3x^2$

39. $f(x) = \sqrt{x + 6}$

40. $f(x) = \sqrt{x + 3}$

Copy each graph. Sketch a graph of the inverse of each function. Then state whether the inverse is a function.

41.

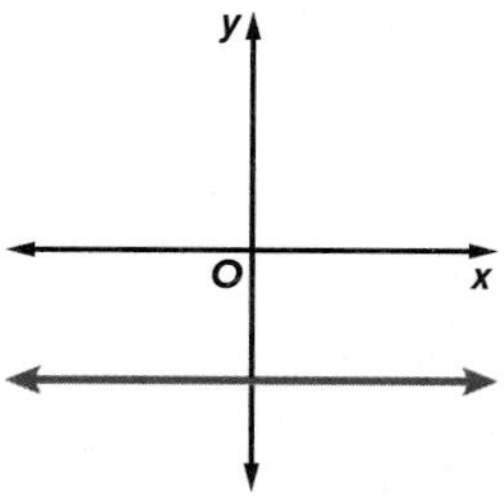

42.

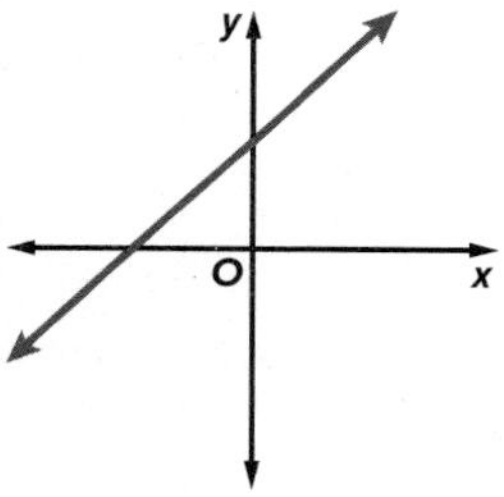

43.

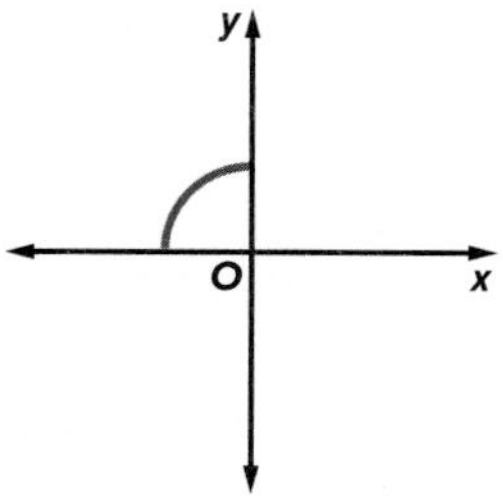

44.

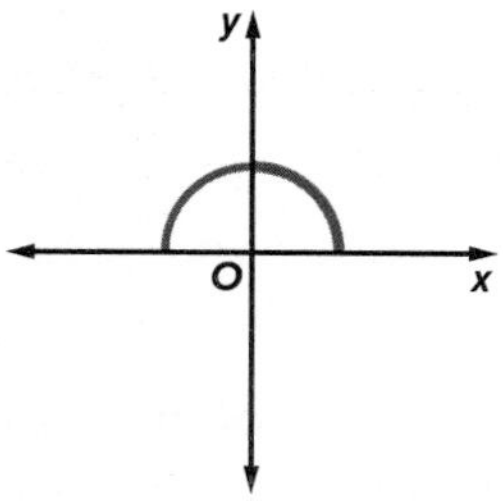

45. FITNESS Alejandro is a personal trainer. For his clients to gain the maximum benefit from their exercise, Alejandro calculates their maximum target heart rate using the function $f(x) = 0.85(220 - x)$, where x represents the age of the client. Find the inverse of this function and interpret its meaning in the context of the situation.

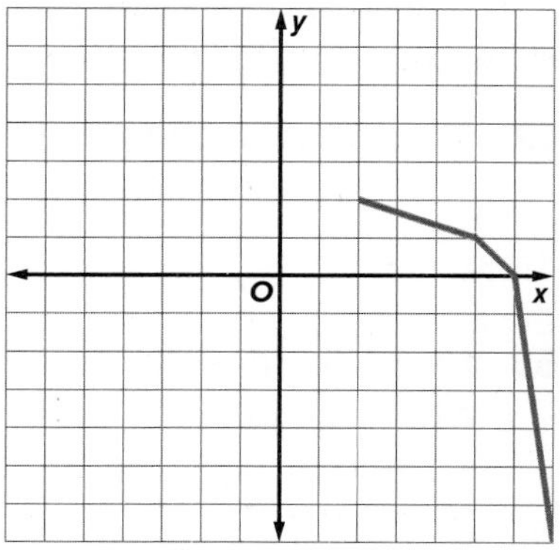

46. Copy the piecewise function shown. Graph the inverse of the piecewise function shown.

47. USE A MODEL The diagram shows a sheet of metal with squares of side length x removed from each corner which can be used to form an open box.

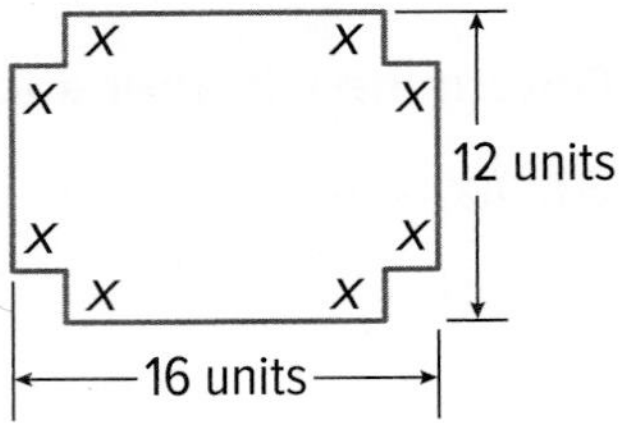

a. Write and graph the function $V(x)$ which gives the volume of the open box. Explain how the domain of this function must be restricted within the context of this scenario.

b. Does restricting the domain in **part a** allow for the inverse to also be a function? Explain your reasoning.

48. STRUCTURE Use the table to find the relationship between $(f + g)^{-1}(x)$ and $f^{-1}(x) + g^{-1}(x)$.

x	0	1	2	3
$f(x)$	0	3	1	4
$g(x)$	1	0	4	3

a. Suppose that functions $f(x)$, $g(x)$, and $(f + g)(x)$ all have inverse functions defined on the domain [0, 3]. Calculate the following values.

i. $f^{-1}(3) + g^{-1}(3)$ **ii.** $f^{-1}(1) + g^{-1}(1)$

b. Use the value of $(f + g)(1)$ to find $(f + g)^{-1}(3)$. Use the value of $(f + g)(0)$ to find $(f + g)^{-1}(1)$.

c. Joyce claims that $(f + g)^{-1}(x) = f^{-1}(x) + g^{-1}(x)$. Determine whether she is correct. Explain your reasoning.

d. Consider the functions $f(x) = 2x + 1$ and $g(x) = 2x - 1$. Compare $(f + g)^{-1}(x)$ and $f^{-1}(x) + g^{-1}(x)$.

Higher-Order Thinking Skills

49. ANALYZE If a relation is not a function, then its inverse is *sometimes*, *always*, or *never* a function. Justify your argument.

50. CREATE Give an example of a function and its inverse. Verify that the two functions are inverses.

51. PERSEVERE Give an example of a function that is its own inverse.

52. ANALYZE Can the graphs of a linear function with a slope $\neq 0$ and its inverse ever be perpendicular? Justify your answer.

53. WRITE Suppose you have a composition of two functions that are inverses. When you put in a value of 5 for x, why is the result always 5?

nth Roots and Rational Exponents

Today's Goals

- Simplify expressions involving radicals and rational exponents.
- Simplify expressions in exponential or radical form.

Today's Vocabulary

nth root
index
radicand
principal root
rational exponent

Explore Inverses of Rational Functions

Online Activity Use a calculator to complete the Explore.

INQUIRY What conjectures can you make about $f(x) = x^n$ and $g(x) = \sqrt[n]{x}$ for all odd positive values of n?

Learn nth Roots

Finding the square root of a number and squaring a number are inverse operations. To find the square root of a, you must find a number with a square of a. The inverse of raising a number to the nth power is finding the ***n*th root** of a number. The symbol $\sqrt[n]{}$ indicates an nth root.

For any real numbers a and b and any positive integer n, if $a^n = b$, then a is an nth root of b. For example, because $(-2)^6 = 64$, -2 is a sixth root of 64 and 2 is a principal root.

An example of an nth root is $\sqrt[n]{36}$, which is read as *the nth root of 36*. In this example, n is the **index** and 36 is the **radicand**, or the expression under the radical symbol.

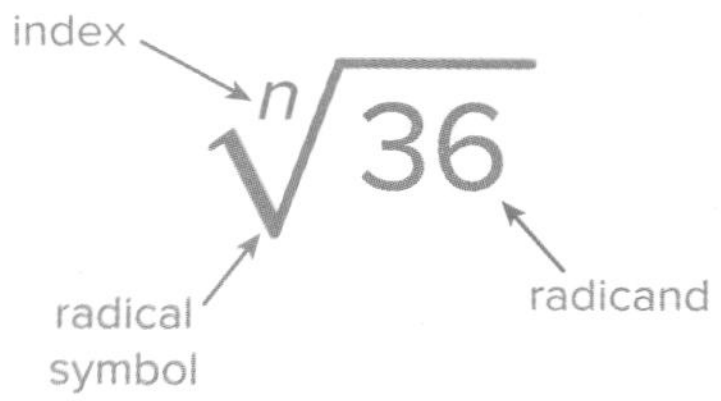

Some numbers have more than one real nth root. For example, 16 has two square roots, 4 and -4, because 4^2 and $(-4)^2$ both equal 16. When there is more than one real root and n is even, the nonnegative root is called the **principal root**.

Key Concept • Real nth Roots

Suppose n is an integer greater than 1, b is a real number, and a is an nth root of b.

b	n is even.	n is odd.
$b > 0$	1 unique positive and 1 unique negative real root: $\pm\sqrt[n]{b}$	1 unique positive and 0 negative real root: $\sqrt[n]{b}$
$b < 0$	0 real roots	0 positive and 1 negative real root: $\sqrt[n]{b}$
$b = 0$	1 real root: $\sqrt[n]{0} = 0$	1 real root: $\sqrt[n]{0} = 0$

The radicand contains no fractions and no radicals appear in the denominator.

Think About It!

Lorena says she can tell that $\sqrt[3]{-64}$ will have a real root without graphing. Do you agree or disagree? Explain your reasoning.

Example 1 Find Roots

Simplify.

a. $\pm\sqrt{25x^4}$

$$\pm\sqrt{25x^4} = \pm\sqrt{(5x^2)^2}$$
$$= \pm 5x^2$$

b. $-\sqrt{(y^2+7)^{12}}$

$$-\sqrt{(y^2+7)^{12}} = -\sqrt{[(y^2+7)^6]^2}$$
$$= -(y^2+7)^6$$

c. $\sqrt[3]{343a^{18}b^6}$

$$\sqrt[3]{343a^{18}b^6} = \sqrt[3]{(7a^6b^2)^3}$$
$$= 7a^6b^2$$

d. $\sqrt{-289c^8d^4}$

There are no real roots of $\sqrt{-289}$. However, there are two imaginary roots, $17i$ and $-17i$. Because we are only finding the principal square root, use $17i$.

$$\sqrt{-289c^8d^4} = \sqrt{-1} \cdot \sqrt{289c^8d^4}$$
$$= i \cdot \sqrt{289c^8d^4}$$
$$= 17ic^4d^2$$

Watch Out!

Principal Roots

Because $5^2 = 25$, 25 has two square roots, 5 and -5. However, the value of $\sqrt{25}$ is 5 only. To indicate both square roots and not just the principal square root, the expression must be written as $\pm\sqrt{25}$.

Check

Write the simplified form of each expression.

a. $\pm\sqrt{196x^4}$ **b.** $-\sqrt{196x^4}$ **c.** $\sqrt{-196x^4}$

When you find an even root of an even power and the result is an odd power, you must use the absolute value of the result to ensure that the answer is nonnegative.

Talk About It!

Compare the simplified expressions in the previous example with the ones in this example. Explain why the simplified expressions in this example require absolute value bars when the simplified expressions in the previous example did not.

Example 2 Simplify Using Absolute Value

Simplify.

a. $\sqrt[4]{81x^4}$

$$\sqrt[4]{81x^4} = \sqrt[4]{(3x)^4}$$
$$= 3|x|$$

Since x could be negative, you must use the absolute value of x to ensure that the principal square root is nonnegative.

b. $\sqrt[8]{256(y^2-2)^{24}}$

$$\sqrt[8]{256(y^2-2)^{24}} = \sqrt[8]{256} \cdot \sqrt[8]{(y^2-2)^{24}}$$
$$= 2|(y^2-2)^3|$$

Since $(y^2-2)^3$ could be negative, you must use the absolute value of $(y^2-2)^3$ to ensure that the principal square root is nonnegative.

Learn Rational Exponents

You can use the properties of exponents to translate expressions from exponential form to radical form or from radical form to exponential form. An expression is in **exponential form** if it is in the form x^n, where n is an exponent. An expression is in **radical form** if it contains a radical symbol.

For any real number b and a positive integer n, $b^{\frac{1}{n}} = \sqrt[n]{b}$, except where $b < 0$ and n is even. When $b < 0$ and n is even, a complex root may exist.

Examples: $125^{\frac{1}{3}} = \sqrt[3]{125}$ or 5 $\qquad (-49)^{\frac{1}{2}} = \sqrt{-49}$ or $7i$

The expression $b^{\frac{1}{n}}$ has a **rational exponent**. The rules for exponents also apply to rational exponents.

Math History Minute:

Christoff Rudolff (1499–1543) wrote the first German algebra textbook. It is believed that he introduced the radical symbol $\sqrt{}$ in 1525 in his book *Die Coss*. Some feel that this symbol was used because it resembled a small *r*, the first letter in the Latin word *radix* or root.

Key Concept • Rational Exponents

For any nonzero number b and any integers x and y, with $y > 1$, $b^{\frac{x}{y}} = \sqrt[y]{b^x} = \left(\sqrt[y]{b}\right)^x$, except when $b < 0$ and y is even. When $b < 0$ and y is even, a complex root may exist.

Examples: $125^{\frac{2}{3}} = \left(\sqrt[3]{125}\right)^2 = 5^2$ or 25

$(-49)^{\frac{3}{2}} = \left(\sqrt{-49}\right)^3 = (7i)^3$ or $-343i$

Key Concept • Simplest Form of Expressions with Rational Exponents

An expression with rational exponents is in simplest form when all of the following conditions are met.

- It has no negative exponents.
- It has no exponents that are not positive integers in the denominator.
- It is not a complex fraction.
- The index of any remaining radical is the least number possible.

Example 3 Radical and Exponential Forms

Simplify.

a. Write $x^{\frac{4}{3}}$ in radical form.

$x^{\frac{4}{3}} = \sqrt[3]{x^4}$

b. Write $\sqrt[5]{x^2}$ in exponential form.

$\sqrt[5]{x^2} = x^{\frac{2}{5}}$

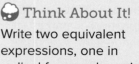

Go Online You can complete an Extra Example online.

Think About It!

Write two equivalent expressions, one in radical form and one in exponential form.

Think About It!

Why did you set t equal to $\frac{1}{4}$?

Think About It!

How can you tell if $x^{\frac{y}{z}}$ will simplify to an integer?

Go Online to see more examples of evaluating expressions with rational exponents.

Watch Out!

Exponents Recall that when you multiply powers, the exponents are added, and when you raise a power to a power, the exponents are multiplied.

Think About It!

How would the expression in part **c** change if the exponents were $\frac{1}{3}$ and $-\frac{3}{4}$?

Example 4 Use Rational Exponents

FINANCIAL LITERACY The expression $c(1 + r)^t$ can be used to estimate the future cost of an item due to inflation, where c represents the current cost of the item, r represents the annual rate of inflation, and t represents the time in years. Write the expression in radical form for the future cost of an item 3 months from now if the annual rate of inflation is 4.7%.

$c(1 + r)^t = c(1 + 0.047)^{\frac{1}{4}}$ — $r = 0.047, t = \frac{1}{4}$

$= c(1.047)^{\frac{1}{4}}$ — Add.

$= c\sqrt[4]{1.047}$ — $b^{\frac{1}{n}} = \sqrt[n]{b}$

Example 5 Evaluate Expressions with Rational Exponents

Evaluate $32^{-\frac{2}{5}}$.

$32^{-\frac{2}{5}} = \frac{1}{32^{\frac{2}{5}}}$ — $b^{-n} = \frac{1}{b^n}$

$= \frac{1}{(2^5)^{\frac{2}{5}}}$ — $32 = 2^5$

$= \frac{1}{2^{5 \cdot \frac{2}{5}}}$ — Power of a Power

$= \frac{1}{2^2}$ or $\frac{1}{4}$ — Multiply the exponents.

Example 6 Simplify Expressions with Rational Exponents

Simplify each expression.

a. $x^{\frac{2}{3}} \cdot x^{\frac{1}{6}}$

$x^{\frac{2}{3}} \cdot x^{\frac{1}{6}} = x^{\frac{2}{3} + \frac{1}{6}}$ — Add powers.

$= x^{\frac{4}{6} + \frac{1}{6}}$ — $\frac{2}{3} = \frac{4}{6}$

$= x^{\frac{5}{6}}$ — Add the exponents.

b. $y^{-\frac{2}{3}}$

$y^{-\frac{2}{3}} = \frac{1}{y^{\frac{2}{3}}}$ — $b^{-n} = \frac{1}{b^n}$

$= \frac{1}{y^{\frac{2}{3}}} \cdot \frac{y^{\frac{1}{3}}}{y^{\frac{1}{3}}}$ — $\frac{y^{\frac{1}{3}}}{y^{\frac{1}{3}}} = 1$

$= \frac{y^{\frac{1}{3}}}{y^{\frac{3}{3}}}$ or $\frac{y^{\frac{1}{3}}}{y}$ — $y^{\frac{2}{3}} \cdot y^{\frac{1}{3}} = y^{\frac{2}{3} + \frac{1}{3}}$

c. $z^{-\frac{1}{3}} \cdot z^{\frac{3}{4}}$

$z^{-\frac{1}{3}} \cdot z^{\frac{3}{4}} = z^{-\frac{1}{3} + \frac{3}{4}}$ — Add powers.

$= z^{-\frac{4}{12} + \frac{9}{12}}$ or $z^{\frac{5}{12}}$ — $-\frac{1}{3} = \frac{4}{12}, \frac{3}{4} = \frac{9}{12}$

Go Online You can complete an Extra Example online.

Practice

Go Online You can complete your homework online.

Examples 1 and 2

Simplify.

1. $\pm\sqrt{121x^4y^{16}}$
2. $\pm\sqrt{225a^{16}b^{36}}$
3. $\pm\sqrt{49x^4}$
4. $-\sqrt{16c^4d^2}$
5. $-\sqrt{81a^{16}b^{20}c^{12}}$
6. $-\sqrt{400x^{32}y^{40}}$
7. $\sqrt[4]{16(x-3)^{12}}$
8. $\sqrt[8]{x^{16}y^8}$
9. $\sqrt[4]{81(x-4)^4}$
10. $\sqrt[6]{x^{18}}$
11. $\sqrt[4]{a^{12}}$
12. $\sqrt[3]{a^{12}}$

Examples 3

Write each expression in radical form, or write each radical in exponential form.

13. $8^{\frac{1}{5}}$
14. $4^{\frac{2}{7}}$
15. $(x^3)^{\frac{3}{2}}$
16. $\sqrt{17}$
17. $\sqrt[3]{5xy^2}$
18. $\sqrt[4]{625x^2}$

Examples 4

19. ORBITING The distance in millions of miles a planet is from the Sun in terms of t, the number of Earth days it takes for the planet to orbit the Sun, can be modeled by the expression $\sqrt[3]{6t^2}$. Write the expression in exponential form.

20. DEPRECIATION The depreciation rate is calculated by the expression $1-\left(\frac{T}{P}\right)^{\frac{1}{n}}$, where n is the age of the item in years, T is the resale price in dollars, and P is the original price in dollars. Write the expression in radical form for an 8 year old car that was originally purchased for \$52,425.

Example 5

Evaluate each expression.

21. $27^{\frac{1}{3}}$
22. $256^{\frac{1}{4}}$
23. $16^{-\frac{3}{2}}$
24. $81^{-\frac{1}{4}}$
25. $1024^{\frac{3}{5}}$
26. $16^{-\frac{5}{4}}$

Example 6

Simplify each expression.

27. $x^{\frac{1}{3}} \cdot x^{\frac{2}{5}}$

28. $a^{\frac{4}{9}} \cdot a^{\frac{1}{4}}$

29. $b^{-\frac{3}{4}}$

30. $y^{-\frac{4}{5}}$

Mixed Exercises

Simplify.

31. $\sqrt[3]{27b^{18}c^{12}}$

32. $-\sqrt{(2x + 1)^6}$

33. $\sqrt[4]{81(x + 4)^4}$

34. $\sqrt[3]{(4x - 7)^{24}}$

35. $\sqrt[3]{(y^3 + 5)^{18}}$

36. $\sqrt[4]{256(5x - 2)^{12}}$

37. $\sqrt{196c^6d^4}$

38. $\sqrt{-64y^8z^6}$

39. $\sqrt[3]{-27a^{15}b^9}$

40. $\sqrt[4]{-16x^{16}y^8}$

41. $a^{\frac{7}{4}} \cdot a^{\frac{5}{4}}$

42. $x^{\frac{2}{3}} \cdot x^{\frac{8}{3}}$

43. $\left(b^{\frac{3}{4}}\right)^{\frac{1}{3}}$

44. $\left(y^{-\frac{3}{5}}\right)^{-\frac{1}{4}}$

45. $d^{-\frac{5}{6}}$

46. $w^{-\frac{7}{8}}$

47. GEOMETRY The volume V of a regular octahedron with edge length ℓ is given by $V = \frac{\ell^3\sqrt{2}}{3}$. Write the volume in simplest form for an octahedron with the given edge lengths.

a. $\sqrt{15}$ cm

b. $\sqrt{24}$ cm

c. $3\sqrt{8}$ cm

Simplify.

48. $\dfrac{f^{-\frac{1}{4}}}{4f^{\frac{1}{2}} \cdot f^{-\frac{1}{3}}}$

49. $\dfrac{c^{\frac{2}{3}}}{c^{\frac{1}{6}}}$

50. $\dfrac{z^{\frac{4}{5}}}{z^{\frac{1}{2}}}$

51. $\sqrt[8]{36h^4j^4}$

52. $\dfrac{ab}{\sqrt{c}}$

53. $\dfrac{xy}{\sqrt[3]{z}}$

54. SPORTS A volleyball has a volume of 864π cm^3. A tennis ball has a volume of 32π cm^3. By how much does the radius of the volleyball exceed that of the tennis ball? Write your answer using rational exponents.

55. WATER TOWER One of the largest sphere water towers in the country is located in Edmond, Oklahoma. It is 218 feet tall and holds 500,000 gallons, or about 66,840 cubic feet of water. Another town is planning to build a similar water tower. However, the new water tower will hold $\frac{3}{5}$ as much water as the tower in Edmond. Determine the radius of the new water tower to the nearest foot.

56. CELLS The number of cells in a cell culture grows exponentially. The number of cells in the culture as a function of time is given by the expression $N\left(\frac{6}{5}\right)^t$, where t is measured in hours and N is the initial size of the culture. Write the following expressions in radical form.

a. the number of cells after 20 minutes with N initial cells

b. the number of cells after 44 minutes with N initial cells

c. the number of cells after 1 hour and 15 minutes with 4000 initial cells

57. REASONING Simplify $\sqrt[b]{m^{3b}}$, where $b > 0$. Explain your reasoning.

58. REGULARITY There are no real nth roots of a number w. What can you conclude about the index and the number w?

59. CONSTRUCT ARGUMENTS Determine the values of x for which $\sqrt{x^2} \neq x$. Explain your answer.

60. STRUCTURE Which of the following functions are equivalent? Justify your answer.

a. $f(x) = \sqrt[3]{x^9}$ **b.** $g(x) = \sqrt{x^6}$ **c.** $r(x) = (\sqrt[3]{x})^9$ **d.** $s(x) = (\sqrt{x})^6$

61. WRITE Explain how it might be easier to simplify an expression using rational exponents rather than using radicals.

62. FIND THE ERROR Destiny and Kimi are simplifying $\sqrt[4]{16x^4y^8}$. Is either of them correct? Explain your reasoning.

Destiny

$$\sqrt[4]{16x^4y^8} = \sqrt[4]{(2xy^2)^4}$$
$$= 2|xy^2|$$

Kimi

$$\sqrt[4]{16x^4y^8} = \sqrt[4]{(2xy^2)^4}$$
$$= 2y^2|x|$$

63. PERSEVERE Under what conditions is $\sqrt{x^2 + y^2} = x + y$ true?

64. ANALYZE Determine whether the statement $\sqrt[4]{(-x)^4} = x$ is *sometimes, always,* or *never* true. Justify your argument.

65. PERSEVERE For what real values of x is $\sqrt[3]{x} > x$?

66. WRITE Explain when and why absolute value symbols are needed when taking an nth root.

67. PERSEVERE Write an equivalent expression for $\sqrt[3]{2x} \cdot \sqrt[3]{8y}$. Simplify the radical.

68. CREATE Find two different expressions that equal 2 in the form $x^{\frac{1}{a}}$.

Graphing Radical Functions

Explore Using Technology to Analyze Square Root Functions

Online Activity Use graphing technology to complete the Explore.

INQUIRY How does adding, subtracting, or multiplying a constant to a function affect the graph of the function?

Learn Graphing Square Root Functions

A **radical function** is a function that contains radicals with variables in the radicand. One type of radical function is a **square root function**, which is a function that contains the square root of a variable expression.

Key Concept • Parent Function of Square Root Functions

The parent function of the square root functions is $f(x) = \sqrt{x}$.

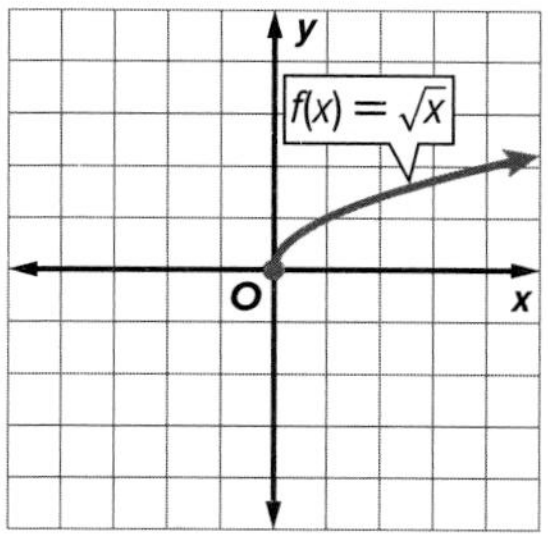

Domain:	$\{x \mid x \geq 0\}$
Range:	$\{f(x) \mid f(x) \geq 0\}$
Intercepts:	$x = 0, f(x) = 0$
End behavior:	As $x \to 0$, $f(x) \to 0$, and as $x \to \infty$, $f(x) \to \infty$.
Increasing/ decreasing:	increasing when $x > 0$
Positive/ negative:	positive for $x > 0$
Symmetry:	no symmetry

A square root function can be written in the form $g(x) = a\sqrt{x - h} + k$. Each constant in the equation affects the parent graph.

- The value of $|a|$ stretches or compresses (dilates) the parent graph.
- When the value of a is negative, the graph is reflected in the x-axis.
- The value of h shifts (translates) the parent graph left or right.
- The value of k shifts (translates) the parent graph up or down.

Today's Goals

- Graph and analyze square root functions.
- Graph and analyze cube root functions.

Today's Vocabulary

radical function
square root function
cube root function

Think About It!

Why is the domain limited to $x \geq 0$ for the parent of the square root function?

Go Online

You may want to complete the Concept Check to check your understanding.

Example 1 Identify Domain and Range Algebraically

Identify the domain and range of $f(x) = \sqrt{2x - 6} + 1$.

The domain is restricted to values for which the radicand is nonnegative.

$2x - 6 \geq 0$	Write an inequality using the radicand.
$2x \geq 6$	Add 6 to each side.
$x \geq 3$	Divide each side by 2.

The domain is $\{x \mid x \geq 3\}$.

Find $f(3)$ to determine the lower limit of the range.

$f(3) = \sqrt{2(3) - 6} + 1$ or 1

The range is $\{f(x) \mid f(x) \geq 1\}$.

Example 2 Graph a Transformed Square Root Function

Graph $g(x) = -3\sqrt{x + 1} + 2$, and identify the domain and range. Then describe how it is related to the graph of the parent function.

Step 1 Determine the minimum domain value.

$x + 1 \geq 0$	Write an inequality using the radicand.
$x \geq -1$	Simplify.

Step 2 Make a table and graph.

Use x-values determined from **Step 1** to make a table.

x	$g(x)$
−1	2
0	−1
1	≈ −2.2
2	≈ −3.2
3	−4

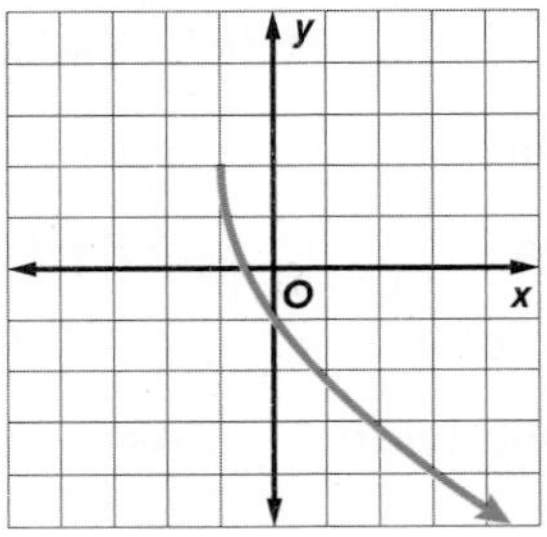

The domain is $\{x \mid x \geq -1\}$ and the range is $\{g(x) \mid g(x) \leq 2\}$.

Step 3 Compare $g(x)$ to the parent function.

The maximum is $(-1, 2)$.

Because $f(x) = \sqrt{x}$, $g(x) = a\sqrt{x - h} + k$ where $a = -3$, $h = -1$, and $k = 2$.

$a < 0$ and $|a| > 1$, so the graph of $f(x) = \sqrt{x}$ is reflected in the x-axis and stretched vertically by a factor of $|a|$, or 3.

$h < 0$, so the graph is then translated left $|h|$ units, or 1 unit.

$k > 0$, so the graph is then translated up k units, or 2 units.

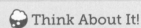

Go Online You can complete an Extra Example online.

Go Online
You can learn how to graph radical functions by watching the video online.

Think About It!
How are the values of h and k related to the domain and range in this example?

Example 3 Analyze the Graph of a Square Root Function

BLOOD DONATION **When blood is donated, medical professionals use a centrifuge to separate it. The centrifuge spins the blood, causing it to separate into three components, which are red cells, platelets, and plasma. In order to efficiently separate the blood, the centrifuge must spin at a specified rate, measured in rotations per minute (RPM), for the required gravitation force, or g-force, exerted on the blood. For a centrifuge with a radius of 7.8 centimeters, the RPM setting of the centrifuge is determined by the product of 104.23 and the square root of the g-force required.**

Part A Write and graph the function.

Complete the table to write the function.

Words	The RPM setting of the centrifuge	is	the product of 104.23	and the square root of the g-force required.
Variables	Let g represent the force and r represent the RPM setting.			
Function	r	$=$	$104.23 \cdot$	$\sqrt{g}$

Make a table to graph the function.

g	r
0	0
400	2085
800	2948
1200	3611
1600	4422

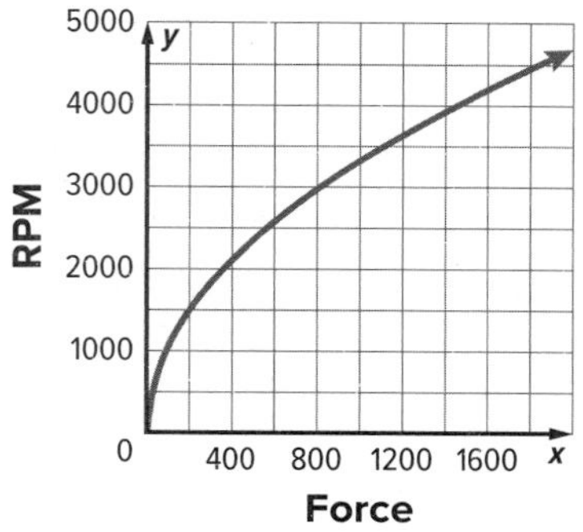

Part B Describe key features of the function.

Domain: $\{g \mid g \geq 0\}$

Range: $\{r \mid r \geq 0\}$

x-intercept: 0

y-intercept: 0

Increasing/Decreasing: increasing as $g \rightarrow \infty$

Positive/Negative: positive for $g > 0$

End Behavior: As $g \rightarrow \infty$, $r \rightarrow \infty$.

Go Online You can complete an Extra Example online.

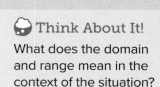

Think About It!
What does the domain and range mean in the context of the situation?

Example 4 Graph a Square Root Inequality

Graph $y < \sqrt{2x + 5}$.

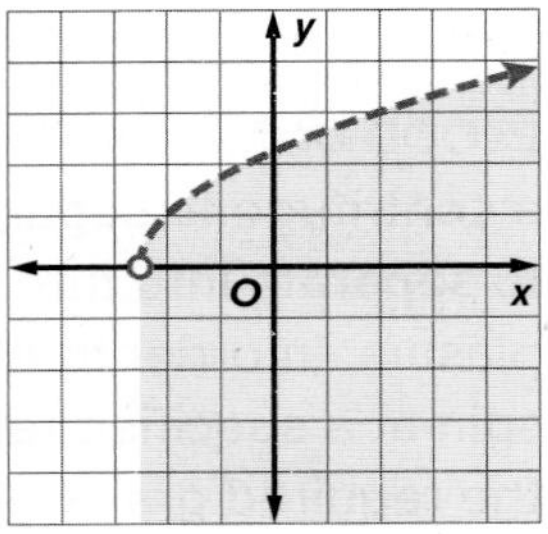

Step 1 Graph the related function.

Graph the boundary $y = \sqrt{2x + 5}$, using a dashed line because the inequality is $<$.

Step 2 Shade.

The domain is $\{x \mid x > -2.5\}$. Because the inequality is less than, shade the region below the boundary and within the domain. Select a test point in the shaded region to verify the solution.

Watch Out!

Test Point When selecting a test point, make sure the point is within the domain of the related function.

Learn Graphing Cube Root Functions

A **cube root function** is a radical function that contains the cube root of a variable expression.

Key Concept • Parent Function of Cube Root Functions

The parent function of the cube root functions is $f(x) = \sqrt[3]{x}$.

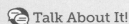

Domain:	all real numbers
Range:	all real numbers
Intercepts:	$x = 0$, $f(x) = 0$
End behavior:	As $x \to -\infty$, $f(x) \to -\infty$, and as $x \to \infty$, $f(x) \to \infty$.
Increasing/ decreasing:	increasing as $x \to \infty$
Positive/ negative:	positive for $x > 0$ negative for $x < 0$
Symmetry:	symmetric about the origin

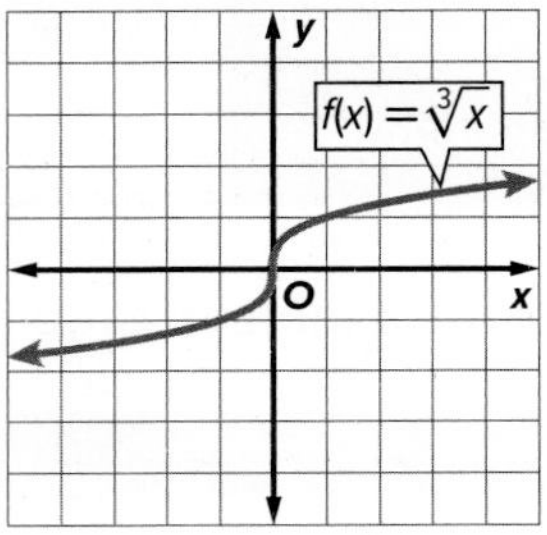

A cube root function can be written in the form $g(x) = a\sqrt[3]{x - h} + k$.

Talk About It!

Describe how the domain and range differ for a radical function with an odd index and a radical function with an even index.

Example 5 Graph Cube Root Functions

Graph each function. State the domain and range.

a. $g(x) = \frac{1}{3}\sqrt[3]{x}$

In $g(x) = a\sqrt[3]{x - h} + k$, $a = \frac{1}{3}$, $h = 0$, and $k = 0$. So the function is centered at the origin and vertically compressed.

x	$g(x)$
-2	≈ -0.42
-1	$-0.\overline{33}$
0	0
1	$0.\overline{33}$
2	≈ 0.42

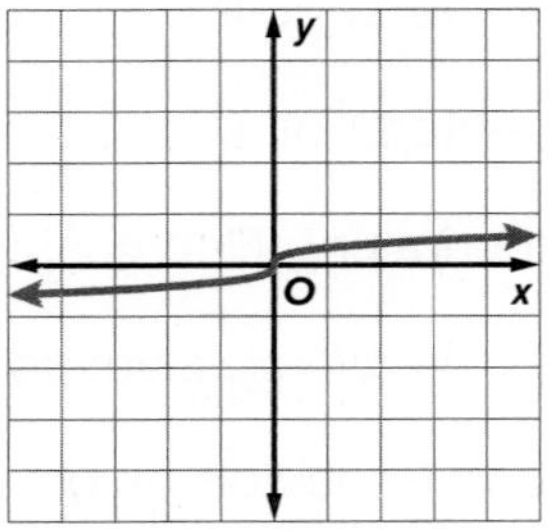

The domain is all real numbers, and the range is all real numbers.

Go Online

You can learn how to graph radical functions on a graphing calculator by watching the video online.

b. $p(x) = \sqrt[3]{x + 5}$

In $p(x) = a\sqrt[3]{x - h} + k$, $a = 1$, $h = -5$, and $k = 0$. So the function is translated 5 units left from the parent graph.

The domain is all real numbers, and the range is all real numbers.

x	y
−7	≈ −1.26
−6	1
−5	0
−4	−1
−3	≈ 1.26

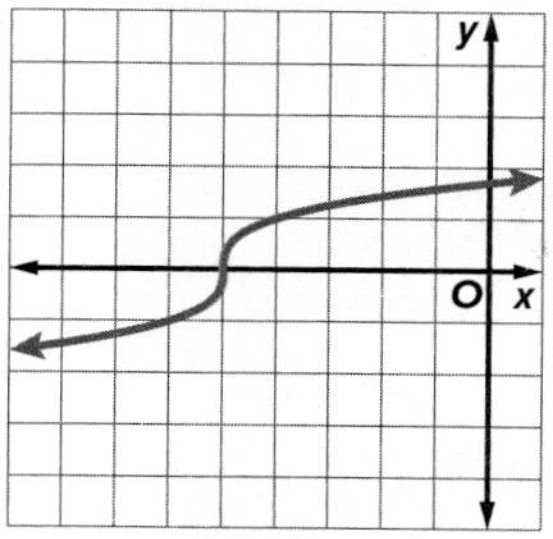

c. $q(x) = \sqrt[3]{4 - x} + 1$

The function can be written as $q(x) = \sqrt[3]{-(x - 4)} + 1$. So the function is reflected and translated 4 units right and 1 unit up from the parent graph.

x	y
2	≈ 2.26
3	2
4	1
5	0
6	≈ −0.26

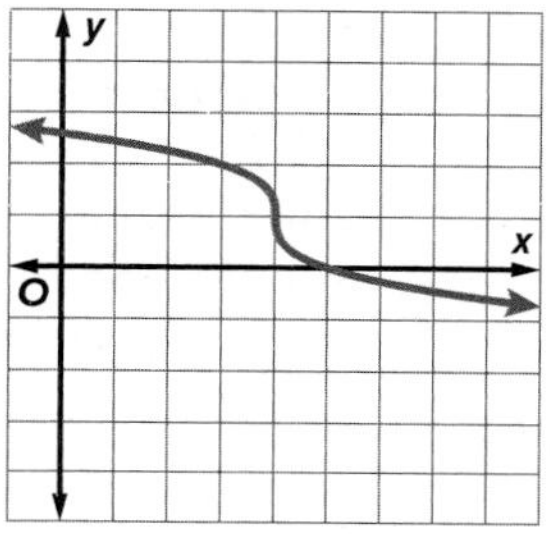

The domain is all real numbers, and the range is all real numbers.

Example 6 Compare Radical Functions

Examine $p(x) = -2\sqrt[3]{x - 6}$ and $q(x)$ shown in the graph.

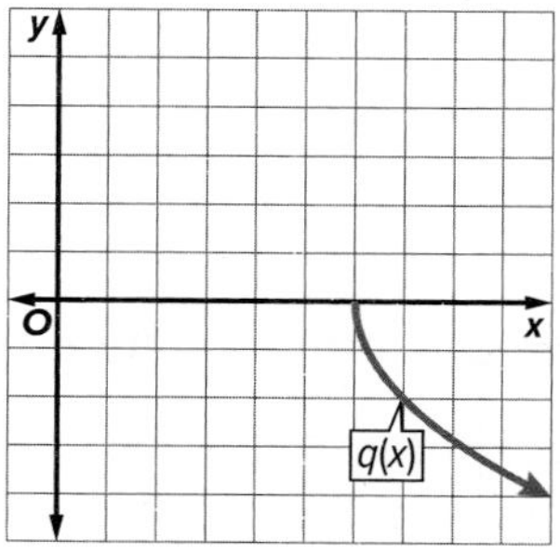

Part A Graph $p(x)$.

Make a table of values for $p(x)$. Then, graph the function.

x	y
0	≈ 3.63
2	≈ 3.17
4	≈ 2.52
6	0
8	≈ −2.52
10	≈ −3.18

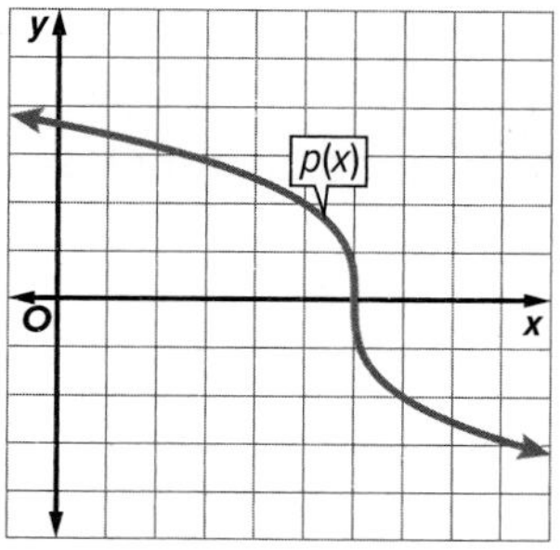

(continued on the next page)

Study Tip

Tables When making a table of values for a radical function, first determine h. Then make a table using values that are greater than and less than h.

Think About It!

Describe the end behavior of the function in part **c**, $y = \sqrt[3]{4 - x} + 1$.

Think About It!

Determine whether the indexes of $p(x)$ and $q(x)$ are even or odd. Justify your response.

Part B Compare key features.

	$p(x)$	$q(x)$
Domain and Range	D: all real numbers, R: all real numbers	D: $\{x \mid x \geq 6\}$, R: $\{q(x) \mid q(x) \leq 0\}$
Intercepts	x-intercept: 6; y-intercept: 3.63	x-intercept: 6; y-intercept: none
Increasing/Decreasing	decreasing as $x \to \infty$	decreasing as $x \to \infty$
Positive/Negative	positive for $x < 6$; negative for $x > 6$	negative for $x > 6$
End Behavior	As $x \to -\infty$, $p(x) \to \infty$, and as $x \to \infty$, $p(x) \to -\infty$.	As $x \to 6$, $q(x) \to 0$, and as $x \to \infty$, $q(x) \to -\infty$.

Example 7 Write a Radical Function

Write a radical function for the graph of $g(x)$.

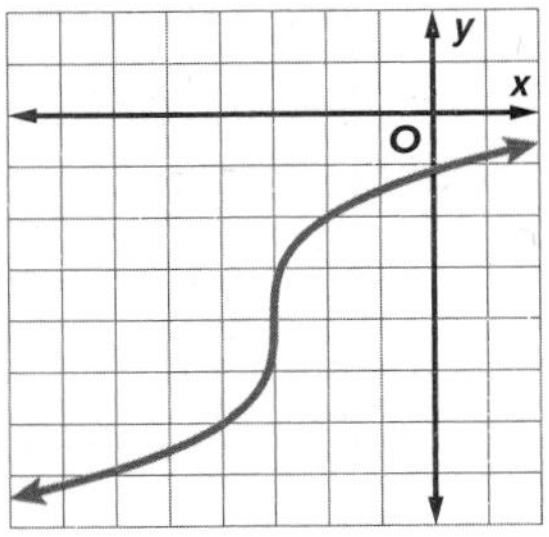

Step 1 Identify the index.

Because the domain and range for $g(x)$ is all real numbers, the index is odd. This function can be represented by $g(x) = a\sqrt[3]{x - h} + k$.

Step 2 Identify any transformations.

The function has been translated 3 units left and 4 units down.

Therefore, $h = -3$ and $k = -4$. To find the value of a, use a point as well as the values of h and k.

$g(x) = a\sqrt[3]{x - h} + k$	Cube root function
$-2 = a\sqrt[3]{-2 - (-3)} + (-4)$	$h = -3$, $k = -4$, $(x, g(x)) = (-2, -2)$
$-2 = a \cdot 1 - 4$	Simplify.
$a = 2$	Simplify.

Step 3 Write the function.

Substitute the values of a, h, and k to write the function. The graph is represented by $g(x) = 2\sqrt[3]{x + 3} - 4$.

Check

Write a radical function for the graph of $g(x)$.

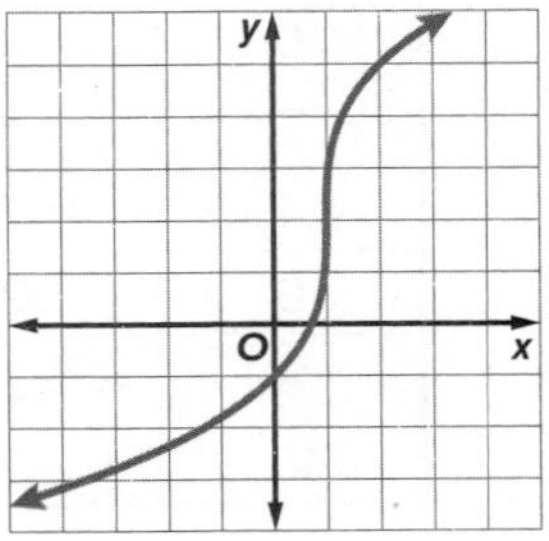

Go Online You can complete an Extra Example online.

Study Tip

Index While one reasonable guess is that the index is 3, the index could be any odd positive integer.

Think About It!

Write a cube root function for a graph that has been reflected in the x-axis and translated 7 units right and 4 units down from the parent function.

Practice

Go Online You can complete your homework online.

Example 1

Identify the domain and range of each function.

1. $y = \sqrt{x - 9}$

2. $y = \sqrt{x + 7}$

3. $y = -\sqrt{6x}$

4. $y = 5\sqrt{x + 2} - 1$

5. $y = \sqrt{3x - 4}$

6. $y = -\sqrt{x - 2} + 2$

Example 2

Graph each function. State the domain and range of each function. Then describe how it is related to the graph of the parent function.

7. $y = -\frac{1}{3}\sqrt{x + 1}$

8. $y = -\sqrt{x - 2} + 3$

9. $y = 2\sqrt{x}$

10. $y = \sqrt{x + 3}$

11. $y = 3\sqrt{x} - 5$

12. $y = \sqrt{x + 4} - 2$

Example 3

13. REFLEXES Raquel and Ashley are testing one another's reflexes. Raquel drops a ruler from a given height so that it falls between Ashley's thumb and index finger. Ashley tries to catch the ruler before it falls through her hand. The time, in seconds, required to catch the ruler is given by the product of $\frac{1}{4}$ and the square root of the distance the ruler falls in inches.

    a. Write and graph a function, where *d* is the distance in inches, and *t* is the time, in seconds.

    b. Describe the key features of the function.

    c. Graph the parent function on the same coordinate grid. How does the function you wrote in **part a** compare to the parent function?

Example 4

Graph each inequality.

14. $y < \sqrt{x - 5}$

15. $y > \sqrt{x + 6}$

16. $y \geq -4\sqrt{x + 3}$

17. $y \leq -2\sqrt{x - 6}$

18. $y > 2\sqrt{x + 7} - 5$

19. $y \geq 4\sqrt{x - 2} - 12$

Example 5

Graph each function. State the domain and range of each function.

20. $f(x) = \sqrt[3]{x + 1}$

21. $f(x) = 3\sqrt[3]{x - 2}$

22. $f(x) = \sqrt[3]{x + 7} - 1$

23. $f(x) = -\sqrt[3]{x - 2} + 9$

Example 6

24. Examine $p(x) = -3\sqrt{x + 2}$ and $q(x)$ shown in the graph.

a. Graph $p(x)$.

b. Compare the key features of the functions.

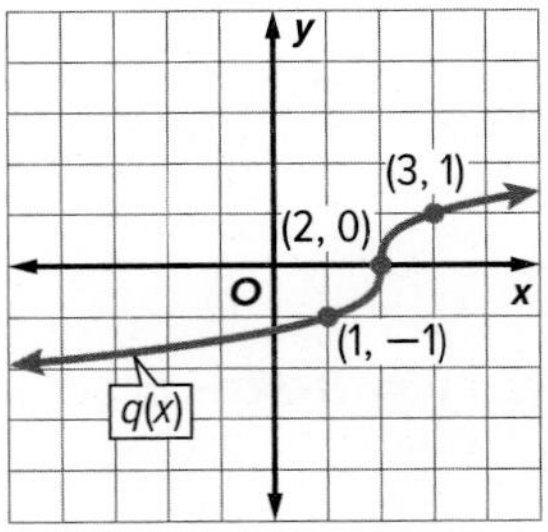

25. Examine $p(x)$, which is 2 less than the cube root of x, and $q(x)$ shown in the graph.

a. Graph $p(x)$.

b. Compare the key features of the functions.

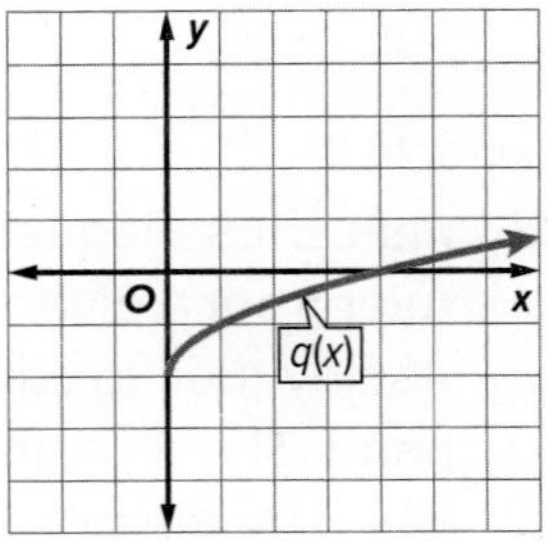

26. Examine $p(x) = \sqrt{x - 2} + 5$ and $q(x)$ shown in the graph.

a. Graph $p(x)$.

b. Compare the key features of the functions.

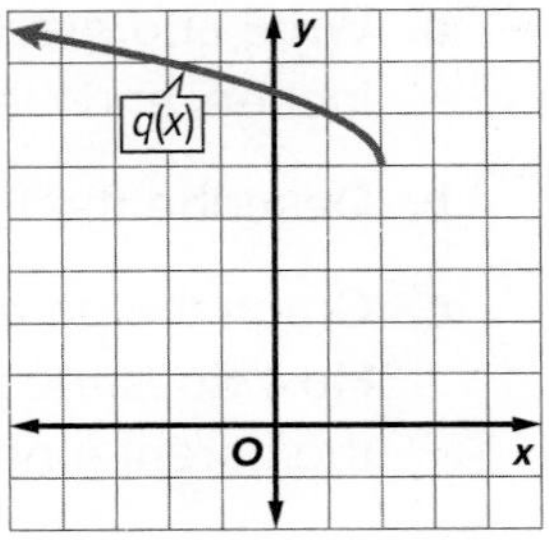

Example 7

Write a radical function for each graph.

27.

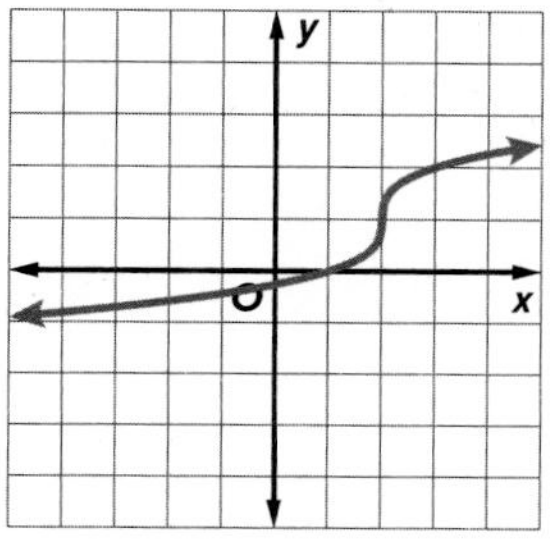

28.

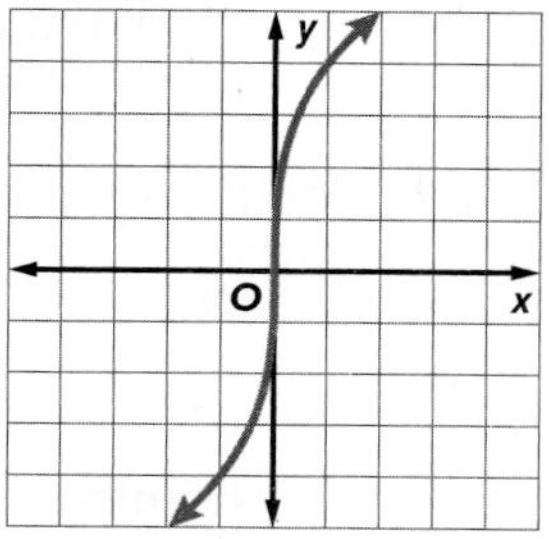

Mixed Exercises

Graph each function and state the domain and range. Then describe how it is related to the graph of the parent function.

29. $f(x) = 2\sqrt{x-5} - 6$

30. $f(x) = \frac{3}{4}\sqrt{x+12} + 3$

31. $f(x) = -\frac{1}{5}\sqrt{x-1} - 4$

32. $f(x) = -3\sqrt{x+7} + 9$

33. $f(x) = -\frac{1}{3}\sqrt[3]{x+2} - 3$

34. $f(x) = -\frac{1}{2}\sqrt[3]{2x-1} + 3$

Graph each inequality.

35. $y \le 6 - 3\sqrt{x-4}$

36. $y < \sqrt{4x-12} + 8$

Write a radical function for each graph.

37.

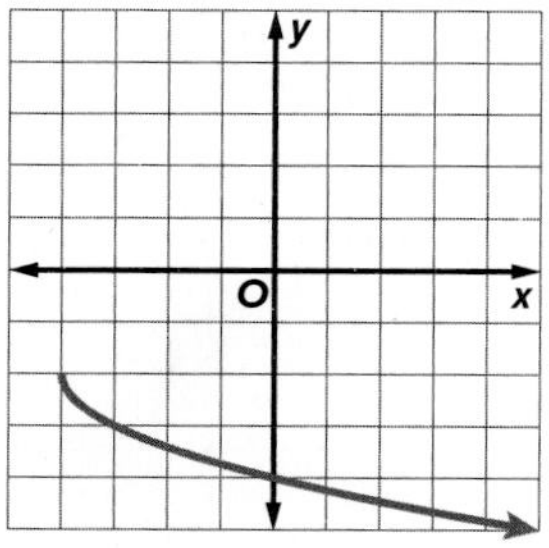

38.

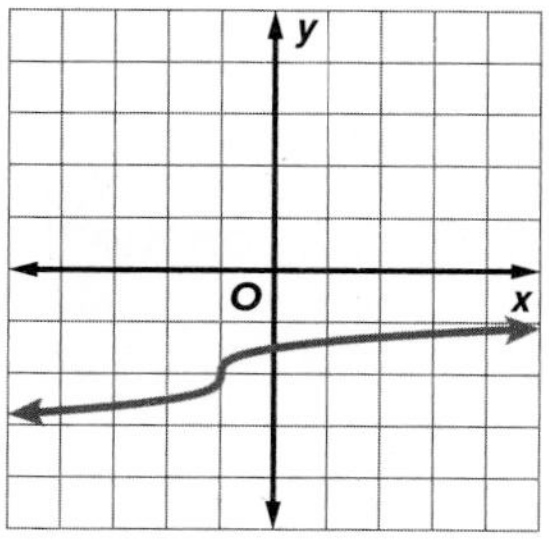

39. STRUCTURE Consider the function $f(x) = -\sqrt{x+3} + \frac{13}{2}$ and the function $g(x)$ shown in the graph.

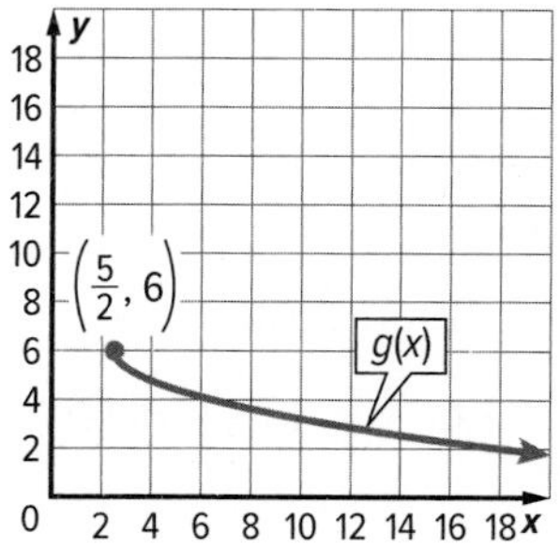

a. Determine which function has the greater maximum value. Explain your reasoning.

b. Compare the domains of the two functions.

c. Compare the average rates of change of the two functions over the interval [6, 13].

40. **DISTANCE** LaRez is standing at the side of a road watching a cyclist. The distance, in feet, between LaRez and the cyclist as a function of time is given by the square root of the sum of 9 and the product of 36 and the time squared.

a. Write and graph a function, where t is the time, in seconds, and d is the distance, in feet.

b. Describe the key features of the function.

41. **GEOMETRY** The length of the radius of a hemisphere can be found using the formula $r = \sqrt[3]{\frac{3V}{2\pi}}$, given the volume V of the hemisphere. Graph the function. State the key features of the graph.

42. **STRUCTURE** Graph $f(x) > \sqrt{-x + 2} - 3$ and its inverse on the same coordinate plane as well as $y = x$. Write the inequality that defines the graph of the inverse. Determine any restrictions that must be placed on the domain of the inverse.

Higher-Order Thinking Skills

43. **PERSEVERE** Write an equation for a square root function with a domain of $\{x \mid x \geq -4\}$, a range of $\{y \mid y \leq 6\}$, and that passes through (5, 3).

44. **ANALYZE** For what positive values of a are the domain and range of $f(x) = \sqrt[a]{x}$ the set of real numbers? Justify your argument.

45. **WRITE** Explain why there are limitations on the domain and range of square root functions.

46. **WRITE** Explain why $y = \pm\sqrt{x}$ is not a function.

47. **CREATE** Write an equation of a relation that contains a radical in its inverse such that:

a. the original relation is a function, and its inverse is not a function.

b. the original relation is not a function, and its inverse is a function.

Operations with Radical Expressions

Learn Properties of Radicals

The properties used to simplify radical expressions involving square roots can be extended to radical expressions involving *n*th roots.

Key Concept • Product Property of Radicals

Words: For any real numbers *a* and *b* and any integer $n > 1$,

$$\sqrt[n]{ab} = \sqrt[n]{a} \cdot \sqrt[n]{b}, \text{ if } n \text{ is even and } a, b \geq 0, \text{ or if } n \text{ is odd.}$$

Examples: $\sqrt{12} \cdot \sqrt{3} = \sqrt{36}$ or 6 and $\sqrt[3]{4} \cdot \sqrt[3]{16} = \sqrt[3]{64}$ or 4

Key Concept • Quotient Property of Radicals

Words: For any real numbers *a* and $b \neq 0$ and any integer $n > 1$,

$$\sqrt[n]{\frac{a}{b}} = \frac{\sqrt[n]{a}}{\sqrt[n]{b}}, \text{ if all roots are defined.}$$

Examples: $\sqrt{\frac{x^4}{25}} = \frac{\sqrt{x^4}}{\sqrt{25}} = \frac{x^2}{5}$ or $\frac{1}{5}x^2$ and $\sqrt[3]{\frac{27}{8}} = \frac{\sqrt[3]{27}}{\sqrt[3]{8}} = \frac{3}{2}$

The properties of radicals and process of rationalizing the denominator can be used to write radical expressions in simplest form.

Key Concept • Simplest Form of Radical Expressions

A radical expression is in simplest form when the following conditions are met.

- The index *n* is as small as possible.
- The radicand contains no factors (other than 1) that are *n*th powers of an integer or polynomial.
- The radicand contains no fractions.
- No radicals appear in the denominator.

Today's Goals

- Simplify radical expressions.
- Add, subtract, and multiply radicals.
- Divide and simplify radical expressions by rationalizing the denominator.

Today's Vocabulary

like radical expressions
conjugates

Example 1 Simplify Expressions with the Product Property

Simplify each expression

a. $\sqrt[3]{-27a^6b^{14}}$

$= \sqrt[3]{(-3)^3 \cdot (a^2)^3 \cdot (b^4)^3 \cdot b^2}$ — Factor into cubes.

$= \sqrt[3]{(-3)^3} \cdot \sqrt[3]{(a^2)^3} \cdot \sqrt[3]{(b^4)^3} \cdot \sqrt[3]{b^2}$ — Product Property of Radicals

$= -3a^2b^4\sqrt[3]{b^2}$ — Simplify.

b. $\sqrt{75x^{12}y^7}$

$= \sqrt{5^2 \cdot 3 \cdot (x^6)^2 \cdot (y^3)^2 \cdot y}$ — Factor into squares.

$= \sqrt{5^2} \cdot \sqrt{3} \cdot \sqrt{(x^6)^2} \cdot \sqrt{(y^3)^2} \cdot \sqrt{y}$ — Product Property of Radicals

$= 5x^6y^3\sqrt{3y}$ — Simplifty.

Think About It!

Why are absolute value symbols not necessary around y^3 in part **b** even though it is an odd power and the result of finding the even root of an even power?

Check

Simplify each expression.

a. $\sqrt[4]{4x^5y^{20}}$

b. $\sqrt{60a^3b^{22}}$

Example 2 Simplify Expressions with the Quotient Property

Simplify each expression.

a. $\sqrt[3]{\dfrac{24a^6}{125}}$

$$\sqrt[3]{\frac{24a^6}{125}} = \frac{\sqrt[3]{24a^6}}{\sqrt[3]{125}}$$ Quotient Property of Radicals

$$= \frac{\sqrt[3]{2^3 \cdot 3 \cdot (a^2)^3}}{\sqrt[3]{5^3}}$$ Factor into cubes.

$$= \frac{\sqrt[3]{2^3} \cdot \sqrt[3]{3} \cdot \sqrt[3]{(a^2)^3}}{\sqrt[3]{5^3}}$$ Product Property of Radicals

$$= \frac{2a^2\sqrt[3]{3}}{5}$$ Simplify.

b. $\sqrt[4]{\dfrac{80y^{14}}{256z^4}}$

$$\sqrt[4]{\frac{80y^{14}}{256z^4}} = \frac{\sqrt[4]{80y^{14}}}{\sqrt[4]{256z^4}}$$ Quotient Property of Radicals

$$= \frac{\sqrt[4]{2^4 \cdot 5 \cdot (y^3)^4 \cdot y^2}}{\sqrt[4]{4^4 \cdot z^4}}$$ Factor into fourth powers.

$$= \frac{\sqrt[4]{2^4} \cdot \sqrt[4]{5} \cdot \sqrt[4]{(y^3)^4} \cdot \sqrt[4]{y^2}}{\sqrt[4]{4^4} \cdot \sqrt[4]{z^4}}$$ Product Property of Radicals

$$= \frac{2|y^3|\sqrt[4]{5y^2}}{4z}$$ Simplify radicals.

$$= \frac{|y^3|\sqrt[4]{5y^2}}{2z}$$ Simplify.

Check

Write the simplified form of $\sqrt{\dfrac{9x^8y^{13}}{25x^2}}$.

Learn Adding and Subtracting Radical Expressions

Radicals can be added and subtracted in the same manner as monomials. In order to add or subtract, the radicals must be like terms. Radicals are **like radical expressions** if *both* the index and the radicand are the same.

Go Online You can complete an Extra Example online.

Think About It!

Jon says that you cannot combine $\sqrt[3]{40}$ and $6\sqrt[3]{5}$ because they do not have the same radicand. Is he correct? Why or why not?

Although two radicals, such as $\sqrt{18}$ and $\sqrt{32}$, may not appear to be like radicals, if you simplify each radical, you can see that $\sqrt{18} = 3\sqrt{2}$ and $\sqrt{32} = 4\sqrt{2}$. These simplified expressions are like radical expressions and can be combined.

Radicals with the same index can be multiplied by using the Product Property of Radicals. If the radicals have coefficients before the radical symbol, multiply the coefficients. Then, multiply the radicands of each expression. To multiply radical expressions with more than one term, you can use the Distributive Property or FOIL method.

$$\left(\sqrt[3]{6} + 1\right)\left(\sqrt[3]{2} - \sqrt[3]{7}\right) = \sqrt[3]{12} - \sqrt[3]{42} + \sqrt[3]{2} - \sqrt[3]{7}$$

Think About It!

Complete the statement to write a general method for multiplying radicals with coefficients.

For any real numbers *a*, *b*, *c*, and *d* and any integer $n > 1$, $c\sqrt{a} \cdot d\sqrt{b} =$ ___?___.

Example 3 Add and Subtract Radicals

Simplify $6\sqrt{45x} + \sqrt{12} - 3\sqrt{20x}$.

$= 6\sqrt{3^2 \cdot 5x} + \sqrt{2^2 \cdot 3} - 3\sqrt{2^2 \cdot 5x}$ — Factor using squares.

$= 6\left(\sqrt{3^2} \cdot \sqrt{5x}\right) + \left(\sqrt{2^2} \cdot \sqrt{3}\right) - 3\left(\sqrt{2^2} \cdot \sqrt{5x}\right)$ — Product Property

$= 6(3\sqrt{5x}) + (2\sqrt{3}) - 3(2\sqrt{5x})$ — Simplify radicals.

$= 18\sqrt{5x} + 2\sqrt{3} - 6\sqrt{5x}$ — Multiply.

$= 12\sqrt{5x} + 2\sqrt{3}$ — Simplify.

Check

Write the simplified form of $\sqrt{18x} - 5\sqrt{28} - 3\sqrt{98x} + 3\sqrt{7x}$.

Think About It!

In Example 4, $\sqrt[5]{-1}$ simplifies to -1 because 5 is an odd root. What would happen if the example used an even root?

Example 4 Multiply Radicals

Simplify $4\sqrt[5]{-10x^2y^6} \cdot 3\sqrt[5]{16x^4y^4}$.

$= 4 \cdot 3 \cdot \sqrt[5]{-10x^2y^6 \cdot 16x^4y^4}$ — Product Property of Radicals

$= 12 \cdot \sqrt[5]{-1 \cdot 2 \cdot 5 \cdot x^2y^6 \cdot 2^4 \cdot x^4y^4}$ — Factor the constants.

$= 12 \cdot \sqrt[5]{-1 \cdot 2^5 \cdot 5 \cdot x^5 \cdot x \cdot y^{10}}$ — Group into powers of 5.

$= 12 \cdot \sqrt[5]{-1} \cdot \sqrt[5]{2^5} \cdot \sqrt[5]{5} \cdot \sqrt[5]{x^5} \cdot \sqrt[5]{x} \cdot \sqrt[5]{y^{10}}$ — Product Power of Radicals

$= 12 \cdot (-1) \cdot 2 \cdot x \cdot y^2 \cdot \sqrt[5]{5} \cdot \sqrt[5]{x}$ — Simplify.

$= -24xy^2\sqrt[5]{5x}$ — Multiply.

Check

Write the simplified form of $5\sqrt[5]{-9x^3y^5} \cdot 3\sqrt[5]{27x^4y^5}$.

Study Tip

Negative Radicands If a radicand has a negative constant, it may be helpful to use -1 as a factor. Then you can simplify the *n*th root of -1, which will be -1 if *n* is odd and *i* if *n* is even.

Go Online You can complete an Extra Example online.

Example 5 Use the Distributive Property to Multiply Radicals

SPORTS **A sports pennant has the dimensions shown. Find the area, in square inches.**

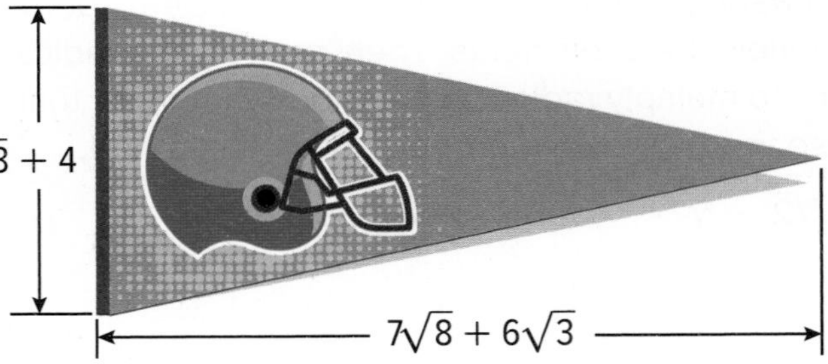

Area $= \frac{1}{2} \cdot$ base $\cdot$ height, so the area is $\frac{1}{2}(3\sqrt{8} + 4)(7\sqrt{8} + 6\sqrt{3})$.

$= \frac{1}{2} \cdot [3\sqrt{8} \cdot 7\sqrt{8} + 3\sqrt{8} \cdot 6\sqrt{3} + 4 \cdot 7\sqrt{8} + 4 \cdot 6\sqrt{3}]$

$= \frac{1}{2} \cdot [21\sqrt{8^2} + 18\sqrt{24} + 28\sqrt{8} + 24\sqrt{3}]$

$= \frac{1}{2} \cdot [21\sqrt{8^2} + 18\sqrt{2^2 \cdot 6} + 28\sqrt{2^2 \cdot 2} + 24\sqrt{3}]$

$= \frac{1}{2} \cdot [21\sqrt{8^2} + 18 \cdot \sqrt{2^2} \cdot \sqrt{6} + 28 \cdot \sqrt{2^2} \cdot \sqrt{2} + 24\sqrt{3}]$

$= \frac{1}{2} \cdot [168 + 36\sqrt{6} + 56\sqrt{2} + 24\sqrt{3}]$

$= 84 + 18\sqrt{6} + 28\sqrt{2} + 12\sqrt{3}$

The area of the pennant is $84 + 18\sqrt{6} + 28\sqrt{2} + 12\sqrt{3}$ in^2, or about 188.5 in^2.

Watch Out!

Simplest Form Do not forget to check that your result is in simplest form. Make sure that none of the individual radicals can be further simplified or combined.

Check

POOLS A rectangular pool safety cover has a length of $7\sqrt{10} - 4$ feet and a width of $6\sqrt{10} + 8\sqrt{5}$ feet. Which expression represents the area of the pool cover in simplest form?

A. $420 + 280\sqrt{2} + 24\sqrt{10} + 32\sqrt{5}$ ft^2

B. $42\sqrt{100} + 280\sqrt{2} - 24\sqrt{10} - 32\sqrt{5}$ ft^2

C. $420 + 280\sqrt{2} - 24\sqrt{10} - 32\sqrt{5}$ ft^2

D. $420 + 56\sqrt{50} - 24\sqrt{10} - 32\sqrt{5}$ ft^2

Learn Rationalizing the Denominator

If a radical expression contains a radical in the denominator, you can rationalize the denominator to simplify the expression. Recall that to rationalize a denominator, you should multiply the numerator and denominator by a quantity so that the radicand has an exact root.

If the denominator is:	Multiply the numerator and denominator by:	Examples
$\sqrt{b}$	$\sqrt{b}$	$\frac{4}{\sqrt{7}} = \frac{4}{\sqrt{7}} \cdot \frac{\sqrt{7}}{\sqrt{7}}$ or $\frac{4\sqrt{7}}{7}$
$\sqrt[n]{b^x}$	$\sqrt[n]{b^{n-x}}$	$\frac{3}{\sqrt[5]{2}} = \frac{3}{\sqrt[5]{2}} \cdot \frac{\sqrt[5]{2^4}}{\sqrt[5]{2^4}}$ or $\frac{3\sqrt[5]{16}}{2}$

Binomials of the form $a\sqrt{b} + c\sqrt{d}$ and $a\sqrt{b} - c\sqrt{d}$, where a, b, c, and d are rational numbers, are called **conjugates** of each other. Multiplying the numerator and denominator by the conjugate of the denominator will eliminate the radical from the denominator of the expression.

Example 6 Rationalize the Denominator

Simplify $\sqrt[3]{\frac{250a^6}{7a}}$.

$$\sqrt[3]{\frac{250a^6}{7a}} = \frac{\sqrt[3]{250a^6}}{\sqrt[3]{7a}}$$ Quotient Property of Radicals

$$= \frac{\sqrt[3]{5^3 \cdot 2 \cdot (a^2)^3}}{\sqrt[3]{7a}}$$ Factor into cubes.

$$= \frac{\sqrt[3]{5^3} \cdot \sqrt[3]{2} \cdot \sqrt[3]{(a^2)^3}}{\sqrt[3]{7a}}$$ Product Property of Radicals

$$= \frac{5a^2\sqrt[3]{2}}{\sqrt[3]{7a}}$$ Simplify.

$$= \frac{5a^2\sqrt[3]{2}}{\sqrt[3]{7a}} \cdot \frac{\sqrt[3]{7^2a^2}}{\sqrt[3]{7^2a^2}}$$ Rationalize the denominator.

$$= \frac{5a^2\sqrt[3]{2 \cdot 7^2a^2}}{\sqrt[3]{7a \cdot 7^2a^2}}$$ Product Property of Radicals

$$= \frac{5a^2\sqrt[3]{98a^2}}{\sqrt[3]{7^3a^3}}$$ Multiply.

$$= \frac{5a^2\sqrt[3]{98a^2}}{7a}$$ $\sqrt[3]{7^3a^3} = 7a$

$$= \frac{5a\sqrt[3]{98a^2}}{7}$$ Simplify.

Check

Write the simplified form of $\sqrt{\frac{20b}{3b^5}}$.

Go Online You can complete an Extra Example online.

Think About It!
Why is the product of $\sqrt[n]{b^x}$ and $\sqrt[n]{b^{n-x}}$ an exact root?

Think About It!
How does multiplying conjugates relate to the difference of squares identity that can be used when multiplying binomials?

Watch Out!
Rationalizing the Denominator When determining the quantity to multiply by when rationalizing the denominator, make sure you raise the entire term under the radical to the power of $n - x$.

Example 7 Use Conjugates to Rationalize the Denominator

Simplify $\frac{4x}{2\sqrt{7} - 5}$.

To rationalize the denominator, multiply the numerator and denominator by the conjugate of $2\sqrt{7} - 5$.

$$\frac{4x}{2\sqrt{7} - 5} = \frac{4x}{2\sqrt{7} - 5} \cdot \frac{2\sqrt{7} + 5}{2\sqrt{7} + 5}$$

$2\sqrt{7} + 5$ is the conjugate of $2\sqrt{7} - 5$.

$$= \frac{4x(2\sqrt{7}) + 4x(5)}{(2\sqrt{7})^2 + 5(2\sqrt{7}) - 5(2\sqrt{7}) - 5(5)}$$

Multiply.

$$= \frac{8x\sqrt{7} + 20x}{28 + 10\sqrt{7} - 10\sqrt{7} - 25}$$

Simplify.

$$= \frac{8x\sqrt{7} + 20x}{3}$$

Subtract.

Check

Write the simplified form of $\frac{6x}{4\sqrt{5} - 4}$.

Go Online You can complete an Extra Example online.

Practice

Go Online You can complete your homework online.

Examples 1 and 2

Simplify.

1. $\sqrt{72a^8b^5}$

2. $\sqrt{9a^{15}b^3}$

3. $\sqrt{24a^{16}b^8c}$

4. $\sqrt{18a^6b^3c^5}$

5. $\sqrt[4]{64a^4b^4}$

6. $\sqrt[3]{-8d^2f^5}$

7. $\sqrt{\frac{25}{36}r^2t}$

8. $\sqrt{\frac{192k^4}{64}}$

9. $\sqrt[5]{\frac{3072h^8}{243f^5}}$

10. $\sqrt[3]{\frac{432n^{12}}{64q^6}}$

Example 3

Simplify.

11. $\sqrt{2} + \sqrt{8} + \sqrt{50}$

12. $\sqrt{12} - 2\sqrt{3} + \sqrt{108}$

13. $8\sqrt{5} - \sqrt{45} - \sqrt{80}$

14. $2\sqrt{48} - \sqrt{75} - \sqrt{12}$

15. $\sqrt{28x} - \sqrt{14} + \sqrt{63x}$

16. $\sqrt{135} + 5\sqrt{10d} - 3\sqrt{60}$

Example 4

Simplify.

17. $3\sqrt{5y} \cdot 8\sqrt{10yz}$

18. $2\sqrt{32a^3b^5} \cdot \sqrt{8a^7b^2}$

19. $6\sqrt{3ab} \cdot 4\sqrt{24ab^3}$

20. $5\sqrt{x^8y^3} \cdot 5\sqrt{2x^5y^4}$

21. $5\sqrt{2x} \cdot 3\sqrt{7x^2y^3}$

22. $3\sqrt{a^5b^7} \cdot 2\sqrt{5a^7b^3}$

Example 5

23. **TRAMPOLINE** There are two trampoline runways at a gymnastics practice facility. Both runways are $\sqrt{3}$ meters wide. One is $6\sqrt{3}$ meters long and the other is $5\sqrt{2}$ meters long. What is the total area of the trampoline runways?

24. **DISTANCE** Jayla walks 5 blocks north, then 8 blocks east to get to the library. Each block is $5\sqrt{10}$ yards long. If Jayla could walk in a straight line to the library instead, how far would the walk be, in yards?

Simplify.

25. $(7\sqrt{2} - 3\sqrt{3})(4\sqrt{6} + 3\sqrt{12})$

26. $(8\sqrt{5} - 6\sqrt{3})(8\sqrt{5} + 6\sqrt{3})$

27. $(12\sqrt{10} - 6\sqrt{5})(12\sqrt{10} + 6\sqrt{5})$

28. $(6\sqrt{3} + 5\sqrt{2})(2\sqrt{6} + 3\sqrt{8})$

Examples 6 and 7

Simplify.

29. $\frac{\sqrt{5a^5}}{\sqrt{b^{13}}}$

30. $\frac{\sqrt{7x}}{\sqrt{10x^3}}$

31. $\frac{3\sqrt[3]{6x^2}}{3\sqrt[3]{5y}}$

32. $\sqrt[4]{\frac{7x^3}{4b^2}}$

33. $\frac{6}{\sqrt{3} - \sqrt{2}}$

34. $\frac{\sqrt{2}}{\sqrt{5} - \sqrt{3}}$

35. $\frac{9 - 2\sqrt{3}}{\sqrt{3} + 6}$

36. $\frac{2\sqrt{2} + 2\sqrt{5}}{\sqrt{5} + \sqrt{2}}$

37. $\frac{3\sqrt{7}}{\sqrt{5} - 1}$

38. $\frac{7x}{3 - \sqrt{2}}$

Mixed Exercises

Simplify.

39. $\sqrt[3]{16y^4z^{12}}$

40. $\sqrt[3]{-54x^6y^{11}}$

41. $\frac{x + 1}{\sqrt{x} - 1}$

42. $\frac{x - 2}{\sqrt{x^2 - 4}}$

43. $3\sqrt{24x} - 2\sqrt{54x} + \sqrt{48}$

44. $5\sqrt{18c} + 3\sqrt{72c} + 6\sqrt{76}$

45. $10\sqrt{175a} - 4\sqrt{112a} - 2\sqrt{63a}$

46. $7\sqrt{204y} + 4\sqrt{459y} - 8\sqrt{140y}$

47. VOLUME McKenzie has a rectangular prism with dimensions 20 inches by 35 inches by 40 inches. She would like to replace it with a cube with the same volume. What should the length of a side of the cube be? Express your answer as a radical expression in simplest form.

48. MUSIC Traditionally, musical instruments are tuned so that the note A has a frequency of 440 Hertz. With each note higher on the instrument, the frequency of the pitch is multiplied by a factor of $\sqrt[12]{2}$. What is the ratio of the frequencies of two notes that are 6 steps apart on the instrument? What is the ratio of the frequencies of two notes that are 9 steps apart on the instrument? Express your answers in simplest form.

49. **PHYSICS** The speed of a wave traveling over a string is given by $\frac{\sqrt{t}}{\sqrt{u}}$, where t is the tension of the string and u is the density. Simplify the expression.

50. **LIGHT** Suppose a light has a brightness intensity of I_1 when it is at a distance of d_1 and a brightness intensity of I_2 when it is at a distance of d_2. These quantities are related by the equation $\frac{d_2}{d_1} = \sqrt{\frac{I_1}{I_2}}$. If $I_1 = 50$ units and $I_2 = 24$ units, find $\frac{d_2}{d_1}$. Express your answer in simplest form.

51. **RACING** Jay likes to race his younger brother while running. To make the race fair, John and Jay start at different locations, but finish at the same point. The diagram shows their running paths. Both of them finished the race in exactly 4 minutes.

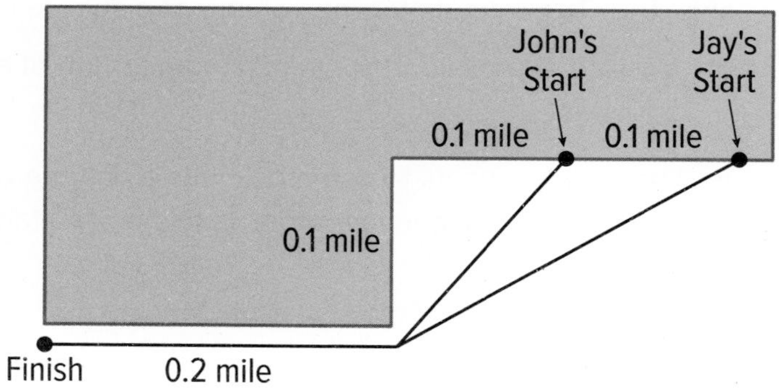

a. If John and Jay continued at their average paces during the race, exactly how many minutes would it take them each to run a mile? Express your answer as a radical expression in simplest form.

b. Exactly how many times faster is Jay compared to John? Express your answer as a radical expression in simplest form.

52. **STRUCTURE** Write the ratio of the side lengths of the two cubes described in simplest form.

a. The volumes of the two cubes are $270x$ cubic inches and $32x^2$ cubic inches.

b. The surface areas of the two cubes are $6x^4$ square feet and $6(x + 1)$ square feet.

53. **REGULARITY** Rewrite each of the following expressions as a single expression in the form ax^m for appropriate choices of a and m. Show your work.

a. $\sqrt{x}(\sqrt{x} + \sqrt{4x})$

b. $\sqrt{x}(\sqrt{x} + \sqrt{4x} + \sqrt{9x})$

c. $\sqrt{x}(\sqrt{x} + \sqrt{4x} + \sqrt{9x} + \sqrt{16x})$

d. More generally, simplify $\sqrt{x}(\sqrt{x} + \sqrt{4x} + \cdots + \sqrt{n^2x})$ for any positive integer n. Use the fact that $1 + 2 + \cdots + n = \frac{n(n + 1)}{2}$.

54. **USE A MODEL** If the area of the trapezoid shown is 200 square feet, what is the height h of the trapezoid?

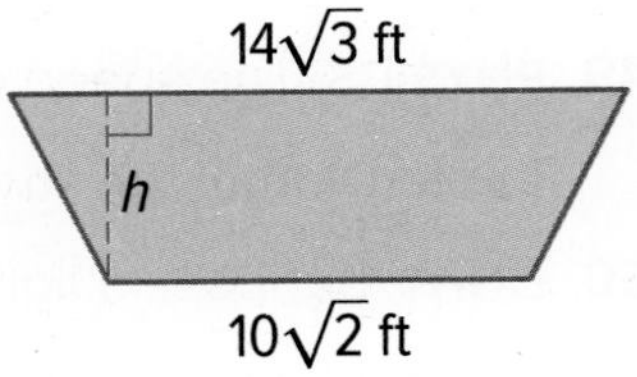

55. **CONSTRUCT ARGUMENTS** A spherical paperweight with a volume of 72π cubic centimeters is to be packaged in a gift box that is a cube. There must be at least 2 centimeters of packing material around the paperweight to protect it during shipping. The formula for the volume of a sphere is $V = \frac{4}{3}\pi r^3$.

a. Write an expression for the minimum length of a side of the gift box. Show your work.

b. The shipper wants to use a box with a volume of 384 cubic centimeters that they already have in inventory. Is this box suitable? Justify your argument.

56. **FIND THE ERROR** Twyla and Brandon are simplifying $4\sqrt{32} + 6\sqrt{18}$. Is either of them correct? Explain your reasoning.

Twyla	Brandon
$4\sqrt{32} + 6\sqrt{18}$	$4\sqrt{32} + 6\sqrt{18}$
$= 4\sqrt{4^2 \cdot 2} + 6\sqrt{3^2 \cdot 2}$	$= 4\sqrt{16 \cdot 2} + 6\sqrt{9 \cdot 2}$
$= 16\sqrt{2} + 18\sqrt{2}$	$= 64\sqrt{2} + 54\sqrt{2}$
$= 34\sqrt{2}$	$= 118\sqrt{2}$

57. **PERSEVERE** Find four combinations of whole numbers that satisfy $\sqrt[a]{256} = b$.

58. **PERSEVERE** Show that $\frac{-1 - i\sqrt{3}}{2}$ is a cube root of 1.

59. **WRITE** Explain why absolute values may be unnecessary when an nth root of an even power results in an odd power.

60. **WHICH ONE DOESN'T BELONG?** Determine which of the radical expressions doesn't belong. Justify your conclusion.

$\sqrt[8]{256g^4h^{16}}$ $\sqrt[4]{16g^8h^2}$ $\sqrt[6]{64g^{12}h^3}$

Solving Radical Equations

Explore Solutions of Radical Equations

Online Activity Use graphing technology to complete the Explore.

INQUIRY When will a radical equation have a solution? When will it have no solution?

Learn Solving Radical Equations Algebraically

A **radical equation** has a variable in a radicand. When solving a radical equation, the result may be an extraneous solution.

Key Concept • Solving Radical Equations

Step 1 Isolate the radical on one side of the equation.

Step 2 To eliminate the radical, raise each side of the equation to a power equal to the index of the radical.

Step 3 Solve the resulting polynomial equation. Check your results.

Example 1 Solve a Square Root Equation

Solve $\sqrt{3x-5}+2=6$.

$\sqrt{3x-5}+2=6$	Original equation
$\sqrt{3x-5}=4$	Subtract 2 from each side.
$3x-5=16$	Square each side to eliminate the radical.
$x=7$	Simplify.

Check that the result satisfies the original equation.

Example 2 Solve a Cube Root Equation

Solve $4(2x+6)^{\frac{1}{3}}-9=3$.

To remove the $\frac{1}{3}$ power, or cube root, you must first isolate it and then raise each side of the equation to the third power.

$4(2x+6)^{\frac{1}{3}}-9=3$	Original equation
$4(2x+6)^{\frac{1}{3}}=12$	Add 9 to each side.
$(2x+6)^{\frac{1}{3}}=3$	Divide each side by 4.
$2x+6=27$	Cube each side.
$2x=21$	Subtract 6 from each side.
$x=\frac{21}{2}$	Divide each side by 2.

(continued on the next page)

Today's Goals

- Solve radical equations in one variable and identify extraneous solutions.
- Solve radical equations by graphing systems of equations.

Today's Vocabulary

radical equation

Think About It!

How could you change the equation so that there is no solution?

Go Online

You can learn how to solve radical equations by watching the video online.

Check

$4(2x + 6)^{\frac{1}{3}} - 9 = 3$	Original equation
$4\left(2 \cdot \underline{\ ?\ } + 6\right)^{\frac{1}{3}} - 9 \stackrel{?}{=} 3$	Replace x with $\frac{21}{2}$.
$4(\underline{\ ?\ })^{\frac{1}{3}} - 9 \stackrel{?}{=} 3$	Simplify.
$4(\underline{\ ?\ }) - 9 \stackrel{?}{=} 3$	The cube root of 27 is 3.
$\underline{\ ?\ } = 3$	True

Example 3 Identify Extraneous Solutions

Talk About It!
In Example 3, could you tell that 4 was an extraneous solution before checking the result? Explain your reasoning.

Solve $\sqrt{x + 21} = 3 - \sqrt{x}$.

$\sqrt{x + 21} = 3 - \sqrt{x}$	Original equation
$x + 21 = 9 - 6\sqrt{x} + x$	Square each side.
$12 = -6\sqrt{x}$	Isolate the radical.
$-2 = \sqrt{x}$	Divide each side by −6.
$4 = x$	Square each side.

Check

$\sqrt{x + 21} = 3 - \sqrt{x}$	Original equation
$\sqrt{\underline{?} + 21} \stackrel{?}{=} 3 - \sqrt{\underline{?}}$	Replace x with 4.
$\sqrt{25} \stackrel{?}{=} 3 - 2$	Simplify.
$\underline{\ ?\ } \neq \underline{\ ?\ }$	False

The result does not satisfy the original equation, so it is an ____?____ solution. Therefore, there is ____?____

Example 4 Solve a Radical Equation

Solve $\frac{2}{3}(11x + 14)^{\frac{1}{6}} + 8 = 10$.

$\frac{2}{3}(11x + 14)^{\frac{1}{6}} + 8 = 10$	Original equation
$\frac{2}{3}(11x + 14)^{\frac{1}{6}} = 2$	Subtract 8 from each side.
$(11x + 14)^{\frac{1}{6}} = 3$	Multiply each side by $\frac{3}{2}$.
$11x + 14 = 729$	Raise each side to the sixth power.
$11x = 715$	Subtract 14 from each side.
$x = 65$	Divide each side by 11.

The value of 65 does make the equation true.

Go Online You can complete an Extra Example online.

Learn Solving Radical Equations by Graphing

To solve a radical equation using the graph of a related function, rewrite the equation with 0 on one side and then replace 0 with $f(x)$.

Equation: $\sqrt{2x+5}+1=4$

Related Function: $f(x)=\sqrt{2x+5}-3$ or $y=\sqrt{2x+5}-3$

The values of x for which $f(x)=0$ are the zeros of the function and occur at the x-intercepts of its graph. The solutions or roots of an equation are the zeros or x-intercepts of its related function.

You can also solve a radical equation by writing and solving a system of equations based on the equation. Set the expressions on each side of the equation equal to y to create the system of equations.

Equation: $\sqrt{2x+5}+1=4$

System of Equations: $y=\sqrt{2x+5}+1$ $\qquad y=4$

The x-coordinate of the intersection of the system of equations is the value of x where the two equations are equal. Thus, the x-coordinate of the point of intersection is the solution of the radical equation.

Think About It!

What would the graph of the related function of a radical equation with no solution look like?

Example 5 Solve a Radical Equation by Graphing

Use a graphing calculator to solve $2\sqrt[3]{3x-4}+10=9$ by graphing.

Step 1 Find a related function. Rewrite equation with 0 on right side.

$2\sqrt[3]{3x-4}+10=9$	Original equation
$2\sqrt[3]{3x-4}+1=0$	Subtract 9 from each side.

Replacing 0 with $f(x)$ gives the related function $f(x)=2\sqrt[3]{3x-4}+1$.

Step 2 Graph the related function.

Use the **Y=** list to graph.

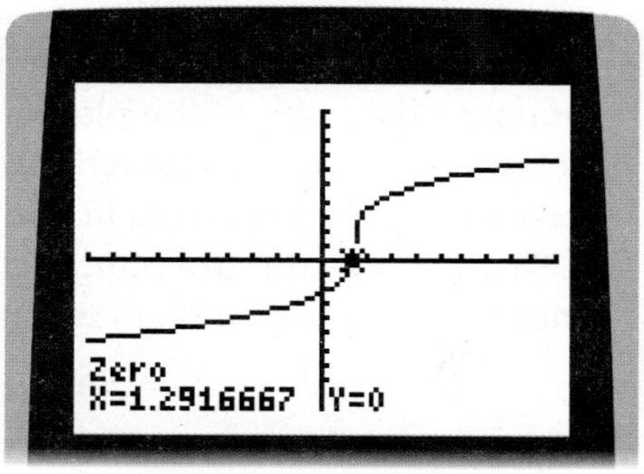

[−10, 10] scl: 1 by [−10, 10] scl: 1

Step 3 Use a table.

You can use the **TABLE** feature to find the interval where the zero lies.

The function changes sign between $x=1$ and $x=2$ which indicates that there is a zero between 1 and 2.

Step 4 Find the zero.

Use the **zero** feature from the **CALC** menu to find the zero of the function.

The zero is about 1.29. This is between 1 and 2, which is consistent with the interval we found using the table.

Watch Out!

Misleading Graphs Although the TI-84 may show what appears to be a discontinuity in the graphs of radical functions with odd roots, these functions are in fact continuous for all real numbers.

Go Online You can complete an Extra Example online.

Think About It!

How can you use the table feature on your calculator to find the intersection?

Example 6 Solve a Radical Equation by Using a System

Use a graphing calculator to solve $\sqrt{x+6} - 5 = -\sqrt{2x} + 1$ by using a system of equations.

Step 1 **Write a system.** Set each side of $\sqrt{x+6} - 5 = -\sqrt{2x} + 1$ equal to y to create a system of equations.

$y = \sqrt{x+6} - 5$ First equation

$y = -\sqrt{2x} + 1$ Second equation

Step 2 **Graph the system.** Enter the equations in the **Y=** list and graph in the standard viewing window.

Step 3 **Find the intersection.**

Use the **intersect** feature from the **CALC** menu to find the coordinates of the point of intersection.

The solution is the x-coordinate of the intersection, which is about 4.02.

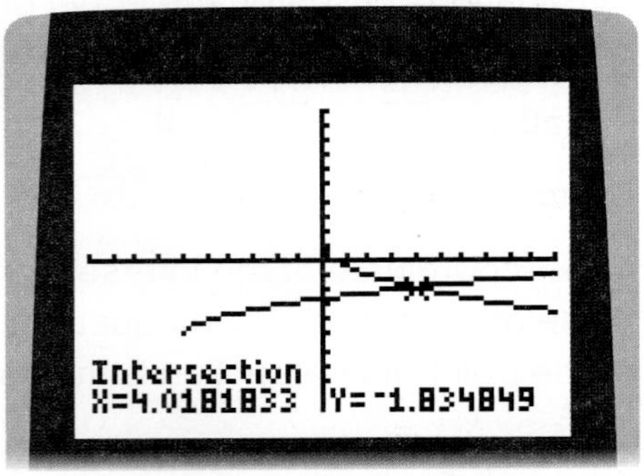

[−10, 10] scl: 1 by [−10, 10] scl: 1

Check

Use a graphing calculator to solve $-4(\sqrt[3]{x-2}) = \sqrt[4]{x-3} - 6$ by using a system of equations. Round to the nearest hundredth if necessary.

$x \approx$ __?__

Example 7 Confirm Solutions by Using Technology

SPACE **The square of the time it takes a planet to orbit the Sun T is equal to the cube of the planet's mean distance from the Sun a. This relationship can also be written as $T = \sqrt{a^3}$, where T is measured in years and a is measured in astronomical units (AU). If it takes Mars 1.88 years to orbit the Sun, use a graphing calculator to find the mean distance from Mars to the Sun.**

$1.88 = \sqrt{a^3}$ $T = 1.88$

$3.5344 = a^3$ Square each side.

$1.5233 \approx a$ Take the cube root of each side.

So, the mean distance from Mars to the Sun is about 1.5233 AU. Use a graphing calculator to confirm this solution by graphing.

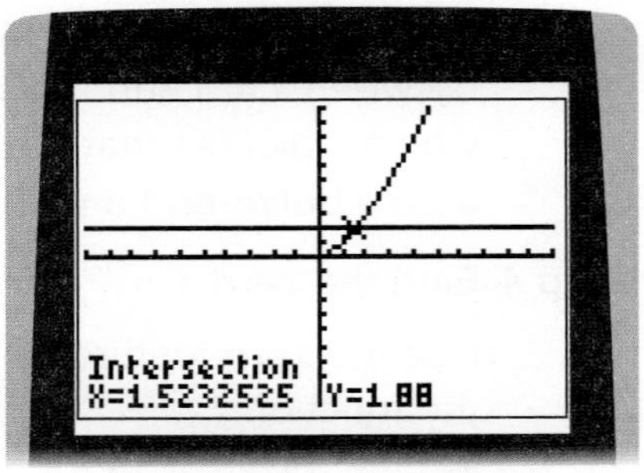

[−10, 10] scl: 1 by [−10, 10] scl: 1

Use a Source

Research the time it takes another planet to orbit the Sun. Write and solve a radical equation to find that planet's mean distance from the Sun.

Go Online You can complete an Extra Example online.

Practice

Go Online You can complete your homework online.

Example 1

Solve each equation.

1. $5\sqrt{j} = 1$

2. $\sqrt{b - 5} = 4$

3. $\sqrt{3n + 1} = 5$

4. $2 + \sqrt{3p + 7} = 6$

5. $\sqrt{k - 4} - 1 = 5$

6. $5 = \sqrt{2g - 7}$

Example 2

Solve each equation.

7. $\sqrt[3]{3r - 6} = 3$

8. $(2d + 3)^{\frac{1}{3}} = 2$

9. $(t - 3)^{\frac{1}{3}} = 2$

10. $4 - (1 - 7u)^{\frac{1}{3}} = 0$

11. $\sqrt[3]{2v - 7} = -2$

12. $4(5n - 1)^{\frac{1}{3}} - 1 = 0$

Examples 3 and 4

Solve each equation. Identify any extraneous solutions.

13. $\sqrt{x - 15} = 3 - \sqrt{x}$

14. $(5q + 1)^{\frac{1}{4}} + 7 = 5$

15. $(3x + 7)^{\frac{1}{4}} - 3 = 1$

16. $(3y - 2)^{\frac{1}{5}} + 5 = 6$

17. $(4z - 1)^{\frac{1}{5}} - 1 = 2$

18. $\sqrt{x - 10} = 1 - \sqrt{x}$

19. $\sqrt[6]{y + 2} + 9 = 14$

20. $(2x - 1)^{\frac{1}{4}} - 2 = 1$

Example 5

Use a graphing calculator to solve each equation by graphing.

21. $\sqrt{x - 7} + 2 = 8$

22. $5 + \sqrt{3m + 9} = 10$

23. $\sqrt[3]{5b - 4} = 2$

24. $\sqrt[3]{2v + 3} = -2$

Example 6

Use a graphing calculator to solve each equation by using a system of equations.

25. $2\left(\sqrt[3]{2k-3}\right) = \left(\sqrt[4]{k+14}\right)$

26. $3\left(\sqrt[3]{d-5}\right) = \left(\sqrt[4]{3d+2}\right)$

27. $\sqrt{2x+7} = 5 - \sqrt{3x}$

28. $\sqrt{n+8} = \sqrt{4n-9}$

Example 7

29. GEOMETRY Heron's Formula states that the area of a triangle whose sides have lengths a, b, and c is $A = \sqrt{s(s-a)(s-b)(s-c)}$ where $s = \frac{1}{2}(a+b+c)$. If the area of the triangle is 270 cm^2, $s = 45$ cm, $a = 15$ cm, and $c = 39$ cm, what is the length of side b?

30. TWINE The largest ball of twine was started in 1953 by Frank Stoeber. In 4 years, the ball had a volume of approximately 268 ft^3. What was the radius of the ball of twine at that time? Round your answer to the nearest tenth.

Mixed Exercises

Solve each equation. Identify any extraneous solutions.

31. $6 + \sqrt{4x+8} = 9$

32. $\sqrt{7a-2} = \sqrt{a+3}$

33. $\sqrt{x-5} - \sqrt{x} = -2$

34. $\sqrt{b-6} + \sqrt{b} = 3$

35. $2(x-10)^{\frac{1}{3}} + 4 = 0$

36. $3(x+5)^{\frac{1}{3}} - 6 = 0$

37. $\frac{1}{7}(14a)^{\frac{1}{3}} = 1$

38. $\frac{1}{4}(32b)^{\frac{1}{3}} = 1$

39. $\sqrt{x-3} = 3 - x$

40. $\sqrt{x-2} = 22 - x$

41. $\sqrt{x+30} = x$

42. $\sqrt{x+22} = x + 2$

43. GEOMETRY The lateral area of a cone with base radius r and height h is given by the formula $L = \pi r\sqrt{r^2 + h^2}$. A cone has a lateral area of 65π square units and a base radius of 5 units. What is the height of the cone?

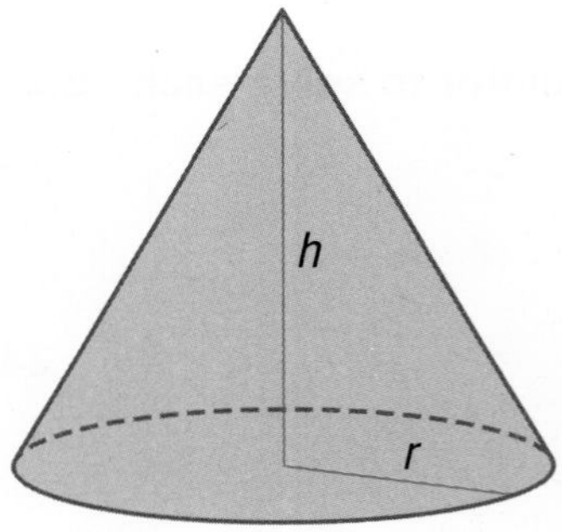

44. **TETHERS** A tether of length y secures a telephone pole at 25 feet off the ground. The distance from the tether to the pole along the ground is represented by x. By the Pythagorean Theorem, the length of the tether is given by $y = \sqrt{x^2 + 25^2}$. If $x + y = 50$ feet, what is the measure of x?

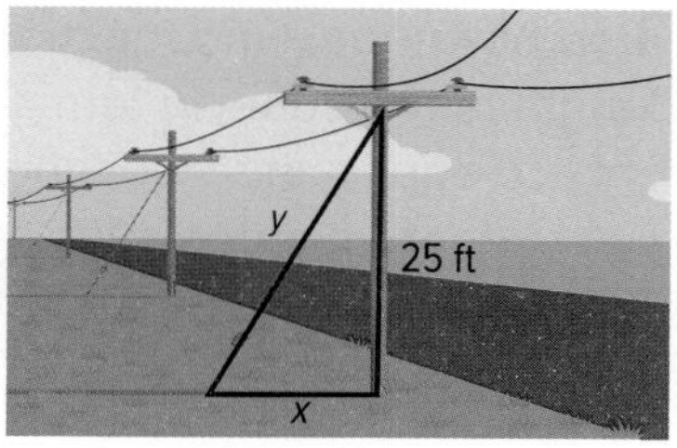

45. **SPACE** NASA's Near-Earth Asteroid Tracking project tracked more than 300 asteroids. An asteroid is passing near Earth. If Earth is located at the origin of a coordinate plane, the path that the asteroid will trace out is given by $y = \frac{17}{x}$, $x > 0$, where unit corresponds to one million miles. One asteroid will be visible by telescope when it is within $\frac{145}{12}$ million miles of Earth.

a. Write an expression that gives the distance of the asteroid from Earth as a function of x.

b. For what values of x will the asteroid be in range of a telescope?

46. **DRIVING** To determine the speed of a car when it begins to skid to a stop, the formula $s = \sqrt{30fd}$ can be used, where s is the speed of the car, f is the coefficient of friction, and d is the length of the skid marks in feet. If the speed limit is 25 mph and the coefficient of friction is 0.6, what is the length of the skid marks if the driver is driving the speed limit?

47. **CHEMISTRY** The nuclear radius of an element can be approximated by $r = (1.2 \times 10^{-15})A^{\frac{1}{3}}$ where r is the length of the radius in meters and A is the molecular mass of the element.

a. The nuclear radius of neon is about 3.267×10^{-15} meter. Find its molecular mass.

b. Which element has a molecular radius of approximately 5.713×10^{-15} meter? Justify your conclusion.

Element	Molecular Mass
copper (Cu)	63.5
gold (Au)	197.0
magnesium (Mg)	24.3
neon (Ne)	?
silver (Ag)	107.9
titanium (Ti)	47.9

48. **USE A MODEL** Explain how to find the solutions to $\sqrt[4]{10x + 11} - \sqrt{x + 2} = 0$ graphically and confirm your results algebraically. What are the solutions?

49. USE TOOLS The surface area of a sphere is 20 cm^2 greater than the surface area of a cube. Find functions to represent the radius of the sphere and the side length of the cube, in terms of the surface area of the cube. If the radius of the sphere equals the side length of cube, describe how to use a graphing calculator to find the surface area of the cube and sphere. Sketch the graph. Find the surface area of the cube and the sphere.

50. CONSTRUCT ARGUMENTS Explain how we know that the equation $\sqrt{x-5}+1=\sqrt{(2-x)}$ has no solutions without having to actually solve it. Confirm this by graphing the two sides of the equation.

Higher-Order Thinking Skills

51. WHICH ONE DOESN'T BELONG? Determine which of the equations doesn't belong. Justify your conclusion.

$\sqrt{x-1}+3=4$	$\sqrt{x+1}+3=4$	$\sqrt{x-2}+7=10$	$\sqrt{x+2}-7=-10$

52. PERSEVERE Haruko is working to solve $(x+5)^{\frac{1}{4}}=-4$. He said that he could tell there was no real solution without even working the problem. Is Haruko correct? Explain your reasoning.

53. ANALYZE Determine whether $\frac{\sqrt{(x^2)^2}}{-x}=x$ is *sometimes*, *always*, or *never* true when x is a real number. Justify your argument.

54. CREATE Write an equation that can be solved by raising each side of the equation to the given power.

a. $\frac{3}{2}$ power **b.** $\frac{5}{4}$ power **c.** $\frac{7}{8}$ power

ANALYZE **Determine whether the following statements are *sometimes*, *always*, or *never* true for $x^{\frac{1}{n}}=a$. Explain your reasoning.**

55. If n is odd, there will be extraneous solutions.

56. If n is even, there will be extraneous solutions.

Review

Essential Question

How can the inverse of a function be used to help interpret a real-world event or solve a problem?

Module Summary

Lesson 6-1

Operations on Functions

- $(f + g)(x) = f(x) + g(x)$
- $(f - g)(x) = f(x) - g(x)$
- $(f \cdot g)(x) = f(x) \cdot g(x)$
- $\left(\frac{f}{g}\right)(x) = \frac{f(x)}{g(x)}, g(x) \neq 0$
- $[f \circ g](x) = f[g(x)]$

Lesson 6-2

Inverse Relations and Functions

- If f and f^{-1} are inverses, then $f(a) = b$ if and only if $f^{-1}(b) = a$.
- $f(x)$ and $g(x)$ are inverses if and only if $[f \circ g](x) = x$ and $[g \circ f](x) = x$.

Lesson 6-3

*n*th Roots and Rational Exponents

- For any real numbers a and b and any positive integer n, if $a^n = b$, then a is an nth root of b.
- When there is more than one real root and n is even, the nonnegative root is called the principal root.
- An expression with rational exponents is in simplest form when it has no negative exponents, it has no exponents that are not positive integers in the denominator, it is not a complex fraction, and the index of any remaining radical is the least number possible.

Lesson 6-4

Graphing Radical Functions

- The parent function of the square root functions is $f(x) = \sqrt{x}$. The parent function of the cube root functions is $f(x) = \sqrt[3]{x}$.
- A square root function can be written in the form $g(x) = a\sqrt{x - h} + k$. A cube root function can be written in the form $g(x) = a\sqrt[3]{x - h} + k$.

Lessons 6-5 and 6-6

Radical Expressions and Equations

- For any real numbers a and b and any integer $n > 1$, $\sqrt[n]{ab} = \sqrt[n]{a} \cdot \sqrt[n]{b}$, if n is even and $a, b \geq 0$ or if n is odd.
- For any real numbers a and $b \neq 0$ and any integer $n > 1$, $\sqrt[n]{\frac{a}{b}} = \frac{\sqrt[n]{a}}{\sqrt[n]{b}}$, if all roots are defined.
- $a\sqrt{b} + c\sqrt{d}$ and $a\sqrt{b} - c\sqrt{d}$ are conjugates of each other.
- To solve a radical equation, isolate the radical on one side of the equation. Raise each side of the equation to a power equal to the index of the radical. Solve the resulting polynomial equation. Check your results.

Study Organizer

Foldables

Use your Foldable to review this module. Working with a partner can be helpful. Ask for clarification of concepts as needed.

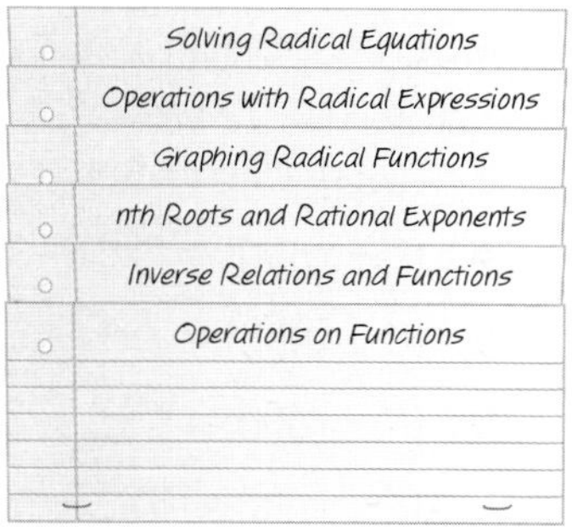

Test Practice

1. **MULTIPLE CHOICE** Given $f(x) = 4x^3 - 5x^2 + 8$ and $g(x) = 2x^3 - 9x^2 - 7x$, find $(f - g)(x)$. (Lesson 6-1)

   A. $(f - g)(x) = 2x^3 - 14x^2 + x$

   B. $(f - g)(x) = 2x^3 + 4x^2 + 15x$

   C. $(f - g)(x) = 2x^3 - 14x^2 - 7x + 8$

   D. $(f - g)(x) = 2x^3 + 4x^2 + 7x + 8$

2. **OPEN RESPONSE** Given $f(x) = 2x^3 + 7x^2 - 7x - 30$ and $g(x) = x - 2$, find $\left(\frac{f}{g}\right)(x)$. (Lesson 6-1)

3. **MULTIPLE CHOICE** Given $f(x) = 4x^3 - 2x^2$ and $g(x) = -x^2 + 3x - 2$, find $(f \cdot g)(x)$. (Lesson 6-1)

   A. $(f \cdot g)(x) = 4x^3 - x^2 - 3x + 2$

   B. $(f \cdot g)(x) = 4x^3 - x^2 + 3x - 2$

   C. $(f \cdot g)(x) = -4x^5 + 14x^4 - 14x^3 + 4x^2$

   D. $(f \cdot g)(x) = 4x^5 - 14x^4 + 14x^3 - 4x^2$

4. **MULTIPLE CHOICE** Given $f(x) = 14x^3 - x^2 + x + 5$ and $g(x) = 7x^3 + 4x^2 - 2x - 1$, find $(f - g)(x)$. (Lesson 6-1)

   A. $(f - g)(x) = 7x^3 + 3x^2 + x + 4$

   B. $(f - g)(x) = 7x^3 - 5x^2 + 3x + 6$

   C. $(f - g)(x) = 7x^3 + 3x^2 + 3x + 6$

   D. $(f - g)(x) = 7x^3 - 5x^2 + x + 4$

5. **OPEN RESPONSE** Given $f(x) = 3x - 7$ and $g(x) = 4x + 5$, find $(f \circ g)(x)$. (Lesson 6-1)

6. **MULTIPLE CHOICE** The graph shows $f(x)$. Which of the following represents $f^{-1}(x)$? (Lesson 6-2)

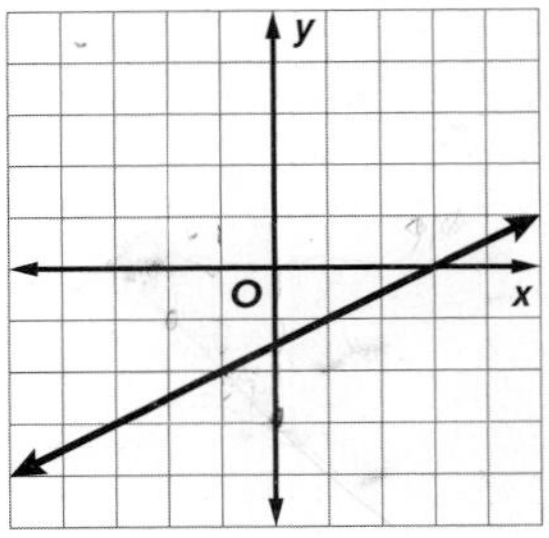

   A. $f^{-1}(x) = 2x + 3$

   B. $f^{-1}(x) = 3x + 2$

   C. $f^{-1}(x) = x + 4$

   D. $f^{-1}(x) = 4x$

7. **OPEN RESPONSE** Find the inverse of $f(x) = x^2 + 4x + 3$. Restrict the domain, if necessary. (Lesson 6-2)

8. **MULTIPLE CHOICE** Evaluate $256^{\frac{3}{8}} + 100{,}000^{-\frac{2}{5}}$. (Lesson 6-3)

A. 7.9

B. 7.99

C. 8.01

D. 8.1

9. **MULTIPLE CHOICE** Which of the following is the simplified form of $\sqrt[3]{729(x-7)^6}$? (Lesson 6-3)

A. $27(x-7)^3$

B. $27(x-7)^2$

C. $9(x-7)^3$

D. $9(x-7)^2$

10. **MULTIPLE CHOICE** The lung volume for mammals can be modeled using the expression $170x^{\frac{4}{5}}$, where x is the mass of the mammal. How can this expression be rewritten using radicals? (Lesson 6-3)

A. $170\sqrt[5]{x^4}$

B. $170\sqrt[4]{x^5}$

C. $\sqrt[5]{170x^4}$

D. $\sqrt[4]{170x^5}$

11. **OPEN RESPONSE** Describe how the graph of $g(x) = -2\sqrt{x+5} - 3$ is related to the graph of the parent function. (Lesson 6-4)

12. **OPEN RESPONSE** Determine the values of a and b. (Lesson 6-3)

$x^{\frac{2}{3}} \cdot x^{-\frac{1}{4}} = x^{\frac{a}{b}}$

13. **MULTIPLE CHOICE** Which of the following is the graph of $g(x) = -\sqrt{x+6} - 2$? (Lesson 6-4)

A.

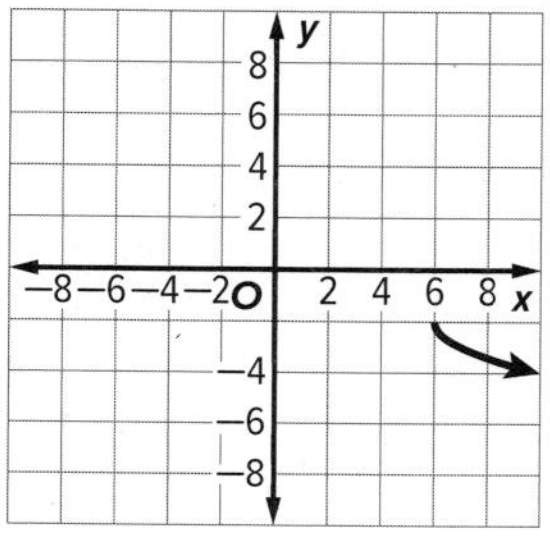

B.

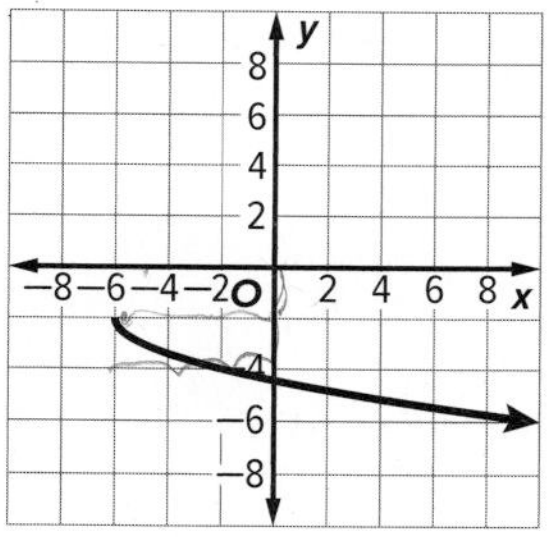

C.

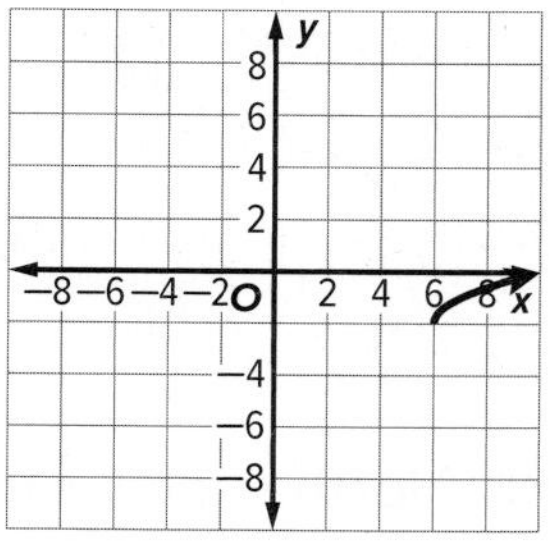

D.

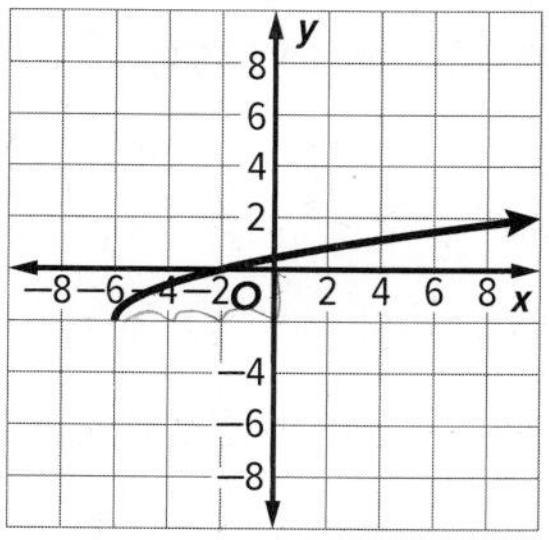

14. MULTIPLE CHOICE Which function is shown on the graph? (Lesson 6-4)

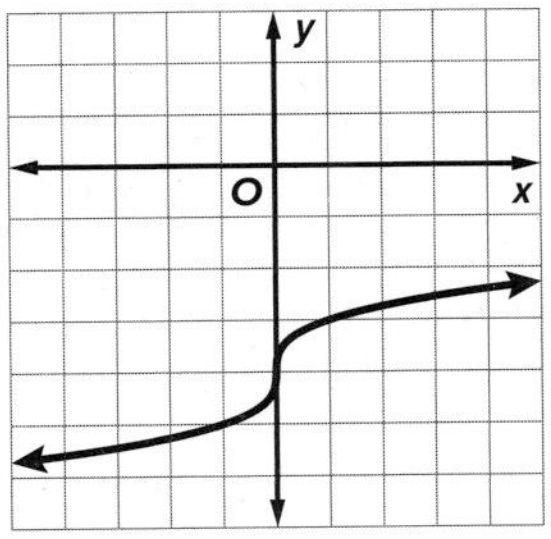

A. $g(x) = (x + 4)^{\frac{1}{3}}$

B. $g(x) = (x - 4)^{\frac{1}{3}}$

C. $g(x) = x^{\frac{1}{3}} - 4$

D. $g(x) = x^{\frac{1}{3}} + 4$

15. OPEN RESPONSE Determine the sum. $7\sqrt{12} + 2\sqrt{48} - 4\sqrt{75}$ (Lesson 6-5)

16. MULTIPLE CHOICE Heather drew a rectangle and labeled the length as $3\sqrt[4]{162x^4y^7}$ and the width as $2\sqrt[4]{40x^3y^5}$. Which is the simplified form of the area of the rectangle? (Lesson 6-5)

A. $36\sqrt[3]{5x^{12}y^{35}}$

B. $36xy^3\sqrt[4]{5x^3}$

C. $36xy^3\sqrt[4]{5x^3y}$

D. $36\sqrt[4]{5y^{12}}$

17. MULTIPLE CHOICE Latasia wants to rationalize the denominator of $\frac{7x^4y^5}{\sqrt[3]{z}}$. What should she multiply the numerator and denominator by? (Lesson 6-5)

A $\sqrt[3]{z}$

B. $\sqrt[3]{z^2}$

C. z^2

D. z^3

18. MULTIPLE CHOICE Solve for x, $3(6x - 8)^{\frac{1}{4}} + 3 = 9$. (Lesson 6-6)

A. $x = 2$

B. $x = 4$

C. $x = 6$

D. $x = 8$

19. MULTIPLE CHOICE Select the solution(s) of the equation $\sqrt{3x + 7} = x - 1$. (Lesson 6-6)

A. 6

B. 3

C. 2 and 3

D. 6 and −1

20. MULTIPLE CHOICE The distance in miles, d, a pilot can see to the horizon is 123% of the square root of the altitude in feet above sea level, a, of the plane. How many miles to the horizon can a pilot flying a plane at an altitude of 30,000 feet see? Round your answer to the nearest mile. (Lesson 6-6)

A. 213 miles

B. 369 miles

C. 21,304 miles

D. 36,900 miles

Exponential Functions

e Essential Question

How are real-world situations involving quantities that grow or decline rapidly modeled mathematically?

What Will You Learn?

How much do you already know about each topic **before** starting this module?

KEY

— I don't know. — I've heard of it. — I know it!

	Before			After		
graph exponential growth functions						
graph exponential decay functions						
solve exponential equations						
solve exponential inequalities						
understand the natural base *e*, and use it to solve problems						
understand and use geometric sequences						
derive formulas for geometric series						
find sums of geometric series						

Foldables Make this Foldable to help you organize your notes about exponential functions. Begin with two sheets of grid paper.

1. **Fold** in half along the width.
2. **On** the first sheet, cut 5 cm along the fold at the ends.
3. **On** the second sheet, cut in the center, stopping 5 cm from the ends.
4. **Insert** the first sheet through the second sheet, and align the folds. Label the front Exponential Functions, and decorate. Label each of the inside pages with a lesson number, using the final page for vocabulary.

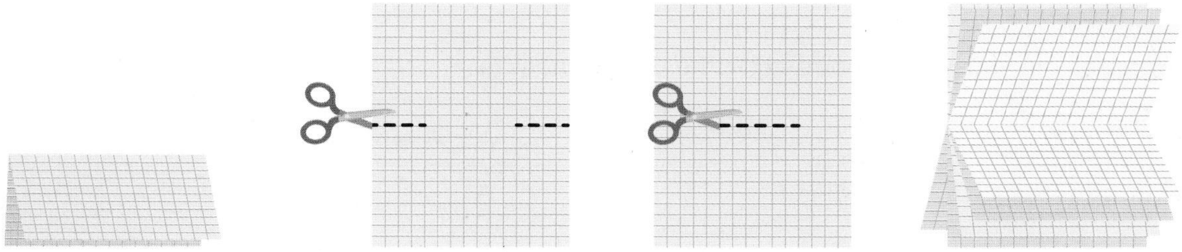

What Vocabulary Will You Learn?

- asymptote
- coefficient of determination
- common ratio
- compound interest
- decay factor
- e
- explicit formula
- exponential decay
- exponential equation
- exponential function
- exponential growth
- exponential inequality
- finite sequence
- geometric means
- geometric sequence
- geometric series
- growth factor
- infinite sequence
- recursive formula
- regression function
- sequence
- series
- sigma notation
- term of a sequence

Are You Ready?

Complete the Quick Review to see if you are ready to start this module. Then complete the Quick Check.

Quick Review

Example 1

Simplify $\frac{(a^3bc^2)^2}{a^4a^2b^2bc^5c^3}$. Assume that no variable equals zero.

$\frac{(a^3bc^2)^2}{a^4a^2b^2bc^5c^3}$

$= \frac{a^6b^2c^4}{a^4a^2b^2bc^5c^3}$ — Power of a Power Rule

$= \frac{a^6b^2c^4}{a^6b^3c^8}$ — Product of Powers Rule

$= \frac{1}{bc^4}$ or $b^{-1}c^{-4}$ — Quotient of Powers Rule

Example 2

Find the inverse of $f(x) = 3x - 1$.

Step 1 Replace $f(x)$ with y in the original equation: $f(x) = 3x - 1 \rightarrow y = 3x - 1$.

Step 2 Interchange x and y: $x = 3y - 1$.

Step 3 Solve for y.

$x = 3y - 1$

$x + 1 = 3y$ — Add 1 to each side.

$\frac{x+1}{3} = y$ — Divide each side by 3.

$\frac{1}{3}x + \frac{1}{3} = y$ — Simplify.

Step 4 Replace y with $f^{-1}(x)$.

$y = \frac{1}{3}x + \frac{1}{3} \rightarrow f^{-1}(x) = \frac{1}{3}x + \frac{1}{3}$

Quick Check

Simplify. Assume that no variable equals zero.

1. $a^4a^3a^5$
2. $(2xy^3z^2)^3$
3. $\frac{-24x^8y^5z}{16x^2y^8z^6}$
4. $\left(\frac{-8r^2n}{36n^3t}\right)^2$

Find the inverse of each function.

5. $f(x) = 2x + 5$
6. $f(x) = x - 3$
7. $f(x) = -4x$
8. $f(x) = \frac{1}{4}x - 3$

How Did You Do?

Which exercises did you answer correctly in the Quick Check?

Graphing Exponential Functions

Explore Using Technology to Analyze Graphs of Exponential Functions

Online Activity Use graphing technology to ocmplete the Explore.

INQUIRY How does performing an operation on an exponential function affect its graph?

Learn Graphing Exponential Growth Functions

In an **exponential function,** the independent variable is an exponent. An exponential function has the form $f(x) = b^x$, where the base b is a constant and the independent variable x is the exponent. For an exponential growth function, $b > 1$. **Exponential growth** occurs when an initial amount increases by the same percent over a given period of time.

Graphs of exponential functions have asymptotes. An **asymptote** is a line that a graph approaches.

Example 1 Graph Exponential Growth Functions

Graph $f(x) = 2^x$. Find the domain, range, y-intercept, asymptote, and end behavior.

Make a table of values. Then plot the points and sketch the graph.

x	$f(x) = 2^x$
−3	0.125
−2	0.25
−1	0.5
0	1
1	2
2	4
3	8

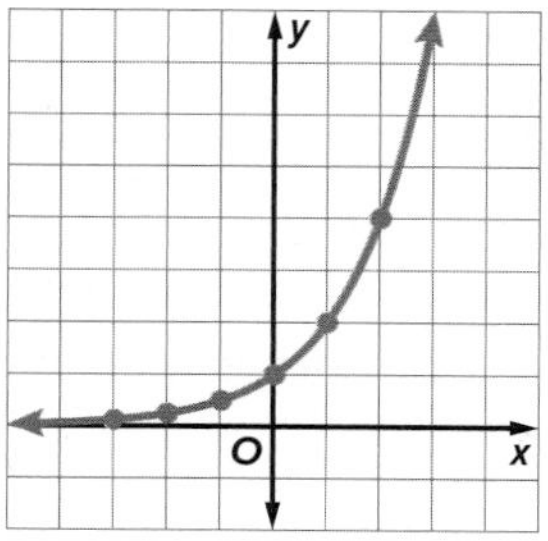

domain: all real numbers

range: all positive real numbers

y-intercept: (0, 1)

asymptote: $y = 0$

end behavior: As $x \to -\infty$, $f(x) \to 0$ and as $x \to \infty$, $f(x) \to \infty$.

Go Online You can complete an Extra Example online.

Today's Goals
- Graph exponential growth functions.
- Graph exponential decay functions.

Today's Vocabulary

exponential function

exponential growth

asymptote

growth factor

exponential decay

decay factor

Think About It!

Why is the range of $f(x)$ all positive real numbers instead of all real numbers?

Study Tip

Asymptotes You can find the asymptote of an exponential function by using a graph or a table. Both tools allow you to identify the line that the function approaches.

> **Go Online** to watch a video to learn how to graph transformations of an exponential function using a graphing calculator.

Example 2 Graph Transformations of Exponential Growth Functions

Graph $g(x) = -\frac{1}{2} \cdot 3^{x+4} + 1$.

Transform the graph of $g(x) = 3^x$.

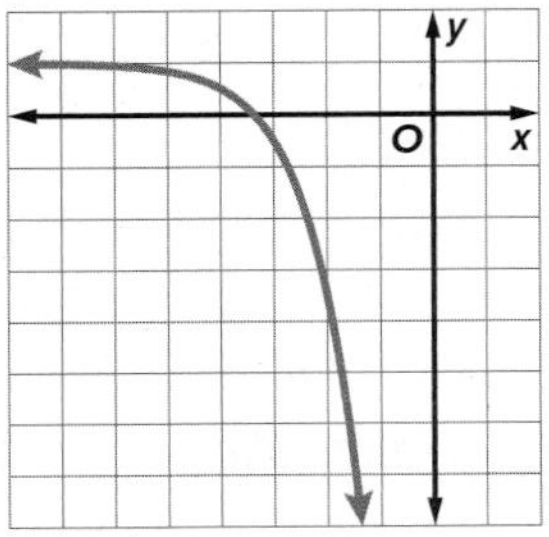

$a = -\frac{1}{2}$; Reflect in the x-axis and compress vertically.

$h = -4$; Translate 4 units left.

$k = 1$; Translate 1 unit up.

> **Go Online** to see other examples about analyzing graphs of exponential functions.

Example 3 Analyze Graphs of Exponential Functions

Identify the value of k and write a function for the graph of $j(x) = f(x) + k$ as it relates to $f(x) = 3.5^x$.

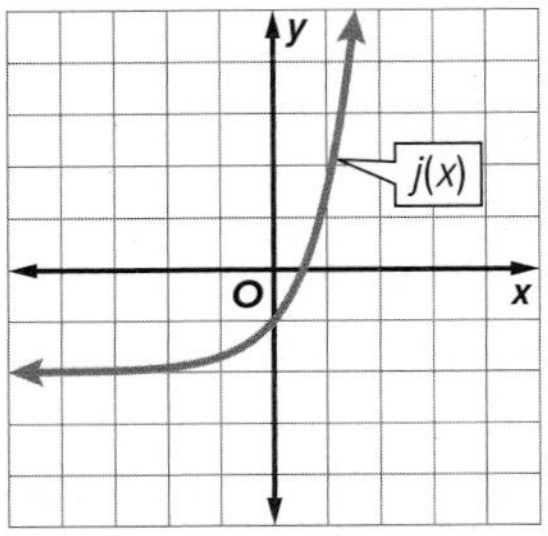

The graph has been translated 2 units down, so $k = -2$ and the function is $j(x) = 3.5^x - 2$.

Exponential growth can be modeled by $A(t) = a(1 + r)^t$, where $A(t)$ represents the amount after t time periods, a is the initial amount, and r is the percent of increase per time period. The **growth factor** is $1 + r$.

Example 4 Use Exponential Growth Functions

LOTTERY **Mr. Lopez recently won the lottery. Suppose he takes the lump-sum payment, and he invests $50 million into an account that yields 5% interest annually. Graph a function that models the amount in his account. Then estimate the amount in the account after 20 years.**

Use the function $A(t) = a(1 + r)^t$ to model the amount of money in his account. Let a be the amount of investment in millions of dollars, and let r be the annual interest rate written as a decimal.

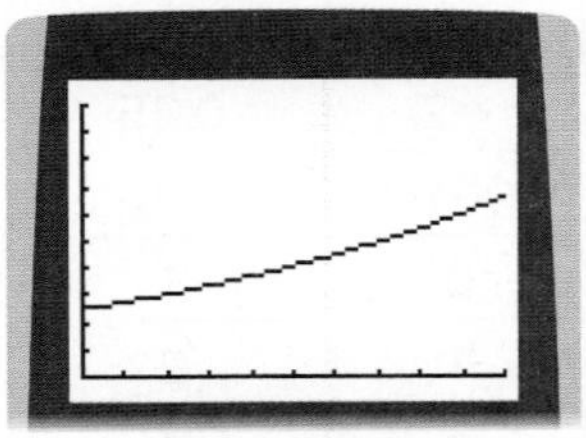

[0, 20] scl: 2 by [0, 200] scl: 20

$A(t) = 50(1.05)^t$

Use a graphing calculator to graph the function.

Use the table feature to estimate the amount of money in the account after 20 years.

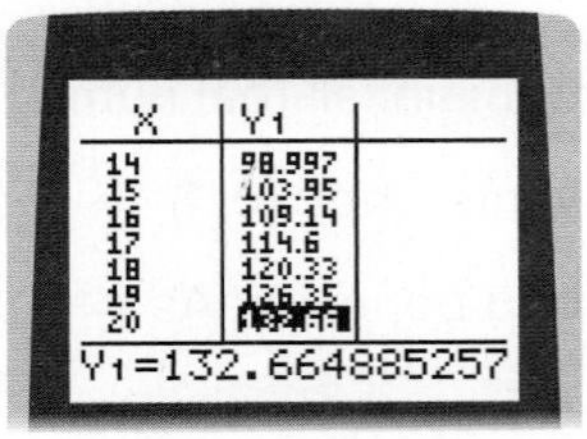

There will be approximately $132,664,885.26 in the account after 20 years.

> **Watch Out!**
> **Labeling Axes** The horizontal axis represents the number of years and the vertical axis represents the amount of money in the account. Be sure to keep units in mind because the amount of money in the account is measured in millions of dollars.

Go Online You can complete an Extra Example online.

Check

FINANCE On New Year's Day, Alma begins a savings plan. She saves \$1 the first day of January and increases the amount she saves each day by 1%. Find the growth factor. Then use a graphing calculator to estimate the amount of money she should save on Day 180 of her savings plan.

growth factor: ___?___ amount saved 180 days later: \$ ___?___

Learn Graphing Exponential Decay Functions

Exponential decay occurs when an initial amount decreases by the same percent over a given period of time. So, for an exponential function of the form $f(x) = b^x$, exponential decay occurs when b is between 0 and 1.

Like exponential growth, exponential decay can be modeled by $A(t) = a(1 - r)^t$, where r is the percent of decrease per time period. The **decay factor** is $1 - r$.

Think About It!

Use a graphing calculator to graph $f(x) = 10^{-x}$ and $g(x) = \left(\frac{1}{10}\right)^x$. Compare $f(x)$ and $g(x)$, and identify whether each function represents exponential growth or decay.

Example 5 Interpret Exponential Functions

Determine whether each function represents *exponential growth* or *exponential decay*.

a. $f(x) = 5^x$ Because $5 > 1$, $f(x)$ is an exponential growth function.

b. $g(x) = \left(\frac{2}{7}\right)^x$ Because $0 < \frac{2}{7} < 1$, $g(x)$ is an exponential decay function.

c. $h(x) = \left(\frac{4}{3}\right)^x$ Because $\frac{4}{3} > 1$, $h(x)$ is an exponential growth function.

d. $j(x) = 1.05^x$ Because $1.05 > 1$, $j(x)$ is an exponential growth function.

e. $k(x) = 0.85^x$ Because $0 < 0.85 < 1$, $k(x)$ is an exponential decay function.

Talk About It!

Explain how you can use the end behavior of the graph of an exponential function to determine whether it represents exponential growth or decay.

Example 6 Graph Exponential Decay Functions

Graph $f(x) = \left(\frac{1}{2}\right)^x$. Find the domain, range, y-intercept, asymptote, and end behavior.

Make a table of values. Then plot the points and sketch the graph.

domain: all real numbers

range: all positive real numbers

y-intercept: (0, 1) asymptote: $y = 0$

end behavior: as $x \to -\infty$, $f(x) \to \infty$ and as $x \to \infty$, $f(x) \to 0$

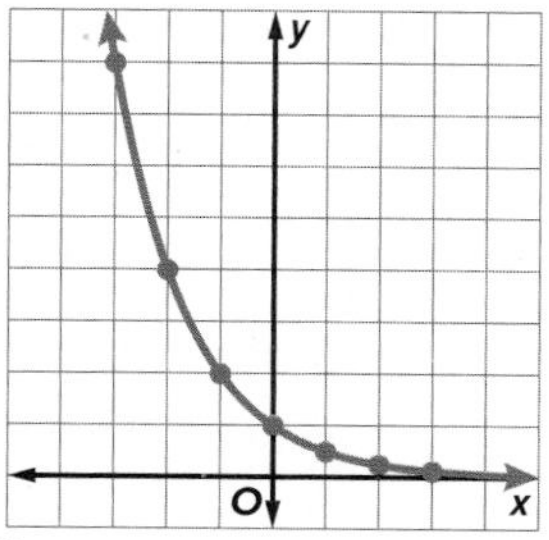

Go Online You can complete an Extra Example online.

Math History Minute

Mathematical biologist **Trachette Jackson (1973—)** uses many different approaches, including continuous and discrete mathematical models, numerical simulations, and experiments, to study tumor growth and treatment. In 2003, Jackson became only the second African-American woman to become a Sloan Fellow in mathematics.

Think About It!

Examine the functions in Example 8. Find the domain and range of $f(x)$. Does $g(x)$ have the same domain and range? Explain.

Example 7 Graph Transformations of Exponential Decay Functions

Graph $g(x) = -2\left(\frac{1}{4}\right)^{x-4} + 3$.

$g(x)$ is a transformation of $f(x) = \left(\frac{1}{4}\right)^x$.

$a = -2$; Reflect in the x-axis and stretch vertically.

$h = 4$; Translate 4 units right.

$k = 3$; Translate 3 units up.

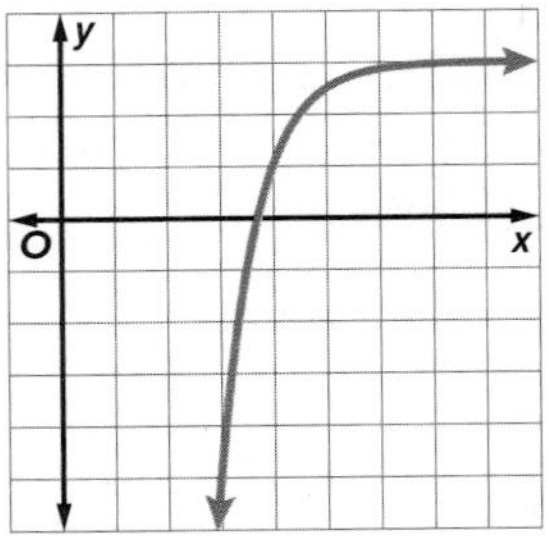

Example 8 Compare Exponential Functions

Consider $f(x) = \begin{cases} \left(\frac{1}{2}\right)^x & \text{for } x < -1 \\ 2x + 4 & \text{for } x \geq -1 \end{cases}$ and $g(x)$ shown in the graph.

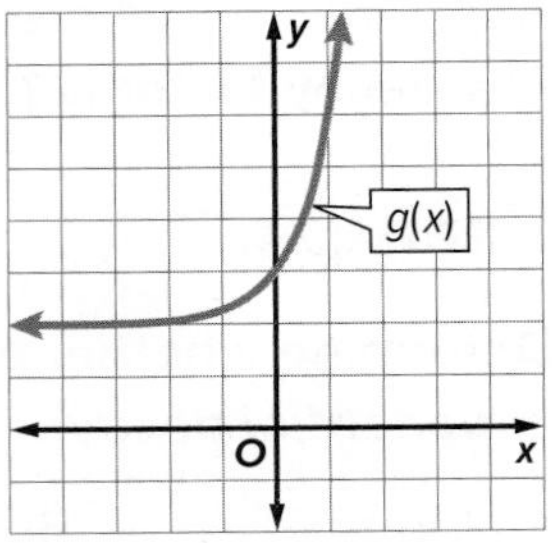

Part A Graph $f(x)$.

First create a table of values to graph the exponential piece. Then graph the linear piece.

x	$f(x)$
−5	32
−4	16
−3	8
−2	4
−1	2

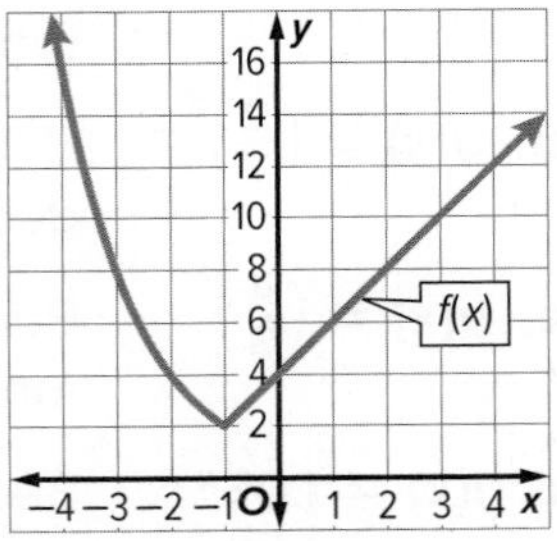

Part B Which function has the lesser relative minimum?

$f(x)$ has a relative minimum of 2. $g(x)$ has no relative minimum. It appears to have an asymptote at $y = 2$, so all of the function values are greater than 2.

Part C Compare the y-intercepts and end behavior of $f(x)$ and $g(x)$.

y-intercepts

$f(x)$: 4

$g(x)$: 3

end behavior

$f(x)$: As $x \longrightarrow -\infty$, $f(x) \longrightarrow \infty$, and as $x \longrightarrow \infty$, $f(x) \longrightarrow \infty$.

$g(x)$: As $x \longrightarrow -\infty$, $g(x) \longrightarrow 2$, and as $x \longrightarrow \infty$, $g(x) \longrightarrow \infty$.

Go Online You can complete an Extra Example online.

Practice

Go Online You can complete your homework online.

Example 1

Graph each function. Find the domain, range, y-intercept, asymptote, and end behavior.

1. $f(x) = 3^x$
2. $f(x) = 5^x$
3. $f(x) = 1.5^x$
4. $f(x) = \left(\frac{5}{2}\right)^x$

Example 2

Graph each function.

5. $f(x) = 2(3)^x$
6. $f(x) = -2(4)^x$
7. $f(x) = 4^{x+1} - 5$
8. $f(x) = 3^{2x} + 1$
9. $f(x) = -0.4(3)^{x+2} + 4$
10. $f(x) = 1.5(2)^x + 6$

Example 3

Identify the value of k and write a function $g(x)$ for each graph as it relates to $f(x)$.

11. $f(x) = 2^x$; $g(x) = f(x) + k$

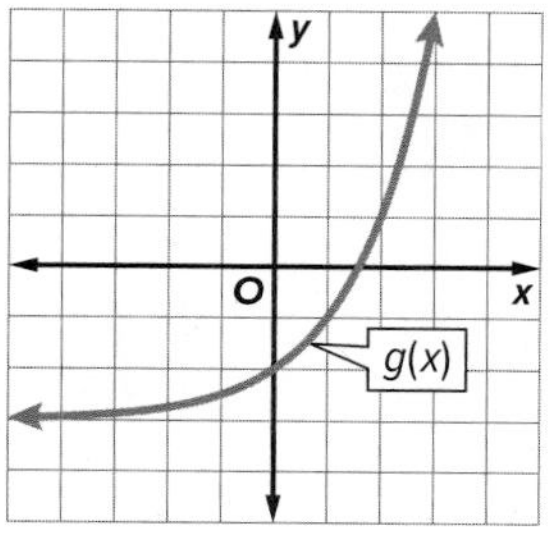

12. $f(x) = 3^x$; $g(x) = f(x) + k$

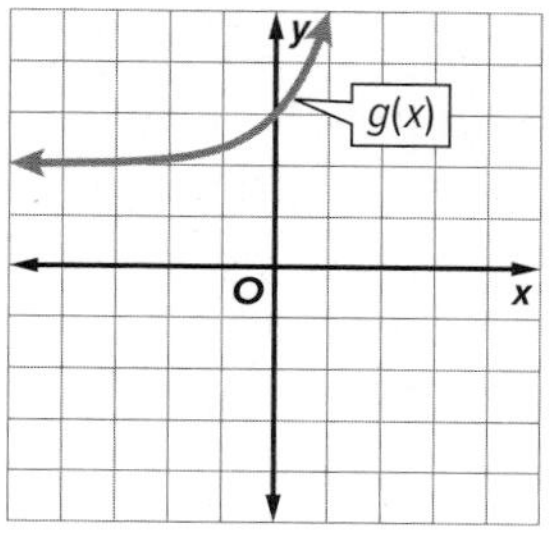

13. $f(x) = \left(\frac{3}{2}\right)^x$; $g(x) = k \cdot f(x)$

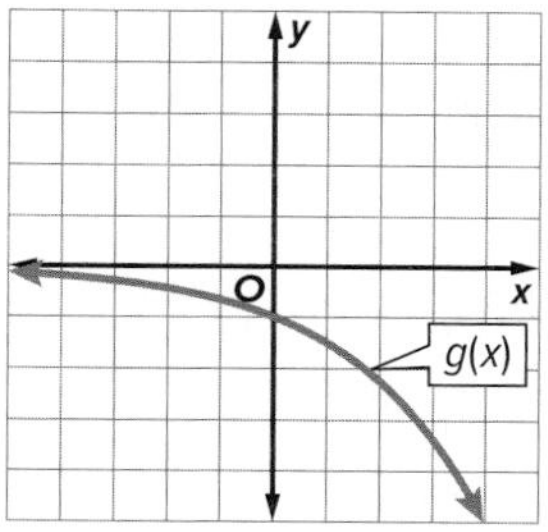

Example 4

14. SUBSCRIPTIONS Subscriptions to an online arts and crafts club have been increasing by 20% every year. The club began with 40 members. Make a graph of the number of subscribers over the first 5 years of the club's existence. About how many subscriptions are there after Year 4?

15. SHOES The cost of a pair of athletic shoes increases about 5.1% every year. In 2018 the average price for a pair of athletic shoes was \$58.17. Graph a function that models the cost of an average pair of athletic shoes. Then estimate the cost of an average pair of shoes in 25 years.

16. MONEY Sunil opened a savings account that compounds interest at a rate of 3% annually. Let P be the initial amount Sunil deposited, and let t be the number of years the account has been open.

    a. Write an equation to find A, the amount of money in the account after t years. Assume that Sunil made no more deposits and no withdrawals.

    b. If Sunil opened the account with \$500 and made no deposits or withdrawals, graph a function to represent the money in his savings account. Then estimate the amount of money in the account 10 years after opening the account.

    c. Estimate the number of years it would take for such an account to double in value.

Example 5

Determine whether each function represents *exponential growth* or *exponential decay*.

17. $f(x) = 7^x$

18. $g(x) = 0.99^x$

19. $h(x) = \left(\frac{2}{3}\right)^x$

20. $j(x) = \left(\frac{5}{4}\right)^x$

21. $k(x) = 0.75^x$

22. $m(x) = 1.02^x$

Example 6

Graph each function. Find the domain, range, *y*-intercept, asymptote, and end behavior.

23. $f(x) = 0.25^x$

24. $f(x) = 0.8^x$

25. $f(x) = \left(\frac{1}{2}\right)^x$

26. $f(x) = \left(\frac{2}{3}\right)^x$

Example 7

Graph each function.

27. $f(x) = -4\left(\frac{3}{5}\right)^{x+4} + 3$

28. $f(x) = 3\left(\frac{2}{5}\right)^{x-3} - 6$

29. $f(x) = \frac{1}{2}\left(\frac{1}{5}\right)^{x+5} + 8$

30. $f(x) = \frac{3}{4}\left(\frac{2}{3}\right)^{x+4} - 2$

31. $f(x) = -\frac{1}{2}\left(\frac{3}{8}\right)^{x+2} + 9$

32. $f(x) = -\frac{5}{4}\left(\frac{4}{5}\right)^{x+4} + 2$

Example 8

33. Consider $f(x) = \begin{cases} \left(\frac{3}{4}\right)^x \text{ for } x < 0 \\ 3x + 1 \text{ for } x \geq 0 \end{cases}$ and $g(x)$ shown in the graph.

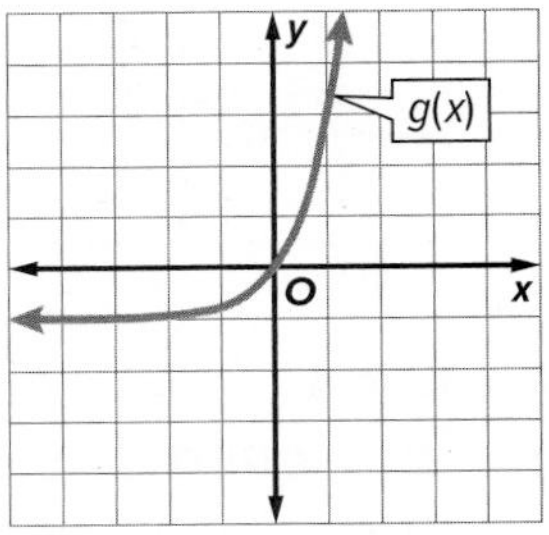

a. Graph $f(x)$.

b. How do the key features, such as domain and range, intercepts, increasing and decreasing, positive and negative, minimum and maximum, symmetry, and end behavior of $f(x)$ and $g(x)$ compare?

34. Consider $f(x) = \begin{cases} -\left(\frac{1}{2}\right)^x + 3 \text{ for } x < -1 \\ -\frac{1}{4}x + \frac{3}{4} \text{ for } x \geq -1 \end{cases}$ and $g(x)$ shown in the graph.

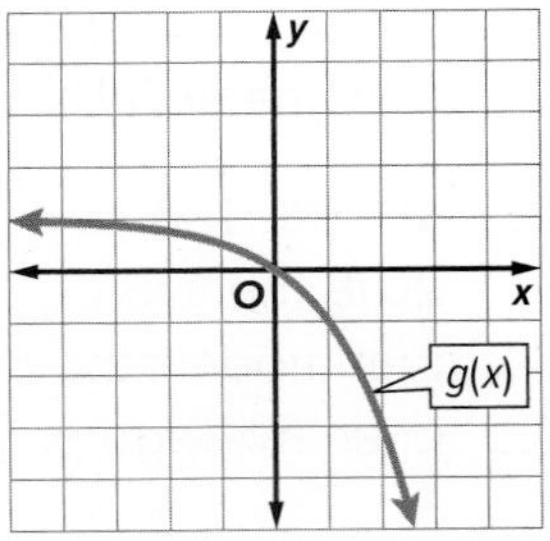

a. Graph $f(x)$.

b. How do the key features, such as domain and range, intercepts, increasing and decreasing, positive and negative, minimum and maximum, symmetry, and end behavior of $f(x)$ and $g(x)$ compare?

Mixed Exercises

Graph each function. State the function's domain and range. Then use the graph to determine whether the function represents an *exponential growth* or *exponential decay* function.

35. $y = 3(2)^x$

36. $y = 2\left(\frac{1}{2}\right)^x$

37. $y = -\frac{3}{2}(1.5)^x$

38. $y = 3\left(\frac{1}{3}\right)^x$

f(x) is the parent function and g(x) is a transformation of f(x). Use the graph to determine g(x).

39. $f(x) = 4^x$

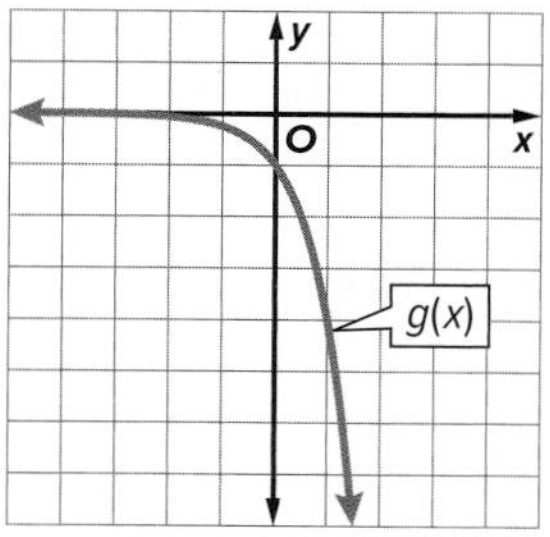

40. $f(x) = \left(\frac{1}{5}\right)^x$

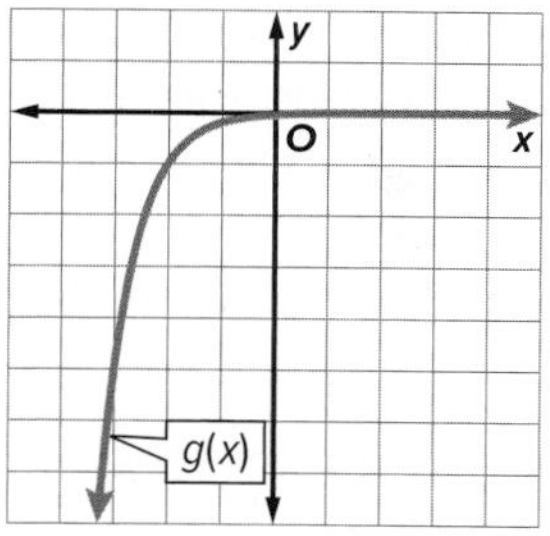

41. REASONING For $y = ab^x$, where $a > 0$, if $b > 1$, the function represents exponential growth. It represents exponential decay if $0 < b < 1$.

a. Choose a positive value for a, and let $b = 1$. Complete the table for these values of a and b. Is $y = ab^x$ an exponential function? Explain your reasoning.

x	−3	−2	−1	0	1	2	3
$y = ab^x$							

b. Choose a positive value for a, and a negative value for b. Complete the table for these values. Is $y = ab^x$ an exponential function? Explain your reasoning.

x	−3	−2	−1	0	1	2	3
$y = ab^x$							

42. INVESTMENTS At age 28, Catalina makes a single \$22,000 investment that earns 5% interest each year.

a. If Catalina leaves the investment untouched until she turns 65, how much will the investment be worth at that time?

b. Catalina's twin brother, Rodrigo, waits 2 years and then makes the same investment as Catalina. If the function that describes Catalina's investment is $C(t)$, what function describes Rodrigo's investment? What will his investment be worth when he turns 65?

43. CARS The value of an automobile depreciates by approximately 15% each year after purchase. Jayden paid \$28,000 when he bought his car 15 years ago.

a. Write and graph a function that models how the value of the car depreciates.

b. How does the average decrease in value during the first five years of ownership compare to the last five years of ownership?

44. STRUCTURE Let $f(x) = 6^x$; $g(x) = -\frac{1}{2}(6)^{x+3} - 1$ and $h(x) = \frac{3}{2}(6)^{x-5}$.

a. What is the asymptote of each function?

b. How could you transform the graph of $f(x)$ to create the graphs of $g(x)$ and $h(x)$?

c. How could you directly transform $g(x)$ to create the graph of $h(x)$?

45. Let $f(x) = (4)^x$ and $g(x) = (4)^{-x} + 1$. What transformations of $f(x)$ will result in the graph of $g(x)$? Graph both functions. For $g(x)$, identify how the y-intercept, intervals where the function is increasing, decreasing, positive, or negative, the asymptote, and the end behavior of $f(x)$ are transformed?

46. If the graph of $f(x) = (0.5)^x$ is reflected in the y-axis, stretched vertically by a factor of 2, and translated 1 unit down to create $g(x)$, find the equation for $g(x)$. What are the domain, range, y-intercept, and zeros of $g(x)$? Graph $g(x)$ and label any intercepts.

47. CHEMISTRY A compound undergoes exponential decay with initial amount 27.3 grams.

a. If the amount of the compound decreases by ten percent each year, define variables and write a function modeling the amount of the compound remaining at a given time.

b. Find the average rate of change for the function over [0, 2] and [3, 5]. Explain the rates of change in terms of the situation.

48. USE TOOLS A population of bacteria grows exponentially with initial population 20,000. After one day the bacteria population grows to 30,000.

a. Write a function $P(t)$ to model the bacteria population after t days.

b. Use a graphing calculator to graph $P(t)$ and sketch the graph. Identify the intercepts, zeros, and the end behavior as $t \to \infty$ and explain these features in the context of the problem.

c. What is an appropriate domain for $P(t)$? Explain your reasoning.

49. ANALYZE Determine whether each statement is *sometimes*, *always*, or *never* true. Explain your reasoning.

a. An exponential function of the form $y = ab^{x-h} + k$ has a y-intercept.

b. An exponential function of the form $y = ab^{x-h} + k$ has an x-intercept.

c. The function $f(x) = |b|^x$ is an exponential growth function if b is an integer.

50. FIND THE ERROR Vince and Grady were asked to graph $f(x)$ and $g(x)$ given the table for $f(x)$ and the description of $g(x)$. Vince thinks they are the same, but Grady disagrees. Who is correct? Explain your reasoning.

x	$f(x)$
0	2
1	1
2	0.5
3	0.25
4	0.125
5	0.0625
6	0.03125

exponential function $g(x)$ with rate of decay of $\frac{1}{2}$ and an initial amount of 2

51. PERSEVERE A substance decays 35% each day. After 8 days, there are 8 milligrams of the substance remaining. How many milligrams were there initially?

52. CREATE Give an example of a value of b for which $f(x) = \left(\frac{8}{b}\right)^x$ represents exponential decay.

53. WRITE Write the procedure for transforming the graph of $g(x) = b^x$ to the graph of $f(x) = ab^{x-h} + k$. Justify each step.

Solving Exponential Equations and Inequalities

Explore Solving Exponential Equations

Online Activity Use the interactive tool to complete the Explore.

INQUIRY How can you rewrite expressions to solve exponential equations?

Today's Goals
- Solve exponential equations in one variable.
- Solve exponential inequalities in one variable.

Today's Vocabulary
exponential equation

compound interest

exponential inequality

Learn Solving Exponential Equations

In an **exponential equation**, the independent variable is an exponent.

Key Concept • Property of Equality for Exponential Equations

If $b > 0$ and $b \neq 1$, then $b^x = b^y$ if and only if $x = y$.

Exponential equations can be solved algebraically or by graphing a system of equations based on the equation.

Equations of exponential functions can be used to calculate compound interest. **Compound interest** is paid on the principal of an investment and any previously earned interest.

Key Concept • Compound Interest

You can calculate compound interest using the formula $A = P\left(1 + \frac{r}{n}\right)^{nt}$, where A is the amount in the account after t years, P is the principal amount invested, r is the annual interest rate, and n is the number of compounding periods each year.

Go Online to watch a video to learn how to solve exponential equations by graphing using a graphing calculator.

Example 1 Solve Exponential Equations Algebraically

Solve each equation.

a. $\mathbf{4^{5x+1} = 64^7}$

$4^{5x+1} = 64^7$	Original equation
$4^{5x+1} = (4^3)^7$	Rewrite 64 as 4^3.
$4^{5x+1} = 4^{21}$	Power of a Power
$5x + 1 = 21$	Property of Equality for Exponential Equations
$5x = 20$	Subtract 1 from each side.
$x = 4$	Divide each side by 5.

(continued on the next page)

Go Online You can complete an Extra Example online.

b. $\left(\frac{1}{2}\right)^{4x-16} = 16^{2x-5}$

$\left(\frac{1}{2}\right)^{4x-16} = 16^{2x-5}$	Original equation
$(2^{-1})^{4x-16} = (2^4)^{2x-5}$	$\frac{1}{2} = 2^{-1}$, $16 = 2^4$
$2^{-4x+16} = 2^{8x-20}$	Power of a Power Property
$-4x + 16 = 8x - 20$	Property of Equality for Exponential Equations
$-12x + 16 = -20$	Subtract $8x$ from each side.
$-12x = -36$	Subtract 16 from each side.
$x = 3$	Divide each side by -12.

Watch Out!

Power of a Power
When using the Power of a Power Property, do not forget to distribute to each term in the exponent.

Example 2 Solve an Exponential Equation by Graphing

SUPER BOWL **Since it began in 1966, the extremely high viewership of the Super Bowl has led companies to pay millions of dollars to advertise for just 30 seconds. The average cost of a 30-second advertisement during the Super Bowl can be approximately modeled by an exponential function where the initial cost is \$58,000 and the average cost increases by 9.9% each Super Bowl.**

Part A Write an exponential function to represent the situation.

Let C be the average cost of a 30-second ad and x be the number of the Super Bowl. The initial cost is \$58,000, or \$0.058 million. The rate of change is 9.9%, or 0.099.

$C = a(1 + r)^x$	Exponential function
$= 0.058(1.099)^x$	$a = 0.58$ and $r = 0.099$

Part B Estimate during which Super Bowl the average cost of a 30-second ad first surpassed \$1 million.

Step 1 Write a system.

To find when the average cost of an ad was \$1 million, let $C = 1$. Then, set each side of $1 = 0.058(1.099)^x$ equal to y to create a system of equations.

$y = 1$	First equation
$y = 0.058(1.099)^x$	Second equation

Step 2 Graph the system.

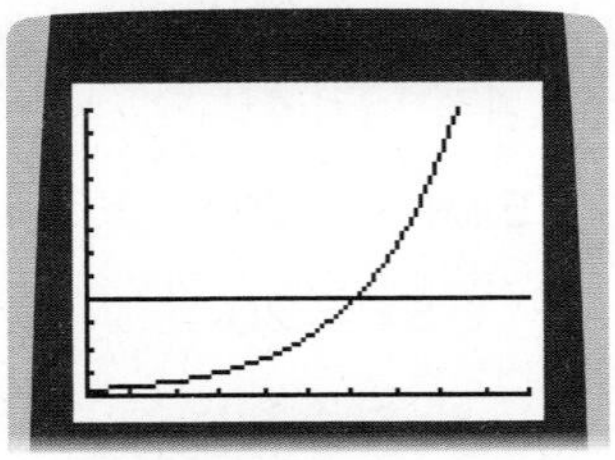

[0, 50] scl: 5 by [0, 3] scl: 0.25

Use a Source
Research during which Super Bowl the average cost of a 30-second ad surpassed \$1 million. Does the equation provide a reasonable estimate of when this occurred?

Go Online You can complete an Extra Example online.

Step 3 Find the intersection.

The solution is the x-coordinate of the point of intersection, which is about 30.16.

Step 4 Interpret the solution.

According to our equation, the average cost of a 30-second ad was \$1 million when $x \approx 30.16$. This means that the average cost was not yet \$1 million during Super Bowl 30. Therefore, it was during Super Bowl 31 that the average cost of a 30-second ad first surpassed \$1 million.

Step 5 Use a table.

You can use the **TABLE** feature to verify the solution. Because the number of the Super Bowl must be a whole number, use 1 as the interval.

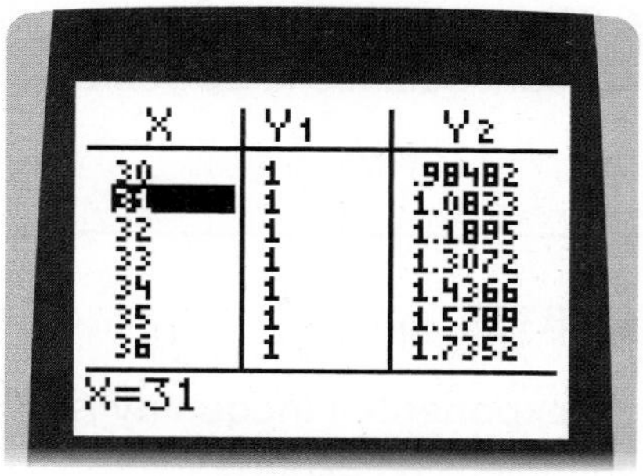

Notice that when $x = 30$, Y_2 was still less than 1 and when $x = 31$, it was more than 1. This confirms that the average cost surpassed \$1 million during Super Bowl 31.

Check

CARS The moment you drive a car off a dealer's lot, it begins to depreciate, or lose its value. In 2017, the average price of a new pickup truck was \$41,000. You can use an exponential function to model the value of a pickup truck over time.

Part A

If the value of a pickup truck depreciates by 9% per year, write a function that represents the value of a new 2017 pickup truck y after x years.

$y =$ ___?___(__?__)^x

Part B

After how many years will a new 2017 pickup truck be worth only \$20,000? Round to the nearest tenth if necessary.

__?__ years

Go Online You can complete an Extra Example online.

Study Tip

Assumptions
Assuming that the average cost of a 30-second ad during the Super Bowl is modeled by the exponential function allows us to make estimations and predictions about the situation. While the actual cost may not follow this trend exactly, it provides a reasonable model.

Example 3 Use the Compound Interest Formula

FINANCE **Luciana deposits $1700 into a savings account that pays 1.8% annual interest compounded monthly. What will be the balance after 5 years?**

$A = P\left(1 + \frac{r}{n}\right)^{nt}$	Compound Interest Formula
$= 1700\left(1 + \frac{0.018}{12}\right)^{12 \cdot 5}$	$P = 1700$, $r = 0.018$, $n = 12$, and $t = 5$
≈ 1859.97	Use a calculator.

After 5 years, there will be about $1859.97 in Luciana's savings account.

Check

FINANCE If Aisha deposits $500 in a checking account that pays 0.9% annual interest compounded twice a month, what will be her account balance in 25 years? Round to the nearest cent.

$__?__

Learn Solving Exponential Inequalities

An **exponential inequality** is an inequality in which the independent variable is an exponent.

Key Concept • Property of Inequality for Exponential Equations

If $b > 1$, then $b^x > b^y$ if and only if $x > y$, and
$b^x < b^y$ if and only if $x < y$.

Example 4 Solve Exponential Inequalities Algebraically

Solve $27^{2x+6} \geq 81^{x-5}$.

$27^{2x+6} \geq 81^{x-5}$	Original inequality
$(3^3)^{2x+6} \geq (3^4)^{x-5}$	$27 = 3^3$, $81 = 3^4$
$3^{6x+18} \geq 3^{4x-20}$	Power of a Power
$6x + 18 \geq 4x - 20$	Property of Inequality for Exponential Equations
$2x + 18 \geq -20$	Subtract $4x$ from each side.
$2x \geq -38$	Subtract 18 from each side.
$x \geq -19$	Divide each side by 2.

The solution set is $\{x \mid x \geq -19\}$.

Check

Solve $125^{x+2} \leq 25^{4x-7}$.

$\{x \mid x$ __?__ $\}$.

Go Online You can complete an Extra Example online.

Practice

Go Online You can complete your homework online.

Example 1

Solve each equation.

1. $25^{2x+3} = 25^{5x-9}$

2. $9^{8x-4} = 81^{3x+6}$

3. $4^{x-5} = 16^{2x-31}$

4. $4^{3x-3} = 8^{4x-4}$

5. $9^{-x+5} = 27^{6x-10}$

6. $125^{3x-4} = 25^{4x+2}$

Example 2

7. INTEREST Bianca invested \$5000 in an account that pays 5% annual interest.

a. Write a function that represents the value in Bianca's account y after x years.

b. After how many years will the value in Bianca's account be \$25,000? Round to the nearest tenth if necessary.

8. POPULATION In 2000, the world population was calculated to be 6,071,675,206. The world population increases at a rate of about 1.2% annually.

a. Write a function that represents the world population y after x years.

b. In what year will the world population be 8,000,000,000? Round to the nearest year if necessary.

9. BUSINESS Ahmed's consulting firm began with 23 clients. The number of clients decreases at an annual rate of 0.5%.

a. Write a function that represents the number of clients at Ahmed's consulting firm y after x years.

b. After how many years will Ahmed's consulting firm have 15 clients? Round to the nearest year if necessary.

10. BATTERY LIFE The battery life of a certain cell phone starts at 8 hours. The battery life decreases at an annual rate of 30%.

a. Write a function that represents the battery life y after x years.

b. After how many years will the battery life of a cell phone be 2 hours? Round to the nearest tenth if necessary.

Example 3

11. BANKING Siobhan deposits \$1200 into a savings account that pays 5.2% annual interest compounded monthly. What will be the balance after 4 years? Round to the nearest cent.

12. INVESTING Nancy deposits \$2500 into an investing account that pays 6.1% annual interest compounded quarterly. What will be the balance after 10 years? Round to the nearest cent.

13. CHECKING Maya deposits $5000 into a checking account that pays 0.75% annual interest compounded monthly. What will be the balance after 8 years? Round to the nearest cent.

14. FINANCE Mr. Fernandez deposits $60,000 into an account that pays 2.5% annual interest compounded quarterly. What will be the balance after 20 years? Round to the nearest cent.

Example 4

Solve each inequality.

15. $\left(\frac{1}{36}\right)^{6x-3} > 6^{3x-9}$

16. $64^{4x-8} < 256^{2x+6}$

17. $\left(\frac{1}{27}\right)^{3x+13} \leq 9^{5x-\frac{1}{2}}$

18. $\left(\frac{1}{9}\right)^{2x+7} \leq 27^{6x-12}$

19. $\left(\frac{1}{8}\right)^{-2x-6} > \left(\frac{1}{32}\right)^{-x+11}$

20. $9^{9x+1} < \left(\frac{1}{243}\right)^{-3x+5}$

Mixed Exercises

Solve each equation or inequality.

21. $8^{4x+2} = 64$

22. $5^{x-6} = 125$

23. $625 \geq 5^{a+8}$

24. $256^{b+2} = 4^{2-2b}$

25. $9^{3c+1} = 27^{3c-1}$

26. $\left(\frac{1}{27}\right)^{2d-2} \leq 81^{d+4}$

27. $\left(\frac{1}{2}\right)^{4x+1} = 8^{2x+1}$

28. $\left(\frac{1}{5}\right)^{x-5} = 25^{3x+2}$

29. $\left(\frac{1}{64}\right)^{c-2} < 32^{2c}$

30. $\left(\frac{1}{8}\right)^{3x+4} = \left(\frac{1}{4}\right)^{-2x+4}$

31. $\left(\frac{2}{3}\right)^{5x+1} = \left(\frac{27}{8}\right)^{x-4}$

32. $\left(\frac{1}{9}\right)^{3t+5} \leq \left(\frac{1}{243}\right)^{t-6}$

33. $81^{a+2} = 3^{3a+1}$

34. $10^{5b+2} > 1000$

35. $216 = \left(\frac{1}{6}\right)^{x+3}$

36. $8^{2y+4} = 16^{y+1}$

37. $\left(\frac{25}{81}\right)^{2x+1} = \left(\frac{729}{125}\right)^{-3x+1}$

38. $\left(\frac{1}{36}\right)^{w+2} \leq \left(\frac{1}{216}\right)^{4w}$

39. $4^{2x} = 8^{x+4}$

40. $\left(\frac{1}{3}\right)^{m} = 27^{m+2}$

41. $25^{4t+1} \geq 125^{2t}$

42. $6^{2x-1} = 36^{-x}$

43. BUSINESS Ingrid and Alberto each opened a business in 2010. Ingrid started with 2 employees, and in 2013 she had 50 employees. Alberto began with 32 employees, and in 2017 he had 310 employees. Since 2010, each company has experienced exponential growth.

a. Write an exponential equation representing the growth for each business.

b. Calculate the number of employees each company had in 2015.

c. Is it reasonable to expect that a business will continue to experience exponential growth? Explain your answer.

Write an exponential function with a graph that passes through the given points.

44. (0, 3) and (3, 375)

45. (0, −1) and (6, −64)

46. (0, 7) and (−2, 28)

47. $\left(0, \frac{1}{2}\right)$ and (2, 40.5)

48. (0, 15) and (1, 12)

49. (0, −6) and (−4, −1536)

50. $\left(0, \frac{1}{3}\right)$ and (3, 9)

51. (0, 1) and (6, 4096)

52. (0, −2) and (−1, −4)

53. USE A MODEL Josiah invested $2000 in an account that pays at least 4% annual interest. He wants to see how much money he will have over the next few years. Graph the inequality $y \geq 2000(1 + 0.04)^x$ to show his potential earnings.

54. ECONOMICS The Jones Corporation estimates that its annual profit could be modeled by $y = 10(0.99)^t$, while the Davis Company's annual profit is modeled by $y = 8(1.01)^t$. For both equations, profit is given in millions of dollars, and t is the number of years since 2015.

a. Find each company's estimated annual profit for the years 2015 and 2025 to the nearest dollar.

b. In which company would you prefer to own stock? Explain your reasoning.

55. MEDICINE After a patient is given a dose of medicine, the concentration in the bloodstream is 3.0 mg/mL. The concentration decays exponentially, and drops to 1.5 mg/mL after 2 hours. The medicine is ineffective at concentrations less than 0.6 mg/mL. If the patient is given a dose at 10 A.M., could the next dose be given at 3 P.M. without the level of medication dropping below the effective concentration? Use a graph to justify your answer.

56. STRUCTURE Dale and Xavier both invest for retirement. Their initial deposits, annual interest rates, and number of compounding periods are shown in the table. Who will have more money in their account after 30 years? How much more?

	Initial Deposit	Annual Interest Rate	Compounding Periods
Dale	$6000	2.5%	quarterly
Xavier	$3250	4.75%	monthly

57. USE A SOURCE Research the average college professor's current salary.

a. Suppose a professor making the average salary receives a 2% raise each year. What will be the professor's annual salary after working 15 more years?

b. Would it be better to start at $85,000 and receive a 3% raise each year? Explain.

58. ANALYZE Tom uses a graphing calculator to graph $f(x) = 5(1.25)^x$ and $g(x) = 1.5^x$. He notices that the gap between the curves increases as x increases, and concludes that $f(x) > g(x)$ for $x > 0$. Is he correct? Use a graph to support your reasoning. Could you have written and solved an equation to come to the same conclusion?

59. FIND THE ERROR Beth and Liz are solving $6^{x-3} > 36^{-x-1}$. Is either correct? Explain.

Beth

$6^{x-3} > 36^{-x-1}$

$6^{x-3} > (6^2)^{-x-1}$

$6^{x-3} > 6^{-2x-2}$

$x - 3 > -2x - 2$

$3x > 1$

$x > \frac{1}{3}$

Liz

$6^{x-3} > 36^{-x-1}$

$6^{x-3} > (6^2)^{-x-1}$

$6^{x-3} > 6^{-x+1}$

$x - 3 > -x + 1$

$2x > 4$

$x > 2$

60. PERSEVERE Solve $16^{18} + 16^{18} + 16^{18} + 16^{18} + 16^{18} = 4^x$.

61. ANALYZE What would be a more beneficial change to a 5-year loan at 8% interest compounded monthly: reducing the term to 4 years or reducing the interest rate to 6.5%? Justify your reasoning.

62. ANALYZE Determine whether the following statements are *sometimes*, *always*, or *never* true. Explain your reasoning.

a. $2^x > -8^{20x}$ for all values of x.

b. The graph of an exponential growth equation $y = ab^x$ is increasing.

c. The graph of an exponential decay equation $y = ab^x$ is increasing.

63. CREATE Write an exponential inequality with a solution of $x \leq 2$.

64. PERSEVERE Show that $27^{2x} \cdot 81^{x+1} = 3^{2x+2} \cdot 9^{4x+1}$.

65. WRITE If you were given the initial and final amounts of a radioactive substance and the amount of time that passes, how would you determine the rate at which the amount was increasing or decreasing in order to write an equation?

Special Exponential Functions

Today's Goals
- Analyze expressions and functions involving the natural base e.

Today's Vocabulary
e

Explore Finding the Value of e

Online Activity Use a table to complete the Explore.

INQUIRY How can you best approximate the value of *e*?

Learn Exponential Functions with Base e

The constant **e** has certain mathematical properties that make it a convenient base for exponential functions. e is the irrational number that $\left(1 + \frac{1}{n}\right)^n$ approaches as n approaches ∞. This value is approximately equal to 2.7182818...

Graphs of exponential functions with base e display the same general characteristics as other exponential functions.

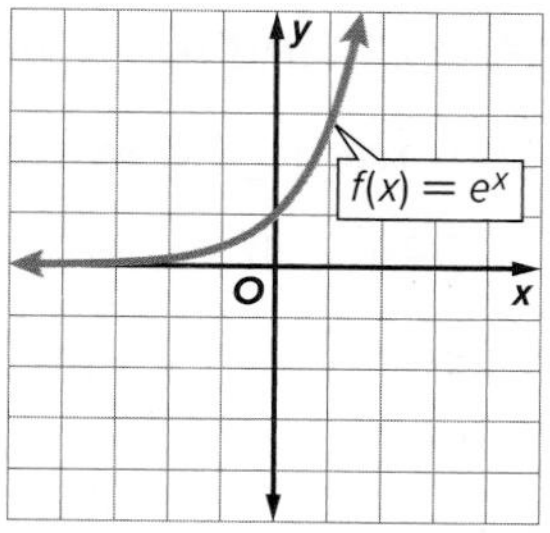

- $f(x)$ approaches an asymptote of $y = 0$ as x approaches $-\infty$.
- $f(x)$ approaches ∞ as x approaches ∞.
- The y-intercept is 1.

You can calculate continuously compounding interest by using the formula $A = Pe^{rt}$, where A is the amount in the account after t years, P is the principal amount invested, and r is the annual interest rate.

Example 1 Expressions with Base e

Simplify each expression.

a. $e^7 \cdot e^{-2}$

$e^7 \cdot e^{-2} = e^{7 + (-2)}$ Product of Powers

$= e^5$ Simplify.

b. $(-3e^{2x})^4$

$(-3e^{2x})^4 = (-3)^4(e^{2x})^4$ Power of a Product

$= 81e^{8x}$ Power of a Power

c. $\frac{42e^7}{14e^5}$

$\frac{42e^7}{14e^5} = 3e^{7-5}$ Quotient of Powers

$= 3e^2$ Simplify.

Go Online You can complete an Extra Example online.

Watch Out!

Calculators
Most calculators have an e button. However, you should use the properties of exponents to evaluate expressions so that your answers are exact.

Check

Simplify each expression.

$e^4 \cdot e^3 \cdot e$ $\quad$ $(6e^2)^4$ $\quad$ $\frac{60e^6}{12e^2}$

Example 2 Graph Functions with Base *e*

Consider $g(x) = -2e^{x-3} + 2$.

Part A Graph the function.

The function $g(x) = -2e^{x-3} + 2$ represents a transformation of the graph $f(x) = e^x$.

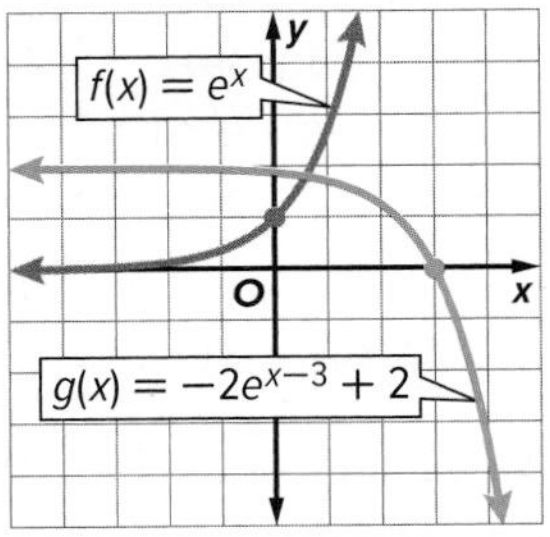

$a = -2$; The graph is reflected in the *x*-axis and stretched vertically.

$h = 3$; The graph is translated 3 units right.

$k = 2$; The graph is translated 2 units up.

Part B Determine the domain and range.

The domain is all real numbers. The range is all real numbers less than 2.

Part C Find the average rate of change.

Determine the average rate of change of g(x) over each interval.

a. [−4, −1]

Based on the graph, the graph from −4 to −1 appears approximately horizontal. So, the average rate of change should be close to 0.

$\frac{g(-1) - g(-4)}{-1 - (-4)} \approx \frac{1.963 - 1.998}{3}$ Evaluate $g(-1)$ and $g(-4)$.

$\approx \frac{-0.035}{3}$ or -0.012 Simplify.

b. [0, 3]

Based on the graph, the curve appears to pass through approximately (0,2) and (3,0). So, the average rate of change for the interval [0, 3] should be about $\frac{0-2}{3-0}$ or $-\frac{2}{3}$.

$\frac{g(3) - g(0)}{3 - 0} \approx \frac{0 - 1.900}{3}$ Evaluate $g(3)$ and $g(0)$.

$\approx \frac{-1.900}{3}$ or -0.633 Simplify.

c. [4, 8]

The graph is very steep for values greater than 4. So, the average rate of change from 4 to 8 should be a negative number with a large absolute value.

$\frac{g(8) - g(4)}{8 - 4} \approx \frac{-294.826 - (-3.437)}{4}$ Evaluate $g(8)$ and $g(4)$.

$\approx \frac{-291.389}{4}$ or -72.847 Simplify.

Go Online You can complete an Extra Example online.

Problem-Solving Tip

Use a Table For Parts A and C, you may find it easier to make a table of values for the function. You can use the values in the table to graph the function and find the values of *g*(*x*) for the given intervals.

Think About It!

Does an exponential function with base *e* behave in the same way as other exponential functions? Explain.

Check

Consider $g(x) = -e^{x+1} + 2$.

Part A

Graph the function.

Part B

Determine the domain and range of

$g(x) = -e^{x+1} + 2$.

Domain: ____?____

Range: ___?___

Part C

Find the average rate of change of $g(x) = -e^{x+1} + 2$ over each interval.

a. $[-6, -3] \approx$ __?__

b. $[-1, 1] \approx$ __?__

c. $[1, 4] \approx$ __?__

Example 3 Apply Functions with Base *e*

COMPOUND INTEREST **LaShawndra deposited $1500 in a savings account that pays 2.5% annual interest compounded continuously. After 4 years, what will be the balance of LaShawndra's account?**

To find the balance of LaShawndra's account after 4 years, use the formula for continuous exponential growth.

$A = Pe^{rt}$	Continuous Compounding Formula
$= 1500\, e^{0.025(4)}$	$P = 1500$, $r = 2.5\% = 0.025$, and $t = 4$
$= 1657.76$	Simplify.

After 4 years, LaShawndra will have $1657.76 in her savings account.

Check

COMPOUND INTEREST Alejandro invested $4500 in a startup company. His investment has been growing continuously at an annual rate of 12.5%.

Part A

Write the function that represents the situation, where A is the value of his investment after t years.

Part B

Evaluate the function.

After 5 years, if it continues to grow in this way Alejandro's investment will be worth $__?__.

After 12 years, it will be worth $___?___.

Go Online You can complete an Extra Example online.

Watch Out!

Rounding When using a calculator to evaluate an exponential function with base e, use the e or *exp* button on the calculator. Using a rounded value of *e* may result in an incorrect solution.

Think About It!

Suppose LaShawndra is also considering a savings account that offers 2.5% annual interest compounded monthly. Which savings account should she choose? Explain.

Example 4 Solve an Exponential Equation by Using Technology

Use a graphing calculator to solve $5 = 2e^x$ by using a system of equations.

Step 1 Write a system.

Set each side of $5 = 2e^x$ equal to y to create a system of equations.

$y = 5$	First equation
$y = 2e^x$	Second equation

Step 2 Graph the system.

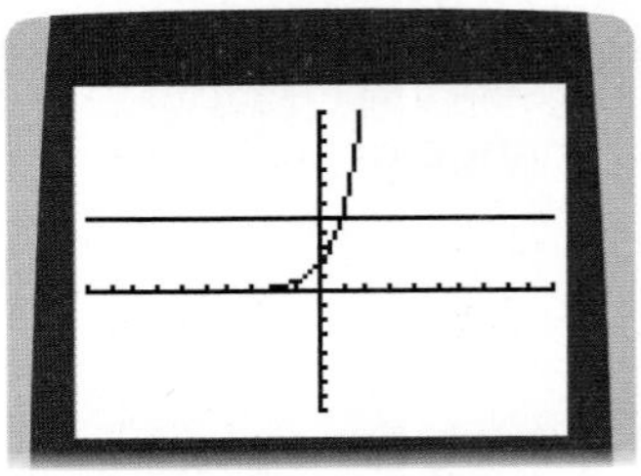

[−10, 10] scl: 1 by [−10, 10] scl: 1

Step 3 Find the intersection.

Use the **intersect** feature from the **CALC** menu to find the coordinates of the point of intersection. When prompted, select each graph. Press **ENTER** again to see the intersection.

The solution is the x-coordinate of the intersection, which is about 0.916.

Step 4 Use a table.

You can use the **TABLE** feature to verify the solution. First enter the starting value and the interval for the table. Because the graph shows the intersection at $x \approx 0.916$, use 0.915 as the starting value and 0.001 as the interval. Scroll through the table.

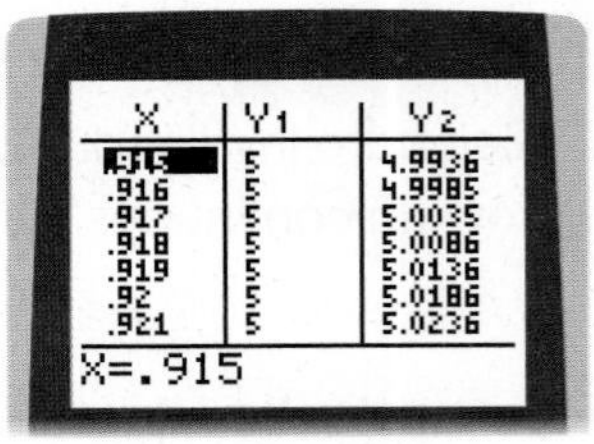

X	Y1	Y2
.915	5	4.9936
.916	5	4.9985
.917	5	5.0035
.918	5	5.0086
.919	5	5.0136
.92	5	5.0186
.921	5	5.0236

X=.915

Notice that when $x \approx 0.916$, **Y1** and **Y2** are approximately equal. This confirms the solution $x \approx 0.916$.

Check

Use a graphing calculator to solve $-6 = -1.5e^{x-3}$ by using a system of equations. Round to the nearest hundredth. $x \approx$ ___?___

Go Online You can complete an Extra Example online.

Study Tip

Estimation Before using the intersection feature on a graphing calculator, analyze the graph and estimate the value of the intersection. This can help you detect potential errors while using the calculator.

Practice

Go Online You can complete your homework online.

Example 1

Simplify each expression.

1. $e^8 \cdot e^3$

2. $e^4 \cdot e^{-1}$

3. $e^9 \cdot e^{-7}$

4. $(2e^{3x})^2$

5. $(3e^{4x})^3$

6. $(-4e^{5x})^2$

7. $\frac{26e^4}{13e}$

8. $\frac{-39e^7}{13e^2}$

9. $\frac{-16e^9}{2e^3}$

Example 2

10. Consider the function $f(x) = 3e^{x-1} + 3$.

a. Graph the function.

b. Determine domain and range.

c. Find the average rate of change over the interval $[-5, -2]$.

11. Consider the function $f(x) = 4e^{2x} - 1$.

a. Graph the function.

b. Determine domain and range.

c. Find the average rate of change over the interval $[-3, -1]$.

12. Consider the function $f(x) = -2e^{x+3} + 2$.

a. Graph the function.

b. Determine domain and range.

c. Find the average rate of change over the interval $[-7, -4]$.

Example 3

13. COMPOUND INTEREST Ryan invested \$5000 in an account that grows continuously at an annual rate of 2.5%.

a. Write the function that represents the situation, where A is the value of Ryan's investment after t years.

b. What will Ryan's investment will be worth after 7 years?

14. SAVINGS Jariah invested \$6500 in a savings account that grows continuously at an annual rate of 3.25%.

a. Write the function that represents the situation, where A is the value of Jariah's investment after t years.

b. What will Jariah's investment will be worth after 18 years?

15. INVESTMENTS Marcella invested \$12,750 in a company. Her investment has been growing continuously at an annual rate of 5.5%.

a. Write the function that represents the situation, where A is the value of Marcella's investment after t years.

b. What will Marcella's investment will be worth after 9 years?

Example 4

Use a graphing calculator to solve each equation by using a system of equations. Round to the nearest hundredth if necessary.

16. $1 = e^{3x}$

17. $-2 = -e^{4x}$

18. $3 = \frac{1}{2}e^{x}$

19. $5 = \frac{1}{2}e^{x+2} - 4$

20. $-3 = -2e^{x} - 1$

21. $-\frac{11}{2} = \frac{2}{3}e^{x+1} - 6$

Mixed Exercises

Simplify each expression.

22. $e^{\frac{1}{8}} \cdot e^{\frac{5}{8}}$

23. $e^{\frac{1}{2}} \cdot e^{-\frac{3}{4}}$

24. $\left(\frac{1}{3}e^{2x}\right)^2$

25. $\left(-\frac{2}{5}e^{\frac{1}{4}x}\right)^3$

26. $\frac{100e^{\frac{9}{10}}}{4e^{\frac{1}{5}}}$

27. $\frac{-216\left(e^{\frac{1}{2}}\right)^5}{\left(3e^{\frac{1}{4}}\right)^2}$

USE TOOLS **Solve each equation. Round to the nearest hundredth if necessary.**

28. $e^{x} = -2e^{x-1} + 3$

29. $-3e^{x} + 1 = 4e^{x+2} - 5$

30. $\frac{1}{2}e^{x+2} - 3 = -\frac{1}{4}e^{2x-4} + 2$

31. REASONING Hailey put \$900 in a savings account that earns 1.2% annual interest compounded continuously. How much will her account be worth in 5 years? Write an equation to represent the problem, then solve your equation.

32. CONSTRUCT ARGUMENTS Justify the conclusion that all functions of the form $f(x) = ae^{x}$ have the same domain.

33. USE TOOLS Graph the function $g(x) = 7e^{x} + 8$. Identify its domain and range, intercepts, zeros, asymptote, and end behavior.

34. USE TOOLS Find the zero of the function $f(x) = 0.8e^{2x} - 9$. Round to the nearest hundredth if necessary.

35. WRITE The compound interest formula can also be used to show population growth, using final and initial populations, growth rate, and time of growth. How do those variables correspond to the variables in the interest formula? Write a paragraph to explain your reasoning.

36. CREATE Write a problem using the formula for compound interest that solves for the initial amount; then solve your problem.

37. ANALYZE Make a conjecture about why you should use the properties of exponents to simplify exponential functions with base *e* before evaluating them on a calculator.

38. FIND THE ERROR Aza is investing \$2000 in an account that grows at 0.7% annual interest compounded continuously. He predicts that he will have a balance of \$16,332.34 after 3 years. Is Aza correct? Explain your reasoning.

39. WHICH ONE DOESN'T BELONG? Which of the functions doesn't belong? Explain your reasoning.

$g(x) = 3e^{2x}$

$g(x) = 5e^{x-4}$

$g(x) = 2e^{x} + 9$

Geometric Sequences and Series

Today's Goals
- Generate geometric sequences.
- Find sums of geometric series.

Today's Vocabulary
sequence
term of a sequence
finite sequence
infinite sequence
geometric sequence
common ratio
explicit formula
recursive formula
geometric means
series
geometric series
sigma notation

Explore Explicit and Recursive Formulas

Online Activity Use a concrete model to complete the Explore.

INQUIRY How can a geometric sequence be defined?

Learn Sequences

A **sequence** is a set of numbers in a particular order or pattern. Each number in a sequence is called a **term**. The first term of a sequence is denoted a_1, the second term is a_2, and so on. A **finite sequence** contains a limited number of terms, while an **infinite sequence** continues without end.

A sequence can be defined as a function.

Key Concept • Sequences as Functions		
Words	A sequence is a function in which the domain consists of natural numbers, and the range consists of real numbers.	
Symbols	Domain: 1 2 3 ... n Range: a_1 a_2 a_3 ... a_n	the position of a term the terms of the sequence

In a **geometric sequence**, each term is determined by multiplying a nonzero constant by the previous term. The constant value is called the **common ratio**.

Key Concept • *n*th Term of a Geometric Sequence

The nth term a_n of a geometric sequence in which the first term is a_1, and the common ratio is r is given by the formula $a_n = a_1r^{n-1}$, where n is any natural number.

An **explicit formula** like the one above allows you to find any term a_n of a sequence by using a formula written in terms of n. A **recursive formula** allows you to find the nth term of a sequence by performing operations to one or more of the preceding terms. The recursive formula for a geometric sequence is $a_n = r \cdot a_{n-1}$, where a_{n-1} is the term in the sequence before the nth term.

Geometric means are the terms between two nonconsecutive terms of a geometric sequence.

Go Online to see another example about identifying geometric sequences.

Example 1 Identify Geometric Sequences

Determine whether the sequence 2, 6, 15, 30, ... is geometric.

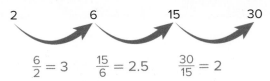

2 → 6 → 15 → 30

$\frac{6}{2} = 3$ $\frac{15}{6} = 2.5$ $\frac{30}{15} = 2$

The ratios are not the same, so the sequence is not geometric.

Talk About It!
The domain of the six terms you plotted is {1, 2, 3, 4, 5, 6}. Is this also the domain of the original sequence? Explain your reasoning.

Example 2 Graph Geometric Sequences

Find and graph the first six terms of the geometric sequence −1, −2, −4, ...

Step 1 Find the common ratio. Using the first two terms, $\frac{-2}{-1} = 2$.

Step 2 Find the next three terms.

Multiply the previous term, −4, by the common ratio. Continue multiplying by 2 to find the following terms.

−4 ×2 → **−8** ×2 → **−16** ×2 → **−32** ...

Step 3 Graph the sequence.

Domain: {1, 2, 3, 4, 5, 6}

Range: {−1, −2, −4, −8, −16, −32}

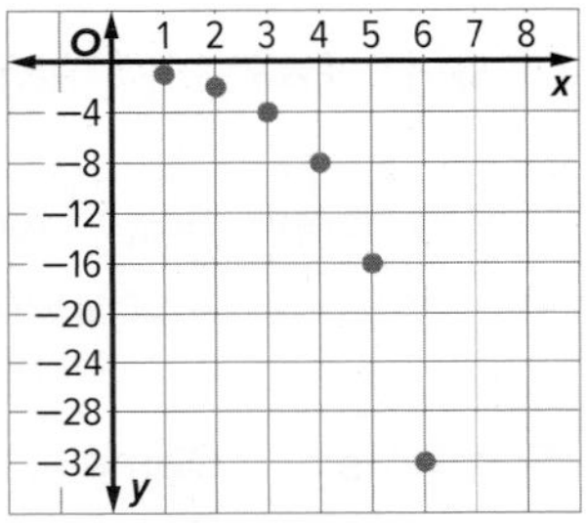

Watch Out!
First Term
Although Javier starts with a 50-gram chocolate bar, the amount he eats on the first day is 25 grams.

Apply Example 3 Find the *n*th Term

CHOCOLATE **Javier wants to eat less chocolate each day. To achieve this, he decides to start with a 50-gram bar of chocolate and eat half of it the first day, then half of what is left of the bar the second day, then half of that the third day, and so on. If Javier starts on a Monday, what is the mass of the piece that he will eat on Friday?**

1 What is the task?

Describe the task in your own words. Then list any questions that you may have. How can you find answers to your questions?

I need to find how much chocolate Javier eats on Friday. What term of the sequence represents Friday?

Go Online You can complete an Extra Example online.

2 How will you approach the task? What have you learned that you can use to help you complete the task?

I know the formula for the *n*th term of a geometric sequence. I will evaluate it for the term that represents Friday.

3 What is your solution?

Use your strategy to solve the problem.

What formula represents the *n*th term of the sequence? $a_n = 25\left(\frac{1}{2}\right)^{n-1}$
Which term represents Friday? a_5
How much chocolate will Javier eat on Friday? about 1.6 g

4 How can you know that your solution is reasonable?

Write About It! Write an argument that can be used to defend your solution.

The amount of chocolate each day is half of the amount the day before. The sequence is 25, 12.5, 6.25, 3.125, 1.5625, ... , so on Friday, Javier eats about 1.6 grams of chocolate.

Think About It!

If you let $a_1 = 50$, then what would a_n represent?

Example 4 Write an Equation for the *n*th Term

Write an equation for the *n*th term of each geometric sequence.

a. −1, 3, −9, 27, ...

Step 1 Find *r*.

$r = \frac{a_3}{a_2}$ — Divide two consecutive terms.

$= \frac{-9}{3}$ or -3 — $a_2 = 3$ and $a_3 = -9$

Step 2 Write the equation.

$a_n = a_1 r^{n-1}$ — *n*th term of a geometric sequence

$= -1(-3)^{n-1}$ or $-(-3)^{n-1}$ — $a_1 = -1$ and $r = -3$

b. $a_8 = 5$ and $r = \frac{1}{2}$

Step 1 Find a_1.

$a_n = a_1 r^{n-1}$ — *n*th term of a geometric sequence

$5 = a_1\left(\frac{1}{2}\right)^{8-1}$ — $n = 8$, $a_8 = 5$, and $r = \frac{1}{2}$

$5 = a_1\left(\frac{1}{2}\right)^{7}$ — Simplify the exponent.

$5 = a_1\left(\frac{1}{128}\right)$ — Evaluate the exponent.

$a_1 = 640$ — Solve for a_1.

Step 2 Write the equation.

$a_n = a_1 r^{n-1}$ — *n*th term of a geometric sequence

$= 640\left(\frac{1}{2}\right)^{n-1}$ — $a_1 = 640$ and $r = \frac{1}{2}$

Go Online You can complete an Extra Example online.

Example 5 Recursive and Explicit Formulas

Given a formula for a geometric sequence in recursive or explicit form, translate it to the other form.

a. $a_1 = 5, a_n = \frac{1}{5}a_{n-1}, n \geq 2$

Because a_n is defined in terms of the previous term, $a_n = \frac{1}{5}a_{n-1}$ is a recursive formula of the form $a_n = r \cdot a_{n-1}$. Thus, $r = \frac{1}{5}$. Now, write the explicit formula.

$a_n = a_1r^{n-1}$	Explicit formula for a geometric sequence
$= 5\left(\frac{1}{5}\right)^{n-1}$	$a_1 = 5$ and $r = \frac{1}{5}$

The explicit formula for $a_1 = 5, a_n = \frac{1}{5}a_{n-1}, n \geq 2$ is $a_n = 5\left(\frac{1}{5}\right)^{n-1}$.

b. $a_n = \frac{2}{3}(7)^{n-1}$

Because a_n is defined in terms of n, $a_n = \frac{2}{3}(7)^{n-1}$ is an explicit formula of the form $a_n = a_1r^{n-1}$.

Thus, $a_1 = \frac{2}{3}$ and $r = 7$. Now, write the recursive formula.

$a_n = ra_{n-1}$	Recursive formula for a geometric sequence
$= 7a_{n-1}$	$r = 7$

The recursive formula for $a_n = \frac{2}{3}(7)^{n-1}$ is $a_1 = \frac{2}{3}, a_n = 7a_{n-1}, n \geq 2$.

Watch Out!

Signs Keep in mind that when r is raised to an even power, r may be positive or negative.

Example 6 Find Geometric Means

Find four geometric means between 243 and 32.

Step 1 Find the total number of terms.

Because there are four terms between the first and last term, there are 4 + 2 or 6 total terms, so $n = 6$.

Step 2 Find *r*.

$a_n = a_1r^{n-1}$	nth term of a geometric sequence
$32 = 243r^{6-1}$	$n = 6$, $a_6 = 32$, and $a_1 = 243$
$\frac{32}{243} = r^5$	Divide each side by 243.
$r = \frac{2}{3}$	Take the 5th root of each side.

Think About It!

Would the geometric means change if the sequence started at 32 and ended at 243? Explain your reasoning.

Step 3 Use *r* to find four geometric means.

243 162 108 72 48 32

$\times\frac{2}{3}$ $\times\frac{2}{3}$ $\times\frac{2}{3}$ $\times\frac{2}{3}$ $\times\frac{2}{3}$

The geometric means are 162, 108, 72, and 48.

Learn Geometric Series

A **series** is the sum of the terms in a sequence. The sum of the first n terms of a series is denoted $\mathbf{S_n}$. A **geometric series** is the sum of the terms of a geometric sequence.

Key Concept • Partial Sums of a Geometric Series

Given	The sum of S_n of the first n terms is:
a_1, r, and n	$S_n = \frac{a_1 - a_1r^n}{1 - r}, r \neq 1$
a_1, r, and a_n	$S_n = \frac{a_1 - a_nr}{1 - r}, r \neq 1$

Go Online Derive the formula for the sum of a finite geometric series in Expand 7-4.

Think About It!

Why does $r \neq 1$ for S_n?

The sum of a series can be written in shorthand by using **sigma notation**, which uses the Greek uppercase letter S to indicate that you should find a sum.

Key Concept • Sigma Notation

Symbols

$$\sum_{k=1}^{n} f(k)$$

last value of k → n; first value of k → $k=1$; $f(k)$ ← formula for the terms of the series

Examples

Arithmetic series:

$$\sum_{k=1}^{6} (3k + 2) = [3(1) + 2] + [3(2) + 2] + [3(3) + 2] + ... + [3(6) + 2]$$
$$= 5 + 8 + 11 + ... + 20$$
$$= 75$$

Geometric series:

$$\sum_{k=1}^{8} 5(3)^{k-1} = 5(3)^{1-1} + 5(3)^{2-1} + ... + 5(3)^{8-1}$$
$$= 5(1) + 5(3) + ... + 5(2187)$$
$$= 5 + 15 + ... + 10{,}935$$
$$= 16{,}400$$

Example 7 Find the Sum of a Geometric Series

DOMINOS **Kateri wants to set up some dominos so she can knock over one, which knocks over two more, each of which knocks over two more, and so on. If she wants to make 6 rows of dominos, how many will she need in total?**

The first row has one domino. So, $a_1 = 1$, $r = 2$, and $n = 6$.

$S_n = \frac{a_1 - a_1r^n}{1 - r}$ Sum formula

$S_n = \frac{1 - 1(2)^6}{1 - 2}$ $a_1 = 1$, $r = 2$, and $n = 6$

$S_n = \frac{-63}{-1}$ Simplify the numerator and the denominator.

$S_n = 63$ Divide.

Go Online You can complete an Extra Example online.

Study Tip

Viability When determining the sum of a geometric series, consider whether the solution makes sense in the context of the situation. For example, if S_n were negative for this problem, that would mean Kateri needs a negative number of dominos, which would not make sense in this context.

Example 8 Find the First Term in a Series

Find a_1 in a geometric series for which $S_n = 21.3125$, $n = 5$, and $r = \frac{1}{2}$. Round to the nearest tenth if necessary.

$S_n = \dfrac{a_1 - a_1r^n}{1 - r}$	Sum formula
$21.3125 = \dfrac{a_1 - a_1\left(\frac{1}{2}\right)^5}{1 - \frac{1}{2}}$	$S_n = 21.3125$, $r = \frac{1}{2}$, and $n = 5$
$21.3125 = \dfrac{a_1 - \frac{1}{32}a_1}{\frac{1}{2}}$	Simplify.
$21.3125 = \dfrac{\frac{31}{32}a_1}{\frac{1}{2}}$	Subtract.
$21.3125 = \frac{31}{16}a_1$	Divide.
$a_1 = 11$	Multiply each side by $\frac{16}{31}$.

So, the first term is 11.

Watch Out!

Sigma Notation
Because the series starts when $k = 5$, $a_1 = 3(-2)^{5-1}$ or 48, rather than $a_1 = 3(-2)^{1-1}$ or 3.

Example 9 Sum in Sigma Notation

Find $\sum_{k=5}^{12} 3(-2)^{k-1}$. If necessary, round your answer to the nearest tenth.

Find a_1, r, and n.

In the first term, $k = 5$, and $a_1 = 3(-2)^{5-1}$ or 48.

The base of the exponential function is r, so $r = -2$.

There are $12 - 5 + 1$ or 8 terms, so $n = 8$.

$S_n = \dfrac{a_1 - a_1r^n}{1 - r}$	Sum formula
$= \dfrac{48 - 48(-2)^8}{1 - (-2)}$	$a_1 = 48$, $r = -2$, and $n = 8$
$= \dfrac{48 - 48\,(256)}{3}$	Simplify.
$= -4080$	Subtract and divide.

So, S_n is -4080.

Check

Find $\sum_{k=2}^{6} 23(2.5)^{k-1}$. Round to the nearest tenth if necessary.

$S_n =$ ___?___

Go Online to learn how to derive the formula for the sum of a finite geometric series in Expand 7-4.

Go Online You can complete an Extra Example online.

Practice

Go Online You can complete your homework online.

Example 1

Determine whether each sequence is geometric.

1. 5, 7, 9, 11, 13, ...

2. 3, 6, 12, 24, 48, ...

3. 2, −4, 8, −16, 32, ...

4. 400, 200, 100, 50, 25, ...

5. 60, 48, 36, 24, 12, ...

6. 8000, −2000, 500, −125, ...

Example 2

Find and graph the first six terms of each geometric sequence.

7. 1, 2, 4, ...

8. 2, 6, 18, ...

9. −1, −3, −9, ...

10. 800, −400, 200, ...

11. 512, −128, 32, ...

12. 160, 80, 40, ...

Example 3

13. FAMILY Amanda is researching her ancestry. She records names and birth dates for her parents, their parents, and so on, in an online research tool. If she can locate all of the information, how many names will Amanda record in the generation that is 5 generations before her?

14. MOORE'S LAW Gordon Moore, co-founder of Intel, suggested that the number of transistors on a square inch of integrated circuit in a computer chip would double every 18 months. Assuming Moore's law is true, how many times as many transistors would you expect on a square inch of integrated circuit in year 6?

Example 4

Write an equation for the *n*th term of each geometric sequence.

15. 3, 9, 27, ...

16. −1, −3, −9, ...

17. 2, −6, 18, ...

18. 5, 10, 20, ...

19. $a_4 = 324$ and $r = 3$

20. $a_3 = 512$ and $r = \frac{1}{8}$

Example 5

Given a formula for a geometric sequence in recursive or explicit form, translate it to the other form.

21. $a_1 = 3, a_n = 0.6a_{n-1}, n \geq 2$

22. $a_n = 0.8(2)^{n-1}$

23. $a_1 = -1, a_n = \frac{1}{2}a_{n-1}, n \geq 2$

24. $a_n = -\frac{2}{3}(6)^{n-1}$

Example 6

Find the geometric means of each sequence.

25. 4, $\underline{\ ?\ }$, $\underline{\ ?\ }$, $\underline{\ ?\ }$, 64

26. 1, $\underline{\ ?\ }$, $\underline{\ ?\ }$, $\underline{\ ?\ }$, 81

27. 38; 228; $\underline{\ ?\ }$; 8208; 49,248; ...

28. 51; $\underline{\ ?\ }$; 4131; $\underline{\ ?\ }$; 334,611; ...

Example 7

29. SCIENTIFIC RESEARCH Scientific balloons carry equipment to observe or conduct experiments. The NASA Balloon Program generally tries to fly balloons above 80,000 to 90,000 feet. Suppose a balloon rises 1000 feet in the first minute after it is launched. For the next hour, each minute it rises 1% more than it rose in the previous minute.

a. Copy and complete the table to show the height of the balloon at various times after launch.

Time (min)	1	2	3	4	5	6
Height (ft)						

b. After an hour will the balloon have reached its target height of 80,000 – 90,000 feet? Explain.

Example 8

Find a_1 for each geometric series described.

30. $S_n = 1295$, $r = 6$, $n = 4$

31. $S_n = 1640$, $r = 3$, $n = 8$

32. $S_n = 218\frac{2}{5}$, $a_n = 1\frac{2}{5}$, $r = \frac{1}{5}$

33. $S_n = -342$, $a_n = -512$, $r = -2$

Example 9

Find the sum of each geometric series.

34. $\sum_{k=1}^{7} 4(-3)^{k-1}$

35. $\sum_{k=1}^{8} (-3)(-2)^{k-1}$

36. $\sum_{k=1}^{9} (-1)(4)^{k-1}$

37. $\sum_{k=1}^{10} 5(-1)^{k-1}$

Mixed Exercises

Find a_n for each geometric sequence.

38. $a_1 = 5$, $r = 2$, $n = 6$

39. $a_1 = 18$, $r = 3$, $n = 6$

40. $a_1 = -3$, $r = -2$, $n = 5$

41. $a_1 = -20$, $r = -2$, $n = 9$

42. $a_1 = 65{,}536$, $r = \frac{1}{4}$, $n = 6$

43. $a_1 = -78{,}125$, $r = \frac{1}{5}$, $n = 9$

Determine whether each sequence is geometric. If a sequence is geometric, write an equation for the *n*th term.

44. 4, 7, 13, 25, ...

45. 25, 175, 1225, 8575, ...

46. −16,384; −8192; −4096; −2048; ...

47. −15, 30, −60, 120, −240, ...

Find the sum of each geometric series to the nearest ten-thousandth.

48. $a_1 = 36, r = \frac{1}{3}, n = 8$

49. $a_1 = 16, r = \frac{1}{2}, n = 9$

50. $a_1 = 240, r = \frac{3}{4}, n = 7$

51. $a_1 = 360, r = \frac{4}{3}, n = 8$

52. Find the missing term in the geometric sequence 64, 96, 144, 216, __?__.

53. Find the first five terms of the geometric sequence for which $a_1 = 3$ and $r = -2$.

54. Find the sum of a geometric series for which $a_1 = 3125$, $a_n = 1$, and $r = \frac{1}{5}$.

55. Find a_1 in a geometric series for which $S_n = 3045$, $r = \frac{2}{5}$ and $a_n = 120$.

56. Find a_1 in a geometric series for which $S_n = -728$, $r = 3$, and $n = 6$.

57. Write an equation to find the *n*th term of the sequence shown in the graph.

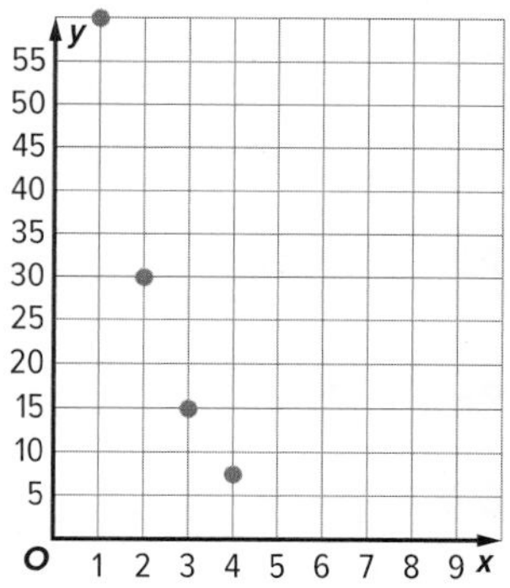

58. Write an equation to find the *n*th term of the geometric sequence $\frac{1}{3}, \frac{2}{9}, \frac{4}{27}, \ldots$.

59. Find the missing terms in the geometric sequence $\frac{729}{64}$, __?__, __?__, __?__, $\frac{324}{9}$.

60. Find two geometric means between 3 and 375.

61. Find two geometric means between 16 and −2.

62. BIOLOGY Mitosis is a process of cell division that results in two identical daughter cells from a single parent cell. The table illustrates the number of cells produced after each of the first 5 cell divisions.

Division Number	0	1	2	3	4	5
Number of Cells	1	2	4	8	16	32

a. Do the entries in the “Number of Cells” row form a geometric sequence? If so, find *r*.

b. Write formulas in explicit and recursive form to find the *n*th term of the sequence. *Hint*: Because the domain of the real-world situation starts at 0, the initial term of the sequence is a_0 instead of a_1.

c. Find the number of cells after 100 divisions.

63. CONSTRUCT ARGUMENTS A grand prize winner has two choices for receiving their winnings. They can choose an \$80 million immediate payout, or 20 annual payments that begin at \$4 million and increase 2% each year. Which option would you advise the winner to take? Explain your reasoning.

64. STRUCTURE An athlete makes a running plan using the formula $a_n = 20(1.1)^{n-1}$, where a_n is the target number of miles for the nth week of training.

a. Interpret the formula in terms of the situation.

b. How many total miles will the athlete have run after training for 10 weeks?

c. The athlete hopes to run a total of 1000 miles in preparation for a race at the end of 20 weeks of training. Will she reach this goal? Explain your reasoning.

d. Another athlete increased running mileage by the same ratio each week but reached 1000 miles in 15 weeks. Approximately how many miles did he run for the first week of training? Explain your reasoning.

65. USE A MODEL A geometric series can be combined with compound interest to find a formula for payments on a loan or mortgage. For a mortgage with a principal P, a monthly interest rate r, and consisting of n monthly payments, each monthly payment is $\frac{Pr}{1-(1+r)^{-n}}$.

a. For a 30-year \$300,000 mortgage with a 6% annual interest rate, how much interest is collected over the life of the loan? Hint: Use r = annual interest rate ÷ 12.

b. Suppose a family can afford to make monthly payments of up to \$1250. If annual interest rate is 4%, what is the greatest 20-year mortgage they can afford? Round your answer to the nearest \$1000.

Higher-Order Thinking Skills

66. PROOF Derive the explicit formula for a geometric sequence using the recursive formula.

67. CREATE Write a geometric series for which $r = \frac{3}{4}$ and $n = 6$.

68. ANALYZE Explain how $\sum_{k=1}^{10} 3(2)^{k-1}$ needs to be altered to refer to the same series if $k = 1$ changes to $k = 0$. Explain your reasoning.

69. PROOF Prove the formula for the nth term of a geometric sequence.

70. PERSEVERE The fifth term of a geometric sequence is $\frac{1}{27}$th of the eighth term. If the ninth term is 702, what is the eighth term?

71. PERSEVERE Use the fact that h is the geometric mean between x and y in the figure at the right to find h^4 in terms of x and y.

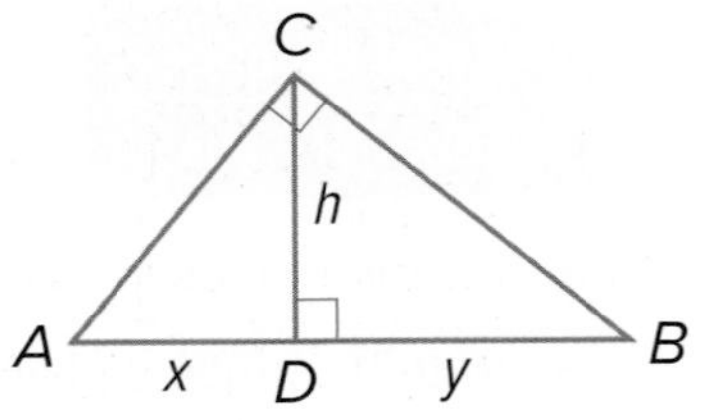

72. CREATE Write a geometric series with 6 terms and a sum of 252.

73. WRITE How can you classify a sequence as arithmetic, geometric, or neither? Explain your reasoning.

Modeling Data

Explore Modeling Exponential Decay

Online Activity Use a concrete model to complete the Explore.

> **INQUIRY** How can experimental data be used to predict outcomes?

Today's Goals
- Choose the best function type to model sets of data.

Today's Vocabulary
regression function
coefficient of determination

Learn Choosing the Best Model

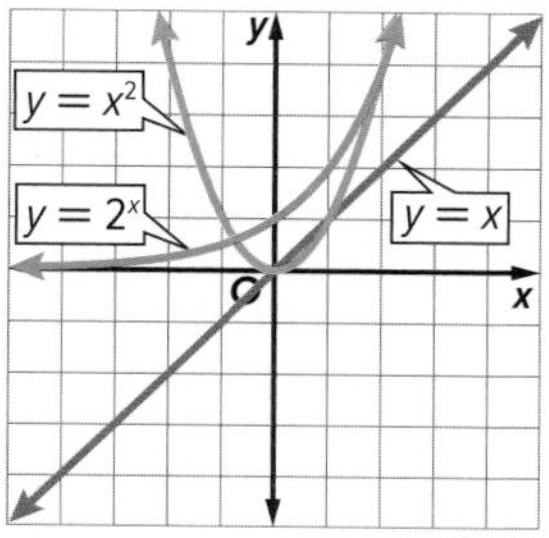

Recall that linear, quadratic, and exponential functions grow differently. Linear functions have linear graphs. Quadratic and exponential functions are nonlinear, with exponential functions growing faster than quadratic functions.

You can use a graphing calculator to find a **regression function**, which is a function generated by an algorithm to find a line or curve that fits a set of data. Calculators may also compute a number r^2, called the **coefficient of determination**. This measure shows how well data are modeled by the regression function. An r^2-value close to 1 is an indication of a strong fit to the data.

Think About It!
Describe the reliability of using a regression function to make predictions for domain values much greater or less than the values in the data set.

Example 1 Examine Scatter Plots

Use a graphing calculator to make a scatter plot of the data in the table. Then determine whether the data are best modeled by a *linear, quadratic*, or *exponential* function.

x	−5	−4	−3	−2	−1	0	1	2	3	4
y	−3	−2.7	−2.4	−2	−1	0	2.8	8	17.2	31

Step 1 Enter the data.

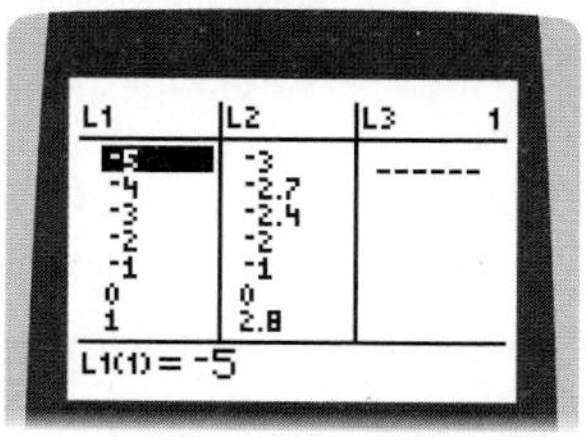

Step 2 Make a scatter plot.

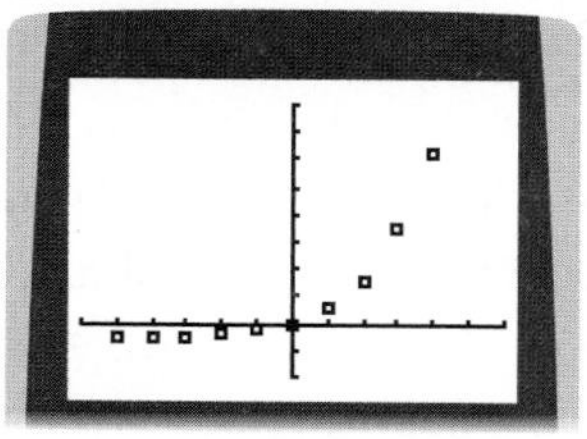

[−6, 6] scl: 1 by [−10, 40] scl: 5

(continued on the next page)

Go Online You can complete an Extra Example online.

Step 3 Examine the scatter plot.

The scatter plot appears to have an asymptote at $y = -3$, and it increases rapidly as $x \to \infty$. The data are likely best modeled by an exponential function.

Talk About It!

Based on the scatter plot, is it reasonable to think that these data may be modeled by another type of function? Justify your argument.

Check

Use a graphing calculator to make a scatter plot of the data. Then determine whether the data are best modeled by a *linear*, *quadratic*, or *exponential* function.

x	−2	−1	0	1	2	3	4	5	6
y	3.8	4.1	4.0	5.2	8.8	13.0	21.3	37.0	66.3

The data are likely best modeled by a(n) ____?____ function.

Example 2 Model Data by Using Technology

COFFEE **Forty years after opening their first shop, a popular coffee company grew to over 17,000 locations worldwide. The table shows the number of shops in the years since the company began. Use the data to determine a best-fit model. Then, use the model to make predictions.**

Years Since Opening	Shops	Years Since Opening	Shops
0	1	28	2498
16	17	30	4709
18	55	32	7225
20	116	34	10,241
22	272	36	15,011
24	677	38	16,635
26	1412	40	17,003

Steps 1 and 2 Enter the data and make a scatter plot.

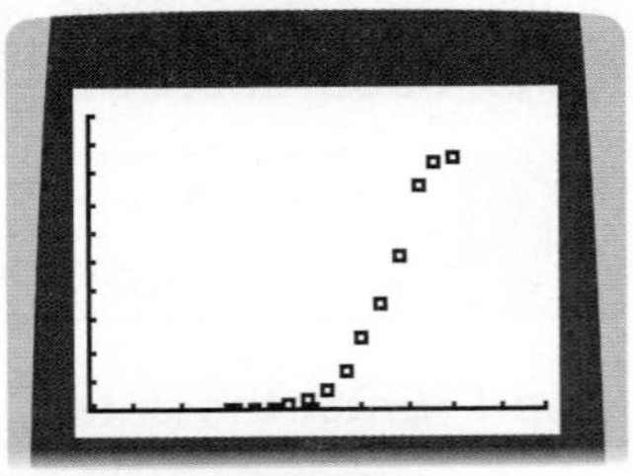
[0, 50] scl: 5 by [0, 20,000] scl: 2000

Step 3 Determine the function of best fit.

To determine the model that best fits the data, perform quadratic and exponential regressions, and compare the coefficients of determination, r^2. The function with a coefficient of determination closest to 1 will fit the data best. The exponential regression has the coefficient of determination closer to 1 for this data. $r^2 \approx 0.964$, which is a strong fit for the data. The regression equation with coefficients rounded to the nearest thousandth is $y \approx (0.655)1.317^x$.

Think About It!

What approximations were made when determining the number of shops at a given year?

Step 4 Evaluate using the regression function.

10 years = 10 shops   42 years = 70,116 shops

Practice

Go Online You can complete your homework online.

Example 1

Use a graphing calculator to make a scatter plot of the data in the table. Then determine whether the data are best modeled by a *linear*, *quadratic*, or *exponential* function.

1.

x	0	−1.4	3	−2.8	5	6.2	8.3	12.1	−4.5
y	2	0.9	5.5	−0.4	7.5	8.3	10.9	14.5	−2

2.

x	0	1	4	0.5	0.8	−0.5	0.4	0.2	2
y	0.4	1.7	8	0.9	1	0.3	0.6	0.05	4.5

3.

x	0	2	−3	1	−0.5	2.5	−2	−1	0.5
y	8	1	−1	4	7	1.5	1.5	4.5	6

Example 2

4. SAVINGS Brittany deposited $1000 into an account. She made no further deposits or withdrawals. The table shows the account balance for a period of 10 years.

Time (years)	Account Balance ($)
0	1000
2	1068.50
4	1155.55
6	1265.25
8	1400.35
10	1560.85

a. Use a graphing calculator to draw a scatterplot of the data.

b. Use the regression feature of the graphing calculator to determine the best model. Round coefficients and constants to the nearest hundredth.

c. About how much will be in the account after 15 years?

5. TEST GRADES A professor conducted a survey by asking 12 students how many hours they spent preparing for the final exam. The professor then matched the students' responses with their final exam grades.

Grade	Hours	Grade	Hours
96	8	93	7.5
84	6	86	6.5
89	5.5	77	5
55	1.5	60	2.25
52	1.25	77	5
60	2	83	6

a. Graph and analyze the data set. State if it is *linear* or *exponential*. Write an equation for the best-fit model.

b. Based on your graph, what is the minimum number of hours that you should spend studying if you want to get a passing grade of 60% or better on the final exam?

6. SCIENCE EXPERIMENTS Shawnte devised an experiment for measuring the average time before a soap bubble pops. She creates 64 identical bubbles and documents the number that have not popped at various times.

Minutes (x)	0	4	8	12	16	20
Population (y)	64	42	19	8	2	1

a. Graph the data.

b. What is the best model? What is the regression function?

c. When would you expect there to be 32 unpopped bubbles?

Mixed Exercises

USE TOOLS **Use a graphing calculator to make a scatter plot of the data. Then determine whether the data are best modeled by a *linear*, *quadratic*, or *exponential* function.**

7. (0, 1), (−1, −3), (10, 41), (2, 9), (7, 29), (−9, −35), (5, 21)

8. (0, 1), (6, 64), (2, 4), (4, 16), (3, 8), (5, 32), (1, 2)

Describe the correlation for each value of *r*.

9. $r = 0.965$

10. $r = -1$

11. $r = 0.22$

12. $r = -0.39$

13. POPULATION The World Bank and the Food and Agriculture Organization track world population trends. The table shows the number of people per square kilometer for several years. Use a graphing calculator to graph the scatter plot that models the data. Then use the scatter plot to describe the relationship between the variables.

World Population Density

Year	People Per km²	Year	People Per km²
2000	47.2	2009	52.8
2001	47.8	2010	53.4
2002	48.4	2011	54.1
2003	49.0	2012	54.7
2004	49.6	2013	55.4
2005	50.2	2014	56.0
2006	50.9	2015	56.7
2007	51.5	2016	57.4
2008	52.2	2017	58.0

14. CREATE Make a table of values where the data in the table is best modeled by an exponential function.

15. WRITE Explain how to use a graphing calculator to determine whether data in a table are best modeled by a linear, quadratic, or exponential function.

16. ANALYZE The path of a stream of water in a fountain is shown in the graph. Should you select an exponential function or quadratic function to model this data? Explain.

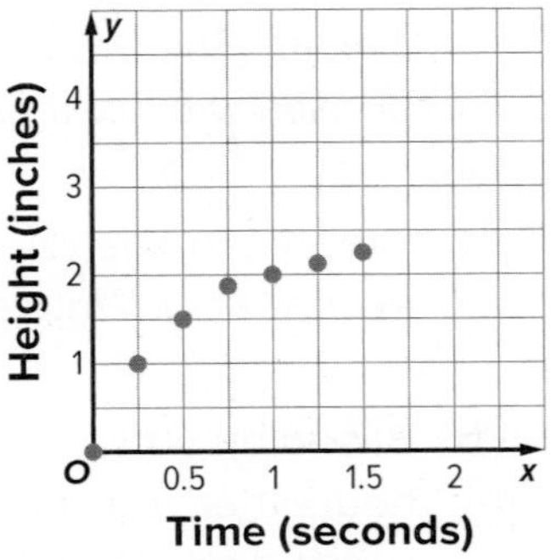

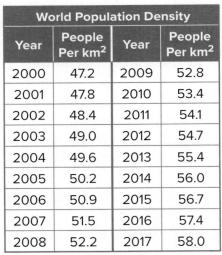

17. PERSEVERE The surface areas *SA* of hemispheres with radius *r* are shown in the table. What is the best model? What is the equation of the regression function?

Radius, r	1	2	3	4	5
Surface Area, SA	3π	12π	27π	48π	75π

Review

Essential Question

How are real-world situations involving quantities that grow or decline rapidly modeled mathematically?

Module Summary

Lesson 7-1

Graphs of Exponential Functions

- An exponential function has the form $f(x) = b^x$, where the base b is a constant and the independent variable x is the exponent.
- Exponential growth occurs when an initial amount increases by the same percent over a given period of time. For an exponential growth function, $b > 1$.
- Exponential decay occurs when an initial amount decreases by the same percent over a given period of time. For an exponential decay function, $0 < b < 1$.
- An asymptote is a line that a graph approaches.

Lesson 7-2

Exponential Equations and Inequalities

- If $b > 0$ and $b \neq 1$, then $b^x = b^y$ if and only if $x = y$.
- If $b > 1$, then $b^x > b^y$ if and only if $x > y$, and $b^x < b^y$ if and only if $x < y$.

Lesson 7-3

Base e Exponential Functions

- The constant e is an irrational number that equals $\left(1 + \frac{1}{n}\right)^n$ as n approaches ∞. This value is approximately equal to 2.7182818...
- Continuously compounded interest: $A = Pe^{rt}$, where A is the amount in the account after t years, P is the principal amount invested, and r is the annual interest rate.

Lesson 7-4

Geometric Sequences and Series

- The nth term a_n of a geometric sequence in which the first term is a_1 and the common ratio is r is given by the formula $a_n = a_1 r^{n-1}$, where n is any natural number.
- Partial sum of a geometric series: $S_n = \frac{a_1 - a_1 r^n}{1 - r}, r \neq 1$ or $S_n = \frac{a_1 - a_n r}{1 - r}, r \neq 1$

Lesson 7-5

Modeling Data

- Calculators may also compute a number r^2, called the coefficient of determination. This measure shows how well data are modeled by the regression function.
- An r^2-value close to 1 is an indication of a strong fit to the data.

Study Organizer

Foldables

Use your Foldable to review this module. Working with a partner can be helpful. Ask for clarification of concepts as needed.

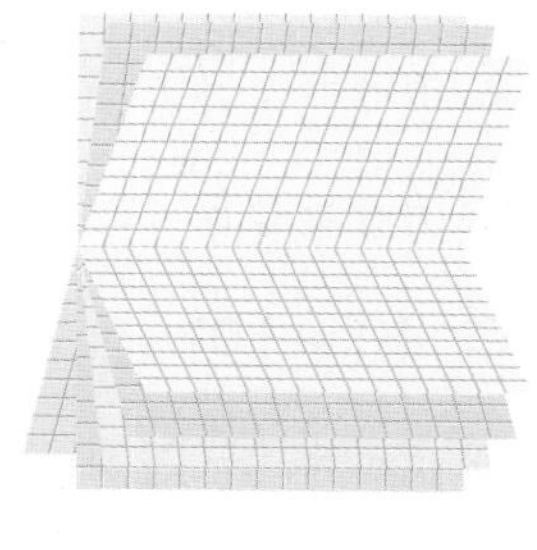

Test Practice

1. MULTIPLE CHOICE Select the graph of $f(x) = 4^x$. (Lesson 7-1)

A.
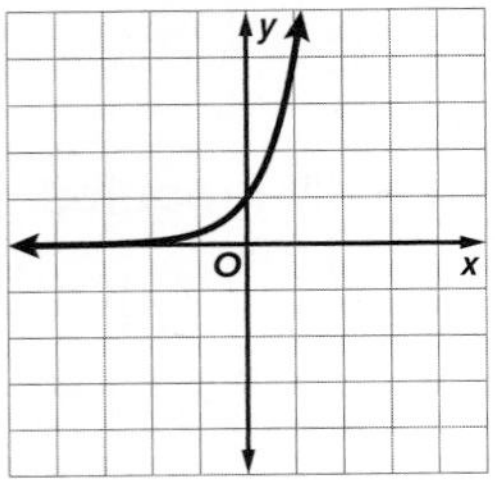

B.
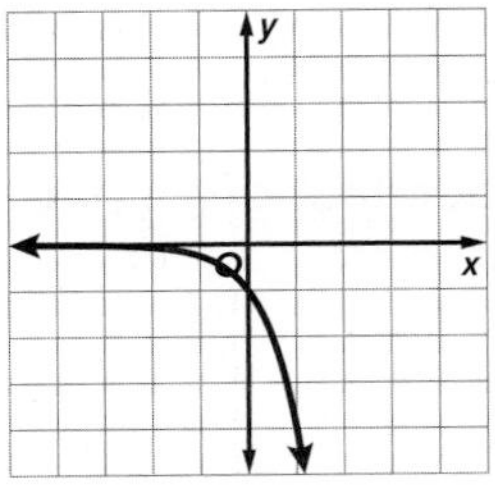

C.
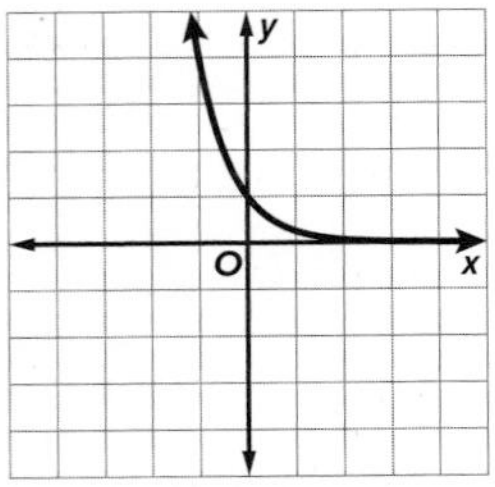

D.
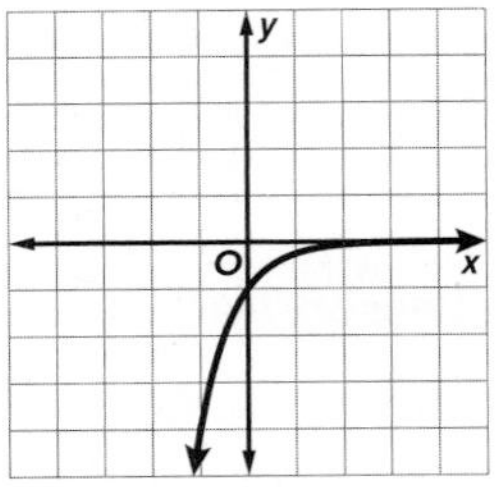

2. OPEN RESPONSE Find the domain, range, y-intercept, and asymptote of the function $f(x) = 6^x$. (Lesson 7-1)

3. MULTI-SELECT Identify the functions that represent exponential decay. Select all that apply. (Lesson 7-1)

A. $f(x) = \left(\frac{10}{11}\right)^x$

B. $f(x) = 0.5^x$

C. $f(x) = 1.01^x$

D. $f(x) = \left(\frac{3}{2}\right)^x$

E. $f(x) = -2^x$

F. $f(x) = 10(0.5)^x$

4. OPEN RESPONSE Eric is preparing for a long-distance race. He currently runs 20 miles each week, and he plans to increase the total distance he runs by 5% each week until race day.

Eric wants to write an exponential function to predict the number of miles he should run each week. What growth or decay factor should he use? Explain your reasoning. (Lesson 7-1)

5. MULTIPLE CHOICE Solve $4^{3x-1} = 8^{x+4}$. (Lesson 7-2)

A. $x = \frac{5}{2}$

B. $x = \frac{5}{3}$

C. $x = \frac{3}{2}$

D. $x = \frac{14}{3}$

6. OPEN RESPONSE The formula $A = 5000\left(1 + \frac{0.021}{12}\right)^{12t}$ represents the amount in dollars A in Padma's college savings account after t years. What does 12 represent? (Lesson 7-2)

7. OPEN RESPONSE The graphs of the functions $y = 2^{x+1}$ and $y = \left(\frac{1}{2}\right)^x$ are shown.

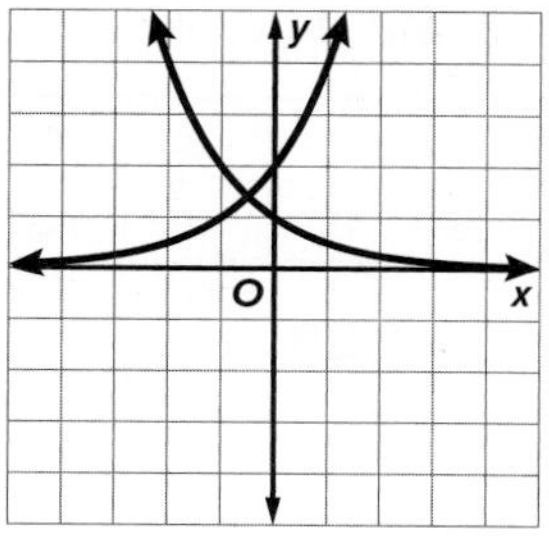

Explain how to use the graph to find the solution of the equation $2^{x+1} = \left(\frac{1}{2}\right)^x$.

(Lesson 7-2)

8. MULTIPLE CHOICE Solve $5^{6x+3} \leq 125^{x-4}$.

(Lesson 7-2)

A. $x \leq -5$

B. $x \geq \frac{2}{5}$

C. $x \leq \frac{7}{5}$

D. $x \geq 5$

9. MULTIPLE CHOICE What is the average rate of change of the function $f(x) = -e^{x+1}$ over the interval $[-1, 2]$? (Lesson 7-3)

A. -3.19

B. -4.93

C. -6.36

D. -12.70

10. MULTIPLE CHOICE Consider the function $f(x) = e^{x-2} + 1$. What is the average rate of change over the interval $[1, 4]$? (Lesson 7-3)

A. 2.34

B. 1.47

C. 0.92

D. 0.07

11. OPEN RESPONSE What is the average rate of change of the function $f(x) = \frac{1}{2}e^{x-1} + 2$ over the interval $[2, 6]$? Round to the nearest hundredth. (Lesson 7-3)

12. OPEN RESPONSE An initial investment of \$2500 grows continuously at an average rate of 8.5% per year. Write a function that will give the value of the investment $f(t)$ after t years. (Lesson 7-3)

13. OPEN RESPONSE Tai wants to watch less television. To achieve this, he starts by watching 10 hours of television during the first week, then watches half the amount of time as the first week during the second week, then watches television half the amount of time as the second week during the third week, and so on. Write a sequence to find the number of hours Tai will watch television during the nth week. (Lesson 7-4)

14. MULTIPLE CHOICE A coach shows two athletes how to do a speed training exercise. Each of these athletes shows two more athletes how to do the exercise. Each of these athletes shows two more athletes how to do the exercise, and so on. How many athletes will know how to do the exercise after 7 generations? (Lesson 7-4)

A. 14

B. 126

C. 254

D. 268

15. MULTIPLE CHOICE On the first day, three friends show a video to two students. The second day, each of those students shows two more students. The third day, each of these students shows two more students. Each day, the students from the previous day continue to show the video to two more students. What is the total number of students who will have seen the video by the end of the eighth day? (Lesson 7-4)

A. 9

B. 384

C. 765

D. 868

16. MULTIPLE CHOICE The table shows the deer population in a certain park in thousands t years after 2010. Which function most closely models the data? (Lesson 7-5)

t	0	1	2	3	4	5	6
$f(t)$	0.6	0.8	1.1	1.4	1.8	2.4	3.1

A. $f(t) = 0.41t + 0.38$

B. $f(t) = 0.61(1.31)^t$

C. $f(t) = 0.5t^2 + 0.11t + 0.63$

D. $f(t) = 0.006t^3 + 0.218t + 0.595$

17. MULTIPLE CHOICE The table shows the annual profit in thousands of dollars y that a marketing company earns x years after it started. Which of these functions most closely models the data? (Lesson 7-5)

x	1	2	3	4	5	6
y	4.4	6.1	9.1	18.2	20.5	35.9

A. $y = 2.77(1.53)^x$

B. $y = 3.44x^{1.15}$

C. $y = 1.17x^2 - 2.22x + 5.67$

D. $y = 5.99x - 5.28$

18. OPEN RESPONSE The table shows the number of employees, in hundreds, y, at an amusement park x years after it opened. Write an exponential function $f(x)$ to model the data. About how many employees were there at the amusement park after 11 years? Round your answer to the nearest whole number. (Lesson 7-5)

x	2	4	6	8	10	12
y	7.3	13.9	33.4	70.7	123.2	247.8

Module 8

Logarithmic Functions

Essential Question

How are logarithms defined and used to model situations in the real world?

What Will You Learn?

How much do you already know about each topic **before** starting this module?

KEY

— I don't know. — I've heard of it. — I know it!

	Before			After		
write and evaluate logarithms						
graph logarithmic functions						
simplify logarithmic expressions						
solve logarithmic equations						
solve exponential equations by using common logarithms						
solve exponential equations by using natural logarithms						
write and solve exponential growth and decay equations using logarithms						
model datasets with natural logarithmic functions using technology						

Foldables Begin with two sheets of grid paper folded along the width.

1. **On** the first sheet, cut 5 centimeters along the fold at the ends.
2. **On** the second sheet, cut in the center, stopping 5 centimeters from the ends.
3. **Insert** the first sheet through the second sheet and align the folds.

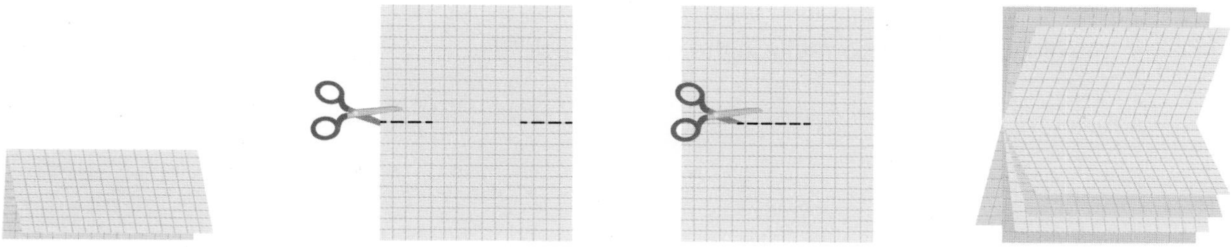

What Vocabulary Will You Learn?

- common logarithms
- logarithm
- logarithmic equation
- logarithmic function
- natural base exponential function
- natural logarithm

Are You Ready?

Complete the Quick Review to see if you are ready to start this module.
Then complete the Quick Check.

Quick Review

Example 1

Find the simple interest on an investment of $4000 at 4% for 3 years using the formula, $I = Prt$.

$I = 4000(0.04)(3)$	Substitute 4000 for principal, 0.04 for interest, and 3 for time.
$I = \$480$	Multiply.

Example 2

Solve $y = 2x + 4$ for x.

$y = 2x + 4$	Original equation
$y - 4 = 2x$	Subtract 4 from each side.
$\frac{y-4}{2} = x$	Divide each side by 2.

Quick Check

Use the simple interest formula, $I = Prt$, to solve each problem.

1. Find the simple interest on an investment of $10,000 at 3.5% for 8 years.

2. A savings account of $3500 earned $875 interest in 5 years. Assuming it remained constant, what interest rate did the account earn?

3. Robin borrowed $1500 for 4 years at a rate of 2.9 %. How much did Robin have to pay back in total?

4. An investment of $2500 earned $2400 in interest at an interest rate of 8%. For how long was the money invested?

Solve each equation for x.

5. $\frac{x}{4} = 2y + 5$

6. $y = \frac{2}{3}x - \frac{5}{6}$

7. $y = -4x + 9$

8. $7 = 2x - y$

How Did You Do?

Which exercises did you answer correctly in the Quick Check?

Logarithms and Logarithmic Functions

Learn Logarithmic Functions

Recall the exponential function $f(x) = 2^x$. You can graph its inverse by interchanging the x- and y-values and graphing the ordered pairs.

$y = 2^x$	
x	y
-3	$\frac{1}{8}$
-2	$\frac{1}{4}$
-1	$\frac{1}{2}$
0	1
1	2
2	4
3	8

$x = 2^y$	
x	y
$\frac{1}{8}$	-3
$\frac{1}{4}$	-2
$\frac{1}{2}$	-1
1	0
2	1
4	2
8	3

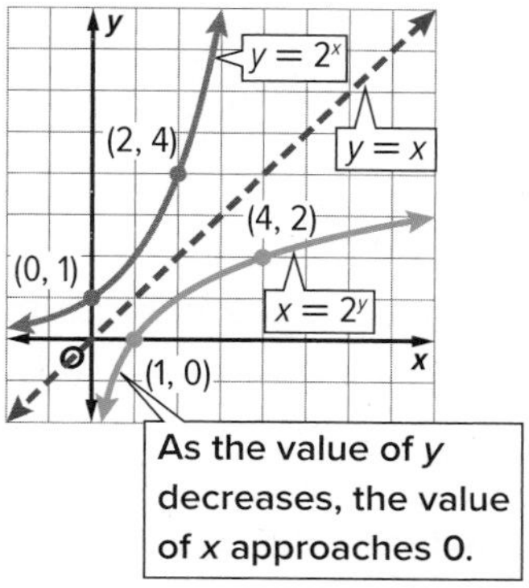

The inverse of $y = 2^x$ can be defined as $x = 2^y$. In general, the inverse of $y = b^x$ is $x = b^y$. In $x = b^y$, y is called the **logarithm**, base b, of x. This is usually written as $y = \log_b x$ and is read as *log base b of x*.

Key Concept • Logarithms with Base b	
Words	Let b and x be positive numbers, $b \neq 1$. The *logarithm* of x with base b is denoted $\log_b x$ and is defined as the exponent y that makes the equation $b^y = x$ true.
Symbols	Suppose $b > 0$ and $b \neq 1$. For $x > 0$, there is a number y such that $\log_b x = y$ if and only if $b^y = x$.
Example	If $\log_2 8 = y$, then $2^y = 8$.

In the expression $\log_b x$, x is called the *argument*.

Example 1 Logarithmic to Exponential Form

Write each equation in exponential form.

a. $\log_4 64 = 3$

$\log_4 64 = 3 \rightarrow 64 = 4^3$

b. $\log_3 \frac{1}{243} = -5$

$\log_3 \frac{1}{243} = -5 \rightarrow \frac{1}{243} = 3^{-5}$

c. $\log_9 3 = \frac{1}{2}$

$\log_9 3 = \frac{1}{2} \rightarrow 3 = 9^{\frac{1}{2}}$

Check

Write $\log_{125} 5 = \frac{1}{3}$ in exponential form.

Today's Goals

- Write logarithmic expressions in exponential form and write exponential expressions in logarithmic form.
- Graph and analyze logarithmic functions.

Today's Vocabulary

logarithm

logarithmic function

Think About It!

If $\log_z 16 = y$, find z^y.

Math History Minute

Scottish mathematician **John Napier (1550–1617)** is known as the discoverer of logarithms. His book, *Mirifici Logarithmorum Cononis Descriptio*, written in 1614, contained a wealth of material related to natural logarithms. He also invented Napier's bones, a numbering rod system that was a mechanical method for facilitating multiplication.

Think About It!

How could you write $(2z + 5)^3 = 8$ in logarithmic form?

Example 2 Exponential to Logarithmic Form

Write each equation in logarithmic form.

a. $7^6 = 117{,}649$

$7^6 = 117{,}649 \rightarrow \log_7 117{,}649 = 6$

b. $8^{-3} = \frac{1}{512}$

$8^{-3} = \frac{1}{512} \rightarrow \log_8 \frac{1}{512} = -3$

c. $8^{\frac{2}{3}} = 4$

$8^{\frac{2}{3}} = 4 \rightarrow \log_8 4 = \frac{2}{3}$

Check

Write $2^9 = 512$ in logarithmic form.

Think About It!

Without evaluating the expression, how can you determine that the value of $\log_{512} 8$ will be less than 1?

Example 3 Evaluate Logarithmic Expressions

Evaluate $\log_{216} 6$.

$\log_{216} 6 = y$	Let the logarithm equal y.
$6 = 216^y$	Definition of logarithm
$6^1 = (6^3)^y$	$6^3 = 216$
$6^1 = 6^{3y}$	Power of a Power
$1 = 3y$	Property of Equality for Exponential Functions
$y = \frac{1}{3}$	Divide each side by 3.

Therefore, $\log_{216} 6 = \frac{1}{3}$.

Study Tip

Exponential and Logarithmic Functions Recall that exponential form is $x = b^y$ and logarithmic form is $y = \log_b x$. In the equation $\frac{h}{14.7} = 10^{-\frac{9P}{100}}$, $x = \frac{h}{14.7}$, $b = 10$, and $y = -\frac{9P}{100}$.

Example 4 Find Inverses of Exponential Functions

AIR PRESSURE **The air pressure outside of an aircraft can be determined by the equation $P = B\left(10^{\frac{-9h}{100}}\right)$, where B is the air pressure at sea level and h is the altitude in miles. If the air pressure at sea level is 14.7 pounds per square inch, write an equation to find the height of an aircraft when the air pressure is known.**

Robert Shealy/Flickr RF/Getty Images

$P = B\left(10^{\frac{-9h}{100}}\right)$	Original equation
$P = 14.7\left(10^{\frac{-9h}{100}}\right)$	Let $B = 14.7$.
$\frac{P}{14.7} = 10^{\frac{-9h}{100}}$	Divide each side by 14.7.
$\log_{10} \frac{P}{14.7} = \frac{-9h}{100}$	Definition of logarithm
$\frac{100}{-9} \log_{10} \frac{P}{14.7} = h$	Multiply each side by $\frac{100}{-9}$.

Go Online

You can complete an Extra Example online.

Check

TORNADOS The distance in miles that a tornado travels is $y = 10^{\frac{w-65}{93}}$, where w represents the speed of the wind in miles per hour.

Part A Write the inverse of the function in logarithmic form.
$w =$ ___?___

Part B Use the inverse function to estimate the speed of the wind, to the nearest mile per hour, of a tornado that travels 50 miles.
___?___ mph

Explore Transforming Logarithmic Functions

Online Activity Use graphing technology to complete the Explore.

INQUIRY How does performing an operation on a logarithmic function affect its graph?

Learn Graphing Logarithmic Functions

A function of the form $f(x) = \log_b x$, where $b > 0$ and $b \neq 1$, is a **logarithmic function**. The graph of $f(x) = \log_b x$ represents the parent graph of the logarithmic functions.

Key Concept • Parent Function of Logarithmic Functions

Parent function	$f(x) = \log_b x$
Type of graph	continuous, one-to-one
Domain	$(0, \infty)$, $\{x \mid x > 0\}$, or all positive real numbers
Range	$(-\infty, \infty)$, $\{f(x) \mid -\infty < f(x) < \infty\}$, or all real numbers
Asymptote	y-axis
x-intercept	(1, 0)
Symmetry	none
Extrema	none

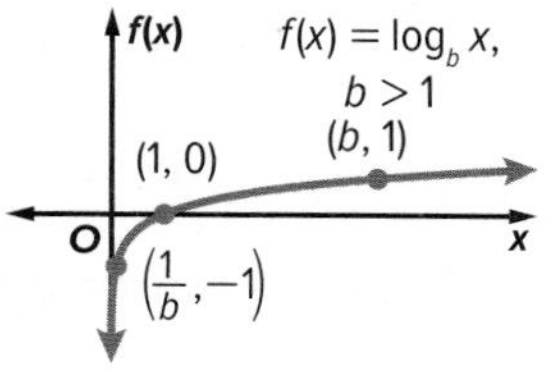

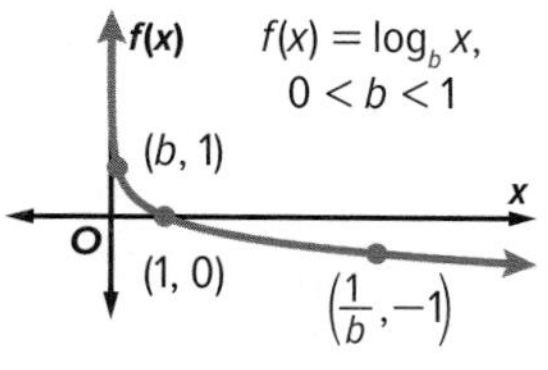

For a function $f(x)$ that is undefined at a vertical asymptote $x = c$, the end behavior of $f(x)$ is described as x approaches c from the right (c^+) or as x approaches c from the left (c^-). For the parent function $f(x) = \log_b x$ where $b > 1$, the end behavior is: As $x \to 0^+$, $f(x) \to -\infty$, and as $x \to \infty$, $f(x) \to \infty$.

Go Online You can complete an Extra Example online.

Watch Out!

Graphing Calculator The **LOG** button on a graphing calculator allows you to only enter logarithmic functions of base 10.

Go Online
You can watch a video to see how to graph transformations of a logarithmic function by using a graphing calculator.

The same techniques used to transform the graphs of other functions can be applied to the graphs of logarithmic functions.

Key Concept • Transformations of Logarithmic Functions	
$g(x) = a \log_b (x - h) + k$	
h – horizontal translation	If $h > 0$, the graph of $f(x)$ is translated h units right. If $h < 0$, the graph of $f(x)$ is translated $\|h\|$ units left.
k – vertical translation	If $k > 0$, the graph of $f(x)$ is translated k units up. If $k < 0$, the graph of $f(x)$ is translated $\|k\|$ units down.
a – reflection and dilation	If $a < 0$, the graph is reflected in the x-axis. If $\|a\| > 1$, the graph is stretched vertically. If $0 < \|a\| < 1$, the graph is compressed vertically.

Example 5 Graph Logarithmic Functions

Graph each function. Then find the intercepts, domain, range, and end behavior.

a. $f(x) = \log_6 x$

Step 1 Identify the base.

$b = 6$

Step 2 Identify ordered pairs.

Use the points $\left(\frac{1}{b}, -1\right)$, (1, 0), and ($b$, 1).

$\left(\frac{1}{b}, -1\right) \rightarrow \left(\frac{1}{6}, -1\right)$

(1, 0)

$(b, 1) \rightarrow (6, 1)$

Step 3 Graph the ordered pairs.

Step 4 Draw a smooth curve through the points.

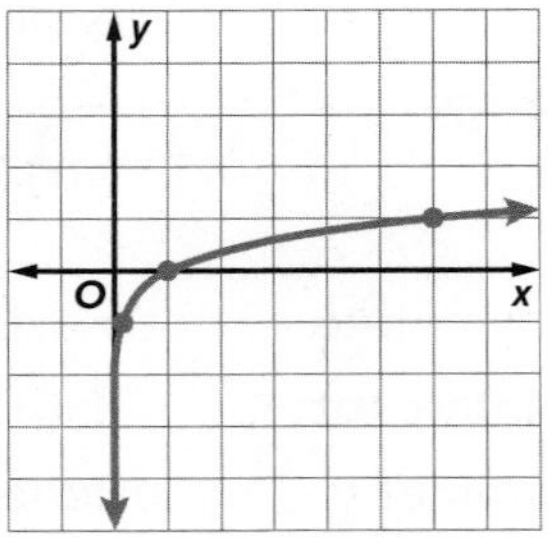

Step 5 Find the intercepts, domain, range, and end behavior.

x-intercept: 1; no y-intercept

domain: all positive real numbers

range: all real numbers

end behavior: As $x \rightarrow 0$, $f(x) \rightarrow -\infty$, and as $x \rightarrow \infty$, $f(x) \rightarrow \infty$.

Go Online
You can watch a video to see how to graph logarithmic functions.

Go Online You can complete an Extra Example online.

b. $g(x) = \log_{\frac{1}{4}} x$

Step 1 **Identify the base.**

$b = \frac{1}{4}$

Step 2 **Identify ordered pairs.**

Use the points $\left(\frac{1}{b}, -1\right)$, (1, 0), and ($b$, 1).

$\left(\frac{1}{b}, -1\right) \rightarrow \left(\frac{1}{\frac{1}{4}}, -1\right)$ or (4, −1)

(1, 0)

$(b, 1) \rightarrow \left(\frac{1}{4}, 1\right)$

Step 3 **Graph the ordered pairs.**

Step 4 **Draw a smooth curve through the points.**

Step 5 **Find the intercepts, domain, range, and end behavior.**

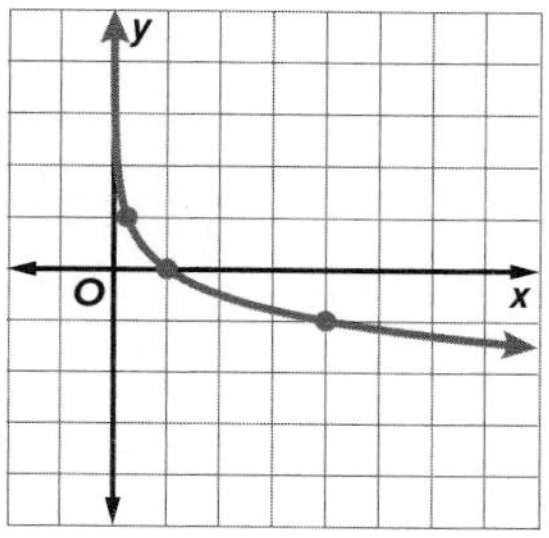

x-intercept: 1; no y-intercept

domain: all positive real numbers

range: all real numbers

end behavior: As $x \rightarrow 0$, $f(x) \rightarrow \infty$, and as $x \rightarrow \infty$, $f(x) \rightarrow -\infty$.

Think About It!

Compare the graphs in parts **a** and **b**.

Example 6 Graph Transformations of Logarithmic Functions

Graph $g(x) = 2\log_{10}(x + 3) - 1$.

$g(x) = 2\log_{10}(x + 3) - 1$ represents a transformation of the graph of $f(x) = \log_{10} x$.

$|a| = 2$ Because $|a| > 1$, the graph is stretched vertically.

$h = -3$ Because $h < 0$, the graph is translated 3 units left.

$k = -1$ Because $k < 0$, the graph is translated 1 unit down.

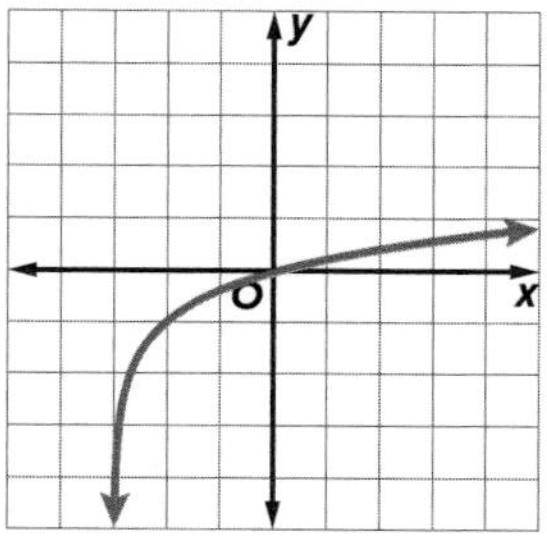

Think About It!

Find the domain and range of $g(x)$.

Alternative Method

Step 1 Identify ordered pairs of the parent function, $f(x) = \log_{10} x$.

$\left(\frac{1}{b}, -1\right)$, (1, 0), and $(b, 1) \rightarrow \left(\frac{1}{10}, -1\right)$, (1, 0), and (10, 1)

Step 2 Transform each point from the parent function.

Because $a = 2$ stretches the graph vertically, each y-value is multiplied by 2.

$\left(\frac{1}{10}, -1\right)$, (1, 0), and $(10, 1) \rightarrow \left(\frac{1}{10}, -2\right)$, (1, 0) and (10, 2)

Because $h = -3$ and $k = -1$, 3 is subtracted from each x-value and 1 is subtracted from each y-value.

$\left(\frac{1}{10}, -2\right)$, (1, 0), and $(10, 2) \rightarrow \left(-\frac{29}{10}, -3\right)$, (−2, −1), and (7, 1)

Step 3 Plot the points of $g(x)$ and draw a smooth curve through the points.

Think About It!

Compare and contrast the methods.

Go Online You can complete an Extra Example online.

Think About It!

Find the domain and range of $g(x)$. Does $j(x)$ have the same domain and range? Explain.

Go Online
An alternate method is available for this example.

Go Online
to see more examples about writing logarithmic functions from graphs.

Example 7 Compare Logarithmic Functions

Consider $g(x) = \log_{10} (x - 2)$ and $j(x)$ shown in the graph.

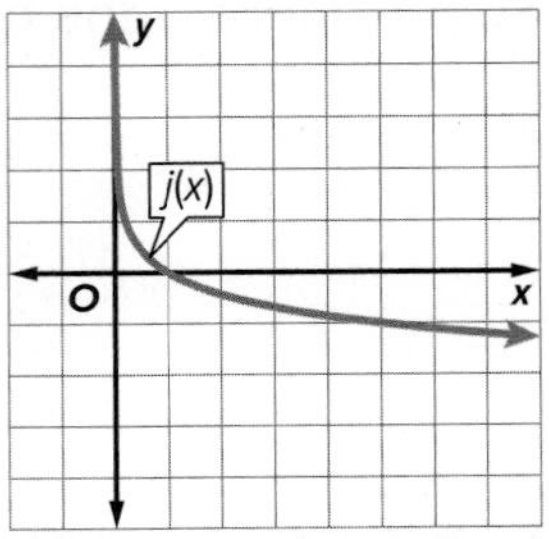

Part A Graph $g(x)$.

The graph of $g(x) = \log_{10} (x - 2)$ is a transformation of $f(x) = \log_{10} x$.

Graph the parent function and transform the graph. For $g(x) = \log_{10} (x - 2)$, $h = 2$.
Because h is positive, the parent graph is translated 2 units right.

Part B Compare the end behavior of $g(x)$ and $j(x)$.

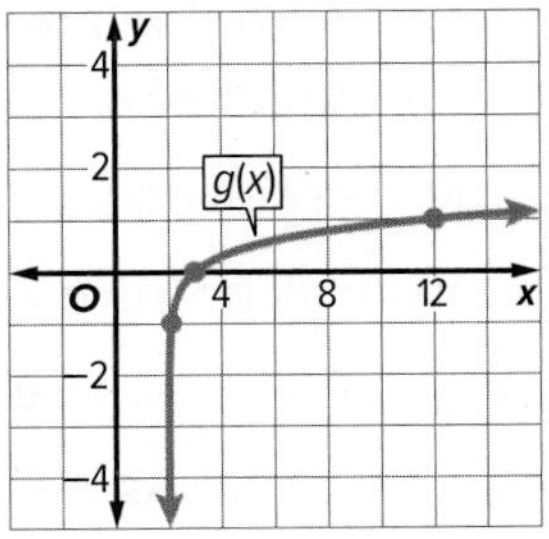

$g(x)$: As $x \to 2$, $g(x) \to -\infty$,
and as $x \to \infty$, $g(x) \to \infty$.

$j(x)$: As $x \to 0$, $j(x) \to \infty$,
and as $x \to \infty$, $j(x) \to -\infty$.

Example 8 Write Logarithmic Functions From Graphs

Identify the value of k, and write a function for the graph as it relates to $f(x) = \log_4 x$.

The graph has been translated 5 units up, so $k = 5$ and the function is $g(x) = \log_4 x + 5$.

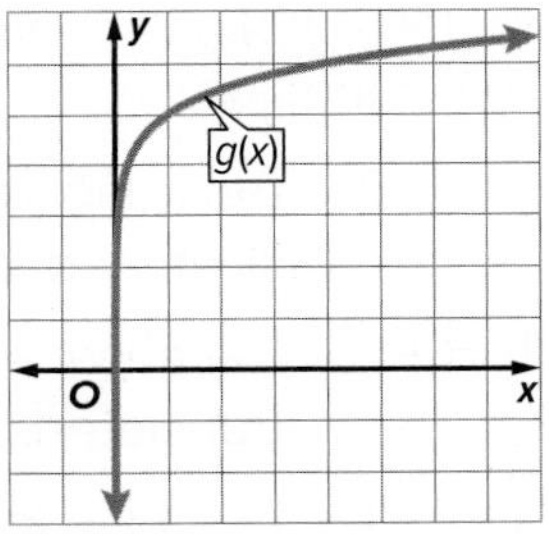

Check

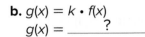

Write a function for each graph as it relates to $f(x) = \log_3 x$.

a. $g(x) = f(x) + k$
$g(x) =$ ___?___

b. $g(x) = k \cdot f(x)$
$g(x) =$ ___?___

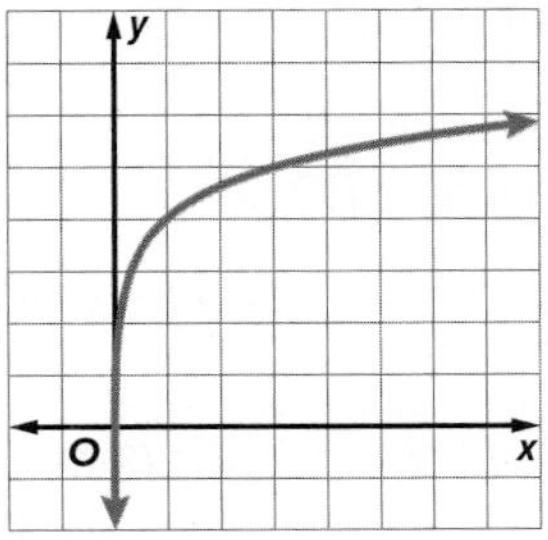

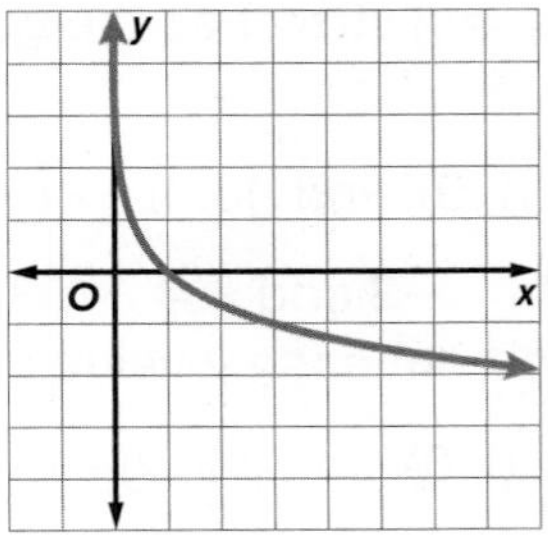

Go Online You can complete an Extra Example online.

Practice

Go Online You can complete your homework online.

Example 1

Write each equation in exponential form.

1. $\log_{15} 225 = 2$

2. $\log_3 \frac{1}{27} = -3$

3. $\log_5 \frac{1}{25} = -2$

4. $\log_3 243 = 5$

5. $\log_4 64 = 3$

6. $\log_4 32 = \frac{5}{2}$

Example 2

Write each equation in logarithmic form.

7. $2^7 = 128$

8. $3^{-4} = \frac{1}{81}$

9. $7^{-2} = \frac{1}{49}$

10. $\left(\frac{1}{7}\right)^3 = \frac{1}{343}$

11. $2^9 = 512$

12. $64^{\frac{2}{3}} = 16$

Example 3

Evaluate each expression.

13. $\log_2 64$

14. $\log_{100} 100{,}000$

15. $\log_5 625$

16. $\log_{27} 81$

17. $\log_4 \frac{1}{32}$

18. $\log_{10} 0.00001$

19. $\log_8 512$

20. $\log_9 1$

21. $\log_8 4$

Example 4

22. STRUCTURE The electric current I in amperes in a particular circuit can be represented by $\log_2 I = -t$, where t is given in seconds. Write an equation to find the current when time is known.

23. STRUCTURE The value of a guitar in dollars after x years can be modeled by the equation $y = g(1.0065)^x$, where g is the initial cost of the guitar. If a guitar costs $400, write an equation to find the number of years it takes for a guitar to reach a certain value.

Example 5

Graph each function. Then find the intercepts, domain, range, and end behavior.

24. $f(x) = \log_{\frac{1}{9}} x$

25. $f(x) = \log_{\frac{1}{5}} x$

26. $f(x) = \log_2 x$

27. $f(x) = \log_8 x$

Example 6

Graph each function.

28. $f(x) = 2 \log_4 (x + 3) - 2$

29. $f(x) = -3 \log_{10} (x - 2) + 1$

30. $f(x) = \log_3 (x + 1) - 4$

31. $f(x) = -\log_5 x + 2.5$

Example 7

32. Consider $g(x) = \log_{10} (x - 4)$ and $p(x)$ shown in the graph.

a. Graph $g(x)$.

b. Compare the end behavior of $g(x)$ and $p(x)$.

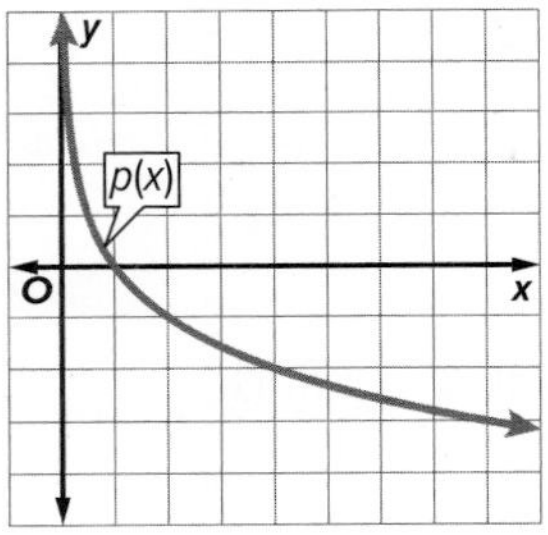

33. Consider $f(x) = \log_{10} (x + 2)$ and the logarithmic function $g(x)$ shown in the table.

a. Graph $f(x)$ and $g(x)$.

b. Compare the end behavior of $f(x)$ and $g(x)$.

x	$\frac{1}{3}$	1	3	9	27
$g(x)$	−1	0	1	2	3

Example 8

Identify the value of k. Write a function for each graph as it relates to $f(x) = \log_5 x$.

34. $g(x) = f(x) + k$

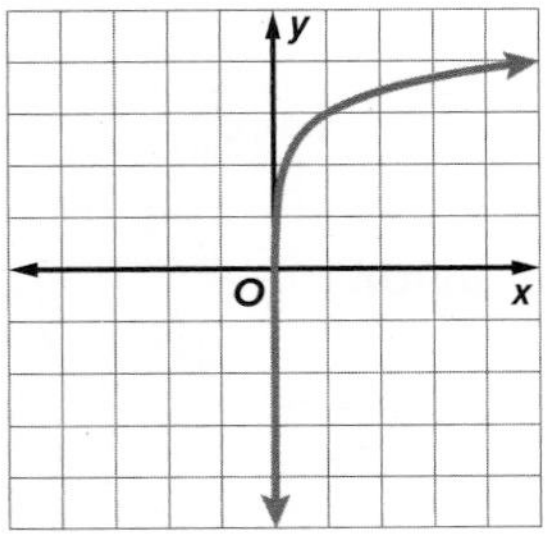

35. $h(x) = k \cdot f(x)$

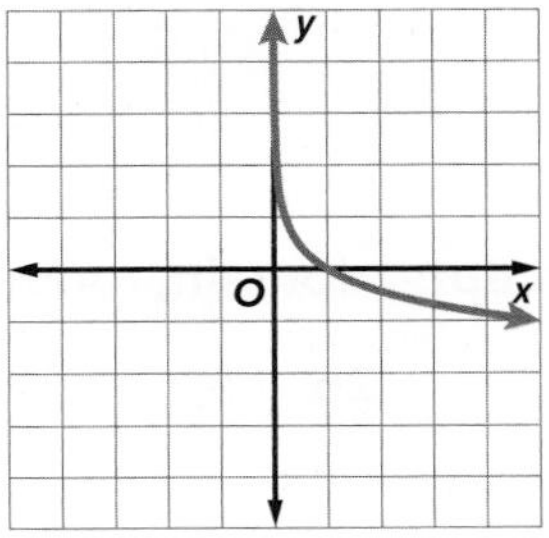

36. $j(x) = f(x) + k$

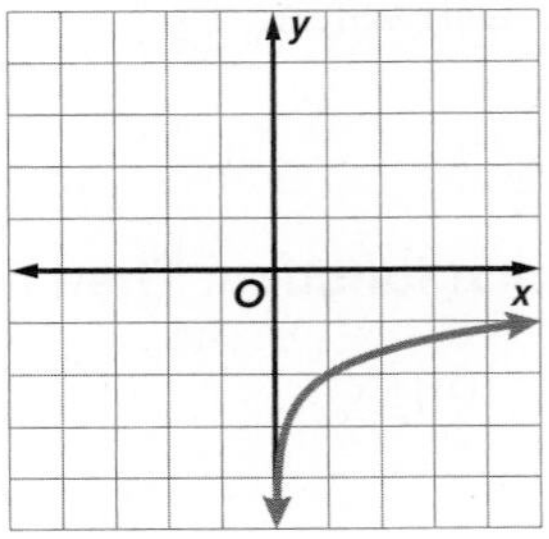

Mixed Exercises

Write each logarithmic equation in exponential form and each exponential equation in logarithmic form.

37. $\log_6 216 = 3$

38. $\log_2 64 = 6$

39. $\log_3 \frac{1}{81} = -4$

40. $3^{-5} = \frac{1}{243}$

41. $\left(\frac{1}{4}\right)^3 = \frac{1}{64}$

42. $7776^{\frac{1}{5}} = 6$

43. EARTHQUAKES The intensity of an earthquake can be measured on the Richter scale using the formula $y = 10^{R-1}$, where y is the absolute intensity of the earthquake and R is its Richter scale measurement. In 1906, an earthquake in San Francisco had an absolute intensity of 6,000,000. What is the Richter scale measurement of the earthquake?

Richter Scale Number	Absolute Intensity
1	1
2	10
3	100
4	1000
5	10,000

44. INVESTING Maria invests $1000 in a savings account that pays 4% interest compounded annually. The value of the account A after x years can be determined from the equation $A = 1000(1.04^x)$. Solve the equation for x.

45. CONSTRUCT ARGUMENTS Julio and Natalia are competing in a math competition. They each select a logarithmic function and compare their functions to see which has a greater value. Julio selected $f(x) = 10 \log_2 x$, and Natalia selected $g(x) = 2 \log_{10} x$.

a. Which function has a greater value when $x = 7$? Explain your reasoning.

b. Which function has a greater value when $x = 1$? Explain your reasoning.

46. BIOLOGY An amoeba divides into two amoebas every hour. The number N of amoebas after t hours can be represented by $t = \log_2 N$. How many hours will it take for a single amoeba to divide into 2048 amoebas?

47. **STRUCTURE** Graph $f(x) = 3^x$ and $g(x) = \log_3 x$ on the same coordinate plane. Define the domain, range, intercepts, and asymptotes of each. Compare the two graphs.

48. **WRITE** What should you consider when using exponential and logarithmic models to make decisions?

49. **ANALYZE** Let $f(x) = \log_{10}(x)$ and $g(x) = 10^x$. Find $f(g(x))$ and $g(f(x))$ and justify your conclusion.

50. **FIND THE ERROR** Betsy says that the graphs of all logarithmic functions cross the y-axis at (0, 1) because any number to the zero power equals 1. Tyrone disagrees because he says log 0 is undefined. Is either of them correct? Explain your reasoning.

51. **PERSEVERE** Without using a calculator, compare $\log_7 51$, $\log_8 61$, and $\log_9 71$. Which of these is the greatest? Explain your reasoning.

52. **CREATE** Write a logarithmic equation of the form $y = \log_b x$ for each of the following conditions.

a. y is equal to 25.

b. y is negative.

c. y is between 0 and 1.

d. x is 1.

e. x is 0.

53. **FIND THE ERROR** Elisa and Matthew are evaluating $\log_{\frac{1}{7}} 49$. Is either of them correct? Explain your reasoning.

Elisa

$\log_{\frac{1}{7}} 49 = y$

$\frac{1^y}{7} = 49$

$(7^{-1})^y = 7^2$

$(7)^{-y} = 7^2$

$y = 2$

Matthew

$\log_{\frac{1}{7}} 49 = y$

$49^y = \frac{1}{7}$

$(7^2)^y = (7)^{-1}$

$7^{2y} = (7)^{-1}$

$2y = -1$

$y = -\frac{1}{2}$

Properties of Logarithms

Explore Logarithmic Expressions and Equations

Online Activity Use the interactive tool to complete the Explore.

INQUIRY How can you determine $\log_b xy$ and $\log_b \frac{x}{y}$ given $\log_b x$ and $\log_b y$?

Learn Logarithmic Equations

A **logarithmic equation** contains one or more logarithms.

Key Concept • Property of Equality for Logarithmic Equations	
Symbols	If b is a positive number other than 1, then $\log_b x = \log_b y$ if and only if $x = y$.
Example	If $\log_3 x = \log_3 7$, then $x = 7$. If $x = 7$, then $\log_3 x = \log_3 7$.
This property also holds true for inequalities.	

Example 1 Solve a Logarithmic Equation by Using Definitions

Solve $\log_4 x = \frac{5}{2}$.

$\log_4 x = \frac{5}{2}$	Original equation
$x = 4^{\frac{5}{2}}$	Definition of logarithm
$x = (2^2)^{\frac{5}{2}}$	$4 = 2^2$
$x = 2^5$ or 32	Power of a Power

Example 2 Solve a Logarithmic Equation by Using Properties of Equality

Solve $\log_5 (2x^2 - 6) = \log_5 4x$.

Step 1 Solve for x.

$\log_5 (2x^2 - 6) = \log_5 4x$	Original equation
$2x^2 - 6 = 4x$	Property of Equality for Logarithmic Equations
$2x^2 - 4x - 6 = 0$	Subtract $4x$ from each side.
$x^2 - 2x - 3 = 0$	Divide each side by 2.
$(x - 3)(x + 1) = 0$	Factor.
$x - 3 = 0$ or $x + 1 = 0$	Zero Product Property
$x = 3$ $\quad$ $x = -1$	Solve each equation.

(continued on the next page)

Today's Goals
- Solve logarithmic equations using properties of equality.
- Simplify and evaluate expressions by using the properties of logarithms.

Today's Vocabulary
logarithmic equation

Go Online
You may want to complete the Concept Check to check your understanding.

Think About It!
Why does the example include the statement, "If $x = 7$, then $\log_3 x = \log_3 7$"?

Go Online
You can complete an Extra Example online.

Step 2 Check for extraneous solutions.

$x = 3$	$x = -1$
$\log_5 (2 \cdot 3^2 - 6) \stackrel{?}{=} \log_5 (4 \cdot 3)$	$\log_5 [2 \cdot (-1)^2 - 6] \stackrel{?}{=} \log_5 [4 \cdot (-1)]$
$\log_5 (12) \stackrel{?}{=} \log_5 (12)$ ✓	$\log_5 (-4) \stackrel{?}{=} \log_5 (-4)$

Because the logarithm of a negative number is undefined, -1 is an extraneous solution. So, $x = 3$.

Check

Solve $\log_{13} (-5x) = \log_{13} (-2x^2 + 3)$.

$x =$ ___?___

Learn Properties of Logarithms

Because logarithms are exponents, the properties of logarithms can be derived from the properties of exponents. For example, the Product Property of Logarithms can be derived from the Product of Powers Property of Exponents.

Key Concept • Product Property of Logarithms	
Words	The logarithm of a product is the sum of the logarithms of its factors.
Symbols	For all positive numbers b, m, and n, where $b \neq 1$, $\log_b mn = \log_b m + \log_b n$.
Example	$\log_5 8(4) = \log_5 8 + \log_5 4$

Go Online to derive the properties of logarithms in Lesson 8-2.

You can use the Product Property of Logarithms to approximate logarithmic expressions.

Key Concept • Quotient Property of Logarithms	
Words	The logarithm of a quotient is the difference of the logarithms of the numerator and the denominator.
Symbols	For all positive numbers b, m, and n, where $b \neq 1$ and $n \neq 0$, $\log_b \frac{m}{n} = \log_b m - \log_b n$.
Example	$\log_8 \frac{2}{3} = \log_8 2 - \log_8 3$

Key Concept • Power Property of Logarithms	
Words	The logarithm of a power is the product of the logarithm and the exponent.
Symbols	For any real number n, and positive numbers m and b, where $b \neq 1$, $\log_b m^n = n \log_b m$.
Example	$\log_2 3^7 = 7 \log_2 3$

Go Online You can complete an Extra Example online.

Example 3 Product Property of Logarithms

Use $\log_3 5 \approx 1.465$ to approximate the value of $\log_3 405$.

$\log_3 405 = \log_3 (3^4 \cdot 5)$	$405 = 3^4 \cdot 5$
$= \log_3 3^4 + \log_3 5$	Product Property of Logarithms
$= 4 + \log_3 5$	Inverse Property of Exponents and Logarithms
$\approx 4 + 1.465$ or 5.465	Replace $\log_3 5$ with 1.465.

Check

Use $\log_8 5 \approx 0.7740$ to approximate the value of $\log_8 320$.

Study Tip

Products and Bases Because the base of the logarithm is 3 and you are given an approximation for $\log_3 5$, the first step in simplifying should be to look for how to write 405 as a product of a power of 3 and 5.

Example 4 Quotient Property of Logarithms

Use $\log_3 5 \approx 1.465$ to approximate the value of $\log_3 \frac{9}{5}$.

$\log_3 \frac{9}{5} = \log_3 \frac{3^2}{5}$	$\frac{9}{5} = \frac{3^2}{5}$
$= \log_3 3^2 - \log_3 5$	Quotient Property of Logarithms
$= 2 - \log_3 5$	Inverse Property of Exponents and Logarithms
$\approx 2 - 1.465$ or 0.535	Replace $\log_3 5$ with 1.465.

Check

Use $\log_6 5 \approx 0.8982$ to approximate the value of $\log_6 \frac{5}{1296}$.

Example 5 Power Property of Logarithms

Use $\log_2 6 \approx 2.585$ to approximate the value of $\log_2 1296$.

$\log_2 1296 = \log_2 6^4$	$1296 = 6^4$
$= 4 \log_2 6$	Power Property of Logarithms
$\approx 4(2.585)$ or 10.34	Replace $\log_2 6$ with 2.585.

Check

Use $\log_2 3 \approx 1.5850$ to approximate the value of $\log_2 243$.

Go Online You can complete an Extra Example online.

Apply Example 6 Solve a Logarithmic Equation by Using Properties

SOUND **The loudness of a sound L in decibels is defined by $L = 10 \log_{10} R$, where R is the relative intensity of the sound. A choir director wants to determine how many members could sing while maintaining a safe level of sound, about 80 decibels. If one person has a relative intensity of 10^6 when singing, then how many people could sing with the same relative intensity to achieve a loudness of 80 decibels?**

1 What is the task?

Describe the task in your own words. Then list any questions that you may have. How can you find answers to your questions?

I need to find the number of people to reach 80 decibels. How can I represent the relative intensity of sound when I do not know the number of choir members? Which properties will I need to use? I can reference the definitions of the properties of logarithms.

2 How will you approach the task? What have you learned that you can use to help you complete the task?

I will interpret the situation to write R in terms of the number of choir members x. Then, I will write an equation, and solve for x. Finally, I will interpret the solution in context. I have learned how to solve logarithmic equations and check for extraneous solutions.

3 What is your solution?

Use your strategy to solve the problem.

What expression represents the relative intensity R of a choir with x members? $x \cdot 10^6$

How many members should the choir have to reach a relative intensity of 80 decibels?

100 members

4 How can you know that your solution is reasonable?

Write About It! Write an argument that can be used to defend your solution.

I can simplify R for $x = 100$ and solve for L. So, $R = 100(10^6) = (10)^8$.

Therefore, $80 = 10 \log_{10} 10^8 = 10(8) = 80$, which checks.

Go Online You can complete an Extra Example online.

Practice

Go Online You can complete your homework online.

Example 1

Solve each equation.

1. $\log_4 x = \frac{3}{2}$
2. $\log_{16} x = \frac{3}{2}$
3. $\log_{25} x = \frac{5}{2}$
4. $\log_9 x = \frac{3}{2}$
5. $\log_{27} x = \frac{2}{3}$
6. $\log_8 x = \frac{2}{3}$

Example 2

Solve each equation.

7. $\log_4 (2x^2 - 4) = \log_4 2x$
8. $\log_5 (x^2 - 6) = \log_5 x$
9. $\log_3 (x^2 - 8) = \log_3 2x$
10. $\log_4 (2x^2 - 20) = \log_4 6x$
11. $\log_2 (6x^2 + 1) = \log_2 5x$
12. $\log_6 (6x^2 - 3) = \log_6 7x$

Examples 3 and 4

Use $\log_4 2 = 0.5$, $\log_4 3 \approx 0.7925$, and $\log_4 5 \approx 1.1610$ to approximate the value of each expression.

13. $\log_4 30$
14. $\log_4 20$
15. $\log_4 \frac{2}{3}$
16. $\log_4 \frac{4}{3}$
17. $\log_4 9$
18. $\log_4 8$

Example 5

Use $\log_2 3 \approx 1.5850$ and $\log_2 5 \approx 2.3219$ to approximate the value of each expression.

19. $\log_2 25$
20. $\log_2 27$
21. $\log_2 125$
22. $\log_2 625$
23. $\log_2 81$
24. $\log_2 243$

Example 6

25. SOUND Recall that the loudness L of a sound in decibels is given by $L = 10 \log_{10} R$, where R is the sound's relative intensity. If the relative intensity of a sound is multiplied by 10 and results in a loudness of 120 decibels, what was the relative intensity of the original sound?

26. EARTHQUAKES The Richter scale magnitude reading m is given by $m = \log_{10} x$, where x represents the amplitude of the seismic wave causing ground motion. Determine the reading of an earthquake that is 10 times less intense than an earthquake that measures 4.5 on the Richter scale.

Mixed Exercises

Solve each equation. Check your solution.

27. $\log_3 56 - \log_3 n = \log_3 7$

28. $\log_2 (4x) + \log_2 5 = \log_2 40$

29. $5 \log_2 x = \log_2 32$

30. $\log_{10} a + \log_{10} (a + 21) = \log_{10} 100$

31. $\log_2 x + \log_2 (x + 2) = \log_2 8$

32. $\log_4 (x^2 + 2x + 1) = \log_4 (11 - x)$

33. $\log_3 \frac{x^2}{4} = \log_3 25$

34. $\log_3 3d = \log_3 9$

35. $\log_{10} (3x^2 - 5x) = \log_{10} 2$

36. $\log_4 (2x^2 - 3x) = \log_4 2$

Use $\log_5 3 \approx 0.6826$ and $\log_5 4 \approx 0.8614$ to approximate the value of each expression.

37. $\log_5 40$

38. $\log_5 30$

39. $\log_5 \frac{3}{4}$

40. $\log_5 \frac{4}{3}$

41. $\log_5 9$

42. $\log_5 16$

43. REASONING A student mistakenly writes the following in his notes:

$$\log_2 a + \log_2 b = \log_2 (a + b).$$

However, after substituting the values for a and b in a problem, he still gets the right answer. If the value of a is 11, what must be the value of b?

44. CHEMISTRY The pH of a solution is given by $\log_{10} \frac{1}{H^+}$, where H^+ is the concentration of hydrogen ions in moles. The pH of a popular soft drink is 2.5. How many moles of hydrogen ions are in a liter of the soft drink? Write your answer in scientific notation.

45. PHYSICS Sir Isaac Newton determined that any two particles exert a gravitational force on each other. The magnitude of the gravitational force of attraction between a particle of mass m_1 and another of mass m_2 is given by $F = k\left(\frac{m_1 m_2}{d^2}\right)$, where d is the distance between the particles and k is a constant. Use the properties of logarithms to find the expanded form of $\log F$.

46. MOTION The period p of simple harmonic motion, such as the swing of a pendulum, can be modeled by $p = 2\pi\sqrt{\frac{m}{k}}$, where m is mass and k is a constant. Use the properties of logarithms to find the expanded form of $\log p$.

47. BIOLOGY Energy is required to transport substances across a cell membrane. The formula $E = 1.4(\log_{10} C_2 - \log_{10} C_1)$ represents the needed energy E in kilocalories per mole for transporting a substance with a concentration of C_1 outside the cell to a concentration of C_2 inside the cell.

a. Express E as a single logarithm of base 10.

b. Given that $\log_{10} 3 \approx 0.4771$, determine the energy needed to transport a substance from outside a cell to inside a cell if the concentration of the substance inside the cell is three times greater than the concentration outside the cell.

c. Given that $\log_{10} 2 \approx 0.3010$, determine the energy needed to transport a substance from outside a cell to inside a cell if the concentration of the substance inside the cell is twice the concentration outside the cell.

48. CONSTRUCT ARGUMENTS Are $\log \frac{y}{3}$ and $\frac{\log y}{\log 3}$ equivalent expressions? Justify your response.

49. Explain how to determine the values of a and b if $\log_2 a^3b^2 = 19$ and $\log_2 \frac{a^4}{b^5} = 10$.

State whether each equation is *true* or *false*.

50. $\log_8 (x - 3) = \log_8 x - \log_8 3$

51. $\log_5 22x = \log_5 22 + \log_5 x$

52. $\log_{10} 19k = 19 \log_{10} k$

53. $\log_2 y^5 = 5 \log_2 y$

54. $\log_7 \frac{x}{3} = \log_7 x - \log_7 3$

55. $\log_4 (z + 2) = \log_4 z + \log_4 2$

56. $\log_8 p^4 = (\log_8 p)^4$

57. $\log_9 \frac{x^2y^3}{z^4} = 2 \log_9 x + 3 \log_9 y - 4 \log_9 z$

58. CREATE Write a logarithmic expression for each condition. Then write the expanded expression.

a. a product and a quotient

b. a product and a power

c. a product, a quotient, and a power

59. PERSEVERE Simplify $\log_{\sqrt{a}}(a^2)$ to find an exact numerical value.

60. WHICH ONE DOESN'T BELONG? Find the expression that does not belong. Justify your conclusion.

$\log_b 24 = \log_b 2 + \log_b 12$	$\log_b 24 = \log_b 20 + \log_b 4$
$\log_b 24 = \log_b 8 + \log_b 3$	$\log_b 24 = \log_b 4 + \log_b 6$

61. PERSEVERE Simplify $x^{3 \log_x 2 - \log_x 5}$ to find an exact numerical value.

62. WRITE Explain how the properties of exponents and logarithms are related. Include examples like the one shown at the beginning of the lesson illustrating the Product Property, but with the Quotient Property and Power Property of Logarithms.

Common Logarithms

Learn Common Logarithms

Base 10 logarithms are called **common logarithms** and can be used in many applications. Common logarithms are usually written without the subscript 10.

$\log_{10} x = \log x, x > 0$

Most scientific calculators have a **log** key for evaluating common logarithms. The graph of the common logarithm function is shown.

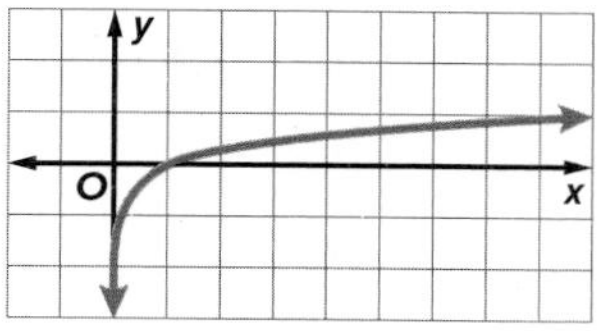

The common logarithms of numbers that differ by integral powers of ten are closely related. Remember that a logarithm represents an exponent. For example, in the equation $y = \log x$, y is the power to which 10 is raised to obtain the value of x.

Complete the table to examine the relationship between common logarithms and exponential form.

$\log x = y$	means	$10^y = x$
$\log 1 = 0$	because	$10^0 = 1$
$\log 10 = 1$	because	$10^1 = 10$
$\log 100 = 2$	because	$10^2 = 100$
$\log 10^m = m$	because	$10^m = 10^m$

Today's Goals
- Solve exponential equations by using common logarithms.
- Evaluate logarithmic expressions by using the Change of Base Formula.

Today's Vocabulary
common logarithms

Example 1 Find Common Logarithms by Using Technology

Use a calculator to evaluate log 8 to the nearest ten-thousandth.

Press **log**, **8**, **)**, and **enter**. The result is 0.903089987, so $\log 8 \approx 0.9031$.

Think About It!
Describe the meaning of $\log 8 \approx 0.9031$ in the context of exponents and the definition of a common logarithm.

Go Online You can complete an Extra Example online.

Example 2 Solve a Logarithmic Equation by Using Exponential Form

SCIENCE The amount of energy E in ergs that is released by an earthquake is related to its Richter scale magnitude M by the equation $\log E = 11.8 + 1.5M$. Although the scale was created in the 1930s, earthquakes that occurred before its invention have been estimated using the Richter scale. For example, an earthquake in Cyprus in 1222 is estimated to have measured 7 on the Richter scale. How much energy was released?

$\log E = 11.8 + 1.5M$	Original equation
$\log E = 11.8 + 1.5(7)$	$M = 7$
$\log E = 22.3$	Simplify.
$E = 10^{22.3}$	Exponential form
$E \approx 2 \times 10^{22}$	Use a calculator.

The earthquake released approximately 2×10^{22} ergs of energy.

Talk About It!

What approximations were made when determining the energy released by the Cyprus earthquake?

Example 3 Solve an Exponential Equation by Using Logarithms

Solve $11^x = 101$. Round to the nearest ten-thousandth.

$11^x = 101$	Original equation
$\log 11^x = \log 101$	Property of Equality for Logarithms
$x \log 11 = \log 101$	Power Property of Logarithms
$x = \dfrac{\log 101}{\log 11}$	Divide each side by log 11.
$x \approx 1.9247$	Use a calculator.

The solution is approximately 1.9247.

Check

You can check this answer graphically by using a graphing calculator.

Enter each side of the equation in the **Y=** list and press **graph**. Then use the **intersect** feature from the **CALC** menu to find the intersection of the two graphs. The intersection is very close to the answer that was obtained algebraically.

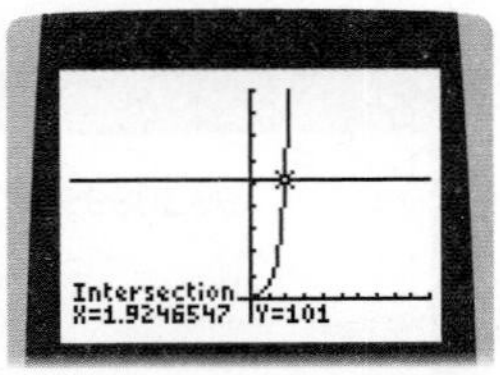

[−10, 10] scl: 1 by [−20, 180] scl: 20

Problem-Solving Tip

Logical Reasoning You can also check to see whether your answer makes sense by considering what x means in the context of the original equation. For example, you know that $11^1 = 11$ and $11^2 = 121$. So, the value of x should be between 1 and 2 because 101 is between 11 and 121.

Go Online You can complete an Extra Example online.

Example 4 Solve an Exponential Inequality by Using Logarithms

Solve $6^{2y-5} < 5^{3y}$. Round to the nearest ten-thousandth.

$6^{2y-5} < 5^{3y}$	Original inequality
$\log 6^{2y-5} < \log 5^{3y}$	Property of Inequality for Logarithmic Functions
$(2y-5)\log 6 < 3y \log 5$	Power Property of Logarithms
$2y \log 6 - 5 \log 6 < 3y \log 5$	Distributive Property
$-5 \log 6 < 3y \log 5 - 2y \log 6$	Subtract $2y \log 6$ from each side.
$-5 \log 6 < y(3 \log 5 - 2 \log 6)$	Distributive Property
$\frac{-5 \log 6}{3 \log 5 - 2 \log 6} < y$	Divide each side by $3 \log 5 - 2 \log 6$.
$\{y \mid y > -7.1970\}$	Use a calculator.

Check

Test $y = 0$.

$6^{2y-5} < 5^{3y}$	Original inequality
$6^{2(0)-5} < 5^{3(0)}$	Replace y with 0.
$6^{-5} < 5^0$	Simplify.
$\frac{1}{7776} < 1$ True ✓	Negative Exponent Property

Learn Change of Base Formula

The Change of Base Formula allows you to write equivalent logarithmic expressions that have different bases.

Key Concept • Change of Base Formula	
Symbols	For all positive numbers a, b, and n, where $a \neq 1$ and $b \neq 1$, $\log_a n = \frac{\log_b n}{\log_b a}$ ← log base b of the original number (numerator); ← log base b of the original base (denominator)
Example	$\log_8 17 = \frac{\log_{10} 17}{\log_{10} 8}$

Example 5 Change of Base Formula

Evaluate $\log_2 11$ by writing it in terms of common logarithms. Round to the nearest ten-thousandth.

$\log_2 11 = \frac{\log_{10} 11}{\log_{10} 2}$	Change of Base Formula
≈ 3.4594	Use a calculator.

Go Online You can complete an Extra Example online.

Watch Out!

Negatives When multiplying or dividing an inequality by a negative value, the inequality sign reverses. In this case $3 \log 5 - 2 \log 6$ is a positive value, so the sign did not change when both sides of the inequality were divided by the expression. While solving inequalities, use your calculator to quickly evaluate logarithmic expressions to determine whether the inequality sign should be reversed.

Go Online to see a proof of the Change of Base Formula.

Talk About It!

Is it possible to evaluate $\log_\pi 5$? Explain.

Check

Evaluate $\log_8 30$. Round to the nearest ten-thousandth.

Example 6 Use the Change of Base Formula

MUSIC The musical cent is a unit of relative pitch. One octave consists of 1200 cents. The formula to determine the difference n in cents between two notes with beginning frequency a and ending frequency b is $n = 1200\left(\log_2 \frac{a}{b}\right)$. Find the frequency of pitch a if pitch b is 1661.22 and the difference between the pitches is 1600 cents.

Think About It!

Describe how to solve $\frac{4}{3} = \frac{\log \frac{a}{1661.22}}{\log 2}$ algebraically.

Step 1 Write the equation in terms of common logarithms.

$n = 1200\left(\log_2 \frac{a}{b}\right)$	Original equation
$1600 = 1200\left(\log_2 \frac{a}{1661.22}\right)$	$n = 1600$ and $b = 1661.22$
$\frac{4}{3} = \log_2\left(\frac{a}{1661.22}\right)$	Divide each side by 1200.
$\frac{4}{3} = \frac{\log \frac{a}{1661.22}}{\log 2}$	Change of Base Formula

Step 2 Use a calculator to solve for a.

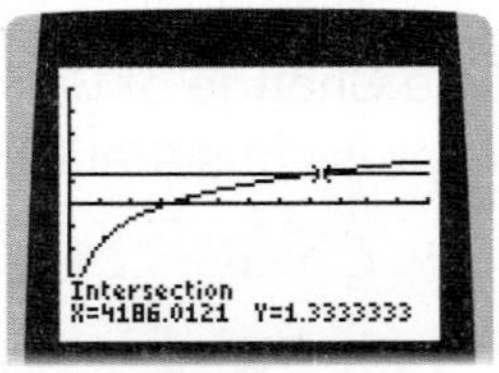

[0, 6000] scl: 500 by [−5, 5] scl: 1

Enter each side of the equation as a function in the **Y=** list. Then, use the **intersect** feature to find the value of a.

The functions intersect at $(4186.0121, 1.\overline{333})$.

Pitch a has a frequency of about 4186.01 Hz.

Go Online

You can watch a video to see how to solve logarithmic equations by graphing using a graphing calculator.

Check

ADVERTISING The revenue of a company is $r = 227 + 44 \log_5 x$, where r is the annual revenue of the company in thousands and x is the amount spent on advertising in thousands. Use a graphing calculator to determine the amount spent on advertising if the revenue for the company is \$325,000. Round to the nearest dollar.

Go Online You can complete an Extra Example online.

Lukas Gojda/123RF

Practice

Go Online You can complete your homework online.

Example 1

Use a calculator to evaluate each expression to the nearest ten-thousandth.

1. log 18

2. log 39

3. log 120

4. log 5.8

5. log 42.3

6. log 0.003

Example 2

7. BIOLOGY There are initially 1000 bacteria in a culture. The number of bacteria doubles each hour. The number of bacteria N present after t hours is given by $N = 1000(2)^t$. How long will it take the culture to increase to 50,000 bacteria?

8. SOUND An equation for loudness L in decibels is given by $L = 10 \log R$, where R is the relative intensity of the sound compared to the minimum threshold of human hearing. One city's emergency weather siren is 138 decibels loud. How many times greater than the minimum threshold of hearing is the siren?

Example 3

Solve each equation. Round to the nearest ten-thousandth.

9. $4^{5k} = 37$

10. $8^p = 50$

11. $7^y = 15$

12. $5^{4x-2} = 120$

13. $6^{x+2} = 18$

14. $2.4^{x+4} = 30$

Example 4

Solve each inequality. Round to the nearest ten-thousandth.

15. $7^{3x-1} \geq 21$

16. $6.5^{2x} \geq 200$

17. $3^x > 243$

18. $16^v \leq \frac{1}{4}$

19. $8^{y+4} > 15$

20. $2^x < 25$

Example 5

Express each logarithm in terms of common logarithms. Then approximate its value to the nearest ten-thousandth.

21. $\log_4 22$

22. $\log_{12} 200$

23. $\log_2 50$

24. $\log_8 15$

25. $\log_3 2$

26. $\log_5 0.4$

Example 6

27. USE ESTIMATION The revenue of a restaurant is $r = 180 + 35 \log_4 x$, where r is the annual revenue in thousands of dollars and x is the amount spent on advertising in thousands of dollars. Use a graphing calculator to determine the amount spent on advertising if the annual revenue for the restaurant is $250,000. Round to the nearest dollar.

28. USE ESTIMATION The function $f(x) = 2 + 10 \log_9 (x + 2)$ gives the estimated number of deer in a county, where x is the number of years after 2012 and $f(x)$ is the deer population in hundreds. A natural resources officer wants to estimate when the population will reach 1500. Write the equation in terms of common logarithms. Then, use a graphing calculator to determine the year.

Mixed Exercises

Solve each equation or inequality. Round to the nearest ten-thousandth.

29. $5^{x^2 - 3} = 72$

30. $4^{2x} = 9^{x + 1}$

31. $2^{n + 1} = 5^{2n - 1}$

32. $7^{2n} > 52^{4n + 3}$

33. $6^p \leq 13^{5 - p}$

34. $2^{y + 3} \geq 8^{3y}$

Express each logarithm in terms of common logarithms. Then approximate its value to the nearest ten-thousandth.

35. $\log_3 (20)^2$

36. $\log_6 (5)^4$

37. $\log_8 (4)^5$

38. $\log_5 (8)^3$

39. $\log_2 (3.6)^6$

40. $\log_{12} (10.5)^4$

41. $\log_3 \sqrt{150}$

42. $\log_4 \sqrt[3]{39}$

43. $\log_5 \sqrt[4]{1600}$

44. **STRUCTURE** Given that $\log_{10} 2 \approx 0.3010$ and $\log_{10} 3 \approx 0.4771$, explain how to find $\log_2 3$.

45. **REASONING** The graph of $y = \log_{10} x$ is shown. Copy the graph. Given that $\frac{1}{\log_{10} 2} \approx 3.32$, sketch a graph of $y = \log_2 x$ on the same coordinate plane.

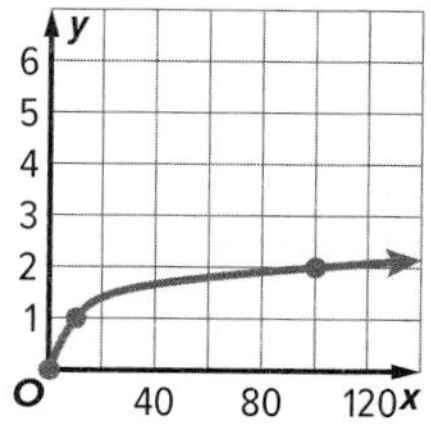

46. **CONSTRUCT ARGUMENTS** When a number n is written in scientific notation, it has the form $n = s \times 10^p$, where $1 \leq s < 10$. Prove that $p \leq \log_{10} n < p + 1$.

47. **STRUCTURE** Use the table shown and the properties of logarithms to determine the following.

a. Determine the missing entries in the table. Justify your reasoning.

b. How can you use the table to determine $\log_{10} 1.5$?

Common Logarithms (approximations)	
x	$\log_{10} x$
2	0.3010
3	0.4771
4	?
5	0.6989
6	?

48. **USE TOOLS** The equation $t = \frac{1}{\log(1 + r)}$ gives the number of years t it takes \$1000 to grow to \$10,000 in a savings account earning an annual interest rate r, compounded annually. If Jo wants to grow her \$1000 investment to \$10,000 in 45 years, use a graphing calculator to determine the interest rate needed for the investment.

49. **STRUCTURE** Rewrite each equation using common logarithms and solve. Show your work.

a. $3 \log_2 (2x) - \log_4 (3x) = 1$

b. $\log_6(x^2) + \log_3(x) = 3$

c. $2 \log_2(3x) = -8 \log_3 (x)$

50. **STRUCTURE** Rewrite each inequality using common logarithms and solve. Show your work.

a. $\log_7 (2x) < 2 \log_6 (x)$

b. $3 \log_2 (4x) \leq \log (8x)$

51. **FIND THE ERROR** Sam and Rosamaria are solving $4^{3p} = 10$. Is either of them correct? Explain your reasoning.

Sam

$4^{3p} = 10$

$\log 4^{3p} = \log 10$

$p \log 4 = \log 10$

$p = \frac{\log 10}{\log 4}$

Rosamaria

$4^{3p} = 10$

$\log 4^{3p} = \log 10$

$3p \log 4 = \log 10$

$p = \frac{\log 10}{3 \log 4}$

52. **PERSEVERE** Solve $\log_{\sqrt{a}} 3 = \log_a x$ for x and explain each step.

53. **CONSTRUCT ARGUMENTS** Find the values of $\log_3 27$ and $\log_{27} 3$. Make a conjecture about the relationship between $\log_a b$ and $\log_b a$. Justify your argument.

54. **FIND THE ERROR** Claudia wants to find the value of $\log_7 8$ to the nearest thousandth. She uses the Change of Base Formula and a calculator to find the approximation. The screen at right shows what she entered, and the result.

a. Explain the error Claudia made and how she can correct it.

b. Consider the relationship between log 8 and log 7. Explain why the result displayed on the calculator should alert Claudia that an error has occurred.

Natural Logarithms

Learn Simplifying Expressions with Natural Logarithms

An exponential function with base e, written as $y = e^x$, is called a **natural base exponential function.** Recall that e is an irrational number with an approximate value of 2.7182818...

The inverse of a natural base exponential function is a **natural logarithm,** which can be written as $\log_e x$, but is more often abbreviated as $\ln x$. Every point on the graph of $y = e^x$ is reflected in the line $y = x$, resulting in the graph of $y = \ln x$.

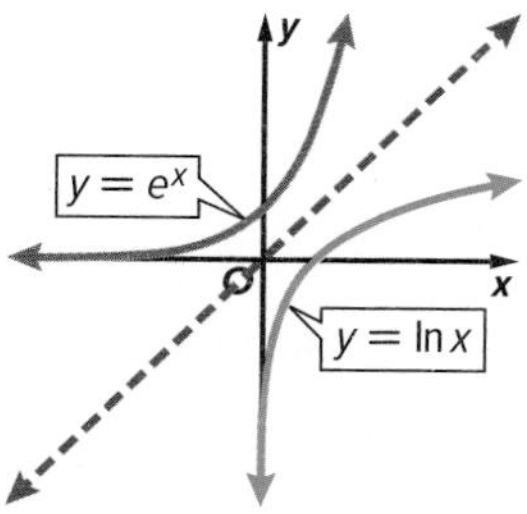

You can write equivalent exponential equations by using the definition of a natural logarithm.

$\ln 6 = x \rightarrow \log_e 6 = x \rightarrow e^x = 6$

Similarly, you can write an equivalent logarithmic equation for an exponential equation with base e.

$e^x = 0.5 \rightarrow \log_e 0.5 = x \rightarrow \ln 0.5 = x$

The same properties that apply to logarithms with a base of b also apply to natural logarithms.

Example 1 Write Equivalent Logarithmic Equations

Write each exponential equation in logarithmic form.

a. $e^x = 14$

$e^x = 14$	Original equation
$\log_e 14 = x$	Definition of natural logarithm
$\ln 14 = x$	Rewrite.

b. $e^7 = x$

$e^7 = x$	Original equation
$\log_e x = 7$	Definition of natural logarithm
$\ln x = 7$	Rewrite.

Check

Write each exponential equation in logarithmic form.

a. $e^3 = 4x$

b. $e^{4x} = 3$

Go Online You can complete an Extra Example online.

Today's Goals

- Simplify expressions with natural logarithms.
- Solve exponential equations by using natural logarithms.

Today's Vocabulary

natural base exponential function

natural logarithm

Think About It!

Why do the same properties that apply to logarithms with a base of b also apply to natural logarithms?

Think About It!

How can you use a calculator to check part **b**?

Example 2 Write Equivalent Exponential Equations

Write each logarithmic equation in exponential form.

a. $\ln 9 = x$

$\ln 9 = x$	Original equation
$\log_e 9 = x$	$\ln 9 = \log_e 9$
$9 = e^x$	Definition of natural logarithm

b. $\ln x \approx 1.7347$

$\ln x \approx 1.7347$	Original equation
$\log_e x \approx 1.7347$	$\ln x = \log_e x$
$x = e^{1.7347}$	Definition of natural logarithm

Think About It!

Elliott says that $\ln e^{3x}$ equals $3x$. Is he correct? Explain.

Check

Write each logarithmic equation in exponential form.

a. $\ln 3x = 4$ **b.** $\ln 4 = 3x$

Example 3 Simplify Logarithmic Expressions

Write $\frac{1}{3} \ln 8 - \ln 3 + \ln 9$ as a single logarithm.

$\frac{1}{3} \ln 8 - \ln 3 + \ln 9$	Original expression
$= \ln 8^{\frac{1}{3}} - \ln 3 + \ln 9$	Power Property of Logarithms
$= \ln 2 - \ln 3 + \ln 9$	$8^{\frac{1}{3}} = \sqrt[3]{8}$ or 2
$= \ln \frac{2}{3} + \ln 9$	Quotient Property of Logarithms
$= \ln 6$	Product Property of Logarithms

Think About It!

In what order should you apply the properties of logarithms to simplify a logarithmic expression?

Check

Use a calculator to verify the solution.

Because both entries result in the same value, $\frac{1}{3} \ln 8 - \ln 3 + \ln 9 = \ln 6$.

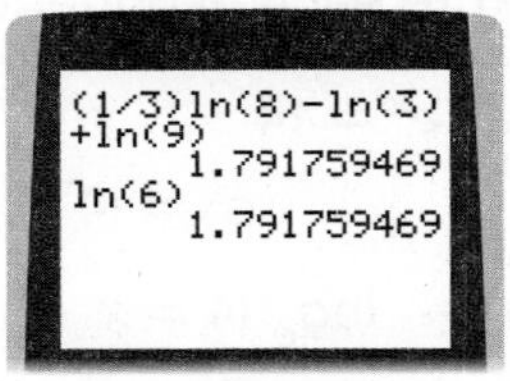

Check

Write $\ln 7 - \frac{1}{3} \ln 27 + \ln 6$ as a single logarithm.

Learn Solving Exponential Equations by Using Natural Logarithms

Because an exponential function with base e is the inverse of the natural logarithm function, they can be used to eliminate each other.

$$e^{\ln x} = x \qquad \ln e^x = x$$

Exponential equations with base e can be solved using natural logarithms because $\ln e = \ln e^1 = 1$.

Go Online You can complete an Extra Example online.

Example 4 Solve Exponential Equations with Base e

Solve $-2e^{x+3} + 8 = -14$. Round to the nearest ten-thousandth.

$-2e^{x+3} + 8 = -14$	Original equation
$-2e^{x+3} = -22$	Subtract 8 from each side.
$e^{x+3} = 11$	Divide each side by −2.
$\ln e^{x+3} = \ln 11$	Property of Equality for Logarithms
$x + 3 = \ln 11$	$\ln e^x = x$
$x + 3 \approx 2.3979$	Use a calculator.
$x \approx -0.6021$	Subtract 3 from each side.

Go Online to see a common error to avoid.

Check

Solve $6e^{0.25x+3} = 26$. Round to the nearest ten-thousandth.

$x \approx$ ___?___

Example 5 Solve Natural Logarithmic Equations

Solve $\frac{1}{6} \ln 5x = 2$. Round to the nearest ten-thousandth.

$\frac{1}{6} \ln 5x = 2$	Original equation
$\ln 5x = 12$	Multiply each side by 6.
$e^{\ln 5x} = e^{12}$	Property of Equality for Exponential Equations
$5x = e^{12}$	$e^{\ln x} = x$
$x = \frac{e^{12}}{5}$	Divide each side by 5.
$x \approx 32{,}550.9583$	Use a calculator.

Example 6 Apply Functions with Base e

TEMPERATURE **Newton's Law of Cooling relates the temperature of an object over time in relation to the initial temperature of the object and the temperature of its surroundings. The temperature in degrees Fahrenheit of a bowl of soup over time is modeled by the function $T(t) = T_a + (T_0 - T_a)e^{-0.032t}$, where T_0 is the temperature of the soup when it is taken off the heat source, T_a is the ambient temperature of the room, and *t* is time in minutes.**

Part A Evaluate the function.

Find the temperature of soup that is initially 200°F in a room that is 72°F after 30 minutes.

$T(t) = T_a + (T_0 - T_a)e^{-0.032t}$	Original function
$= 72 + (200 - 72)e^{-0.032(30)}$	$T_a = 72$, $T_0 = 200$, and $t = 30$
≈ 121	Use a calculator.

After 30 minutes, the soup will be about 121°F.

(continued on the next page)

Go Online You can complete an Extra Example online.

Think About It!

Considering the function given for the situation, will the soup ever become colder than the temperature of the restaurant? Explain.

Part B Evaluate the function for t.

Suppose a restaurant manager wants to make sure that the soup is at least 180°F when it is served to customers. If the soup is initially 200°F and the temperature of the restaurant is 72°F, how long from the time the soup is taken off the heat source should the server serve the soup?

$T(t) \leq T_a + (T_0 - T_a)e^{-0.032t}$	Original inequality
$180 \leq 72 + (200 - 72)e^{-0.032t}$	$T(t) = 180$, $T_a = 72$, and $T_0 = 200$
$108 \leq 128e^{-0.032t}$	Subtract 72 from each side and simplify.
$0.84375 \leq e^{-0.032t}$	Divide each side by 128.
$\ln 0.84375 \leq \ln e^{-0.032t}$	Property of Equality of Logarithms
$\ln 0.84375 \leq -0.032t$	$\ln e^x = x$
$\frac{\ln 0.84375}{-0.032} \geq t$	Divide each side by −0.032.
$5.3 \geq t$	Use a calculator.

The server should serve the soup within 5.3 minutes off taking it off the heat source.

Problem-Solving Tip

Use Reasoning

Examining the structure of an expression can help you interpret the situation. For example, in the expression $T_a + (T_0 - T_a)e^{-0.032t}$, as t increases, $e^{-0.032t}$ will approach 0. So, $(T_0 - T_a)e^{-0.032t}$ will approach 0 and $T_a + (T_0 - T_a)e^{-0.032t}$ will approach T_a. The temperature will approach the ambient temperature as time passes.

Part C Evaluate the function for T_0.

Suppose a restaurant manager increases the temperature of the restaurant to 75°F and wants the servers to have at least 7 minutes to serve the soup to the customers while the soup is 180°F or warmer. At what temperature will the soup have to be when it is initially taken off the heat source?

$T(t) \leq T_a + (T_0 - T_a)e^{-0.032t}$	Original inequality
$180 \leq 75 + (T_0 - 75)e^{-0.032(7)}$	$T(t) = 180$, $T_a = 75$, and $t = 7$
$105 \leq (T_0 - 75)e^{-0.224}$	Subtract 75 from each side.
$105 \leq T_0(e^{-0.224}) - 75(e^{-0.224})$	Distributive Property
$105 + 75(e^{-0.224}) \leq T_0(e^{-0.224})$	Add $75(e^{-0.224})$ to each side.
$\frac{105 + 75(e^{-0.224})}{e^{-0.224}} \leq T_0$	Divide each side by $e^{-0.224}$.
$206.4 \leq T_0$	Use a calculator.

The initial temperature of the soup must be at least 206.4°F so that the servers have 7 minutes to serve it.

Go Online You can complete an Extra Example online.

Explore Using a Scatter Plot to Analyze Data

Online Activity Use the video to complete the Explore.

> **INQUIRY** How can you use a scatter plot to determine the type of function that best models a set of data?

Example 7 Examine Logarithmic Data

GAMING **The table shows the percentage of people in the U.S. who play games on their mobile phones in the years since 2010. Use a graphing calculator to analyze the data.**

Years Since 2010	1	2	3	4	5	6	7	8	9	10
Population (%)	25.9	33.9	40.9	46.3	51.3	55.7	58.9	61.6	63.1	63.7

Source: eMarketer; *projected values for years 2016 and beyond*

Step 1 Enter the data.

Before you begin, make sure that your Diagnostic setting is on. You can find this in the **[CATALOG]** menu. Press **[D]** and scroll down to **DiagnosticOn**. Then press **enter**. Enter the data by pressing **stat** and selecting the **Edit** option.

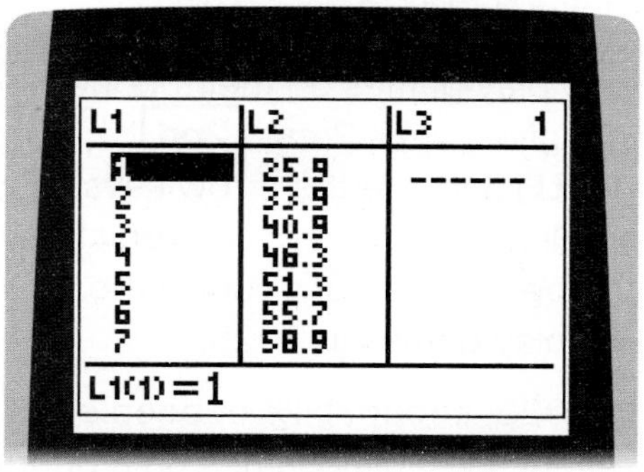

Enter the years since 2010 into List 1 (**L1**) and the percentage of the population in List 2 (**L2**).

Step 2 Make a scatter plot.

Press **2nd**, **[STAT PLOT]**, and then **enter** to turn on the plot. Choose the scatter plot icon. Use **L1** for the **Xlist** and **L2** for the **Ylist**.

Press **window** and adjust the parameters to fit the data.

Press **graph** to see the scatter plot.

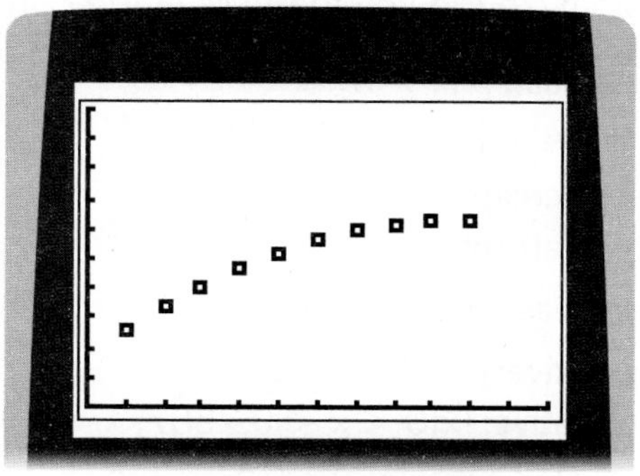

[0, 12] scl: 1 by [0, 100] scl: 10

From the scatter plot, the data appear to have a curved trend that approximates a natural logarithmic function.

(*continued on the next page*)

Go Online
You can complete an Extra Example online.

Step 3 Determine the function of best fit.

To determine the function that best fits the data, perform a natural logarithmic regression. Check that the coefficient of determination is close to 1, indicating a strong fit of the data.

Press **stat** and access the **CALC** menu. Select a natural logarithmic regression **LnReg** from the list. Select **L1** and **L2** as the data set and store the regression equation in the **Y=** list.

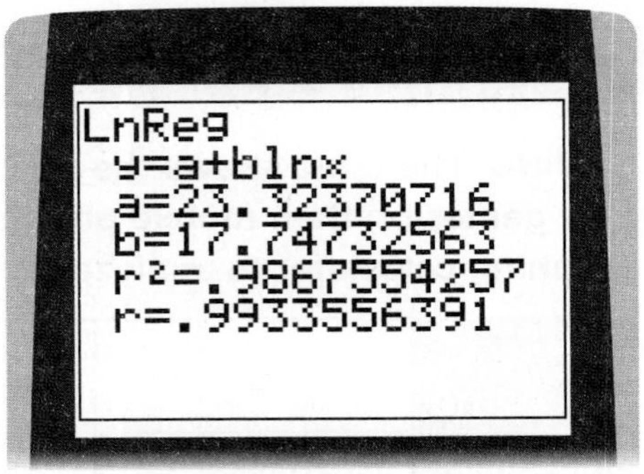

The regression has a coefficient of determination of $r \approx 0.9934$, which is a very strong fit of the data. Because the function fits the data well and makes sense in the context of the situation, the data can be represented by a logarithmic function.

The regression equation is $y = 23.3237 + 17.7473 \ln x$.

Step 4 Evaluate using the regression function.

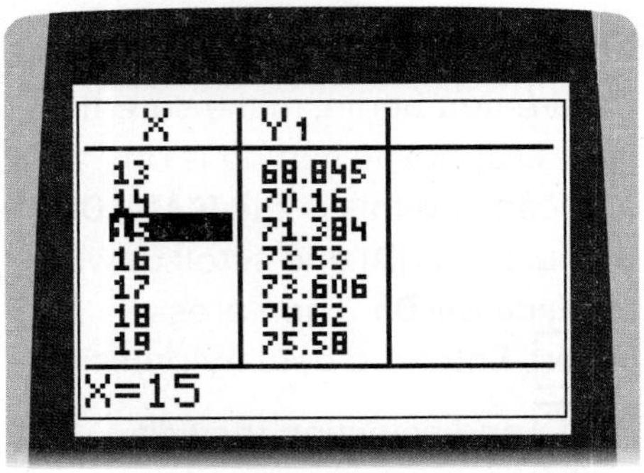

You can use the table generated by the graphing calculator to predict the percentage of people who play games on their phones in a given year. Press **2nd** **[TABLE]** and use the arrow keys to scroll down to $x = 15$ to predict the percentage of people who will play games in 2025.

In 2025, about 71.4% of people in America will play games on a phone.

Step 5 Analyze the graph.

Press **graph**. You can use the functionality of the calculator to help analyze the key features in the context of the situation.

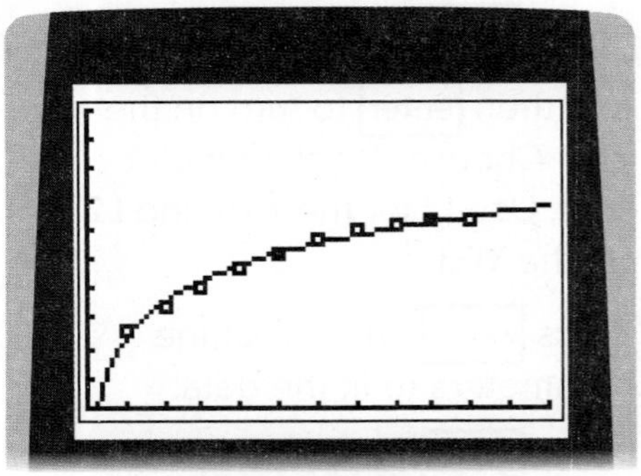

[0, 12] scl: 1 by [0, 100] scl: 10

domain: $\{x | x > 0\}$
range: $\{y | y > 0\}$

***x*-intercept:** 0.2687
***y*-intercept:** none

increasing: $\{x | x > 0\}$

positive: $\{x | x > 0.2687\}$
negative: $\{x | 0 < x < 0.2687\}$

end behavior: As $x \to 0$, $y \to 0$ and as $x \to \infty$, $y \to \infty$.

The function $y = 23.3237 + 17.7473 \ln x$ includes y-values less than 0. However, in the context of the situation, y represents a percent of the population and must be positive.

Watch Out!

Range The function for the percentage of people who play games on their phone has a range of all real numbers. In the context of the situation, only positive values of y no greater than 100 make sense because the percentage of the population cannot be negative or greater than 100%.

Use a Source

Use the data you collected in the Explore activity to write a logarithmic function to model the temperature of the coffee over time. Then, use the function to predict the temperature of the coffee after 45 minutes.

Practice

Go Online You can complete your homework online.

Example 1

Write each exponential equation in logarithmic form.

1. $e^{15} = x$

2. $e^{3x} = 45$

3. $e^{-5x} = 0.2$

4. $e^{8.2} = 10x$

5. $e^{x} = 3$

6. $e^{4} = 8x$

7. $e^{4} = x - 3$

8. $e^{x} = 8$

9. $e^{5} = 10x$

Example 2

Write each logarithmic equation in exponential form.

10. $\ln 50 = x$

11. $\ln 36 = 2x$

12. $\ln 20 = x$

13. $\ln x = 8$

14. $\ln (4x) = 9.6$

15. $\ln 0.0002 = x$

16. $\ln x \approx 0.3345$

17. $\ln 15 = x$

18. $\ln 5.34 = 2x$

Example 3

Write each expression as a single logarithm.

19. $3 \ln 3 - \ln 9$

20. $4 \ln 16 - \ln 256$

21. $2 \ln x + 2 \ln 4$

22. $3 \ln 4 + 3 \ln 3$

23. $\ln 125 - 2 \ln 5$

24. $3 \ln 10 + 2 \ln 100$

25. $4 \ln \frac{1}{3} - 6 \ln \frac{1}{9}$

26. $7 \ln \frac{1}{2} + 5 \ln 2$

27. $8 \ln x - 4 \ln 5$

Example 4

Solve each equation. Round to the nearest ten-thousandth.

28. $2e^{x} - 1 = 11$

29. $5e^{x} + 3 = 18$

30. $e^{3x} = 30$

31. $e^{x} = 5.8$

32. $2e^{x} - 3 = 1$

33. $4 + e^{x} = 19$

Example 5

Solve each equation. Round to the nearest ten-thousandth.

34. $\ln(x - 6) = 1$

35. $\ln(x + 2) = 3$

36. $\ln(x + 3) = 5$

37. $\ln 3x + \ln 2x = 9$

38. $\ln 5x + \ln x = 7$

39. $\ln 3x = 2$

Example 6

40. USE A MODEL Monique wants to invest $4000 in a savings account that pays 3.4% annual interest compounded continuously. The formula $A = Pe^{rt}$ is used to find the amount in the account, where A is the amount in the account after t years, P is the principal amount invested, and r is the annual interest rate.

a. What is the balance of Monique's account after 5 years?

b. Suppose Monique wants to wait until there is at least $6000 in her account before withdrawing any money. How long must she keep her money in the savings account?

c. Suppose Monique wants to have at least $10,000 in her account after 10 years. If the interest rate remains 3.4% compounded continuously, what is the principal she should invest?

41. USE A MODEL Sarita deposits $750 in an account paying 2.7% annual interest compounded continuously. Use the formula for continuously compounded interest, $A = Pe^{rt}$, where A is the amount in the account after t years, P is the principal amount invested, and r is the annual interest rate.

a. What is the balance of Sarita's account after 3 years?

b. How long will it take the balance in Sarita's account to reach at least $2000?

c. Suppose Sarita wants to have at least $1000 in her account after 5 years. If the interest rate remains 2.7% compounded continuously, what is the principal she should invest?

Example 7

42. USE TOOLS The table shows the number of people y who attended the annual neighborhood festival since it began in 2005. Use a graphing calculator to find a logarithmic function which models the data. Then use the model to predict how many people will attend in 2025.

Years Since 2005	1	2	3	4	5	6	7	8	9	10	11
People Attending	150	223	267	290	319	339	354	367	380	402	411

43. USE TOOLS The table shows the mass y of a decaying compound x years since the compound was made. Use a graphing calculator to find a logarithmic function to model the data. Then use the model to predict the mass of the compound after 30 years.

Years	1	2	3	4	5	6	7	8	9	10
Mass (g)	800	775	742	734	719	710	703	692	689	680

44. USE TOOLS The table shows the number of cell colonies y that remain after each minute x of an experiment. Use a graphing calculator to find a logarithmic function to model the data. Then use the model to predict the number of cell colonies remaining after 15 minutes.

Time (min)	1	2	3	4	5	6	7	8	9	10
Number of cells	1500	1385	1350	1315	1300	1280	1270	1258	1243	1232

45. USE TOOLS The table shows the number of visitors y in hundreds to a wildlife sanctuary's hawk exhibit since it opened in 2012. Use a graphing calculator to find a logarithmic function to model the data. Then use the model to predict how many visitors the sanctuary can expect in 2028.

Years Since 2012	1	2	3	4	5	6	7	8
Visitors (hundreds)	2100	2110	2150	2180	2200	2220	2232	2245

Mixed Exercises

Write the expression as a sum or difference of logarithms or multiples of logarithms.

46. $\ln \frac{16}{125}$

47. $\ln \sqrt[5]{x^3}$

Use the natural logarithm to solve each equation.

48. $3^x = 0.4$

49. $2^{3x} = 18$

50. POPULATION The growth of the world's population can be represented as $A = A_0e^{rt}$, where A is the population at time t, A_0 is the population at $t = 0$, and r is the annual growth rate. The world's population at the beginning of 2008 was estimated at 6,641,000,000. If the annual growth rate is 1.2%, when will the world population reach 9 billion?

51. FINANCE Janie's bank pays 2.8% annual interest compounded continuously on her savings account. She invested \$2000 in the account. Using the formula $A = Pe^{rt}$, determine how long it will take for her initial deposit to double in value. Assume that she makes no additional deposits and no withdrawals. Round your answer to the nearest quarter year.

STRUCTURE Given $\ln 5 \approx 1.6094$ and $\ln 8 \approx 2.0794$, evaluate each expression without using a calculator. Explain your reasoning.

52. $\ln 200$

53. $\ln 3.125$

54. $\ln 10$

55. USE ESTIMATION Over time, the amount of Carbon-14 present in non-living organic material decreases. Radiocarbon dating is a technique that measures the amount of carbon-14 that is present in these materials to estimate age. The exponential function $y = ae^{kt}$, represents the relationship between the current amount y of carbon-14 in organic material and the initial amount a of Carbon-14 over time t with rate of decay k.

a. The amount of time it takes for the amount of Carbon-14 present in organic material to decrease by half is 5730 years. This is called the half-life. Use this information to find the rate of decay k associated with Carbon-14. Write your answer as an exact value.

b. Suppose a fossil initially contained 200 milligrams of Carbon-14. Write an equation that represents the relationship between the age of the fossil and the amount of carbon-14 currently in the sample.

c. Use a graphing calculator to estimate how long it will take the fossil to contain 40% or less of the carbon-14 it initially contained. Then, solve the inequality algebraically and compare your results.

56. PRECISION If \$5000 is invested into an account that earns 2% compounded continuously and at the same time \$3000 is invested into an account that earns 4% compounded continuously, after how long will the two accounts contain the same amount of money?

57. ANALYZE Nevaeh says that $e^{e^x} = x$. Do you agree or disagree? Justify your answer.

58. PERSEVERE Solve $4^x - 2^{x+1} = 15$ for x.

59. CONSTRUCT ARGUMENTS Prove $\ln ab = \ln a + \ln b$ for natural logarithms.

60. ANALYZE Determine whether $x > \ln x$ is *sometimes*, *always*, or *never* true. Explain your reasoning.

Using Exponential and Logarithmic Functions

Learn Using Logarithms to Solve Exponential Growth Problems

Exponential functions with base *e* are frequently used in an alternate form of the Exponential Growth Formula to represent situations involving continuous growth.

Key Concept • Continuous Exponential Growth

Exponential growth can be modeled by the function

$$f(x) = ae^{kt},$$

where a is the initial value, t is time in years, and k is a constant representing the rate of continuous growth.

The Continuous Exponential Growth equation $y = ae^{kt}$ is often used to represent population growth and is the same as the continuously compounded interest formula.

Continuously Compounded Interest	Continuous Exponential Growth
$A = Pe^{rt}$	$y = ae^{kt}$
P = initial amount	a = initial population
A = amount at time t	y = population at time t
r = interest rate	k = rate of continuous growth

Today's Goals

- Write and solve exponential growth equations and inequalities.
- Write and solve exponential decay equations.

Think About It!

Write an exponential growth equation to represent a population that grows at a rate of 6% per year and has an initial population of 250.

Example 1 Continuous Exponential Growth

POPULATION **In 2016, the population of Florida was 20.61 million people. In 2000, it was 15.98 million.**

Part A Write an exponential growth equation.

Because the population is growing as time passes, use the year as the independent variable and the population as the dependent variable. Find the value of k, Florida's relative growth rate. Then write an equation that represents the population of Florida t years after 2000.

Since t is the number of years after 2000, 15.98 represents the initial population, and 20.61 represents the population 16 years after 2000, or at $t = 16$.

(continued on the next page)

$$y = ae^{kt}$$ Formula for continuous exponential growth

$$20.61 = 15.98e^{k(16)}$$ $y = 20.61$, $a = 15.98$, and $t = 2016 - 2000$ or 16

$$\frac{20.61}{15.98} = e^{16k}$$ Divide each side by 15.98.

$$\ln \frac{20.61}{15.98} = \ln e^{16k}$$ Property of Equality for Logarithmic Equations

$$\ln \frac{20.61}{15.98} = 16k$$ $\ln e^x = x$

$$\frac{\ln \frac{20.61}{15.98}}{16} = k$$ Divide each side by 16.

$$0.0159 \approx k$$ Use a calculator.

Florida's relative rate of growth is about, 0.0159, or about 1.59%.

$$y = ae^{kt}$$ Formula for continuous exponential growth

$$y = 15.98e^{0.0159t}$$ $a = 15.98$ and $k = 0.0159$

So, the population of Florida t years after 2000 can be modeled by the equation $y = 15.98e^{0.0159t}$.

Study Tip

Assumptions We know that populations cannot grow infinitely and may be affected by other factors such as available land area, resources, diseases, and so on. However, assuming that a population will continue to grow at the same rate allows us to represent the situation with an exponential growth equation.

Part B Predict when the population will reach 25 million people.

Use your equation from **Part A** to predict the population.

$$y = 15.98e^{0.0159t}$$ Original equation

$$25 = 15.98e^{0.0159t}$$ $y = 25$

$$1.564 = e^{0.0159t}$$ Divide each side by 15.98.

$$\ln 1.564 = \ln e^{0.0159t}$$ Property of Equality for Logarithmic Equations

$$\ln 1.564 = 0.0159t$$ $\ln e^x = x$

$$\frac{\ln 1.564}{0.0159} = t$$ Divide each side by 0.0159.

$$28.129 \approx t$$ Use a calculator.

Florida's population will reach 25 million people approximately 28.129 years after 2000, or in 2028.

Think About It!

Based on the given models for the populations of Florida and California, does it make sense that Florida's population will eventually surpass California's? Explain your reasoning.

Part C Compare the populations of Florida and California.

California's population in 2000 was 33.9 million, and its population growth can be modeled by $y = 33.9e^{0.0092t}$. Determine when Florida's population will surpass California's.

$$15.98e^{0.0159t} > 33.9e^{0.0092t}$$ Formula for exponential growth

$$\ln 15.98e^{0.0159t} > \ln 33.9e^{0.0092t}$$ Prop. of Ineq. for Logarithms

$$\ln 15.98 + \ln e^{0.0159t} > \ln 33.9 + \ln e^{0.0092t}$$ Product Property of Logarithms

$$\ln 15.98 + 0.0159t > \ln 33.9 + 0.0092t$$ $\ln e^x = x$

$$\ln 15.98 + 0.0067t > \ln 33.9$$ Subtract $0.0092t$ from each side.

$$0.0067t > \ln 33.9 - \ln 15.98$$ Subtract $\ln 15.98$ from each side.

$$t > \frac{\ln 33.9 - \ln 15.98}{0.0067}$$ Divide each side by 0.0067.

$$t > 112.25$$ Use a calculator.

Assuming that this trend continues, Florida's population is on track to surpass California's in 2112.

Go Online You can complete an Extra Example online.

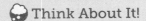

Check

SCIENCE An experiment starts with 20 bacteria A cells. After 45 minutes, there are 710 bacteria A cells.

Part A

Write the equation that models the number of bacteria A cells y after t minutes. Round the value of k to the nearest thousandth.

Part B

After how many minutes will there be 1000 bacteria A cells? Round to the nearest tenth if necessary.

___?___ minutes

Part C

Bacteria B grows exponentially according to the model $y = 52e^{0.064t}$. After how many minutes will there be more bacteria A cells than bacteria B cells? Round to the nearest tenth if necessary.

___?___ minutes

Learn Using Logarithms to Solve Exponential Decay Problems

Exponential functions with base e are frequently used by researchers and scientists to represent situations involving continuous decay.

Key Concept • Continuous Exponential Decay

Exponential decay can be modeled by the function

$$f(x) = ae^{-kt},$$

where a is the initial value, t is time in years, and k is a positive constant representing the rate of continuous decay.

Think About It!

Given time in months and a rate per year, how would you find the value of t to substitute into the continuous exponential decay function?

Example 2 Continuous Exponential Decay

SCIENCE **The half-life of a radioactive substance is the time it takes for half of the atoms of the substance to disintegrate. The radioactive substance Thorium-230 is used to determine the ages of cave formations and coral. The half-life of Thorium-230 is 75,381 years.**

(continued on the next page)

Use a Source

Research the half-life of another radioactive substance. Find the value of k and the equation of decay for the substance.

Watch Out!

Rate and Time Some radioactive substances have half-life periods that are more easily recorded in a unit of time other than years. Make sure that time t and the time period of the rate k use the same unit. For example, the half-life of Phosphorus-32 is 14.29 days, so its decay constant of 0.0485 represents a decay rate of 4.85% per day, not per year.

Part A

Determine the value of k and the equation of decay for Thorium-230.

If a is the initial amount of the substance, then the amount y that remains after one half-life period, or 75,381 years, is $\frac{1}{2}a$ or $0.5a$.

$y = ae^{-kt}$	Formula for continuous exponential decay
$0.5a = ae^{-k(75,381)}$	$y = 0.5a$ and $t = 75,381$
$0.5 = e^{-75,381k}$	Divide each side by a.
$\ln 0.5 = \ln e^{-75,381k}$	Property of Equality for Logarithmic Equations
$\ln 0.5 = -75,381k$	$\ln e^x = x$
$\frac{\ln 0.5}{-75,381} = k$	Divide each side by $-75,381$.
$0.0000092 \approx k$	Use a calculator.

The rate of decay of Thorium-230 is about 0.0000092, or about 0.00092% per year. Thus, the equation for the decay of Thorium-230 is $y = ae^{-0.0000092t}$.

Part B

How much of a 2-gram sample of Thorium-230 should be left after 1500 years?

$y = ae^{-0.0000092t}$	Formula for the decay of Thorium-230
$= 2e^{-0.0000092(1500)}$	$a = 2$ and $t = 1500$
$= 2e^{-0.0138}$	Simplify.
≈ 1.97	Use a calculator.

After 1500 years, there will be about 1.97 grams of Thorium-230 remaining. Since Thorium-230 has such a long half-life, it is reasonable that after 1500 years only a small amount, 0.03 gram, of the original sample will have decayed.

Check

HALF-LIFE Iodine-131, a radioactive isotope commonly used to treat thyroid cancer, has a half-life of 8.02 days.

Part A

Write the continuous exponential decay equation for Iodine-131. Round the value of k to the nearest thousandth.

Part B

How much of a 15-gram sample will be left after 20 days? Round to the nearest tenth if necessary.

Go Online You can complete an Extra Example online.

Apply Example 3 Radiocarbon Dating

ARCHAEOLOGY **Carbon-14 is used to date objects in a method called radiocarbon dating. Carbon-14 has as half-life of 5730 years and a decay constant k of 0.00012. In 2016, charcoal flakes from the excavation site of an ancient human settlement in Canada were sent for radiocarbon dating. If the charcoal flakes were found to contain about 18.6% as much Carbon-14 as they would have originally contained, how old are the charcoal flakes to the nearest year?**

1 What is the task?

Describe the task in your own words. Then list any questions that you may have. How can you find answers to your questions?

I need to use the information about the half-life of Carbon-14 to find the age of the charcoal flakes. Which values should I use in the exponential decay equation? When I solve the equation, what will the solution mean? I can use what I know about half-life to estimate a solution.

2 How will you approach the task? What have you learned that you can use to help you complete the task?

I will estimate the solution. Then I will write and solve an exponential decay equation for Carbon-14. I will interpret the solution in the context of the situation. I have learned how to use the exponential decay equation.

3 What is your solution?

Use your strategy to solve the problem.

Based on the half-life of Carbon-14, how could you estimate the age of the charcoal flakes?

After two half-life periods, there would be 25% of the original amount of Carbon-14 remaining in the charcoal flake because 100% ÷ 2 = 50% and 50% ÷ 2 = 25%. After three half-life periods, there would be 12.5% left. If the flake contains 18.6% as much Carbon-14 as it originally did, then between two and three half-life periods have passed. So, the charcoal flake is between 11,460 and 17,190 years old.

What equation can be used to find the age of the charcoal flakes?

$0.186 = e^{-0.00012t}$

What does the solution of the equation represent in the context of the situation?

The charcoal flakes are about 14,017 years old. This means that the ancient human settlement existed about 14,000 years ago.

(*continued on the next page*)

Study Tip

Radiocarbon Dating When given a percent or fraction of decay, use an original amount of 1 for *a*.

4 How can you know that your solution is reasonable?

Write About It! Write an argument that can be used to defend your solution.

The solution falls within the initial estimate of 11,460 to 17,190 years. I can also substitute 14,017 into the formula for the decay of Carbon-14 to check that the remaining amount is about 18.6%.
For $y = ae^{-0.00012t}$, let $a = 1$ and $t = 14{,}017$. Then $y = 1 \cdot e^{-0.00012(14{,}017)}$ ≈ 0.186 or 18.6%.

Check

ARCHAEOLOGY An archaeologist examining a prehistoric painting estimates that it contains about 2.34% as much Carbon-14 as it would have contained when it was painted. How long ago was the painting created? Round to the nearest year.

[*Hint*: The formula for the decay of Carbon-14 is $y = ae^{-0.00012t}$.]

Practice

Go Online You can complete your homework online.

Example 1

1. POPULATION In 2000, the world population was estimated to be 6.124 billion people. In 2005, it was 6.515 billion.

   a. Write an exponential growth equation to represent the population y in billions t years after 2000.

   b. Use the equation to predict the year in which the world population reached 7.5 billion people.

2. CONSUMER AWARENESS Jason wants to buy a new HD television but he thinks that if he waits, the quality of HD televisions will improve. The television he wants to buy costs $2500 now, and based on pricing trends, Jason thinks that the price will increase by 4% each year.

   a. Write an exponential growth equation to represent the price y of a new HD television t years from now.

   b. Use the equation to predict when a new HD television will cost $3000.

   c. Jason decides to wait to buy a new television and saves his money. He puts $2200 in a savings account with 4.7% annual interest compounded continuously. Determine when the amount in his savings will exceed the cost of a new television.

Example 2

3. REASONING A radioactive substance has a half-life of 32 years.

   a. Determine the value of k and the equation of decay for this radioactive substance.

   b. How much of a 5-gram sample of the radioactive substance should be left after 100 years?

Example 3

4. CARBON DATING Carbon-14 has a decay constant k of 0.00012. Use this information to determine the age of the objects based on the amount of Carbon-14.

   a. a fossil that has lost 95% of its Carbon-14

   b. an animal skeleton that has 95% of its Carbon-14 remaining

5. HALF-LIFE Archeologists uncover an ancient wooden tool. They analyze the tool and find that it has 22% as much Carbon-14 compared to the likely amount that it contained when it was made. Given that the decay constant of Carbon-14 is 0.00012, about how old is the artifact?

Mixed Exercises

6. RADIOACTIVE DECAY The half-life of Rubidium-87 is about 48.8 billion years.

a. Determine the value of k and the equation of decay for Rubidium-87.

b. A specimen currently contains 50 milligrams of Rubidium-87. How long will it take the specimen to decay to only 18 milligrams of Rubidium-87?

c. How many milligrams of the 50-milligram sample will be left after 800 million years?

d. How long will it take Rubidium-87 to decay to one-sixteenth its original amount?

7. USE A MODEL Consider a certain bacteria which is undergoing continuous exponential growth.

a. If there are 80 cells initially and 675 cells after 30 minutes, determine the value of k.

b. When will the bacteria reach a population of 6000 cells?

c. If a second type of bacteria is growing exponentially according to the model $y = 35e^{0.0978t}$, determine how long before the number of cells of this bacteria exceed the number of cells in the original bacteria.

8. NUCLEAR POWER The element plutonium-239 is highly radioactive. Nuclear reactors can produce and also use this element. The heat that plutonium-239 emits has helped to power equipment on the moon. If the half-life of Plutonium-239 is 24,360 years, what is the value of k for this element?

9. DEPRECIATION A Global Positioning Satellite (GPS) system uses satellite information to locate ground position. Abu's surveying firm bought a GPS system for $12,500. The GPS is now worth $8600. If the value of the system depreciates at a rate of 6.2% annually, how many years ago did Abu buy the GPS system?

10. WHALES Modern whales appeared between 5 and 10 million years ago. A paleontologist claims to have discovered a whale vertebrae which contains 80% less Carbon-14 than it originally contained. Is it possible for this vertebrae to be from a modern whale? Use the decay constant of 0.00012 for Carbon-14. Justify your response.

11. REGULARITY Luisa read that the population of her town has increased exponentially. The current population of her town is 68,735. One year ago, the population was 67,387.

a. Based on this information, write an exponential growth equation. Let y represent the population after t years.

b. Use the equation to estimate the population 100 years ago.

12. RADIOACTIVE DECAY Cobalt, an element used to make alloys, has several isotopes. One of these, Cobalt-60, is radioactive and has a half-life of 5.7 years. What is the rate of decay for Cobalt-60?

13. WILDLIFE The initial population of rabbits in an area is 8000 and the population grows continuously at a rate of 26% each year. Write an equation to represent the rabbit population P in thousands after t years. Then, determine how long it will take for the population to reach 25,000.

14. STATE YOUR ASSUMPTIONS A population is growing continuously at a rate of 3%. If the population is now 5 million, when will the population reach 8.3 million? State an assumption needed to solve the problem.

15. ORGANISMS The table shows the amount of Carbon-14 left in a 1000-milligram sample over time. Use the data to verify that the decay constant is approximately −0.00012.

Time (years)	Carbon-14 (mg)
0	1000
1	999.876
2	999.752
3	999.628

16. SCIENCE The number of bacteria in a colony is growing exponentially. At 10:00 A.M. the number of bacteria was 20, and the colony population has continuously increased at a rate of 8% each hour.

a. Write an equation to represent the number of bacteria y after t hours.

b. If this trend continues, determine the time when the number of bacteria in the colony will reach 50. Round to the nearest minute.

17. USE A MODEL A biology experiment starts with 1,000,000 cells and 30% of the cells are dying every minute. The biologist wants to determine when there will be less than 1000 cells.

a. Copy and complete the table of values and graph the points. Determine what kind of mathematical model best describes the points.

t (min)	Surviving Cells After *t* Minutes	(t, $f(t)$)
0	initial amount	(0, 1,000,000)
1	(0.70)(1,000,000) = 700,000 survive	(1, 700,000)
2	(0.70)(700,000) =	
3	(0.70)(490,000) =	

b. Write an equation to represent the situation and define each variable.

c. Find the value of the constant k to 6 decimal places, and tell whether it indicates growth or decay. Write the exponential equation to represent this experiment.

d. How long will it take to have less than 1000 cells?

18. USE A MODEL The population of a city is modeled by $f(t) = 250e^{0.01753t}$, where $f(t)$ is the population in thousands t years after 2012.

a. Based on the equation, what information do you know about the population of the city?

b. In what year will the city's population reach 500,000?

19. PERSEVERE Solve $\frac{120{,}000}{1 + 48e^{-0.015t}} = 24e^{0.055t}$ for t.

20. ARGUMENTS Explain mathematically why $f(t) = \frac{c}{1 + 60e^{-0.5t}}$ approaches, but never reaches the value of c as $t \rightarrow \infty$.

21. WRITE How are exponential and continuous exponential functions used to model different real-world situations?

Review

Essential Question

How are logarithms defined and used to model situations in the real world?

Module Summary

Lesson 8-1

Logarithmic Functions

- Let b and x be positive numbers, $b \neq 1$. The *logarithm* of x with base b is denoted $\log_b x$ and is defined as the exponent y that makes the equation $b^y = x$ true.
- The same techniques used to transform the graphs of other functions can be applied to the graphs of logarithmic functions in the form $g(x) = a \log_b (x - h) + k$.

Lesson 8-2

Properties of Logarithms

- If b is a positive number other than 1, then $\log_b x = \log_b y$ if and only if $x = y$.
- For all positive numbers b, m, and n, where $b \neq 1$, $\log_b mn = \log_b m + \log_b n$.
- For all positive numbers b, m, and n, where $b \neq 1$ and $n \neq 0$, $\log_b \frac{m}{n} = \log_b m - \log_b n$.
- For any real number n, and positive numbers m and b, where $b \neq 1$, $\log_b m^n = n \log_b m$.

Lessons 8-3 and 8-4

Common Logarithms and Natural Logarithms

- Base 10 logarithms are called common logarithms.
- For all positive numbers a, b, and n, where $a \neq 1$ and $b \neq 1$, $\log_a n = \frac{\log_b n}{\log_b a}$.
- An exponential function with base e, written as $y = e^x$, is called a natural base exponential function.
- The inverse of a natural base exponential function is a natural logarithm, which can be written as $\log_e x$, but is more often abbreviated as $\ln x$.
- The same properties that apply to logarithms with a base of b also apply to natural logarithms.

Lesson 8-5

Using Exponential and Logarithmic Functions

- Exponential growth can be modeled by the function $f(x) = ae^{kt}$, where a is the initial value, t is time in years, and k is a constant representing the rate of continuous growth.
- The Continuous Exponential Growth equation $y = ae^{kt}$ is often used to represent population growth and is the same as the continuously compounded interest formula.
- Exponential decay can be modeled by the function $f(x) = ae^{-kt}$, where a is the initial value, t is time in years, and k is a positive constant representing the rate of continuous decay.

Study Organizer

Foldables

Use your Foldable to review this module. Working with a partner can be helpful. Ask for clarification of concepts as needed.

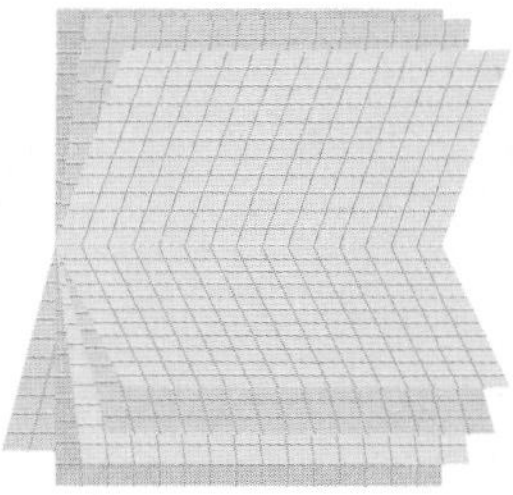

Test Practice

1. OPEN RESPONSE Write each equation in logarithmic form. (Lesson 8-1)

a) $6^4 = 1296$

b) $24^{\frac{3}{5}} = 27$

c) $8^{-2} = \frac{1}{64}$

2. GRAPH Graph $f(x) = \log_7 x - 3$. (Lesson 8-1)

3. GRAPH Graph $g(x) = -\log_4 x + 2$. (Lesson 8-1)

4. MULTIPLE CHOICE Which of the following is equivalent to $\log_{81} 3$? (Lesson 8-1)

A. -4

B. $-\frac{1}{4}$

C. $\frac{1}{4}$

D. 4

5. OPEN RESPONSE Explain how you can use $\log_3 7 \approx 1.7712$ to approximate the value of $\log_3 15{,}309$. (Lesson 8-2)

6. MULTIPLE CHOICE Find the value of x that makes the equation true. (Lesson 8-2)

$\log_{11}(x^2 + 5x - 108) = \log_{11} 28x$

A. -7

B. -4

C. 18

D. 27

7. **MULTIPLE CHOICE** Solve $\log_{32} x = \frac{7}{5}$. (Lesson 8-2)

A. 2

B. 14

C. 44.8

D. 128

8. **OPEN RESPONSE** Some stars appear brighter than others because they are closer to Earth. The equation $0.58 = 5.03 - 5 \log_{10} x$ can be used to determine the distance x in parsecs that the star, Vega, is from Earth. Explain how to use a graphing calculator to find the distance Vega is from Earth. (Lesson 8-3)

9. **MULTI-SELECT** Select all the solutions of $14^{x-5} = 8^x$. (Lesson 8-3)

A. −23.5792

B. 11.6667

C. 23.5792

D. $\dfrac{5 \log 14}{\log 8 - \log 14}$

E. $\dfrac{5 \log 14}{\log 14 - \log 8}$

F. $\log 14(x - 5) = \log 8x$

10. **OPEN RESPONSE** Solve $6^{k-3} > 11^k$. (Lesson 8-3)

11. **MULTIPLE CHOICE** Which equation is equivalent to $12 \log_x 8 = 15$? (Lesson 8-3)

A. $\frac{5}{4} = \frac{\log x}{\log 8}$

B. $\frac{5}{4} = \frac{\log 8}{\log x}$

C. $3 = \frac{\log x}{\log 8}$

D. $3 = \frac{\log 8}{\log x}$

12. **MULTIPLE CHOICE** Each key on a piano has a certain frequency, related to its position on the keyboard. The relationship can be described using the equation $n = 1 + 12 \log_2 \frac{f}{27.5}$, where n is the number of the key and f is the frequency in hertz. If middle C has a frequency of 261.1 Hz, what number key is it on a regular keyboard? Round to the nearest whole number, if necessary. (Lesson 8-3)

A. 3

B. 10

C. 39

D. 40

13. **OPEN RESPONSE** Write each equation in logarithmic form. (Lesson 8-4)

a) $e^7 = 8x$

b) $e^{2x} = 17$

14. MULTI-SELECT Select all expressions that show 8 ln 81 − 10 ln 27 as a single logarithm. (Lesson 8-4)

A. $\ln \frac{1}{3}$

B. ln 6

C. ln 9

D. 2 ln 3

E. −2 ln 53

15. MULTIPLE CHOICE Cho started a savings account with $2800. The account pays 3.8% compounded continuously. Cho wanted to withdraw the money after 5 years, but her friend says she should wait to withdraw it after 10 years.

Use the continuously compounded interest formula, $A = Pe^{rt}$, to determine how much more money will be in the savings account if Cho waits 10 years instead of 5 years to withdraw the money? Round to the nearest cent. (Lesson 8-4)

A. $691.66

B. $708.50

C. $3385.90

D. $4094.40

16. OPEN RESPONSE Solve the inequality. Round to the nearest ten-thousandth. (Lesson 8-4)

$\ln (2x + 5)^3 < 6$

17. MULTIPLE CHOICE Alaska ranks as the 48th state when comparing population sizes. In 1980, the population was 410,851 and in 2010 the population was 713,985. Which equation models the population of Alaska *t* years after 1980? (Lesson 8-5)

A. $y = 410{,}851e^{0.0184t}$

B. $y = 410{,}851e^{-0.0184t}$

C. $y = 713{,}985e^{0.0184t}$

D. $y = 713{,}985e^{-0.0184t}$

18. OPEN RESPONSE Suppose the population of Alaska continues to grow at a continuous rate. Use the equation found in Exercise 17 to explain how to predict when the population of Alaska will reach 1,600,000 people. (Lesson 8-5)

19. OPEN RESPONSE Satellites in space are powered by radioisotopes, or radioactive elements. The amount of power the radioisotope generates over time can be represented by the equation $P = 53e^{-0.0042t}$, where *P* is the output in watts and *t* is the time in days. After how many days will the remaining power be 11.44 watts? Round to the nearest whole number. (Lesson 8-5)

Module 9

Rational Functions

Essential Question

How are the rules for operations with rational numbers applied to operations with rational expressions and equations?

What Will You Learn?

How much do you already know about each topic **before** starting this module?

KEY

— I don't know. — I've heard of it. — I know it!

	Before			After		
multiply and divide rational expressions						
add and subtract rational expressions						
graph and analyze reciprocal functions						
graph and analyze rational functions						
recognize and solve direct variation equations						
recognize and solve joint variation equations						
recognize and solve inverse variation equations						
solve rational equations and inequalities algebraically and by graphing						

Foldables Make this Foldable to help you organize your notes about rational functions and relations. Begin with an $8\frac{1}{2}'' \times 11''$ sheet of grid paper.

1. **Fold** in thirds along the height.
2. **Fold** the top edge down, making a 2″ tab at the top. Cut along the folds.
3. **Label** the outside tabs *Expressions, Functions*, and *Equations*. Use the inside tabs for definitions and notes.
4. **Write** examples of each topic in the space below each tab.

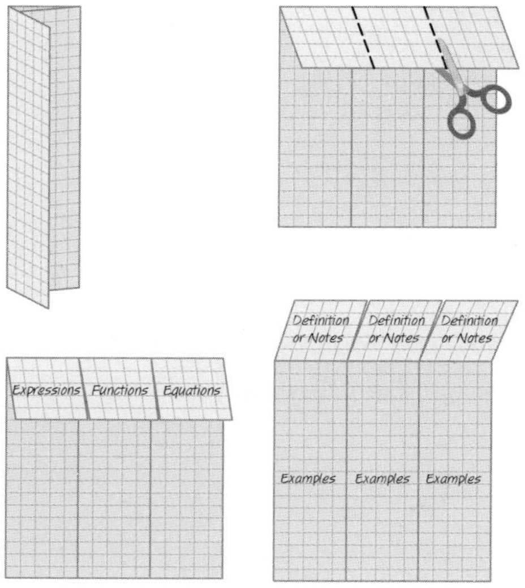

What Vocabulary Will You Learn?

- combined variation
- complex fraction
- constant of variation
- direct variation
- excluded values
- horizontal asymptote
- hyperbola
- inverse variation
- joint variation
- oblique asymptote
- point discontinuity
- rational equation
- rational expression
- rational function
- rational inequality
- reciprocal function
- vertical asymptote

Are You Ready?

Complete the Quick Review to see if you are ready to start this module. Then complete the Quick Check.

Quick Review

Example 1

Solve $\frac{9}{11} = \frac{7}{8}r$. Write in simplest form.

$\frac{9}{11} = \frac{7}{8}r$	Original equation
$\frac{72}{11} = 7r$	Multiply each side by 8.
$\frac{72}{77} = r$	Divide each side by 7.

Since the GCF of 72 and 77 is 1, the solution is in simplest form.

Example 2

Simplify $\frac{1}{3} + \frac{3}{4} - \frac{5}{6}$.

$\frac{1}{3} + \frac{3}{4} - \frac{5}{6}$	
$= \frac{1}{3}\left(\frac{4}{4}\right) + \frac{3}{4}\left(\frac{3}{3}\right) - \frac{5}{6}\left(\frac{2}{2}\right)$	The LCM of 3, 4, and 6 is 12.
$= \frac{4}{12} + \frac{9}{12} - \frac{10}{12}$	Simplify.
$= \frac{3}{12}$	Add and subtract.
$= \frac{3 \div 3}{12 \div 3}$ or $\frac{1}{4}$	Simplify.

Quick Check

Solve each equation. Write in simplest form.

1. $\frac{5}{14} = \frac{1}{3}x$

2. $\frac{8}{5} = \frac{1}{4}k$

3. $\frac{1}{8}m = \frac{7}{3}$

4. $\frac{10}{9}p = 7$

Simplify each expression.

5. $\frac{3}{4} - \frac{7}{8}$

6. $\frac{8}{9} - \frac{7}{6} + \frac{1}{3}$

7. $\frac{9}{10} - \frac{4}{15} + \frac{1}{3}$

8. $\frac{10}{3} + \frac{5}{6} + 3$

How Did You Do?

Which exercises did you answer correctly in the Quick Check?

Lesson 9-1

Multiplying and Dividing Rational Expressions

Learn Simplifying Rational Expressions

A **rational expression** is a ratio of two polynomial expressions.

Because variables in algebra often represent real numbers, operations with rational numbers and rational expressions are similar. For example, when you write a fraction in simplest form, you divide the numerator and denominator by the greatest common factor (GCF).

$$\frac{35}{40} = \frac{\cancel{5} \cdot 7}{\cancel{5} \cdot 8} = \frac{7}{8} \qquad \text{GCF} = 5$$

You use the same process to simplify a rational expression.

$$\frac{x^2 + 7x + 10}{x^2 - x - 6} = \frac{(x + 5)\cancel{(x + 2)}}{(x - 3)\cancel{(x + 2)}} = \frac{(x + 5)}{(x - 3)} \qquad \text{GCF} = x + 2$$

Sometimes, you can also factor out −1 in the numerator or denominator to help simplify a rational expression.

Example 1 Simplify a Rational Expression

Simplify $\frac{x^2 - 2x - 24}{2x^3 + 6x^2 - 8x}$, and state when the original expression is undefined.

Part A Simplify.

$$\frac{x^2 - 2x - 24}{2x^3 + 6x^2 - 8x} = \frac{x^2 - 2x - 24}{2x(x^2 + 3x - 4)} \qquad \text{Distributive Property}$$

$$= \frac{(x + 4)(x - 6)}{2x(x + 4)(x - 1)} \qquad \text{Factor the numerator and denominator.}$$

$$= \frac{(x - 6)}{2x(x - 1)} \cdot \frac{\overset{1}{\cancel{(x + 4)}}}{\underset{1}{\cancel{(x + 4)}}} \qquad \text{Eliminate common factors.}$$

$$= \frac{x - 6}{2x(x - 1)} \qquad \text{Simplify.}$$

Part B State when the expression is undefined.

The expression is undefined when the denominator is equal to 0.

Because the question asks when the original expression is undefined, consider the denominator before the common factors are eliminated. Thus, the expression is undefined when $2x(x + 4)(x - 1) = 0$. By the Zero Product Property, the expression is undefined when $2x = 0$ or $x = 0$, when $x + 4 = 0$ or $x = -4$, and when $x - 1 = 0$ or $x = 1$.

Go Online You can complete an Extra Example online.

Today's Goals

- Simplify rational expressions.
- Simplify rational expressions by multiplying and dividing.

Today's Vocabulary

rational expression

complex fraction

Think About It!

Are all rational expressions defined for all values of x? Justify your argument.

Study Tip

Checking Because you are simplifying the expression, you can check your answer by testing the original expression and your answer for various values of x.

Check

Simplify $\frac{x^2 + 2x + 1}{4x^2 + 3x - 1}$, and state when the original expression is undefined.

Part A Select the simplified expression.

A. 1

B. $\frac{x + 1}{4x - 1}$

C. $\frac{(x + 1)^2}{(4x + 1)(x - 1)}$

D. $\frac{1}{4x - 1}$

Part B State when the expression is undefined.

$x =$ __?__, __?__

Think About It!

How would your answer differ if you factored out −1 in the denominator rather than the numerator? Would the expression be equivalent?

Example 2 Simplify by Using −1

Simplify $\frac{(6x^2 - 5xy)(x + 2y)}{(x + y)(5y - 6x)}$.

$$\frac{(6x^2 - 5xy)(x + 2y)}{(x + y)(5y - 6x)} = \frac{x(6x - 5y)(x + 2y)}{(x + y)(5y - 6x)} \quad \text{Factor.}$$

$$= \frac{x(-1)\overset{1}{\cancel{(5y - 6x)}}(x + 2y)}{(x + y)\underset{1}{\cancel{(5y - 6x)}}} \quad 6x - 5y = -1(5y - 6x)$$

$$= \frac{(-x)(x + 2y)}{x + y} \text{ or } -\frac{x(x + 2y)}{x + y} \quad \text{Simplify.}$$

Check

Select the simplified form of $\frac{(7y - 3x)(5x - 1)}{(5x^3 + x^2)(3x - 7y)}$.

A. $-\frac{5x - 1}{x^2(5x + 1)}$

B. $\frac{(7y - 3x)(5x - 1)}{x^2(5x + 1)(3x - 7y)}$

C. $\frac{1}{x^2}$

D. $\frac{5x - 1}{5x^3 + x^2}$

Explore Simplifying Complex Fractions

Online Activity Use the interactive tool to complete the Explore.

INQUIRY Can you simplify complex fractions that contain polynomials in the numerator or denominator?

Go Online

You can complete an Extra Example online.

Learn Multiplying and Dividing Rational Expressions

The method for multiplying and dividing fractions also works with rational expressions.

Key Concept • Multiplying Rational Expressions

Words: To multiply rational expressions, multiply the numerators and the denominators.

Symbols: For all rational expressions $\frac{a}{b}$ and $\frac{c}{d}$ with $b \neq 0$ and $d \neq 0$, $\frac{a}{b} \cdot \frac{c}{d} = \frac{ac}{bd}$.

Key Concept • Dividing Rational Expressions

Words: To divide rational expressions, multiply the dividend by the reciprocal of the divisor.

Symbols: For all rational expressions $\frac{a}{b}$ and $\frac{c}{d}$ with $b \neq 0$, $c \neq 0$, and $d \neq 0$, $\frac{a}{b} \div \frac{c}{d} = \frac{a}{b} \cdot \frac{d}{c} = \frac{ad}{bc}$.

A **complex fraction** is a rational expression with a numerator and/or denominator that is also a rational expression. To simplify a complex fraction, first rewrite it as a division expression.

Think About It!

Why is it that for multiplying rational expressions we require that $d \neq 0$, but for dividing rational expressions we require that $c \neq 0$ and $d \neq 0$?

Example 3 Multiply and Divide Rational Expressions

Simplify each expression.

a. $\frac{3x}{8y} \cdot \frac{12x^2y}{9xy^3}$

$$\frac{3x}{8y} \cdot \frac{12x^2y}{9xy^3} = \frac{3 \cdot x \cdot 2 \cdot 2 \cdot 3 \cdot x \cdot x \cdot y}{2 \cdot 2 \cdot 2 \cdot y \cdot 3 \cdot 3 \cdot x \cdot y \cdot y \cdot y}$$ Factor.

$$= \frac{\overset{1}{\cancel{3}} \cdot \overset{1}{\cancel{x}} \cdot \overset{1}{\cancel{2}} \cdot \overset{1}{\cancel{2}} \cdot \overset{1}{\cancel{3}} \cdot x \cdot x \cdot \overset{1}{\cancel{y}}}{\underset{1}{\cancel{2}} \cdot \underset{1}{\cancel{2}} \cdot 2 \cdot \underset{1}{\cancel{y}} \cdot \underset{1}{\cancel{3}} \cdot \underset{1}{\cancel{3}} \cdot \underset{1}{\cancel{x}} \cdot y \cdot y \cdot y}$$ Eliminate common factors.

$$= \frac{x \cdot x}{2 \cdot y \cdot y \cdot y} \text{ or } \frac{x^2}{2y^3}$$ Simplify. Write using exponents.

b. $\frac{10d^5}{6cd} \div \frac{30c^3d^2}{4c}$

$$\frac{10d^5}{6cd} \div \frac{30c^3d^2}{4c} = \frac{10d^5}{6cd}\left(\frac{4c}{30c^3d^2}\right)$$ Multiply by the reciprocal of the divisor.

$$= \frac{2 \cdot 5 \cdot d \cdot d \cdot d \cdot d \cdot d \cdot 2 \cdot 2 \cdot c}{2 \cdot 3 \cdot c \cdot d \cdot 2 \cdot 3 \cdot 5 \cdot c \cdot c \cdot c \cdot d \cdot d}$$ Factor.

$$= \frac{\overset{1}{\cancel{2}} \cdot \overset{1}{\cancel{5}} \cdot \overset{1}{\cancel{d}} \cdot \overset{1}{\cancel{d}} \cdot \overset{1}{\cancel{d}} \cdot d \cdot d \cdot \overset{1}{\cancel{2}} \cdot 2 \cdot \overset{1}{\cancel{c}}}{\underset{1}{\cancel{2}} \cdot 3 \cdot \underset{1}{\cancel{c}} \cdot \underset{1}{\cancel{d}} \cdot \underset{1}{\cancel{2}} \cdot 3 \cdot \underset{1}{\cancel{5}} \cdot c \cdot c \cdot c \cdot \underset{1}{\cancel{d}} \cdot \underset{1}{\cancel{d}}}$$ Eliminate common factors.

$$= \frac{d \cdot d \cdot 2}{3 \cdot 3 \cdot c \cdot c \cdot c} \text{ or } \frac{2d^2}{9c^3}$$ Simplify. Write using exponents.

Go Online You can complete an Extra Example online.

Example 4 Multiply and Divide Polynomial Expressions

GEOMETRY A manufacturer that creates and sells rectangular planters wants to compare the areas of the top sections of two potential sizes of planters. For some given measure x, the area for the first planter is represented by $\frac{x^2 + 15x + 50}{x + 2}$, and the area for the second planter is represented by $\frac{x^2 + 30x + 200}{x + 2}$. Write and simplify an expression that represents the ratio of the area for the first planter to that of the second.

Because the question calls for a ratio of the area of the top of the first planter to that of the second, find $\frac{x^2 + 15x + 50}{x + 2} \div \frac{x^2 + 30x + 200}{x + 2}$.

$$\frac{x^2 + 15x + 50}{x + 2} \div \frac{x^3 + 30x + 200}{x + 2} = \frac{x^2 + 15x + 50}{x + 2} \cdot \frac{x + 2}{x^3 + 30x + 200}$$

$$= \frac{(x + 5)\overset{1}{\cancel{(x + 10)}}}{\underset{1}{\cancel{x + 2}}} \cdot \frac{\overset{1}{\cancel{x + 2}}}{\underset{1}{\cancel{(x + 10)}}(x + 20)}$$

$$= \frac{x + 5}{x + 20}$$

Example 5 Simplify Complex Fractions

Simplify.

$$\frac{\frac{3x}{x - y}}{\frac{6xy}{4x^2 - 4y^2}} = \frac{3x}{x - y} \div \frac{6xy}{4x^2 - 4y^2}$$ Express as a division expression.

$$= \frac{3x}{x - y} \cdot \frac{4x^2 - 4y^2}{6xy}$$ Multiply by the reciprocal.

$$= \frac{3x}{x - y} \cdot \frac{4(x^2 - y^2)}{6xy}$$ Distributive Property

$$= \frac{\overset{1\,1}{\cancel{3x}}}{\underset{1}{\cancel{x - y}}} \cdot \frac{\overset{2}{\cancel{4}}(x + y)\overset{1}{\cancel{(x - y)}}}{\underset{1\,1}{\cancel{6xy}}}$$ Factor and eliminate factors.

$$= \frac{2(x + y)}{y}$$ Simplify.

Check

Simplify $\frac{\frac{x^2 - 9y^2}{xy}}{\frac{2x + 6y}{x^2}}$.

Go Online You can complete an Extra Example online.

Talk About It!

If you simplify a ratio of two rational expressions that both contain polynomials in their numerators and denominators, will the simplified expression *always*, *sometimes*, or *never* be a rational expression with polynomials? Explain your reasoning.

Practice

Go Online You can complete your homework online.

Example 1

Simplify each expression, and state when the original expression is undefined.

1. $\frac{x(x-3)(x+6)}{x^2+x-12}$

2. $\frac{y^2(y^2+3y+2)}{2y(y-4)(y+2)}$

3. $\frac{(x^2-9)(x^2-z^2)}{4(x+z)(x-3)}$

4. $\frac{(x^2-16x+64)(x+2)}{(x^2-64)(x^2-6x-16)}$

5. $\frac{x^2(x+2)(x-4)}{6x(x^2+x-20)}$

6. $\frac{3y(y-8)(y^2+2y-24)}{15y^2(y^2-12y+32)}$

Example 2

Simplify each expression.

7. $\frac{x^2-5x-14}{28+3x-x^2}$

8. $\frac{9x^2-x^3}{x^2-3x-54}$

9. $\frac{(x-4)(x^2+2x-48)}{(36-x^2)(x^2+4x-32)}$

10. $\frac{16-c^2}{c^2+c-20}$

Example 3

Simplify each expression.

11. $\frac{3ac^3f^3}{8a^2bcf^4} \cdot \frac{12ab^2c}{18ab^3c^2f}$

12. $\frac{14xy^2z^3}{21w^4x^2z} \cdot \frac{7wxyz}{12w^2y^3z}$

13. $\frac{64a^2b^5}{35b^2c^3f^4} \div \frac{12a^4b^3c}{70abcf^2}$

14. $\frac{9x^2yz}{5z^4} \div \frac{12x^4y^2}{50xy^4z^2}$

15. $\frac{15a^2b^2}{21ac} \cdot \frac{14a^4c^2}{6ab^3}$

16. $\frac{14c^2f^5}{9a^2} \div \frac{35cf^4}{18ab^3}$

Example 4

17. **BEAUTY** A producer of beauty care products wants to compare the areas of two face cream containers. For some measure a, the area for the first container is represented by $\frac{16a^2+40a+25}{3a^2-10a-8}$. The area of the second container is represented by $\frac{4a+5}{a^2-8a+16}$. Write and simplify an expression that represents the ratio of the area of the first container to the area of the second container.

18. **PACKAGING** A packaging plant quality assurance agent compares the surface area and volume of packages to ensure the packages will be transported economically. The most popular package size is a cylinder with a height of 18 inches. Define a variable, then write and simplify an expression that represents the ratio of the surface area to the volume of the cylinder.

19. **LANDSCAPING** Rashard is building a rectangular patio. If $\frac{t^2+19t+84}{4t-4}$ represents the length of the patio and $\frac{2t-2}{t^2+9t+14}$ represents the width, write and simplify an expression that represents the area of the patio.

Example 5

Simplify each expression.

20. $\dfrac{\frac{x^2-9}{6x-12}}{\frac{x^2+10x+21}{x^2-x-2}}$

21. $\dfrac{\frac{y-x}{z^3}}{\frac{x-y}{6z^2}}$

22. $\dfrac{\frac{a^2-b^2}{b^3}}{\frac{b^2-ab}{a^2}}$

23. $\dfrac{\frac{x-y}{a+b}}{\frac{x^2-y^2}{b^2-a^2}}$

Mixed Exercises

Simplify each expression.

24. $\dfrac{y^2+8y+15}{y-6} \cdot \dfrac{y^2-9y+18}{y^2-9}$

25. $\dfrac{c^2-6c-16}{c^2-d^2} \div \dfrac{c^2-8c}{c+d}$

26. $\dfrac{x^2+9x+20}{8x+16} \cdot \dfrac{4x^2+16x+16}{x^2-25}$

27. $\dfrac{3a^2+6a+3}{a^2-3a-10} \div \dfrac{12a^2-12}{a^2-4}$

28. $\dfrac{9-x^2}{x^2-4x-21} \cdot \left(\dfrac{2x^2+7x+3}{2x^2-15x+7}\right)^{-1}$

29. $\left(\dfrac{2x^2+2x-12}{x^2+4x-5}\right)^{-1} \cdot \dfrac{2x^3-8x}{x^2-2x-35}$

30. $\left(\dfrac{3xy^3z}{2a^2bc^2}\right)^3 \cdot \dfrac{16a^4b^3c^5}{15x^7yz^3}$

31. $\dfrac{20x^2y^6z^{-2}}{3a^3c^2} \cdot \left(\dfrac{16x^3y^3}{9acz}\right)^{-1}$

32. $\dfrac{\frac{8x^2-10x-3}{10x^2+35x-20}}{\frac{2x^2+x-6}{4x^2+18x+8}}$

33. $\dfrac{\frac{2x^2+7x-30}{-6x^2+13x+5}}{\frac{4x^2+12x-72}{3x^2-11x-4}}$

34. $\dfrac{x^2+4x-32}{2x^2+9x-5} \cdot \dfrac{3x^2-75}{3x^2-11x-4} \div \dfrac{6x^2-18x-60}{x^3-4x}$

35. $\dfrac{8x^2+10x-3}{3x^2-12x-36} \div \dfrac{2x^2-5x-12}{3x^2-17x-6} \cdot \dfrac{4x^2+3x-1}{4x^2-40x+24}$

36. JELLY BEANS A large jar contains B blue jelly beans and R red jelly beans. A bag of 100 red and 100 blue jelly beans is added to the jar. What is the ratio of red to blue jelly beans in the jar?

37. MILEAGE Martina drives a hybrid car that gets 45 miles per gallon when driving in the city and 48 miles per gallon when driving on the highway. If Martina uses C gallons of gas in the city and H gallons of gas on the highway, write an expression for the average number of miles per gallon that Martina gets with her car in terms of C and H.

38. HEIGHT The front face of a Nordic house is triangular. The area of the face is $x^2 + 3x + 10$, where x is the base of the triangle. What is the height of the triangle in terms of x?

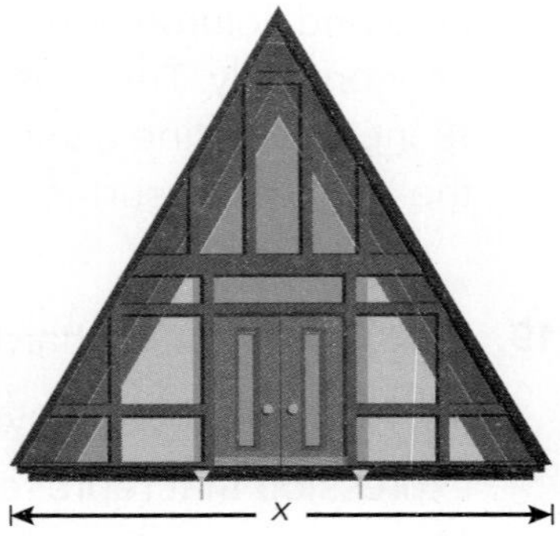

39. CONTAINERS David is designing a cylindrical container. He wants the height of the container to be twice the length of the radius. Write an expression for the ratio of the volume of the container to the surface area of the container, where h is the height and r is the radius.

40. ARCHERY Harold and Hinto are on the archery team. So far at practice, Harold has hit the target 7 out of 10 times for an average of 70%, and Hinto has hit the target 9 out of 10 times for an average of 90%. Harold and Hinto each hit the target on their next x consecutive attempts.

a. Write an expression for Harold and for Hinto that shows their average of hitting the target after hitting the target on x consecutive attempts.

b. Write a simplified expression for the ratio of the percentage Harold hit the target to the percentage Hinto hit the target if they both hit the target on their next x consecutive attempts.

41. GEOMETRY The volume of the rectangular box shown is given by $V = (2x^3 + 26x^2 + 60x)$ cubic inches.

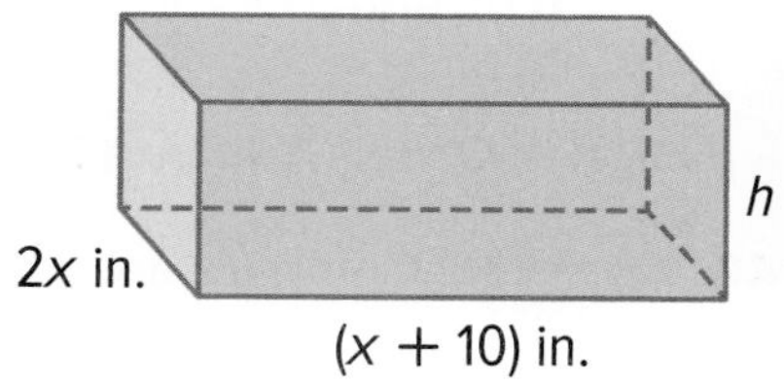

a. Explain how to find an expression in terms of x for the height h of the box.

b. In terms of x, what is h in simplest form?

c. Explain how you could check the expression you found in **part b**. Then check your expression.

42. ROWING The time it takes Irfan to row 9 miles upstream is $\frac{9}{5 - c}$. The time it takes him to row 9 miles downstream is $\frac{9}{5 + c}$, where c is the speed of the current in miles per hour and $0 \leq c < 5$.

a. Which way does Irfan travel faster, downstream or upstream? Give two reasons for your answer.

b. Determine how many times faster Irfan travels in the faster direction than in the slower direction. Explain your method.

43. REASONING During her first week of running, Rafaela takes t minutes to run a mile. In miles per hour, her average speed is given by $\frac{60}{t}$. Rafaela hopes to get her average time to $t - 2$ minutes to run a mile during her second week of running. On average, how many times faster does Rafaela hope to run a mile during her second week than during her first week of running? Explain how you know.

44. CONSTRUCTING ARGUMENTS Saquita graphs the function $y = \frac{5(x + 1)(x + 2)(x + 3)}{(x + 1)(x + 2)(x + 3)}$ on a graphing calculator and claims that the result is a horizontal line because of the display shown. Is Saquita correct? Justify your argument.

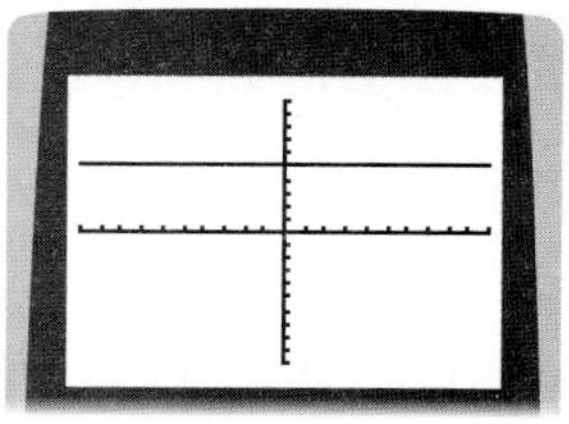

45. USE A MODEL Anita's yard is being professionally landscaped. The final design will consist of a circular fountain x feet in diameter in square A surrounded by a grassy area in square B and a gravel pathway in square C that borders the grassy area. The square areas will be centered on each other as shown in the diagram. Square A will have a side length of $2x$ feet.

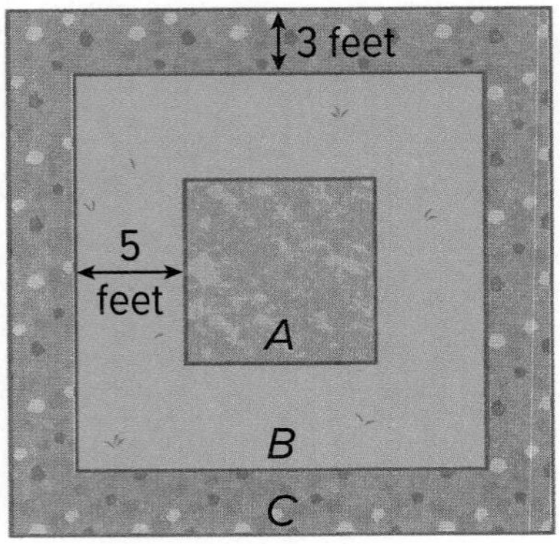

a. Anita would like the lengths of the sides to be proportional. For what values of x will the ratio of the lengths of a side of square C to a side of square B equal the ratio of the lengths of a side of square B to a side of square A? Explain your reasoning. What diameter could the fountain have?

b. If the landscape architect changed the width of the gravel pathway to 4 feet and the width of the grassy area to 2 feet, is there a value for x that would make the ratios equal? Explain your reasoning. What diameter could the fountain have?

Higher-Order Thinking Skills

46. ANALYZE Compare and contrast $\frac{(x-6)(x+2)(x+3)}{x+3}$ and $(x-6)(x+2)$.

47. FIND THE ERROR Troy and Beverly are simplifying $\frac{x+y}{x-y} \div \frac{4}{y-x}$. Is either of them correct? Explain your reasoning.

Troy

$$\frac{x+y}{x-y} \div \frac{4}{y-x} = \frac{x-y}{x+y} \cdot \frac{4}{y-x}$$
$$= \frac{-4}{x+y}$$

Beverly

$$\frac{x+y}{x-y} \div \frac{4}{y-x} = \frac{x+y}{x-y} \cdot \frac{y-x}{4}$$
$$= -\frac{x+y}{4}$$

48. PERSEVERE Find the expression that makes the following statement true for all values of x within the domain.

$$\frac{x-6}{x+3} \cdot \frac{?}{x-6} = x-2$$

49. WHICH ONE DOESN'T BELONG? Identify the expression that does not belong with the other three. Justify your conclusion.

$\frac{1}{x-1}$	$\frac{x^2+3x+2}{x-5}$	$\frac{x+1}{\sqrt{x}+3}$	$\frac{x^2+1}{3}$

50. ANALYZE Determine whether the following statement is *sometimes*, *always*, or *never* true. Justify your argument.

A rational function that has a variable in the denominator is defined for all real values of x.

51. CREATE Write a rational expression that simplifies to $\frac{x-1}{x+4}$.

52. WRITE The rational expression $\frac{x^2+3x}{4x}$ is simplified to $\frac{x+3}{4}$. Explain why this new expression is not defined for all values of x.

53. CREATE Write three different rational expressions that are equivalent to the expression $\frac{a}{a-5}$, $a \neq 5$.

Lesson 9-2

Adding and Subtracting Rational Expressions

Today's Goals
- Simplify rational expressions by adding and subtracting.
- Simplify complex fractions.

Explore Closure of Rational Expressions

Online Activity Use the interactive tool to complete the Explore.

INQUIRY If you multiply, divide, add, or subtract two rational expressions, is the result also a rational expression?

Learn Adding and Subtracting Rational Expressions

Just as with rational numbers in fractional form, to add or subtract two rational expressions that have unlike denominators, you must first find the least common denominator (LCD). The LCD is the least common multiple (LCM) of the two denominators.

To find the LCM of two or more numbers or polynomials, factor them. The LCM contains each factor the greatest number of times it appears as a factor.

Key Concept • Adding Rational Expressions

Words: To add rational expressions, find the least common denominator. Rewrite each expression with the LCD. Then add.

Symbols: For all rational expressions $\frac{a}{b}$ and $\frac{c}{d}$ with $b \neq 0$ and $d \neq 0$, $\frac{a}{b} + \frac{c}{d} = \frac{ad}{bd} + \frac{bc}{bd} = \frac{ad + bc}{bd}$.

Key Concept • Subtracting Rational Expressions

Words: To subtract rational expressions, find the least common denominator. Rewrite each expression with the LCD. Then subtract.

Symbols: For all rational expressions $\frac{a}{b}$ and $\frac{c}{d}$ with $b \neq 0$ and $d \neq 0$, $\frac{a}{b} - \frac{c}{d} = \frac{ad}{bd} - \frac{bc}{bd} = \frac{ad - bc}{bd}$.

Example 1 Add and Subtract Rational Expressions with Monomial Denominators

Simplify each expression.

a. $\mathbf{\frac{7a}{4b} + \frac{4c^2}{10}}$

$\frac{7a}{4b} + \frac{4c^2}{10} = \frac{7a}{4b} \cdot \frac{5}{5} + \frac{4c^2}{10} \cdot \frac{2b}{2b}$ — The LCD is $20b$.

$= \frac{35a}{20b} + \frac{8bc^2}{20b}$ — Multiply.

$= \frac{35a + 8bc^2}{20b}$ — Add.

Go Online You can complete an Extra Example online.

Talk About It!

For part **b**, if you identified the LCD as $9y \cdot 6z = 54yz$, multiplied, and simplified the expression, would your answer still be correct? Explain your reasoning.

Think About It!

Why is each rational expression not multiplied by $\frac{x-3}{x-3}$?

b. $\frac{2x}{9y} - \frac{7y}{6z}$

$$\frac{2x}{9y} - \frac{7y}{6z} = \frac{2x}{9y} \cdot \frac{2z}{2z} - \frac{7y}{6z} \cdot \frac{3y}{3y}$$ The LCD is $18yz$.

$$= \frac{4xz}{18yz} - \frac{21y^2}{18yz}$$ Multiply.

$$= \frac{4xz - 21y^2}{18yz}$$ Subtract.

Example 2 Add and Subtract Rational Expressions with Polynomial Denominators

Simplify.

$$\frac{2x+1}{x^2+2x-15} - \frac{7}{5x-15}$$

$$= \frac{2x+1}{(x-3)(x+5)} - \frac{7}{5(x-3)}$$ Factor the denominators.

$$= \frac{5(2x+1)}{5(x-3)(x+5)} - \frac{7(x+5)}{5(x-3)(x+5)}$$ The LCD is $5(x-3)(x+5)$.

$$= \frac{10x+5-7x-35}{5(x-3)(x+5)}$$ Distributive Property

$$= \frac{3x-30}{5(x-3)(x+5)}$$ Subtract.

Check

Simplify $\frac{3x}{4x^2+4} - \frac{2x^2}{x^4-1}$.

Example 3 Use Addition and Subtraction of Rational Expressions

PRODUCTION **The rate at which some oil wells pump oil in thousands of barrels, given *x* years of pumping, can be given by $\frac{20}{x} + \frac{200x}{3x^2+20}$. Simplify the expression.**

$$\frac{20}{x} + \frac{200x}{3x^2+20}$$ Original expression

$$= \frac{20(3x^2+20)}{x(3x^2+20)} + \frac{200x(x)}{(3x^2+20)(x)}$$ Multiply by the missing factors.

$$= \frac{60x^2+400+200x^2}{3x^3+20x}$$ Distributive Property

$$= \frac{260x^2+400}{3x^3+20x}$$ Add.

Go Online You can complete an Extra Example online.

Learn Simplifying Complex Fractions

Complex fractions can be simplified by simplifying the numerator and denominator separately and then simplifying the resulting expression. You can also simplify a complex fraction by finding the LCD of all of the denominators. Then, the denominators can all be eliminated by multiplying by the LCD.

> **Talk About It!**
> In the complex fraction $\frac{\frac{1}{2}+\frac{1}{3}}{\frac{1}{4}-\frac{1}{2}}$, what is the LCD of the numerator, the denominator, and of all denominators in the fraction?

Example 4 Simplify Complex Fractions by Using Different LCDs

Simplify $\frac{\frac{x}{y}+1}{\frac{y}{x}-\frac{1}{y}}$.

Step 1 Determine the LCD.

LCD of the Numerator:

The numerator is $\frac{x}{y}+1$, and the LCD is y.

LCD of the Denominator:

The denominator is $\frac{y}{x}-\frac{1}{y}$, and the LCD is xy.

Step 2 Simplify.

$\frac{\frac{x}{y}+1}{\frac{y}{x}-\frac{1}{y}} = \frac{\frac{x}{y}+\frac{y}{y}}{\frac{y^2}{xy}-\frac{x}{xy}}$ The LCD of the numerator is y. The LCD of the denominator is xy.

$= \frac{\frac{x+y}{y}}{\frac{y^2-x}{xy}}$ Simplify.

$= \frac{x+y}{y} \div \frac{y^2-x}{xy}$ Write as a division expression.

$= \frac{x+y}{y} \cdot \frac{xy}{y^2-x}$ Multiply by the reciprocal of the divisor.

$= \frac{x+y}{\cancel{y}_1} \cdot \frac{x\cancel{y}^1}{y^2-x}$ Eliminate common factors.

$= \frac{x+y}{1} \cdot \frac{x}{y^2-x}$ Simplify.

$= \frac{x(x+y)}{y^2-x}$ Multiply.

$= \frac{x^2+xy}{y^2-x}$ Distributive Property

Go Online You can complete an Extra Example online.

Check

Simplify $\dfrac{\frac{x}{xy} - 2}{2xy - \frac{3}{y}}$.

A. $\dfrac{2xy^2 - 4xy^3 + 6y - 3}{y^2}$

B. $\dfrac{1 - 2y}{2xy^2 - 3}$

C. $\dfrac{4xy^3 - 2xy^2 - 6y + 3}{6xy^2}$

D. 0

Example 5 Simplify Complex Fractions by Using the Same LCD

Simplify $\dfrac{\frac{x}{y} + \frac{2}{x}}{\frac{x}{2} - \frac{y}{x}}$.

Step 1 Determine the LCD.

The denominators of the terms in the numerator and denominator are 2, x, and y.

So, the LCM of all of the denominators is $2xy$.

Step 2 Simplify.

$$\frac{\frac{x}{y} + \frac{2}{x}}{\frac{x}{2} - \frac{y}{x}} = \frac{\left(\frac{x}{y} + \frac{2}{x}\right)}{\left(\frac{x}{2} - \frac{y}{x}\right)} \cdot \frac{2xy}{2xy}$$ The LCD is $2xy$.

$$= \frac{\frac{x}{y} \cdot 2xy + \frac{2}{x} \cdot 2xy}{\frac{x}{2} \cdot 2xy - \frac{y}{x} \cdot 2xy}$$ Distributive Property

$$= \frac{2x^2 + 4y}{x^2y - 2y^2}$$ Multiply.

Check

Simplify $\dfrac{1 + \frac{b^2}{a}}{\frac{a}{2} - \frac{a}{b}}$.

A. $\dfrac{2ab + 2b^3}{a^2b - 2a^2}$

B. $\dfrac{ab - 2a + b^3 - 2b^2}{2b}$

C. $-\dfrac{ab - 2a + b^3 - 2b^2}{a^2}$

D. $\dfrac{2 + b^3}{a - a^2}$

Go Online to practice what you've learned in Lessons 9-1 and 9-2.

Go Online You can complete an Extra Example online.

Practice

Go Online You can complete your homework online.

Examples 1 and 2

Simplify each expression.

1. $\frac{3}{x} + \frac{5}{y}$

2. $\frac{3}{8p^2r} + \frac{5}{4p^2r}$

3. $\frac{2c - 7}{3} + 4$

4. $\frac{2}{m^2p} + \frac{5}{p}$

5. $\frac{12}{5y^2} - \frac{2}{5yz}$

6. $\frac{7}{4gh} + \frac{3}{4h^2}$

7. $\frac{3}{w - 3} - \frac{2}{w^2 - 9}$

8. $\frac{3t}{2 - x} + \frac{5}{x - 2}$

9. $\frac{k}{k - n} - \frac{k}{n - k}$

10. $\frac{4z}{z - 4} + \frac{z + 4}{z + 1}$

11. $\frac{n}{n - 3} + \frac{2n + 2}{n^2 - 2n - 3}$

12. $\frac{3}{y^2 + y - 12} - \frac{2}{y^2 + 6y + 8}$

Example 3

13. ELECTRIC POTENTIAL The electric potential between two electrons is given by a formula that has the form $\frac{1}{r} + \frac{1}{1 - r}$. Simplify this expression.

14. GEOMETRY The cross section of a solid consists of two trapezoids stacked one on top of the other. The total area of the cross section is x^2 square units. Assuming the trapezoids have the same height, write an expression for the height of the solid in terms of x. Write your answer in simplest form. Recall that the area of a trapezoid with height h and bases b_1 and b_2 is given by $\frac{1}{2}h(b_1 + b_2)$.

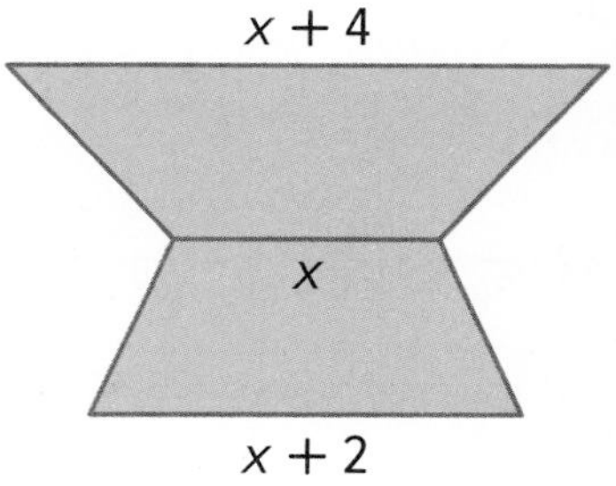

15. TRACK Morgan, Connell, Zack, and Moses run the 400-meter relay together. Each of them runs 100 meters. Their average speeds were r, $r + 0.5$, $r - 0.5$, and $r - 1$ meters per second, respectively.

a. Write an expression for each individual's time for their leg of the race.

b. Write an expression for their time as a team. Write your answer as a ratio of two polynomials.

c. The world record for the 100-meter relay is 37.4 seconds. What will r equal if the team ties the world record?

Examples 4 and 5

Simplify each expression.

16. $\dfrac{\frac{2}{x - 3} + \frac{3x}{x^2 - 9}}{\frac{3}{x + 3} - \frac{4x}{x^2 - 9}}$

17. $\dfrac{\frac{4}{x + 5} + \frac{9}{x - 6}}{\frac{5}{x - 6} - \frac{8}{x + 5}}$

18. $\dfrac{\frac{5}{x + 6} - \frac{2x}{2x - 1}}{\frac{x}{2x - 1} + \frac{4}{x + 6}}$

19. $\dfrac{\frac{8}{x - 9} - \frac{x}{3x + 2}}{\frac{3}{3x + 2} + \frac{4x}{x - 9}}$

Mixed Exercises

Simplify each expression.

20. $\frac{2}{a+2} - \frac{3}{2a}$

21. $\frac{5}{3b+d} - \frac{2}{3bd}$

22. $\frac{1}{x^2+2x+1} + \frac{x}{x+1}$

23. $\frac{2x+1}{x-5} - \frac{4}{x^2-3x-10}$

24. $\frac{5a}{24cf^4} + \frac{a}{36bc^4f^3}$

25. $\frac{4b}{15x^3y^2} - \frac{3b}{35x^2y^4z}$

26. $\frac{5b}{6a} + \frac{3b}{10a^2} + \frac{2}{ab^2}$

27. $\frac{4}{3x} + \frac{8}{x^3} + \frac{2}{5xy}$

28. $\frac{8}{3y} + \frac{2}{9} - \frac{3}{10y^2}$

29. $\frac{1}{16a} + \frac{5}{12b} - \frac{9}{10b^3}$

30. $\frac{8}{x^2-6x-16} + \frac{9}{x^2-3x-40}$

31. $\frac{6}{y^2-2y-35} + \frac{4}{y^2+9y+20}$

32. $\frac{12}{3y^2-10y-8} - \frac{3}{y^2-6y+8}$

33. $\frac{6}{2x^2+11x-6} - \frac{8}{x^2+3x-18}$

34. $\frac{2x}{4x^2+9x+2} + \frac{3}{2x^2-8x-24}$

35. $\frac{4x}{3x^2+3x-18} - \frac{2x}{2x^2+11x+15}$

36. LENSES The focal length of a lens f is related to the distance from the subject to the lens p and the distance from the lens to the image on the sensor q by the formula $\frac{1}{p} + \frac{1}{q} = \frac{1}{f}$. Express $\frac{1}{p} + \frac{1}{q}$ as a single fraction.

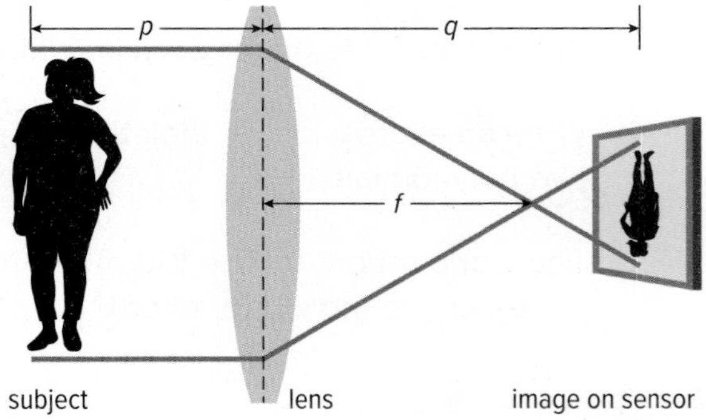

37. USE A SOURCE In the seventeenth century, Lord William Brouncker conducted research in mathematics and music and served as the first president of the Royal Society in London, England. Research Lord Brouncker's work in the area of continued fractions. Describe the discovery for which he is best known.

Simplify each expression.

38. $\frac{1}{12a} + 6 - \frac{3}{5a^2}$

39. $\frac{5}{16y^2} - 4 - \frac{8}{3x^2y}$

40. $\frac{5}{6x^2 + 46x - 16} + \frac{2}{6x^2 + 57x + 72}$

41. $\frac{1}{8x^2 - 20x - 12} + \frac{4}{6x^2 + 27x + 12}$

42. $\frac{x^2 + y^2}{x^2 - y^2} + \frac{y}{x + y} - \frac{x}{x - y}$

43. $\frac{x^2 + x}{x^2 - 9x + 8} + \frac{4}{x - 1} - \frac{3}{x - 8}$

44. $\dfrac{\frac{2}{a-1} + \frac{3}{a-4}}{\frac{6}{a^2 - 5a + 4}}$

45. $\dfrac{\frac{1}{x} + \frac{1}{y}}{\left(\frac{1}{x} - \frac{1}{y}\right)(x + y)}$

Find the slope of the line that passes through each pair of points.

46. $A\left(\frac{2}{p}, \frac{1}{2}\right)$ and $B\left(\frac{1}{3}, \frac{3}{p}\right)$

47. $C\left(\frac{1}{4}, \frac{4}{q}\right)$ and $D\left(\frac{5}{q}, \frac{1}{5}\right)$

48. USE A MODEL Hachi needs to buy fencing for her rectangular garden.

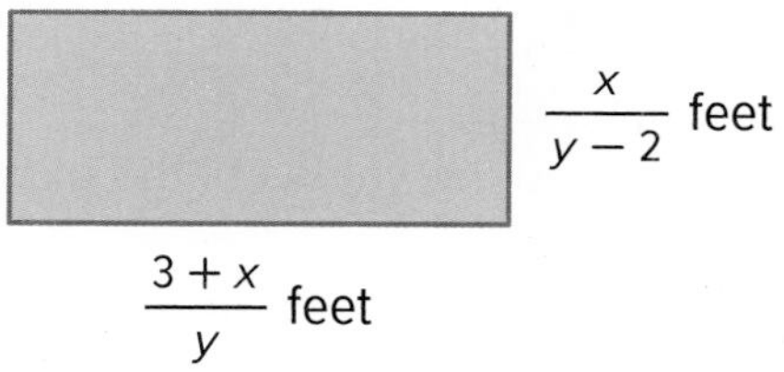

a. Write an expression, in simplest form, that represents the number of feet of fencing Hachi needs. Are there any restrictions on the variables? Explain.

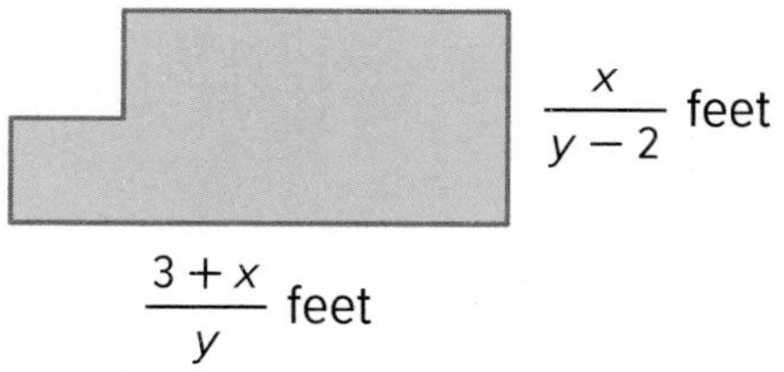

b. Hachi wants to remove a square corner from her garden. The square section removed will have sides the length of half the width of the original garden. What expression represents the perimeter of the new garden? Explain.

49. STRUCTURE Determine the average of three rational numbers represented by these rational expressions: $\frac{1}{x}$, $\frac{1}{x-3}$, and $\frac{1}{2x}$ for $x \neq 3$, and $x \neq 0$. Explain how you found the average.

50. ELECTRONICS A resistor is an electrical component that reduces the flow of electrical current through a circuit. A resistor is connected in parallel when both of its terminals are connected to both terminals of an adjacent resistor. When three resistors are connected in parallel, the total resistance, R_T, is given by $R_T = \dfrac{1}{\frac{1}{R_1} + \frac{1}{R_2} + \frac{1}{R_3}}$.

a. Simplify the complex fraction. Explain how you know your result is simplified as much as possible.

b. Timothy found this formula for total resistance, $\frac{1}{R_T} = \frac{1}{R_1} + \frac{1}{R_2} + \frac{1}{R_3}$. He said that this formula is equivalent to the original formula. Is Timothy correct? Explain.

Find and simplify the sum or difference.

51. $\frac{2(x+5)}{2x^2-1} + \frac{4x}{x+1}$

52. $\frac{x-1}{x^2+x-12} - 2$

53. WORK The ecology club is landscaping a local park. Dell can plant a flower bed in 3 hours. Max can plant a flower bed of the same size in 4 hours. Using t, time in hours, write an expression representing how many flower beds they will complete in t hours working together. Explain your reasoning.

54. REASONING Determine three real numbers that divide the real number line between $\frac{x}{6}$ and $\frac{x}{2}$ into four equal parts. Assume that $x \neq 0$. Explain your reasoning.

55. STRUCTURE The Fibonacci numbers are a famous sequence in which each term is the sum of the previous two. So, the Fibonacci sequence is 1, 1, 2, 3, 5, 8, 13, 21, 34, ... The sequence is related to a sequence of continued fractions. Simplify each continued fraction.

a. $\frac{1}{1+\frac{1}{x}}$

b. $\frac{1}{1+\frac{1}{1+\frac{1}{x}}}$

c. $\frac{1}{1+\frac{1}{1+\frac{1}{1+\frac{1}{x}}}}$

d. $\frac{1}{1+\frac{1}{1+\frac{1}{1+\frac{1}{1+\frac{1}{x}}}}}$

e. Find the values of the expressions in **parts a–d** for $x = 1$.

f. How are the terms of the Fibonacci sequence related to the values you found in **part e**? Make a conjecture about the next value in the pattern.

56. PERSEVERE Simplify $\frac{5x^{-2} - \frac{x+1}{x}}{\frac{4}{3-x^{-1}} + 6x^{-1}}$.

57. ANALYZE The sum of any two rational numbers is always a rational number. So, the set of rational numbers is said to be closed under addition. Determine whether the set of rational expressions is closed under addition, subtraction, multiplication, and division by a nonzero rational expression. Justify your argument.

58. CREATE Write three monomials with an LCM of $180a^4b^6c$.

59. WRITE Explain how to add rational expressions that have unlike denominators. How does this compare to adding rational numbers?

60. CREATE Write a rational expression by adding and subtracting three terms that each have a monomial denominator and a sum of $\frac{4w-9}{20w}$.

Graphing Reciprocal Functions

Learn Graphing Reciprocal Functions

A **reciprocal function** has an equation of the form $f(x) = \frac{n}{b(x)}$, where n is a real number and $b(x)$ is a linear expression that cannot equal 0.

The parent function of a reciprocal function is $f(x) = \frac{1}{x}$. A **vertical asymptote** is a vertical line that a graph approaches. A **horizontal asymptote** is a horizontal line that a graph approaches. Reciprocal functions have vertical asymptotes when the denominator equals 0, so the function $f(x) = \frac{1}{x}$ has a vertical asymptote at $x = 0$. The type of graph formed by a reciprocal function is called a **hyperbola**.

The domain of a function is limited to values for which the function is defined. Values for which the function is not defined are called **excluded values**.

Key Concept • Reciprocal Functions

Parent function	$f(x) = \frac{1}{x}$
Type of graph	hyperbola
Domain and range	all nonzero real numbers
Asymptotes	$x = 0$ and $f(x) = 0$
Intercepts	none
Not defined	$x = 0$

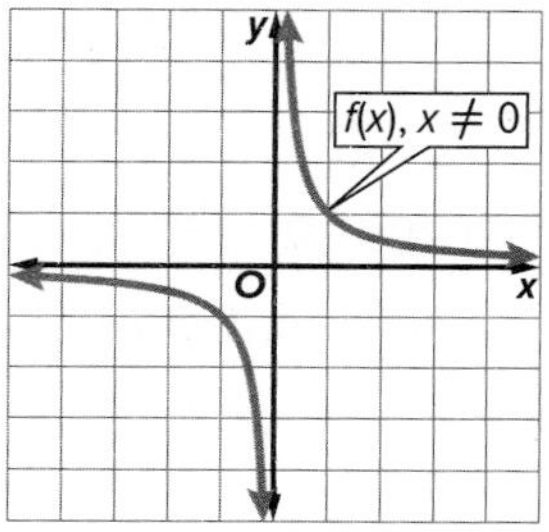

A reciprocal function has two asymptotes, which are lines that a graph approaches. The vertical asymptote is determined by the excluded value of x, and the horizontal asymptote is determined by the value that is undefined for $f(x)$.

For a reciprocal function in the form $f(x) = \frac{n}{b(x)}$, the horizontal asymptote is $f(x) = 0$ because there is no value of x that will result in $f(x) = 0$. For a translated reciprocal function of the form $f(x) = \frac{n}{b(x)} + k$, where k is a constant, the horizontal asymptote is $f(x) = k$.

Example 1 Limitations on the Domains of Reciprocal Functions

Determine the excluded value of x for each function.

a. $g(x) = \frac{6}{x}$

The function is undefined for $x = 0$.

(*continued on the next page*)

Today's Goal

- Graph reciprocal functions by making tables of values.
- Graph and write reciprocal functions by using transformations.

Today's Vocabulary

reciprocal function
vertical asymptote
horizontal asymptote
hyperbola
excluded values

Talk About It!

How many vertical asymptotes would you expect a function with a quadratic expression in the denominator to have? Justify your response.

Study Tip

Alternative Notation The parent function of the reciprocal function can also be written as $f(x) = x^{-1}$.

Think About It!

Explain why reciprocal functions have a restricted domain.

Think About It!

Explain the relationship between the asymptotes and the domain and range.

b. $g(x) = \frac{2}{x-7}$

$x - 7 = 0$ Set the denominator of the expression equal to 0.

$x = 7$ Add 7 each side.

The function is undefined for $x = 7$.

c. $g(x) = \frac{-5}{3x+4}$

$3x + 4 = 0$ Set the denominator of the expression equal to 0.

$3x = -4$ Subtract 4 from each side.

$x = -\frac{4}{3}$ Divide each side by 3.

The function is undefined for $x = -\frac{4}{3}$.

Check

Determine the excluded value of x for each function.

a. $g(x) = \frac{9}{17x}$ is undefined when $x =$ ___?___.

b. $g(x) = \frac{3}{x-11}$ is undefined when $x =$ ___?___.

c. $g(x) = \frac{1}{2x+8}$ is undefined when $x =$ ___?___.

Example 2 Graph a Reciprocal Function by Using a Table

Consider $g(x) = \frac{1}{2x-5} + 2$.

Part A Identify key features.

Identify the asymptotes, domain, and range of $g(x)$.

To determine the vertical asymptote, find the excluded value for x.

$2x - 5 = 0$ Set the denominator of the rational expression equal to 0.

$x = \frac{5}{2}$ Simplify.

The vertical asymptote is $x = \frac{5}{2}$, and the domain is $\left\{x \mid x \neq \frac{5}{2}\right\}$.

The function is of the form $f(x) = \frac{n}{b(x)} + k$, where $k = 2$. So, the graph has a horizontal asymptote of $g(x) = 2$, and the range is $\{g(x) \mid g(x) \neq 2\}$.

Part B Graph the function, and identify the intercepts.

Graph the vertical and horizontal asymptotes with dashed lines.

Use a table to find values of $g(x)$. Include values that are less than and greater than the excluded value. Then, plot the points.

Connect the points with smooth curves without crossing the asymptotes.

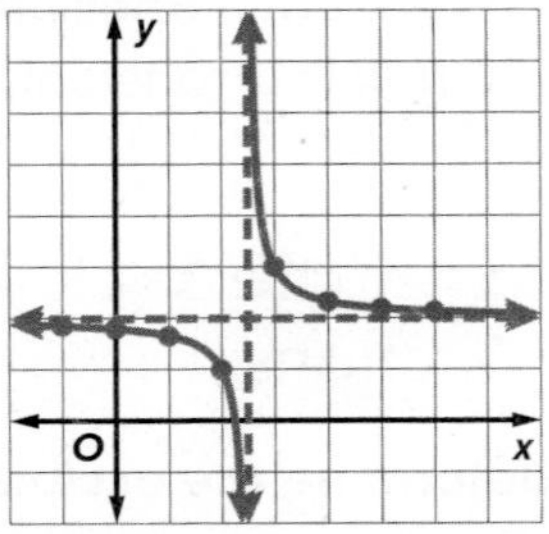

x	−1	0	1	2	3	4	5	6
g(x)	1.86	1.8	1.67	1	3	2.33	2.2	2.14

Go Online You can complete an Extra Example online.

Find the x-intercept by substituting 0 for $g(x)$.

$g(x) = \frac{1}{2x - 5} + 2$	Original function
$0 = \frac{1}{2x - 5} + 2$	Substitute 0 for $g(x)$.
$-2 = \frac{1}{2x - 5}$	Subtract 2 from each side.
$-2(2x - 5) = 1$	Multiply each side by $2x - 5$.
$-4x + 10 = 1$	Distributive Property
$x = 2.25$	Solve for x.

Find the y-intercept by substituting 0 for x.

$g(x) = \frac{1}{2x - 5} + 2$	Original function
$g(x) = \frac{1}{2(0) - 5} + 2$	Substitute 0 for x.
$g(x) = 1.8$	Solve for $g(x)$.

The x-intercept is 2.25 and the y-intercept is 1.8.

Check

Consider $g(x) = \frac{3}{x} + 1$.

Identify features of $g(x)$.

vertical asymptote: $x =$ __?__ horizontal asymptote: $y =$ __?__

domain: $\{x \mid x \neq$ __?__$\}$ range: $\{g(x) \mid g(x) \neq$ __?__$\}$

x-intercept: __?__ y-intercept: __?__

Example 3 Analyze a Reciprocal Function

ELECTRICITY **Ohm's law states that electric circuits operate according to the equation $I = \frac{V}{R}$, where I is the electrical current in amperes, V is the electromotive force in volts, and R is the resistance of the circuit in ohms. Analyze the electrical current of a 110-volt curling iron in relation to the resistance of its circuit.**

Part A Write and graph a function.

Because $V = 110$, the electrical current is given by $I = \frac{110}{R}$.

The vertical asymptote is $R = 0$ and the horizontal asymptote is $I = 0$. Make a table of values and plot ordered pairs on a coordinate plane. Then, connect the points with smooth curves.

(*continued on the next page*)

Go Online You can complete an Extra Example online.

Watch Out!

Symmetry When graphing, keep in mind that the pieces of the graph of the reciprocal function will be in opposite regions. For example, a function with asymptotes of $y = 0$ and $x = 0$ will either be graphed in Quadrants I and III or Quadrants II and IV.

R	I
−30	−3.67
−20	−5.5
−10	−11
10	11
20	5.5
30	3.67

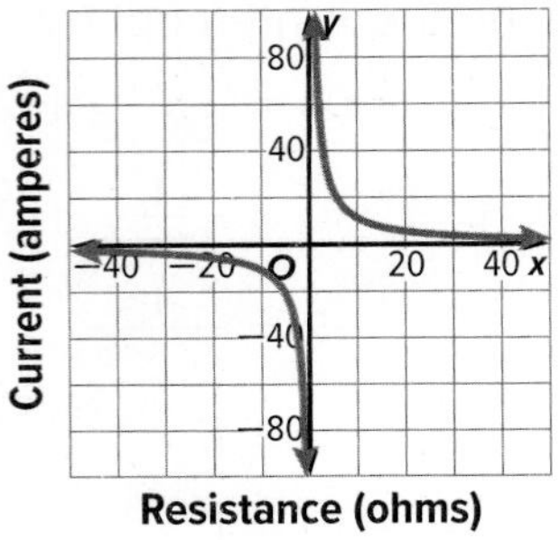

Part B Analyze the key features.

domain: $\{R \mid R \neq 0\}$

range: $\{I \mid I \neq 0\}$

intercepts: none

positive: when $R > 0$

negative: when $R < 0$

symmetry: symmetric about the origin

end behavior: As $R \to -\infty$, $I \to 0$ and as $R \to \infty$, $I \to 0$.

Use a Source

In the United States, a voltage of 110 is standard for most small appliances, like curling irons. Find the standard voltage for another country and determine how it would affect the graph.

Check

CABIN RENTAL A group of friends plans to rent a cabin at Red Pine Ski Lodge. The cabin costs \$750 to rent for the weekend. Let x be the number of friends who are sharing the cost, and let y be the cost for each person.

Graph a function representing the average cost per person to rent the cabin. Analyze the key features of the graph.

domain: $\{x \mid x$ __?__$\}$

range: $\{y \mid y$ __?__$\}$

x-intercept: __?__

y-intercept: __?__

positive: when x __?__

negative: when x __?__

end behavior: As $x \to -\infty$, $y \to$ __?__ and as $x \to \infty$, $y \to$ __?__.

Given the context of the situation, the domain is limited to __?__.

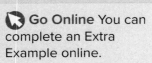

Go Online You can complete an Extra Example online.

Explore Transforming Reciprocal Functions

Online Activity Use graphing technology to complete the Explore.

> **INQUIRY** How does performing an operation on a reciprocal function affect its graph?

Go Online You can watch the video to see how to graph transformations of a reciprocal function on a TI-84.

Learn Transformations of Reciprocal Functions

The same techniques used to transform the graphs of other functions can be applied to the graphs of reciprocal functions.

Go Online You may want to complete the Concept Check to check your understanding.

Key Concept • Transformations of Reciprocal Functions	
	$g(x) = \frac{a}{x-h} + k$
h – horizontal translation	If $h > 0$, the graph of $f(x)$ is translated h units right. If $h < 0$, the graph of $f(x)$ is translated $\lvert h \rvert$ units left. The *vertical asymptote* is at $x = h$.
k – vertical translation	If $k > 0$, the graph of $f(x)$ is translated k units up. If $k < 0$, the graph of $f(x)$ is translated $\lvert k \rvert$ units down. The *horizontal asymptote* is at $f(x) = k$.
a – orientation and shape	If $\lvert a \rvert > 1$, the graph is stretched vertically. If $0 < \lvert a \rvert < 1$, the graph is compressed vertically.

Example 4 Graph a Transformation of a Reciprocal Function

Graph $g(x) = \frac{-4}{x+1} - 2$. State the domain and range.

$g(x) = \frac{-4}{x+1} - 2$ represents a transformation of the graph of $f(x) = \frac{1}{x}$.

$a = -4$ The graph is reflected in the x-axis and stretched vertically.

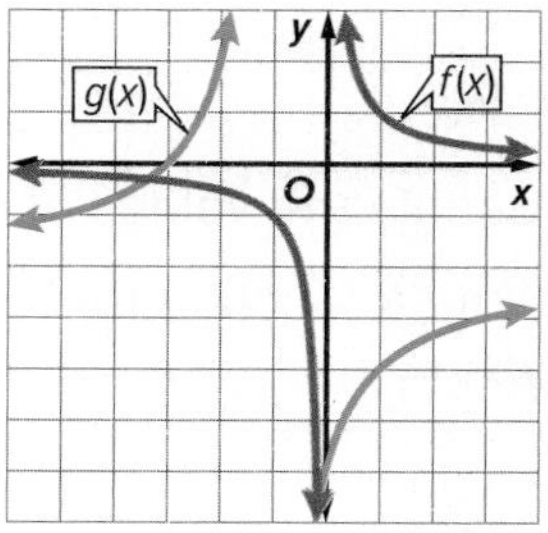

$h = -1$ The graph is translated left 1 unit. The vertical asymptote is $x = -1$.

$k = -2$ The graph is translated down 2 units. The horizontal asymptote is $g(x) = -2$.

The domain is $\{x \mid x \neq -1\}$.

The range is $\{g(x) \mid g(x) \neq -2\}$.

Go Online You can complete an Extra Example online.

Think About It!

Describe the transformations in $g(x) = \frac{-4}{x+1} - 2$ as it relates to the parent function.

Watch Out!

Scaling When graphing transformations of reciprocal functions, you may need to use a larger coordinate plane or scale the axes so you can get a better view of the function. Using a larger coordinate plane will make asymptotes more obvious.

Check

Graph $g(x) = \frac{1}{x-1} + 3$. Then state the domain and range.

Think About It!

Determine the domain and range of $g(x) = \frac{5}{x+4} - 2$.

Example 5 Write a Reciprocal Function from a Graph

Identify the values of *a*, *h*, and *k*. Then write a function for the graph $g(x) = \frac{a}{x-h} + k$.

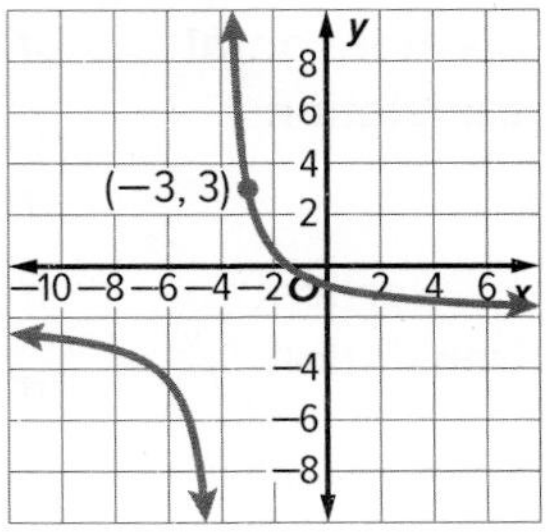

The asymptotes are $x = -4$ and $y = -2$. So, $h = -4$ and $k = -2$. Use a point on the graph to solve for a.

$g(x) = \frac{a}{x-h} + k$	General form of reciprocal function
$3 = \frac{a}{(-3)-(-4)} + (-2)$	$h = -4$, $k = -2$, and $(x, g(x)) = (-3, 3)$
$3 = a - 2$	Simplify.
$5 = a$	Add 2 to each side.

Use the values of a, h, and k to write the function.

$g(x) = \frac{5}{x+4} - 2$ $\quad$ $a = 5$, $h = -4$, and $k = -2$

Check

Identify the values of a, h, and k for the graph.

Write a function for the graph $g(x) = \frac{a}{x-h} + k$.

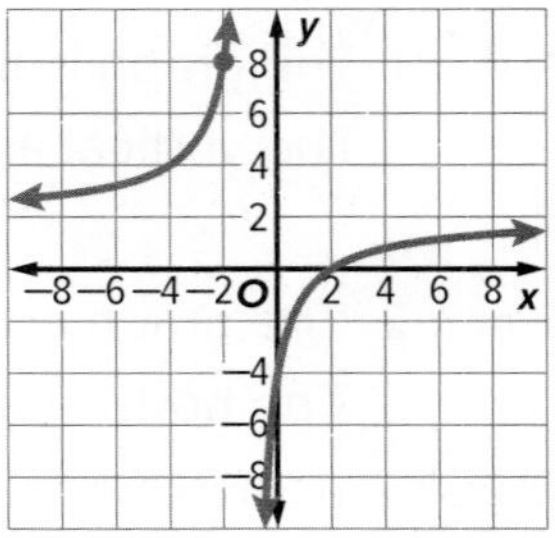

Go Online You can complete an Extra Example online.

Practice

Go Online You can complete your homework online.

Example 1

Determine the excluded value of x for each function.

1. $f(x) = \frac{5}{x}$
2. $g(x) = \frac{-2}{x + 2}$
3. $f(x) = \frac{10}{x - 3}$
4. $g(x) = \frac{5}{-6x}$
5. $f(x) = \frac{5}{2x + 3}$
6. $g(x) = \frac{5}{7x - 9}$

Example 2

Identify the asymptotes, domain, and range of each function. Then graph the function and identify its intercepts.

7. $f(x) = \frac{1}{x - 1}$
8. $f(x) = -\frac{1}{x} + 4$
9. $f(x) = \frac{5}{x + 4}$
10. $f(x) = \frac{6}{x} - 3$

Example 3

11. PLANES A plane is scheduled to leave Dallas for an 800-mile flight to Chicago's O'Hare airport. However, the departure is delayed for two hours.
    a. If $t = 0$ represents the scheduled departure time, write a function that represents the plane's average speed r on the vertical axis as a function of travel time t, which is based on the travel from the scheduled departure time to the destination. Graph the function.
    b. Analyze the key features of the graph in the context of the situation.

12. COMPUTERS To manufacture a specific model of computer, a company pays \$5000 for rent and overhead and \$435 per computer for parts.
    a. Write the function relating the average cost to make a computer C to how many computers n are being made. Graph the function.
    b. Analyze the key features of the graph.

Example 4

Graph each function. State the domain and range.

13. $f(x) = \frac{1}{x + 3} - 3$
14. $f(x) = \frac{-1}{x + 5} - 6$
15. $f(x) = \frac{-1}{x + 1} + 3$
16. $f(x) = \frac{1}{x + 4} - 2$

Example 5

Identify the values of a, h, and k. Then write a function for the graph $g(x) = \frac{a}{x-h} + k$.

17.

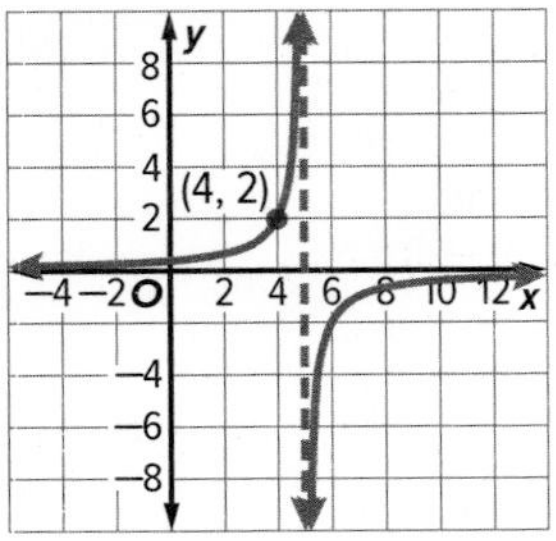

18.

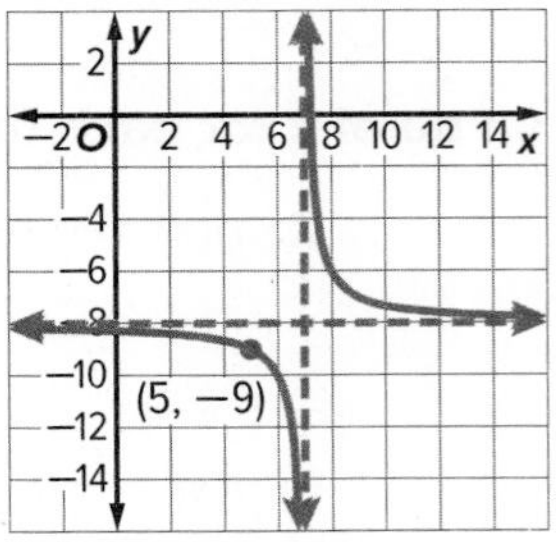

19.

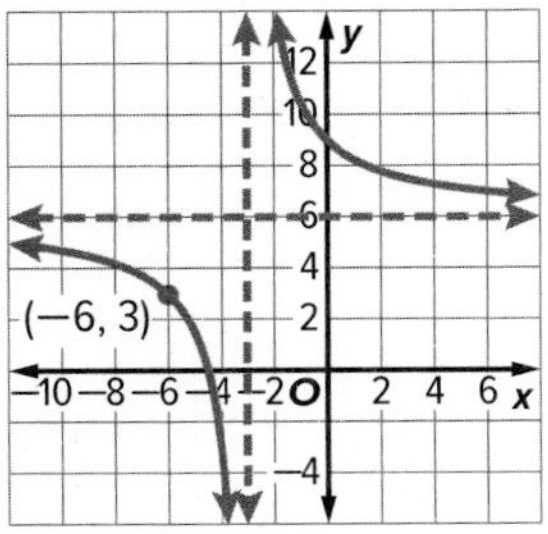

20.

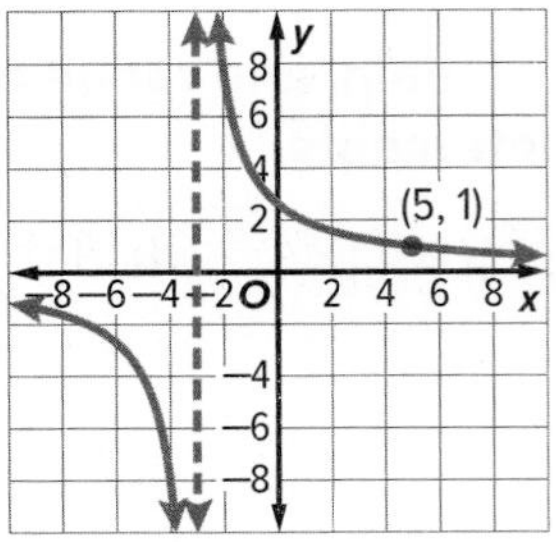

21.

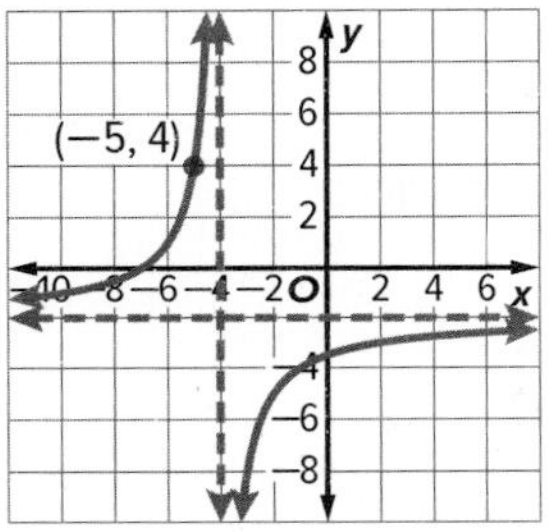

22.

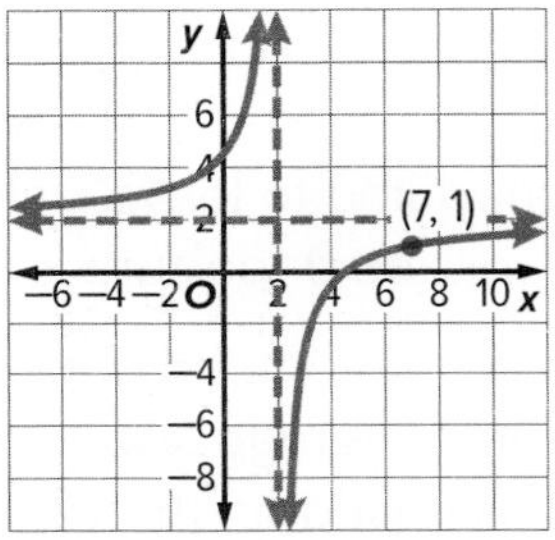

Mixed Exercises

Determine the values of x for which $f(x)$ is undefined.

23. $f(x) = \frac{2}{-2x + 5}$

24. $f(x) = \frac{-12}{-3x - 7}$

25. $f(x) = \frac{4}{x^2 - 2x - 3}$

26. $f(x) = \frac{6}{x^2 + 4x - 12}$

27. $f(x) = \frac{-3}{x^2 - 2x - 15}$

28. $f(x) = \frac{-2}{(x + 4)(x^2 - 9)}$

29. What are the domain and range for the function $f(x) = \frac{3}{x + 1}$?

30. Determine the equations of any asymptotes of the graph of $f(x) = \frac{3}{x^2 - x - 12}$.

Graph each function. State the domain and range, and identify the asymptotes.

31. $f(x) = \frac{3}{2x - 4}$

32. $f(x) = \frac{5}{3x}$

33. $f(x) = \frac{2}{4x + 1}$

34. $f(x) = \frac{1}{2x + 3}$

35. $f(x) = \frac{-3}{x + 7} - 1$

36. $f(x) = \frac{-4}{x + 2} - 5$

37. $f(x) = \frac{6}{x - 1} + 2$

38. $f(x) = \frac{2}{x - 4} + 3$

39. $f(x) = \frac{-7}{x - 8} - 9$

40. $f(x) = \frac{-6}{x - 7} - 8$

Write a function for each graph $f(x) = \frac{a}{x - h} + k$.

41.

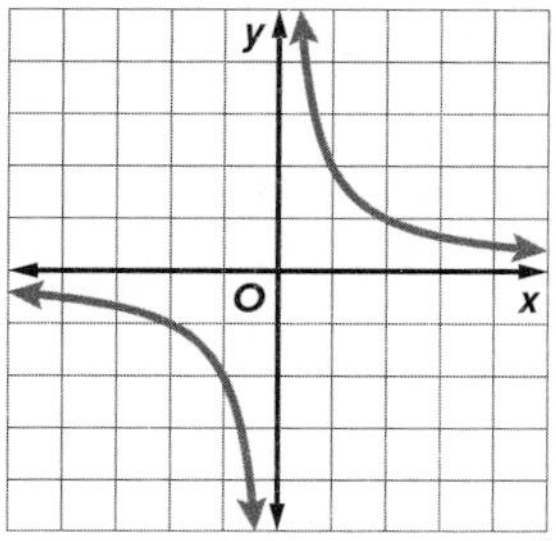

42.

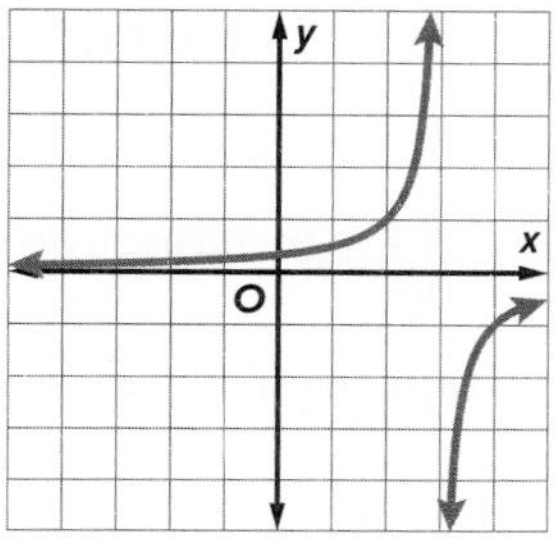

43.

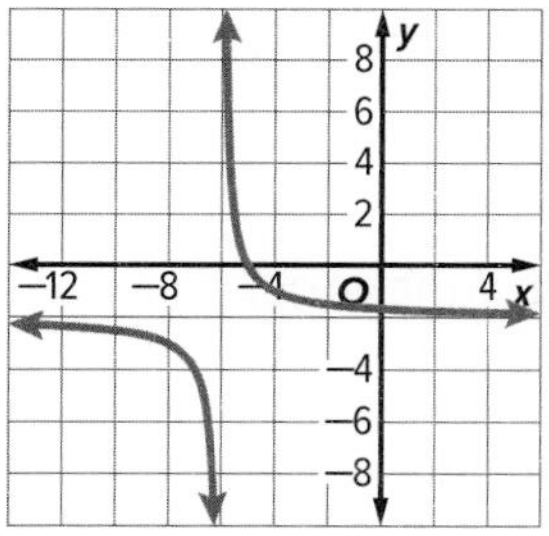

44.

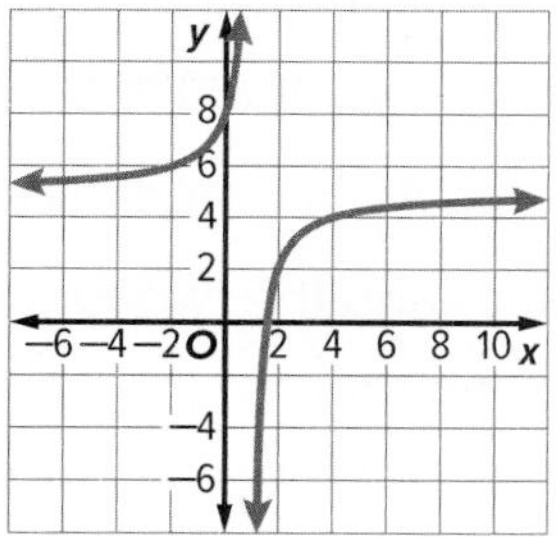

45. VACATION The Castellanos family is planning to rent a car for a trip. The rental costs \$125 plus \$0.30 per mile.

a. Write the equation that relates the cost per mile, C, to the number of miles traveled, m.

b. Explain any limitations to the range or domain in this situation.

46. BIOLOGY The population of a certain bacteria can be approximated by the function $P(t) = \frac{40}{t + 2} + 10$, where $P(t)$ is equal to the number of bacteria after t minutes. Eventually, what will happen to the bacteria population?

STRUCTURE Graph each reciprocal function and identify any transformations from the graph of the parent function $f(x) = \frac{1}{x}$.

47. $g(x) = \frac{-1}{x+2} + 1$

48. $h(x) = \frac{2}{x-4} - 5$

49. PRECISION The figure shows the graph of the parent reciprocal function, $f(x)$, and the graph of a reciprocal function, $k(x)$. Explain how to use transformations to write the equation for $k(x)$.

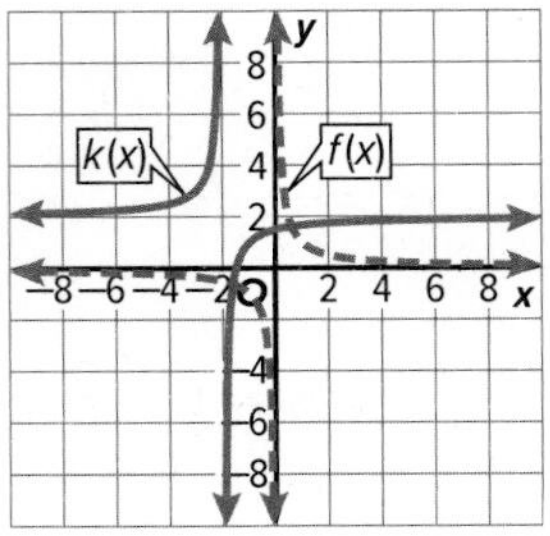

50. MONEY Some students are renting a bus for a trip to an aquarium. The bus costs \$200, and the aquarium tickets are \$25 per person.

a. Write a function that gives the total cost per student y, assuming that there are x students and they share the cost of the bus equally. Then graph the function on a coordinate plane.

b. What is an appropriate domain for the function? Explain.

c. Find the average rate of change on the interval $x = 10$ to $x = 20$. Interpret the result.

51. CREATE Write a reciprocal function for which the graph has a vertical asymptote at $x = -4$ and a horizontal asymptote at $f(x) = 6$.

52. ANALYZE Consider the functions $f(x) = \frac{1}{x}$ and $g(x) = \frac{1}{x^2}$.

a. Make a table of values comparing the two functions. Then graph both functions.

b. Compare and contrast the two graphs.

c. Make a conjecture about the difference between the graphs of reciprocal functions with an even exponent in the denominator and those with an odd exponent in the denominator.

53. WHICH ONE DOESN'T BELONG? Find the function that does not belong. Justify your conclusion.

$f(x) = \frac{3}{x+1}$	$g(x) = \frac{x+2}{x^2+1}$	$h(x) = \frac{5}{x^2+2x+1}$	$j(x) = \frac{20}{x-7}$

54. PERSEVERE Write two different reciprocal functions with graphs having the same vertical and horizontal asymptotes. Then graph the functions.

55. PERSEVERE Graph $f(x) = \frac{4}{(x+2)^2}$. What are the asymptotes of the graph?

56. WRITE Explain why only part of the graph of a rational function may be meaningful in a real-world situation in the context of the problem.

Graphing Rational Functions

Explore Analyzing Rational Functions

Online Activity Use graphing technology to complete the Explore.

INQUIRY How can you use a graphing calculator to analyze a rational function?

Today's Goals

- Graph and analyze rational functions with vertical and horizontal asymptotes.
- Graph and analyze rational functions with oblique asymptotes.

Today's Vocabulary

rational function

oblique asymptote

point discontinuity

Learn Graphing Rational Functions with Vertical and Horizontal Asymptotes

A **rational function** has an equation of the form $f(x) = \frac{a(x)}{b(x)}$, where $a(x)$ and $b(x)$ are polynomial functions and $b(x) \neq 0$.

Key Concept • Vertical and Horizontal Asymptotes

If $f(x) = \frac{a(x)}{b(x)}$, $a(x)$ and $b(x)$ are polynomial functions with no common factors other than 1, and $b(x) \neq 0$, then:

- $f(x)$ has a vertical asymptote whenever $b(x) = 0$.
- $f(x)$ has at most one horizontal asymptote.
  - If the degree of $a(x)$ is greater than the degree of $b(x)$, there is no horizontal asymptote.
  - If the degree of $a(x)$ is less than the degree of $b(x)$, the horizontal asymptote is the line $y = 0$.
  - If the degree of $a(x)$ equals the degree of $b(x)$, the horizontal asymptote is the line $y = \frac{\text{leading coefficient of } a(x)}{\text{leading coefficient of } b(x)}$.

Think About It!

Finding the degree of a polynomial is important to remember when finding the asymptotes of a rational function. Describe how to find the degree of a polynomial.

Example 1 Graph with No Horizontal Asymptotes

Graph $f(x) = \frac{x^3}{x + \frac{2}{3}}$.

Step 1 Find the zeros.

Set $a(x) = 0$.

$x^3 = 0$

$x = 0$

There is a zero at $x = 0$.

(continued on the next page)

Go Online You can complete an Extra Example online.

Go Online

You can watch a video to see how to graph rational functions with vertical and horizontal asymptotes.

Talk About It!

Why is it helpful to graph the asymptote(s) first when graphing a rational function?

Watch Out!

Zeros vs. Vertical Asymptotes Zeros of rational functions occur at the values that make the numerator zero. Vertical asymptotes occur at the values that make the denominator zero.

Step 2 Find the asymptotes.

Find the vertical asymptote. Set $b(x) = 0$.

$$x + \frac{2}{3} = 0$$
$$x = -\frac{2}{3}$$

There is a vertical asymptote at $x = -\frac{2}{3}$.

Because the degree of the numerator is greater than the degree of the denominator, there is no horizontal asymptote.

Step 3 Draw the graph.

Graph the asymptote. Then make a table of values, and graph.

x	$f(x)$
−2	6
−1	3
0	0
1	0.6
2	3

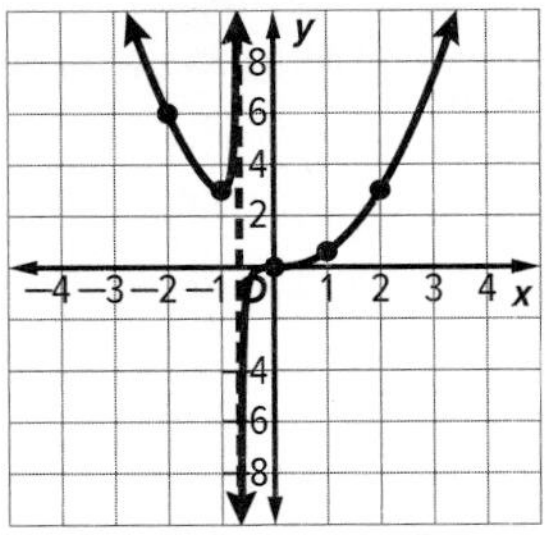

Check

Consider $g(x) = \frac{(0.5x - 1)^3}{x}$.

Part A There is a zero at $x =$ __?__.

Part B There is a vertical asymptote at $x =$ __?__.

Part C Graph the function.

Example 2 Use Graphs of Rational Functions

PHARMACY **Young's Rule can be used to estimate the dosage of medicine to give a child if you know the adult dosage of the same medicine. Young's rule can be written as $y = \frac{Ax}{x + 12}$, where A is the adult dosage in milligrams and x is the age of the child in years.**

Part A Graph the function.

Graph the function for an adult dose of 200 mg.

The function is $y = \frac{200x}{x + 12}$.

The vertical asymptote is at $x = -12$.

Because the degree of $a(x)$ = the degree of $b(x)$, the horizontal asymptote is the line $y = \frac{\text{leading coefficient of } a(x)}{\text{leading coefficient of } b(x)}$, so $y = \frac{200}{1}$ or $y = 200$.

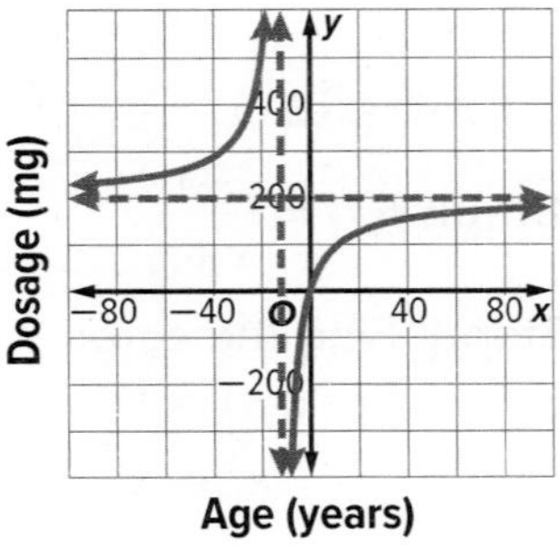

Go Online You can complete an Extra Example online.

Part B Find key features.

Find the x- and y-intercepts and end behavior of the function.

x-intercept: 0

y-intercept: 0

end behavior: As $x \to \infty$, $y \to 200$, and as $x \to -\infty$, $y \to 200$.

Part C Use the graph.

Find the dosage for a 12-year-old child if the adult dosage is 200 mg.

From the graph, it appears that when x is 12, y is approximately 100. So, the dosage for a 12-year old child is approximately 100 mg.

CHECK

Check your answer by substituting in the original equation.

$y = \frac{200x}{x + 12}$ Original equation

$y = \frac{200(12)}{12 + 12}$ $x = 12$

$y = 100$ Simplify.

Example 3 Compare Rational Functions

Consider $g(x) = \frac{x - 2}{2x + 2}$ and $h(x)$ shown in the graph.

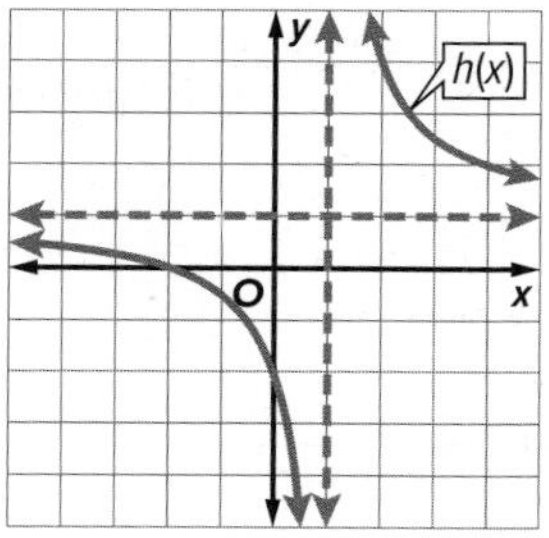

Part A Graph $g(x)$.

Step 1 Find the zeros.

$x - 2 = 0$ Set $a(x) = 0$.

$x = 2$ Add 2 to each side.

There is a zero at $x = 2$.

Step 2 Find the asymptotes.

Find the vertical asymptote.

$2x + 2 = 0$ Set $b(x) = 0$.

$2x = -2$ Subtract 2 from each side.

$x = -1$ Divide each side by 2.

There is a vertical asymptote at $x = -1$.

(*continued on the next page*)

Go Online You can complete an Extra Example online.

Think About It!
Describe the domain in the context of the situation.

Think About It!

What do you know about the degrees of the numerator and denominator of $h(x)$? Explain your reasoning.

Because the degree of the numerator equals the degree of the denominator, the horizontal asymptote is the line

$y = \frac{\text{leading coefficient of } a(x)}{\text{leading coefficient of } b(x)}$ or $y = \frac{1}{2}$.

Step 3 Draw the graph.

Draw the asymptotes. Then make a table of values, and graph the ordered pairs.

Part B Which function has the greater *y*-intercept?

$g(x)$ has a y-intercept of -1. $h(x)$ appears to have a y-intercept at $x = -2$, so its y-intercept is less than -1.

Part C Compare the asymptotes of $g(x)$ and $h(x)$.

vertical asymptotes	$g(x)$: $x = -1$	$h(x)$: $x = 1$
horizontal asymptotes	$g(x)$: $y = \frac{1}{2}$	$h(x)$: $y = 1$

Check

Consider $g(x) = \frac{-6x + 4}{2x - 3}$ and the graph of $h(x)$. Which function has the higher horizontal asymptote?

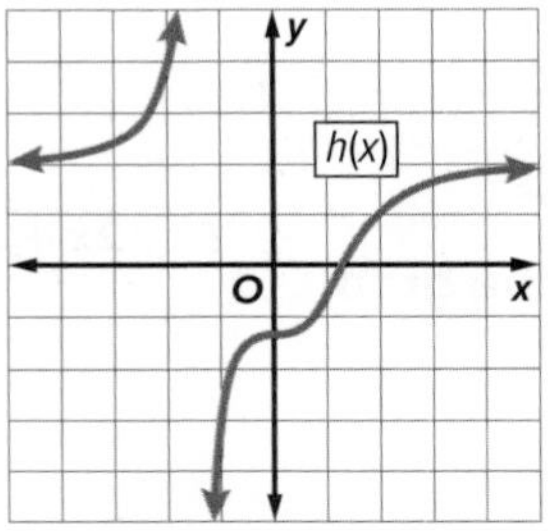

___?___ has the higher horizontal asymptote.

Learn Graphing Rational Functions with Oblique Asymptotes

An **oblique asymptote**, or slant asymptote, is neither horizontal nor vertical.

Key Concept • Oblique Asymptotes

If $f(x) = \frac{a(x)}{b(x)}$, where $a(x)$ and $b(x)$ are polynomial functions with no common factors other than 1 and $b(x) \neq 0$, then $f(x)$ has an oblique asymptote if the degree of $a(x)$ minus the degree of $b(x)$ equals 1. The equation of the asymptote is $f(x) = \frac{a(x)}{b(x)}$ with no remainder.

Go Online You can complete an Extra Example online.

Math History Minute

Charlotte Scott (1858–1931) was the first British woman to receive a doctorate in mathematics. In 1894, Scott published a book titled *Projective Methods in Plane Analytical Geometry,* which had a chapter devoted to Rational Curves and the Tracing of Curves. Her mathematical specialty was the study of algebraic functions of degree higher than two.

In some cases, graphs of rational functions may have **point discontinuity**, which looks like a hole in the graph. This is because the function is undefined at that point. If the original function is undefined for $x = a$ but the related rational expression of the function in simplest form is defined for $x = a$, then there is a point discontinuity or hole in the graph at $x = a$.

Key Concept • Point Discontinuity

If $f(x) = \frac{a(x)}{b(x)}$, $b(x) \neq 0$, and $x - c$ is a factor of both $a(x)$ and $b(x)$, then there is a point discontinuity at $x = c$.

Go Online You can watch a video to see how to graph rational functions with oblique asymptotes.

Example 4 Graph with Oblique Asymptotes

Consider $f(x) = \frac{x^2 + 2x + 1}{x + 4}$.

Part A Find the zeros.

$x^2 + 2x + 1 = 0$	Set $a(x) = 0$.
$(x + 1)^2 = 0$	Factor.
$x + 1 = 0$	Take the square root of each side.
$x = -1$	Subtract 1 from each side.

There is a zero at $x = -1$.

Part B Find the asymptotes.

Find the vertical asymptote.

$x + 4 = 0$	Set $b(x) = 0$.
$x = -4$	Subtract 4 from each side.

There is a vertical asymptote at $x = -4$.

Because the degree of the numerator is greater than the degree of the denominator, there is no horizontal asymptote.

The difference between the degree of the numerator and the degree of the denominator is 1, so there is an oblique asymptote. To find the oblique asymptote, divide the numerator by the denominator.

$$\begin{array}{r} x - 2 \\ x + 4 \overline{)\, x^2 + 2x + 1} \\ \underline{(-)x^2 + 4x} \\ -2x + 1 \\ \underline{(-)-2x - 8} \\ 9 \end{array}$$

The equation of the asymptote is the quotient excluding the remainder. So, there is an oblique asymptote at $f(x) = x - 2$.

Go Online You can complete an Extra Example online.

Study Tip

Oblique Asymptotes To determine whether a rational function has an oblique asymptote, find the degrees of the numerator and denominator. If the difference between the degree of the numerator and the degree of the denominator is 1, then the rational function has an oblique asymptote.

Part C Graph $f(x) = \frac{x^2 + 2x + 1}{x + 4}$.

Graph the asymptotes. Then, make a table of values and graph the ordered pairs.

x	$f(x)$
−7	−12
−6	−12.5
−5	−16
−3	4
−2	0.5
−1	0
0	0.25
1	0.8
2	1.5

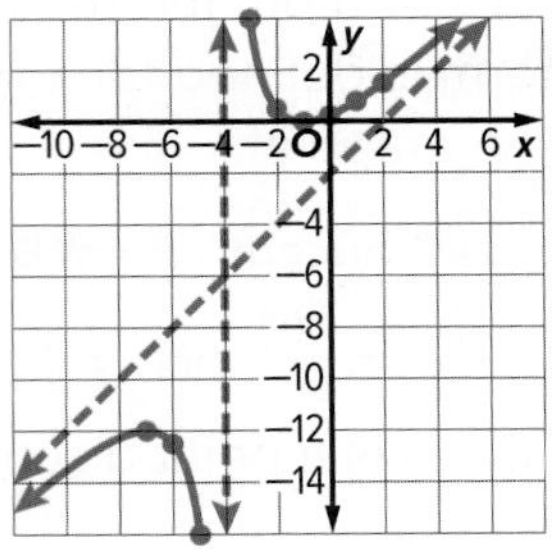

Think About It!

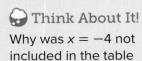

Why was $x = -4$ not included in the table of values?

Check

Find the asymptotes of $g(x) = \frac{2x^2 + 3x + 2}{2x + 1}$.

Then graph the function.

There is a vertical asymptote at $x =$ __?__.

There is an oblique asymptote at $f(x) = x +$ __?__.

Example 5 Graph with Point Discontinuity

Graph $f(x) = \frac{x^2 - 4}{x + 2}$. Find the point discontinuity.

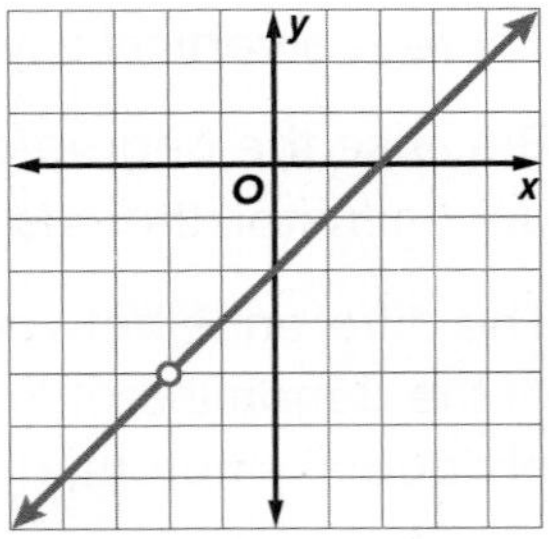

Notice that $\frac{x^2 - 4}{x + 2} = \frac{(x + 2)(x - 2)}{x + 2}$ or $x - 2$.

However, because the denominator of the original function cannot be 0, there is a discontinuity at $x + 2 = 0$ or $x = -2$.

Therefore, the graph of $f(x) = \frac{x^2 - 4}{x + 2}$ is the graph of $f(x) = x - 2$ with a hole or point discontinuity at $x = -2$.

Think About It!

Create a rational function with a point of discontinuity at $x = 2$.

Check

Consider $f(x) = \frac{(x + 3)^2}{x + 3}$. Find the point discontinuity. Then graph the function.

$f(x) = \frac{(x + 3)^2}{x + 3}$ has a point discontinuity at $x =$ __?__.

Practice

Go Online You can complete your homework online.

Example 1

Graph each function.

1. $f(x) = \frac{x^4}{6x + 12}$

2. $f(x) = \frac{x^3}{8x - 4}$

3. $f(x) = \frac{x^4 - 16}{x^2 - 1}$

4. $f(x) = \frac{x^3 + 64}{16x - 24}$

Example 2

5. **INTERNET** An Internet service provider charges customers a \$60 installation fee plus \$30 per month for Internet service. A function that models the average monthly cost is $f(x) = \frac{60 + 30x}{x}$, where x is the number of months.
   - **a.** Graph the function.
   - **b.** Find the x- and y-intercepts and end behavior of the graph.
   - **c.** Find the average monthly cost to a customer that has Internet service for 8 months.

6. **SALES** The quantity of a certain product sold in week x is approximated by the function $f(x) = \frac{80x}{x^2 + 40}$.
   - **a.** Graph the function.
   - **b.** Find the x- and y-intercepts and the end behavior of the graph.
   - **c.** During which week(s) did 5 of the products sell?

7. **FACTORY** The cost in cents to create a certain part of a small engine is modeled by $f(x) = \frac{18x - 12}{6x}$, where x is the number of parts made.
   - **a.** Graph the function.
   - **b.** Find the x- and y-intercepts and the end behavior of the graph.
   - **c.** About how much does the 6th part cost to make?

Example 3

For Exercises 8-10, consider the given function and the function shown in the graph.

- **a.** Copy the graph. Graph the given function.
- **b.** Which function has the greater y-intercept?
- **c.** Compare the asymptotes of the two functions.

8. $f(x) = \frac{x - 5}{3x + 5}$ and $g(x)$ shown in the graph

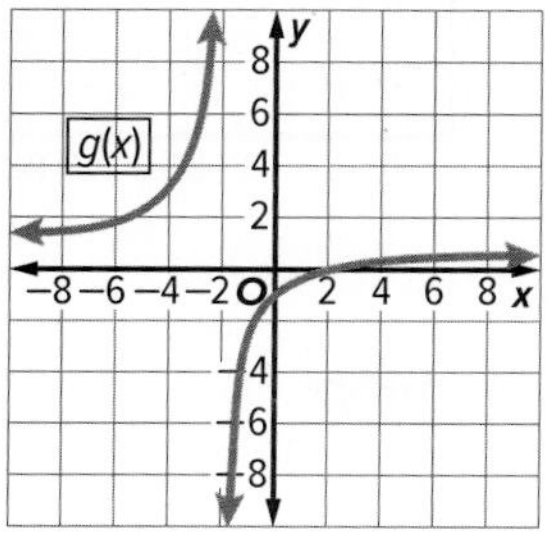

9. $h(x) = \frac{x + 1}{4x - 4}$ and $j(x)$ shown in the graph

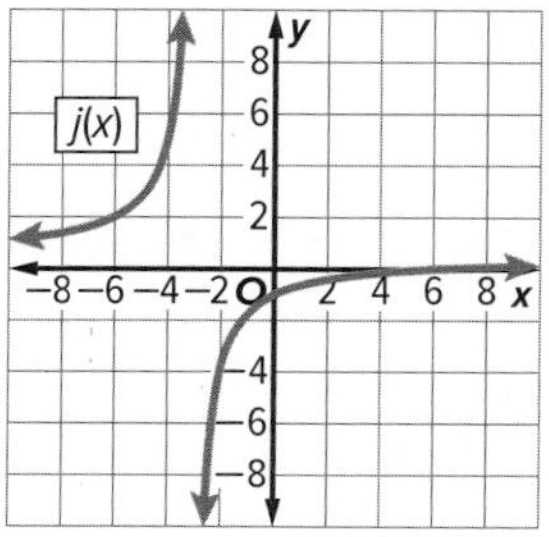

10. $f(x) = \frac{x - 3}{2x + 7}$ and $g(x)$ shown in the graph

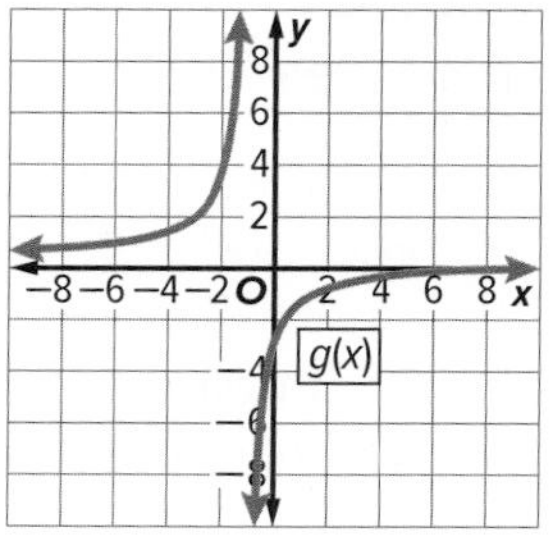

Example 4

Find the zeros and asymptotes of each function. Then graph each function.

11. $f(x) = \frac{(x-4)^2}{x+2}$

12. $f(x) = \frac{(x+3)^2}{x-5}$

13. $f(x) = \frac{6x^2+4x+2}{x+2}$

14. $f(x) = \frac{2x^2+7x}{x-2}$

15. $f(x) = \frac{3x^2+8}{2x-1}$

16. $f(x) = \frac{2x^2+5}{3x+4}$

Example 5

Graph each function. Find the point discontinuity.

17. $f(x) = \frac{x^2-2x-8}{x-4}$

18. $f(x) = \frac{x^2+4x-12}{x-2}$

19. $f(x) = \frac{x^2-25}{x+5}$

20. $f(x) = \frac{x^2-64}{x-8}$

21. $f(x) = \frac{(x-4)(x^2-4)}{x^2-6x+8}$

22. $f(x) = \frac{(x+5)(x^2+2x-3)}{x^2+8x+15}$

Mixed Exercises

Graph each function.

23. $f(x) = \frac{x}{x+2}$

24. $f(x) = \frac{x^2-4}{x-2}$

25. $f(x) = \frac{x^2+x-12}{x-3}$

26. $f(x) = \frac{x-1}{x^2-4x+3}$

27. $f(x) = \frac{3}{x^2-2x-8}$

28. $f(x) = \frac{x^3}{2x+2}$

29. $f(x) = \frac{2x^3+4x^2-10x-12}{2x^2+8x+6}$

30. $f(x) = \frac{(x+1)^2}{2x-1}$

31. Consider $f(x) = \frac{(x+2)(x-3)}{x+2}$ and $g(x) = \frac{(x+2)(x-3)}{x(x+2)}$.

a. Graph $f(x)$ and $g(x)$.

b. Which function has a y-intercept of -3?

c. Compare the asymptotes and point discontinuity of $f(x)$ and $g(x)$.

32. BACTERIA The graph shows the cost in millions of dollars y to remove x percent of the bacteria from drinking water.

a. What does the x-coordinate of the vertical asymptote represent?

b. What happens to the cost of removing bacteria from drinking water as p approaches 100?

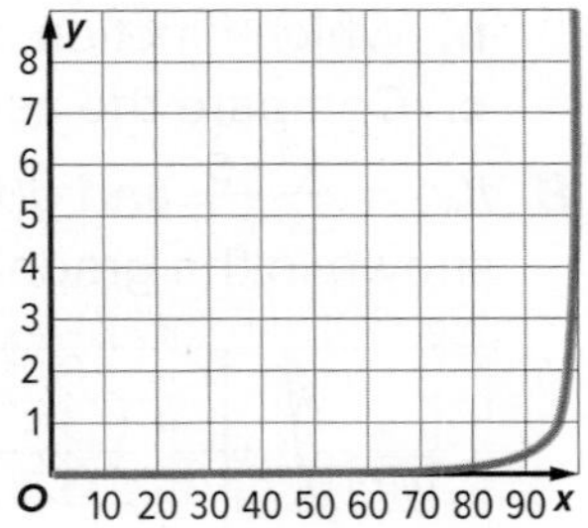

33. Sir Isaac Newton studied the rational function $f(x) = \frac{ax^3+bx^2+cx+d}{x}$. Assuming that $d \neq 0$, where will the graph of this function have a vertical asymptote?

Graph each function.

34. $f(x) = \frac{5}{(x-1)(x+4)}$

35. $f(x) = \frac{4}{(x-2)^2}$

36. $f(x) = \frac{x-3}{x+1}$

37. $f(x) = \frac{1}{(x+4)^2}$

38. $f(x) = \frac{2x}{(x+2)(x-5)}$

39. $f(x) = \frac{x^4 - 2x^2 + 1}{x^3 + 2}$

40. $f(x) = \frac{x^4 - x^2 - 12}{x^3 - 6}$

41. $f(x) = \frac{3x^4 + 6x^3 + 3x^2}{x^2 + 2x + 1}$

42. $f(x) = \frac{2x^4 + 10x^3 + 12x^2}{x^2 + 5x + 6}$

43. $f(x) = \frac{x+1}{x^2 + 6x + 5}$

44. $f(x) = \frac{x^2 - 10x - 24}{x+2}$

45. $f(x) = \frac{x^3 + 1}{x^2 - 4}$

46. CONSTRUCT ARGUMENTS Alina graphed the function $f(x) = \frac{x^2 - 4x}{x - 4}$, as shown. Did she correctly graph the function? If so, justify your argument. If not, explain how to correct the graph.

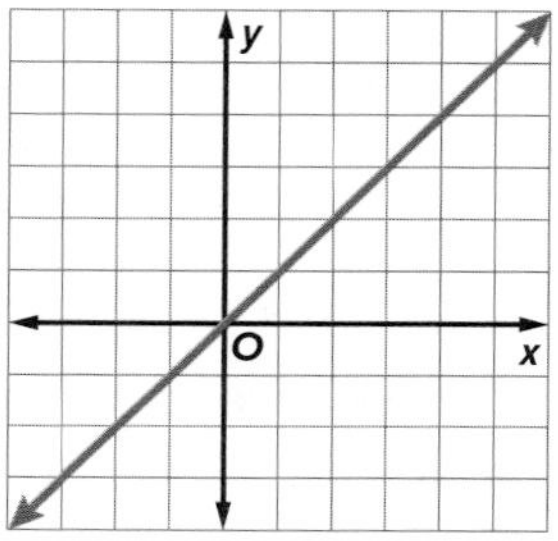

47. BATTING AVERAGES A major league baseball player had a lifetime batting average of 0.305 at the beginning of the 2017 season with 2067 hits out of 6767 at bats. During the 2017 season, the player had 183 hits.

a. Write an equation describing the baseball player's batting average y at the end of the 2017 season using x to represent the number of at bats the player had during the season.

b. Determine the location of the horizontal and vertical asymptotes for the graph of the equation.

c. What is the meaning of the horizontal asymptote for the graph of this equation?

48. REASONING A music studio uses the function $f(x) = \frac{300x}{x^2 + 4}$ to estimate the number of downloads per hour in thousands in the hours after the release of a new song. Graph the function. Then restrict the domain of the function as required by the context, and graph the function with the restricted domain. Explain the shape of the graph in the context of the situation.

49. STRUCTURE Identify the domain, zeros, intercepts, and asymptotes of the graph, and determine a function that corresponds to the graph.

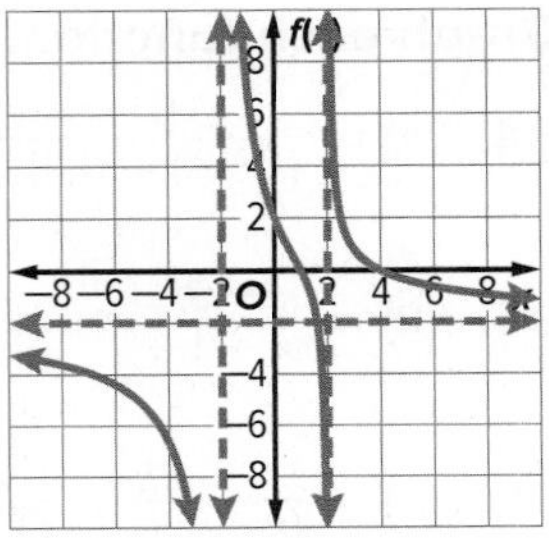

50. STRUCTURE Consider the functions $y = x + 1$, $y = \frac{(x + 1)(x - 1)}{x - 1}$, $y = \frac{(x + 1)(x - 1)^2}{x - 1}$, and $y = \frac{(x + 1)(x - 1)^2}{(x - 1)^2}$. Which, if any, are equivalent?

51. SKETCH A GRAPH Graph a rational function that has a y-intercept at 5, a vertical asymptote at $x = 3$, and a horizontal asymptote at $y = 6.5$.

52. PRECISION Analyze the graph of the function shown. Over what intervals of x is the function positive? Over what intervals of x is the function negative? Over what intervals is the function increasing? Over what intervals is the function decreasing?

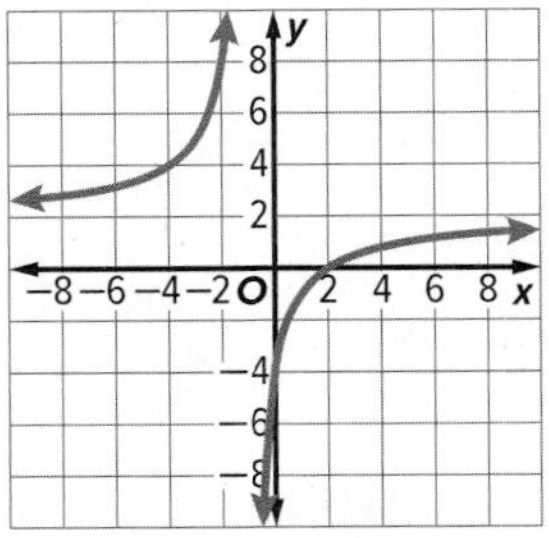

53. REGULARITY Describe how you use the degrees of the numerator and denominator to learn about a function's asymptotes.

54. PERSEVERE On the drive to visit a nearby college, the Marshall family averages 40 miles per hour.

a. Define variables and write a function for the average speed for the entire trip, in terms of the average speed for the drive home. (Hint: Write an expression for the average speed in terms of distance and times for the outgoing trip and return trip. Then express time in terms of speed and distance.)

b. If the family averages 60 miles per hour driving home, is the average speed for the entire trip equal to 50 miles per hour? Explain. What is the horizontal asymptote, and what does it represent?

55. CREATE Sketch the graph of a rational function with a horizontal asymptote $y = 1$ and a vertical asymptote $x = -2$.

56. PERSEVERE Compare and contrast $g(x) = \frac{x^2 - 1}{x(x^2 - 2)}$ and $f(x)$ shown in the graph.

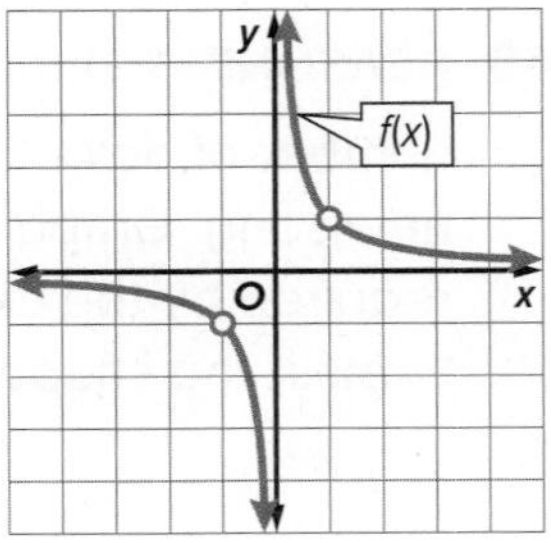

57. ANALYZE Describe the difference between the graphs of $f(x) = x - 2$ and $g(x) = \frac{(x + 3)(x - 2)}{x + 3}$.

58. PERSEVERE A rational function has an equation of the form $f(x) = \frac{a(x)}{b(x)}$, where $a(x)$ and $b(x)$ are polynomial functions and $b(x) \neq 0$. Show that $f(x) = \frac{x}{a - b} + c$ is a rational function.

59. WRITE How can factoring be used to determine the vertical asymptotes or point discontinuity of a rational function?

Variation

Explore Variation

Online Activity Use the interactive tool to complete the Explore.

INQUIRY How can you relate the dimensions of a rectangle with its area?

Learn Direct Variation and Joint Variation

Two quantities x and y are related by a **direct variation** if y is equal to a constant k times x. The constant k is called the **constant of variation**.

Key Concept • Direct Variation

y varies directly as x if there is some nonzero constant k such that $y = kx$.

The graph of a direct variation function, such as $f(x) = 4x$, is a line through the origin. A direct variation is a special case of an equation in slope-intercept form, $y = mx + b$, where $m = k$ and $b = 0$. The slope of a direct variation equation is its constant of variation.

If you know that y varies directly as x and one set of values, you can use a proportion to find another set of corresponding values.

Joint variation occurs when one quantity varies directly as the product of two or more other quantities.

Key Concept • Joint Variation

y varies jointly as x and z if there is some nonzero constant k such that $y = kxz$.

Example 1 Direct Variation

If y varies directly as x and $y = -3$ when $x = 24$, find y when $x = -16$.

Use a proportion that relates the values.

$$\frac{y_1}{x_1} = \frac{y_2}{x_2} \quad \text{Direct variation}$$

$$\frac{-3}{24} = \frac{y_2}{-16} \quad y_1 = -3,\ x_1 = 24,\ \text{and } x_2 = -16$$

$$-16\left(\frac{-3}{24}\right) = -16\left(\frac{y_2}{-16}\right) \quad \text{Multiply each side by } -16.$$

$$2 = y_2 \quad \text{Simplify.}$$

Go Online You can complete an Extra Example online.

Today's Goals
- Recognize and solve direct and joint variation equations.
- Recognize and solve inverse and combined variation equations.

Today's Vocabulary
direct variation
constant of variation
joint variation
inverse variation
combined variation

Talk About It!
Explain why the graph of a direct variation function always passes through the origin.

Think About It!
If y varies jointly as x and z, what must be true when $y = 0$?

Check

If y varies directly as x and $y = -9$ when $x = 6$, find y when $x = -7$.

$y_2 =$ ___?___

Example 2 Joint Variation

Suppose y varies jointly as x and z. Find y when $x = 4$ and $z = -3$, if $y = -15$ when $x = -6$ and $z = 1$.

Use a proportion that relates the values.

$\frac{y_1}{x_1 z_1} = \frac{y_2}{x_2 z_2}$	Joint variation
$\frac{y_1}{4(-3)} = \frac{-15}{-6(1)}$	$x_1 = 4$, $z_1 = -3$, $y_2 = -15$, $x_2 = -6$, and $z_2 = 1$
$\frac{y_1}{-12} = \frac{-15}{-6}$	Simplify.
$-12\left(\frac{y_1}{-12}\right) = -12\left(\frac{-15}{-6}\right)$	Multiply by the LCD, −12.
$y_1 = -30$	Simplify.

Check

Suppose y varies jointly as x and z. Find y when $x = -4$ and $z = -5$, if $y = -135$ when $x = 5$ and $z = -9$. $y_2 = 60$

Learn Inverse Variation and Combined Variation

Two quantities, x and y, are related by an **inverse variation** if their product is equal to a constant k.

Study Tip

Proportions Several different proportions can be used to solve an inverse variation problem with the same result.

Key Concept • Inverse Variation

Words: y varies inversely as x if there is some nonzero constant k such that $xy = k$ or $y = \frac{k}{x}$, where $x \neq 0$ and $y \neq 0$.

Example: If $xy = 8$ and $x = 12$, then $y = \frac{8}{12}$ or $\frac{2}{3}$.

When y varies inversely as x and the constant of proportionality k is positive, one quantity increases while the other decreases. You can use a proportion such as $\frac{x_1}{y_2} = \frac{x_2}{y_1}$ to solve inverse variation problems in which some quantities are known.

Combined variation occurs when one quantity varies directly and/or inversely as two or more other quantities.

If you know that y varies directly as x, that y varies inversely as z, and one set of values, you can use a proportion to find another set of corresponding values.

If $y_1 = \frac{kx_1}{z_1}$ and $y_2 = \frac{kx_2}{z_2}$, then $\frac{y_1 z_1}{x_1} = k$ and $\frac{y_2 z_2}{x_2} = k$. So, $\frac{y_1 z_1}{x_1} = \frac{y_2 z_2}{x_2}$.

Go Online You can complete an Extra Example online.

Example 3 Inverse Variation

If m varies inversely as n, and $m = -4$ when $n = 6$, find m when $n = -10$.

Use a proportion that relates the values.

$\frac{m_1}{n_2} = \frac{m_2}{n_1}$	Inverse variation
$\frac{-4}{-10} = \frac{m_2}{6}$	$m_1 = -4$, $n_1 = 6$, and $n_2 = -10$
$6\left(\frac{-4}{-10}\right) = 6\left(\frac{m_2}{6}\right)$	Cross multiply.
$2.4 = m_2$	Simplify.

Check

If y varies inversely as x, and $x = -3$ when $y = 3$, find y when $x = 18$.

$y_2 =$ ___?___

Example 4 Combined Variation

Suppose a varies directly as b, and a varies inversely as c. Find b when $a = 6$ and $c = 28$, if $b = 7$ when $a = -49$ and $c = 3$.

Use a proportion that relates the given values.

$a_1 = \frac{kb_1}{c_1}$ and $a_2 = \frac{kb_2}{c_2}$	b varies directly as a, so b is in the numerator. c varies inversely as a, so c is in the denominator.
$k = \frac{a_1c_1}{b_1}$ and $k = \frac{a_2c_2}{b_2}$	Solve for k.
$\frac{a_1c_1}{b_1} = \frac{a_2c_2}{b_2}$	Substitute for k.
$\frac{6(28)}{b_1} = \frac{-49(3)}{7}$	$a_1 = 6$, $c_1 = 28$, $a_2 = -49$, $b_2 = 7$, and $c_2 = 3$
$\frac{168}{b_1} = \frac{-147}{7}$	Simplify.
$7b_1\left(\frac{168}{b_1}\right) = 7b_1\left(\frac{-147}{7}\right)$	Multiply by the LCD, $7b_1$.
$-147b_1 = 1176$	Simplify.
$b_1 = -8$	Divide each side by -147.

When $a = 6$ and $c = 28$, the value of b is -8.

Check

Suppose m varies directly as n, and m varies inversely as p. Find m when $n = 20$ and $p = 25$, if $m = -\frac{4}{3}$ when $n = 6$ and $p = 9$.

$m_2 =$ ___?___

Go Online You can complete an Extra Example online.

Study Tip

Subscripts When writing the initial proportion, use subscripts with the variables. This will help you substitute the correct value for each of the variables.

Example 5 Write and Solve a Combined Variation

BICYCLES A rider's speed in miles per hour varies jointly with the revolutions per minute (RPM) of the pedals and the number of teeth on the front gear and inversely with the number of teeth on the rear gear. On a flat stretch of road, a rider is traveling at 17.1 miles per hour while using a front gear with 50 teeth and a rear gear with 16 teeth and pedaling at 70 RPM. Find the speed of the rider when she bikes uphill using a front gear with 34 teeth and a rear gear with 23 teeth while pedaling at 70 RPM.

Let x = speed, p = revolutions of pedals, f = number of teeth on front gear, and r = number of teeth on rear gear.

$x_1 = \frac{kp_1f_1}{r_1}$ and $x_2 = \frac{kp_2f_2}{r_2}$	x varies jointly as p and f and inversely as r.
$k = \frac{x_1r_1}{p_1f_1}$ and $k = \frac{x_2r_2}{p_2f_2}$	Solve for k.
$\frac{x_1r_1}{p_1f_1} = \frac{x_2r_2}{p_2f_2}$	Substitute.
$\frac{17.1(16)}{70(50)} = \frac{x_2(23)}{70(34)}$	$x_1 = 17.1$, $p_1 = 70$, $f_1 = 50$, $r_1 = 16$, $f_2 = 34$, $p_2 = 70$, and $r_2 = 23$
$\frac{273.6}{3500} = \frac{23x_2}{2380}$	Simplify.
$59{,}500\left(\frac{273.6}{3500}\right) = 59{,}500\left(\frac{23x_2}{2380}\right)$	Multiply by the LCD, 59,500.
$4651.2 = 575x_2$	Simplify.
$8.1 \approx x_2$	Divide each side by 575.

The speed of the rider pedaling at 70 RPM using a front gear with 34 teeth and a rear gear with 23 teeth is about 8.1 mph.

Check

BASEBALL The earned run average (ERA) a of a baseball pitcher varies directly as the number of earned runs r and inversely as the number of pitched innings p. In one season, a pitcher has an ERA of 2.475 with 80 innings pitched and 22 earned runs. Find the number of earned runs if the pitcher has an ERA of 2.25 after pitching 112 innings in the next season.

Part A Write a proportion you can use to solve the problem.

Part B Find the number of earned runs when the pitcher has an ERA of 2.25 after pitching 112 innings in the next season.

Go Online You can complete an Extra Example online.

Think About It!

What assumptions were made when determining the speed of the rider?

Practice

Go Online You can complete your homework online.

Example 1

If x varies directly as y, find x when $y = 8$.

1. $x = 6$ when $y = 32$
2. $x = 11$ when $y = -3$
3. $x = 14$ when $y = -2$
4. $x = -4$ when $y = 10$
5. If y varies directly as x, and $y = 35$ when $x = 7$, find y when $x = 11$.
6. If y varies directly as x, and $y = 360$ when $x = 180$, find y when $x = 270$.

Example 2

If a varies jointly as b and c, find a when $b = 4$ and $c = -3$.

7. $a = -96$ when $b = 3$ and $c = -8$
8. $a = -60$ when $b = -5$ and $c = 4$
9. $a = -108$ when $b = 2$ and $c = 9$
10. $a = 24$ when $b = 8$ and $c = 12$
11. If y varies jointly as x and z, and $y = 18$ when $x = 2$ and $z = 3$, find y when x is 5 and z is 6.
12. If y varies jointly as x and z, and $y = -16$ when $x = 4$ and $z = 2$, find y when x is -1 and z is 7.

Example 3

If f varies inversely as g, find f when $g = -6$.

13. $f = 15$ when $g = 9$
14. $f = 4$ when $g = 28$
15. $f = -12$ when $g = 19$
16. $f = 0.6$ when $g = -21$
17. If y varies inversely as x, and $y = 2$ when $x = 2$, find y when $x = 1$.
18. If y varies inversely as x, and $y = 6$ when $x = 5$, find y when $x = 10$.

Example 4

19. Suppose a varies directly as b, and a varies inversely as c. Find b when $a = 5$ and $c = -4$, if $b = 12$ when $c = 3$ and $a = 8$.
20. Suppose x varies directly as y, and x varies inversely as z. Find z when $x = 10$ and $y = -7$, if $z = 20$ when $x = 6$ and $y = 14$.
21. Suppose a varies directly as b, and a varies inversely as c. Find b when $a = 2.5$ and $c = 18$, if $b = 6$ when $c = 4$ and $a = 96$.
22. Suppose x varies directly as y, and x varies inversely as z. Find z when $x = 32$ and $y = 9$, if $z = 16$ when $x = 12$ and $y = 4$.

Example 5

23. ELECTRICITY The resistance R of a piece of wire varies directly as its length L and inversely as its cross-sectional area A. A piece of copper wire 3 meters long and 2 millimeters in diameter has a resistance of 0.0107 ohms (Ω). Find the resistance of a second piece of copper wire 2 meters long and 6 millimeters in diameter. (*Hint*: The cross-sectional area is πr^2.)

24. PLANTS The size of plant cells before cell division is directly related to the water available for nutrition and inversely proportional to the bacteria present in the soil. If a given plant's cells grow to 1.5 mm when 0.25 liter of water is available and the bacteria level is restricted to 500 parts per million, what would the bacteria level be for the same plant whose cells grow to 2 mm with 0.4 liter of water available?

25. PRESSURE The volume of a gas varies directly with the temperature and inversely with the pressure. A certain gas has a volume of 15 L, a temperature of 290 K, and a pressure of 1 atm. If the gas is compressed to a volume of 12 L and is heated to 310 K, what will the new pressure be? Round your answer to the nearest thousandth.

Mixed Exercises

State whether each equation represents a *direct, joint, inverse,* or *combined* variation for the given variable. Then name the constant of variation.

26. $c = 12m$; c

27. $p = \frac{4}{q}$; p

28. $A = \frac{1}{2}bh$; A

29. $rw = 15$; r

30. $y = 2rgt$; y

31. $f = 5280m$; f

32. $y = 0.2d$; y

33. $vz = -25$; z

34. $t = 16rh$; t

35. $R = \frac{8}{w}$; R

36. $b = \frac{1}{3}a$; b

37. $C = 2\pi r$; C

38. $\frac{x}{y} = 2.75$; x

39. $fg = -2$; f

40. $a = 3bc$; a

41. $10 = \frac{xy^2}{z}$; x

42. $y = -11x$; y

43. $\frac{n}{p} = 4$; n

44. $9n = pr$; p

45. $-2y = z$; z

46. $a = 27b$; a

47. $c = \frac{7}{d}$; c

48. $-10 = gh$; g

49. $m = 20cd$; m

50. If y varies directly as x, and $y = 540$ when $x = 10$, find x when $y = 1080$.

51. If y varies directly as x, and $y = 12$ when $x = 72$, find x when $y = 9$.

52. If y varies jointly as x and z, and $y = 120$ when $x = 4$ and $z = 6$, find y when x is 3 and z is 2.

53. If y varies inversely as x, and $y = 3$ when $x = 14$, find x when $y = 6$.

54. **STRUCTURE** Bulan works at a car wash. The amount Bulan earns varies directly with the number of hours she works. Bulan earns $144 for working 12 hours. Write the direct variation equation for the amount Bulan earns y for working x hours. How much will Bulan earn for working 18 hours?

55. **DESIGN** As a general rule, the number of parking spaces in a parking lot for a movie theater complex varies directly with the number of theaters in the complex. A typical complex has 30 parking spaces for each theater. A developer wants to build a new cinema complex on a lot that has enough space for 210 parking spaces. Write the direct variation equation for the number of parking spaces y for x theaters. How many theaters should the developer build in his complex?

Determine whether each relation shows *direct* or *inverse* variation, or *neither*.

56.

x	y
4	12
8	24
16	48
32	96

57.

x	y
8	2
4	4
−2	−8
−8	−2

58.

x	y
2	4
3	9
4	16
5	25

59. **ART** Premade art canvas is available in the standard sizes shown.

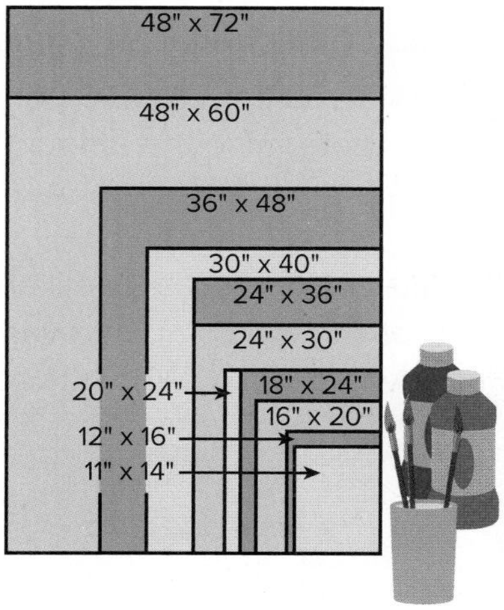

a. Analyze the relationship between the width and length of each canvas size. For each size, write a function that relates the length ℓ to the width w.

b. For which sizes is the relationship between the width and the length the same? Explain your reasoning.

c. Explain what it means in the context of the situation if the relationship between the width and length is the same for two canvas sizes.

60. **PRECISION** The cost of painting a wall varies directly with the area of the wall. Write a formula for the cost of painting a rectangular wall with dimensions ℓ by w. With respect to ℓ and w, does the cost vary *directly*, *jointly*, or *inversely*? Explain your reasoning.

61. **RENT** An apartment rents for m dollars per month.

a. If n students share the rent equally, how much would each student have to pay?

b. How does the cost per student vary with the number of students?

c. If 2 students have to pay $700 each, how much money would each student have to pay if there were 5 students sharing the rent?

62. **COMMUNICATIONS** On average, the number of calls c each day between two cities is directly proportional to the product of the populations P_1 and P_2 of the cities and inversely proportional to the square of the distance between them. That is, $c = \frac{kP_1P_2}{d^2}$. Use the function and the population and distance information in the table below.

	Birmingham, AL (pop. 210,710)	Indianapolis, IN (pop. 872,680)	Tallahassee, FL (pop. 191,049)	San Francisco, CA (pop. 884,363)
Birmingham	—	479 mi	301 mi	2327 mi
Indianapolis	479 mi	—	778 mi	2274 mi
Tallahassee	301 mi	778 mi	—	2637 mi
San Francisco	2327 mi	2274 mi	2637 mi	—

a. The average number of daily calls between Indianapolis and Birmingham is about 16,000. Find the value of *k*. Round to the nearest hundredth.

b. Find the average number of daily calls between Tallahassee and each other city listed.

c. Could you use this formula to find the average number of phone calls made within a city? Explain.

63. **INVESTING** You decide to invest 10% of your before-tax income in a retirement fund, so your employer deducts this money from your weekly paycheck.

a. Write an equation to represent the amount deducted from your paycheck *d* for investment in your retirement fund for a week during which you worked *h* hours and are paid *r* dollars per hour.

b. Is your equation a *direct, joint,* or *inverse* variation? Explain your reasoning.

c. If you earn $19.50 per hour and worked 36 hours last week, explain how to determine the amount deducted last week for your retirement fund.

64. **FIND THE ERROR** Jamil and Savannah are setting up a proportion to begin solving the combined variation in which *z* varies directly as *x*, and *z* varies inversely as *y*. Who has set up the correct proportion? Explain your reasoning.

Jamil

$z_1 = \frac{kx_1}{y_1}$ and $z_2 = \frac{kx_2}{y_2}$

$k = \frac{z_1y_1}{x_1}$ and $k = \frac{z_2y_2}{x_2}$

$\frac{z_1y_1}{x_1} = \frac{z_2y_2}{x_2}$

Savannah

$z_1 = \frac{kx_1}{y_1}$ and $z_2 = \frac{kx_2}{y_2}$

$k = \frac{z_1x_1}{y_1}$ and $k = \frac{z_2x_2}{y_2}$

$\frac{z_1x_1}{y_1} = \frac{z_2x_2}{y_2}$

65. **PERSEVERE** If *a* varies inversely as *b*, *c* varies jointly as *b* and *f*, and *f* varies directly as *g*, how are *a* and *g* related?

66. **ANALYZE** Explain why some mathematicians consider every joint variation a combined variation, but not every combined variation a joint variation.

67. **CREATE** Describe three real-life quantities that vary jointly with each other.

68. **WRITE** Determine the type(s) of variation(s) for which 0 cannot be one of the values. Explain your reasoning.

Solving Rational Equations and Inequalities

Explore Solving Rational Equations

Online Activity Use a real-world situation to complete the Explore.

> INQUIRY How can you solve rational equations by graphing?

Today's Goals
- Solve rational equations in one variable.
- Solve rational inequalities in one variable.

Today's Vocabulary

rational equation

rational inequality

Learn Solving Rational Equations

A **rational equation** contains at least one rational expression. To solve these equations, it is often easier to first eliminate the fractions. You can eliminate the fractions by multiplying each side of the equation by the least common denominator (LCD). Solving rational equations in this way can yield results that are not solutions of the original equation. You can identify these extraneous solutions by substituting each result into the original equation to see if it makes the equation true.

There are three types of problems that are commonly solved by using rational equations: mixture problems, uniform motion problems, and work problems.

Example 1 Solve a Rational Equation

Solve $\frac{7}{12} + \frac{9}{x-4} = \frac{55}{48}$.

The LCD for the terms is $48(x - 4)$.

$$\frac{7}{12} + \frac{9}{x-4} = \frac{55}{48}$$

$$48(x-4)\left(\frac{7}{12}\right) + 48(x-4)\left(\frac{9}{x-4}\right) = 48(x-4)\left(\frac{55}{48}\right)$$

$$\underset{4}{\cancel{48}}(x-4)\left(\frac{7}{\underset{1}{\cancel{12}}}\right) + 48\,\underset{1}{\cancel{(x-4)}}\left(\frac{9}{\underset{1}{\cancel{x-4}}}\right) = \underset{1}{\cancel{48}}\,(x-4)\left(\frac{55}{\underset{1}{\cancel{48}}}\right)$$

$$28x - 112 + 432 = 55x - 220$$

$$28x + 320 = 55x - 220$$

$$540 = 27x$$

$$20 = x$$

Go Online You can complete an Extra Example online.

Check

Select the solution(s) of $\frac{5}{6} - \frac{2}{4x+1} = \frac{x}{3}$.

A. $-\frac{67}{4}$ B. $-\frac{1}{4}$

C. $\frac{1}{4}$ D. $\frac{1}{2}$

E. $\frac{7}{4}$ F. 3

Go Online
You can watch a video to see how to solve rational equations.

Example 2 Solve a Rational Equation with an Extraneous Solution

Solve $\mathbf{\frac{2m}{m-4} - \frac{m^2+7m+4}{3m^2-18m+24} = \frac{4m}{3m-6}}$.

The LCD for the terms is $(m-4)(3m-6)$.

$$\frac{2m}{m-4} - \frac{m^2+7m+4}{3m^2-18m+24} = \frac{4m}{3m-6}$$

$$\frac{(m-4)(3m-6)(2m)}{(m-4)} - \frac{(m-4)(3m-6)(m^2+7m+4)}{3m^2-18m+24} = \frac{(m-4)(3m-6)(4m)}{3m-6}$$

$$\frac{\overset{1}{\cancel{(m-4)}}(3m-6)(2m)}{\underset{1}{\cancel{m-4}}} - \frac{\overset{1}{\cancel{(m-4)}}\overset{1}{\cancel{(3m-6)}}(m^2+7m+4)}{\underset{1}{\cancel{3m^2-18m+24}}} = \frac{(m-4)\overset{1}{\cancel{(3m-6)}}(4m)}{\underset{1}{\cancel{3m-6}}}$$

$$(3m-6)(2m) - (m^2+7m+4) = (m-4)4m$$

$$6m^2 - 12m - m^2 - 7m - 4 = 4m^2 - 16m$$

$$5m^2 - 19m - 4 = 4m^2 - 16m$$

$$m^2 - 3m - 4 = 0$$

$$(m-4)(m+1) = 0$$

$$m - 4 = 0 \text{ or } m + 1 = 0$$

$$m = 4 \qquad m = -1$$

When solving a rational equation, any possible solution that results in a zero in the denominator must be excluded from the list of solutions.

Check each solution by substituting into the original equation.

Since $m = 4$ results in a zero in the denominator, it is extraneous. So, the solution is -1.

Talk About It!
How could you identify the extraneous solution without substituting each possible solution back into the original equation? Explain.

Check

Solve $\frac{3x}{x-4} - \frac{x^2-7x-4}{x^2-16} = \frac{5}{x+4}$.

Go Online You can complete an Extra Example online.

Example 3 Mixture Problem

FIRST AID **Rubbing alcohol, a commonly used first aid antiseptic, typically contains 70% isopropyl alcohol. Suppose you are adding a 50% isopropyl alcohol liquid to 200 milliliters of a liquid that is 91% isopropyl alcohol. How much of a 50% isopropyl alcohol liquid should be added to create rubbing alcohol that is 70% isopropyl alcohol?**

Step 1 Estimate the solution.

Since $(0.5 + 0.91) \div 2 = 0.705$, or 70.5%, creating a new liquid with 70% isopropyl alcohol should require a similar amount of 50% and 91% isopropyl alcohol liquids. So, around 200 milliliters of the 50% isopropyl alcohol liquid should be added to the 91% liquid.

Step 2 Write an equation for the concentration of the new liquid.

Complete the table. Let x be the amount of 50% isopropyl alcohol liquid that is added.

	Original	Added	New
Isopropyl Alcohol (mL)	$0.91(200)$	$0.5(x)$	$0.91(200) + 0.5x$
Total Amount of Liquid (mL)	200	x	$200 + x$

The percentage of isopropyl alcohol in the new liquid must equal the amount of isopropyl alcohol divided by the total amount of new liquid.

$$\text{Percentage of isopropyl alcohol in liquid} = \frac{\text{amount of isopropyl alcohol}}{\text{total amount of liquid}}$$

$$\frac{70}{100} = \frac{0.91(200) + 0.5x}{200 + x}$$

Step 3 Solve the equation.

$$\frac{70}{100} = \frac{0.91(200) + 0.5x}{200 + x} \quad \text{Original equation}$$

$$\frac{70}{100} = \frac{182 + 0.5x}{200 + x} \quad \text{Simplify the numerator.}$$

$$100(200 + x)\left(\frac{70}{100}\right) = 100(200 + x)\left(\frac{182 + 0.5x}{200 + x}\right) \quad \text{Multiply by LCD, } 100(200 + x).$$

$$\overset{1}{\cancel{100}}(200 + x)\left(\frac{70}{\underset{1}{\cancel{100}}}\right) = 100(\underset{1}{\cancel{200 + x}})\left(\frac{182 + 0.5x}{\underset{1}{\cancel{200 + x}}}\right) \quad \text{Divide common factors.}$$

$$(200 + x)70 = 100(182 + 0.5x) \quad \text{Simplify.}$$

$$14{,}000 + 70x = 18{,}200 + 50x \quad \text{Distributive Property}$$

$$x = 210 \quad \text{Solve.}$$

Step 4 Check for reasonableness and interpret the solution.

So, 210 milliliters of a 50% isopropyl liquid added to the 200 milliliters of 91% liquid creates a 70% isopropyl alcohol liquid. The answer is reasonable because it is close to our estimate.

Problem-Solving Tip

Make a Table When there is more than one element in a problem, it may be helpful to organize the information in a table when writing expressions and equations relating the items. Tables can be useful when solving mixture, work, and distance problems involving rational equations.

Think About It!

How long does it take the bat to travel with the wind and against the wind? Is this reasonable? Justify your argument.

Study Tip

Assumptions Although a bat's speed likely varies throughout its flight, assuming that a bat flies at a constant rate of speed allows us to create a reasonable model for the speed of the bat with and against the wind. We must also assume that the bat is flying directly with or against the wind during its entire flight.

Example 4 Distance Problem

BATS **Mexican free-tailed bats have an average flight speed of 25 miles per hour. Suppose it takes one of the bats 5 hours to fly 121.8 miles round trip one night. Assuming that the bat flew at a constant speed, determine the speed of the wind.**

The bat flies 121.8 miles round trip, or 60.9 miles each way. Use the formula relating distance, rate, and time in the form $t = \frac{d}{r}$ to write an equation for the total time. Let w represent the speed of the wind.

$$\frac{60.9}{25+w} + \frac{60.9}{25-w} = 5$$

$$(25+w)(25-w)\left(\frac{60.9}{25+w}\right) + (25+w)(25-w)\left(\frac{60.9}{25-w}\right) = (25+w)(25-w)5$$

$$\cancel{(25+w)}_1(25-w)\left(\frac{60.9}{\cancel{25+w}_1}\right) + (25+w)\cancel{(25-w)}_1\left(\frac{60.9}{\cancel{25-w}_1}\right) = (25+w)(25-w)5$$

$$(25-w)60.9 + (25+w)60.9 = (625-w^2)5$$

$$1522.5 - 60.9w + 1522.5 + 60.9w = 3125 - 5w^2$$

$$3045 = 3125 - 5w^2$$

$$16 = w^2$$

$$w = 4 \text{ or } -4$$

The only viable solution for the speed of the wind is 4 miles per hour.

Example 5 Work Problem

AGRICULTURE **If it takes a 24-row planter 10 hours to plant a field and 6 hours if a 16-row planter is also used, how long would it take to plant the field if only the 16-row planter were used?**

Let p represent the time for the 16-row planter to plant alone.

$$\frac{1}{10} + \frac{1}{p} = \frac{1}{6}$$ Write the equation.

$$30p\left(\frac{1}{10}\right) + 30p\left(\frac{1}{p}\right) = 30p\left(\frac{1}{6}\right)$$ Multiply by the LCD, 30p.

$$\overset{3}{\cancel{30}}p\left(\frac{1}{\cancel{10}_1}\right) + 30\overset{1}{\cancel{p}}\left(\frac{1}{\cancel{p}_1}\right) = \overset{5}{\cancel{30}}p\left(\frac{1}{\cancel{6}_1}\right)$$ Divide common factors.

$$3p + 30 = 5p$$ Simplify.

$$30 = 2p$$ Subtract from each side.

$$15 = p$$ Divide each side by 2.

It would take the 16-row planter 15 hours to plant the field alone.

Go Online You can complete an Extra Example online.

Learn Solving Rational Inequalities

A **rational inequality** contains at least one rational expression. You can use these steps to solve rational inequalities.

Key Concept • Solving Rational Inequalities

Step 1 State the excluded values. These are the values for which the denominator is 0.

Step 2 Solve the related equation.

Step 3 Use the values determined from the previous steps to divide a number line into intervals.

Step 4 Test a value in each interval to determine which intervals contain values that satisfy the inequality.

Example 6 Solve a Rational Inequality

Solve $\frac{x}{2} - \frac{5}{x+1} > \frac{x-4}{3}$.

Step 1 Find the excluded values.

When $x = -1$, the denominator of $\frac{5}{x+1}$ is 0. The excluded value is -1.

Step 2 Solve the related equation.

$$\frac{x}{2} - \frac{5}{x+1} = \frac{x-4}{3}$$

$$6(x+1)\left(\frac{x}{2}\right) - 6(x+1)\left(\frac{5}{x+1}\right) = 6(x+1)\left(\frac{x-4}{3}\right)$$

$$3x^2 + 3x - 30 = 2x^2 - 6x - 8$$

$$x^2 + 9x - 22 = 0$$

$$(x-2)(x+11) = 0$$

$$x = 2 \text{ or } -11$$

Steps 3 and 4 Divide a number line into intervals. Test each interval.

Draw vertical lines at the excluded value and at the solution points.

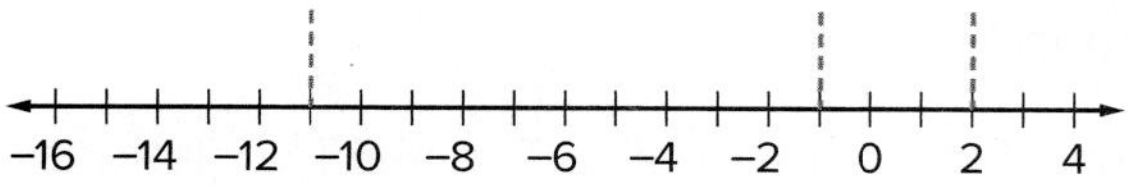

Test $x = -13$. False

Test $x = -4$. True

Test $x = 0$. False

Test $x = 3$. True

The statement is true for $x = -4$ and $x = 3$. Therefore, the solution is $-11 < x < -1$ or $x > 2$.

Check

Solve $-\frac{6}{x} - \frac{x-2}{4} > \frac{3-x}{3}$.

Go Online You can complete an Extra Example online.

Talk About It

Describe how you would graph the solution set on a number line.

Apply Example 7 Write and Solve a Rational Inequality

CLOTHING **Jamila runs an online store that sells custom hats. She spends $378 on an embroidery machine. In addition, it costs Jamila $1.85 to make each hat. How many hats will Jamila have to make so that the average cost per hat is less than or equal to $5?**

1 What is the task?

Describe the task in your own words. Then list any questions that you may have. How can you find answers to your questions?

I need to find the number of hats that Jamila needs to make so that the average cost to produce each hat is less than $5. I know that the fixed cost of buying the embroidery machine is $378 and the cost to make each hat is $1.85. How can I represent the situation with an inequality?

2 How will you approach the task? What have you learned that you can use to help you complete the task?

I will write an inequality and identify the excluded values. Then I will solve the related equation, divide the number line into intervals, and determine which intervals satisfy the inequality.

3 What is your solution?

Use your strategy to solve the problem.

What inequality represents the situation?

$$\frac{1.85h + 378}{h} \leq 5$$

Are there any excluded values? If so, identify them.

Yes; 0 is excluded.

What are the solutions to the inequality? What solutions are viable in the context of the situation?

$x < 0$ or $x \geq 120$ is the solution to the inequality. The viable solutions are limited to $x \geq 120$.

4 How can you know that your solution is reasonable?

Write About It! Write an argument that can be used to defend your solution.

Because the embroidery machine was so expensive, it would require a large number of hats to be made for the average cost to decrease. As Jamila makes more hats, the average cost per hat goes down. Therefore, it makes sense that it would require 120 hats or more to get the average cost down to $5 or less.

Practice

Go Online You can complete your homework online.

Example 1

Solve each equation. Check your solutions.

1. $\frac{2x+3}{x+1} = \frac{3}{2}$

2. $\frac{-12}{y} = y - 7$

3. $\frac{9}{x-7} - \frac{7}{x-6} = \frac{13}{x^2 - 13x + 42}$

4. $\frac{13}{y+3} - \frac{12}{y+4} = \frac{18}{y^2 + 7y + 12}$

5. $\frac{14}{x-2} - \frac{18}{x+1} = \frac{22}{x^2 - x - 2}$

6. $\frac{2}{a+2} + \frac{10}{a+5} = \frac{36}{a^2 + 7a + 10}$

Example 2

7. $\frac{x}{2x-1} + \frac{3}{x+4} = \frac{21}{2x^2 + 7x - 4}$

8. $\frac{2}{y-5} + \frac{y-1}{2y+1} = \frac{2}{2y^2 - 9y - 5}$

9. $\frac{x-8}{2x+2} + \frac{x}{2x+2} = \frac{2x-3}{x+1}$

10. $\frac{12p+19}{p^2 + 7p + 12} - \frac{3}{p+3} = \frac{5}{p+4}$

11. $\frac{2f}{f^2 - 4} + \frac{1}{f-2} = \frac{2}{f+2}$

12. $\frac{8}{t^2 - 9} + \frac{4}{t+3} = 1$

Examples 3–5

13. SOLUTION Evita adds a 75% acid solution to 8 milliliters of solution that is 15% acid.
    a. Write a function that represents the percent of acid in the resulting solution, where x is the amount of 75% acid solution added.
    b. How much 75% acid solution should be added to create a solution that is 50% acid?

14. FLIGHT TIME The distance between John F. Kennedy International Airport and Los Angeles International Airport is about 2500 miles. Let S be the airspeed of a jet, which is the speed at which the jet would be traveling in still air. The ground speed, which indicates how fast the jet is traveling over the land surface, is affected by the airspeed and the wind. Traveling in the same direction as the wind increases the ground speed; and traveling in the opposite direction decreases ground speed.
    a. Suppose the current wind speed is 100 miles per hour at the altitude at which the jet will fly. Write an equation for S if it takes 2 hours and 5 minutes longer to fly between New York and Los Angeles against the wind versus flying with the wind.
    b. Solve the equation in **part a** for S.
    c. Write and solve an equation and find how much longer it would take to fly against the wind between New York and Los Angeles if the wind speed increases to 150 miles per hour and the airspeed of the jet is 525 miles per hour.

15. CONSTRUCTION It takes Rosita 32 hours to drywall a basement by herself and 18 hours if Paola helps her. How long would it take Paola to drywall the basement by herself? Round your answer to the nearest hour.

16. BASEBALL A major league baseball team plays 162 games a season. Halfway through the season, one team has won 60% of their games. What percent of their remaining games would the team need to win to end the season winning 75% of their games?

17. CURRENT The distance for the Valleyview River Cruise trip is 28 miles round trip. The boat travels 10 miles per hour in still water. Because of the current, it takes longer to cruise upstream than downstream.

a. Define a variable and write an equation to represent the situation if it takes 30 minutes longer to cruise upstream versus downstream.

b. What is the speed of the current?

18. YARDWORK Each week Imani and Demond must mow their 4-acre yard. When they use both their 36-inch mower and 42-inch mower, it takes them 2 hours. When the 36-inch mower is out for repairs, it takes them $3\frac{1}{4}$ hours. How long would the job take if the 42-inch mower were broken?

Example 6

Solve each inequality. Check your solutions.

19. $3 - \frac{4}{x} > \frac{5}{4x}$

20. $\frac{5}{3a} - \frac{3}{4a} > \frac{5}{6}$

21. $\frac{x-2}{x+2} + \frac{1}{x-2} > \frac{x-4}{x-2}$

22. $\frac{3}{4} - \frac{1}{x-3} > \frac{x}{x+4}$

23. $\frac{x}{5} + \frac{2}{3} < \frac{3}{x-4}$

24. $\frac{x}{x+2} + \frac{1}{x-1} < \frac{3}{2}$

Example 7

25. ORIGAMI For prom, Muna wants to fold 1000 origami cranes. She is asking volunteers to help and does not want to make anyone fold more than 15 cranes.

a. Write an inequality to represent this situation, if N is the number of people enlisted to fold cranes.

b. What is the minimum number of people that will satisfy the inequality in **part a**?

Le Do/istock/Getty Images

26. PROM Caleb manages the budget for his school's junior prom. His class has spent $1250 for the prom venue, $625 for a DJ, and $1470 for decorations. They will also serve dinner before the dance, which costs $12 per student. If he wants to keep the cost of prom tickets less than $20, how many students will need to buy tickets?

Mixed Exercises

27. HEIGHT Fabiana is 8 inches shorter than her sister Pilar, or 12.5% shorter than Pilar. How tall is Fabiana?

Solve each equation or inequality. Check your solutions.

28. $\frac{x-2}{x+4} > \frac{x+1}{x+10}$

29. $\frac{3}{k} - \frac{4}{3k} = 0$

30. $2 - \frac{3}{v} = \frac{5}{v}$

31. $n + \frac{3}{n} < \frac{12}{n}$

32. $\frac{1}{2m} - \frac{3}{m} < -\frac{5}{2}$

33. $\frac{1}{2x} < \frac{2}{x} - 1$

34. $\frac{6}{x+2} = \frac{x-7}{x+2} + \frac{1}{4}$

35. $\frac{t-5}{t-3} = \frac{t-3}{t+3} + \frac{1}{t-3}$

36. $3 + \frac{2}{t} > \frac{8}{t}$

37. $\frac{6}{m+5} > 2$

38. NUMBER THEORY The ratio of two less than a number to six more than that number is 2 to 3. What is the number?

39. BUSINESS The Franklin Electronics Company has determined that, after its first 50 wireless speakers are produced, the average cost of producing one speaker can be approximated by $C(x) = \frac{60x + 17{,}000}{x - 50}$, where x represents the number of speakers produced. Consumer research has indicated that the company should charge $80 per speaker in order to maximize its profit.

a. The company wants to determine how many speakers must be produced and sold in order to ensure that the revenue from each one is greater than the average cost of producing each one. Write an inequality whose solution represents this information.

b. Solve your inequality and interpret your solution in the context of the situation.

40. ANALYZE Let $f(x) = -x - 2$ and $g(x) = -\frac{3}{x}$.

a. Find any x values for which $f(x)$ or $g(x)$ is undefined and any solutions to $f(x) = g(x)$.

b. Write the solution set to the inequality $f(x) \leq g(x)$. Graph your results on a number line.

c. Explain why each value from **part a** should be included in or excluded from the solution set of the inequality.

41. USE TOOLS Graph $f(x) = \frac{2}{x^2 - x}$ and $g(x) = \frac{1}{x-1}$ on a graphing calculator. Then use the graph to solve $\frac{2}{x^2 - x} = \frac{1}{x-1}$. Explain why this method works.

42. USE A MODEL It takes one fuel line 3 hours to fill an oil tanker. How fast must a second fuel line be able to fill the oil tanker so that, when used together, the two lines will fill the tanker in 45 minutes?

43. The prom committee is spending \$1000 for the venue and DJ plus \$35 per person for the catered meal. How many students must attend the prom to keep the ticket price at or below \$60?

44. STATISTICS The harmonic mean is often the most accurate average of numbers in situations involving rates or ratios. The number x is the harmonic mean of y and z if $\frac{1}{x}$ is the average of $\frac{1}{y}$ and $\frac{1}{z}$.

a. Write an equation for the harmonic mean of 50 and 75.

b. What is the harmonic mean of 50 and 75?

45. CONSTRUCT ARGUMENTS Hideki used the rational function $f(x) = \frac{x^2 + x - 12}{x^2 - 2x - 3}$ to determine the solutions of $f(x) = 0$. He found that x could be 3 or -4. Do you agree? Justify your argument.

46. ANALYZE Given $\frac{5x}{x-2} = 7 + \frac{10}{x-2}$.

a. Solve the equation for x. Is your solution extraneous? Explain how you know.

b. How can you use a graphical method to check your answer for **part a**?

47. FIND THE ERROR A student is trying to solve the inequality $\frac{1}{x} < 2$. Refer to the work shown at the right. Do you agree? Explain your reasoning. What is one method this student could have used to check their work?

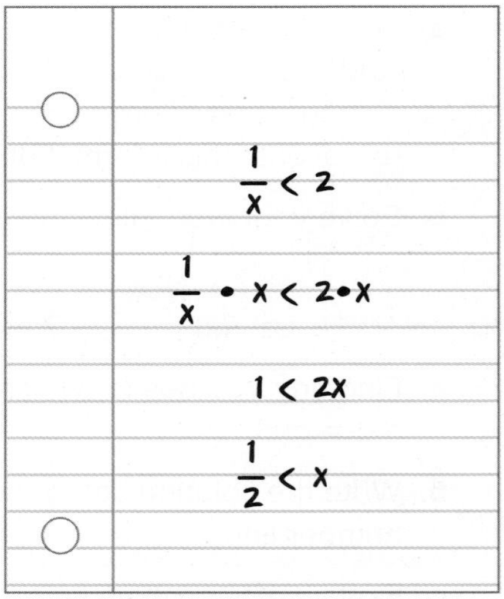

$\frac{1}{x} < 2$
$\frac{1}{x} \cdot x < 2 \cdot x$
$1 < 2x$
$\frac{1}{2} < x$

48. CREATE Give an example of a rational equation that can be solved by multiplying each side of the equation by $4(x + 3)(x - 4)$.

49. PERSEVERE Solve $\frac{1 + \frac{9}{x} + \frac{20}{x^2}}{\frac{x^2 - 25}{x^2}} = \frac{x + 4}{x - 5}$.

50. ANALYZE While using the table feature on a graphing calculator to explore $f(x) = \frac{1}{x^2 - x - 6}$, the values -2 and 3 say "ERROR." Explain its meaning.

51. WRITE Why should you check solutions of rational equations and inequalities?

Review

Essential Question

How are the rules for operations with rational numbers applied to operations with rational expressions and equations?

Module Summary

Lessons 9-1 and 9-2

Operations with Rational Expressions

- To multiply rational expressions, multiply the numerators and the denominators.
- To divide rational expressions, multiply by the reciprocal of the divisor.
- To simplify a complex fraction, first rewrite it as a division expression.
- To add or subtract rational expressions, find the least common denominator. Rewrite each expression with the LCD. Then add or subtract.

Lessons 9-3 and 9-4

Graphing Reciprocal and Rational Functions

- A reciprocal function has an equation of the form $f(x) = \frac{n}{b(x)}$, where n is a real number and $b(x)$ is a linear expression that cannot equal 0.
- A rational function has an equation of the form $f(x) = \frac{a(x)}{b(x)}$, where $a(x)$ and $b(x)$ are polynomial functions and $b(x) \neq 0$.

Lesson 9-5

Variation

- y varies directly as x if there is some nonzero constant k such that $y = kx$.
- y varies jointly as x and z if there is some nonzero constant k such that $y = kxz$.
- y varies inversely as x if there is some nonzero constant k such that $xy = k$ or $y = \frac{k}{x}$, where $x \neq 0$ and $y \neq 0$.
- If you know that y varies directly as x, that y varies inversely as z, and one set of values, you can use a proportion to find another set of corresponding values.

Lesson 9-6

Rational Equations and Inequalities

- To solve a rational equation, it is often easier to first eliminate the fractions. You can eliminate the fractions by multiplying each side of the equation by the least common denominator (LCD).
- To solve a rational inequality, state the excluded values. Next, solve the related equation. Use the values determined from the previous steps to divide a number line into intervals. Test a value in each interval to determine which intervals contain values that satisfy the inequality.

Study Organizer

Foldables

Use your Foldable to review this module. Working with a partner can be helpful. Ask for clarification of concepts as needed.

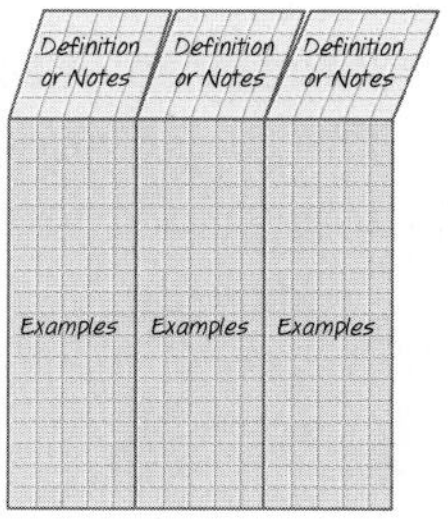

Test Practice

1. **OPEN RESPONSE** Multiply. (Lesson 9-1)

$$\frac{42x^2}{15} \cdot \frac{5y^2}{7x^3}$$

2. **OPEN RESPONSE** Simplify $\frac{3x^2 + x - 2}{2x^2 - 3x - 5}$. (Lesson 9-1)

3. **MULTIPLE CHOICE** The volume of a rectangular solid is $x^3 - 9x$ cubic inches.

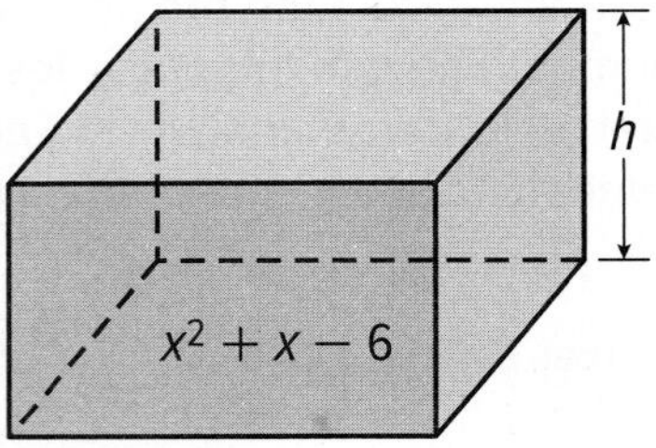

If the area of the base of the solid is $x^2 + x - 6$ square inches, what is the height of the solid in terms of x? (Lesson 9-1)

A. $\frac{x(x - 3)}{x - 2}$

B. $\frac{x(x + 3)}{x + 2}$

C. $\frac{x^2 - 3}{x - 2}$

D. $\frac{x - 9}{-6}$

4. **OPEN RESPONSE** Add. (Lesson 9-2)

$$\frac{5}{x^2 - 9} + \frac{2}{x^2 + 10x + 21}$$

5. **OPEN RESPONSE** Subtract. (Lesson 9-2)

$$\frac{2y}{3x^2} - \frac{5}{y}$$

6. **MULTIPLE CHOICE** When two electrical resistors with resistance R_1 and R_2 are connected in parallel, their combined resistance is $\frac{1}{\frac{1}{R_1} + \frac{1}{R_2}}$. (Lesson 9-2)

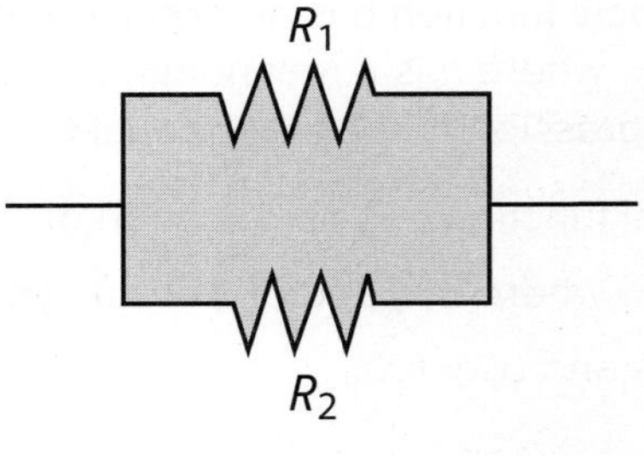

Which of these expressions is equivalent?

A. $\frac{R_1 + R_2}{2}$

B. $\frac{R_1R_2}{R_1 + R_2}$

C. $\frac{R_1 + R_2}{R_1R_2}$

D. $\frac{2}{R_1 + R_2}$

7. MULTIPLE CHOICE Add. (Lesson 9-2)

$$\frac{x^2-1}{x^2-3x-10}+\frac{x+3}{x+2}$$

A. $\frac{2x+x+2}{(x-4)(x+2)}$

B. $\frac{3x^2+20x+32}{(x+2)(3x+10)}$

C. $\frac{x^2+x+2}{(x+2)(x-5)}$

D. $\frac{2x^2-2x-16}{(x-5)(x+2)}$

8. OPEN RESPONSE What is the excluded value of x for the function $f(x)=\frac{2}{3x-18}$? (Lesson 9-3)

9. GRAPH Graph $f(x)=\frac{3}{x-2}+1$. (Lesson 9-3)

10. OPEN RESPONSE Identify the values of a, h, and k. Then write a function for the graph $g(x)=\frac{a}{x-h}+k$. (Lesson 9-3)

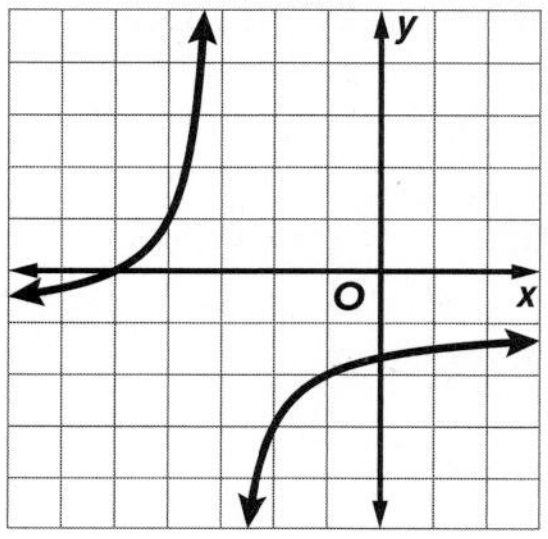

11. OPEN RESPONSE Consider the graph of $f(x)=\frac{x^2-4x+4}{x^2-4}$. (Lesson 9-4)

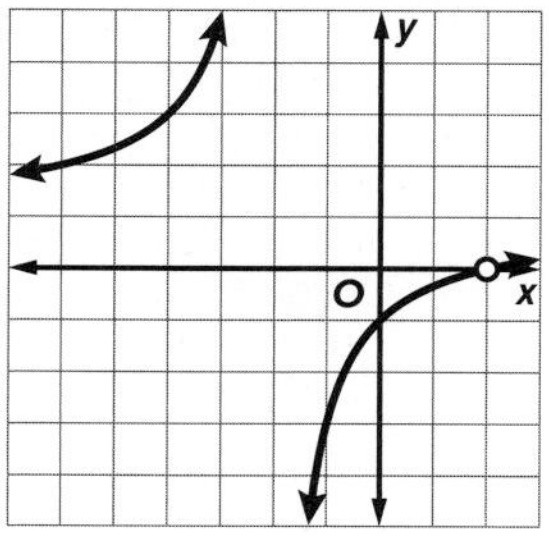

a) Find x-and y-intercepts.

b) Find any asymptotes.

12. OPEN RESPONSE Consider the graph of $f(x)=\frac{2x+6}{x+4}$. (Lesson 9-4)

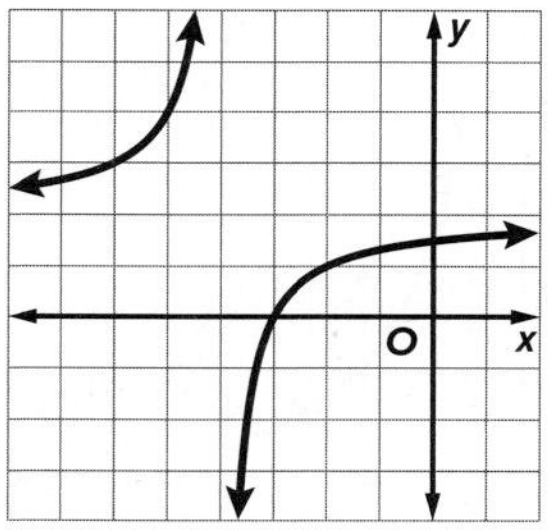

a) Describe the end behavior.

b) Find the domain.

13. MULTIPLE CHOICE Suppose y varies jointly as x and z, and $y = 72$ when $x = 4$ and $z = -6$. What is the value of y when $x = -5$ and $z = 7$? (Lesson 9-5)

A. −105 C. −24

B. −35 D. 105

14. MULTIPLE CHOICE Suppose y varies inversely as x, and $y = 10.2$ when $x = 3$. What is the value of y when $x = 2$? (Lesson 9-5)

A. 3.4

B. 6.8

C. 15.3

D. 30.6

15. OPEN RESPONSE The time required to drive a fixed distance is inversely proportional to the speed. The graph shows the time, in minutes, required to drive from Centerville to Concord, depending on the speed, in miles per hour. (Lesson 9-5)

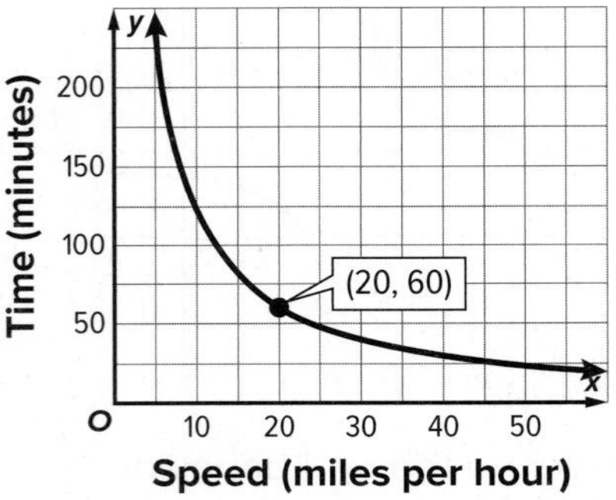

Find the speed for each given time required.

a) 48 minutes

b) 40 minutes

16. MULTI-SELECT Select all solutions of the equation shown. (Lesson 9-6)

$$\frac{x}{x-2} + \frac{1}{x+3} = \frac{3}{2}$$

A. −7

B. −3

C. −2

D. 2

E. 3

F. 7

17. MULTIPLE CHOICE Which number line shows the solution of $\frac{x-3}{x+1} < 5$? (Lesson 9-6)

A. 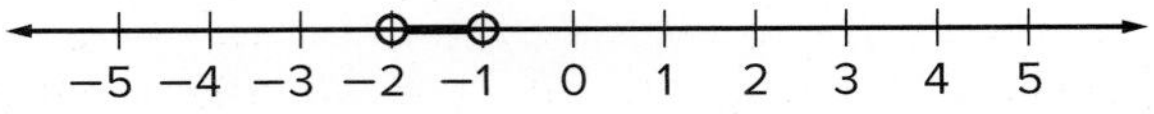

B. 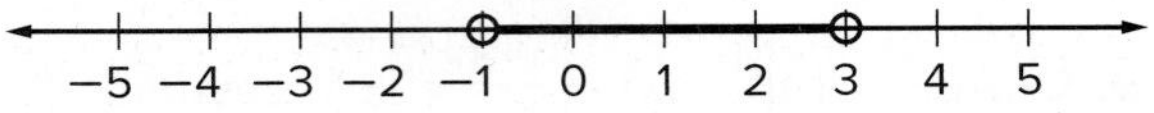

C. 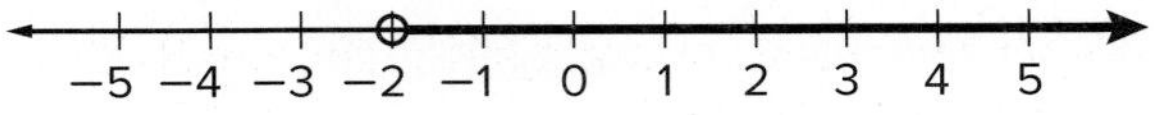

D. 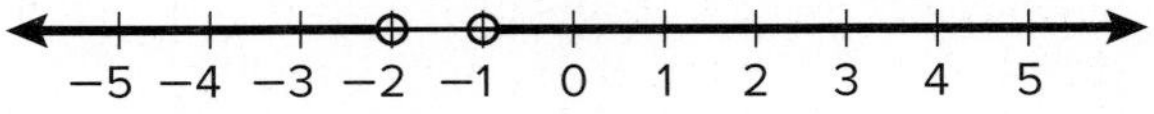

18. MULTIPLE CHOICE Working alone, an experienced painter can paint a room in 7 hours. When an apprentice helps, they can paint the same room in 4 hours. Approximately how many hours would it take the apprentice, working alone, to paint the room? Round to the nearest tenth of an hour. (Lesson 9-6)

A. 3 hours

B. 9.3 hours

C. 11 hours

D. 28.6 hours

Module 10

Inferential Statistics

e Essential Question

How can data be collected and interpreted so that it is useful to a specific audience?

What Will You Learn?

How much do you already know about each topic **before** starting this module?

KEY

— I don't know. — I've heard of it. — I know it!

	Before			After		
classify sampling methods and identify bias in samples and survey questions						
explore experimental probabilities and fair decision making						
compare theoretical and experimental probabilities						
describe and compare distributions by finding their mean and standard deviation						
classify variables and analyze probability distributions to determine expected outcomes						
use statistics and normal distributions to analyze data						
use sample data to make inferences about populations						

Foldable Make this Foldable to help you organize your notes about inferential statistics. Begin with an $8\frac{1}{2}$" × 11" sheet of paper.

1. **Fold** in half lengthwise.
2. **Fold** the top to the bottom.
3. **Open.** Cut along the second fold to make two tabs.
4. **Label** each tab as shown.

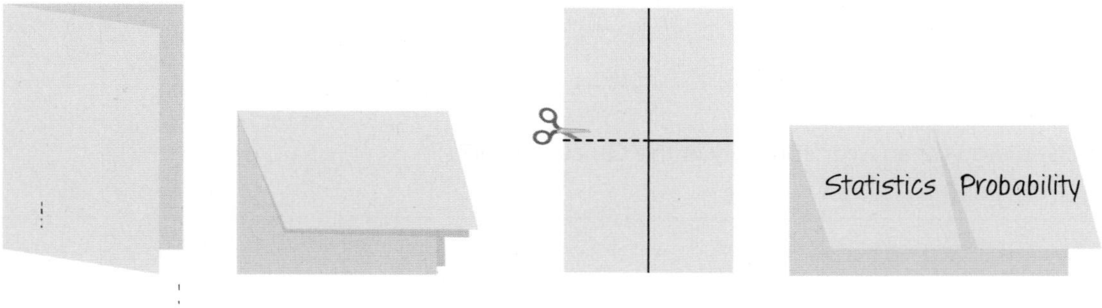

What Vocabulary Will You Learn?

- bias
- confidence interval
- continuous random variable
- descriptive statistics
- discrete random variable
- distribution
- experiment
- experimental probability
- inferential statistics
- maximum error of the estimate
- normal distribution
- observational study
- outcome
- outlier
- parameter
- population proportion
- probability distribution
- probability model
- sample space
- sampling error
- simulation
- standard deviation
- standard error of the mean
- standard normal distribution
- statistic
- survey
- symmetric distribution
- theoretical probability
- variance
- *z*-value

Are You Ready?

Complete the Quick Review to see if you are ready to start this module. Then complete the Quick Check.

Quick Review

Example 1

The number of days it rained in each month over the past year are shown below. Find the mean, median, and mode.

4, 2, 9, 16, 13, 9, 8, 9, 7, 6, 8, 5

Mean $\bar{x} = \frac{4+2+9+16+13+9+8+9+7+6+8+5}{12}$ or 8 days

Median 2, 4, 5, 6, 7, 8, 8, 9, 9, 9, 13, 16

$\frac{8+8}{2}$ or 8 days

Mode The value that occurs most often in the set is 9, so the mode of the data set is 9 days.

Example 2

A die is rolled and a coin is tossed. What is the probability that the die shows a 1 and the coin lands tails up?

$P(1, \text{tails}) = \frac{1}{6} \cdot \frac{1}{2}$ or $\frac{1}{12}$

Quick Check

Find the mean, median, and mode of each data set.

1. number of customers at a store each day: 78, 80, 101, 66, 73, 92, 97, 125, 110, 76, 89, 90, 82, 87
2. quiz scores for the first grading period: 88, 70, 85, 92, 88, 77, 98, 88, 70, 82
3. the number of touchdowns scored by a player over the last 10 seasons: 7, 5, 10, 12, 4, 10, 11, 6, 9, 3

A die is rolled and a coin is tossed. Find each probability.

4. $P(4, \text{heads})$
5. $P(\text{odds}, \text{tails})$
6. $P(2 \text{ or } 4, \text{heads})$
7. $P(2, 4, \text{or } 6, \text{tails})$

How Did You Do?

Which exercises did you answer correctly in the Quick Check?

Random Sampling

Learn Randomness and Bias

Statistics is an area of mathematics that deals with collecting, analyzing, and interpreting data.

A characteristic of a population is called a **parameter**. Because it is not always practical to collect data from a large population, it may be collected from a representative sample. A characteristic of a sample is called a **statistic**.

A **population** consists of all the members of a group of interest. A **bias** is an error that results in a misrepresentation of a population.

Key Concept • Types of Samples	
simple random	Each member has an equal chance of being selected.
systematic	Members are selected according to a specified interval.
self-selected	Members volunteer to be included in the sample.
convenience	Members that are readily available are selected.
stratified	Members are selected from similar, nonoverlapping groups.

Today's Goals
- Classify sampling methods and identify bias in samples and survey questions.
- Distinguish among sample surveys, experiments, and observational studies.

Today's Vocabulary
parameter

statistic

population

bias

survey

experiment

observational study

Example 1 Classify a Random Sample

APPS **The administration of Jefferson High School developed a free app to help their students with studying and organization. At the end of the school year, the students were asked to complete an online questionnaire about their satisfaction with the app. Classify the sample.**

Step 1 Identify the population.

the entire student body represented by the administration

Step 2 Identify the sample.

the students who responded to the online questionnaire

Step 3 Classify the sample.

The sample is self-selected.

Study Tip
Interests When identifying interests served by a survey, ask: Who is performing the survey? What is the purpose of the survey, and who would benefit from the results leaning one way or another?

Check

MEMBERSHIP A club member would like to ask the other members in her state's council about new requirements for first-year members. Classify each sampling method.

a. She asks every 5th member on the council roster.

b. She asks every other club member in her 25-member club.

c. She splits the council into four age groups and randomly asks 20 people from each group.

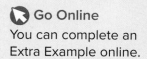

Go Online
You can complete an Extra Example online.

Example 2 Identify Sample Bias

TRAFFIC **The traffic commission took a random sample of travelers using Route 15 and determined the average daily commute on this road during rush hour. After viewing the results, the commission proposed expanding the road from 2 lanes to 4 lanes. Identify bias and potential interests in the sample.**

Route 15 Congestion
Distance: 10.6 miles
Average Commute: 26.3 minutes
Standard Deviation: 4.4 minutes
Sample Size: 10

Step 1 Identify the purpose of the sample.
The purpose is to evaluate the length of the commute in order to determine whether road expansion is necessary.

Step 2 Identify the bias in the sample. While the sample is random, a sample size of 10 is too small to accurately reflect all of the cars that use the road.

Step 3 Identify interests. The bias in the sample size may lead to a sample average that does not reflect the true average commute of the population. This bias could serve the interests of contractors who hope to win the business of expanding the road.

Example 3 Identify Biased Questions

FOOD **A restaurant owner wants to determine which kinds of meals she should add to her menu: vegetarian options or more traditional meat-based dishes. She releases a survey to her customers asking the following question: *Do you prefer a plain salad or a delicious steak?* Identify any bias in the question.**

Step 1 Identify the purpose of the question. The purpose is to determine which dishes are most popular with customers.

Step 2 Identify the bias in the sample. The question provides descriptions of each kind of food, and the description for the steak favors it over the salad.

Learn Types of Studies

In a **survey**, data are collected from responses given by members of a group regarding their characteristics, behaviors, or opinions.

In an **experiment**, the sample is divided into two randomly selected groups, the *experimental group* and the *control group*. The effect on the experimental group is then compared to the control group.

In an **observational study**, members of a sample are measured or observed without being affected by the study.

To determine when to use each type of study, think about how the data will be obtained and how the study will affect the participants. Each method of study relies on random selection of a sample to ensure that it accurately represents the population.

Go Online You can complete an Extra Example online.

Think About It!

How large a sample size do you think would be needed to accurately reflect the population?

Use a Source

Find a biased online survey question. Explain why it is biased, and rephrase the question.

Talk About It!

Describe an example of a biased observational study. Explain how the study is biased.

Example 4 Classify Study Types

UNIFORMS **A research team wants to test new football uniform designs and their appeal to young adults. They randomly select 100 young adults to view the different uniforms. The research team observes and records the reactions to the uniforms.**

Step 1 What is the purpose of the study?

The purpose is to determine if the new uniforms will be appealing to young adults.

Step 2 Does this situation represent a *survey,* an *experiment*, or an *observational study*?

This is a(n) observational study because the participants are observed without being affected by the study.

Step 3 Identify the sample and population.

The sample is the 100 young adults involved in the study. The population is all young adults.

Check

Match each study subject with the corresponding study type that should be used.

___?___ A principal wants to determine the favorite after-school activity of his students.

___?___ A researcher wants to determine whether young adults would be interested in a new line of smartwatches entering the market.

___?___ A teacher wants to determine whether bright colors affect the test-taking abilities of high school students.

A. observational study

B. census

C. survey

D. experiment

Example 5 Design a Survey

EXERCISE **The owner of a company would like to convert a conference room into an exercise room. She would like to survey her employees to see if they would use the exercise room. Design the survey.**

Step 1 State the objective of the survey.

The objective of the survey is to determine the employees' interest in converting a conference room into an exercise room.

Step 2 Identify the population.

The population is all employees of the company that work in the building.

Step 3 Write an unbiased survey question.

Sample answer: "If available, would you use an exercise room in the building if it were created by converting a conference room?"

Think About It!

In Example 5, seven out of the 85 people who were surveyed said they would use the exercise room. What conclusion can you draw from these results? Should the owner proceed with building the exercise room? Support your conclusion.

Go Online

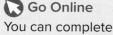

You can complete an Extra Example online.

Example 6 Draw Conclusions from a Study

TECHNOLOGY **Researchers are testing different types of virtual reality goggles. Each randomly selected participant could choose a type of goggle. The researchers observed the number of people who chose each type and the number of positive reactions.**

Type	Number of People	Positive Reactions (%)
A	11	45%
B	17	53%
C	45	91%
D	52	92%
E	6	33%

Step 1 Identify the type of study. The study is a(n) observational study.

Step 2 Draw a conclusion based on the results.

The Types C and D goggles seemed to be preferred by the most people in the study, and received a very high percent of positive reactions. Therefore, the researchers should recommend that those types of goggles be produced.

Think About It!

Suppose all 6 of the people who used the Type E goggles had a positive reaction. Do you think this should cause the researchers to alter their recommendation? Justify your reasoning.

Example 7 Identify Bias in Studies

Identify any flaws in the design of the experiment.

Experiment: A teacher wants to study the impact of music on testing. She gives a test to her Algebra 2 class in a quiet environment. She then gives the same test to her Geometry class, but this time with music playing softly in the room.

Results: She found that the mean score of the tests taken by the Geometry class was higher than the mean score of the tests taken by the Algebra 2 class. She concluded that music is beneficial when testing.

Step 1 Identify the control and experimental groups.

Because the Algebra 2 class takes the test in a quiet environment, and the change the teacher wants to measure is the effect of music on test-taking, the Algebra 2 class is the control group. Because the Geometry class takes the test with the music playing, they are the experimental group.

Step 2 Identify any flaws in the experiment.

The experimental and control groups were not randomly selected, and the two groups have more than one difference: beside the presence of music while taking the test, the classes are different and may be at different skill levels.

Step 3 Correct any flaws.

The teacher could use a sample of Algebra 2 or Geometry students at similar ability levels and randomly select members of the control and experimental groups from this sample.

Go Online

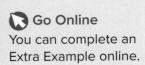

You can complete an Extra Example online.

Practice

Go Online You can complete your homework online.

Example 1

Identify each sample, and suggest a population from which it was selected. Then classify the sample as *simple random, systematic, self-selected, convenience,* or *stratified*. Explain your reasoning.

1. Berton divides his sports T-shirts by team. Then he randomly selects four T-shirts from each team and records the size.

2. The project manager at a new business inspects every tenth smart phone produced to check that it is operating correctly.

3. A grocery store manager asks its customers to submit suggestions for items on the salad bar during the week.

Examples 2 and 3

Identify each sample or question as *biased* or *unbiased*. Explain your reasoning.

4. A random sample of eight people is asked to select their favorite food for a survey about Americans' food preferences.

5. Every tenth student at band camp is asked to name his or her favorite band for a survey about the campers.

6. Every fifth person entering a museum is asked to name his or her favorite type of book to read for a survey about reading interests of people in the city.

7. Do you think that the workout facility needs a new treadmill and racquetball court?

8. Which is your favorite type of music, pop, or country?

9. Are you a member of any after-school clubs?

10. Don't you agree that employees should pack their lunch?

Example 4

Determine whether each situation describes a *survey*, an *experiment*, or an *observational study*. Then identify the sample, and suggest a population from which it may have been selected.

11. An Internet service provider conducts an online study in which customers are randomly selected and asked to provide feedback on their customer service.

12. A research group randomly selects 100 business owners, half of whom started their own businesses, and compares their success.

13. A research group randomly chooses 50 people to participate in a study to determine whether exercising regularly reduces the risk of diabetes in adults.

14. An online video streaming service mails a questionnaire to randomly selected people across the country to determine whether they prefer streaming movies or sports.

Example 5

15. PROM A prom committee wants to conduct a survey to determine how many juniors and seniors are planning to attend prom. State the objective of the survey, suggest a population, and write two unbiased survey questions.

16. RESTAURANTS A restaurant manager wants to conduct a survey to determine what new food items to include on the restaurant's menu. State the objective of the survey, suggest a population, and write two unbiased survey questions.

Example 6

17. SPORTS DRINKS A sports drink company gives out free samples of their new sports drink at the mall. The percent of teens that take the sample is 62%. The percent of adults that take the sample is 20%. Identify the type of study. Draw a conclusion about the study based on the results. Explain your reasoning.

18. EDUCATION Researchers want to determine a teacher's effectiveness. They observe five class periods and record the difficulty of the lesson, students' reactions, and the overall effectiveness of the lesson. The results are shown in the table. Identify the type of study. Draw a conclusion about the study based on the results. Explain your reasoning.

Period	Difficulty of Lesson	Student Reaction	Effectiveness (%)
1	Easy	Engaged	91%
2	Difficult	Engaged	97%
3	Medium	Engaged	93%
4	Difficult	Not Engaged	60%
5	Easy	Not Engaged	55%

Example 7

Identify the control and experimental groups in the experiments. Identify and correct any flaws in the design of the experiment.

19. MORALE A school principal hopes to increase the morale of students. The principal selects some students at random. The principal gives each of the selected students a gift card to a local restaurant. The principal compares the morale of the students in each group.

20. SPORTS The owner of a professional sports team wants to see if new uniforms will improve ticket sales. The team opened the season by playing three away games. The fourth game was played at home, and the team wore the new uniforms. The owner then compares the number of tickets purchased for the home game with the number of tickets purchased for the first three games of the season.

Mixed Exercises

Classify each sample as *simple random, systematic, self-selected, convenience,* or *stratified.* Then determine whether each situation describes a *survey*, an *observational study*, or an *experiment.*

21. To determine the music preferences of their customers, the manager of a music store selected 10 customers in the store to participate in an interview.

22. Administrators at a community library want to know the type of materials patrons are most likely to use. Every Friday, they record the type of media each patron uses.

23. To determine whether the school should purchase new computer software, the technology team divides 50 students into two groups by age. Half of the students from each age group are randomly selected to complete an activity using the current computer software, and the other half of the students from each group complete the same activity using the new computer software. The students' actions are recorded and analyzed.

24. VIDEO GAMES A behavioral scientist studies the influence of video games on students' academic performance.

a. Describe an observational study the scientist can perform to study the influence.

b. Describe an experiment the scientist can perform to test the influence.

25. BAND A band director wants to compare the musical instruments played by males in band class with the musical instruments played by females. What type of study should the band director conduct?

26. USE A SOURCE Research to find an article online, in a magazine, or in a newspaper that shows the results of a survey or study. Analyze the data. Then analyze the summary of the report, or article. Explain whether the data in the report is biased.

CONSTRUCT ARGUMENTS **Explain whether an *experiment*, an *observational study*, or a *survey* is best for achieving the goal of the researcher. If applicable, describe the conditions for which another type of study might be used.**

27. The buyer of a new car wants to determine whether it will be more cost efficient to run the car she chose using E85 flex-fuel or regular gasoline.

28. A pizza shop owner wants to determine how satisfied his customers are with a new special that he has on the menu.

29. STATE YOUR ASSUMPTION A cat rescue group wants to determine whether feeding rescued stray cats a grain-free food helps them gain weight more quickly. They perform an experiment by giving a control group of 10 cats regular food and an experimental group of 10 cats the grain-free food. They record the cats' weight gains after two weeks, as shown in the table. Is the mean weight gain a good measure to determine whether the rescue group should switch food? State any assumptions you made to come to your conclusion.

Weight Gain (oz)			
Control Group		Experimental Group	
6	5	4	4
3	6	7	8
8	4	5	3
5	9	7	6
7	4	9	5

30. FIND THE ERROR Mia and Esteban are describing one way to improve a survey. Are Mia and Esteban correct? Explain your reasoning.

Mia

The survey should include as many people in the population as possible.

Esteban

The sample for the survey should be chosen randomly. Several random samples should be taken.

31. ANALYZE Aaliyah selects 10 students at random from each class in her high school and asks the following survey question. Analyze the strengths and weaknesses of the question.
How do students feel about the new dress code?

32. WRITE Why are accurate studies important to companies?

ANALYZE **Determine whether each statement is *true* or *false*. If false, explain.**

33. To save time and money, population parameters are used to estimate sample statistics.

34. Observational studies and experiments can both be used to study cause-and-effect relationships.

35. CREATE Design an observational study. Identify the objective of the study, define the population and sample, collect and organize the data, and calculate a sample statistic.

36. WRITE What factors should be considered when determining whether a given statistical study is reliable?

Using Statistical Experiments

Explore Fair Decisions

Online Activity Use a real-world situation to complete the Explore.

INQUIRY How can you use probability to make a fair decision?

Learn Theoretical and Experimental Probability

Recall that the probability of an event is the ratio of the number of outcomes in which a specified event occurs to the total number of possible outcomes.

The **theoretical probability** is what is expected to happen, and the **experimental probability** is calculated using data from an actual experiment.

The Law of Large Numbers states that as a sample size increases, the variation in a data set decreases. So, as the sample size increases, experimental probabilities will more closely resemble their associated probabilities.

Example 1 Find Probabilities

A student tossed a fair eight-sided die 200 times and recorded the results. Find the theoretical and experimental probabilities of rolling an 8.

Number on Die	Frequency
1	28
2	19
3	24
4	22
5	21
6	18
7	26
8	42

Theoretical Probability
The theoretical probability is what is expected to happen. Because the die is fair, each side has a $\frac{1}{8}$ chance of being the result. Thus, the theoretical probability is $\frac{1}{8}$, or 12.5%.

Experimental Probability
The experimental probability is based on the data collected from the experiment. Because the die was thrown 200 times and 42 of those throws landed on 8, the experimental probability is $\frac{42}{200}$, or 21%.

Go Online You can complete an Extra Example online.

Today's Goals
- Compare theoretical and experimental probabilities.
- Determine whether models are consistent with results from simulations of real-life situations.

Today's Vocabulary
theoretical probability

experimental probability

probability model

simulation

Talk About It!
A popular gambling expression, *The house always wins,* argues that ultimately a casino wins in the long term, despite possible wins by gamblers in the short term. Explain how this expression relates to the Law of Large Numbers. Use the phrases *experimental probability* and *theoretical probability* in your explanation.

Think About It!

What does it mean for a die to be fair?

Check

A student tossed a fair die 150 times and recorded the results. Find the theoretical and experimental probabilities of rolling a 5. Write your answer as a percentage rounded to the nearest tenth, if necessary.

Number	Frequency
1	51
2	16
3	23
4	19
5	24
6	17

The theoretical probability of rolling a 5 is __?__ %.

The experimental probability of rolling a 5 is __?__ %.

Explore Simulations and Experiments

Online Activity Use the interactive tool to complete the Explore.

INQUIRY How can simulations help you analyze the results of an experiment?

Think About It!

A claim was made that the probability of a specific coin landing on heads cannot be 0.5 because it was flipped 20 times in a row, and it never landed on heads. How would you use a simulation to test this claim?

Learn Simulations

A **probability model** is a mathematical representation of a random event. It consists of the sample space and the probability of each outcome.

A **simulation** uses a probability model to imitate a process or situation so it can be studied. Coins, dice, and random number generators can be used to conduct simulations of experiments that might be difficult or impractical to perform in reality.

Key Concept • Conducting a Simulation	
Step 1	Determine each possible outcome of the situation to be simulated and its theoretical probability.
Step 2	Describe a probability model that accurately represents the theoretical probability of each outcome.
Step 3	Define the trial for the situation, and state the number of trials to be conducted.
Step 4	Conduct the simulation.
Step 5	Analyze the results.

Example 2 Design and Run a Simulation

PROMOTIONS **A company that produces bottles of water runs a promotion in which 1 out of every 8 bottle caps wins the customer a free bottle of water.**

Step 1 Describe the probability model.

There are 2 possible outcomes: winning a free bottle of water or not winning a free bottle of water.

The theoretical probability of winning a free bottle is $\frac{1}{8}$.

The theoretical probability of not winning a free bottle is $\frac{7}{8}$.

Step 2 Define the trial for the situation, and state the number of trials to be conducted.

Of the six available probability tools, identify the tool(s) that can be used for this scenario. The tools that can model theoretical probabilities of $\frac{1}{8}$ and $\frac{7}{8}$ are circled below.

(cards) a single coin toss (marbles) standard die

(random number generator) (spinner)

All the outcomes can be represented by the numbers 1 through 8, so a random number generator is a good tool to use. Because there is a $\frac{1}{8}$ chance of winning and a $\frac{7}{8}$ chance of not winning, the number 1 can be used to represent winning, and the numbers 2 through 8 can be used to represent not winning.

One trial will represent one bottle cap. The simulation can consist of any number of trials. Run the simulation for 100 trials, and the number generator will return 100 random numbers between 1 and 8.

Step 3 Conduct the simulation.

For a simulation using a random number generator, determine the number of trials and the minimum and maximum values. Then conduct a simulation.

Number of trials: 100

Minimum Value: 1 Maximum Value: 8

Step 4 Analyze the results.

What are your results? That is, what is the experimental probability that a random bottle cap is a winner?

Sample answer: In one simulation, the number 1 was generated 12 times in 100 trials, or 12% of the time.

Go Online You can complete an Extra Example online.

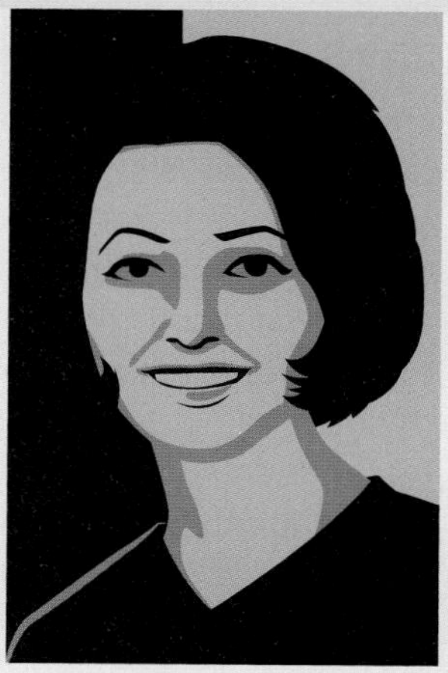

Math History Minute

Statistician **Joan Ginther** has been called "the luckiest woman in the world" after winning multi-million dollar lottery jackpots on four separate occasions for a total of $20.4 million. The odds of winning four major jackpots are about one in eighteen septillion.

Go Online to use a sketch to conduct a simulation.

Study Tip

Modeling Simulations often provide a safe and efficient problem-solving strategy in situations that otherwise may be costly, dangerous, or impossible to solve using theoretical techniques. Simulations should involve data that are easier to obtain than the actual data you are modeling.

Think About It!

Why do you think multiple simulations were needed to evaluate this model?

Go Online to use a sketch to conduct a simulation.

Example 3 Run and Evaluate a Simulation

HEALTH CLUBS **A health club ran a promotion in which they sold tickets to their customers, and 1 in every 10 tickets won free training sessions valued at $150. A customer complained that they bought 20 tickets in a row and did not win a prize. Run and evaluate a simulation, and decide whether the customer has a legitimate complaint.**

Step 1 Describe the probability model.

There are 2 possible outcomes: winning the prize or not winning the prize.

The theoretical probability of winning the prize is $\frac{1}{10}$.

The theoretical probability of not winning the prize is $\frac{9}{10}$.

Step 2 Define the trial for the situation, and state the number of trials to be conducted.

How can a spinner be used to simulate this scenario, where the spinner landing in one sector represents winning and in another represents losing?

The sector that represents winning should contain 10% of the spinner's area.

The sector that represents losing should contain 90% of the spinner's area.

One trial will represent one ticket. Run the simulation for 100 trials.

Step 3 Conduct the simulation.

Spin the spinner 100 times and determine the number of times the spinner lands in the winning sector.

Step 4 Analyze the results and evaluate the model.

What are your results? That is, what is the experimental probability of winning?

In one simulation, the spinner landed in the winning sector 11 times in 100 trials, or 11% of the time.

How can the simulation be evaluated to address the concern of the customer?

Run the simulation multiple times to see whether there are instances where the spinner landed in the losing sector at least 20 times in a row.

Suppose the simulation was conducted 50 times, and the spinner landed in the losing sector at least 20 times in a row in 36 of the 50 simulations, or 72% of the simulations. In this case, the customer's experience is not unlikely, and their complaint is not legitimate.

Go Online You can complete an Extra Example online.

Practice

Go Online You can complete your homework online.

Example 1

1. A student spun a spinner with 4 equal sections 100 times and recorded the results.

Spinner Section	Frequency
Red	35
Blue	38
Green	13
Yellow	14

   a. Find the theoretical probability of spinning blue. Write your answer as a percentage rounded to the nearest tenth, if necessary.

   b. Find the experimental probability of spinning blue. Write your answer as a percentage rounded to the nearest tenth, if necessary.

2. A student flipped a coin 125 times and recorded the results.

Coin Result	Frequency
Heads	73
Tails	52

   a. Find the theoretical probability of the coin landing on heads. Write your answer as a percentage rounded to the nearest tenth, if necessary.

   b. Find the experimental probability of the coin landing on heads. Write your answer as a percentage rounded to the nearest tenth, if necessary.

3. A fair 6-sided die is rolled 150 times.

Number on Die	Frequency
1	32
2	18
3	27
4	16
5	33
6	24

   a. Find the theoretical probability of rolling a 3. Write your answer as a percentage rounded to the nearest tenth, if necessary.

   b. Find the experimental probability of rolling a 3. Write your answer as a percentage rounded to the nearest tenth, if necessary.

Examples 2 and 3

4. INTERNET Tiana sells handmade earrings online. Last month she sold 60% of her inventory. Design and run a simulation that can be used to estimate the probability of selling inventory.

5. PROGRAMMING Lamar designed a soccer computer game. He coded the program such that a player will make a goal on 35% of the attempts. Paola is testing the game and thinks there may be an error in the game's programming. She attempted to make 30 goals and only 4 were successful. Run and evaluate a simulation, and decide whether Paola is correct.

Mixed Exercises

6. Describe a way to simulate the test results for someone who guesses on a multiple choice test with 4 choices for each question.

7. Describe a spinner that could be used to simulate an event with 2 possible outcomes, one with a probability of 60%.

8. How should numbers be assigned to use a random number generator for the following situation?
Probability of choosing blue: 40%
Probability of choosing green: 20%
Probability of choosing red: 30%
Probability of choosing purple: 10%

9. HAIR COLOR A survey of Hartford High School students found that 50% of the students had brown hair, 40% had blonde hair, and 10% had red hair. Design a simulation that can be used to estimate the probability that a randomly selected student will have one of these hair colors. Assume that each student's hair color will fall into one of these categories.

10. DESIGN Marta designed a simulation using a fair coin and a spinner with six equal sections. Describe a situation that Marta may be simulating.

11. TRAVEL According to a survey done by a travel agency, 40% of their cruise clients went to the Bahamas, 25% went to Mexico, 20% went to Alaska, and 15% went to Greece. When Jinhai and Jayme were asked to design a simulation that could be used to estimate the probability of a client going to each of these places, Jinhai designed the spinner with central angles shown. Jayme wants to assign integers to each destination and use the random number generator on her graphing calculator. Which model would allow more trials to be conducted in a shorter period of time?

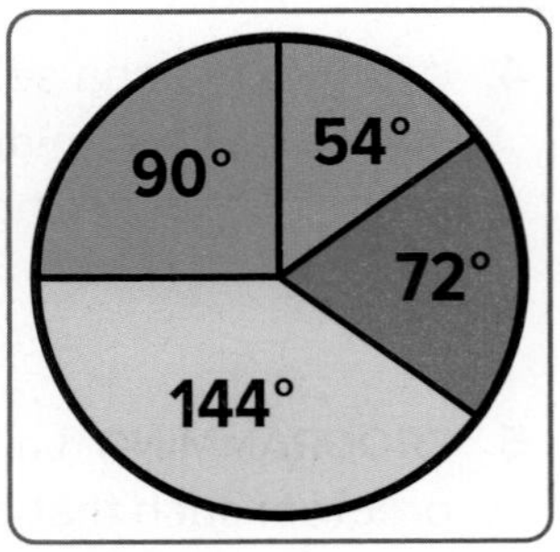

12. EARTHQUAKES Geologists conclude that there is a 62% probability of a magnitude 6.7 or greater quake striking the San Francisco Bay region before 2032. Design a simulation that can be used to estimate the probability that such an earthquake occurs.

13. **STATE YOUR ASSUMPTION** Sarah kept track of the color of each car that passed her house. She recorded the car color for 250 cars.

Car Color	Frequency
White	61
Black	56
Silver	72
Red	41
Blue	20

a. Find the theoretical probability of a blue car passing by Sarah's house. Write your answer as a percentage rounded to the nearest tenth, if necessary.

b. Find the experimental probability of a blue car passing by Sarah's house. Write your answer as a percentage rounded to the nearest tenth, if necessary.

c. What assumption did you make to find the theoretical probability?

14. **USE TOOLS** Describe an event that can be simulated using each method.

a. A coin is tossed.

b. A four-sided die is rolled.

15. **USE A MODEL** Forty-two students participated in a raffle with three prizes. Fifteen of the participants were members of the basketball team. All three winners were members of the basketball team. The other students claim the raffle was unfair.

a. Describe a simulation you could use to test the results of the raffle.

b. Run a simulation and record your results in a table.

c. Based on your results, do you think the raffle was fair? Explain your reasoning.

16. **CARDS** A package of cards has an equal number of red, black, white, blue, green, yellow, orange, and purple cards. Judy kept track of the color card she randomly drew after shuffling the cards. She recorded the card color results for 500 trials.

Card Color	Frequency
Red	63
Black	55
White	39
Blue	62
Green	52
Yellow	78
Orange	56
Purple	95

a. Find the theoretical probability of Judy drawing any color card. Write your answer as a percentage rounded to the nearest tenth, if necessary.

b. Find the experimental probability of Judy drawing a green card. Write your answer as a percentage rounded to the nearest tenth, if necessary.

c. Find the experimental probability of Judy drawing a yellow card. Write your answer as a percentage rounded to the nearest tenth, if necessary.

17. USE A MODEL A volleyball player serves an ace 66% of the time in a set.

a. Design a model or simulation you could use to estimate the probability that the volleyball player will serve an ace on the next serve.

b. Run a simulation and record your results in a table.

c. Examine the data. Is the data from your simulation consistent with the probability model?

18. CONSTRUCT ARGUMENTS There are 2 candidates running for class president. Of the six models discussed in this lesson—cards, coin toss, marbles, dice, random number generator, or spinner—which model would be most appropriate for a simulation to determine how students will vote? Justify your conclusion.

19. ANALYZE Jevon tosses a coin several times and finds that the experimental probability for the coin landing heads up is 25%. Should Jevon be concerned about the fairness of the coin? Explain your reasoning.

Higher-Order Thinking Skills

20. ANALYZE Is the experimental probability of heads when a coin is tossed 15 times *sometimes*, *never*, or *always* equal to the theoretical probability? Justify your argument.

21. PERSEVERE *True* or *false*: If the theoretical probability of an event is 1, the experimental probability of the event cannot be 0. Explain your reasoning.

22. WRITE What should you consider when using the results of a simulation to make a prediction?

23. ANALYZE An experiment has three equally likely outcomes, *A*, *B*, and *C*. Is it possible to use the spinner shown in a simulation to predict the probability of outcome *C*? Explain your reasoning.

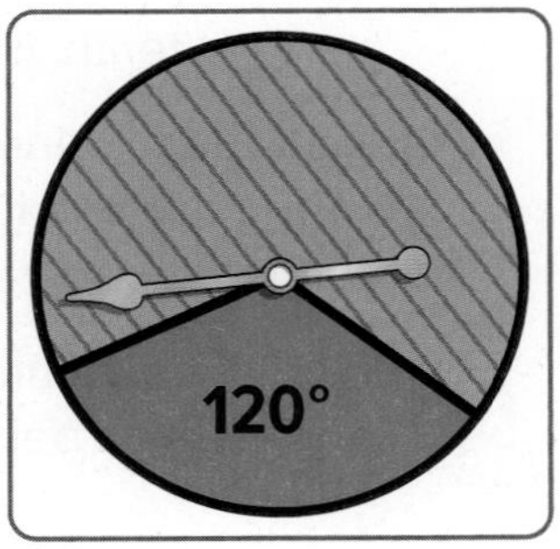

24. WRITE Can tossing a coin *sometimes*, *always*, or *never* be used to simulate an experiment with two possible outcomes? Explain your reasoning.

Analyzing Population Data

Learn Describing Distributions

Measures of center and measures of spread or variation are often called **descriptive statistics**. To analyze data sets, we often look at their **distribution**, which is a graph or table that shows the theoretical frequency of each possible data value.

Symmetric Distribution	Negatively Skewed Distribution	Positively Skewed Distribution
The mean and median are approximately equal, and the data are approximately symmetric about the center.	Typically, the median is greater than the mean and less data is on the left side of the graph.	Typically, the mean is greater than the median and less data is on the right side of the graph.
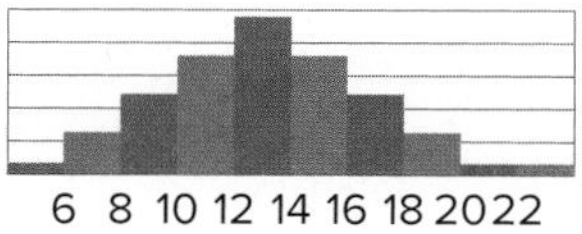 mean = 13.1 median = 13.2	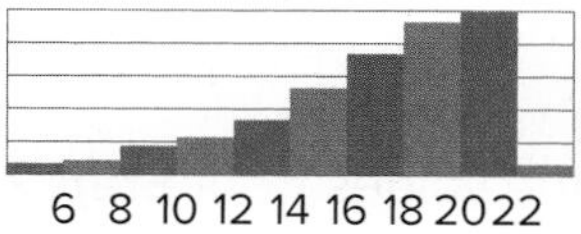 mean = 14.8 median = 16.7	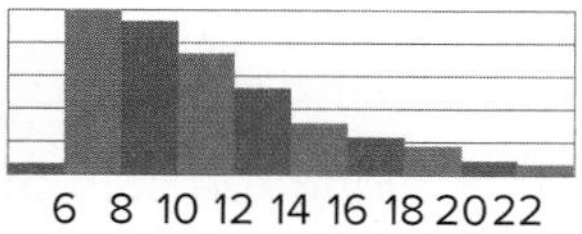 mean = 12.7 median = 11.6

A bimodal distribution has two distinct peaks. When a distribution is symmetric, the mean accurately reflects the center of the data. However, when a distribution is skewed, the mean is not as reliable. **Outliers**, which are extremely high or low values when compared to the rest of the values in a data set, have a strong effect on the mean of a data set. When a distribution is skewed or has outliers, the mean lies away from the majority of the data toward the tail and may not accurately describe of the data.

In a symmetric distribution, the data are distributed symmetrically about the center. The center is measured with the mean, and the spread is measured with **variance** and **standard deviation**, which describe how the data deviate from the mean.

Key Concept • Finding the Standard Deviation

Step 1 Find the mean μ.

Step 2 Find the square of the difference between each data value x_n and the mean, $(x_n - \mu)^2$.

Step 3 Find the sum of all the values in **Step 2.**

Step 4 Divide the sum by the number of values in the set of data n. This value is the variance.

Step 5 Take the square root of the variance.

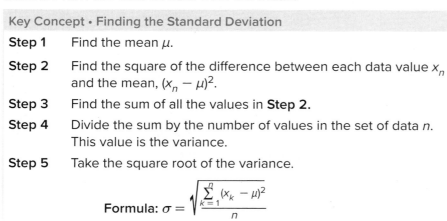

Formula: $\sigma = \sqrt{\dfrac{\sum_{k=1}^{n} (x_k - \mu)^2}{n}}$

Today's Goal

- Describe distributions by finding their mean and standard deviation.

Today's Vocabulary

descriptive statistics

distribution

symmetric distribution

outlier

variance

standard deviation

Watch Out!

Symmetric Distributions
Every set of data has a mean and standard deviation, but they are only accurate measurements for symmetric distributions.

Example 1 Find a Standard Deviation

TRACK **A coach recorded the race times of each track member for a 400-meter race. Find and interpret the standard deviation of the data.**

400m Practice Times (seconds)	
57.1	55.9
59.3	54.9
54.6	50.3
55.2	53.5

Step 1 Find the mean, μ.

$$\mu = \frac{57.1 + 59.3 + 54.6 + 55.2 + 55.9 + 54.9 + 50.3 + 53.5}{8}$$
$$= 55.1$$

The mean running time for the team is 55.1 seconds.

Step 2 Find the squares of the differences, $(\mu - x_n)^2$.

$(55.1 - 57.1)^2 = 4.00$ $(55.1 - 55.9)^2 = 0.64$

$(55.1 - 59.3)^2 = 17.64$ $(55.1 - 54.9)^2 = 0.04$

$(55.1 - 54.6)^2 = 0.25$ $(55.1 - 50.3)^2 = 23.04$

$(55.1 - 55.2)^2 = 0.01$ $(55.1 - 53.5)^2 = 2.56$

Step 3 Find the sum.

Find the sum of the values from **Step 2**.

$4.00 + 17.64 + 0.25 + 0.01 + 0.64 + 0.04 + 23.04 + 2.56$
$= 48.18$

Step 4 Divide by the number of values.

Divide the sum from **Step 3** by the number of running times.

$\frac{48.18}{8} = 6.0225$ This is the variance.

Step 5 Take the square root of the variance.

$\sqrt{6.0225} \approx 2.45$ This is the standard deviation.

The standard deviation is about 2.45. This is small compared with the run times, which means that the majority of the team members' times are close to the mean of 55.1 seconds, and almost all of the times will likely fall within 2 standard deviations of the mean, or between 50.2 and 60.0 seconds.

Check

How do you know that your solution is reasonable? (*Hint:* The majority of the data should be within one standard deviation of the mean, and almost all of the data should be within two.)

Over half of the times are within one standard deviation of the mean, which is the interval [52.65, 57.55]. All of the data are within two standard deviations of the mean, or the interval [50.2, 60].

Go Online You can complete an Extra Example online.

Talk About It!

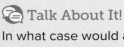

In what case would a standard deviation of 2.45 be considered large? Provide a real-world example.

Check

BREAKFAST The manager of a cafeteria tracked the number of waffles that were sold during breakfast every day. Find and interpret the standard deviation of the data. Round your answers to the nearest tenth.

Number Sold	
36	42
48	40
44	56
57	53

Part A The standard deviation is approximately __?__ waffles.

Part B The majority of the sales were between __?__ and __?__ waffles.

Example 2 Calculate Statistics

Use a graphing calculator to find the mean and standard deviation of the set of data.

26, 12, 15, 20, 17, 19, 18, 16, 14, 23, 13, 18, 19, 20, 22

On a TI-84, enter the data by pressing **stat** and selecting the **Edit** option. Enter all data in List 1 **(L1)**. Then, press **stat**, access the **CALC** menu, select **1-Var Stats**, and press **enter**.

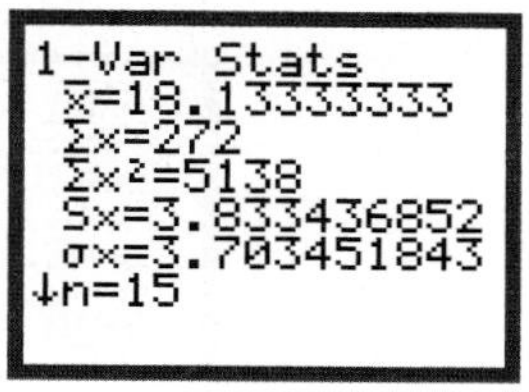

The mean is represented by $\bar{x}$ and standard deviation is represented by σ_x.

The mean is about 18.1, and the standard deviation is about 3.7.

Check

Use a graphing calculator to find the mean and standard deviation of the set of data. Round your answers to the nearest tenth.

68, 55, 52, 63, 54, 47, 56, 56, 58, 62, 63, 52

The mean is about __?__, and the standard deviation is about __?__.

Go Online You can complete an Extra Example online.

Think About It!

How would an outlier affect the statistics you calculated? Consider the case where the outlier is extremely high and the case where the outlier is extremely low.

Study Tip

Terminology On a calculator, some of the menus are labeled **STATS** or **Statistics**. You have learned that while statistics are used to describe a sample, parameters are used to describe a population. So even though you are working with populations in this lesson, you can still use the statistics features of the calculator to analyze the population.

Example 3 Compare Distributions

COURSES **The exam scores for two Algebra 2 classes are shown. Use the mean and standard deviation to compare the distributions.**

Mr. Jackson's Algebra 2 Class				
95	96	95	94	90
98	92	91	94	92
92	92	94	93	93
96	95	88	94	90

Ms. Hettrick's Algebra 2 Class				
73	86	83	79	78
66	82	79	83	77
70	85	84	76	91
68	73	89	92	84

Part A Find the mean and standard deviation of the distributions.

Measures of center and spread are used to compare symmetric distributions of data.

Using a TI-84 calculator, enter the data for Mr. Jackson's class in **L1** and the data for Ms. Hettrick's class in **L2**. To display the statistics, press **stat**, access the **CALC** menu, and select **1-Var Stats**. For Mr. Jackson's class, enter **L1** and press **enter** again. For Ms. Hettrick's class, enter **L2** and press **enter** again instead.

Mr. Jackson's Class

```
1-Var Stats
 x̄=93.2
 Σx=1864
 Σx²=173834
 Sx=2.397366977
 σx=2.336664289
↓n=20
```

Ms. Hettrick's Class

```
1-Var Stats
 x̄=79.9
 Σx=1598
 Σx²=128710
 Sx=7.362064928
 σx=7.17565328
↓n=20
```

Part B Compare the distributions.

Mr. Jackson's class has a mean score of 93.2 and standard deviation of 2.3. Ms. Hettrick's class has a mean score of 79.9 and a standard deviation of 7.2. This means that Mr. Jackson's class scored higher on average, and that the scores in Ms. Hettrick's class are more spread out from the mean.

Check

TRAINING Twenty trainees were provided with a 100-point pretest before their training began. Once their training was complete, they were each given a 100-point posttest that was very similar to the pretest. Use the mean and standard deviation to compare the distributions.

Pretest Scores				
80	60	85	75	90
70	65	70	75	50
55	65	80	95	65
90	60	85	70	75

Posttest Scores				
73	88	84	88	85
83	93	94	94	93
93	97	99	96	81
89	96	93	93	93

The average score was higher for the ___?___. The scores were more varied in the ___?___.

Go Online You can complete an Extra Example online.

Practice

Go Online You can complete your homework online.

Example 1

1. **BARBER** A barber wants to purchase new professional shears from a Web site. The prices of all of the shears are shown in the table. Use the standard deviation formula to find and interpret the standard deviation of the data. Round your answers to the nearest cent.

Cost of Shears ($)			
50	165	55	79
84	68	38	42

2. **READING** Ms. Sanchez keeps track of the total number of books each student in the book club reads during the school year. Use the standard deviation formula to find and interpret the standard deviation of the data. Round your answers to the nearest tenth.

Books Read		
9	6	12
8	9	14
10	13	8

Example 2

Use a graphing calculator to find the mean and standard deviation of each set of data. Round to the nearest tenth.

3. 20, 23, 24, 23, 22, 25, 21, 23, 24, 22, 21, 23, 22, 24

4. 150, 153, 125, 136, 143, 150, 166, 148, 150, 173, 150, 153, 143, 142, 153

5. 9.0, 3.8, 6.2, 7.1, 5.3, 6.2, 7.1, 8.2, 7.1, 4.5, 9.9, 8.2

6. 3350, 2800, 4525, 2150, 2800, 2150, 3350, 1800, 5250, 3975, 580, 2800

Example 3

7. **TRACK** Twenty track members were timed on a one-mile run before their season began and again after their season was completed. Use the mean and standard deviation to compare the distributions.

Mile Time Before Season (min)				
9	8.5	9.5	9	7
7.5	11	10.75	9.25	9
10.25	8	10	8.25	9
10.5	10	8.5	9	9.25

Mile Time After Season (min)				
7	6.5	7.5	6	5.5
8.5	8	7	7.25	7.5
6.75	6.5	7	5	6.5
9	7.5	8.25	6.5	7

8. **GRADES** Mr. Williams recorded students' test scores of his two Geometry classes. Use the mean and standard deviation to compare the distributions.

Geometry Class 1				
58	90	95	70	85
90	100	95	75	85
95	70	90	85	75
90	95	90	98	98

Geometry Class 2				
70	80	95	55	65
40	80	85	85	100
65	85	90	70	80
80	80	85	90	70

Mixed Exercises

Use a graphing calculator to find the mean and standard deviation of each set of data. Round to the nearest thousandth.

9. 0.2, 0.4, 0.35, 0.25, 0.3, 0.45, 0.3, 0.4, 0.65, 0.35, 0.15, 0.35, 0.4, 0.45, 0.35, 0.5, 0.6, 0.3

10. $\frac{1}{4}, \frac{5}{8}, \frac{1}{2}, \frac{3}{8}, \frac{5}{8}, \frac{3}{8}, \frac{1}{2}, 1, \frac{1}{2}, \frac{1}{8}, \frac{5}{8}, \frac{3}{4}, \frac{1}{2}, \frac{3}{4}, \frac{7}{8}, \frac{1}{4}$

11. TESTING A community college offers 10 dates for taking the entrance exam. An administrator records the number of attendees at each session. Find and interpret the standard deviation of the data. Round your answers to the nearest tenth.

Number of Students
98, 72, 92, 72, 102, 55, 99, 76, 117, 107

12. USE A MODEL Juan Carlos is researching the effect of tank size on the development of guppies. He measures each fish in both tanks. How can Juan Carlos use the mean and standard deviation to develop a conclusion in his research?

Length of Guppies in Small Tank (in.)				
2.5	1.8	2	1.5	1.5
1.7	1.6	1.5	1.4	1.5
1.5	1.4	1.2	1.5	2.2
1.4	1.5	1.6	1.5	1.8

Length of Guppies in Large Tank (in.)				
2.5	1.7	1.3	2.2	1.6
1.1	1.9	2.7	2.3	2.6
2.2	2.7	1.9	1.6	1.9
2.1	1.5	1.7	2.6	2.4

13. CONSTRUCT ARGUMENTS The manager of a fast food restaurant is analyzing the hourly wages of the employees. To make the wages more even, she wants to reduce the spread of the data that represent the wages by adding $1 to each employee's hourly wage. Is this an effective method to reduce the spread? Justify your response.

Employee Hourly Wages ($)			
11.20	11.05	9.80	10.88
10.75	10.30	11.15	11.57
9.92	10.65	12.00	10.32
10.15	9.45	10.55	9.95

14. WRITE How can the center and spread of a bimodal distribution, like the one shown, be described?

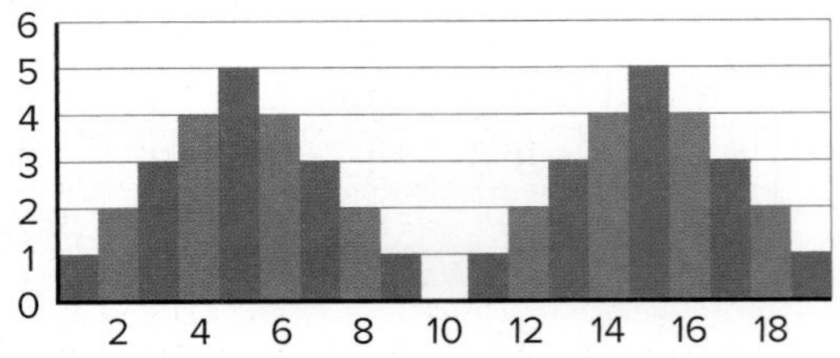

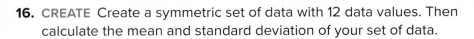

15. ANALYZE Approximate the mean and standard deviation of the distribution of data shown. Explain your reasoning.

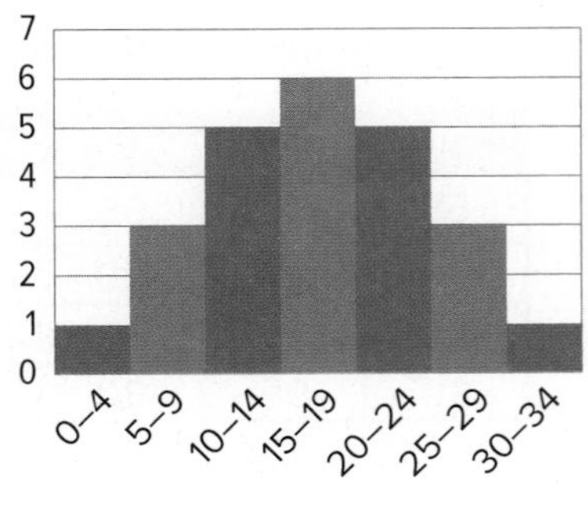

16. CREATE Create a symmetric set of data with 12 data values. Then calculate the mean and standard deviation of your set of data.

17. WRITE Explain how to find the standard deviation of a set of data without using technology.

Normal Distributions

Explore Probability Distributions

Online Activity Use a simulation to complete the Explore.

INQUIRY What is the relationship between the expected value of a discrete random variable and the mean of the distribution of that variable as the sample size increases?

Today's Goals

- Classify variables and analyze probability distributions to determine expected outcomes.
- Analyze normally distributed variables by using the Empirical Rule.
- Analyze standardized data and distributions by using *z*-values.

Today's Vocabulary

probability distribution
discrete random variable
continuous random variable
outcome
sample space
normal distribution
z-value
standard normal distribution

Learn Probability Distributions

A **probability distribution** is a mapping of each outcome of a statistical experiment to its probability of occurrence. It is similar to a frequency distribution, except that the probability of an event occurring is measured rather than the number of times it occurs.

A probability distribution is a statistical function that describes all of the possible values and probabilities that a random variable can take within a given range. These probabilities can be determined theoretically or experimentally.

The value of a random variable is the numerical outcome of a random event. An **outcome** is the result of a single event X. The **sample space** is the set of all possible outcomes in a distribution. The probability $P(X)$ represents the likelihood of the event occurring. A probability distribution for a particular random variable maps the sample space to the probabilities of the outcomes within the sample space.

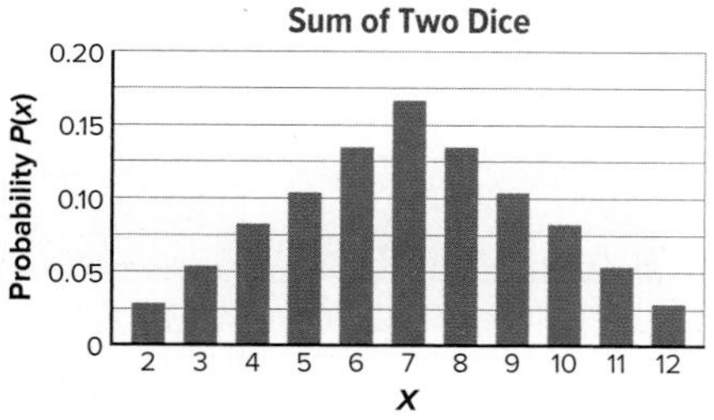

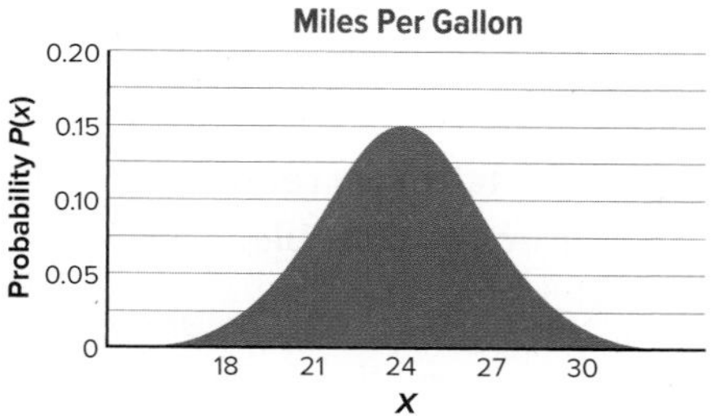

A **discrete random variable** is finite and can be counted. It is graphed with individual bars separated by space because no other values of X are possible. A **continuous random variable** is the numerical outcome of a random event that can take on any value. A continuous distribution is graphed with a histogram or as the area under a curve because any value under the curve is possible.

The probability distribution of a random variable X must satisfy the conditions:

- The probability of each value of X must be between 0 and 1.
- The sum of the probabilities of all of the values of X must equal 1.

Think About It!

Why must the probability of each value of a random variable be between 0 and 1?

Example 1 Classify Random Variables

Classify each random variable as *discrete* or *continuous*.

a. the number of songs on a random selection of smartphones

The number of songs on a random selection of smartphones is discrete because the number of songs is countable.

b. the air pressure in a random selection of basketballs

Air pressure is continuous because it can take on any value within a certain range. The air pressure of a standard basketball can be anywhere between 7.5 and 8.5 pounds per square inch.

Think About It!

Why is there a space between the bars for a graphical distribution of a discrete random variable?

Check

Classify each random variable as *discrete* or *continuous*.

200-meter swim times

number of students in class

individual archery scores

the amount of water in a bottle

Example 2 Analyze a Probability Distribution

MARKETING **A candy company had a promotion in which 4000 of their candy wrappers won the purchaser a cash prize. The frequency table shows the number of winning wrappers for each prize.**

Construct a relative frequency table, and graph the probability distribution.

Prize, *X*	Winners
\$5	1250
\$10	1000
\$20	750
\$50	600
\$100	300
\$500	80
\$1000	16
\$2500	4

Step 1 Construct a relative frequency table.

The relative frequency table converts the frequencies to probabilities. Divide each frequency by the total number of winning wrappers, 4000.

Prize (*X*)	Frequency	Relative Frequency
\$5	1250	0.3125
\$10	1000	0.2500
\$20	750	0.1875
\$50	600	0.1500
\$100	300	0.0750
\$500	80	0.0200
\$1000	16	0.0040
\$2500	4	0.0010

Go Online You can complete an Extra Example online.

Step 2 Graph the probability distribution.

The bars are separated on the graph because the distribution is discrete. Each unique outcome is indicated on the horizontal axis, and the probability of each outcome occurring is indicated on the vertical axis.

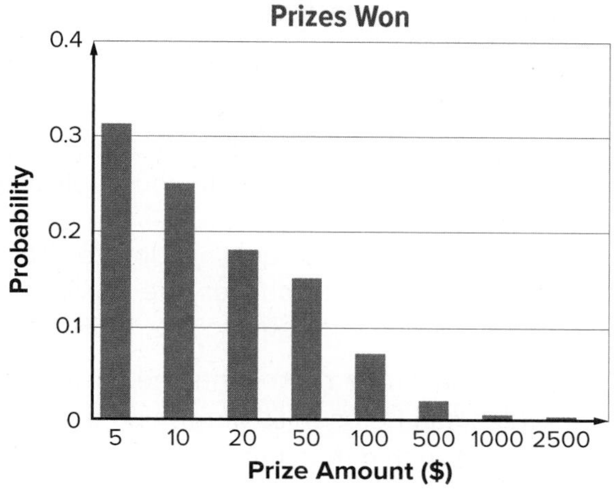

Example 3 Misleading Distributions

SALARY **A business publishes a report saying that most of its employees earn at least $40,000 annually. When asked to show the distribution of salaries, they provide a probability distribution. Identify any flaws in the representation of the probability distribution.**

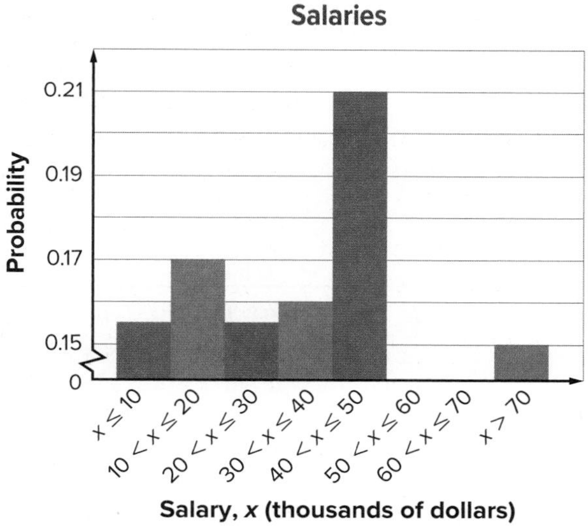

Step 1 Identify interests.

The company provides the distribution as evidence of fair salary to their employees. The company benefits from a distribution that makes the salary of $40,000 to $50,000 appear significantly more likely than the lower amounts.

Step 2 Examine the distribution.

The scale for the x-axis is potentially misleading. While most bins are evenly distributed, the bin for salaries greater than $70,000 may result in outliers because values significantly greater than $70,000 may be included.

The scale for the y-axis is misleading because of the small increments. The scale makes it appear that many more employees earn $40,000 to $50,000 than the lower salaries. However, most of the salary ranges have very similar probability.

Think About It!
What is the probability that a winning wrapper has a value of $500 or greater?

Think About It!
Describe an appropriate scale for the y-axis of the distribution.

Go Online You can complete an Extra Example online.

Learn Normal Distributions

The **normal distribution** is the most common continuous probability distribution of a random variable. A normal distribution occurs as a sample size increases due to the Law of Large Numbers.

Key Concept • Law of Large Numbers

As the number of trials increases, the experimental probability approaches the theoretical probability.

Key Concept • The Normal Distribution

- The graph of a normal distribution is continuous, bell-shaped, and symmetric with respect to the mean.
- The mean, median, and mode are equal and located at the center.
- The curve approaches, but never touches, the *x*-axis.
- The total area under the curve is equal to 1, or 100%.

The area under the normal curve represents the amount of data within a certain interval, which can be used to determine the probability that a random data value falls within that interval. This area is defined in relation to the mean and standard deviation of the data.

When a set of data is normally distributed, or approximately normal, the Empirical Rule can be used.

Study Tip

Normal Distributions
In order for a distribution to be approximately normal, there must be a large number of values in the data set.

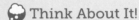

Think About It!

How can you find the area under the normal curve that is more than 2 standard deviations greater than the mean?

Key Concept • The Empirical Rule

In a normal distribution with mean μ and standard deviation σ,

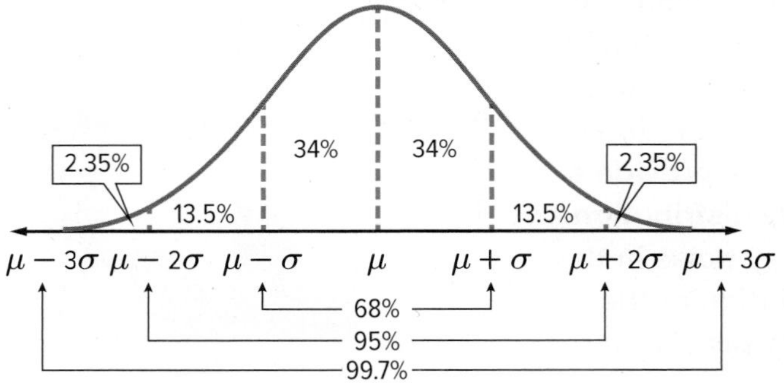

- approximately 68% of the data fall within 1σ of the mean,
- approximately 95% of the data fall within 2σ of the mean, and
- approximately 99.7% of the data fall within 3σ of the mean.

When a set of data is *not* approximately normal, it cannot be represented by the Empirical Rule. Skewed data is one example of a set of data that is not approximately normal.

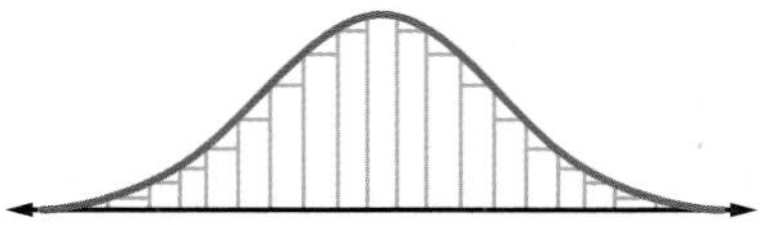

The data are normally distributed, symmetric about the mean, and bell-shaped.

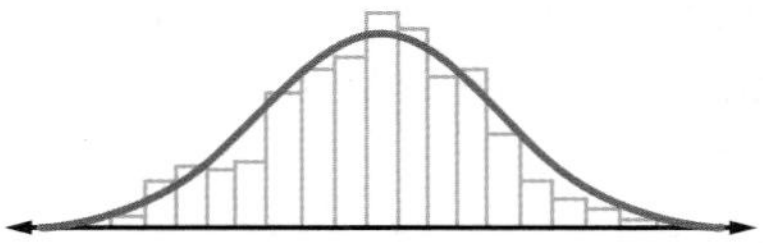

The data are approximately normally distributed. The data can be modeled by the normal distribution.

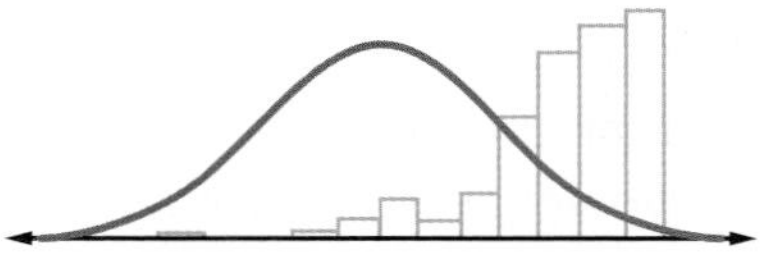

The data are skewed to the left. A normal curve would not be the best curve to model the distribution.

Example 4 Approximate Data by Using a Normal Distribution

GOALS **The total number of goals scored by a soccer team during the season are shown.**

Create a histogram of the set of data. Determine whether the data can be approximated with a normal distribution.

Goals Scored Per Player			
0	2	0	0
0	0	1	1
1	2	0	2
1	3	1	0
0	0	0	3
0	2	5	8
0	9	11	1
2	0	1	0

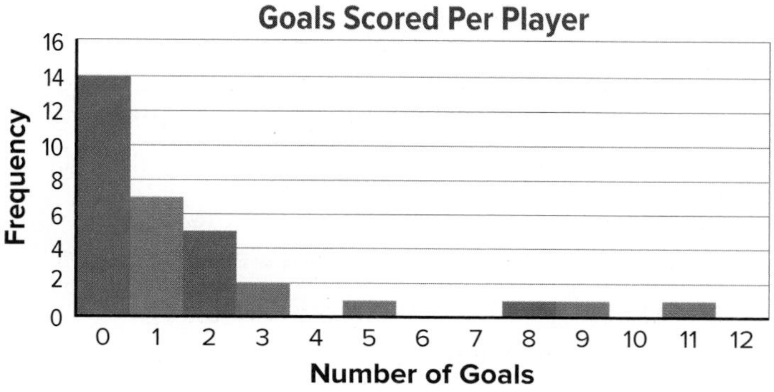

The data set is skewed. Thus, the data cannot be approximated with a normal distribution.

> **Think About It!**
> Use the histogram to estimate the mean and standard deviation of the data. Explain your reasoning.

Example 5 Use the Empirical Rule to Analyze Data

A normal distribution has a mean of 42 and a standard deviation of 6.

a. Find the range of values that represents the middle 95% of the distribution.

By the Empirical Rule, the middle 95% of the data in a normal distribution is under the curve between the interval

$[\mu - 2\sigma, \mu + 2\sigma]$.

$\mu - 2\sigma = 42 - 2(6) = 30$ $\qquad$ $\mu + 2\sigma = 42 + 2(6) = 54$

b. What percent of the data will be greater than 36?

36 is σ less than μ. Therefore, the range of values greater than 36 is $X > \mu - \sigma$.

This range includes:

- All of the data greater than μ, or 50%.
- The data between $\mu - \sigma$ and μ, or 34%.

The percent of data greater than 36 is 84%.

Go Online You can complete an Extra Example online.

> **Think About It!**
> What are the limitations of the Empirical Formula?

> **Go Online**
> You can watch a video to learn how to graph a normal distribution using a graphing calculator online.

Think About It!

Using the formula for the z-value, what is the z-score of a data value equal to $\mu + \sigma$? $\mu - 2\sigma$?

Learn The Standard Normal Distribution

The Empirical Rule is only useful for evaluating specific values. Once the data set is *standardized*, any value can be evaluated. Data are standardized by converting them to z-values. The ***z*-value** of a specific data value is a measure of position, and it represents the number of standard deviations that a given data value is from the mean.

The z-value for a data value X in a set of normally distributed data is $z = \frac{X - \mu}{\sigma}$, where μ is the mean and σ is the standard deviation.

If the distributions are standardized by using z-values, any distribution can be compared to any other distribution.

The **standard normal distribution** is a normal distribution with a mean of 0 and a standard deviation of 1. You can take any normal distribution and convert it to the standard normal distribution by standardizing it.

The standard normal distribution allows you to determine how much of the data falls between any two z-values, or to the left or right of any z-value. The area under the normal curve between two points corresponds to the proportion of the data that falls in a given interval.

Study Tip

Standard Normal Distribution The standard normal distribution is the set of all z-values. z-values are also called *z-scores*.

Because the area under the curve is 1, the area under the curve between two points represents the probability that a data value falls in that interval.

The proportion of values within each interval, or the associated probability of a data point falling within that interval, can also be found by using probability tables. For these values, round to the hundredths place.

To use a Standard Normal Distribution Table to find the probability of data being in the interval [−1.09, 0.69], complete the following steps.

Step 1 Find the area under the curve for $X < -1.09$.

The probability associated with $z = -1.09$ is 0.1379. The probability of a randomly selected value in a normally distributed data set being less than $\mu - 1.09\sigma$ is 0.1379.

z	0.00	⋯	0.09
−3.4	0.0003	⋯	0.0002
⋮	⋯	⋯	⋯
−1.0	0.1587	⋯	0.1379

Step 2 Find the area under the curve for $X < 0.69$.

The probability associated with $z = 0.69$ is 0.7549. The probability of a randomly selected value in a normally distributed data set being less than $\mu + 0.69\sigma$ is 0.7549.

z	0.00	⋯	0.09
0.0	0.5000	⋯	0.5359
⋮	⋯	⋯	⋯
0.6	0.7257	⋯	0.7549

Step 3 Find the area within the interval.

The area under the curve within [−1.09, 0.69], and the associated probability, is $0.7549 - 0.1379 = 0.6170$, or 61.7%.

Go Online You can complete an Extra Example online.

Example 6 Use z-Values to Locate Position

Find z if $X = 24$, $\mu = 19$, and $\sigma = 3.8$.

$z = \frac{X - \mu}{\sigma}$ Formula for z-values

$z = \frac{24 - 19}{3.8}$ $X = 24$, $\mu = 19$, and $\sigma = 3.8$

$z = \frac{5}{3.8}$ Subtract.

$z \approx 1.316$ Divide.

Check

Find z if $X = 106.3$, $\mu = 88.8$, and $\sigma = 9.6$. Round your answer to the nearest thousandth.

Example 7 Find Area Under the Standard Normal Curve by Using a Table

Use a Standard Normal Distribution Table to find the area under the normal curve within the interval $z < -1.18$.

Step 1 Identify the probability for the left endpoint, $z = -1.18$.

Using a Standard Normal Distribution Table, move down the left column until you reach the correct tenths place, −1.1. Then move across the row until you reach the hundredths place, 0.08.

z	0.00	···	0.08	0.09
−3.4	0.0003	···	0.0003	0.0002
⋮	···	···	···	···
−1.1	0.1357	···	0.1190	0.1170

The probability associated with $z = -1.18$ is 0.1190.

Step 2 Identify the area under the curve within the interval.

As shown in the graph, 0.119 represents the area under the curve to the left of $z = -1.18$.

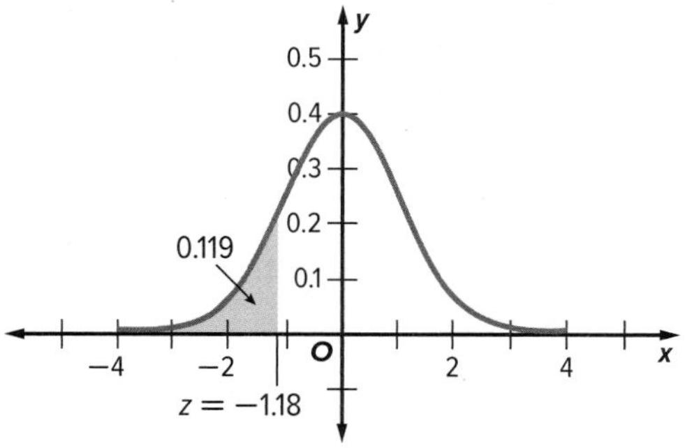

Check

Use a table to find the area under the normal curve within $z < 0.19$. Round your answer to the nearest ten thousandth.

Think About It!

Why is it not necessary to go past $z = 4$ or $z = -4$ when assessing a proportion under the normal curve? Justify your answer.

Study Tip

Standard Normal Distribution Table A table is available online and in the back of your book.

Think About It!

What is the area under the curve to the right of $z = -1.18$?

Study Tip

Percentage, Proportion, Probability, and Area When a problem asks for a percentage, proportion, or probability, it is asking for the same value: the corresponding area under the normal curve.

Example 8 Find Area Under the Standard Normal Curve by Using a Calculator

INTERNET TRAFFIC **The number of daily hits to a local news Web site is normally distributed with $\mu = 98{,}452$ hits and $\sigma = 10{,}325$ hits. Find the probability that the Web site will get at least 100,000 hits on a given day, $P(X > 100{,}000)$.**

Step 1 Find the corresponding *z*-value for $X = 100{,}000$.

$z = \frac{X - \mu}{\sigma}$ Formula for *z*-value

$z = \frac{100{,}00 - 98{,}452}{10{,}325}$ $X = 100{,}000$, $\mu = 98{,}452$, and $\sigma = 10{,}325$

$z \approx 0.150$ Simplify.

Step 2 Find the probability.

The area under the curve when $z > 4$ is negligible. So $z = 4$ can be used as an upper bound for finding area. You can use a graphing calculator to find the area between $z = 0.150$ and $z = 4$.

```
normalcdf(0.150,
4)
        .4403506018
```

Press **2nd** **[DISTR]** and select **normalcdf**. Enter the interval and press **enter** to display the area.

The area is 0.44. Therefore, the probability of the Web site getting at least 100,000 hits in one day is 44.0%.

Check

CUSTOMER SERVICE The length of time that a customer spends waiting to be connected to a customer service representative is normally distributed with a mean of 21.3 seconds and a standard deviation of 3.8 seconds. Find the probability that a wait time will be greater than 30 seconds.

Go Online

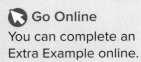

You can complete an Extra Example online.

Practice

Go Online You can complete your homework online.

Example 1

Identify the random variable in each distribution, and classify it as *discrete* or *continuous*. Explain your reasoning.

1. the number of texts received per week
2. the number of "likes" for a Web page
3. the height of a plant after a specific amount of time

Examples 2-4

4. FUNDRAISING At a fundraising dinner, the underside of 200 plates were randomly tagged with a sticker to indicate winning a cash prize. The frequency table shows the number of winning plates for each prize. Construct a relative frequency table, and graph the probability distribution.

Prize, (X)	Frequency
$5	150
$50	40
$100	9
$1000	1

5. BASKETBALL An athletic director made a probability distribution of the heights of her team's basketball players, and distributed a flyer that claimed that the majority of the players on the basketball team are 71 inches or taller. Identify any flaws in the representation of the probability distribution.

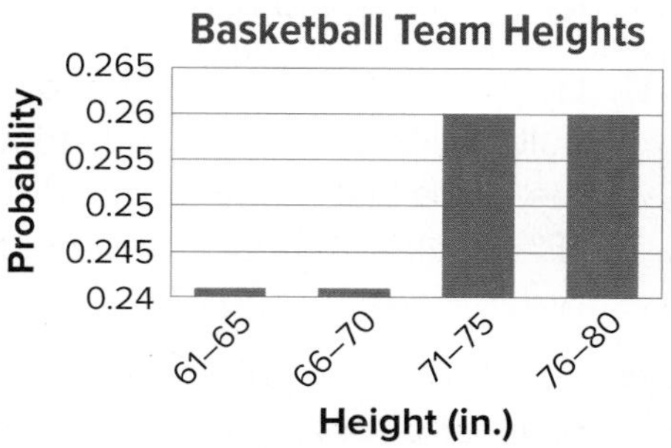

6. TRACK The preliminary times for a 110-meter hurdles race are shown. Create a histogram of the set of data. Determine whether the data can be approximated with a normal distribution.

Times (seconds)
14.75, 14.77, 14.31, 14.83, 14.84, 14.35, 14.69, 14.63, 14.74, 14.82, 14.25, 14.93

Example 5

7. A normal distribution has a mean of 186.4 and a standard deviation of 48.9.

   a. What range of values represents the middle 99.7% of the data?

   b. What percent of data will be greater than 235.3?

   c. What range of values represents the upper 2.5% of the data?

Example 6

Find the *z*-value for each standard normal distribution.

8. $\sigma = 9.8, X = 55.4$, and $\mu = 68.34$
9. $\sigma = 11.6, X = 42.80$, and $\mu = 68.2$
10. $\sigma = 11.9, X = 119.2$, and $\mu = 112.4$

Example 7

Use a table to find the area under the normal curve for each interval.

11. $z > 0.58$
12. $z < -1.56$
13. $-2.29 < z < 2.76$

Example 8

14. TESTING The scores on a test administered to prospective employees are normally distributed with a mean of 100 and a standard deviation of 12.3.

a. What percent of the scores are between 70 and 80?

b. What percent of the scores are over 115?

c. If 75 people take the test, how many would you expect to score lower than 75?

Mixed Exercises

15. USE A MODEL Shawndra conducted a study to find how far people park from parking blocks by measuring the distance from the end of the parking block to the front fender of the parked car. The distribution of the data closely approximated a normal distribution with a mean of 8.5 inches. About 2.5% of cars parked more than 11.5 inches from the parking block. What percentage of cars would you expect to have parked less than 5.5 inches from the parking block?

16. PRECISION Use the normal distribution of data.

a. What is the standard deviation of the data set?

b. What percent of the data will be greater than 60?

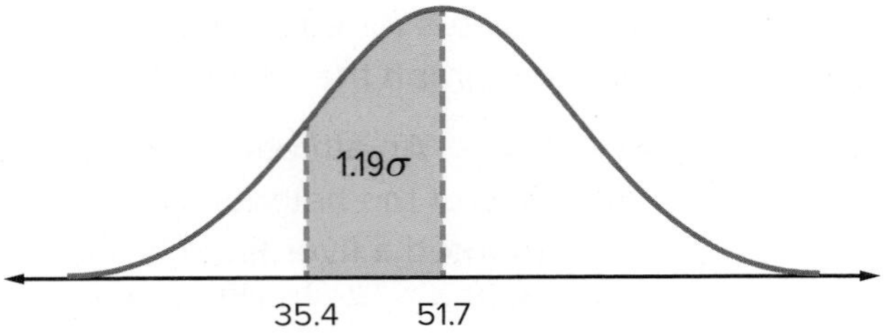

17. LIGHT BULBS The time that a certain brand of light bulb will last is normally distributed. About 2.5% of the bulbs last longer than 6800 hours, and about 16% last longer than 6500 hours. How long does the average bulb last?

18. STRUCTURE Each day, Roberto takes a 20-question practice college entrance exam and records his score. Find the probability distribution for this situation.

Monday	Tuesday	Wednesday	Thursday	Friday
17	16	16	17	18
15	17	16	18	16

Higher-Order Thinking Skills

19. FIND THE ERROR A data set of tree diameters is normally distributed with mean 11.5 centimeters, standard deviation 2.5, and range from 3.6 to 19.8. Monica and Hiroko want to find the range that represents the middle 68% of the data. Is either of them correct? Explain your reasoning.

Monica	Hiroko
The data span 16.2 cm, and 68% of 16.2 is about 11 cm. Center this 11-cm interval at the mean of 11.5 cm. This will range from about 6 cm to 17 cm.	The middle 68% span from $\mu + \sigma$ to $\mu - \sigma$. So we move 2.5 cm above and below 11.5. The 68% group will range from 9 cm to 14 cm.

20. PERSEVERE A shipment of cell phones has an average battery life of 8.2 hours with a standard deviation of 0.7 hour. Eight of the phones have a battery life greater than 9.3 hours. If the sample is normally distributed, how many phones are in the shipment?

21. WRITE Describe the relationship between the *z*-value, the position of an interval of *X* in the normal distribution, the area under the normal curve, and the probability of the interval occurring.

Estimating Population Parameters

Learn Estimating the Population Mean

The data from a sample can be used to make inferences about the corresponding population. This is called **inferential statistics**.

Sampling error is the variation between samples of the same population and is due to randomness. It does not include bias or other errors that may occur when obtaining the sample. As the sample size increases, the size of the sampling error will decrease because the sample increasingly represents the larger population.

The **standard error of the mean** $\sigma_{\overline{x}}$ is equal to $\frac{\sigma}{\sqrt{n}}$, where σ is the standard deviation of the population, $\overline{x}$ is the sample mean, and n is the sample size.

A **confidence interval** is an estimate of the population parameter using a sample statistic, stated as a range with a specific degree of certainty, typically 90%, 95%, or 99%. The confidence interval for a sample mean $\overline{x}$ is $\overline{x} + E$, where E is the maximum error of the estimate. The **maximum error of the estimate** is the maximum difference between the estimate of the population parameter and its actual value. The maximum error of the estimate is also called the margin of error.

Key Concept • Maximum Error of the Estimate for a Sample Mean

The maximum error of the estimate E for a sample mean is given by $E = z \cdot \frac{\sigma}{\sqrt{n}}$, and can be approximated by the formula $E = z \cdot \frac{s}{\sqrt{n}}$, where

- z is a specific value that corresponds to the confidence level,
- s is the standard deviation of the sample, and
- n is the sample size, $n \geq 30$.

The three most common degrees of certainty and their corresponding z-values, called critical values, are given.

- A certainty of 90% has a z-value of 1.645.
- A certainty of 95% has a z-value of 1.960.
- A certainty of 99% has a z-value of 2.576.

Key Concept • Estimating the Population Mean μ

Step 1 Find the mean of a random sample $\overline{x}$.

Step 2 Determine the desired degree of certainty.

Step 3 Find the maximum error of the estimate E.

Step 4 Estimate the mean of the population μ by finding the interval $(\overline{x} - E, \overline{x} + E)$.

Today's Goals

- Use sample data to infer a population mean by using confidence intervals.
- Use sample data to infer a population proportion by using confidence intervals.

Today's Vocabulary

inferential statistics

sampling error

standard error of the mean

confidence interval

maximum error of the estimate

population proportion

Think About It!

How will the confidence level affect the size of the confidence interval? Justify your argument.

Talk About It!

Why do you think confidence levels are 90%, 95%, and 99% instead of 25%, 50%, and 75%? Justify your argument with an example.

Example 1 Find the Maximum Error of the Estimate

CHESS **A poll of 315 randomly selected members of an online chess club showed that they spent an average of 4.6 hours per week playing chess with a standard deviation of 1.2 hours. Use a 90% confidence interval to find the maximum error of the estimate for the time spent playing chess.**

Certainty	*z*-Value
90%	1.645
95%	1.960
99%	2.576

Step 1 Identify the *z*-value.

A 90% confidence interval is equivalent to the area under the normal curve in the range (−1.645, 1.645).

Step 2 Find the maximum error of the estimate.

$E = z \cdot \frac{s}{\sqrt{n}}$ Maximum Error of the Estimate Formula

$= 1.645 \cdot \frac{1.2}{\sqrt{315}}$ $z = 1.645$, $s = 1.2$, and $n = 315$

≈ 0.11 Simplify.

So, we can say with 90% confidence that the mean time a chess club member plays per week is within 0.11 hour of the sample mean time.

Example 2 Estimate a Population Mean

WEB TRAFFIC **A survey of 50 students at East High School asked them how many Web sites they visit at least three times per day. The sample mean is 8.61, and the sample standard deviation is 2.6. Use a 95% confidence level to estimate the mean for the population of East High School students.**

Step 1 Calculate *E*.

Use the *z*-value that corresponds with a 95% confidence interval, 1.960, to calculate *E*.

$E = z \cdot \frac{s}{\sqrt{n}}$ Maximum Error of the Estimate Formula

$= 1.960 \cdot \frac{2.6}{\sqrt{50}}$ $z = 1.960$, $s = 2.6$, and $n = 50$

≈ 0.72 Simplify.

Step 2 Determine the confidence interval.

The confidence interval is the interval defined by $\bar{x} \pm E$.

$\bar{x} - E = 8.61 - 0.72 = 7.89$

$\bar{x} + E = 8.61 + 0.72 = 9.33$

At a 95% confidence interval, the population mean is $7.89 \leq \mu \leq 9.33$. Therefore, we are 95% confident that the number of Web sites students visit at least three times a day is between 7.89 and 9.33.

Problem-Solving Tip

Reference Critical Values The most commonly used confidence levels and their corresponding *z*-values are shown in the table. Have this table handy when solving problems.

Certainty	*z*-Value
90%	1.645
95%	1.960
99%	2.576

Go Online You can complete an Extra Example online.

Check

CAFETERIA A poll of 219 randomly selected high school seniors showed that they used the school cafeteria an average of 2.6 times per week, with a standard deviation of 0.7. Use a 95% confidence level to find the maximum error of the estimate, and then estimate the mean of the population. Round to the nearest hundredth.

At a 95% confidence interval, the maximum error of the estimate is ___?___. The population mean is ___?___ $\leq \mu \leq$ ___?___.

Learn Estimating the Population Proportion

You can also use confidence intervals to estimate a **population proportion *p*,** which is the number of members in the population with a particular characteristic divided by the total number of members in the population.

Like the population mean, the population proportion is often difficult to determine. Instead, the sample proportion $\hat{p}$ (called *p* hat) is used to estimate *p*. The sample proportion is the proportion of successes in a sample. The proportion of failures is represented by $\hat{q}$.

- The proportion of successes $\hat{p}$ is given by $\hat{p} = \frac{x}{n}$, where *x* is the number of successes and *n* is the sample size (the number of trials).
- The proportion of failure $\hat{q}$ is given by $\hat{q} = 1 - \hat{p}$.

Like the sample mean, $\hat{p}$ is an estimate and will likely not equal the population proportion. Therefore, a confidence interval using $\hat{p}$ and a maximum error of the estimate *E* is used to find the range of values within which *p* likely falls.

The maximum error of the estimate *E* for a population proportion is given by $E = z \cdot \sqrt{\frac{\hat{p}\hat{q}}{n}}$, where *z* is the *z*-value that corresponds to the confidence level, $\hat{p}$ is the sample proportion of success, $\hat{q}$ is the sample proportion of failure, and *n* is the sample size, where $n\hat{p} \geq 5$ and $n\hat{q} \geq 5$.

The maximum error of the estimate *E* for a population proportion can be used to estimate the **population proportion**.

To estimate the population proportion:

- Find the proportions of success and failure.
- Find the sample size.
- Identify the corresponding *z*-value.
- Find the maximum error of the estimate *E*.
- The confidence interval for the population proportion is $\hat{p} \pm E$.

Go Online You can complete an Extra Example online.

Apply Example 3 Estimate a Population Proportion

LUNCH PERIOD **A survey of 150 students at West High School asked them if they agree with the administration's plan to split the lunch period into three different periods, with 72 of the respondents saying that they agree. Use a 90% confidence level to estimate the population proportion for all 1334 West High School students.**

1 What is the task?

Describe the task in your own words. Then list any questions that you may have. How can you find answers to your questions?

I need to find $\hat{p}$ and $\hat{q}$, and then use those values to estimate the population proportion with 90% confidence.

Is the sample size large enough to use a confidence interval? What formulas do I need to use?

I can test the sample size. I can use the previous example within the lesson to check that I am using the correct formulas.

2 How will you approach the task? What have you learned that you can use to help you complete the task?

I will find $\hat{p}$ and $\hat{q}$ and check that the sample size is appropriate to use a confidence interval. Next, I will find the maximum error of the estimate. Then I will use the confidence interval to find the population proportion.

3 What is your solution?

Use your strategy to solve the problem.

What are the values of $\hat{p}$ and $\hat{q}$? $\hat{p} = 0.48$; $\hat{q} = 0.52$

How can you ensure that proportions have large enough sample sizes to use a confidence interval?

I can verify that $n\hat{p} \geq 5$ and $n\hat{q} \geq 5$. Because the sample size was 150 students, $n\hat{p} = 150 \cdot 0.48$, or 72 and $n\hat{q} = 150 \cdot 0.52$, or 78. So the sample size is large enough.

What is the confidence interval? $CI = [0.413, 0.547]$

With 90% confidence, what proportion of students agree with the administration's plan? between 41.3% and 54.7%

4 How can you know that your solution is reasonable?

Write About It! Write an argument that can be used to defend your solution.

The proportion of success of the sample falls within the range I found using the confidence interval. Because the administration used a large enough sample size, it is reasonable to assume that proportion of success within the population sample will be representative of the population.

Go Online You can complete an Extra Example online.

Check

PARKS A survey of 175 residents of a particular neighborhood asked if they agreed with the neighborhood board's decision to charge a fee to build a park in the middle of the neighborhood. 42 residents agreed with the decision.

Part A Use a 95% confidence level to estimate the population proportion.

A. $0.187 \leq p \leq 0.293$

B. $0.155 \leq p \leq 0.325$

C. $0.177 \leq p \leq 0.303$

D. $0.587 \leq p \leq 0.713$

Part B We are 95% confident that the proportion of residents who agree with the neighborhood board's decision is between __?__% and __?__%.

Example 4 Misleading Population Estimates

POLLS A lobbying group that supports the passing of a ballot measure releases a report based on a poll they performed. They claim that, with 90% confidence, the proportion of people who will vote for the measure is 50%, but could be as low as 40% due to the margin of error of 5%. The sample size was 200. Identify any misleading representation of the data.

Step 1 Verify that $n\hat{p} \geq 5$ and $n\hat{q} \geq 5$.

The sample size is 200. The report claims that the proportion is 0.5, but the sample proportion $\hat{p}$ is actually 0.45 with a margin of error of 0.05. So, $\hat{p} = 0.45$ and $\hat{q} = 1 - 0.45 = 0.55$.

$n\hat{p} = 200 \cdot 0.45$ Substitute.

$= 90$ Simplify.

$n\hat{q} = 200 \cdot 0.55$ Substitute.

$= 110$ Simplify.

Because $n\hat{p} \geq 5$ and $n\hat{q} \geq 5$ a confidence interval is appropriate.

Step 2 Examine the confidence level and interval.

The confidence level of 90% is acceptable for a poll. However, instead of stating that the population proportion is estimated to be 45% with a margin of error of 5%, they state the highest possible estimate for the population proportion to give the impression the candidate is likely to attain 50% of the vote.

Go Online You can complete an Extra Example online.

Check

COMPUTERS A computer manufacturer has received several complaints about their recent line of hard drives failing too often. Internal policy is that if 20% or more of a product fails, they will perform a recall and refund anyone who purchased that hard drive. The company conducts a survey and finds that, with 99.99% confidence, as few as 10% of hard drives have failed. The sample size was 100. Identify any misleading representation of the data. Select all that apply.

A. The sample size is too small to use a confidence interval.

B. The confidence level is too high.

C. The confidence level is too low.

D. The way the confidence interval is stated is misleading.

E. The internal policy is too lenient.

F. The report is not misleading.

Practice

Go Online You can complete your homework online.

Example 1

1. HOMEWORK A poll of 225 randomly selected high school students showed that they spent an average of 2.75 hours per night doing homework with a standard deviation of 1.5 hours. Use a 90% confidence interval to find the maximum error of the estimate for the time spent doing homework.
2. ATHLETES A poll of 350 randomly selected athletes showed that they spent an average of 10.25 hours per week at practice with a standard deviation of 2.3 hours. Use a 95% confidence interval to find the maximum error of the estimate for the time spent practicing.
3. PETS A poll of 470 randomly selected pet owners showed that they have an average of 1.9 pets per household with a standard deviation of 0.5 pets. Use a 99% confidence interval to find the maximum error of the estimate for the number of pets per household.

Example 2

4. FREIGHT A sample of 200 packages at a shipping company is weighed. The sample mean is 5.8 pounds, and the sample standard deviation is 2.7 pounds. Use a 90% confidence level to estimate the mean for the packages.
5. AIRLINES A survey of 700 airline passengers asks about satisfaction. The passengers must rate the airline on a scale of 1 to 10, where 1 is *not satisfied* and 10 is *extremely satisfied*. The sample mean is 5.5, and the sample standard deviation is 1.5. Use a 99% confidence level to estimate the mean for the passenger ratings.
6. CREDIT HOURS A sample of 650 college students is surveyed about the number of credit hours they are taking this semester. The sample mean is 14.75 credit hours, and the sample standard deviation is 2.25 credit hours. Use a 95% confidence level to estimate the mean for the credit hours.

Example 3

7. PART-TIME JOB A survey of 120 students at South Grove High School asked them if they agree that high school students should have part-time jobs, with 93 of the respondents saying that they agree. Use a 90% confidence level to estimate the population proportion for all 1532 South Grove High School students.
8. BUS A survey of 215 students at Landry High School asked them if they ride the bus to school, with 46 respondents saying that they do. Use a 95% confidence level to estimate the population proportion for all 1109 students.
9. FIELD TRIP A survey of 245 students at Jackson High School asked them if they agree with their principal's decision to have students pack their lunches for a field trip, with 102 of the respondents saying that they agree. Use a 99% confidence level to estimate the population proportion for all 973 students.

Example 4

10. MENU An employee at a restaurant conducts a poll about the satisfaction with their lunch menu. The report claims, with 80% confidence, that the proportion of people who are satisfied with the lunch menu is between 85% and 90%. The sample size was 15. Identify any misleading representation of the data.
11. POLLS A recent poll showed that 120 citizens were in favor of building more parks in their community. A claim was made, with 95% confidence, that the proportion of people who want to build more parks is 51%, but could be as low as 45% due to the margin of error of 3%. The sample size was 250. Identify any misleading representation of the data.

Mixed Exercises

12. CONSTRUCT ARGUMENTS Enrique is assessing his morning commute times for the first two weeks of his new job. For these 10 commutes, Enrique has calculated $\overline{x} = 30.8$ minutes and $s \approx 4.83$ minutes.

a. Find 90%, 95%, and 99% confidence intervals for Enrique's mean commute time. Express these as ranges (from x min to y min).

b. Enrique collects another two weeks' data, which gives him $\overline{x} = 31.1$ min and $s \approx 6.40$ min for the 20-workday period. During this time, Enrique is slightly late for work twice. Which confidence level should Enrique pick if he wants to avoid being late, and why? Calculate this confidence interval for μ, and explain how Enrique should use it with his new estimate of σ to ensure he is on time.

13. PRECISION Karen, an artisan potter, is concerned that her kiln is not heating evenly. She finds it acceptable to throw away no more than 10% of the pieces she fires in the kiln, but recently has had to discard 32 pieces out of a sample of 205. Find a 95% confidence interval for the population proportion of discards. Based on your finding, write a mathematically precise statement to determine whether Karen should buy a new kiln.

14. REASONING A pharmaceutical company is testing a new medication that is supposed to shorten the number of days a patient experieneces symptoms of influenza. It was tested on 45 patients, and the average number of days until symptoms cleared was 5.3 with a standard deviation of 1.1 days.

a. Use the sample size and standard deviation to find the maximum error of estimate for a confidence level of 99%. Interpret the results in the context of the population mean.

b. Write the 99% confidence interval for the population mean number of days until symptoms resolve with the new medication.

c. The company claims that the average number of days until recovery using their new medication can be as low as four days. Is their claim reasonable?

Higher-Order Thinking Skills

15. ANALYZE Determine how the sample size affects the maximum error.

16. WRITE Explain how to estimate the population proportion.

17. CREATE Survey students in your class about the number of hours they watch television each week. Find the sample mean and the sample standard deviation. Use a 90% confidence level to estimate the mean number of hours students in your class watch television each week.

Module 10 • Inferential Statistics

Review

Essential Question

How can data be collected and interpreted so that it is useful to a specific audience?

Module Summary

Lessons 10-1 and 10-2

Samples and Experiments

- Because it is not always practical to collect data from a large population, it may be collected from a representative sample.
- A bias is an error that results in a misrepresentation of a population.
- Surveys, experiments, and observations are used to collect data.
- The probability of an event is the ratio of the number of outcomes in which an event occurs to the total number of possible outcomes.
- A simulation uses a probability model to imitate a situation so it can be studied.

Lessons 10-3 and 10-4

Distributions

- A symmetric distribution has mean and median approximately equal. A negatively skewed distribution typically has a median greater than the mean. There is less data on the left side of the graph. A positively skewed distribution typically has a mean greater than the median. There is less data on the right side of the graph.
- The standard deviation of a data set is found using the formula $\sigma = \sqrt{\frac{\sum_{k=1}^{n}(x_k - \mu)^2}{n}}$.
- The expected value of a discrete random variable is calculated by finding the sum of the products of every possible value of X and its associated probability $P(X)$.
- In a normal distribution with mean μ and standard deviation σ, approximately 68% of the data fall within 1σ of the mean, approximately 95% of the data fall within 2σ of the mean, and approximately 99.7% of the data fall within 3σ of the mean.

Lesson 10-5

Population Parameters

- The maximum error of the estimate E for a sample mean can be approximated by the formula $E = z \cdot \frac{s}{\sqrt{n}}$, where z is a specific value that corresponds to the confidence level, s is the standard deviation of the sample, and n is the sample size, $n \geq 30$.
- The maximum error of the estimate E for a population proportion is given by $E = z \cdot \sqrt{\frac{\hat{p}\hat{q}}{n}}$, where z is the z-value that corresponds to the confidence level, $\hat{p}$ is the sample proportion of success, $\hat{q}$ is the sample proportion of failure, and n is the sample size, where $n\hat{p} \geq 5$ and $n\hat{q} \geq 5$.

Study Organizer

Foldables

Use your Foldable to review this module. Working with a partner can be helpful. Ask for clarification of concepts as needed.

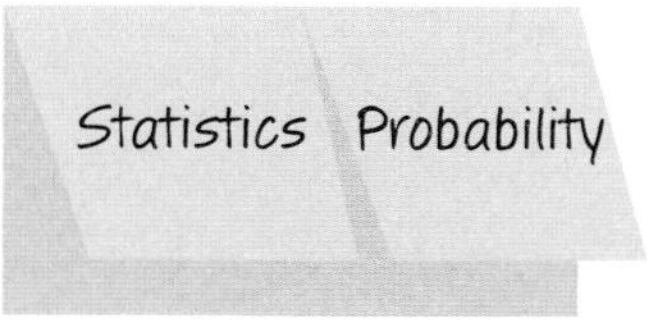

Test Practice

1. **MULTIPLE CHOICE** Blake wants to open an ice cream shop. He is creating a survey for local people near the planned location to determine if there is interest in a new ice cream shop before he completes his plans. Which question from the survey is biased?. (Lesson 10-1)

   A. How many times per month do you go out for ice cream?

   B. What is your favorite ice cream flavor?

   C. What time of the day do you typically visit an ice cream shop?

   D. Would you visit a new, tasty and local ice cream shop instead of an old, dirty shop with few flavors to choose from?

2. **OPEN RESPONSE** Two researchers are discussing ways of determining children's favorite breakfast cereals.

   - Shelly says that the randomly selected participants should be asked a series of questions about their cereal eating.
   - Raul says that they should watch the students selecting from different cereals and write down what they notice.

Indicate what type of study that each of the researchers is suggesting. (Lesson 10-1)

3. **MULTIPLE CHOICE** In an effort to survey the student population about possible themes for the spring prom, the student council separates the students into 4 groups by grade level and then chooses 25 students from each grade. Which sampling method did the student council use? (Lesson 10-1)

   A. Stratified

   B. Convenience

   C. Systematic

   D. Self-selected

4. **MULTIPLE CHOICE** Leonard randomly selected a card from a standard deck of playing cards, recorded the suit, and returned the card. He followed this set of steps 120 times. The results are shown.

Suit	Frequency
Heart	28
Diamond	37
Spade	34
Club	21

Which statement about the results is true? (Lesson 10-2)

   A. The theoretical probability of selecting a heart is less than the experimental probability of selecting a heart.

   B. The theoretical probability of selecting a diamond is less than the experimental probability of selecting a diamond.

   C. The experimental probability of selecting a club is greater than the theoretical probability of selecting a club.

   D. The experimental probability of selecting a heart is equal to the experimental probability of selecting a diamond.

5. **MULTIPLE CHOICE** The quality control manager of a food processing manufacturer estimates that 1 can in every case of 12 canned vegetables will likely become damaged during shipment. Determine a simulation that best represents the situation. (Lesson 10-2)

   A. Use a random number generator with the numbers 1 through 12.

   B. Use 12 play cards including 4 kings, 4 queens, and 4 jacks.

   C. Use two dice, where tossing two ones represents a dented can.

   D. Use a set of 12 marbles with 6 red and 6 blue.

6. OPEN RESPONSE A football player makes 70% of the field goals he attempts. Diana used a random number generator to generate integers 1 through 100. The integers 1–70 represented a made field goal on the next attempt, and the integers 71–100 represented a missed field goal on the next attempt. The simulation consisted of 50 trials, with 42 trials being integers 1–70. Explain the data. Is the data from the simulation consistent with the model? (Lesson 10-2)

7. MULTI-SELECT Zain is analyzing two sets of data. (Lesson 10-3)

Set A				
35	28	35	33	32
32	31	30	36	49
36	33	29	34	37

Set B				
22	26	26	24	27
38	35	31	22	30
24	33	30	25	38

Select all the true statements about each data set.

A. Data set A has a higher average value.

B. Data set B has a higher average value.

C. C. Data set A is more varied.

D. D. Data set B is more varied.

8. OPEN RESPONSE Catherine is growing several tomato plants. She measures the height of each plant. The results are shown.

Tomato Plant Heights (inches)			
13	10	12	14
14	18	16	12
16	14	14	11
12	15	15	10

Find and interpret the standard deviation of the plant heights. (Lesson 10-3)

9. MULTI-SELECT Kent owns a shop and has workers that produce products during two different shifts throughout the day. He kept track of how many products each group completed during their shift for several days. The data are shown. (Lesson 10-3)

A Shift Production		
250	207	232
278	264	247
244	288	258

B Shift Production		
263	245	255
249	264	262
277	270	268

Describe Kent's data. Select all that apply.

A. On average, the A shift produced more products.

B. On average, the B shift produced more products.

C. The number of products produced by the A shift is more varied.

D. The number of products produced by the B shift is more varied.

E. When shifts A and B are combined, they have a higher average and less variation than either shift by itself.

10. **OPEN RESPONSE** A normal distribution has a mean of 347.2 and a standard deviation of 13.9. (Lesson 10-4)

Part A What percent of the data is less than 319.4?

Part B What percent of the data is greater than 361.1?

11. **OPEN RESPONSE** A normal distribution has a mean of 63.4 and a standard deviation of 2.5. Find the range of values that represent the outer 0.3%. (Lesson 10-4)

12. **MULTIPLE CHOICE** Hiroyuki randomly surveyed 325 students asking how much time they spent getting ready for school in the morning. The average time spent was 26 minutes with a standard deviation of 5.8 minutes.

Use a 90% confidence interval to find the maximum error of the estimate of time (in minutes) spent getting ready for school. Round to the nearest hundredth. (Lesson 10-5)

A. 0.22

B. 0.26

C. 0.53

D. 0.63

13. **OPEN RESPONSE** A restaurant randomly surveyed 360 of their patrons asking if they would like the restaurant to stay open later. A total of 273 people responded that they would appreciate the later hours. Using a 95% confidence interval, estimate the population proportion and explain what it means. (Lesson 10-5)

14. **MULTIPLE CHOICE** A national business has randomly surveyed 12 employees asking whether the employees would prefer to work on Thanksgiving. The business reports that with 95% confidence the proportion of people that would prefer to work on Thanksgiving is between 33% and 45%, with a margin of error of 6%. Identify any misleading representations of the data. (Lesson 10-5)

A. The report is not misleading.

B. The confidence level is too low.

C. The confidence level is stated in a misleading way.

D. The sample is too small to use a confidence level.

Module 11

Trigonometric Functions

Essential Question

What are the key features of the graph of a trigonometric function and how do they represent real-world situations?

What Will You Learn?

How much do you already know about each topic **before** starting this module?

KEY

— I don't know. — I've heard of it. — I know it!

	Before			After		
draw angles in standard position						
convert between degree and radian measures of angles						
find values of trigonometric functions using general and reference angles						
find trigonometric values using the unit circle and properties of periodic functions						
graph and analyze sine and cosine functions and their reciprocals						
graph and analyze the tangent function and its reciprocal						
graph translations of trigonometric functions						
find angle measures by using inverse trigonometric functions						
solve trigonometric equations						

Foldables Make this Foldable to help you organize your notes about trigonometric functions. Begin with four sheets of grid paper.

1. **Stack** the paper and measure 2.5 inches from the bottom.
2. **Fold** on the diagonal.
3. **Cut** the extra tabs off, and staple along the diagonal to form a book.
4. **Label** the edge as *Trigonometric Functions*.

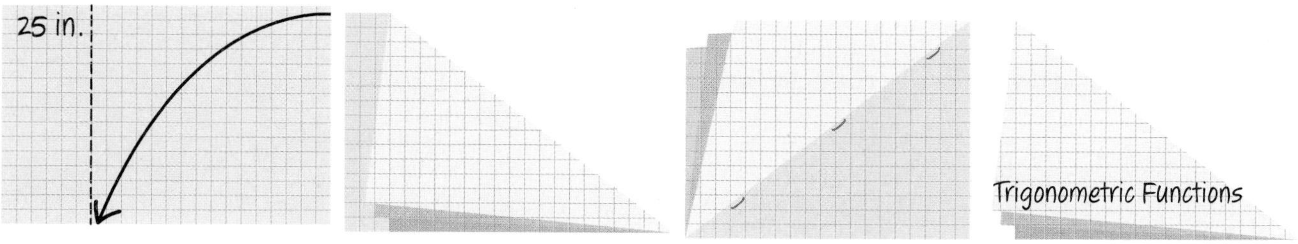

What Vocabulary Will You Learn?

- amplitude
- central angle of a circle
- circular function
- cosecant
- cosine
- cotangent
- coterminal angles
- cycle
- frequency
- initial side
- inverse trigonometric functions
- midline
- oscillation
- period
- periodic function
- phase shift
- principal values
- quadrantal angle
- radian
- reciprocal trigonometric functions
- reference angle
- secant
- sine
- sinusoidal function
- standard position
- tangent
- terminal side
- trigonometric function
- trigonometric ratio
- trigonometry
- unit circle
- vertical shift

Are You Ready?

Complete the Quick Review to see if you are ready to start this module. Then complete the Quick Check.

Quick Review

Example 1

Find the value of a. Round to the nearest tenth if necessary.

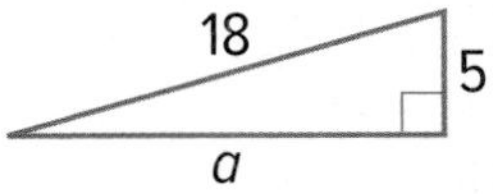

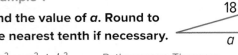

$c^2 = a^2 + b^2$	Pythagorean Theorem
$18^2 = a^2 + 5^2$	Replace c with 18 and b with 5.
$324 = a^2 + 25$	Simplify.
$299 = a^2$	Subtract 25 from each side.
$17.3 \approx a$	Take the positive square root of each side.

Example 2

Find the missing measures. Write all radicals in simplest form.

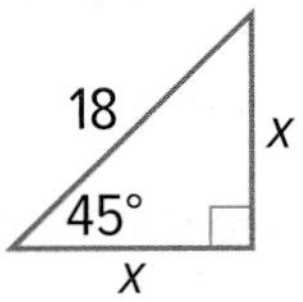

$x^2 + x^2 = 18^2$	Pythagorean Theorem
$2x^2 = 18^2$	Combine like terms.
$2x^2 = 324$	Simplify.
$x^2 = 162$	Divide each side by 2.
$x = \sqrt{162}$	Take the positive square root of each side.
$x = 9\sqrt{2}$	Simplify.

Quick Check

Find the value of x. Round to the nearest tenth if necessary.

1.

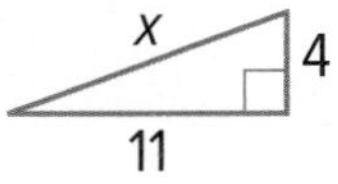

2.

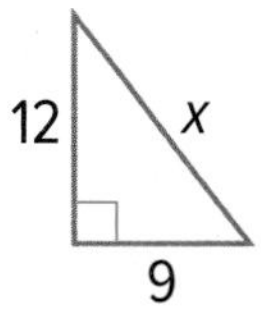

3.

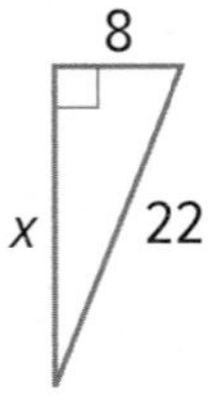

4.

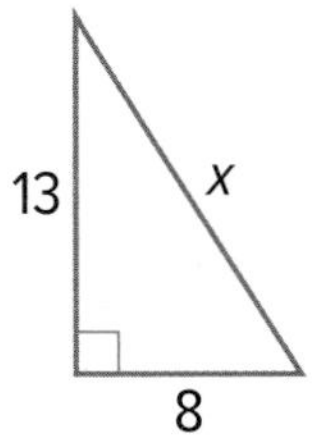

Find the values of x and y. Write all radicals in simplest form.

5.

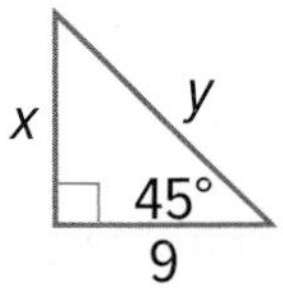

6.

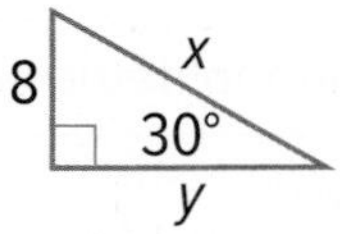

7.

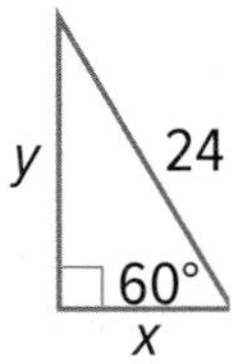

8.

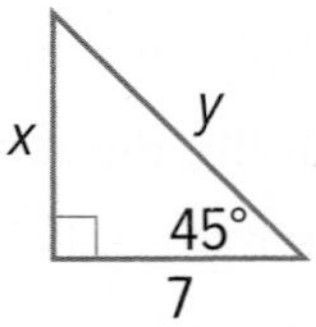

How Did You Do?

Which exercises did you answer correctly in the Quick Check?

Lesson 11-1

Angles and Angle Measure

Learn Angles in Standard Position

An angle on the coordinate plane is in **standard position** if the vertex is at the origin and one ray is on the positive *x*-axis. The **initial side** of the angle is the ray that is fixed on the *x*-axis. The **terminal side** is the ray that rotates about the center.

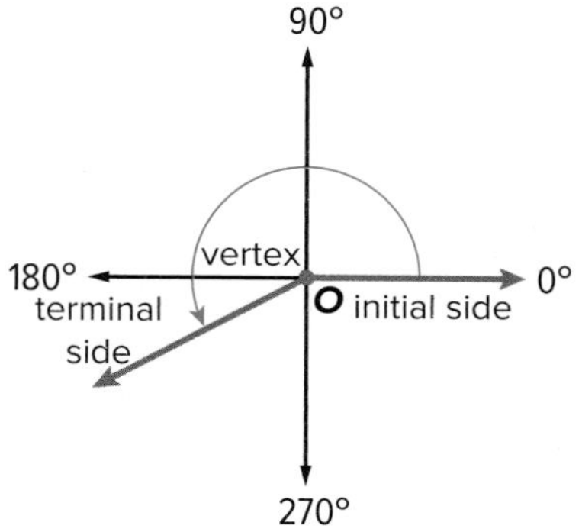

An angle in standard position is measured by the amount of rotation in degrees from the initial side to the terminal side. Angle measures can be positive or negative.

Key Concept • Angle Measures

Positive Angle Measures	Negative Angle Measures
If the measure of an angle is positive, the terminal side is rotated counterclockwise.	If the measure of an angle is negative, the terminal side is rotated clockwise.
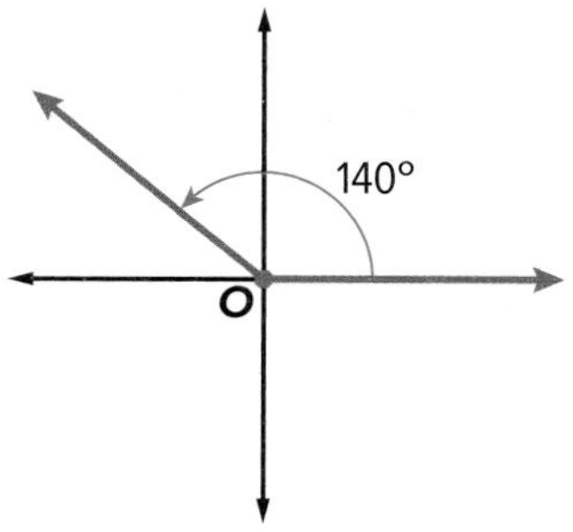	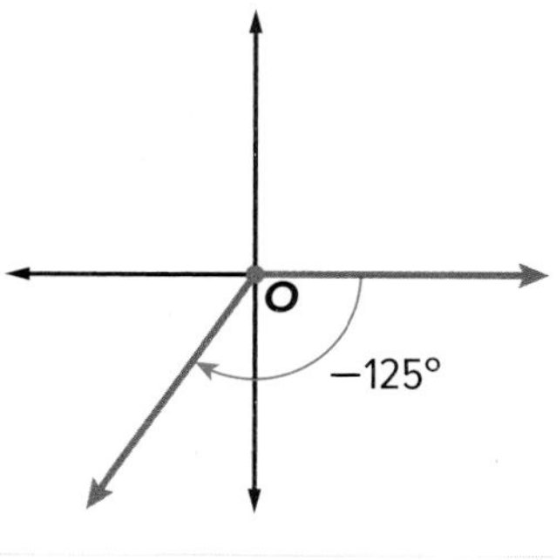

The terminal side of an angle can make more than one complete rotation. If rotating counterclockwise, the angle measures increase past 360°. If rotating clockwise, the angle measures decrease past −360°.

Angles in standard position that have the same terminal side are **coterminal angles**. Coterminal angles can have different measures.

$30° + 360° = 390°$ $\qquad$ $30° - 360° = -330°$

Add or subtract any multiple of 360° to find another coterminal angle. An infinite number of coterminal angles can be found for any angle.

Today's Goals

- Draw angles in standard position and identify coterminal angles.
- Convert between degree and radian measures and find arc lengths by using central angles.

Today's Vocabulary

standard position
initial side
terminal side
coterminal angles
radian
central angle of a circle

Talk About It

Research the meaning of *initial* and *terminal*. How do these definitions relate to the meanings of the *initial side* and *terminal side* of an angle?

Study Tip

Angle of Rotation In trigonometry, an angle is sometimes referred to as an *angle of rotation*.

Example 1 Draw an Angle in Standard Position

Draw an angle in standard position that measures 200°.

Because the measure of the angle is positive, rotate the terminal side counterclockwise.

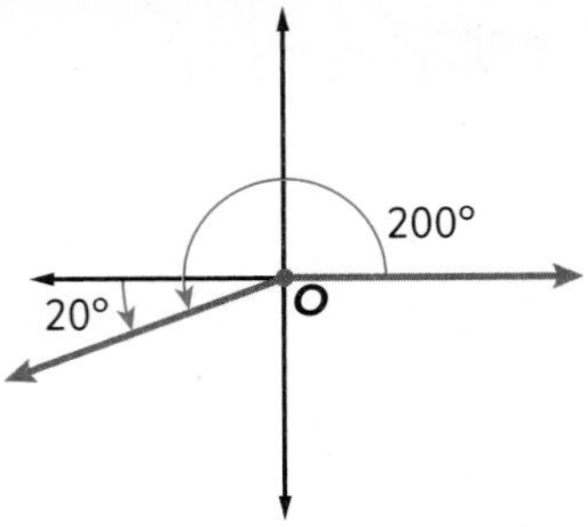

The negative x-axis represents 180°. Because $200° = 180° + 20°$, draw the terminal side 20° counterclockwise past the negative x-axis.

Go Online to see another example about drawing angles in standard position.

Example 2 Draw an Angle with More Than One Rotation

Draw an angle in standard position that measures 475°.

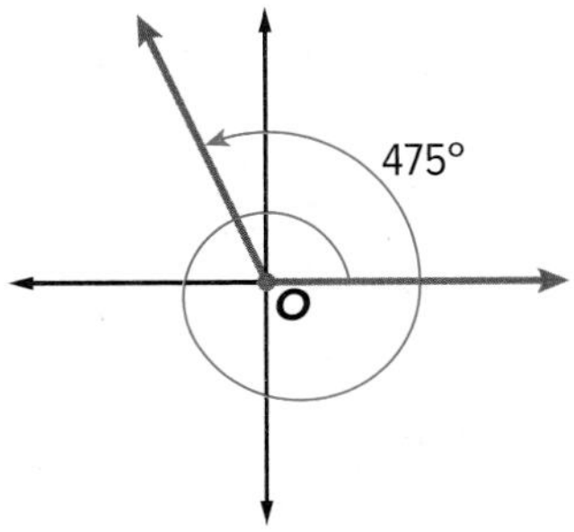

$475° = 360° + 115°$

The terminal side of the angle will make more than one rotation counterclockwise. First, make a full rotation starting at the positive x-axis. Continue the rotation. Draw the terminal side of the angle 115° counterclockwise past the positive x-axis.

Check

Draw an angle in standard position that measures −400°.

Example 3 Identify Coterminal Angles

Find an angle with a positive measure and an angle with a negative measure that are coterminal with a 35° angle.

positive angle: $35° + 360° = 395°$

negative angle: $35° - 360° = -325°$

Check

Find an angle with a positive measure and an angle with a negative measure that are coterminal with a −50° angle.

positive angle: ___?___°

negative angle: ___?___°

Go Online You can complete an Extra Example online.

Study Tip

Axes It may be helpful to use the angle measures represented by the positive and negative x- and y-axes when drawing an angle in standard form. As the terminal side rotates around the coordinate plane, it has rotated an additional 90° each time it reaches an axis. You can use the axes as benchmarks and continue the rotation from the appropriate axis.

Think About It!

Write expressions that can be used to find the measure of four different angles that are all coterminal with an angle that measures $x°$.

Explore Arc Length

Online Activity

INQUIRY How can you use a central angle to determine the length of an arc?

Learn Degrees and Radians

Angles can also be measured in units called **radians**, which are based on arc length. A radian is the measure of an angle θ in standard position with a terminal side that intercepts an arc with the same length as the radius of the circle. The circumference of a circle is $2\pi r$ and an angle that measures 1 radian has an arc length of r. So, one complete revolution around a circle equals 2π radians.

Key Concept • Convert Between Degrees and Radians

Degrees to Radians	Radians to Degrees
To convert a degree measure to radians, multiply the number of degrees by $\frac{\pi \text{ radians}}{180°}$.	To convert a radian measure to degrees, multiply the number of radians by $\frac{180°}{\pi \text{ radians}}$.

There are special angles with equivalent degree and radian measures that may be helpful to memorize. Other special angles are multiples of these angles.

$30° = \frac{\pi}{6}$ radians

$45° = \frac{\pi}{4}$ radians

$60° = \frac{\pi}{3}$ radians

$90° = \frac{\pi}{2}$ radians

A **central angle of a circle** has its vertex at the center of the circle and sides that are radii. Central angles intercept an arc of the circle. The endpoints of the intercepted arc lie on the rays of the central angle. If you know the measure of a central angle and the radius of the circle, you can find the length of the arc that is intercepted by the angle.

Key Concept • Arc Length

Words: For a circle with radius r and central angle θ (in radians), the arc length s equals the product of r and θ.

Symbols: $s = r\theta$

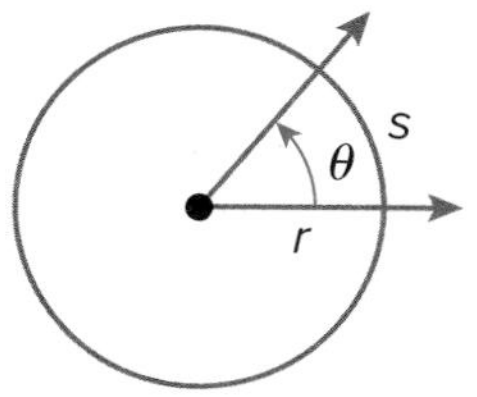

Go Online You can complete an Extra Example online.

Think About It!

Given that 2π radians = 360° and π radians = 180°, what is the equivalent degree measure of 1 radian? Explain.

Study Tip

Positive and Negative Radians As with degrees, radians measure the amount of rotation from the initial side to the terminal side. Radians may also be positive or negative, depending on the direction of the rotation.

Think About It!

How would you draw an angle that measures $-\frac{5\pi}{9}$ radians in standard position? Describe the position of the terminal side.

Example 4 Convert Degrees to Radians

Rewrite −100° in radians.

$-100° = -100° \cdot \frac{\pi \text{ radians}}{180°}$ — Multiply by conversion factor.

$= \frac{-100\pi}{180}$ or $-\frac{5\pi}{9}$ radians — Simplify.

Check

Rewrite −220° in radians. Write the solution as a fraction in simplest form.

___?___ radians

Example 5 Convert Radians to Degrees

Rewrite $\frac{11\pi}{4}$ in degrees.

$\frac{11\pi}{4} = \frac{11\pi}{4} \text{ radians} \cdot \frac{180°}{\pi \text{ radians}}$ — Multiply by conversion factor.

$= \frac{11\cancel{\pi}}{4} \cancel{\text{radians}} \cdot \frac{180°}{\cancel{\pi \text{ radians}}}$ — Cancel units.

$= \frac{1980°}{4}$ or $495°$ — Simplify.

Think About It!

Is it reasonable that the degree measure of the angle in Example 5 is greater than 360°? Justify your conclusion.

Example 6 Find Arc Length

TRAFFIC **A traffic circle, or roundabout, is a circular roadway at the intersection of two or more streets that allows cars to travel through more continuously than a traffic light or stop sign. The diameter of a traffic circle is 160 feet. How far does a car travel in the roundabout if it goes three-fourths of the way around?**

Step 1 Find the central angle in radians.

$\theta = \frac{3}{4} \cdot 2\pi$ or $\frac{3\pi}{2}$ — The angle is $\frac{3}{4}$ of a rotation.

Step 2 Find the arc length.

$s = r\theta$ — Formula for arc length

$= 80 \cdot \frac{3\pi}{2}$ — $r = 0.5d = 0.5(160)$ or 80 and $\theta = \frac{3\pi}{2}$

$= 120\pi$ — Multiply.

≈ 376.99 — Use a calculator.

A car that travels three-fourths of the way around the traffic circle will travel about 377 feet.

Watch Out!

Arc Length Remember to write angle measures in radians, not degrees, when finding arc length.

Check

SURVEYING If a surveyor's wheel with a radius of 15 inches completes $\frac{13}{20}$ of a rotation, what is the total distance traveled in feet? Round to the nearest hundredth if necessary.

___?___ ft

Go Online You can complete an Extra Example online.

Practice

Go Online You can complete your homework online.

Examples 1 and 2

Draw an angle with the given measure in standard position.

1. $160°$ **2.** $280°$ **3.** $400°$

4. $185°$ **5.** $810°$ **6.** $390°$

7. $495°$ **8.** $-50°$ **9.** $-420°$

10. $210°$ **11.** $305°$ **12.** $580°$

13. $135°$ **14.** $-450°$ **15.** $-560°$

Example 3

Find an angle with a positive measure and an angle with a negative measure that are coterminal with each angle.

16. $65°$ **17.** $-75°$ **18.** $230°$

19. $45°$ **20.** $60°$ **21.** $370°$

22. $-90°$ **23.** $420°$ **24.** $30°$

25. $55°$ **26.** $80°$ **27.** $110°$

28. $\frac{2\pi}{5}$ **29.** $\frac{5\pi}{6}$ **30.** $-\frac{3\pi}{2}$

31. $\frac{2\pi}{3}$ **32.** $\frac{5\pi}{2}$ **33.** $-\frac{3\pi}{4}$

Example 4

Rewrite each degree measure in radians.

34. $140°$ **35.** $-260°$ **36.** $-75°$

37. $380°$ **38.** $130°$ **39.** $720°$

40. $210°$ **41.** $90°$ **42.** $-30°$

Example 5

Rewrite each radian measure in degrees.

43. $-\frac{3\pi}{5}$ **44.** $\frac{7\pi}{6}$ **45.** $\frac{\pi}{3}$

46. $\frac{5\pi}{6}$ **47.** $\frac{2\pi}{3}$ **48.** $\frac{5\pi}{4}$

49. $-\frac{3\pi}{4}$ **50.** $-\frac{7\pi}{6}$ **51.** $\frac{\pi}{4}$

Example 6

52. TRANSPORTATION A traffic roundabout has a diameter of 200 meters. How far does an automobile travel in the roundabout if it goes one-fourth of the way around?

53. ANALOG CLOCKS The length of the minute hand of an analog clock is 5 inches. If the minute hand rotates from 12 noon to 12:40 P.M., then how far does its point move?

Mixed Exercises

REGULARITY **Rewrite each degree measure in radians and each radian measure in degrees.**

54. 18° **55.** 6° **56.** −72°

57. −820° **58.** 4π **59.** $\frac{5\pi}{2}$

60. $-\frac{9\pi}{2}$ **61.** $-\frac{7\pi}{12}$ **62.** −270°

Find the length of each arc. Round to the nearest tenth.

63.

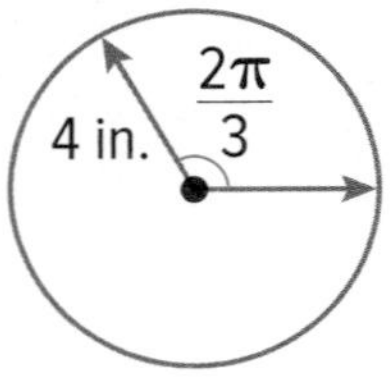

64.

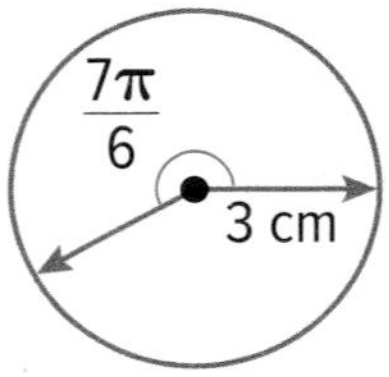

65.

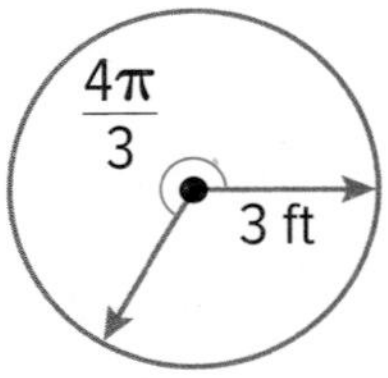

66. TIME Find both the degree and radian measures of the angle through which the hour hand on a clock rotates from 5 A.M. to 10 P.M.

67. ROTATION A truck with 16-inch radius wheels is driven at 77 feet per second (52.5 miles per hour). Find the measure of the angle through which a point on the outside of the wheel travels each second. Round to the nearest degree and nearest radian.

Ian Hubball/Alamy Stock Photo

68. PLANETS Earth makes one full rotation on its axis every 24 hours. How long does it take Earth to rotate through 150°? Neptune makes one full rotation on its axis every 16 hours. How long does it take Neptune to rotate through 150°?

69. SURVEYING If a surveyor's wheel with a diameter of 19 inches completes $\frac{5}{6}$ of a rotation, what is the total distance traveled in inches? Round to the nearest hundredth if necessary.

70. **STRUCTURE** It is convenient to know the measures of some specific angles in both degree and radian measures.

a. Copy and complete the figure at the right by writing the radian measure for each angle.

b. Describe at least one pattern or symmetry that you notice in the completed figure.

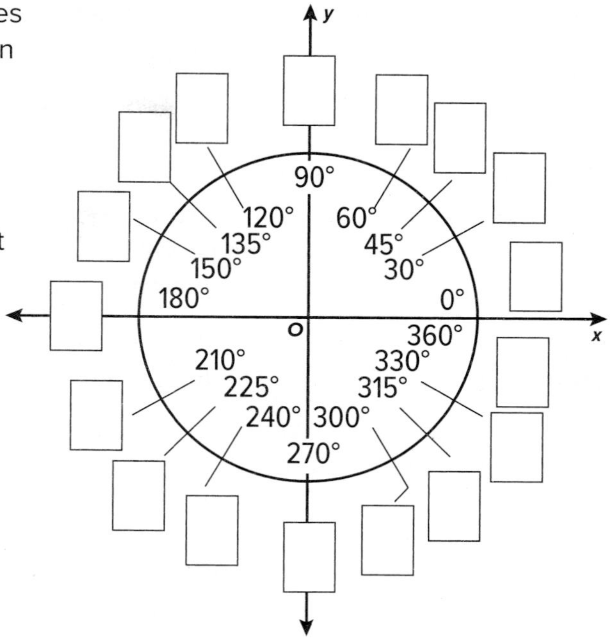

71. **CLOCKS** Through what angle, in degrees and radians, does the hour hand on a clock rotate between 4 P.M. and 7 P.M.? Assuming the length of the hour hand is 6 inches, find the arc length of the circle made by the hour hand during that time.

72. **AMUSEMENT PARKS** The carousel at an amusement park has 20 horses spaced evenly around its circumference. The horses are numbered consecutively from 1 to 20. The carousel completes one rotation about its axis every 40 seconds.

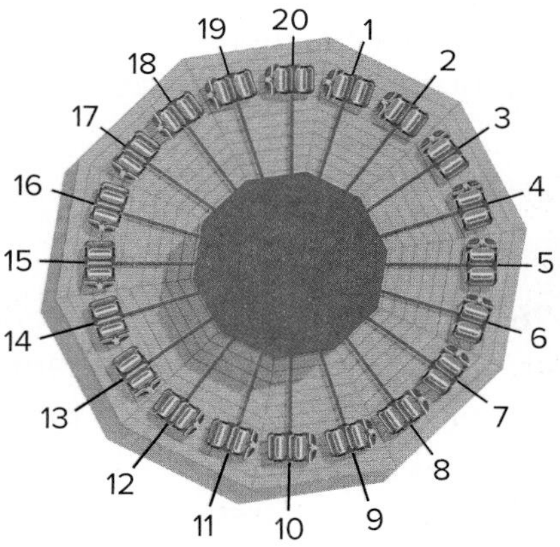

a. What is the central angle, in degrees, formed by horse #1 and horse #8?

b. What is the speed of the carousel in rotations per minute?

c. What is the speed of the carousel in radians per minute?

d. A child rides the carousel for 6 minutes. Through how many radians will the child pass in the course of the carousel ride?

73. **FERRIS WHEELS** The London Eye is one of the world's largest Ferris wheels. The diameter of the wheel is 135 meters, and it makes a complete rotation in 30 minutes. A passenger gets on the ride and travels for 5 minutes before the ride stops. The passenger wants to know how far she traveled during this time.

a. During the 5-minute interval, what is the measure of the central angle of the wheel's rotation in radians? Explain.

b. Explain how to find the distance the passenger traveled to the nearest meter.

c. Explain how you know your answer is reasonable.

74. PIZZA A circular pizza with a diameter of 18 inches is cut into 8 congruent slices. What is the radian measure of the central angle of each slice? Explain.

75. CONSTRUCT ARGUMENTS A lawn sprinkler produces a stream of water that reaches 15 feet from the sprinkler head. The sprinkler rotates to sweep out part of a circle. The area of the lawn that gets watered as the sprinkler moves back and forth is 75π square feet. Melinda says it is possible to determine the radian measure of the angle that is swept out by the sprinkler. Seiji says there is not enough information to determine this. Who is correct? Justify your argument.

76. REASONING Consider the figure at the right.

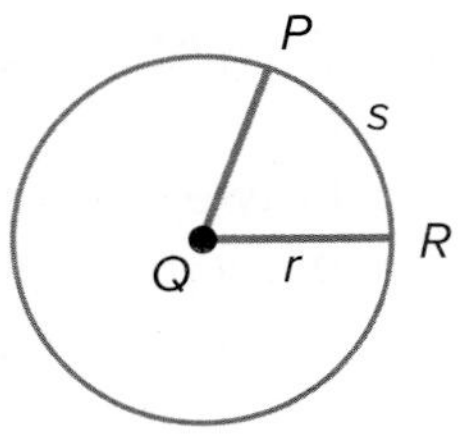

a. Suppose $s = r$. What is $m\angle PQR$ in radians? What is the measure of the angle to the nearest tenth of a degree?

b. Suppose $2s = r$. What is $m\angle PQR$ in radians? What is the measure of the angle to the nearest tenth of a degree?

77. STRUCTURE Circle C has a radius of 5 centimeters. The function $f(x)$ gives the length in centimeters of an arc of circle C that is subtended by a central angle of x radians. Describe the graph of $f(x)$ as precisely as possible, and justify your answer.

78. FIND THE ERROR Tarshia and Alan are writing an expression for the measure of an angle coterminal with the angle shown at the right. Is either of them correct? Explain your reasoning.

Tarshia	Alan
The measure of a coterminal angle is $(x - 360)°$.	The measure of a coterminal angle is $(360 - x)°$.

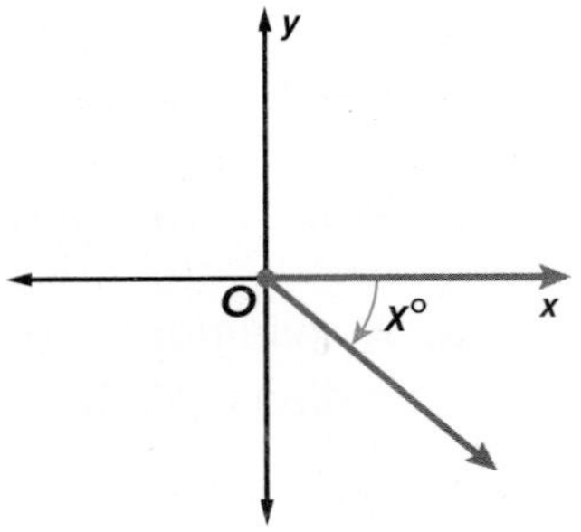

79. PERSEVERE A line makes an angle of $\frac{\pi}{2}$ radians with the positive x-axis at the point (2, 0). Find the equation of the line.

80. ANALYZE Express $\frac{1}{8}$ of a revolution in degrees and in radians. Justify your argument.

81. CREATE Draw and label an acute angle in standard position. Find two angles, one positive and one negative, that are coterminal with the angle.

82. ANALYZE Justify the formula for the length of an arc.

83. WRITE Use a circle with radius r to describe what one degree and one radian represent. Then explain how to convert between the measures.

Trigonometric Functions of General Angles

Learn Trigonometric Functions in Right Triangles

Trigonometry is the study of relationships among the angles and sides of triangles. A **trigonometric ratio** compares the lengths of two sides of a right triangle. A **trigonometric function** relates the angles of a triangle to the lengths of its sides.

Key Concepts • Trigonometric Functions in Right Triangles

If θ is the measure of an acute angle of a right triangle, then the trigonometric functions involving the opposite side *opp*, the adjacent side *adj*, and the hypotenuse *hyp* are defined as follows.

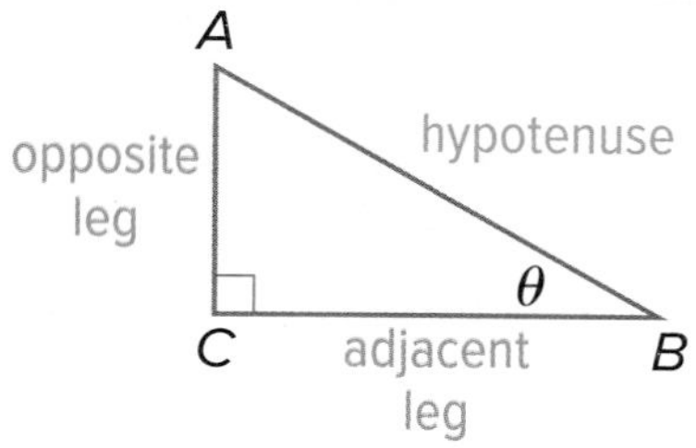

sine: $\sin\theta = \frac{\text{opp}}{\text{hyp}}$

cosine: $\cos\theta = \frac{\text{adj}}{\text{hyp}}$

tangent: $\tan\theta = \frac{\text{opp}}{\text{adj}}$

cosecant: $\csc\theta = \frac{\text{hyp}}{\text{opp}}$

secant: $\sec\theta = \frac{\text{hyp}}{\text{adj}}$

cotangent: $\cot\theta = \frac{\text{adj}}{\text{opp}}$

The Key Concept shows how to define the six trigonometric functions by using ratios of the sides of a triangle. Alternately, once sine and cosine are defined in this way, the remaining trigonometric functions can be defined in terms of sine and/or cosine. For example, since $\sin\theta = \frac{\text{opp}}{\text{hyp}}$, $\frac{1}{\sin\theta} = \frac{\text{hyp}}{\text{opp}}$, which is the same ratio used to define $\csc\theta$. Thus, $\csc\theta = \frac{1}{\sin\theta}$.

Example 1 Evaluate Trigonometric Functions

Find the exact values of the six trigonometric functions for angle θ.

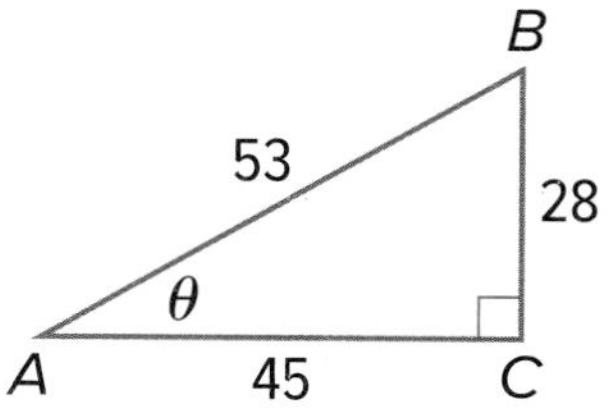

leg opposite θ: $BC = 28$

leg adjacent θ: $AC = 45$

hypotenuse: $AB = 53$

$\sin\theta = \frac{\text{opp}}{\text{hyp}} = \frac{28}{53}$ $\qquad$ $\cos\theta = \frac{\text{adj}}{\text{hyp}} = \frac{45}{53}$

$\tan\theta = \frac{\text{opp}}{\text{adj}} = \frac{28}{45}$ $\qquad$ $\csc\theta = \frac{\text{hyp}}{\text{opp}} = \frac{53}{28}$

$\sec\theta = \frac{\text{hyp}}{\text{adj}} = \frac{53}{45}$ $\qquad$ $\cot\theta = \frac{\text{adj}}{\text{opp}} = \frac{45}{28}$

Go Online You can complete an Extra Example online.

Today's Goals
- Find values of trigonometric functions for acute angles.
- Find values of trigonometric functions of general angles.
- Find values of trigonometric functions by using reference angles.

Today's Vocabulary
trigonometry
trigonometric ratio
trigonometric function
sine
cosine
tangent
cosecant
secant
cotangent
quadrantal angle
reference angle

Go Online
You may want to complete the Concept Check to check your understanding.

Check

Find the exact values of the six trigonometric functions for angle θ.

$\sin \theta =$? $\cos \theta =$?

$\tan \theta =$? $\csc \theta =$?

$\sec \theta =$? $\cot \theta =$?

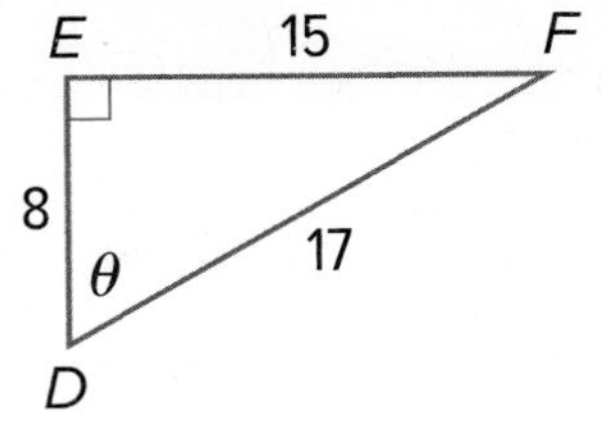

Study Tip

Reciprocal Functions The secant, cosecant, and cotangent functions are called the **reciprocal functions** because their ratios are the reciprocals of the sine, cosine, and tangent functions.

Example 2 Find Trigonometric Ratios

If $\cos A = \frac{9}{13}$, find the exact values of the five remaining trigonometric functions for A.

Step 1 Use the given information to draw a right triangle. Copy the figure and label the sides and vertices.

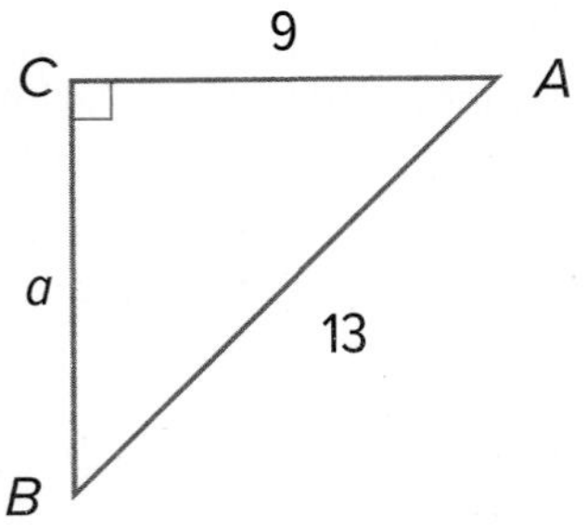

Step 2 Use the Pythagorean Theorem to find a.

To write the remaining trigonometric functions, first find the missing side length, a. Because $\triangle ABC$ is a right triangle, use the Pythagorean Theorem.

$a^2 + b^2 = c^2$	Pythagorean Theorem
$a^2 + 9^2 = 13^2$	$b = 9$ and $c = 13$
$a^2 + 81 = 169$	Simplify.
$a^2 = 88$	Subtract 81 from each side.
$a = \pm\sqrt{88}$	Take the square root of each side.
$a = \sqrt{88}$ or $2\sqrt{22}$	Length cannot be negative.

Step 3 Find the values.

leg opposite $\angle A$: $BC = 2\sqrt{22}$

leg adjacent $\angle A$: $AC = 9$

hypotenuse: $AB = 13$

$\sin A = \frac{\text{opp}}{\text{hyp}} = \frac{2\sqrt{22}}{13}$ $\csc A = \frac{\text{hyp}}{\text{opp}} = \frac{13}{2\sqrt{22}}$ or $\frac{13\sqrt{22}}{44}$

$\cos A = \frac{\text{adj}}{\text{hyp}} = \frac{9}{13}$ $\sec A = \frac{\text{hyp}}{\text{adj}} = \frac{13}{9}$

$\tan A = \frac{\text{opp}}{\text{adj}} = \frac{2\sqrt{22}}{9}$ $\cot A = \frac{\text{adj}}{\text{opp}} = \frac{9}{2\sqrt{22}}$ or $\frac{9\sqrt{22}}{44}$

Study Tip

Labeling Triangles A capital letter is often used to represent both a vertex of a triangle and the measure of the angle at that vertex. The same letter in lowercase is used to represent both the side opposite that angle and the length of that side.

Check

If $\sec B = \frac{11}{3}$, find the exact values of the five remaining trigonometric functions for B.

$\sin B =$?, $\cos B =$?, $\tan B =$?, $\csc B =$?, $\cot B =$?

Go Online You can complete an Extra Example online.

Learn Trigonometric Functions of General Angles

You can find values of trigonometric functions for angles greater than 90° or less than 0° when given a point on the terminal side of the angle.

Key Concept • Trigonometric Functions of General Angles

Let θ be an angle in standard position, and let $P(x, y)$ be a point on its terminal side. By the Pythagorean Theorem, $r = \sqrt{x^2 + y^2}$, where r is the distance from the origin to point P along the terminal side. Using the coordinates of point P and r, the six trigonometric functions of θ are defined below.

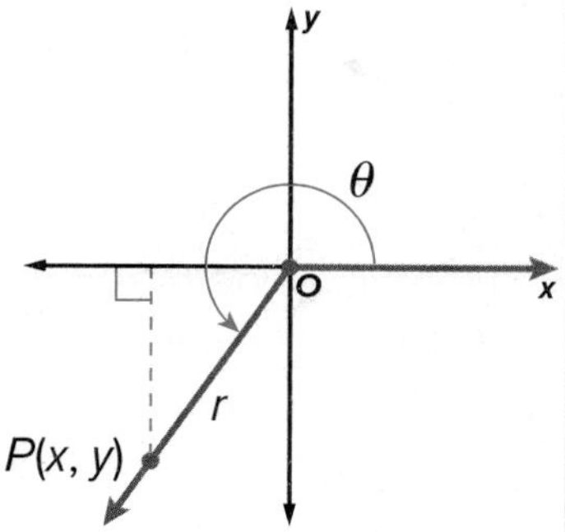

$\sin \theta = \frac{y}{r}$ $\qquad \cos \theta = \frac{x}{r}$

$\tan \theta = \frac{y}{x}, x \neq 0$ $\qquad \csc \theta = \frac{r}{y}, y \neq 0$

$\sec \theta = \frac{r}{x}, x \neq 0$ $\qquad \cot \theta = \frac{x}{y}, y \neq 0$

If the terminal side of angle θ in standard position coincides with one of the axes, the angle is called a **quadrantal angle**.

Since r is the distance from the origin to point $P(x, y)$ along the terminal side:

- if $\theta = 0°$ or $360°$, then $r = x$.
- if $\theta = 90°$, then $r = y$.
- if $\theta = 180°$, then $r = |x|$.
- if $\theta = 270°$, then $r = |y|$.

Think About It!
Why can the Pythagorean Theorem be used to find r?

Think About It!
Why is it necessary to define r using absolute value for $\theta = 180°$ and $\theta = 270°$?

Example 3 Evaluate Trigonometric Functions Given a Point

The terminal side of θ in standard position contains the point (−6, 4). Find the exact values of the six trigonometric functions of θ.

Step 1 Draw the angle.
Draw Point P to draw θ with the terminal side through (−6, 4).

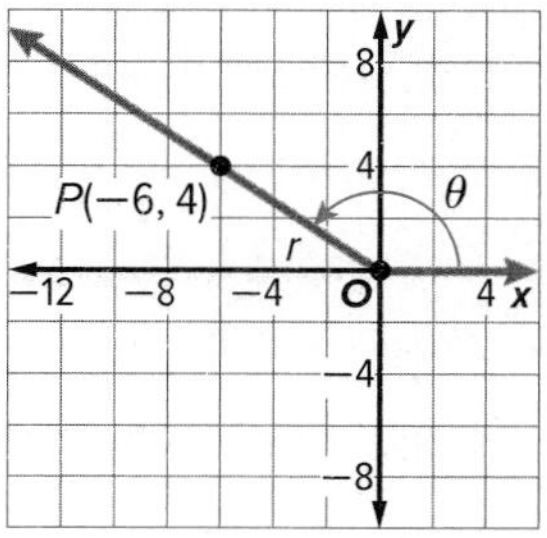

Step 2 Find *r*.
Use the Pythagorean Theorem to find the value of r.

$r = \sqrt{x^2 + y^2}$ — Pythagorean Theorem

$= \sqrt{(-6)^2 + 4^2}$ — $x = -6$ and $y = 4$

$= \sqrt{52}$ or $2\sqrt{13}$ — Simplify.

(continued on the next page)

Step 3 Find the trigonometric functions.

Use $x = -6$, $y = 4$, and $r = 2\sqrt{13}$ to write the trigonometric functions.

$$\sin\theta = \frac{y}{r} = \frac{4}{2\sqrt{13}} \text{ or } \frac{2\sqrt{13}}{13} \qquad \csc\theta = \frac{r}{y} = \frac{2\sqrt{13}}{4} \text{ or } \frac{\sqrt{13}}{2}$$

$$\cos\theta = \frac{x}{r} = \frac{-6}{2\sqrt{13}} \text{ or } -\frac{3\sqrt{13}}{13} \qquad \sec\theta = \frac{r}{x} = \frac{2\sqrt{13}}{-6} \text{ or } -\frac{\sqrt{13}}{3}$$

$$\tan\theta = \frac{y}{x} = \frac{4}{-6} \text{ or } -\frac{2}{3} \qquad \cot\theta = \frac{x}{y} = \frac{-6}{4} \text{ or } -\frac{3}{2}$$

Check

The terminal side of θ in standard position contains the point (2, −8). Find the exact values of the six trigonometric functions of θ.

$\sin\theta =$ ___?___, $\cos\theta =$ ___?___, $\tan\theta =$ ___?___,

$\csc\theta =$ ___?___, $\sec\theta =$ ___?___, $\cot\theta =$ ___?___

Example 4 Evaluate Trigonometric Functions of Quadrantal Angles

The terminal side of θ in standard position contains the point $P(-5, 0)$. Find the exact values of the six trigonometric functions of θ.

The point P lies on the negative x-axis, so the measure of the quadrantal angle θ is 180°.

Use $x = -5$, $y = 0$, and $r = 5$ to write the trigonometric functions.

$$\sin\theta = \frac{y}{r} = \frac{0}{5} \text{ or } 0 \qquad \csc\theta = \frac{r}{y} = \frac{5}{0} \text{ undefined}$$

$$\cos\theta = \frac{x}{r} = \frac{-5}{5} \text{ or } -1 \qquad \sec\theta = \frac{r}{x} = \frac{5}{-5} \text{ or } -1$$

$$\tan\theta = \frac{y}{x} = \frac{0}{-5} \text{ or } 0 \qquad \cot\theta = \frac{x}{y} = \frac{-5}{0} \text{ undefined}$$

Learn Trigonometric Functions with Reference Angles

For a nonquadrantal angle θ in standard position, its **reference angle** is the acute angle θ' formed by the terminal side and the x-axis. The rules for finding the measures of reference angles vary depending on the quadrant in which the terminal side is located.

If the terminal side is in:

- Quadrant I, then $\theta' = \theta$.
- Quadrant II, then $\theta' = 180° - \theta$ or $\theta' = \pi - \theta$.
- Quadrant III, then $\theta' = \theta - 180°$ or $\theta' = \theta - \pi$.
- Quadrant IV, then $\theta' = 360° - \theta$ or $\theta' = 2\pi - \theta$.

You can use reference angles to evaluate trigonometric functions for any angle θ. The sign of the function is determined by the quadrant in which the terminal side of θ lies.

Go Online You can complete an Extra Example online.

Math History Minute

Indian mathematician **Aryabhata (476–550)** was the author of several treatises on mathematics and astronomy. He approximated the value of π and discussed the concept of sine in one of his works, *ardha-jya*, which literally means "half-chord." India's first satellite, launched in 1975, is named in his honor.

Think About It!

For each quadrantal angle, which trigonometric functions are undefined? Justify your reasoning.

Key Concept • Evaluating Trigonometric Functions

Step 1 Find the measure of the reference angle θ'.
Step 2 Evaluate the trigonometric function for θ'.
Step 3 Determine the sign of the trigonometric function values. Use the quadrant in which the terminal side of θ lies.

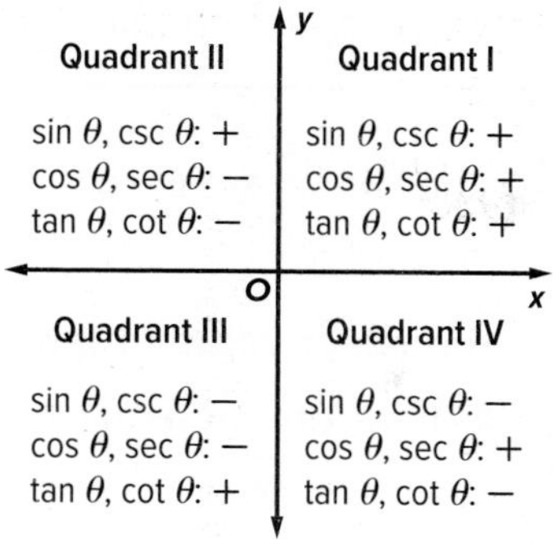

It may be helpful to memorize the trigonometric values for these special angles.

Trigonometric Values for Special Angles

$\sin 30° = \frac{1}{2}$	$\sin 45° = \frac{\sqrt{2}}{2}$	$\sin 60° = \frac{\sqrt{3}}{2}$
$\cos 30° = \frac{\sqrt{3}}{2}$	$\cos 45° = \frac{\sqrt{2}}{2}$	$\cos 60° = \frac{1}{2}$
$\tan 30° = \frac{\sqrt{3}}{3}$	$\tan 45° = 1$	$\tan 60° = \sqrt{3}$
$\csc 30° = 2$	$\csc 45° = \sqrt{2}$	$\csc 60° = \frac{2\sqrt{3}}{3}$
$\sec 30° = \frac{2\sqrt{3}}{3}$	$\sec 45° = \sqrt{2}$	$\sec 60° = 2$
$\cot 30° = \sqrt{3}$	$\cot 45° = 1$	$\cot 60° = \frac{\sqrt{3}}{3}$

Example 5 Find Reference Angles

Sketch each angle. Then find the measure of its reference angle.

a. 155°

The terminal side of 155° lies in Quadrant II. The reference angle θ' is formed by the terminal side and the negative *x*-axis.

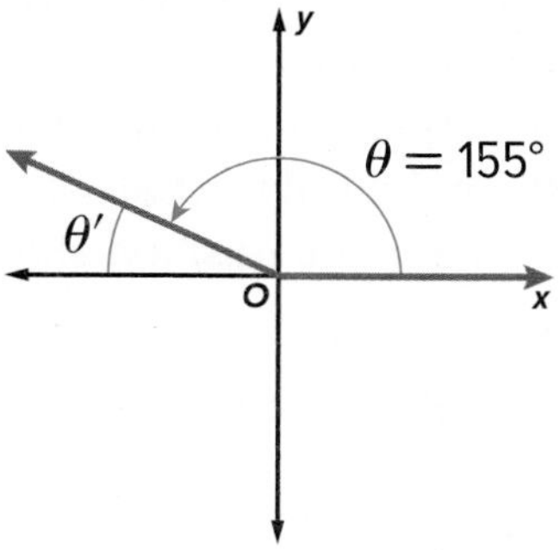

$\theta' = 180 - \theta$

$= 180 - 155$ or $25°$

b. $-\frac{8\pi}{3}$

Because $-\frac{8\pi}{3}$ is not between 0 and 2π, first find a coterminal angle with a positive measure.

coterminal angle: $-\frac{8\pi}{3} + 2(2\pi) = \frac{4\pi}{3}$

The terminal side of $\frac{4\pi}{3}$ lies in Quadrant III. The reference angle θ' is formed by the terminal side and the negative *x*-axis.

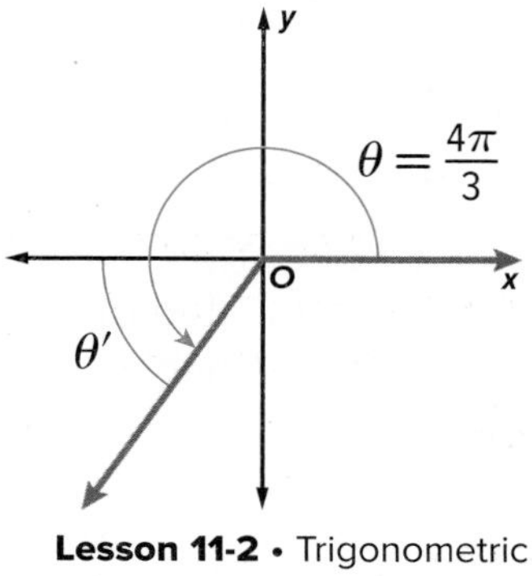

$\theta' = \theta - \pi$

$= \frac{4\pi}{3} - \pi$ or $\frac{\pi}{3}$

Think About It!
What are the possible measures for a reference angle θ'?

Study Tip
Coterminal Angles If the measure of θ is not between 0° and 360°, then use a coterminal angle with a positive measure between 0° and 360° to find the reference angle.

Go Online
You can complete an Extra Example online.

Study Tip

Special Angles If the measure of an angle is a multiple of 30°, 45°, or 60°, its reference angle will be a special angle.

Check

Find the reference angle θ' of an angle that measures $\frac{23\pi}{6}$. ___?___

Example 6 Use a Reference Angle to Find a Trigonometric Value

Find the exact value of tan $\frac{7\pi}{4}$.

Part A Find the measure of the reference angle.

The terminal side of $\frac{7\pi}{4}$ lies in Quadrant IV.

$\theta' = 2\pi - \theta$ — Find the reference angle θ'.

$= 2\pi - \frac{7\pi}{4}$ or $\frac{\pi}{4}$ — $\theta = \frac{7\pi}{4}$

Part B Use the reference angle to evaluate the function.

$\tan \frac{7\pi}{4} = -\tan \frac{\pi}{4}$ — The tangent function is negative in Quadrant IV.

$= -\tan 45°$ — $\frac{\pi}{4}$ radians $= 45°$

$= -1$ — $\tan 45° = 1$

Check

Find the exact value of sec 225°.

Use a Source

Measure the length of your arm from shoulder to hand and your shoulder height. Use the data to find the height of the ball at the same position in the example if you were pitching.

Example 7 Use Trigonometric Functions

SOFTBALL **During a windmill pitch in fastpitch softball, a pitcher's arm makes a complete clockwise rotation before releasing the ball. Suppose a pitcher has an arm length of 28 inches, and the axis from which the pitcher's arm swings is at her shoulder height of 57 inches. What is the height of the ball when the pitcher's arm is in the position shown?**

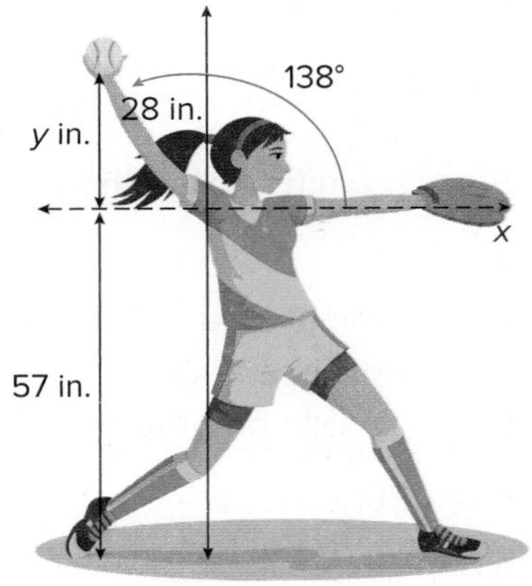

The pitching arm is in Quadrant II, so $\theta = 180° - 138°$ or $42°$.

We know the radius r and need to find y. The sine function relates these values.

$\sin \theta = \frac{y}{r}$ — Sine function of a general angle

$\sin 42° = \frac{y}{28}$ — $\theta = 42°$ and $r = 28$

$28 \sin 42° = y$ — Multiply each side by 28.

$18.7 \approx y$ — Use a calculator to solve for y.

Since $y \approx 18.7$, the total height of the ball is $18.7 + 57$, or about 75.7 inches.

Go Online You can complete an Extra Example online.

Practice

Go Online You can complete your homework online.

Example 1

Find the exact values of the six trigonometric functions for angle θ.

1.

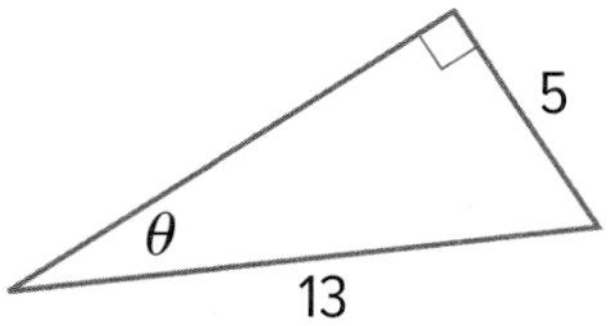

2.

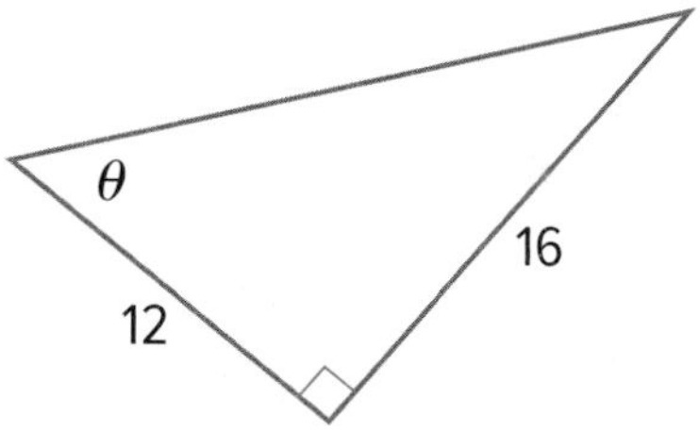

3.

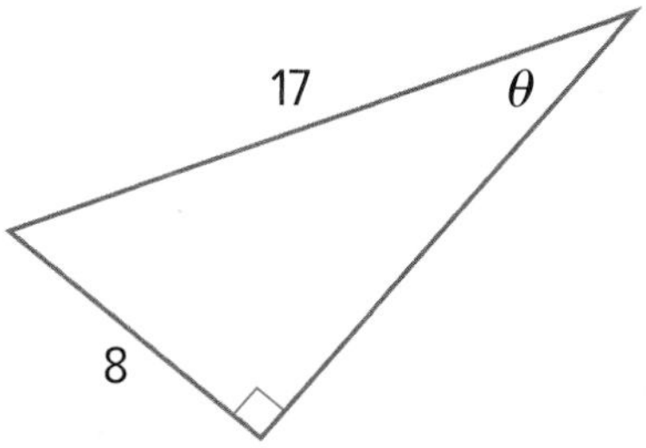

4.

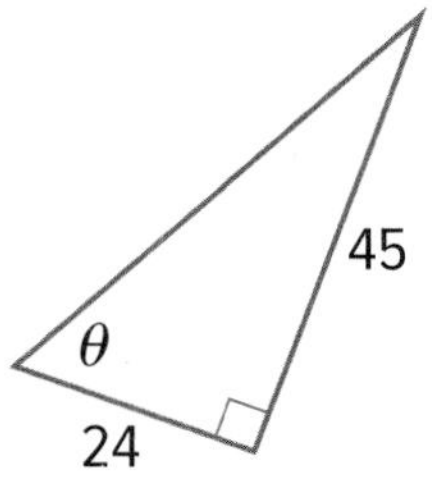

5.

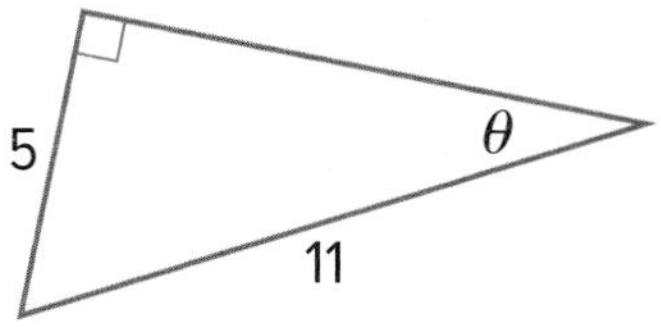

6.

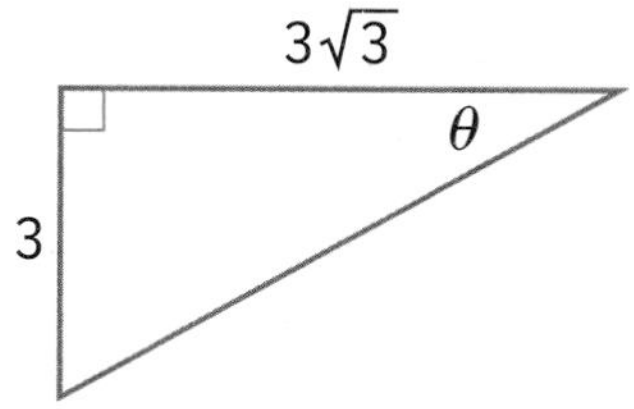

Example 2

In a right triangle, $\angle A$ and $\angle B$ are acute. Find the values of the five remaining trigonometric functions.

7. $\tan A = \frac{8}{15}$

8. $\cos A = \frac{3}{10}$

9. $\tan B = 3$

10. $\sin B = \frac{4}{9}$

11. $\cos A = \frac{1}{2}$

12. $\sin A = \frac{15}{17}$

Examples 3 and 4

The terminal side of θ in standard position contains each point. Find the exact values of the six trigonometric functions of θ.

13. $(5, 12)$

14. $(3, 4)$

15. $(8, -15)$

16. $(-4, 3)$

17. $(-9, -40)$

18. $(1, 2)$

19. $(8, 4)$

20. $(4, 4)$

21. $(6, 2)$

22. $(-5, 5\sqrt{2})$

23. $(3, -9)$

24. $(-8, 12)$

25. $(3, 0)$

26. $(0, -7)$

27. $(0, 4)$

Example 5

Sketch each angle. Then find the measure of each reference angle.

28. $\frac{31\pi}{36}$

29. 230°

30. 205°

31. $\frac{4\pi}{3}$

32. $-\frac{\pi}{6}$

33. $\frac{7\pi}{4}$

34. 135°

35. 200°

36. $\frac{5\pi}{3}$

37. $\frac{13\pi}{8}$

38. −210°

39. $-\frac{7\pi}{4}$

Example 6

PRECISION **Find the exact value of each trigonometric function.**

40. tan 330°

41. $\cos\left(-\frac{11\pi}{4}\right)$

42. cot 30°

43. $\csc\frac{\pi}{4}$

44. sin (−150°)

45. $\tan\left(-\frac{\pi}{4}\right)$

Example 7

46. AMUSEMENT RIDES An amusement park thrill ride swings its riders back and forth on a pendulum that spins. Suppose the swing arm of the ride is 62 feet in length, and the axis from which the arm swings is about 64 feet above the ground. What is the height of the riders above the ground at the peak of the arc? Round to the nearest foot if necessary.

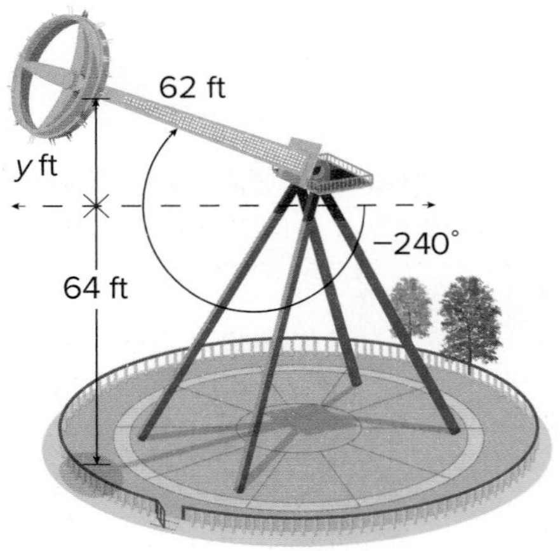

47. ROOFING A roofer rests a ladder at a height of 12 feet against a building so that the base of the ladder is *x* feet from the bottom of the side of the building, forming a 71.6° angle with the ladder and ground. Find the distance from the bottom of the ladder to the side of the building.

a. Write an equation that can be used to find the distance from the bottom of the ladder to the side of the building.

b. How far is the bottom of the ladder from the side of the building? Round to the nearest tenth if necessary.

Mixed Exercises

In a right triangle, ∠*A* is acute.

48. If $\tan A = 3$, what is $\sin A$?

49. If $\sin A = \frac{1}{16}$, what is $\cos A$?

50. If $\tan A = \frac{7}{12}$, what is $\cos A$?

51. If $\sin B = \frac{3}{8}$, what is $\tan B$?

Find the exact value of each trigonometric function.

52. sin 150°

53. cos 270°

54. cot 135°

55. tan (−30°)

56. tan $\frac{\pi}{4}$

57. cot $(-\pi)$

58. LIGHT Light rays are reflected by a surface. If the surface is partially transparent, some of the light rays are bent, or *refracted*, as they pass from the air through the material. The angles of reflection θ_1 and of refraction θ_2 in the diagram at the right are related by the equation $\sin \theta_1 = n \sin \theta_2$. If $\theta_1 = 60°$ and $n = \sqrt{3}$, find the measure of θ_2.

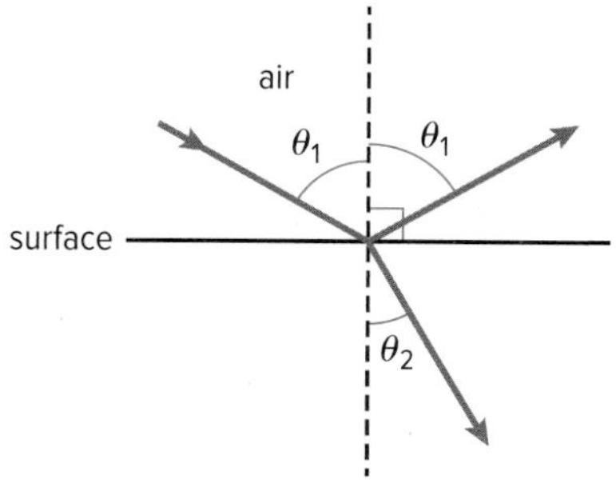

59. RADIOS Two correspondence radios are located 2 kilometers away from a base camp. The angle formed between the first radio, the base camp, and the second radio is 120°. If the first radio has coordinates (2, 0) relative to the base camp, what is the position of the second radio relative to the base camp?

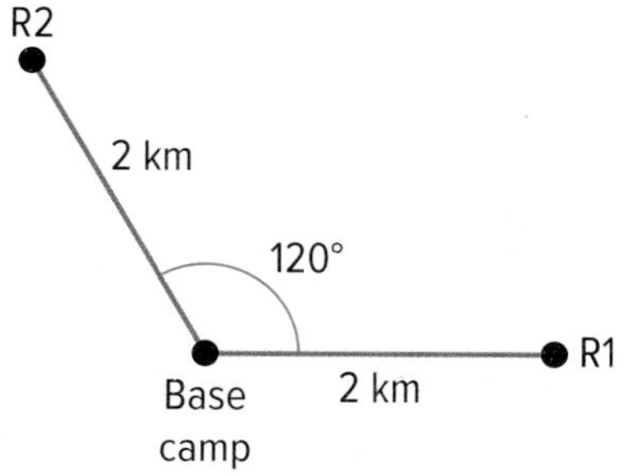

60. PAPER AIRPLANES The formula $R = \frac{V_0^2 \sin 2\theta}{32} + 15 \cos \theta$ gives the distance traveled by a paper airplane that is thrown with an initial velocity of V_0 feet per second at an angle of θ with the ground.

a. If the airplane is thrown with an initial velocity of 15 feet per second at an angle of 25°, how far will the airplane travel?

b. Two airplanes are thrown with an initial velocity of 10 feet per second. One airplane is thrown at an angle of 15° to the ground, and the other airplane is thrown at an angle of 45° to the ground. Which will travel farther?

61. CLOCKS The hands on the clock form an acute angle, θ. The minute hand is about 5 inches long, and the angle formed by the hands at the center of the clock is 75°. If the tips of the hands were connected, it would form a right triangle, with the minute hand being the hypotenuse. Use a trigonometric ratio to find the length of the hour hand to the nearest tenth of an inch.

Peter Horrox/Shutterstock

62. FERRIS WHEELS Luis rides a Ferris wheel in Japan called the Sky Dream Fukuoka, which has a radius of about 60 m and is 5 m off the ground. After he enters the bottom car, the wheel rotates 210.5° counterclockwise before stopping. How high above the ground is Luis when the car has stopped?

REASONING Use the congruent triangles in the figure for Exercises 63–66.

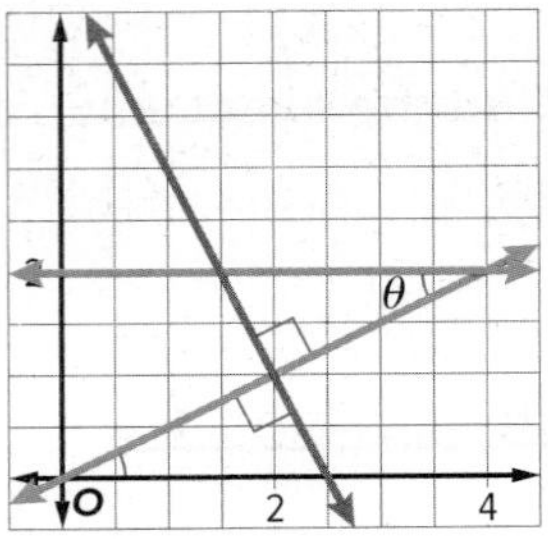

63. What is the cosecant of θ?

64. What is the cosine of θ?

65. What is the cotangent of θ?

66. What is the secant of θ?

67. STRUCTURE There are many angles that share the same reference angle. Give 3 angles that have a reference angle of 60°, including at least one angle with a measure greater than 360°. Explain your method.

68. SHADOWS A tree is $15\frac{1}{3}$ feet tall and casts a shadow at a right angle from the base of the tree. The length of the shadow depends on the angle at which the sunlight hits the tree, θ.

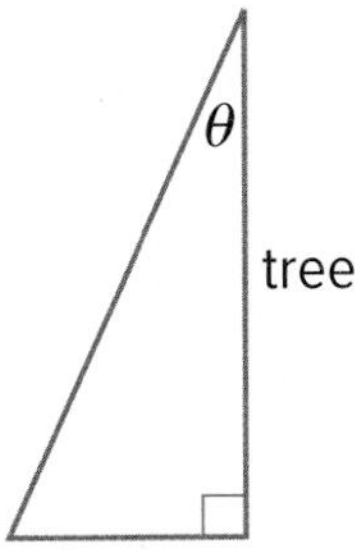

a. Find the distance from the top of the tree to the end of the tree's shadow, in feet, when the sun is at an angle of 13°. Round your answer to the nearest tenth.

b. How long is the tree's shadow, in feet, when the Sun is at an angle of 58°? Round your answer to the nearest tenth.

Higher-Order Thinking Skills

69. ANALYZE Determine whether the following statement is *true* or *false*. Justify your argument.
For any acute angle, the sine function will never have a negative value.

70. CREATE In right triangle *ABC*, sin *A* = sin *C*. What can you conclude about $\triangle ABC$? Justify your reasoning.

71. PERSEVERE For an angle θ in standard position, $\sin \theta = \frac{\sqrt{2}}{2}$ and $\tan \theta = -1$. Can the value of θ be 225°? Justify your reasoning.

72. ANALYZE Determine whether $3 \sin 60° = \sin 180°$ is *true* or *false*. Justify your argument.

73. WRITE Use the sine and cosine functions to explain why cot 180° is undefined.

74. CREATE Give an example of a negative angle θ for which $\sin \theta > 0$ and $\cos \theta < 0$.

75. WRITE Describe the steps for evaluating a trigonometric function for an angle θ that is greater than 90°. Include a description of a reference angle.

76. PERSEVERE When will all six trigonometric functions have a rational value? Justify your argument.

Circular and Periodic Functions

Explore Trigonometric Functions of Special Angles

Online Activity

INQUIRY How can you use special right triangles and the unit circle to find the exact trigonometric values of special angles?

Today's Goals

- Find values of trigonometric functions given a point on a unit circle or the measure of a special angle.
- Find values of trigonometric functions that model periodic events.

Today's Vocabulary

unit circle
circular function
periodic function
cycle
period

Learn Circular Functions

A **unit circle** is a circle with a radius of 1 unit centered at the origin on the coordinate plane. Notice that on a unit circle, the radian measure of a central angle $\theta = \frac{s}{1}$ or s, so the radian measure of an angle is the length of the arc on the unit circle subtended by the angle.

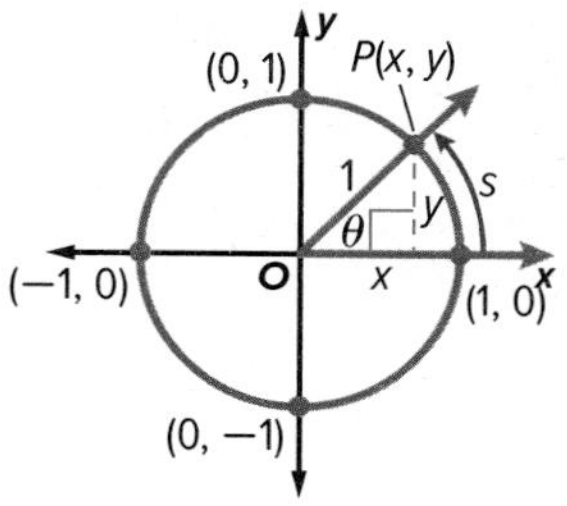

You can use a point P on the unit circle to generalize sine and cosine functions by applying the definitions of trigonometric functions in right triangles.

Key Concept • Functions on a Unit Circle

Words: If the terminal side of an angle θ in standard position intersects the unit circle at $P(x, y)$, then $\cos \theta = x$ and $\sin \theta = y$.

Symbols: $P(x, y) = P(\cos \theta, \sin \theta)$

Example:

If $\theta = \frac{5\pi}{4}$,

$P(x, y) = P\left(\cos \frac{5\pi}{4}, \sin \frac{5\pi}{4}\right)$.

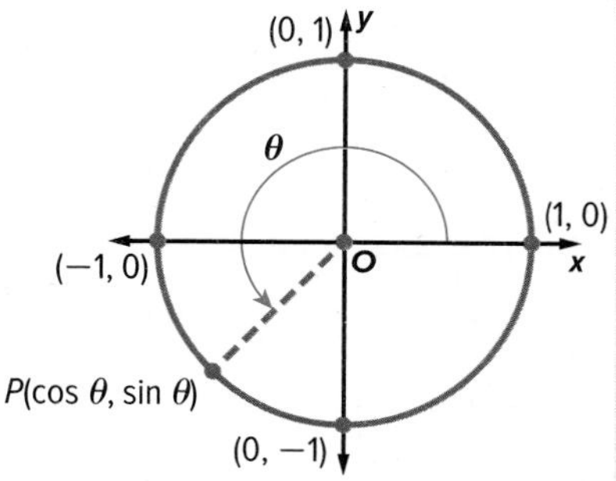

Study Tip

Sine and Cosine To help you remember that for (x, y) on the unit circle, $x = \cos \theta$ and $y = \sin \theta$, notice that alphabetically x comes before y and *cosine* comes before *sine*.

Both $\cos \theta = x$ and $\sin \theta = y$ are functions of θ. Because they are defined using a unit circle, they are **circular functions**, which describe a point on a circle as the function of an angle defined in radians.

Study Tip

Unit Circle Angles in the unit circle are in standard position. The terminal side travels counterclockwise around the unit circle for positive angle measures and clockwise for negative angle measures.

Study Tip

Trigonometric Functions Because tan θ, csc θ, sec θ, and cot θ can be defined in terms of sin θ and cos θ, you can use the coordinates of the point at which an angle intersects the unit circle to find the remaining trigonometric functions as well.

The unit circle is commonly used to show the exact values of cos θ and sin θ for special angles. The cosine values are the x-coordinates of the points where the terminal sides of the angles intersect the unit circle, and the sine values are the y-coordinates.

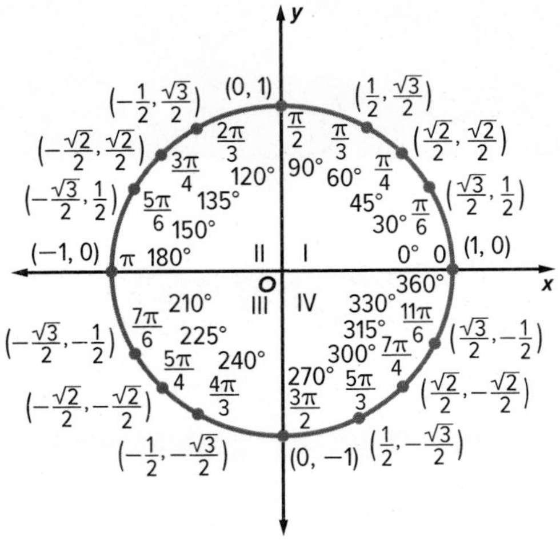

Example 1 Find Sine and Cosine Given a Point on the Unit Circle

The terminal side of θ in standard position intersects the unit circle at $P\left(-\frac{12}{13}, \frac{5}{13}\right)$. Find cos θ and sin θ.

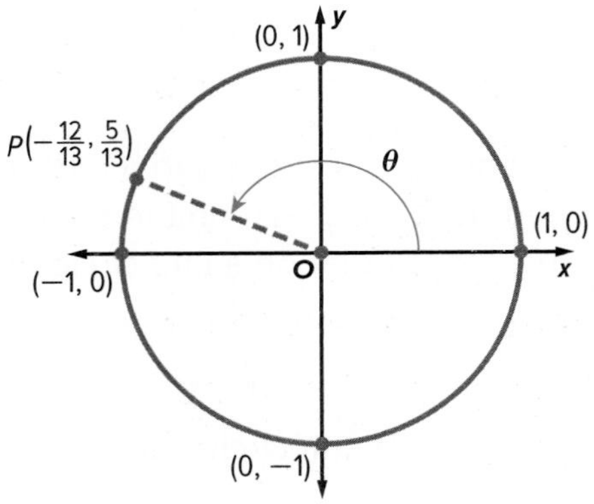

$P\left(-\frac{12}{13}, \frac{5}{13}\right) = P(\cos\theta, \sin\theta)$

$\cos\theta = -\frac{12}{13}$

$\sin\theta = \frac{5}{13}$

Check

The terminal side of θ in standard position intersects the unit circle at $P\left(-\frac{4}{5}, -\frac{3}{5}\right)$. Find cos θ and sin θ. Write the solutions as decimals.

cos θ = __?__ sin θ = __?__

Example 2 Find Trigonometric Values of Special Angles

Find the exact values of the six trigonometric functions for an angle that measures $\frac{5\pi}{4}$ radians.

Using the unit circle, we know that the special angle $\frac{5\pi}{4}$ intersects the unit circle in Quadrant III at $P\left(-\frac{\sqrt{2}}{2}, -\frac{\sqrt{2}}{2}\right)$.

So, $\cos\theta = -\frac{\sqrt{2}}{2}$ and $\sin\theta = -\frac{\sqrt{2}}{2}$.

Go Online You can complete an Extra Example online.

Use the values of $\cos\theta$ and $\sin\theta$ to find the remaining trigonometric values.

$\tan\theta = \frac{\sin\theta}{\cos\theta} = 1$　　　$\csc\theta = \frac{1}{\sin\theta} = -\sqrt{2}$

$\sec\theta = \frac{1}{\cos\theta} = -\sqrt{2}$　　　$\cot\theta = \frac{\cos\theta}{\sin\theta} = 1$

Check

Find the exact values of the six trigonometric functions for an angle that measures $\frac{4\pi}{3}$ radians.

$\sin\theta =$ ___?___　　$\cos\theta =$ ___?___　　$\tan\theta =$ ___?___

$\csc\theta =$ ___?___　　$\sec\theta =$ ___?___　　$\cot\theta =$ ___?___

Think About It!

How could you use the sine and cosine values of an angle that measures $\frac{\pi}{4}$ radians to determine $\sin\frac{5\pi}{4}$ and $\cos\frac{5\pi}{4}$?

Learn Periodic Functions

A **periodic function** has *y*-values that repeat at regular intervals. One complete pattern of a periodic function is called a **cycle,** and the horizontal length of one cycle is called the **period**.

The values of the sine and cosine functions can be found by using the unit circle. As you move around the unit circle, the values of these functions repeat every 360° or 2π. So, the sine and cosine functions are periodic functions where $\sin(x + 2\pi) = \sin x$ and $\cos(x + 2\pi) = \cos x$. Because tangent, cosecant, secant, and cotangent can be defined in terms of sine and cosine, they are also periodic functions.

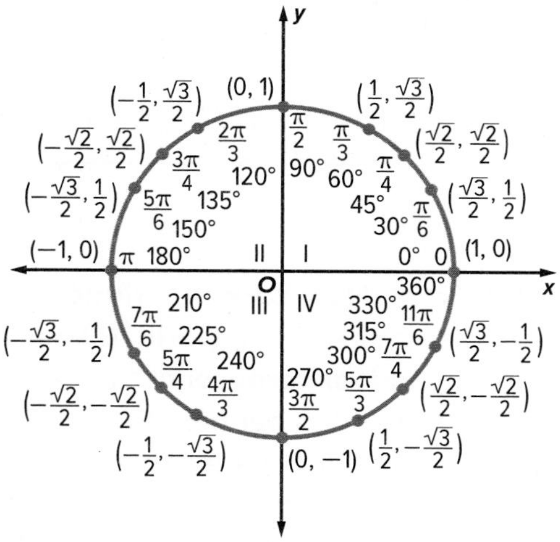

Recall that the values of $\sin\theta$ and $\cos\theta$ are the *y*- and *x*-coordinates, respectively, of the point $P(x, y)$ where the terminal side of an angle θ intersects the unit circle. Thus, you can use the values of $\sin\theta$ and $\cos\theta$ shown on the unit circle to graph the sine and cosine functions, where the *x*-axis represents the values of θ and the *y*-axis represents the values of $\sin\theta$ or $\cos\theta$.

Each point on the graph of $y = \sin x$ is given by $(x, \sin x)$. Using the unit circle, *x* is the measure of the angle, and $\sin x$ is the *y*-coordinate of the corresponding point on the unit circle.

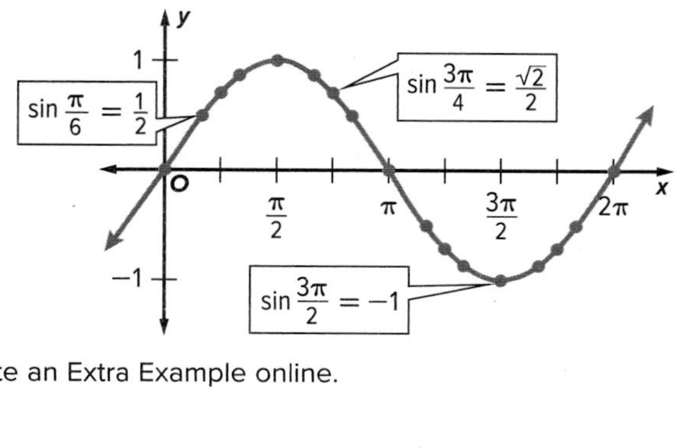

Go Online You can complete an Extra Example online.

Study Tip

Radian and Degrees The sine and cosine functions can also be graphed using degrees as the units of the *x*-axis.

Similarly, each point on the graph of $y = \cos x$ is given by $(x, \cos x)$. Using the unit circle, x is the measure of the angle, and $\cos x$ is the x-coordinate of the corresponding point on the unit circle.

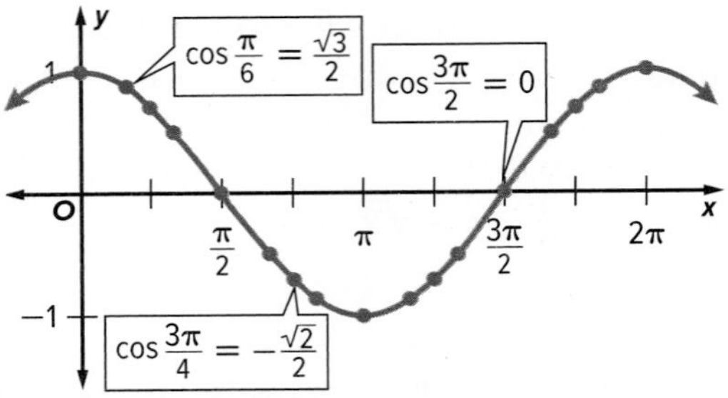

Study Tip

Cycles A cycle can begin at any point on the graph of a periodic function. In the example, if the beginning of the cycle is at $\frac{\pi}{2}$, then the pattern repeats at 2π. The period is still $2\pi - \frac{\pi}{2}$ or $\frac{3\pi}{2}$.

Example 3 Identify the Period of a Function

Determine the period of the function.

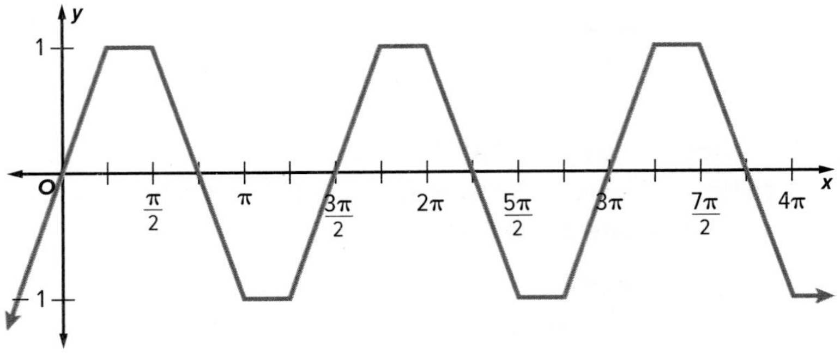

The pattern repeats at $\frac{3\pi}{2}$, 3π, and so on. So, the period is $\frac{3\pi}{2}$.

Apply Example 4 Graph Periodic Functions

CAROUSELS **New York's Eldridge Park Carousel is considered the fastest carousel in the world, taking riders for a spin at 18 miles per hour. The indoor carousel has a diameter of 50 feet and can complete about 10 rotations per minute. The distance of a rider *d* from the front wall of the building varies periodically as a function of time *t*. Identify the period of the function. Then graph the function. Assume that a rider begins at the point closest to the wall, 20 feet from the wall.**

1 What is the task?

Describe the task in your own words. Then list any questions that you may have. How can you find answers to your questions?

I need to find the period and graph the function. I know that the diameter of the carousel is 50 feet and that the carousel completes 10 rotations per minute. How are the period and the graph affected by the diameter and the number of rotations per minute? I can relate the function to the unit circle to answer my questions.

2 How will you approach the task? What have you learned that you can use to help you complete the task?

I will use the key features to analyze the function. I have learned about key features of the sine and cosine functions.

Go Online You can complete an Extra Example online.

3 What is your solution?

Use your strategy to solve the problem.

When is the function increasing and when is it decreasing? What are the extrema?

The distance the rider is from the wall will increase until the rider reaches the maximum distance halfway through the period; then the distance will decrease until the rider is back at the starting position. The closest the rider is to the wall is 20 feet. The carousel has a diameter of 50 feet, so the farthest is 70 feet.

What are a cycle and the period in terms of the situation? What is the period of the function?

One full rotation of the carousel is one cycle. The time it takes to complete one rotation is the period. The carousel can complete 10 rotations per minute, so the period is 60 seconds ÷ 10 or 6 seconds.

The diagram shows the distance of the rider from the wall during one rotation.

Use the diagram to complete a table showing the distance of a rider from the wall.

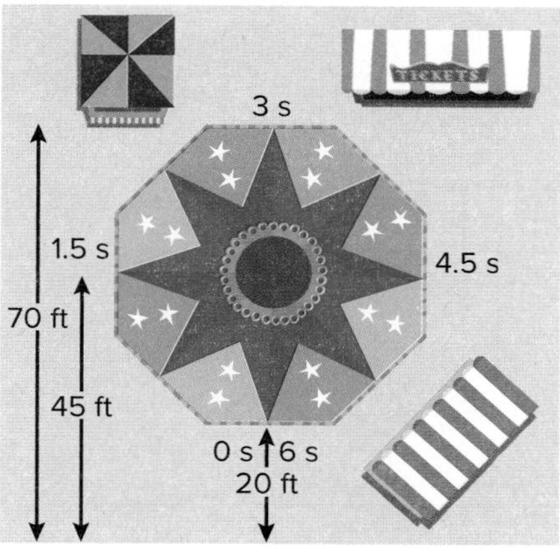

Time (s)	Distance (ft)
0	20
1.5	45
3	70
4.5	45
6	20

Graph and interpret the function.

What are the minimum and maximum distances? How often does the graph repeat?

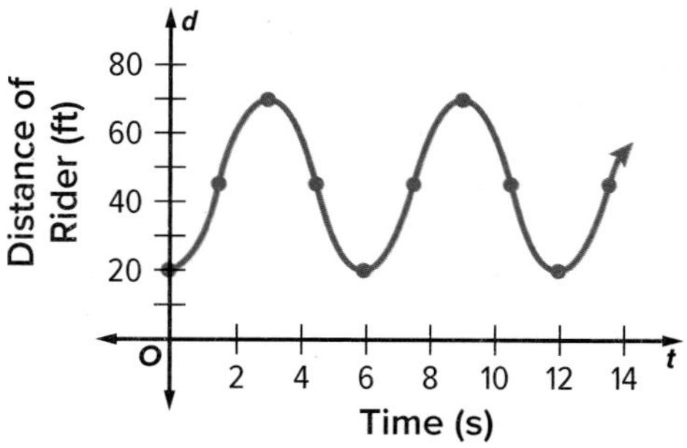

The minimum distance is 20 feet and the maximum distance is 70 feet. The graph repeats every 6 seconds.

4 How can you know that your solution is reasonable?

Write About It! Write an argument that can be used to defend your solution.

The graph shows a repeating pattern as a periodic function should. The key features we predicted match the points in the table and the key features of the graph.

Talk About It!

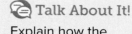

Explain how the assumption that the rider starts at the point closest to the wall affects the graph of the function. What would happen to the graph if the rider started at a different point on the carousel?

Go Online

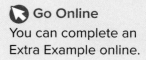

You can complete an Extra Example online.

Check

POGO STICK In 2016, Henry Cabelus set the record for the fewest pogo stick jumps in one minute, when he jumped up and down only 38 times in one minute. Cabelus's height off the ground h while jumping is a function of time t. Suppose that at the highest point of each jump, Cabelus was 6 feet off the ground. Select the graph of the function.

A.

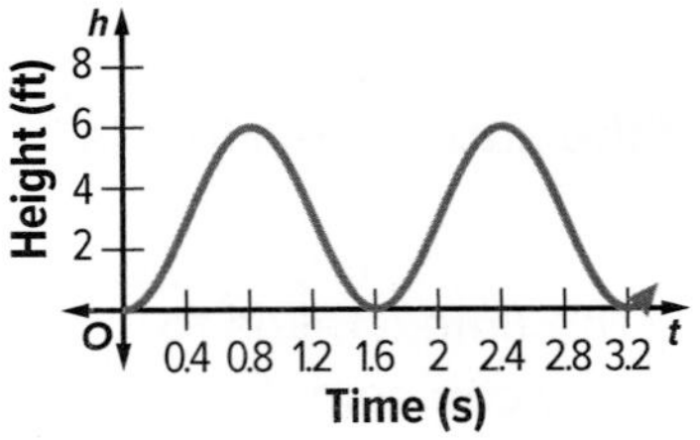

B.

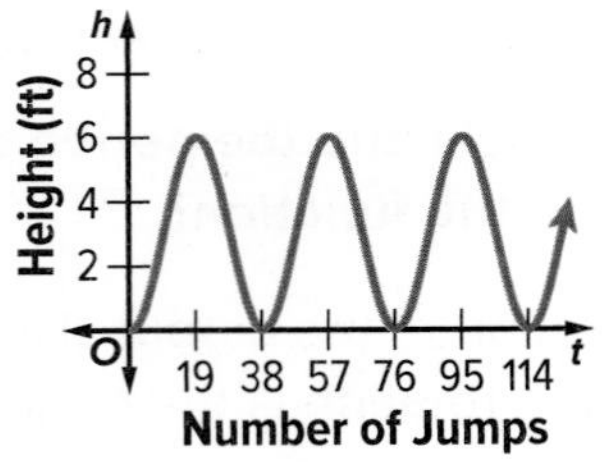

C.

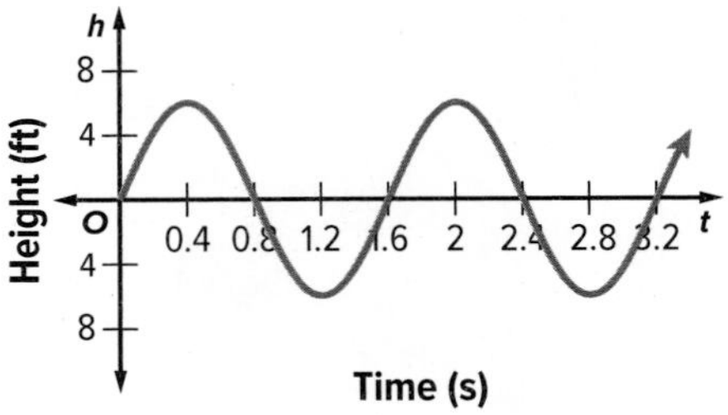

D.

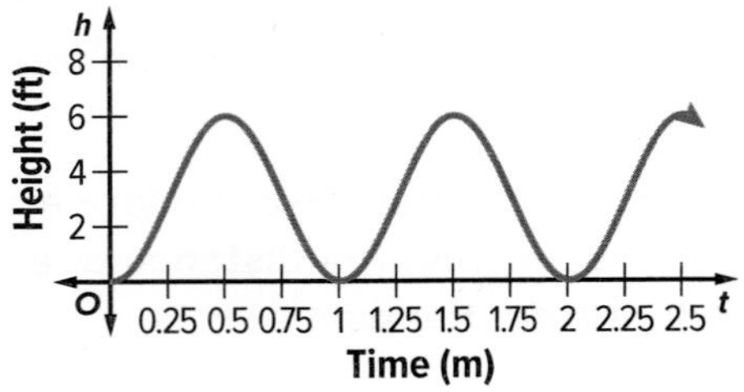

Think About It!

Tanisha claims that because $\frac{10\pi}{3}$ contains three multiples of π, $\cos \frac{10\pi}{3} = \cos\left(\frac{\pi}{3} + \frac{9\pi}{3}\right) = \cos \frac{\pi}{3}$. Is she correct? Justify your argument.

Example 5 Evaluate Trigonometric Expressions

Find the exact value of $\cos \frac{10\pi}{3}$.

$\cos \frac{10\pi}{3} = \cos\left(\frac{4\pi}{3} + \frac{6\pi}{3}\right)$ $\quad$ $\frac{4\pi}{3} + \frac{6\pi}{3} = \frac{10\pi}{3}$

$= \cos \frac{4\pi}{3}$ $\quad$ $\cos(x + 2\pi) = \cos x$

$= -\frac{1}{2}$ $\quad$ Use the unit circle.

Check

Find the exact value of each expression.

$\sin \frac{8\pi}{3} =$ ___?___ $\qquad$ $\cos \frac{5\pi}{6} =$ ___?___

$\sin \frac{21\pi}{4} =$ ___?___ $\qquad$ $\cos \frac{11\pi}{3} =$ ___?___

$\cos 60° =$ ___?___ $\qquad$ $\sin 600° =$ ___?___

Practice

Go Online You can complete your homework online.

Example 1

The terminal side of angle θ in standard position intersects the unit circle at each point *P*. Find cos θ and sin θ.

1. $P\left(-\frac{\sqrt{3}}{2}, \frac{1}{2}\right)$

2. $P(0, -1)$

3. $P\left(-\frac{2}{3}, \frac{\sqrt{5}}{3}\right)$

4. $P\left(-\frac{4}{5}, -\frac{3}{5}\right)$

5. $P\left(\frac{1}{6}, -\frac{\sqrt{35}}{6}\right)$

6. $P\left(\frac{\sqrt{7}}{4}, \frac{3}{4}\right)$

Example 2

Find the exact values of the six trigonometric functions for each given angle measure.

7. $\frac{\pi}{6}$ radians

8. $\frac{3\pi}{4}$ radians

9. $\frac{\pi}{2}$ radians

10. $\frac{5\pi}{3}$ radians

11. π radians

12. $\frac{7\pi}{6}$ radians

Example 3

Determine the period of the function.

13.

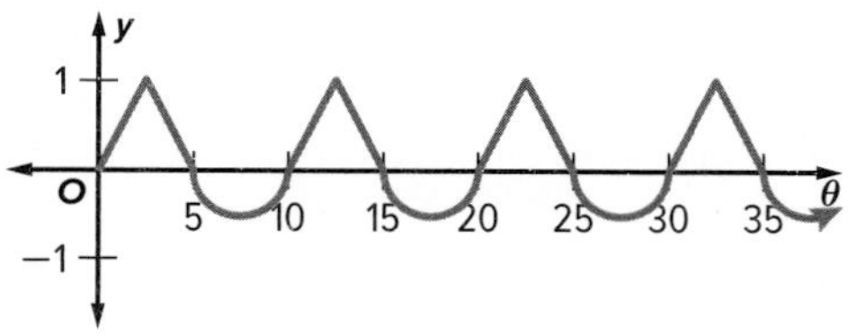

14.

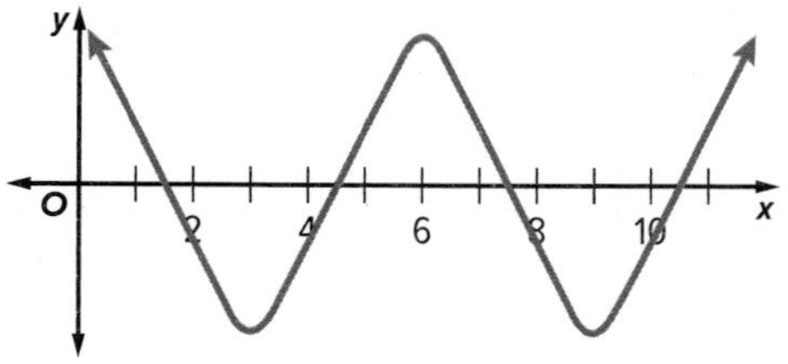

15.

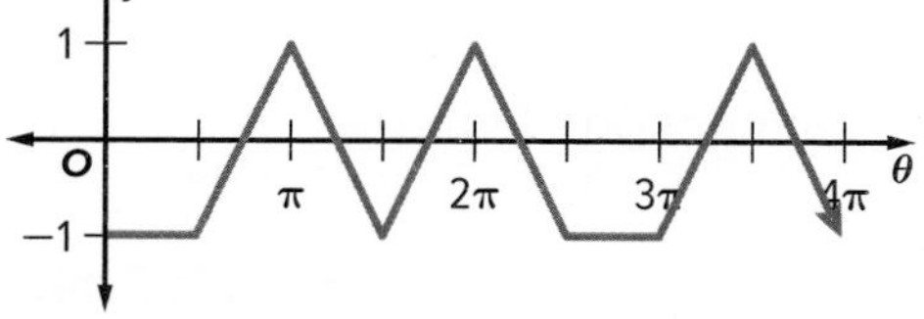

16.

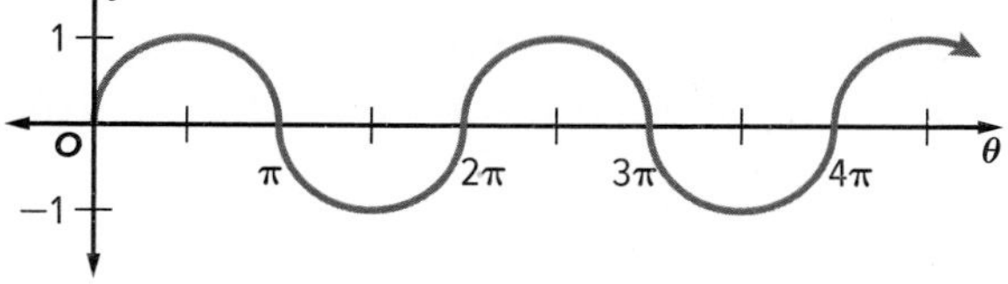

17.

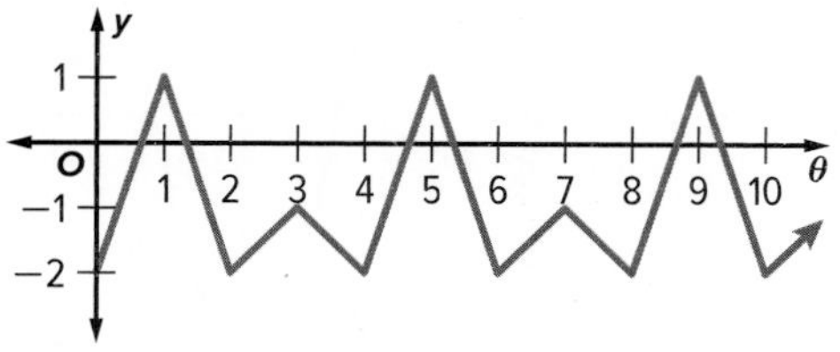

18.

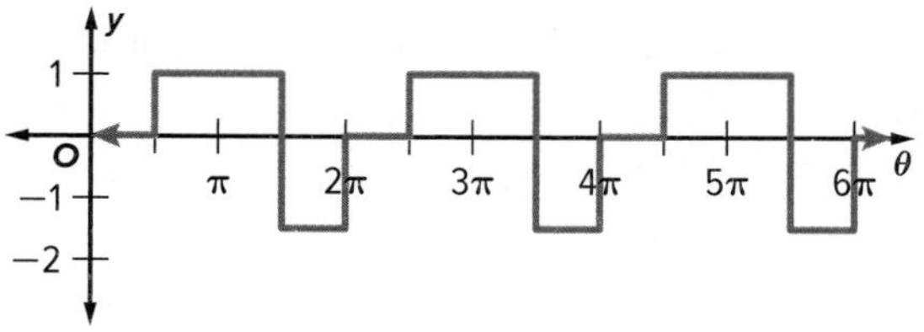

Example 4

19. GEARS A gear at a factory has a diameter of 24 inches and can complete about 15 rotations per minute. One spot on the gear has a specific notch. The distance of the notch *d* from the edge of the machine to which the gear is attached varies periodically as a function of time *t*. Identify the period of the function. Then graph the function. Assume that the notch begins at the point closest to the edge of the machine, 10 inches from the edge.

20. PENDULUM The height of a pendulum varies periodically as a function of time. The pendulum swings in one direction and reaches its high point of 3 feet. It then swings the opposite direction and reaches 3 feet again. Its lowest point is 1.5 feet. The time it takes for the pendulum to swing from its low point to one of its high points is 3 seconds. Identify the period of the function. Then graph the function. Assume that the pendulum begins at the low point, 1.5 feet above the ground.

Example 5

Find the exact value of each expression.

21. $\sin(-510°)$

22. $\sin 495°$

23. $\cos\left(-\frac{5\pi}{2}\right)$

24. $\sin \frac{5\pi}{3}$

25. $\cos \frac{11\pi}{4}$

26. $\sin\left(-\frac{3\pi}{4}\right)$

Mixed Exercises

Find the exact value of each expression.

27. $\cos 45° - \cos 30°$

28. $6(\sin 30°)(\sin 60°)$

29. $2 \sin \frac{4\pi}{3} - 3 \cos \frac{11\pi}{6}$

30. $\cos\left(-\frac{2\pi}{3}\right) + \frac{1}{3} \sin 3\pi$

31. $(\sin 45°)^2 + (\cos 45°)^2$

32. $\frac{(\cos 30°)(\cos 150°)}{\sin 315°}$

33. FERRIS WHEELS A Ferris wheel with a diameter of 100 feet completes 2.5 revolutions per minute. What is the period of the function that describes the height of a seat on the outside edge of the Ferris wheel as a function of time?

34. MOON The Moon's period of revolution is the number of days it takes for the Moon to revolve around Earth. The period can be determined by graphing the percentage of sunlight reflected by the Moon each day, as seen by an observer on Earth. Use the graph to determine the Moon's period of revolution.

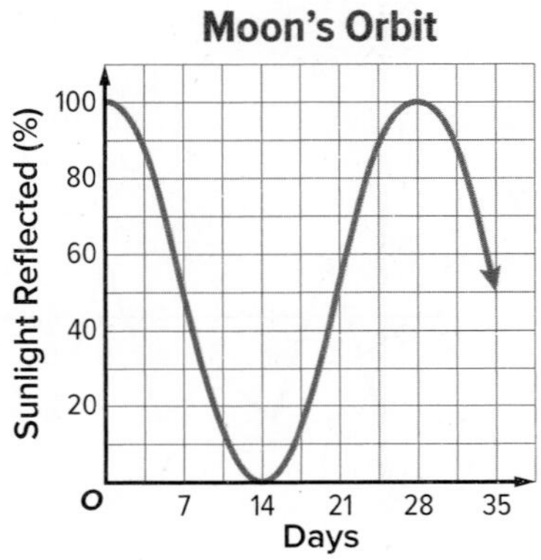

35. CONSTRUCT ARGUMENTS Determine whether each statement is *always*, *sometimes*, or *never* true. Justify your argument.

a. If k is a real number, then there is a value of θ such that $\cos \theta = k$.

b. $\sin \theta = \sin (\theta + 2\pi)$

c. If $\theta = n\pi$, where n is a whole number, then $\cos \theta = 1$.

d. If θ is an angle in standard position in which the terminal side lies in Quadrant IV, then $\sin \theta$ is positive.

36. REASONING Point P lies on the unit circle and on the line $y = x$. If θ is an angle in standard position in which the terminal side contains P, what can you conclude about $\sin \theta$ and $\cos \theta$? Explain.

37. USE A MODEL The wheel at a water park has a radius of 1 meter. As the water flows, the wheel turns counterclockwise, as shown. A point P on the edge of the wheel begins at the surface of the water. The function $f(x) = \sin x$ represents the height of P above or below the surface of the water as the wheel rotates through an angle of x radians.

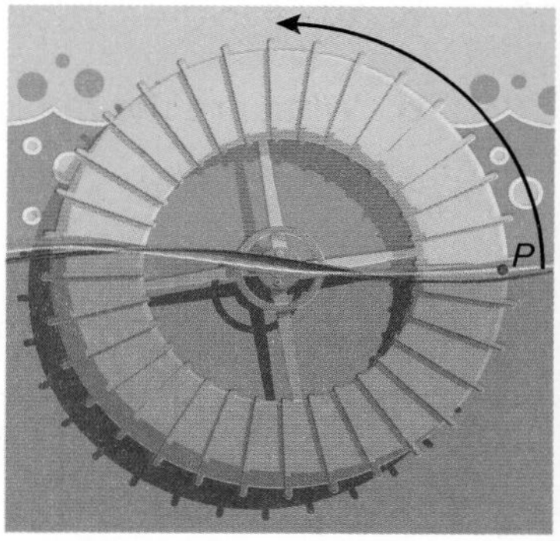

a. How far does point P travel as the wheel rotates through an angle of $\frac{3\pi}{4}$ radians? Explain.

b. Graph $f(x) = \sin x$ on the coordinate plane.

c. What is the period of the function? Explain how you know, and explain how the period is shown in the graph. What does the period tell you about point P?

d. What are the x-intercepts? What do these represent?

e. Identify an interval where the function is decreasing. What does this represent?

38. TIRES A point on the edge of a car tire is marked with paint. As the car moves slowly, the marked point on the tire varies in distance from the surface of the road. The height in inches of the point is given by the function $h = -8 \cos t + 8$, where t is the time in seconds.

a. What is the maximum height above ground that the point on the tire reaches?

b. What is the minimum height above ground that the point on the tire reaches?

c. How many rotations does the tire make per second?

d. How far does the marked point travel in 30 seconds? How far does the marked point travel in one hour?

39. **TEMPERATURES** The temperature T in degrees Fahrenheit of a city t months into the year is approximated by the formula $T = 42 + 30 \sin \frac{\pi}{6}t$.

a. What is the highest monthly temperature for the city?

b. In what month does the highest temperature occur?

c. What is the lowest monthly temperature for the city?

d. In what month does the lowest temperature occur?

40. **FACTORIES** A machine in a factory has a gear with a radius of 1 foot. A point P on the edge of the gear begins at the furthest point from a wall, and then the gear begins to rotate counterclockwise. The function $f(x) = \cos x + 2$ represents the distance of P from the wall as the gear rotates through an angle of x radians.

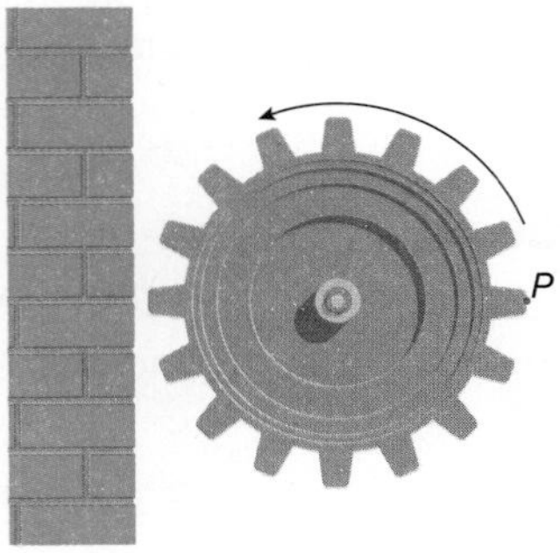

a. What is $f\left(\frac{\pi}{2}\right)$? What does it represent?

b. Graph $f(x)$ on a coordinate plane.

c. What is the period of the function? What does this tell you about P?

d. What are the maximum and minimum values of the function?

41. **FIND THE ERROR** Francis and Benita are finding the exact value of $\cos\left(-\frac{\pi}{3}\right)$. Is either of them correct? Explain your reasoning.

Francis	Benita
$\cos\left(-\frac{\pi}{3}\right) = -\cos\frac{\pi}{3}$ $= -0.5$	$\cos\left(-\frac{\pi}{3}\right) = \cos\left(-\frac{\pi}{3}\right) + 2\pi$ $= \cos\frac{5\pi}{3}$ $= 0.5$

42. **PERSEVERE** A ray has its endpoint at the origin of the coordinate plane, and point $P\left(\frac{1}{2}, -\frac{\sqrt{3}}{2}\right)$ lies on the ray. Find the angle θ formed by the positive x-axis and the ray.

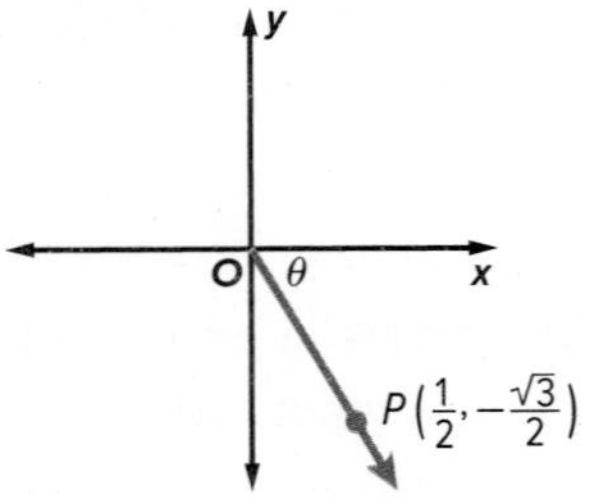

43. **ANALYZE** Is the period of a sine curve *sometimes*, *always*, or *never* an integer multiple of π? Justify your argument.

44. **CREATE** Draw the graph of a periodic function that has a maximum value of 10 and a minimum value of −10. Describe the period of the function.

45. **WRITE** Explain how to determine the period of a periodic function from its graph. Include a description of a cycle.

Graphing Sine and Cosine Functions

Learn Graphing Sine and Cosine Functions

Like other functions you have studied, trigonometric functions can be graphed on the coordinate plane. Sine and cosine functions are oscillating functions. The **oscillation** of a function refers to how much the graph of the function varies between its extreme values as it approaches positive or negative infinity. The **midline** is the line about which a graph oscillates, so it is halfway between the maximum and minimum. The **amplitude** of the graph of a sine or cosine function equals half the difference between the maximum and minimum values of the function.

Key Concept • Sine and Cosine Functions

Parent	$y = \sin x$	$y = \cos x$
Graph	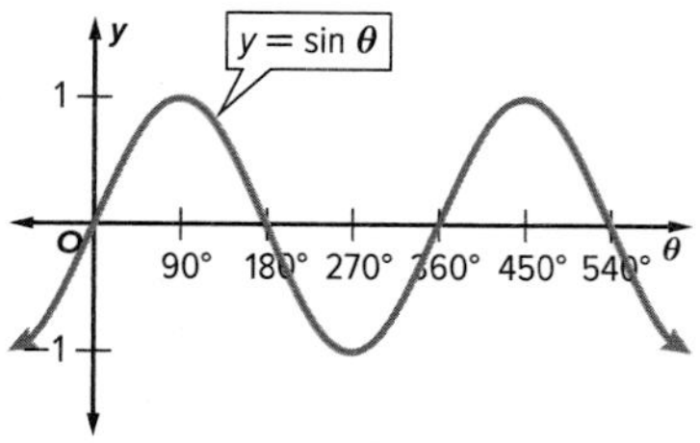	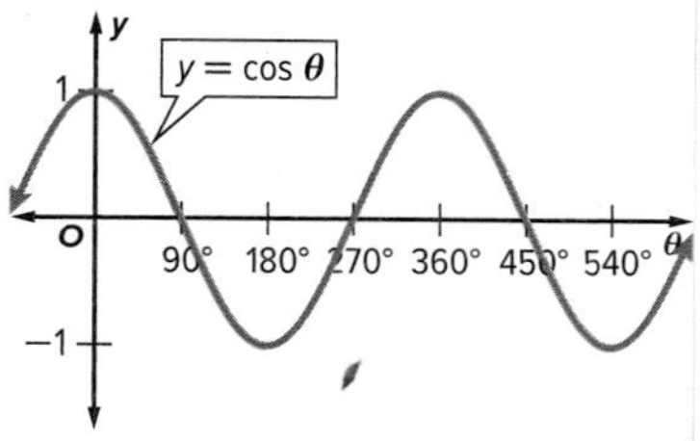
Domain	all real numbers	all real numbers
Range	$\{y \mid -1 \leq y \leq 1\}$	$\{y \mid -1 \leq y \leq 1\}$
Amplitude	1	1
Midline	$y = 0$	$y = 0$
Period	360°	360°
Oscillation	between −1 and 1	between −1 and 1

Sine and cosine functions, like other functions, can be transformed. You can use the graphs of the parent functions to graph $y = a \sin bx$ and $y = a \cos bx$. For functions of the form $y = a \sin bx$ and $y = a \cos bx$, $|a|$ is the amplitude and $\frac{360°}{|b|}$ is the period.

You can also use x-intercepts to help graph the functions. The x-intercepts for one cycle of the sine and cosine functions are:

$y = a \sin bx$	$y = a \cos bx$
$(0, 0), \left(\frac{1}{2} \cdot \frac{360°}{b}, 0\right), \left(\frac{360°}{b}, 0\right)$	$\left(\frac{1}{4} \cdot \frac{360°}{b}, 0\right), \left(\frac{3}{4} \cdot \frac{360°}{b}, 0\right)$

Go Online You can complete an Extra Example online.

Today's Goals
- Graph and analyze sine and cosine functions.
- Model periodic real-world situations with sine and cosine functions.

Today's Vocabulary
oscillation
midline
amplitude
sinusoidal function
frequency

Go Online You can watch a video to see how to graph sine and cosine functions.

Think About It!
Describe the relationships between the values of a and b in $y = a \sin bx$, and the amplitude, period, and dilation of the function.

Example 1 Identify the Amplitude and Period from a Graph

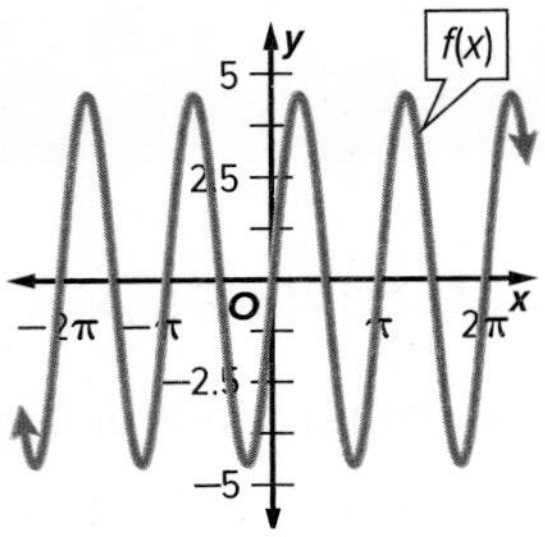

Identify the amplitude, midline, and period of *f*(*x*).

The maximum is 4. The minimum is −4.

Therefore, the amplitude is $\frac{4-(-4)}{2}$ or 4.

The midline is at $y = 0$. The pattern in the graph repeats every π radians. Therefore, the period is π.

Check

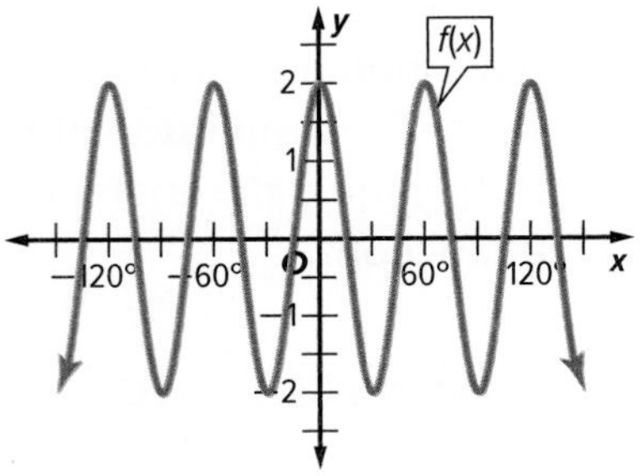

Identify the amplitude, midline, and period of *f*(*x*).

The amplitude is __?__.

The midline is at $y =$ __?__.

The period is __?__°.

Think About It!

Why are absolute value symbols used for *a* and *b* when finding the amplitude and period?

Example 2 Identify the Amplitude and Period from an Equation

Identify the amplitude and period of $f(x) = 3 \cos 5x$.

The function is written as $f(x) = a \cos bx$.

a is used to find the amplitude. *b* is used to find the period.

amplitude: $|a| = |3|$ or 3   period: $\frac{360°}{|b|} = \frac{360°}{|5|}$ or 72°

Check

Identify the amplitude and period of $f(x) = 7 \sin 8x$.

The amplitude is __?__.   The period is __?__°.

Example 3 Graph a Sine Function

Graph $y = 0.5 \sin 4x$.

Step 1 Find the amplitude.

amplitude: $|a| = 0.5$

This is a vertical dilation. Therefore, the graph is compressed vertically. The maximum value is 0.5, and the minimum value is −0.5.

Go Online You can complete an Extra Example online.

Step 2 Find the period.

period: $\frac{360°}{|b|} = \frac{360°}{|4|}$ or 90°

One cycle has a length of 90°.

Step 3 Find *x*-intercepts.

There are *x*-intercepts at:

$(0, 0) \qquad \left(\frac{1}{2} \cdot \frac{360°}{b}, 0\right) = (45°, 0) \qquad \left(\frac{360°}{b}, 0\right) = (90°, 0)$

The sine curve goes up from (0, 0), so a maximum is located halfway between (0, 0) and (45, 0) or at (22.5, 0.5).

The sine curve goes down from (45, 0), so a minimum is located halfway between (45, 0) and (90, 0) or at (67.5, −0.5).

Step 4 Graph the function.

Plot (0, 0), (45, 0), (90, 0), (22.5, 0.5), and (67.5, −0.5) on the coordinate plane, and sketch the curve through the points.

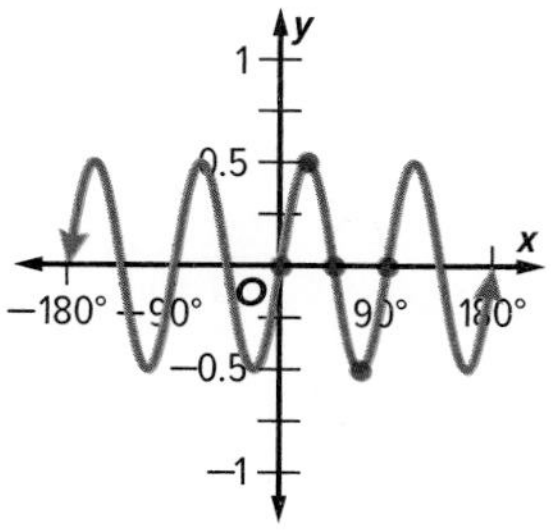

Check

Select the graph of $y = -2 \sin 3x$.

A.

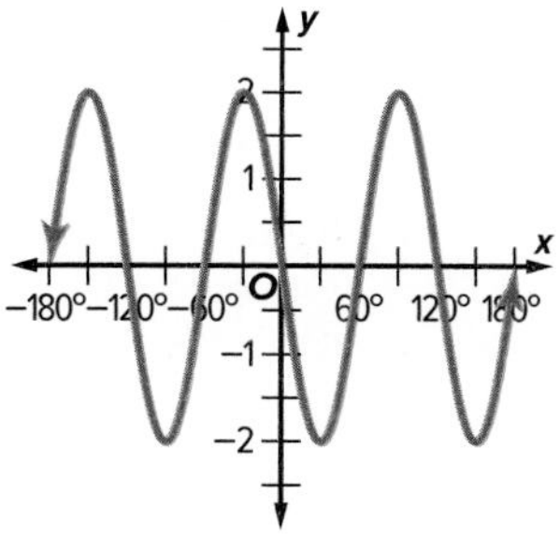

B.

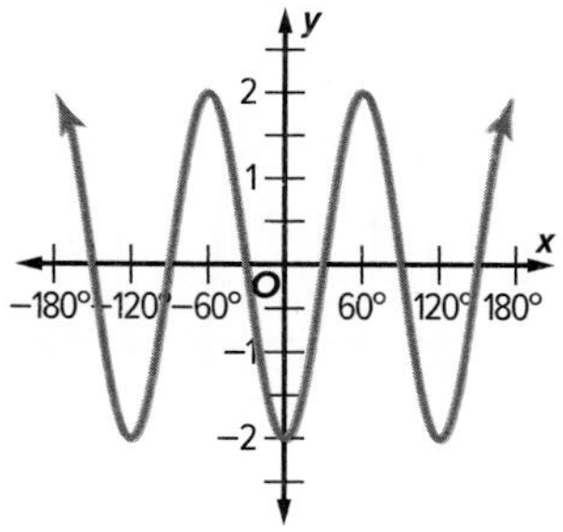

C.

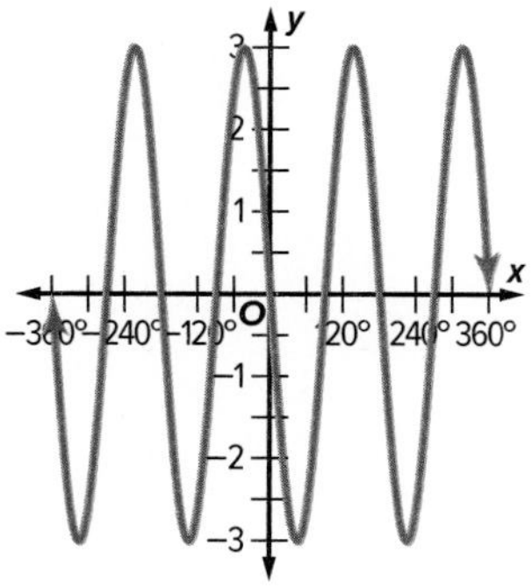

D.

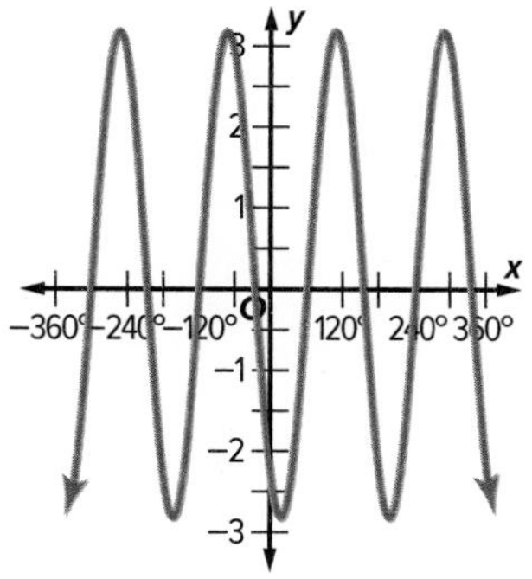

Go Online You can complete an Extra Example online.

Think About It!

How would you graph $y = -0.5 \sin 4x$?

Example 4 Graph a Cosine Function

Graph $y = 2 \cos 3x$.

Step 1 Find the amplitude.

amplitude: $|a| = 2$
This is a vertical dilation. Therefore, the graph is stretched vertically. The maximum value is 2, and the minimum value is -2.

Step 2 Find the period.

period: $\frac{360°}{|b|} = \frac{360°}{|3|}$ or $120°$
One cycle has a length of 120°.

Step 3 Find *x*-intercepts.

There are *x*-intercepts at:

$\left(\frac{1}{4} \cdot \frac{360°}{b}, 0\right) = (30°, 0)$ $\quad$ $\left(\frac{3}{4} \cdot \frac{360°}{b}, 0\right) = (90°, 0)$

A minimum is located halfway between (30, 0) and (90, 0) or at (60, −2).

A maximum is located at the *y*-intercept or (0, 2).

Step 4 Graph the function.

Plot (30, 0), (90, 0), (60, −2), and (0, 2) on the coordinate plane, and sketch the curve through the points.

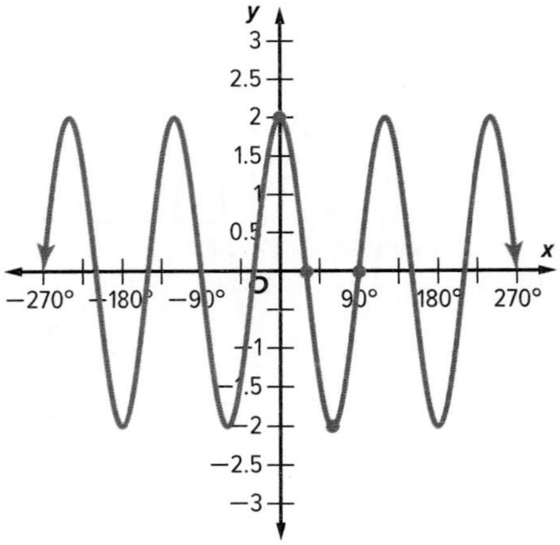

Check

Graph $y = -\cos 2x$.

Watch Out!

Amplitude The graphs of $y = a \sin b\theta$ and $y = a \cos b\theta$ with amplitude $|a|$ have maxima at $y = a$ and minima at $y = -a$.

Go Online You can complete an Extra Example online.

Learn Modeling with Sine and Cosine Functions

The sine and cosine functions are sometimes referred to as sinusoidal functions. A **sinusoidal function** is a function that can be produced by translating, reflecting, or dilating the sine function. The **frequency** of a sinusoidal function is the number of cycles in a given unit of time. This frequency is the reciprocal of the period of the function. So, if the period is $\frac{1}{100}$ of a second, then the frequency is 100 cycles per second.

Think About It!
When modeling periodic phenomena, should the *x*-axis be labeled in degrees, radians, or seconds? Should you use degree or radian mode when using a graphing calculator?

Key Concept • Modeling with $y = a \sin bx$ and $y = a \cos bx$

Step 1 Use the amplitude to find a.

Step 2 Use the period to find b. The period is the reciprocal of the frequency. It can be written as $\frac{360°}{|b|}$ or $\frac{2\pi}{|b|}$.

Step 3 Write the function.

Example 5 Characteristics of the Sine and Cosine Functions

SPRINGS **An object on a spring oscillates according to the function $y = 40 \cos \pi t$, where y is the distance in centimeters above its equilibrium position at time t in seconds.**

Part A Find the period and frequency, and describe them in the context of the situation.

The period is $\frac{2\pi}{|b|}$. Since $b = \pi$, the period is $\frac{2\pi}{\pi}$ or 2.

The frequency is $\frac{1}{\text{period}}$ or $\frac{1}{2}$.

Therefore, the object completes $\frac{1}{2}$ of a cycle per second, and it will reach the maximum distance above the equilibrium point every 2 seconds.

Part B Identify the domain and range in the context of the situation.

The domain of $y = 40 \cos \pi t$ is all real numbers. Because time cannot be negative, the relevant domain in the context of the situation is $[0, \infty)$.

The range is $[-40, 40]$. This references the farthest away that the object can get from its point of equilibrium. With the equilibrium at the center, the object can be as much as 40 centimeters from the center in either direction. This is verified by the amplitude.

Watch Out!
Assumptions
Assuming that the object continues to oscillate indefinitely allows us to use a periodic function to model its path.

Go Online You can complete an Extra Example online.

Check

ELECTRONICS The electric power delivered for household use in the United States is 110 volts. The function $y = 100\sqrt{2} \sin 120\pi t$ represents the effective current, where t is the time in seconds.

Part A Find the period and frequency, and describe them in the context of the situation.

period = __?__ frequency = __?__

The current completes __?__ cycle(s) per second, and will reach the maximum every __?__ second(s).

Part B Graph the function.

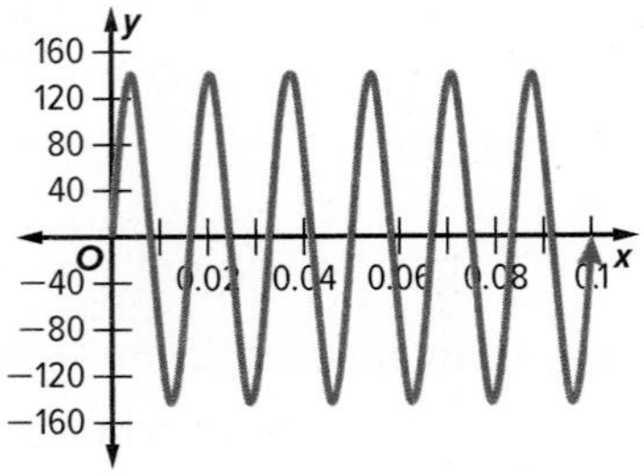

Think About It!
Describe the domain, end behavior, and intercepts of the function in Example 6 in the context of the situation.

Example 6 Model Periodic Situations

ELECTRICITY **The voltage supplied by an electrical outlet can be represented by a periodic function. The voltage oscillates between –120 and 120 volts, with a frequency of 50 cycles per second. Write and graph a function for the voltage v as a function of time t.**

Step 1 The voltage oscillates from −120 to 120.

The maximum is 120. The minimum is −120.

The amplitude is 120. $a = 120$

Step 2 The frequency is 50 cycles per second.

The period is $\frac{1}{\text{frequency}}$ or $\frac{1}{50}$.

$\text{period} = \frac{2\pi}{b}$ Period in terms of b

$\frac{1}{50} = \frac{2\pi}{b}$ Substitute.

$b = 100\pi$ Solve.

Step 3 Use t instead of x to write the function as a function of time.

$y = a \sin bt$ General equation

$y = 120 \sin 100\pi t$ Substitute.

Step 4 Use a graphing calculator to graph the function.

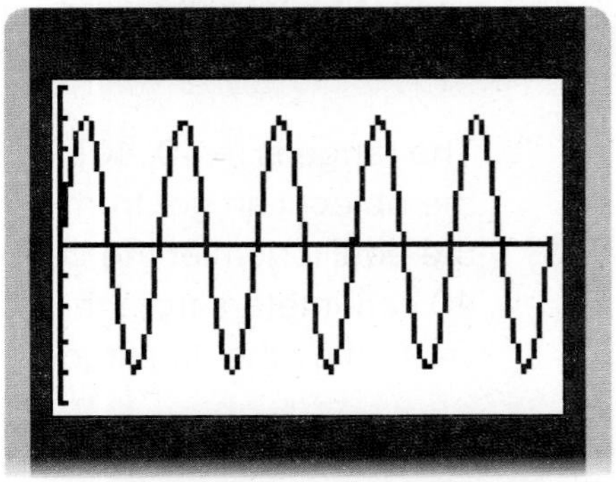

[0, 0.1] scl: 0.01 by [−150, 150] scl: 30

Think About It!
Could cosine have been used to model the situation? If so, provide the cosine function, and explain the difference between the two functions.

Go Online You can complete an Extra Example online.

Practice

Go Online You can complete your homework online.

Example 1

Find the amplitude, midline, and period of each function.

1.

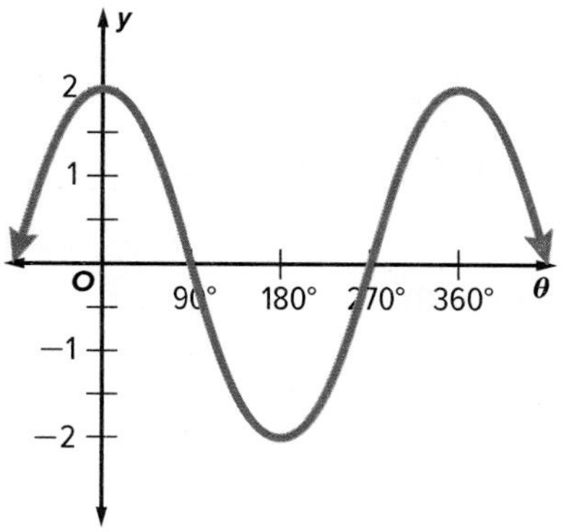

2.

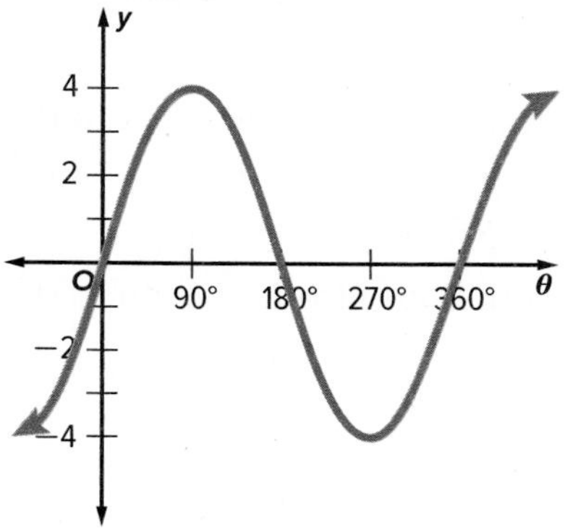

3.

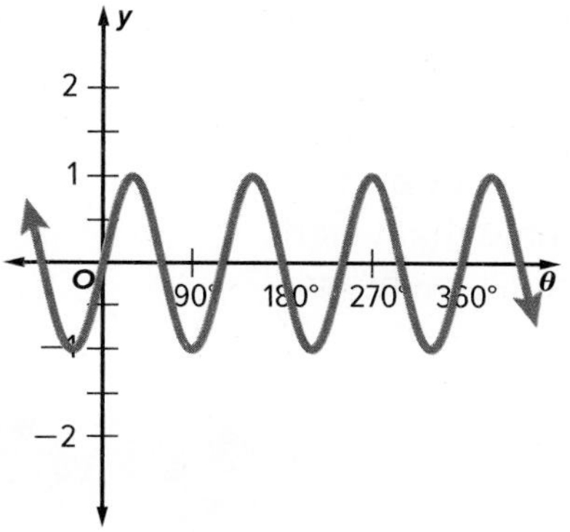

4.

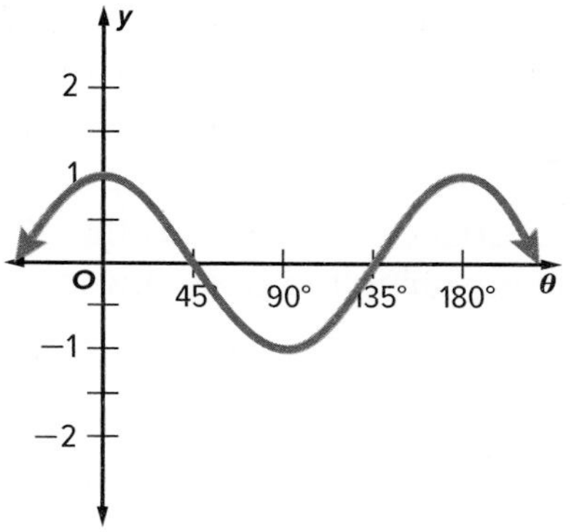

Example 2

Find the amplitude and period of each function.

5. $y = 2\cos\theta$

6. $y = 2\sin\theta$

7. $y = \cos\frac{1}{2}\theta$

8. $y = \frac{3}{4}\cos\theta$

9. $y = \frac{1}{2}\sin 2\theta$

10. $y = 3\cos 2\theta$

Examples 3 and 4

Find the amplitude and period of each function. Then graph the function.

11. $y = 3\sin\theta$

12. $y = \cos 3\theta$

13. $y = \sin 4\theta$

14. $y = \frac{3}{2}\sin\theta$

15. $y = 4\cos 2\theta$

16. $y = 5\sin\frac{2}{3}\theta$

Example 5

17. USE A MODEL An object on a spring oscillates using the function $y = 25 \cos \pi t$, where y is the distance in inches from its equilibrium position in t seconds.

a. Find and describe the period and frequency.

b. Identify the domain and range in the context of the situation.

18. SWINGS Suppose a tire swing is rotated $\frac{\pi}{5}$ radians and released. The function $y = \frac{\pi}{5} \cos 2t$ represents the displacement of the swing at time t for a frequency of radians per second.

a. Find the period and frequency, and describe them in the context of the situation.

b. Identify the domain and range in the context of the situation.

Example 6

19. REASONING A boat that is tied to a dock moves vertically up and down with the waves. Delray watches the boat for 30 seconds and notes that the boat moves up and down a total of 6 times. The difference between the boat's highest point and lowest point is 3 feet. Write and graph a trigonometric function that models the boat's vertical position x seconds after she began watching. Assume that when Delray began watching the boat, it was at its highest point and that its average vertical position was 0 feet.

20. FERRIS WHEELS A Ferris wheel at a state fair has a diameter of 65 feet and makes 4 complete revolutions each minute. Santiago boards a car of the Ferris wheel at the car's lowest point, and he rides for 2 minutes. Write and graph a trigonometric function that models his height above or below the axle of the Ferris wheel θ minutes after the ride starts.

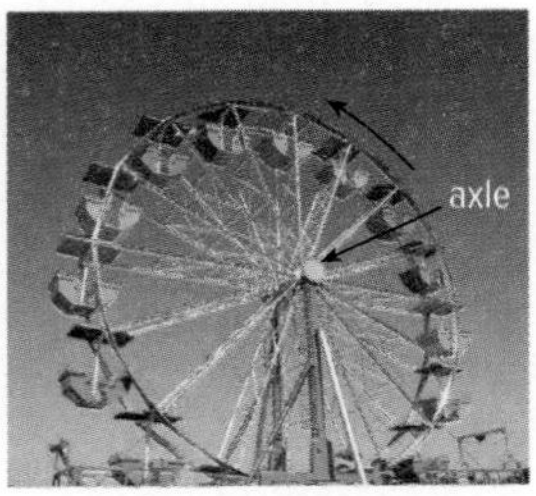

Mixed Exercises

Find the amplitude and period of each function. Then graph the function.

21. $y = 3 \sin \frac{2}{3}\theta$

22. $y = \frac{1}{2} \cos \frac{3}{4}\theta$

23. $y = 4 \cos \frac{\theta}{3}$

REGULARITY **For each graph, identify the period and write an equation.**

24.

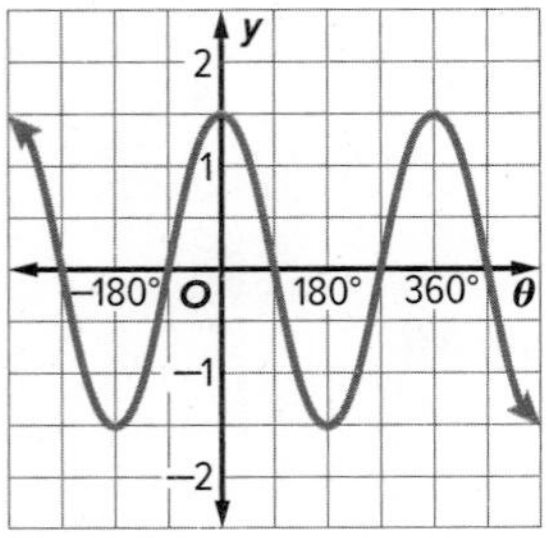

25.

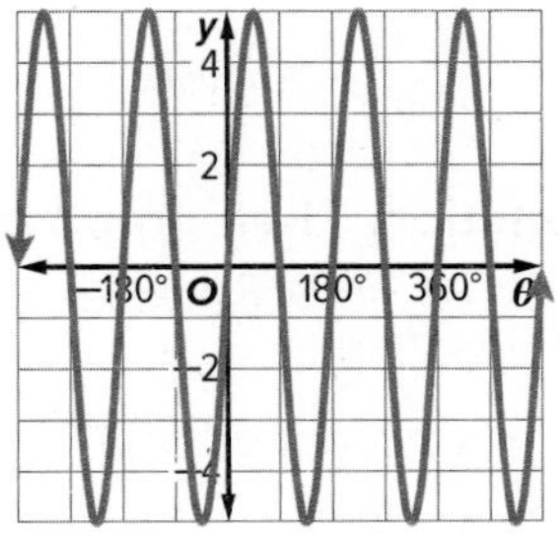

26.

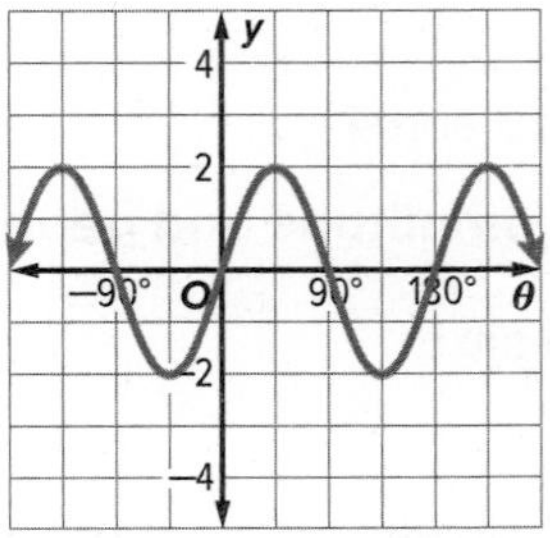

Ivan Cholakov/123RF

27. USE A SOURCE Research how the sine function can be used to model ocean waves. Explain.

28. PHYSICS An anchoring cable exerts a force of 500 Newtons on a pole. The force has the horizontal and vertical components F_x and F_y.

a. The function $F_x = 500 \cos \theta$ describes the relationship between the angle θ and the horizontal force. What are the amplitude and period of this function?

b. The function $F_y = 500 \sin \theta$ describes the relationship between the angle θ and the vertical force. What are the amplitude and period of this function?

29. WEATHER The function $y = 60 + 25 \sin \frac{\pi}{6}t$, where t is in months and $t = 0$ corresponds to April 15, models the average high temperature in degrees Fahrenheit.

a. Determine the period of this function. What does this period represent?

b. What is the maximum high temperature and when does this occur?

30. MODELING A cyclist pedals at a rate of 6 rotations every 5 seconds. The motion of the pedals is circular with a radius of 7 inches. The closest the pedals get to the ground is four inches away. Write a function $h(t)$ that models the height of a pedal in inches as a function of the time t in seconds. Assume the pedal starts at its highest point.

31. STRUCTURE Functions can be used to model the wave patterns of different colors of light emitted from a particular source, where y is the height of the wave in nanometers and t is the length from the start of the wave in nanometers.

Color	Function
Red	$y = 300 \sin\left(\frac{\pi}{350}t\right)$
Orange	$y = 125 \sin\left(\frac{\pi}{305}t\right)$
Yellow	$y = 460 \sin\left(\frac{\pi}{290}t\right)$
Green	$y = 200 \sin\left(\frac{\pi}{260}t\right)$
Blue	$y = 40 \sin\left(\frac{\pi}{235}t\right)$
Violet	$y = 80 \sin\left(\frac{\pi}{210}t\right)$

a. What are the amplitude and period of the function describing green light waves?

b. The intensity of a light wave corresponds directly to its amplitude. Which color emitted from the source is the most intense?

c. The color of light depends on the period of the wave. Which color has the shortest period? The longest period?

32. SWIMMING As Kazuo swims a 25-meter sprint, the position of his right hand relative to the water surface can be modeled by the graph shown, where y is the height of the hand in inches from the water level and t is the time in seconds past the start of the sprint. What function describes this graph?

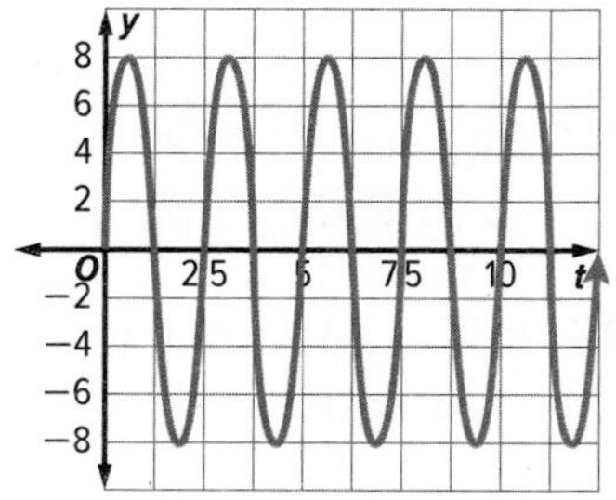

33. **REASONING** Maribel sets up an experiment with a spring at a physics lab. She hangs the spring from a hook and attaches a weight to the bottom of the spring. She records the length of the spring when is it fully compressed and fully extended, as shown. When she releases the spring from the fully-compressed position, she finds that it takes 2 seconds to come back to this position. Maribel wants to write a function $f(t)$ that models the length of the spring, in centimeters, t seconds after it has been released.

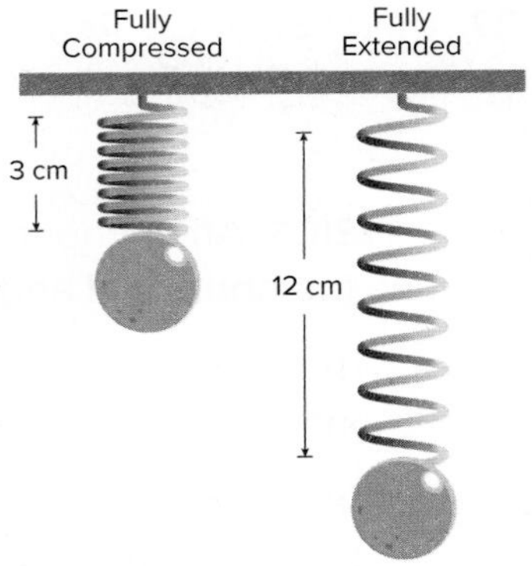

a. What is the amplitude of $f(t)$? What is the period?

b. Write and graph a function $f(t)$ that models the length of the spring.

34. **AMUSEMENT RIDES** An amusement park ride consists of two 16-foot arms that each have a car on one end of the arm and a counterweight on the other end. The arms rotate in opposite directions. Each car has a minimum height of 3 feet above the ground. It takes each car 3 seconds to make a complete rotation. The function $f(x)$ models the height of one car above the ground x seconds after the ride starts. (Assume the cars make complete rotations as soon as the ride begins.)

a. Write and graph a function $f(x)$ that models the motion of a car.

b. What is the first interval of your graph in which the function is increasing? What does it represent?

c. Explain how your graph shows the period of the ride.

35. **CREATE** Write a sine function in which the amplitude is 2 and its graph has 3 complete cycles on the interval $0 \leq \theta \leq \pi$. Justify your answer.

36. **PERSEVERE** The function $h(x) = 6 \sin 30\pi x + 10$ models the height above ground in inches of a point P at the tip of a blade of a floor fan x seconds after the fan is turned on. What is the speed of the fan in rotations per minute? Explain.

37. **ANALYZE** Compare and contrast the graphs of $y = \frac{1}{2} \sin \theta$ and $y = \sin \frac{1}{2}\theta$.

38. **CREATE** Write a trigonometric function that has an amplitude of 3 and a period of 180°. Then graph the function.

39. **WRITE** How can you use the characteristics of a trigonometric function to sketch its graph?

Graphing Other Trigonometric Functions

Learn Graphing Tangent Functions

Because $\tan x = \frac{\sin x}{\cos x}$, the tangent function is undefined when $\cos x = 0$, which occurs at $x = (90 + 180n)°$, where n is an integer. As a result, the graph of the tangent function has asymptotes where $x = (90 + 180n)°$.

Key Concept • Graphs of Tangent Functions

Parent Function	$y = \tan x$
Domain	$\{x \mid x \neq (90 + 180n)°$, n is an integer$\}$
Range	all real numbers
Amplitude	undefined
Period	180°
Number of x-Intercepts in One Cycle	1
Midline	$y = 0$

Graph

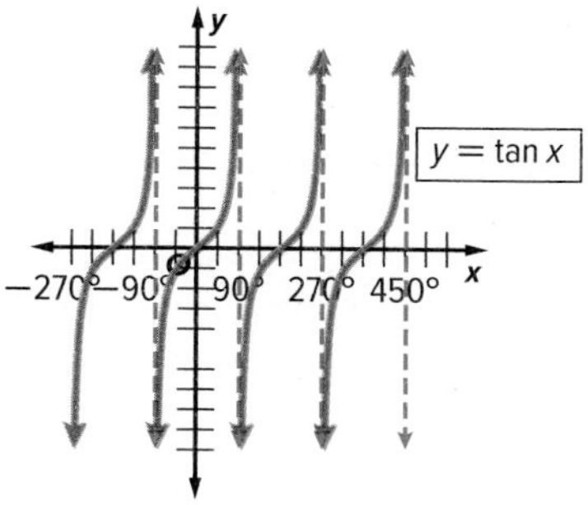

For the graph of $y = a \tan bx$, the period is $\frac{180°}{|b|}$, there is no amplitude, and the asymptotes are at $x = \frac{(90 + 180n)°}{|b|}$, where n is an integer.

The value of a determines the vertical dilation, where $|a| > 1$ results in a vertical stretch and $0 < |a| < 1$ results in a vertical compression. The function is reflected in the x-axis when $a < 0$. The value of b determines the horizontal dilation, where $|b| > 1$ results in a horizontal compression and $0 < |b| < 1$ results in a horizontal stretch.

Example 1 Graph a Tangent Function with a Dilation

Find the period, asymptotes, x-intercepts, midline, and transformations of $y = \tan 3x$. Then graph the function.

Step 1 Analyze the function.

- For a tangent function in the form $y = a \tan bx$, $\frac{180°}{|b|}$ represents the period. Because $b = 3$, the period is $\frac{180°}{|3|}$ or 60°.
- The asymptotes occur at $x = \frac{(90 + 180n)°}{|3|}$ or $(60n + 30)°$, where n is an integer.

(continued on the next page)

Today's Goals

- Graph and analyze tangent functions.
- Graph and analyze reciprocal trigonometric functions.

Today's Vocabulary

reciprocal trigonometric functions

Talk About It!

Explain why the tangent function does not have an amplitude.

Go Online

You can learn how to graph tangent functions by watching the video.

Think About It!

Compare the period, asymptotes, and x-intercepts of $y = \tan 3x$ and $y = -\tan 3x$.

- x-intercepts occur at integer multiples of the period. The period of this function is 60°, so the x-intercepts are 0°, 60°, 120°, 180°, ...
- The midline of $y = \tan 3x$ is $y = 0$.
- Because $a = 1$, the function is not vertically dilated in relation to the parent function. Because $b = 3$, the function is compressed horizontally in relation to the parent function.

Step 2 Graph the function.

Use the period, asymptotes, and x-intercepts to graph the function. Notice that $y = \tan 3x$ is compressed horizontally in relation to the parent function.

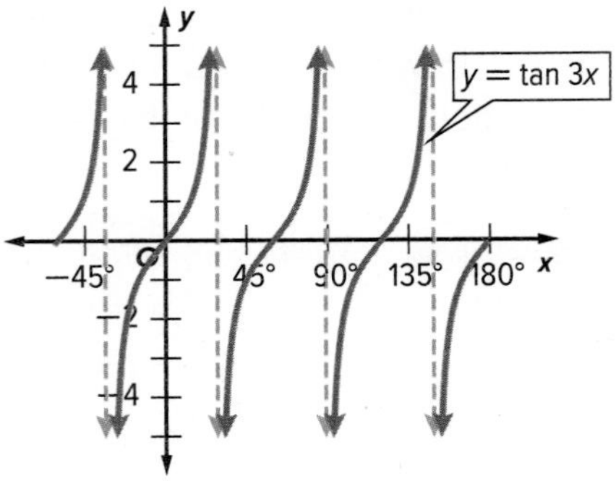

Check

Consider $y = \tan 0.25x$.

Part A Identify the period, asymptotes, x-intercepts, and transformations of $y = \tan 0.25x$ for $-2\pi \leq x \leq 6\pi$.

period: ___?___

asymptotes: ___?___

x-intercepts: ___?___

The function is ___?___ in relation to the parent function.

Part B Select the graph of $y = \tan 0.25x$.

A.

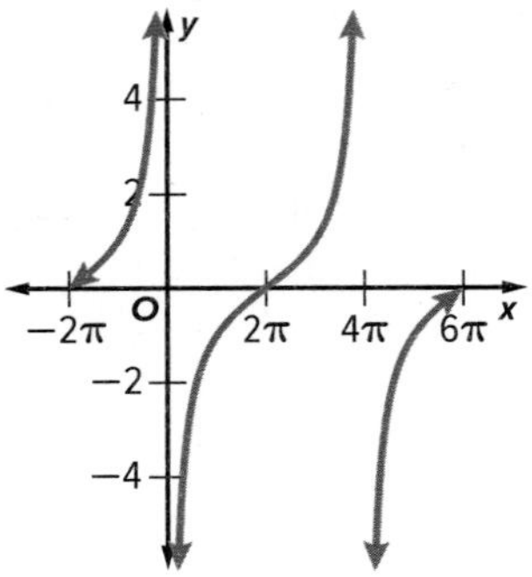

B.

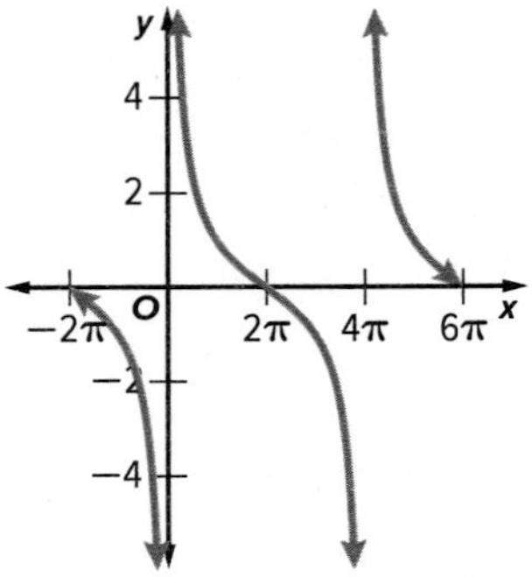

C.

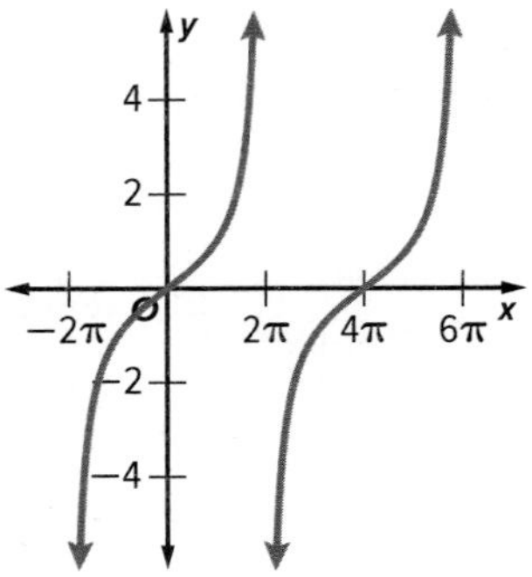

D.

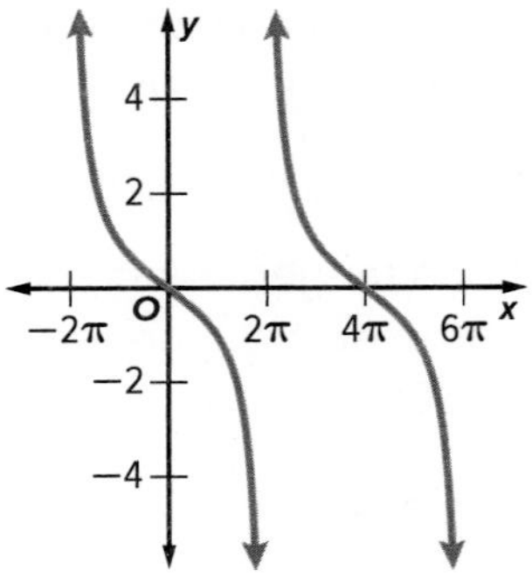

Go Online You can complete an Extra Example online.

Example 2 Graph a Tangent Function with a Dilation and a Reflection

Find the period, asymptotes, x-intercepts, midline, and transformations of $y = -\frac{1}{3}\tan 2x$. Then graph the function.

Step 1 Analyze the function.

- The period is $\frac{\pi}{|2|}$ or $\frac{\pi}{2}$.
- The asymptotes occur at $\frac{\frac{\pi}{2} + \pi n}{|b|}$, where n is an integer. The asymptotes are at $x = \frac{\frac{\pi}{2} + \pi n}{|2|}$ or $\frac{\pi}{4} + \frac{\pi n}{2}$, where n is an integer.
- x-intercepts occur at integer multiples of the period. The period of this function is $\frac{\pi}{2}$, so the x-intercepts are $-\frac{\pi}{2}$, 0, $\frac{\pi}{2}$, π, $\frac{3\pi}{2}$, ...
- The midline of $y = -\frac{1}{3}\tan 2x$ is $y = 0$.
- Because $a = -\frac{1}{3}$, the function is compressed vertically and reflected in the x-axis in relation to the parent function. Because $b = 2$, the function is compressed horizontally in relation to the parent function.

Think About It!

Are $y = \frac{1}{2}\tan 2x$ and $y = \tan x$ equivalent functions? Justify your answer.

Step 2 Graph the function.

Use the period, asymptotes, and x-intercepts to graph the function.

Notice that $y = -\frac{1}{3}\tan 2x$ is compressed vertically, compressed horizontally, and reflected in the x-axis in relation to the parent function.

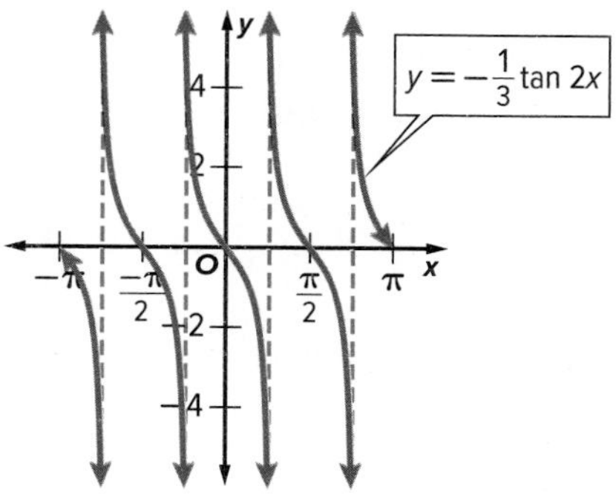

Check

Consider $y = -2\tan\frac{3}{2}x$.

Part A

Identify the period, asymptotes, x-intercepts, and transformations of $y = -2\tan\frac{3}{2}x$ for $-60° \leq x \leq 180°$.

period: __?__

asymptotes: ______?______

x-intercepts: __?__

The function is ___?___ vertically, ___?___ horizontally and reflected in the __?__ in relation to the parent function.

Part B

Graph $y = -2\tan\frac{3}{2}x$.

Learn Graphing Reciprocal Trigonometric Functions

The **reciprocal trigonometric functions**, cosecant, secant, and cotangent, can be expressed as $\frac{1}{\sin x}$, $\frac{1}{\cos x}$, and $\frac{1}{\tan x}$, respectively. As a result, the graphs of the reciprocal trigonometric functions have asymptotes when the corresponding sine, cosine, or tangent function equals 0 and the reciprocal function is undefined.

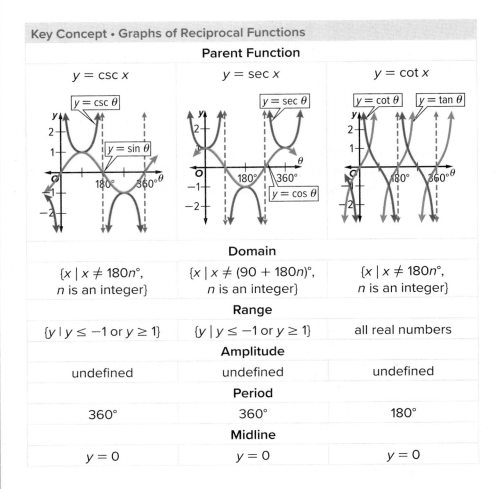

<table>
<tr><th colspan="3">Key Concept • Graphs of Reciprocal Functions</th></tr>
<tr><td colspan="3">Parent Function</td></tr>
<tr><td>$y = \csc x$</td><td>$y = \sec x$</td><td>$y = \cot x$</td></tr>
<tr><td>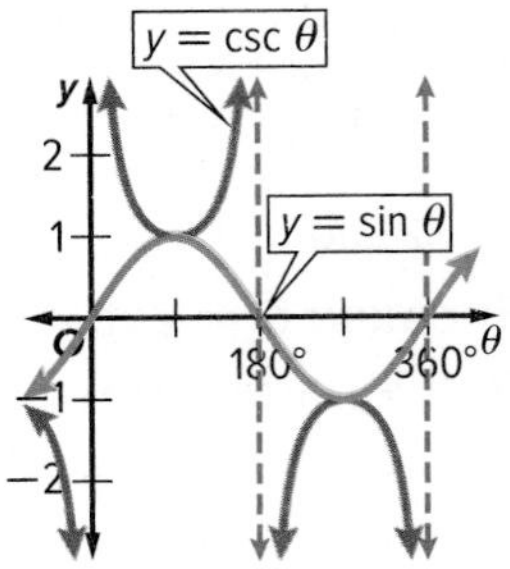
</td><td>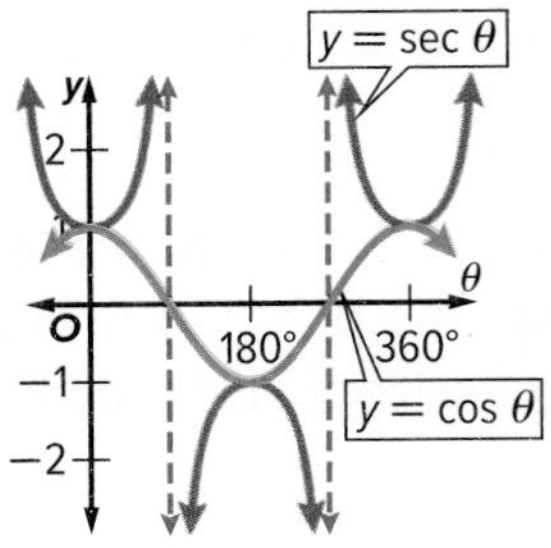
</td><td>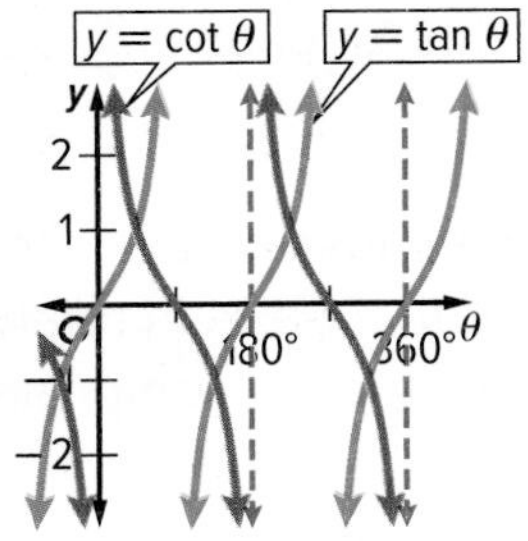
</td></tr>
<tr><td colspan="3">Domain</td></tr>
<tr><td>$\{x \mid x \neq 180n°$, n is an integer$\}$</td><td>$\{x \mid x \neq (90 + 180n)°$, n is an integer$\}$</td><td>$\{x \mid x \neq 180n°$, n is an integer$\}$</td></tr>
<tr><td colspan="3">Range</td></tr>
<tr><td>$\{y \mid y \leq -1 \text{ or } y \geq 1\}$</td><td>$\{y \mid y \leq -1 \text{ or } y \geq 1\}$</td><td>all real numbers</td></tr>
<tr><td colspan="3">Amplitude</td></tr>
<tr><td>undefined</td><td>undefined</td><td>undefined</td></tr>
<tr><td colspan="3">Period</td></tr>
<tr><td>360°</td><td>360°</td><td>180°</td></tr>
<tr><td colspan="3">Midline</td></tr>
<tr><td>$y = 0$</td><td>$y = 0$</td><td>$y = 0$</td></tr>
</table>

Think About It!

Compare the range of $y = 3 \cos x$ and the range of $y = 3 \sec x$.

To graph a reciprocal trigonometric function,

- find the period of the corresponding reciprocal function.
- determine the vertical asymptotes by finding when the corresponding reciprocal function equals 0.
- determine the relative maxima and minima for secant and cosecant functions.
- determine x-intercepts for cotangent functions.
- plot the corresponding reciprocal function as a guide.

Example 3 Graph a Cosecant Function

Find the period, asymptotes, relative extrema, and midline of $y = \csc 0.5x$. Then graph the function.

Step 1 Analyze the function.

Since $y = \csc 0.5x$ is the reciprocal of $y = \sin 0.5x$, the graphs have the same period of $\frac{2\pi}{0.5}$, or 4π. The vertical asymptotes occur when $\sin 0.5x = 0$. So, the asymptotes are $x = 0$, $x = 2\pi$, $x = 4\pi$, ... or $x = 2\pi n$, where n is an integer.

The relative maxima and minima of $y = \csc 0.5x$ occur at the same points as the relative minima and maxima of $y = \sin 0.5x$. So, $y = \csc 0.5x$ has relative maxima at $x = -\pi$, $x = 3\pi$, $x = 7\pi$, ..., and $y = \csc 0.5x$ has relative minima at $x = -3\pi$, $x = \pi$, $x = 5\pi$, ... The midline is $y = 0$.

Step 2 Graph the function.

Use the period, asymptotes, and relative extrema to graph $y = \csc 0.5x$.

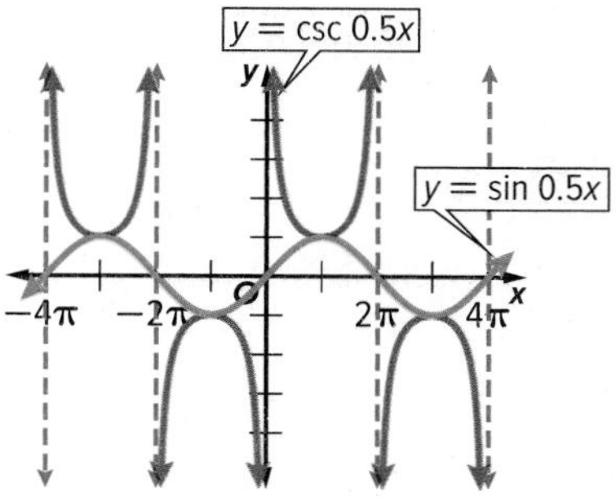

Think About It!

Use reflections to identify another cosecant function with a graph identical to $y = \csc 0.5x$.

Example 4 Graph a Cotangent Function

Find the period, asymptotes, x-intercepts, and midline of $y = -4 \cot 2x$. Then graph the function.

Step 1 Analyze the function.

Since cotangent functions are reciprocals of tangent functions, $y = -4 \cot 2x$ has the same period as $y = 4 \tan 2x$, $\frac{180°}{2}$ or $90°$.

The vertical asymptotes occur when $-4 \tan 2x = 0$. So, the asymptotes are $x = 0°$, $x = 90°$, $x = 180°$, ... or $x = 90n°$, where n is an integer.

The x-intercepts of $y = -4 \cot 2x$ occur at the asymptotes of $y = -4 \tan 2x$. Therefore, the x-intercepts are odd multiples of $\frac{180°}{2|b|}$ or $45°$. The midline is $y = 0$.

Step 2 Graph the function.

Use the period, asymptotes, and x-intercepts to graph $y = -4 \cot 2x$.

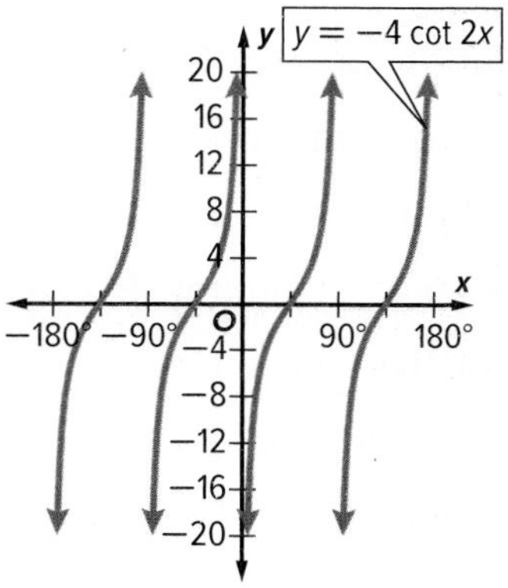

Go Online You can complete an Extra Example online.

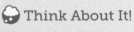

Example 5 Apply a Reciprocal Trigonometric Function

ADVERTISING Suppose a banner towing plane is approaching a music festival at an elevation of 1200 feet above the crowd at the festival. The plane will eventually fly directly over the crowd. Let *d* be the distance in feet the banner is from the music festival, and let *x* be the angle of elevation to the banner from the perspective of the crowd. Write a function that relates the distance as a function of an angle *x*. Then, graph the function and analyze its key features.

Part A Write a function.

Begin by making a sketch. Because the side opposite x is known and the hypotenuse is d, use the cosecant function.

$$\sin x = \frac{1200}{d}$$

$$d = \frac{1200}{\sin x}$$

$$d = 1200 \csc x$$

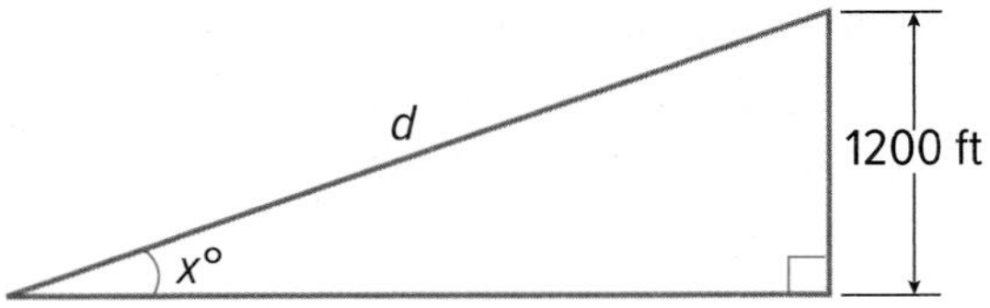

Study Tip

Assumptions To write a function to represent this situation, you must assume that the plane is flying in a constant direction without varying altitude.

Part B Graph the function.

The period of the function is $\frac{360°}{b}$ or 360°.

The vertical asymptotes occur when $1200 \sin x = 0$. The asymptotes are $x = 180n°$, where n is an integer.

Use the graph of $y = 1200 \sin x$ to help graph $y = 1200 \csc x$.

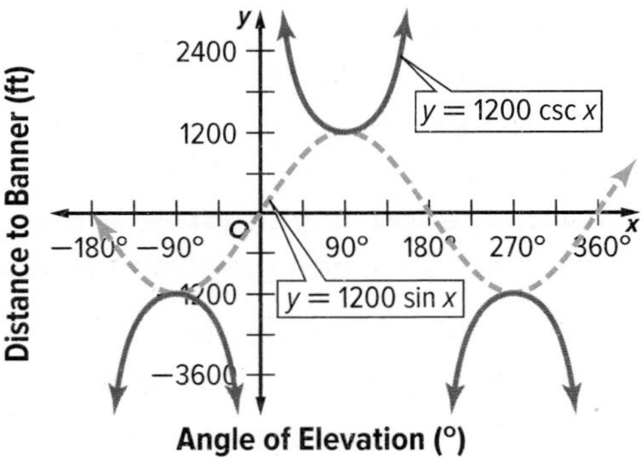

Part C Analyze the graph.

In the context of the situation, only the x-values between 0° and 180° are relevant, so analyze that portion of the graph.

Domain: $\{x \mid 0 < x < 180\}$

Range: $\{d \mid d \geq 1200\}$

x-Intercept: none

y-Intercept: none

Relative minimum: (90, 1200)

Relative maximum: none

Increasing: $\{x \mid 90 < x < 180\}$

Decreasing: $\{x \mid 0 < x < 90\}$

Think About It!

What is the distance of the banner when the angle of elevation is 55°? Round to the nearest foot.

Go Online You can complete an Extra Example online.

Practice

Go Online You can complete your homework online.

Examples 1 and 2

Find the period, asymptotes, *x*-intercepts, midline, and transformations of each tangent function. Then graph the function.

1. $y = \tan 5x$

2. $y = \tan 4x$

3. $y = \tan 2x$

4. $y = \frac{1}{2} \tan x$

5. $y = 2 \tan \frac{1}{2}x$

6. $y = -\frac{1}{2} \tan 2x$

Examples 3 and 4

Find the period and asymptotes of each function, the relative extrema of each secant and cosecant function, and the *x*-intercepts of each cotangent function. Then graph each function.

7. $y = \csc \frac{3}{4}x$

8. $y = \csc 3x$

9. $y = \frac{1}{2} \csc x$

10. $y = \cot 2x$

11. $y = \cot \frac{1}{2}x$

12. $y = -2 \cot x$

13. $y = 3 \sec x$

14. $y = \sec \frac{1}{3}x$

15. $y = \sec 4x$

Example 5

16. ACROBATS Suppose an acrobat is walking from a high wire that is attached on its ends to two different towers at a height of 200 feet above the floor, as shown. The acrobat will eventually walk directly over the location of a camera in the floor. Let *d* be the distance in feet the acrobat is from the camera.

a. Write a function that relates the distance as a function of an angle *x*.

b. Graph the function.

c. Analyze the graph.

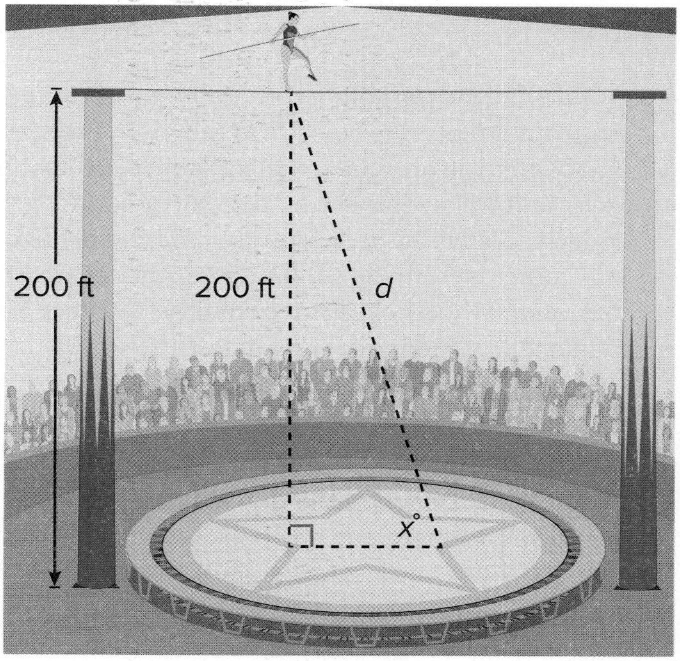

17. AIR TRAFFIC CONTROL Ground-based air traffic controllers direct aircraft on the ground and in controlled airspace from airport control towers. Let *y* be the length of the shadow of a 300-foot control tower as the Sun moves across the sky at angle *x*.

a. Write the function that relates the length of the shadow to the of angle.

b. Graph the function.

c. Analyze the graph.

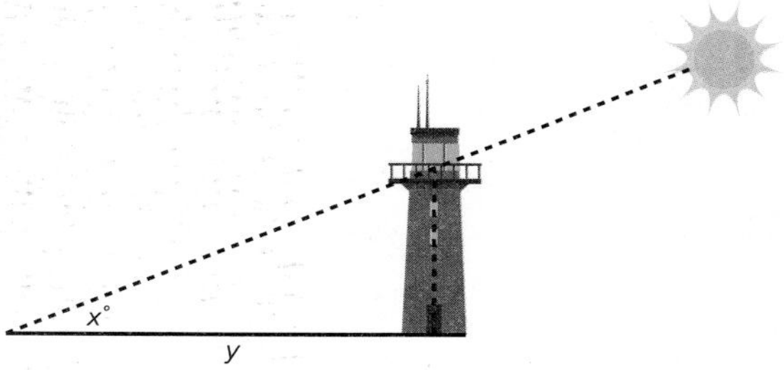

Mixed Exercises

Find the period of each function. Then graph the function.

18. $y = 2 \tan \frac{1}{4}\theta$

19. $y = 2 \sec \frac{4}{5}\theta$

20. $y = 5 \csc 3\theta$

21. $y = 2 \cot 6\theta$

22. $y = \sec 3\theta$

23. $y = 2 \sec \theta$

24. ATTRACTIONS A replica of the Eiffel Tower at an amusement park has an elevator that lifts riders vertically up the tower to an observation deck. Suppose Susana is waiting 20 meters away from the base of the tower while her sister Elva is on the elevator. Let d be the distance in meters from Susana to Elva as she is moving up the tower, and let x be the angle of elevation to Susana from Elva's perspective. Write the function that relates the distance d as a function of an angle x.

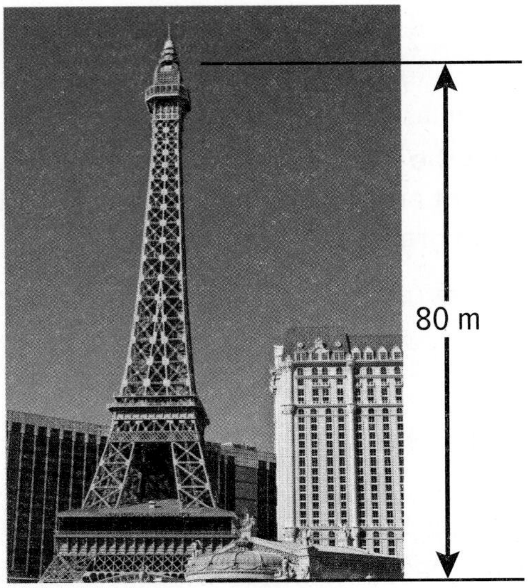

25. FOOTBALL As part of the pre-game celebration a military helicopter flies over the crowd of fans in the stadium. The helicopter will approach the stadium 2000 feet above field level and will eventually fly directly over the marching band located on the 50-yard line. Let d be the distance in feet the helicopter is from the 50-yard line, and let x be the angle of elevation to the helicopter from there. Write a function that relates the distance to the angle.

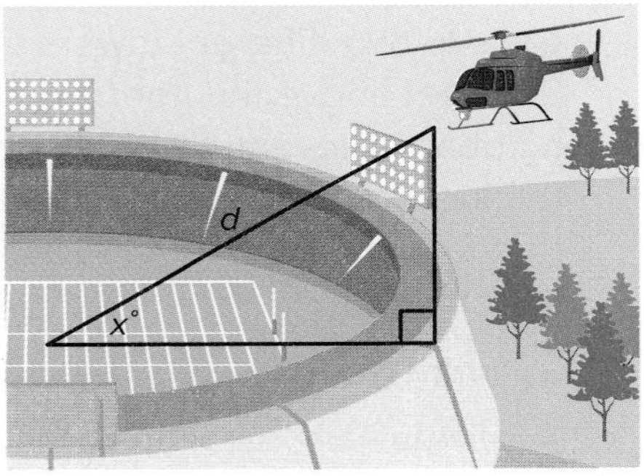

Higher-Order Thinking Skills

26. PERSEVERE Describe the domain and range of $y = a \cos \theta$ and $y = a \sec \theta$, where a is any positive real number.

27. FIND THE ERROR Tyler says the period of the function $y = \tan \frac{1}{2}\theta$ is 360°. Lacretia says it is 180°. Is either of them correct? Explain your reasoning.

28. CREATE Write a trigonometric function that has a domain of $\{\theta | \theta \neq 90° + 180n$, n is an integer$\}$ and a range of $\{y | y \leq -3$ or $y \geq 3\}$.

29. WRITE How can you use the key features of a trigonometric function to sketch its graph?

Translations of Trigonometric Graphs

Explore Analyzing Trigonometric Functions by Using Technology

Online Activity

INQUIRY How does adding a constant to, subtracting a constant from, or multiplying a constant by a function affect the graph of a trigonometric function?

Learn Horizontal Translations of Trigonometric Functions

A horizontal translation of the graph of a trigonometric function is called a **phase shift**.

Key Concept • Phase Shift

The phase shift of the functions $y = a \sin b(x - h)$, $y = a \cos b(x - h)$, and $y = a \tan b(x - h)$ is h, where $b > 0$.

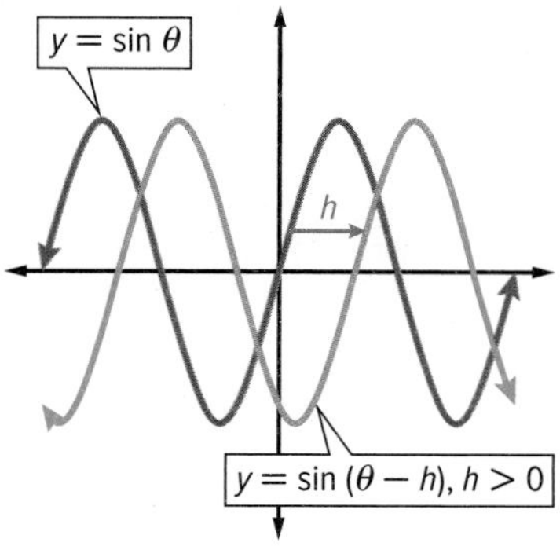

If $h > 0$, the parent function is translated right h units.

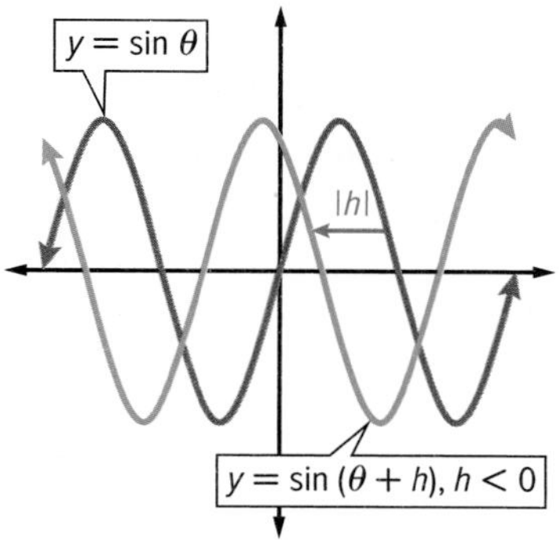

If $h < 0$, the parent function is translated left $|h|$ units.

Example 1 Graph a Phase Shift

State the amplitude, period, and phase shift for $y = \cos (x - 270°)$. Then graph the function and state the domain and range.

Step 1 Analyze the function.

For $y = a \cos b(x - h)$, a represents the amplitude, $\frac{360°}{|b|}$ represents the period, and h represents the phase shift.

For $y = \cos (x - 270°)$, the amplitude is 1, the period is $\frac{360°}{|1|}$ or 360°, and the phase shift is 270° right in relation to the parent function.

(continued on the next page)

Go Online You can complete an Extra Example online.

Today's Goals

- Graph horizontal translations of trigonometric functions.
- Graph vertical translations of trigonometric functions.

Today's Vocabulary

phase shift
vertical shift

Go Online
You may want to complete the Concept Check to check your understanding.

Talk About It!
Find a phase shift to the left that will produce a graph that is identical to $y = \cos (x - 270°)$. Explain your reasoning.

Step 2 Graph the function.

Translate the graph right 270° in relation to $y = \cos x$. The domain is all real numbers, and the range is $\{y \mid -1 \leq y \leq 1\}$.

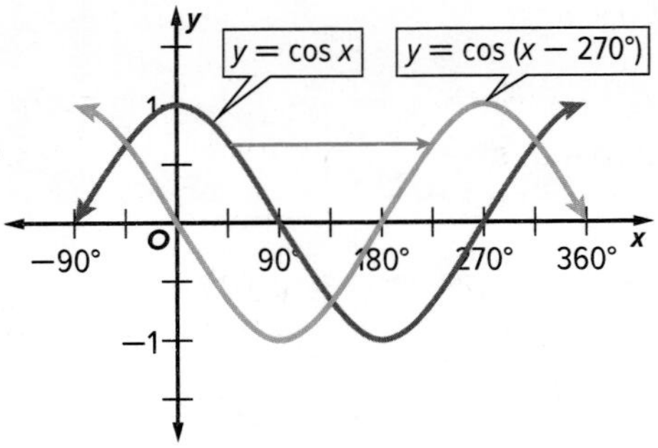

Example 2 Graph a Transformation of a Trigonometric Function

State the amplitude, period, and phase shift for $y = 3 \sin\left(x + \frac{\pi}{4}\right)$. Then graph the function and state the domain and range.

Step 1 Analyze the function.

For $y = 3 \sin\left(x + \frac{\pi}{4}\right)$, $a = 3$, $b = 1$, and $h = -\frac{\pi}{4}$. So the amplitude is 3, the period is $\frac{2\pi}{|1|}$ or 2π radians, and the phase shift is left $\frac{\pi}{4}$ radians in relation to the parent function.

Step 2 Graph the function.

Use the amplitude and period to graph $y = 3 \sin x$. Translate the function left $\frac{\pi}{4}$ radians.

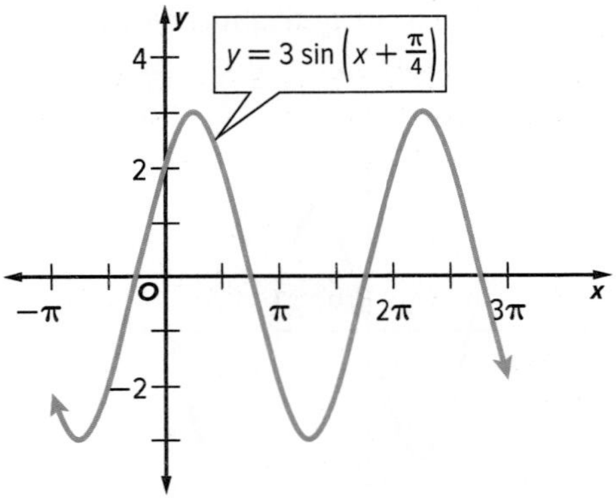

The graph of $y = 3 \sin\left(x + \frac{\pi}{4}\right)$ has been stretched vertically and shifted left $\frac{\pi}{4}$ radians in relation to the parent function. The domain is all real numbers, and the range is $\{y \mid -3 \leq y \leq 3\}$.

Think About It!

Compare the graphs of $y = 3 \sin\left(x + \frac{\pi}{4}\right)$ and $y = 3 \sin(x + 45°)$.

Check

Identify the key features of $y = 2 \tan\left(x - \frac{\pi}{3}\right)$. Then graph the function.

amplitude: _?_

period: _?_

phase shift: _?_

domain: $\{x \mid x \neq$ _?_ n, where n is an integer$\}$

range: $\{y \mid$ _?_ $\leq y \leq$ _?_$\}$

Go Online You can complete an Extra Example online.

Learn Vertical Translations of Trigonometric Functions

A vertical translation of the graph of a trigonometric function is called a **vertical shift**. By adding or subtracting a positive constant k, the graph is translated up or down in relation to the parent function.

> **Think About It!**
> What is the midline for $y = \tan x$? How does this relate to its vertical shift and the value of k?

Key Concept • Vertical Shift

The vertical shift of the functions $y = a \sin b\theta + k$, $y = a \cos b\theta + k$, and $y = a \tan b\theta + k$ is k, where $b > 0$.

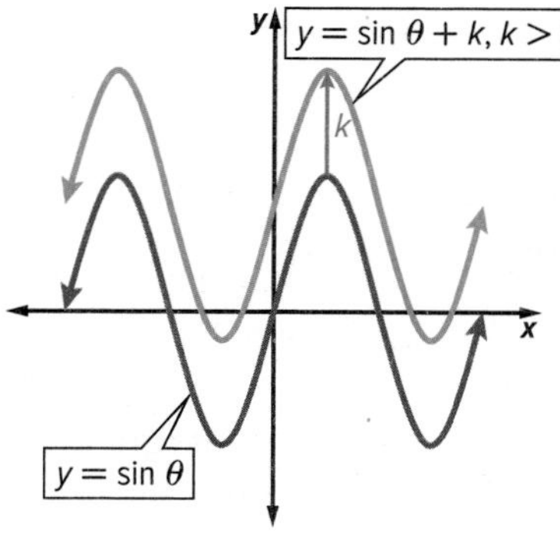

If $k > 0$, the parent function is translated up k units.

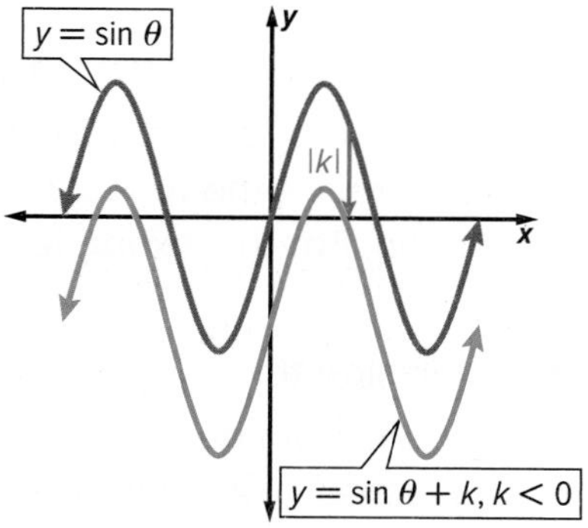

If $k < 0$, the parent function is translated down $|k|$ units.

When a trigonometric function is shifted vertically k units, the midline is $y = k$, which is the line about which the graph of a function oscillates. You can use the midline to help sketch graphs of functions with vertical shifts.

Example 3 Graph a Vertical Shift

State the amplitude, period, phase shift, vertical shift, and midline equation of $y = \tan \frac{1}{2}x - 1$. Then graph the function and state the domain and range.

> **Think About It!**
> Compare and contrast the graphs of $y = \tan \frac{1}{2}x - 1$ and $y = \tan x$.

Step 1 Analyze the function.

For $y = \tan \frac{1}{2}x - 1$, $a = 1$, $b = \frac{1}{2}$, and $k = -1$. So the period is $\frac{180°}{\left|\frac{1}{2}\right|}$ or 360°, the vertical shift is down 1 unit, and the midline is $y = -1$.

Step 2 Graph the function.

For $y = \tan \frac{1}{2}x$, the vertical asymptotes are at $x = (360n + 180)°$, where n is an integer. Translate the function down 1 unit. The graph of $y = \tan \frac{1}{2}x - 1$ is stretched horizontally and shifted down 1 unit in relation to the parent function. The domain is $\{x \mid x \neq 180n°$, where n is an integer$\}$ and the range is all real numbers.

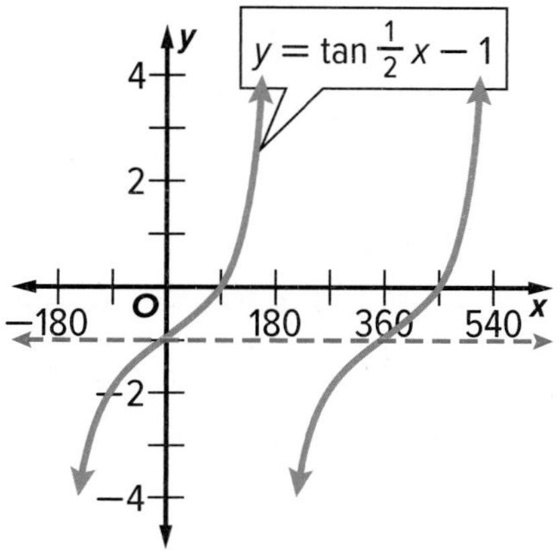

Go Online You can complete an Extra Example online.

Example 4 Model Translations of Trigonometric Functions

EXERCISE **Suppose as a person jumps rope, the height of the rope oscillates between a maximum of 108 inches and a minimum of 0 inches and hits the ground two times per second. Write a trigonometric function that represents the height of the rope *y* at time *x* seconds if the rope begins at a height of 0 inches. Then graph the function.**

Step 1 Choose which function to use.

Because the function oscillates between 0 and 108 inches continuously, the function can be modeled by a sine or cosine function. For this example, we will use cosine.

Step 2 Determine the parameters of the function.

The minimum height is 0 inches and the maximum height is 108 inches. The midline is $y = \frac{0 + 108}{2}$ or $y = 54$. So, the vertical shift is $k = 54$.

Because the amplitude is half the difference between the maximum and minimum values, $a = \frac{108 - 0}{2}$ or $a = 54$.

The period is determined by how often the jump rope completes one cycle. The rope reaches the minimum twice every second, so the period is 0.5 second. Solve for the parameter b using the value of the period.

$0.5 = \frac{2\pi}{\lvert b\rvert}$	Period $= \frac{2\pi}{\lvert b\rvert}$
$0.5\lvert b\rvert = 2\pi$	Multiply each side by $\lvert b\rvert$.
$b = \pm 4\pi$	Simplify.

Use the positive value of b to represent the period.

Find the phase shift by first considering the parent function. The maximum of the parent function $y = \cos x$ occurs at $x = 0$. Because the rope starts at a height of 0 and the period of the function is 0.5, the maximum height of the jump rope occurs at $x = 0.25$. So the phase shift h is $0.25 - 0$ or 0.25.

Step 3 Write the function.

Write the function relating height y and time x.

$y = a \cos b(x - h) + k$	Standard cosine function
$y = 54 \cos 4\pi(x - 0.25) + 54$	Substitute $a = 54$, $b = 4\pi$, $h = 0.25$, and $k = 54$.

Go Online You can complete an Extra Example online.

Problem-Solving Tip

Make a Sketch Making a sketch of the graph might be useful when identifying the parameters of the function. Because you are given maximums, minimums, and the period, you can plot several points and draw a smooth curve between them. The sketch may help you identify which trigonometric function to use and the parameters of the function.

Step 4 Graph and analyze the function.

Use the amplitude, period, and midline to graph $y = 54 \cos 4\pi(x - 0.25) + 54$.

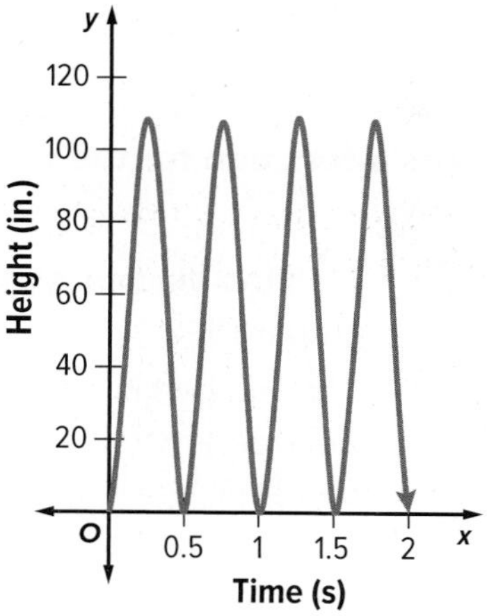

Notice that the graph oscillates between a maximum of 108 inches and a minimum of 0 inches, and is positive for all values in the domain.

The graph has a *y*-intercept of 0 and *x*-intercepts of $0.5n$, where *n* is an integer.

Because *x* represents time and *y* represents height, both values must be positive in the context of the situation.

The domain is $\{x \mid x \geq 0\}$ and $\{y \mid 0 \leq y \leq 180\}$.

Check

TIDES The tides at Florida's Cape Canaveral reached a maximum height of 3.4 feet at 2:00 A.M., minimum height of 0.4 feet at 8:30 A.M., and another maximum at 3:00 P.M. the next day.

Select the trigonometric function that represents the height *y* of the tide *x* hours since midnight. Then graph the function.

A. $y = 1.5 \cos 13(x - 2) + 1.9$

B. $y = 1.5 \cos \frac{2\pi}{13}(x - 2) + 1.9$

C. $y = 1.5 \cos 13(x - 2) + 1.5$

D. $y = 1.9 \cos \frac{2\pi}{13}(x + 2) - 1.5$

Think About It!

Use the function to find the height of the jump rope at 2.2 seconds. Round to the nearest inch.

Go Online

You can complete an Extra Example online.

Example 5 Write a Trigonometric Function from a Graph

Write a function for the cosine graph shown.

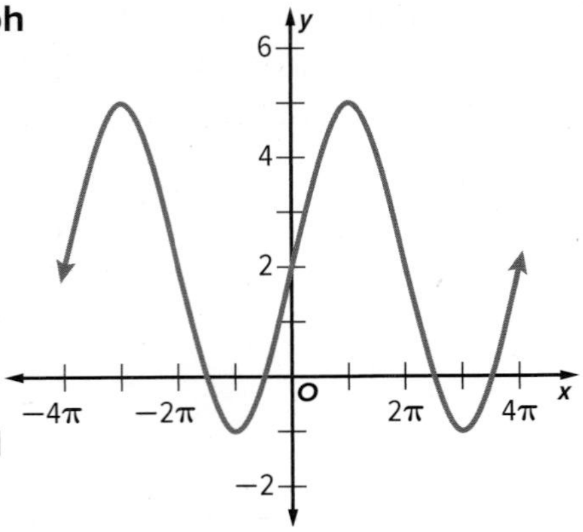

Step 1 Find the vertical shift and amplitude.

The midline is halfway between the relative extrema, $y = -1$ and $y = 5$. So the equation of the midline is $y = \frac{-1 + 5}{2}$ or 2 and the vertical shift k is 2.

The amplitude is half the difference between the maximum and minimum values, $a = \frac{5 - (-1)}{2}$ or 3.

Step 2 Find the phase shift.

$y = \cos x$ has a maximum at $x = 0$. The maximum has been translated right π radians. So the phase shift h is π.

Step 3 Find the period.

The period of the function is the distance between any two consecutive sets of repeating points on the graph.

Use the value of the period, 4π, to find b.

$4\pi = \frac{2\pi}{\lvert b \rvert}$	Period $= \frac{2\pi}{\lvert b \rvert}$
$4\pi \lvert b \rvert = 2\pi$	Multiply each side by $\lvert b \rvert$.
$b = \pm 0.5$	Simplify.

Use the positive value of b to represent the period.

Step 4 Write the function.

Use the parameters from the graph to write the function.

$y = a \cos b(x - h) + k$	Cosine function
$y = 3 \cos 0.5(x - \pi) + 2$	$a = 3$, $b = 0.5$, $h = \pi$, and $k = 2$

The graph is represented by $y = 3 \cos 0.5(x - \pi) + 2$.

Use a graphing calculator to check the solution. You can find the extrema or trace along the function to check that it has the same features as the original function.

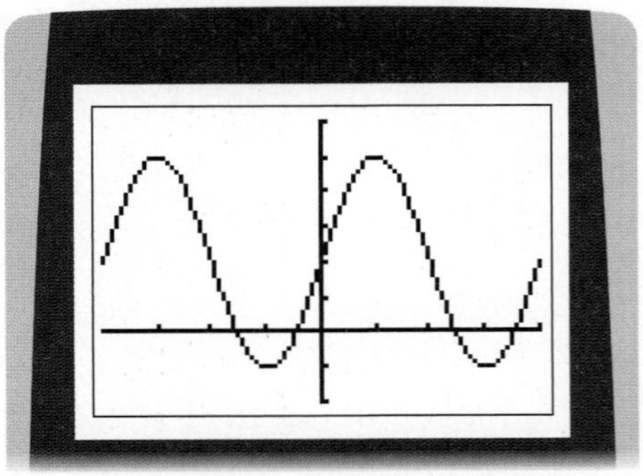

$[-4\pi, 4\pi]$ scl: π by $[-2, 6]$ scl: 1

Think About It!

Write another function for the cosine graph where $h < 0$.

Study Tip

Finding the Period The maximum or minimum values of the graph are often convenient points to use when determining the period.

Go Online

You can complete an Extra Example online.

Practice

Go Online You can complete your homework online.

Example 1

State the amplitude, period, and phase shift for each function. Then graph the function and state the domain and range.

1. $y = \sin(\theta + \pi)$
2. $y = \tan\left(\theta - \frac{\pi}{2}\right)$
3. $y = \sin(\theta + 90°)$
4. $y = \cos(\theta - 45°)$
5. $y = \tan\left(\theta + \frac{\pi}{6}\right)$
6. $y = \cos(\theta + 180°)$

Example 2

State the amplitude, period, and phase shift for each function. Then graph the function and state the domain and range.

7. $y = 3\cos(\theta - 45°)$
8. $y = 2\sin(\theta + 60°)$
9. $y = \frac{1}{2}\sin 3\left(\theta - \frac{\pi}{3}\right)$
10. $y = \frac{1}{2}\cos 3\left(\theta - \frac{\pi}{2}\right)$
11. $y = 2\cos(\theta + 90°)$
12. $y = \frac{1}{4}\sin(\theta - 30°)$

Example 3

State the amplitude, period, phase shift, vertical shift, and midline equation of each function. Then graph the function and state the domain and range.

13. $y = \cos\theta + 3$
14. $y = \tan\theta - 1$
15. $y = \tan\theta + \frac{1}{2}$
16. $y = 2\cos\theta - 5$
17. $y = 2\sin\theta - 4$
18. $y = \frac{1}{3}\sin\theta + 7$

Example 4

19. **REASONING** An office building has a large clock on its exterior. The center of the clock is located 79 feet above ground level, and the minute hand of the clock is 3 feet long. The function $h(t)$ gives the height of the tip of the minute hand above ground level in feet at any time t minutes after 12:00 noon.

   a. What is the amplitude of $h(t)$? What is the midline? What is the period?

   b. Write the function $h(t)$ as a cosine function.

   c. Graph $h(t)$.

   d. On the interval $t = 0$ to $t = 120$, how many times does the function attain its minimum value? What does this represent in the real-world situation?

20. MODELING As the tide moves in and out of a bay, the depth of the water in the bay varies from a low of 15 meters to a high of 18 meters. It takes 6.2 hours for the tide to come in and 6.2 hours for the tide to go out. A marine biologist starts recording the depth of the water in the bay at high tide. She would like to develop a model that gives the depth of the water in the bay at any time *t* hours after she starts recording the data.

a. What are the maximum and minimum depths of the water in the bay?

b. Determine the amplitude and the midline of the function.

c. Choose a trigonometric function that can be used to represent the depth of the water in the bay at any time *t* hours after high tide when the marine biologist starts recording the data. Justify your choice. Then write a function to represent the situation.

d. Graph and analyze the function.

e. Use your graph to estimate the depth of the water 9 hours after high tide.

Example 5

21. Write a function for the cosine graph shown.

a. Find the vertical shift and amplitude.

b. Find the phase shift, if any.

c. Find the period.

d. Write the function.

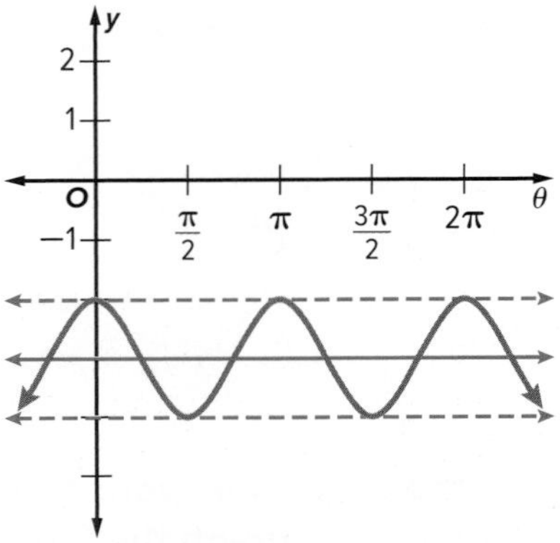

22. Write a function for the sine graph shown.

a. Find the vertical shift and amplitude.

b. Find the phase shift.

c. Find the period.

d. Write the function.

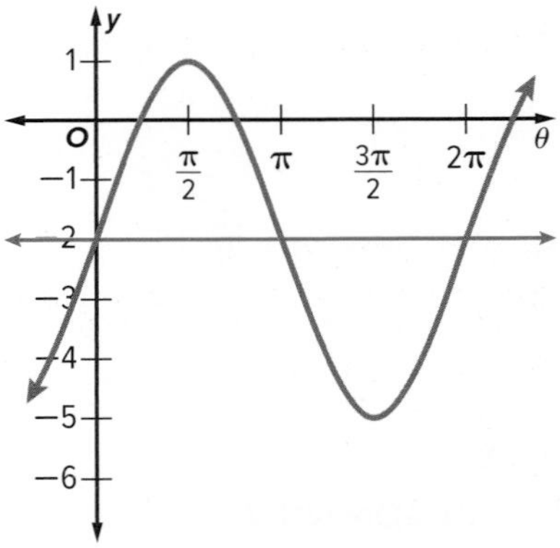

Mixed Exercises

Write an equation for each translation.

23. $y = \sin x$, translated 4 units to the right and 3 units up

24. $y = \cos x$, translated 5 units to the left and 2 units down

25. $y = \tan x$, translated π units to the right and 2.5 units up

State the amplitude, period, phase shift, and vertical shift for each function. Then graph the function.

26. $y = -3 + 2 \sin 2\left(\theta + \frac{\pi}{4}\right)$ 27. $y = 3 \cos 2(\theta + 45°) + 1$ 28. $y = -1 + 4 \tan (\theta + \pi)$

29. **USE A MODEL** What trigonometric function can best be used to represent the height of a cart on a Ferris wheel as a function of time? Explain.

30. **ECOLOGY** The population of an insect species in a stand of trees follows the growth cycle of a particular tree species. The insect population can be modeled by $y = 40 + 30 \sin 6t$, where t is the number of years since November, 1920.

a. Evaluate the function in degrees to determine how often the insect population reaches its maximum level.

b. When did the population last reach its maximum?

31. **POPULATION** The population of predators and prey in a closed ecological system tends to vary periodically over time. In a certain system, the population of snakes P can be represented by $P = 100 + 20 \sin \frac{\pi}{5}t$, where t is the number of years since January 1, 2000. In that same system, the population of rats can be represented by $R = 200 + 75 \sin\left(\frac{\pi}{5}t + \frac{\pi}{10}\right)$.

a. What is the maximum snake population? When is this population first reached?

b. What is the minimum rat population? When is this population first reached?

32. **USE A MODEL** An ambulance stops outside a hospital 10 feet from the outer wall. The rotating light at the top of the ambulance projects a beam of light $\overline{AB}$ on the wall. In the figure, c represents the length of the beam and θ represents the measure of the angle in radians that the beam makes with the perpendicular segment to the wall. The light makes one complete rotation every second. What trigonometric function can best be used to represent the length c of the beam of light, in feet, as a function of time in x seconds? Explain.

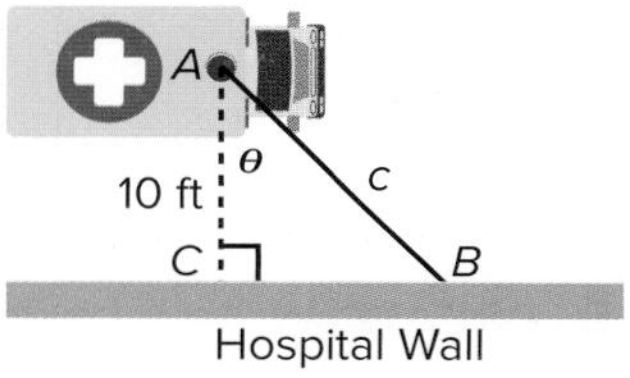

Find the coordinates of a point that represents a maximum for each graph.

33. $y = -2 \cos\left(x - \frac{\pi}{2}\right)$

34. $y = 4 \sin\left(x + \frac{\pi}{3}\right)$

35. $y = 3 \tan\left(x + \frac{\pi}{2}\right) + 2$

36. $y = -3 \sin\left(x - \frac{\pi}{4}\right) - 4$

Compare each pair of graphs.

37. $y = -\cos 3\theta$ and $y = \sin 3(\theta - 90°)$

38. $y = 2 + 0.5 \tan \theta$ and $y = 2 + 0.5 \tan(\theta + \pi)$

39. $y = 2 \sin\left(\theta - \frac{\pi}{6}\right)$ and $y = -2 \sin\left(\theta + \frac{5\pi}{6}\right)$

40. **STRUCTURE** Let $f(\theta) = 2.7 \cos 3(\theta - \pi) + 1$. Write a sine function $g(\theta)$ that has the same graph as $f(\theta)$.

Identify the period of each function. Then write an equation for the graph using the given trigonometric function.

41. sine

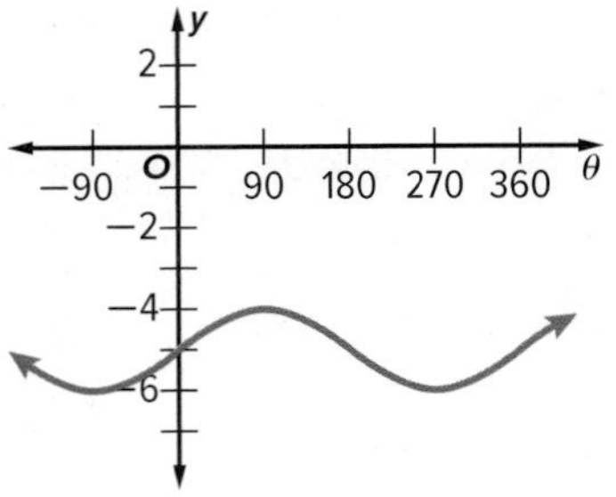

42. cosine

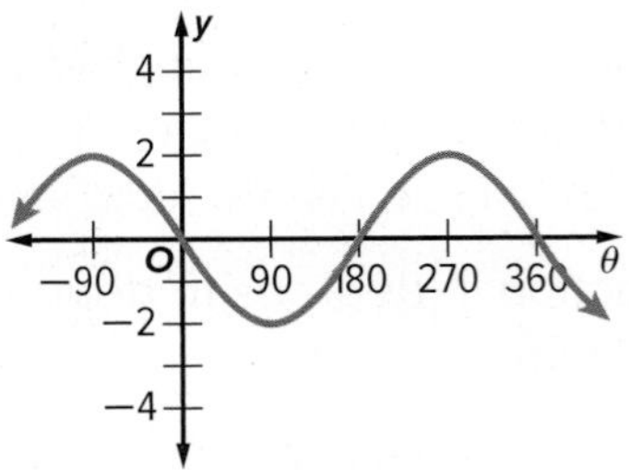

43. cosine

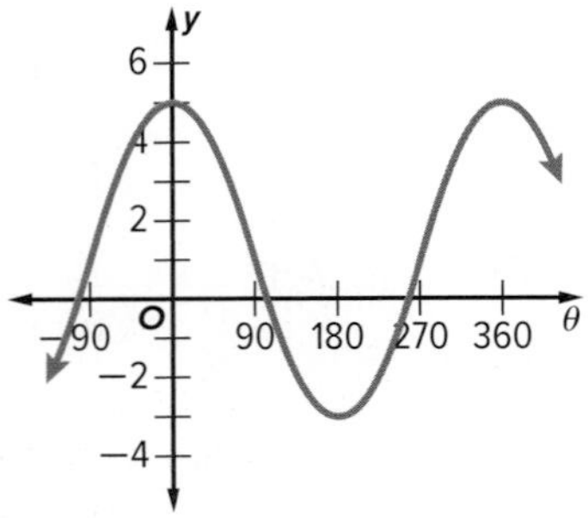

44. sine

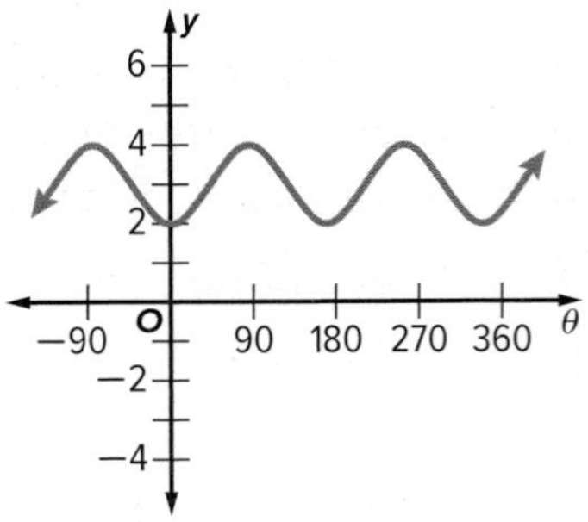

45. CREATE Write a cosine function $h(\theta)$ with midline $y = -2$ and period π, that has no θ-intercepts.

46. ANALYZE If you are given the amplitude and period of a cosine function, is it *sometimes*, *always*, or *never* possible to find the maximum and minimum values of the function? Explain your reasoning.

47. PERSEVERE Describe how the graph of $y = 3 \sin 2\theta + 1$ is different from $y = \sin \theta$.

48. WRITE Describe two different phase shifts that will translate the sine curve onto the cosine curve shown at the right. Then write an equation for the new sine curve using each phase shift.

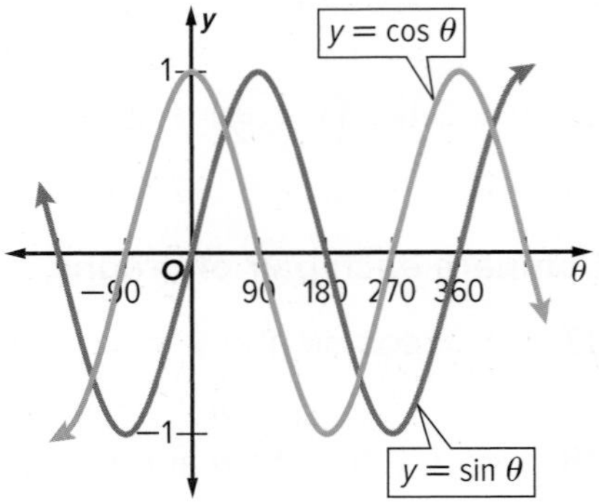

49. CREATE Write a periodic function that has an amplitude of 2 and midline at $y = -3$. Then graph the function.

50. ANALYZE How many different sine graphs pass through the point $(n\pi, 0)$? Justify your argument.

51. PERSEVERE Find a sine function equivalent to $y = \cos \theta - 3$.

52. FIND THE ERROR Alex claimed that $y = 4 \sin \frac{1}{4}\theta$ and $y = \sin \theta$ are equivalent because $4 \cdot \frac{1}{4} = 1$. Is Alex correct? Explain your reasoning.

Inverse Trigonometric Functions

Learn Inverse Trigonometric Functions

If you know the value of a trigonometric function for an angle, you can use the inverse function to find the angle measure. Because the inverse of a function transposes the *x*- and *y*-values, the inverse of $y = \sin x$ is $x = \sin y$.

The relation $x = \sin y$ is not a function because there are many values of *y* for each value of *x*. Therefore, the domain must be restricted in order for the inverse to be a function. Domains for the inverses of the cosine and tangent functions must also be restricted. The values in the restricted domain are called **principal values**. Trigonometric functions with domains restricted to the principal values are denoted with a capital letter. **Inverse trigonometric functions** are inverses of the trigonometric functions with restricted domains.

Key Concept • Inverse Trigonometric Functions

Inverse Function	Symbols	Domain	Range
Arcsine	$y = \text{Arcsin}\, x$	$-1 \leq x \leq 1$	$-\frac{\pi}{2} \leq y \leq \frac{\pi}{2}$ or
	$y = \text{Sin}^{-1} x$		$-90° \leq y \leq 90°$
Arccosine	$y = \text{Arccos}\, x$	$-1 \leq x \leq 1$	$0 \leq y \leq \pi$ or
	$y = \text{Cos}^{-1} x$		$0° \leq y \leq 180°$
Arctangent	$y = \text{Arctan}\, x$	all real numbers	$-\frac{\pi}{2} \leq y \leq \frac{\pi}{2}$ or
	$y = \text{Tan}^{-1} x$		$-90° \leq y \leq 90°$

Example 1 Evaluate Inverse Trigonometric Functions

Find $\text{Tan}^{-1} \sqrt{3}$. Write angle measures in degrees and radians.

Find the angle θ for $-90° \leq \theta \leq 90°$ with a tangent value of $\sqrt{3}$.

Because $\tan \theta = \frac{\sin \theta}{\cos \theta}$, use the unit circle to find a point where the ratio of sine to cosine is $\sqrt{3}$ to 1. Every special angle on the unit circle can be written as a value over 2. A point on the unit circle at which $\sin \theta = \frac{\sqrt{3}}{2}$ and $\cos \theta = \frac{1}{2}$ has a tangent value of $\frac{\frac{\sqrt{3}}{2}}{\frac{1}{2}}$ or $\sqrt{3}$.

When $\theta = 60°$, $\tan \theta = \sqrt{3}$. So, $\text{Tan}^{-1} \sqrt{3} = 60°$ or $\frac{\pi}{3}$.

Check

Find $\text{Cos}^{-1}\left(\frac{\sqrt{2}}{2}\right)$. Write the angle measure in degrees.

$\text{Cos}^{-1}\left(\frac{\sqrt{2}}{2}\right) = \underline{\ ?\ }°$

Go Online You can complete an Extra Example online.

Today's Goal

- Find values of angle measures by using inverse trigonometric functions.

Today's Vocabulary

inverse trigonometric functions

principal values

Think About It!

Describe the relationship between the restricted domain of $y = \sin x$ and the range of $y = \sin^{-1} x$.

Example 2 Find a Trigonometric Value by Using a Calculator

Find $\sin(\text{Cos}^{-1} 0.36)$. Round to the nearest hundredth.

Use a TI-84 Plus Family calculator to evaluate the expression.

Keystrokes: [sin] [2nd] [cos^{-1}] 0.36 [)] [)] [enter]

$\sin(\text{Cos}^{-1} 0.36) \approx 0.93$

Watch Out!

Angle Measure Remember that when evaluating an inverse trigonometric function, the result is an angle measure.

Example 3 Find an Angle Measure by Using a Graphing Calculator

If $\text{Sin}\ \theta = -0.17$, find θ in degrees. Round to the nearest hundredth.

You can use the inverse of sine to solve for θ. The sine of angle θ is -0.17, so $\text{Arcsin}(-0.17) = \theta$.

Use a TI-84 Plus Family calculator to evaluate the expression.

Keystrokes: [2nd] [sin^{-1}] -0.17 [enter]

So, $\theta \approx -9.79°$.

Talk About It!

Given the restrictions on the $\text{Sin}\ \theta$ function, does the solution of Example 3 make sense? Explain.

Example 4 Use Inverse Trigonometric Functions

PLANES **Suppose a pilot has 30 miles to land a plane at the Santa Barbara airport from an elevation of 15,000 feet. Find the angle in degrees at which the airplane should descend.**

Step 1 Draw and label a diagram.

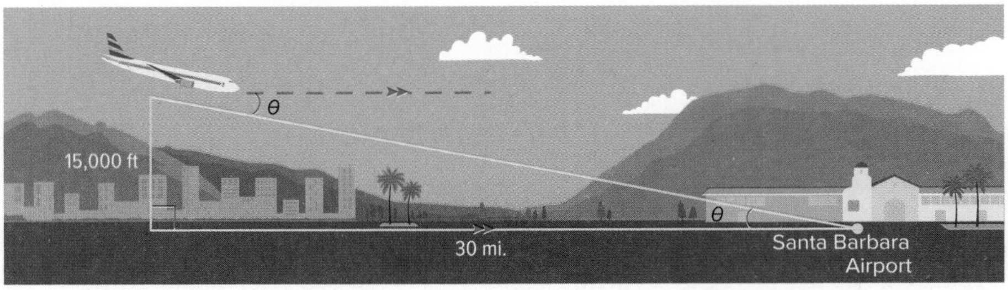

Step 2 Write and solve the trigonometric equation.

$\tan \theta = \frac{15{,}000 \text{ ft}}{30 \text{ mi}}$ — Tangent function

$\tan \theta = \frac{15{,}000 \text{ ft}}{30 \text{ mi}} \cdot \frac{1 \text{ mi}}{5280 \text{ ft}}$ — Convert miles to feet, 1 mile = 5280 feet.

$\tan \theta = \frac{15{,}000}{158{,}400}$ — Simplify.

$\theta = \text{Tan}^{-1}\left(\frac{15{,}000}{158{,}400}\right)$ — Inverse tangent function

$\theta \approx 5.4°$ — Simplify.

The angle of descent is about 5.4°.

Problem-Solving Tip

Draw a Diagram If you are unsure how to begin a problem, try drawing a diagram. For many trigonometric problems, it helps to begin by drawing a right triangle and labeling any sides or angles given in the problem. Continue labeling the triangle as you solve for unknowns.

Go Online You can complete an Extra Example online.

Practice

Go Online You can complete your homework online.

Example 1

Find each value. Write angle measures in degrees and radians.

1. $\text{Cos}^{-1}\left(\frac{\sqrt{3}}{2}\right)$

2. $\text{Sin}^{-1}\left(-\frac{\sqrt{3}}{2}\right)$

3. $\text{Arccos}\left(-\frac{1}{2}\right)$

4. $\text{Arctan}\ \sqrt{3}$

5. $\text{Arccos}\left(-\frac{\sqrt{2}}{2}\right)$

6. $\text{Tan}^{-1}(-1)$

7. $\text{Sin}^{-1}\frac{\sqrt{2}}{2}$

8. $\text{Cos}^{-1}\left(-\frac{\sqrt{3}}{2}\right)$

9. $\text{Arcsin}\ 1$

Example 2

USE TOOLS **Find each value. Round to the nearest hundredth if necessary.**

10. $\cos\left[\text{Sin}^{-1}\left(-\frac{\sqrt{2}}{2}\right)\right]$

11. $\tan\left[\text{Arcsin}\left(-\frac{5}{7}\right)\right]$

12. $\sin\left(\text{Tan}^{-1}\frac{5}{12}\right)$

13. $\text{Cos}\ [\text{Arcsin}\ (-0.7)]$

14. $\cos\ (\text{Arctan}\ 5)$

15. $\sin\ (\text{Cos}^{-1}\ 0.3)$

16. $\sin\ (\text{Cos}^{-1}\ 1)$

17. $\sin\left(\text{Sin}^{-1}\frac{1}{2}\right)$

18. $\tan\left(\text{Arcsin}\ \frac{\sqrt{3}}{2}\right)$

Example 3

USE TOOLS **Find θ in degrees. Round to the nearest hundredth if necessary.**

19. $\text{Sin}\ \theta = 0.8$

20. $\text{Tan}\ \theta = 4.5$

21. $\text{Cos}\ \theta = 0.5$

22. $\text{Cos}\ \theta = -0.95$

23. $\text{Sin}\ \theta = -0.1$

24. $\text{Tan}\ \theta = -1$

25. $\text{Cos}\ \theta = 0.52$

26. $\text{Cos}\ \theta = -0.2$

27. $\text{Sin}\ \theta = 0.35$

28. $\text{Tan}\ \theta = 8$

29. $\text{Cos}\ \theta = 0.25$

30. $\text{Sin}\ \theta = -0.57$

Example 4

31. KITCHEN The exit from a restaurant kitchen has two pairs of swinging doors that meet in the middle of the doorway. Each door is three feet wide. A waiter needs to take a cart of plates into the dining area from the kitchen. The cart is two feet wide.

    a. What is the minimum angle θ through which the doors must each be opened to prevent the cart from hitting either door? Round to the nearest tenth.

    b. What assumption did you make when solving for θ in part **a**?

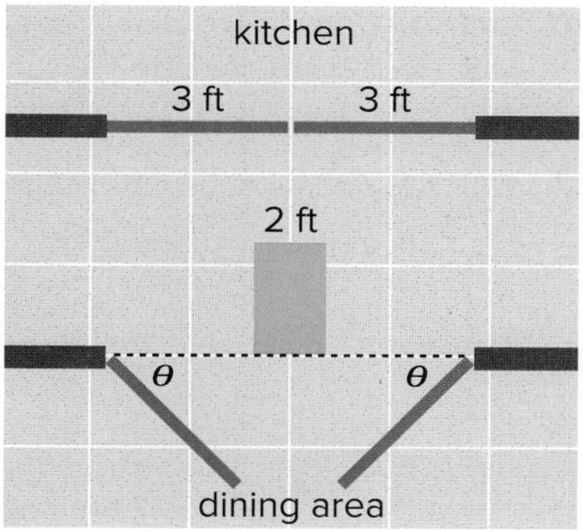

32. **BIKING** Compass directions are given as angles compared to North, South, East, or West. For example, northeast could be described as 45° east of north. Bong-Cha rides her bike two miles east and four miles south to get to her friend Marco's house. If Bong-Cha could have traveled directly from her house to Marco's, in what direction would she have traveled? Round to the nearest tenth.

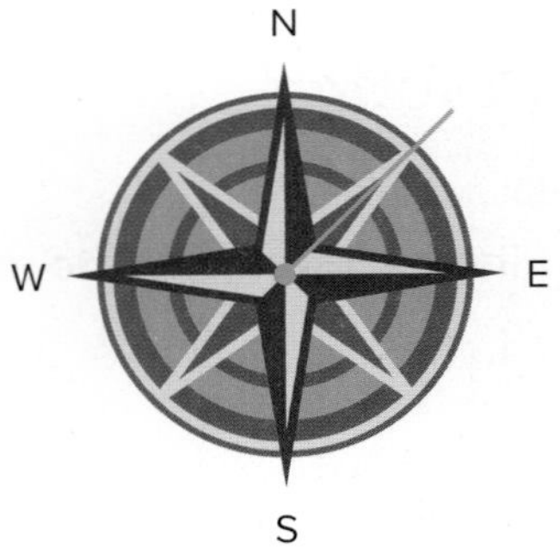

Mixed Exercises

Solve each equation for $0 \leq \theta \leq 2\pi$.

33. $\csc \theta = 1$

34. $\sec \theta = -1$

35. $\sec \theta = 1$

36. $\csc \theta = \frac{1}{2}$

37. $\cot \theta = 1$

38. $\sec \theta = 2$

39. **SURVEYING** In ancient times, it was known that a triangle with side lengths of 3, 4, and 5 units was a right triangle. Surveyors used ropes with knots at each unit of length to make sure that an angle was a right angle. Such a rope was placed on the ground so that one leg of the triangle had three knots and the other had four. This guaranteed that the triangle formed was a right triangle, meaning that the surveyor had formed a right angle. To the nearest degree, what are the angle measures of a triangle formed this way?

40. **PERSEVERE** Determine whether *cos (Arccosx)* $= x$ *for all values of* x is *true* or *false*. If false, give a counterexample.

41. **FIND THE ERROR** Desiree and Oscar are solving $\cos \theta = 0.3$ where $90° < \theta < 180.°$ Is either of them correct? Explain your reasoning.

Desiree	Oscar
$\cos \theta = 0.3$	$\cos \theta = 0.3$
$\cos^{-1} 0.3 = 162.5^0$	$\cos^{-1} 0.3 = 72.5^0$

42. **CREATE** Write an equation with an Arcsine function and an equation with a Sine function that both involve the same angle measure.

43. **WRITE** Compare and contrast the relations $y = \tan^{-1}x$ and $y = \mathrm{Tan}^{-1}x$. Include information about the domains and ranges.

44. **ANALYZE** Explain why $\mathrm{Sin}^{-1}8$ and $\mathrm{Cos}^{-1}8$ are undefined while $\mathrm{Tan}^{-1}8$ is defined.

45. **CREATE** Write an inverse trigonometric function with domain $\{-2 \leq \theta \leq 2\}$ and range $\{1 \leq y \leq \pi + 1\}$. Justify your answer and graph the function.

Module 11 • Trigonometric Functions

Review

Essential Question

What are the key features of the graph of a trigonometric function and how do they represent real-world situations?

Module Summary

Lessons 11-1 and 11-2

Trigonometric Functions and Angles

- An angle that is coterminal with another angle can be found by adding or subtracting a multiple of 360°.
- To convert a degree measure to radians, multiply the number of degrees by $\frac{\pi \text{ radians}}{180°}$. To convert a radian measure to degrees, multiply the number of radians by $\frac{180°}{\pi \text{ radians}}$.
- For a circle with radius r and central angle θ (in radians), the arc length s equals the product of r and θ.
- $\sin \theta = \frac{\text{opp}}{\text{hyp}}$, $\cos \theta = \frac{\text{adj}}{\text{hyp}}$, and $\tan \theta = \frac{\text{opp}}{\text{adj}}$.
- $\csc \theta = \frac{\text{hyp}}{\text{opp}}$, $\sec \theta = \frac{\text{hyp}}{\text{adj}}$, and $\cot \theta = \frac{\text{adj}}{\text{opp}}$.

Lesson 11-3

Circular and Periodic Functions

- If the terminal side of an angle θ in standard position intersects the unit circle at $P(x, y)$, then $\cos \theta = x$ and $\sin \theta = y$.
- The sine and cosine functions are periodic functions where $\sin (x + 2\pi) = \sin x$ and $\cos (x + 2\pi) = \cos x$. Tangent, cosecant, secant, and cotangent can be defined in terms of sine and cosine, so they are periodic functions.

Lessons 11-4 through 11-6

Graphing Trigonometric Functions

- For functions of the form $y = a \sin bx$ and $y = a \cos bx$, $|a|$ is the amplitude and $\frac{360°}{|b|}$ is the period.
- The frequency of a function is the reciprocal of the period of the function.
- The graph of the tangent function has asymptotes at $x = (90 + 180n)°$, where n is an integer.
- The reciprocal trigonometric functions, cosecant, secant, and cotangent, are $\frac{1}{\sin x}$, $\frac{1}{\cos x}$, and $\frac{1}{\tan x}$, respectively.
- The phase shift of $y = a \sin b(x - h)$, $y = a \cos b(x - h)$, and $y = a \tan b(x - h)$ is h, where $b > 0$.
- The vertical shift of $y = a \sin b\theta + k$, $y = a \cos b\theta + k$, and $y = a \tan b\theta + k$ is k, where $b > 0$.

Lesson 11-7

Inverse Trigonometric Functions

- Because the inverse of a function transposes the x- and y-values, the inverse of $y = \sin x$ is $x = \sin y$.

Study Organizer

Foldables

Use your Foldable to review this module. Working with a partner can be helpful. Ask for clarification of concepts as needed.

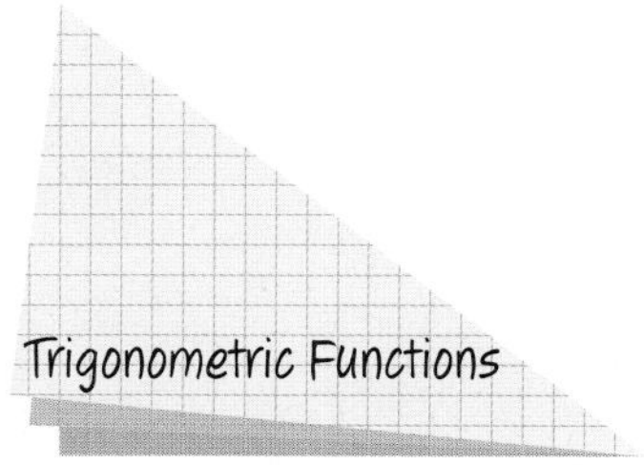

Test Practice

1. **OPEN RESPONSE** Convert each measure from degrees to radians. (Lesson 11-1)

   a) 20°

   b) 80°

   c) 160°

2. **MULTIPLE CHOICE** Convert $\frac{\pi}{3}$ radians to degrees. (Lesson 11-1)

   A. 30°

   B. 45°

   C. 60°

   D. 90°

3. **OPEN RESPONSE** In a central pivot irrigation system, sprinklers mounted on a long pipe are moved around a field in a circular pattern.

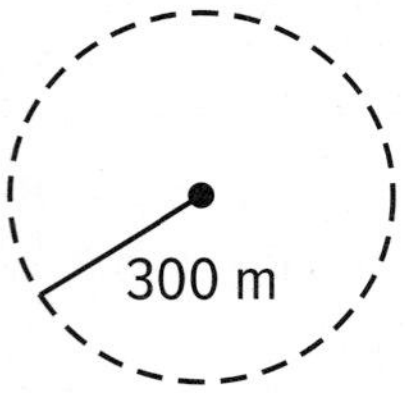

   If it takes 12 hours for the system to complete one pass around the circle, how many meters will the outer edge of the pipe travel in 1 hour? Round to the nearest meter. (Lesson 11-1)

4. **OPEN RESPONSE** Given that $\cos \theta = \frac{3}{4}$ and $0 < \theta < \frac{\pi}{2}$, find the values of the other five trigonometric functions at θ. (Lesson 11-2)

5. **OPEN RESPONSE** If $\sin \theta = \frac{7}{25}$, what is $\tan \theta$? (Lesson 11-2)

6. **MULTIPLE CHOICE** A door is 24 inches wide and can swing open 140°. (Lesson 11-2)

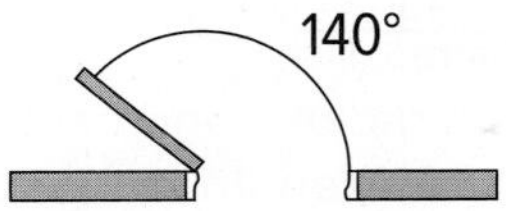

   What is the distance, in inches, from the wall to the edge of the door? Round to the nearest tenth of an inch.

   A. 11.8 inches

   B. 15.4 inches

   C. 20.1 inches

   D. 40 inches

7. **OPEN RESPONSE** Angle θ is drawn in standard position. At which point does the terminal side of θ intersect the unit circle if $\theta = \frac{5\pi}{6}$? (Lesson 11-3)

8. **OPEN RESPONSE** The terminal side of θ in standard position intersects the unit circle at $(x, 0.25)$. Find the approximate values of each of the trigonometric functions of θ given that $x > 0$. (Lesson 11-3)

9. **OPEN RESPONSE** Determine the period of each function. (Lesson 11-3)

a)

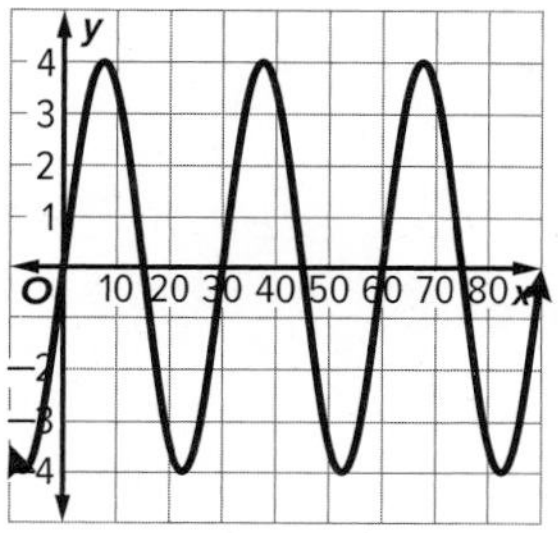

b)

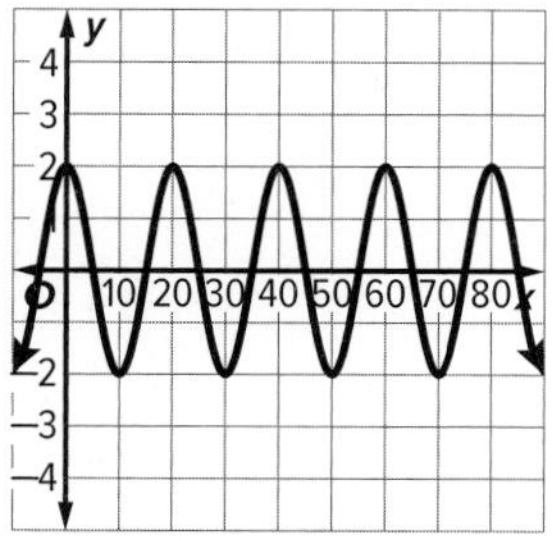

10. **OPEN RESPONSE** Use this function to identify the following. (Lesson 11-4)

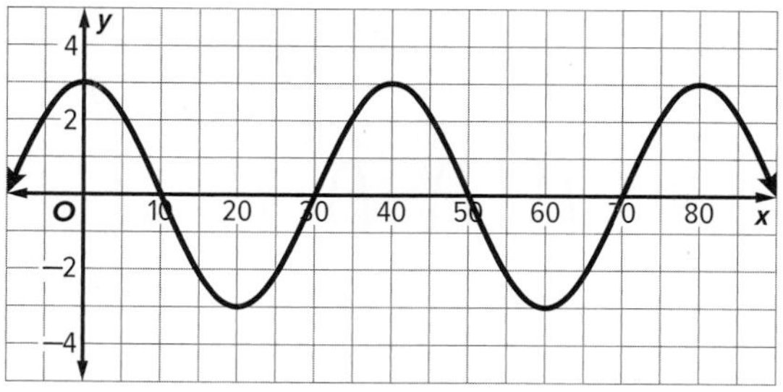

a) amplitude

b) midline

11. **MULTIPLE CHOICE** A spring oscillates with a frequency of 1 cycle per second. The distance between the maximum and minimum points of the oscillation is 3 centimeters. Which of these functions can be used to model the oscillation if y represents the distance in centimeters from the equilibrium position and t is given in seconds. (Lesson 11-4)

A. $y = 1.5 \sin 2\pi t$

B. $y = 1.5 \sin \pi t$

C. $y = 3 \sin 2\pi t$

D. $y = 3 \sin \pi t$

12. **MULTI-SELECT** For which of these values of x does the function $f(x) = \tan 2x$ have a vertical asymptote? Select all that apply. (Lesson 11-5)

A. $-\frac{\pi}{2}$

B. $-\frac{\pi}{4}$

C. 0

D. $\frac{\pi}{4}$

E. $\frac{\pi}{2}$

13. MULTIPLE CHOICE Which function is graphed below? (Lesson 11-5)

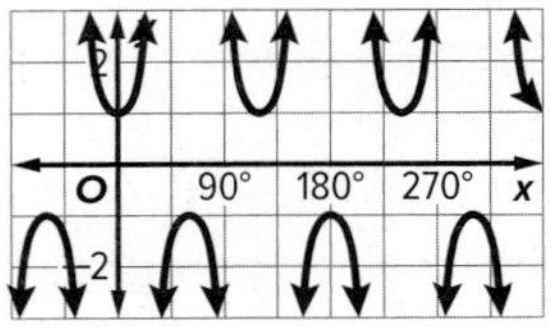

A. $f(x) = \sec 3x$

B. $f(x) = \cos 3x$

C. $f(x) = 3 \sec x$

D. $f(x) = 3 \cos x$

14. MULTI-SELECT A rotating beacon is positioned 10 meters from a wall. The beacon projects a dot of light onto the wall. The location of the dot can be modeled by the function $f(t) = 10 \tan 0.628t$, where $f(t)$ is the position left or right of the centerline in meters and t is given in seconds. How is the graph of $f(t)$ related to the graph of its parent function? Select all that apply. (Lesson 11-6)

A. The graph has been compressed vertically.

B. The graph has been stretched vertically.

C. The graph has been compressed horizontally.

D. The graph has been stretched horizontally.

E. The graph has been stretched vertically and compressed horizontally.

15. MULTI-SELECT Select the key features of the graph of $y = \cos(x + 45°)$. (Lesson 11-6)

A. The amplitude is 1.

B. The period is 315°.

C. The phase shift is 45° right.

D. The domain is all real numbers.

E. The range is [−1, 1].

16. OPEN RESPONSE The graph shows the average temperature in Haleiwa, Hawaii. The data can be approximated by the function $f(t) = a \sin\left[\frac{\pi}{26}(x - h)\right] + k$. (Lesson 11-6)

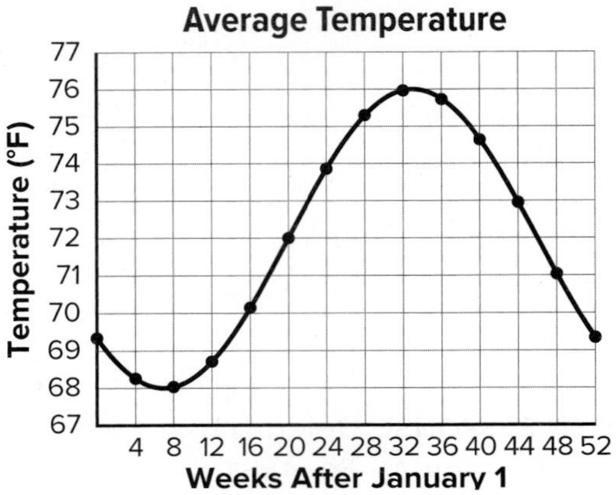

a) Find the value of a.

b) Find the value of h.

c) Find the value of k.

17. MULTIPLE CHOICE Find Arccos $\left(-\frac{\sqrt{3}}{2}\right)$. (Lesson 11-7)

A. $-\frac{5\pi}{6}$

B. $-\frac{\pi}{6}$

C. $\frac{5\pi}{6}$

D. $\frac{\pi}{6}$

18. OPEN RESPONSE Fala is 64 inches tall. Her shadow measures 40 inches long. What is the angle of elevation of the Sun to the nearest degree? (Lesson 11-7)

Module 12

Trigonometric Identities and Equations

Essential Question

How are trigonometric identities similar to and different from other equations?

What Will You Learn?

How much do you already know about each topic **before** starting this module?

KEY

— I don't know. — I've heard of it. — I know it!

	Before			After		
	I don't know.	I've heard of it.	I know it!	I don't know.	I've heard of it.	I know it!
find trigonometric values using trigonometric identities						
simplify trigonometric expressions using trigonometric identities						
verify trigonometric identities by transforming equations						
find trigonometric values using sum and difference identities						
verify identities using sum and difference identities						
find values of sine and cosine using double-angle and half-angle identities						
solve equations and determine extraneous solutions using trigonometric identities						

Foldables Make this Foldable to help you organize your notes about trigonometric identities and equations. Begin with one sheet of 11" × 17" paper and four sheets of grid paper.

1. **Fold** the short sides of the 11" × 17" paper to meet in the middle.
2. **Cut** each tab in half as shown.
3. **Cut** four sheets of grid paper in half and fold the half-sheets in half.
4. **Insert** two folded half-sheets under each of the four tabs and staple along the fold.
5. **Label** each tab as shown.

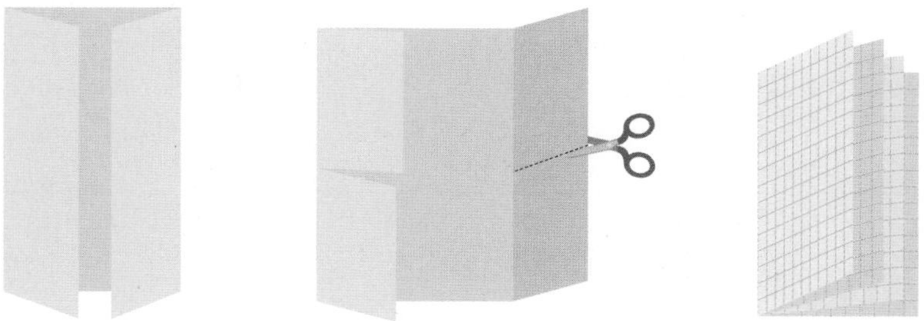

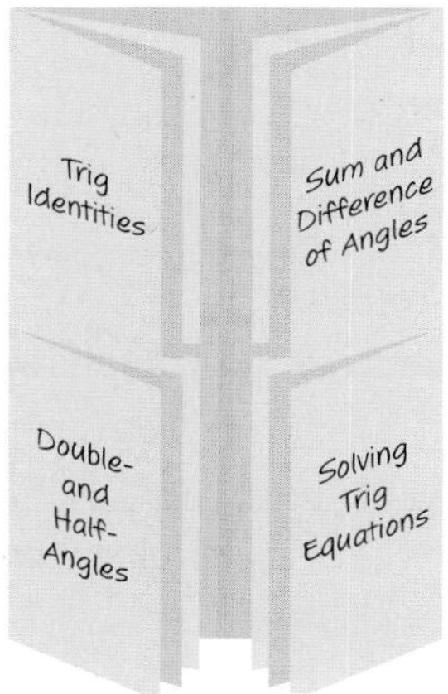

What Vocabulary Will You Learn?

- cofunction identities
- Pythagorean identities
- trigonometric equation
- trigonometric identity

Are You Ready?

Complete the Quick Review to see if you are ready to start this module.

Then complete the Quick Check.

Quick Review

Example 1

Factor $x^3 + 2x^2 - 24x$ completely.

$x^3 + 2x^2 - 24x = x(x^2 + 2x - 24)$

The product of the coefficients of the x-terms in factors of $x^2 + 2x - 24$ must be -24, and their sum must be 2. The product of 6 and -4 is -24, and their sum is 2.

$x(x^2 + 2x - 24) = x(x + 6)(x - 4)$

Example 2

Find the exact value of cos 135°.

The reference angle is 180° − 135°, or 45°.

$\cos 45°$ is $\frac{\sqrt{2}}{2}$. Since 135° is in the second quadrant, $\cos 135° = -\frac{\sqrt{2}}{2}$.

Quick Check

Factor. Assume that no variable equals zero.

1. $-16a^2 + 4a$

2. $5x^2 - 20$

3. $x^3 + 9$

4. $2y^2 - y - 15$

Find the exact value of each trigonometric function.

5. sin 45°

6. cos 225°

7. tan 150°

8. sin 120°

How Did You Do?

Which exercises did you answer correctly in the Quick Check?

Trigonometric Identities

Explore Pythagorean Identity

Online Activity Use graphing technology to complete the Explore.

INQUIRY How can the unit circle be used to justify trigonometric identities?

Learn Using Trigonometric Identities to Find Values

A **trigonometric identity** is an equation involving trigonometric functions that is true for all values for which every expression in the equation is defined. A *counterexample* can be used to show that an equation is false and, therefore, is not an identity.

Key Concept • Quotient and Reciprocal Identities

Quotient Identities	
$\tan\theta = \frac{\sin\theta}{\cos\theta}$; $\cos\theta \neq 0$	$\cot\theta = \frac{\cos\theta}{\sin\theta}$; $\sin\theta \neq 0$

Reciprocal Identities	
$\csc\theta = \frac{1}{\sin\theta}$; $\sin\theta \neq 0$	$\sin\theta = \frac{1}{\csc\theta}$; $\csc\theta \neq 0$
$\sec\theta = \frac{1}{\cos\theta}$; $\cos\theta \neq 0$	$\cos\theta = \frac{1}{\sec\theta}$; $\sec\theta \neq 0$
$\cot\theta = \frac{1}{\tan\theta}$; $\tan\theta \neq 0$	$\tan\theta = \frac{1}{\cot\theta}$; $\cot\theta \neq 0$

The identity $\tan\theta = \frac{\sin\theta}{\cos\theta}$ is true except for angle measures 90°, 270°, . . . , 90° + k180°, where k is an integer. The cosine of each of these angle measures is 0, so tan θ is not defined when $\cos\theta = 0$. Similarly, $\cot\theta = \frac{\cos\theta}{\sin\theta}$ is undefined when $\sin\theta = 0$ for angle measures 0°, 180°, . . . , k180°, where k is an integer.

The **Pythagorean identities** express the Pythagorean Theorem in terms of the trigonometric functions.

Key Concept • Pythagorean Identities

$\cos^2\theta + \sin^2\theta = 1$	$\tan^2\theta + 1 = \sec^2\theta$	$\cot^2\theta + 1 = \csc^2\theta$

Today's Goals
- Find trigonometric values by using trigonometric identities.
- Simplify trigonometric expressions by using trigonometric identities.

Today's Vocabulary
trigonometric identity

Pythagorean identities

cofunction identities

Study Tip

Reading Trigonometric Functions $\sin^2\theta$ is read as *sine squared theta*. It has the same value and meaning as the square of the quantity sin θ, or $(\sin\theta)^2$.

Go Online to see a proof of the Pythagorean identity $\cos^2\theta + \sin^2\theta = 1$.

Math History Minute

Swiss clockmaker **Jost Bürgi (1552–1632)** was able to calculate sines using several algorithms. He explained them in his work *Fundamentum Astronomiae* (1592). Bürgi is also known for constructing a table of logarithms around 1600. Few copies of this work were saved, and it went virtually unnoticed by the mathematics community.

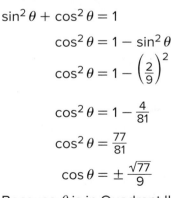

A trigonometric function f is a cofunction of another trigonometric function g if $f(A) = g(B)$ when A and B are complementary angles. The **cofunction identities** show the relationships between sine and cosine, tangent and cotangent, and secant and cosecant.

All six of the trigonometric functions are either odd or even. Recall that a function is even if $f(-x) = f(x)$ is true for every value in the domain. A function is odd if $f(-x) = -f(x)$ is true for every value in the domain. These relationships are given in the negative-angle identities, which are also sometimes called odd-even identities.

Key Concept • Cofunction Identities and Negative-Angle Identities		
Cofunction Identities		
$\sin\left(\frac{\pi}{2} - \theta\right) = \cos\theta$	$\cos\left(\frac{\pi}{2} - \theta\right) = \sin\theta$	$\tan\left(\frac{\pi}{2} - \theta\right) = \cot\theta$
Negative-Angle Identities		
$\sin(-\theta) = -\sin\theta$	$\cos(-\theta) = \cos\theta$	$\tan(-\theta) = -\tan\theta$

Example 1 Use the Pythagorean Identities

Find the exact value of $\cos\theta$ if $\sin\theta = \frac{2}{9}$ and $90° < \theta < 180°$.

You can use the unit circle to estimate the value of θ and $\cos\theta$.

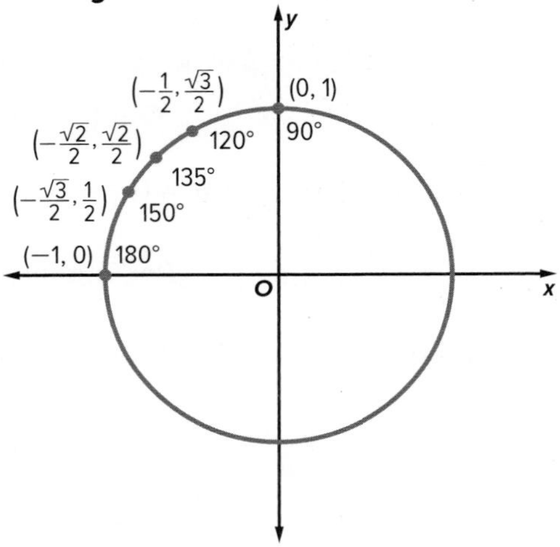

Because $90° < \theta < 180°$, θ is in Quadrant II. $\sin\theta$ is $\frac{2}{9}$, which is between 0 and $\frac{1}{2}$, so θ will be between 150° and 180°. Therefore, $\cos\theta$ is between -1 and $-\frac{\sqrt{3}}{2}$.

To find the exact value of $\cos\theta$, use a Pythagorean identity.

$\sin^2\theta + \cos^2\theta = 1$	Pythagorean Identity
$\cos^2\theta = 1 - \sin^2\theta$	Subtract $\sin^2\theta$ from each side.
$\cos^2\theta = 1 - \left(\frac{2}{9}\right)^2$	Substitute $\frac{2}{9}$ for $\sin\theta$.
$\cos^2\theta = 1 - \frac{4}{81}$	Square $\frac{2}{9}$.
$\cos^2\theta = \frac{77}{81}$	Subtract.
$\cos\theta = \pm\frac{\sqrt{77}}{9}$	Take the square root of each side.

Because θ is in Quadrant II, $\cos\theta$ is negative. Therefore, $\cos\theta = -\frac{\sqrt{77}}{9}$.

Using the unit circle, the estimated value of $\cos\theta$ was between -1 and $-\frac{\sqrt{3}}{2}$. The solution lies within this interval.

Go Online You can complete an Extra Example online.

You can use a graphing calculator to approximate values and check the solution. Find $\cos\left[\sin^{-1}\left(\frac{2}{9}\right)\right]$ using a calculator and compare it to $-\frac{\sqrt{77}}{9}$. Because the two values are approximately equal, our solution is correct.

Check

Find the exact value of $\sin\theta$ if $\cot\theta = 2\sqrt{6}$ and $0° < \theta < 90°$.

Example 2 Use the Cofunction and Negative-Angle Identities

Find the exact value of $\cot\left(\theta - \frac{\pi}{2}\right)$ if $\tan\theta = 1.79$.

$$\cot\left(\theta - \frac{\pi}{2}\right) = \frac{\cos\left(\theta - \frac{\pi}{2}\right)}{\sin\left(\theta - \frac{\pi}{2}\right)}$$ Quotient Identity

$$= \frac{\cos\left[-\left(\frac{\pi}{2} - \theta\right)\right]}{\sin\left[-\left(\frac{\pi}{2} - \theta\right)\right]}$$ Factor −1 from each angle measure expression.

$$= \frac{\cos\left(\frac{\pi}{2} - \theta\right)}{-\sin\left(\frac{\pi}{2} - \theta\right)}$$ Negative-Angle Identity

$$= -\frac{\sin\theta}{\cos\theta}$$ Cofunction Identity

$$= -\tan\theta$$ Quotient Identity

$$= -1.79$$ $\tan\theta = 1.79$

Check

Find the exact value of $-\sin\theta$ if $\cos\left(\theta - \frac{\pi}{2}\right) = 1.4$ and $0 < \theta < \frac{\pi}{2}$.

Think About It!

Why did we not need to know the quadrant in which θ was located in order to solve the problem?

Explore Negative-Angle Identity

Online Activity Use graphing technology to complete the Explore.

INQUIRY How can the graphs of trigonometric functions be used to justify trigonometric identities?

Learn Using Trigonometric Identities to Simplify

You can use the basic trigonometric identities along with algebra to simplify trigonometric expressions. In particular, you should familiarize yourself with the Quotient and Reciprocal Identities and the Pythagorean Identities.

Go Online You can complete an Extra Example online.

Think About It!

Why do algebraic methods like factoring work with trigonometric expressions?

To simplify trigonometric expressions, you will want to use these strategies.

- Recognize and use the Pythagorean Identities. Because $\sin^2 x + \cos^2 x = 1$, $\cos^2 x = 1 - \sin^2 x$ is also true.
- Factor. For example, given an expression like $\sin x - \sin x \cos^2 x$, you can factor and then use a Pythagorean Identity to simplify.
- Rewrite. When an expression includes tangent or one of the reciprocal trigonometric functions, it may help to rewrite the expression in terms of sine or cosine.
- Separate or combine. Writing two fractions over a common denominator or splitting one fraction into two fractions, like $\frac{a-b}{c} = \frac{a}{c} - \frac{b}{c}$, may also help.

Simplifying an expression that contains trigonometric functions means that the expression is written as a numerical value or in terms of a single trigonometric function, if possible.

Think About It!

How can you graphically check that your simplification is correct and is equivalent to the original expression?

Example 3 Simplify a Trigonometric Expression

Simplify $\frac{\csc\theta\cos\theta}{\sin\theta\cot\theta}$.

$\frac{\csc\theta\cos\theta}{\sin\theta\cot\theta} = \frac{\frac{1}{\sin\theta}\cdot\cos\theta}{\sin\theta\cot\theta}$ — Reciprocal Identity, $\csc\theta = \frac{1}{\sin\theta}$

$= \frac{\frac{\cos\theta}{\sin\theta}}{\sin\theta\cot\theta}$ — Simplify the numerator.

$= \frac{\frac{\cos\theta}{\sin\theta}}{\sin\theta\cdot\frac{\cos\theta}{\sin\theta}}$ — $\cot\theta = \frac{\cos\theta}{\sin\theta}$

$= \frac{\frac{\cos\theta}{\sin\theta}}{\cos\theta}$ — Simplify.

$= \frac{\frac{\cos\theta}{\sin\theta}\cdot\frac{1}{\cos\theta}}{\cos\theta\cdot\frac{1}{\cos\theta}}$ — Multiply by $\frac{\frac{1}{\cos\theta}}{\frac{1}{\cos\theta}}$.

$= \frac{1}{\sin\theta}$ — Simplify.

$= \csc\theta$ — Reciprocal Identity, $\frac{1}{\sin\theta} = \csc\theta$

Check

Simplify $\frac{\tan^2 x\csc^2 x - 1}{\sec^2 x}$.

Study Tip

Simplifying Expressing all trigonometric functions in terms of sine and cosine often helps with simplifying complicated trigonometric expressions.

Example 4 Simplify a Trigonometric Expression by Factoring

Simplify $-\sec\left(\frac{\pi}{2} - \theta\right) - \cot^2\theta \csc\theta$.

$-\sec\left(\frac{\pi}{2} - \theta\right) - \cot^2\theta \csc\theta$	Original expression
$= -\frac{1}{\cos\left(\frac{\pi}{2} - \theta\right)} - \cot^2\pi \csc\theta$	$\sec\theta = \frac{1}{\cos\theta}$
$= -\frac{1}{\sin\theta} - \cot^2\theta \csc\theta$	$\cos\left(\frac{\pi}{2} - \theta\right) = \sin\theta$
$= -\csc\theta - \cot^2\theta \csc\theta$	$\frac{1}{\sin\theta} = \csc\theta$
$= -\csc\theta\,(1 + \cot^2\theta)$	Factor $-\csc\theta$ from each term.
$= -\csc\theta\,(\csc^2\theta)$	$\cot^2\theta + 1 = \csc^2\theta$
$= -\csc^3\theta$	Simplify.

Check

Simplify $\cos x \tan^2 x - \cos x \sec^2 x$.

Go Online You can complete an Extra Example online.

Talk About It!

Without calculating the answer, would you expect $-\csc^3 \frac{8\pi}{7}$ to be positive or negative? Explain.

Example 5 Use a Trigonometric Expression

ART GALLERY **An art gallery is rearranging its large canvas murals. One of the gallery's hallways that is 8 feet wide meets another hallway that is 10 feet wide at a right angle. The length L of the longest canvas mural that can be maneuvered around the corner, without tipping the canvas, can be represented by $L = \frac{8 \cot \theta \cos \theta}{1 - \sin^2 \theta} + 10 \csc\left(\frac{\pi}{2} - \theta\right)$, where θ is the angle between the canvas and the wall of the narrower hallway. Rewrite the equation in terms of sine and cosine.**

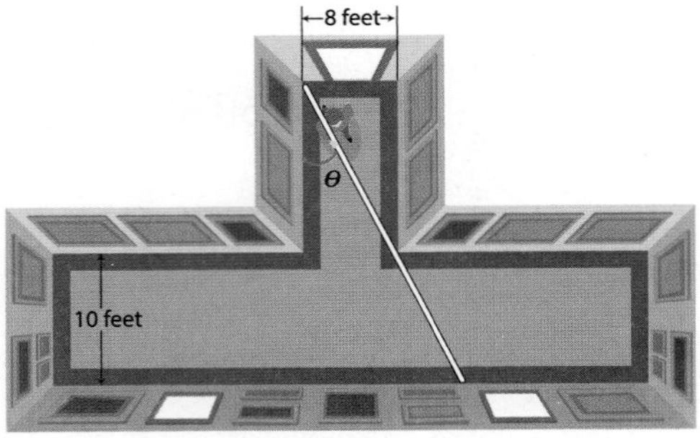

$L = \frac{8 \cot \theta \cos \theta}{1 - \sin^2 \theta} + 10 \csc\left(\frac{\pi}{2} - \theta\right)$ Original equation

$= \frac{8 \cot \theta \cos \theta}{1 - \sin^2 \theta} + 10 \cdot \frac{1}{\sin\left(\frac{\pi}{2} - \theta\right)}$ $\csc \theta = \frac{1}{\sin \theta}$

$= \frac{8 \cot \theta \cos \theta}{1 - \sin^2 \theta} + 10 \cdot \frac{1}{\cos \theta}$ $\sin\left(\frac{\pi}{2} - \theta\right) = \cos \theta$

$= \frac{8 \cot \theta \cos \theta}{1 - \sin^2 \theta} + \frac{10}{\cos \theta}$ Multiply.

$= \frac{8 \cot \theta \cos \theta}{\cos^2 \theta} + \frac{10}{\cos \theta}$ $\cos^2 \theta + \sin^2 \theta = 1$

$= \frac{8 \cot \theta}{\cos \theta} + \frac{10}{\cos \theta}$ $\frac{\cos \theta}{\cos \theta} = 1$

$= \frac{8\left(\frac{\cos \theta}{\sin \theta}\right)}{\cos \theta} + \frac{10}{\cos \theta}$ $\cot \theta = \frac{\cos \theta}{\sin \theta}$

$= \frac{8\left(\frac{\cos \theta}{\sin \theta}\right)}{\cos \theta} \cdot \frac{\frac{1}{\cos \theta}}{\frac{1}{\cos \theta}} + \frac{10}{\cos \theta}$ Multiply the numerator and denominator by $\frac{1}{\cos \theta}$.

$= \frac{8}{\sin \theta} + \frac{10}{\cos \theta}$ Simplify.

The length of the longest mural canvas can be expressed as $L = \frac{8}{\sin \theta} + \frac{10}{\cos \theta}$.

Check

FOOTBALL A football player kicks a field goal at an initial velocity of 55 feet per second. The maximum height h in feet of the football is given by the equation $h = \frac{47 \tan^2 x}{\sec^2 x}$, where x is the degree measure of the angle at which the ball is kicked. Write the simplified form of the expression of the equation.

$h = \underline{\quad ? \quad}$

Go Online You can complete an Extra Example online.

Think About It!

Find L if the angle between the canvas and the wall of the narrower hallway is 30°. Round to the nearest tenth.

Problem-Solving Tip

Identify Subgoals
Often there are multiple ways to simplify a trigonometric expression. Before beginning a problem, it may help to identify the subgoals and possible identities that can be used to simplify the expression. This will help you determine the shortest and clearest way to simplify.

Practice

Go Online You can complete your homework online.

Examples 1 and 2

Find the exact value of each expression if $0° < \theta < 90°$.

1. If $\sin\theta = \frac{3}{5}$, find $\cos\theta$.
2. If $\sin\theta = \frac{1}{2}$, find $\tan\theta$.
3. If $\cos\theta = \frac{3}{5}$, find $\csc\theta$.

Find the exact value of each expression if $180° < \theta < 270°$.

4. If $\sin\theta = -\frac{1}{2}$, find $\cos\theta$.
5. If $\cos\theta = -\frac{3}{5}$, find $\csc\theta$.
6. If $\sec\theta = -3$, find $\tan\theta$.

Find the exact value of each expression if $270° < \theta < 360°$.

7. If $\csc\theta = -\frac{5}{3}$, find $\cos\theta$.
8. If $\cos\theta = \frac{5}{13}$, find $\sin\theta$.
9. If $\tan\theta = -1$, find $\sec\theta$.

10. Find the exact value of $\tan\left(\theta - \frac{\pi}{2}\right)$ if $\cot\theta = 1.53$.

11. Find the exact value of $\cos\theta$ if $\sin\left(\theta - \frac{\pi}{2}\right) = 2.5$ and $0 < \theta < \frac{\pi}{2}$.

12. Find the exact value of $\cos\theta$ if $\sin\theta = \frac{2}{3}$ and $90° < \theta < 180°$.

13. Find $\sin\theta$ if $\cos\theta = \frac{3}{4}$ and $0° \leq \theta < 90°$.

14. Find $\cos\theta$ if $\sin\theta = \frac{1}{2}$ and $90° \leq \theta < 180°$.

15. Find $\cot\theta$ if $\tan\theta = 2$ and $180° < \theta < 270°$.

Examples 3 and 4

Simplify each expression.

16. $\sec\theta\tan^2\theta + \sec\theta$
17. $\cos\left(\frac{\pi}{2} - \theta\right)\cot\theta$
18. $\sin\theta\sec\theta$

19. $\cot\theta\sec\theta$
20. $(\sin\theta)(1 + \cot^2\theta)$
21. $\csc\theta\sin\theta$

22. $\sin\left(\frac{\pi}{2} - \theta\right)\sec\theta$
23. $\frac{\cos(-\theta)}{\sin(-\theta)}$
24. $\frac{\cos\theta}{\sec\theta}$

25. $4(\tan^2\theta - \sec^2\theta)$
26. $\frac{1 + \tan^2\theta}{\csc^2\theta}$
27. $\csc\theta\tan\theta - \tan\theta\sin\theta$

Example 5

28. WAVES The path P of a wave in the ocean is given by the equation $P = \frac{1 + \sin^2\theta\sec^2\theta}{\sec^2\theta} - \cos^2\theta$, where θ is the angle between sea level and the wave. Simplify the equation.

29. LIGHT WAVE The distance in feet a light wave d is from its source is given by the equation $d = \sin^2\theta + \tan^2\theta + \cos^2\theta$, where θ is the angle between the source and the light wave. Simplify the equation.

Mixed Exercises

STRUCTURE **Simplify each expression.**

30. $\dfrac{1 - \sin^2\theta}{\sin\theta + 1}$

31. $\csc\theta + \cot\theta$

32. $\dfrac{1 - \sin^2\theta}{\sin^2\theta}$

33. $\tan\theta\csc\theta$

34. $\dfrac{1}{\sin^2\theta} - \dfrac{\cos^2\theta}{\sin^2\theta}$

35. $2(\csc^2\theta - \cot^2\theta)$

36. $(1 + \sin\theta)(1 - \sin\theta)$

37. $2 - 2\sin^2\theta$

38. $\dfrac{\tan\left(\frac{\pi}{2} - \theta\right)\sec\theta}{1 - \csc^2\theta}$

39. $\dfrac{\cos\left(\frac{\pi}{2} - \theta\right) - 1}{1 + \sin(-\theta)}$

40. $\dfrac{\sec\theta\sin\theta + \cos\left(\frac{\pi}{2} - \theta\right)}{1 + \sec\theta}$

41. $\dfrac{\cot\theta\cos\theta}{\tan(-\theta)\sin\left(\frac{\pi}{2} - \theta\right)}$

42. PERSEVERE Find a counterexample to show that $1 - \sin x = \cos x$ is not an identity.

43. WRITE Pythagoras is most famous for the Pythagorean Theorem. The identity $\cos^2\theta + \sin^2\theta = 1$ is an example of a Pythagorean identity. Why do you think that this identity is classified in this way?

44. PERSEVERE Prove that $\tan(-a) = -\tan a$ by using the quotient and negative angle identities.

45. CREATE Write two expressions that are equivalent to $\tan\theta\sin\theta$.

46. ANALYZE Explain how you can use division to rewrite $\sin^2\theta + \cos^2\theta = 1$ as $1 + \cot^2\theta = \csc^2\theta$.

47. FIND THE ERROR Jordan and Ebony are simplifying $\dfrac{\sin^2\theta}{\cos^2\theta + \sin^2\theta}$. Is either of them correct? Explain your reasoning.

Jordan

$$\frac{\sin^2\theta}{\cos^2\theta + \sin^2\theta} = \frac{\sin^2\theta}{\cos^2\theta}\ \frac{\sin^2\theta}{\sin^2\theta}$$
$$= \tan^2\theta + 1$$
$$= \sec^2\theta$$

Ebony

$$\frac{\sin^2\theta}{\cos^2\theta + \sin^2\theta} = \frac{\sin^2\theta}{1}$$
$$= \sin^2\theta$$

48. PERSEVERE Prove that $\tan^2\theta + 1 = \sec^2\theta$ and $\cot^2\theta + 1 = \csc^2\theta$.

Verifying Trigonometric Identities

Learn Verify Trigonometric Identities by Transforming One Side

You can use the basic trigonometric identities along with the definitions of the trigonometric functions to verify identities. By proving that both sides of the equation are equal for all defined values of the variable, you can show that the identity is true.

Key Concept • Strategies for Verifying Trigonometric Identities

- First, try to verify the identity by simplifying one side. It is usually best to start with the more complicated side and simplify it to match the less complicated side.
- Try substituting basic identities. Rewriting everything on the more complicated side in terms of sine and cosine may make the steps easier.
- Use what you have learned from algebra to simplify. Factor, multiply, add, simplify, or combine fractions as necessary.
- When a term includes $1 + \sin \theta$ or $1 + \cos \theta$, think about multiplying the numerator and denominator by the conjugate. Then you can use a Pythagorean Identity.

Example 1 Verify a Trigonometric Identity by Transforming One Side

Verify that $\cot \theta (\cot \theta + \tan \theta) = \csc^2 \theta$ is an identity.

$\cot \theta (\cot \theta + \tan \theta) \stackrel{?}{=} \csc^2 \theta$	Original equation
$\cot^2 \theta + \cot \theta \tan \theta \stackrel{?}{=} \csc^2 \theta$	Distributive Property
$\cot^2 \theta + \frac{\sin \theta}{\cos \theta} \cdot \frac{\cos \theta}{\sin \theta} \stackrel{?}{=} \csc^2 \theta$	Quotient Identities
$\cot^2 \theta + \frac{\cos\theta \sin\theta}{\sin\theta \cos\theta} \stackrel{?}{=} \csc^2 \theta$	Multiply.
$\cot^2 \theta + 1 \stackrel{?}{=} \csc^2 \theta$	Simplify.
$\csc^2 \theta = \csc^2 \theta$	Pythagorean Identity

Because $\csc^2 \theta = \csc^2 \theta$ is true for all values of θ, $\cot \theta (\cot \theta + \tan \theta) = \csc^2 \theta$ is an identity.

Go Online You can complete an Extra Example online.

Today's Goal
- Verify trigonometric identities.

Think About It!
Why should every step be given a reason, usually another verified trigonometric identity, algebraic operation, or definition, when verifying an identity?

Think About It!
As you are verifying an identity, what are some ways to quickly check that the simplified identity is equivalent to the original identity?

Watch Out!

Operations Because you cannot assume that both sides of the equation are equal, you cannot perform operations to both sides like when solving an equation. You must work each side of the equation separately to verify an identity.

Alternate Method

$$\cot\theta(\cot\theta + \tan\theta) \stackrel{?}{=} \csc^2\theta \quad \text{Original equation}$$

$$\cot\theta\left(\frac{\cos\theta}{\sin\theta} + \frac{\sin\theta}{\cos\theta}\right) \stackrel{?}{=} \csc^2\theta \quad \text{Quotient Identities}$$

$$\cot\theta\left(\frac{\cos\theta}{\sin\theta}\cdot\frac{\cos\theta}{\cos\theta} + \frac{\sin\theta}{\cos\theta}\cdot\frac{\sin\theta}{\sin\theta}\right) \stackrel{?}{=} \csc^2\theta \quad \text{Find LCD.}$$

$$\cot\theta\left(\frac{\cos^2\theta}{\sin\theta\cos\theta} + \frac{\sin^2\theta}{\sin\theta\cos\theta}\right) \stackrel{?}{=} \csc^2\theta \quad \text{Simplify.}$$

$$\cot\theta\left(\frac{\cos^2\theta + \sin^2\theta}{\sin\theta\cos\theta}\right) \stackrel{?}{=} \csc^2\theta \quad \text{Combine fractions.}$$

$$\cot\theta\left(\frac{1}{\sin\theta\cos\theta}\right) \stackrel{?}{=} \csc^2\theta \quad \text{Pythagorean Identity}$$

$$\left(\frac{\cancel{\cos\theta}}{\sin\theta}\right)\left(\frac{1}{\sin\theta\cancel{\cos\theta}}\right) \stackrel{?}{=} \csc^2\theta \quad \text{Quotient Identity}$$

$$\frac{1}{\sin^2\theta} \stackrel{?}{=} \csc^2\theta \quad \text{Simplify.}$$

$$\csc^2\theta = \csc^2\theta \quad \text{Reciprocal Identity}$$

Examine the Alternate Method. Compare and contrast the methods.

The alternate method simplified the expression within the parentheses instead of distributing the cotangent function. The alternate method required more steps, but both methods proved that the equation is an identity.

Check

Write the explanations to verify that $\csc\theta \cdot \sec\theta = \tan\theta + \cot\theta$ is an identity.

$$\csc\theta\cdot\sec\theta \stackrel{?}{=} \tan\theta + \cot\theta \quad \text{Original equation}$$

$$\csc\theta\cdot\sec\theta \stackrel{?}{=} \frac{\sin\theta}{\cos\theta} + \frac{\cos\theta}{\sin\theta} \quad \underline{\quad ? \quad}$$

$$\csc\theta\cdot\sec\theta \stackrel{?}{=} \frac{\sin\theta}{\cos\theta}\cdot\frac{\sin\theta}{\sin\theta} + \frac{\cos\theta}{\sin\theta}\cdot\frac{\cos\theta}{\cos\theta} \quad \text{Find common denominator.}$$

$$\csc\theta\cdot\sec\theta \stackrel{?}{=} \frac{\sin^2\theta}{\cos\theta\sin\theta} + \frac{\cos^2\theta}{\cos\theta\sin\theta} \quad \text{Simplify.}$$

$$\csc\theta\cdot\sec\theta \stackrel{?}{=} \frac{1}{\cos\theta\sin\theta} \quad \underline{\quad ? \quad}$$

$$\csc\theta\cdot\sec\theta = \csc\theta\cdot\sec\theta \quad \underline{\quad ? \quad}$$

Example 2 Verify a Trigonometric Identity by Transforming One Side

Verify that $2\sec^2 x = \frac{1}{1-\sin x} + \frac{1}{1+\sin x}$ is an identity.

$2\sec^2 x \stackrel{?}{=} \frac{1}{1-\sin x} + \frac{1}{1+\sin x}$	Original equation
$2\sec^2 x \stackrel{?}{=} \frac{1}{1-\sin x} \cdot \frac{1+\sin x}{1+\sin x} + \frac{1}{1+\sin x} \cdot \frac{1-\sin x}{1-\sin x}$	Find common denominators.
$2\sec^2 x \stackrel{?}{=} \frac{1+\sin x}{1-\sin^2 x} + \frac{1-\sin x}{1-\sin^2 x}$	Simplify.
$2\sec^2 x \stackrel{?}{=} \frac{1+\sin x + 1 - \sin x}{1-\sin^2 x}$	Combine fractions.
$2\sec^2 x \stackrel{?}{=} \frac{2}{1-\sin^2 x}$	Simplify the numerator.
$2\sec^2 x \stackrel{?}{=} \frac{2}{\cos^2 x}$	Pythagorean Identity
$2\sec^2 x = 2\sec^2 x$	Reciprocal Identity

$2\sec^2 x = \frac{1}{1-\sin x} + \frac{1}{1+\sin x}$ is an identity.

Check

Complete each statement or reason to verify that $\frac{\sin\theta}{1-\cos\theta} = \csc\theta + \cot\theta$ is an identity.

$\frac{\sin\theta}{1-\cos\theta} \stackrel{?}{=} \csc\theta + \cot\theta$	Original equation
___?___ $\stackrel{?}{=} \csc\theta + \cot\theta$	Multiply the numerator and denominator by $1 + \cos\theta$.
$\frac{\sin\theta + \sin\theta\cos\theta}{1-\cos^2\theta} \stackrel{?}{=} \csc\theta + \cot\theta$	Multiply.
___?___ $\stackrel{?}{=} \csc\theta + \cot\theta$	___?___
___?___ $\stackrel{?}{=} \csc\theta + \cot\theta$	___?___
___?___ $\stackrel{?}{=} \csc\theta + \cot\theta$	Divide by the common factor.
$\csc\theta + \cot\theta = \csc\theta + \cot\theta$	___?___

Talk About It!

In the first step, the fractions on the right side of the equation were multiplied by two different fractions, $\frac{1+\sin x}{1+\sin x}$ and $\frac{1-\sin x}{1-\sin x}$, while the left side of the equation was not multiplied by anything. Justify how equality was maintained.

Study Tip

Transforming When transforming an equation to verify an identity, try to simplify the more complicated side of the equation until both sides are the same.

Study Tip

Additional Steps Because there are often many alternative methods to verify an identity, you may notice that the same identity can be verified using fewer or more steps. If each step is mathematically sound and the final equation is true, then the verification should be correct, regardless of the number of steps.

Example 3 Verify a Trigonometric Identity by Transforming Each Side

Verify that $\frac{\sin\theta\tan\theta}{1-\cos\theta} = (1+\cos\theta)\sec\theta$ is an identity.

$\frac{\sin\theta\tan\theta}{1-\cos\theta} \stackrel{?}{=} (1+\cos\theta)\sec\theta$		Original equation
$\frac{\sin\theta\cdot\frac{\sin\theta}{\cos\theta}}{1-\cos\theta} \stackrel{?}{=} (1+\cos\theta)\sec\theta$		Reciprocal Identity
$\frac{\frac{\sin^2\theta}{\cos\theta}}{1-\cos\theta}\cdot\frac{1+\cos\theta}{1+\cos\theta} \stackrel{?}{=} (1+\cos\theta)\sec\theta$		Multiply by $\frac{1+\cos\theta}{1+\cos\theta} = 1$.
$\frac{\frac{\sin^2\theta}{\cos\theta}+\sin^2\theta}{1-\cos^2\theta} \stackrel{?}{=} (1+\cos\theta)\sec\theta$		Distributive Property
$\frac{\frac{\sin^2\theta}{\cos\theta}+\sin^2\theta}{\sin^2\theta} \stackrel{?}{=} (1+\cos\theta)\sec\theta$		Pythagorean Identity
$\frac{\sin^2\theta}{\cos\theta}\cdot\frac{1}{\sin^2\theta}+\sin^2\theta\cdot\frac{1}{\sin^2\theta} \stackrel{?}{=} (1+\cos\theta)\sec\theta$		Simplify the complex fraction.
$\frac{1}{\cos\theta}+1 \stackrel{?}{=} (1+\cos\theta)\sec\theta$		Simplify.
$\sec\theta+1 \stackrel{?}{=} (1+\cos\theta)\sec\theta$		Reciprocal Identity

Isolate the right side of the equation, $(1+\cos\theta)\sec\theta$, with the goal of simplifying the expression to $\sec\theta+1$.

$\frac{\sin\theta\tan\theta}{1-\cos\theta} \stackrel{?}{=}$	$(1+\cos\theta)\sec\theta$	Original equation
$\sec\theta+1 \stackrel{?}{=}$	$\sec\theta+\cos\theta\sec\theta$	Distribute.
$\sec\theta+1 \stackrel{?}{=}$	$\sec\theta+\cos\theta\cdot\frac{1}{\cos\theta}$	Reciprocal Identity
$\sec\theta+1 =$	$\sec\theta+1$	Simplify.

Because both sides of the equation can be simplified to $\sec\theta+1$, $\frac{\sin\theta\tan\theta}{1-\cos\theta} = (1+\cos\theta)\sec\theta$ is an identity.

Check

Verify that $\sin^2\theta+\cos^2\theta-\sec^2\theta = \frac{1+\tan^2\theta}{\csc^2\theta}$ is an identity.

$\sin^2\theta+\cos^2\theta-\sec^2\theta \stackrel{?}{=} \frac{1+\tan^2\theta}{\csc^2\theta}$	Original equation
___?___ $\stackrel{?}{=} \frac{1+\tan^2\theta}{\csc^2\theta}$	___?___
$\tan^2\theta \stackrel{?}{=} \frac{1+\tan^2\theta}{\csc^2\theta}$	___?___
$\tan^2\theta \stackrel{?}{=} \frac{\sec^2\theta}{\csc^2\theta}$	___?___
$\tan^2\theta \stackrel{?}{=} \frac{\frac{1}{\cos^2\theta}}{\frac{1}{\sin^2\theta}}$	___?___
$\tan^2\theta \stackrel{?}{=} \frac{\sin^2\theta}{\cos^2\theta}$	Simplify the complex fraction.
$\tan^2\theta = \tan^2\theta$	___?___

Go Online You can complete an Extra Example online.

Practice

Go Online You can complete your homework online.

Examples 1 and 2

Verify that each equation is an identity by transforming one side.

1. $\cos^2\theta + \tan^2\theta\cos^2\theta = 1$
2. $\cot\theta(\cot\theta + \tan\theta) = \csc^2\theta$
3. $1 + \sec^2\theta\sin^2\theta = \sec^2\theta$
4. $\sin\theta\sec\theta\cot\theta = 1$
5. $\frac{1-\cos\theta}{1+\cos\theta} = (\csc\theta - \cot\theta)^2$
6. $\frac{1-2\cos^2\theta}{\sin\theta\cos\theta} = \tan\theta - \cot\theta$
7. $(\sin\theta - 1)(\tan\theta + \sec\theta) = -\cos\theta$
8. $\cos\theta\cos(-\theta) - \sin\theta\sin(-\theta) = 1$
9. $\sec\theta - \tan\theta = \frac{1-\sin\theta}{\cos\theta}$
10. $\frac{1+\tan\theta}{\sin\theta+\cos\theta} = \sec\theta$

Example 3

Verify that each equation is an identity by transforming each side.

11. $\left(\sin\theta + \frac{\cot\theta}{\csc\theta}\right)^2 = \frac{2+\sec\theta\csc\theta}{\sec\theta\csc\theta}$
12. $\frac{\cos\theta}{1-\sin\theta} = \frac{1+\sin\theta}{\cos\theta}$
13. $\csc^2\theta - 1 = \frac{\cot^2\theta}{\csc\theta\sin\theta}$
14. $\cos\theta\cot\theta = \csc\theta - \sin\theta$
15. $\csc^2\theta = \cot^2\theta + \sin\theta\csc\theta$
16. $\frac{\sec\theta - \csc\theta}{\csc\theta\sec\theta} = \frac{\sin\theta - \cos\theta}{\sin^2\theta + \cos^2\theta}$

Mixed Exercises

Verify that each equation is an identity.

17. $\tan\theta\cos\theta = \sin\theta$
18. $\cot\theta\tan\theta = 1$
19. $(\tan\theta)(1 - \sin^2\theta) = \sin\theta\cos\theta$
20. $\frac{\csc\theta}{\sec\theta} = \cot\theta$
21. $\frac{\sin^2\theta}{1-\sin^2\theta} = \tan^2\theta$
22. $\frac{\cos^2\theta}{1-\sin\theta} = 1 + \sin\theta$
23. **OPTICS** The polarizing angle for any substance can be found using Brewster's Law. It states that the relationship between the two indices of fractions n_1 and n_2 and the polarizing angle θ_p is $\tan\theta_p = \frac{n_2}{n_1}$. Use the law of refraction $n_1\sin\theta_p = n_2\sin\theta_r$ to prove Brewster's Law.

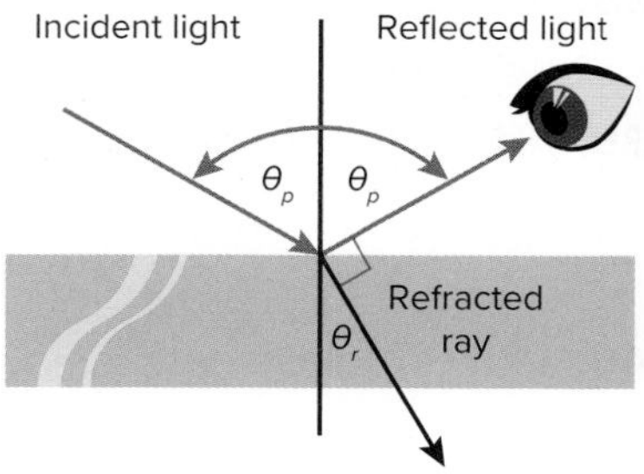

24. STRUCTURE The graph of $y = \frac{\cos^2 x}{1 - \sin x}$ is shown below on the left. The graph of $y = \sin x$ is shown below on the right. Use the graphs to write an identity involving $\frac{\cos^2 x}{1 - \sin x}$ and $\sin x$. Then verify the identity.

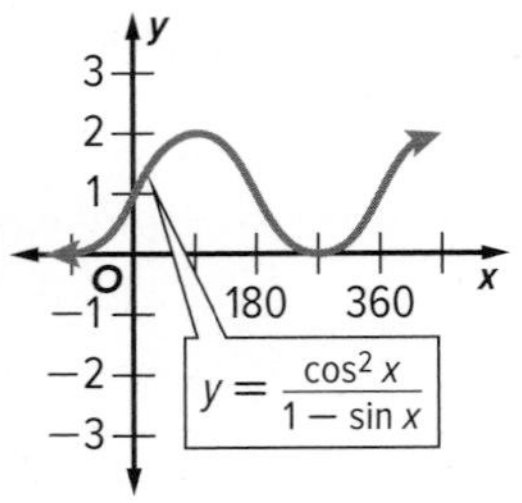

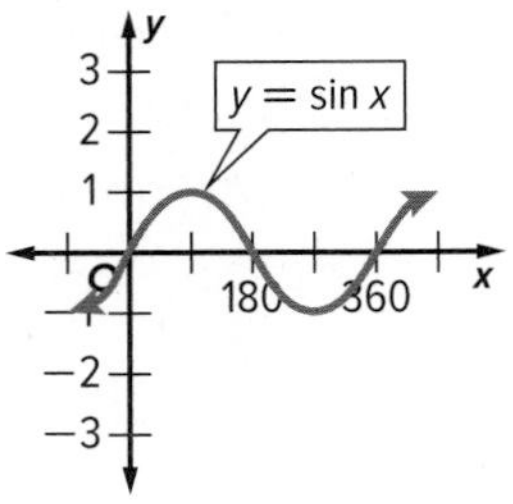

25. CONSTRUCT ARGUMENTS Show two different methods of verifying that $\frac{1}{1 - \sin^2 \theta} = \tan^2 \theta + 1$ is a trigonometric identity.

Determine whether each equation is an identity. Justify your argument.

26. $\frac{\tan\left(\frac{\pi}{2} - \theta\right) \csc \theta}{\csc^2 \theta} = \cos \theta$

27. $\frac{1 + \tan \theta}{1 + \cot \theta} = \tan \theta$

28. $\sin \theta \csc (-\theta) = 1$

29. $\frac{\sec^2 \theta - \tan^2 \theta}{\cos^2 \theta + \sin^2 \theta} = 1$

30. $\tan \theta + \cos \theta = \sin \theta$

31. $\cot (-\theta) \cot \left(\frac{\pi}{2} - \theta\right) = 1$

32. $\sec \theta \sin \left(\frac{\pi}{2} - \theta\right) = 1$

33. $\frac{1 + \tan^2 \theta}{\csc^2 \theta} = \sin^2 \theta$

34. $\sec^2 (-\theta) - \tan^2 (-\theta) = 1$

35. REASONING Diego decides that if $\sin^2 A + \cos^2 B = 1$, and A and B both have measures between 0° and 180°, then $A = B$. Is he correct? Explain your reasoning.

36. USE TOOLS How can you use your calculator to show that $\sin^2 \theta + \csc^2 \theta = 1$ is not an identity? Explain your reasoning.

37. WHICH ONE DOESN'T BELONG? Identify the equation that does not belong with the other three. Justify your conclusion.

$\sin^2\theta + \cos^2\theta = 1$	$1 + \cot^2\theta = \csc^2\theta$
$\sin^2\theta - \cos^2\theta = 2\sin^2\theta$	$\tan^2\theta + 1 = \sec^2\theta$

38. PERSEVERE Transform the right side of $\tan^2 \theta = \frac{\sin^2 \theta}{\cos^2 \theta}$ to show that $\tan^2 \theta = \sec^2 \theta - 1$.

39. ANALYZE Explain why $\sin^2 \theta + \cos^2 \theta = 1$ is an identity, but $\sin \theta = \sqrt{1 - \cos^2 \theta}$ is not.

40. WRITE A classmate is having trouble trying to verify a trigonometric identity. Write a question you could ask to help her work through the problem.

41. CREATE Let $x = \frac{1}{2} \tan \theta$, where $-\frac{\pi}{2} < \theta < \frac{\pi}{2}$. Write $f(x) = \frac{x}{\sqrt{1 + 4x^2}}$ in terms of a single trigonometric function of θ.

42. CREATE A statement such as $\cos \theta = 2$ is a *contradiction*. Write two contradictions involving the sine or tangent function.

Sum and Difference Identities

Learn Use Sum and Difference Identities to Find Trigonometric Values

By writing angle measures as the sums or differences of more familiar angle measures, you can use these sum and difference identities to find exact values of trigonometric functions for angles that are less common.

Key Concept • Sum and Difference Identities
$\sin(A + B) = \sin A \cos B + \cos A \sin B$
$\cos(A + B) = \cos A \cos B - \sin A \sin B$
$\tan(A + B) = \frac{\tan A + \tan B}{1 - \tan A \tan B}$, $A, B \neq 90° + 180n°$, where n is any integer
$\sin(A - B) = \sin A \cos B - \cos A \sin B$
$\cos(A - B) = \cos A \cos B + \sin A \sin B$
$\tan(A - B) = \frac{\tan A - \tan B}{1 + \tan A \tan B}$, $A, B \neq 90° + 180n°$, where n is any integer

Today's Goals

- Find values of sine and cosine by using sum and difference identities.
- Verify trigonometric identities by using sum and difference identities.

Talk About It!

Which of the sum and difference identities are undefined for certain angle measures? Explain your reasoning.

Example 1 Use a Sum Identity

Find the exact value of sin 165°.

$\sin 165° = \sin(45° + 120°)$	$A = 45°, B = 120°$
$= \sin 45° \cos 120° + \cos 45° \sin 120°$	Sum Identity for Sine
$= \frac{\sqrt{2}}{2} \cdot \left(-\frac{1}{2}\right) + \frac{\sqrt{2}}{2} \cdot \frac{\sqrt{3}}{2}$	Evaluate each expression.
$= -\frac{\sqrt{2}}{4} + \frac{\sqrt{6}}{4}$	Multiply.
$= \frac{\sqrt{6} - \sqrt{2}}{4}$	Combine the fractions.

Go Online

to see a common error to avoid.

Think About It!

How can you check your solution?

Check

Find the exact value of tan 255°.

A. undefined

B. $\sqrt{3} + 2$

C. $\frac{2}{3}$

D. $\sqrt{3} - 2$

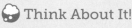

Go Online You can complete an Extra Example online.

Example 2 Use a Difference Identity

Find the exact value of $\cos \frac{5\pi}{12}$.

$$
\begin{aligned}
\cos \frac{5\pi}{12} &= \cos\left(\frac{2\pi}{3} - \frac{\pi}{4}\right) && A = \frac{2\pi}{3}, B = \frac{\pi}{4}\\
&= \cos \frac{2\pi}{3} \cos \frac{\pi}{4} - \sin \frac{2\pi}{3} \sin \frac{\pi}{4} && \text{Difference Identity for Cosine}\\
&= \left(-\frac{1}{2} \cdot \frac{\sqrt{2}}{2}\right) - \left(\frac{\sqrt{3}}{2} \cdot \frac{\sqrt{2}}{2}\right) && \text{Evaluate each expression.}\\
&= -\frac{\sqrt{2}}{4} - \frac{\sqrt{6}}{4} && \text{Multiply.}\\
&= -\frac{\sqrt{2} + \sqrt{6}}{4} && \text{Combine the fractions.}
\end{aligned}
$$

Think About It!
How would you find the exact value of csc 105°?

Check

Find the exact value of $\sin\left(-\frac{7\pi}{12}\right)$.

Example 3 Use Sum and Difference Identities

ELECTRICITY **For a certain circuit carrying alternating current, the formula $c = 4 \sin 25t°$ can be used to find the current c after t seconds. Find the exact current in amperes at $t = 3$ seconds.**

Think About It!
Why is the degree symbol necessary in $c = 4 \sin 25t°$ in the context of this problem?

$$
\begin{aligned}
c &= 4 \sin 25t° && \text{Original equation}\\
&= 4 \sin (25 \cdot 3)° && t = 3\\
&= 4 \sin 75° && \text{Multiply.}\\
&= 4 \sin (30° + 45°) && 30° + 45° = 75°\\
&= 4 (\sin 30° \cos 45° + \cos 30° \sin 45°) && \text{Sum Identity for Sine}\\
&= 4\left(\frac{1}{2} \cos 45° + \cos 30° \cdot \frac{\sqrt{2}}{2}\right) && \text{Evaluate sin 30° and sin 45°.}\\
&= 4\left(\frac{1}{2} \cdot \frac{\sqrt{2}}{2} + \frac{\sqrt{3}}{2} \cdot \frac{\sqrt{2}}{2}\right) && \text{Evaluate cos 30° and cos 45°.}\\
&= 4\left(\frac{\sqrt{2}}{4} + \frac{\sqrt{6}}{4}\right) && \text{Multiply.}\\
&= 4\left(\frac{\sqrt{2} + \sqrt{6}}{4}\right) && \text{Add.}\\
&= \sqrt{2} + \sqrt{6} && \text{Distribute.}
\end{aligned}
$$

The exact current after 3 seconds is $\sqrt{2} + \sqrt{6}$ amperes.

Check

ART A digital artist uses tessellations of basic shape patterns to create fractal art. To make one shape, he combines two right triangles as shown, with lengths measured in pixels. In order to complete his work, he needs to find the exact value of the sine of angle BAC. What is it?

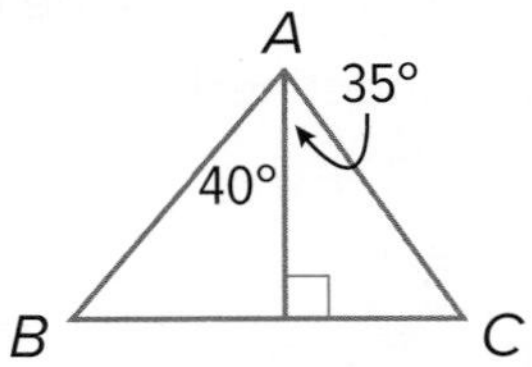

Go Online You can complete an Extra Example online.

Learn Use Sum and Difference Identities to Verify Trigonometric Identities

Sum and difference identities can be used to rewrite trigonometric expressions in which one of the angles is a multiple of 90°. The resulting identity is called a reduction identity because it reduces the complexity of the expression.

> **Think About It!**
> Why is it easier to simplify an expression when one of the angles is a multiple of 90°?

Example 4 Verify a Cofunction Identity

Verify that $\tan(90° - \theta) = \cot\theta$ is an identity.

$\tan(90° - \theta) \stackrel{?}{=} \cot\theta$	Original equation
$\frac{\sin(90° - \theta)}{\cos(90° - \theta)} \stackrel{?}{=} \cot\theta$	Quotient Identity
$\frac{\sin 90° \cos\theta - \cos 90° \sin\theta}{\cos 90° \cos\theta + \sin 90° \sin\theta} \stackrel{?}{=} \cot\theta$	Difference Identities
$\frac{\sin 90° \cos\theta - 0 \cdot \sin\theta}{0 \cdot \cos\theta + \sin 90° \sin\theta} \stackrel{?}{=} \cot\theta$	$\cos 90° = 0$
$\frac{1 \cdot \cos\theta - 0 \cdot \sin\theta}{0 \cdot \cos\theta + 1 \cdot \sin\theta} \stackrel{?}{=} \cot\theta$	$\sin 90° = 1$
$\frac{\cos\theta}{\sin\theta} \stackrel{?}{=} \cot\theta$	Simplify.
$\cot\theta = \cot\theta$	Quotient Identity

> **Think About It!**
> Why can't the tangent difference identity be used to verify $\tan(90° - \theta) = \cot\theta$? Justify your argument.

Check

Complete the statements to verify that $\cos(90° - \theta) = \sin\theta$ is an identity.

$\cos(90° - \theta) \stackrel{?}{=} \sin\theta$	Original equation
___?___ $\stackrel{?}{=} \sin\theta$	___?___
___?___ $\stackrel{?}{=} \sin\theta$	$\cos 90° = 0$; $\sin 90° = 1$
___?___ $= \sin\theta$	Simplify.

Go Online You can complete an Extra Example online.

Example 5 Verify a Reduction Identity

Verify that $\tan(\theta - 180°) = \tan\theta$.

$\tan(\theta - 180°) \stackrel{?}{=} \tan\theta$	Original equation
$\dfrac{\tan\theta - \tan 180°}{1 + \tan\theta \tan 180°} \stackrel{?}{=} \tan\theta$	Difference Identity
$\dfrac{\tan\theta - 0}{1 + \tan\theta \cdot 0} \stackrel{?}{=} \tan\theta$	$\tan 180° = 0$
$\dfrac{\tan\theta}{1} \stackrel{?}{=} \tan\theta$	Simplify the numerator and denominator.
$\tan\theta = \tan\theta$	Simplify.

Check

Complete the statements to verify that $\cot(\theta - 180°) = \cot\theta$.

$\cot(\theta - 180°) \stackrel{?}{=} \cot\theta$	Original equation
$\dfrac{1}{\tan(\theta - 180°)} \stackrel{?}{=} \cot\theta$	___?___
___?___ $\stackrel{?}{=} \cot\theta$	Difference Identity
___?___ $\stackrel{?}{=} \cot\theta$	Simplify.
$\dfrac{1 + \tan\theta \cdot 0}{\tan\theta - 0} \stackrel{?}{=} \cot\theta$	$\tan 180° = 0$
$\dfrac{?}{} \stackrel{?}{=} \cot\theta$	Simplify.
$\cot\theta = \cot\theta$	___?___

Go Online You can complete an Extra Example online.

Practice

Go Online You can complete your homework online.

Examples 1 and 2

Find the exact value of each expression.

1. $\sin 135°$

2. $\cos 165°$

3. $\cos \frac{7\pi}{12}$

4. $\sin \frac{\pi}{12}$

5. $\tan 195°$

6. $\cos \left(-\frac{\pi}{12}\right)$

Example 3

7. ART As part of a mosaic that an artist is making, she places two right triangular tiles together to make a new triangular piece. One tile has lengths of 3 inches, 4 inches, and 5 inches. The other tile has lengths 4 inches, $4\sqrt{3}$ inches, and 8 inches. The pieces are placed with the sides of 4 inches against each other as shown in the figure.

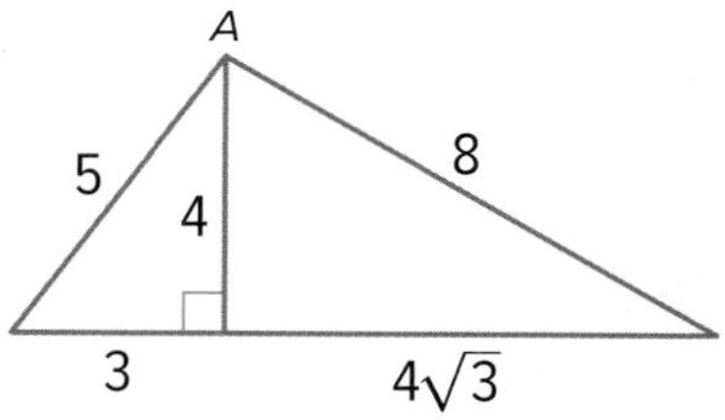

a. What are the exact values of $\sin A$ and $\cos A$?

b. What is the measure of angle A?

c. Is the new triangle formed from the two triangular tiles also a right triangle? Explain.

8. CAMERAS Security cameras are being installed in the Community Center. One camera will be placed on the wall 4 yards above the pool deck. If the pool deck is 5 yards wide, through what angle θ must the camera rotate to view the entire length of the pool? Round to the nearest tenth of a degree.

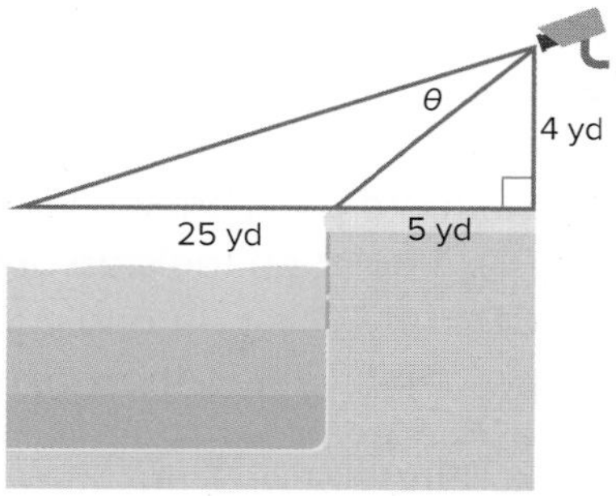

Examples 4 and 5

Verify that each equation is an identity.

9. $\cos\left(\frac{\pi}{2} + \theta\right) = -\sin \theta$

10. $\cos(60° + \theta) = \sin(30° - \theta)$

11. $\cos(180° + \theta) = -\cos \theta$

12. $\tan(\theta + 45°) = \frac{1 + \tan \theta}{1 - \tan \theta}$

Mixed Exercises

Find the exact value of each expression.

13. sin 330°

14. cos (−165°)

15. sin (−225°)

16. cos 135°

17. sin (−45)°

18. cos 210°

19. cos (−135°)

20. tan 75°

21. sin (−195°)

22. sin 75°

23. cos (−225°)

24. tan 210°

25. $\sin \frac{4\pi}{3}$

26. $\sin \frac{23\pi}{12}$

Verify that each equation is an identity.

27. $\sin (90° + \theta) = \cos \theta$

28. $\sin (180° + \theta) = -\sin \theta$

29. $\cos (270° - \theta) = -\sin \theta$

30. $\cos (\theta - 90°) = \sin \theta$

31. $\sin \left(\theta - \frac{\pi}{2}\right) = -\cos \theta$

32. $\cos (\pi + \theta) = -\cos \theta$

33. CONSTRUCT ARGUMENTS You can use the figure to prove that $\cos (A + B) = \cos A \cos B - \sin A \sin B$. Angle A was drawn in standard position and $\angle B$ shares a side with the $\angle A$, as shown. P is a point on the side of the angle with measure $A + B$ so that $OP = 1$. $\overline{PQ}$ is perpendicular to the terminal side of $\angle A$. $\overline{QS}$ is perpendicular to the x-axis. $\overline{QT}$ is perpendicular to $\overline{PR}$.

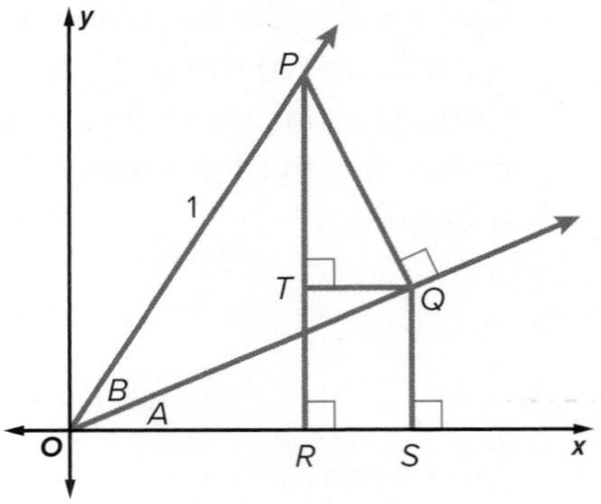

a. Find $m\angle TPQ$. Justify your answer.

b. Explain how to write PQ and OQ in terms of sin B and/or cos B. (*Hint*: Focus on $\triangle POQ$.)

c. Use the side lengths of $\triangle QOS$ to write a ratio for cos A. Then use this, and the expressions for PQ and OQ from part **b**, to write an expression for OS that involves sines and/or cosines of A and/or B.

d. Use the side lengths of $\triangle TPQ$ to write a ratio for sin A. Then use this, and the expressions for PQ and OQ from part **b**, to write an expression for QT that involves sines and/or cosines of A and/or B.

e. Prove the sum formula for cosines by using your work in the previous steps.

Find the exact value of each expression.

34. $\tan 165°$

35. $\sec 1275°$

36. $\sin 735°$

37. $\tan \frac{23\pi}{12}$

38. $\csc \frac{5\pi}{12}$

39. $\cot \frac{113\pi}{12}$

Verify that each equation is an identity.

40. $\sin (A + B) = \frac{\tan A + \tan B}{\sec A \sec B}$

41. $\cos (A + B) = \frac{1 - \tan A \tan B}{\sec A \sec B}$

42. $\sec (A - B) = \frac{\sec A \sec B}{1 + \tan A \tan B}$

43. $\sin (A + B) \sin (A - B) = \sin^2 A - \sin^2 B$

44. CONSTRUCT ARGUMENTS Explain how to use the sum formula for cosines to prove the difference formula for cosines.

45. CONSTRUCT ARGUMENTS You can use the sum and difference formulas for sine and cosine to prove the sum and difference formulas for tangent.

a. Prove the sum formula for tangent.

b. Prove the difference formula for tangent.

46. REASONING The figure shows the graphs of $y = \sin \theta$ and $y = \cos \theta$.

a. Explain how to use the graphs to find the value of h in the equation $\sin (\theta - h) = \cos \theta$.

b. Use one or more sum or difference formulas to prove that you found the correct value of h.

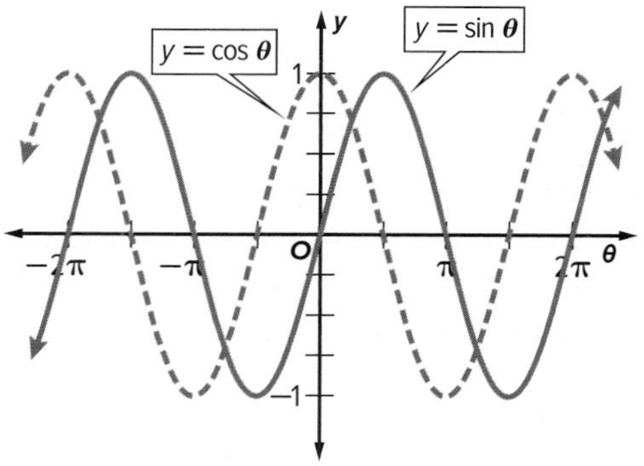

47. STRUCTURE Show how to find the exact value of $\sin \frac{13\pi}{12}$.

48. USE A MODEL Demetri stands 6 feet from the base of a flagpole and sights the top of the pole with an angle of elevation of 75°. His eyes are 5 feet above the ground. Find the exact height of the flagpole. Then find the height to the nearest tenth of a foot.

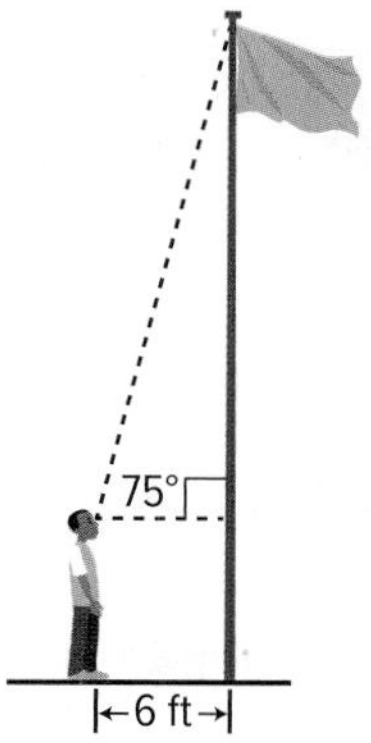

49. STRUCTURE Solve $\cos (\theta + \pi) = \sin (\theta - \pi)$ algebraically, given that $0 \leq \theta \leq \pi$. Explain your steps.

50. PRECISION Use sum or difference identities to prove that the equation $\cos \left(\frac{\pi}{2} - \theta\right) = \sin \theta$ is an identity.

STRUCTURE **Use sum or difference identities to write each expression as a single trigonometric expression. Then find the exact value of each expression.**

51. $\sin 19° \cos 11° + \cos 19° \sin 11°$

52. $\dfrac{\tan 144° - \tan 9°}{1 + \tan 144° \tan 9°}$

53. $\cos 111° \cos 21° + \sin 111° \sin 21°$

54. $\sin 108° \cos 18° - \cos 108° \sin 18°$

55. $\dfrac{\tan \frac{\pi}{16} + \tan \frac{3\pi}{16}}{1 - \tan \frac{\pi}{16} \tan \frac{3\pi}{16}}$

56. $\cos \frac{\pi}{3} \cos \frac{2\pi}{3} - \sin \frac{\pi}{3} \sin \frac{2\pi}{3}$

Higher-Order Thinking Skills

57. ANALYZE Simplify the following expression without expanding any of the sums or differences. $\sin\left(\frac{\pi}{3} - \theta\right) \cos\left(\frac{\pi}{3} + \theta\right) - \cos\left(\frac{\pi}{3} - \theta\right) \sin\left(\frac{\pi}{3} + \theta\right)$

58. WRITE You may have experienced a wireless Internet provider temporarily losing the signal. Waves that pass through the same place at the same time cause interference. Interference occurs when two waves combine to have a greater, or smaller, amplitude than either of the component waves. *Constructive interference occurs* when two waves combine to have a greater amplitude than either of the component waves. *Destructive interference* occurs when the component waves combine to have a smaller amplitude. Explain how the sum and difference identities are used to describe wireless Internet interference. Include an explanation of the difference between constructive and destructive interference.

59. PERSEVERE Derive an identity for $\cot(A + B)$ in terms of $\cot A$ and $\cot B$.

60. ANALYZE The top figure shows two angles A and B in standard position on the unit circle. In the bottom figure, an angle with measure $A - B$ is in standard position.

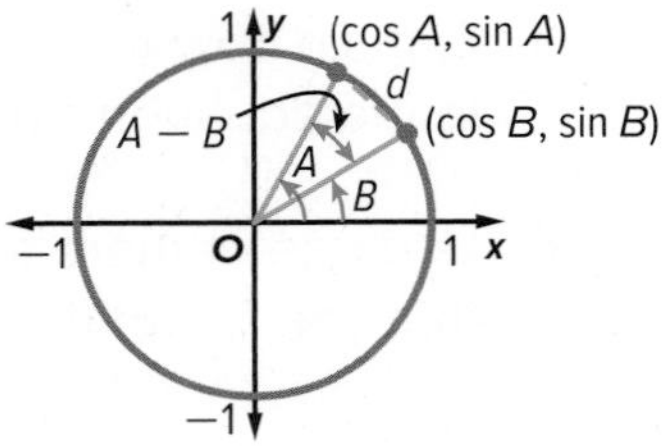

a. Find d, where $(x_1, y_1) = (\cos B, \sin B)$ and $(x_2, y_2) = (\cos A, \sin A)$.

b. Find d when an angle with measure $A - B$ is in standard position.

c. Find and simplify d^2 for each expression. Then equate the values of d^2 to derive a formula for $\cos(A - B)$.

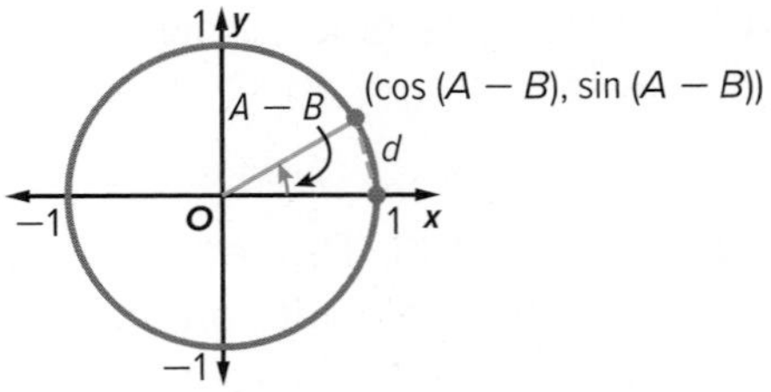

61. CREATE Consider the following theorem. If A, B, and C are the angles of an oblique triangle, then $\tan A + \tan B + \tan C = \tan A \tan B \tan C$. Choose values for A, B, and C. Verify that the conclusion is true for your specific values.

62. CREATE Write an expression that has an exact value of $\dfrac{\sqrt{6} - \sqrt{2}}{4}$.

63. PERSEVERE Verify that $\sin(90° - \theta) = \cos \theta$ is an identity using the Difference Identity.

Double-Angle and Half-Angle Identities

Explore Proving the Double-Angle Identity for Cosine

Online Activity

INQUIRY How can you use an Angle Sum Identity to find a Double-Angle Identity?

Today's Goals
- Find values of sine and cosine by using double-angle identities.
- Find values of sine and cosine by using half-angle identities.

Learn Double-Angle Identities

Double-angle formulas can help you find the value of a function of twice an angle measure. The double-angle formulas are identities, so they are true for all real numbers.

Key Concept • Double-Angle Identities

The following identities hold true for all values of θ.

$\sin 2\theta = 2 \sin \theta \cos \theta$ $\qquad$ $\cos 2\theta = 2\cos^2 \theta - 1$

$\cos 2\theta = \cos^2 \theta - \sin^2 \theta$ $\qquad$ $\cos 2\theta = 1 - 2\sin^2 \theta$

$\tan 2\theta = \frac{2 \tan \theta}{1 - \tan^2 \theta}$, $\theta \neq 45° + 90n°$, where n is any integer

Study Tip

Multiple Identities Note that three variations of the cosine double-angle identity are provided because all three can be used to derive other identities.

Example 1 Use a Double-Angle Identity to Find an Exact Value

Find the exact value of $\sin 2\theta$ if $\sin \theta = -\frac{4}{9}$ and $180° < \theta < 270°$.

You can use the double-angle identity for sine.

Step 1 Find $\cos \theta$.

$\cos^2 \theta = 1 - \sin^2 \theta$	Pythagorean Identity
$\cos^2 \theta = 1 - \left(-\frac{4}{9}\right)^2$	Substitute.
$\cos^2 \theta = 1 - \frac{16}{81}$	Square $-\frac{4}{9}$.
$\cos^2 \theta = \frac{65}{81}$	Subtract.
$\cos \theta = \pm \frac{\sqrt{65}}{9}$	Take the square root.

Step 2 Determine the sign.

Because $180° < \theta < 270°$, $360° < 2\theta < 540°$.

Because 2θ is in Quadrant I or II, $\sin 2\theta$ is positive.

(continued on the next page)

Go Online You can complete an Extra Example online.

Step 3 Find sin 2θ.

$$\sin 2\theta = 2 \sin \theta \cos \theta$$ Double-Angle Identity

$$= 2\left(-\frac{4}{9}\right) \cdot \left(-\frac{\sqrt{65}}{9}\right)$$ $\sin \theta = -\frac{4}{9}$ and $\cos \theta = -\frac{\sqrt{65}}{9}$

$$= \frac{8\sqrt{65}}{81}$$ Multiply.

Example 2 Use a Double-Angle Identity to Rewrite an Identity

BASEBALL **The distance d in meters that a baseball travels from the time it leaves the bat to the time it returns to the batted height is represented by $d = \frac{v_0^2 \sin \theta \cos \theta}{0.5g}$, where v_0 is the initial velocity, and g is the acceleration due to gravity. Rewrite this formula in terms of d and 2θ.**

Step 1 Determine the correct identity.

Note that the numerator has $\sin \theta \cos \theta$ as a factor, which is exactly half of $\sin 2\theta = 2 \sin \theta \cos \theta$. Thus, you can use the identity if you multiply each side of the equation by 2.

Step 2 Rewrite the equation.

$$d = \frac{v_0^2 \sin \theta \cos \theta}{0.5g}$$ Original equation

$$2d = 2\left(\frac{v_0^2 \sin \theta \cos \theta}{0.5g}\right)$$ Multiply each side by 2.

$$2d = \frac{v_0^2(2 \sin \theta \cos \theta)}{0.5g}$$ Commutative Property

$$2d = \frac{v_0^2 \sin 2\theta}{0.5g}$$ $\sin 2\theta = 2 \sin \theta \cos \theta$

$$d = \frac{v_0^2 \sin 2\theta}{g}$$ Multiply each side by $\frac{1}{2}$.

Thus, $d = \frac{v_0^2 \sin 2\theta}{g}$.

Check

ELECTRICITY The power P delivered to a resistor in a certain AC circuit two seconds after it has been started is given by the equation $P = R \sin^2 2\theta$, where R is the resistance. Select the formula that is rewritten in terms of P and θ.

A. $P = \frac{R}{\tan^2\theta}$

B. $P = 4R \sin^2\theta \cos^2\theta$

C. $P = R \sin^2\theta \cos^2 \theta$

D. $P = 2R \sin \theta \cos \theta$

Go Online You can complete an Extra Example online.

Watch Out!

Calculator Check You can check your answer by using a calculator. Find $\sin^{-1}\left(-\frac{4}{9}\right)$ to get $-26.39°$, but θ is in Quadrant III. Using the symmetry of the unit circle, $\sin 206.39°$ also equals $-\frac{4}{9}$, so $\theta = 206.39°$, $2\theta = 412.78°$, and $\sin 2\theta = 0.796$. This is equal to $\frac{8\sqrt{65}}{81}$.

Study Tip

Identities Write down the identities so you can examine the equation you want to rewrite and find which identity is best to use.

Think About It!

What are the units for g and v_0 in $d = \frac{v_0^2 \sin 2\theta}{g}$?

Learn Half-Angle Identities

Half-angle formulas can help you find the value of a function of half an angle measure. The half-angle formulas are identities, so they are true for all real numbers.

Key Concept • Half-Angle Identities

The following identities hold true for all values of θ.

$\sin \frac{\theta}{2} = \pm\sqrt{\frac{1-\cos\theta}{2}}$ $\qquad$ $\cos \frac{\theta}{2} = \pm\sqrt{\frac{1+\cos\theta}{2}}$

$\tan \frac{\theta}{2} = \pm\sqrt{\frac{1-\cos\theta}{1+\cos\theta}}$, $\theta \neq 180° + 360n°$, where n is an integer

Example 3 Use a Half-Angle Identity to Find an Exact Value

Find the exact value of $\cos \frac{\theta}{2}$ if $\sin \theta = \frac{12}{13}$ and $90° < \theta < 180°$.

You can use the half-angle identity for sine.

Step 1 Find $\cos \theta$.

$\cos^2 \theta = 1 - \sin^2 \theta$	Pythagorean Identity
$\cos^2 \theta = 1 - \left(\frac{12}{13}\right)^2$	Substitute.
$\cos^2 \theta = 1 - \frac{144}{169}$	Square $\frac{12}{13}$.
$\cos^2 \theta = \frac{25}{169}$	Subtract.
$\cos \theta = \pm\frac{5}{13}$	Take the square root.

Step 2 Determine the signs.

Because $90° < \theta < 180°$, $45° < \frac{\theta}{2} < 90°$. Because $\frac{\theta}{2}$ is in Quadrant I and θ is in Quadrant II, $\cos \frac{\theta}{2}$ is positive and $\cos \theta = -\frac{5}{13}$.

Step 3 Find $\cos \frac{\theta}{2}$.

$\cos \frac{\theta}{2} = \pm\sqrt{\frac{1+\cos\theta}{2}}$	Half-Angle Identity
$= \pm\sqrt{\frac{1+\left(-\frac{5}{13}\right)}{2}}$	$\cos \theta = -\frac{5}{13}$
$= \pm\sqrt{\frac{\frac{8}{13}}{2}}$	Add.
$= \pm\sqrt{\frac{4}{13}}$	Simplify the fraction.
$= \pm\frac{2}{\sqrt{13}}$	Simplify the numerator.
$= \pm\frac{2\sqrt{13}}{13}$	Rationalize the denominator.
$= \frac{2\sqrt{13}}{13}$	$\frac{\theta}{2}$ is in Quadrant I.

Go Online You can complete an Extra Example online.

Talk About It!

Why is it necessary to know the quadrant in which θ lies? Provide an example where a different quadrant for θ can lead to a different answer.

Study Tip

Signs To determine which sign is appropriate using a half-angle identity, check quadrant $\frac{\theta}{2}$ and not θ.

Check

Find the exact value of $\tan \frac{\theta}{2}$ if $\sin \theta = -\frac{7}{11}$ and $180° < \theta < 270°$.

Example 4 Use a Half-Angle Identity to Evaluate an Expression

Find the exact value of sin 22.5°.

Step 1 Find cos θ.

To use the half-angle identity for sine, the expression must be of the form $\sin \frac{\theta}{2}$. Because $22.5 = \frac{45}{2}$, $\theta = 45$.

$\sin \frac{\theta}{2} = \pm\sqrt{\frac{1-\cos\theta}{2}}$ Half-Angle Identity

$\sin \frac{45°}{2} = \pm\sqrt{\frac{1-\cos 45°}{2}}$ $\theta = 45$

Step 2 Determine the signs.

Because $\frac{\theta}{2}$ and θ, or 22.5° and 45°, are in Quadrant I, $\sin \frac{\theta}{2}$ is positive and $\cos \theta$ is positive.

Step 3 Use the half-angle identity.

$\sin 22.5° = \pm\sqrt{\frac{1-\cos 45°}{2}}$ Half-Angle Identity; θ is in Quadrant I, so $\sin \theta$ is positive.

$= \sqrt{\frac{1-\frac{\sqrt{2}}{2}}{2}}$ Evaluate cos 45°.

$= \sqrt{\frac{\frac{2}{2}-\frac{\sqrt{2}}{2}}{2}}$ The least common denominator is 2.

$= \sqrt{\frac{\frac{2-\sqrt{2}}{2}}{2}}$ Combine the fractions.

$= \sqrt{\frac{2-\sqrt{2}}{4}}$ Divide.

$= \frac{\sqrt{2-\sqrt{2}}}{\sqrt{4}}$ Quotient Property of Radicals

$= \frac{\sqrt{2-\sqrt{2}}}{2}$ Simplify.

Check

Use a calculator to check your answer.

$\sin 22.5° = \frac{\sqrt{2-\sqrt{2}}}{2}$, so the answer is correct.

Check

Find the exact value of cos 22.5°.

Think About It!
Identify two different trigonometric identities that can be used to evaluate cos 15°.

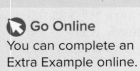

Go Online
You can complete an Extra Example online.

Practice

Go Online You can complete your homework online.

Examples 1 and 3

Find the exact values of sin 2θ, cos 2θ, sin $\frac{\theta}{2}$, and cos $\frac{\theta}{2}$.

1. $\sin \theta = \frac{2}{3}$; $90° < \theta < 180°$

2. $\sin \theta = -\frac{15}{17}$; $\pi < \theta < \frac{3\pi}{2}$

3. $\cos \theta = \frac{3}{5}$; $\frac{3\pi}{2} < \theta < 2\pi$

4. $\cos \theta = \frac{1}{5}$; $270° < \theta < 360°$

5. $\tan \theta = \frac{4}{3}$; $180° < \theta < 270°$

6. $\tan \theta = \frac{3}{5}$; $\frac{\pi}{2} < \theta < \pi$

7. $\cos \theta = \frac{7}{25}$, $0° < \theta < 90°$

8. $\sin \theta = -\frac{4}{5}$, $180° < \theta < 270°$

9. $\sin \theta = \frac{40}{41}$, $90° < \theta < 180°$

10. $\cos \theta = \frac{3}{7}$, $270° < \theta < 360°$

Example 2

11. SOUND WAVES The sound waves produced by vibrating a tuning fork is represented by $S = 2 \sin 2\theta$. Rewrite this formula in terms of S and θ.

12. MONUMENTS The World War II Memorial consists of 56 granite pillars arranged around a plaza with two triumphal arches on opposite sides. When the shadow from one of the pillars is 16.1 meters long, the angle of elevation to the Sun is θ. When the shadow is 7.2 meters long, the angle of elevation is 2θ. What is the height of the pillar?

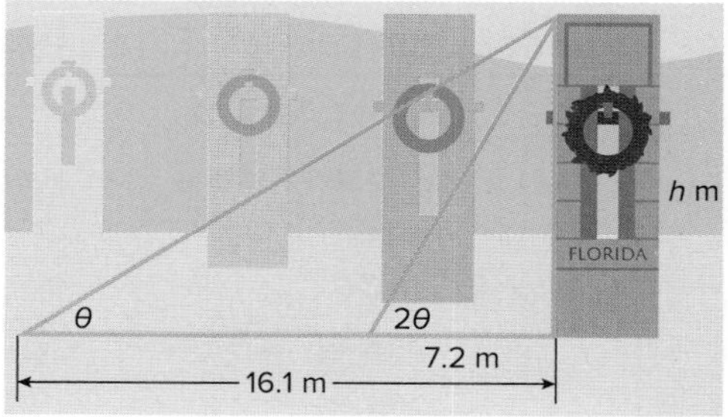

Example 4

Find the exact value of each expression.

13. $\sin 75°$

14. $\sin \frac{3\pi}{8}$

15. $\sin \frac{7\pi}{12}$

16. $\tan 165°$

17. $\tan \frac{5\pi}{12}$

18. $\tan 22.5°$

19. $\cos 22.5°$

20. $\sin 165°$

21. $\cos 105°$

22. $\sin \frac{\pi}{8}$

23. $\sin \frac{15\pi}{8}$

24. $\cos 75°$

Mixed Exercises

Find the exact values of sin 2θ, cos 2θ, and tan 2θ.

25. $\cos \theta = \frac{4}{5}$, $0° < \theta < 90°$

26. $\sin \theta = \frac{1}{3}$, $0 < \theta < \frac{\pi}{2}$

27. $\tan \theta = -3$, $90° < \theta < 180°$

28. $\sec \theta = -\frac{4}{3}$, $90° < \theta < 180°$

29. $\csc \theta = -\frac{5}{2}$, $\frac{3\pi}{2} < \theta < 2\pi$

30. $\cot \theta = \frac{3}{2}$, $180° < \theta < 270°$

31. **STRUCTURE** The large triangle is an isosceles right triangle. The small triangle inside the large triangle was formed by bisecting each of the acute angles of the right triangle.

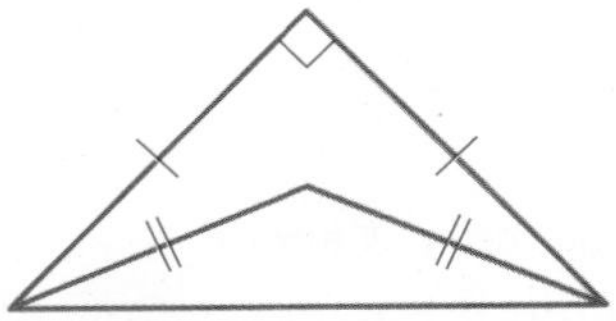

a. What are the exact values of sine and cosine for either of the congruent angles of the small triangle?

b. What are the exact values of sine and cosine for the obtuse angle of the small triangle?

32. **RAMPS** A ramp for loading goods onto a truck was mistakenly built with the dimensions shown. The degree measure of the angle the ramp makes with the ground should have been twice the degree measure of the angle used.

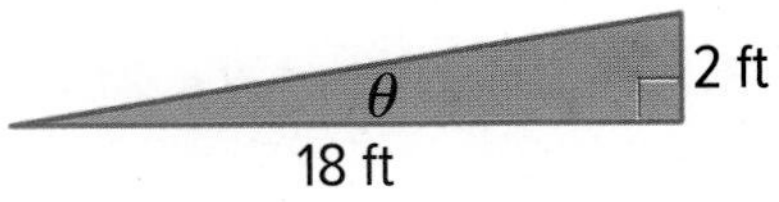

a. Find the exact values of the sine and cosine of the angle the ramp should have made with the ground.

b. If the ramp had been built properly, what would the degree measures of the two acute angles have been?

33. Show how to find the exact value of sin 240° by each method indicated.

a. using a sum of angles formula

b. using a difference of angles formula

c. using a double-angle formula

d. using a half-angle formula

34. **FIND THE ERROR** Teresa and Armando are calculating the exact value of sin 15°. Is either of them correct? Explain your reasoning.

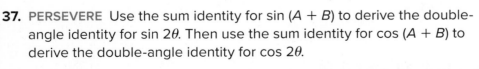

Teresa

$\sin(A - B) = \sin A \cos B - \cos A \sin B$

$\sin(45° - 30°) = \sin 45° \cos 30° -$

$\cos 45° \sin 30°$

$= \frac{\sqrt{2}}{2} \cdot \frac{\sqrt{3}}{2} - \frac{\sqrt{2}}{2} \cdot \frac{1}{2}$

$= \frac{\sqrt{4}}{4}$

Armando

$\sin \frac{A}{2} = \pm \frac{\sqrt{1 - \cos A}}{2}$

$\sin \frac{30°}{2} = \pm \frac{\sqrt{1 - \frac{1}{2}}}{2}$

$= 0.5$

35. **PERSEVERE** Circle O is a unit circle. Use the figure to prove that $\tan \frac{1}{2}\theta = \frac{\sin \theta}{1 + \cos \theta}$.

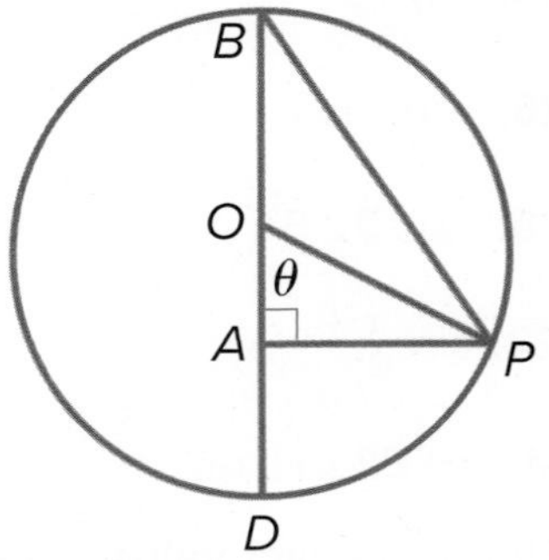

36. **WRITE** Write a short paragraph about the conditions under which you would use each of the three identities for $\cos 2\theta$.

37. **PERSEVERE** Use the sum identity for $\sin(A + B)$ to derive the double-angle identity for $\sin 2\theta$. Then use the sum identity for $\cos(A + B)$ to derive the double-angle identity for $\cos 2\theta$.

38. **ANALYZE** Derive the half-angle identities from the double-angle identities.

39. **PERSEVERE** Suppose a golfer consistently hits the ball so that it leaves the tee with an initial velocity of 115 feet per second. If $d = \frac{2v^2 \sin \theta \cos \theta}{g}$, explain why the maximum distance is attained when $\theta = 45°$.

40. **CREATE** Choose an integer n greater than 2 and write an identity for a trigonometric ratio involving $n\theta$.

Solving Trigonometric Equations

Learn Solving Trigonometric Equations

Trigonometric identities are equations that are true for all values of the variable for which both sides are defined. Typically, **trigonometric equations** are true for only certain values of the variable. Solving these equations resembles solving algebraic equations.

Trigonometric equations are usually solved for values of the variable between 0° and 360° or between 0 and 2π radians. There may be solutions outside that interval. These other solutions differ by integral multiples of the period of the function.

Key Concept • Some Approaches to Solving Trigonometric Equations

- Factor and use the Zero Product Property.
- Use identities to write in terms of one common angle.
- Use identities to write in terms of one common trigonometric function.
- Take the square root of both sides to reduce the power of a function.
- Square both sides to convert to one common trigonometric function.
- Graph a system of equations to approximate the solutions, or check a solution algebraically.

Today's Goal
- Solve trigonometric equations.

Today's Vocabulary
trigonometric equation

Study Tip

Check Your Answers You can check your answers algebraically by entering the solutions into the original equation and graphically by graphing each side of the original equation and noting the intersections.

Example 1 Solve an Equation Over a Given Interval

Solve $2 \sin \theta \cos^2 \theta - \cos^2 \theta = 0$ if $0 \leq \theta < 360°$.

Step 1 Factor.

$2 \sin \theta \cos^2 \theta - \cos^2 \theta = 0$ Original equation

$\cos^2 \theta (2 \sin \theta - 1) = 0$ Factor.

Step 2 Solve.

By the Zero Product Property, θ is a solution when $\cos^2 \theta = 0$ or $2 \sin \theta - 1 = 0$.

Find the values of θ such that $\cos^2 \theta = 0$.

$\cos^2 \theta = 0$ Zero Product Property

$\cos \theta = 0$ Take the square root of each side.

$\theta = 90°$ or $270°$ $\cos \theta = 0$ for [0, 360)

Find the values of θ such that $2 \sin \theta - 1 = 0$.

$2 \sin \theta - 1 = 0$ Zero Product Property

$2 \sin \theta = 1$ Add.

$\sin \theta = \frac{1}{2}$ Divide.

$\theta = 30°$ or $150°$ $\sin \theta = \frac{1}{2}$ for [0, 360)

The solutions are 30°, 90°, 150°, and 270°.

Go Online You can complete an Extra Example online.

Go Online to watch a video to see how to solve trigonometric equations by using a graphing calculator.

Think About It!

Why are there infinitely many solutions of $2 \sin \theta \cos^2 \theta - \cos^2 \theta = 0$ when θ is not restricted?

Check

Solve $\tan \theta \sin 2\theta - \sin \theta = 0$ for $0 \leq \theta < 360°$.

$\theta = \underline{\ ?\ }°, \underline{\ ?\ }°, \underline{\ ?\ }°, \underline{\ ?\ }°$

Study Tip

Expressing Solutions as Multiples The expression $\pi + 2k\pi$ includes 2π and its multiples, so it is not necessary to list them separately.

Think About It!

After graphing $y = \sin 2\theta + \sin \theta$ on a graphing calculator and selecting the zero function, the calculator returns a value of (4.1887902, 0) as one of the zeros. This value approximates $\frac{4\pi}{3}$. How can the calculator be used to confirm that $\frac{4\pi}{3}$ is a zero?

Example 2 Solve an Equation for All Values of θ

Solve $\sin 2\theta + \sin \theta = 0$ for all values of θ with θ measured in radians.

Step 1 Factor.

$\sin 2\theta + \sin \theta = 0$	Original equation
$2 \sin \theta \cos \theta + \sin \theta = 0$	Double-Angle Identity
$\sin \theta (2 \cos \theta + 1) = 0$	Factor.

Step 2 Solve.

By the Zero Product Property, θ is a solution when $\sin \theta = 0$ or $2 \cos \theta + 1 = 0$.

Solve each equation.

The period of $\sin \theta$ is 2π. Find the solutions within $[0, 2\pi]$.

$\sin \theta = 0$	Zero Product Property
$\theta = 0, \pi, 2\pi$	$\sin \theta = 0$ for $[0, 2\pi]$

The solutions will repeat every period, so the solutions can be written as $0 + 2k\pi$ and $\pi + 2k\pi$, where k is any integer. This can be simplified to $k\pi$, where k is any integer.

The period of $2 \cos \theta + 1$ is 2π. Find the solutions within $[0, 2\pi]$.

$2 \cos \theta + 1 = 0$	Zero Product Property
$2 \cos \theta = -1$	Subtract.
$\cos \theta = -\frac{1}{2}$	Divide.
$\theta = \frac{2\pi}{3}$ or $\frac{4\pi}{3}$	$\cos \theta = -\frac{1}{2}$ for $[0, 2\pi]$

The solutions will repeat every period, so the solutions can be written as $\frac{2\pi}{3} + 2k\pi$ and $\frac{4\pi}{3} + 2k\pi$, where k is any integer.

The solutions of $\sin 2\theta + \sin \theta = 0$ are $k\pi$, $\frac{2\pi}{3} + 2k\pi$ and $\frac{4\pi}{3} + 2k\pi$, where k is any integer.

Check

Solve $\frac{2\sqrt{3}}{3} \sin \theta = 1$ for all values of θ. Assume that k is an integer. Select all that apply.

A. $\theta = 2k\pi$

B. $\theta = \frac{\pi}{3} + 2k\pi$

C. $\theta = \frac{\pi}{2} + 2k\pi$

D. $\theta = \frac{2\pi}{3} + 2k\pi$

E. $\theta = \pi + 2k\pi$

F. $\theta = \frac{3\pi}{2} + 2k\pi$

Go Online You can complete an Extra Example online.

Apply Example 3 Use a Trigonometric Equation

JUMP ROPE **During a jump rope activity, the height of the center of the rope can be modeled by $y = 4 \sin 3\pi\left(t - \frac{1}{6}\right) + 4$, where t is the time in seconds. How often does the rope hit the ground? How many times will it hit the ground in one minute?**

1 What is the task?

Describe the task in your own words. Then list any questions that you may have. How can you find answers to your questions?

I need to use the equation that models the height of the jump rope to find when the rope hits the ground and the number of times the rope hits the ground in one minute.

What height represents the rope hitting the ground? How many times does it hit in one period? How many periods are in one minute?

2 How will you approach the task? What have you learned that you can use to help you complete the task?

I will find the period of the equation. Then I will find height and times at which the rope hits the ground. I will use that information to find all the times during which the rope hits the ground.

3 What is your solution?

Use your strategy to solve the problem.

How long is one period, or one revolution of the jump rope?
$\frac{2}{3}$ second

At what height will the jump rope hit the ground? 0 feet

What equation can be used to find the times the rope hits the ground during one period? $4 \sin 3\pi\left(t - \frac{1}{6}\right) + 4 = 0$

In one minute, how many times will the jump rope hit the ground?
90 times

4 How can you know that your solution is reasonable?

Write About It! Write an argument that can be used to defend your solution.

Ground height is represented by $y = 0$. By substituting $y = 0$ in the given equation and solving, I know that the rope hits the ground at $t = \frac{2}{3}$ in the interval $\left(0, \frac{2}{3}\right]$. That means that the rope hits the ground once each period. Because there are 60 seconds in one minute and a period is $\frac{2}{3}$ second, I found $60 \div \frac{2}{3} = 90$, which is the number of times the rope hits the ground in one minute.

Think About It!
What assumption did you make while solving this problem?

Problem-Solving Tip
Determine Reasonable Answers When you have solved a problem, check your answer for reasonableness. In this example, the rope hits the ground 90 times per minute, which is a reasonable number of times.

Example 4 Solve a Trigonometric Inequality

Solve $2 \sec \theta \leq 4$ for $0° \leq \theta < 360°$.

Step 1 Isolate the trigonometric expression.

$2 \sec \theta \leq 4$ Original inequality

$\sec \theta \leq 2$ Divide each side by 2.

Step 2 Identify the solutions of the related equation.

Because $\sec \theta = \frac{1}{\cos \theta}$, $\sec \theta = 2$ when $\cos \theta = \frac{1}{2}$.

So, the two solutions within [0, 360) are 60° and 300°.

Step 3 Identify where $\sec \theta$ is undefined.

For $0° \leq \theta < 360°$, $\sec \theta$ is undefined when $\cos \theta = 0$, or when $\theta = 90°$ or $270°$.

Step 4 Create a sign chart.

Label solutions of the related equation and undefined values.

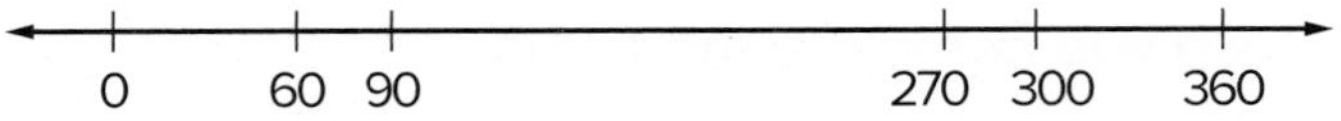

Because secant is continuous on its domain, $\sec \theta$ will be a constant sign between each interval.

Step 5 Test values within each interval.

Because $\sec \theta \leq 2$, the inequality is valid when $\sec \theta - 2$ is 0 or negative.

$\sec 30° - 2 \approx -0.845$ $\sec 75° - 2 \approx 1.864$ $\sec 180° - 2 = -3$

$\sec 280° - 2 \approx 3.759$ $\sec 330° - 2 \approx -0.845$

Step 6 Complete the sign chart and identify the solutions.

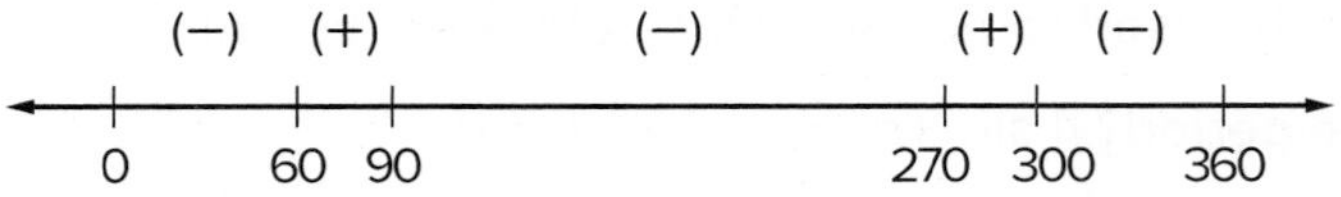

$\sec \theta - 2 \leq 0$, so the solution set includes intervals where $\sec \theta - 2$ is negative. The solution set is [0, 60°], (90°, 270°), [300°, 360°).

Step 7 Check the solutions with a graphing calculator.

Graph $y = \sec \theta - 2$ and note where it is negative within [0, 360°).

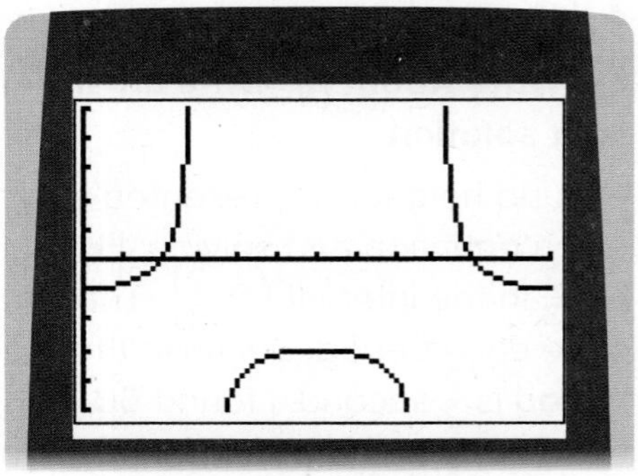

[0, 360] scl: 30 by [−5, 5] scl:1

Check

Solve $\csc \theta < \sqrt{2}$ for $45° < \theta < 135°$ and $180° < \theta < 360°$. Select all that apply.

A. (0, 45°]
B. [45°, 90°)
C. (90°, 270°)
D. [135°, 180°)
E. (180°, 360°)
F. [180°, 360°)
G. (270°, 315°]

Go Online
You can complete an Extra Example online.

Some trigonometric equations have no solution. Other times, a value may appear to be a solution, but is extraneous. Check all of the solutions within a period to identify extraneous solutions.

Example 5 Determine Whether a Solution Exists Over a Given Interval

Solve $\cos \theta - 1 = \sin \theta$ for [0, 360°).

Step 1 Factor.

$\cos \theta - 1 = \sin \theta$	Original equation
$(\cos \theta - 1)^2 = \sin^2 \theta$	Square each side.
$\cos^2 \theta - 2 \cos \theta + 1 = \sin^2 \theta$	Multiply.
$\cos^2 \theta - 2 \cos \theta + 1 = 1 - \cos^2 \theta$	$\sin^2 \theta = 1 - \cos^2 \theta$
$2 \cos^2 \theta - 2 \cos \theta + 1 = 1$	Add $\cos^2 \theta$ to each side.
$2 \cos^2 \theta - 2 \cos \theta = 0$	Subtract 1 from each side.
$2 \cos \theta (\cos \theta - 1) = 0$	Factor.

Step 2 Solve.

By the Zero Product Property, θ is a solution when $2 \cos \theta = 0$ or $\cos \theta - 1 = 0$.

Solve each equation.

Case 1		Case 2
$2 \cos \theta = 0$	Zero Product Property	$\cos \theta - 1 = 0$
$\cos \theta = 0$	Simplify.	$\cos \theta = 1$
$\theta = 90°$ or $270°$	Solve.	$\theta = 0°$

Step 3 Check for extraneous solutions.

Check each solution by substituting it into the original equation.

$\theta = 0°$

$\cos \theta - 1 = \sin \theta$	Original equation
$\cos 0° - 1 \stackrel{?}{=} \sin 0°$	$\theta = 0°$
$1 - 1 \stackrel{?}{=} 0$	Evaluate cos 0° and sin 0°.
$0 = 0$	Subtract.

$\theta = 90°$

$\cos \theta - 1 = \sin \theta$	Original equation
$\cos 90° - 1 \stackrel{?}{=} \sin 90°$	$\theta = 90°$
$0 - 1 \stackrel{?}{=} 1$	Evaluate cos 90° and sin 90°.
$-1 \neq 1$	Subtract.

$\theta = 270°$

$\cos \theta - 1 = \sin \theta$	Original equation
$\cos 270° - 1 \stackrel{?}{=} \sin 270°$	$\theta = 270°$
$0 - 1 \stackrel{?}{=} -1$	Evaluate cos 270° and sin 270°.
$-1 = -1$	Subtract.

The solutions are 0° and 270°. 90° is an extraneous solution.

(continued on the next page)

Go Online You can complete an Extra Example online.

Talk About It!

How do you think an extraneous solution was introduced during the process of solving this problem? Explain your reasoning.

Check Confirm with a graph.

The graphs of $y = \cos\theta - 1$ and $y = \sin\theta$ intersect at two locations within [0°, 360°). Those intersections, 0° and 270°, are the solutions of $\cos\theta - 1 = \sin\theta$ on [0°, 360°).

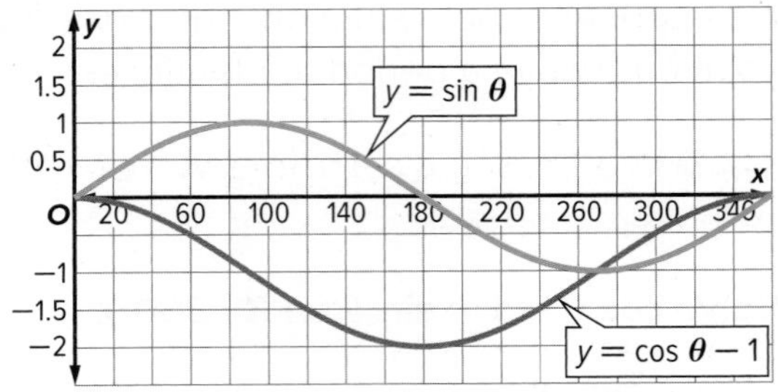

Example 6 Determine Whether a Solution Exists for All Values of θ

Solve $2\sin^4\theta - \cos^2\theta = 0$ for all values of θ with θ measured in degrees.

Step 1 Factor.

$2\sin^4\theta - \cos^2\theta = 0$	Original equation
$2\sin^4\theta - (1 - \sin^2\theta) = 0$	$\cos^2\theta = 1 - \sin^2\theta$
$2\sin^4\theta + \sin^2\theta - 1 = 0$	Distributive Property
$(2\sin^2\theta - 1)(\sin^2\theta + 1) = 0$	Factor.

Think About It!

Why do θ = 135°, 225°, and 315° not need to be checked once 45° was confirmed as a solution? Explain your reasoning.

Step 2 Solve.

By the Zero Product Property, θ is a solution when $2\sin^2\theta - 1 = 0$ or $\sin^2\theta + 1 = 0$. Solve each equation.

$2\sin^2\theta - 1 = 0$	Zero Product Property
$2\sin^2\theta = 1$	Add 1 to each side.
$\sin^2\theta = \frac{1}{2}$	Divide each side by 2.
$\sin\theta = \pm\frac{\sqrt{2}}{2}$	Take the square root of each side.
$\theta = 45°, 135°, 225°, \text{ or } 315°$	$\sin\theta = \pm\frac{\sqrt{2}}{2}$ for [0, 360°)
$\sin^2\theta + 1 = 0$	Zero Product Property
$\sin^2\theta = -1$	Subtract 1 from each side.

$\sin^2\theta$ is never negative, so there are no real solutions for $\sin^2\theta + 1 = 0$.

Study Tip

Patterns Look for patterns in your solutions. Look for pairs of solutions that differ by exactly π or 2π and write your solutions with the simplest possible pattern.

Step 3 Check the solutions.

Because all four values check, 45°, 135°, 225°, and 315° are all solutions. Thus, the solutions are $45° + 90k°$, where k is any integer. You can also confirm the solutions with a graph and check that the zeros are $45° + 90k°$.

Go Online You can complete an Extra Example online.

Practice

Go Online You can complete your homework online.

Example 1

Solve each equation for the given interval.

1. $\cos^2 \theta = \frac{1}{4}$; $0° \leq \theta \leq 360°$

2. $2 \sin^2 \theta = 1$; $90° \leq \theta \leq 270°$

3. $\sin 2\theta - \cos \theta = 0$; $0 \leq \theta \leq 2\pi$

4. $3 \sin^2 \theta = \cos^2 \theta$; $0 \leq \theta \leq \frac{\pi}{2}$

5. $2 \sin \theta + \sqrt{3} = 0$; $180° \leq \theta \leq 360°$

6. $4 \sin^2 \theta - 1 = 0$; $180° \leq \theta \leq 360°$

Example 2

Solve each equation for all values of θ if θ is measured in radians.

7. $\cos 2\theta + 3 \cos \theta = 1$

8. $2 \sin^2 \theta = \cos \theta + 1$

9. $\cos^2 \theta - \frac{3}{2} = \frac{5}{2} \cos \theta$

10. $3 \cos \theta - \cos \theta = 2$

Solve each equation for all values of θ if θ is measured in degrees.

11. $\sin \theta - \cos \theta = 0$

12. $\tan \theta - \sin \theta = 0$

13. $\sin^2 \theta = 2 \sin \theta + 3$

14. $4 \sin^2 \theta = 4 \sin \theta - 1$

Example 3

15. SANDCASTLES The water level on Sunset Beach can be modeled by the function $y = 7 + 7 \sin \frac{\pi}{6}t$, where y is the distance in feet of the waterline above the low tide mark and t is the number of hours past 6 A.M. At 2 P.M., Nashiko built her sandcastle 10 feet above the low tide mark. At what time will the waterline reach Nashiko's sandcastle?

16. BATTERY The light of a battery indicator pulses while the battery is charging. This can be modeled by the equation $y = 60 + 60 \sin \frac{\pi}{4}t$, where y is the lumens emitted from the bulb and t is the number of seconds since the beginning of a pulse. At what time will the amount of light emitted be equal to 110 lumens?

Example 4

17. Solve $3 \csc \theta \leq 2$ for $0° \leq \theta \leq 360°$.

18. Solve $\sqrt{3} \sec \theta \leq 2$ for $0° \leq \theta \leq 360°$.

Example 5

Solve each equation over the given interval.

19. $\cos\theta + 1 = \sin\theta$; $[0, 2\pi)$

20. $\cos\theta + \sin\theta = 1$; $[0, 2\pi)$

21. $1 - \cos\theta = \sqrt{3}\sin\theta$; $[0, 2\pi)$

22. $\tan\frac{\theta}{2} - 1 = 0$; $[0, 2\pi)$.

Example 6

Solve each equation for θ with θ measured in degrees.

23. $\sin\theta + \sqrt{2} = -\sin\theta$

24. $\tan 3\theta = 1$

Mixed Exercises

Solve each equation for the given interval.

25. $\sin\theta = \frac{\sqrt{2}}{2}, 0° \leq \theta \leq 360°$

26. $2\cos\theta = -\sqrt{3}, 90° < \theta < 180°$

27. $\tan^2\theta = 1, 180° < \theta < 360°$

28. $2\sin\theta = 1, 0 \leq \theta < \pi$

29. $\sin^2\theta + \sin\theta = 0, \pi \leq \theta < 2\pi$

30. $2\cos^2\theta + \cos\theta = 0, 0 \leq \theta < \pi$

Solve each equation for all values of θ if θ is measured in radians.

31. $2\cos^2\theta - \cos\theta = 1$

32. $\sin^2\theta - 2\sin\theta + 1 = 0$

33. $\sin\theta + \sin\theta\cos\theta = 0$

34. $\sin^2\theta = 1$

35. $4\cos\theta = -1 + 2\cos\theta$

36. $\tan\theta\cos\theta = \frac{1}{2}$

Solve each equation for all values of θ if θ is measured in degrees.

37. $2\sin\theta + 1 = 0$

38. $2\cos\theta + \sqrt{3} = 0$

39. $\sqrt{2}\sin\theta + 1 = 0$

40. $4\sin^2\theta = 3$

41. $2\cos^2\theta = 1$

42. $\cos 2\theta = -1$

Solve each equation.

43. $3\cos^2\theta - \sin^2\theta = 0$

44. $\sin\theta + \sin 2\theta = 0$

45. $2\sin^2\theta = \sin\theta + 1$

46. $\cos\theta + \sec\theta = 2$

47. Find all solutions for $\sin \theta = \cos 2\theta$ if $0° \leq \theta < 360°$.

48. Find all solutions for $4 \cos^2 \theta = 1$ if $0 \leq \theta < 2\pi$.

49. Solve $\cos 2\theta = \cos \theta$ for all values of θ if θ is measured in degrees.

50. Solve $\cos 2\theta = 3 \sin \theta - 1$ for all values of θ if θ is measured in radians.

51. SOUND The sound from stringed instruments is generated by waves that travel along the strings. The wave traveling on a guitar string can be modeled by $D = 0.5 \sin (6.5x)° \sin (2500t)°$, where D is the displacement in millimeters at the position x millimeters from the bridge of the guitar at time t seconds. Find the first positive time when the point 50 centimeters from the bridge has a displacement of 0.01 millimeter.

52. RIDES The original Ferris wheel had a diameter of about 260 feet and would make a complete revolution in 9 minutes. The motion of the wheel can be modeled by $h = 134 - 130 \cos \frac{2\pi t}{9}$, where h is the height of the rider in feet above the ground t minutes after they begin the ride. When is the rider about 175 feet above the ground?

53. USE A MODEL The table shows the number of hours of daylight on various days of the year in Seattle, Washington. The function $f(x) = 3.725 \sin (0.016x - 1.180) + 11.932$ models the number of hours of daylight in Seattle on day x of the calendar year.

Number of Hours of Daylight in Seattle						
Date	Jan 1	Feb 11	May 3	Aug 22	Oct 15	Nov 29
Day of Year	1	42	123	234	288	333
Hours of Daylight	8.5	10.1	14.6	13.9	10.9	8.8

a. What is the maximum value of $f(x)$? What does this tell you?

b. During what period of the year does Seattle get more than 14 hours of daylight per day? Explain.

54. USE A SOURCE Research the outside temperature several times during the day. Make a table of the data, where x is the number of hours since midnight and y is the temperature in degrees Fahrenheit. Use the data and your calculator to write a sine function $T(x)$ that models the temperature. Approximate the high temperature for the day. At approximately what time did the high temperature occur? Justify your reasoning.

55. BUSINESS Carissa has a small business installing and repairing air conditioners. She has kept track of her monthly profit since she started the business. The function $P(x) = 1808.831 \sin (0.543x - 2.455) + 1942.476$ models the monthly profit in dollars, x months since Carissa started the business.

a. Determine the approximate period of the function. What does this tell you about Carissa's business?

b. Next month will mark the 6th anniversary of the day that Carissa started the business. What profit should she expect to make next month? Explain.

56. STRUCTURE A garden sprinkler is attached to a spigot by a straight hose that is 3 meters long. The sprinkler makes 6 complete rotations every minute. As it rotates, it sprays a stream of water that hits the wall, as shown. Let d be the length of the stream of water from the sprinkler to the wall, and assume the stream of water hits the spigot when the sprinkler is first turned on. The function $d(x) = 3 \sec \frac{\pi}{5}x$ models the length of the stream to strike the wall in meters as a function of the time x in seconds. After the sprinkler is turned on, what is the first time when the stream of water that strikes the wall is 5 meters long?

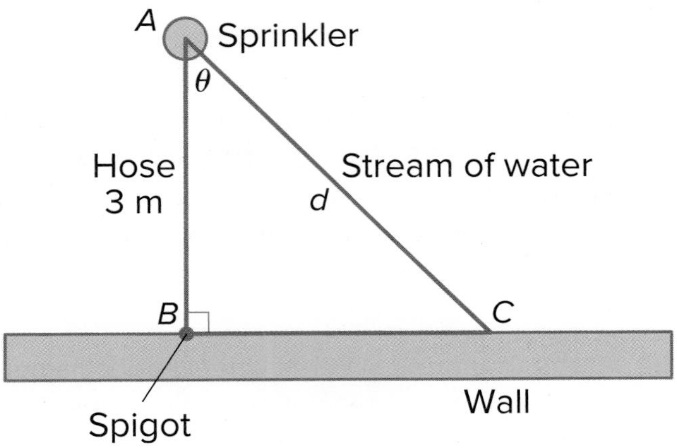

57. PERSEVERE Solve $\sin 2x < \sin x$ for $0 \leq x \leq 2\pi$ without a calculator.

58. ANALYZE Compare and contrast solving trigonometric equations with solving linear and quadratic equations. What techniques are the same? What techniques are different? How many solutions do you expect?

59. WRITE Why do trigonometric equations often have infinitely many solutions?

60. CREATE Write an example of a trigonometric equation that has exactly two solutions if $0° \leq \theta \leq 360°$.

61. PERSEVERE How many solutions in the interval $0° \leq \theta < 360°$ should you expect for $a \sin (b\theta + c) = d$, if $a \neq 0$ and b is a positive integer?

62. FIND THE ERROR Ms. Rollins divided her students into four groups, asking each to solve the equation $\sin \theta \cot \theta = \cos^2 \theta$. Do any of the groups have the correct solution? Explain your reasoning.

Group A:	$0° + k \cdot 360°, 90° + k \cdot 360°, 270° + k \cdot 360°$
Group B:	$0° + k \cdot 360°, 90° + k \cdot 180°$
Group C:	$90° + k \cdot 180°$
Group D:	$90° + k \cdot 360°, 270° + k \cdot 360°$

Review

Essential Question

How are trigonometric identities similar to and different from other equations?

Module Summary

Lessons 12-1 and 12-2

Trigonometric Identities

- A trigonometric identity holds true for all values for which every expression in the equation is defined.
- The Pythagorean identities are $\cos^2\theta + \sin^2\theta = 1$, $\tan^2\theta + 1 = \sec^2\theta$, and $\cot^2\theta + 1 = \csc^2\theta$.
- To verify a trigonometric identity:
  - First, try to verify the identity by simplifying one side.
  - Try substituting basic identities.
  - Use what you have learned from algebra to simplify.
  - When a term includes $1 + \sin\theta$ or $1 + \cos\theta$, think about multiplying the numerator and denominator by the conjugate. Then you can use a Pythagorean Identity.

Lesson 12-3

Sum and Difference Identities

- $\sin(A + B) = \sin A\cos B + \cos A\sin B$
- $\cos(A + B) = \cos A\cos B - \sin A\sin B$
- $\tan(A + B) = \frac{\tan A + \tan B}{1 - \tan A\tan B}$, $A, B \neq 90° + 180n°$, where n is any integer
- $\sin(A - B) = \sin A\cos B - \cos A\sin B$
- $\cos(A - B) = \cos A\cos B + \sin A\sin B$
- $\tan(A - B) = \frac{\tan A - \tan B}{1 + \tan A\tan B}$, $A, B \neq 90° + 180n°$, where n is any integer

Lesson 12-4

Double-Angle and Half-Angle Identities

- $\sin 2\theta = 2\sin\theta\cos\theta$
- $\cos 2\theta = \cos^2\theta - \sin^2\theta = 2\cos^2\theta - 1 = 1 - 2\sin^2\theta$
- $\tan 2\theta = \frac{2\tan\theta}{1 - \tan^2\theta}$, $\theta \neq 45° + 90n°$, where n is any integer
- $\sin\frac{\theta}{2} = \pm\sqrt{\frac{1 - \cos\theta}{2}}$, $\cos\frac{\theta}{2} = \pm\sqrt{\frac{1 + \cos\theta}{2}}$, $\tan\frac{\theta}{2} = \pm\sqrt{\frac{1 - \cos\theta}{1 + \cos\theta}}$, $\theta \neq 180° + 360n°$, where n is any integer

Lesson 12-5

Solving Trigonometric Equations

- To solve a trigonometric equation:
  - Factor and use the Zero Product Property.
  - Use identities to write in terms of one common angle or trig function.
  - Take the square root of both sides to reduce the power of a function.
  - Square both sides to convert to one common trigonometric function.
  - Graph a system of equations to approximate the solutions or check a solution algebraically.

Study Organizer

Foldables

Use your Foldable to review this module. Working with a partner can be helpful. Ask for clarification of concepts as needed.

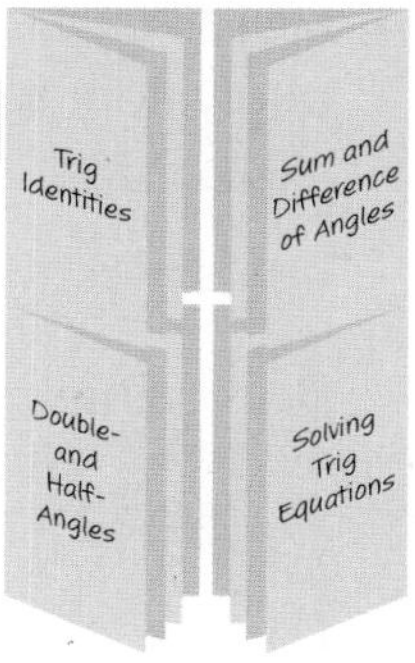

Test Practice

1. OPEN RESPONSE What is the exact value of $\cot \theta$ given $\cos\left(\frac{\pi}{2} - \theta\right) = -\frac{8}{9}$ and $90° < \theta < 180°$? (Lesson 12-1)

2. MULTIPLE CHOICE Which expression is equivalent to $\sin x\,(\cot x - \csc x)$? (Lesson 12-1)

A. $\sin x$

B. $-\sin x$

C. $\cos x - \tan x$

D. $\cos x - 1$

3. MULTIPLE CHOICE If $\tan \theta = \frac{8}{15}$, then what is the value of $\csc \theta$ when $0° < \theta < 90°$? (Lesson 12-1)

A. 0.471

B. 0.533

C. 1.133

D. 2.125

4. MULTIPLE CHOICE Which expression is equivalent to $\left(\tan\left(\frac{\pi}{2} - x\right)\right)(\cos x)(\sin x) - 1$? (Lesson 12-1)

A. $\sin^2 x$

B. $-\sin^2 x$

C. $\sec^2 x$

D. $-\sec^2 x$

5. MULTI-SELECT Select the reasons that justify each step. (Lesson 12-2)

$\tan^2 \theta \cos^2 \theta \stackrel{?}{=} 1 - \cos^2 \theta$	Original equation
$\frac{\sin^2 \theta}{\cos^2 \theta} \cdot \cos^2 \theta \stackrel{?}{=} 1 - \cos^2 \theta$	*Step 2 Reason*
$\sin^2 \theta \stackrel{?}{=} 1 - \cos^2 \theta$	*Step 3 Reason*
$1 - \cos^2 \theta = 1 - \cos^2 \theta$	Pythagorean Identity

A. Step 2: Cofunction Identity

B. Step 2: Quotient Identity

C. Step 2: Reciprocal Identity

D. Step 3: Simplify.

E. Step 3: Quotient Identity

F. Step 3: Pythagorean Identity

6. OPEN RESPONSE The verification of $\frac{\cos \theta}{1 + \sin \theta} + \tan \theta = \sec \theta$ is shown.

$$\frac{\cos \theta}{1 + \sin \theta} + \tan \theta \stackrel{?}{=} \sec \theta$$

$$\textit{Missing Step} \stackrel{?}{=} \sec \theta$$

$$\frac{\cos^2 \theta}{\cos \theta\,(1 + \sin \theta)} + \frac{\sin \theta(1 + \sin \theta)}{\cos \theta(1 + \sin \theta)} \stackrel{?}{=} \sec \theta$$

$$\frac{\cos^2 \theta + \sin \theta + \sin^2 \theta}{\cos \theta\,(1 + \sin \theta)} \stackrel{?}{=} \sec \theta$$

$$\frac{1 + \sin \theta}{\cos \theta\,(1 + \sin \theta)} \stackrel{?}{=} \sec \theta$$

$$\frac{1}{\cos \theta} \stackrel{?}{=} \sec \theta$$

$$\sec \theta = \sec \theta$$

What is the missing step? (Lesson 12-2)

7. OPEN RESPONSE The verification of $\frac{\cot^2\theta}{1+\csc\theta} = \frac{1-\sin\theta}{\sin\theta}$ is shown. Give the reason that justifies each step. (Lesson 12-2)

Step 1: $\frac{\cot^2\theta}{1+\csc\theta} \stackrel{?}{=} \frac{1-\sin\theta}{\sin\theta}$

Step 2: $\frac{\csc^2\theta - 1}{1+\csc\theta} \stackrel{?}{=} \frac{1-\sin\theta}{\sin\theta}$

Step 3: $\frac{(\csc\theta+1)(\csc\theta-1)}{1+\csc\theta} \stackrel{?}{=} \frac{1-\sin\theta}{\sin\theta}$

Step 4: $\csc\theta - 1 \stackrel{?}{=} \frac{1-\sin\theta}{\sin\theta}$

Step 5: $\csc\theta - 1 \stackrel{?}{=} \frac{1}{\sin\theta} - \frac{\sin\theta}{\sin\theta}$

Step 6: $\csc\theta - 1 \stackrel{?}{=} \frac{1}{\sin\theta} - 1$

Step 7: $\csc\theta - 1 = \csc\theta - 1$

8. MULTI-SELECT Which expression(s) could be used to determine cos 75°? Select all that apply. (Lesson 12-3)

A. cos 45° cos 30° + sin 45° sin 30°

B. cos 45° cos 30° − sin 45° sin 30°

C. cos 90° cos 15° + sin 90° sin 15°

D. cos 90° cos 15° − sin 90° sin 15°

E. sin 45° cos 30° + cos 45° sin 30°

F. sin 45° cos 30° − cos 45° sin 30°

G. sin 90° cos 15° + cos 90° sin 15°

H. sin 90° cos 15° − cos 90° sin 15°

9. MULTIPLE CHOICE What is the exact value of $\tan \frac{7\pi}{12}$? (Lesson 12-3)

A. $-2+\sqrt{3}$

B. $-2-\sqrt{3}$

C. $1+\sqrt{3}$

D. $1-\sqrt{3}$

10. OPEN RESPONSE Verify that $\tan(2\pi - x) = -\tan x$ is an identity. (Lesson 12-3)

11. MULTIPLE CHOICE Find the exact value of $\cos 2\theta$. (Lesson 12-4)

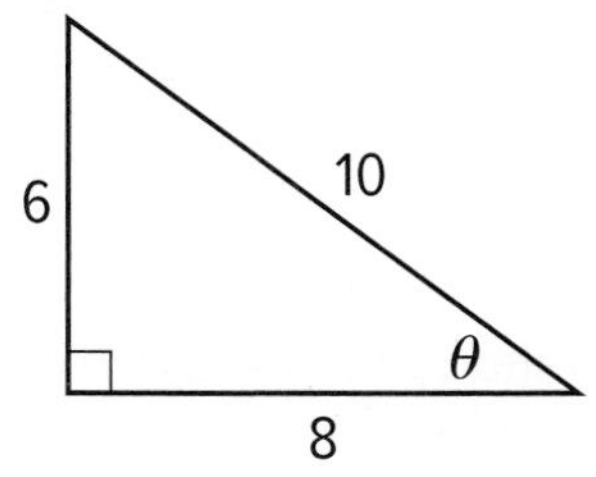

A. 0.28

B. 0.32

C. 0.56

D. 0.64

12. **MULTIPLE CHOICE** What is the value of $\sin 2\theta$ if $\sin \theta = \frac{8}{17}$ and $0^0 < \theta < 90°$? (Lesson 12-4)

A. $-\frac{161}{289}$

B. $\frac{161}{289}$

C. $\frac{240}{289}$

D. $\frac{16}{17}$

13. **OPEN RESPONSE** What is the exact value of $\sin 157.5°$? (Lesson 12-4)

14. **MULTIPLE CHOICE** Monisha drew the triangle shown. She then drew a line bisecting θ. What is the value of the tangent of the new angle formed? (Lesson 12-4)

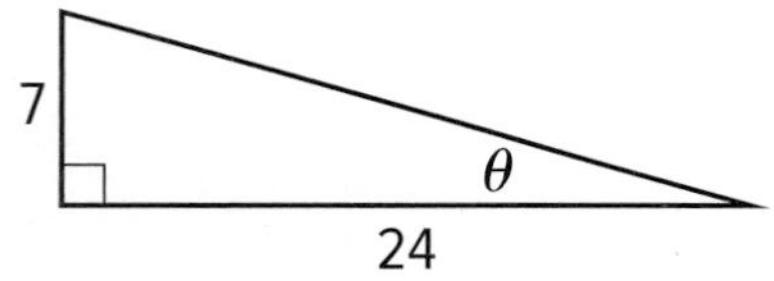

A. $\frac{\sqrt{2}}{10}$

B. $\frac{7\sqrt{2}}{10}$

C. $\frac{1}{7}$

D. $\frac{7}{48}$

15. **MULTI-SELECT** Which of the given values is a solution to the equation $\tan^2 \theta = \sin \theta \sec \theta$? Select all that apply. (Lesson 12-5)

A. 45°

B. 135°

C. 225°

D. 315°

16. **MULTIPLE CHOICE** The distance a baseball travels after being thrown can be modeled by $d = \frac{v^2}{16} \sin \theta \cos \theta$, where d is is the horizontal distance in feet and v is the initial velocity. If the ball was thrown with an initial velocity of 72 feet per second and traveled 81 feet, at what angle (in degrees) was the ball thrown? (Lesson 12-5)

A. 15°

B. 30°

C. 45°

D. 50°

17. **MULTI-SELECT** Solve $\tan^2 x + \tan x = 0$ for $0° \leq x < 360°$. Select all that apply. (Lesson 12-5)

A. 0°

B. 45°

C. 90°

D. 135°

E. 150°

F. 180°

G. 225°

H. 315°

18. **OPEN RESPONSE** Solve $\sin \theta \cos \theta = \frac{1}{2}$ for all values of θ, if θ is measured in radians. (Lesson 12-5)

Selected Answers

Module 1

Quick Check

1. −15 **3.** 10 **5.** $b = \frac{a}{3} - 3$ **7.** $x = \frac{4}{3}y + \frac{8}{3}$

Lesson 1-1

1. D = {all real numbers}; R = {all real numbers}; Codomain = {all real numbers}; onto **3.** D = {all real numbers}, R = $\{y \mid y \geq 0\}$, Codomain = {all real numbers}; not onto
5. D = {1, 2, 3, 4, 5, 6, 7}; R = {56, 52, 44, 41, 43, 46, 53}; one-to-one
7. neither **9.** both
11. continuous; D = {all real numbers}, R = {all real numbers}
13. discrete; D = {1, 2, 3, 4, 5, 6}, R = {3, 4, 5, 6}
15. continuous; D = all positive real numbers, R = $\{y \mid y \geq 0\}$
17. D = $\{x \mid x \in \mathbb{R}\}$ or $(-\infty, \infty)$; R = $\{y \mid y \leq 0\}$ or $(-\infty, 0]$
19. D = $\{x \mid x \leq -1 \text{ or } x \geq 1\}$ or $(-\infty, -1] \cup [1, \infty)$; R = $\{y \mid y \in \mathbb{R}\}$ or $[-\infty, \infty)$
21. D = $\{x \mid x \leq -2 \text{ or } x \geq 1\}$ or $(-\infty, -2] \cup [1, \infty)$; R = $\{y \mid y \geq -2\}$ or $[-2, \infty)$
23. D = $\{x \mid x \in \mathbb{R}\}$ or $(-\infty, \infty)$; R = $\{y \mid y \geq -4\}$ or $[-4, \infty)$; neither one-to-one nor onto; continuous
25. D = $\{x \mid x \in \mathbb{R}\}$ or $(-\infty, \infty)$; R = $\{y \mid y \in \mathbb{R}\}$ or $[-\infty, \infty)$; both; continuous
27. D = $\{x \mid x \in \mathbb{R}\}$ or $(-\infty, \infty)$; R = $\{y \mid y \in \mathbb{R}\}$ or $[-\infty, \infty)$; onto; continuous
29. D = $\{w \mid 0 \leq w \leq 15\}$; R = $\{L \mid 4 \leq L \leq 11.5\}$; one-to-one; continuous
31. neither one-to-one nor onto; discrete
33. D = $\{n \mid n \geq 0\}$; R = $\{T(n) \mid T(n) \geq 0\}$; both (within the restrictions of the domain and codomain); continuous
35.

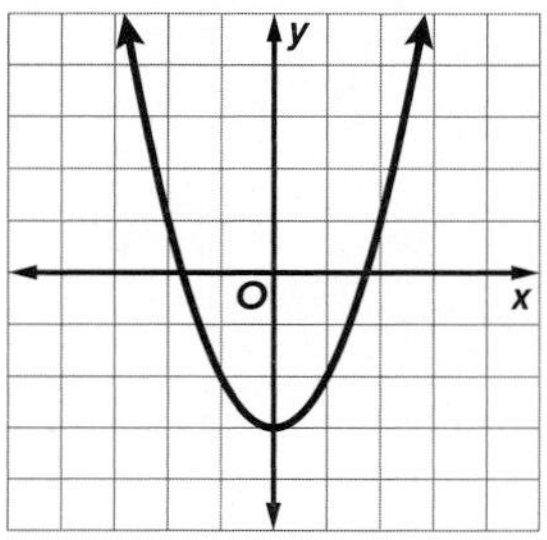

37. D = $\{x \mid x \neq 0\}$ or $(-\infty, 0) \cup (0, \infty)$; R = $\{y \mid y \neq 0\}$ or $(-\infty, 0) \cup (0, \infty)$; neither; continuous

39. Sample answer: The vertical line test is used to determine whether a relation is a function. If no vertical line intersects a graph in more than one point, the graph represents a function. The horizontal line test is used to determine whether a function is one-to-one. If no horizontal line intersects the graph more than once, then the function is one-to one. The horizontal line test can also be used to determine whether a function is onto. If every horizontal line intersects the graph at least once, then the function is onto.

Lesson 1-2

1. Yes; it can be written in $y = mx + b$ form.
3. Yes; it can be written in $y = mx + b$ form.
5. nonlinear **7.** nonlinear **9.** The number of inches and corresponding number of feet is a linear function because when graphed, a line contains all of the points shown in the table.
11. *x*-int: 12; *y*-int: −18 **13.** *x*-int: 6; *y*-int: −18
15. *x*-int: 0, 4; *y*-int: 0 **17a.** *x*-int: 5; *y*-int: 20
17b. The *x*-intercept represents the number of days until Aksa will run out of money. The *y*-intercept represents the total amount Aksa had in her lunch account at the beginning of the week. **19.** point symmetry
21. point symmetry **23.** neither; $f(-x) = (-x)^3 + (-x)^2 = -x^3 + x^2 \neq f(x)$ and $\neq -f(-x)$.
25. No; *x* is in a denominator. The equation is neither even nor odd. **27.** Yes; it can be written in $y = mx + b$ form. The equation is even. **29.** No; there is an x^2 term. The equation is even. **31.** linear; *x*-int: $\frac{10}{3}$; *y*-int: $\frac{30}{7}$; point symmetry **33.** nonlinear; *x*-int: −3, −2, −1, 1, 2; *y*-int: 12; neither point nor line symmetry **35.** line symmetry; $x = 0.4$
37a. $2x + 2y + 10 = 500$

37b. Yes; it can be written in $y = mx + b$ form.

37c. point symmetry;

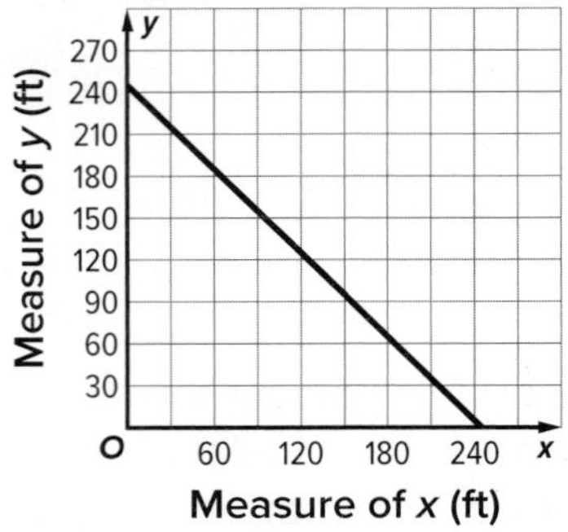

39. Odd; $f(-r) = -f(r)$

41. No; sample answer: $f(x) = x^3 + 2x - 5$

43. Never; sample answer: the graph of $x = a$ is a vertical line, so it is not a function.

Lesson 1-3

1. rel. max. at $(-2, -2)$ rel. min. at $(-1, -3)$

3. rel. max. at $(0, -2)$, min. at $(-1, -3)$ and $(1, -3)$

5. The relative maxima occur at $x = -3.7$ and $x = 4.5$, and the relative minimum occurs at $x = 0$. The relative maxima at $x = -3.7$ and $x = 4.5$ represents the top of two hills. The relative minimum at $x = 0$ represents a valley between the hills.

7. As $x \to -\infty$, $y \to -\infty$ and as $x \to \infty$, $y \to -\infty$.

9. As $x \to -\infty$, $y \to \infty$ and as $x \to \infty$, $y \to -\infty$.

11a. $t = 1.5$; The fish reaches its maximum height 1.5 seconds after it is thrown.

11b. As $t \to -\infty$, $h(t) \to -\infty$ and as $t \to \infty$, $h(t) \to -\infty$; The height cannot be negative because we are considering the path of the fish above the surface of the water, $h = 0$. Time cannot be negative because we're measuring from the initial time the fish was thrown at $t = 0$.

13. rel. max. at $x = 2.7$, rel. min. at $x = -1.2$; As $x \to -\infty$, $f(x) \to \infty$ and as $x \to \infty$, $f(x) \to -\infty$.

15. rel. max. at $x = -2$, rel. min. at $x = -1$; As $x \to -\infty$, $f(x) \to \infty$ and as $x \to \infty$, $f(x) \to \infty$.

17. no rel. max, min: $x = 0$; As $x \to -\infty$, $f(x) \to \infty$ and as $x \to \infty$, $f(x) \to \infty$.

19. As temperature increases, density decreases.

21. rel. max $(-2.8, 6)$, $(1.8, 3)$; rel. min. $(0, 2)$, $(5, -6)$; As $x \to -\infty$, $y \to -\infty$ and as $x \to \infty$, $y \to \infty$.

23. no relative max or min

25. rel. max: $x = -1.87$; rel. min: $x = 1.52$

27. The dynamic pressure would approach ∞.

29. as $r \to \infty$, $V \to \infty$

31. Sample answer: The end behavior of a graph describes the output values the input values approach negative and positive infinity. It can be determined by examining the graph.

33. As the concentration of the catalyst is increased, the reaction rate approaches 0.5.

35. Joshua switched the $f(x)$ values. He read the graph from right to left instead of left to right.

Lesson 1-4

1.

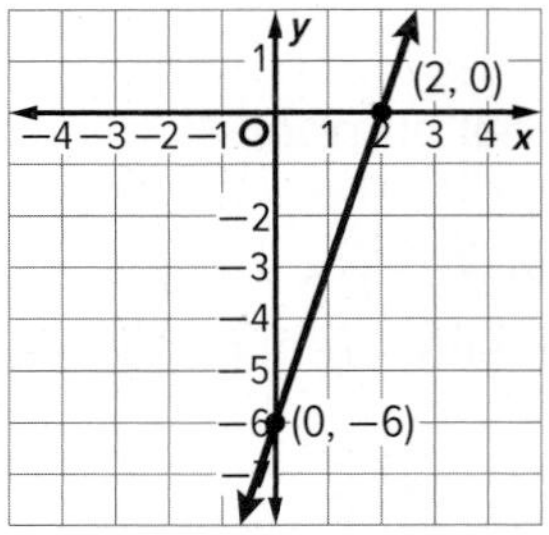

3.

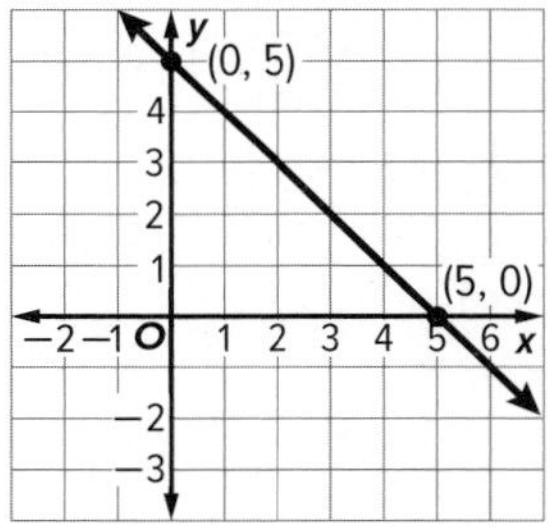

5.

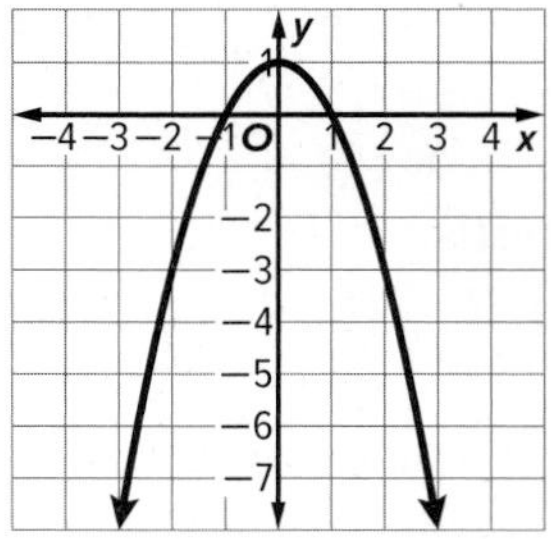

7.

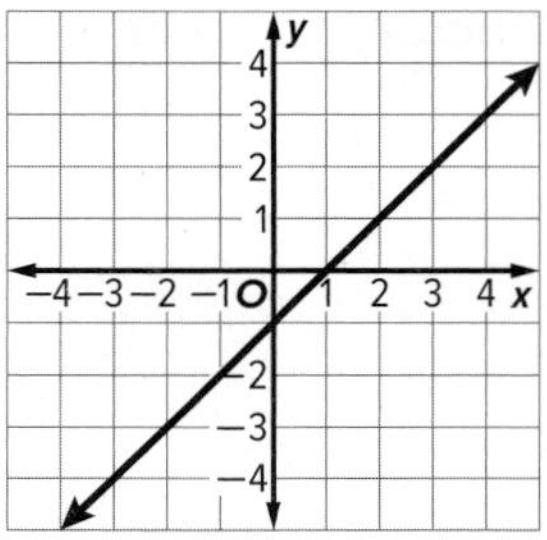

9.

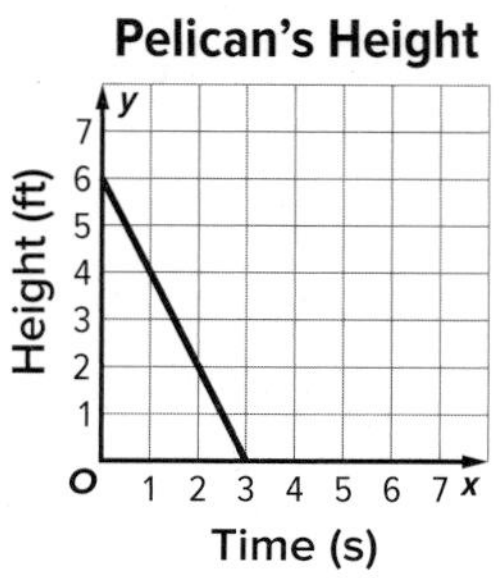

11. The x-intercept of $f(x)$ is 2, and the x-intercept of $g(x)$ is $-\frac{2}{3}$. The x-intercept of $f(x)$ is greater than the x-intercept of $g(x)$. So, $f(x)$ intersects the x-axis at a point farther to the right than $g(x)$. The y-intercept of $f(x)$ is -1, and the y-intercept of $g(x)$ is 2. The y-intercept of $g(x)$ is greater than the y-intercept of $f(x)$. So, $g(x)$ intersects the y-axis at a higher point than $f(x)$. The slope of $f(x)$ is $\frac{1}{2}$ and the slope of $g(x)$ is 3. Each function is increasing, but the slope of $g(x)$ is greater than the slope of $f(x)$. So, $g(x)$ increases faster than $f(x)$. **13.** Both x-intercepts of $f(x)$ are less than the x-intercept of $g(x)$. The graph of $f(x)$ intersects the x-axis more times than $g(x)$. The y-intercept of $g(x)$ is less than the y-intercept of $f(x)$. So, $f(x)$ intersects the y-axis at a higher point than $g(x)$. Neither function has a relative minimum or relative maximum. $f(x)$ has a minimum at $(-2, -4)$. The two functions have the opposite end behaviors as $x \rightarrow -\infty$. The two functions have the same end behavior as $x \rightarrow \infty$.

15. The x-intercept of $f(x)$ is $\frac{2}{3}$, and the x-intercept of $g(x)$ is $\frac{3}{8}$. The x-intercept of $f(x)$ is greater than the x-intercept of $g(x)$. So, $f(x)$ intersects the x-axis at a point farther to the right than $g(x)$. The y-intercept of $f(x)$ and $g(x)$ is $\frac{1}{2}$. So, $f(x)$ and $g(x)$ intersect the y-axis at the same point. The slope of $f(x)$ is $\frac{3}{4}$ and the slope of $g(x)$ is $-\frac{4}{3}$. Each function is decreasing, but the slope of $g(x)$ is less than the slope of $f(x)$. So, $g(x)$ decreases faster than $f(x)$.

17.

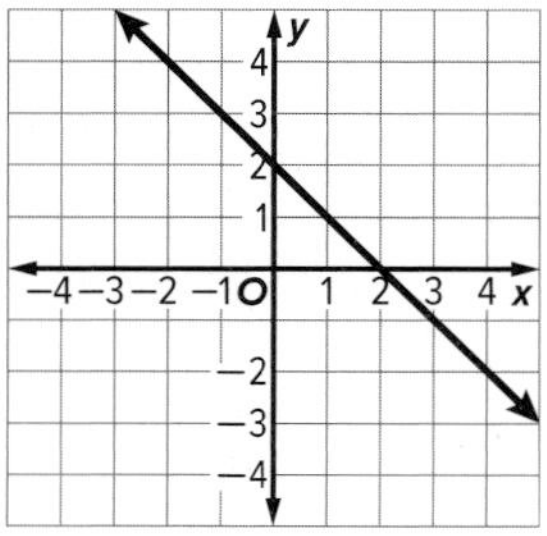

19.

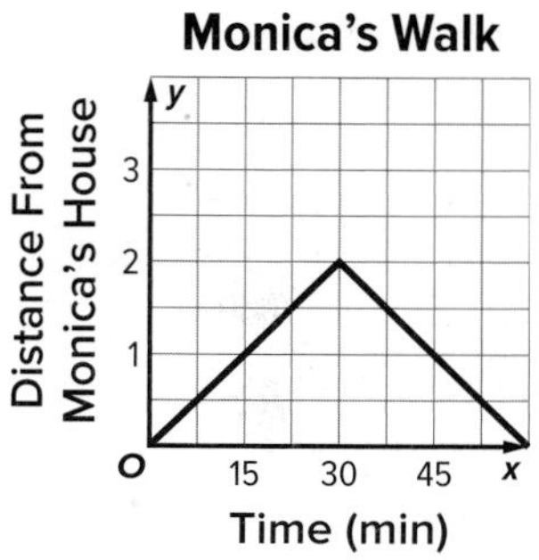

21a. linear; Sample answer: It is linear because it makes no stops along the way, and it descends at a steady pace, which indicates a constant rate of change, or slope.

21b.

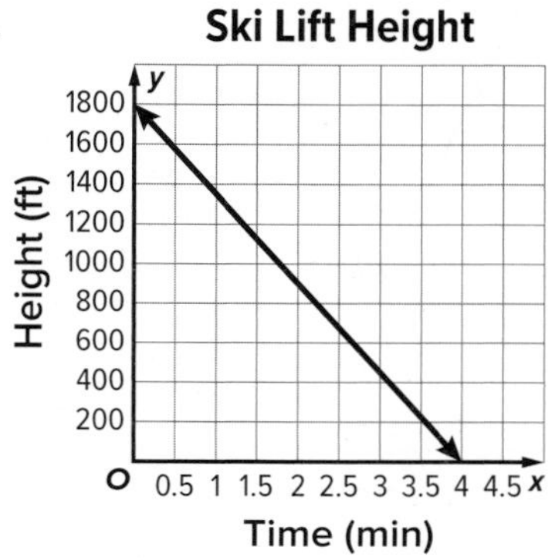

23. Sample answer: The function is continuous. The function has a y-intercept at -3. The function has a maximum at $(-3, 1)$. The function has a minimum at $(1.4, -4)$. As $x \rightarrow -\infty$, $f(x) \rightarrow -\infty$ and as $x \rightarrow \infty$, $f(x) \rightarrow \infty$.

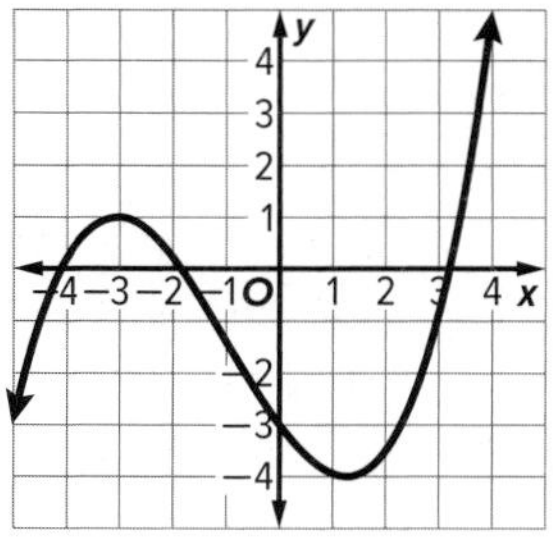

25. Always; Sample answer: A linear function cannot cross the x-axis more than once. So, if a function has more than one x-intercept, it is a nonlinear function.

27. Both Linda and Rubio sketched correct graphs. Both graphs have an x-intercept at 2, a y-intercept at -9, are positive for $x > 2$, and have an end behavior of as $x \rightarrow -\infty$, $f(x) \rightarrow -\infty$ and as $x \rightarrow \infty$, $f(x) \rightarrow \infty$.

Lesson 1-5

1.
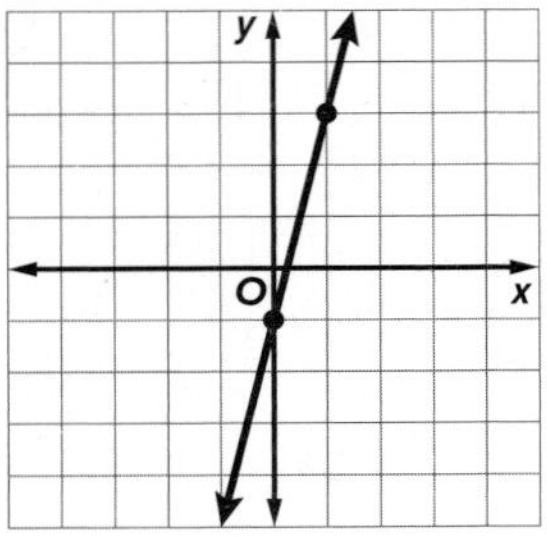

3.
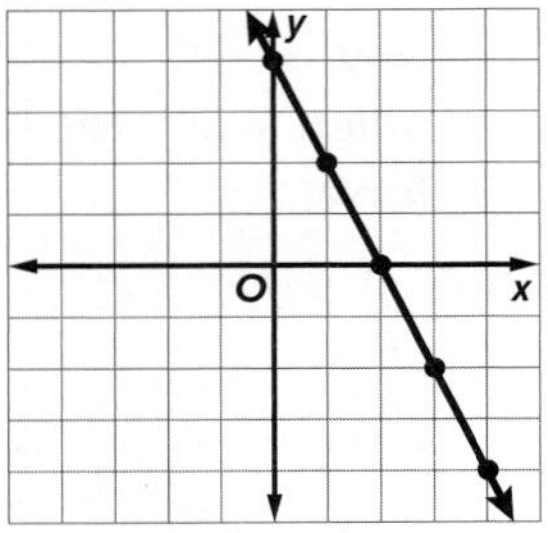

5.
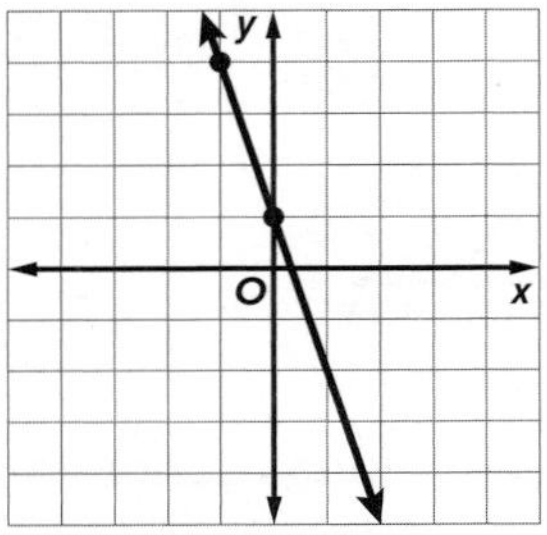

7.
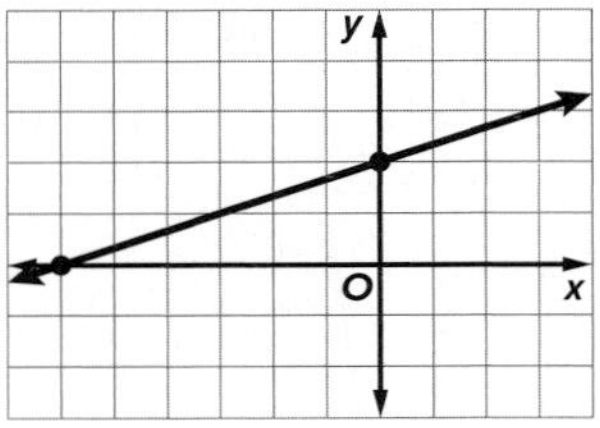

9.
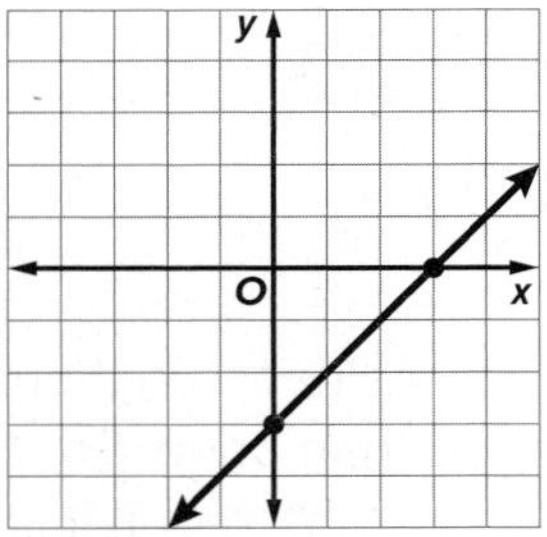

11.
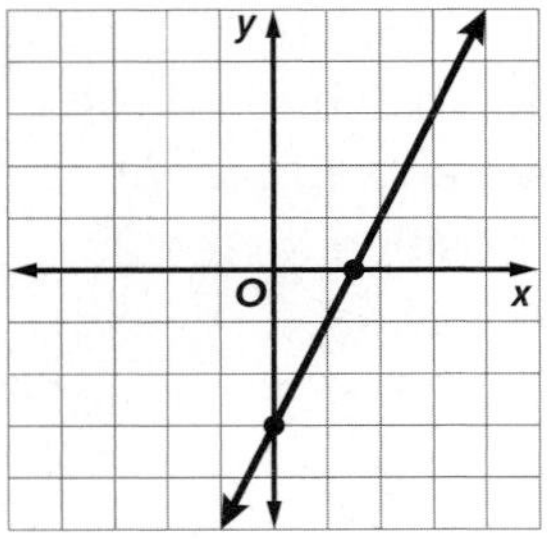

13.
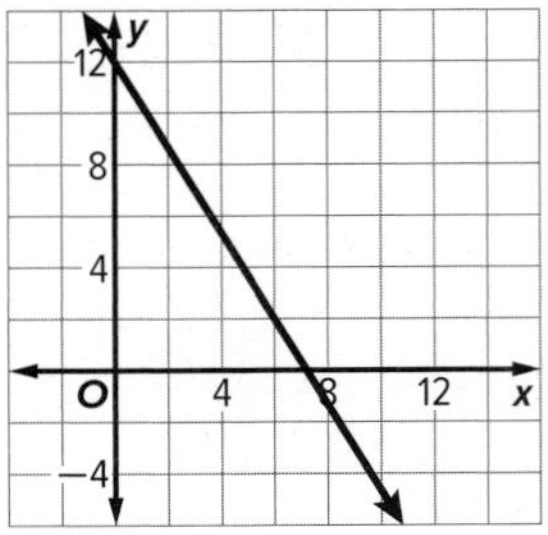

15.
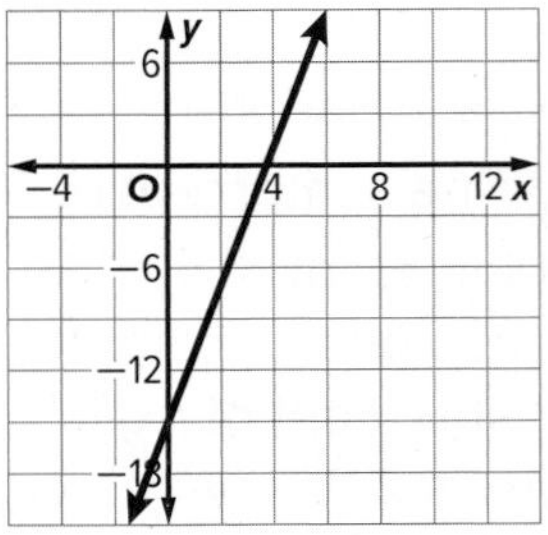

17.
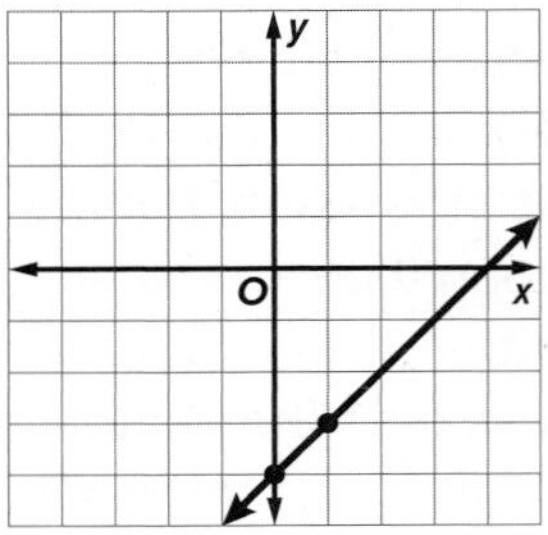

19.
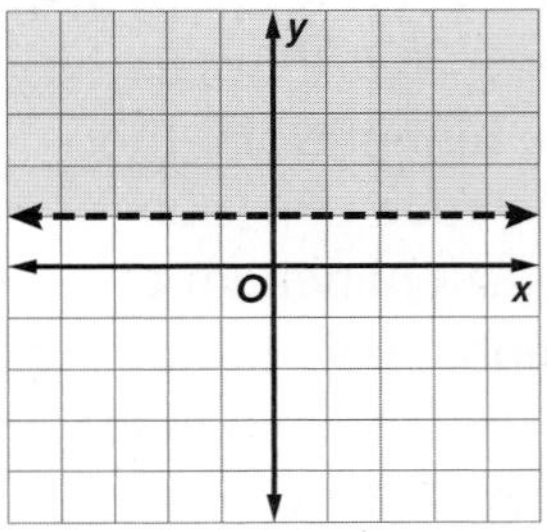

21.
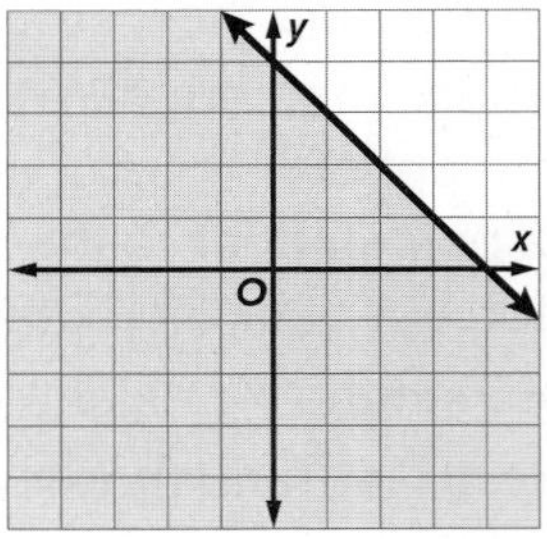

23.
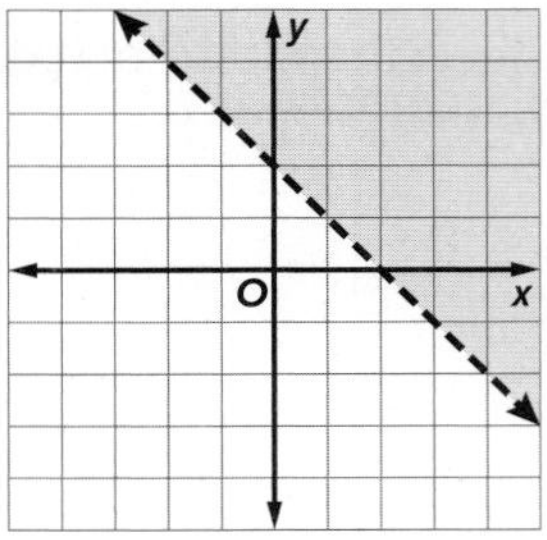

25.
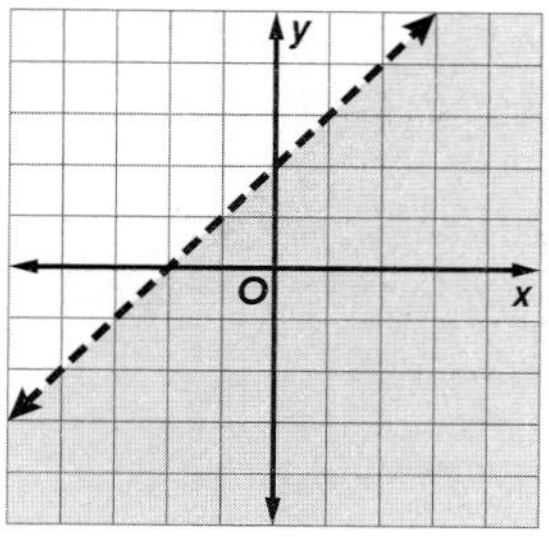

27.
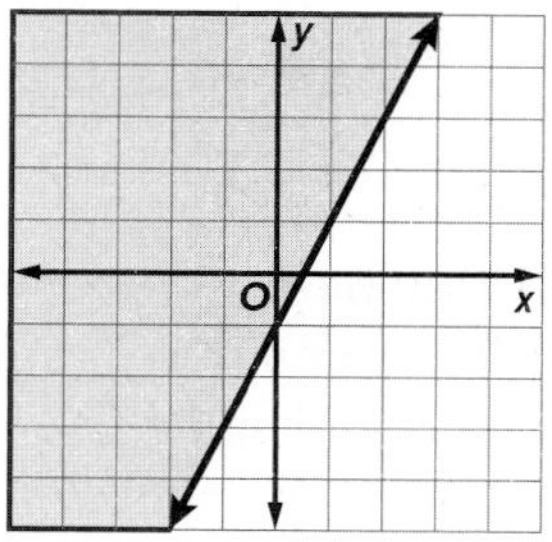

29.
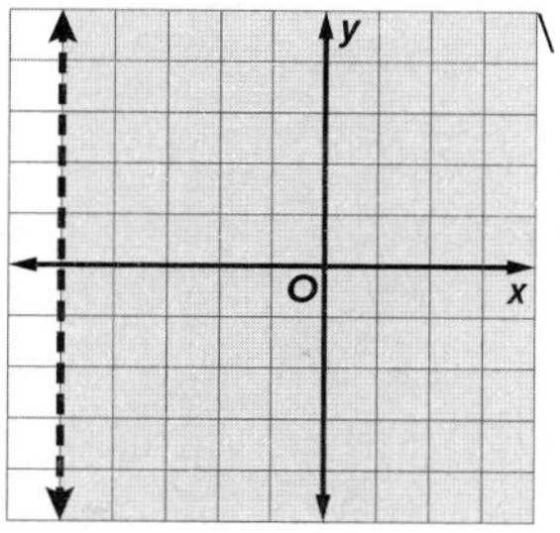

31a. $x + y \geq 400$

31b.
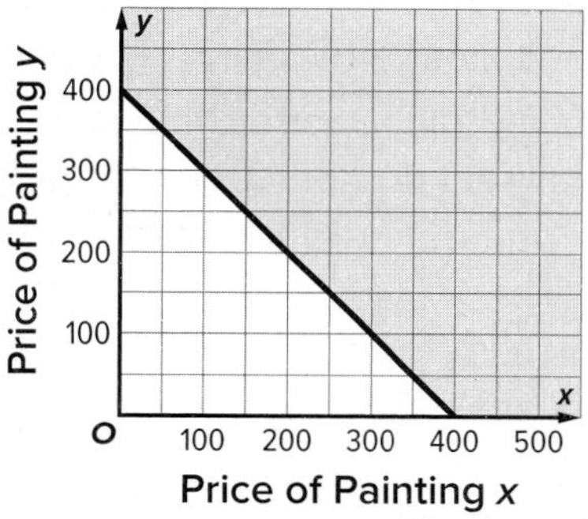

33.
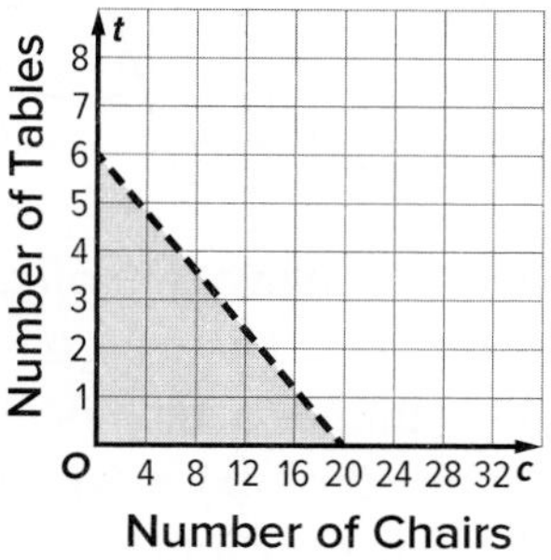

35.
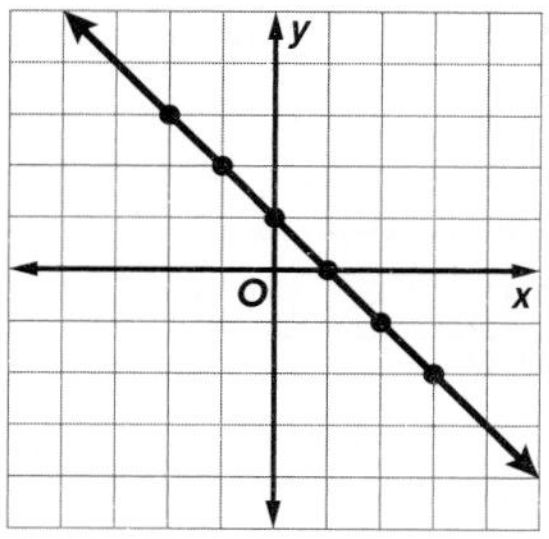

37.
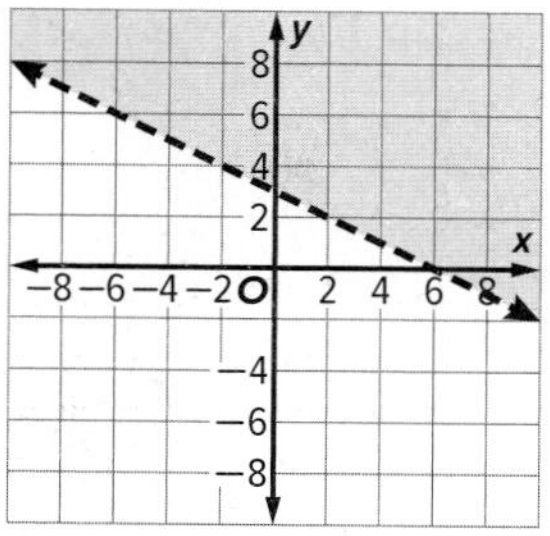

39.
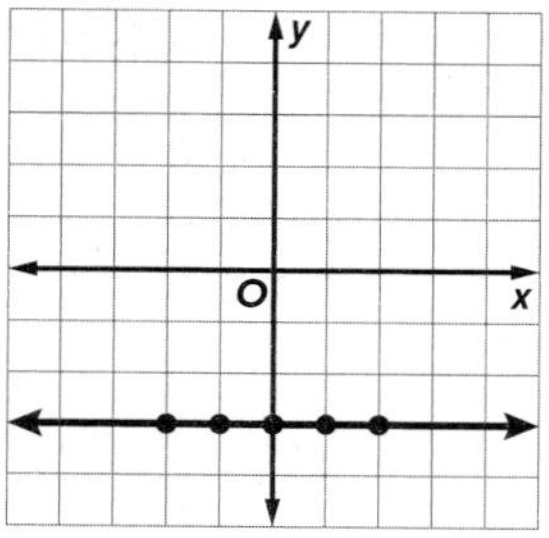

41.
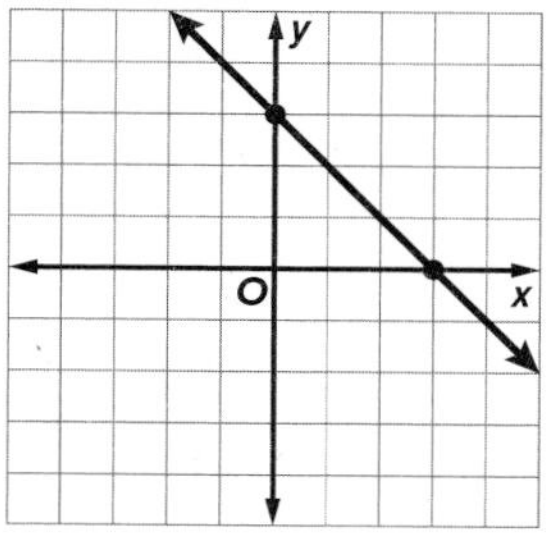

43.
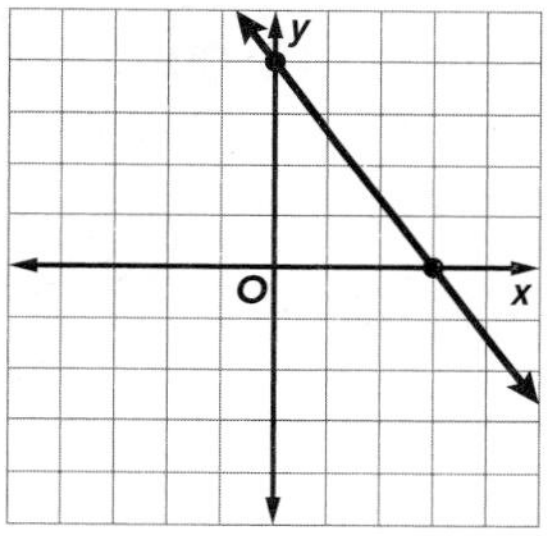

45.
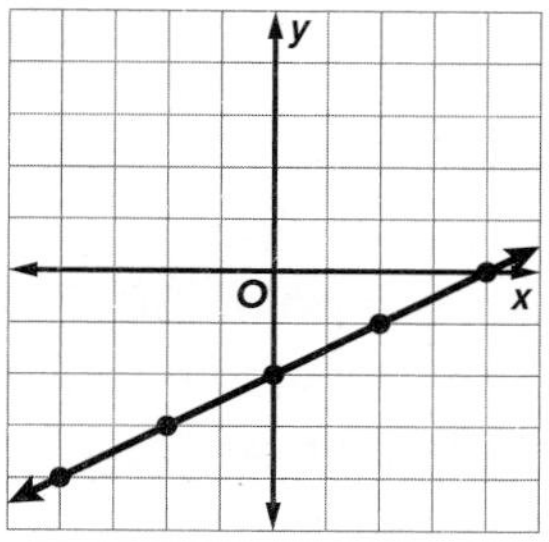

47.
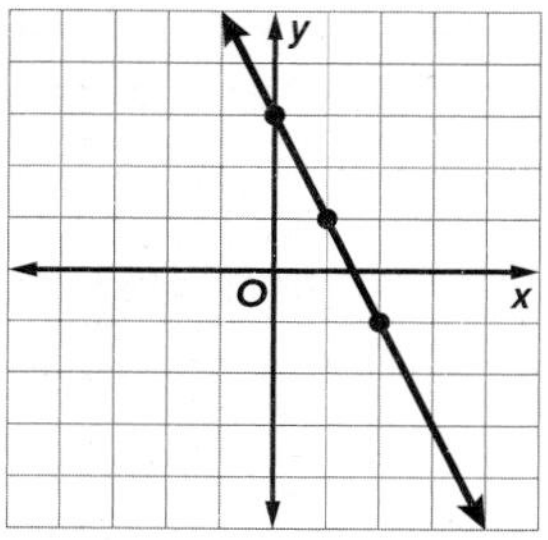

49.

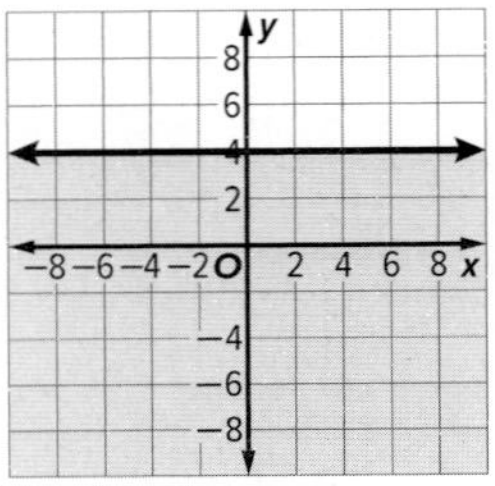

51.

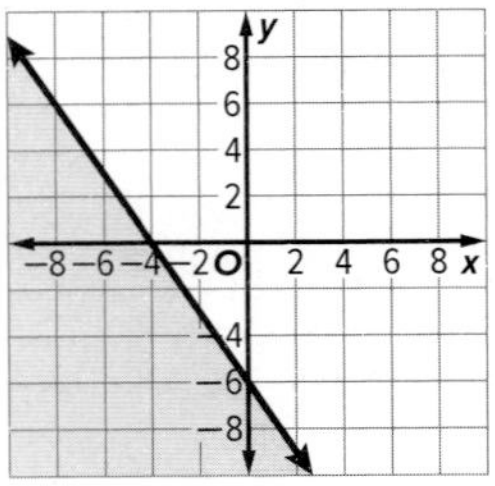

53. x-int = 2, y-int = −6; graph is increasing;

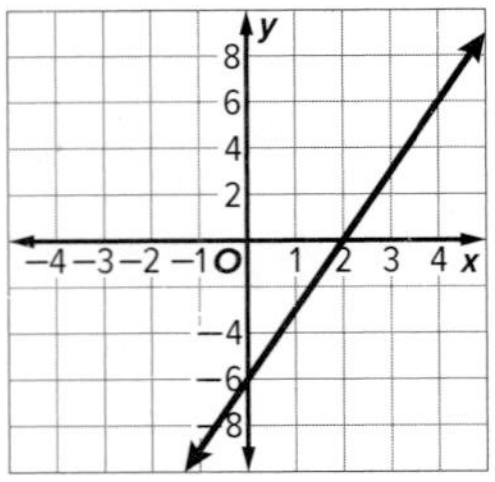

55a. Let x = number of desktops; let y = number of notebooks; $1000x + 1200y \leq 80{,}000$.

55b.

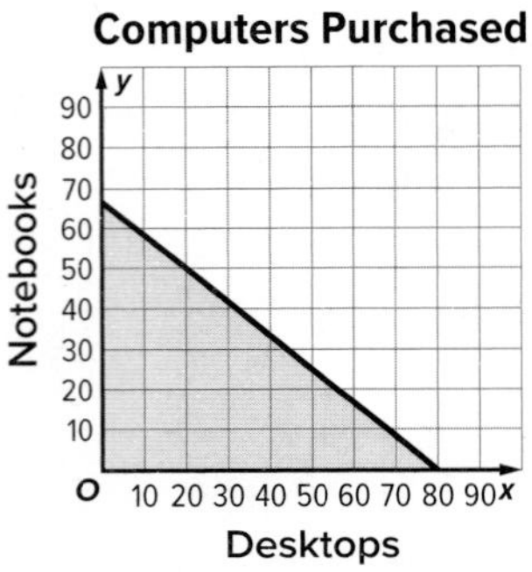

55c. Yes; Sample answer: the point (50, 25) is on the line, which is part of the viable region. **57a.** Let x = cost of a student ticket; let y = cost of an adult ticket; $300x + 150y = 1800$. **57b.** Sample answer: $x = \$2.00$, $y = \$8.00$; $x = \$3.00$, $y = \$6.00$; $x = \$4.60$, $y = \$2.80$; $x = \$6.00$, $y = \$0.00$

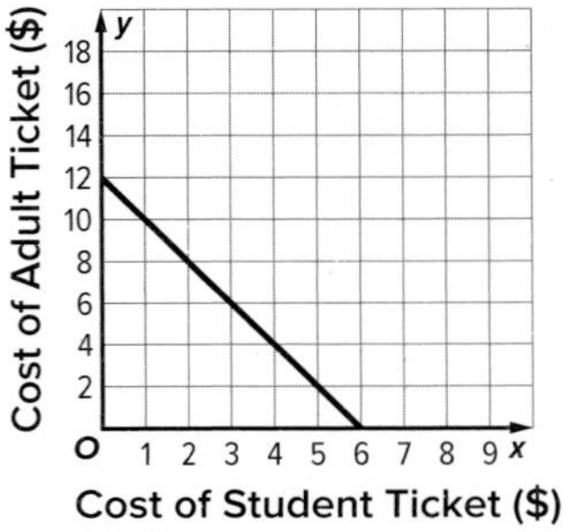

59a. Let x = long-sleeved shirts; let y = short-sleeved shirts; $7x + 4y \geq 280$;

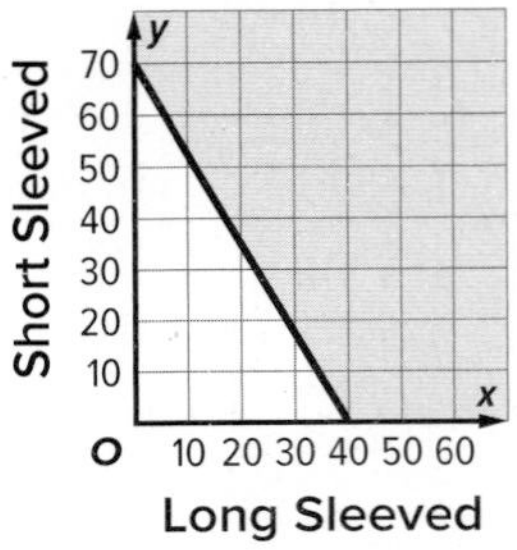

59b. Sample answer: 30 long-sleeved and 50 short-sleeved shirts; 60 long-sleeved and 40 short-sleeved shirts. **59c.** Domain and range values must be positive integers since you cannot buy a negative number of shirts or a portion of a shirt. **59d.** No, you cannot buy −10 long-sleeved shirts. **61.** Paulo; Janette shaded the incorrect region. **63.** Sample answer: If given the x- and y-intercepts of a linear function, I already know two points on the graph. To graph the equation, I only need to graph those two points and connect them with a straight line. **65.** $y = \frac{1}{4}x + 5$

Lesson 1-6

1. The function is defined for all values of x, so the domain is all real numbers. The range is −1 and all real numbers greater than or equal to 0 and less than or equal to 6, which is also represented as $\{f(x) \mid f(x) = -1 \text{ or } 0 \leq f(x) \leq 6\}$. The y-intercept is 0, and the x-intercept is 0. The function is increasing when $0 \leq x \leq 3$.

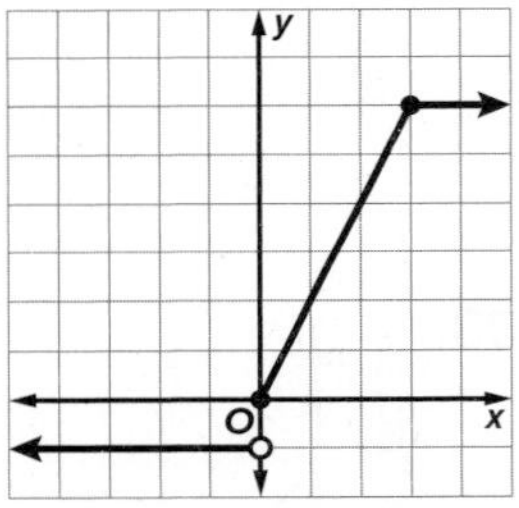

3. The function is defined for all values of x, so the domain is all real numbers. The range is 2 and all real numbers less than 0, which is also represented as $\{f(x) \mid f(x) = 0 \text{ or } f(x) < 0\}$. The y-intercept is 2, and there is no x-intercept. The function is decreasing when $x < 0$.

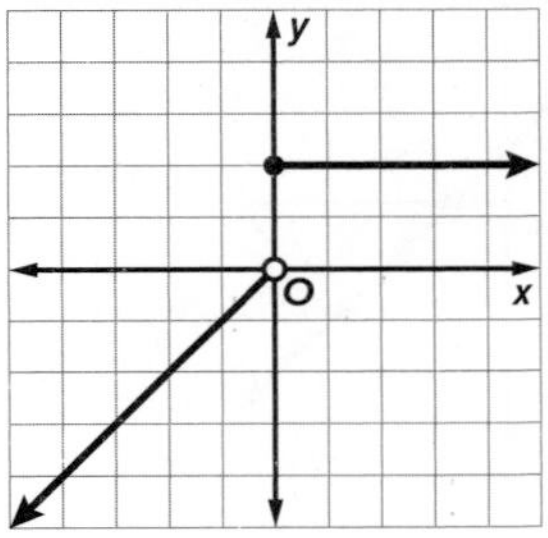

5a. $f(x) = \begin{cases} 48x & \text{if } 0 < x \leq 3 \\ 45x & \text{if } 3 < x \leq 8 \\ 42x & \text{if } 8 < x \leq 19 \\ 38x & \text{if } x > 19 \end{cases}$

5b. \$225; \$798

7. D = {all real numbers}; R = {all integers}

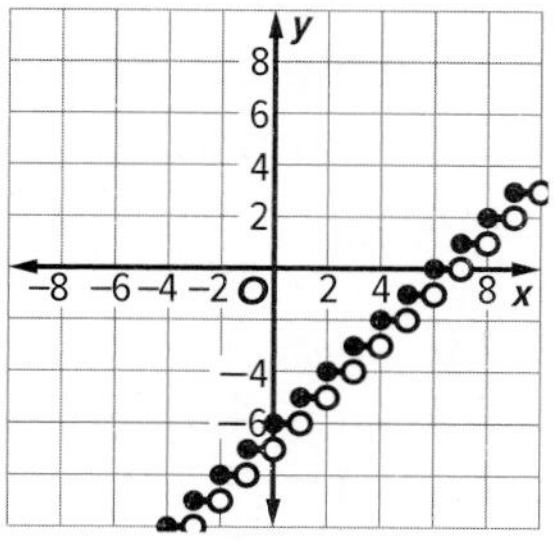

9. D = {all real numbers}; R = {all integers}

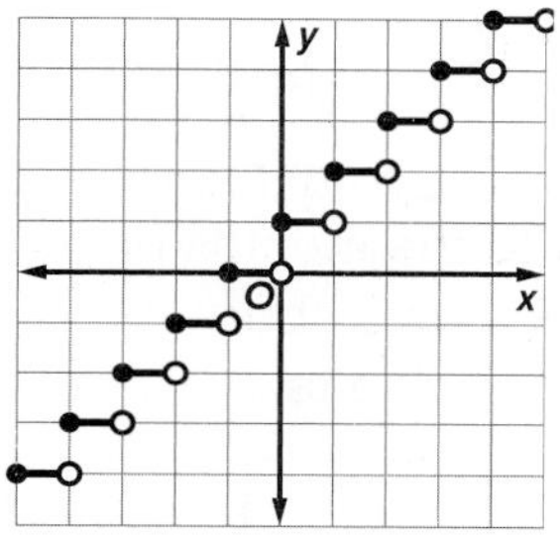

11. D = $\{x \mid 0 < x \leq 8\}$;
R = $\{y \mid y = 5.00, 10.00, 15.00, 20.00\}$

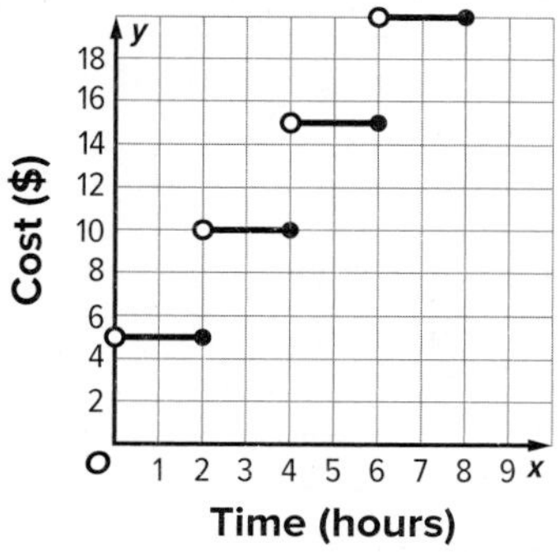

13. D = {all real numbers}; R = $\{f(x) \mid f(x) \geq 0\}$

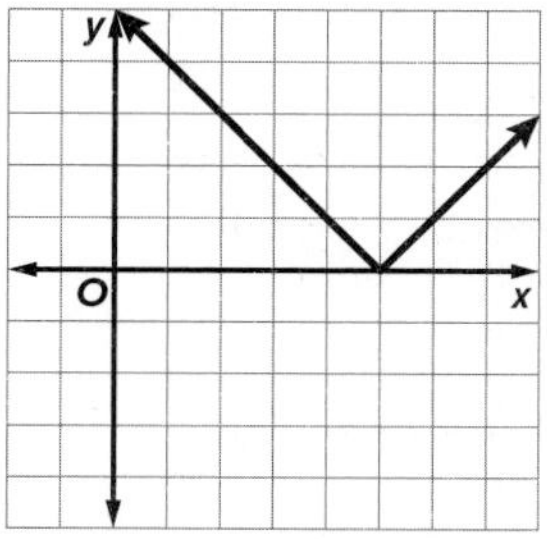

15. D = {all real numbers}; R = $\{h(x) \mid h(x) \geq -8\}$

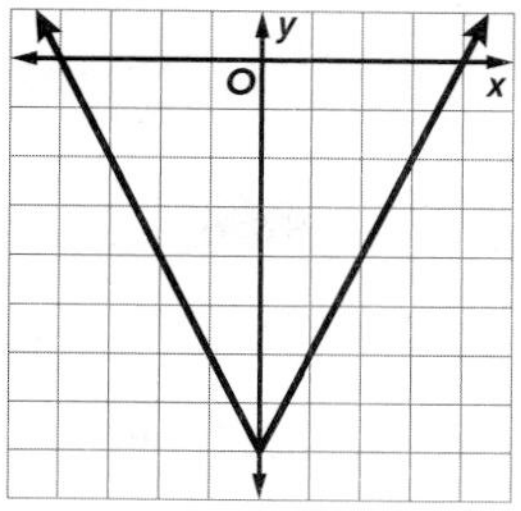

17. D = {all real numbers}; R = $\{f(x) \mid f(x) \geq 6\}$

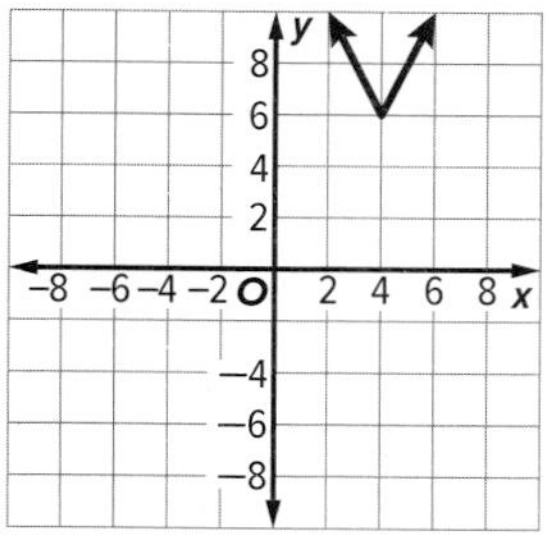

19. D = {all real numbers}; R = $\{g(x) \mid g(x) \geq 0\}$

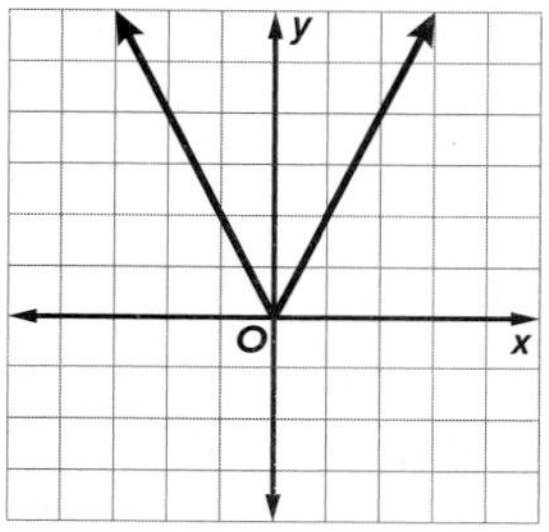

21. D = $\{x \mid x \leq -4 \text{ or } 0 < x\}$;
R = $\{f(x) \mid f(x) \geq 12, f(x) = 8, \text{ or } 0 < f(x) \leq 3\}$

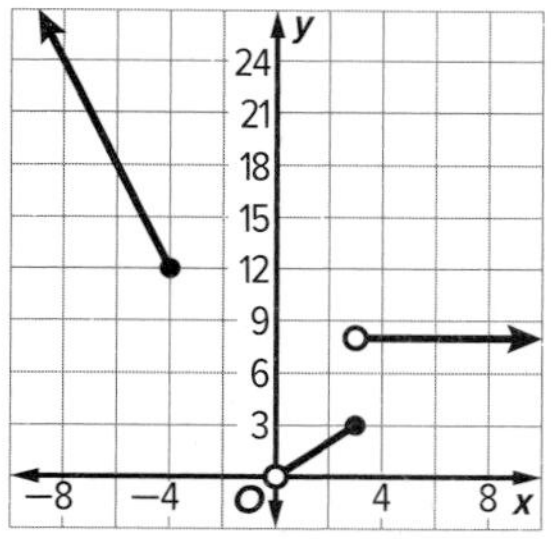

23. D = {all real numbers};
R = $\{g(x) \mid g(x) < -10 \text{ or } -6 \leq g(x) \leq 2\}$

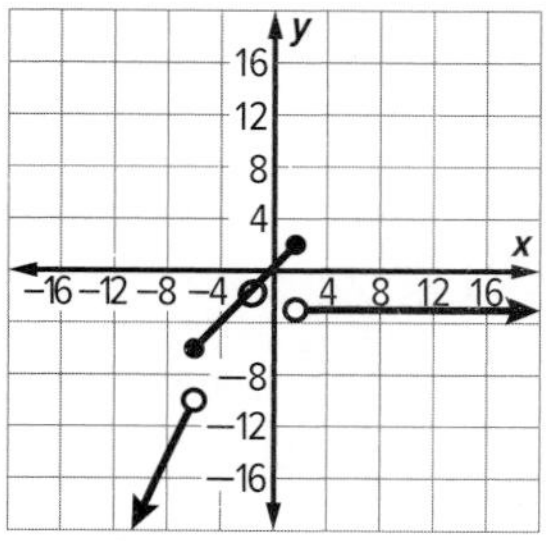

25. D = {all real numbers}; R = $\{f(x) \mid f(x) > -3\}$

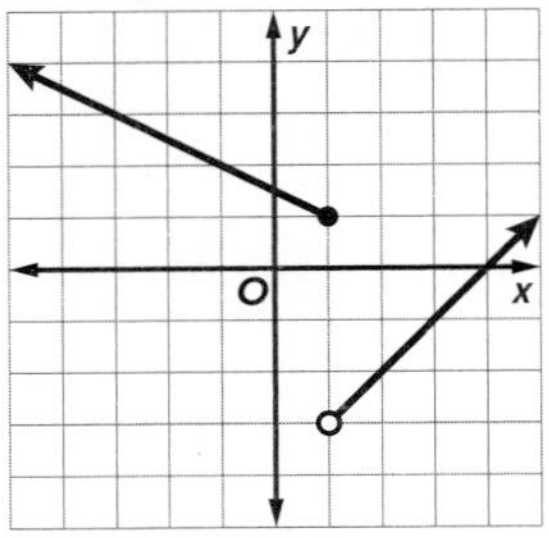

27. D = {all real numbers}; R = {all integers}

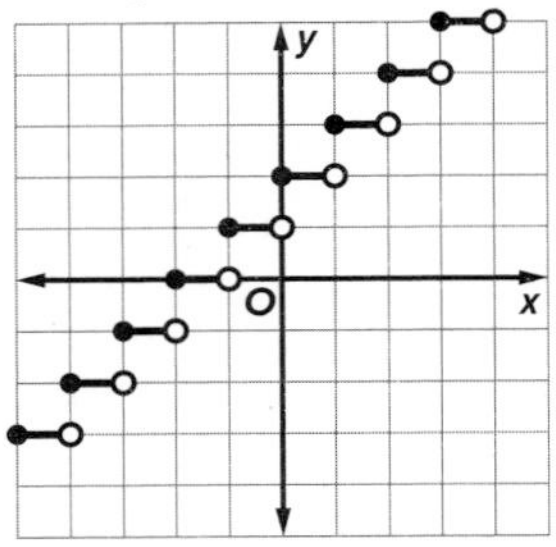

29. D = {all real numbers};
R = {all whole numbers}

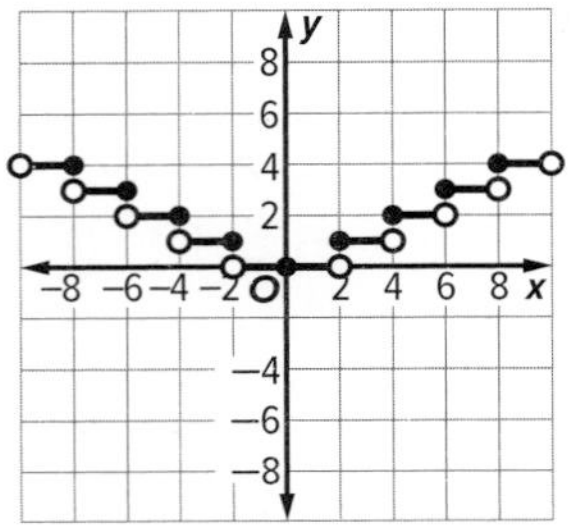

31. D = {all real numbers}; R = $\{g(x) \mid g(x) \leq 4\}$

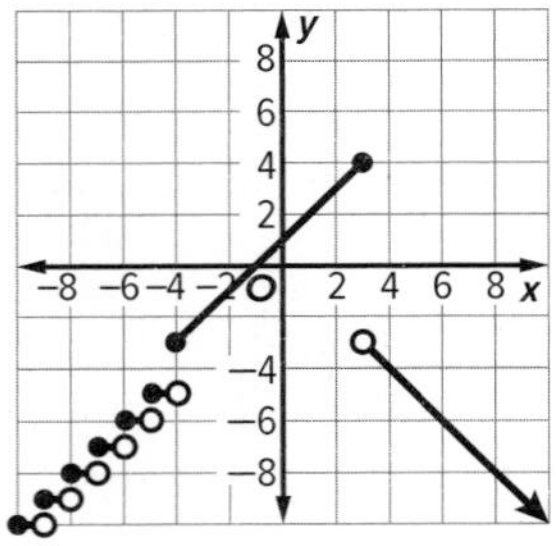

33. D = {all real numbers}; R = $\{h(x) \mid h(x) \geq 2\}$

35. $C(x) = \begin{cases} 5 & \text{if } 0 \leq x \leq 1 \\ 7.5 & \text{if } 1 < x \leq 2 \\ 10 & \text{if } 2 < x \leq 3 \\ 12.5 & \text{if } 3 < x \leq 4 \\ 15 & \text{if } 4 < x \leq 24 \end{cases}$

37a. $C(x) = \begin{cases} 500 + 17.50x \text{ if } 0 \leq x \leq 40 \\ 1200 + 14.75(x - 40) \text{ if } x \geq 41 \end{cases}$

37b.

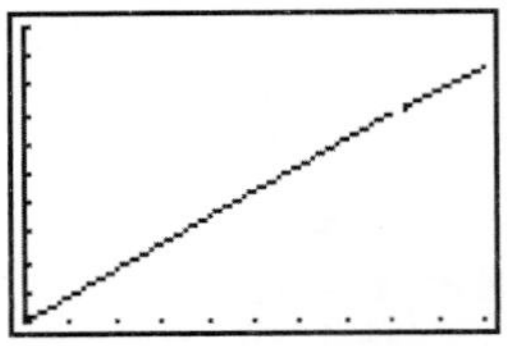

[0, 50] scl: 5 by [500, 1500] scl: 100

37c. Because it costs \$1200 for 40 guests to attend, use the first expression in the function $C(x)$. Solve the equation $500 + 17.50x = 900$ to obtain about 22.9. Because there cannot be a fraction of a guest, at most 22 guests can be invited to the reunion.

39a. $R(t) = \begin{cases} \frac{20}{3}t + 30 & \text{if } 0 \leq t \leq 3 \\ 60 & \text{if } 3 < t < 4 \\ 80 & \text{if } 4 \leq t \leq 5 \\ 60 & \text{if } 5 < t < 6 \\ -\frac{50}{3} + 160 & \text{if } 6 \leq t \leq 9 \end{cases}$

Range = [10, 60] ∪ {80}

39b. The graph is increasing from $t = 0$ to $t = 3$. This corresponds to the months of September, October, and November.

41. Because the absolute value takes negative f(x)-values and makes them positive, the graph retains the step-like nature of the greatest integer function, but it also has the "v" shape of the absolute value.

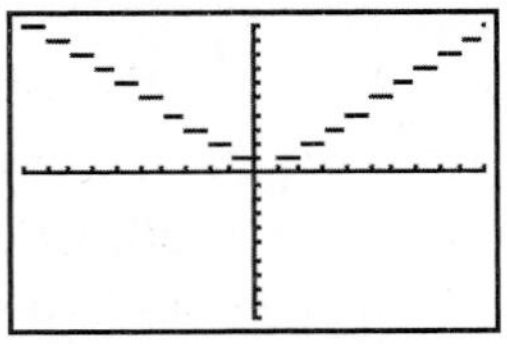

[−10, 10] scl: 1 by [−10, 10] scl: 1

43. $f(x) = \begin{cases} -x + 2 & \text{if } x \leq 0 \\ -x - 2 & \text{if } x > 0 \end{cases}$

45. $f(x) = \begin{cases} \frac{1}{2}x + 1 & \text{if } x < 2 \\ x - 4 & \text{if } x > 2 \end{cases}$

47. D = $\{m \mid 0 < m \leq 6\}$; R = {C | C = 2.00, 4.00, 6.00, 8.00, 10.00, 12.00}

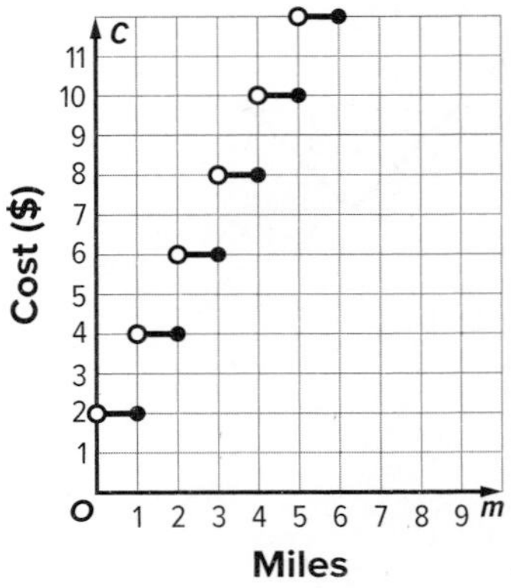

49. Sample answer: $|y| = x$ **51.** Sample answer: 8.6; The greatest integer function asks for the greatest integer less than or equal to the given value; thus 8 is the greatest integer. If we were to round this value to the nearest integer, we would round up to 9. **53.** Sample answer: Piecewise functions can be used to represent the cost of items when purchased in quantities, such as a dozen eggs.

Lesson 1-7

1. translation of the graph of $y = x^2$ up 4 units
3. translation of the graph of $y = x$ down 1 unit
5. translation of the graph of $y = x^2$ right 5 units
7. $y = x^2 - 2$ **9.** $y = x + 1$ **11.** $y = |x + 3| + 1$
13. compressed horizontally and reflected in the y-axis **15.** stretched vertically and reflected in the x-axis **17.** compressed vertically and reflected in the x-axis
19. stretched horizontally and reflected in the y-axis **21.** reflected in the x-axis, stretched vertically, and translated down 4 units
23. compressed vertically and translated down 2 units **25.** reflected in the x-axis, compressed vertically, and translated left 3 units **27.** stretched horizontally by a factor of 2; The absolute value function shows the ball bouncing off the edge of the pool table and the stretch shows the wide angle.
29. compressed vertically by a factor of 0.75 and translated up 25; There is a \$25 fixed cost, plus \$0.75 per mile, regardless of direction.

31. $y = 2|x + 2| + 5$ **33.** $y = -|x| - 3$
35. $y = -(x - 4)^2$

37. translation of $y = |x|$ down 2 units

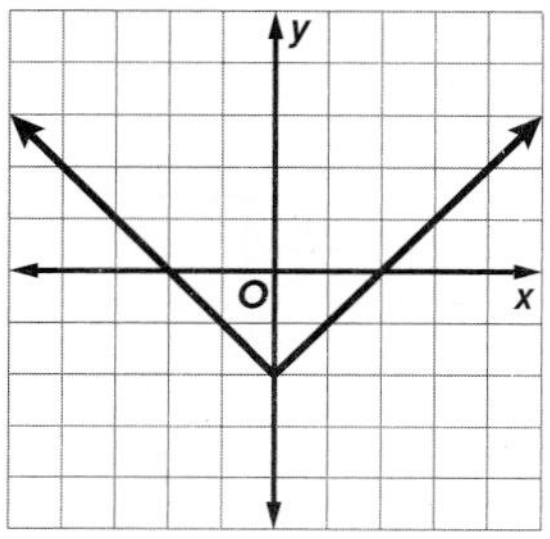

39. reflection of $y = x$ in the x-axis

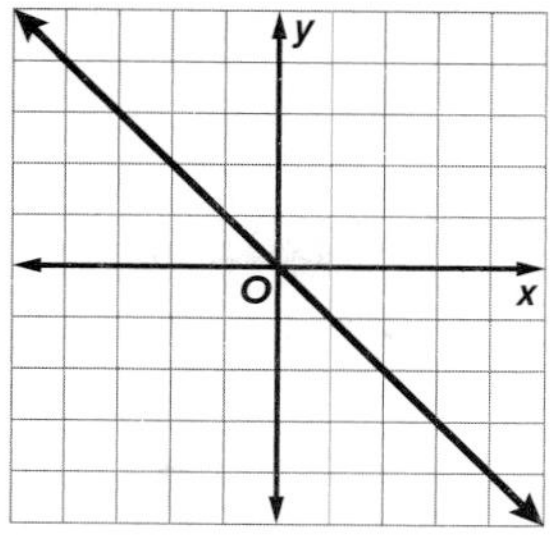

41. vertical stretch of $y = x$

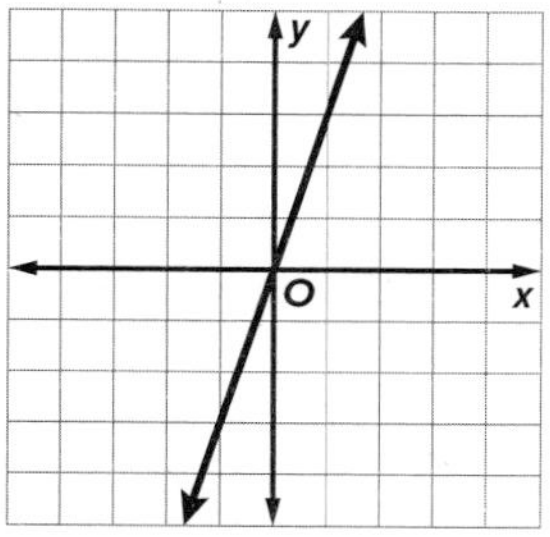

43. translation of $y = x^2$ down 4 units
45. horizontal compression of the graph of $y = x^2$

47a. translation of $y = |x|$ right 8 units

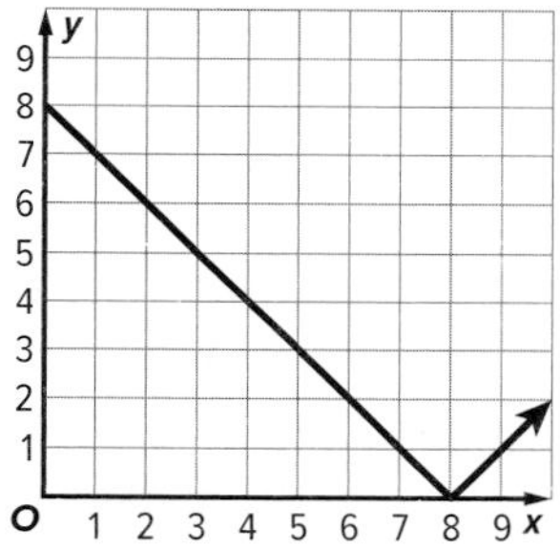

47b. Sample answer: The speedometer is stuck at 8 mph.

49. stretched vertically by a factor of 4
51. Maria stretched the function vertically by a factor of 10. **53a.** quadratic **53b.** x-axis
53c. right 25 units and up 81 units
53d. $y = -(x - 25)^2 + 81$

55a.

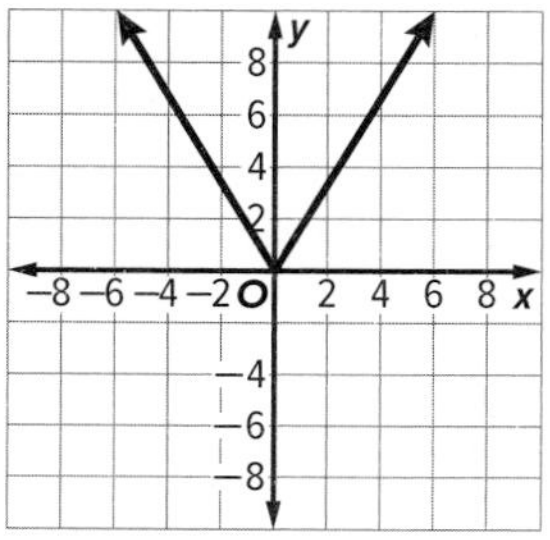

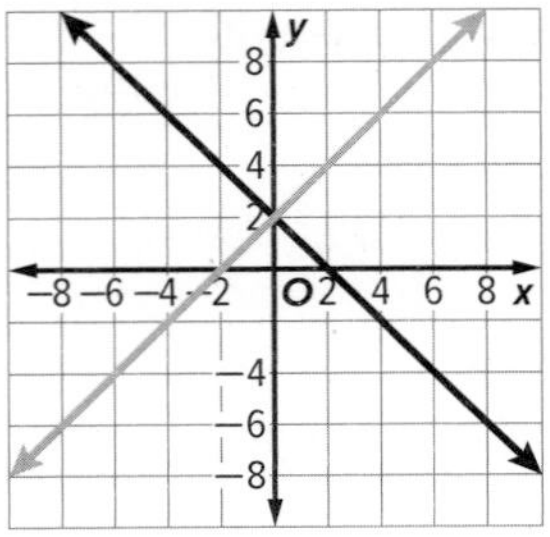

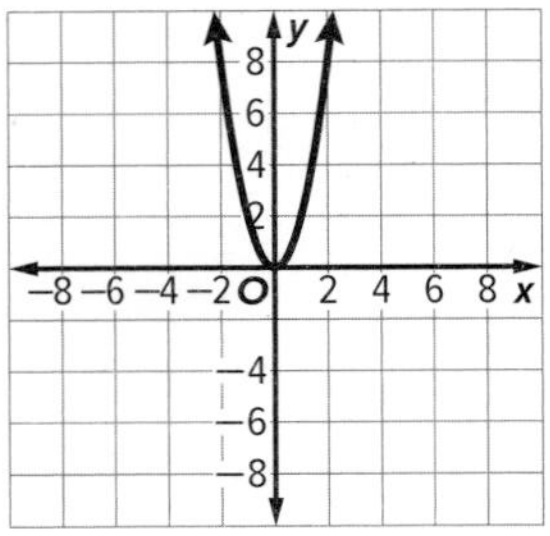

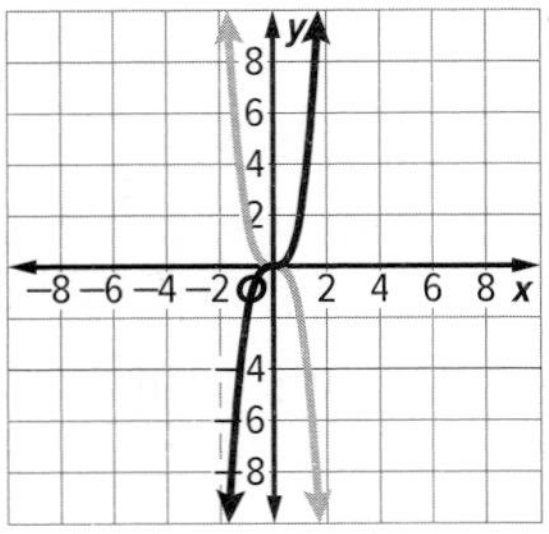

55b. $f(x)$ and $h(x)$ are even, $g(x)$ is neither, and $k(x)$ is odd. **55c.** Even functions are symmetric in the y-axis. If $f(-x) = f(x)$, then the graphs of $f(-x)$ and $f(x)$ coincide. If the graph of a function coincides with its own reflection in the y-axis, then the graph is symmetric in the y-axis. Odd functions are symmetric in the origin, which means that the graph of an odd function coincides with its rotation of 180° about the origin. A rotation of 180° is equivalent to reflection in two perpendicular lines. $f(-x)$ is a reflection in the y-axis and $-f(x)$ is a reflection in the x-axis. Thus if the graphs of $f(-x)$ and $-f(x)$ coincide, $f(x)$ is symmetric about the origin.

57. Sample answer: The graph in Quadrant II has been reflected in the x-axis and moved right 10 units.

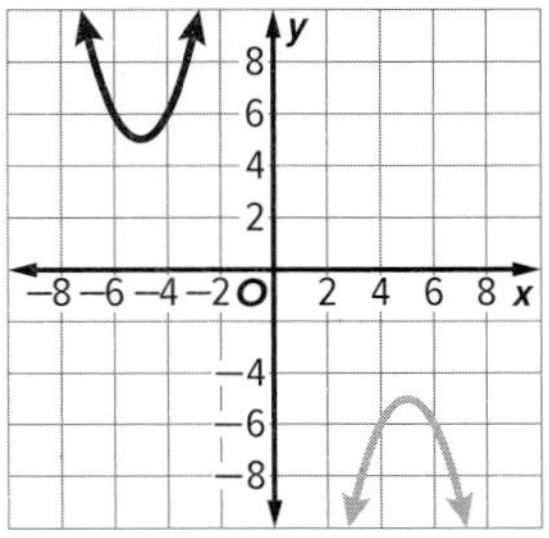

59. Sample answer: Because the graph of $g(x)$ is symmetric about the y-axis, reflecting in the y-axis results in a graph that appears the same. It is not true for all quadratic functions. When the axis of symmetry of the parabola is not along the y-axis, the graph and the graph reflected in the y-axis will be different.

Module 1 Review

1. C **3.** Sample answer: The x-intercept is (6, 0). This means that after 6 weeks, Tia owes her friend $0. The y-intercept is (0, 80). This means that Tia initially owed her friend $80, or that Tia borrowed $80 from her friend to go to a theme park.

5.

x	Relative Maximum	Relative Minimum	Neither
−5	X		
−4			X
−2		X	
0			X
1	X		
5		X	

7. The x-intercept of $f(x)$ is 2, and the x-intercept of $g(x)$ is 1. The x-intercept of $f(x)$ is greater than the x-intercept of $g(x)$. So, $f(x)$ intersects the x-axis at a point farther to the right than $g(x)$. The y-intercept of $f(x)$ is 4, and the y-intercept of $g(x)$ is −5. The y-intercept of $f(x)$ is greater than the y-intercept of $g(x)$. So, $f(x)$ intersects the y-axis at a higher point than $g(x)$. The slope of $f(x)$ is −2 and the slope of $g(x)$ is 5. $f(x)$ is decreasing and $g(x)$ is increasing. The slope of $g(x)$ is greater than the slope of $f(x)$. **9.** A

11.

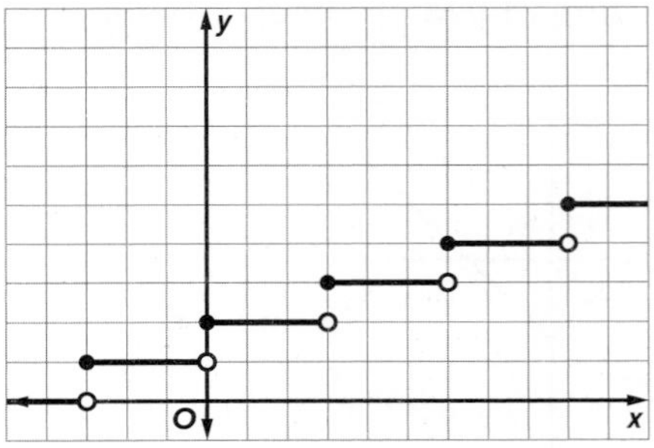

13. Sample answer: The graph of the parent function $f(x) = x^2$ has been stretched vertically, translated 2 units to the right, and translated up 9 units. **15.** C

Module 2

Quick Check

1.

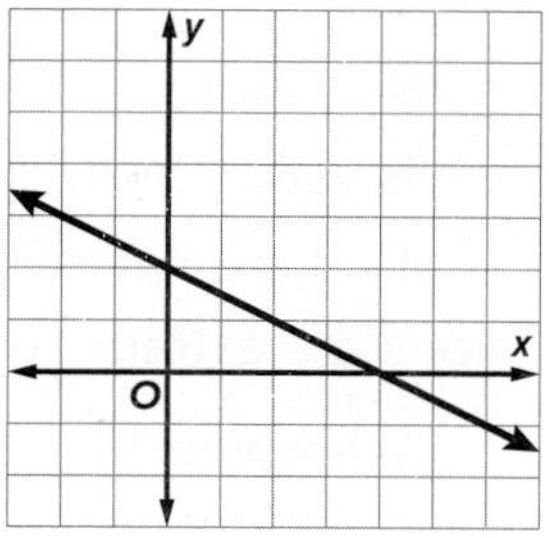

3.

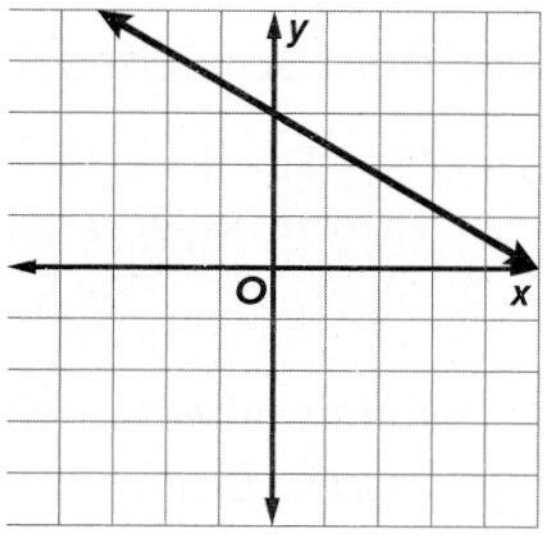

5.

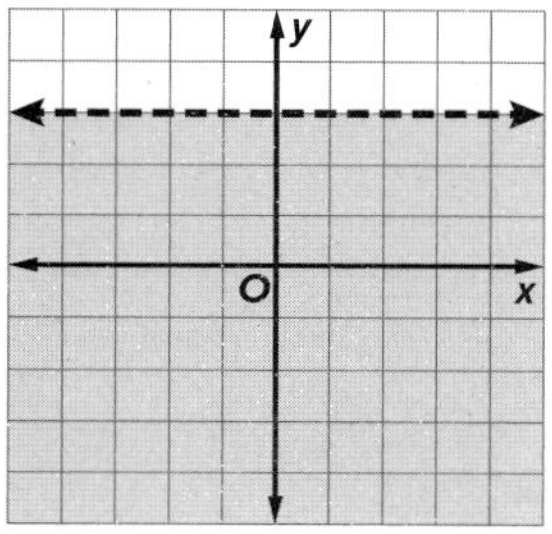

7.

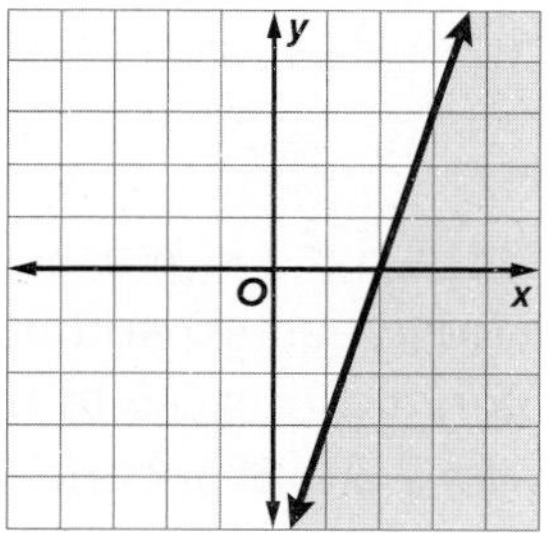

Lesson 2-1

1. $\frac{4}{5}$ **3.** $\frac{1}{4}$ **5.** 5 **7.** 0 **9.** −3

11. g = green fees per person; $6(2) + 4g = 76$; $16

13. $y = 2A - x$

15. $h = \frac{A - 2\pi r^2}{2\pi r}$

17. $f(x) = 2x + 12$ The solution is −6.

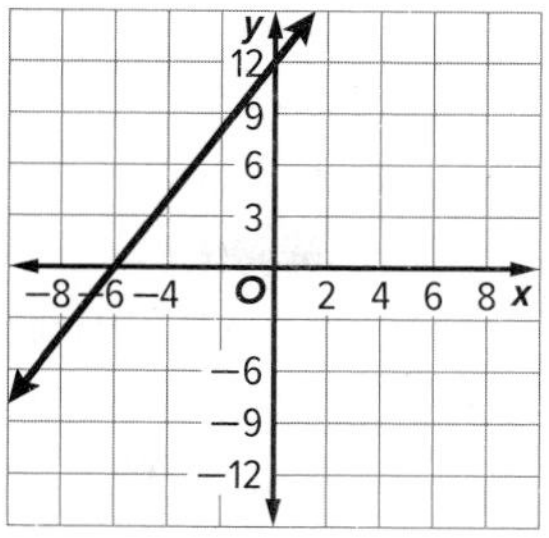

19. $f(x) = \frac{1}{2}x - 6$ The solution is 12.

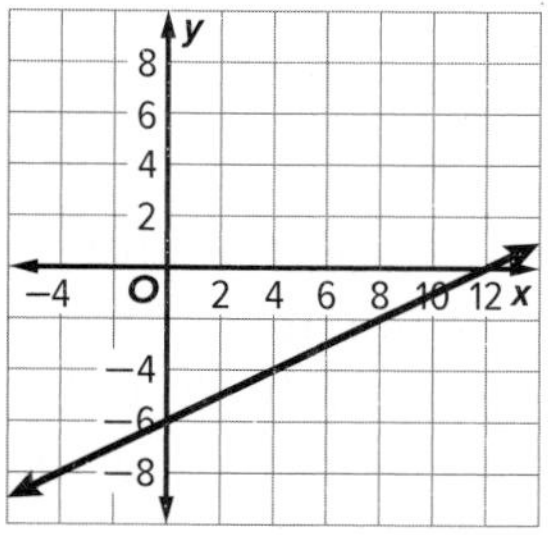

21. $f(x) = -3x - 2$ The solution is $x = -\frac{2}{3}$.

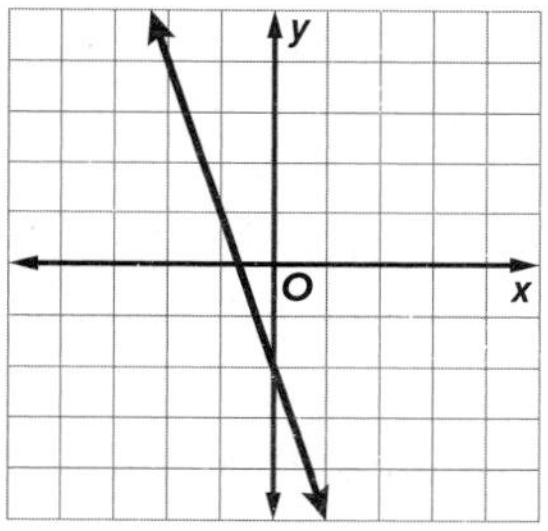

23a. Sample answer: about 5.5 weeks

23b. $w \approx 5.56$

25. $\{z \mid z \leq -8\}$

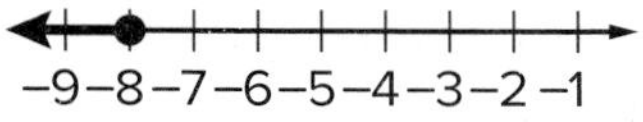

27. $\{n \mid n < 2\}$

29. $\{m \mid m \leq 4\}$

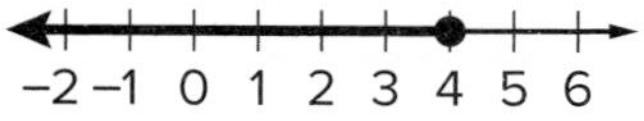

31. $15P + 300 \geq 1500$; $P \geq 80$; Manuel must translate at least 80 pages.

33. $\frac{22}{5}$ **35.** 5

37. $\{x \mid x \leq 6\}$

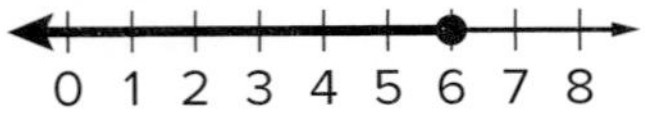

39. Sample answer: Let x represent the number of skating session. With a membership: $6x + 60 \leq 90$; $x \leq 5$. Without a membership: $10x \leq 90$; $x \leq 9$. She should not buy a membership.

41. Jade; Sample answer: in the last step, when Steven subtracted b_1 from each side, he mistakenly put b_1 in the numerator instead of subtracting it from the fraction.

43. $y_1 = y_2 - \sqrt{d^2 - (x_2 - x_1)^2}$

45. Sample answer: When one number is greater than another number, it is either more positive or less negative than that number. When these numbers are multiplied by a negative value, their roles are reversed. That is, the number that was more positive is now more negative than the other number. Thus, it is now *less than* that number and the inequality symbol needs to be reversed.

Lesson 2-2

1. $\{11\}$ **3.** $\{3, 4\}$ **5.** $\{-8\}$ **7.** $\{-2, -1\}$

9. $|x - 87.4| \leq 1.5$; 85.9° F; 88.9° F

11. $\{x \mid -2 \leq x \leq 0\}$

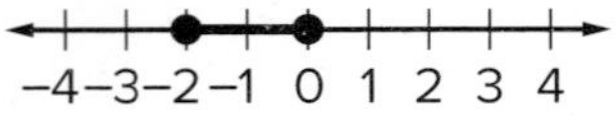

13. ∅

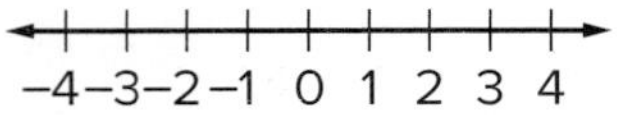

15. $\{x \mid -1.8 < x < 3.4\}$

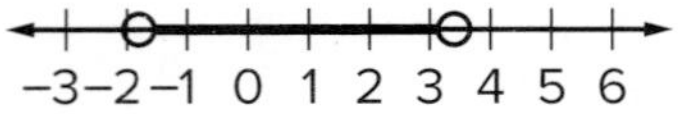

17. $\left\{x \mid x < -1 \text{ or } x > \frac{1}{3}\right\}$

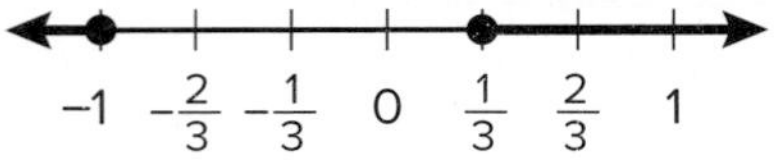

19. $|r - 24| > 6.5$; $\{r \mid r < 17.5 \text{ or } r > 30.5\}$

21. $\left\{1, \frac{1}{5}\right\}$

23. $\left\{z \mid z > -\frac{8}{3}\right\}$

25. $|4x + 7| = 2x + 3$; $x = -2$, $x = -\frac{5}{3}$; The absolute value equation is valid when $2x + 3 \geq 0$, so the equation is valid when $x \geq -\frac{3}{2}$. Since neither value of x is greater than or equal to $-\frac{3}{2}$, both solutions are extraneous.

27a. $|x - 36| \leq 0.125$; Sample answer: The inequality shows that the length of the lumber x could be as much as 0.125 inch greater than 36 inches or 0.125 inch less than 36 inches

27b. $\{x \mid 35.875 \leq x \leq 36.125\}$; The length of the lumber can range from 35.875 inches to 36.125 inches

29. Yuki; Sample answer: Yuki is correct because if $|a| = |b|$, then either $a = b$ or $a = -b$. They will get the same answers because $a = -b$ and $b = -a$ and $a = b$ and $-a = -b$ are equivalent equations.

31. $\left\{p \mid -\frac{9}{4} < p < \frac{5}{4}\right\}$

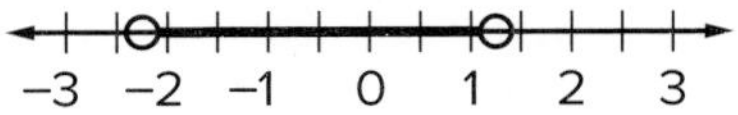

33. $\left\{w \mid w \leq -\frac{23}{2} \text{ or } w \geq \frac{7}{2}\right\}$

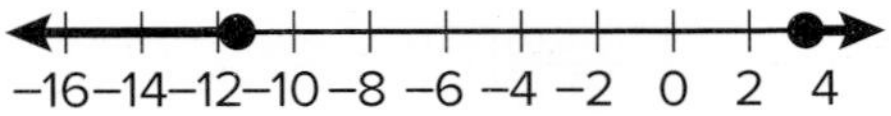

35. ∅

37. $40 < |200 - 32t| < 88$; $3.5 < t < 5$ or $7.5 < t < 9$; The speed is between 40 and 88 feet per second in the intervals from 3.5 to 5 seconds going up and from 7.5 to 9 seconds coming down.

39. Roberto is correct. Sample answer: The solution set for each inequality is all real numbers. For any value of c (positive, negative, or zero), each inequality will be true.

41. $x > 5$ and $x < 1$; Sample answer: Each of these has a non-empty solution set except for $x > 5$ and $x < 1$. There are no values of x that are simultaneously greater than 5 and less than 1.

43. The 4 potential solutions are:

1. $(2x - 1) \geq 0$ and $(5 - x) \geq 0$
2. $(2x - 1) \geq 0$ and $(5 - x) < 0$
3. $(2x - 1) < 0$ and $(5 - x) \geq 0$
4. $(2x - 1) < 0$ and $(5 - x) < 0$

The resulting equations corresponding to these cases are:

1. $2x - 1 + 3 = 5 - x : x = 1$
2. $2x - 1 + 3 = x - 5 : x = -7$
3. $1 - 2x + 3 = 5 - x : x = -1$
4. $1 - 2x + 3 = x - 5 : x = 3$

The solutions from case 1 and case 3 work. The others are extraneous. The solution set is $\{-1, 1\}$.

45. Always; if $|x| < 3$, then x is between -3 or 3. Adding 3 to the absolute value of any of the numbers in this set will produce a positive number.

47. Sample answer: $\left|x - \frac{a+b}{2}\right| \leq b - \frac{a+b}{2}$

Lesson 2-3

1. $7x + 5y = -35$; $A = 7$, $B = 5$, $C = -35$

3. $3x - 10y = -5$; $A = 3$, $B = -10$, $C = -5$

5. $5x + 32y = 160$; $A = 5$, $B = 32$, $C = 160$

7. $y = -2x + 4$; $m = -2$, $b = 4$

9. $y = -4x + 12$; $m = -4$, $b = 12$

11. $y = -\frac{2}{3}x + \frac{5}{3}$; $m = -\frac{2}{3}$, $b = \frac{5}{3}$

13. $y = 20x + 83$; There were 83 shirts collected before noon. There were 20 shirts collected each hour after noon.

15a. Let x represent the number of hours the plumber spends working at a job site, and y represent the total cost for the services.

15b. slope: 42; y-intercept: 65; $y = 42x + 65$

15c. $275

17. $y + 8 = -5(x + 3)$

19. $y + 8 = -\frac{2}{3}(x - 6)$

21. $y + 3 = -8(x - 2)$ or $y - 5 = -8(x - 1)$

23. $y + 2 = -\frac{3}{2}(x + 1)$ or $y - 1 = -\frac{3}{2}(x + 3)$

25. $y - 5.919 = 0.856(x - 1)$ or $y - 11.055 = 0.856(x - 7)$

27. $2x + y = 5$; $y = -2x + 5$; $y + 7 = -2(x - 6)$

29. $x - y = 5$; $y = x - 5$; $y + 1 = 1(x - 4)$ or $y - 3 = 1(x - 8)$

31. $y = -0.25x + 648$; Sample answer: I assumed that the water level continues to drop at a constant rate.

33. $16x - 19y = -41$

35a. Sample answer: $(8, -9)$; I used the given two points to write an equation. Then, I substituted 8 for x and solved for y.

35b. Sample answer: The equations are equivalent when simplified.

37. Never; sample answer: The graph of $x = a$ is a vertical line.

39. Sample answer: $y - 0 = 2(x - 3)$

41. No; Sample answer: You can choose points on the graph and show on a coordinate plane that they do not fall on a single line. For instance, the points (0, 2), (1, 10), (2, 24), and (3, 49) do not lie on a straight line.

43. Sample answer: Depending on what information is given and what the problem is, it might be easier to represent a linear equation in one form over another. For example, if you are given the slope and the y-intercept, you could represent the equation in slope-intercept form. If you are given a point and the slope, you could represent the equation in point-slope form. If you are trying to graph an equation using the x- and y-intercepts, you could represent the equation in standard form.

Lesson 2-4

1. 1; consistent and independent

3. 1; consistent and independent

5. infinitely many; consistent and dependent

7. (2, 1)

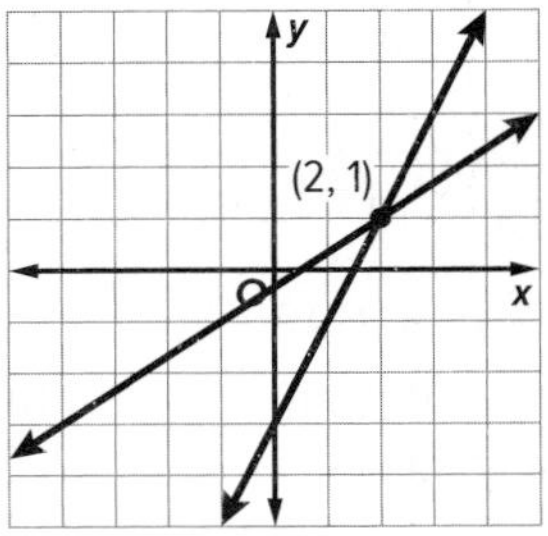

9. (3, −3)

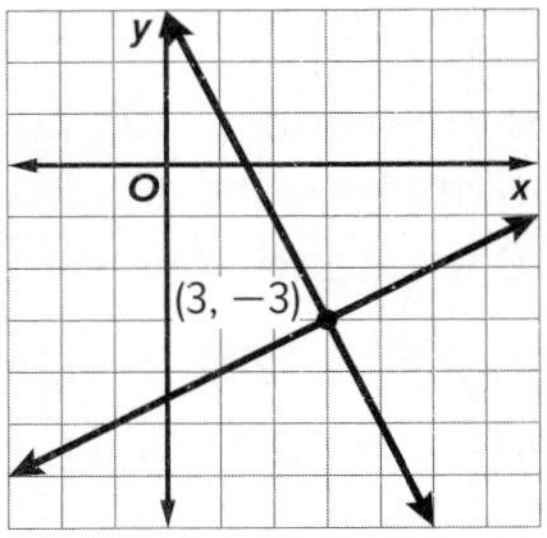

11. no solution

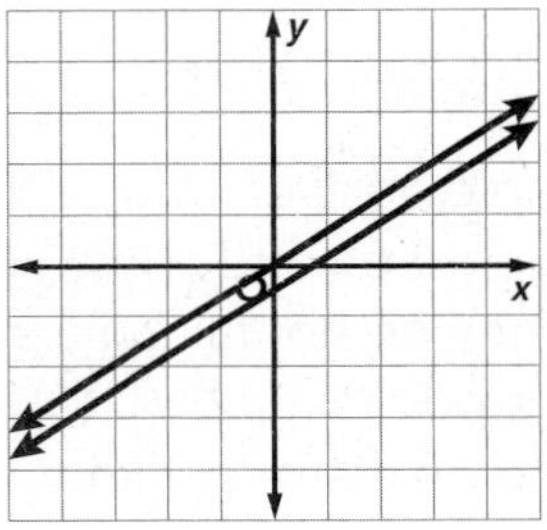

13a. Company A: $y = 24x + 42$;
Company B: $y = 28x + 25$

13b. about 4 containers

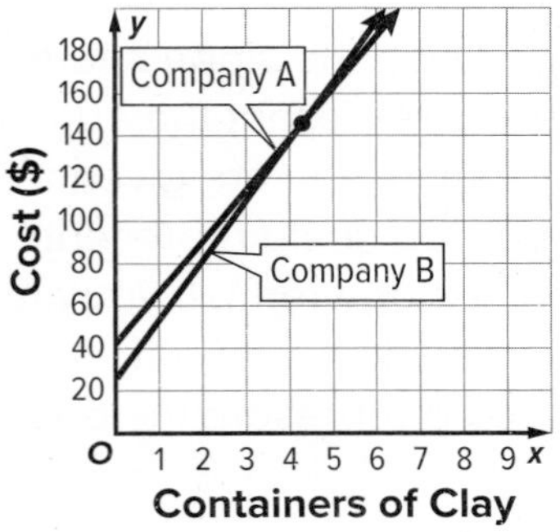

13c. Sample answer: I estimated that the cost would be the same when ordering 4 containers. By substituting $x = 4$ in the equations, the cost at Company A is \$138 and the cost at Company B is \$137. These values are approximately equal, so the estimate is reasonable.

15. (2.07, −0.39)

17. (15.03, 10.98)

19. 2.76

21. −0.99

23. (2, 1)

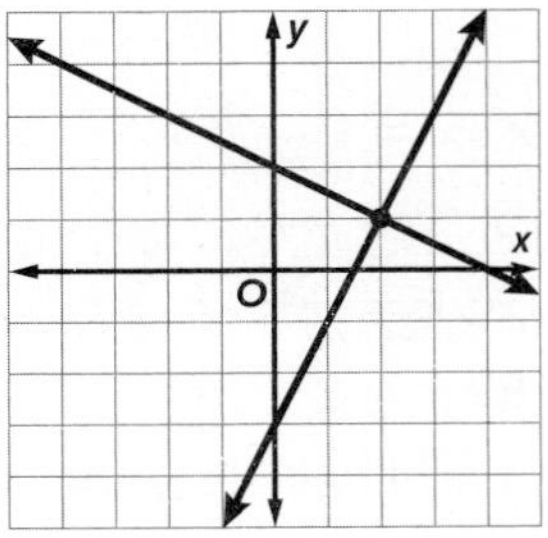

25. (3.78, 5.04)

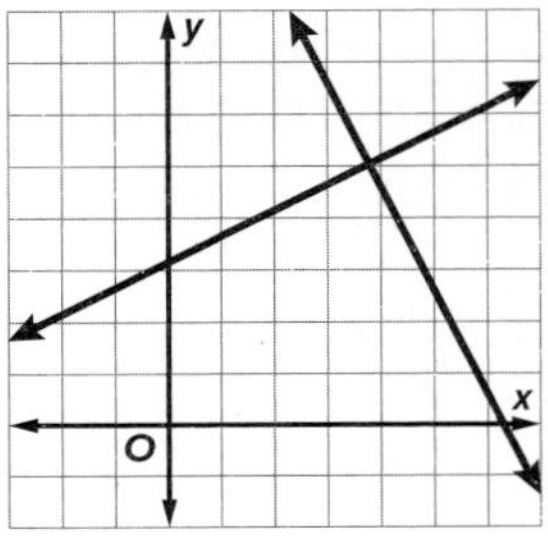

27. Always; Sample answer: *a* and *b* are the same line. *b* is parallel to *c*, so *a* is also parallel to *c*. Since *c* and *d* are consistent and independent, then *c* is not parallel to *d* and, thus, intersects *d*. Since *a* and *c* are parallel, then *a* cannot be parallel to *d*, so, *a* must intersect *d* and must be consistent and independent with *d*.

29. Never; Sample answer: Lines cannot intersect at exactly two distinct points. Lines intersect once (one solution), coincide (infinite solutions), or never intersect (no solution).

Lesson 2-5

1. (3, −3) **3.** (2, 1) **5.** no solution

7a. Cassandra: $3x + 14y = 203$;
Alberto: $11x + 11y = 220$; $x = 7$, $y = 13$

7b. The cost of each small pie is \$7. The cost of each large pie is \$13.

7c. Sample answer: By substituting the solution into each equation in the system, you can verify that it is correct. $3(7) + 14(13) = 203$, and $11(7) + 11(13) = 220$.

9. $(-2, -5)$ **11.** no solution **13.** $(3, 5)$
15. $(-2, 3)$ **17.** $(1, 6)$ **19.** $(6, -5)$ **21.** 4, 8
23. adult ticket \$5.50; student ticket \$2.75.
25. Gloria is correct.; Sample answer: Syreeta subtracted 26 from 17 instead of 17 from 26 and got $3x = -9$ instead of $3x = 9$. She proceeded to get a value of -11 for y. She would have found her error if she had substituted the solution into the original equations.
27. Sample answer: It is more helpful to use substitution when one of the variables has a coefficient of 1 or if a coefficient can easily be reduced to 1.

Lesson 2-6

1.

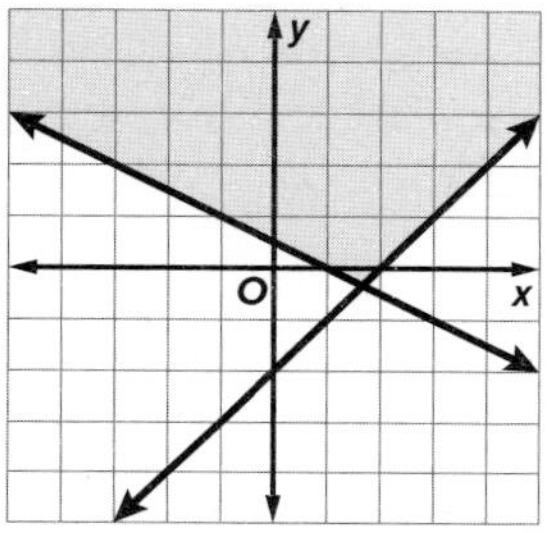

3.

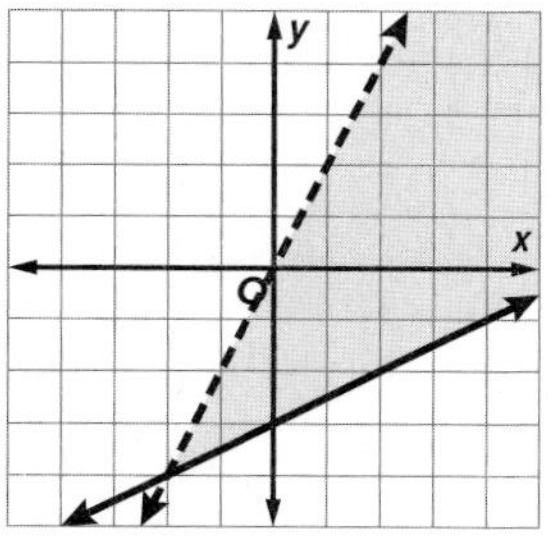

5.

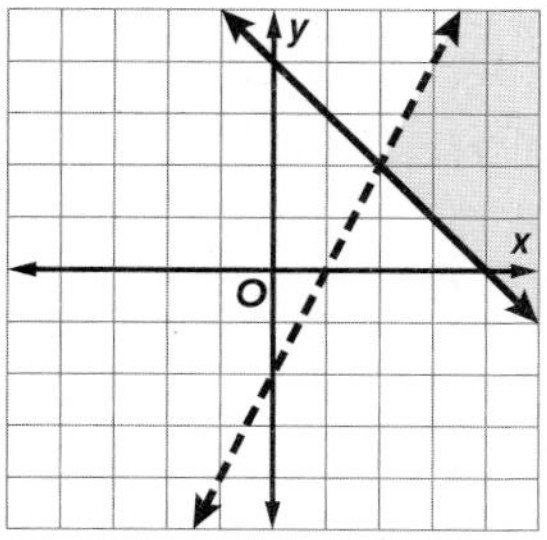

7.

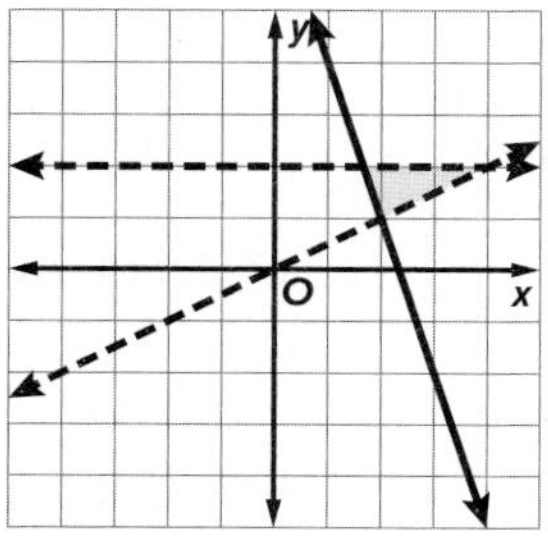

9.

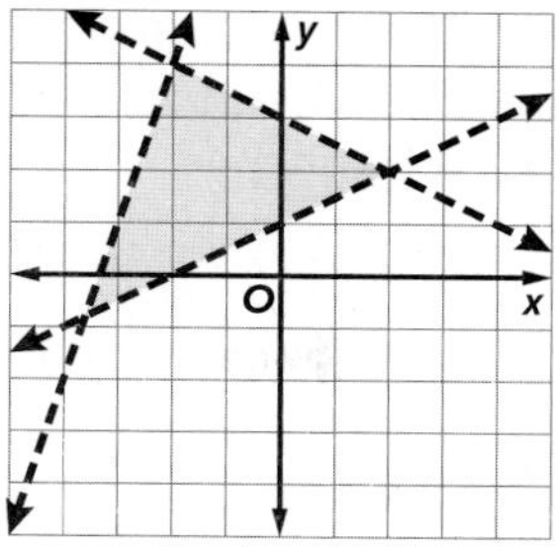

11.

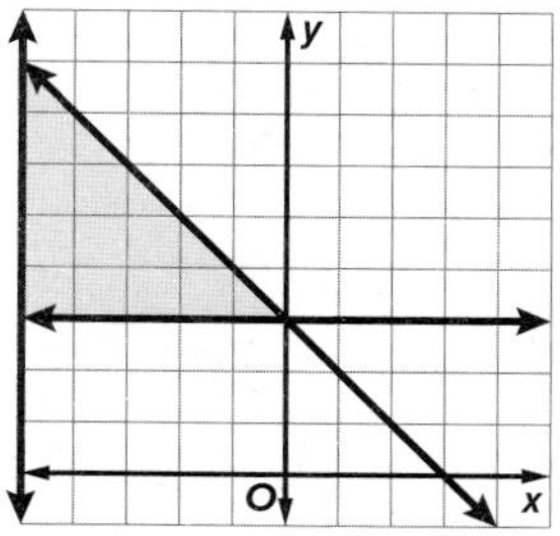

13a. Let x be student tickets and y be adult tickets. $x + y \leq 800$; $4x + 7y \geq 3400$

13b. Quadrant I

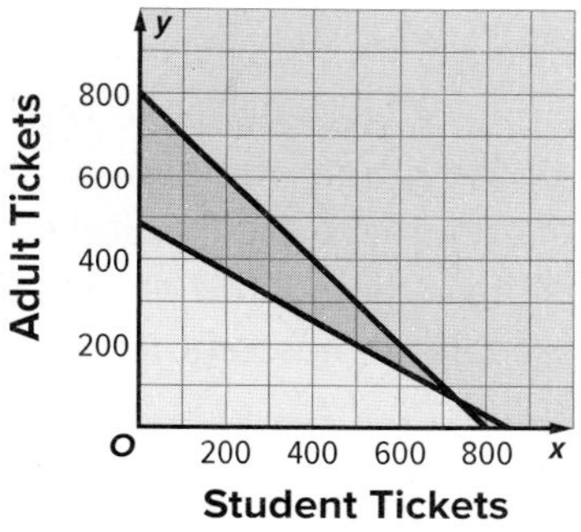

13c. No; they would only make \$3300.

15.

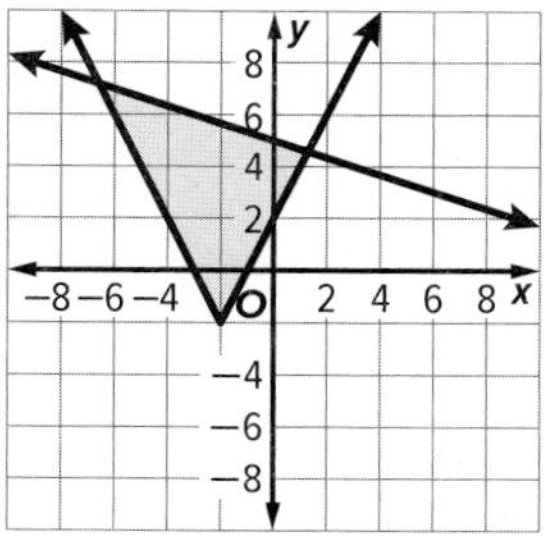

17.

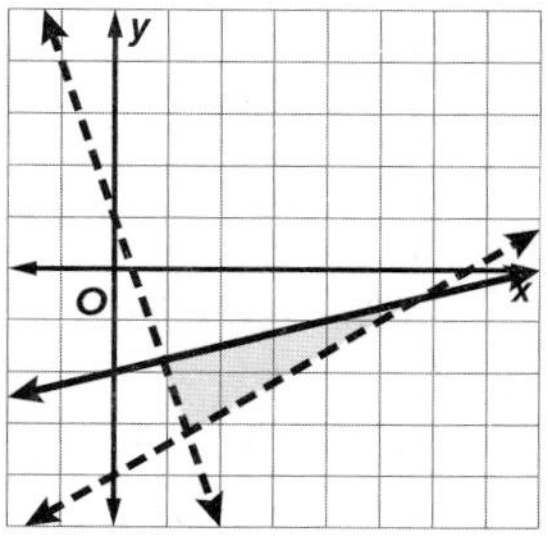

19a. Let x represent the low risk investment and y represent the high risk investment.
$x + y \leq 2000$; $0.03x + 0.12y \geq 150$

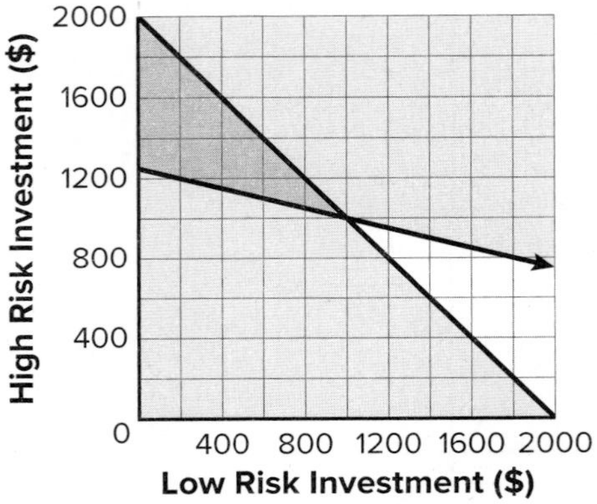

19b. Sample answer: Because Sheila cannot invest a negative amount of money, the graph is limited to positive values of x and y.

19c. Sample answer: $200 in low risk and $1700 in high risk; $400 in low risk and $1600 in high risk; $1000 in low risk and $1000 in high risk

21. 75 units^2

23. True; sample answer: The feasible region is the intersection of the graph of the inequalities. If the graphs intersect, there are infinitely many points in the feasible region. If the graphs do not intersect, they contain no common points and there is no solution.

Lesson 2-7

1. vertices: (1, 2), (1, 4), (5, 8), (5, 2);
max: 11; min: -5

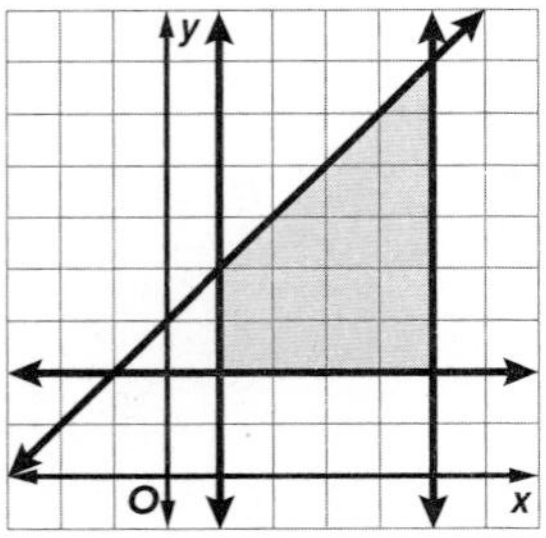

3. vertices: (0, 2), (4, 3), $\left(\frac{7}{3}, -\frac{1}{3}\right)$;
max: 25; min: 6

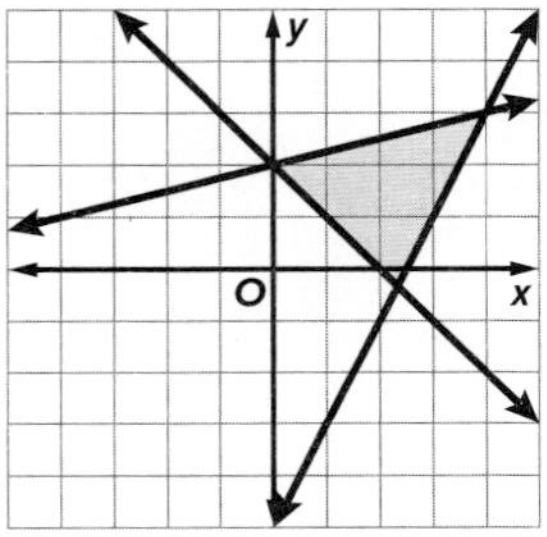

5. vertices: (1, -1), (1, 6), (8, 6);
max: 2; min: -5

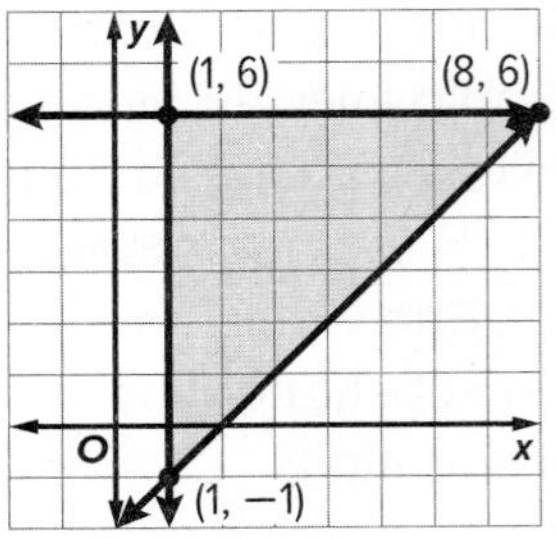

7. vertex: (-1, 7);
max: 13; no min.

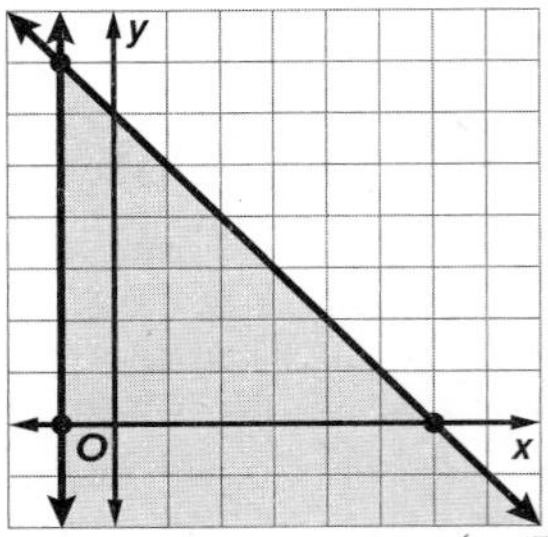

9. vertices: (-2, 0); $\left(-\frac{7}{5}, \frac{9}{5}\right)$;
max: $\frac{34}{5}$; no min.

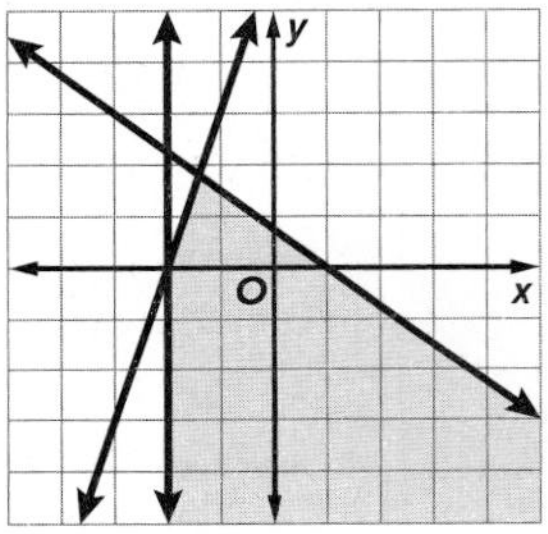

11a. Let x represent clay beads and y represent glass beads.; $0 \leq x \leq 10$; $y \geq 4$; $4y \leq 2x + 8$
$C = 0.20x + 0.40y$; The total cost equals 0.20 times the number of clay beads plus 0.40 times the number of glass beads.

11b. (4, 4), (10, 4), and (10, 7)

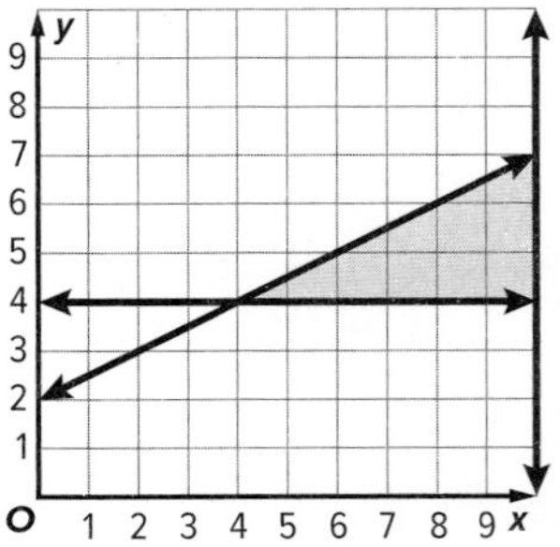

11c. Substitute into $C = 0.20x + 0.40y$: (4, 4) yields $2.40, (10, 4) yields $3.60, and (10, 7) yields $4.80. The minimum cost would be $2.40 at (4, 4), which represents 4 clay beads and 4 glass beads.

13. (5, 2); 19 feet

15. Always; Sample answer: if a point on the unbounded region forms a minimum, then a maximum cannot also be formed because of the unbounded region. There will always be a value in the solution that will produce a higher value than any projected maximum.

17. Sample answer: Even though the region is bounded, multiple maximums occur at A and B and all of the points on the boundary of the feasible region containing both A and B. This happened because that boundary of the region has the same slope as the function.

Lesson 2-8

1. $(4, -3, -1)$ **3.** infinitely many solutions

5. no solution **7.** $(5, -5, -20)$

9. $(-3, 2, 1)$ **11.** $(2, -1, 3)$

13. 60 grams of Mix A, 50 grams of Mix B, and 40 grams of Mix C

15. 3 oz of apples, 7 oz of raisins, 6 oz of peanuts

17. 9, 6, 3 **19.** fastest press: 1700 papers; slower press: 1000 papers; slowest press: 800 papers

21. orchestra ticket: \$10; mezzanine ticket: \$8; balcony ticket: \$7

23. $a = -3, b = 4, c = -6; y = -3x^2 + 4x - 6$

25. Sample answer:

$$\begin{aligned} 3x + 4y + z &= -17 \\ 2x - 5y - 3z &= -18 \\ -x + 3y + 8z &= 47 \end{aligned}$$

$$\begin{aligned} 3x + 4y + z &= -17 \\ 3(-5) + 4(-2) + 6 &= -17 \\ -15 + (-8) + 6 &= -17 \\ -17 &= -17 \checkmark \end{aligned}$$

$$\begin{aligned} 2x - 5y - 3z &= -18 \\ 2(-5) - 5(-2) - 3(6) &= -18 \\ -10 + 10 - 18 &= -18 \\ -18 &= -18 \checkmark \end{aligned}$$

$$\begin{aligned} -x + 3y + 8z &= 47 \\ -(-5) + 3(-2) + 8(6) &= 47 \\ 5 - 6 + 48 &= 47 \\ 47 &= 47 \checkmark \end{aligned}$$

Lesson 2-9

1. $\{-1, 9\}$

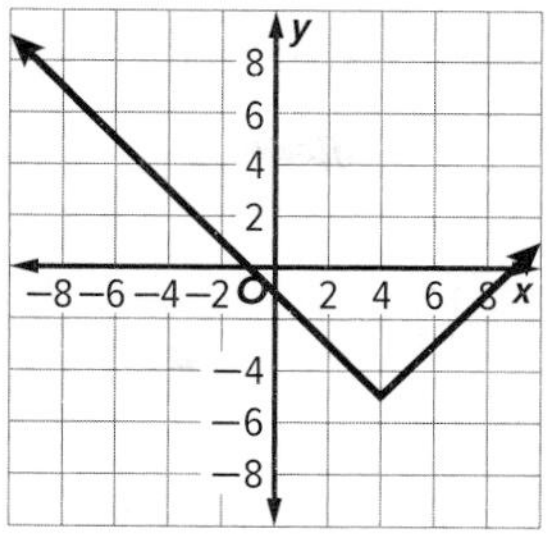

3. $\left\{-\frac{1}{2}\right\}$

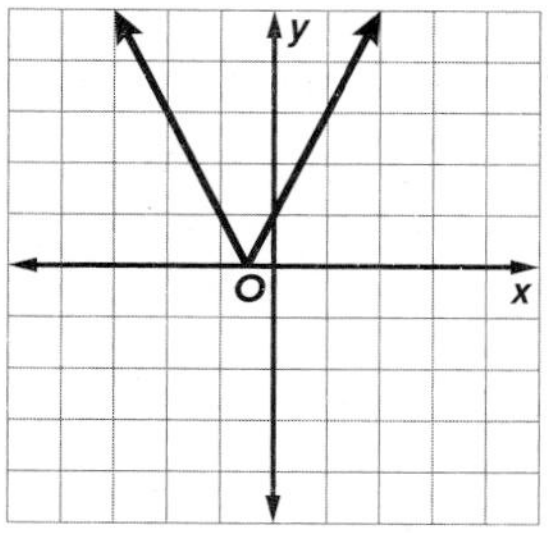

5. $\left\{\frac{1}{3}\right\}$

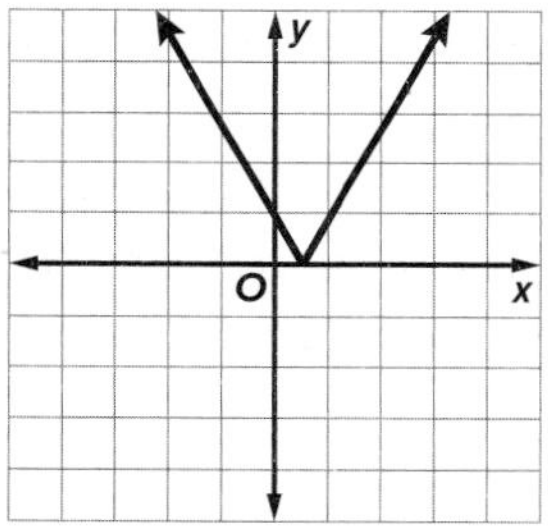

7. $\{-7, 9\}$

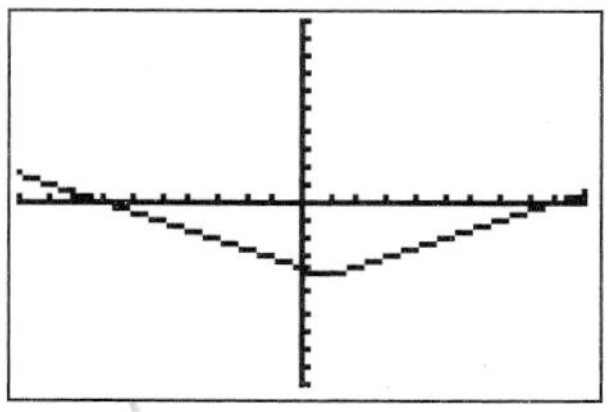

[−10, 10] scl: 1 by [−10, 10] scl: 1

9. $\{-10, 14\}$

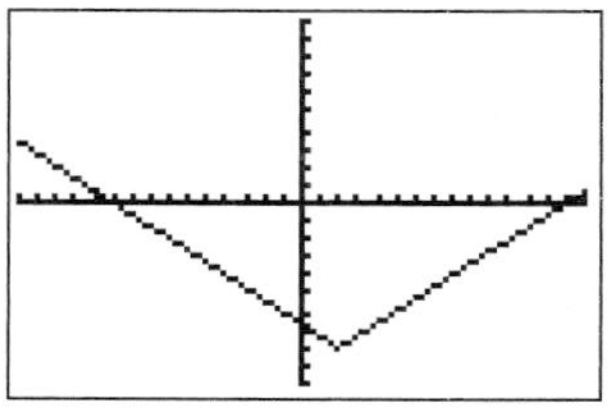

[−15, 15] scl: 1 by [−10, 10] scl: 1

11. $\{-56, 44\}$

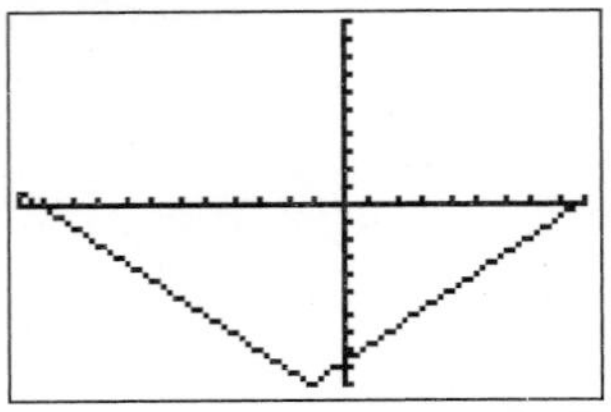

[−60, 45] scl: 5 by [−10, 10] scl: 1

13. $\{-12, 16\}$ **15.** $\varnothing$ **17.** $\{-2, -1\}$

19. $\{x \mid 1 \leq x \leq 5\}$

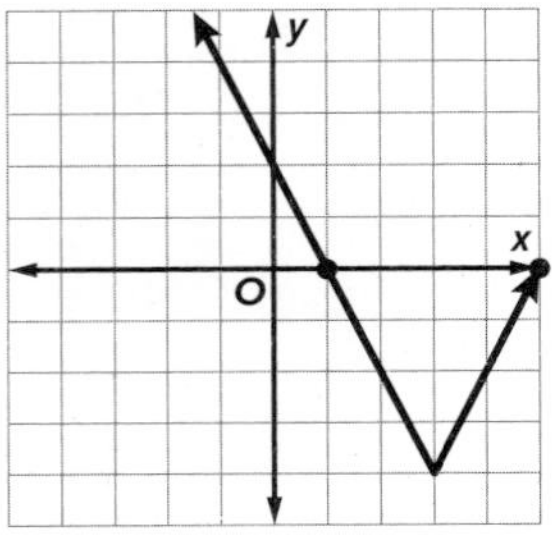

21. $\left\{ x \mid x \leq -\frac{3}{2} \text{ or } x \geq \frac{5}{2} \right\}$

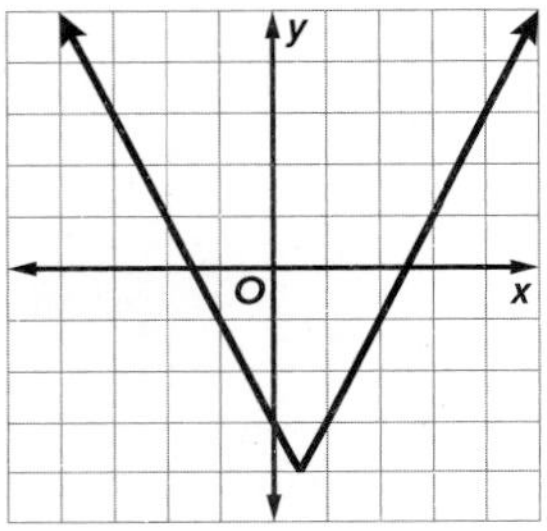

23. $\{x \mid -6 < x < 2\}$

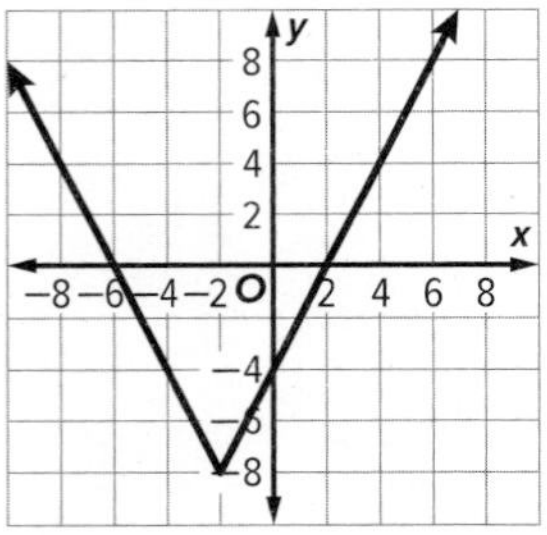

25. $\{x \mid x < -20 \text{ or } x > 12\}$

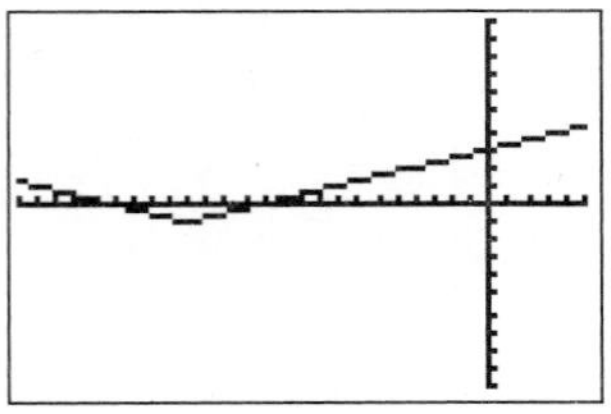

[−25, 5] scl: 2 by [−10, 10] scl: 1

27. $\left\{x \mid -\frac{1}{3} < x < 1\right\}$

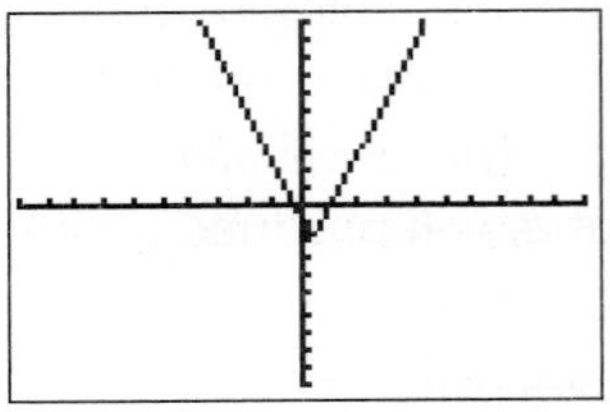

[−10, 10] scl: 1 by [−10, 10] scl: 1

29. $\varnothing$

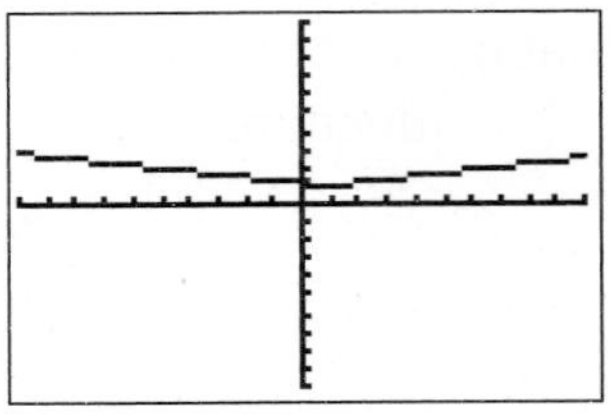

[−10, 10] scl: 1 by [−10, 10] scl: 1

31. $\{0.5, 1.5\}$

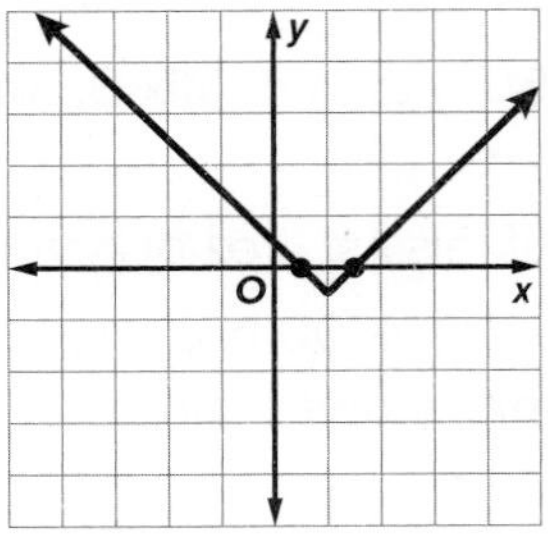

33. $\varnothing$

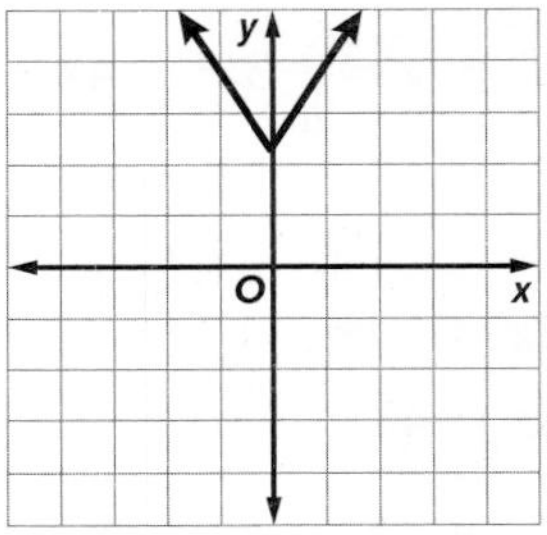

35. $\{x \mid 2 \leq x \leq 4\}$

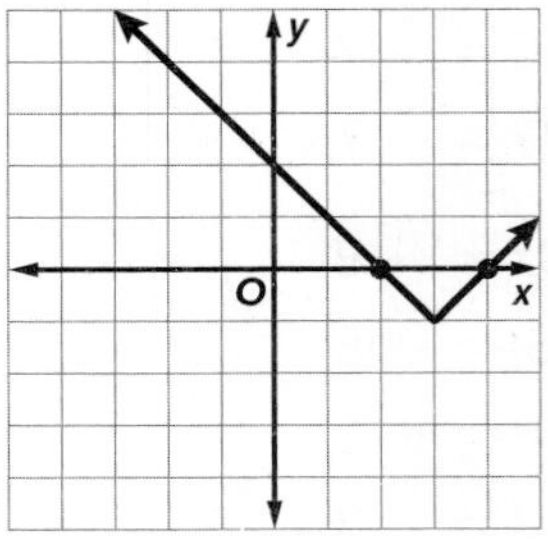

37. $\left\{x \mid x < -10\frac{4}{5} \text{ or } x > 7\frac{1}{5}\right\}$

39. $\{-85, 95\}$

41. $|x - 1.524| \leq 0.147$; $\{x \mid 1.377 \leq x \leq 1.671\}$

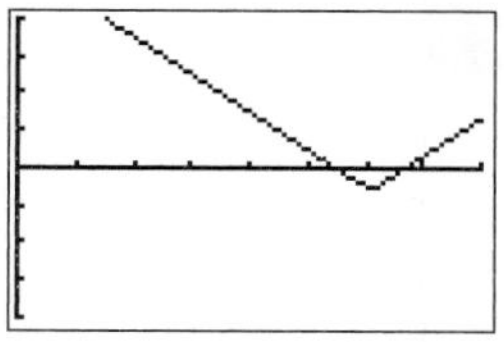

[0, 2] scl: 0.25 by [−1, 1] scl: 0.25

43. Sample answer: The process by which the equation or inequality is set to zero to represent $f(x)$ and making a table of values is the same. However, the graph of an absolute value equation is restricted to having either 0, 1, or 2 solutions; whereas the absolute value inequalities can have infinitely many solutions.

45. Sample answer: $|x - 10| = 1$

Module 2 Review

1. Sample answer: First, use the Distributive Property: $6x + 27 + 2x - 4 = 55$. Combine like terms: $8x + 23 = 55$. Then use the Subtraction Property of Equality: $8x = 32$. Finally, use the Division Property of Equality: $x = 4$.

3. $C = \frac{5}{9}(F - 32)$; 77°F

5. $y = -2x + 10$; -2 represents the miles Allie is running each day. 10 represents that Allie's goal at the beginning of the week is to jog 10 miles.

7. B **9.** (0.5, 0.75)

11. No; Sample answer: Jazmine would only earn $210.

13. A, D, G

15. $2x + 3y + z = 285$

$x + 5y + 2z = 400$

$3x + 2y + 4z = 440$

Selected Answers

Module 3

Quick Check

1. 22 **3.** − 21 **5.** $(x - 3)(x - 7)$

7. $(2x + 3)(x - 5)$

Lesson 3-1

1. D = {all real numbers}, R = $\{y \mid y \geq -1\}$

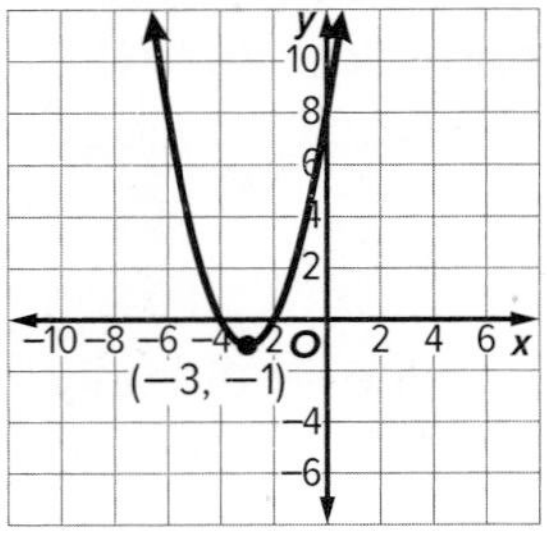

3. D = {all real numbers}, R = $\{y \mid y \geq 1\}$

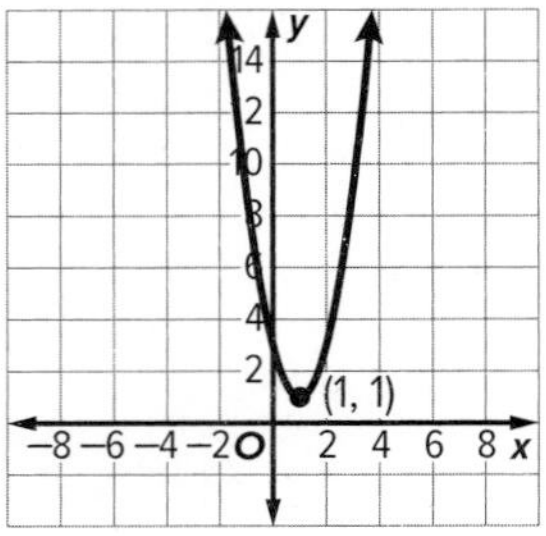

5. D = {all real numbers}, R = $\{y \mid y \geq 0\}$

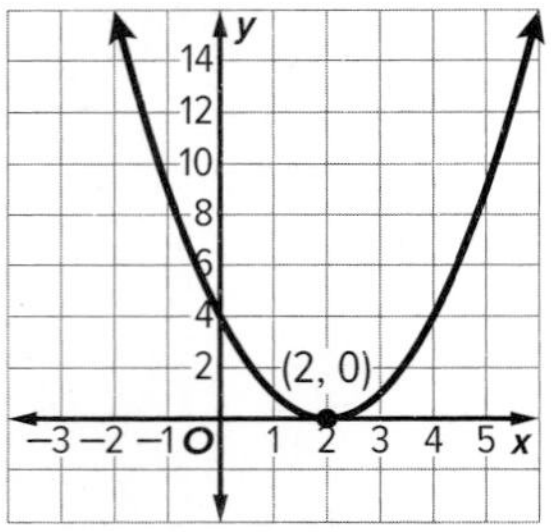

7. $g(x)$; Sample answer: Its vertex is a maximum point at (1, 3), which is 4 units above the vertex of $f(x)$ which is (1, −1).

9. $f(x)$; Sample answer: Its vertex is a minimum point at (5, −20), which is 10 units below the vertex of $g(x)$ which is (5, −10).

11a. For x = the number of price increases and y = revenue, $y = (80 + 4x)(480 - 16x)$.

11b. $100 per permit

11c. D = $\{x \mid 0 \leq x \leq 30\}$, R = $\{y \mid 0 \leq y \leq 40{,}000\}$

13. −10 **15.** −3 **17.** 0 **19.** 2 **21.** 3 **23.** 0

25a. $\approx \frac{12.5}{5}$ or 2.5; $\frac{55.35 - 42.97}{5} = 2.476$

25b. Sample answer: From 2011–2016, the number of people in the U.S. who consumed between 8 and 11 bags of potato chips annually increased by 2.476 million people per year.

27a. y-int = −9; axis of symmetry: $x = 1.5$; x-coordinate of vertex = 1.5

27b.

x	$f(x)$
0	9
1	−13
1.5	−13.5
2	−13
3	−9

27c.

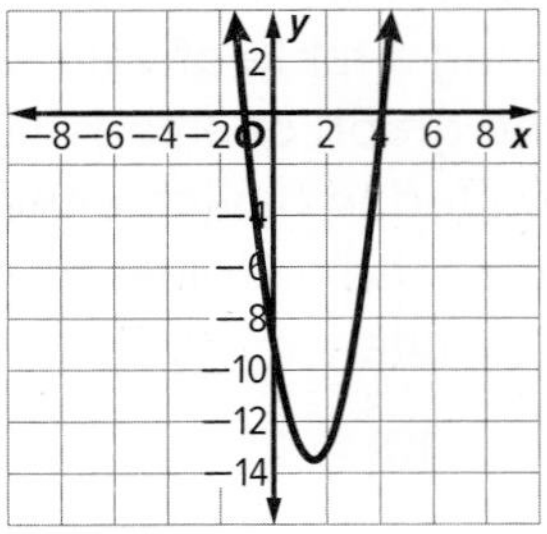

29a. y-int = 0; axis of symmetry: $x = \frac{5}{8}$; x-coordinate of vertex = $\frac{5}{8}$

29b.

x	$f(x)$
$-\frac{3}{4}$	−6
$\frac{1}{4}$	1
$\frac{5}{8}$	1.5625
1	1
2	−6

29c.

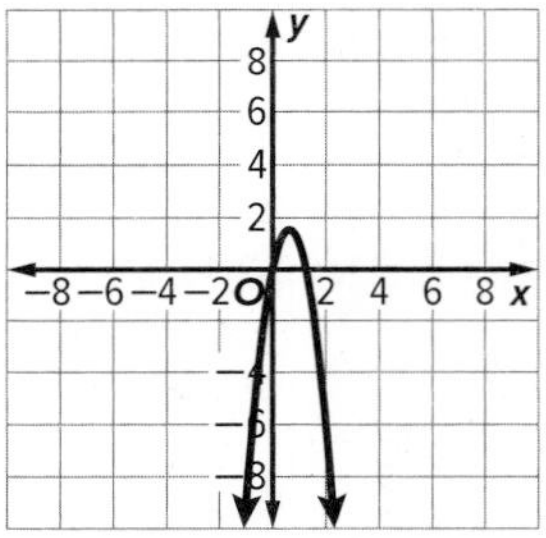

31a. y-int = 4; axis of symmetry: $x = -6$; x-coordinate of vertex = −6

31b.

x	$f(x)$
−10	−1
−8	−4
−6	−5
−4	−4
−2	−1

31c.

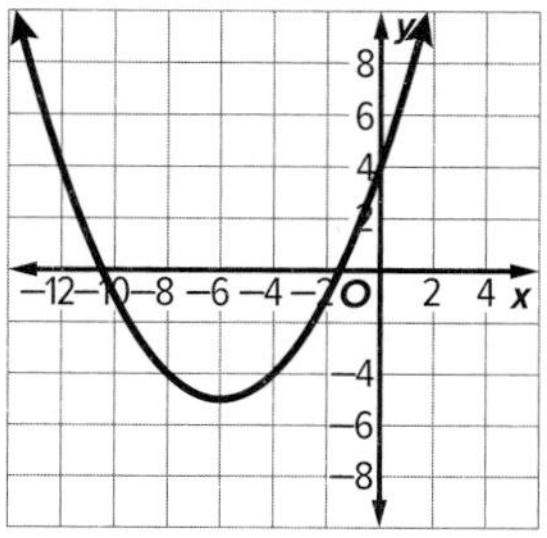

33. max; $\frac{2}{3}$; 23

35. max; 1.4; 3.8 **37.** max ≈ −4.11

39a. Let x represent the number of \$1 price increases. Let y represent the income. Then $y = (70 - x)(20 + x)$.

x	y
0	1400
5	1625
10	1800
15	1925
20	2000
25	2025
30	2000
35	1925
40	1800
45	1625
50	1400

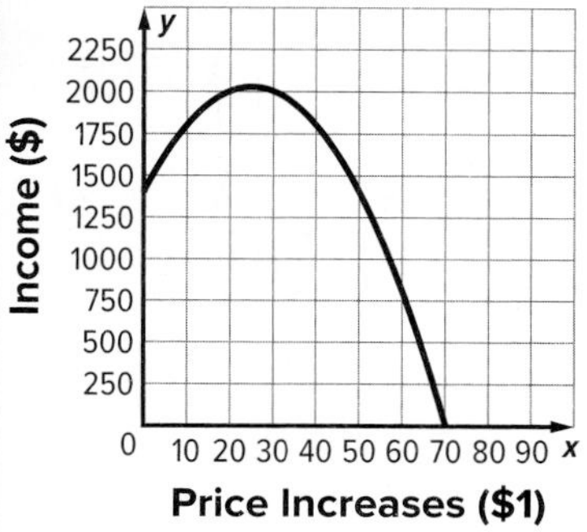

39b. (25, 2025); The club should increase the price by \$25 to make a maximum profit of \$2025. Sample answer: Increasing the price by \$25 is more than twice the current price, so this may be unreasonable. The club may want to revisit their assumptions.

41a. If x = number of price increases, the revenue is $R(x) = -5x^2 + 40x + 1200$. Because the price changes and the revenue will not be negative, the domain is $\{x \mid 0 \leq x \leq 20\}$.

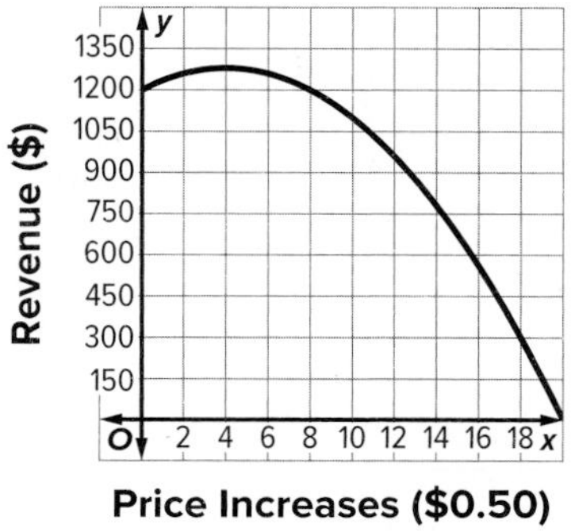

41b. The graph shows that the maximum revenue value occurs at $x = 4$, which corresponds to a ticket price of \$6.00 + 4(\$0.50) or \$8.00. The maximum revenue is $R(4)$ = \$1280.00.

41c. As price increases continue, the demand for tickets will decrease. The x-intercept indicates a price that is too high, and no one is estimated buy tickets for the cinema.

43. 20 ft

45. Madison; sample answer: $f(x)$ has a maximum of −2. $g(x)$ has a maximum of 1.

47. Sample answer: A function is quadratic if it has no other terms than a quadratic term, linear term, and constant term. The function has a maximum if the coefficient of the quadratic term is negative and has a minimum if the coefficient of the quadratic term is positive.

Lesson 3-2

1. no real solution **3.** −4

5. $-\frac{1}{2}$

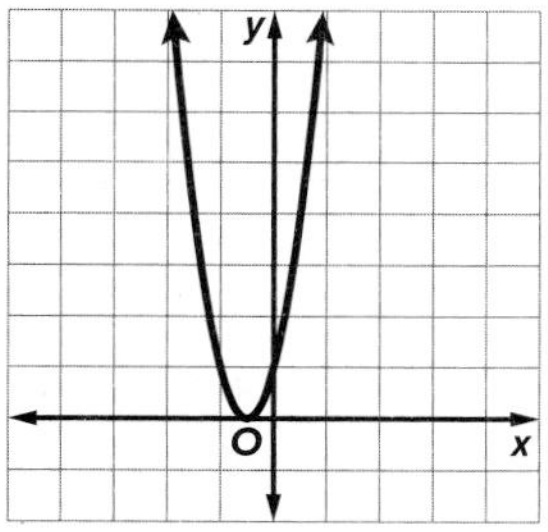

7. −3, 1

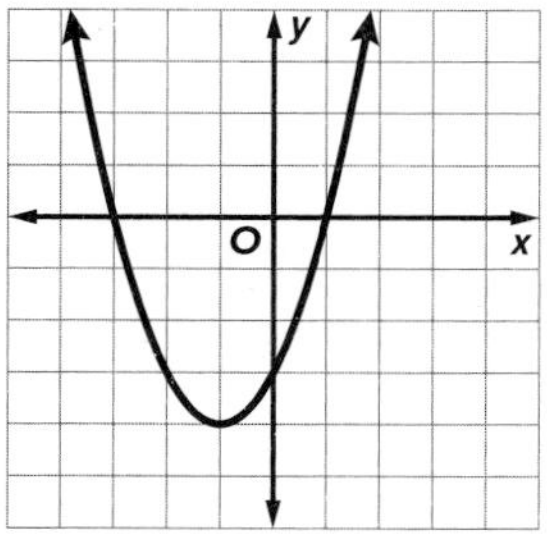

9. 1, 5

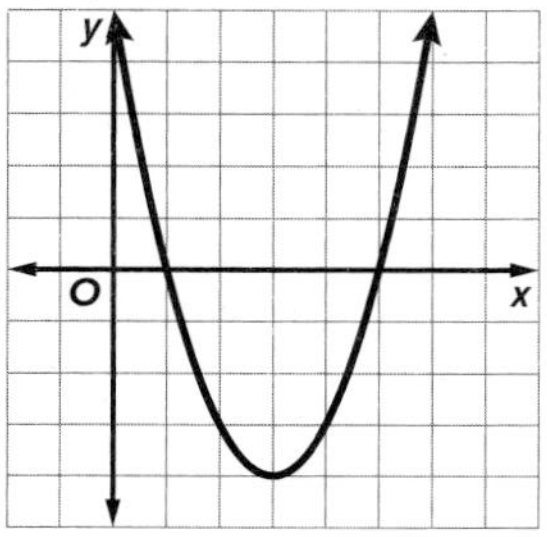

11. −2, 5

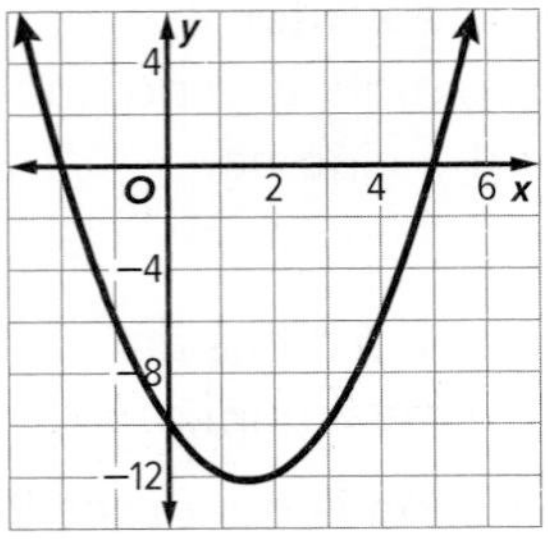

13. 6 and −4

15. between 0 and 1; between 3 and 4

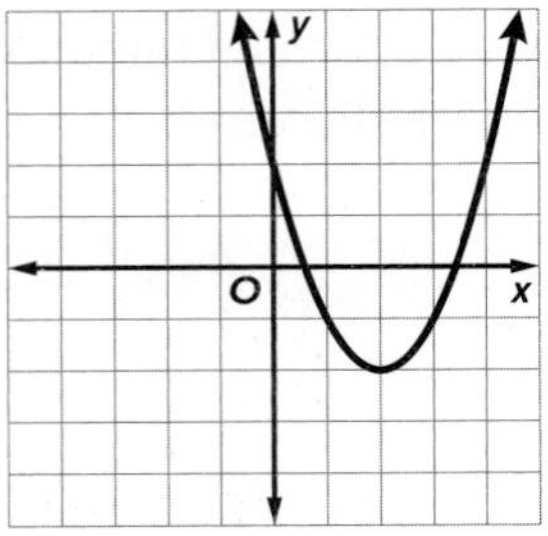

17. between −1 and 0; between −4 and −3

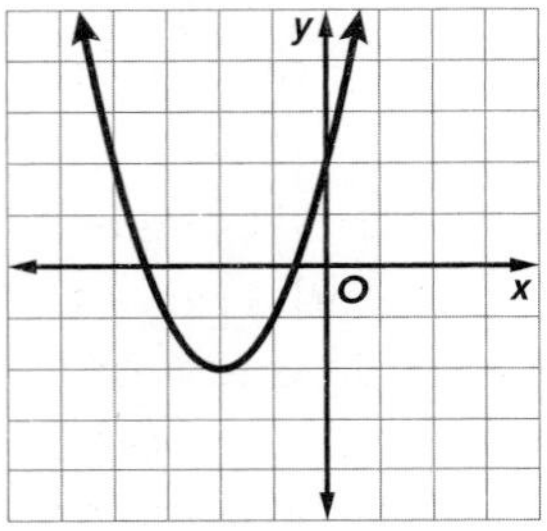

19. −6, 6

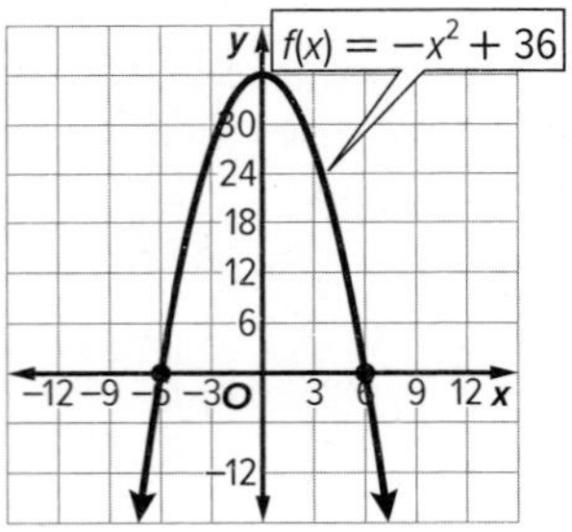

21. between −1 and 0; between −5 and −4

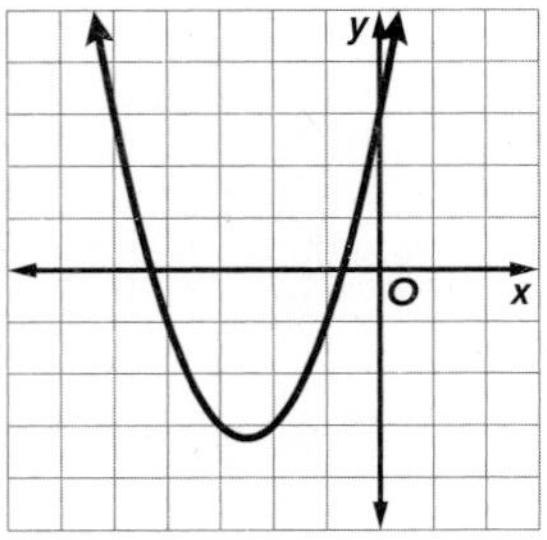

23. no real solution

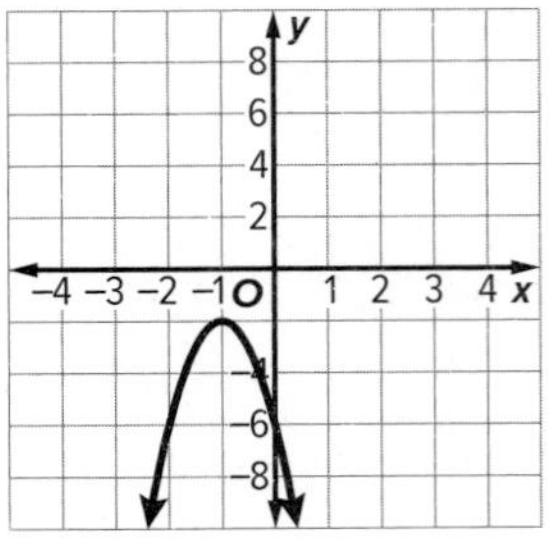

25. between 0 and 1; between 2 and 3

27. −1, 1 **29.** ≈−1.45, ≈3.45

31. ≈3.24, ≈−1.24 **33.** 2.1 seconds

35. between −3 and −2, between 1 and 2

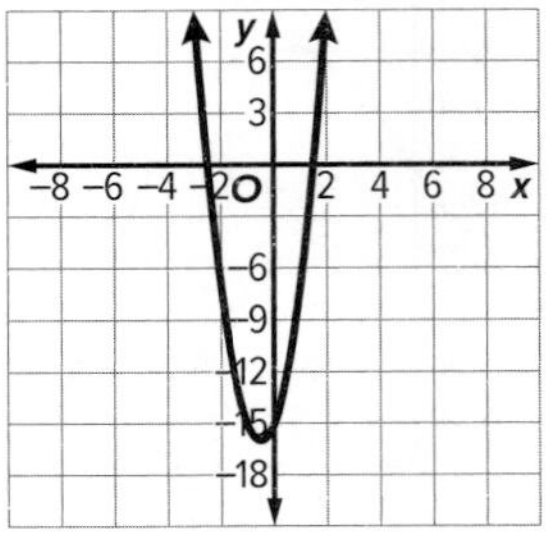

37. between −1 and 0, between 4 and 5

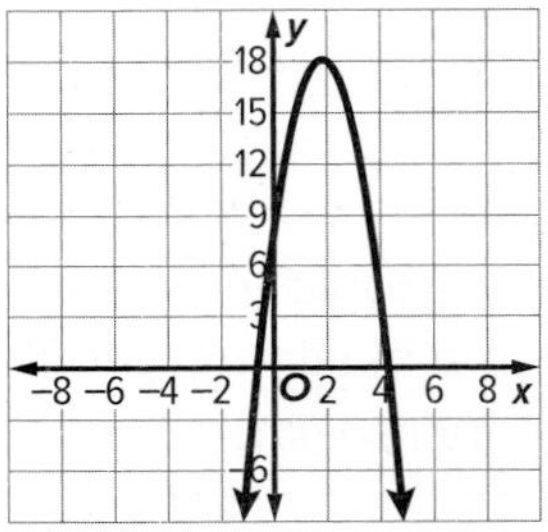

39. between 3 and 4, between 8 and 9

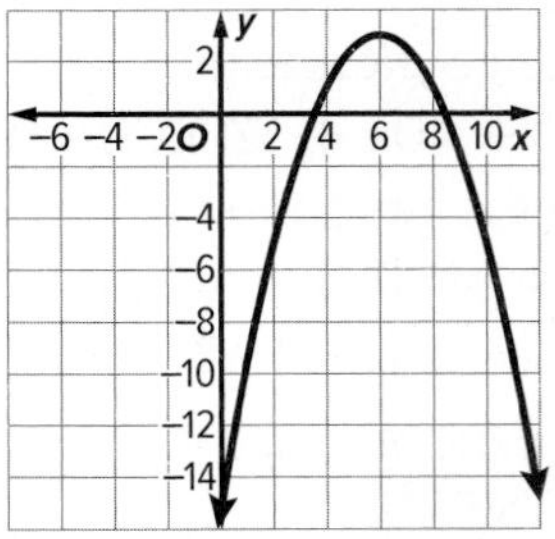

41. between −3 and −2; 0

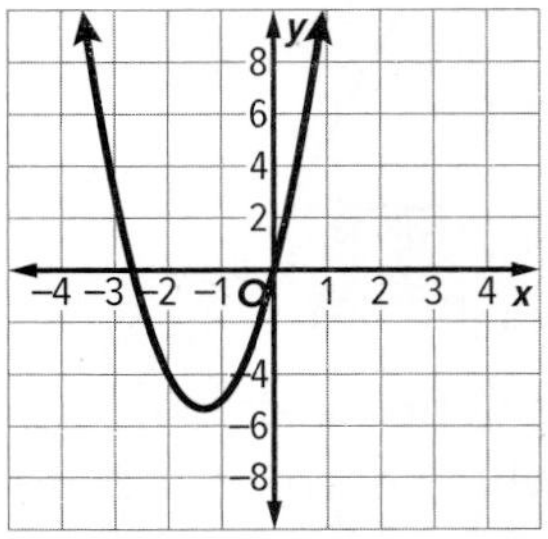

43. between −8 and −7; between 12 and 13

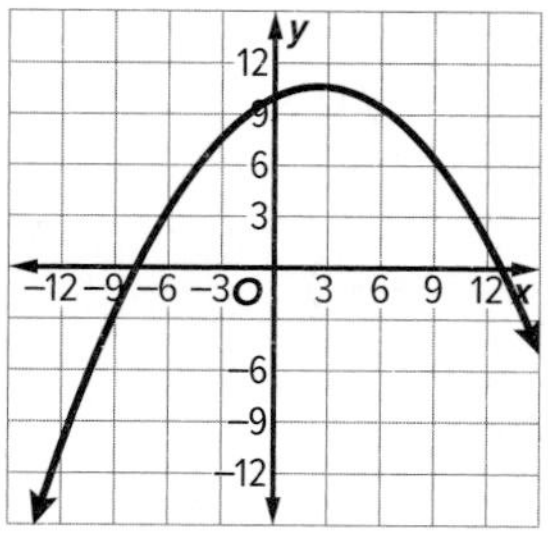

45. no real solution **47.** −2

49. between −3 and −2, between 4 and 5

51. about −5 and 17 **53.** 11 and −19

55. 64 m

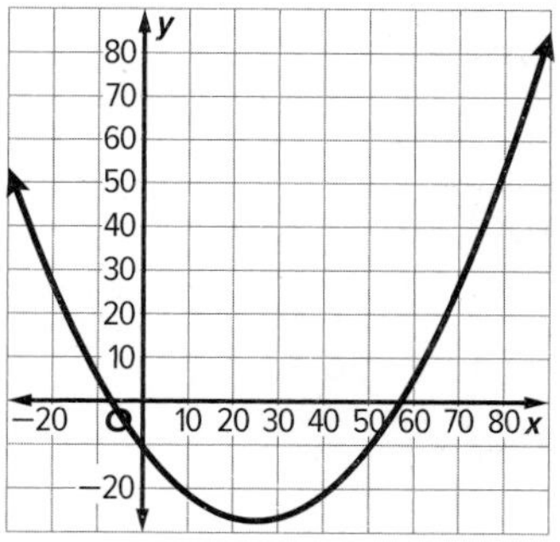

57. 5 seconds

59. No; sample answer: Hakeem is right about the location of one of the roots, but his reason is not accurate. The roots are located where $f(x)$ changes signs.

61. 5; Sample answer: The intercepts are equidistant from the axis of symmetry.

63. Sample answer: Graph the function using the axis of symmetry. Determine where the graph intersects the x-axis. The x-coordinates of those points are solutions of the quadratic equation.

Lesson 3-3

1. $4i\sqrt{3}$ **3.** $6i\sqrt{2}$ **5.** $2i\sqrt{21}$ **7.** $-23\sqrt{2}$

9. $-5\sqrt{2}$ **11.** $-i$ **13.** $\pm 3i$ **15.** $\pm i$ **17.** $\pm 5i$

19. 3, 3 **21.** $-\frac{11}{2}, -3$ **23.** 4, −3 **25.** $10 - 4i$

27. $2 + i$ **29.** $10 - 5i$ **31.** $7 + i$ **33.** $-10i$

35. $\frac{3}{2} - \frac{1}{2}i$ **37.** $-\frac{5}{3} - 2i$ **39.** $\frac{7}{9} - \frac{4i\sqrt{2}}{9}$

41. $(3 + i)x^2 + (-2 + i)x - 8i + 7$

43. $10 + 10j$ volts **45.** $8 - 2j$ ohms

47. Zoe; $i^3 = -i$, not −1.

49. Always; Sample answer: the value of 5 can be represented by $5 + 0i$, and the value of $3i$ can be represented by $0 + 3i$.

51. Some quadratic equations have complex solutions and cannot be solved using only the real numbers.

Lesson 3-4

1. $0, \frac{1}{3}$ **3.** $0, -\frac{5}{4}$ **5.** −11, −3 **7.** 7 ft by 9 ft

9. $\frac{3}{2}, -1$ **11.** $\frac{2}{5}, -6$ **13.** $8, -\frac{5}{2}$

15. −8, 8 **17.** −17, 17 **19.** −13, 13 **21.** $\frac{7}{2}$

23. $\frac{3}{4}$ **25.** $-\frac{8}{5}$ **27.** $5i, -5i$

29. $15i, -15i$ **31.** $\frac{5}{6}i, -\frac{5}{6}i$ **33.** $-3, \frac{1}{2}$

35. $9i, -9i$ **37.** $\frac{1}{5}i, -\frac{1}{5}i$

39. 13 inches by 16 inches

41. 24, 26

43. Neither; both students made a mistake in Step 3.

45. Sample answer: 3 and 6 → $x^2 - 9x + 18 = 0$. −3 and −6 → $x^2 + 9x + 18 = 0$. The linear term changes sign.

47. Sample answer: Standard form is $ax^2 + bx + c$. Multiply a and c. Then find a pair of integers, g and h, that multiply to equal ac and add to equal b. Then write the quadratic expression, substituting the middle term, bx, with $gx + hx$. The expression is now $ax^2 + gx + hx + c$. Then factor the GCF from the first two terms and factor the GCF from the second two terms. So, the expression becomes $\text{GCF}(x - q) + \text{GCF}_2(x - q)$. Simplify to get $(\text{GCF} + \text{GCF}_2)(x - q)$ or $(x - p)(x - q)$.

Lesson 3-5

1. 2, 16 **3.** $0, \frac{4}{3}$ **5.** $\frac{15}{2}, -\frac{1}{2}$

7. $\frac{-1 \pm 3\sqrt{2}}{6}$ **9.** $\frac{-2 \pm 5\sqrt{3}}{5}$ **11.** $\frac{3 \pm 4\sqrt{6}}{5}$

13. $8 \pm 8i$ **15.** $-7 \pm 10i$

17. $-4 \pm 6i$ **19.** 25; $(x + 5)^2$

21. 144; $(x + 12)^2$ **23.** $\frac{81}{4}$; $\left(x - \frac{9}{2}\right)^2$

25. 4, 9 **27.** $2 \pm \sqrt{17}$ **29.** $\frac{1 \pm \sqrt{13}}{2}$

31. 16 in. by 16 in. by 16 in.

33. $1, \frac{1}{2}$ **35.** $\frac{1}{5}, -\frac{9}{5}$ **37.** $\frac{1}{3}, -1$

39. $\frac{3 \pm i\sqrt{31}}{4}$ **41.** $1 \pm i\sqrt{2}$ **43.** $1 \pm i\sqrt{3}$

45. $y = (x + 3)^2 - 8$; $x = -3$; $(h, k) = (-3, -8)$; minimum

47. $y = -(x + 4)^2 + 11$; $x = -4$; $(h, k) = (-4, 11)$; maximum

49. $y = 3(x + 1)^2 - 4$; $x = -1$; $(h, k) = (-1, -4)$; minimum

51a. $h(t) = -4.9(t - 0.427)^2 + 8.931$

51b. $t = 0.427$; points equidistant from the axis of symmetry represent the times when the diver will be at the same height during his dive. Vertex = (0.427, 8.391); Malik reaches a maximum height of about 8.391 meters approximately 0.427 second after he begins his dive.

53. −2.77, −0.56 **55.** −1.44, 0.24

57. $1.6 \pm 0.9i$ **59.** $-1.3 \pm 4.09i$

61. $h(t) = -4.9t^2 + 25.8$; 2.3 s **63.** 5%

65. $y = (x - 5)^2 + 3$; (5, 3)

67. $y = (x - 10)^2 + 4$; (10, 4)

69a. $n^2 + 90n$ **69b.** 10

71a. w = width, $V(w)$ = the volume; $V(w) = 6w^2 - 32w$

71b. 20.7 by 62.1 in.

73. Alsonso; Aika did not add 16 to each side; she added it only to the left side.

75a. 2; rational; 16 is a perfect square, so $x + 2$ and x are rational.

75b. 2; rational; 16 is a perfect square, so $x - 2$ and x are rational.

75c. 2; complex; if the opposite of a square is positive, the square is negative. The square root of a negative number is complex.

75d. 2; real; the square must equal 20. Since that is positive but not a perfect square, the solutions will be real but not rational.

75e. 1; rational; the expression must be equal to 0 and only −2 makes the expression equal to 0.

75f. 1; rational; the expressions $(x + 4)$ and $(x + 6)$ must either be equal or opposites. No value makes them equal, −5 makes them opposites. The only solution is −5.

77. Sample answer: Completing the square is rewriting one side of a quadratic equation in the form of a perfect square. Once in this form, the equation can be solved by using the Square Root Property.

Lesson 3-6

1. −3, −5 **3.** $\frac{3}{2}, \frac{1}{3}$ **5.** $-4 \pm \sqrt{11}$

7. 2.9 s

9. $\frac{1}{4}, -5$ **11.** $-2, \frac{1}{3}$ **13.** $\frac{-5 \pm \sqrt{57}}{16}$

15. $\frac{3 \pm \sqrt{34}}{4}$ **17.** $\pm 5i$ **19.** $\frac{1 \pm i}{4}$

21. $7 \pm 2i$ **23.** $\frac{3 \pm 2i\sqrt{6}}{3}$

25. 225; 2 rational roots

27. 289; 2 rational roots

29. 24; 2 irrational roots

31. 21; 2 irrational roots

33. −196; 2 complex roots

35. −7; 2 complex roots

37. $b^2 - 4ac > 0$; 2 real rational or irrational roots

39. 2 complex roots; $\frac{1 \pm 4i}{2}$

41. 2 rational roots; 0, $\frac{5}{7}$

43. 2 rational roots; 3, 8

45. 2 rational roots; 4, $\frac{4}{3}$

47. 2 irrational roots; $\frac{-5 \pm \sqrt{3}}{2}$

49. about 2.9 seconds

51. x = the length of a side of the base, $(x + 2)^2(3x + 4) = x^2(3x + 1) + 531$; 5 in. by 5 in. by 16 in.

53. $7x^2 + 6x + 2 = 0$ is different from the other 3 equations because it has 2 complex roots, where the other 3 equations each have 2 rational roots.

55a. Sample answer: Always; when a and c are opposite signs, then ac will always be negative and $-4ac$ will always be positive. Because b^2 will also always be positive, then $b^2 - 4ac$ represents the addition of two positive values, which will never be negative. Hence, the discriminant can never be negative and the solutions can never be imaginary.

55b. Sample answer: Sometimes; the roots will only be irrational if $b^2 - 4ac$ is not a perfect square.

57. −0.75

Lesson 3-7

1.

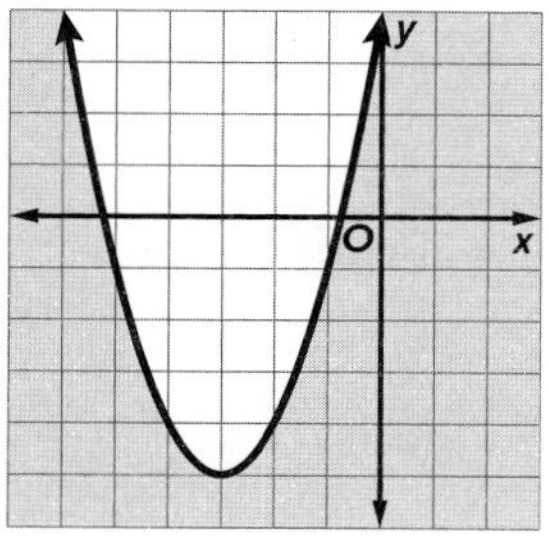

3.

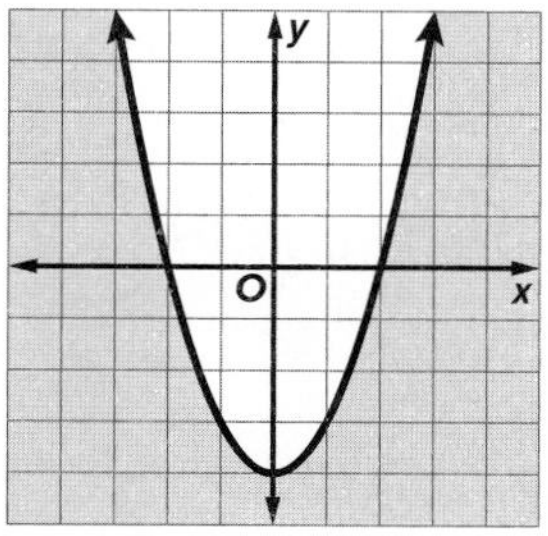

5.

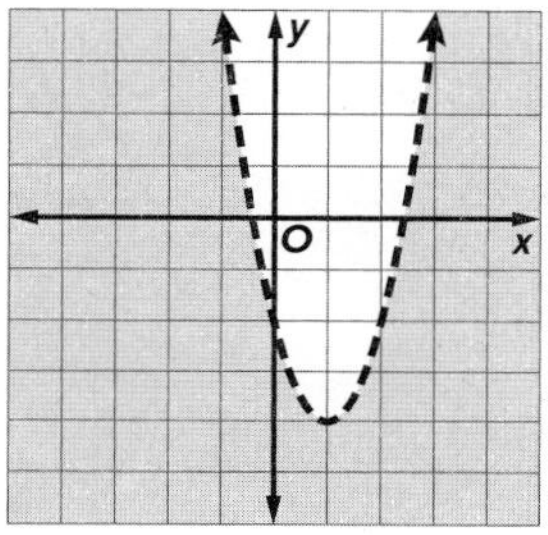

7.

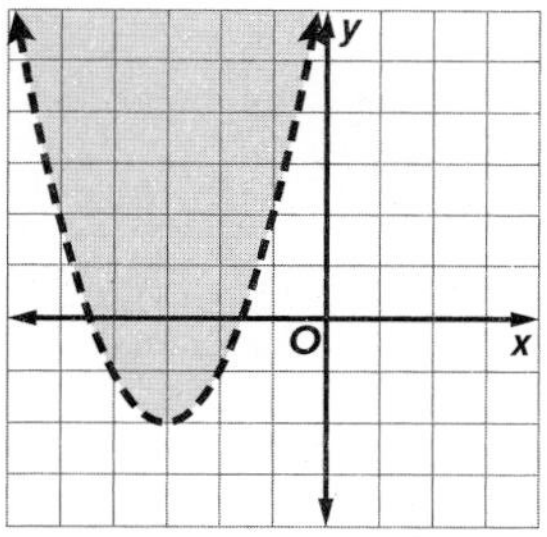

9.

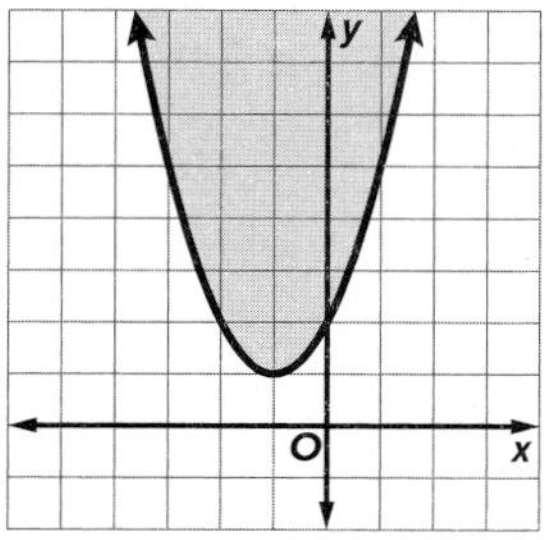

11.

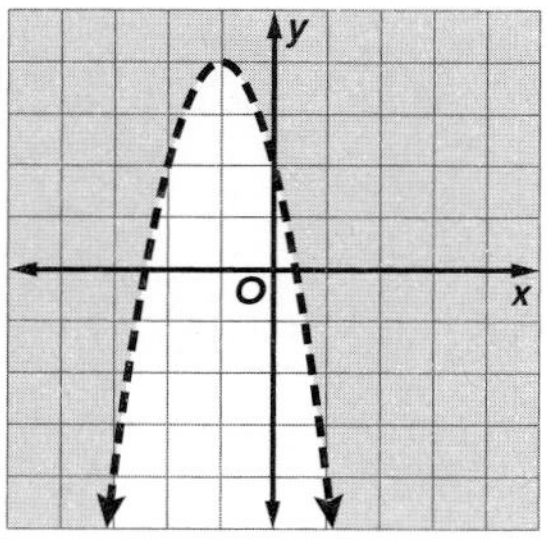

13. $\{x \mid x = 3\}$

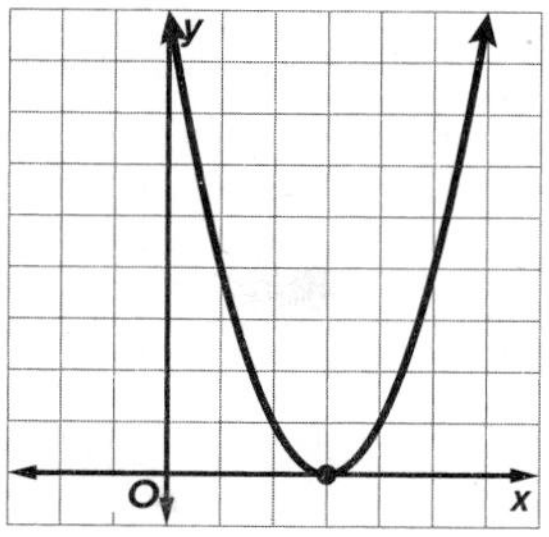

15. $\{x \mid x < -5 \text{ or } x > 4\}$

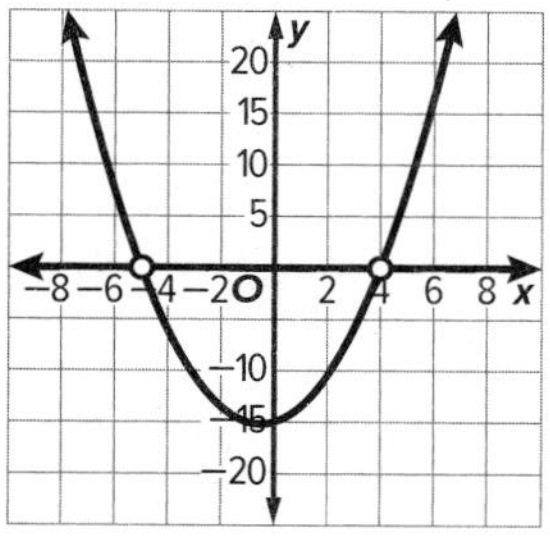

17. {all real numbers}

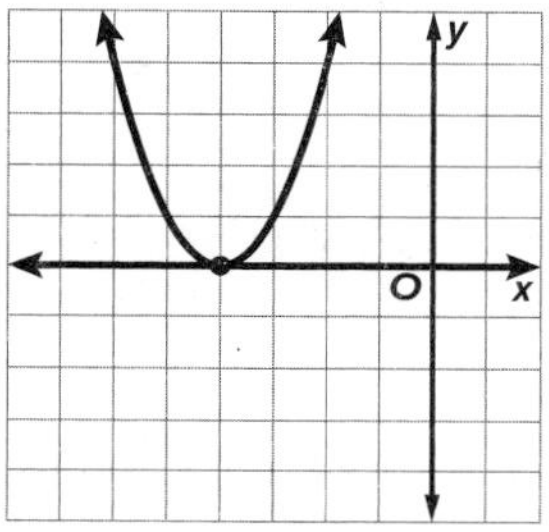

19. 30 ft to 60 ft **21.** $\{x \mid x < -1.06 \text{ or } x > 7.06\}$

23. $\{x \mid x \leq -2.75 \text{ or } x \geq 1\}$

25. $\{x \mid x < 0.61 \text{ or } x > 2.72\}$

27. {all real numbers}

29. $\{x \mid x \leq -9.24 \text{ or } x \geq -0.76\}$

31. $\{x \mid -5 \leq x \leq -3\}$

33. $\left\{x \mid x \neq -\frac{1}{2}\right\}$ **35.** $\left\{x \mid -3 \leq x \leq -\frac{4}{9}\right\}$

37. $y > x^2 - 4x - 6$ **39.** $y > -0.25x^2 - 4x + 2$

41. No; the graphs of the inequalities intersect the x-axis at the same points.

43a. Sample answer: $x^2 + 2x + 1 \geq 0$

43b. Sample answer: $x^2 - 4x + 6 < 0$

45.

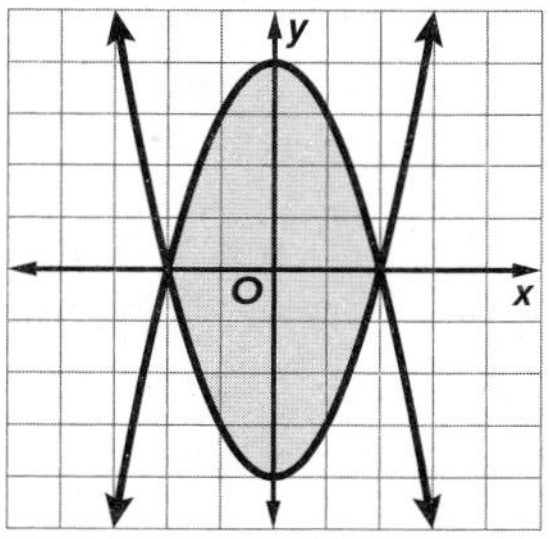

Lesson 3-8

1. (2, −1), (−1, −4) **3.** (−1, 2), (1.5, 4.5)

5. (2, −2), $\left(\frac{1}{2}, 1\right)$ **7.** (1, 2)

9. $\left(\frac{-1 + \sqrt{33}}{4}, \frac{-1 + \sqrt{33}}{4}\right), \left(\frac{-1 - \sqrt{33}}{4}, \frac{-1 - \sqrt{33}}{4}\right)$

11. −2, 1

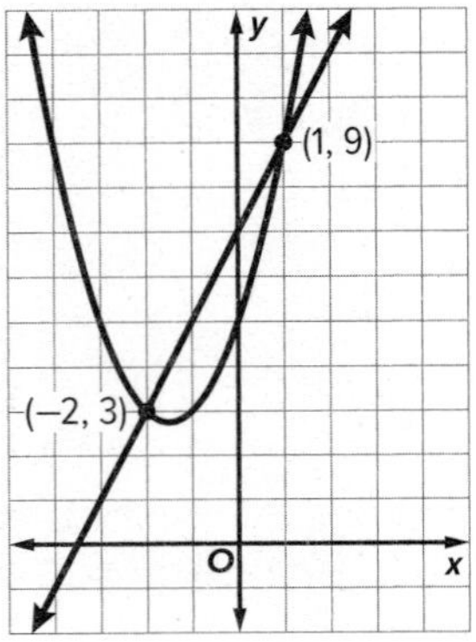

13. $-\frac{1}{4}$, 1

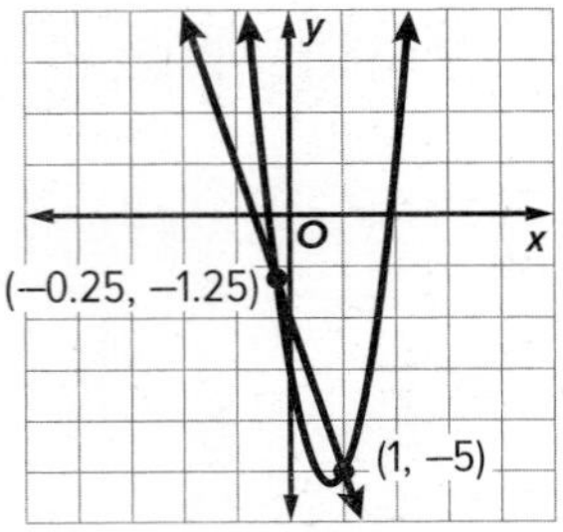

15. −1, 7

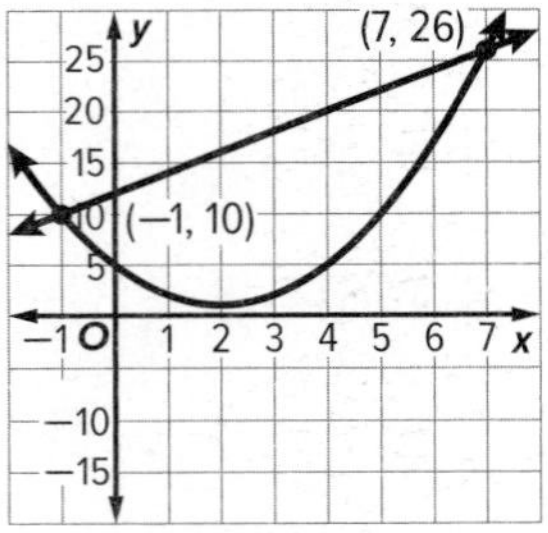

17a. $P = 2x^2 + 30x$ and $P = 50{,}000$

17b. $x \approx -165.79$, $x \approx \$150.79$

17c. Sample answer: Because the selling price cannot be negative, (150.79, 50,000) is the only viable solution in the context of the situation. This means that the business can earn a $50,000 profit when the selling price is about $150.79 per item.

19. no solution

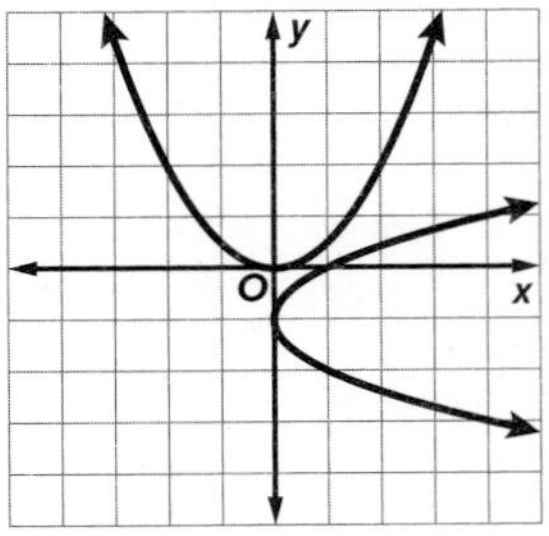

21. (0, 2) and (1, 3)

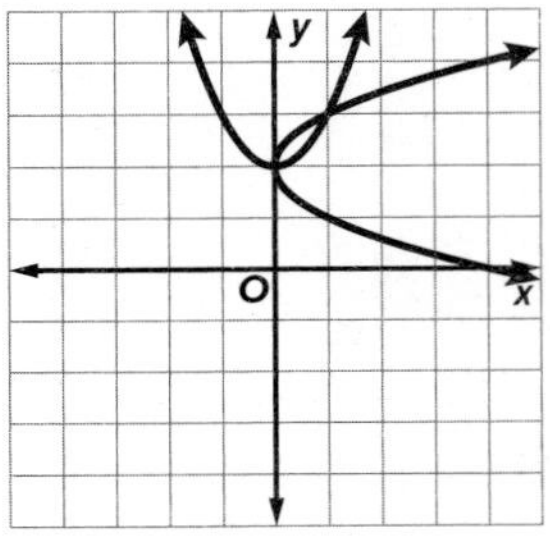

23. (1, 1) and (5, 5)

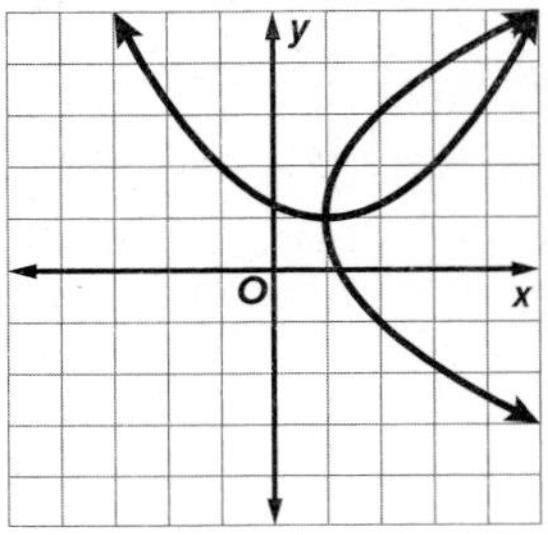

25. no solution **27.** (2, 2) and (2, −2)

29. (−1, −7) and (4, 23)

31. $-\frac{3}{2}$, 2

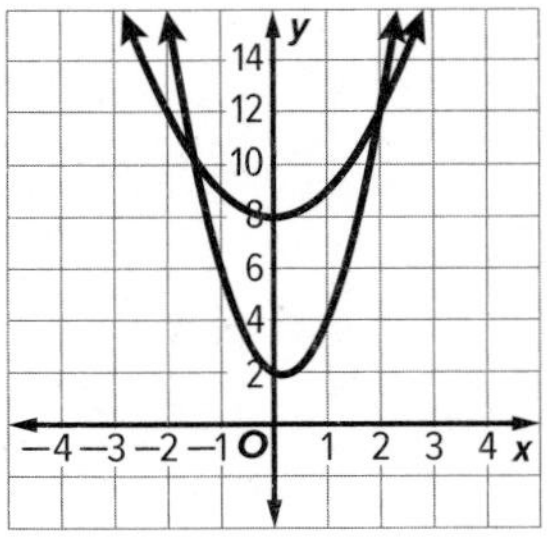

33. −2

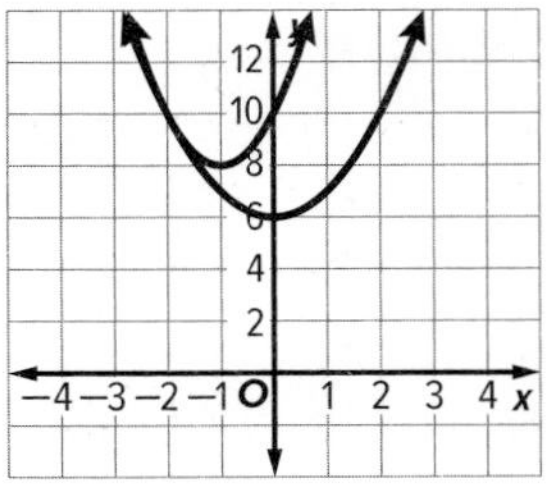

35. $-\frac{1}{2}, 3$

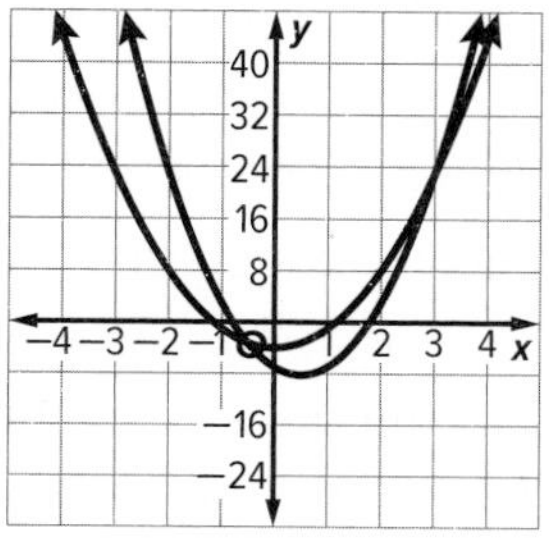

37. -2 **39.** $(-2, -6), (0, 8)$

41. $(-1, 2), (2, 5)$ **43.** $(0.42, 0.352)$ and $(1, 2)$

45. $(-1.4, 2)$ and $(1.4, 2)$

47. $(-1.8, 3.2)$ and $(1.1, 1.3)$

49. $(2, -2), (0, 0)$ **51.** no solution

53. no solution **55.** $(-\sqrt{31}, 7)$ and $(\sqrt{31}, 7)$

57. $-6\frac{1}{2}, 1$ **59.** $-3, 1$ **61.** $-\frac{1}{2}, -1$

63. -3 **65.** $-\frac{1}{2}, 3$

67. $-6, -3$ **69a.** 0.7 s

69b. about 20.9 m; The faster rocket lands after about 9.96 seconds. Solving $y = -4.9t^2 + 46.7t + c$ when $y = 0$ and $t = 9.96$, shows that $c \approx 20.9$.

71a. Yes; the maximum height of the ball is about 12.77 feet, which is higher than the net.

71b. Yes; the player's hands could strike the ball at a height of about 9 feet. Because that is above the height of the net, the ball may be blocked.

71c. Sample answer: I assumed that the student bumping the ball was far enough away from the net that she wouldn't hit it. I also assumed that the player attempting to block the ball is in the correct position for the path of the ball to intersect the path of her hands.

73. Sample answer: $y = x^2$ and $y = -x^2$

75. Danny is correct. Carol incorrectly solved for y in the second equation of the system before using the substitution method.

Module 3 Review

1. A **3.** 10 and 21 **5.** C **7.** $x = -3, x = 5$

9. D **11.** A, D **13.** C

15. between 0 and 5 feet

17. $(2, 1), (7, 6)$

Module 4

Quick Check

1. $-5 + (-13)$ **3.** $5mr + (-7mp)$

5. $-4a - 20$ **7.** $-m + \frac{5}{2}$

Lesson 4-1

1. As $x \to -\infty$, $f(x) \to \infty$ and as $x \to \infty$ and $f(x) \to \infty$; D = $(-\infty, \infty)$, R = $[0, \infty)$

3. As $x \to -\infty$, $f(x) \to \infty$ and as $x \to \infty$ and $f(x) \to -\infty$; D = $(-\infty, \infty)$, R = $(-\infty, \infty)$

5. D = $(-\infty, \infty)$, R = $[0, \infty)$

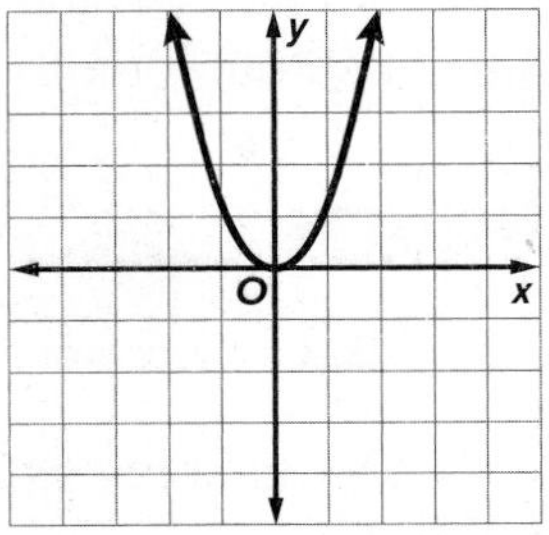

7. degree = 1, leading coefficient = 1

9. degree = 5, leading coefficient = -5

11. not in one variable because there are two variables, u and t

13a. 455

13b.

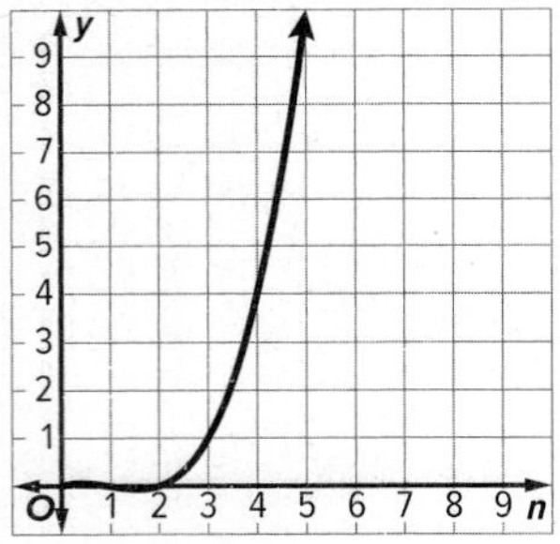

15. 1 **17.** 3 **19.** 3 **21a.** $f(x)$

21b. zeros: $f(x)$: -1.39, 0.23, 3.16; $g(x)$: -2.75, -0.5, 0.25

x-intercepts: $f(x)$: -1.39, 0.23, 3.16; $g(x)$: -2.75, -0.5, 0.25 y-intercept: $f(x)$: 1; $g(x)$: 1 end behavior: $f(x)$: As $x \to -\infty$, $f(x) \to -\infty$, and as $x \to \infty$, $f(x) \to \infty$; $g(x)$: As $x \to -\infty$, $g(x) \to \infty$, and as $x \to \infty$, $g(x) \to -\infty$.

23. As $x \to -\infty$, $f(x) \to -\infty$ and as $x \to \infty$, $f(x) \to -\infty$. degree = 4; leading coefficient = -5

25. As $x \to -\infty$, $g(x) \to -\infty$ and as $x \to \infty$, $g(x) \to \infty$. degree = 5; leading coefficient = 5

27. This is not a polynomial because there is a negative exponent.

29. As $x \to -\infty$, $h(x) \to \infty$ and as $x \to \infty$, $h(x) \to \infty$. degree = 2, leading coefficient = 3

31. As $x \to \infty$, $y \to -\infty$, and as $x \to -\infty$, $y \to \infty$; degree = 3; the leading coefficient is negative.

33. $f(1 - x) = a(1 - x)^3 - b(1 - x)^2 + (1 - x)$. So, $f(1 - x) = -ax^3 + (3a - b)x^2 + (-3a + 2b - 1)x + (a - b + 1)$. The function $f(1 - x)$ has the opposite leading coefficient, representing a reflection in the y-axis. So, it has the opposite end behavior.

35. Sample answer: The volume of the new box is modeled by the function $V(x) = (10 - x)^2(4 + 2x)$. The graph appears to have a relative maximum at $x = 2$ and $V(2) = 512$. So, the dimensions of the box with the greatest volume will be 8 centimeters by 8 centimeters by 8 centimeters.

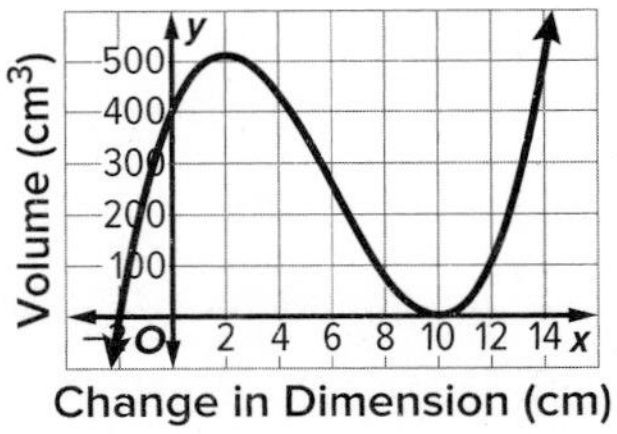

37. $f(x)$; $f(x)$ has potential for 5 or more real zeros and a degree of 5 or more. $g(x)$ has potential for 4 real zeros and a degree of 4.

39. Sample answer:

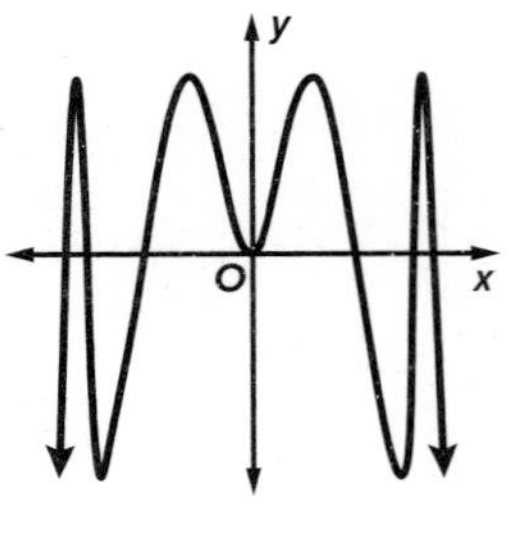

Lesson 4-2

1. zeros between $x = -4$ and $x = -3$, and between $x = 0$ and $x = 1$

x	$f(x)$
−4	3
−3	−1
−2	−3
−1	−3
0	−1
1	3
2	9

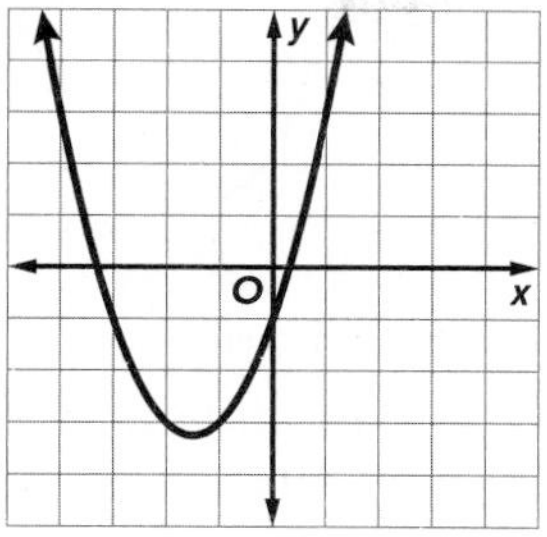

3a. zero between $x = -6$ and $x = -5$

x	$f(x)$
−6	37
−5	5
−4	25
−3	29
−2	23
−1	13
0	5
1	5
2	19
3	53

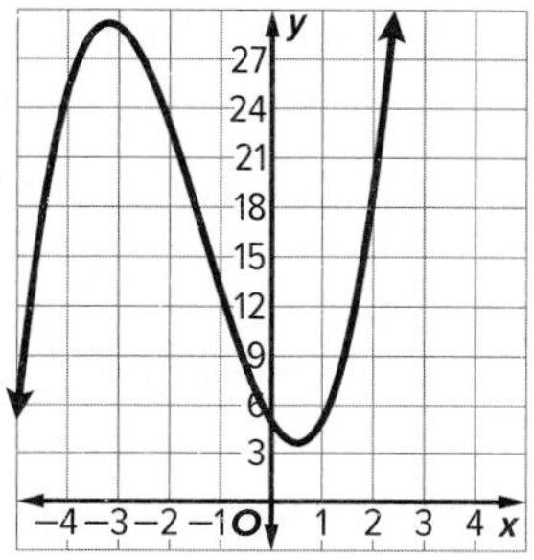

5. relative minimum between $x = 0$ and $x = 1$; relative maximum near $x = 4$

x	$f(x)$
−1	22
0	0
1	2
2	16
3	30
4	32
5	10
6	−48
7	−154

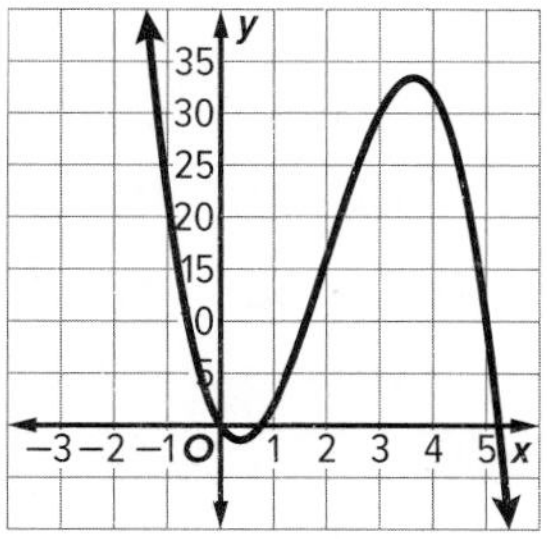

7. relative minimum near $x = -1$; no relative maximum

x	$f(x)$
−4	247
−3	74
−2	11
−1	−2
0	−1
1	2
2	19
3	86
4	263

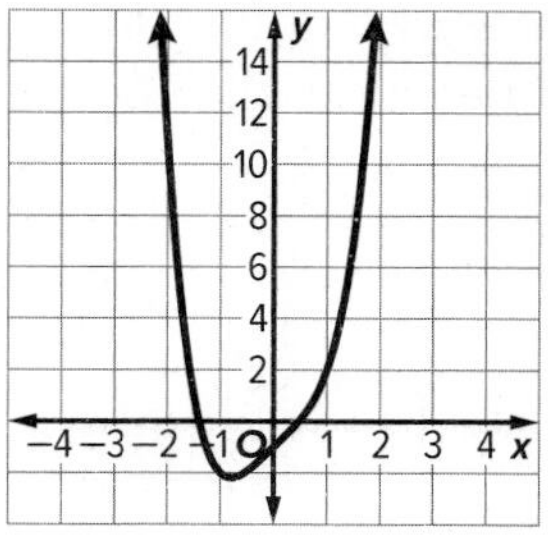

9. Domain and Range: The domain and range of the function are all real numbers. Because the function models years, the relevant domain is $\{n \mid n \geq 0\}$ and the relevant range is $\{v \mid v \geq -4\}$.

Extrema: There is a relative minimum at $n = 2$ in the relevant domain.

End Behavior: As $x \to \infty$, $v \to \infty$.

Intercepts: In the relevant domain, the v-intercept is at (0, 0). The n-intercept, or zero, is at about (3, 0).

Symmetry: In the relevant domain, the graph does not have symmetry.

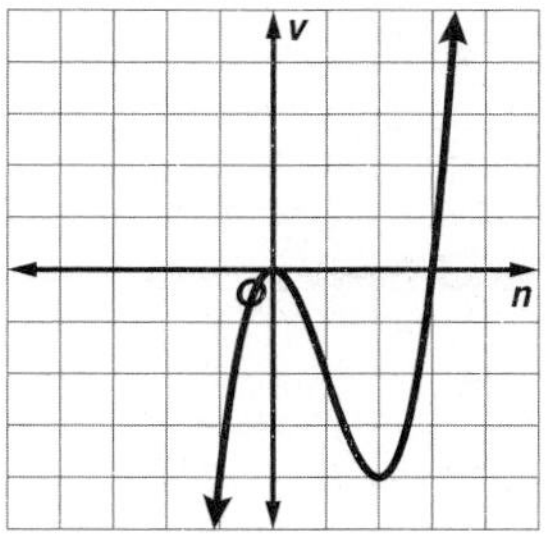

11. Curve of best fit: $y = -0.012x^4 + 0.217x^3 - 1.116x^2 + 1.486x + 7.552$; car sales in 2017: 9.991 million

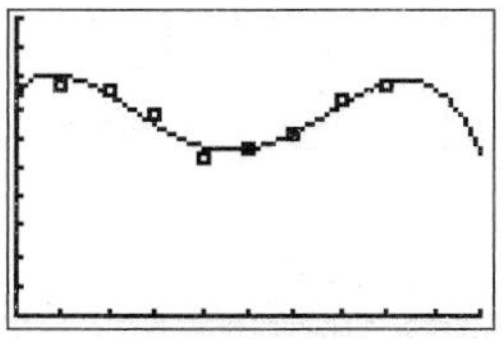

[0, 10] scl: 1 by [0, 10] scl: 1

Selected Answers

13. Curve of best fit: $y = -0.122x^4 + 1.565x^3 - 6.325x^2 + 20.321x + 33.125$, where x is the number of months since January; average volunteer hours for September: 44

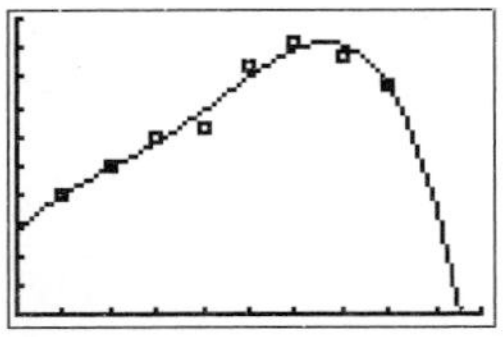

[0, 10] scl: 1 by [0, 120] scl: 12

15. average rate of change: 4500; From 2012 to 2017, the salesman's average rate of change in salary was an increase of $4500 per year.

17.

x	$f(x)$
−3	−7
−2	−2
−1	−3
0	2
1	25

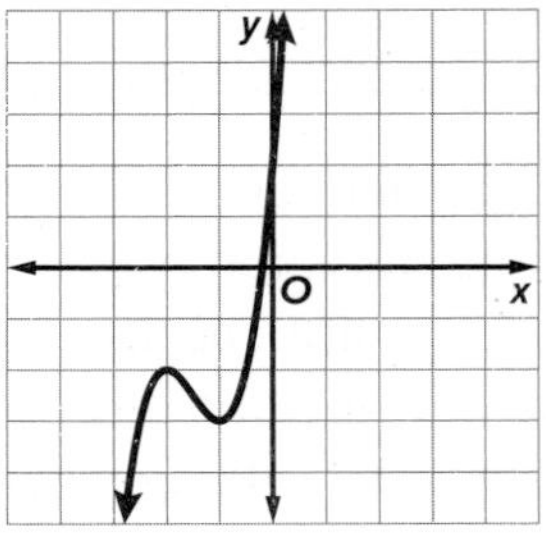

zero between $x = -1$ and $x = 0$;

rel. max. at $x = -2$, rel. min. at $x = -1$;

D = all real numbers; R = all real numbers

19.

x	$f(x)$
−3	61
−2	6
−1	−3
0	−2
1	−3
2	6
3	61

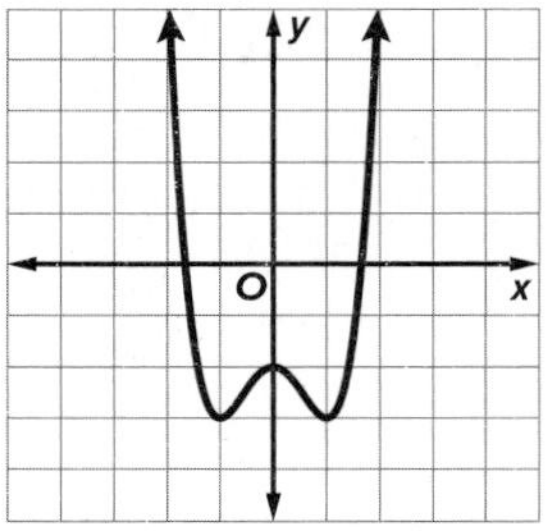

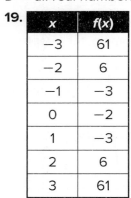

zeros between $x = -2$ and $x = -1$, and between $x = 1$ and $x = 2$;

rel. max. at $x = 0$, min. at $x = -1$, and $x = 1$;

D = all real numbers; R = $\{f(x) \mid f(x) \geq -3\}$

21. rel. max: $x \approx -2.73$; rel. min: $x \approx 0.73$

23. max: $x \approx 1.34$; no rel. min

25. Sample answer: $f(x) \to -\infty$ as $x \to \infty$ and $f(x) \to \infty$ as $x \to -\infty$; the leading coefficient is negative.

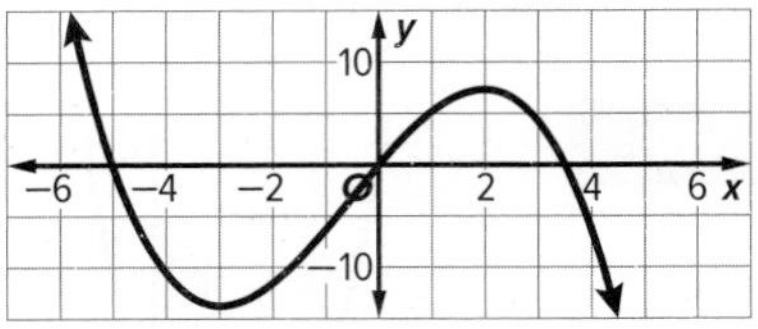

27. The type of polynomial function that should be used to model the graph is a function with an even degree with a negative leading coefficient. This is based on the fact that the graph is reflected in the x-axis and as $x \to -\infty$, $f(x) \to -\infty$ and as $x \to \infty$, $f(x) \to -\infty$.

29. As the x-values approach large positive or negative numbers, the term with the largest degree becomes more and more dominant in determining the value of $f(x)$.

31. Sample answer:

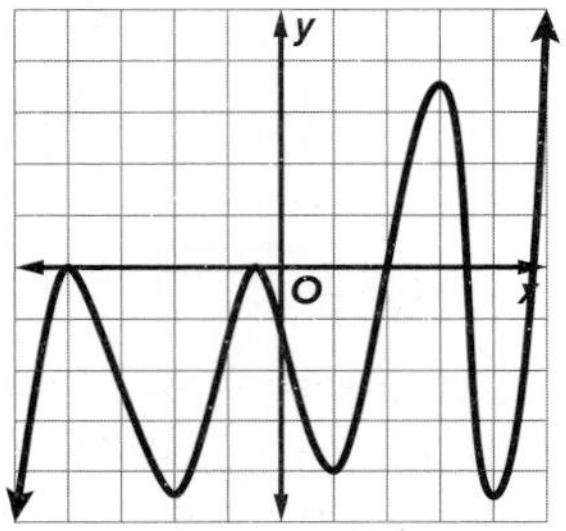

33. Sample answer: No; $f(x) = x^2 + x$ is an even degree, but $f(1) \neq f(-1)$.

Lesson 4-3

1. yes; 2 **3.** no **5.** $2a^2 - a - 2$ **7.** $3g + 12$
9. $3x^2 + 4x - 5$ **11.** $-3x + 3$ **13.** $3np^2 - 3pz$
15. $-10c^2 + 5d^2$ **17.** $a^2 - 10a + 25$
19. $x^3 + x^2y - xy^2 - y^3$
21. $2x^3 + x^2y - 2xy^2 - y^3$ **23.** $r^2 - 4t^2$
25. $2x^5 - 7x^4 + 5x^3 - 4x^2 - x + 2$
27. $s^2 + 11s + 15$ **29a.** $4x^2 + 8x + 3$
29b. $8x^2 + 16x + 6$
31. $12a^2b + 8a^2b^2 - 15ab^2 + 4b^2$

33. $2n^4 - 3n^3p + 6n^4p^4$

35. $2n^5 - 14n^3 + 4n^2 - 28$

37. $64n^3 - 240n^2 + 300n - 125$

39. $7x - 2y$

41a. $f(x)g(x) = (3x^2 - 1)(x + 2) = 3x^3 + 6x^2 - x - 2$; 3rd degree

41b. $h(x)f(x) = (-x^2 - x)(3x^2 - 1) = -3x^4 + x^2 - 3x^3 + x$; 4th degree

41c. $[f(x)]^2 = (3x^2 - 1)^2 = 9x^4 - 6x^2 + 1$; 4th degree

43.

	Addition and Subtraction	Multiplication	Division
Integers	yes	yes	No; $3 \div 2 = 1.5$
Rational Numbers	yes	yes	yes
Polynomials	yes	yes	No; $\frac{3}{x^2} = 3x^{-2}$

45. $-0.022x + 1590$

47a. Area of sidewalk = area of larger circle − area of smaller circle, so $384\pi = \pi(r + 12)^2 - \pi r^2 = \pi(r^2 + 24r + 144) - \pi r^2 = 24\pi r + 144\pi$. $24\pi r = 240\pi$; $r = 10$; the radius of the smaller circle is 10 feet, and the radius of the larger circle is $r + 12 = 10 + 12$ or 22 feet.

47b.

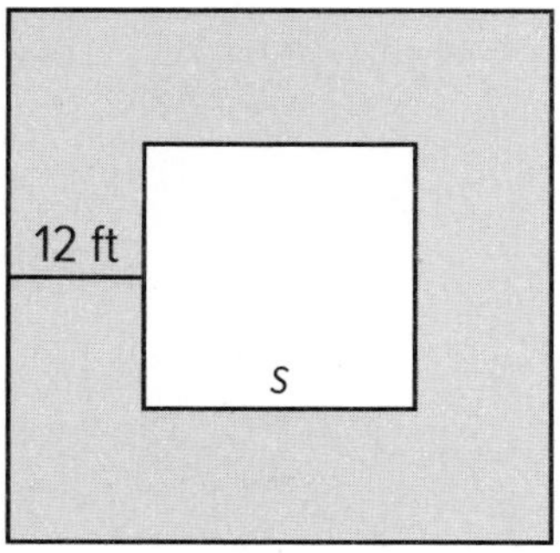

47c. $48s + 576$ ft^2

49a. $m \cdot n$; each term of one polynomial multiplies each term of the other one time.

49b. 2; adding like terms may result in a sum of 0; but the first and last terms are unique.

51. $(3 - 2b)(3 + 2b)$

Lesson 4-4

1. $5y^2 + 2y + 1$ **3.** $2j - 3k$ **5.** $n + 2$

7. $2t + 1 + \frac{9}{t + 6}$ **9.** $2g - 3$ **11.** $3v + 5 + \frac{10}{v - 4}$

13. $y^2 - 2y + 4 - \frac{2}{y + 2}$

15. $\frac{4}{3}p^2 + \frac{p}{9} + \frac{19}{27} + \frac{19}{27(3x - 1)}$

17. $m - 3 + \frac{6}{m + 4}$ **19.** $2x^2 - 3x + 1$

21. $x^2 - 2x + 4$ **23.** $4c^2d - \frac{3}{2}d^2$

25. $n^2 - n - 1$

27. $3z^4 - z^3 + 2z^2 - 4z + 9 - \frac{13}{z + 2}$

29. A is x^2; B is $6x$; C is 11. **31.** $\pi(x^2 - 8x + 16)$

33. Yes; Because $3x$ times the divisor is $9x^2 + 3x$, the divisor must be $3x + 1$. The second and third terms of the dividend must be $0x + 5$ because the first difference is $-3x + 5$.

35. $8x^3 - 8x^2 - 2x - 10$; I multiplied the divisor and the quotient and added the remainder: $(2x^2 + x + 1)(4x - 6) + (-4) = 8x^3 - 8x^2 - 2x - 10$.

37a. No; the degrees of $f(x)$ and $d(x)$ may be equal. For example, $\frac{3x^2 + 6}{x^2} = 3 + \frac{6}{x^2}$.

37b. Yes; if the degree of $r(x)$ is greater than or equal to the degree of $d(x)$, then the expression $\frac{r(x)}{d(x)}$ may be simplified by division. For example, if $\frac{r(x)}{d(x)} = \frac{8x + 1}{x}$, then $8x + 1$ may be divided by x to get $8 + \frac{1}{x}$.

37c. Yes; because $\frac{f(x)}{d(x)} = q(x) + \frac{r(x)}{d(x)}$, the degree of $\frac{f(x)}{d(x)}$ must equal the degree of $q(x) + \frac{r(x)}{d(x)}$. The degree of $r(x)$ is less than the degree of $d(x)$, so the degree of $q(x) + \frac{r(x)}{d(x)}$ equals the degree of $q(x)$. This means the degree of $\frac{f(x)}{d(x)}$ equals the degree of $q(x)$, and the degree of $\frac{f(x)}{d(x)}$ is the degree of $f(x)$ minus the degree of $d(x)$. For example, in $\frac{2x^3 - 1}{x^2 + 3} = 2x + \frac{-6x - 1}{x^2 + 3}$, the degree of $q(x)$ is the degree of $f(x)$ minus the degree of $d(x)$.

39. The binomial is a factor of the polynomial.

41. Sample answer: $\frac{x^2 + 5x + 9}{x + 2}$

43a. $B^2 + 1$

43b. $B + \frac{1}{B^2 + B + 1}$

Lesson 4-5

1. $x^3 - 3x^2y + 3xy^2 - y^3$

3. $g^4 - 4g^3h + 6g^2h^2 - 4gh^3 + h^4$

5. $y^3 - 21y^2 + 147y - 343$

7. 38%

9. $243x^5 + 1620x^4y + 4320x^3y^2 + 5760x^2y^3 + 3840xy^4 + 1024y^5$

11. $4096h^4 - 6144h^3j + 3456h^2j^2 - 864hj^3 + 81j^4$

13. $x^5 + \frac{5}{2}x^4 + \frac{5}{2}x^3 + \frac{5}{4}x^2 + \frac{5}{16}x + \frac{1}{32}$

15. $32b^5 + 20b^4 + 5b^3 + \frac{5}{8}b^2 + \frac{5}{128}b + \frac{1}{1024}$

17. 23%

19. If C represents a correct circuit board and N represents an incorrect circuit board, then the number of ways for the robot to produce 5 of 7 circuit boards accurately is given by the coefficient of C^5N^2 in the expansion of $(C + N)^7$. Using the Binomial Theorem, there are 21 ways for the robot to produce a correct circuit board out of 128 possibilities. So, the probability that 5 of 7 are correct is $\frac{21}{128}$ or about 16% probability.

21. If c represents a correct answer and w represents a wrong answer, then the coefficients expansion of $(c + w)^{10}$ can be used to represent situation. Using Pascal's triangle, there are 45 ways to get 8 questions correct, 10 ways to get 9 questions correct, and 1 way to get all correct. So there are 56 ways to get 8 or more correct. By adding all of the values in this row of Pascal's triangle, I found that there are 1024 different ways he could answer the questions in the quiz. Matthew has a $\frac{56}{1024}$ or about a 5.5% chance of getting 8 or more correct.

23. There are 9 judges on the Supreme Court. The majority could be 5, 6, 7, 8, or 9 votes. So, there are ${}_9C_5 + {}_9C_6 + {}_9C_7 + {}_9C_8 + {}_9C_9 = 256$ combinations.

25. 1.21896; $(1.02)^{10}$ + 1.21899442; the approximation differs from the value given by a calculator by 0.00003442.

27. Sample answer: While they have the same terms, the signs for $(x + y)^n$ will all be positive, while the signs for $(x - y)^n$ will alternate.

29. Sample answer: $\left(x + \frac{6}{5}y\right)^5$

Module 4 Review

1. B **3.** B **5.** $g(x) \to -\infty$ **7.** C

9. $x^6 + 7x^5 - 6x^3 + 8x^2 + 2x$

11. C **13.** $5x^2 + 7x - 3$ **15.** B

Module 5

Quick Check

1. $24x^3 + 8x^2 + 6x + 4$

3. $14x^3 - 40x^2 + 12x + 24$

5. $-4, 2$ **7.** $-\frac{4}{3}, \frac{1}{2}$

Lesson 5-1

1. -4.12 **3.** $-0.47, 0.54, 3.94$ **5.** -1.27

7a. $6x^3 + 110x^2 - 200x = 15{,}000$

7b. $y = 6x^3 + 110x^2 - 200x$, $y = 15{,}000$; $x = 10$ cm

7c. 25 cm by 20 cm by 30 cm

9a. $\pi x^3 + 3\pi x^2 = 628$

9b. $y = \pi x^3 + 3\pi x^2$, $y = 628$; $x = 5$

9c. radius = 5 in., height = 8 in.

11. $-3.63, -1.35, 1.35, 3.63$ **13.** no solution

15. \$0.29, 0.88; sample answer: graph $f(x) = 70{,}000(x - x^4)$ and $f(x) = 20{,}000$ and find the x-values of the points of intersections.

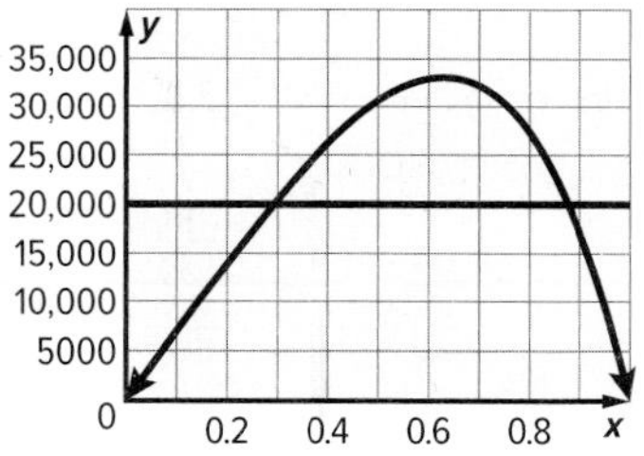

17. Sample answer: A function with an even degree reverses direction, so both ends extend in the same direction. That means that if the function opens up and the vertex is above the x-axis or if the function opens down and the vertex is below the x-axis, there are no real solutions. A function with an odd degree does not reverse direction, so the ends extend in opposite directions. Therefore, it must cross the x-axis at least once.

19. Sometimes; sample answer: The positive solution is often correct when a negative value doesn't make sense, such as for distance or time; however, sometimes a negative solution is reasonable, such as for temperature or position problems. Also, sometimes there are two positive solutions and one may be unreasonable.

21. $5x^2 = -2x - 11$; It is the only one that has no real solutions.

Lesson 5-2

1. $(2c - 3d)(4c^2 + 6cd + 9d^2)$

3. $a^2(a - b)(a^2 + ab + b^2)(a + b)(a^2 - ab + b^2)$

5. prime **7.** $(x + y)(x - y)(6f + g - 3h)$

9. $(a - b)(a^2 + ab + b^2)(x - 8)^2$ **11.** $0, 7, 2$

13. $0, -5, 8$

15a. Let x represent the length of a side of the leg and y represent the length of the notch; $x^3 - y^3$.

15b. 15 in.

17. $-15(x^2)^2 + 18(x^2) - 4$ **19.** not possible

21. $4(2x^5)^2 + 1(2x^5) + 6$ **23.** $\pm\sqrt{5}, \pm i\sqrt{2}$

25. $\pm\frac{2\sqrt{3}}{3}, \pm\frac{\sqrt{15}}{3}$ **27.** $\pm\frac{\sqrt{6}}{6}, \pm i\frac{\sqrt{3}}{2}$

29. $(x + 2)(x^2 - 2x + 4)(x - 2)(x^2 + 2x + 4)$

31. $x^2y^2(2x + 3y)(4x^2 + 6xy + 9y^2)$

33. $(y - 1)^3(y^2 + y + 1)^3$

35. $3, -3, \pm i\sqrt{10}$ **37.** $\pm\sqrt{11}, \pm 2i$

39. $-6, 3 \pm 3i\sqrt{3}$

41. $\pm\frac{1}{2}, \pm i\frac{\sqrt{6}}{2}$ **43.** $\pm\frac{3}{2}, \pm\frac{\sqrt{10}}{5}$

45. $\pm\frac{1}{2}, \pm\sqrt{2}$ **47.** $1, -2, \frac{-1 \pm i\sqrt{3}}{2}, 1 \pm i\sqrt{3}$

49. $-5, \frac{1}{2}, \frac{-1 \pm i\sqrt{3}}{2}, \frac{5 \pm 5i\sqrt{3}}{2}$ **51.** $\pm 3, \pm 1, \pm i$

53a. Sample answer: $\frac{1}{2} = 2\left(\frac{1}{4}\right)$

53b. $u = x^{\frac{1}{4}}$; $u^2 - 8u + 15 = 0$

53c. 81, 625

55. prime **57.** $(2a + 1)(k - 3)$ **59.** prime

61. $(d - 6)^2$ **63.** $(y + 9)^2$ **65.** $19x^2(x - 2)$

67. $(m^2 + 1)(m - 1)(m + 1)$ **69.** 2.4 in.

71. The solutions are $x = \pm\sqrt{m}$ and $\pm\sqrt{n}$. because both equations have the same coefficients, take the solutions for the quadratic equation and substitute x^2 for x.

73. $\frac{16}{3}, \frac{3}{4}$

75. Sample answer: $12x^6 + 6x^4 + 8x^2 + 4 = 12(x^2)^3 + 6(x^2)^2 + 8(x^2) + 4$

Lesson 5-3

1. $(x - y)^2 = x^2 - 2xy + y^2$ (Original equation)
$x^2 - 2xy + y^2 = x^2 - 2xy + y^2$ (Distributive Property)

3. $4(x - 7)^2 = 4x^2 - 56x + 196$ (Original equation)
$4(x^2 - 14x + 49) = 4x^2 - 56x + 196$
(Distributive Property)
$4x^2 - 56x + 196 = 4x^2 - 56x + 196$
(Distributive Property)

5. $a^2 - b^2 = (a + b)(a - b)$ (Original equation)
$= a^2 - ab + ab - b^2$ (Distributive Property)
$= a^2 - b^2$ (Simplify.)

7. $p^4 - q^4 = (p - q)(p + q)(p^2 - q^2)$ (Original equation)
$= (p^2 + pq - pq - q^2)(p^2 + q^2)$ (Distributive Property)
$= (p^2 - q^2)(p^2 + q^2)$ (Simplify.)
$= p^4 + p^2q^2 - p^2q^2 - q^4$ (Distributive Property)
$= p^4 - q^4$ (Simplify.)

9. $(3x + y)^2 = 9x^2 + 6xy + y^2$ (Original equation)
$9x^2 + 6xy + y^2 = 9x^2 + 6xy + y^2$ (Square the left side.)
$9x^2 + 6xy + y^2 = 9x^2 + 6xy + y^2$ (True)
Because the identity is true, this proves that Aponi is correct. Her process for finding the area of a square will always work.

11. identity **13.** not an identity **15.** not an identity

17. $g^6 + h^6$
$= (g^2 + h^2)(g^4 - g^2h^2 + h^4)$ (Original equation)
$= g^6 + g^4h^2 + g^2h^4 + g^4h^2 - g^2h^4 + h^6$ (Distributive Property)
$= g^6 + h^6$ (Simplify.)

19. $u^6 - w^6$
$= (u + w)(u - w)(u^2 + uw + w^2)(u^2 - uw + w^2)$ (Original equation)
$= (u^2 + w^2)(u^2 + uw + w^2)(u^2 - uw + w^2)$ (FOIL)
$= (u^4 + u^3w - uw^3 - w^4)(u^2 - uw + w^2)$ (Distributive Property)
$= u^6 + u^5w - u^3w^3 - u^2w^4 - u^5w - u^4w^2 + u^2w^4 + uw^5 + u^4w^2 + u^3w^3 - uw^5 - w^6$ (Distributive Property)
$= u^6 - w^6$ (Simplify.)

21. A polynomial identity is a polynomial equation that is satisfied for any values that are substituted for the variables. To prove that a polynomial equation is an identity you begin with the more complicated side of the equation and use algebra properties to transform that side of the equation until it is simplified to look like the other side.

23. $x^2 - y^2 = 3$, $2xy = 4$, $x^2 + y^2 = 5$; $x = 2$, $y = 1$

25. When George multiplied b and a^2 in the second line he mistakenly made the term negative so it did not cancel when he simplified.

Lesson 5-4

1. −59; 11 **3.** 1707; 62 **5.** −98; 0 **7.** 13,190; 2
9. 21, 0 **11.** −2, 1 **13.** 3, 3 **15.** −4, −7
17. 15, −6 **19.** −22, 20
21a. \$20.4 billion **21b.** \$144.16 billion
23. $(x - 1)^2$ **25.** $x - 4$, $x + 1$
27. $x + 6$, $2x + 7$ **29.** $x + 1$, $x^2 + 2x + 3$
31. $x - 1$, $x + 2$ **33.** $x - 2$, $x + 2$
35. $x + 3$, $x - 6$ **37.** $x + 1$, $x - 4$
39. $2x + 1$, $x - 1$ **41.** $x - 1$, $x + 2$
43. $x - 4$, $3x - 2$ **45.** 11; 764

47.

0.1	$\frac{1}{6}$	$\frac{1}{2}$	1	1
		$\frac{1}{60}$	$\frac{31}{600}$	$\frac{631}{6000}$
	$\frac{1}{6}$	$\frac{31}{60}$	$\frac{631}{600}$	$\frac{6631}{6000}$

$1\frac{631}{6000}$

49. 8 **51.** −3

53. $(x - 1)$, $(5x - 2)$, $(2x + 3)$; Sample answer: By the Factor Theorem, $(x - 1)$ is a factor. Use synthetic division with $x = 1$. The remainder is $k - 3$. For $(x - 1)$ to be a factor, $k - 3 = 0$, so $k = 3$. The quotient is $10x^2 + 13x - 3$, which factors as $(5x - 1)(2x + 3)$. $P(x) = (x - 1)(5x - 1)(2x + 3)$. The cubic has a positive leading coefficient and zeros at −1.5, 0.2, and 1.

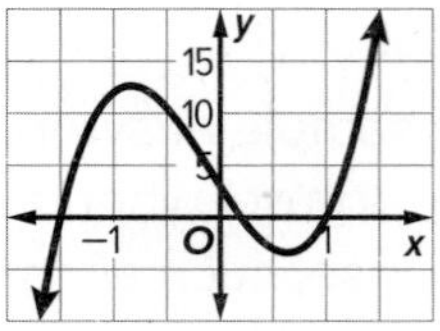

55a. (2003, 621), (2012, 1197), (2021, 1740), (2025, −15,255)

55b. The model still represents the situation after 25 years.; Sample answer: No, the average value is unlikely to fall so quickly.

57. −5; Sample answer: If $P(x)$ is symmetric to the y-axis, it must also contain the point $(-2, -5)$. According to the Remainder Theorem, the remainder when polynomial $P(x)$ is divided by $(x - r)$ is $P(r)$. Since $P(-2), = -5$, the remainder when $P(x)$ is divided by $(x + 2)$ is −5.

59. Use synthetic substitution with $a = -6$. The remainder must be zero, so $432 - 216k = 0$, or $k = 2$. The depressed polynomial is $2x^2 + 3x - 5$, which factors to $(2x + 5)(x - 1)$. Therefore, we can write $kx^3 + 15x^2 + 13x - 30 = (x + 6)(2x + 5)(x - 1)$.

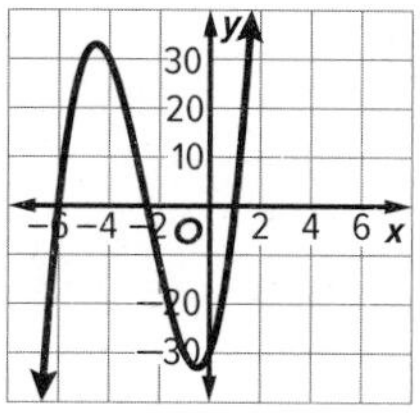

61. $f(x) = 0.1(x - 1)^2(x + 5)^2$

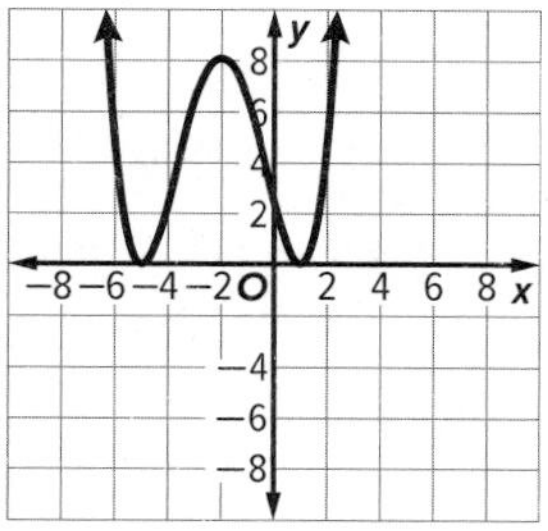

63. If $x - a$ is a factor of $f(x)$, then $f(a)$ has a factor of $(a - a)$ or 0. Since a factor of $f(a)$ is 0, $f(a) = 0$. Now assume that $f(a) = 0$. If $f(a) = 0$, then the Remainder Theorem states that the remainder is 0 when $f(x)$ is divided by $x - a$. This means that $x - a$ is a factor of $f(x)$. This proves the Factor Theorem.

65. Sample answer: When $x = 1$, $f(1)$ is the sum of all of the coefficients and constants in $f(x)$, in this case, a, b, c, d, and e. The sum of a, b, c, d, and e is 0, so however the coefficients are arranged, $f(1)$ will always equal 0, and $f(x)$ will have a rational root.

67. Tyrone; Sample answer: By the Factor Theorem, $(x - r)$ is a factor when $P(r) = 0$. $P(r) = 0$ when $x = -1$ and $x = 2$.

Lesson 5-5

1. $-\frac{12}{5}$; 1 real

3. 0, 0, 0, $2i$, $-2i$; 3 real, 2 imaginary

5. $\frac{1 \pm \sqrt{2}}{2}$; 2 real

7. −2, $\frac{3}{2}$; 2 real

9. −1, $\frac{1 \pm i\sqrt{3}}{2}$; 1 real, 2 imaginary

11. $-\frac{8}{3}$, 1; 2 real

13. $-\frac{5}{2}, \frac{5}{2}, -\frac{5}{2}i, \frac{5}{2}i$; 2 real, 2 imaginary

15. −2, −2, 0, 2, 2; 5 real

17. 2 or 0; 1; 2 or 0

19. 3 or 1; 0; 2 or 0

21. 1; 1; 2

23. 0 or 2; 0 or 2; 0, 2, or 4

25. 0 or 2; 1; 2 or 4

27. 0 or 2; 0 or 2; 2, 4, or 6

29. 3, $1 + \sqrt{2}$, $1 - \sqrt{2}$

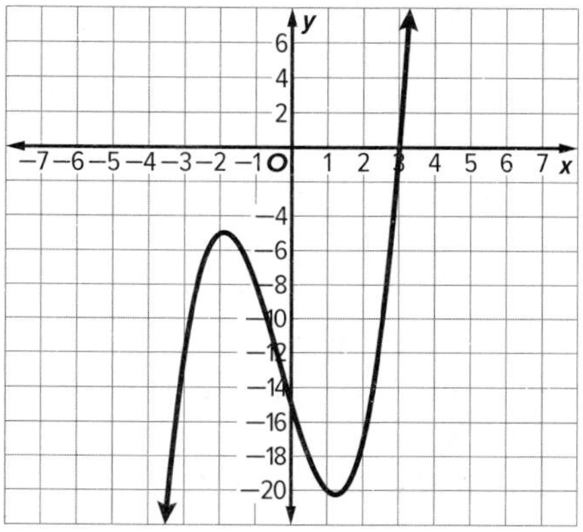

31. 1, −2, −3

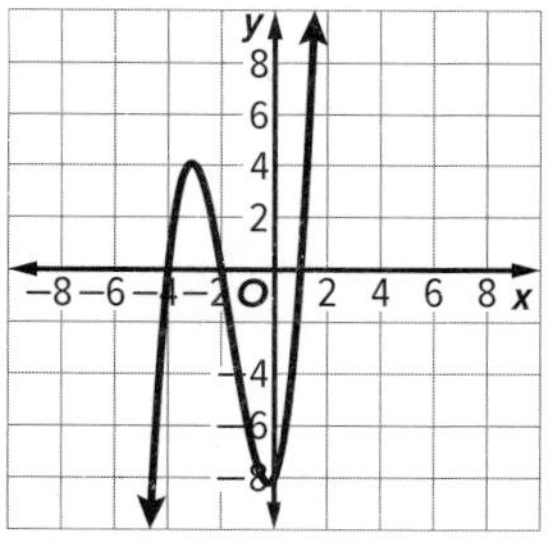

33. −1, −1, 1, 4

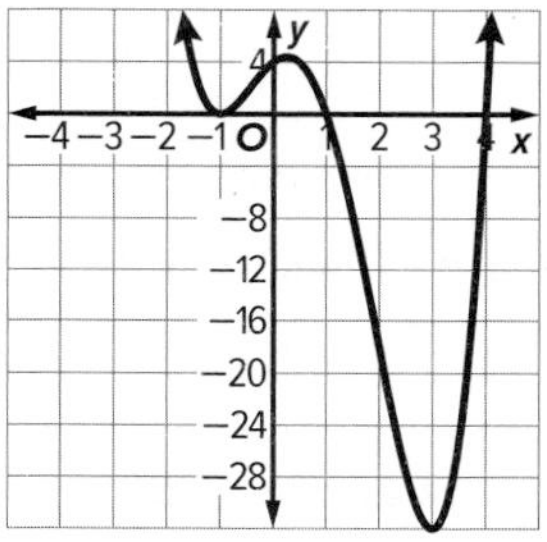

35. −6, −2, 1

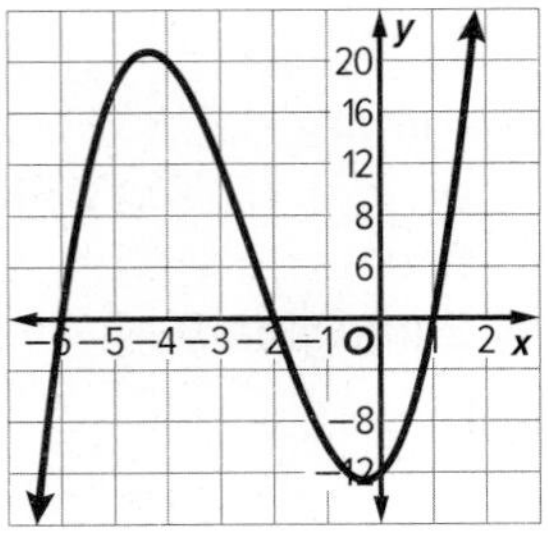

37. −4, 7, −5*i*, 5*i*

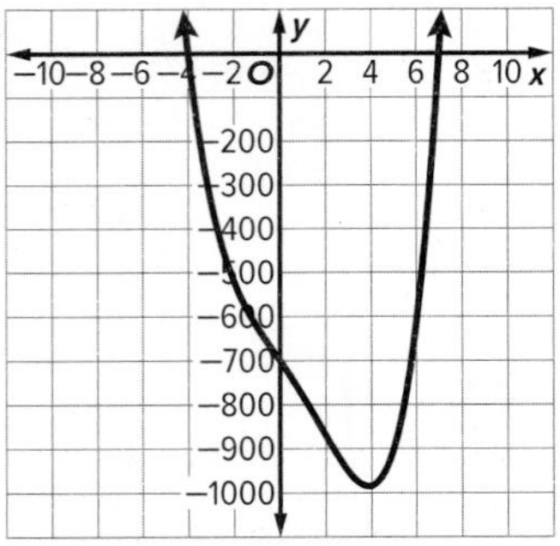

39. 4, 4, −2*i*, 2*i*

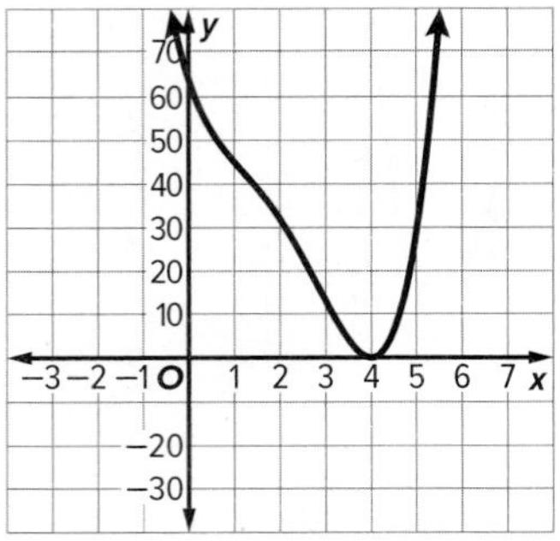

41. $y = x^2 + x - 6$

43. $y = x^4 - 6x^3 + 7x^2 + 6x - 8$

45. Sample answer:

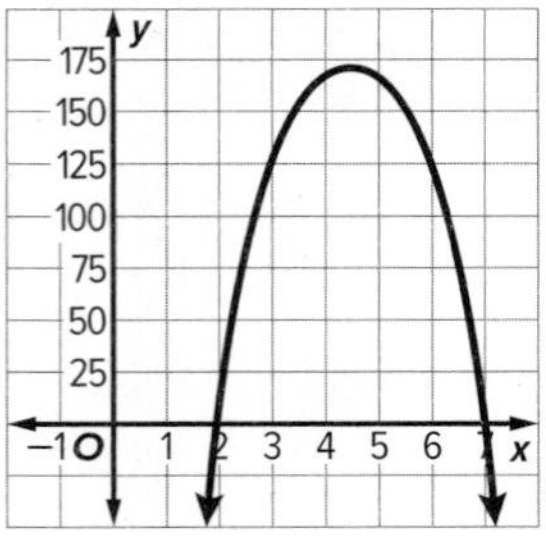

47. Sample answer: $f(x) = x^3 + 2x^2 - 23x - 60$

49. Sample answer: $f(x) = x^4 + 2x^3 + 6x^2 + 18x - 27$

51. Sample answer: $f(x) = x^4 - 3x^3 - 9x^2 + 77x + 150$

53.

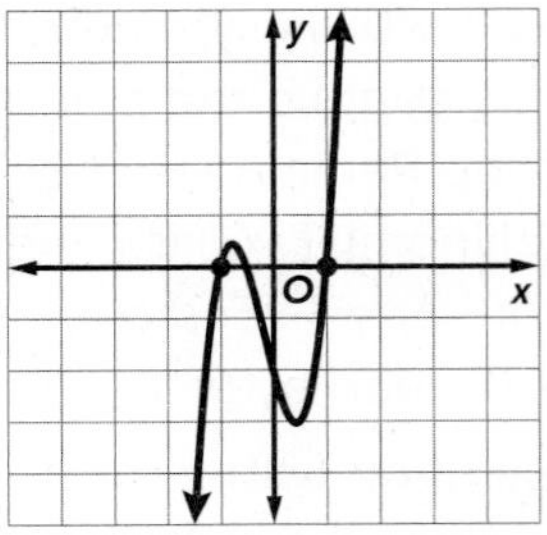

55.

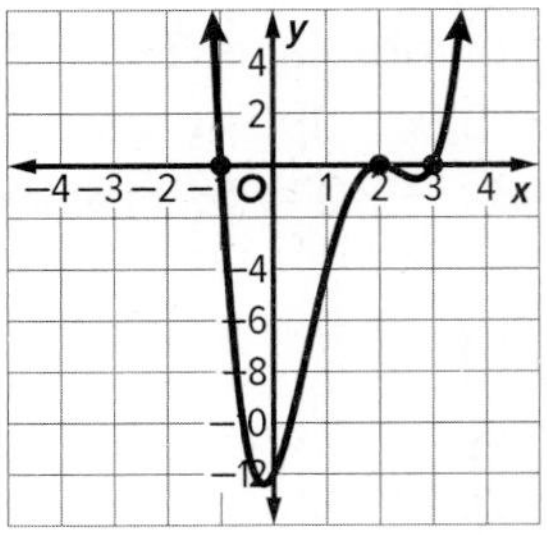

57a. $(x - 5)^2(7x + 9.5)\pi = 234\pi$; 7, $\frac{23 \pm 3\sqrt{65}}{28}$

57b. height 41.5 m, radius 2 m; $x = 7$ is the only reasonable solution in the context of the situation. The other possible values of x result in negative measures.

59. The graph is accurate as far as the location of the zeros, but does not consider any vertical dilation. It could be improved by finding more points between the roots.

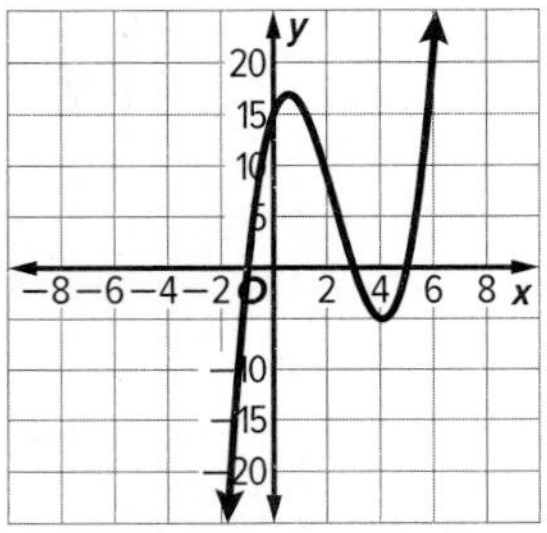

61a.

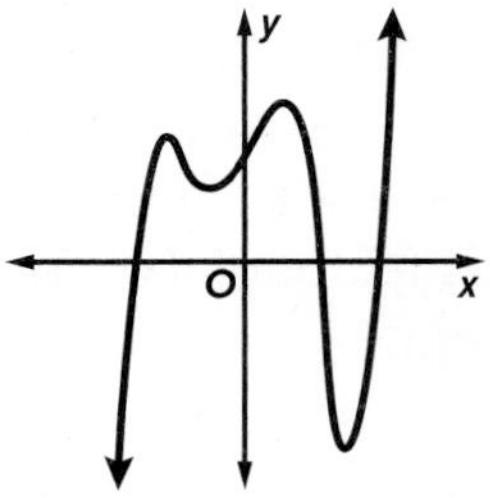

61b.

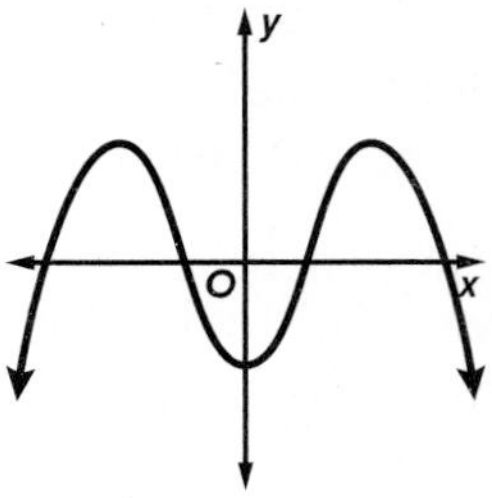

61c. 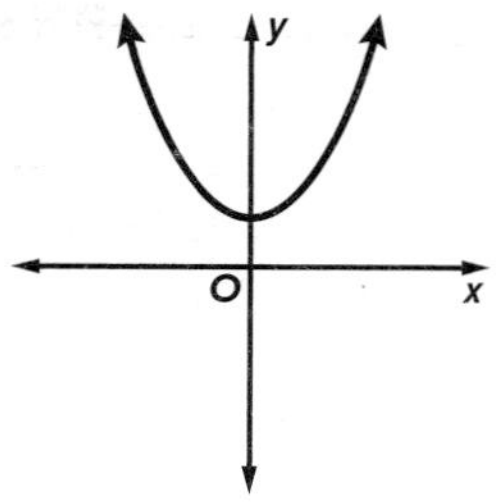

63. $r^4 + 1 = 0$; Sample answer: The equation has imaginary solutions and all of the others have real solutions.

65. Sample answer: To determine the number of positive real roots, determine how many time the signs change in the polynomial as you move from left to right. In this function there are 3 changes in sign. Therefore, there may be 3 or 1 positive real roots. To determine the number of negative real roots, I would first evaluate the polynomial for $-x$. All of the terms with an odd-degree variable would change signs. Then I would again count the number of sign changes as I move from left to right. There would be only one change. Therefore there may be 1 negative root.

Module 5 Review

1. A **3.** A, B, D

5. Set the expression for the volume of the first figure equal to the expression for the surface area of the second figure. Solve the equation $x^3 - 9x = 8x^2$. Solving the equation gives the solutions -1, 0, and 9. The value of x cannot be 0 because when substituting 0 for x in the original expressions, the volume of the first figure is $(0)3 - 9(0) = 0 - 0 = 0$ and surface area of the second figure is $8(0)^2 = 8(0) = 0$.

Volume and surface area cannot be 0, so the value of x is -1 or 9.

7. $x = 6$

9. B **11.** C **13.** B

15.

x	Real Root	Imaginary Root	Not a Root
-2			X
-1	X		
1	X		
2			X
$-2i$		X	
$-i$			X
i			X
$2i$		X	

17. A

Selected Answers

Module 6

Quick Check

1. between 0 and 1, and between 3 and 4

3. between −1 and 0, and between 1 and 2

5. $5x + 3$ **7.** $2x^2 + 5x - 6$

Lesson 6-1

1. $(f + g)(x) = -2x + 5$; $(f - g)(x) = 6x - 5$; $(f \cdot g)(x) = -8x^2 + 10x$; $\left(\frac{f}{g}\right)(x) = \frac{2x}{-4x + 5}, x \neq \frac{5}{4}$

3. $(f + g)(x) = 3x - 9$; $(f - g)(x) = -x + 5$; $(f \cdot g)(x) = 2x^2 - 11x + 14$; $\left(\frac{f}{g}\right)(x) = \frac{x - 2}{2x - 7}, x \neq \frac{7}{2}$

5. $(f + g)(x) = x^2 + 3x + 1$; $(f - g)(x) = -3x^2 - 3x + 11$; $(f \cdot g)(x) = -2x^4 - 3x^3 + 17x^2 + 18x - 30$; $\left(\frac{f}{g}\right)(x) = \frac{-x^2 + 6}{2x^2 + 3x - 5}, x \neq 1 \text{ or } -\frac{5}{2}$

7a. $(a - b)(x)$; $(a - b)(x) = 85x + 1750$

7b.

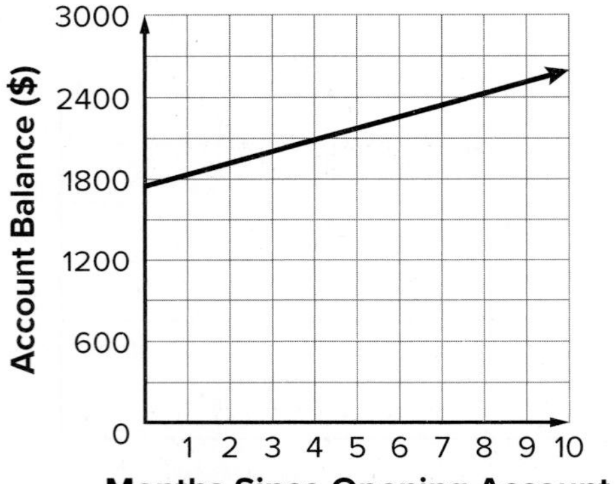

9. $f \circ g = \{(-4, 4)\}$, D = $\{-4\}$, R = $\{4\}$; $g \circ f = \{(-8, 0), (0, -4), (2, -5), (-6, -1)\}$, D = $\{-6, 0, 2\}$, R = $\{-5, -4, -1, 0\}$

11. $f \circ g$ is undefined, D = ∅, R = ∅; $g \circ f$ is undefined, D = ∅, R = ∅

13. $[f \circ g](x) = 2x + 10$, D = {all real numbers}, R = {all even numbers}; $[g \circ f](x) = 2x + 5$, D = {all real numbers}, R = {all odd numbers}

15. $[f \circ g](x) = x^2 - 6x - 2$, D = {all real numbers}, R = $\{y \mid y \geq -11\}$; $[g \circ f](x) = x^2 + 6x - 8$, D = {all real numbers}, R = $\{y \mid y \geq -17\}$

17. $p(x) = 0.85x$; $t(x) = 1.065x$; $t[p(x)] = 1.065(0.85x)$; $1222.09

19. Sample answer: The order of the discounts does not matter. Either composition results in a final cost of $54.91.

21. 15 **23.** 1 **25.** 12 **27.** 189 **29.** 21

31. −3 **33.** 9 **35.** 440

37a. $V[r(t)] = \frac{\pi t^3}{6} + 2\pi t^2 + 8\pi t + \frac{32}{3}\pi$

37b. 2145 m^3

39a. $(f + g)(-1) = 4$

39b. $(h - g)(0) = 8$

39c. $(f \cdot h)(4) = 5$

39d. $\left(\frac{f}{g}\right)(3) = 4$

39e. $\left(\frac{g}{h}\right)(2) = 0$

39f. $\left(\frac{g}{f}\right)(1) =$ undefined

41. Sample answer: $f(x) = x - 9$, $g(x) = x + 5$

43a. D = {all real numbers}

43b. D = $\{x \mid x \geq 0\}$

Lesson 6-2

1. {(6, −8), (−2, 6), (−3, 7)}

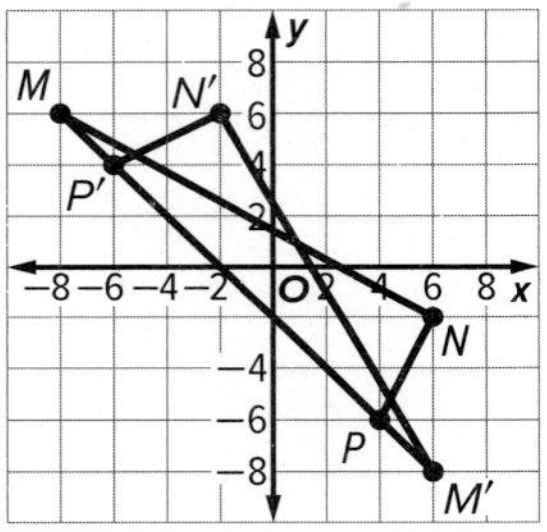

3. {(−1, 8), (−1, −8), (−8, −2), (−8, 2)}

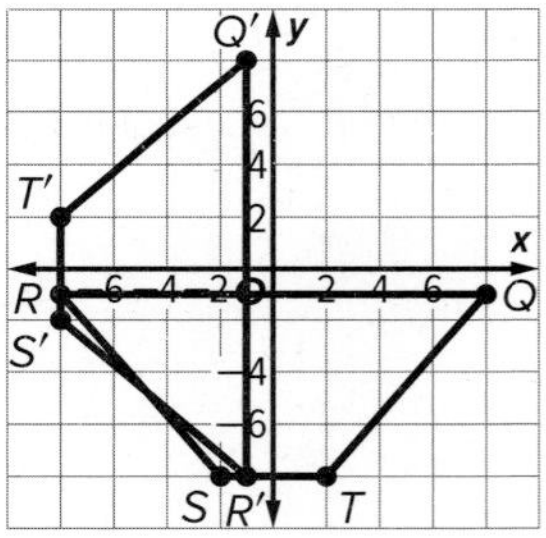

5. $f^{-1}(x) = x - 2$

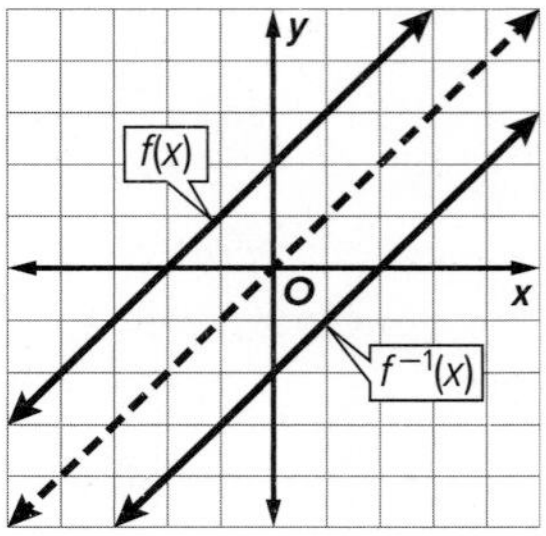

7. $f^{-1}(x) = -\frac{x}{2} + \frac{1}{2}$

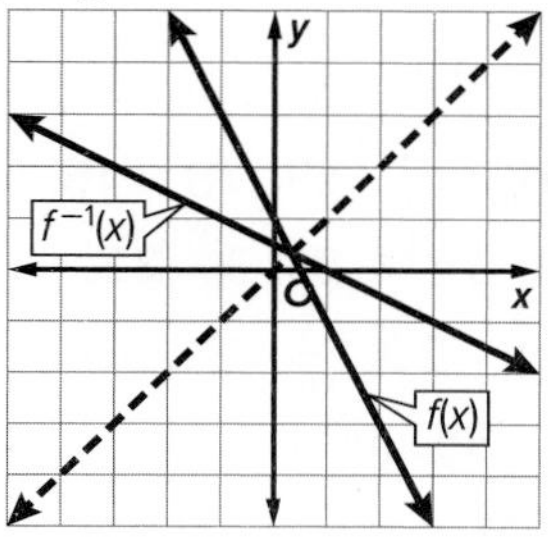

9. $f^{-1}(x) = -\frac{3}{5}(x + 8)$

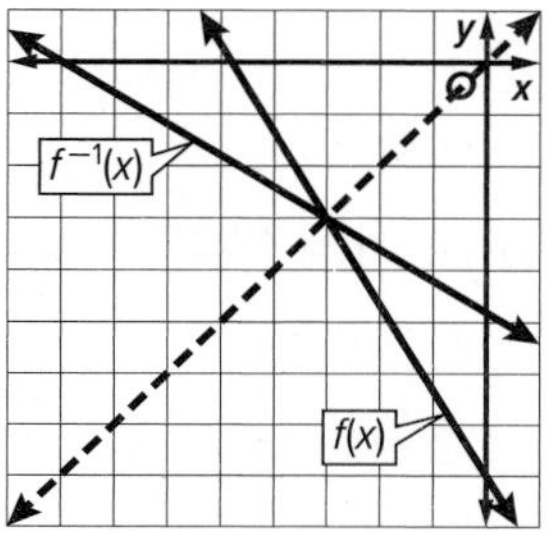

11. $f^{-1}(x) = \frac{1}{4}x$

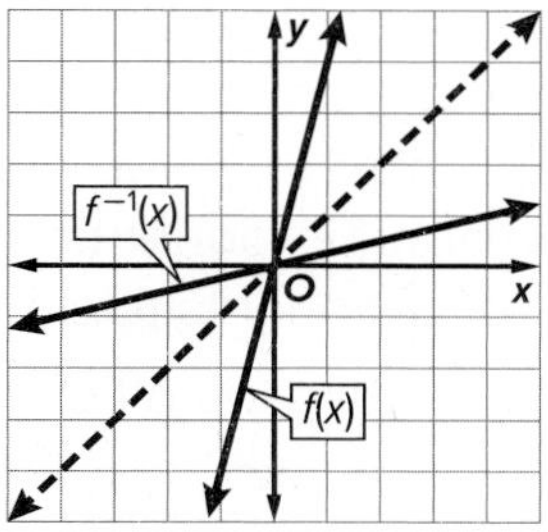

13. $f^{-1}(x) = \pm\frac{\sqrt{5x}}{5}$

If the domain of $f(x)$ is restricted to $(-\infty, 0]$, then the inverse is $f^{-1}(x) = -\frac{\sqrt{5x}}{5}$.
If the domain of $f(x)$ is restricted to $[0, \infty)$, then the inverse is $f^{-1}(x) = \frac{\sqrt{5x}}{5}$.

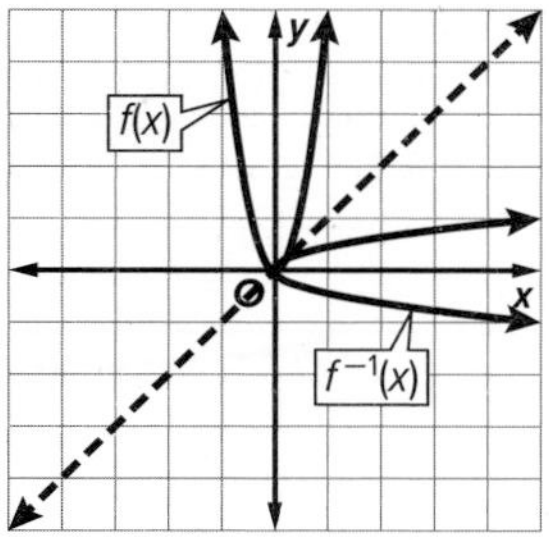

15a. $p^{-1}(x) = 14x$; Sample answer: The inverse converts stones to pounds, where x is the weight in stones.

15b. About 35 pounds

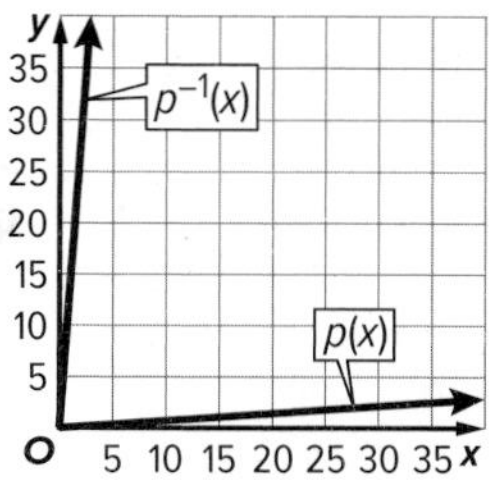

17. no **19.** yes **21.** no **23.** yes

25. $x = 4$

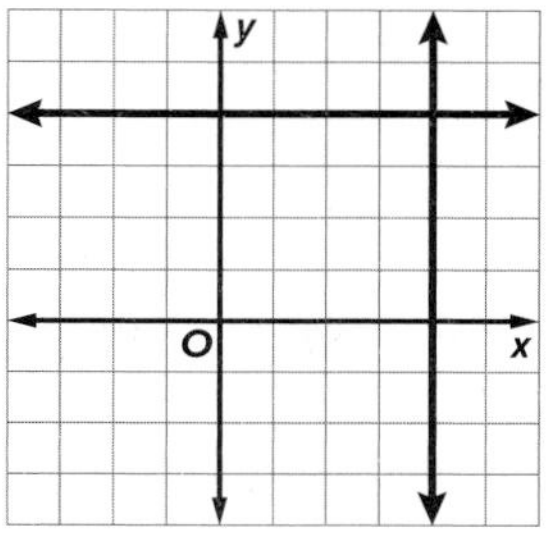

27. $f^{-1}(x) = x - 2$

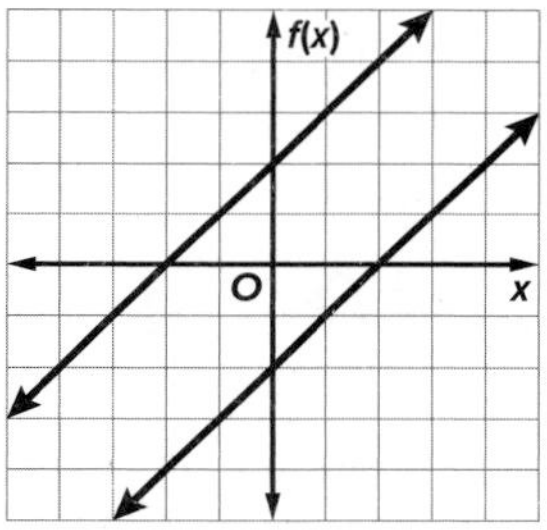

29. $f^{-1}(x) = \pm\sqrt{2x + 2}$
If the domain of $f(x)$ is restricted to $(-\infty, 0]$, then the inverse is $f^{-1}(x) = \sqrt{2x + 2}$. If the domain of $f(x)$ is restricted to $[0, \infty)$, then the inverse is $f^{-1}(x) = -\sqrt{2x + 2}$.

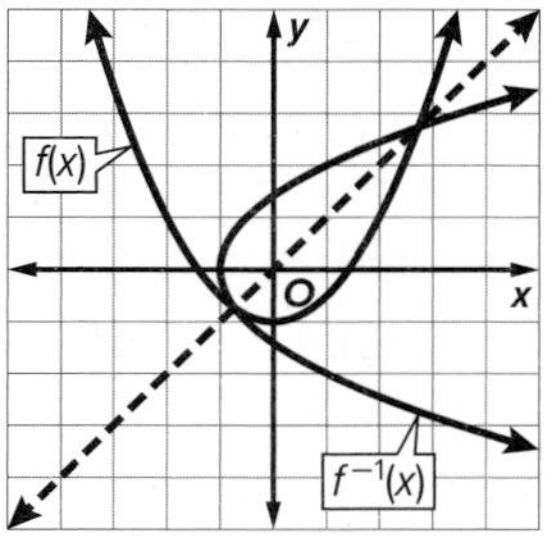

31. yes **33.** no **35.** yes

37. $\{x \mid x \geq 0\}$ or $\{x \mid x \leq 0\}$; $\{x \mid x \geq 5\}$

39. $\{x \mid x \geq -6\}$; $\{x \mid x \geq 0\}$

41.

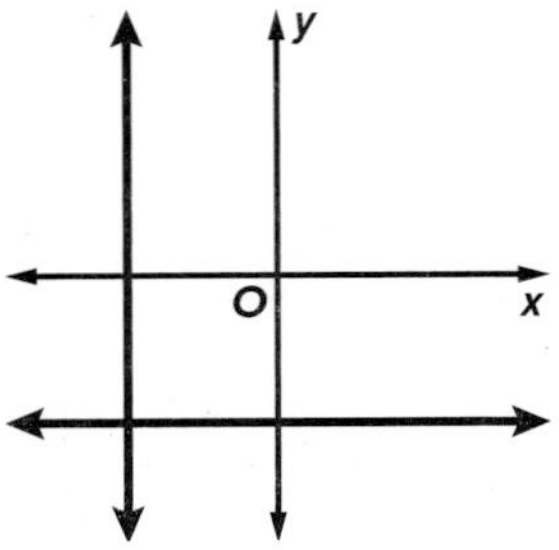

43.

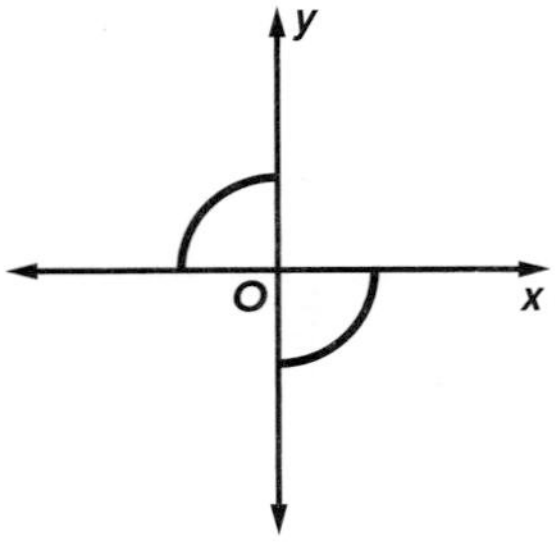

45. $f^{-1}(x) = 220 - \frac{x}{0.85}$; Sample answer: The inverse function gives the age of the clients based on their maximum target heart rate.

47a. $V(x) = x(12 - 2x)(16 - 2x)$; Sample answer: The value of x must be between 0 and 6 because the length of x must be positive, but the length of the side of the box $(12 - 2x)$ cannot be less than 0.

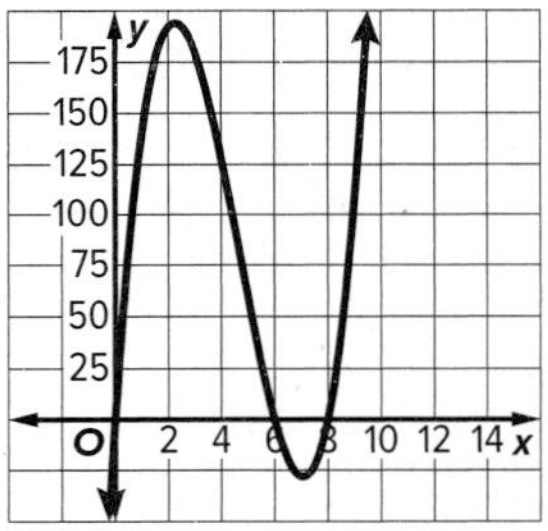

47b. No; sample answer: even with the restriction, the function does not pass the horizontal line test, so its inverse will not be a function.

49. Sample answer: Sometimes; $y = \pm\sqrt{x}$ is an example of a relation that is not a function, with an inverse being a function. A circle is an example of a relation that is not a function with an inverse not being a function.

51. Sample answer: $f(x) = x$ and $f^{-1}(x) = x$ or $f(x) = -x$ and $f^{-1}(x) = -x$

53. Sample answer: One of the functions carries out an operation on 5. Then the second function that is an inverse of the first function reverses the operation on 5. Thus, the result is 5.

Lesson 6-3

1. $\pm 11x^2y^8$ **3.** $\pm 7x^2$ **5.** $-9a^8b^{10}c^6$ **7.** $2|(x-3)^3|$
9. $3|x-4|$ **11.** $|a^3|$ **13.** $\sqrt[5]{8}$ **15.** $\sqrt{x^9}$
17. $5^{\frac{1}{3}}x^{\frac{1}{3}}y^{\frac{2}{3}}$ **19.** $6^{\frac{1}{3}}t^{\frac{2}{3}}$ **21.** 3 **23.** $\frac{1}{64}$
25. 64 **27.** $x^{\frac{11}{15}}$ **29.** $\frac{b^{\frac{1}{4}}}{b}$ **31.** $3b^6c^4$
33. $3|(x+4)|$ **35.** $(y^3+5)^6$
37. $14|c^3|d^2$ **39.** $-3a^5b^3$ **41.** a^3
43. $b^{\frac{1}{4}}$ **45.** $\frac{d^{\frac{1}{6}}}{d}$
47a. $5\sqrt{30}$ cm^3
47b. $32\sqrt{3}$ cm^3
47c. 288 cm^3
49. $c^{\frac{1}{2}}$ **51.** $6^{\frac{1}{4}}h^{\frac{1}{2}}j^{\frac{1}{2}}$ **53.** $\frac{xy\sqrt[3]{z^2}}{z}$ **55.** 21 feet
57. $\sqrt[b]{m^{3b}} = m^3$
59. $x < 0$; If $x < 0$, then $\sqrt{x^2} = -x$.

61. Sample answer: It may be easier to simplify an expression when it has rational exponents because all the properties of exponents apply. We do not have as many properties dealing directly with radicals. However, we can convert all radicals to rational exponents, and then use the properties of exponents to simplify.

63. when x or $y = 0$ and the other variable is ≥ 0

65. $0 < x < 1, x < -1$ **67.** $2\sqrt[3]{2xy}$

Lesson 6-4

1. $D = \{x \mid x \geq 9\}$; $R = \{y \mid y \geq 0\}$
3. $D = \{x \mid x \geq 0\}$; $R = \{y \mid y \leq 0\}$
5. $D = \left\{x \mid x \geq \frac{4}{3}\right\}$; $R = \{y \mid y \geq 0\}$
7. $D = \{x \mid x \geq -1\}$; $R = \{y \mid y \leq 0\}$; reflected in the x-axis, compressed vertically, and translated left 1 unit

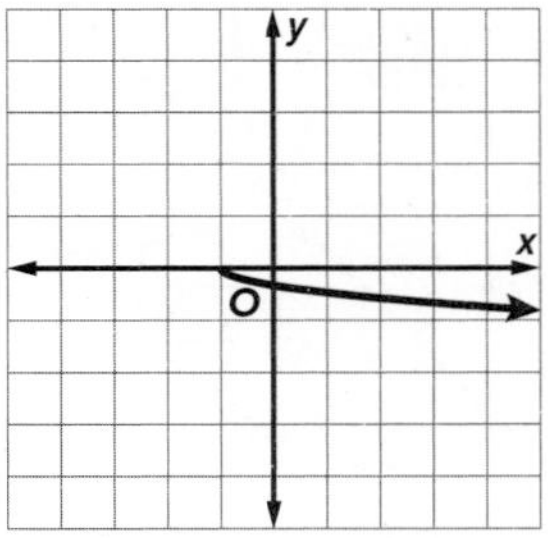

9. D = {$x \mid x \geq 0$}; R = {$y \mid y \geq 0$}
stretched vertically

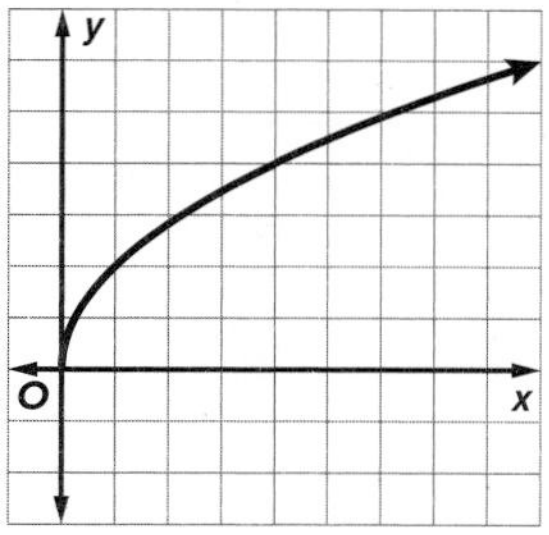

11. D = {$x \mid x \geq 0$}; R = {$y \mid y \geq -5$}; stretched vertically and translated down 5 units

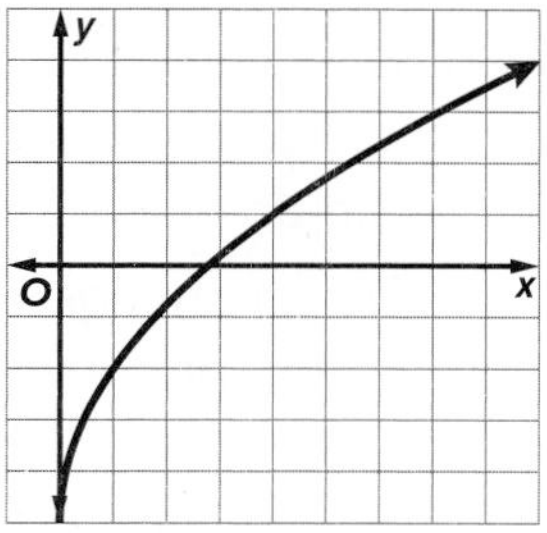

13a. $t = \frac{1}{4}\sqrt{d}$

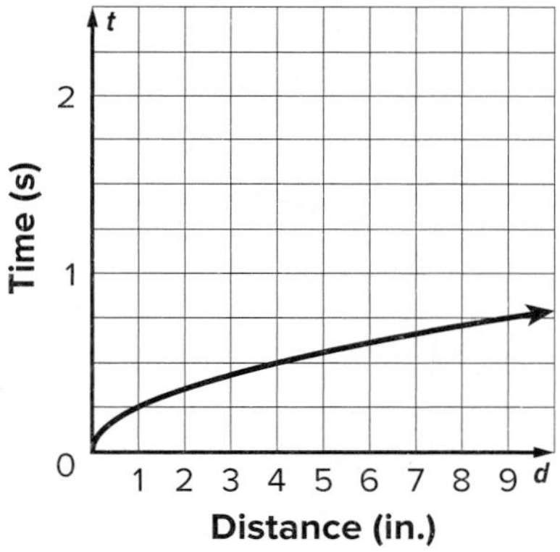

13b. D = {$d \mid d \geq 0$}; R = {$t \mid t \geq 0$}; increasing as $x \to \infty$; positive for $x > 0$; As $x \to \infty$, $y \to \infty$.

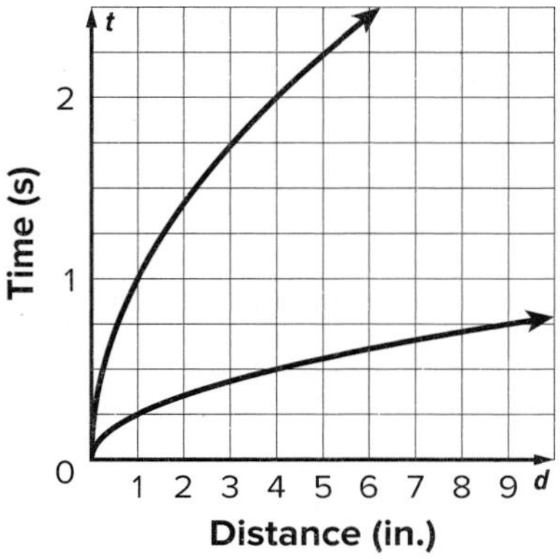

13c. compressed vertically

15.

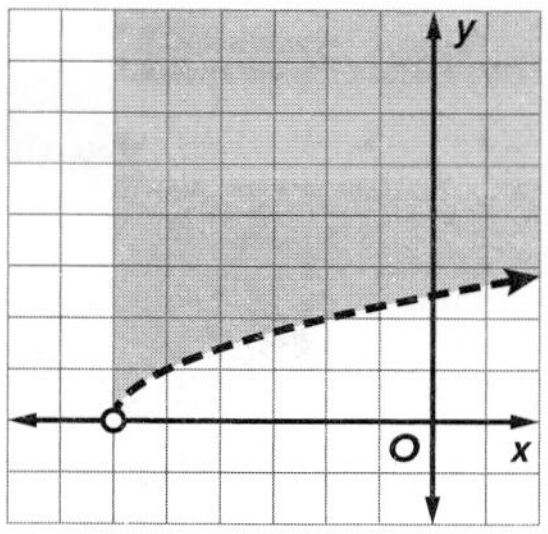

17.

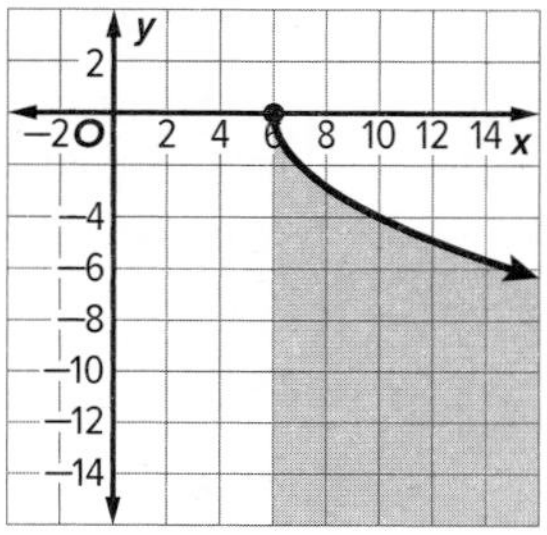

19.

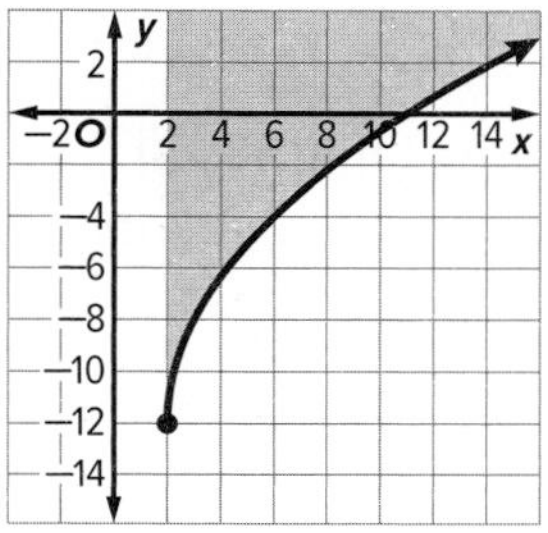

21. D = $(-\infty, \infty)$; R = $(-\infty, \infty)$

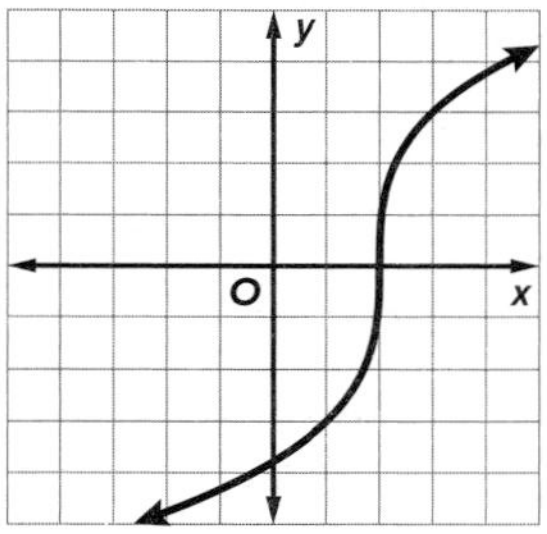

23. D = $(-\infty, \infty)$; R = $(-\infty, \infty)$

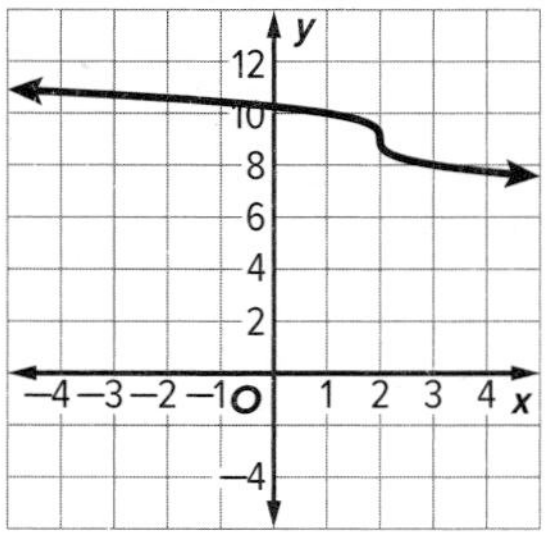

25a.

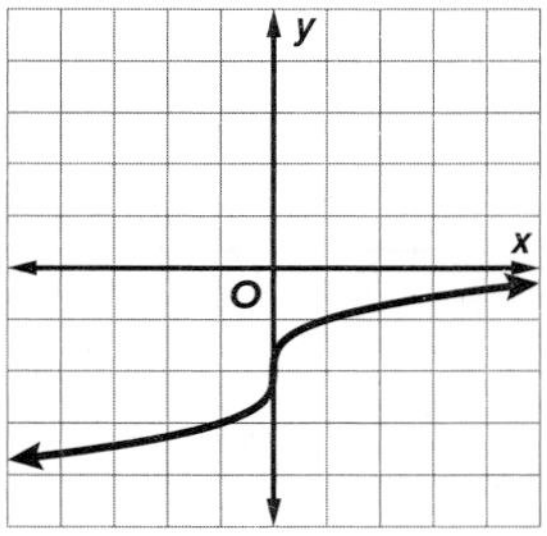

25b.

	$p(x) = \sqrt[3]{x} - 2$	$q(x)$
Domain	all real numbers	$\{x \mid x \geq 0\}$
Range	all real numbers	$\{y \mid y \leq -3\}$
Intercepts	x-int: 8; y-int: -2	x-int: 9; y-int: -3
Increasing/ Decreasing	increasing as $x \longrightarrow \infty$	increasing as $x \longrightarrow \infty$
Positive/ Negative	negative for $x < 8$; positive for $x > 8$	negative for $x < 9$; positive for $x > 9$
End Behavior	as $x \longrightarrow -\infty$, $p(x) \longrightarrow -\infty$ as $x \longrightarrow \infty$, $p(x) \longrightarrow \infty$	as $x \longrightarrow -\infty$, $q(x) \longrightarrow -2$ as $x \longrightarrow \infty$, $q(x) \longrightarrow \infty$

27. $f(x) = \sqrt[3]{x - 2} + 1$

29. D = $\{x \mid x \geq 5\}$; R = $\{f(x) \mid f(x) \geq -6\}$ stretched vertically translated 5 units right and 6 units down

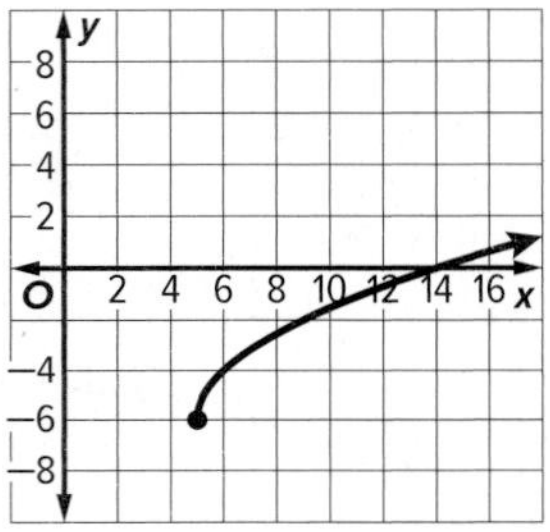

31. D = $\{x \mid x \geq 1\}$; R = $\{f(x) \mid f(x) \leq -4\}$ compressed vertically, translated 1 unit right and 4 units down, reflected across the x-axis

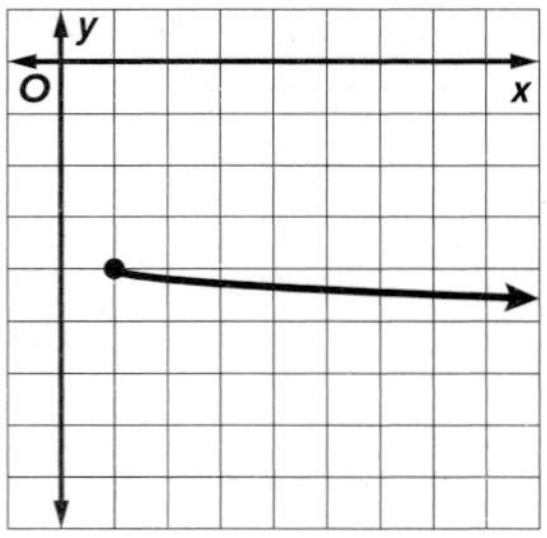

33. D = $(-\infty, \infty)$; R = $(-\infty, \infty)$

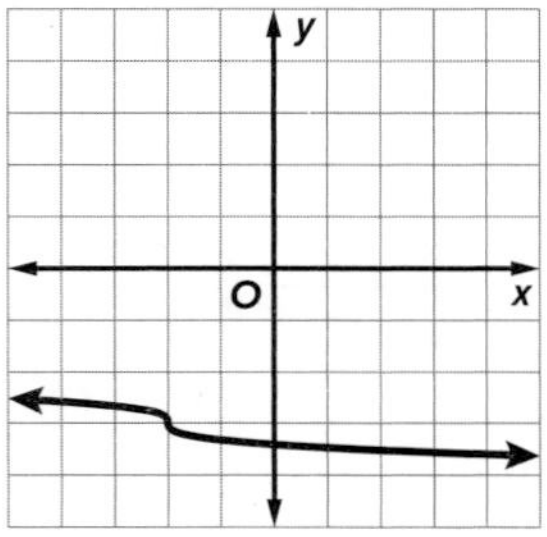

35.

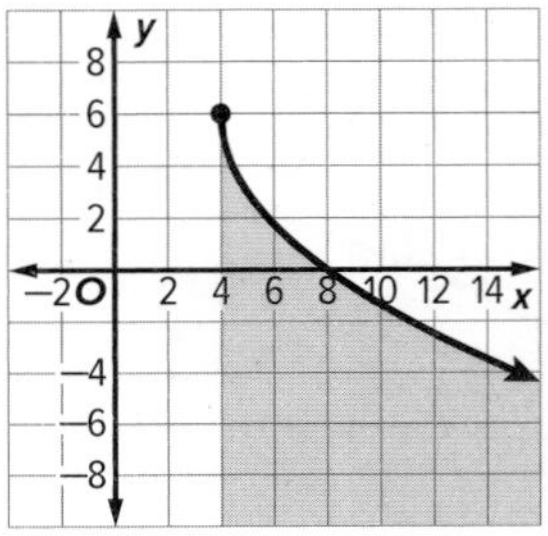

37. $f(x) = -\sqrt{x + 4} - 2$

39a. $f(x)$ has the greater maximum value because its maximum, $\frac{13}{2}$, is greater than 6, the maximum value of $g(x)$.

39b. The domain of $f(x)$ is $x \geq -3$, since any values less than -3 produce a negative value under the radical. The domain of $g(x)$ is $x \geq \frac{5}{2}$.

39c. The average rate of change over the interval is $-\frac{1}{7}$ for $f(x)$. It appears that the rate of change for $g(x)$ is the same.

41. D = $\{V \mid -\infty < V < \infty\}$ or $(-\infty, \infty)$; R = $\{r(V) \mid -\infty < r(V) < \infty\}$ or $(-\infty, \infty)$; End behavior: The values of r increase as the values of V increase.

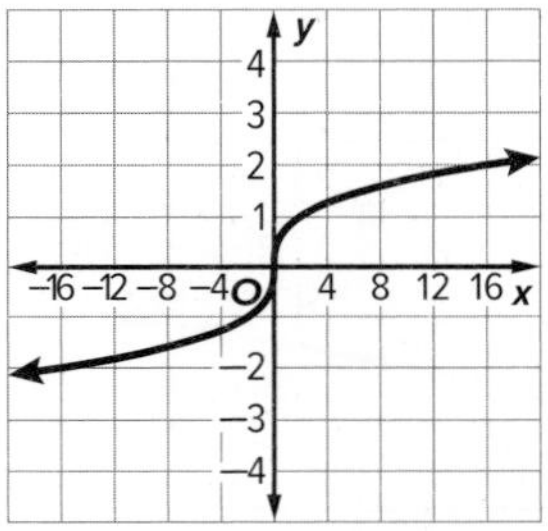

43. Sample answer: $y = -\sqrt{x + 4} + 6$

45. Sample answer: The domain is limited because square roots of negative numbers are imaginary. The range is limited due to the limitation of the domain.

47a. Sample answer: The original is $y = x^2 + 2$ and inverse is $y = \pm\sqrt{x - 2}$.

47b. Sample answer: The original is $y = \pm\sqrt{x + 4}$ and inverse is $y = (x - 4)^2$.

Lesson 6-5

1. $6a^4b^2\sqrt{2b}$ **3.** $2a^8b^4\sqrt{6cb}$ **5.** $2|ab|\sqrt[4]{4}$
7. $\frac{5}{6}|r|\sqrt{t}$ **9.** $\frac{4h \cdot \sqrt[5]{3h^3}}{3f}$ **11.** $8\sqrt{2}$ **13.** $\sqrt{5}$
15. $5\sqrt{7x} - \sqrt{14}$ **17.** $120|y|\sqrt{2z}$ **19.** $144|a|b^2\sqrt{2}$

21. $15xy\sqrt{14xy}$

23. $18 + 5\sqrt{6}\text{m}^2$

25. $56\sqrt{3} + 42\sqrt{6} - 36\sqrt{2} - 54$ **27.** 1260

29. $\frac{a^2\sqrt{5ab}}{b^7}$ **31.** $\frac{\sqrt[3]{150x^2y^2}}{5y}$

33. $6\sqrt{3} + 6\sqrt{2}$ **35.** $\frac{20 - 7\sqrt{3}}{11}$ **37.** $\frac{3\sqrt{7} + 3\sqrt{35}}{4}$

39. $2yz^4\sqrt[3]{2y}$

41. $\frac{(x + 1)(\sqrt{x} + 1)}{x - 1}$ or $\frac{x\sqrt{x} + \sqrt{x} + x + 1}{x - 1}$

43. $4\sqrt{3}$ **45.** $28\sqrt{7a}$ **47.** $10\sqrt[3]{28}$ in. **49.** $\frac{\sqrt{tu}}{u}$

51a. John: $\frac{0.8 - 4\sqrt{0.02}}{0.02}$ or $40 - 20\sqrt{2}$ min;
Jay: $\frac{0.8 - 4\sqrt{0.05}}{-0.01}$ or $40\sqrt{5} - 80$ min

51b. $2 - \sqrt{2} + \sqrt{5} + \frac{\sqrt{10}}{2}$

53a. $3x$; $(\sqrt{x} \cdot x) + (\sqrt{x} \cdot \sqrt{4x}) = (\sqrt{x^2} + \sqrt{4x^2}) = x + 2x = 3x$

53b. $6x$; $(\sqrt{x} \cdot \sqrt{x}) + (\sqrt{x} \cdot \sqrt{4x}) + (\sqrt{x} + \sqrt{9x}) = (\sqrt{x^2} + \sqrt{4x^2} + \sqrt{9x^2}) = x + 2x + 3x = 6x$

53c. $10x$; $(\sqrt{x} \cdot \sqrt{x}) + (\sqrt{x} \cdot \sqrt{4x}) + (\sqrt{x} \cdot \sqrt{9x}) + (\sqrt{x} \cdot \sqrt{16x}) = (\sqrt{x^2} + \sqrt{4x^2} + \sqrt{9x^2} + \sqrt{16x^2}) = x + 2x + 3x + 4x = 10x$

53d. $(\sqrt{x} \cdot \sqrt{x}) + (\sqrt{x} \cdot \sqrt{4x}) + \ldots + (\sqrt{x} + \sqrt{n^2x}) = x + 2x + \ldots + nx = (1 + 2 + \ldots + n)x = \left(\frac{n(n + 1)}{2}\right)x$

55a. $6\sqrt[3]{2} + 1$; $r = \sqrt[3]{\frac{3}{4\pi} \cdot 72\pi} = \sqrt[3]{54} = 3\sqrt[3]{2}$;
$S = 3\sqrt[3]{2} + 2(2)$

55b. No; sample answer: a box with a volume of 384 cm^3 has a side that is $\sqrt[3]{384}$, or about 7.27 cm; $s = 6\sqrt[3]{2} + 4 \approx 11.55$ is the least possible value for a side of the gift box.

57. $a = 1, b = 256$; $a = 2, b = 16$; $a = 4, b = 4$; $a = 8, b = 2$

59. Sample answer: It is only necessary to use absolute values when it is possible that n could be odd or even and still be defined. It is when the radicand must be nonnegative in order for the root to be defined that the absolute values are not necessary.

Lesson 6-6

1. $\frac{1}{25}$ **3.** 8 **5.** 40 **7.** 11 **9.** 11 **11.** $-\frac{1}{2}$

13. no real solution; 16 is an extraneous solution.

15. 83 **17.** 61 **19.** 15,623 **21.** 43

23. 2.4 **25.** 2 **27.** 1.23 **29.** 36 cm

31. $\frac{1}{4}$ **33.** $\frac{81}{16}$

35. 2 **37.** 24.5

39. 3, 4 (extraneous solution)

41. 6, −5 (extraneous solution)

43. 12 units

45a. $\sqrt{x^2 + \frac{289}{x^2}}$

45b. $\frac{17}{12} \leq x \leq 12$

47a. 20.2

47b. Silver; $5.713 \times 10^{-15} \approx (1.2 \times 10^{-15})(A)^{\frac{1}{3}}$; $4.761 \approx (A)^{\frac{1}{3}}$; $107.917 \approx A$

49. Graph functions $f(x) = \sqrt{\frac{x}{6}}$ and $g(x) = \sqrt{\frac{x + 20}{4\pi}}$, where x is surface area of the cube. The graphs intersect at $x \approx 18.27$. The surface areas of the cube and sphere are approximately 18.27 cm^2 and 38.27 cm^2.

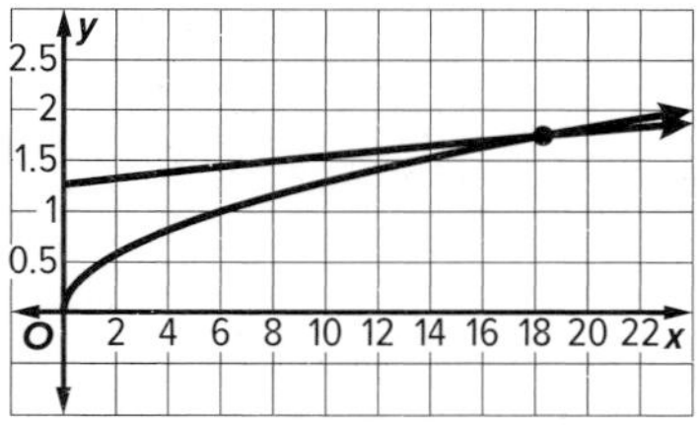

51. $\sqrt{x + 2} - 7 = -10$; Sample answer: This equation does not have a real solution while the other three equations do.

53. never;

$$\frac{\sqrt{(x^2)^2}}{-x} = x$$
$$\frac{x^2}{-x} = x$$
$$x^2 = (x)(-x)$$
$$x^2 \neq -x^2$$

55. Never; sample answer: The radicand can be negative.

Module 6 Review

1. D

3. C

5. $(f \circ g)(x) = 12x + 8$

7. If the domain of $f(x)$ is restricted to $(-\infty, -2]$, then the inverse is $f^{-1}(x) = -2 - \sqrt{x+1}$. If the domain of $f(x)$ is restricted to $[-2, \infty)$, then the inverse is $f^{-1}(x) = -2 + \sqrt{x+1}$.

9. D

11. The parent function is $f(x) = \sqrt{x}$ and $g(x) = a\sqrt{x-h} + k$ where $a = -2$, $h = -5$, and $k = -3$. Because $a < 0$ and $|a| > 1$, the graph of $g(x)$ is the graph of the parent function reflected across the x-axis and stretched vertically by a factor of $|a|$, or 2. Also, because $h < 0$, the graph of $g(x)$ is the graph of the parent function translated left $|h|$ units, or 5 units. Finally, because $k < 0$, the graph of $g(x)$ is the graph of the parent function translated down $|k|$ units, or 3 units.

13. B **15.** $2\sqrt{3}$ **17.** B **19.** A

Module 7

Module 7 Opener

1. a^{12} **3.** $\frac{-3x^6}{2y^3z^5}$

5. $f^{-1}(x) = \frac{1}{2}x - \frac{5}{2}$

7. $f^{-1}(x) = -\frac{1}{4}x$

Lesson 7-1

1. Domain: all real numbers; Range: all positive real numbers; y-intercept: (0, 1); Asymptote: $y = 0$; End Behavior: as $x \to -\infty$, $f(x) \to 0$ and as $x \to \infty$, $f(x) \to \infty$

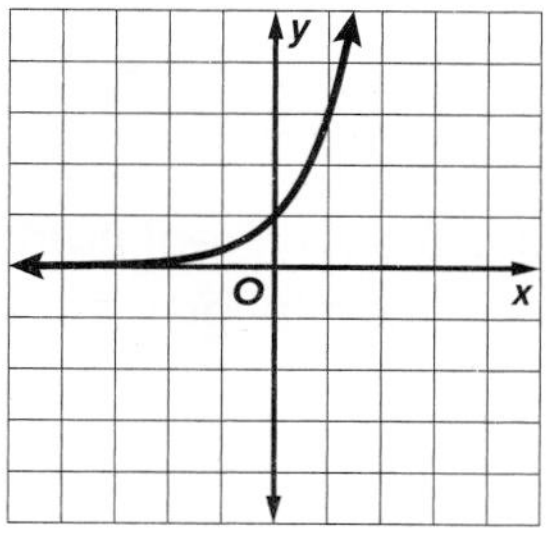

3. Domain: all real numbers; Range: all positive real numbers; y-intercept: (0, 1); Asymptote: $y = 0$; End Behavior: as $x \to -\infty$, $f(x) \to 0$ and as $x \to \infty$, $f(x) \to \infty$

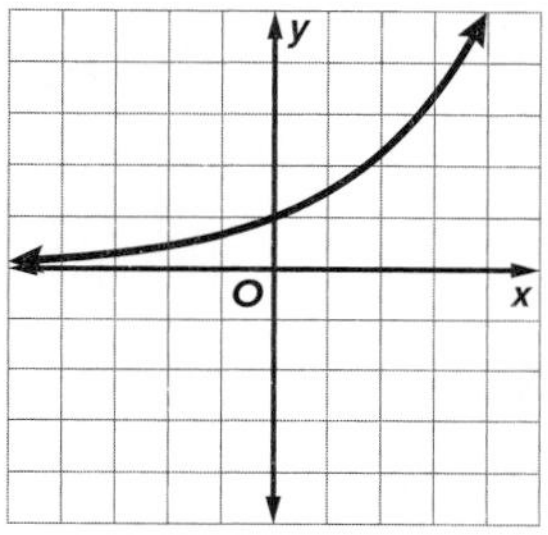

5.

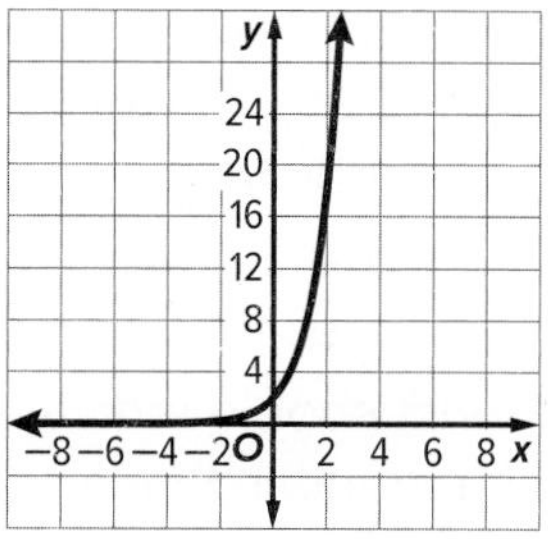

7.

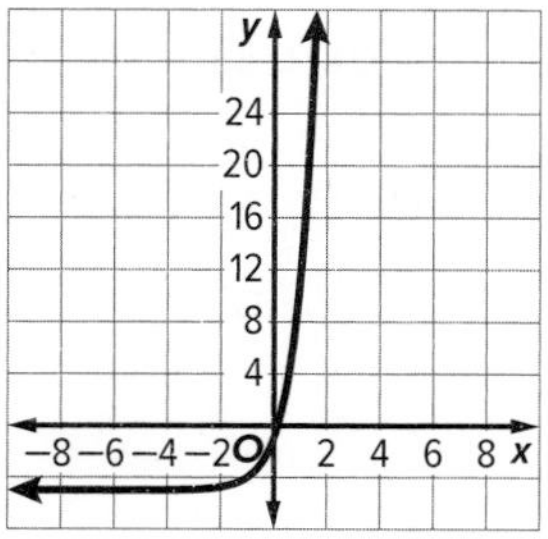

9.

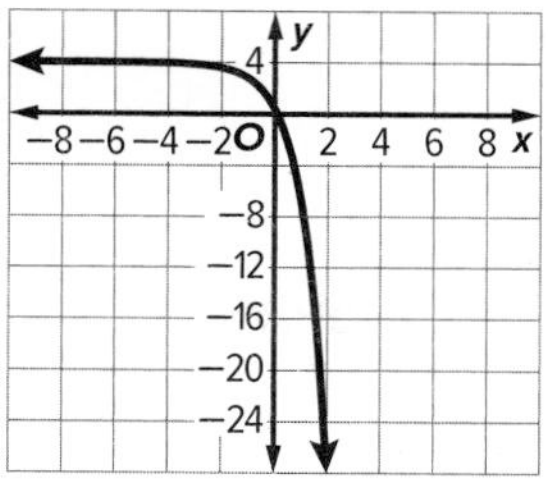

11. $k = -3$; $g(x) = 2^x - 3$

13. $k = -1$; $g(x) = -\left(\frac{3}{2}\right)^x$

15. about $201.73

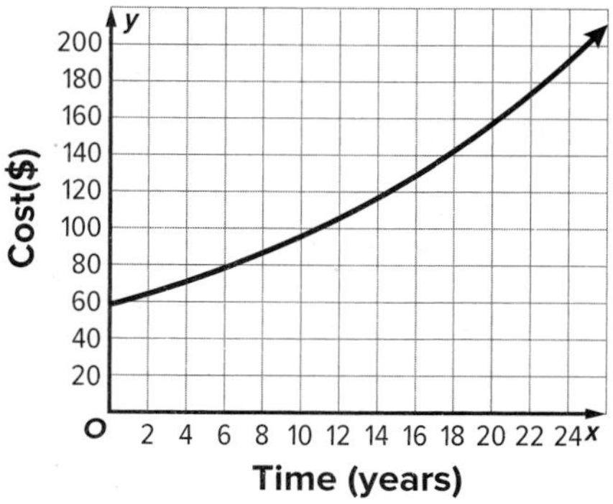

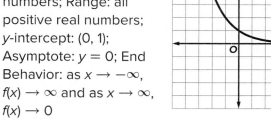

17. exponential growth

19. exponential decay

21. exponential decay

23. Domain: all real numbers; Range: all positive real numbers; y-intercept: (0, 1); Asymptote: $y = 0$; End Behavior: as $x \to -\infty$, $f(x) \to \infty$ and as $x \to \infty$, $f(x) \to 0$

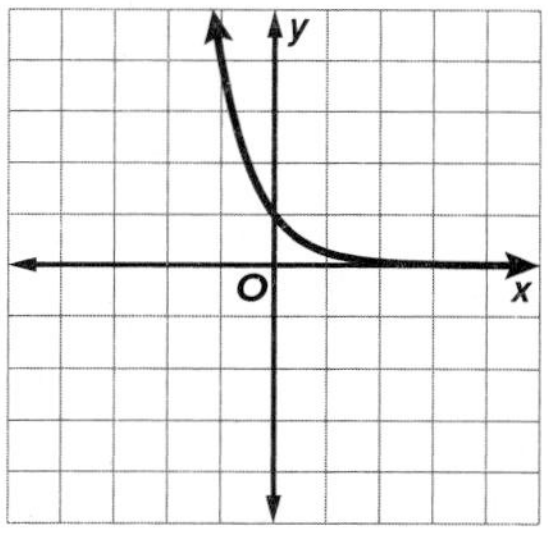

25. Domain: all real numbers; Range: all positive real numbers; y-intercept: (0, 1); Asymptote: $y = 0$; End Behavior: as $x \to -\infty$, $f(x) \to \infty$ and as $x \to \infty$, $f(x) \to 0$

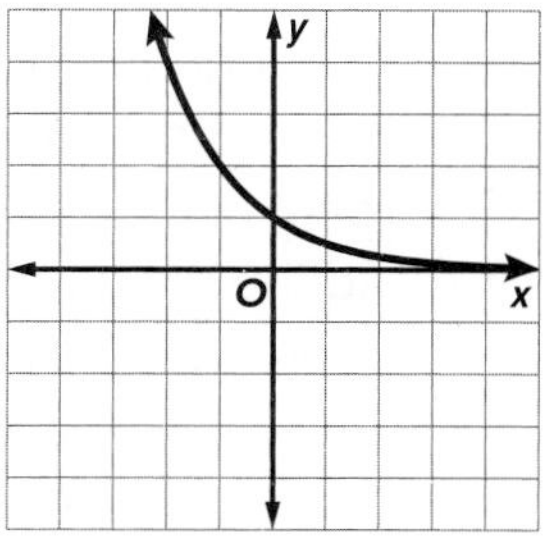

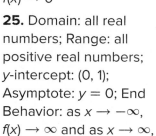

27.

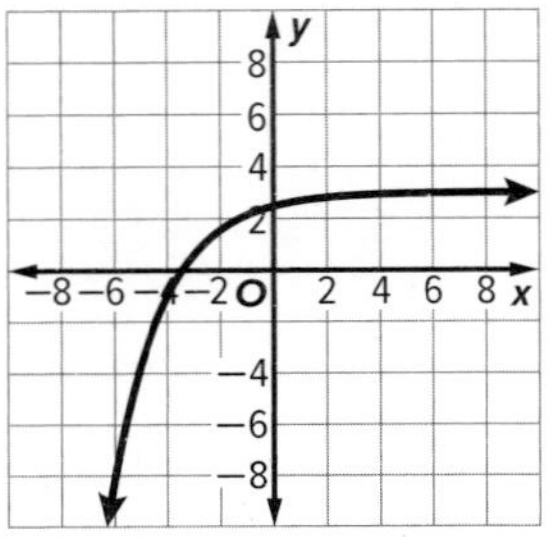

29.

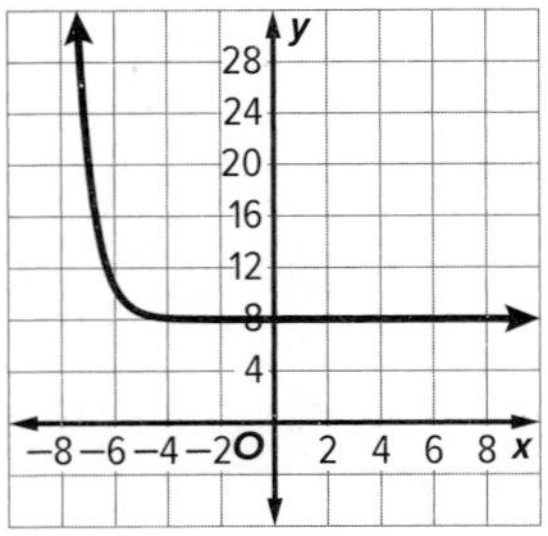

31.

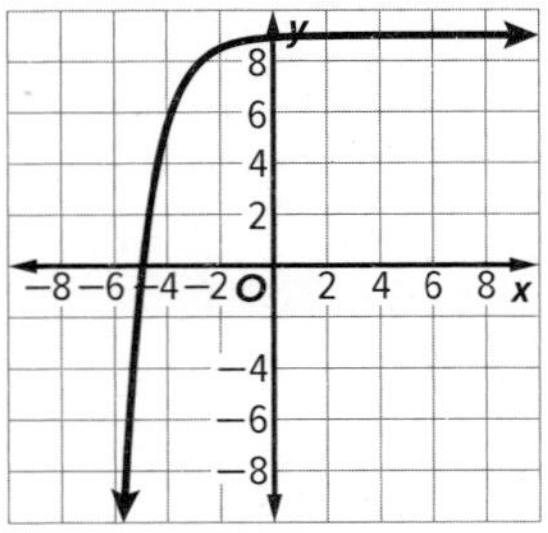

33a.

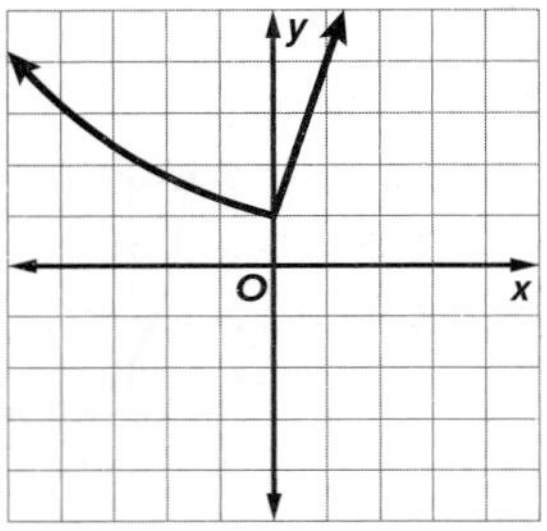

33b. Both $f(x)$ and $g(x)$ have a domain of all real numbers. $f(x)$ has a range of $y \geq 1$ and $g(x)$ has a range of $y > -1$, so the range of $f(x)$ is greater than the range of $g(x)$. $f(x)$ has no x-intercept and $g(x)$ has an x-intercept at (0, 0). $f(x)$ has a y-intercept at (0, 1) and $g(x)$ has a y-intercept at (0, 0), so the y-intercept of $f(x)$ is greater than the y-intercept of $g(x)$. $f(x)$ is increasing when $x > 0$ and $g(x)$ is increasing for all values of x. $f(x)$ is decreasing when $x < 0$ and $g(x)$ is never decreasing. $f(x)$ is positive for all values of x and $g(x)$ is positive when $x > 0$. $f(x)$ is never negative and $g(x)$ is negative when $x < 0$. $f(x)$ has a minimum at (0, 1) and $g(x)$ has a minimum slightly greater than −1, so the minimum of $f(x)$ is greater than the minimum of $g(x)$. Neither $f(x)$ nor $g(x)$ are symmetric. The end behavior of $f(x)$ is: As $x \to -\infty$, $f(x) \to \infty$, and as $x \to \infty$, $f(x) \to \infty$. The end behavior of $g(x)$ is: As $x \to -\infty$, $g(x) \to -1$ and as $x \to \infty$, $g(x) \to \infty$.

35. D = {all real numbers}, R = $\{y \mid y > 0\}$; exponential growth

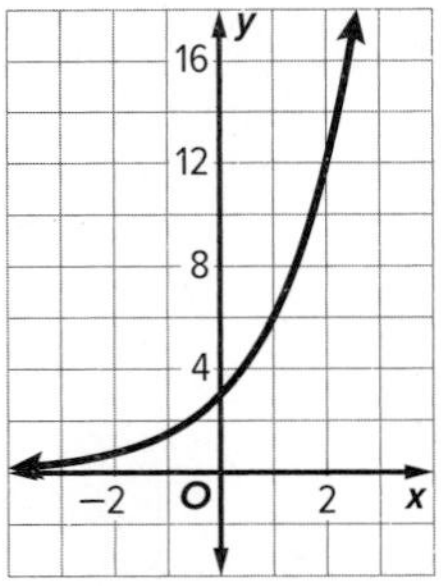

37. D = {all real numbers}, R = $\{y \mid y < 0\}$; exponential growth

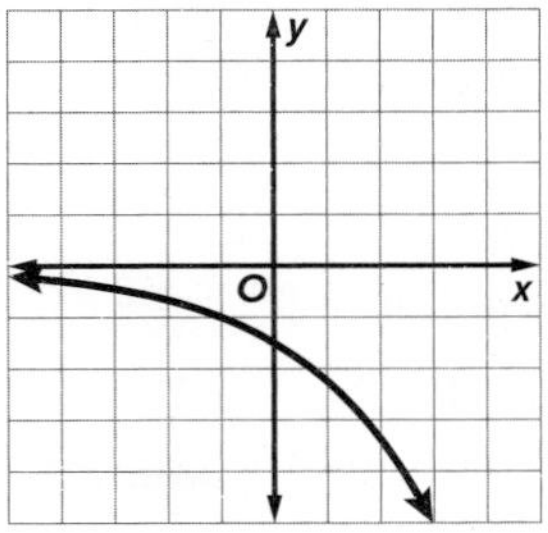

39. $g(x) = -4^x$

41a. Sample answer: Choose $a = 2$, so $y = 2(1)^x$. All values for y are 2. Students should indicate that $y = a$ (regardless of their choice of a) is a constant function, not an exponential one, since the value of y never changes.

41b. Sample answer: Choose $a = 2$ and $b = -2$, so $y = 2(-2)^x$. The table of values for y are $-\frac{1}{4}$, $\frac{1}{2}$, −1, 2, −4, 8, −16. Students should indicate that this function is not exponential. Since values of y alternate signs, the function does not continuously increase or decrease.

43a. $V(t) = 28{,}000(0.85)^t$;

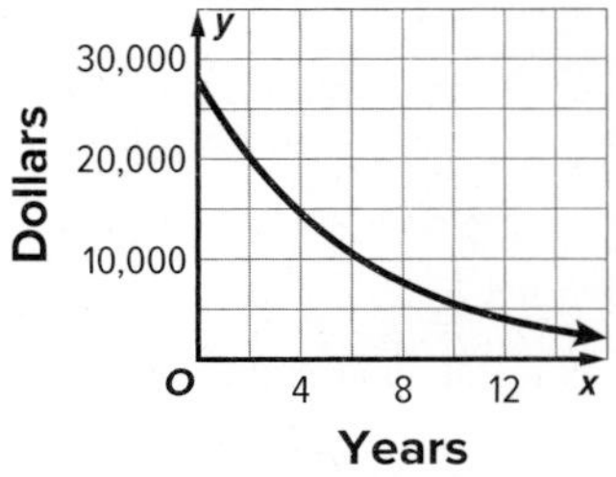

43b. Average annual value lost during Years 0–4 = $\frac{28{,}000(0.85)^0 - 28{,}000(0.85)^4}{5}$ = \$2676.77; Average annual value lost during Years 11–15: $\frac{28{,}000(0.85)^{11} - 28{,}000(0.85)^{15}}{5}$ = \$447.94

45. Reflect $f(x)$ in the y-axis and translate the result one unit up to obtain the graph of $g(x)$. The y-intercept of $g(x)$ is (0, 2), which was translated up 1 unit from the y-intercept of $f(x)$, (0, 1). $g(x)$ is decreasing on its entire domain, $f(x)$ is increasing on its entire domain. Both $g(x)$ and $f(x)$ are positive on its entire domain. The asymptote of $g(x)$ is $y = 1$, which was translated up one unit from the asymptote of $f(x)$, $y = 0$. For $g(x)$ the end behavior is as $x \to -\infty$, $g(x) \to \infty$, and as $x \to \infty$, $g(x) \to 1$. The end behavior of $f(x)$ is as $x \to -\infty$, $f(x) \to 0$, and as $x \to \infty$, $f(x) \to \infty$.

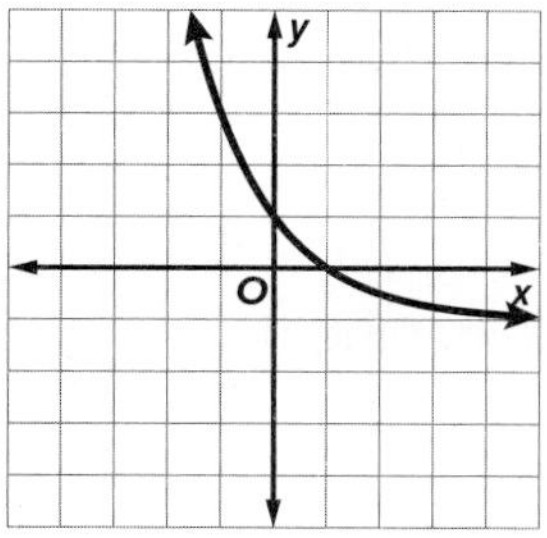

47a. A is the amount in grams, t is time in years, $A(t) = 27.3(0.9)^t$

47b. rate of change for [0, 2] = −2.5935, rate of change for [3, 5] ≈ −1.891; The amount of the compound decreases quickly at first and then more slowly as time passes.

49a. Always; sample answer: The domain of exponential functions is all real numbers, so (0, y) always exists.

49b. Sometimes; sample answer: The graph of an exponential function crosses the x-axis when $k < 0$.

49c. Sometimes; Sample answer: The function is not exponential if $b = 1$ or -1.

51. about 251 mg

53. Sample answer: The parent function, $g(x) = b^x$, is stretched if $|a|$ is greater than 1 or compressed if $|a|$ is less than 1. The graph is reflected in the x-axis when a is negative. The parent function is translated up k units if k is positive and down $|k|$ units if k is negative. The graph is reflected in the x-axis when a is negative. The parent function is translated h units to the right if h is positive and $|h|$ units to the left if h is negative.

Lesson 7-2

1. 4 **3.** 19

5. 2 **7a.** $y = 5000(1.05)^x$

7b. 33.0 years **9a.** $y = 23(0.995)^x$

9b. 85 years **11.** \$1476.79

13. \$5309.08 **15.** $x < 1$

17. $x \geq -2$ **19.** $x > -73$

21. 0 **23.** $a \leq -4$

25. $\frac{5}{3}$ **27.** $-\frac{2}{5}$

29. $c > \frac{3}{4}$ **31.** $\frac{11}{8}$

33. − 7 **35.** − 6

37. 1 **39.** 12

41. $t \geq -1$

43a. Ingrid: $y = 2(2.924)^x$; Alberto: $y = 32(1.383)^x$

43b. Ingrid: 427; Alberto: 162

43c. No; sample answer: A business cannot grow exponentially indefinitely.

45. $y = -1(2)^x$ **47.** $y = \frac{1}{2}(9)^x$

49. $y = -6\left(\frac{1}{4}\right)^x$ **51.** $y = (4)^x$

53.

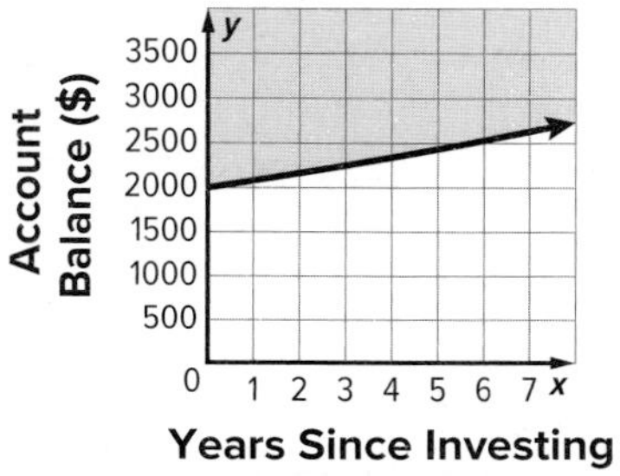

55. No; A function to model the concentration is $f(t) = 3(0.5)^{0.5t}$. The intersection with $g(t) = 0.6$ occurs at $t \approx 4.6$, meaning the concentration drops below the effective level before 3 P.M.

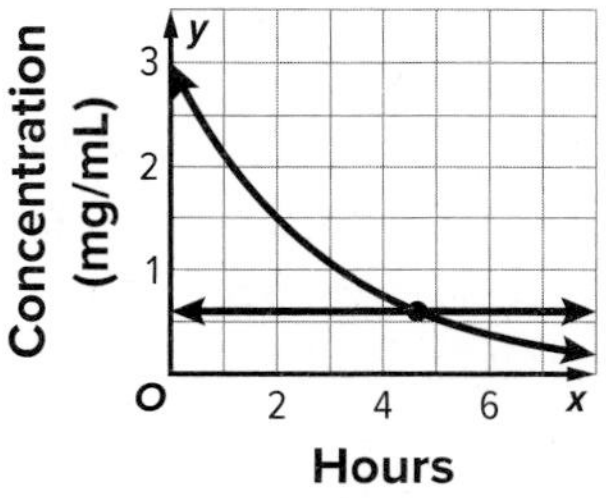

57a. Sample answer: According to the US Census Bureau and the American Association of University Professors, the average salary for a professor in 2017 was \$100,100. After 15 years, the professor's annual salary would be $100{,}100(1.02)^{15}$ = \$134,721.42.

57b. Sample answer: It depends on how long the professor intends to work. For years 1–16, the average salary with a 2% raise earns more. After year 17, the lower starting salary with a 3% raise earns more. The professor should consider the cumulative salary over the time he or she intends to work to determine which is better.

59. Beth; sample answer: Liz added the exponents instead of multiplying them when taking the power of a power.

61. Reducing the term will be more beneficial. The multiplier is 1.3756 for the 4-year and 1.3828 for the 6.5%.

63. Sample answer: $4^x \leq 4^2$

65. Sample answer: Divide the final amount by the initial amount. If n is the number of time intervals that pass, take the nth root of the answer.

Lesson 7-3

1. e^{11} **3.** e^2

5. $27e^{12x}$ **7.** $2e^3$ **9.** $-8e^6$

11a.

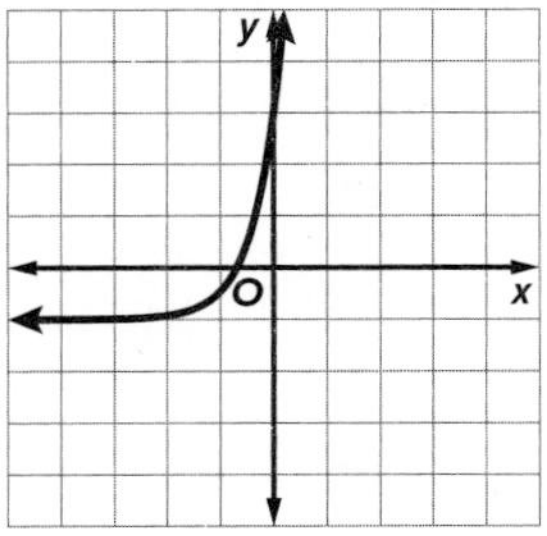

11b. The domain is all real numbers. The range is all real numbers greater than −1.

11c. ≈ 0.266 **13a.** $A = 5000e^{0.025t}$

13b. \$5956.23 **15a.** $A = 12{,}750e^{0.055t}$

15b. \$20,916.35 **17.** 0.17

19. 0.89 **21.** −1.29

23. $e^{-\frac{1}{4}}$ **25.** $-\frac{8}{125}e^{\frac{3}{4}x}$

27. $-24e^2$ **29.** −1.69

31. $f(x) = 900e^{0.012(5)}$; \$955.65

33.

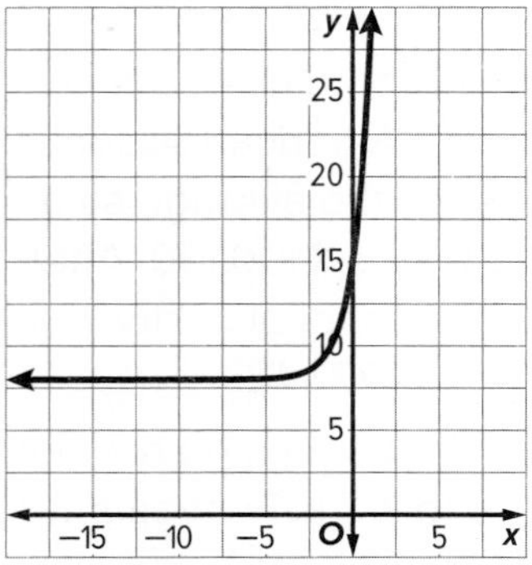

Domain: all real numbers; Range: all real numbers greater than 8; y-intercept: 15; no zeros; asymptote: $y = 8$; end behavior: as $x \to -\infty$, $f(x) \to 8$, as $x \to \infty$, $f(x) \to \infty$

35. Final amount corresponds to final population, initial amount corresponds to initial population, interest rate corresponds to growth rate, and time of growth is the same for both equations.

37. Sample answer: Since e is irrational, evaluating any expression with e will approximate its value, which introduces error. To minimize the error, do as much work as possible with the exact value (e) before evaluating.

39. $g(x) = 2e^x + 9$ because it has a range of all real numbers greater than 9 and the other two functions have a range of all real numbers greater than 0.

Lesson 7-4

1. not geometric **3.** geometric

5. not geometric

7. 1, 2, 4, 8, 16, 32

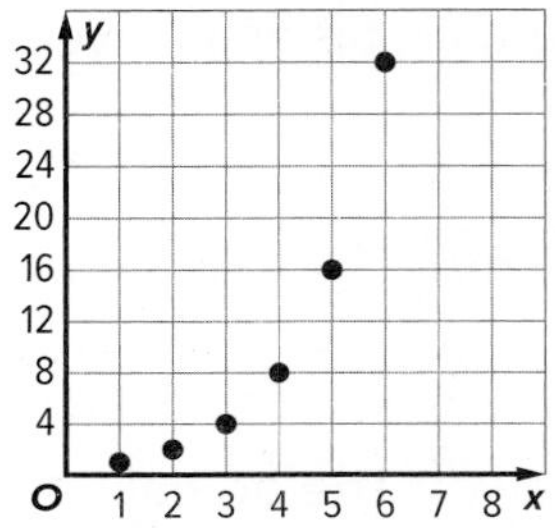

9. −1, −3, −9, −27, −81, −243

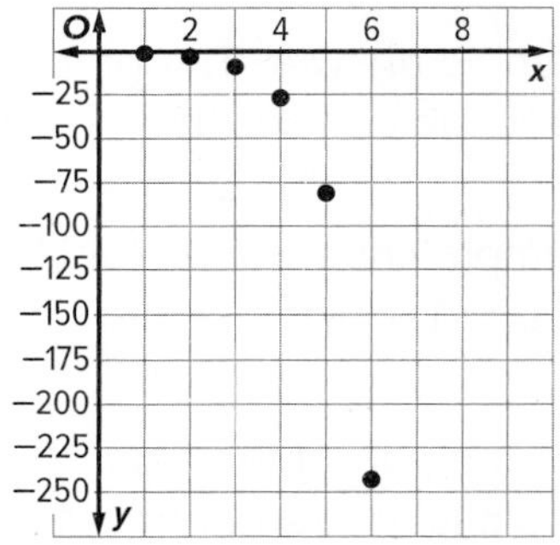

11. 512, −128, 32, −8, 2, −0.5

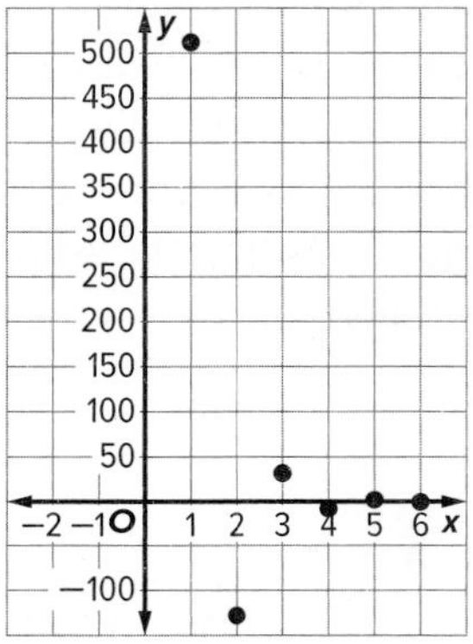

13. 32

15. $a_n = 3(3)^{n-1}$

17. $a_n = 2(-3)^{n-1}$

19. $a_n = 12(3)^{n-1}$

21. $a_n = 3(0.6)^{n-1}$

23. $a_n = -\left(\frac{1}{2}\right)^{n-1}$

25. ±8, 16, ±32

27. 1368

29a. The heights in feet after the first 6 seconds are 1000, 2010, 3030.1, 4060.401, 5101.00501, and 6152.01506.

29b. Yes; the height after 60 minutes is about 81,669.7 ft.

29c. 6.34×10^{29}

31. $\frac{1}{2}$

33. −2

35. 255

37. 0

39. 4374

41. −5120

43. −0.2

45. geometric; $a_n = 25(7)^{n-1}$

47. geometric; $a^n = -15(-2)^{n-1}$

49. 31.9375

51. 9707.8189

53. 3, −6, 12, −24, 48

55. 1875

57. $a_n = 60\left(\frac{1}{2}\right)^{n-1}$

59. $\pm\frac{243}{16}, \frac{81}{4}, \pm 27$

61. −8, 4

63. Sample answer: The second option; receiving 20 annual payments represents a geometric series, and $S = 4{,}000{,}000\frac{1-(1.02)^{20}}{1-(1.02)}$ or $97,189,479.20. By choosing the annual payments, the winner will receive approximately $17 million more than by choosing the single payout.

65a. $347,514.00

65b. $206,000

67. Sample answer: $256 + 192 + 144 + 108 + 81 + \frac{243}{4}$

69. Sample answer:

Let a_n = the nth term of the sequence and r = the common ratio.

$a_2 = a_1 \cdot r$	Definition of the second term of a geometric sequence
$a_3 = a_2 \cdot r$	Definition of the third term of a geometric sequence
$a_3 = a_1 \cdot r \cdot r$	Substitution
$a_3 = a_1 \cdot r^2$	Associative Property of Multiplication
$a_3 = a_1 \cdot r^{3-1}$	$3 - 1 = 2$
$a_n = a_1 \cdot r^{n-1}$	$n = 3$

71. x^2y^2

73. Sample answer: A series is arithmetic if every pair of consecutive terms shares a common difference. A series is geometric if every pair of consecutive terms shares a common ratio. If the series displays both qualities, then it is both arithmetic and geometric. If the series displays neither quality, then it is neither geometric nor arithmetic.

Lesson 7-5

1.

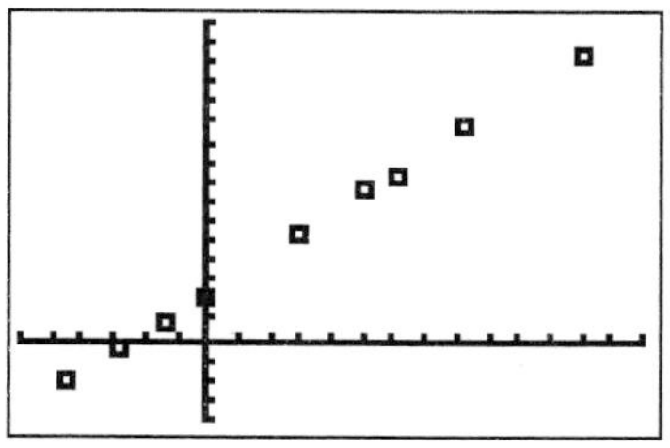

[−6, 14] scl: 1 by [−4, 16] scl: 1

linear

3.

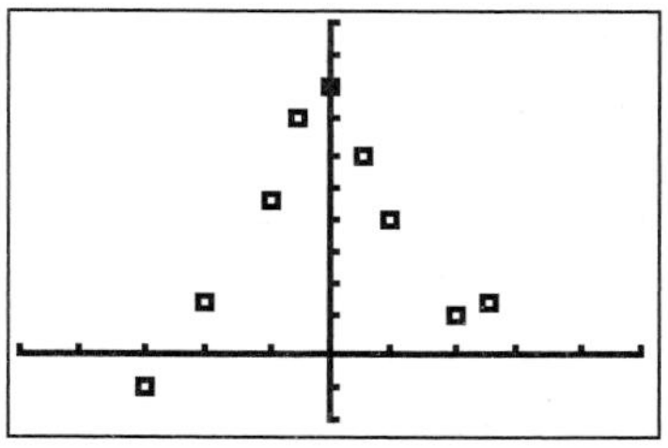

[−5, 5] scl: 1 by [−2, 10] scl: 1

quadratic

5a.

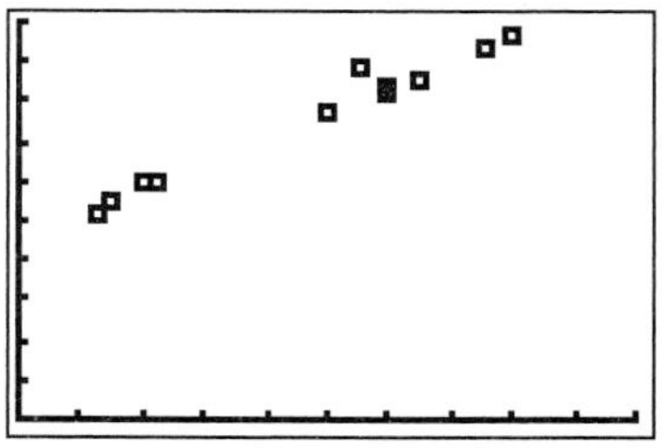

[0, 10] scl: 1 by [0, 100] scl: 10

linear; $y = 6.4x + 45.85$

5b. ≈2.2 hours

7.

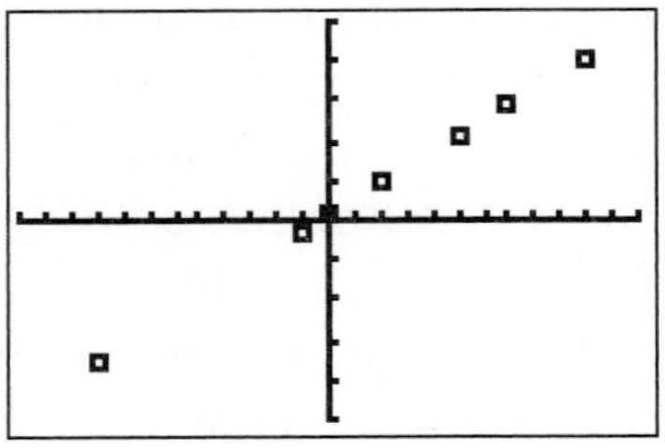

[−12, 12] scl: 1 by [−50, 50] scl: 10

linear

9. strong, positive

11. weak, positive

13.

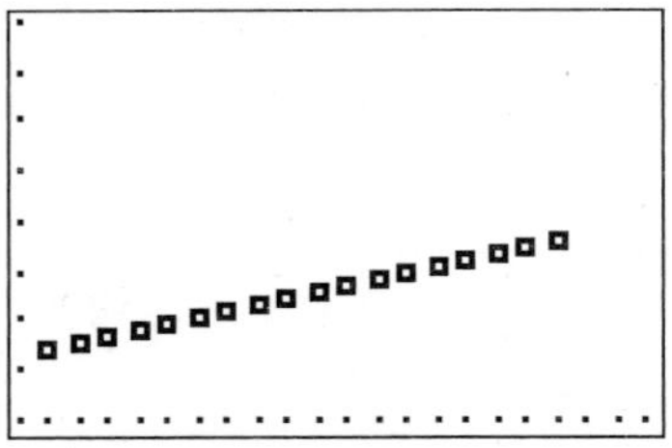

[1999, 2020] scl: 1 by [40, 80] scl: 5

The scatter plot is linear, with a positive direction, and a strong correlation.

15. Enter the *x*-values in the table in L1 in a calculator. Enter the *y*-values in the table in L2 in a calculator. Perform linear, quadratic, and exponential regressions. Then compare the coefficients of determination, r^2. The function with a coefficient of determination closest to 1 will fit the data best.

17. quadratic; $SA = 3\pi r^2$

Module 7 Review

1. A

3. A, B, F

5. D

7. Sample answer: The *x*-coordinate of the intersection of the graphs will give an approximate solution of the equation.

9. C

11. 18.21

13. $a_n = 10\left(\frac{1}{2}\right)^{n-1}$

15. C

17. A

Module 8

Module 8 Opener

1. \$2800 **3.** \$1674 **5.** $x = 8y + 20$
7. $x = \frac{y - 9}{-4}$

Lesson 8-1

1. $15^2 = 225$ **3.** $5^{-2} = \frac{1}{25}$ **5.** $4^3 = 64$
7. $\log_2 128 = 7$ **9.** $\log_7 \frac{1}{49} = -2$
11. $\log_2 512 = 9$
13. 6 **15.** 4 **17.** $-\frac{5}{2}$ **19.** 3 **21.** $\frac{2}{3}$
23. $\log_{1.0065} \frac{y}{400} = x$
25.

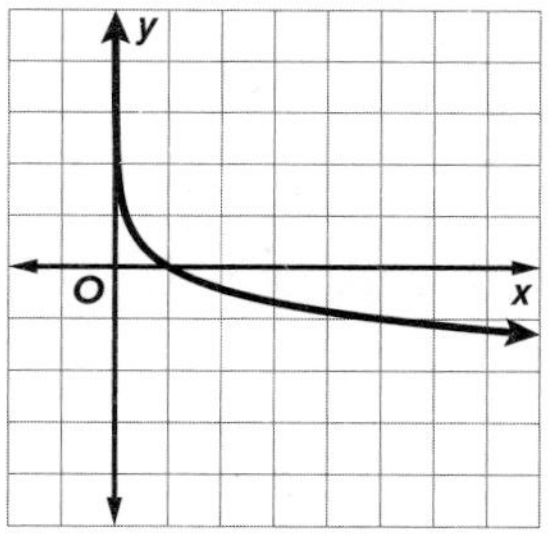

x-intercept: 1; no y-intercept; D: $(0, \infty)$; R: $(-\infty, \infty)$; end behavior: As $x \to 0^+$, $f(x) \to \infty$, and as $x \to \infty$, $f(x) \to -\infty$.

27.

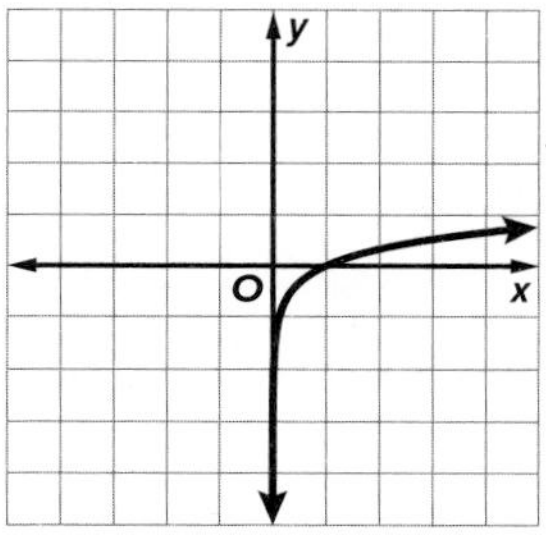

x-intercept: 1; no y-intercept; D: $(0, \infty)$; R: $(-\infty, \infty)$; end behavior: As $x \to 0^+$, $f(x) \to -\infty$, and as $x \to \infty$, $f(x) \to \infty$.

29.

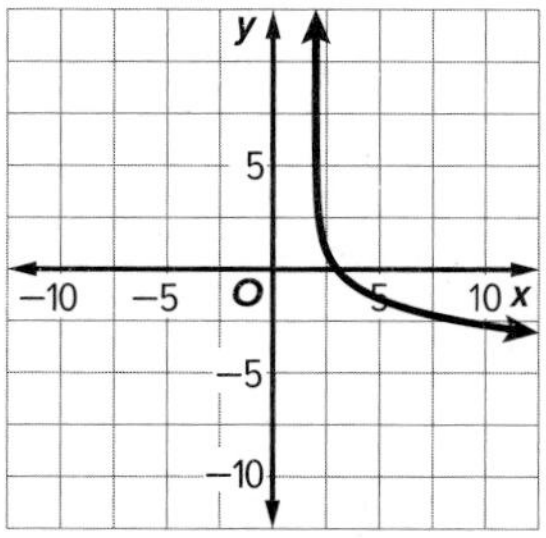

31.

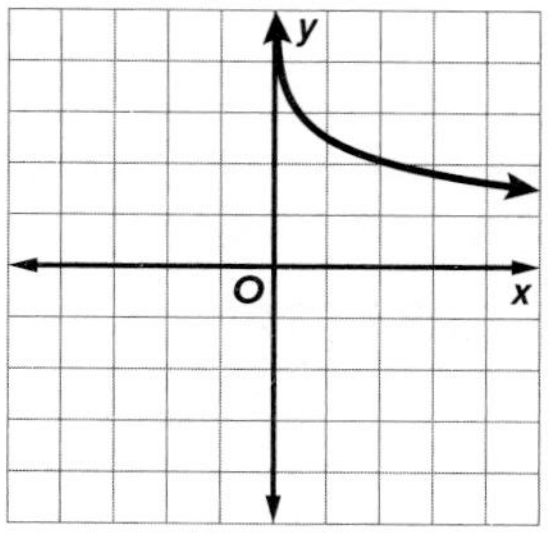

33a.

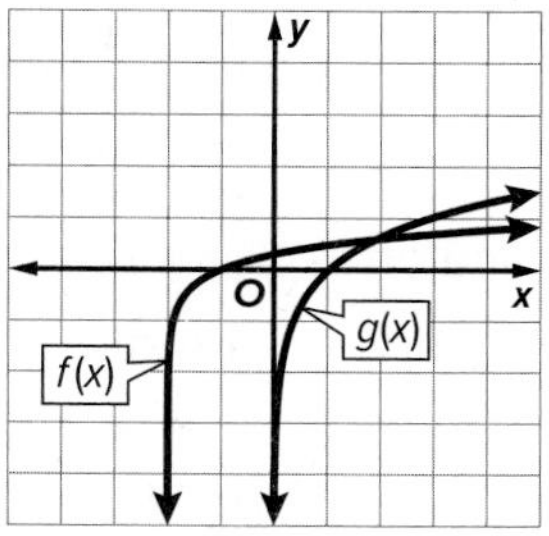

33b. $f(x)$: As $x \to -2$, $f(x) \to -\infty$, and as $x \to \infty$, $f(x) \to \infty$.
$g(x)$: As $x \to 0$, $g(x) \to -\infty$, and as $x \to \infty$, $g(x) \to \infty$.

35. $k = -1$; $g(x) = -(\log_5 x)$ **37.** $6^3 = 216$

39. $3^{-4} = \frac{1}{81}$ **41.** $\log_{\frac{1}{4}} \frac{1}{64} = 3$

43. 7.8 **45a.** Julio's; sample answer: I can use the graph to determine that $f(7) \approx 28$ and $g(7) \approx 1.7$. **45b.** Neither; sample answer: both $f(1)$ and $g(1)$ equal 0.

47.

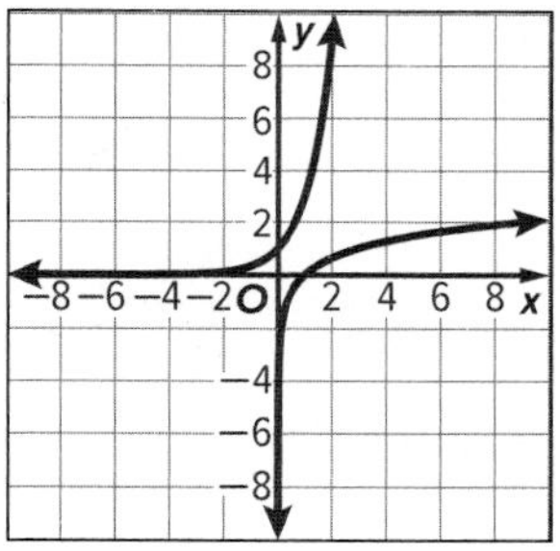

$f(x)$: D: $(-\infty, \infty)$ R: $(0, \infty)$; y-int: $(0, 1)$; asymptote: $y = 0$; $g(x)$: D: $(0, \infty)$ R: $(-\infty, \infty)$; x-int: $(1, 0)$; asymptote: $x = 0$; The graphs have an inverse relationship with one another. They are symmetric about the line $y = x$.

49. $f(g(x)) = x$ and $g(f(x)) = x$, so $f(x) = \log_{10}(x)$ and $g(x) = 10^x$ are inverse of each other.

51. $\log_7 51$; sample answer: $\log_7 51$ equals a little more than 2. $\log_8 61$ equals a little less than 2. $\log_9 71$ equals a little less than 2. Therefore, $\log_7 51$ is the greatest.

53. No; sample answer: Elisa was closer. She should have $-y = 2$ or $y = -2$ instead of $y = 2$. Matthew used the definition of logarithms incorrectly.

Lesson 8-2

1. 8 **3.** 3125 **5.** 9 **7.** 2 **9.** 4 **11.** $\frac{1}{3}$ or $\frac{1}{2}$
13. 2.4535 **15.** −0.2925 **17.** 1.5850

19. 4.6438 **21.** 6.9657 **23.** 6.34
25. 10^{11} **27.** 8 **29.** 2 **31.** 2
33. −10 or 10 **35.** $-\frac{1}{3}$ or 2 **37.** 2.2921
39. −0.1788 **41.** 1.3652 **43.** 1.1

45. $\log F = \log k + \log m_1 + \log m_2 - 2 \log d$

47a. $E = 1.4 \log_{10} \frac{C_2}{C_1}$

47b. about 0.6679 kilocalories per mole

47c. about 0.4214 kilocalories per mole

49. Rewrite the equations as $3 \cdot \log_2 a + 2 \cdot \log_2 b = 19$ and $4 \cdot \log_2 a - 5 \cdot \log_2 b = 10$. Solve the system of equations to get $\log_2 a = 5$ and $\log_2 b = 2$. Then $a = 32$ and $b = 4$.

51. true **53.** true **55.** false **57.** true **59.** 4
61. $\frac{8}{5}$

Lesson 8-3

1. 1.2553 **3.** 2.0792 **5.** 1.6263 **7.** about 5.6 h

9. 0.5209 **11.** 1.3917 **13.** −0.3869
15. $\{x \mid x \geq 0.8549\}$ **17.** $\{x \mid x > 5\}$
19. $\{y \mid y > -2.6977\}$

21. $\frac{\log 22}{\log 4}$, 2.2297

23. $\frac{\log 50}{\log 2}$, 5.6439

25. $\frac{\log 2}{\log 3}$, 0.6309 **27.** $16,000 **29.** ±2.3785

31. 0.9177 **33.** $\{p \mid p \leq 2.9437\}$

35. $\frac{2 \log 20}{2}$, 5.4537

37. $\frac{5 \log 4}{\log 8}$, 3.3333 **39.** $\frac{6 \log 3.6}{2}$, 11.0880

41. $\frac{\log 150}{2 \log 3}$, 2.2804 **43.** $\frac{\log 1600}{4 \log 5}$, 1.1460

45.

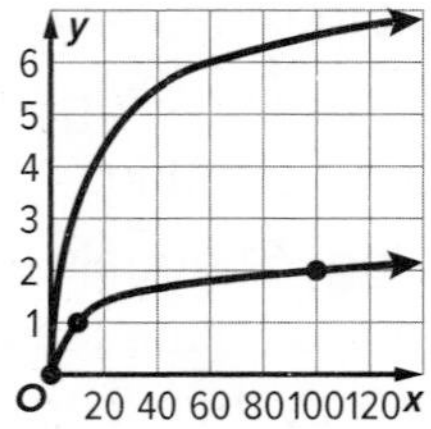

47a. $\log 4 = \log 2^2 = 2 \log 2 \approx 2 \cdot 0.3010 \approx 0.6020$, $\log 6 = \log (2 \cdot 3) = \log 2 + \log 3 \approx 0.3010 + 0.4771 \approx 0.7781$

47b. Sample answer:

$$\log 1.5 = \log 3 + \log 5 - \log 10$$
$$\approx 0.4771 + 0.6989 - 1$$
$$\approx 0.1760$$

49a. $3 \log_2 (2x) - \log_4 (3x) = 1$

$$3\left[\frac{\log(2x)}{\log(2)}\right] - \frac{\log(3x)}{\log(4)} = 1$$

$$\frac{3 \log(2x)}{\log(2)} - \frac{\log(3x)}{2 \log(2)} = 1$$

$$\frac{6 \log(2x) - \log(3x)}{2 \log(2)} = 1$$

$6(\log (2) + \log (x)) - \log (3) - \log (x) = 2 \log (2)$

$5 \log (x) = \log (3) - 4 \log (2)$

So, $x \approx 0.715$.

49b. $\log_6 (x^2) + \log_3 (x) = 3$

$$\frac{\log(x^2)}{\log(6)} + \frac{\log(x)}{\log(3)} = 3$$

$$\frac{\log(3) \log(x^2) + \log(6) \log(x)}{\log(3) \log(6)} = 3$$

$2 \log (3) \log (x) + \log (6) \log (x) = 3 \log (3) \log (6)$

$\log (x) [2 \log (3) + \log (6)] = 3 \log (3) \log (6)$

So, $x \approx 4.39$.

49c. $2 \log_2 (3x) = -8 \log_3 (x)$

$$2\left[\frac{\log(3x)}{\log(2)}\right] = -8\left[\frac{\log(x)}{\log(3)}\right]$$

$2 \log (3) \log (3x) = -8 \log (2) \log (x)$

$2 \log (3)[\log (3) + \log (x)] = -8 \log (2) \log (x)$

$2 \log (x)[2 \log (3) + 8 \log (2)] = -2[\log (3)]^2$

So, $x \approx 0.732$.

51. Rosamaria is correct. Sample answer: Sam forgot to bring the 3 down from the exponent when he applied the Property of Equality for Logarithms.

53. $\log_3 27 = 3$ and $\log_{27} 3 = \frac{1}{3}$;
Conjecture: $\log_a b = \frac{1}{\log_b a}$

Proof:

$\log_a b \stackrel{?}{=} \frac{1}{\log_b a}$	Original statement
$\frac{\log_b b}{\log_b a} \stackrel{?}{=} \frac{1}{\log_b a}$	Change of Base Formula
$\frac{1}{\log_b a} = \frac{1}{\log_b a}$	Inverse Property of Exponents and Logarithms

Lesson 8-4

1. $\ln x = 15$ **3.** $-5x = \ln 0.2$ **5.** $x = \ln 3$
7. $\ln (x - 3) = 4$ **9.** $5 = \ln 10x$ **11.** $e^{2x} = 36$
13. $x = e^8$ **15.** $e^x = 0.0002$ **17.** $e^x = 15$
19. $\ln 3$ **21.** $\ln 16x^2$ or $2 \ln 4x$ **23.** $\ln 5$
25. $-8 \ln \frac{1}{3}$ **27.** $\ln \frac{x^8}{625}$ **29.** 1.0986
31. 1.7579 **33.** 2.7081 **35.** 18.0855
37. 36.7493 **39.** 2.4630
41a. \$813.28 **41b.** about 36.3 years
41c. \$873.72 **43.** $y = 804.30 - 52.90 \ln x$; about 624 grams **45.** $y = 2079.08 + 75.85 \ln x$; about 228,940 visitors **47.** $\frac{3}{5} \ln x$
49. 1.3900 **51.** about 24.75 years
53. $\ln(5 \cdot 5 \div 8) = \ln 5 + \ln 5 - \ln 8 = 1.1394$
55a. $P = P_0 e^{kt}$; $0.5P_0 = P_0 e^{5730k}$; $0.5 = e^{5730k}$; $\ln (0.5) = 5730k (\ln e)$; $k = \frac{\ln 0.5}{5730}$
55b. $P = 200e^{\left(\frac{\ln 0.5}{5730}\right)t}$

Amount of Carbon-14 in a Sample Over Time

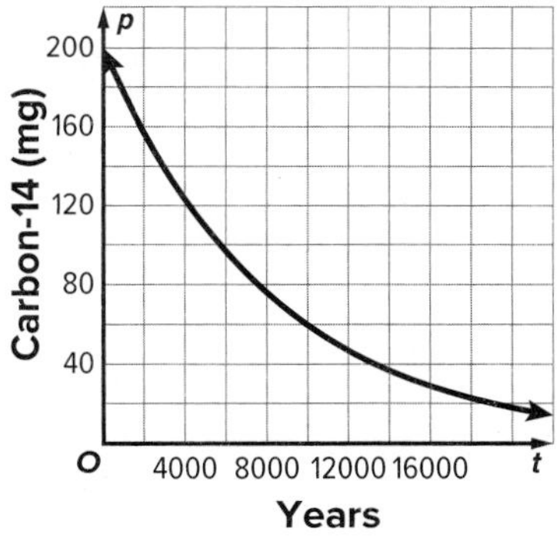

55c. Sample answer: at least about 7500 years. Solve inequality: $P \leq 200e^{\frac{\ln 0.5}{5730}t}$; $80 \leq 200e^{\frac{\ln 0.5}{5730}t}$; $0.4 \leq e^{\frac{\ln 0.5}{5730}t}$; $\ln (0.4) \leq \frac{\ln 0.5}{5730} t \cdot \ln (e)$; $t \geq \frac{5730 \cdot \ln 0.4}{\ln 0.5}$; ≈ 7575 years

57. Disagree; $e^{e^x} = x$ can be rewritten as $e^x(\ln e) = \ln x$, which can be written as $e^x = \ln x$. The functions $y = e^x$ and $y = \ln x$ are inverse functions which have no point of intersection so there are no values of x for which $e^{e^x} = x$.

59. Let $p = \ln a$ and $q = \ln b$. That means that $e^p = a$ and $e^q = b$.

$ab = e^p \times e^q$

$ab = e^{p+q}$

$\ln(ab) = (p + q)$

$\ln(ab) = \ln a + \ln b$

Lesson 8-5

1a. $y = 6.124e^{0.01238t}$ **1b.** 2016
3a. about 0.02166 or about 2.166%; $y = ae^{-0.02166t}$
3b. about 0.57 g **5.** about 12,618 years old
7a. $k \approx 0.071$ **7b.** about 60.8 min **7c.** about 30.85 min **9.** about 6.0 yr
11a. $y = 67{,}387e^{0.0198t}$ **11b.** about 9304
13. $P = 8e^{0.26t}$; about 4.4 years
15. Sample answer: Based on the data in the table, the decay constant is about −0.000124, which is approximately equal to the given constant. $f(t) = 1000e^{-0.000124t}$; Use the exponential model: $f(t) = ae^{kt}$; $999.628 = 1000^{k(3)}$; $0.999628 = e^{3k}$; $\ln 0.999628 = \ln e^{3k}$; $\ln 0.999628 = 3k$; $k = -0.000124$; Verify: $f(2) = 1000e^{-0.000124(2)} = 999.752$

17a.

t (min)	Surviving Cells After t Minutes	$(t, f(t))$
0	initial amount	(0, 1,000,000)
1	(0.70)(1,000,000) = 700,000 survive	(1, 700,000)
2	(0.70)(700,000) = 490,000	(2, 490,000)
3	(0.70)(490,000) = 343,000	(3, 343,000)

Sample answer: An exponential model best describes the points because the number of cells decreases by the same percentage every minute.

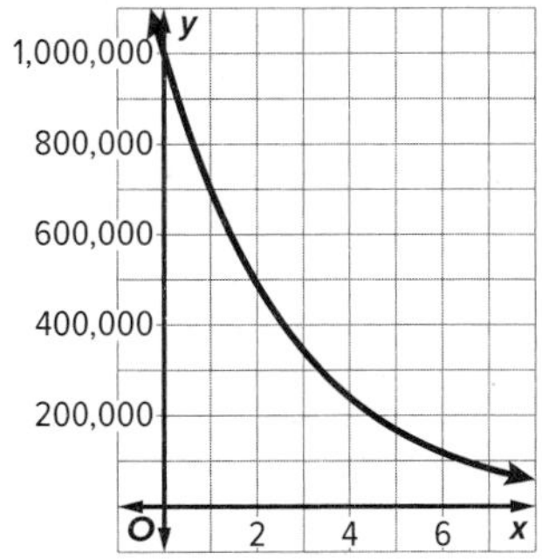

17b. In the model $f(t) = ae^{kt}$, $f(t)$ represents the number of cells remaining after t minutes, a represents the number of cells at the start of the experiment, t represents the number of minutes since the experiment began, and k is the growth or decay constant. $f(t) = 1{,}000{,}000e^{kt}$.

Selected Answers

17c. $k \approx -0.356675$; Because k is negative, it is exponential decay. Substitute known values into the formula: $700{,}000 = 1{,}000{,}000e^{k(1)}$; $0.7 = e^k$; $\ln 0.7 = k$; $k \approx -0.356675$; The exponential decay model for this experiment is $f(t) = 1{,}000{,}000e^{-0.356675t}$.

17d. It will take approximately 19.367 minutes to have less than 1000 cells. Write the model and solve for t; $1000 = 1{,}000{,}000e^{-0.356675t}$; $0.001 = e^{-0.356675t}$; $\ln 0.001 = -0.356675t$; $t \approx 19.367$.

19. $t \approx 113.45$

21. Sample answer: Exponential functions can be used to model situations that incorporate a percentage of growth or decay for a specific number of times per year. Continuous exponential functions can be used to model situations that incorporate a percentage of growth or decay continuously.

Module 8 Review

1a. $\log_6 1296 = 4$ **1b.** $\log_{24} 27 = \frac{3}{5}$

1c. $\log_8 \frac{1}{64} = -2$

3.

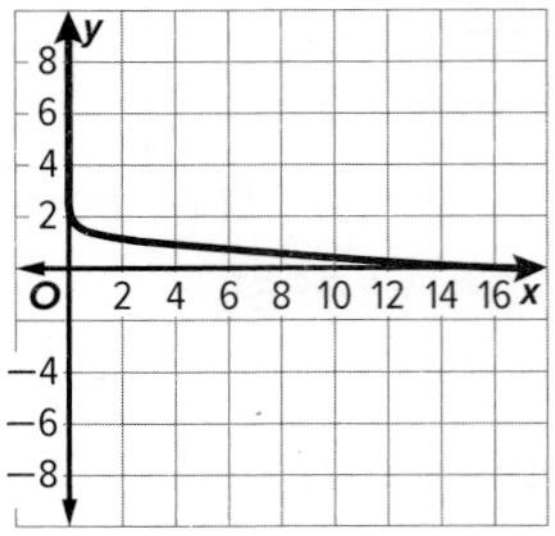

5. Sample answer: If I divide 15,309 by 7, the result is 2187, or 37. So, the expression that needs to be approximated can be written as $\log_3 7 \cdot 3^7$. Then, using the Product Property of Logarithms, the expression can be rewritten again as $\log_3 7 + \log_3 3^7$. I know that $\log_3 3^7 = 7$, so the approximate value of $\log_3 15{,}309$ is about $1.7712 + 7$ or 8.7712.

7. D **9.** C, E **11.** B **13a.** $\ln 8x = 7$

13b. $\ln 17 = 2x$ **15.** B **17.** A **19.** 365 days

Module 9

Module 9 Opener

1. $x = \frac{15}{14}$ **3.** $M = \frac{56}{3}$ **5.** $-\frac{1}{8}$ **7.** $\frac{29}{30}$

Lesson 9-1

1. $\frac{x(x+6)}{x+4}$; $x = -4, 3$ **3.** $\frac{(x+3)(x-z)}{4}$; $x = -z, 3$

5. $\frac{x(x+2)}{6(x+5)}$; $x = -5, 0, 4$ **7.** $-\frac{x+2}{x+4}$ **9.** $-\frac{1}{x+6}$

11. $\frac{c}{4ab^2f^2}$ **13.** $\frac{32b}{3ac^3f^2}$ **15.** $\frac{5a^4c}{3b}$ **17.** $\frac{(4a+5)(a-4)}{3a+2}$

19. $\frac{t+12}{2(t+2)}$ **21.** $-\frac{6}{z}$ **23.** $\frac{b-a}{x+y}$ **25.** $\frac{c+2}{c(c-d)}$

27. $\frac{(a+1)(a-2)}{4(a-5)(a-1)}$ **29.** $\frac{x(x+2)(x-1)}{(x+3)(x-7)}$ **31.** $\frac{15y^3}{4a^2cxz}$

33. $\frac{x-4}{-4(x-3)}$ **35.** $\frac{(4x-1)^2(3x+1)(x+1)}{12(x+2)(x-4)(x^2-10x+6)}$

37. $\frac{48H+45c}{c+H}$ **39.** $\frac{r}{3}$

41a. The height can be found by dividing the volume by the product of the length and width of the box.

41b. $(x + 3)$ in.

41c. Sample answer: Substitute a value for x in each of the given expressions for the length, width, and volume, and the same value for x in the expression found for h, and then check that $V = \ell wh$.

CHECK: For $x = 5$,
length $= (5) + 10 = 15$ in.
width $= 2(5) = 10$ in.
volume $= 2(5)^3 + 26(5)^2 + 60(5) = 1200$ in^3
height $= (5) + 3 = 8$ in.
Verify $V = \ell wh$: $1200 = (15)(10)(8)$

43. By dividing the second week's speed by the first week's speed, you can determine how much faster she hopes to run on average. $\frac{60}{t-2} \div \frac{60}{t} = \frac{60}{t-2} \cdot \frac{t}{60} = \frac{t}{t-2}$. She hopes to run $\frac{t}{t-2}$ times faster during her second week than during her first week.

45a. $C : B = \frac{(x+8)}{(x+5)}$, $B : A = \frac{(x+5)}{x}$; For these two ratios to be equal, $\frac{C}{B} \cdot \frac{A}{B}$ must equal 1. To equal 1, their product must have the numerator equal to the denominator. So, $\frac{(x+8)}{(x+5)} \cdot \frac{x}{(x+5)} = \frac{x^2+8x}{(x+5)^2}$ and $x^2 + 8x = (x+5)^2$. Solving for x results in $x = -12.5$. Since x cannot be a negative value, there are no values of x that will make these ratios equal.

45b. Yes; $C : B = \frac{(x+6)}{(x+2)}$, $B : A = \frac{(x+2)}{x}$ and $\frac{(x+6)}{(x+2)} \cdot \frac{x}{(x+2)} = \frac{x^2+6x}{(x+2)^2}$. When $x^2 + 6x = (x+2)^2$, $x = 2$. The length of a side of square A would need to be 4 feet, and the fountain would need to be 2 feet in diameter.

47. Beverly; sample answer: Troy's mistake was multiplying by the reciprocal of the dividend instead of the divisor.

49. $\frac{x+1}{\sqrt{x+3}}$; The other three expressions are rational expressions. Since the denominator of $\frac{x+1}{\sqrt{x+3}}$ is not a polynomial, $\frac{x+1}{\sqrt{x+3}}$ is not a rational expression.

51. Sample answer: $\frac{x^2-1}{x^2+5x+4}$

53. Sample answer: $\frac{3a}{3a-15}$, $a \neq 5$; $\frac{a^2}{a^2-5a}$, $a \neq 5$ or 0; $\frac{a(a+1)}{(a-5)(a+1)}$, $a \neq 5$ or -1

Lesson 9-2

1. $\frac{5x+3y}{xy}$ **3.** $\frac{2c+5}{3}$ **5.** $\frac{12z-2y}{5y^2z}$

7. $\frac{3w+7}{(w-3)(w+3)}$ **9.** $\frac{2k}{k-n}$ **11.** $\frac{n+2}{n-3}$ **13.** $\frac{1}{r(1-r)}$

15a. $\frac{100}{r}$, $\frac{100}{r+\frac{1}{2}}$, $\frac{100}{r-\frac{1}{2}}$, $\frac{100}{r-1}$

15b. $\frac{100(16r^3-12r^2-2r+1)}{4r^4-4r^3-r^2+r}$ **15c.** 10.97 m/s

17. $\frac{13x+21}{-3x+73}$ **19.** $\frac{-x^2+33x+16}{12x^2+11x-27}$

21. $\frac{15bd-6b-2d}{3bd(3b+d)}$ **23.** $\frac{2x^2+5x-2}{(x-5)(x+2)}$

25. $\frac{28by^2z-9bx}{105x^3y^4z}$ **27.** $\frac{2(10x^2y+60y+3x^2)}{15x^3y}$

29. $\frac{15b^3+100ab^2-216a}{240ab^3}$ **31.** $\frac{2(5y-2)}{(y-7)(y+5)(y+4)}$

33. $\frac{-10(x+1)}{(2x-1)(x+6)(x-3)}$ **35.** $\frac{2x(x+16)}{3(x-2)(x+3)(2x+5)}$

37. Sample answer: Lord Brouncker is best known for the continued fraction approximation of π, $\pi = \cfrac{4}{1+\cfrac{1^2}{2+\cfrac{3^2}{2+\cfrac{5^2}{2+\cfrac{7^2}{\cdots}}}}}$.

39. $\frac{15x^2-192x^2y^2-128y}{48x^2y^2}$ **41.** $\frac{19x-36}{12(2x+1)(x-3)(x+4)}$

43. $\frac{x^2+2x-29}{(x-1)(x-8)}$ **45.** $\frac{1}{y-x}$ **47.** $-\frac{4}{5}$

Selected Answers

49. The average is the sum of the given numbers divided by the number of given numbers. So, the average of $\frac{1}{x}$, $\frac{1}{x-3}$, and $\frac{1}{2x}$ is

$$\frac{\frac{1}{x}+\frac{1}{x-3}+\frac{1}{2x}}{3} = \frac{\frac{1(2)(x-3)}{x(2)(x-3)}+\frac{1(2)(x)}{(x-3)(2)(x)}+\frac{1(x-3)}{2x(x-3)}}{3} =$$

$$\frac{\frac{2(x-3)}{2x(x-3)}+\frac{2x}{2x(x-3)}+\frac{x-3}{2x(x-3)}}{3} = \frac{\frac{2(x-3)+2x+x-3}{2x(x-3)}}{3} =$$

$$\frac{2(x-3)+2x+x-3}{6x(x-3)} = \frac{5x-9}{6x(x-3)} \text{ for } x \neq 0$$

and $x \neq 3$.

51. $\frac{8x^3+2x^2+8x+10}{(x+1)(2x^2-1)}$

53. $\frac{7t}{12}$; Sample answer: In 1 hour, Dell will plant $\frac{1}{3}$ of a flower bed, or $\frac{t}{3}$ flower beds in t hours. In 1 hour, Max will plant $\frac{1}{4}$ of a flower bed, or $\frac{t}{4}$ flower beds in t hours. Adding the two expressions, $\frac{t}{3}+\frac{t}{4}=\frac{7t}{12}$. So, $\frac{7t}{12}$ represents how many flowers beds will be planted in t hours.

55a. $\frac{x}{x+1}$, $x \neq 0, -1$ **55b.** $\frac{x+1}{2x+1}$, $x \neq 0, -1, -\frac{1}{2}$

55c. $\frac{2x+1}{3x+2}$, $x \neq 0, -1, -\frac{1}{2}, -\frac{2}{3}$ **55d.** $\frac{3x+2}{5x+3}$, $x \neq 0, -1, -\frac{1}{2}, -\frac{2}{3}, -\frac{3}{5}$ **55e.** $\frac{1}{2}; \frac{2}{3}; \frac{3}{5}; \frac{5}{8}$

55f. Sample answer: The values are ratios of two consecutive Fibonacci numbers. The next value should be $\frac{8}{13}$.

57. Sample answer: The set of rational expressions is closed under all of these operations because the sum, difference, product, and quotient of two rational expressions is a rational expression.

59. Sample answer: First, factor the denominators of all of the expressions. Find the LCD of the denominators. Convert each expression so they all have the LCD. Add or subtract the numerators. Then simplify. It is the same.

Lesson 9-3

1. $x \neq 0$ **3.** $x \neq 3$ **5.** $x \neq -\frac{3}{2}$

7. asymptotes: $x = 1$, $f(x) = 0$; $D = \{x \mid x \neq 1\}$; $R = \{f(x) \mid f(x) \neq 0\}$ The y-intercept is -1.

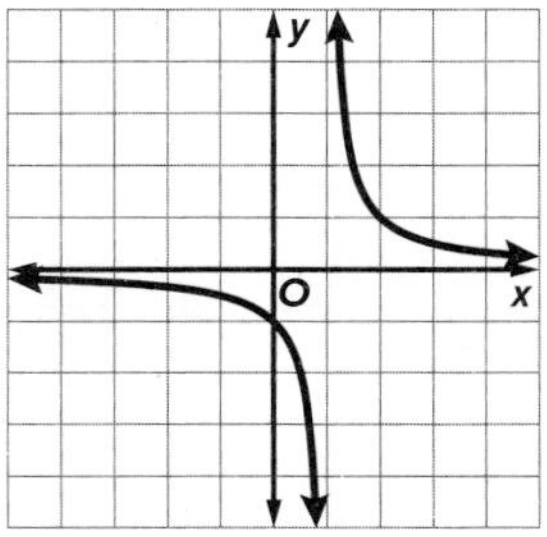

9. asymptotes: $x = -4$, $f(x) = 0$; $D = \{x \mid x \neq -4\}$; $R = \{f(x) \mid f(x) \neq 0\}$

The y-intercept is 1.25.

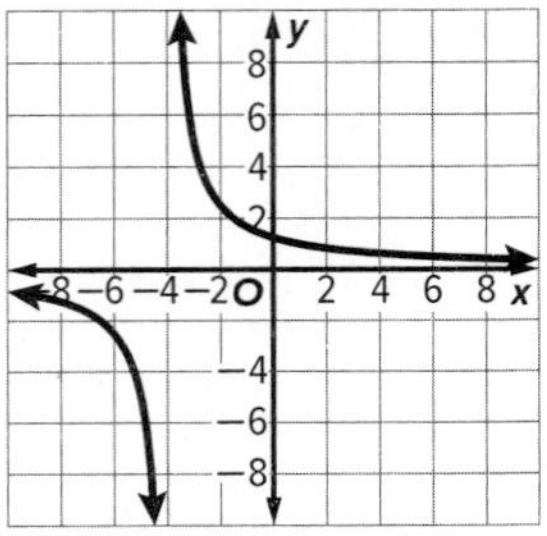

11a. $r = \frac{800}{t-2}$

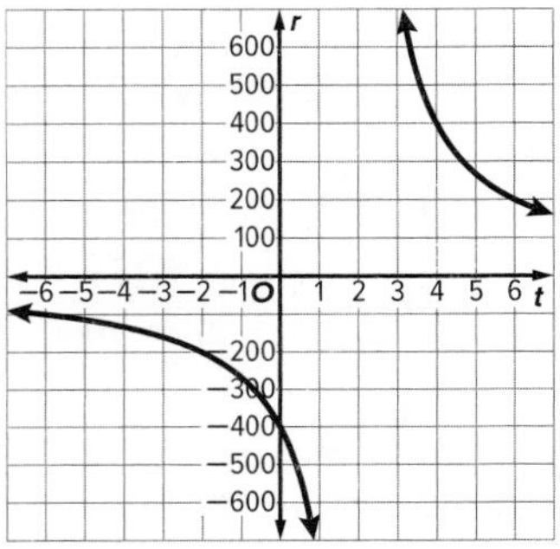

11b. Sample answer: $D = \{t \mid t > 2\}$; $R = \{r(t) \mid r(t) > 0\}$; intercepts: none; positive: when $t > 2$; negative: none; symmetry: symmetric about (2, 0); end behavior: As $t \to -\infty$, $r \to 0$, and as $t \to \infty$, $r \to 0$. Because the plane's speed and travel time cannot be negative, only values in the domain $\{t \mid t > 2\}$ makes sense in the context of this situation.

13. $D = \{x \mid x \neq -3\}$; $R = \{f(x) \mid f(x) \neq -3\}$

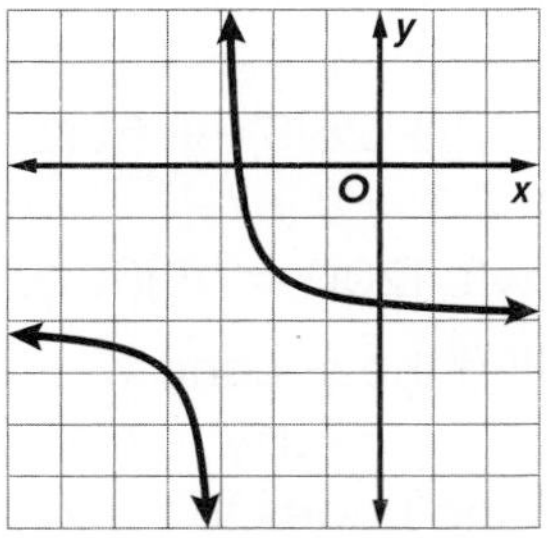

15. $D = \{x \mid x \neq -1\}$; $R = \{f(x) \mid f(x) \neq 3\}$

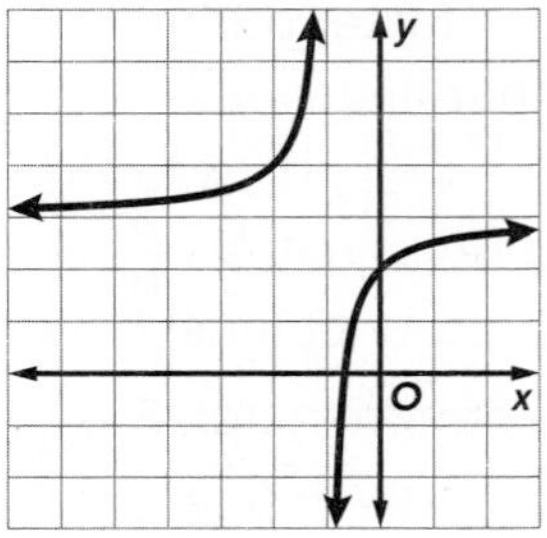

17. $a = -2, h = 5, k = 0; g(x) = \frac{-2}{x - 5}$

19. $a = 9, h = -3, k = 6; g(x) = \frac{9}{x + 3} + 6$

21. $a = -6, h = -4, k = -2; g(x) = \frac{-6}{x + 4} - 2$

23. $\frac{5}{2}$ **25.** $-1, 3$

27. $-3, 5$

29. $\{x \mid x \neq -1\}$ and $\{f(x) \mid f(x) \neq 0\}$

31. asymptotes: $x = 2, f(x) = 0$; $D = \{x \mid x \neq 2\}$; $R = \{f(x) \mid f(x) \neq 0\}$

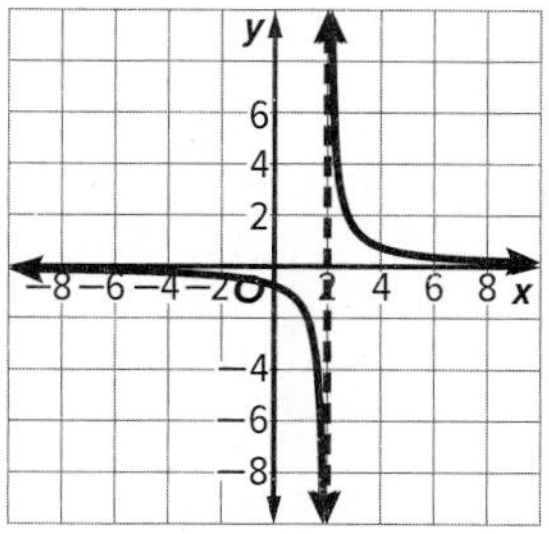

33. asymptotes: $x = \frac{1}{4}, f(x) = 0$; $D = \left\{x \mid x \neq -\frac{1}{4}\right\}$; $R = \{f(x) \mid f(x) \neq 0\}$

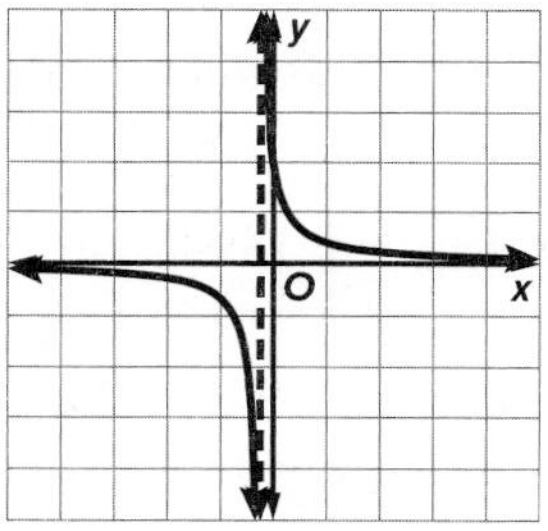

35. asymptotes: $x = -7, f(x) = -1$; $D = \{x \mid x \neq -7\}$; $R = \{f(x) \mid f(x) \neq -1\}$

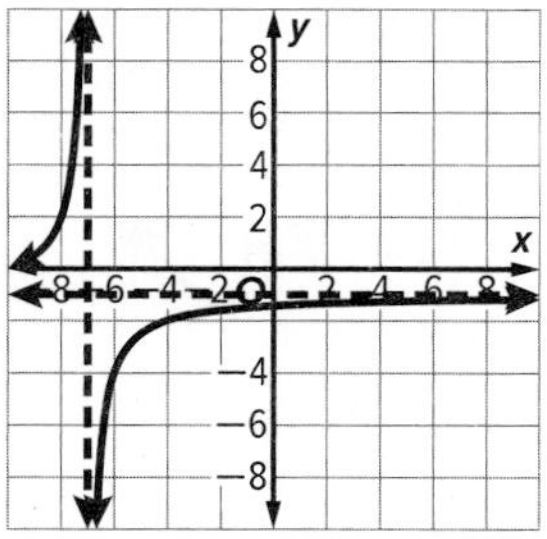

37. asymptotes: $x = 1, f(x) = 2$; $D = \{x \mid x \neq 1\}$; $R = \{f(x) \mid f(x) \neq 2\}$

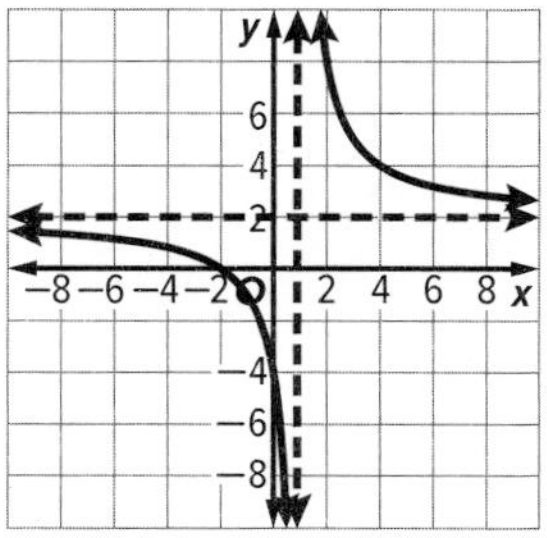

39. asymptotes: $x = 8, f(x) = -9$; $D = \{x \mid x \neq 8\}$; $R = \{f(x) \mid f(x) \neq -9\}$

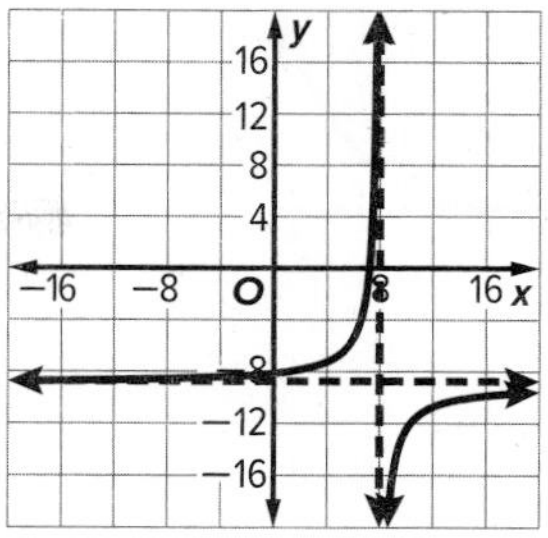

41. $f(x) = \frac{2}{x}$ **43.** $f(x) = \frac{2}{x + 6} - 2$

45a. $C = \frac{125}{m} + 0.3$

45b. They cannot travel zero miles or a negative number of miles.

47. The graph of $g(x)$ is a reflection in the x-axis, a translation 2 units left, and a translation 1 unit up of the graph of $f(x)$.

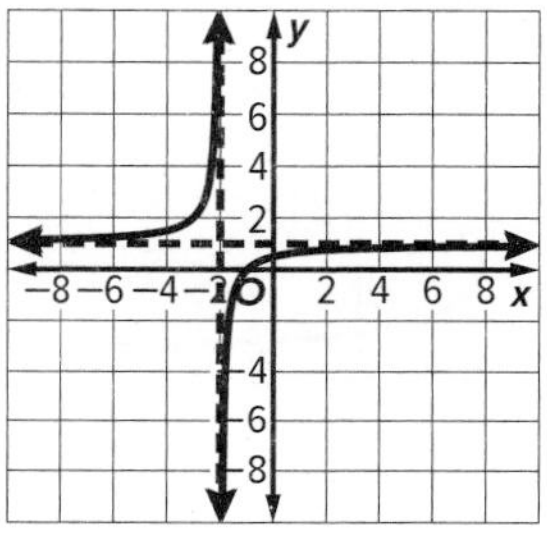

49. The graph of $k(x)$ is a reflection in the x-axis, a translation 2 units left, and a translation 2 units up of the graph of $f(x)$; so, $k(x) = -f(x + 2) + 2 = -\frac{1}{x + 2} + 2$.

51. Sample answer: $f(x) = \frac{1}{x + 4} + 6$

53. Sample answer: $g(x)$; all other choices have unknowns only in the denominator.

55. asymptotes: $x = -2, y = 0$

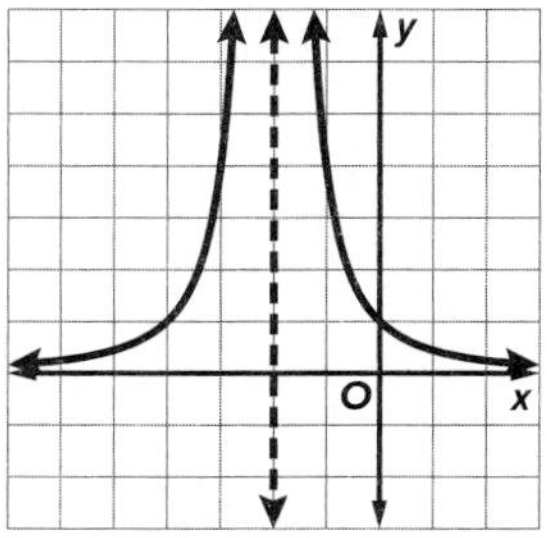

Lesson 9-4

1.

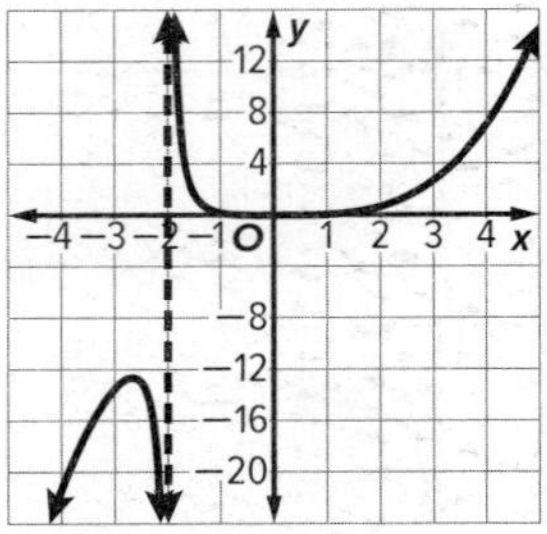

3.

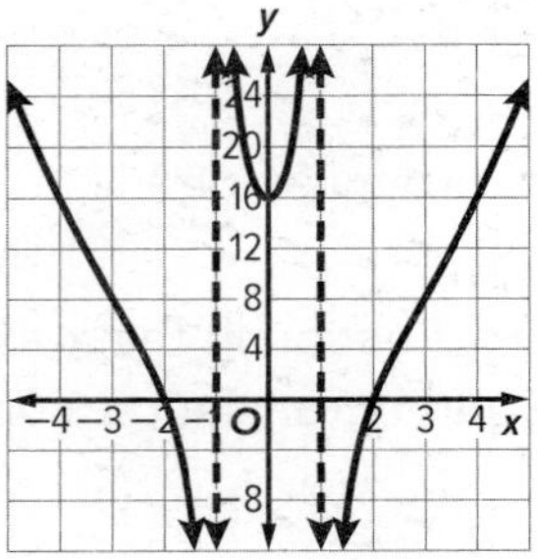

5a.

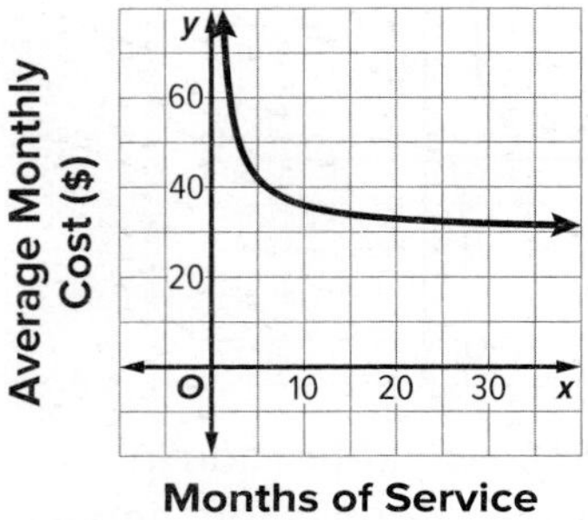

5b. x-intercept: -2; y-intercept: none; end behavior: As $x \to -\infty$, $f(x) \to 30$ and as $x \to -\infty$, $f(x) \to 30$.

5c. \$37.50

7a.

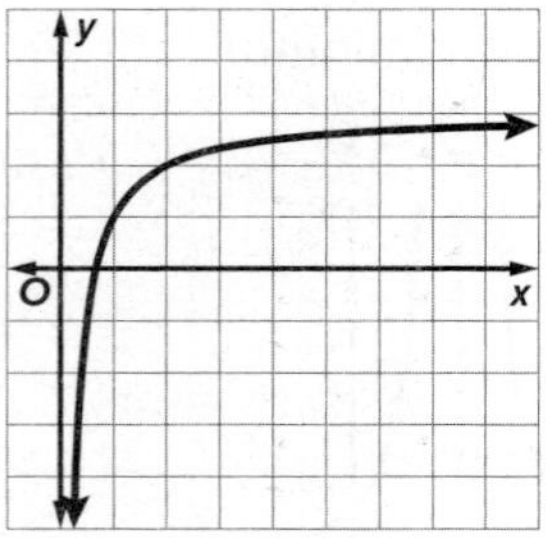

7b. x-intercept: $\frac{2}{3}$; y-intercept: none; end behavior: As $x \to -\infty$, $f(x) \to 3$ and as $x \to \infty$, $f(x) \to 3$.

7c. ≈ 2.7 cents

9a.

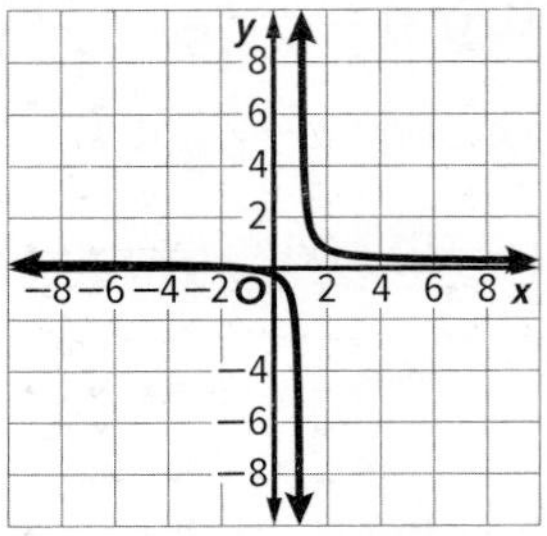

9b. $h(x)$, since $-\frac{1}{4} > -1$

9c. Vertical asymptotes $h(x)$: $x = 1$ $j(x)$: $x = -3$
Horizontal asymptotes $h(x)$: $y = \frac{1}{4}$ $j(x)$: $y = \frac{1}{2}$

11. zero: $x = 4$; vertical asymptote: $x = -2$; oblique asymptote: $f(x) = x - 10$

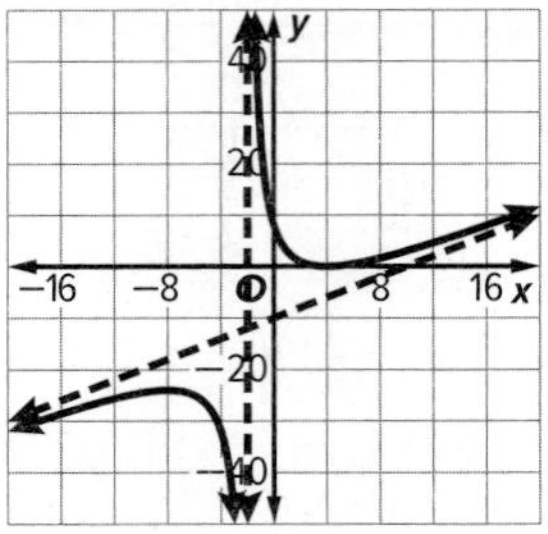

13. zero: none; vertical asymptote: $x = -2$; oblique asymptote: $f(x) = 6x - 8$

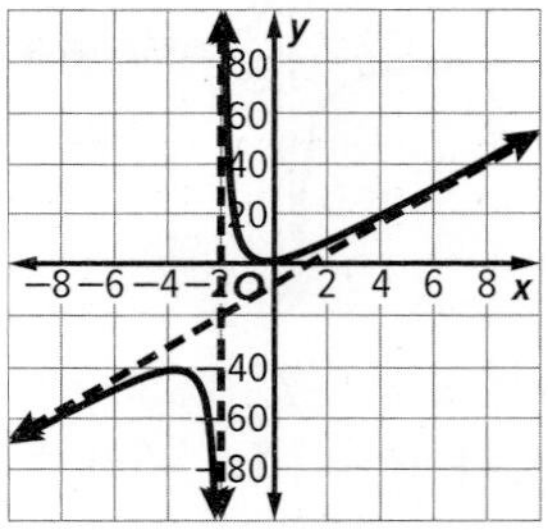

15. zero: none; vertical asymptote: $x = \frac{1}{2}$; oblique asymptote: $f(x) = \frac{3}{2}x + \frac{3}{4}$

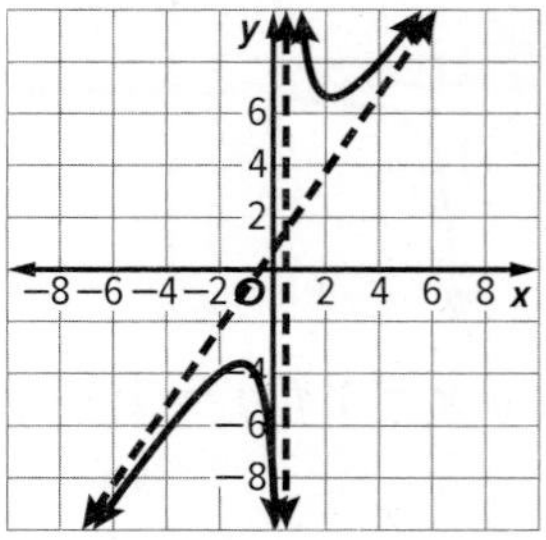

17. point discontinuity at $x = 4$

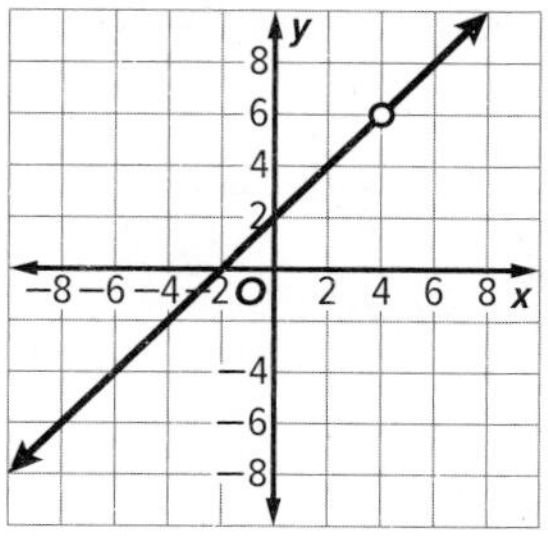

19. point discontinuity at $x = -5$

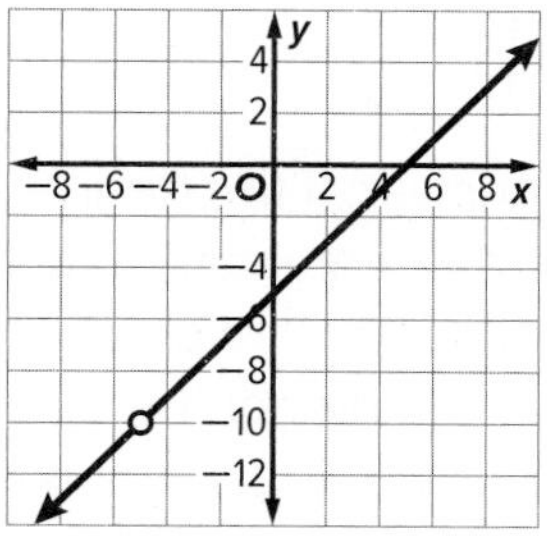

21. point discontinuity at $x = 2$ and $x = 4$

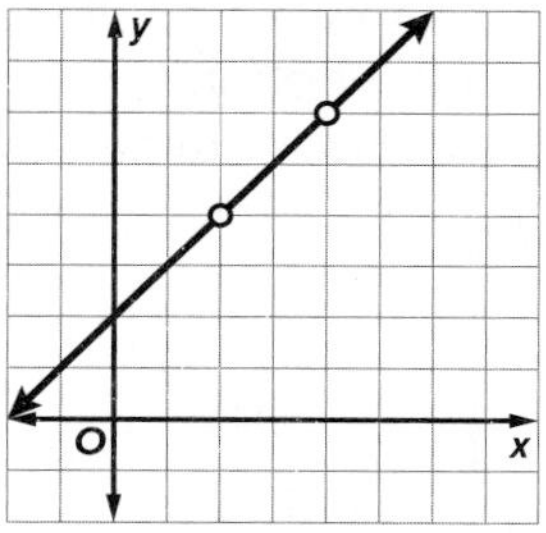

23.

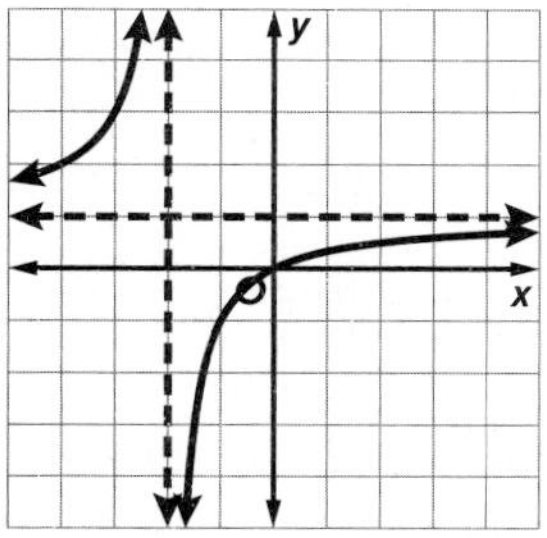

25.

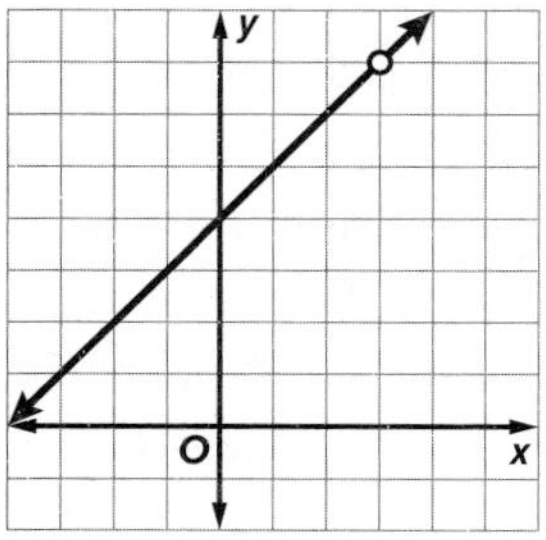

27.

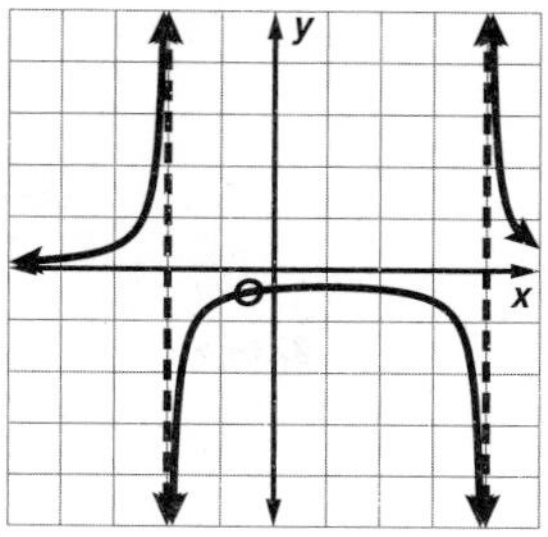

29.

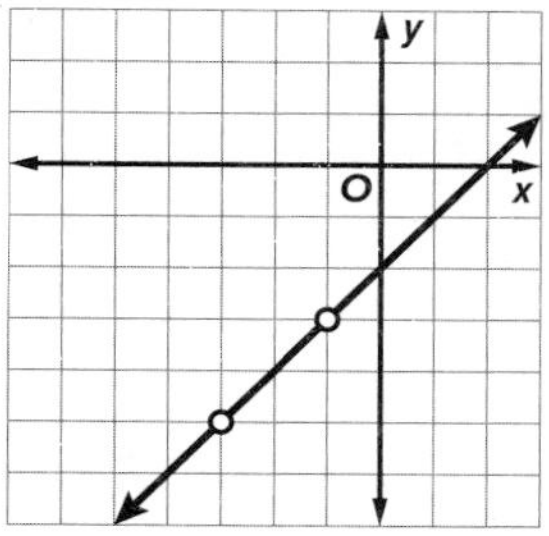

31a.

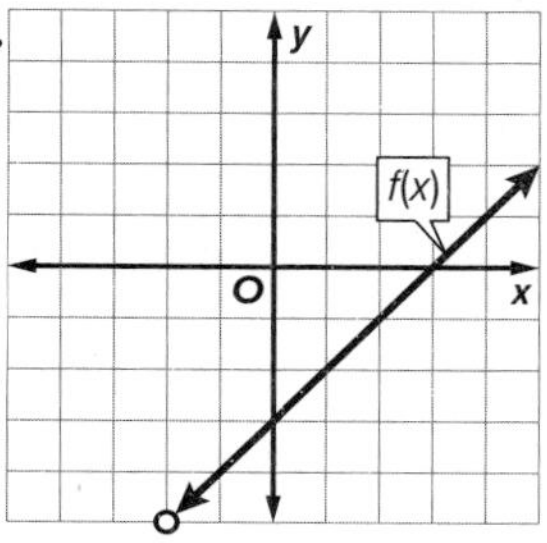

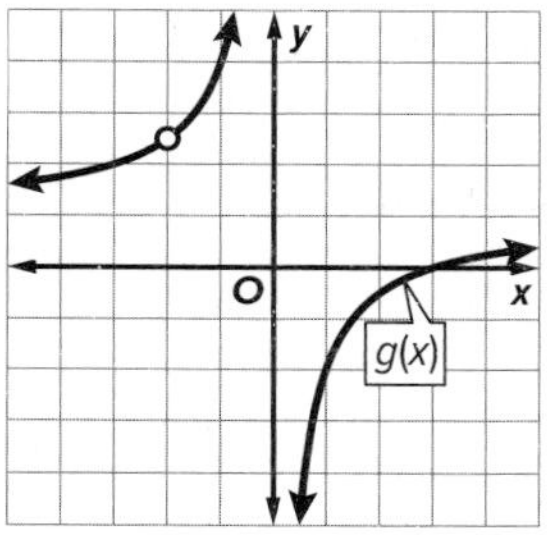

31b. $f(x)$

31c. The graph of $f(x)$ has a hole at $x = -2$, but no vertical asymptote. Its graph is a straight line with a hole in it at $(-2, -5)$. The graph of $g(x)$ also has a hole at $x = -2$, but has a vertical asymptote at $x = 0$. Its graph is not a straight line, but two curves having a hole in the graph at $\left(-2, \frac{5}{2}\right)$.

33. $x = 0$

35.

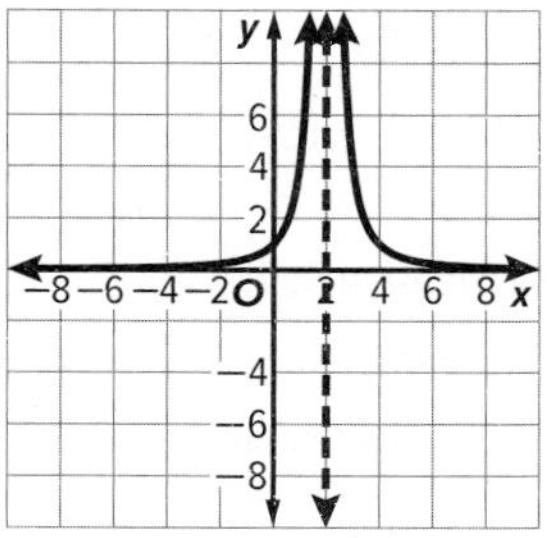

37.

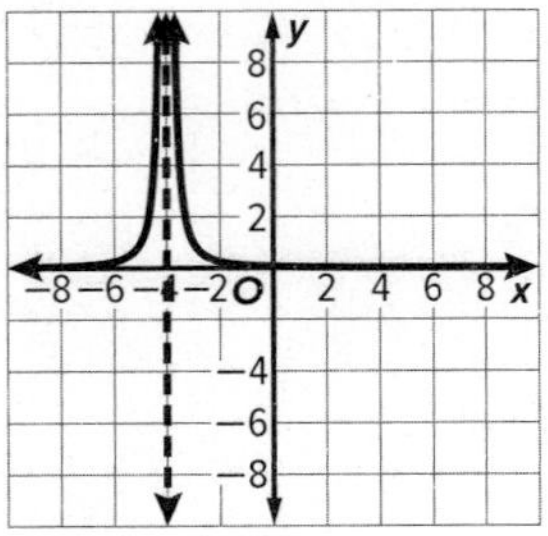

39.

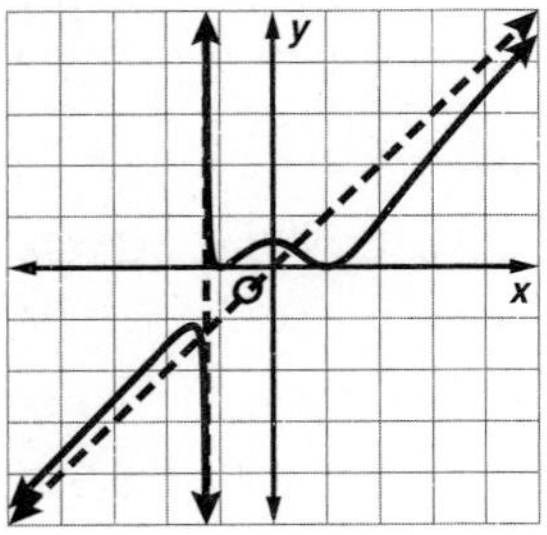

41.

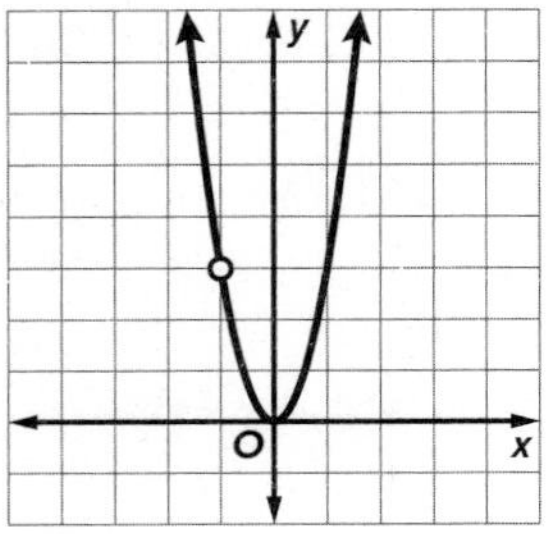

43.

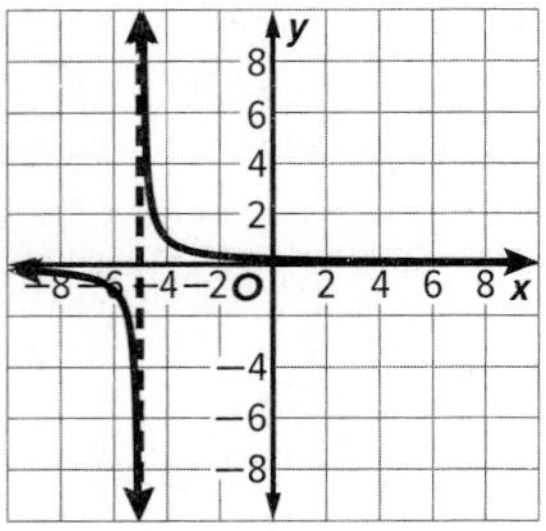

45.

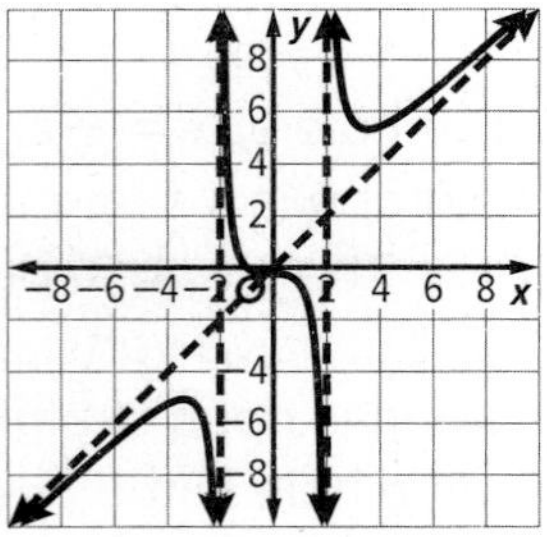

47a. $y = \frac{2067 + 6767}{6767 + x}$

47b. vertical asymptote at $x = -6767$; horizontal asymptote at $y = 0$

47c. As the number of at bats increases, the ratio grows closer and closer to zero, since the number of hits remains constant.

49. Sample answer: The domain is $\{x \mid x \neq -2 \text{ and } x \neq 2\}$. The graph has zeros at $x = 1$ and $x = 4$ and a y-intercept at (0, 1). The graph has vertical asymptotes at $x = -2$ and $x = 2$ and a horizontal asymptote at $f(x) = -1$ for $2 \leq x$ or $x \leq -2$. A function with these features is $f(x) = -\frac{(x-1)(x-4)}{(x+2)(x-2)} = -\frac{x^2 - 5x + 4}{x^2 - 4}$.

51. Sample answer:

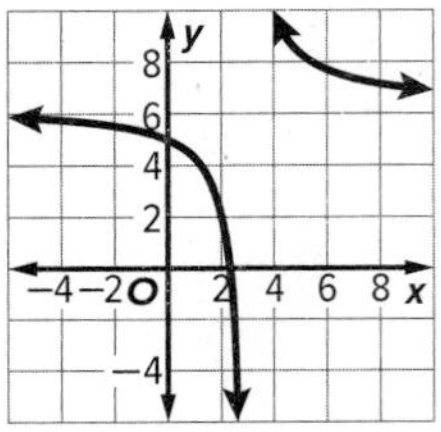

53. Sample answer: If the degrees of the numerator and denominator are equal, there is a horizontal asymptote determined by the coefficients of the numerator and denominator. If the degree of the numerator is greater, there is no horizontal asymptote. If the degree of the numerator is exactly 1 greater than the degree of the denominator, there is an oblique asymptote. If the degree of the denominator is greater, the horizontal asymptote is the line $y = 0$.

55. Sample Answer:

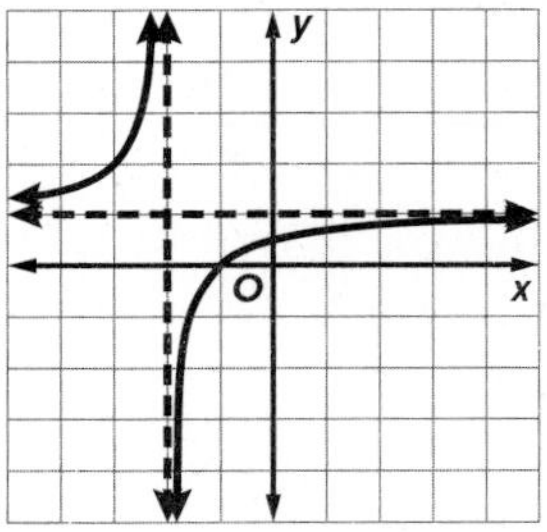

57. The graph of $g(x)$ has a hole at $x = -3$.

59. Sample answer: By factoring the denominator of the expression in a rational function and determining the the values that cause each factor to equal zero, you can find the asymptotes or discontinuity of the function. If the denominator has a factor $x - c$ that does not appear in the numerator, then there is a vertical asymptote at $x = c$. If the numerator and denominator have a common factor of $x - c$, then there is point discontinuity at $x = c$.

Lesson 9-5

1. 1.5 **3.** −56 **5.** 55 **7.** −48 **9.** 72

11. 90 **13.** −22.5 **15.** 38

17. 4 **19.** -10 **21.** about 0.7 **23.** 0.000793 Ω
25. 1.336 atm **27.** inverse; 4 **29.** inverse; 15
31. direct; 5280 **33.** inverse; -25
35. inverse; 8 **37.** direct; 2π **39.** inverse; -2
41. combined; 10 **43.** direct; 4 **45.** direct; -2
47. inverse; 7 **49.** joint; 20 **51.** 54 **53.** 7
55. $y = 30x$; 7 theaters **57.** inverse
59a. 11×14: $\ell = \frac{14}{11}w$; 12×16: $\ell = \frac{4}{3}w$; 16×20: $\ell = \frac{5}{4}w$; 18×24: $\ell = \frac{4}{3}w$; 20×24: $\ell = \frac{6}{5}w$; 24×30: $\ell = \frac{5}{4}w$; 24×36: $\ell = \frac{3}{2}w$; 30×40: $\ell = \frac{4}{3}w$; 36×48: $\ell = \frac{4}{3}w$; 48×60: $\ell = \frac{5}{4}w$; 48×72: $\ell = \frac{3}{2}w$ **59b.** 12×16, 18×24, 30×40, 36×48 all have the relationship $\ell = \frac{4}{3}w$. 16×20, 24×30, and 48×60 have the relationship $\ell = \frac{5}{4}w$. 24×36 and 48×72 have the relationship $\ell = \frac{3}{2}w$.
59c. Sample answer: One canvas is an enlargement of the other.
61a. $\frac{m}{n}$ dollars **61b.** inversely **61c.** \$280
63a. $d = 0.10hr$ **63b.** Joint variation; the amount deducted varies directly as the product of two quantities, the hourly wage and the number of hours worked. **63c.** Substitute $r = \$19.50$ and $h = 36$ in the formula for part a. The amount deducted was \$70.20.
65. a and g are directly related.
67. Sample answer: The force of an object varies jointly as its mass and acceleration.

Lesson 9-6

1. -3 **3.** 9 **5.** 7 **7.** $-12, 2$ **9.** ∅ **11.** 6
13a. $f(x) = \frac{8(0.15) + x(0.75)}{8 + x}$ **13b.** 11.2 mL
15. 41 hours
17a. $\frac{14}{10 - c} - \frac{14}{10 + c} = \frac{1}{2}$ **17b.** 1.7 mph
19. $x < 0$ or $x > 1.75$ **21.** $x < -2$ or $2 < x < 14$
23. $x < -5$ or $4 < x < \frac{17}{3}$ **25a.** $\frac{1000}{N} \leq 15$
25b. 67 **27.** 56 in. **29.** no solution
31. $n < -3$ or $0 < n < 3$ **33.** $0 < x < \frac{3}{2}$
35. 9 **37.** $-5 < m < -2$
39a. $80 > \frac{60x + 17{,}000}{x - 50}$
39b. $x > 1050$; the company must produce and sell at least 1050 speakers in order to ensure that the revenue from each one is greater than the average cost of producing each one.
41. $x = 2$; this method works because the x-coordinate of the point of intersection will make both sides of the equation equal to the same value, and no extraneous roots are shown.
43. at least 40 students
45. Sample answer: 3 must be excluded as a solution, as it will make the denominator of $f(x)$ equal to 0. Only -4 is a solution.
47. Sample answer: I do not agree with the solution set. Multiplying both sides of the inequality by x does not result in an equivalent inequality if x is negative. The work is only valid for positive x-values. The student could have graphed the two equations $f(x) = \frac{1}{x}$ and $g(x) = 2$ and looked at x-values for which the graph of $f(x)$ is below the graph of $g(x)$.
49. all real numbers except 5, -5, 0
51. Sample answer: Multiplying each side of a rational equation or inequality by the LCD can result in extraneous solutions. Therefore, you should check all solutions to make sure that they satisfy the original equation or inequality.

Module 9 Review

1. $\frac{2y^2}{x}$ **3.** A **5.** $\frac{2y^2 - 15x^2}{3x^2y}$ **7.** D
9.
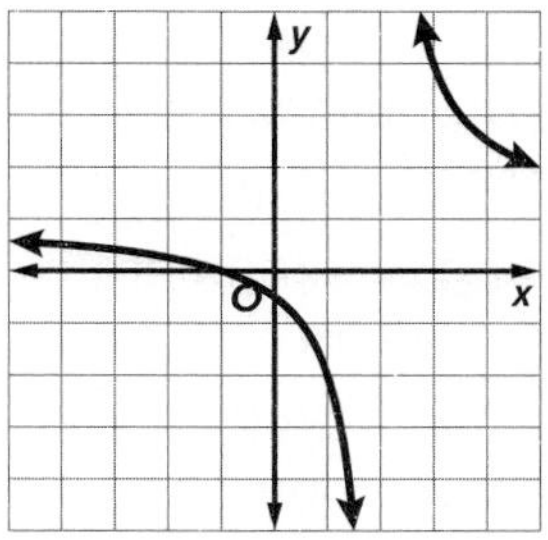

11a. x-intercept: undefined; y-intercept: -1
11b. horizontal, $y = 1$, vertical, $x = -2$
13. D **15a.** 25 mph **15b.** 30 mph **17.** D

Module 10

Module 10 Opener

1. 89 customers, 88 customers, no mode

3. 7.7, 8, 10

5. $\frac{1}{4}$

7. $\frac{1}{4}$

Lesson 10-1

1. Sample: the T-shirts Berton selects; population: all of Berton's sports T-shirts; stratified: Berton divides the T-shirts by team before the sample is selected.

3. Sample: the customers who submit suggestions; population: all customers; self-selected: the customers voluntarily submit suggestions.

5. unbiased; Sample answer: The students are randomly selected. The fact that the sample is selected at band camp does not influence the response of the selected sample.

7. biased; Sample answer: The question is asking about two issues: whether the workout facility needs a new treadmill and whether the workout facility needs a new racquetball court.

9. unbiased; Sample answer: The question does not influence participants.

11. survey; sample: customers that take the online survey; population: all customers

13. experiment; sample: the 50 adults participating in the study; population: all adults

15. objective: to determine the number of juniors and seniors planning to attend prom; population: all juniors and seniors at the school; sample survey questions: What grade are you in? Do you plan on attending prom?

17. observational study; Sample answer: Teenagers are more willing than adults to try a new sports drink.

19. control group: students who do not receive gift cards; experimental group: students receiving gift cards; Sample answer: The students may not have any interest in the gift cards they are receiving. The principal should conduct a survey to find what gifts or activities might motivate the students.

21. convenience; survey

23. stratified; experiment

25. observational study

27. Sample answer: She should use an observational study to determine the average miles per gallon that identical cars are able to go on each type of fuel. She can then divide the current price of each fuel by the number of miles to find the average cost per mile. The group with the lower cost per mile would be the one that is more cost efficient. She might also look for the results of a survey by a motor club or magazine stating the average gas mileage for cars similar to her new car.

29. No; sample answer: the difference between the means of the experimental group and control group is only 0.1 ounce. Both groups had cats that gained significant weight and others that gained little. To compare the groups, I assumed that the cats in each group were similar in age, starting weight, and health. I also assumed that each cat was given the same amount of food.

31. Sample answer: This method of selecting a sample is valid. Each student has an equally likely chance of being selected for the sample. A weakness may be that this would not reflect that one grade may feel strongly about the dress code than another.

33. False; sample answer: A sample statistic is used to estimate a population parameter.

35. Sample answer:
objective: Determine the average amount of time that students spend studying at the library.
population: All students that study at the library.
sample: 25 randomly selected students studying at the library during a given week.

Study Time (minutes)				
38	16	45	41	63
18	20	17	8	15
41	28	55	19	15
30	11	20	79	24
78	24	26	32	19

mean: ≈31.3 min

Lesson 10-2

1a. 25%

1b. 38%

3a. 16.7%

3b. 18%

5. Sample answer: The theoretical probability of making a goal is $\frac{7}{20}$ and the theoretical probability of missing the goal is $\frac{13}{20}$. I can use a random number generator to create a simulation. For numbers 1–20, let 1–7 represent a goal and 8–20 represent a miss. Based on 100 trials, the experimental probability of making a goal is 41%. Because Paola's success rate was only about 13%, it is likely that there is an error in the programming.

7. spinner with 2 regions, one central angle 216°, and one 144°

9. Sample answer: Use a random number generator where 0, 1, 2, 3, and 4 represent brown hair, 5, 6, 7, and 8 represent blonde hair, and 9 represents red hair.

11. Jayme's model

13a. 20% **13b.** 8% **13c.** Sample answer: I assumed that each car color was equally likely.

15a. Sample answer: Use a random number generator to generate a set of three numbers from 1 to 42. Numbers 1–15 will represent players on the basketball team and numbers 16–42 will represent the other students. Discard any trial that repeats any of the same number. Run the trial fifty times.

15b.

35, 13, 36	15, 32, 9	6, 14, 12	16, 26, 27	4, 34, 20	16, 32, 24	12, 9, 18	10, 17, 26
15, 13, 23	18, 15, 41	13, 9, 20	33, 38, 30	34, 40, 35	37, 18, 36	22, 16, 36	23, 5, 28
29, 7, 2	21, 22, 1	33, 31, 22	30, 39, 18	31, 7, 17	35, 40, 36	35, 9, 29	32, 20, 21
21, 3, 23	16, 36, 5	30, 36, 11	28, 35, 10	5, 42, 21	11, 18, 40	26, 10, 2	28, 35, 41
13, 11, 23	12, 7, 24	5, 3, 7	15, 39, 2	30, 21, 24	32, 25, 7	18, 9, 34	29, 20, 2
18, 26, 6	32, 6, 39	17, 3, 21	31, 18, 27	24, 29, 31	25, 37, 21	32, 23, 13	21, 15, 28
3, 34, 15	12, 2, 20						

15c. Sample answer: Only 2 of the 50 trials, or 4%, resulted in only team members winning the raffle. According to this simulation it is highly unlikely that all the winners are members of the basketball team. There is not enough data to determine if the raffle is unfair. If it happens again for another raffle, then the fairness should be examined more closely.

17a. The theoretical probability that she serves an ace on her next serve is 66%, and the theoretical probability that she doesn't is 34%. Use a random number generator to generate integers 1 through 100. The integers 1–66 will represent an ace, and the integers 67–100 will represent any other outcome of her serve. The simulation will consist of 30 trials.

17b. Sample data is shown:

14	16	8	94	9	87
41	41	68	42	66	11
74	38	86	41	26	92
28	36	4	12	25	68
24	4	76	3	61	35

17c. Sample answer: 22 of the 30 trials, or 73.3%, resulted in an ace on the next serve. According to this simulation it is highly likely that her next serve will be an ace. The data is not completely consistent with the model, but if a larger trial were done it might correlate more closely.

19. Sample answer: Jevon's concern about the fairness of the coin should be dependent on the number of times he tossed it. If he completed only 4, or even 20, trials, then the sample size is not large enough to warrant concern. However, if he completed 100 or more trials, then he should be concerned since the experimental probability should be closer to the theoretical probability of $\frac{1}{2}$, or 50%.

21. True; sample answer: The event will always occur when the theoretical probability is 1. Therefore, the experimental probability can never be 0. For example, the theoretical probability of rolling a die and getting 1, 2, 3, 4, 5, or 6 is 1 since it will always show one of these six numbers. So, when a die is rolled, the result will be one of these six numbers, making the experimental probability always greater than 0.

23. Yes; sample answer: If the spinner were going to be divided equally into three outcomes, each sector would measure 120. Because you only want to know the probability of outcome *C*, you can record spins that end in the red area as a success, or the occurrence of outcome *C*, and spins that end in the blue area as a failure, or an outcome of *A* or *B*.

Lesson 10-3

1. 38.22; Sample answer: Because the standard deviation is great compared to the mean of $72.63, the data are spread farther from the mean. Based on the standard deviation, most of the shears cost between $34.41 and $110.85.

3. 22.6; 1.3

5. 6.9; 1.7

7. Before the season the mean time was 9.2 minutes and the standard deviation was 1.0 minute. After the season the mean time was 7.0 minutes and the standard deviation was 0.9 minutes. This means that before the season the mile times were higher on average, but that the mile times were generally spread the same before and after the season.

9. 0.375; 0.123

11. 18.4; Sample answer: The standard deviation is small compared to the mean of 89. Most of the exam sessions will have between 71 and 107 attendees.

13. No; sample answer: adding $1 to each employee's hourly wage has no effect on the standard deviation of the data. The standard deviation both before and after the wage change is 0.67.

15. Sample answer: Assume that each data value falls in the center of each bar of the histogram. Using technology, the mean is 17 and the standard deviation is about 7.4.

17. Find the mean of the set of data. Then, find the square of the different between each data value and the mean. Next, find the sum of all of the squared differences. Then, divide the sum by the number of values in the set of data, which is the variance. Finally, take the square root of the quotient, or variance.

Lesson 10-4

1. texts; discrete; The number of texts can only be represented as a whole number.

3. height of plant; continuous; The height of a plant may be any positive value.

5. Sample answer: The scale of the *y*-axis misleadingly shows the differences in probabilities.

7a. $39.7 < X < 333.1$ **7b.** 16% **7c.** $X > 284.2$

9. −2.19 **11.** 0.281 **13.** 0.9861

15. 2.5% **17.** 6200 hours

19. Hiroko; Sample answer: Monica's solution would work with a uniform distribution.

21. Sample answer: The *z*-value represents the position of a value *X* in a normal distribution. The area under a curve to the left of a *z*-value represents the probability that a value from the distribution will be less than the given value *X*.

Lesson 10-5

1. $E = 0.16$

3. $E = 0.06$

5. At a 99% confidence interval, the population mean is $5.35 \leq \mu \leq 5.65$. Therefore, we are

99% confident that the rating of the airline is between 5.35 and 5.65.

7. At a 90% confidence interval, the population proportion is $0.737 \le p \le 0.813$. Therefore, we are 90% confident that the proportion of students who agree that high school students should have a part-time job is between 73.7% and 81.3%.

9. At a 99% confidence interval, the population proportion is $0.326 \le p \le 0.506$. Therefore, we are 99% confident that the proportion of students who agree with the principal's plan is between 32.6% and 50.6%.

11. Instead of stating that the population proportion is estimated at $\frac{120}{250} = 0.48 = 48\%$ with a margin of error of 3%, the claim uses the highest possible estimate for the population proportion to give the impression that the proportion of people who want parks to be built is greater than 50%.

13. $\hat{p} = \frac{32}{205} \approx 0.156$ and $ME = 1.96 \sqrt{\frac{0.156(1 - 0.156)}{205}} \approx 0.050$, so $CI = 15.6\% \pm 5.0\%$; Sample answer: With 95% confidence, the mean proportion of discards for the population of all pieces fired in her kiln is between 10.6% and 20.6%, so Karen should probably buy a new kiln on the basis that the discard rate is most likely higher than 10%.

15. As the sample size increase, the maximum error decreases.

17. Sample answer: sample mean: 18.8 hours; sample standard deviation: 8.8; the mean number of hours with a 90% confidence level is $15.4 \le \mu \le 22.2$.

Hours of Television Watched Each Week					
12	8	19	27	20	18
0	33	25	14	9	19
23	6	20	26	30	29

Module 10 Review

1. D **3.** A **5.** A

7.

Data Set	has a higher average value	is more varied
A	X	
B		X

9. B, C **11.** $x < 55.9$ and $x > 70.9$

13. $E = 1.96 \sqrt{\frac{\left(\frac{273}{360}\right)\left(\frac{87}{360}\right)}{360}}$

$E \approx 0.044$

$\hat{p} = \frac{273}{360} \approx 0.7583$

$\hat{q} = \frac{87}{360} \approx 0.2417$

Sample answer: At a 95% confidence level, I am confident that the proportion of patrons that would like to see the restaurant stay open later is between 71.41% and 80.26%.

Selected Answers

Module 11 Opener

1. 11.7 **3.** 20.5 **5.** $x = 9, y = 9\sqrt{2}$
7. $x = 12, y = 12\sqrt{3}$

Lesson 11-1

1.

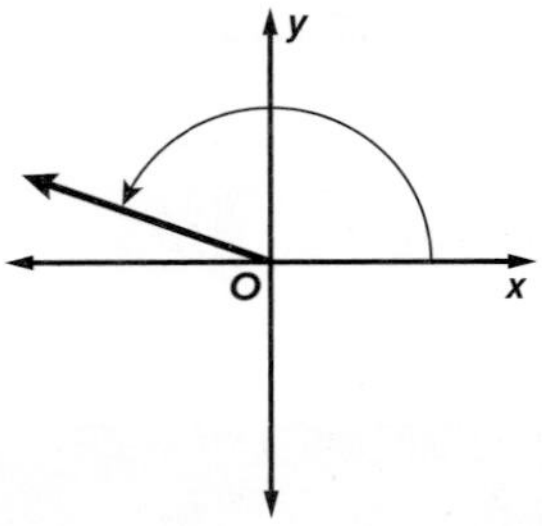

3.

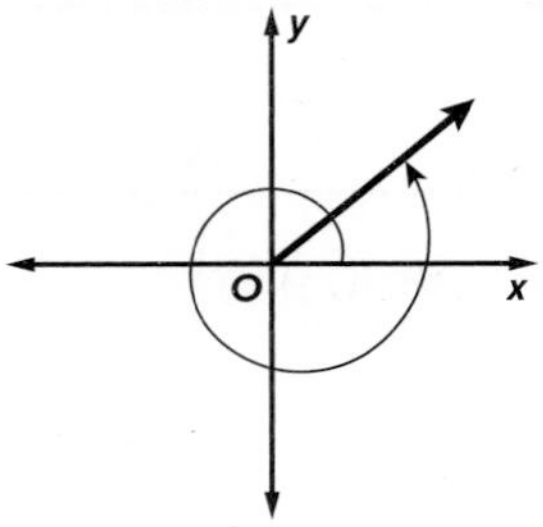

5.

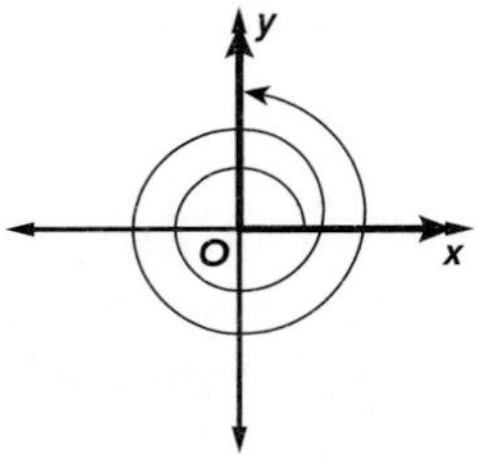

7.

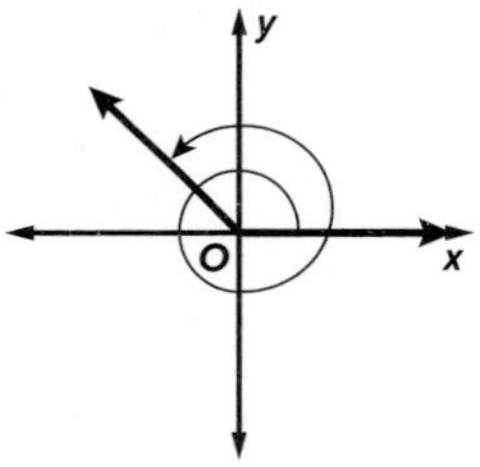

9.

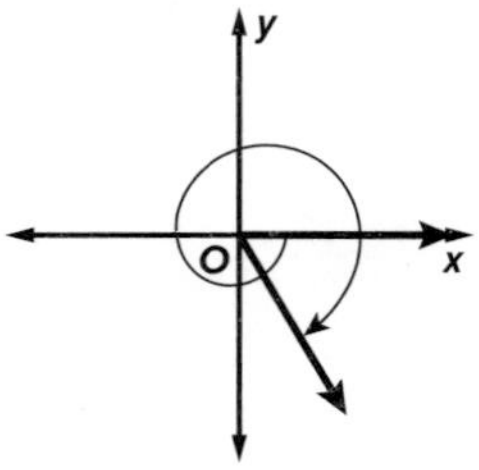

11.

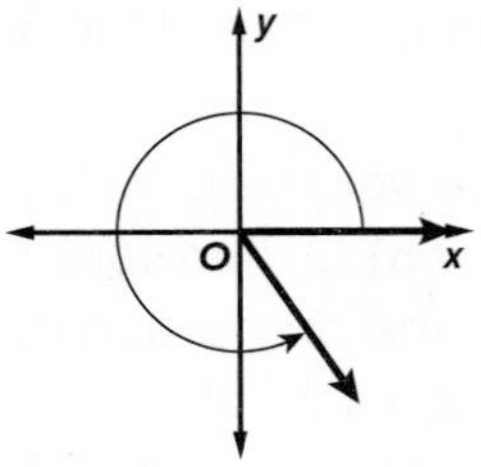

13.

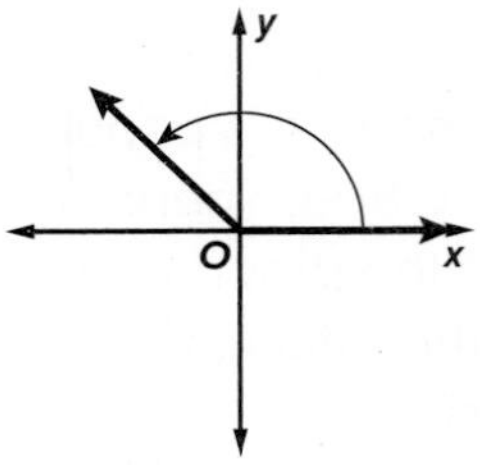

15.

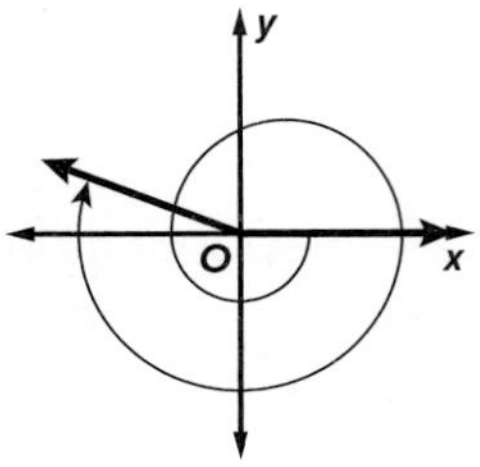

17. 285°, −435° **19.** 405°, −315°
21. 10°, −350° **23.** 60°, −300°
25. 425°, −295° **27.** 470°, −250°
29. $\frac{17\pi}{6}, -\frac{7\pi}{6}$ **31.** $\frac{8\pi}{3}, -\frac{4\pi}{3}$ **33.** $\frac{5\pi}{4}, -\frac{11\pi}{4}$
35. $-\frac{13\pi}{9}$ **37.** $\frac{19\pi}{9}$ **39.** 4π **41.** $\frac{\pi}{2}$
43. −108° **45.** 60° **47.** 120° **49.** −135°
51. 45° **53.** $\frac{20\pi}{3}$ or approximately 20.9 in.
55. $\frac{\pi}{30}$ **57.** $-\frac{41\pi}{9}$ **59.** 450°
61. −105° **63.** 8.4 in. **65.** 12.6 ft
67. 3309°/s; 58 radians/s **69.** 49.74 in.
71. −90°, $-\frac{\pi}{2}$ radians; 3π inches
73a. $\frac{\pi}{3}$; Because a complete rotation takes 30 minutes, 5 minutes represents $\frac{5}{30}$ or $\frac{1}{6}$ of a complete rotation, and $\frac{1}{6} \cdot 2\pi = \frac{\pi}{3}$.
73b. The radius r of the wheel is 67.5 m. The distance the passenger traveled is the arc length, so $s = r\theta = 67.5\left(\frac{\pi}{3}\right) \approx 71$ m.
73c. The circumference is πd, which is approximately $3.14d = 3.14(135) = 423.9$ m and $\frac{1}{6}(423.9) = 70.65$, which is close to 71, so the answer is reasonable.

75. Melinda is correct. The area of the complete circle is $\pi r^2 = \pi(15)^2 = 225\pi$ ft^2. Because the area that gets watered by the sprinkler is 75π ft^2, this area is $\frac{75\pi}{225\pi} = \frac{1}{3}$ of the circle. Therefore, the measure of the central angle is $\frac{1}{3}(2\pi) = \frac{2\pi}{3}$.

77. Because $s = r\theta$ and $r = 5$, the function may be written as $f(x) = 5x$. This means the graph is a straight line with a slope of 5 that passes through the origin.

79. $x = 2$

81. 440° and −280°

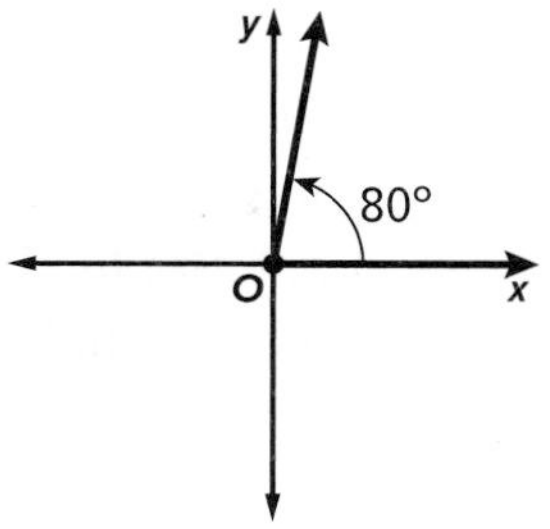

83. One degree represents an angle measure that equals $\frac{1}{360}$ rotation around a circle. One radian represents the measure of an angle in standard position that intercepts an arc of length r. To convert from degrees to radians, multiply the number of degrees by $\frac{\pi \text{ radians}}{180°}$. To convert from radians to degrees, multiply the number of radians by $\frac{180°}{\pi \text{ radians}}$.

Lesson 11-2

1. $\sin\theta = \frac{5}{13}$, $\cos\theta = \frac{12}{13}$, $\tan\theta = \frac{5}{12}$, $\csc\theta = \frac{13}{5}$, $\sec\theta = \frac{13}{12}$, $\cot\theta = \frac{12}{5}$

3. $\sin\theta = \frac{8}{17}$, $\cos\theta = \frac{15}{17}$, $\tan\theta = \frac{8}{15}$, $\csc\theta = \frac{17}{8}$, $\sec\theta = \frac{17}{15}$, $\cot\theta = \frac{15}{8}$

5. $\sin\theta = \frac{5}{11}$, $\cos\theta = \frac{4\sqrt{6}}{11}$, $\tan\theta = \frac{5\sqrt{6}}{25}$, $\csc\theta = \frac{11}{5}$, $\sec\theta = \frac{11\sqrt{6}}{24}$, $\cot\theta = \frac{4\sqrt{6}}{5}$

7. $\sin A = \frac{8}{17}$, $\cos A = \frac{15}{17}$, $\csc A = \frac{17}{8}$, $\sec A = \frac{17}{15}$, $\cot A = \frac{15}{8}$

9. $\sin B = \frac{3\sqrt{10}}{10}$, $\cos B = \frac{\sqrt{10}}{10}$, $\csc B = \frac{\sqrt{10}}{3}$, $\sec B = \sqrt{10}$, $\cot B = \frac{1}{3}$

11. $\sin A = \frac{\sqrt{3}}{2}$, $\tan A = \sqrt{3}$, $\csc A = \frac{2\sqrt{3}}{3}$, $\sec A = 2$, $\cot A = \frac{\sqrt{3}}{3}$

13. $\sin\theta = \frac{12}{13}$, $\cos\theta = \frac{5}{13}$, $\tan\theta = \frac{12}{5}$, $\csc\theta = \frac{13}{12}$, $\sec\theta = \frac{13}{5}$, $\cot\theta = \frac{5}{12}$

15. $\sin\theta = -\frac{15}{17}$, $\cos\theta = \frac{8}{17}$, $\tan\theta = -\frac{15}{8}$, $\csc\theta = -\frac{17}{15}$, $\sec\theta = \frac{17}{8}$, $\cot\theta = -\frac{8}{15}$

17. $\sin\theta = -\frac{40}{41}$, $\cos\theta = -\frac{9}{41}$, $\tan\theta = \frac{40}{9}$, $\csc\theta = -\frac{41}{40}$, $\sec\theta = -\frac{41}{9}$, $\cot\theta = \frac{9}{40}$

19. $\sin\theta = \frac{\sqrt{5}}{5}$, $\cos\theta = \frac{2\sqrt{5}}{5}$, $\tan\theta = \frac{1}{2}$, $\csc\theta = \sqrt{5}$, $\sec\theta = \frac{\sqrt{5}}{2}$, $\cot\theta = 2$

21. $\sin\theta = \frac{\sqrt{10}}{10}$, $\cos\theta = \frac{3\sqrt{10}}{10}$, $\tan\theta = \frac{1}{3}$, $\csc\theta = \sqrt{10}$, $\sec\theta = \frac{\sqrt{10}}{3}$, $\cot\theta = 3$

23. $\sin\theta = -\frac{3\sqrt{10}}{10}$, $\cos\theta = \frac{\sqrt{10}}{10}$, $\tan\theta = -3$, $\csc\theta = -\frac{\sqrt{10}}{3}$, $\sec\theta = \sqrt{10}$, $\cot\theta = -\frac{1}{3}$

25. $\sin\theta = 0$, $\cos\theta = 1$, $\tan\theta = 0$, $\csc\theta =$ undefined, $\sec\theta = 1$, $\cot\theta =$ undefined

27. $\sin\theta = 1$, $\cos\theta = 0$, $\tan\theta =$ undefined, $\csc\theta = 1$, $\sec\theta =$ undefined, $\cot\theta = 0$

29. 50°

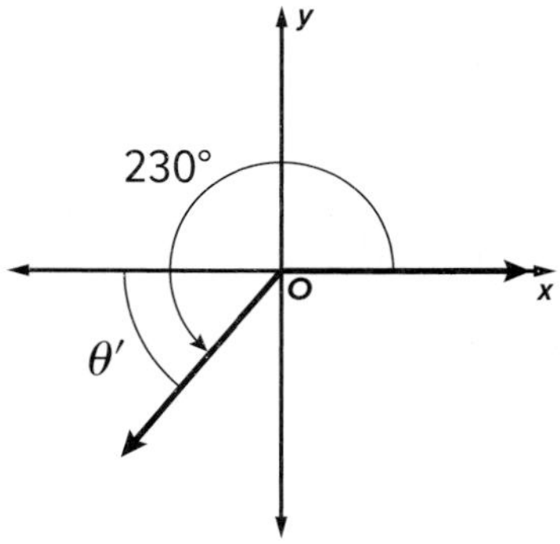

31. $\frac{\pi}{3}$

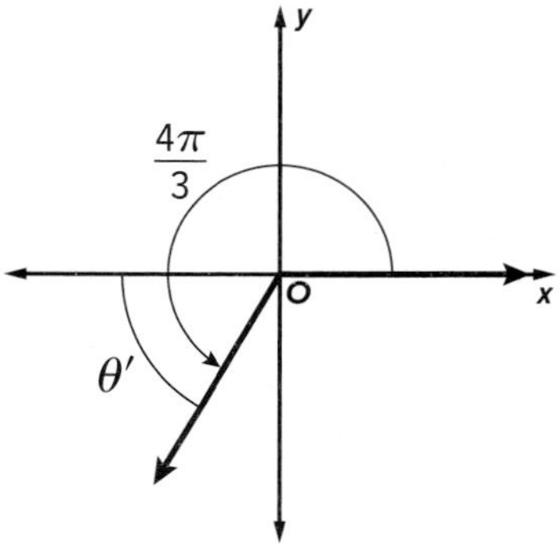

33. $\frac{\pi}{4}$

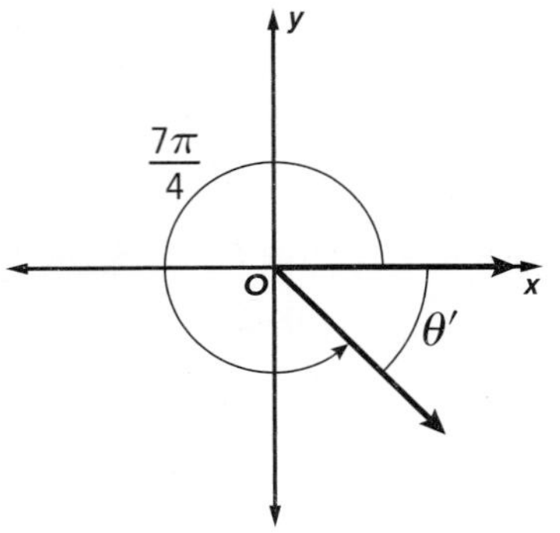

35. 20°

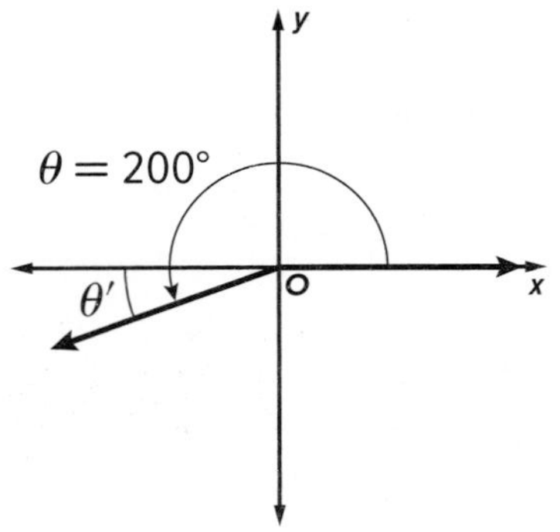

37. $\frac{3\pi}{8}$

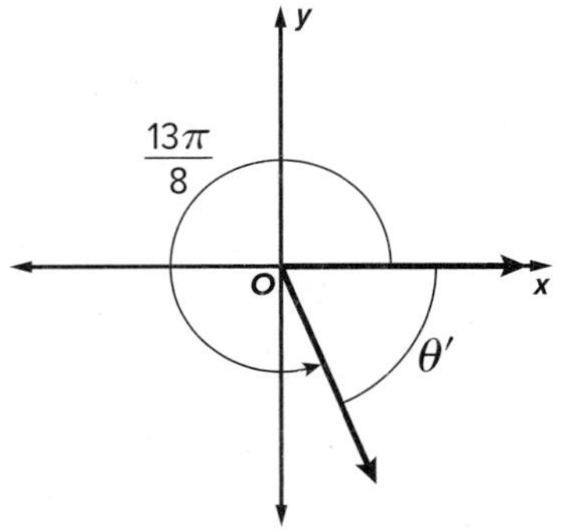

39. $\frac{\pi}{4}$

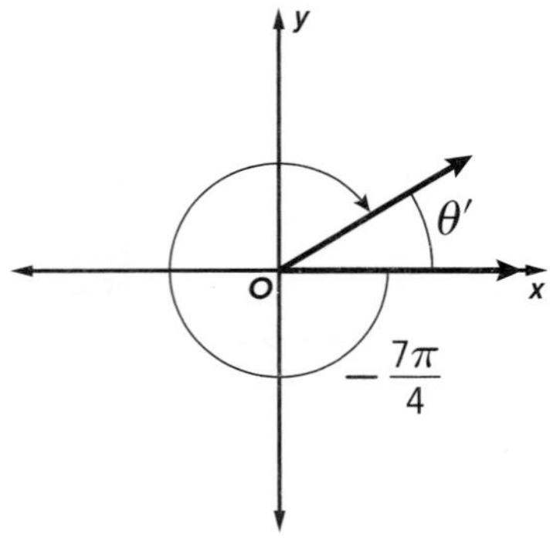

41. $-\frac{\sqrt{2}}{2}$ **43.** $\sqrt{2}$ **45.** −1 **47a.** $\tan 71.6 = \frac{12}{x}$

47b. about 4.0 ft **49.** $\frac{\sqrt{255}}{16}$ or ≈ 0.998

51. $\frac{3\sqrt{55}}{55}$ or ≈ 0.405

53. 0 **55.** $-\frac{\sqrt{3}}{3}$ **57.** undefined **59.** $(-1, \sqrt{3})$

61. 1.3 in. **63.** $\sqrt{5}$ **65.** 2

67. Sample answer: 120°, 240°, 420°; For an angle in Quadrant II, $180° - \theta = 60°$, so $\theta = 120°$; for an angle in Quadrant III, $\theta - 180° = 60°$, so $\theta = 240°$; 60° plus any multiple of 360° will have a reference angle of 60°, so $60° + 360° = 420°$.

69. True; $\sin \theta = \frac{\text{opp}}{\text{hyp}}$ and the values of the opposite side and the hypotenuse of an acute triangle are positive, so the value of the sine function is positive.

71. No; for $\sin \theta = \frac{\sqrt{2}}{2}$ and $\tan \theta = -1$, the reference angle is 45°. However, for $\sin \theta$ to be positive and $\tan \theta$ to be negative, the reference angle must be in the second quadrant. So, the value of θ must be 135° or an angle coterminal with 135°.

73. We know that $\cot \theta = \frac{x}{y}$, $\sin \theta = \frac{y}{r}$, and $\cos \theta = \frac{x}{r}$. Since $\sin 180° = 0$, it must be true that $\cot 180° = \frac{x}{0}$, which is undefined.

75. First, sketch the angle and determine in which quadrant it is located. Then use the appropriate rule for finding its reference angle θ'. A reference angle is the acute angle formed by the terminal side of θ and the x-axis. Next, find the value of the trigonometric function for θ'. Finally, use the quadrant location to determine the sign of the trigonometric function value of θ.

Lesson 11-3

1. $\cos \theta = -\frac{\sqrt{3}}{2}$, $\sin \theta = \frac{1}{2}$

3. $\cos \theta = -\frac{2}{3}$, $\sin \theta = \frac{\sqrt{5}}{3}$

5. $\cos \theta = \frac{1}{6}$, $\sin \theta = -\frac{\sqrt{35}}{6}$

7. $\sin \theta = \frac{1}{2}$; $\cos \theta = \frac{\sqrt{3}}{2}$, $\tan \theta = \frac{\sqrt{3}}{3}$, $\csc \theta = 2$, $\sec \theta = \frac{2\sqrt{3}}{3}$, $\cot \theta = \sqrt{3}$

9. $\sin \theta = 1$; $\cos \theta = 0$, $\tan \theta =$ undefined, $\csc \theta = 1$, $\sec \theta =$ undefined, $\cot \theta = 0$

11. $\sin \theta = 0$; $\cos \theta = -1$, $\tan \theta = 0$, $\csc \theta =$ undefined, $\sec \theta = -1$, $\cot \theta =$ undefined

13. 10 **15.** $\frac{5\pi}{2}$ **17.** 4

19. period = 4 seconds

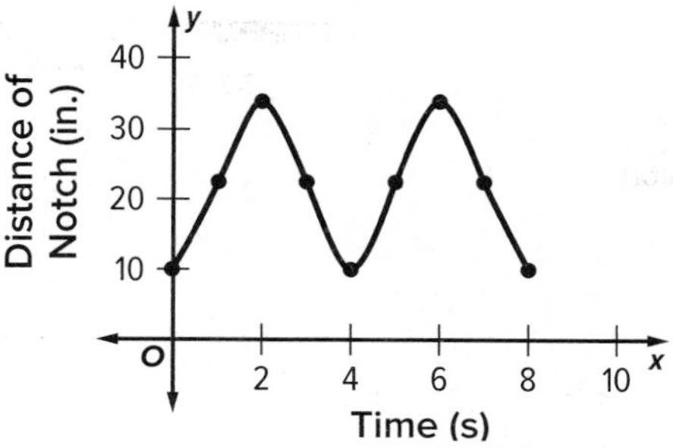

21. $-\frac{1}{2}$ **23.** 0 **25.** $-\frac{\sqrt{2}}{2}$ **27.** $\frac{\sqrt{2}-\sqrt{3}}{2}$

29. $-\frac{5\sqrt{3}}{2}$ **31.** 1 **33.** 24 s

35a. Sometimes; the cosine function can only result in values between −1 and 1, inclusive.

35b. Always; the sine function has a period of 2π.

35c. Sometimes; $\cos \theta = 1$ when n is even and $\cos \theta = -1$ when n is odd.

35d. Never; the y-coordinate of the corresponding point on the unit circle is negative.

37a. $\frac{3\pi}{4}$ meters; Sample answer: The radian measure of the angle is the length of the arc on the unit circle subtended by the angle.

37b.

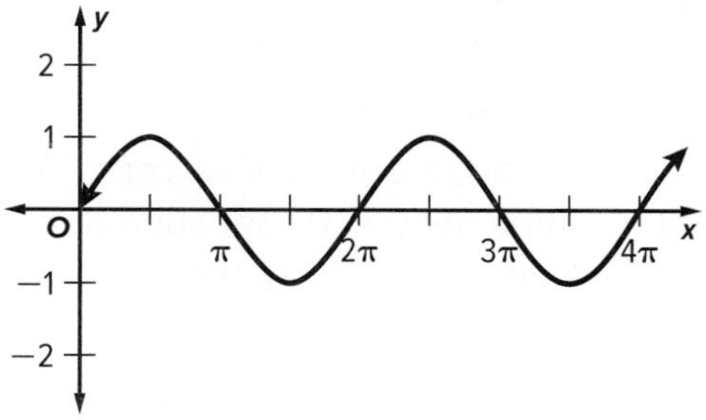

37c. Sample answer: The period is 2π because the values of the function repeat every 2π radians. This is shown in the graph when the shape of the curve from 0 to 2π is repeated from 2π to 4π. The heights of point P above or below the surface of the water repeat once as the wheel makes a complete rotation (2π radians).

37d. The x-intercepts are whole-number multiples of π. These are the angles of rotation for which P is on the surface of the water.

37e. Sample answer: $\left(\frac{\pi}{2}, \frac{3\pi}{2}\right)$; as the wheel rotates through an angle from $\frac{\pi}{2}$ radians to $\frac{3\pi}{2}$ radians, P moves downward.

39a. 72°F **39b.** March **39c.** 12°F

39d. September

41. Benita; Francis incorrectly wrote $\cos\left(-\frac{\pi}{3}\right) = -\cos\frac{\pi}{3}$.

43. Sometimes; the period of a sine curve could be $\frac{\pi}{2}$, which is not an integer multiple of π.

45. The period of a periodic function is the horizontal distance of the part of the graph that is nonrepeating. Each nonrepeating part of the graph is one cycle.

Lesson 11-4

1. 2; $y = 0$; 360° **3.** 1; $y = 0$; 120°

5. 2; 360° **7.** 1; 720° **9.** $\frac{1}{2}$; 180°

11. 3; 360°

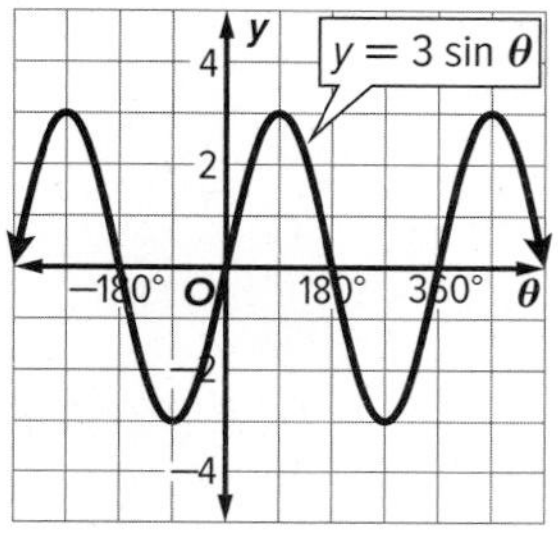

13. 1; 90°

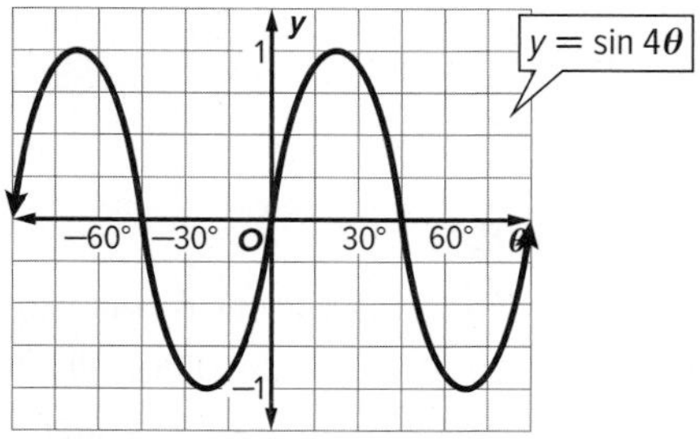

15. 4; 180°

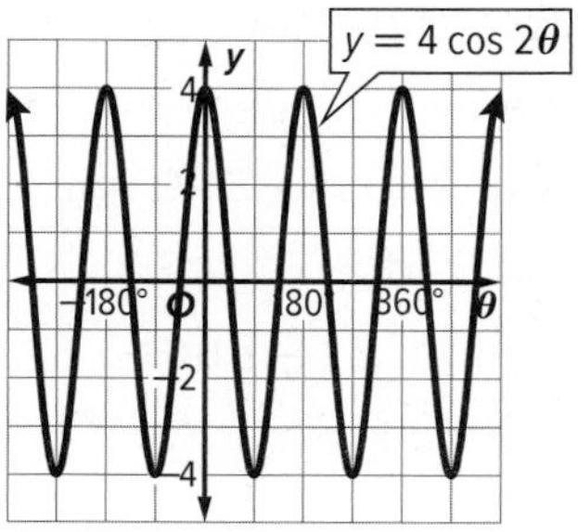

17a. period: 2; frequency: $\frac{1}{2}$ The object completes $\frac{1}{2}$ of a cycle per second, and it will reach maximum distance from the equilibrium point every 2 seconds.

17b. Domain: all real numbers; sample answer: Because time cannot be negative, the relevant domain in the context of the situation is $[0, \infty)$. The range is $[-25, 25]$.

19. $h(x) = 1.5 \cos \frac{2\pi}{5}x$

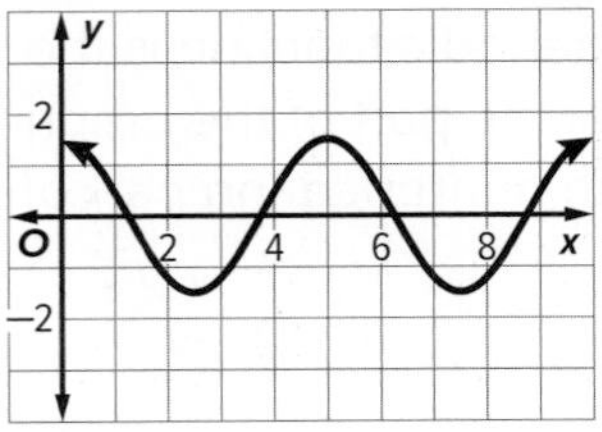

21. 3; 540°

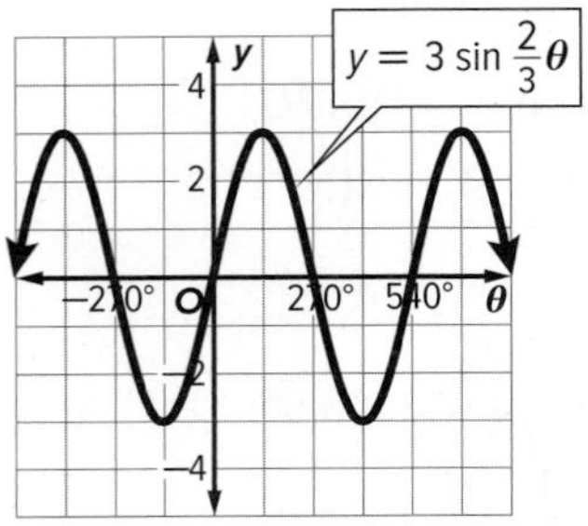

23. 4; 1080°

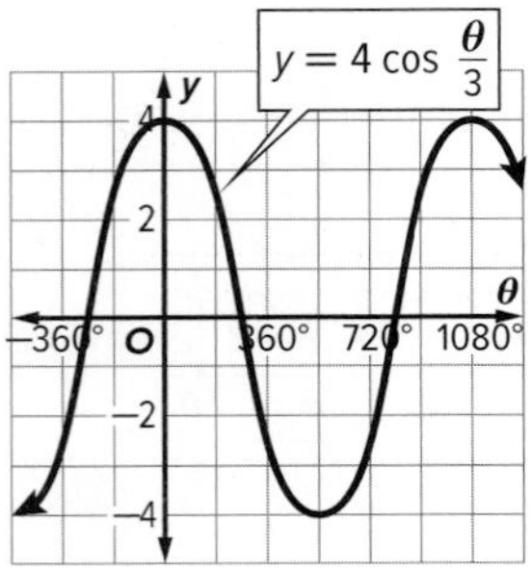

25. 180°; $y = 5 \sin 2\theta$

27. Sample answer: The sine function has the same shape as an ocean wave. Additionally, the sine function and ocean waves share similar behaviors, as they both repeat themselves over a consistent interval of time and distance.

29a. 12; a calendar year

29b. 85°F; July 15 **31a.** 200 nm, 520 nm

31b. yellow **31c.** violet; red

33a. amplitude: 4.5; period: 2 seconds

33b. $f(t) = -4.5 \cos \pi t + 7.5$

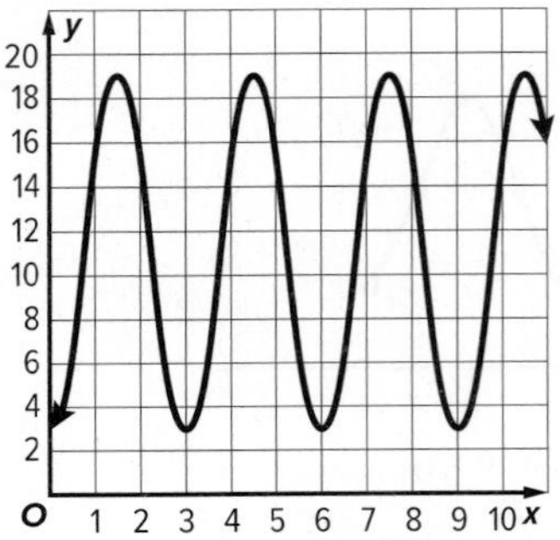

35. Sample answer: $y = 2 \sin 6\theta$; the amplitude is $|a| = 2$. The period is $\frac{2\pi}{|b|} = \frac{2\pi}{6} = \frac{\pi}{3}$, so there are 3 complete cycles between $\theta = 0$ and $\theta = \pi$.

37. The graph of $y = \frac{1}{2} \sin \theta$ has an amplitude of $\frac{1}{2}$ and a period of 360°. The graph of $y = \sin \frac{1}{2}\theta$ has an amplitude of 1 and a period of 720°.

39. Sample answer: Determine the amplitude and period of the function; find and graph and x-intercepts and extrema; use the parent function to sketch the graph.

Lesson 11-5

1. period = 36°; asymptotes: $(36n + 18)°$, where n is an odd integer; x-intercepts: 0°, 36°, 72°, 108°, ...; midline: $y = 0$; Because $a = 1$, the function is not vertically dilated in relation to the parent function. Because $b = 5$, the function is compressed horizontally in relation to the parent function.

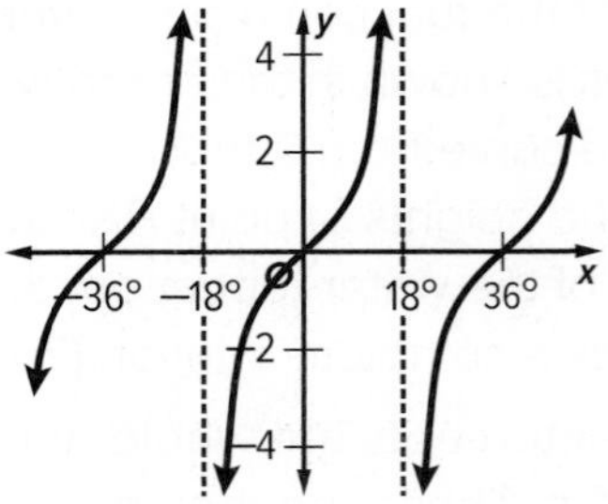

3. period = 90°; asymptotes: (90n + 45)°, where n is an odd integer; x-intercepts: 0°, 90°, 180°, 270°, ...; midline: $y = 0$; Because $a = 1$, the function is not vertically dilated in relation to the parent function. Because $b = 2$, the function is compressed horizontally in relation to the parent function.

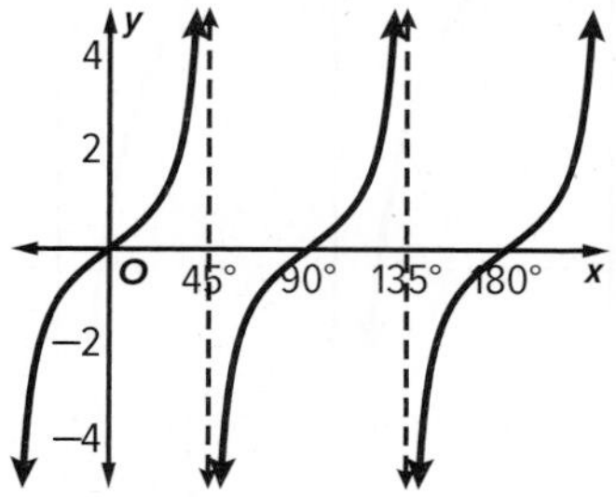

5. period = 360°; asymptotes: (360n + 180)°, where n is an odd integer; x-intercepts: 0°, 360°, 720°, 1080°, ...; midline: $y = 0$; Because $a = 2$, the function is stretched vertically in relation to the parent function. Because $b = \frac{1}{2}$, the function is stretched horizontally in relation to the parent function.

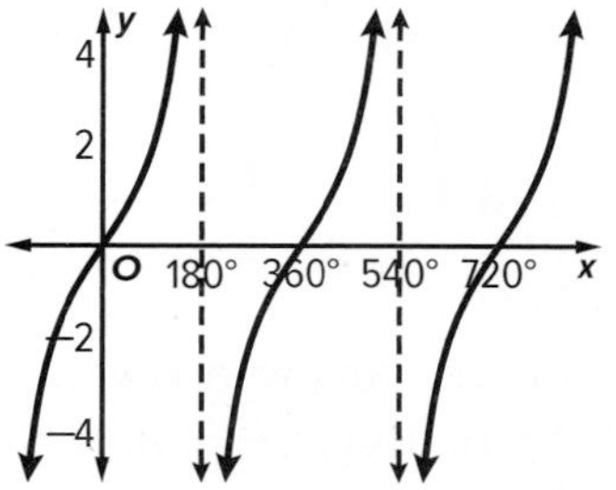

7. period = $\frac{8\pi}{3}$ asymptotes: $x = \frac{4\pi}{3} + \frac{4\pi}{3}n$, where n is an integer; relative minima at $x = \frac{2\pi}{3}$, $x = \frac{10\pi}{3}$, or $x = \frac{2\pi}{3} + \frac{8\pi}{3}n$, where n is an integer; relative maxima at $x = 2\pi$, $x = \frac{14\pi}{3}$, or $x = 2\pi + \frac{8\pi}{3}n$, where n is an integer.

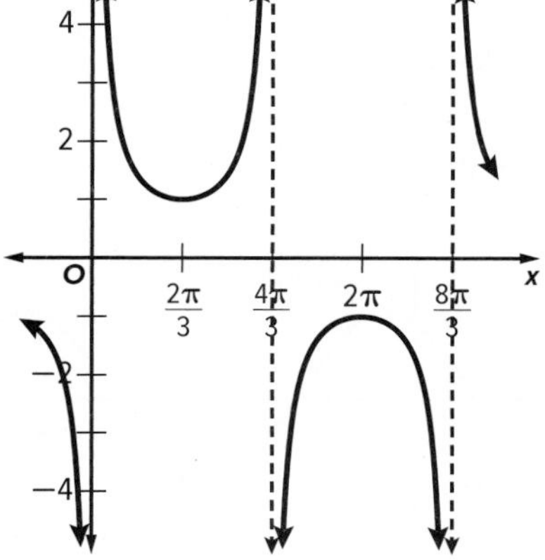

9. period = 2π; asymptotes: $x = 2\pi + 2\pi n$, where n is an integer; relative minima at $x = \frac{\pi}{2}$, $x = \frac{5\pi}{2}$, or $x = \frac{\pi}{2} + 2\pi n$, where n is an integer; relative maxima at $\frac{3\pi}{2}, \frac{7\pi}{2}$, or $x = \frac{3\pi}{2} + 2\pi n$, where n is an integer.

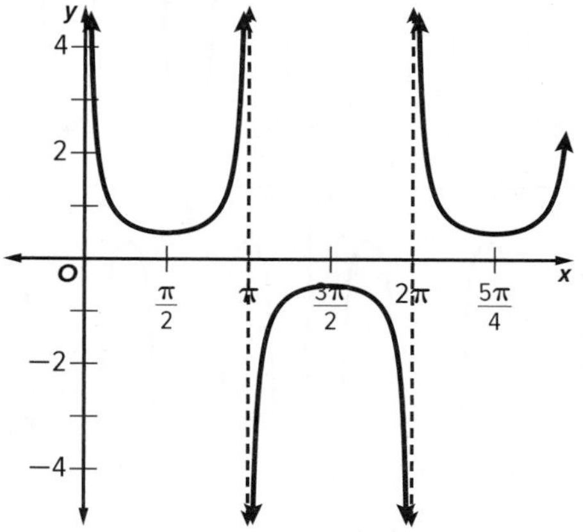

11. period = 360°; asymptotes: $x = 360n°$, where n is an integer; x-intercepts: 180°, 540°, ...; (odd multiples of 180°).

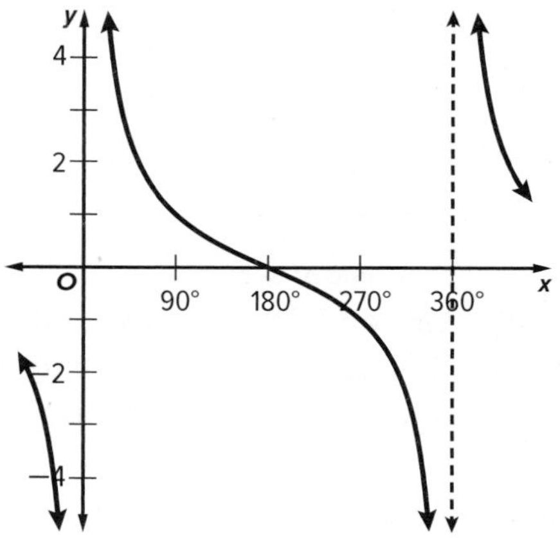

13. period = 360°; asymptotes: $x = (90 + 180n)°$, relative maxima at $x = (180 + 360n)°$, where n is an integer, relative minima at $x = 360n°$, where n is an integer.

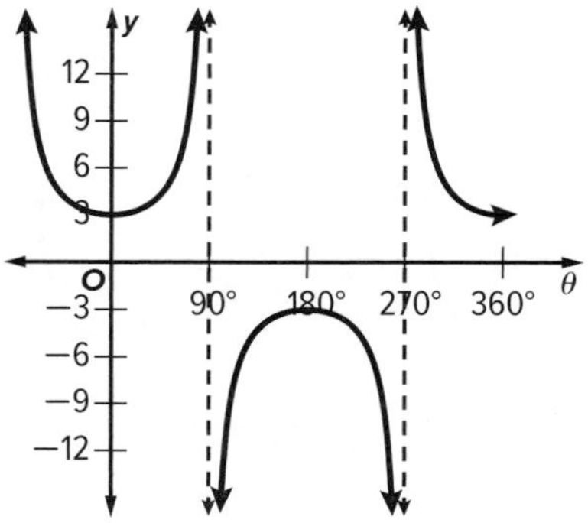

15. period $= \frac{\pi}{2}$; asymptotes: $x = \frac{\pi}{8} + \frac{\pi}{4}n$, relative maxima at $x = \frac{\pi}{4} + \frac{\pi}{2}n$ where n is an integer, relative minima at $x = \left(\frac{\pi}{2}\right)n$ where n is an integer.

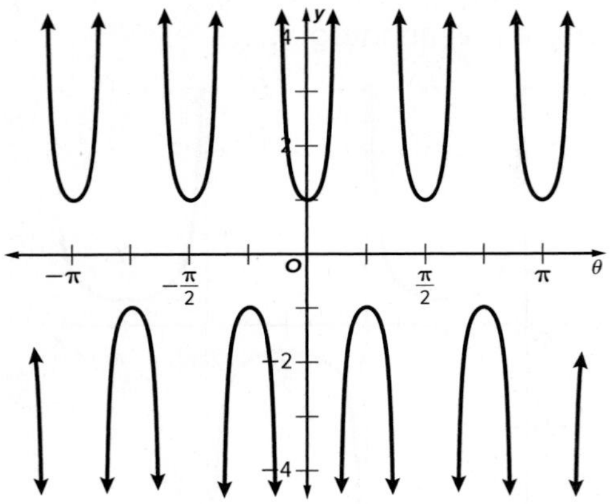

17a. $y = 300 \cot x$

17b.

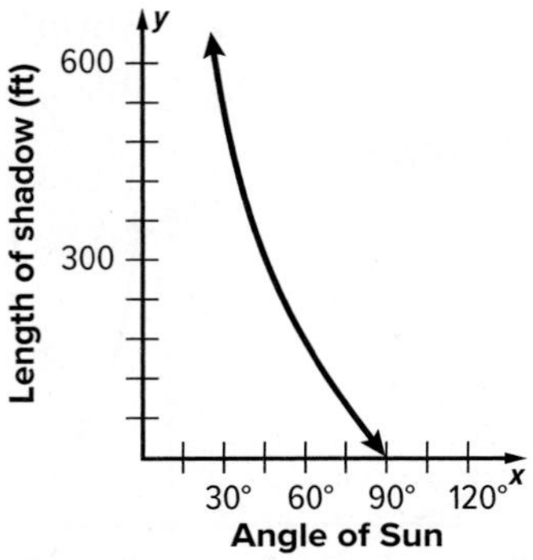

17c. Sample answer: In the context of the situation, only x-values between 0° and 90° are relevant, so focus on analyzing that portion of the graph.

domain: $\{x \mid 0 < x < 90\}$
range: $\{y \mid 0 \leq y \leq \infty\}$
x-intercept: 90
y-intercept: none
relative minimum: none
relative maximum: none
The period of $y = 300 \cot x$ is 180°.

19. period 450°

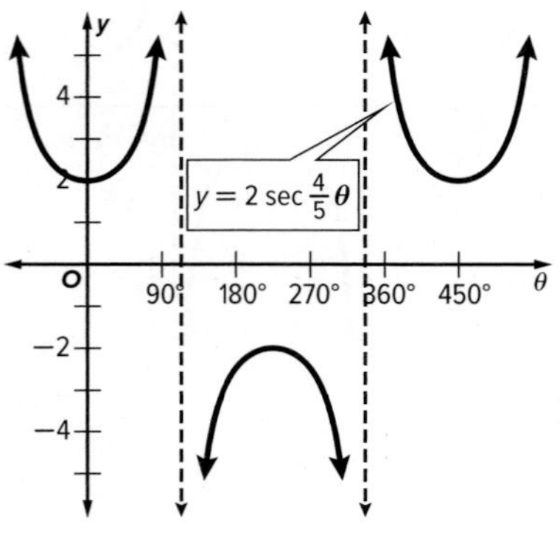

21. period: 30°

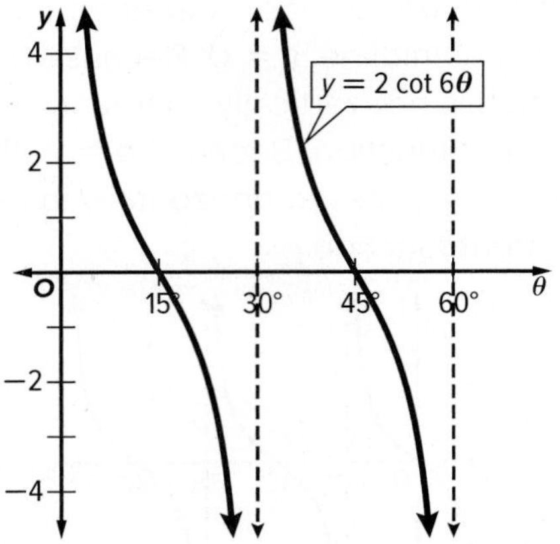

23. period: 360°

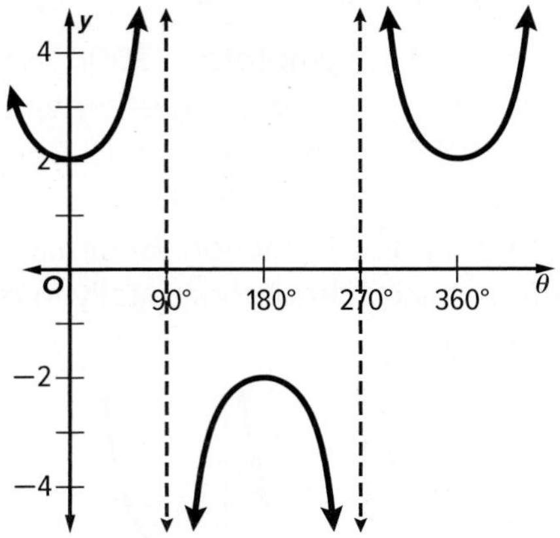

25. $d = 2000 \csc x$

27. Tyler is correct; Sample answer: The period of the tangent function is $\frac{180°}{|b|}$, and b is $\frac{1}{2}$. So the period is $\frac{180°}{\frac{1}{2}}$, or 360°.

29. Sample answer: Determine the amplitude and period of the function, then find and graph any x-intercepts, extrema, and asymptotes. Sketch the parent function and transform the graph to graph the given function.

Lesson 11-6

1. 1; 2π; π to the left

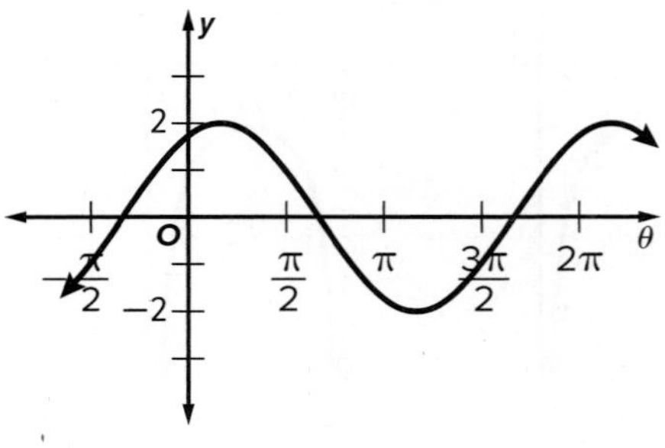

domain: $-\infty < \theta < \infty$; range: $-2 \leq f(\theta) \leq 2$

3. 1; 360°; 90° to the left

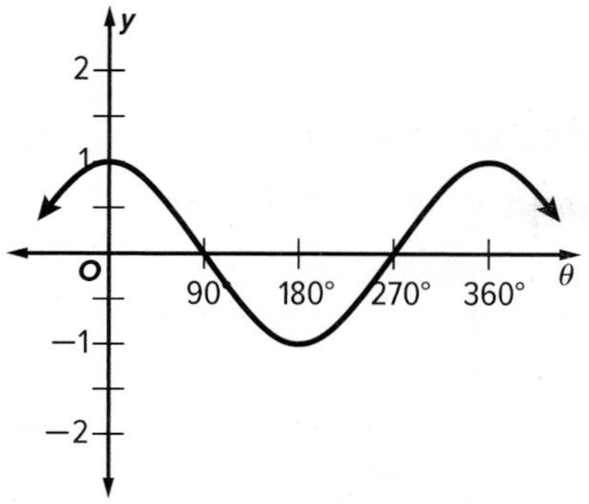

domain: $-\infty < \theta < \infty$; range: $-1 \leq f(\theta) \leq 1$

5. no amplitude; π; $\frac{\pi}{6}$ to the left

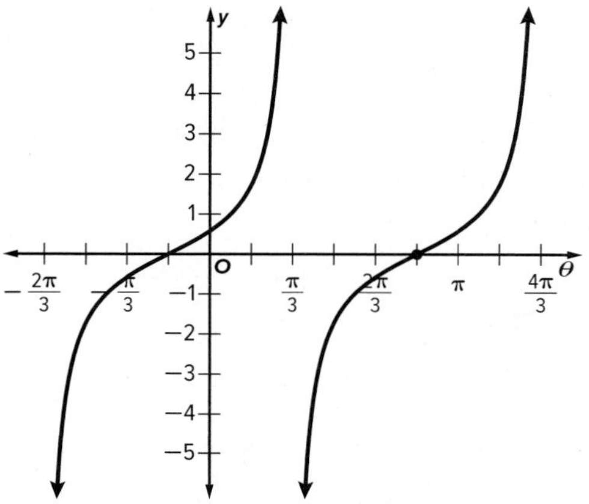

domain: $\frac{\pi}{3} + \pi n < \theta < \frac{\pi}{3} + \pi(n+1)$, where n is an integer; range: $-\infty < f(\theta) < \infty$

7. 3; 360°; 45° to the right

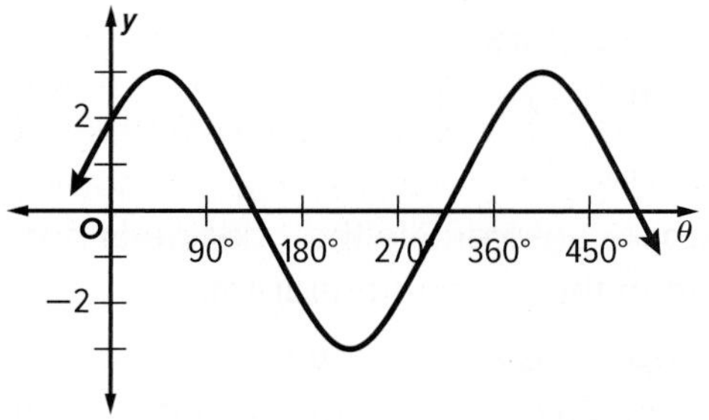

domain: $-\infty < \theta < \infty$; range: $-3 \leq f(\theta) \leq 3$

9. $\frac{1}{2}$; $\frac{2\pi}{3}$; $\frac{\pi}{3}$ to the right

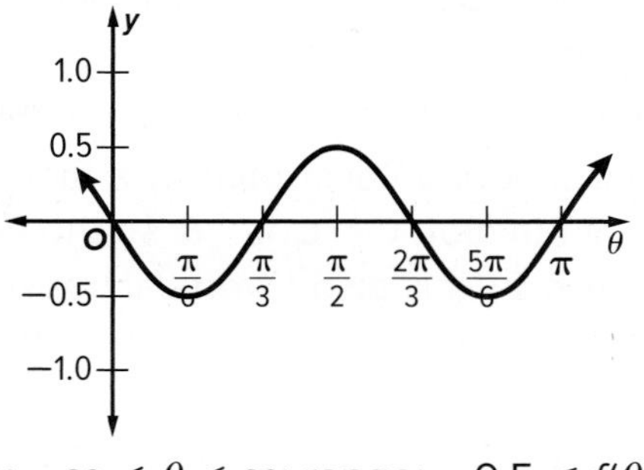

domain: $-\infty < \theta < \infty$; range: $-0.5 \leq f(\theta) \leq 0.5$

11. 2; 360°; 90° to the left

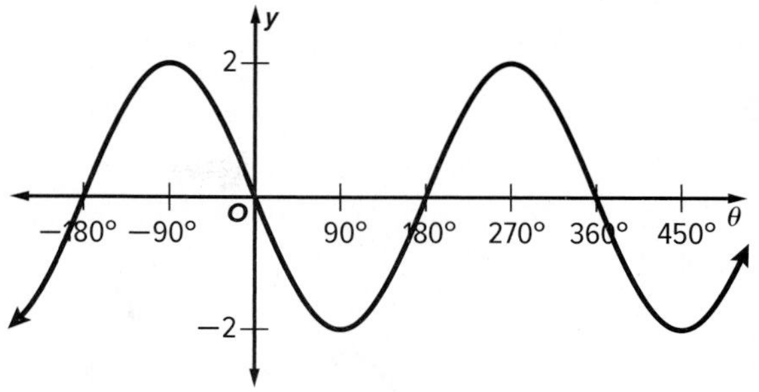

domain: $-\infty < \theta < \infty$; range: $-2 \leq f(\theta) \leq 2$

13. 1; 360°; $k = 3$; $y = 3$

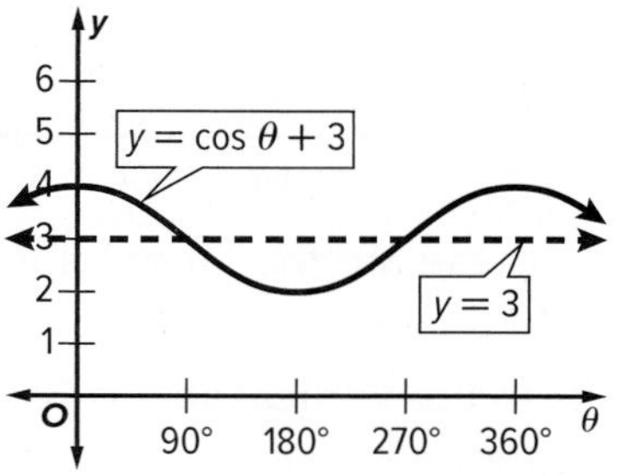

domain: $-\infty < \theta < \infty$; range: $2 \leq f(\theta) \leq 4$

15. no amplitude; 180° $k = \frac{1}{2}$; $y = \frac{1}{2}$

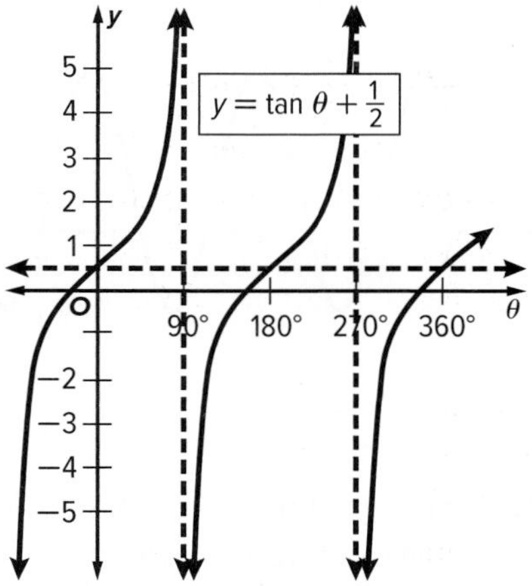

domain: $90 + 180n < \theta < 270 + 180n$, where n is an integer; range: $-\infty < f(\theta) < \infty$

17. 2; 360°; $k = -4$; $y = -4$

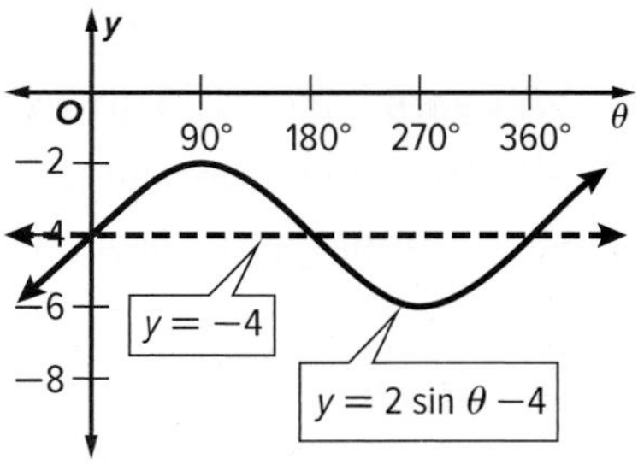

domain: $-\infty < \theta < \infty$; range: $-6 \leq f(\theta) \leq -2$

19a. The amplitude is 3; the midline is $y = 79$; the period is 60 minutes.

19b. $h(t) = 3\cos\left(\frac{\pi}{30}t\right) + 79$

19c.

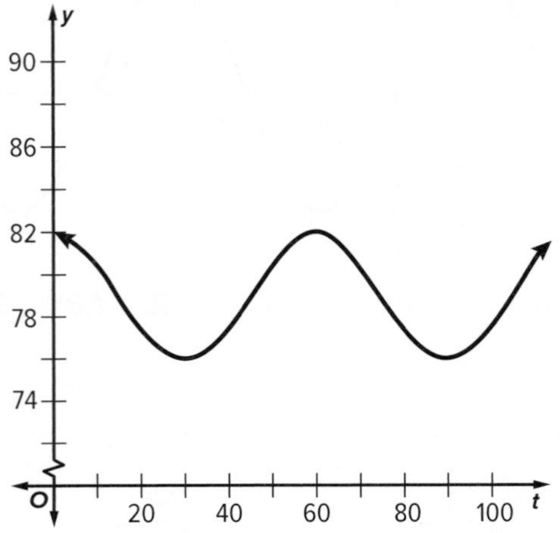

19d. 2 times; these are the times when the minute hand is on the 6 (i.e., 12:30 and 1:30).

21a. $k = -3$, so the vertical shift is 3 units down; amplitude = 1 **21b.** none **21c.** π

21d. $y = \cos 2\theta - 3$ **23.** $y = \sin(x - 4) + 3$

25. $y = \tan(x - \pi) + 2.5$ **27.** 3; 180°; −45°; 1

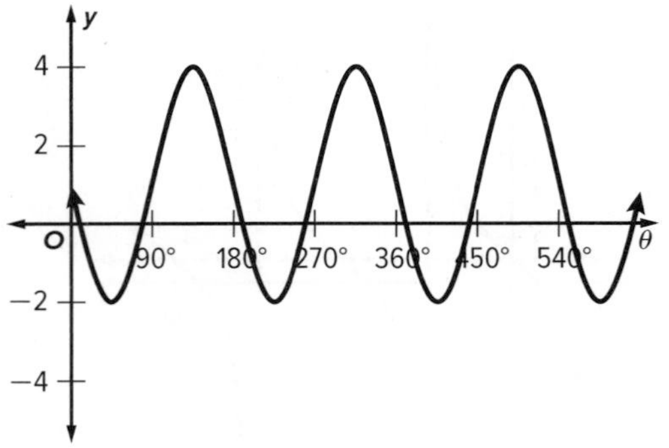

29. Sample answer: sine; Because as a Ferris wheel turns, the distance a rider is above the ground varies sinusoidally with time. Note: any phenomenon that can be modeled with a sine function can also be modeled with a cosine function using the appropriate horizontal shift and/or reflection about the horizontal axis.

31a. 120; in 2.5 years **31b.** 125; in 7 years

33. $\left(\frac{3\pi}{2}, 2\right)$ **35.** no maximum values

37. The graphs are reflections of each other over the *x*-axis.

39. The graphs are identical.

41. 360°; Sample answer: $y = \sin\theta - 5$

43. 360°; Sample answer: $y = 4\cos\theta + 1$

45. Sample answer: $h(\theta) = 1.5\cos 2\theta - 2$

47. The graph of $y = 3\sin 2\theta + 1$ has an amplitude of 3 rather than an amplitude of 1. It is shifted up 1 unit from the parent graph and is compressed so that it has a period of 180°.

49. Sample answer: $y = 2\sin\theta - 3$

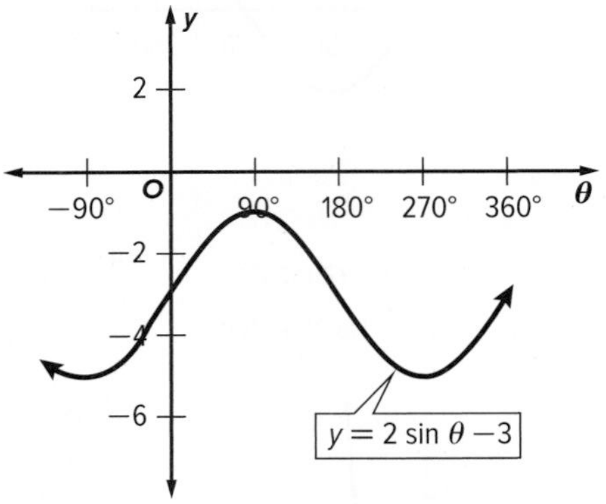

51. Sample answer: $y = \sin\left(\theta + \frac{\pi}{2}\right) - 3$ is equivalent to $y = \cos(\theta) - 3$.

Lesson 11-7

1. 30°, $\frac{\pi}{6}$ **3.** 120°, $\frac{2\pi}{3}$ **5.** 135°, $\frac{3\pi}{4}$ **7.** 45°; $\frac{\pi}{4}$

9. 90°; $\frac{\pi}{2}$ **11.** −1.02 **13.** 0.71 **15.** 0.95

17. 0.5 **19.** 53.13° **21.** 60° **23.** −5.74°

25. 58.67° **27.** 20.49° **29.** 75.52°

31a. 48.2°

31b. Sample answer: I assumed that the angle at which the doors would be opened would be equal in measure.

33. $\frac{\pi}{2}$ **35.** 0, 2π **37.** $\frac{\pi}{4}, \frac{5\pi}{4}$

39. 37°, 53°, 90°

41. Sample answer: Neither; cosine is not positive in the second quadrant.

43. Sample answer: $y = \tan^{-1} x$ is a relation that has a domain of all real numbers and a range of all real numbers except odd multiples of $\frac{\pi}{2}$. The relation is not a function. $y = \text{Tan}^{-1} x$ is a function that has a domain of all real numbers and a range of $-\frac{\pi}{2} < y < \frac{\pi}{2}$.

45. Sample answer: $y = \text{Cos}^{-1}\frac{1}{2}\theta + 1$; this function is a horizontal stretch by a factor 2 of the parent function $y = \text{Cos}^{-1}\theta$, so the domain is $\{-2 \le \theta \le 2\}$. It is also a vertical translation 1 unit up, so the range is $\{1 \le y \le \pi + 1\}$.

Module 11 Review

1a. $\frac{\pi}{9}$ **1b.** $\frac{4\pi}{9}$ **1c.** $\frac{8\pi}{9}$
3. 157 meters **5.** $\frac{7}{24}$ **7.** $\left(-\frac{\sqrt{3}}{2}, \frac{1}{2}\right)$ **9a.** 30
9b. 20 **11.** A **13.** A **15.** A, D, E **17.** C

Selected Answers

Module 12

Module 12 Opener

1. $-4a(4a-1)$ **3.** prime **5.** $\frac{\sqrt{2}}{2}$ **7.** $-\frac{\sqrt{3}}{3}$

Lesson 12-1

1. $\frac{4}{5}$ **3.** $\frac{5}{4}$ **5.** $-\frac{5}{4}$ **7.** $\frac{4}{5}$ **9.** $\sqrt{2}$

11. -2.5 **13.** $\frac{\sqrt{7}}{4}$ **15.** $\frac{1}{2}$ **17.** $\cos\theta$ **19.** $\csc\theta$

21. 1 **23.** $-\cot\theta$ **25.** -4 **27.** $\cos\theta$

29. $d=\sec^2\theta$ **31.** $\frac{1+\cos\theta}{\sin\theta}$ **33.** $\sec\theta$

35. 2 **37.** $2\cos^2\theta$ **39.** -1 **41.** $-\cot^2\theta$

43. Sample answer: The functions $\cos\theta$ and $\sin\theta$ can be thought of as the lengths of the legs of a right triangle, and the number 1 can be thought of as the measure of the corresponding hypotenuse.

45. Sample answer: $\frac{\sin\theta}{\cos\theta}\cdot\sin\theta$ and $\frac{\sin^2\theta}{\cos\theta}$.

47. Ebony; Jordan did not use the identity that $\sin^2\theta+\cos^2\theta=1$ and made an error adding rational expressions.

Lesson 12-2

1.
$$\cos^2\theta+\tan^2\theta\cos^2\theta\stackrel{?}{=}1$$
$$\cos^2\theta+\frac{\sin^2\theta}{\cos^2\theta}\cdot\cos^2\theta\stackrel{?}{=}1$$
$$\cos^2\theta+\sin^2\theta\stackrel{?}{=}1$$
$$1=1\checkmark$$

3.
$$1+\sec^2\theta\sin^2\theta\stackrel{?}{=}\sec^2\theta$$
$$1+\frac{1}{\cos^2\theta}\cdot\sin^2\theta\stackrel{?}{=}\sec^2\theta$$
$$1+\tan^2\theta\stackrel{?}{=}\sec^2\theta$$
$$\sec^2\theta=\sec^2\theta\checkmark$$

5.
$$\frac{1-\cos\theta}{1+\cos\theta}\stackrel{?}{=}(\csc\theta-\cot\theta)^2$$
$$\frac{1-\cos\theta}{1+\cos\theta}\stackrel{?}{=}\csc^2\theta-2\cot\theta\csc\theta+\cot^2\theta$$
$$\frac{1-\cos\theta}{1+\cos\theta}\stackrel{?}{=}\frac{1}{\sin^2\theta}-2\cdot\frac{\cos\theta}{\sin\theta}\cdot\frac{1}{\sin\theta}+\frac{\cos^2\theta}{\sin^2\theta}$$
$$\frac{1-\cos\theta}{1+\cos\theta}\stackrel{?}{=}\frac{1}{\sin^2\theta}-\frac{2\cos\theta}{\sin^2\theta}+\frac{\cos^2\theta}{\sin^2\theta}$$
$$\frac{1-\cos\theta}{1+\cos\theta}\stackrel{?}{=}\frac{1-2\cos\theta+\cos^2\theta}{\sin^2\theta}$$
$$\frac{1-\cos\theta}{1+\cos\theta}\stackrel{?}{=}\frac{(1-\cos\theta)(1-\cos\theta)}{(1-\cos^2\theta)}$$
$$\frac{1-\cos\theta}{1+\cos\theta}\stackrel{?}{=}\frac{(1-\cos\theta)(1-\cos\theta)}{(1-\cos\theta)(1+\cos\theta)}$$
$$\frac{1-\cos\theta}{1+\cos\theta}=\frac{1-\cos\theta}{1+\cos\theta}\checkmark$$

7.
$$(\sin\theta-1)(\tan\theta+\sec\theta)\stackrel{?}{=}-\cos\theta$$
$$\sin\theta\tan\theta+\sin\theta\sec\theta-\tan\theta-\sec\theta\stackrel{?}{=}-\cos\theta$$
$$\frac{\sin^2\theta}{\cos\theta}+\frac{\sin\theta}{\cos\theta}-\frac{\sin\theta}{\cos\theta}-\frac{1}{\cos\theta}\stackrel{?}{=}-\cos\theta$$
$$\frac{\sin^2\theta}{\cos\theta}-\frac{1}{\cos\theta}\stackrel{?}{=}-\cos\theta$$
$$\frac{\sin^2\theta-1}{\cos\theta}\stackrel{?}{=}-\cos\theta$$
$$\frac{-\cos^2\theta}{\cos\theta}\stackrel{?}{=}-\cos\theta$$
$$-\cos\theta=-\cos\theta\checkmark$$

9.
$$\sec^2\theta-\tan^2\theta\stackrel{?}{=}\frac{1-\sin\theta}{\cos\theta}$$
$$\frac{1}{\cos\theta}-\frac{\sin\theta}{\cos\theta}\stackrel{?}{=}\frac{1-\sin\theta}{\cos\theta}$$
$$\frac{1-\sin\theta}{\cos\theta}=\frac{1-\sin\theta}{\cos\theta}\checkmark$$

11.
$$\left(\sin\theta+\frac{\cot\theta}{\csc\theta}\right)^2\stackrel{?}{=}\frac{2+\sec\theta\csc\theta}{\sec\theta\csc\theta}$$
$$\left(\sin\theta+\frac{\frac{\cos\theta}{\sin\theta}}{\frac{1}{\sin\theta}}\right)^2\stackrel{?}{=}\frac{2+\sec\theta\csc\theta}{\sec\theta\csc\theta}$$
$$(\sin\theta+\cos\theta)^2\stackrel{?}{=}\frac{2+\sec\theta\csc\theta}{\sec\theta\csc\theta}$$
$$(\sin\theta+\cos\theta)^2\stackrel{?}{=}\frac{2+\frac{1}{\cos\theta}\cdot\frac{1}{\sin\theta}}{\frac{1}{\cos\theta}\cdot\frac{1}{\sin\theta}}$$
$$(\sin\theta+\cos\theta)^2\stackrel{?}{=}\left(2+\frac{1}{\cos\theta\sin\theta}\right)\cdot\frac{\cos\theta\sin\theta}{1}$$
$$(\sin\theta+\cos\theta)^2\stackrel{?}{=}2\cos\theta\sin\theta+\cos^2\theta+\sin^2\theta$$
$$(\sin\theta+\cos\theta)^2\stackrel{?}{=}(\sin\theta+\cos\theta)^2$$

13.
$$\csc^2\theta-1\stackrel{?}{=}\frac{\cot^2\theta}{\csc\theta+\sin\theta}$$
$$\cot^2\theta\frac{1}{\cos\theta-\frac{\sin\theta}{\cos\theta}}\stackrel{?}{=}\cot^2\theta\frac{1}{\cos\theta-\frac{\sin\theta}{\cos\theta}}$$
$$\cot^2\theta=\cot^2\theta\checkmark$$

15.
$$\csc^2\theta\stackrel{?}{=}\cot^2\theta+\sin\theta\csc\theta$$
$$\cot^2\theta+1\stackrel{?}{=}\cot^2\theta+\sin\theta\cdot\frac{1}{\sin\theta}$$
$$\cot^2\theta+1=\cot^2\theta+1\checkmark$$

17.
$$\tan\theta\cos\theta\stackrel{?}{=}\sin\theta$$
$$\frac{\sin\theta}{\cos\theta}\cdot\cos\theta\stackrel{?}{=}\sin\theta$$
$$\sin\theta=\sin\theta\checkmark$$

19.
$$(\tan\theta)(1-\sin^2\theta)\stackrel{?}{=}\sin\theta\cos\theta$$
$$\tan\theta\cos^2\theta\stackrel{?}{=}\sin\theta\cos\theta$$
$$\frac{\sin\theta}{\cos\theta}\cdot\cos^2\theta\stackrel{?}{=}\sin\theta\cos\theta$$
$$\sin\theta\cos\theta=\sin\theta\cos\theta\checkmark$$

21.
$$\frac{\sin^2\theta}{1-\sin^2\theta}\stackrel{?}{=}\tan^2\theta$$
$$\frac{\sin^2\theta}{\cos^2\theta}\stackrel{?}{=}\tan^2\theta$$
$$\left(\frac{\sin\theta}{\cos\theta}\right)^2\stackrel{?}{=}\tan^2\theta$$
$$\tan^2\theta=\tan^2\theta\checkmark$$

23. $n_1 \sin \theta_p = n_2 \sin \theta_r$

$n_1 \sin \theta_p = n_2 \sin (90 - \theta_p)$

$n_1 \sin \theta_p = n_2 \cos \theta_p$

$\frac{\sin \theta_p}{\cos \theta_p} = \frac{n_2}{n_1}$

$\tan \theta_p = \frac{n_2}{n_1}$

25. An identity can be verified by transforming one side of the equation to form of the other side, or by transforming both sides of the equation separately so that both sides form the same expression.

Sample answer by transforming one side of the equation:

$$\frac{1}{1 - \sin^2 \theta} \stackrel{?}{=} \tan^2 \theta + 1$$
$$\frac{1}{1 - (1 - \cos^2 \theta)} \stackrel{?}{=} \tan^2 \theta + 1$$
$$\frac{1}{\cos^2 \theta} \stackrel{?}{=} \tan^2 \theta + 1$$
$$\sec^2 \theta \stackrel{?}{=} \tan^2 \theta + 1$$
$$\tan^2 \theta + 1 = \tan^2 \theta + 1 \checkmark$$

Sample answer by transforming both sides of the equation:

$$\frac{1}{1 - \sin^2 \theta} \stackrel{?}{=} \tan^2 \theta + 1$$
$$\frac{1}{1 - (1 - \cos^2 \theta)} \stackrel{?}{=} \frac{\sin^2 \theta}{\cos^2 \theta} + 1$$
$$\frac{1}{\cos^2 \theta} \stackrel{?}{=} \frac{\sin^2 \theta}{\cos^2 \theta} + \frac{\sin^2 \theta}{\cos^2 \theta}$$
$$\frac{1}{\cos^2 \theta} \stackrel{?}{=} \frac{\sin^2 \theta + \cos^2 \theta}{\cos^2 \theta}$$
$$\frac{1}{\cos^2 \theta} = \frac{1}{\cos^2 \theta} \checkmark$$

27.
$$\frac{1 + \tan \theta}{1 + \cot \theta} \stackrel{?}{=} \tan \theta$$
$$\frac{1 + \tan \theta}{1 + \frac{1}{\tan \theta}} \stackrel{?}{=} \tan \theta$$
$$\left(\frac{\tan \theta}{\tan \theta}\right)\left(\frac{1 + \tan \theta}{1 + \frac{1}{\tan \theta}}\right) \stackrel{?}{=} \tan \theta$$
$$\frac{\tan \theta(1 + \tan \theta)}{\tan \theta + 1} \stackrel{?}{=} \tan \theta$$
$$\tan \theta = \tan \theta \checkmark$$

29.
$$\frac{\sec^2 \theta - \tan^2 \theta}{\cos^2 \theta + \sin^2 \theta} \stackrel{?}{=} 1$$
$$\frac{1}{\cos^2 \theta + \sin^2 \theta} \stackrel{?}{=} 1$$
$$\frac{1}{1} \stackrel{?}{=} 1$$
$$1 = 1 \checkmark$$

31.
$$\cot (-\theta) \cot \left(\frac{\pi}{2} - \theta\right) \stackrel{?}{=} 1$$
$$\left(\frac{\cos (-\theta)}{\sin (-\theta)}\right)\left(\frac{\cos \left(\frac{\pi}{2} - \theta\right)}{\sin \left(\frac{\pi}{2} - \theta\right)}\right) \stackrel{?}{=} 1$$
$$\left(\frac{\cos \theta}{-\sin \theta}\right)\left(\frac{\sin \theta}{\cos \theta}\right) \stackrel{?}{=} 1$$
$$-1 \neq 1$$

35. No; use the Pythagorean Identity to rewrite the equation as $\sin^2 A + (1 - \sin^2 B) = 1$. This simplifies to $\sin^2 A = \sin^2 B$, which is true if $\sin A = \sin B$ or $\sin A = -\sin B$. Because A and B are both less than 180°, $\sin A > 0$ and $\sin B > 0$. So, the second equation cannot be true. The first equation is true if $A = B$ or if $A = 180° - B$.

37. $\sin^2 \theta - \cos^2 \theta = 2 \sin^2 \theta$; the other three are Pythagorean identities, but this is not.

39. Sample answer: $\sin^2 45° + \cos^2 45° = 1$, but $\sin 45° \neq \sqrt{1 - \cos^2 45°}$ and $\sin^2 30° + \cos^2 30° = 1$, but $\sin 30° \neq \sqrt{1 - \cos^2 30°}$. Because sine is a function, it does not behave the same way as variables or constants when verifying an identity.

41. $f(\theta) = \frac{1}{2} \sin \theta$

Lesson 12-3

1. $\frac{\sqrt{2}}{2}$ **3.** $\frac{\sqrt{2} - \sqrt{6}}{4}$ **5.** $2 - \sqrt{3}$ **7a.** $\frac{3 + 4\sqrt{3}}{10}$; $\frac{4 - 3\sqrt{3}}{10}$ **7b.** 96.9° **7c.** No; sample answer: $\sin 90° = 1$, and $\cos 90° = 0$. Because the exact values of $\sin A$ and $\cos A$ do not equal 1 and 0, respectively, the measure of A cannot be 90°.

9.
$$\cos \left(\frac{\pi}{2} + \theta\right) \stackrel{?}{=} -\sin \theta$$
$$\cos \frac{\pi}{2} \cos \theta - \sin \frac{\pi}{2} \sin \theta \stackrel{?}{=} -\sin \theta$$
$$(0)\cos \theta - (1)\sin \theta \stackrel{?}{=} -\sin \theta$$
$$-\sin \theta = -\sin \theta \checkmark$$

11.
$$\cos (180° + \theta) \stackrel{?}{=} -\cos \theta$$
$$\cos 180° \cos \theta - \sin 180 \sin \theta \stackrel{?}{=} -\cos \theta$$
$$-1 \cdot \cos \theta - 0 \cdot \sin \theta \stackrel{?}{=} -\cos \theta$$
$$-\cos \theta = -\cos \theta \checkmark$$

13. $-\frac{1}{2}$ **15.** $\frac{\sqrt{2}}{2}$ **17.** $-\frac{\sqrt{2}}{2}$ **19.** $-\frac{\sqrt{2}}{2}$

21. $\frac{\sqrt{6} - \sqrt{2}}{4}$ **23.** $-\frac{\sqrt{2}}{2}$ **25.** $-\frac{\sqrt{3}}{2}$

27.
$$\sin (90° + \theta) \stackrel{?}{=} \cos \theta$$
$$\sin 90° \cos \theta + \cos 90° \sin \theta \stackrel{?}{=} \cos \theta$$
$$1 \cdot \cos \theta + 0 \cdot \sin \theta \stackrel{?}{=} \cos \theta$$
$$\cos \theta = \cos \theta \checkmark$$

29.
$$\cos (270° - \theta) \stackrel{?}{=} -\sin \theta$$
$$\cos 270° \cos \theta + \sin 270° \sin \theta \stackrel{?}{=} -\sin \theta$$
$$0 \cdot \cos \theta + (-1) \cdot \sin \theta \stackrel{?}{=} -\sin \theta$$
$$-\sin \theta = -\sin \theta \checkmark$$

31.
$$\sin\left(\theta - \frac{\pi}{2}\right) \stackrel{?}{=} -\cos \theta$$
$$\sin \theta \cos \frac{\pi}{2} - \cos \theta \sin \frac{\pi}{2} \stackrel{?}{=} -\cos \theta$$
$$(\sin \theta)(0) - (\cos \theta)(1) \stackrel{?}{=} -\cos \theta$$
$$-\cos \theta = -\cos \theta \checkmark$$

33a. $m\angle TPQ = A$; Sample answer: Because $\overline{QT}$ is parallel to the x-axis, $m\angle OQT = A$ since this angle is an alternate interior angle with $\angle QOS$. $\angle OQT$ is complementary to $\angle TQP$ because $\angle OQP$ is a right angle. $\angle TPQ$ is complementary to $\angle TQP$ because these are the acute angles in a right triangle. So, $\angle OQT \cong \angle TPQ$ because these angles are both complementary to the same angle. Therefore, $m\angle TPQ = A$.

33b. In right triangle POQ, $\sin B = \frac{PQ}{OP}$, but $OP = 1$, so $PQ = \sin B$. Also in right triangle POQ, $\cos B = \frac{OQ}{OP}$, and $OP = 1$, so $OQ = \cos B$.

33c. In $\triangle QOS$, $\cos A = \frac{QS}{OQ}$. Therefore, $OS = OQ \cos A$. But from part b, $OQ = \cos B$, so by substitution, $OS = \cos B \cos A$.

33d. In $\triangle TPQ$, $\sin A = \frac{QT}{PQ}$. Therefore, $QT = PQ \sin A$. But from part b, $PQ = \sin B$, so by substitution, $QT = \sin B \sin A$.

33e. In $\triangle POR$, $\cos (A + B) = \frac{OR}{OP} = OR = OS - RS$, by the Segment Addition Postulate. From part c, $OS = \cos B \cos A$. Also, $RS = QT$ because $TQRS$ is a rectangle, and from part d, $QT = \sin B \sin A$. So, $\cos (A + B) = \cos B \cos A - \sin B \sin A = \cos A \cos B - \sin A \sin B$.

35. $\sqrt{2} - \sqrt{6}$ **37.** $-2 + \sqrt{3}$ **39.** $2 - \sqrt{3}$

41. $\cos (A + B) \stackrel{?}{=} \dfrac{1 - \tan A \tan B}{\sec A \sec B}$

$$\cos (A + B) \stackrel{?}{=} \frac{1 - \frac{\sin A}{\cos A} \cdot \frac{\sin B}{\cos B}}{\frac{1}{\cos A} \cdot \frac{1}{\cos B}}$$

$$\cos (A + B) \stackrel{?}{=} \frac{1 - \frac{\sin A}{\cos A} \cdot \frac{\sin B}{\cos B}}{\frac{1}{\cos A} \cdot \frac{1}{\cos B}} \cdot \frac{\cos A \cos B}{\cos A \cos B}$$

$$\cos (A + B) \stackrel{?}{=} \frac{\cos A \cos B - \sin A \sin B}{1}$$

$$\cos (A + B) = \cos (A + B) \checkmark$$

43. $\sin (A + B) \sin (A - B) \stackrel{?}{=} \sin^2 A - \sin^2 B$

$(\sin A \cos B + \cos A \sin B)(\sin A \cos B - \cos A \sin B) \stackrel{?}{=} \sin^2 A - \sin^2 B$

$\sin^2 A \cos^2 B - \cos^2 A \sin^2 B \stackrel{?}{=} \sin^2 A - \sin^2 B$

$\sin^2 A \cos^2 B + \sin^2 A \sin^2 B - \sin^2 A \sin^2 B - \cos^2 A \sin^2 B \stackrel{?}{=} \sin^2 A - \sin^2 B$

$\sin^2 A(\cos^2 B + \sin^2 B) - \sin^2 B(\sin^2 A + \cos^2 A) \stackrel{?}{=} \sin^2 A - \sin^2 B$

$\sin^2 A(1) - \sin^2 B(1) \stackrel{?}{=} \sin^2 A - \sin^2 B$

$\sin^2 A - \sin^2 B = \sin^2 A - \sin^2 B \checkmark$

45a. $\tan (A + B) = \dfrac{\sin (A + B)}{\cos (A + B)} = \dfrac{\sin A \cos B + \cos A \sin B}{\cos A \cos B - \sin A \sin B}$; divide the numerator and denominator of this fraction by $\cos A \cos B$. This gives $\dfrac{\frac{\sin A \cos B}{\cos A \cos B} + \frac{\cos A \sin B}{\cos A \cos B}}{\frac{\cos A \cos B}{\cos A \cos B} - \frac{\sin A \sin B}{\sin A \sin B}} = \dfrac{\tan A + \tan B}{1 - \tan A \tan B}$.

45b. $\tan (A - B) = \tan (A + (-B)) = \dfrac{\tan A + \tan (-B)}{1 - \tan A \tan (-B)}$, but $\tan (-B) = -\tan B$, so $\tan (A - B) = \dfrac{\tan A - \tan B}{1 + \tan A \tan B}$.

47. Sample answer: $\sin \frac{13\pi}{12} = \sin\left(\frac{5\pi}{6} + \frac{\pi}{4}\right) = \sin \frac{5\pi}{6} \cos \frac{\pi}{4} + \cos \frac{5\pi}{6} \sin \frac{\pi}{4} = \frac{1}{2} \cdot \frac{\sqrt{2}}{2} + \left(-\frac{\sqrt{3}}{2}\right)\left(\frac{\sqrt{2}}{2}\right) = \frac{\sqrt{2} - \sqrt{6}}{4}$

49. $\theta = \frac{\pi}{4}$; by the sum and difference formulas, the equation can be written as $\cos \theta \cos \pi - \sin \theta \sin \pi = \sin \theta \cos (-\pi) - \cos \theta \sin (-\pi)$, or $-\cos \theta = -\sin \theta$, which shows that $\theta = \frac{\pi}{4}$.

51. $\sin (19° + 11°)$; $\frac{1}{2}$ **53.** $\cos (111° - 21°)$; 0

55. $\tan \frac{\pi}{16} + \frac{3\pi}{16}$; 1 **57.** $\sin (-2\theta)$

59. $\cot (A + B) = \dfrac{1}{\tan (A + B)}$

$$\cot (A + B) = \frac{1}{\frac{\tan A + \tan B}{1 - \tan A \tan B}}$$

$$\cot (A + B) = \frac{1 - \tan A \tan B}{\tan A + \tan B}$$

$$\cot (A + B) = \frac{1 - \frac{1}{\cot A} \cdot \frac{1}{\cot B}}{\frac{1}{\cot A} + \frac{1}{\cot B}} \cdot \frac{\cot A \cot B}{\cot A \cot B}$$

$$\cot (A + B) = \frac{\cot A \cot B - 1}{\cot A + \cot B}$$

61. Sample answer: $A = 35°$, $B = 60°$, $C = 85°$; $0.7002 + 1.7321 + 11.4301 \stackrel{?}{=} (0.7002)(1.7321)(11.4301)$; $13.86 = 13.86$ ✓

63.

$\sin (90° - \theta) \stackrel{?}{=} \cos \theta$	Original equation
$\sin 90° \cos \theta - \cos 90° \sin \theta \stackrel{?}{=} \cos \theta$	Difference Identity
$1 \cdot \cos \theta - 0 \cdot \sin \theta \stackrel{?}{=} \cos \theta$	$\cos 90° = 0$; $\sin 90° = 1$
$\cos \theta = \sin \theta$ ✓	Simplify.

Lesson 12-4

1. $-\frac{4\sqrt{5}}{9}, \frac{1}{9}, \sqrt{\frac{3 - \sqrt{5}}{6}}, \sqrt{\frac{3 + \sqrt{5}}{6}}$

3. $-\frac{24}{25}, -\frac{7}{25}, \frac{\sqrt{5}}{5}, -\frac{2\sqrt{5}}{5}$ **5.** $\frac{24}{25}, -\frac{7}{25}, \frac{2\sqrt{5}}{5}, \frac{\sqrt{5}}{5}$

7. $\frac{336}{625}, -\frac{527}{625}, \frac{3}{5}, \frac{4}{5}$ **9.** $-\frac{720}{1681}, -\frac{1519}{1681}, \frac{5\sqrt{41}}{41}, \frac{4\sqrt{41}}{41}$

11. $S = 4 \sin \theta \cos \theta$ **13.** $\frac{\sqrt{2+\sqrt{3}}}{3}$

15. $-\frac{\sqrt{2-\sqrt{3}}}{2}$ **17.** $2 + \sqrt{3}$ **19.** $\frac{\sqrt{2+\sqrt{2}}}{2}$

21. $-\frac{\sqrt{2-\sqrt{3}}}{2}$ **23.** $-\frac{\sqrt{2-\sqrt{2}}}{2}$ **25.** $\frac{24}{25}, \frac{7}{25}, \frac{24}{7}$

27. $-\frac{3}{5}, -\frac{4}{5}, \frac{3}{4}$ **29.** $-\frac{4\sqrt{21}}{25}, \frac{17}{25}, -\frac{4\sqrt{21}}{17}$

31a. $\frac{\sqrt{2-\sqrt{2}}}{2}; \frac{\sqrt{2+\sqrt{2}}}{2}$ **31b.** $\frac{\sqrt{2}}{2}; -\frac{\sqrt{2}}{2}$

33a. Sample answer: $\sin 240° = \sin (180° + 60°)$

$= \sin 180° \cos 60° + \cos 180° \sin 60°$

$= -\frac{\sqrt{3}}{2}$

33b. Sample answer: $\sin 240° = \sin (270° - 30°)$

$= \sin 270° \cos 30° - \cos 270° \sin 30°$

$= -\frac{\sqrt{3}}{2}$

33c. $\sin 240° = \sin (2 \cdot 120°)$

$= 2 \sin 120° \cos 120°$

$= -\frac{\sqrt{3}}{2}$

33d. $\sin 240° = \sin \frac{480°}{2}$

$= -\sqrt{\frac{1 - \cos 480°}{2}}$

$= -\frac{\sqrt{3}}{2}$

35. $\angle PBD$ is an inscribed angle that subtends the same arc as the central angle $\angle POD$, so $m\angle PBD = \frac{1}{2}\theta$. By right triangle trigonometry, $\tan \frac{1}{2}\theta = \frac{PA}{BA} = \frac{PA}{1 + OA} = \frac{\sin \theta}{1 + \cos \theta}$.

37. $\sin 2\theta = \sin(\theta + \theta)$

$= \sin \theta \cos \theta + \cos \theta \sin \theta$

$= 2 \sin \theta \cos \theta$

$\cos 2\theta = \cos(\theta + \theta)$

$= \cos \theta \cos \theta - \sin \theta \sin \theta$

$= \cos^2\theta - \sin^2\theta$

You can find alternate forms for $\cos 2\theta$ by making substitutions into the expression $\cos^2\theta - \sin^2\theta$.

$\cos^2\theta - \sin^2\theta = (1 - \sin^2\theta) - \sin^2\theta$ (Substitute $1 - \sin^2\theta$ for $\cos^2\theta$.)

$= 1 - 2 \sin^2\theta$ (Simplify.)

$\cos^2\theta - \sin^2\theta = \cos^2\theta - (1 - \cos^2\theta)$ (Substitute $1 - \cos^2\theta$ for $\sin^2\theta$.)

$= 2 \cos^2\theta - 1$ (Simplify.)

39. Sample answer: Since $d = \frac{v^2 \sin 2\theta}{b}$, d is at a maximum when $\sin 2\theta = 1$, that is, when $2\theta = 90°$ or $\theta = 45°$.

Lesson 12-5

1. 60°, 120°, 240°, 300°

3. $\frac{\pi}{6}, \frac{\pi}{2}, \frac{5\pi}{6}, \frac{3\pi}{2}$ **5.** 240°, 300°

7. $\frac{\pi}{3} + 2k\pi, \frac{5\pi}{3} + 2k\pi$

9. $\frac{2\pi}{3} + 2k\pi, \frac{4\pi}{3} + 2k\pi$ **11.** $45 + k \cdot 180°$

13. $270° + k \cdot 360°$ **15.** approximately 7 P.M.

17. (180°, 360°)

19. $\frac{\pi}{2}, \pi$

21. $0, \frac{2\pi}{3}$ **23.** $225° + 360k°, 315° + 360k°$

25. 45°, 135° **27.** 225°, 315° **29.** $\pi, \frac{3\pi}{2}$

31. $0 + 2k\pi, \frac{2\pi}{3} + 2k\pi$, and $\frac{4\pi}{3} + 2k\pi$

33. $k\pi$ **35.** $\frac{2\pi}{3} + 2k\pi, \frac{4\pi}{3} + 2k\pi$

37. $210° + k \cdot 360°$ and $330° + k \cdot 360°$

39. $225° + k \cdot 360°$ and $315° + k \cdot 360°$

41. $45° + k \cdot 90°$

43. $\frac{\pi}{3} + k\pi$ and $\frac{2\pi}{3} + k\pi$, or $60° + k \cdot 180°$ and $120° + k \cdot 180°$

45. $\frac{\pi}{2} + 2\pi k, \frac{7\pi}{6} + 2\pi k, \frac{11\pi}{6} + 2\pi k$ or $90° + k \cdot 360°, 210° + k \cdot 360°, 330° + k \cdot 360°$

47. 30°, 150°, 270°

49. $k \cdot 120°$, where k is an integer **51.** 0.0026 s

53a. The maximum value is approximately 15.66, so there are about 15 h 40 min of daylight in Seattle on the longest day of the year.

53b. The graph of $y = f(x)$ intersects the graph of $y = 14$ at $x \approx 111$ and $x \approx 233$; this represents the period from April 21 to August 21.

55a. The period is about 11.6 months, which shows that the business has a yearly cycle and it is likely that there is more business during the summer months.

55b. Approximately \$366; 6 years is 72 months; Evaluating the function for $x = 72$ shows that $P(72) \approx 366$.

57. $\frac{\pi}{3} < x < \pi$ or $\frac{5\pi}{3} < x < 2\pi$

59. Sample answer: All trigonometric functions are periodic. Therefore, once one or more solutions are found for a certain interval, there will be additional solutions that can be found by adding integral multiples of the period of the function to those solutions.

61. 0, b, or $2b$

Module 12 Review

1. $\frac{\sqrt{17}}{8}$ **3.** D **5.** B, D

7.

	1	2	3	4	5	6	7
Factor.			X				
Pythagorean Identity		X					
Original Equation	X						
Divide.						X	
Reciprocal Identity							X
Simplify.				X			
Write as two fractions.					X		

9. B **11.** A **13.** $\frac{\sqrt{2-\sqrt{2}}}{2}$ **15.** A, C

17. A, D, F, H

Glossary

English	Español

A

30°-60°-90° triangle A right triangle with two acute angles that measure 30° and 60°.

triángulo 30°-60°-90° Un triángulo rectángulo con dos ángulos agudos que miden 30° y 60°.

45°-45°-90° triangle A right triangle with two acute angles that measure 45°.

triángulo 45°-45°-90° Un triángulo rectángulo con dos ángulos agudos que miden 45°.

absolute value The distance a number is from zero on the number line.

valor absoluto La distancia que un número es de cero en la línea numérica.

absolute value function A function written as $f(x) = |x|$, in which $f(x) \geq 0$ for all values of x.

función del valor absoluto Una función que se escribe $f(x) = |x|$, donde $f(x) \geq 0$, para todos los valores de x.

accuracy The nearness of a measurement to the true value of the measure.

exactitud La proximidad de una medida al valor verdadero de la medida.

additive identity Because the sum of any number a and 0 is equal to a, 0 is the additive identity.

identidad aditiva Debido a que la suma de cualquier número a y 0 es igual a, 0 es la identidad aditiva.

additive inverses Two numbers with a sum of 0.

inverso aditivos Dos números con una suma de 0.

adjacent angles Two angles that lie in the same plane and have a common vertex and a common side but have no common interior points.

ángulos adyacentes Dos ángulos que se encuentran en el mismo plano y tienen un vértice común y un lado común, pero no tienen puntos comunes en el interior.

adjacent arcs Arcs in a circle that have exactly one point in common.

arcos adyacentes Arcos en un circulo que tienen un solo punto en común.

algebraic expression A mathematical expression that contains at least one variable.

expresión algebraica Una expresión matemática que contiene al menos una variable.

algebraic notation Mathematical notation that describes a set by using algebraic expressions.

notación algebraica Notación matemática que describe un conjunto usando expresiones algebraicas.

alternate exterior angles When two lines are cut by a transversal, nonadjacent exterior angles that lie on opposite sides of the transversal.

ángulos alternos externos Cuando dos líneas son cortadas por un ángulo transversal, no adyacente exterior que se encuentran en lados opuestos de la transversal.

alternate interior angles When two lines are cut by a transversal, nonadjacent interior angles that lie on opposite sides of the transversal.

ángulos alternos internos Cuando dos líneas son cortadas por un ángulo transversal, no adyacente interior que se encuentran en lados opuestos de la transversal.

altitude of a parallelogram A perpendicular segment between any two parallel bases.

altitud de un paralelogramo Un segmento perpendicular entre dos bases paralelas.

altitude of a prism or cylinder A segment perpendicular to the bases that joins the planes of the bases.

altitud de un prisma o cilindro Un segmento perpendicular a las bases que une los planos de las bases.

altitude of a pyramid or cone A segment perpendicular to the base that has the vertex as one endpoint and a point in the plane of the base as the other endpoint.

altitud de una pirámide o cono Un segmento perpendicular a la base que tiene el vértice como un punto final y un punto en el plano de la base como el otro punto final.

altitude of a triangle A segment from a vertex of the triangle to the line containing the opposite side and perpendicular to that side.

altitud de triángulo Un segmento de un vértice del triángulo a la línea que contiene el lado opuesto y perpendicular a ese lado.

ambiguous case When two different triangles could be created or described using the given information.

caso ambiguo Cuando dos triángulos diferentes pueden ser creados o descritos usando la información dada.

amplitude For functions of the form $y = a \sin b\theta$ or $y = a \cos b\theta$, the amplitude is $|a|$.

amplitud Para funciones de la forma $y = a$ sen $b\theta$ o $y = a \cos b\theta$, la amplitud es $|a|$.

analytic geometry The study of geometry that uses the coordinate system.

geometría analítica El estudio de la geometría que utiliza el sistema de coordenadas.

angle The intersection of two noncollinear rays at a common endpoint.

ángulo La intersección de dos rayos no colineales en un extremo común.

angle bisector A ray or segment that divides an angle into two congruent angles.

bisectriz de un ángulo Un rayo o segmento que divide un ángulo en dos ángulos congruentes.

angle of depression The angle formed by a horizontal line and an observer's line of sight to an object below the horizontal line.

ángulo de depresión El ángulo formado por una línea horizontal y la línea de visión de un observador a un objeto por debajo de la línea horizontal.

angle of elevation The angle formed by a horizontal line and an observer's line of sight to an object above the horizontal line.

ángulo de elevación El ángulo formado por una línea horizontal y la línea de visión de un observador a un objeto por encima de la línea horizontal.

angle of rotation The angle through which a figure rotates.

ángulo de rotación El ángulo a través del cual gira una figura.

apothem A perpendicular segment between the center of a regular polygon and a side of the polygon or the length of that line segment.

apotema Un segmento perpendicular entre el centro de un polígono regular y un lado del polígono o la longitud de ese segmento de línea.

approximate error The positive difference between an actual measurement and an approximate or estimated measurement.

error aproximado La diferencia positiva entre una medida real y una medida aproximada o estimada.

arc Part of a circle that is defined by two endpoints.

arco Parte de un círculo que se define por dos puntos finales.

arc length The distance between the endpoints of an arc measured along the arc in linear units.

longitude de arco La distancia entre los extremos de un arco medido a lo largo del arco en unidades lineales.

area The number of square units needed to cover a surface.

área El número de unidades cuadradas para cubrir una superficie.

arithmetic sequence A pattern in which each term after the first is found by adding a constant, the common difference *d*, to the previous term.

secuencia aritmética Un patrón en el cual cada término después del primero se encuentra añadiendo una constante, la diferencia común *d*, al término anterior.

asymptote A line that a graph approaches.

asíntota Una línea que se aproxima a un gráfico.

auxiliary line An extra line or segment drawn in a figure to help analyze geometric relationships.

línea auxiliar Una línea o segmento extra dibujado en una figura para ayudar a analizar las relaciones geométricas.

average rate of change The change in the value of the dependent variable divided by the change in the value of the independent variable.

tasa media de cambio El cambio en el valor de la variable dependiente dividido por el cambio en el valor de la variable independiente.

axiom A statement that is accepted as true without proof.

axioma Una declaración que se acepta como verdadera sin prueba.

axiomatic system A set of axioms from which theorems can be derived.

sistema axiomático Un conjunto de axiomas de los cuales se pueden derivar teoremas.

axis of symmetry A line about which a graph is symmetric.

eje de simetría Una línea sobre la cual un gráfica es simétrico.

axis symmetry If a figure can be mapped onto itself by a rotation between 0° and 360° in a line.

eje simetría Si una figura puede ser asignada sobre sí misma por una rotación entre 0° y 360° en una línea.

B

bar graph A graphical display that compares categories of data using bars of different heights.

gráfico de barra Una pantalla gráfica que compara las categorías de datos usando barras de diferentes alturas.

base In a power, the number being multiplied by itself.

base En un poder, el número se multiplica por sí mismo.

base angles of a trapezoid The two angles formed by the bases and legs of a trapezoid.

ángulos de base de un trapecio Los dos ángulos formados por las bases y patas de un trapecio.

base angles of an isosceles triangle The two angles formed by the base and the congruent sides of an isosceles triangle.

ángulo de la base de un triángulo isosceles Los dos ángulos formados por la base y los lados congruentes de un triángulo isosceles.

base edge The intersection of a lateral face and a base in a solid figure.

arista de la base La intersección de una cara lateral y una base en una figura sólida.

base of a parallelogram Any side of a parallelogram.

base de un paralelogramo Cualquier lado de un paralelogramo.

base of a pyramid or cone The face of the solid opposite the vertex of the solid.

base de una pirámide o cono La cara del sólido opuesta al vértice del sólido.

bases of a prism or cylinder The two parallel congruent faces of the solid.

bases de un prisma o cilindro Las dos caras congruentes paralelas de la figura sólida.

bases of a trapezoid The parallel sides in a trapezoid.

bases de un trapecio Los lados paralelos en un trapecio.

best-fit line The line that most closely approximates the data in a scatter plot.

línea de ajuste óptimo La línea que más se aproxima a los datos en un diagrama de dispersión.

betweenness of points Point C is between A and B if and only if A, B, and C are collinear and $AC + CB = AB$.

intermediación de puntos El punto C está entre A y B si y sólo si A, B, y C son colineales y $AC + CB = AB$.

bias An error that results in a misrepresentation of a population.

sesgo Un error que resulta en una tergiversación de una población.

biconditional statement The conjunction of a conditional and its converse.

declaración bicondicional La conjunción de un condicional y su inverso.

binomial The sum of two monomials.

binomio La suma de dos monomios.

bisect To separate a line segment into two congruent segments.

bisecar Separe un segmento de línea en dos segmentos congruentes.

bivariate data Data that consists of pairs of values.

datos bivariate Datos que constan de pares de valores.

boundary The edge of the graph of an inequality that separates the coordinate plane into regions.

frontera El borde de la gráfica de una desigualdad que separa el plano de coordenadas en regiones.

bounded When the graph of a system of constraints is a polygonal region.

acotada Cuando la gráfica de un sistema de restricciones es una región poligonal.

box plot A graphical representation of the five-number summary of a data set.

diagram de caja Una representación gráfica del resumen de cinco números de un conjunto de datos.

C

categorical data Data that can be organized into different categories.

datos categóricos Datos que pueden organizarse en diferentes categorías.

causation When a change in one variable produces a change in another variable.

causalidad Cuando un cambio en una variable produce un cambio en otra variable.

center of a circle The point from which all points on a circle are the same distance.

centro de un círculo El punto desde el cual todos los puntos de un círculo están a la misma distancia.

center of a regular polygon The center of the circle circumscribed about a regular polygon.

centro de un polígono regular El centro del círculo circunscrito alrededor de un polígono regular.

center of dilation The center point from which dilations are performed.

centro de dilatación Punto fijo en torno al cual se realizan las homotecias.

center of rotation The fixed point about which a figure rotates.

centro de rotación El punto fijo sobre el que gira una figura.

center of symmetry A point in which a figure can be rotated onto itself.

centro de la simetría Un punto en el que una figura se puede girar sobre sí misma.

central angle of a circle An angle with a vertex at the center of a circle and sides that are radii.

ángulo central de un círculo Un ángulo con un vértice en el centro de un círculo y los lados que son radios.

central angle of a regular polygon An angle with its vertex at the center of a regular polygon and sides that pass through consecutive vertices of the polygon.

ángulo central de un polígono regular Un ángulo con su vértice en el centro de un polígono regular y lados que pasan a través de vértices consecutivos del polígono.

centroid The point of concurrency of the medians of a triangle.

baricentro El punto de intersección de las medianas de un triángulo.

chord of a circle or sphere A segment with endpoints on the circle or sphere.

cuerda de un círculo o esfera Un segmento con extremos en el círculo o esfera.

circle The set of all points in a plane that are the same distance from a given point called the center.

círculo El conjunto de todos los puntos en un plano que están a la misma distancia de un punto dado llamado centro.

circular function A function that describes a point on a circle as the function of an angle defined in radians.

función circular Función que describe un punto en un círculo como la función de un ángulo definido en radianes.

circumcenter The point of concurrency of the perpendicular bisectors of the sides of a triangle.

circuncentro El punto de concurrencia de las bisectrices perpendiculares de los lados de un triángulo.

circumference The distance around a circle.

circunferencia La distancia alrededor de un círculo.

circumscribed angle An angle with sides that are tangent to a circle.

ángulo circunscrito Un ángulo con lados que son tangentes a un círculo.

circumscribed polygon A polygon with vertices outside the circle and sides that are tangent to the circle.

poligono circunscrito Un polígono con vértices fuera del círculo y lados que son tangentes al círculo.

closed If for any members in a set, the result of an operation is also in the set.

cerrado Si para cualquier número en el conjunto, el resultado de la operación es también en el conjunto.

closed half-plane The solution of a linear inequality that includes the boundary line.

semi-plano cerrado La solución de una desigualdad linear que incluye la línea de limite.

codomain The set of all the *y*-values that could possibly result from the evaluation of the function.

codominar El conjunto de todos los valores y que podrían resultar de la evaluación de la función.

coefficient The numerical factor of a term.

coeficiente El factor numérico de un término.

coefficient of determination An indicator of how well a function fits a set of data.

coeficiente de determinación Un indicador de lo bien que una función se ajusta a un conjunto de datos.

cofunction identities Identities that show the relationships between sine and cosine, tangent and cotangent, and secant and cosecant.

collinear Lying on the same line.

combination A selection of objects in which order is not important.

combined variation When one quantity varies directly and/or inversely as two or more other quantities.

common difference The difference between consecutive terms in an arithmetic sequence.

common logarithms Logarithms of base 10.

common ratio The ratio of consecutive terms of a geometric sequence.

common tangent A line or segment that is tangent to two circles in the same plane.

complement of *A* All of the outcomes in the sample space that are not included as outcomes of event A.

complementary angles Two angles with measures that have a sum of 90°.

completing the square A process used to make a quadratic expression into a perfect square trinomial.

complex conjugates Two complex numbers of the form $a + bi$ and $a - bi$.

complex fraction A rational expression with a numerator and/or denominator that is also a rational expression.

complex number Any number that can be written in the form $a + bi$, where a and b are real numbers and i is the imaginary unit.

component form A vector written as $<x, y>$, which describes the vector in terms of its horizontal component x and vertical component y.

identidades de cofunción Identidades que muestran las relaciones entre seno y coseno, tangente y cotangente, y secante y cosecante.

colineal Acostado en la misma línea.

combinación Una selección de objetos en los que el orden no es importante.

variación combinada Cuando una cantidad varía directamente y / o inversamente como dos o más cantidades.

diferencia común La diferencia entre términos consecutivos de una secuencia aritmética.

logaritmos comunes Logaritmos de base 10.

razón común El razón de términos consecutivos de una secuencia geométrica.

tangente común Una línea o segmento que es tangente a dos círculos en el mismo plano.

complemento de *A* Todos los resultados en el espacio muestral que no se incluyen como resultados del evento A.

ángulo complementarios Dos ángulos con medidas que tienen una suma de 90°.

completar el cuadrado Un proceso usado para hacer una expresión cuadrática en un trinomio cuadrado perfecto.

conjugados complejos Dos números complejos de la forma $a + bi$ y $a - bi$.

fracción compleja Una expresión racional con un numerador y / o denominador que también es una expresión racional.

número complejo Cualquier número que se puede escribir en la forma $a + bi$, donde a y b son números reales e i es la unidad imaginaria.

forma de componente Un vector escrito como $<x, y>$, que describe el vector en términos de su componente horizontal x y componente vertical y.

composite figure A figure that can be separated into regions that are basic figures, such as triangles, rectangles, trapezoids, and circles.

figura compuesta Una figura que se puede separar en regiones que son figuras básicas, tales como triángulos, rectángulos, trapezoides, y círculos.

composite solid A three-dimensional figure that is composed of simpler solids.

solido compuesta Una figura tridimensional que se compone de figuras más simples.

composition of functions An operation that uses the results of one function to evaluate a second function.

composición de funciones Operación que utiliza los resultados de una función para evaluar una segunda función.

composition of transformations When a transformation is applied to a figure and then another transformation is applied to its image.

composición de transformaciones Cuando una transformación se aplica a una figura y luego se aplica otra transformación a su imagen.

compound event Two or more simple events.

evento compuesto Dos o más eventos simples.

compound inequality Two or more inequalities that are connected by the words *and* or *or*.

desigualdad compuesta Dos o más desigualdades que están unidas por las palabras *y* u *o*.

compound interest Interest calculated on the principal and on the accumulated interest from previous periods.

interés compuesto Intereses calculados sobre el principal y sobre el interés acumulado de períodos anteriores.

compound statement Two or more statements joined by the word *and* or *or*.

enunciado compuesto Dos o más declaraciones unidas por la palabra *y* o *o*.

concave polygon A polygon with one or more interior angles with measures greater than 180°.

polígono cóncavo Un polígono con uno o más ángulos interiores con medidas superiores a 180°.

concentric circles Coplanar circles that have the same center.

círculos concéntricos Círculos coplanarios que tienen el mismo centro.

conclusion The statement that immediately follows the word *then* in a conditional.

conclusión La declaración que inmediatamente sigue la palabra *entonces* en un condicional.

concurrent lines Three or more lines that intersect at a common point.

líneas concurrentes Tres o más líneas que se intersecan en un punto común.

conditional probability The probability that an event will occur given that another event has already occurred.

probabilidad condicional La probabilidad de que un evento ocurra dado que otro evento ya ha ocurrido.

conditional relative frequency The ratio of the joint frequency to the marginal frequency.

frecuencia relativa condicional La relación entre la frecuencia de la articulación y la frecuencia marginal.

conditional statement A compound statement that consists of a premise, or hypothesis, and a conclusion, which is false only when its premise is true and its conclusion is false.

enunciado condicional Una declaración compuesta que consiste en una premisa, o hipótesis, y una conclusión, que es falsa solo cuando su premisa es verdadera y su conclusión es falsa.

cone A solid figure with a circular base connected by a curved surface to a single vertex.

cono Una figura sólida con una base circular conectada por una superficie curvada a un solo vértice.

confidence interval An estimate of the population parameter stated as a range with a specific degree of certainty.

intervalo de confianza Una estimación del parámetro de población se indica como un rango con un grado específico de certeza.

congruent Having the same size and shape.

congruente Tener el mismo tamaño y forma.

congruent angles Two angles that have the same measure.

ángulo congruentes Dos ángulos que tienen la misma medida.

congruent arcs Arcs in the same or congruent circles that have the same measure.

arcos congruentes Arcos en los mismos círculos o congruentes que tienen la misma medida.

congruent polygons All of the parts of one polygon are congruent to the corresponding parts or matching parts of another polygon.

poligonos congruentes Todas las partes de un polígono son congruentes con las partes correspondientes o partes coincidentes de otro polígono.

congruent segments Line segments that are the same length.

segmentos congruentes Línea segmentos que son la misma longitud.

congruent solids Solid figures that have exactly the same shape, size, and a scale factor of 1:1.

sólidos congruentes Figuras sólidas que tienen exactamente la misma forma, tamaño y un factor de escala de 1:1.

conic sections Cross sections of a right circular cone.

secciones cónicas Secciones transversales de un cono circular derecho.

conjecture An educated guess based on known information and specific examples.

conjetura Una suposición educada basada en información conocida y ejemplos específicos.

conjugates Two expressions, each with two terms, in which the second terms are opposites.

conjugados Dos expresiones, cada una con dos términos, en la que los segundos términos son opuestos.

conjunction A compound statement using the word *and*.

conjunción Una declaración compuesta usando la palabra *y*.

consecutive interior angles When two lines are cut by a transversal, interior angles that lie on the same side of the transversal.

ángulos internos consecutivos Cuando dos líneas se cortan por un ángulo transversal, interior que se encuentran en el mismo lado de la transversal.

consistent A system of equations with at least one ordered pair that satisfies both equations.

consistente Una sistema de ecuaciones para el cual existe al menos un par ordenado que satisfice ambas ecuaciones.

constant function A linear function of the form $y = b$; The function $f(x) = a$, where a is any number.

función constante Una función lineal de la forma $y = b$; La función $f(x) = a$, donde a es cualquier número.

constant of variation The constant in a variation function.

constante de variación La constante en una función de variación.

constant term A term that does not contain a variable.	**término constante** Un término que no contiene una variable.
constraint A condition that a solution must satisfy.	**restricción** Una condición que una solución debe satisfacer.
constructions Methods of creating figures without the use of measuring tools.	**construcciones** Métodos de creación de figuras sin el uso de herramientas de medición.
continuous function A function that can be graphed with a line or an unbroken curve.	**función continua** Una función que se puede representar gráficamente con una línea o una curva ininterrumpida.
continuous random variable The numerical outcome of a random event that can take on any value.	**variable aleatoria continua** El resultado numérico de un evento aleatorio que puede tomar cualquier valor.
contrapositive A statement formed by negating both the hypothesis and the conclusion of the converse of a conditional.	**antítesis** Una afirmación formada negando tanto la hipótesis como la conclusión del inverso del condicional.
convenience sample Members that are readily available or easy to reach are selected.	**muestra conveniente** Se seleccionan los miembros que están fácilmente disponibles o de fácil acceso.
converse A statement formed by exchanging the hypothesis and conclusion of a conditional statement.	**recíproco** Una declaración formada por el intercambio de la hipótesis y la conclusión de la declaración condicional.
convex polygon A polygon with all interior angles measuring less than 180°.	**polígono convexo** Un polígono con todos los ángulos interiores que miden menos de 180°.
coordinate proofs Proofs that use figures in the coordinate plane and algebra to prove geometric concepts.	**pruebas de coordenadas** Pruebas que utilizan figuras en el plano de coordenadas y álgebra para probar conceptos geométricos.
coplanar Lying in the same plane.	**coplanar** Acostado en el mismo plano.
corollary A theorem with a proof that follows as a direct result of another theorem.	**corolario** Un teorema con una prueba que sigue como un resultado directo de otro teorema.
correlation coefficient A measure that shows how well data are modeled by a regression function.	**coeficiente de correlación** Una medida que muestra cómo los datos son modelados por una función de regresión.
corresponding angles When two lines are cut by a transversal, angles that lie on the same side of a transversal and on the same side of the two lines.	**ángulos correspondientes** Cuando dos líneas se cortan transversalmente, los ángulos que se encuentran en el mismo lado de una transversal y en el mismo lado de las dos líneas.
corresponding parts Corresponding angles and corresponding sides of two polygons.	**partes correspondientes** Ángulos correspondientes y lados correspondientes de dos polígonos.
cosecant The ratio of the length of a hypotenuse to the length of the leg opposite the angle.	**cosecante** Relación entre la longitud de la hipotenusa y la longitud de la pierna opuesta al ángulo.

cosine The ratio of the length of the leg adjacent to an angle to the length of the hypotenuse.

coseno Relación entre la longitud de la pierna adyacente a un ángulo y la longitud de la hipotenusa.

cotangent The ratio of the length of the leg adjacent to an angle to the length of the leg opposite the angle.

cotangente La relación entre la longitud de la pata adyacente a un ángulo y la longitud de la pata opuesta al ángulo.

coterminal angles Angles in standard position that have the same terminal side.

ángulos coterminales Ángulos en posición estándar que tienen el mismo lado terminal.

counterexample An example that contradicts the conjecture showing that the conjecture is not always true.

contraejemplo Un ejemplo que contradice la conjetura que muestra que la conjetura no siempre es cierta.

critical values The *z*-values corresponding to the most common degrees of certainty.

valores críticos Los valores z correspondientes a los grados de certeza más comunes.

cross section The intersection of a solid and a plane.

sección transversal Intersección de un sólido con un plano.

cube root One of three equal factors of a number.

raíz cúbica Uno de los tres factores iguales de un número.

cube root function A radical function that contains the cube root of a variable expression.

función de la raíz del cubo Función radical que contiene la raíz cúbica de una expresión variable.

curve fitting Finding a regression equation for a set of data that is approximated by a function.

ajuste de curvas Encontrar una ecuación de regresión para un conjunto de datos que es aproximado por una función.

cycle One complete pattern of a periodic function.

ciclo Un patron completo de una función periódica.

cylinder A solid figure with two congruent and parallel circular bases connected by a curved surface.

cilindro Una figura sólida con dos bases circulares congruentes y paralelas conectadas por una superficie curvada.

D

decay factor The base of an exponential expression, or $1 - r$.

factor de decaimiento La base de una expresión exponencial, o $1 - r$.

decomposition Separating a figure into two or more nonoverlapping parts.

descomposición Separar una figura en dos o más partes que no se solapan.

decreasing Where the graph of a function goes down when viewed from left to right.

decreciente Donde la gráfica de una función disminuye cuando se ve de izquierda a derecha.

deductive argument An argument that guarantees the truth of the conclusion provided that its premises are true.

argumento deductivo Un argumento que garantiza la verdad de la conclusión siempre que sus premisas sean verdaderas.

deductive reasoning The process of reaching a specific valid conclusion based on general facts, rules, definitions, or properties.

razonamiento deductivo El proceso de alcanzar una conclusión válida específica basada en hechos generales, reglas, definiciones, o propiedades.

define a variable To choose a variable to represent an unknown value.

definir una variable Para elegir una variable que represente un valor desconocido.

defined term A term that has a definition and can be explained.

término definido Un término que tiene una definición y se puede explicar.

definitions An explanation that assigns properties to a mathematical object.

definiciones Una explicación que asigna propiedades a un objeto matemático.

degree The value of the exponent in a power function; $\frac{1}{360}$ of the circular rotation about a point.

grado Valor del exponente en una función de potencia. $\frac{1}{360}$ de la rotación circular alrededor de un punto.

degree of a monomial The sum of the exponents of all its variables.

grado de un monomio La suma de los exponents de todas sus variables.

degree of a polynomial The greatest degree of any term in the polynomial.

grado de un polinomio El grado mayor de cualquier término del polinomio.

density A measure of the quantity of some physical property per unit of length, area, or volume.

densidad Una medida de la cantidad de alguna propiedad física por unidad de longitud, área o volumen.

dependent A consistent system of equations with an infinite number of solutions.

dependiente Una sistema consistente de ecuaciones con un número infinito de soluciones.

dependent events Two or more events in which the outcome of one event affects the outcome of the other events.

eventos dependientes Dos o más eventos en que el resultado de un evento afecta el resultado de los otros eventos.

dependent variable The variable in a relation, usually y, with values that depend on x.

variable dependiente La variable de una relación, generalmente y, con los valores que depende de x.

depressed polynomial A polynomial resulting from division with a degree one less than the original polynomial.

polinomio reducido Un polinomio resultante de la división con un grado uno menos que el polinomio original.

descriptive modeling A way to mathematically describe real-world situations and the factors that cause them.

modelado descriptivo Una forma de describir matemáticamente las situaciones del mundo real y los factores que las causan.

descriptive statistics The branch of statistics that focuses on collecting, summarizing, and displaying data.

estadística descriptiva Rama de la estadística cuyo enfoque es la recopilación, resumen y demostración de los datos.

diagonal A segment that connects any two nonconsecutive vertices within a polygon.

diagonal Un segmento que conecta cualquier dos vértices no consecutivos dentro de un polígono.

diameter of a circle or sphere A chord that passes through the center of a circle or sphere.

diámetro de un círculo o esfera Un acorde que pasa por el centro de un círculo o esfera.

difference of squares A binomial in which the first and last terms are perfect squares.

diferencia de cuadrados Un binomio en el que los términos primero y último son cuadrados perfectos.

difference of two squares The square of one quantity minus the square of another quantity.

diferencia de dos cuadrados El cuadrado de una cantidad menos el cuadrado de otra cantidad.

dilation A nonrigid motion that enlarges or reduces a geometric figure; A transformation that stretches or compresses the graph of a function.

dilatación Un movimiento no rígido que agranda o reduce una figura geométrica; Una transformación que estira o comprime el gráfico de una función.

dimensional analysis The process of performing operations with units.

análisis dimensional El proceso de realizar operaciones con unidades.

direct variation When one quantity is equal to a constant times another quantity.

variación directa Cuando una cantidad es igual a una constante multiplicada por otra cantidad.

directed line segment A line segment with an initial endpoint and a terminal endpoint.

segment de línea dirigido Un segmento de línea con un punto final inicial y un punto final terminal.

directrix An exterior line perpendicular to the line containing the foci of a curve.

directriz Una línea exterior perpendicular a la línea que contiene los focos de una curva.

discontinuous function A function that is not continuous.

función discontinua Una función que no es continua.

discrete function A function in which the points on the graph are not connected.

función discreta Una función en la que los puntos del gráfico no están conectados.

discrete random variable The numerical outcome of a random event that is finite and can be counted.

variable aleatoria discreta El resultado numérico de un evento aleatorio que es finito y puede ser contado.

discriminant In the Quadratic Formula, the expression under the radical sign that provides information about the roots of the quadratic equation.

discriminante En la Fórmula cuadrática, la expresión bajo el signo radical que proporciona información sobre las raíces de la ecuación cuadrática.

disjunction A compound statement using the word *or*.

disyunción Una declaración compuesta usando la palabra *o*.

distance The length of the line segment between two points.

distancia La longitud del segmento de línea entre dos puntos.

distribution A graph or table that shows the theoretical frequency of each possible data value.

distribución Un gráfico o una table que muestra la frecuencia teórica de cada valor de datos posible.

domain The set of the first numbers of the ordered pairs in a relation; The set of x-values to be evaluated by a function.

dominio El conjunto de los primeros números de los pares ordenados en una relación; El conjunto de valores x para ser evaluados por una función.

dot plot A diagram that shows the frequency of data on a number line.

gráfica de puntos Una diagrama que muestra la frecuencia de los datos en una línea numérica.

double root Two roots of a quadratic equation that are the same number.

raíces dobles Dos raíces de una función cuadrática que son el mismo número.

E

e An irrational number that approximately equals 2.7182818....

e Un número irracional que es aproximadamente igual a 2.7182818 ....

edge of a polyhedron A line segment where the faces of the polyhedron intersect.

arista de un poliedro Un segmento de línea donde las caras del poliedro se cruzan.

elimination A method that involves eliminating a variable by combining the individual equations within a system of equations.

eliminación Un método que consiste en eliminar una variable combinando las ecuaciones individuales dentro de un sistema de ecuaciones.

empty set The set that contains no elements, symbolized by { } or ∅.

conjunto vacio El conjunto que no contiene elementos, simbolizado por { } o ∅.

end behavior The behavior of a graph at the positive and negative extremes in its domain.

comportamiento extremo El comportamiento de un gráfico en los extremos positivo y negativo en su dominio.

enlargement A dilation with a scale factor greater than 1.

ampliación Una dilatación con un factor de escala mayor que 1.

equation A mathematical statement that contains two expressions and an equal sign, =.

ecuación Un enunciado matemático que contiene dos expresiones y un signo igual, =.

equiangular polygon A polygon with all angles congruent.

polígono equiangular Un polígono con todos los ángulos congruentes.

equidistant A point is equidistant from other points if it is the same distance from them.

equidistante Un punto es equidistante de otros puntos si está a la misma distancia de ellos.

equidistant lines Two lines for which the distance between the two lines, measured along a perpendicular line or segment to the two lines, is always the same.

líneas equidistantes Dos líneas para las cuales la distancia entre las dos líneas, medida a lo largo de una línea o segmento perpendicular a las dos líneas, es siempre la misma.

equilateral polygon A polygon with all sides congruent.

polígono equilátero Un polígono con todos los lados congruentes.

equivalent equations Two equations with the same solution.

ecuaciones equivalentes Dos ecuaciones con la misma solución.

equivalent expressions Expressions that represent the same value.

expresiones equivalentes Expresiones que representan el mismo valor.

evaluate To find the value of an expression.

evaluar Calcular el valor de una expresión.

even functions Functions that are symmetric in the y-axis.

incluso funciones Funciones que son simétricas en el eje y.

event A subset of the sample space.

evento Un subconjunto del espacio de muestra.

excluded values Values for which a function is not defined.

valores excluidos Valores para los que no se ha definido una función.

experiment A sample is divided into two groups. The experimental group undergoes a change, while there is no change to the control group. The effects on the groups are then compared; A situation involving chance.

experimento Una muestra se divide en dos grupos. El grupo experimental experimenta un cambio, mientras que no hay cambio en el grupo de control. A continuación se comparan los efectos sobre los grupos; Una situación de riesgo.

experimental probability Probability calculated by using data from an actual experiment.

probabilidad experimental Probabilidad calculada utilizando datos de un experimento real.

exponent When n is a positive integer in the expression x^n, n indicates the number of times x is multiplied by itself.

exponente Cuando n es un entero positivo en la expresión x^n, n indica el número de veces que x se multiplica por sí mismo.

exponential decay Change that occurs when an initial amount decreases by the same percent over a given period of time.

desintegración exponencial Cambio que ocurre cuando una cantidad inicial disminuye en el mismo porcentaje durante un período de tiempo dado.

exponential decay function A function in which the independent variable is an exponent, where $a > 0$ and $0 < b < 1$.

función exponenciales de decaimiento Una ecuación en la que la variable independiente es un exponente, donde $a > 0$ y $0 < b < 1$.

explicit formula A formula that allows you to find any term a_n of a sequence by using a formula written in terms of n.

fórmula explícita Una fórmula que le permite encontrar cualquier término a_n de una secuencia usando una fórmula escrita en términos de n.

exponential equation An equation in which the independent variable is an exponent.

ecuación exponencial Una ecuación en la que la variable independiente es un exponente.

exponential form When an expression is in the form x^n.

forma exponencial Cuando una expresión está en la forma x^n.

exponential function A function in which the independent variable is an exponent.

función exponencial Una función en la que la variable independiente es el exponente.

exponential growth Change that occurs when an initial amount increases by the same percent over a given period of time.

crecimiento exponencial Cambio que ocurre cuando una cantidad inicial aumenta por el mismo porcentaje durante un período de tiempo dado.

exponential growth function A function in which the independent variable is an exponent, where $a > 0$ and $b > 1$.

función de crecimiento exponencial Una función en la que la variable independiente es el exponente, donde $a > 0$ y $b > 1$.

exponential inequality An inequality in which the independent variable is an exponent.

desigualdad exponencial Una desigualdad en la que la variable independiente es un exponente.

exterior angle of a triangle An angle formed by one side of the triangle and the extension of an adjacent side.

ángulo exterior de un triángulo Un ángulo formado por un lado del triángulo y la extensión de un lado adyacente.

exterior angles When two lines are cut by a transversal, any of the four angles that lie outside the region between the two intersected lines.

ángulos externos Cuando dos líneas son cortadas por una transversal, cualquiera de los cuatro ángulos que se encuentran fuera de la región entre las dos líneas intersectadas.

exterior of an angle The area outside of the two rays of an angle.

exterior de un ángulo El área fuera de los dos rayos de un ángulo.

extraneous solution A solution of a simplified form of an equation that does not satisfy the original equation.

solución extraña Una solución de una forma simplificada de una ecuación que no satisface la ecuación original.

extrema Points that are the locations of relatively high or low function values.

extrema Puntos que son las ubicaciones de valores de función relativamente alta o baja.

extreme values The least and greatest values in a set of data.

valores extremos Los valores mínimo y máximo en un conjunto de datos.

F

face of a polyhedron A flat surface of a polyhedron.

cara de un poliedro Superficie plana de un poliedro.

factored form A form of quadratic equation, $0 = a(x - p)(x - q)$, where $a \neq 0$, in which p and q are the x-intercepts of the graph of the related function.

forma factorizada Una forma de ecuación cuadrática, $0 = a(x - p)(x - q)$, donde $a \neq 0$, en la que p y q son las intercepciones x de la gráfica de la función relacionada.

factorial of n The product of the positive integers less than or equal to n.

factorial de n El producto de los enteros positivos inferiores o iguales a n.

factoring The process of expressing a polynomial as the product of monomials and polynomials.

factorización por agrupamiento Utilizando la Propiedad distributiva para factorizar polinomios que possen cuatro o más términos.

factoring by grouping Using the Distributive Property to factor some polynomials having four or more terms.

factorización El proceso de expresar un polinomio como el producto de monomios y polinomios.

family of graphs Graphs and equations of graphs that have at least one characteristic in common.

familia de gráficas Gráficas y ecuaciones de gráficas que tienen al menos una característica común.

feasible region The intersection of the graphs in a system of constraints.

región factible La intersección de los gráficos en un sistema de restricciones.

finite sample space A sample space that contains a countable number of outcomes.

espacio de muestra finito Un espacio de muestra que contiene un número contable de resultados.

finite sequence A sequence that contains a limited number of terms.

secuencia finita Una secuencia que contiene un número limitado de términos.

five-number summary The minimum, quartiles, and maximum of a data set.

resumen de cinco números El mínimo, cuartiles *y* máximo de un conjunto de datos.

flow proof A proof that uses boxes and arrows to show the logical progression of an argument.

demostración de flujo Una prueba que usa cajas y flechas para mostrar la progresión lógica de un argumento.

focus A point inside a parabola having the property that the distances from any point on the parabola to them and to a fixed line have a constant ratio for any points on the parabola.

foco Un punto dentro de una parábola que tiene la propiedad de que las distancias desde cualquier punto de la parábola a ellos ya una línea fija tienen una relación constante para cualquier punto de la parábola.

formula An equation that expresses a relationship between certain quantities.

fórmula Una ecuación que expresa una relación entre ciertas cantidades.

fractional distance An intermediary point some fraction of the length of a line segment.

distancia fraccionaria Un punto intermediario de alguna fracción de la longitud de un segmento de línea.

frequency The number of cycles in a given unit of time.

frecuencia El número de ciclos en una unidad del tiempo dada.

function A relation in which each element of the domain is paired with exactly one element of the range.

función Una relación en que a cada elemento del dominio de corresponde un único elemento del rango.

function notation A way of writing an equation so that $y = f(x)$.

notación functional Una forma de escribir una ecuación para que $y = f(x)$.

G

geometric means The terms between two nonconsecutive terms of a geometric sequence; The nth root, where n is the number of elements in a set of numbers, of the product of the numbers.

medios geométricos Los términos entre dos términos no consecutivos de una secuencia geométrica; La enésima raíz, donde n es el número de elementos de un conjunto de números, del producto de los números.

geometric model A geometric figure that represents a real-life object.

modelo geométrico Una figura geométrica que representa un objeto de la vida real.

geometric probability Probability that involves a geometric measure such as length or area.

probabilidad geométrica Probabilidad que implica una medida geométrica como longitud o área.

geometric sequence A pattern of numbers that begins with a nonzero term and each term after is found by multiplying the previous term by a nonzero constant r.

secuencia geométrica Un patrón de números que comienza con un término distinto de cero y cada término después se encuentra multiplicando el término anterior por una constante no nula r.

geometric series The indicated sum of the terms in a geometric sequence.

series geométricas La suma indicada de los términos en una secuencia geométrica.

glide reflection The composition of a translation followed by a reflection in a line parallel to the translation vector.

reflexión del deslizamiento La composición de una traducción seguida de una reflexión en una línea paralela al vector de traslación.

greatest integer function A step function in which $f(x)$ is the greatest integer less than or equal to x.

función entera más grande Una función del paso en que $f(x)$ es el número más grande menos que o igual a x.

growth factor The base of an exponential expression, or $1 + r$.

factor de crecimiento La base de una expresión exponencial, o $1 + r$.

H

half-plane A region of the graph of an inequality on one side of a boundary.

semi-plano Una región de la gráfica de una desigualdad en un lado de un límite.

height of a parallelogram The length of an altitude of the parallelogram.

altura de un paralelogramo La longitud de la altitud del paralelogramo.

height of a solid The length of the altitude of a solid figure.

altura de un sólido La longitud de la altitud de una figura sólida.

height of a trapezoid The perpendicular distance between the bases of a trapezoid.

altura de un trapecio La distancia perpendicular entre las bases de un trapecio.

histogram A graphical display that uses bars to display numerical data that have been organized in equal intervals.

histograma Una exhibición gráfica que utiliza barras para exhibir los datos numéricos que se han organizado en intervalos iguales.

horizontal asymptote A horizontal line that a graph approaches.

asíntota horizontal Una línea horizontal que se aproxima a un gráfico.

hyperbola The graph of a reciprocal function.

hipérbola La gráfica de una función recíproca.

hypothesis The statement that immediately follows the word *if* in a conditional.

hipótesis La declaración que sigue inmediatamente a la palabra *si* en un condicional.

I

identity An equation that is true for every value of the variable.

identidad Una ecuación que es verdad para cada valor de la variable.

identity function The function $f(x) = x$.

función identidad La función $f(x) = x$.

if-then statement A compound statement of the form *if p, then q*, where p and q are statements.

enunciado si-entonces Enunciado compuesto de la forma *si p, entonces q*, donde p y q son enunciados.

image The new figure in a transformation.

imagen La nueva figura en una transformación.

imaginary unit *i* The principal square root of -1.

unidad imaginaria *i* La raíz cuadrada principal de -1.

incenter The point of concurrency of the angle bisectors of a triangle.

incentro El punto de intersección de las bisectrices interiors de un triángulo.

included angle The interior angle formed by two adjacent sides of a triangle.

ángulo incluido El ángulo interior formado por dos lados adyacentes de un triángulo.

included side The side of a triangle between two angles.

lado incluido El lado de un triángulo entre dos ángulos.

inconsistent A system of equations with no ordered pair that satisfies both equations.

inconsistente Una sistema de ecuaciones para el cual no existe par ordenado alguno que satisfaga ambas ecuaciones.

increasing Where the graph of a function goes up when viewed from left to right.

crecciente Donde la gráfica de una función sube cuando se ve de izquierda a derecha.

independent A consistent system of equations with exactly one solution.

independiente Un sistema consistente de ecuaciones con exactamente una solución.

independent events Two or more events in which the outcome of one event does not affect the outcome of the other events.

eventos independientes Dos o más eventos en los que el resultado de un evento no afecta el resultado de los otros eventos.

independent variable The variable in a relation, usually x, with a value that is subject to choice.

variable independiente La variable de una relación, generalmente x, con el valor que sujeta a elección.

index In nth roots, the value that indicates to what root the value under the radicand is being taken.

índice En enésimas raíces, el valor que indica a qué raíz está el valor bajo la radicand.

indirect measurement Using similar figures and proportions to measure an object.

medición indirecta Usando figuras y proporciones similares para medir un objeto.

indirect proof One assumes that the statement to be proven is false and then uses logical reasoning to deduce that a statement contradicts a postulate, theorem, or one of the assumptions.

demostración indirecta Se supone que la afirmación a ser probada es falsa y luego utiliza el razonamiento lógico para deducir que una afirmación contradice un postulado, teorema o uno de los supuestos.

indirect reasoning Reasoning that eliminates all possible conclusions but one so that the one remaining conclusion must be true.

razonamiento indirecto Razonamiento que elimina todas las posibles conclusiones, pero una de manera que la conclusión que queda una debe ser verdad.

inductive reasoning The process of reaching a conclusion based on a pattern of examples.

razonamiento inductive El proceso de llegar a una conclusión basada en un patrón de ejemplos.

inequality A mathematical sentence that contains $<$, $>$, $\leq$, $\geq$, or $\neq$.

desigualdad Una oración matemática que contiene uno o más de $<$, $>$, $\leq$, $\geq$, o $\neq$.

inferential statistics When the data from a sample is used to make inferences about the corresponding population.

estadísticas inferencial Cuando los datos de una muestra se utilizan para hacer inferencias sobre la población correspondiente.

infinite sample space A sample space with outcomes that cannot be counted.

espacio de muestra infinito Un espacio de muestra con resultados que no pueden ser contados.

infinite sequence A sequence that continues without end.

secuencia infinita Una secuencia que continúa sin fin.

informal proof A paragraph that explains why the conjecture for a given situation is true.

prueba informal Un párrafo que explica por qué la conjetura para una situación dada es verdadera.

initial side The part of an angle that is fixed on the *x*-axis.

lado inicial La parte de un ángulo que se fija en el eje *x*.

inscribed angle An angle with its vertex on a circle and sides that contain chords of the circle.

ángulo inscrito Un ángulo con su vértice en un círculo y lados que contienen acordes del círculo.

inscribed polygon A polygon inside a circle in which all of the vertices of the polygon lie on the circle.

polígono inscrito Un polígono dentro de un círculo en el que todos los vértices del polígono se encuentran en el círculo.

intercept A point at which the graph of a function intersects an axis.

interceptar Un punto en el que la gráfica de una función corta un eje.

intercepted arc The part of a circle that lies between the two lines intersecting it.

arco intersecado La parte de un círculo que se encuentra entre las dos líneas que se cruzan.

interior angle of a triangle An angle at the vertex of a triangle.

ángulo interior de un triángulo Un ángulo en el vértice de un triángulo.

interior angles When two lines are cut by a transversal, any of the four angles that lie inside the region between the two intersected lines.

ángulos interiores Cuando dos líneas son cortadas por una transversal, cualquiera de los cuatro ángulos que se encuentran dentro de la región entre las dos líneas intersectadas.

interior of an angle The area between the two rays of an angle.

interior de un ángulo El área entre los dos rayos de un ángulo.

interquartile range The difference between the upper and lower quartiles of a data set.

rango intercuartil La diferencia entre el cuartil superior *y* el cuartil inferior de un conjunto de datos.

intersection A set of points common to two or more geometric figures; intersection The graph of a compound inequality containing *and*.

intersección Un conjunto de puntos communes a dos o más figuras geométricas; intersección La gráfica de una desigualdad compuesta que contiene la palabra *y*.

intersection of *A* and *B* The set of all outcomes in the sample space of event *A* that are also in the sample space of event *B*.

intersección de *A* y *B* El conjunto de todos los resultados en el espacio muestral del evento *A* que también se encuentran en el espacio muestral del evento *B*.

interval The distance between two numbers on the scale of a graph.

intervalo La distancia entre dos números en la escala de un gráfico.

interval notation Mathematical notation that describes a set by using endpoints with parentheses or brackets.

notación de intervalo Notación matemática que describe un conjunto utilizando puntos finales con paréntesis o soportes.

inverse A statement formed by negating both the hypothesis and conclusion of a conditional statement.

inverso Una declaración formada negando tanto la hipótesis como la conclusión de la declaración condicional.

inverse cosine The ratio of the length of the hypotenuse to the length of the leg adjacent to an angle.

inverso del coseno Relación de la longitud de la hipotenusa con la longitud de la pierna adyacente a un ángulo.

inverse functions Two functions, one of which contains points of the form (a, b) while the other contains points of the form (b, a).

funciones inversas Dos funciones, una de las cuales contiene puntos de la forma (a, b) mientras que la otra contiene puntos de la forma (b, a).

inverse relations Two relations, one of which contains points of the form (a, b) while the other contains points of the form (b, a).

relaciones inversas Dos relaciones, una de las cuales contiene puntos de la forma (a, b) mientras que la otra contiene puntos de la forma (b, a).

inverse sine The ratio of the length of the hypotenuse to the length of the leg opposite an angle.

inverso del seno Relación de la longitud de la hipotenusa con la longitud de la pierna opuesta a un ángulo.

inverse tangent The ratio of the length of the leg adjacent to an angle to the length of the leg opposite the angle.

inverso del tangente Relación de la longitud de la pierna adyacente a un ángulo con la longitud de la pierna opuesta a un ángulo.

inverse trigonometric functions Arcsine, Arccosine, and Arctangent.

funciones trigonométricas inversas Arcsine, Arccosine y Arctangent.

inverse variation When the product of two quantities is equal to a constant k.

variación inversa Cuando el producto de dos cantidades es igual a una constante k.

isosceles trapezoid A quadrilateral in which two sides are parallel and the legs are congruent.

trapecio isósceles Un cuadrilátero en el que dos lados son paralelos y las patas son congruentes.

isosceles triangle A triangle with at least two sides congruent.

triángulo isósceles Un triángulo con al menos dos lados congruentes.

J

joint frequencies Entries in the body of a two-way frequency table. In a two-way frequency table, the frequencies in the interior of the table.

frecuencias articulares Entradas en el cuerpo de una tabla de frecuencias de dos vías. En una tabla de frecuencia bidireccional, las frecuencias en el interior de la tabla.

joint variation When one quantity varies directly as the product of two or more other quantities.

variación conjunta Cuando una cantidad varía directamente como el producto de dos o más cantidades.

K

kite A convex quadrilateral with exactly two distinct pairs of adjacent congruent sides.

cometa Un cuadrilátero convexo con exactamente dos pares distintos de lados congruentes adyacentes.

L

lateral area The sum of the areas of the lateral faces of the figure.

área lateral La suma de las áreas de las caras laterales de la figura.

lateral edges The intersection of two lateral faces.

aristas laterales La intersección de dos caras laterales.

lateral faces The faces that join the bases of a solid.

caras laterales Las caras que unen las bases de un sólido.

lateral surface of a cone The curved surface that joins the base of a cone to the vertex.

superficie lateral de un cono La superficie curvada que une la base de un cono con el vértice.

lateral surface of a cylinder The curved surface that joins the bases of a cylinder.

superficie lateral de un cilindro La superficie curvada que une las bases de un cilindro.

leading coefficient The coefficient of the first term when a polynomial is in standard form.

coeficiente líder El coeficiente del primer término cuando un polinomio está en forma estándar.

legs of a trapezoid The nonparallel sides in a trapezoid.

patas de un trapecio Los lados no paralelos en un trapezoide.

legs of an isosceles triangle The two congruent sides of an isosceles triangle.

patas de un triángulo isósceles Los dos lados congruentes de un triángulo isósceles.

like radical expressions Radicals in which both the index and the radicand are the same.

expresiones radicales semejantes Radicales en los que tanto el índice como el radicand son iguales.

like terms Terms with the same variables, with corresponding variables having the same exponent.

términos semejantes Términos con las mismas variables, con las variables correspondientes que tienen el mismo exponente.

line A line is made up of points, has no thickness or width, and extends indefinitely in both directions.

línea Una línea está formada por puntos, no tiene espesor ni anchura, y se extiende indefinidamente en ambas direcciones.

line of fit A line used to describe the trend of the data in a scatter plot.

línea de ajuste Una línea usada para describir la tendencia de los datos en un diagrama de dispersión.

line of reflection A line midway between a preimage and an image; The line in which a reflection flips the graph of a function.

línea de reflexión Una línea a medio camino entre una preimagen y una imagen; La línea en la que una reflectión voltea la gráfica de una función.

line of symmetry An imaginary line that separates a figure into two congruent parts.

línea de simetría Una línea imaginaria que separa una figura en dos partes congruentes.

line segment A measurable part of a line that consists of two points, called endpoints, and all of the points between them.

segmento de línea Una parte medible de una línea que consta de dos puntos, llamados extremos, y todos los puntos entre ellos.

line symmetry A graph has line symmetry if it can be reflected in a vertical line so that each half of the graph maps exactly to the other half.

simetría de línea Un gráfico tiene simetría de línea si puede reflejarse en una línea vertical, de modo que cada mitad del gráfico se asigna exactamente a la otra mitad.

linear equation An equation that can be written in the form $Ax + By = C$ with a graph that is a straight line.

ecuación lineal Una ecuación que puede escribirse de la forma $Ax + By = C$ con un gráfico que es una línea recta.

linear extrapolation The use of a linear equation to predict values that are outside the range of data.

extrapolación lineal El uso de una ecuación lineal para predecir valores que están fuera del rango de datos.

linear function A function in which no independent variable is raised to a power greater than 1; A function with a graph that is a line.

función lineal Una función en la que ninguna variable independiente se eleva a una potencia mayor que 1; Una función con un gráfico que es una línea.

linear inequality A half-plane with a boundary that is a straight line.

desigualdad lineal Un medio plano con un límite que es una línea recta.

linear interpolation The use of a linear equation to predict values that are inside the range of data.

interpolación lineal El uso de una ecuación lineal para predecir valores que están dentro del rango de datos.

linear pair A pair of adjacent angles with noncommon sides that are opposite rays.

par lineal Un par de ángulos adyacentes con lados no comunes que son rayos opuestos.

linear programming The process of finding the maximum or minimum values of a function for a region defined by a system of linear inequalities.

programación lineal El proceso de encontrar los valores máximos o mínimos de una función para una región definida por un sistema de desigualdades lineales.

linear regression An algorithm used to find a precise line of fit for a set of data.

regresión lineal Un algoritmo utilizado para encontrar una línea precisa de ajuste para un conjunto de datos.

linear transformation One or more operations performed on a set of data that can be written as a linear function.

transformación lineal Una o más operaciones realizadas en un conjunto de datos que se pueden escribir como una función lineal.

literal equation A formula or equation with several variables.

ecuación literal Un formula o ecuación con varias variables.

logarithm In $x = b^y$, y is called the logarithm, base b, of x.

logaritmo En $x = b^y$, y se denomina logaritmo, base b, de x.

logarithmic equation An equation that contains one or more logarithms.

ecuación logarítmica Una ecuación que contiene uno o más logaritmos.

logarithmic function A function of the form $f(x) =$ log base b of x, where $b > 0$ and $b \neq 1$.

función logarítmica Una función de la forma $f(x) =$ base log b de x, donde $b > 0$ y $b \neq 1$.

logically equivalent Statements with the same truth value.

lógicamente equivalentes Declaraciones con el mismo valor de verdad.

lower quartile The median of the lower half of a set of data.

cuartil inferior La mediana de la mitad inferior de un conjunto de datos.

M

magnitude The length of a vector from the initial point to the terminal point.

magnitud La longitud de un vector desde el punto inicial hasta el punto terminal.

magnitude of symmetry The smallest angle through which a figure can be rotated so that it maps onto itself.

magnitud de la simetria El ángulo más pequeño a través del cual una figura se puede girar para que se cargue sobre sí mismo.

major arc An arc with measure greater than 180°.

arco mayor Un arco con una medida superior a 180°.

mapping An illustration that shows how each element of the domain is paired with an element in the range.

cartografía Una ilustración que muestra cómo cada elemento del dominio está emparejado con un elemento del rango.

marginal frequencies In a two-way frequency table, the frequencies in the totals row and column; The totals of each subcategory in a two-way frequency table.

frecuencias marginales En una tabla de frecuencias de dos vías, las frecuencias en los totales de fila y columna; Los totales de cada subcategoría en una tabla de frecuencia bidireccional.

maximum The highest point on the graph of a function.

máximo El punto más alto en la gráfica de una función.

maximum error of the estimate The maximum difference between the estimate of the population mean and its actual value.

error máximo de la estimación La diferencia máxima entre la estimación de la media de la población y su valor real.

measurement data Data that have units and can be measured.

medicion de datos Datos que tienen unidades y que pueden medirse.

measures of center Measures of what is average.

medidas del centro Medidas de lo que es promedio.

measures of spread Measures of how spread out the data are.

medidas de propagación Medidas de cómo se extienden los datos son.

median The beginning of the second quartile that separates the data into upper and lower halves.

mediana El comienzo del segundo cuartil que separa los datos en mitades superior e inferior.

median of a triangle A line segment with endpoints that are a vertex of the triangle and the midpoint of the side opposite the vertex.

mediana de un triángulo Un segmento de línea con extremos que son un vértice del triángulo y el punto medio del lado opuesto al vértice.

metric A rule for assigning a number to some characteristic or attribute.

métrico Una regla para asignar un número a alguna caracteristica o atribuye.

midline The line about which the graph of a function oscillates.

linea media La línea sobre la cual oscila la gráfica de una función periódica.

midpoint The point on a line segment halfway between the endpoints of the segment.

punto medio El punto en un segmento de línea a medio camino entre los extremos del segmento.

midsegment of a trapezoid The segment that connects the midpoints of the legs of a trapezoid.

segment medio de un trapecio El segmento que conecta los puntos medios de las patas de un trapecio.

midsegment of a triangle The segment that connects the midpoints of the legs of a triangle.

segment medio de un triángulo El segmento que conecta los puntos medios de las patas de un triángulo.

minimum The lowest point on the graph of a function.

mínimo El punto más bajo en la gráfica de una función.

minor arc An arc with measure less than 180°.

arco menor Un arco con una medida inferior a 180°.

mixture problems Problems that involve creating a mixture of two or more kinds of things and then determining some quantity of the resulting mixture.

problemas de mezcla Problemas que implican crear una mezcla de dos o más tipos de cosas y luego determinar una cierta cantidad de la mezcla resultante.

monomial A number, a variable, or a product of a number and one or more variables.

monomio Un número, una variable, o un producto de un número y una o más variables.

monomial function A function of the form $f(x) = ax^n$, for which a is a nonzero real number and n is a positive integer.

función monomial Una función de la forma $f(x) = ax^n$, para la cual a es un número real no nulo y n es un entero positivo.

multi-step equation An equation that uses more than one operation to solve it.

ecuaciones de varios pasos Una ecuación que utiliza más de una operación para resolverla.

multiplicative identity Because the product of any number a and 1 is equal to a, 1 is the multiplicative identity.

identidad multiplicativa Dado que el producto de cualquier número a y 1 es igual a, 1 es la identidad multiplicativa.

multiplicative inverses Two numbers with a product of 1.

inversos multiplicativos Dos números con un producto es igual a 1.

multiplicity The number of times a number is a zero for a given polynomial.

multiplicidad El número de veces que un número es cero para un polinomio dado.

mutually exclusive Events that cannot occur at the same time.

mutuamente exclusivos Eventos que no pueden ocurrir al mismo tiempo.

N

natural base exponential function An exponential function with base e, written as $y = e^x$.

función exponencial de base natural Una función exponencial con base e, escrita como $y = e^x$.

natural logarithm The inverse of the natural base exponential function, most often abbreviated as $\ln x$.

logaritmo natural La inversa de la función exponencial de base natural, más a menudo abreviada como $\ln x$.

negation A statement that has the opposite meaning, as well as the opposite truth value, of an original statement.

negación Una declaración que tiene el significado opuesto, así como el valor de verdad opuesto, de una declaración original.

negative Where the graph of a function lies below the x-axis.

negativo Donde la gráfica de una función se encuentra debajo del eje x.

negative correlation Bivariate data in which y decreases as x increases.

correlación negativa Datos bivariate en el cual y disminuye a x aumenta.

negative exponent An exponent that is a negative number.

exponente negativo Un exponente que es un número negativo.

negatively skewed distribution A distribution that typically has a median greater than the mean and less data on the left side of the graph.

distribución negativamente sesgada Una distribución que típicamente tiene una mediana mayor que la media y menos datos en el lado izquierdo del gráfico.

net A two-dimensional figure that forms the surfaces of a three-dimensional object when folded.

red Una figura bidimensional que forma las superficies de un objeto tridimensional cuando se dobla.

no correlation Bivariate data in which x and y are not related.

sin correlación Datos bivariados en los que x e y no están relacionados.

nonlinear function A function in which a set of points cannot all lie on the same line

función no lineal Una función en la que un conjunto de puntos no puede estar en la misma línea

nonrigid motion A transformation that changes the dimensions of a given figure.

movimiento no rígida Una transformación que cambia las dimensiones de una figura dada.

normal distribution A continuous, symmetric, bell-shaped distribution of a random variable.

distribución normal Distribución con forma de campana, simétrica y continua de una variable aleatoria.

*n*th root If $a^n = b$ for a positive integer n, then a is an nth root of b.

raíz enésima Si $a^n = b$ para cualquier entero positive n, entonces a se llama una raíz enésima de b.

*n*th term of an arithmetic sequence The nth term of an arithmetic sequence with first term a_1 and common difference d is given by $a_n = a_1 + (n - 1)d$, where n is a positive integer.

enésimo término de una secuencia aritmética El enésimo término de una secuencia aritmética con el primer término a_1 y la diferencia común d viene dado por $a_n = a_1 + (n - 1)d$, donde n es un número entero positivo.

numerical expression A mathematical phrase involving only numbers and mathematical operations.

expresión numérica Una frase matemática que implica sólo números y operaciones matemáticas.

O

oblique asymptote An asymptote that is neither horizontal nor vertical.

asíntota oblicua Una asíntota que no es ni horizontal ni vertical.

observational study Members of a sample are measured or observed without being affected by the study.

estudio de observación Los miembros de una muestra son medidos o observados sin ser afectados por el estudio.

octant One of the eight divisions of three-dimensional space.

octante Una de las ocho divisiones del espacio tridimensional.

odd functions Functions that are symmetric in the origin.

funciones extrañas Funciones que son simétricas en el origen.

one-to-one function A function for which each element of the range is paired with exactly one element of the domain.

función biunívoca Función para la cual cada elemento del rango está emparejado con exactamente un elemento del dominio.

onto function A function for which the codomain is the same as the range.

sobre la función Función para la cual el codomain es el mismo que el rango.

open half-plane The solution of a linear inequality that does not include the boundary line.

medio plano abierto La solución de una desigualdad linear que no incluye la línea de limite.

opposite rays Two collinear rays with a common endpoint.

rayos opuestos Dos rayos colineales con un punto final común.

optimization The process of seeking the optimal value of a function subject to given constraints.

optimización El proceso de buscar valor óptimo de una función sujeto a restricciones dadas.

order of symmetry The number of times a figure maps onto itself.

orden de la simetría El número de veces que una figura se asigna a sí misma.

ordered triple Three numbers given in a specific order used to locate points in space.

triple ordenado Tres números dados en un orden específico usado para localizar puntos en el espacio.

orthocenter The point of concurrency of the altitudes of a triangle.

ortocentro El punto de concurrencia de las altitudes de un triángulo.

orthographic drawing The two-dimensional views of the top, left, front, and right sides of an object.

dibujo ortográfico Las vistas bidimensionales de los lados superior, izquierdo, frontal y derecho de un objeto.

oscillation How much the graph of a function varies between its extreme values as it approaches positive or negative infinity.

oscilación Cuánto la gráfica de una función varía entre sus valores extremos cuando se acerca al infinito positivo o negativo.

outcome The result of a single event; The result of a single performance or trial of an experiment.

resultado El resultado de un solo evento; El resultado de un solo rendimiento o ensayo de un experimento.

outlier A value that is more than 1.5 times the interquartile range above the third quartile or below the first quartile.

parte aislada Un valor que es más de 1,5 veces el rango intercuartílico por encima del tercer cuartil o por debajo del primer cuartil.

P

parabola A curved shape that results when a cone is cut at an angle by a plane that intersects the base; The graph of a quadratic function.

parábola Forma curvada que resulta cuando un cono es cortado en un ángulo por un plano que interseca la base; La gráfica de una función cuadrática.

paragraph proof A paragraph that explains why the conjecture for a given situation is true.

prueba de párrafo Un párrafo que explica por qué la conjetura para una situación dada es verdadera.

parallel lines Coplanar lines that do not intersect; Nonvertical lines in the same plane that have the same slope.

líneas paralelas Líneas coplanares que no se intersecan; Líneas no verticales en el mismo plano que tienen pendientes iguales.

parallel planes Planes that do not intersect.

planos paralelas Planos que no se intersecan.

parallelogram A quadrilateral with both pairs of opposite sides parallel.

paralelogramo Un cuadrilátero con ambos pares de lados opuestos paralelos.

parameter A measure that describes a characteristic of a population; A value in the equation of a function that can be varied to yield a family of functions.

parámetro Una medida que describe una característica de una población; Un valor en la ecuación de una función que se puede variar para producir una familia de funciones.

parent function The simplest of functions in a family.

función basica La función más fundamental de un familia de funciones.

Pascal's triangle A triangle of numbers in which a row represents the coefficients of an expanded binomial $(a + b)^n$.

triángulo de Pascal Un triángulo de números en el que una fila representa los coeficientes de un binomio expandido $(a + b)^n$.

percent rate of change The percent of increase per time period.

por ciento tasa de cambio El porcentaje de aumento por período de tiempo.

percentile A measure that tells what percent of the total scores were below a given score.

percentil Una medida que indica qué porcentaje de las puntuaciones totales estaban por debajo de una puntuación determinada.

perfect cube A rational number with a cube root that is a rational number.

cubo perfecto Un número racional con un raíz cúbica que es un número racional.

perfect square A rational number with a square root that is a rational number.

cuadrado perfecto Un número racional con un raíz cuadrada que es un número racional.

perfect square trinomials Squares of binomials.

trinomio cuadrado perfecto Cuadrados de los binomios.

perimeter The sum of the lengths of the sides of a polygon.

perimetro La suma de las longitudes de los lados de un polígono.

period The horizontal length of one cycle.

periodo La longitud horizontal de un ciclo.

periodic function A function with y-values that repeat at regular intervals.

función periódica Una función con y-valores aquella repetición con regularidad.

permutation An arrangement of objects in which order is important.

permutación Un arreglo de objetos en el que el orden es importante.

perpendicular Intersecting at right angles.

perpendicular Intersección en ángulo recto.

perpendicular bisector Any line, segment, or ray that passes through the midpoint of a segment and is perpendicular to that segment.

mediatriz Cualquier línea, segmento o rayo que pasa por el punto medio de un segmento y es perpendicular a ese segmento.

perpendicular lines Nonvertical lines in the same plane for which the product of the slopes is –1.

líneas perpendiculares Líneas no verticales en el mismo plano para las que el producto de las pendientes es −1.

phase shift A horizontal translation of the graph of a trigonometric function.

cambio de fase Una traducción horizontal de la gráfica de una función trigonométrica.

pi The ratio $\frac{\text{circumference}}{\text{diameter}}$.

pi Relación $\frac{\text{circunferencia}}{\text{diámetro}}$.

piecewise-defined function A function defined by at least two subfunctions, each of which is defined differently depending on the interval of the domain.

función definida por piezas Una función definida por al menos dos subfunciones, cada una de las cuales se define de manera diferente dependiendo del intervalo del dominio.

piecewise-linear function A function defined by at least two linear subfunctions, each of which is defined differently depending on the interval of the domain.

función lineal por piezas Una función definida por al menos dos subfunciones lineal, cada una de las cuales se define de manera diferente dependiendo del intervalo del dominio.

plane A flat surface made up of points that has no depth and extends indefinitely in all directions.

plano Una superficie plana compuesta de puntos que no tiene profundidad y se extiende indefinidamente en todas las direcciones.

plane symmetry When a plane intersects a three-dimensional figure so one half is the reflected image of the other half.

simetría plana Cuando un plano cruza una figura tridimensional, una mitad es la imagen reflejada de la otra mitad.

Platonic solid One of five regular polyhedra.

sólido platónico Uno de cinco poliedros regulares.

point A location with no size, only position.

punto Una ubicación sin tamaño, solo posición.

point discontinuity An area that appears to be a hole in a graph.

discontinuidad de punto Un área que parece ser un agujero en un gráfico.

point of concurrency The point of intersection of concurrent lines.

punto de concurrencia El punto de intersección de líneas concurrentes.

point of symmetry The point about which a figure is rotated.

punto de simetría El punto sobre el que se gira una figura.

point of tangency For a line that intersects a circle in one point, the point at which they intersect.

punto de tangencia Para una línea que cruza un círculo en un punto, el punto en el que se cruzan.

point symmetry A figure or graph has this when a figure is rotated 180° about a point and maps exactly onto the other part.

simetría de punto Una figura o gráfica tiene esto cuando una figura se gira 180° alrededor de un punto y se mapea exactamente sobre la otra parte.

polygon A closed plane figure with at least three straight sides.

polígono Una figura plana cerrada con al menos tres lados rectos.

polyhedron A closed three-dimensional figure made up of flat polygonal regions.

poliedros Una figura tridimensional cerrada formada por regiones poligonales planas.

polynomial A monomial or the sum of two or more monomials.

polinomio Un monomio o la suma de dos o más monomios.

polynomial function A continuous function that can be described by a polynomial equation in one variable.

función polinómica Función continua que puede describirse mediante una ecuación polinómica en una variable.

polynomial identity A polynomial equation that is true for any values that are substituted for the variables.

identidad polinomial Una ecuación polinómica que es verdadera para cualquier valor que se sustituya por las variables.

population All of the members of a group of interest about which data will be collected.

población Todos los miembros de un grupo de interés sobre cuáles datos serán recopilados.

population proportion The number of members in the population with a particular characteristic divided by the total number of members in the population.

proporción de la población El número de miembros en la población con una característica particular dividida por el número total de miembros en la población.

positive Where the graph of a function lies above the x-axis.

positiva Donde la gráfica de una función se encuentra por encima del eje x.

positive correlation Bivariate data in which y increases as x increases.

correlación positiva Datos bivariate en el cual y aumenta a x disminuye.

positively skewed distribution A distribution that typically has a mean greater than the median.

distribución positivamente sesgada Una distribución que típicamente tiene una media mayor que la mediana.

postulate A statement that is accepted as true without proof.

postulado Una declaración que se acepta como verdadera sin prueba.

power function A function of the form $f(x) = ax^n$, where a and n are nonzero real numbers.

función de potencia Una ecuación polinomial que es verdadera para una función de la forma $f(x) = ax^n$, donde a y n son números reales no nulos.

precision The repeatability, or reproducibility, of a measurement.

precisión La repetibilidad, o reproducibilidad, de una medida.

preimage The original figure in a transformation.

preimagen La figura original en una transformación.

prime polynomial A polynomial that cannot be written as a product of two polynomials with integer coefficients.

polinomio primo Un polinomio que no puede escribirse como producto de dos polinomios con coeficientes enteros.

principal root The nonnegative root of a number.

raíz principal La raíz no negativa de un número.

principal square root The nonnegative square root of a number.

raíz cuadrada principal La raíz cuadrada no negativa de un número.

principal values The values in the restricted domains of trigonometric functions.

valores principales Valores de los dominios restringidos de las functiones trigonométricas.

principle of superposition Two figures are congruent if and only if there is a rigid motion or series of rigid motions that maps one figure exactly onto the other.

principio de superposición Dos figuras son congruentes si y sólo si hay un movimiento rígido o una serie de movimientos rígidos que traza una figura exactamente sobre la otra.

prism A polyhedron with two parallel congruent bases connected by parallelogram faces.

prisma Un poliedro con dos bases congruentes paralelas conectadas por caras de paralelogramo.

probability The number of outcomes in which a specified event occurs to the total number of trials.

probabilidad El número de resultados en los que se produce un evento especificado al número total de ensayos.

probability distribution A function that maps the sample space to the probabilities of the outcomes in the sample space for a particular random variable.

distribución de probabilidad Una función que mapea el espacio de muestra a las probabilidades de los resultados en el espacio de muestra para una variable aleatoria particular.

probability model A mathematical representation of a random event that consists of the sample space and the probability of each outcome.

modelo de probabilidad Una representación matemática de un evento aleatorio que consiste en el espacio muestral y la probabilidad de cada resultado.

projectile motion problems Problems that involve objects being thrown or dropped.

problemas de movimiento del proyectil Problemas que involucran objetos que se lanzan o caen.

proof A logical argument in which each statement is supported by a statement that is accepted as true.

prueba Un argumento lógico en el que cada sentencia está respaldada por una sentencia aceptada como verdadera.

proof by contradiction One assumes that the statement to be proven is false and then uses logical reasoning to deduce that a statement contradicts a postulate, theorem, or one of the assumptions.

prueba por contradicción Se supone que la afirmación a ser probada es falsa y luego utiliza el razonamiento lógico para deducir que una afirmación contradice un postulado, teorema o uno de los supuestos.

proportion A statement that two ratios are equivalent.

proporción Una declaración de que dos proporciones son equivalentes.

pure imaginary number A number of the form bi, where b is a real number and i is the imaginary unit.

número imaginario puro Un número de la forma bi, donde b es un número real e i es la unidad imaginaria.

pyramid A polyhedron with a polygonal base and three or more triangular faces that meet at a common vertex.

pirámide Poliedro con una base poligonal y tres o más caras triangulares que se encuentran en un vértice común.

Pythagorean identities Identities that express the Pythagorean Theorem in terms of the trigonometric functions.

identidades pitagóricas Identidades que expresan el Teorema de Pitágoras en términos de las funciones trigonométricas.

Pythagorean triple A set of three nonzero whole numbers that make the Pythagorean Theorem true.

triplete Pitágorico Un conjunto de tres números enteros distintos de cero que hacen que el Teorema de Pitágoras sea verdadero.

Q

quadrantal angle An angle in standard position with a terminal side that coincides with one of the axes.

ángulo de cuadrante Un ángulo en posición estándar con un lado terminal que coincide con uno de los ejes.

quadratic equation An equation that includes a quadratic expression.

ecuación cuadrática Una ecuación que incluye una expresión cuadrática.

quadratic expression An expression in one variable with a degree of 2.

expresión cuadrática Una expresión en una variable con un grado de 2.

quadratic form A form of polynomial equation, $au^2 + bu + c$, where u is an algebraic expression in x.

forma cuadrática Una forma de ecuación polinomial, $au^2 + bu + c$, donde u es una expresión algebraica en x.

quadratic function A function with an equation of the form $y = ax^2 + bx + c$, where $a \neq 0$.

función cuadrática Una función con una ecuación de la forma $y = ax^2 + bx + c$, donde $a \neq 0$.

quadratic inequality An inequality that includes a quadratic expression.

desigualdad cuadrática Una desigualdad que incluye una expresión cuadrática.

quadratic relations Equations of parabolas with horizontal axes of symmetry that are not functions.

relaciones cuadráticas Ecuaciones de parábolas con ejes horizontales de simetría que no son funciones.

quartic function A fourth-degree function.

función cuartica Una función de cuarto grado.

quartiles Measures of position that divide a data set arranged in ascending order into four groups, each containing about one fourth or 25% of the data.

cuartiles Medidas de posición que dividen un conjunto de datos dispuestos en orden ascendente en cuatro grupos, cada uno de los cuales contiene aproximadamente un cuarto o el 25% de los datos.

quintic function A fifth-degree function.

función quíntica Una función de quinto grado.

R

radian A unit of angular measurement equal to $\frac{180°}{\pi}$ or about 57.296°.

radián Una unidad de medida angular igual o $\frac{180°}{\pi}$ alrededor de 57.296°.

radical equation An equation with a variable in a radicand.

ecuación radical Una ecuación con una variable en un radicand.

radical expression An expression that contains a radical symbol, such as a square root.

expresión radicales Una expresión que contiene un símbolo radical, tal como una raíz cuadrada.

radical form When an expression contains a radical symbol.

forma radical Cuando una expresión contiene un símbolo radical.

radical function A function that contains radicals with variables in the radicand.

función radical Función que contiene radicales con variables en el radicand.

radicand The expression under a radical sign.

radicando La expresión debajo del signo radical.

radius of a circle or sphere A line segment from the center to a point on a circle or sphere.

radio de un círculo o esfera Un segmento de línea desde el centro hasta un punto en un círculo o esfera.

radius of a regular polygon The radius of the circle circumscribed about a regular polygon.

radio de un polígono regular El radio del círculo circunscrito alrededor de un polígono regular.

range The difference between the greatest and least values in a set of data; The set of second numbers of the ordered pairs in a relation; The set of y-values that actually result from the evaluation of the function.

rango La diferencia entre los valores de datos más grande or menos en un sistema de datos; El conjunto de los segundos números de los pares ordenados de una relación; El conjunto de valores y que realmente resultan de la evaluación de la función.

rate of change How a quantity is changing with respect to a change in another quantity.

tasa de cambio Cómo cambia una cantidad con respecto a un cambio en otra cantidad.

rational equation An equation that contains at least one rational expression.

ecuación racional Una ecuación que contiene al menos una expresión racional.

rational exponent An exponent that is expressed as a fraction.

exponente racional Un exponente que se expresa como una fracción.

rational expression A ratio of two polynomial expressions.

expresión racional Una relación de dos expresiones polinomiales.

rational function An equation of the form $f(x) = \frac{a(x)}{b(x)}$, where $a(x)$ and $b(x)$ are polynomial expressions and $b(x) \neq 0$.

función racional Una ecuación de la forma $f(x) = \frac{a(x)}{b(x)}$, donde $a(x)$ y $b(x)$ son expresiones polinomiales y $b(x) \neq 0$.

rational inequality An inequality that contains at least one rational expression.

desigualdad racional Una desigualdad que contiene al menos una expresión racional.

rationalizing the denominator A method used to eliminate radicals from the denominator of a fraction or fractions from a radicand.

racionalizando el denominador Método utilizado para eliminar radicales del denominador de una fracción o fracciones de una radicand.

ray Part of a line that starts at a point and extends to infinity.

rayo Parte de una línea que comienza en un punto y se extiende hasta el infinito.

reciprocal function An equation of the form $f(x) = \frac{n}{b(x)}$, where n is a real number and $b(x)$ is a linear expression that cannot equal 0.

función recíproca Una ecuación de la forma $f(x) = \frac{n}{b(x)}$, donde n es un número real y $b(x)$ es una expresión lineal que no puede ser igual a 0.

reciprocal trigonometric functions Trigonometric functions that are reciprocals of each other.

funciones trigonométricas recíprocas Funciones trigonométricas que son reciprocales entre sí.

reciprocals Two numbers with a product of 1.

recíprocos Dos números con un producto de 1.

rectangle A parallelogram with four right angles.

rectángulo Un paralelogramo con cuatro ángulos rectos.

recursive formula A formula that gives the value of the first term in the sequence and then defines the next term by using the preceding term.

formula recursiva Una fórmula que da el valor del primer término en la secuencia y luego define el siguiente término usando el término anterior.

reduction A dilation with a scale factor between 0 and 1.

reducción Una dilatación con un factor de escala entre 0 y 1.

reference angle The acute angle formed by the terminal side of an angle and the x-axis.

ángulo de referencia El ángulo agudo formado por el lado terminal de un ángulo en posición estándar y el eje x.

reflection A function in which the preimage is reflected in the line of reflection; A transformation in which a figure, line, or curve is flipped across a line.

reflexión Función en la que la preimagen se refleja en la línea de reflexión; Una transformación en la que una figura, línea o curva se voltea a través de una línea.

regression function A function generated by an algorithm to find a line or curve that fits a set of data.

función de regresión Función generada por un algoritmo para encontrar una línea o curva que se ajuste a un conjunto de datos.

regular polygon A convex polygon that is both equilateral and equiangular.

polígono regular Un polígono convexo que es a la vez equilátero y equiangular.

regular polyhedron A polyhedron in which all of its faces are regular congruent polygons and all of the edges are congruent.

poliedro regular Un poliedro en el que todas sus caras son polígonos congruentes regulares y todos los bordes son congruentes.

regular pyramid A pyramid with a base that is a regular polygon.

pirámide regular Una pirámide con una base que es un polígono regular.

regular tessellation A tessellation formed by only one type of regular polygon.

teselado regular Un teselado formado por un solo tipo de polígono regular.

relation A set of ordered pairs.

relación Un conjunto de pares ordenados.

relative frequency In a two-way frequency table, the ratios of the number of observations in a category to the total number of observations; The ratio of the number of observations in a category to the total number of observations.

frecuencia relativa En una tabla de frecuencia bidireccional, las relaciones entre el número de observaciones en una categoría y el número total de observaciones; La relación entre el número de observaciones en una categoría y el número total de observaciones.

relative maximum A point on the graph of a function where no other nearby points have a greater *y*-coordinate.

máximo relativo Un punto en la gráfica de una función donde ningún otro punto cercano tiene una coordenada *y* mayor.

relative minimum A point on the graph of a function where no other nearby points have a lesser *y*-coordinate.

mínimo relativo Un punto en la gráfica de una función donde ningún otro punto cercano tiene una coordenada *y* menor.

remote interior angles Interior angles of a triangle that are not adjacent to an exterior angle.

ángulos internos no adyacentes Ángulos interiores de un triángulo que no están adyacentes a un ángulo exterior.

residual The difference between an observed *y*-value and its predicted *y*-value on a regression line.

residual La diferencia entre un valor de *y* observado y su valor de *y* predicho en una línea de regresión.

rhombus A parallelogram with all four sides congruent.

rombo Un paralelogramo con los cuatro lados congruentes.

rigid motion A transformation that preserves distance and angle measure.

movimiento rígido Una transformación que preserva la distancia y la medida del ángulo.

root A solution of an equation.

raíz Una solución de una ecuación.

rotation A function that moves every point of a preimage through a specified angle and direction about a fixed point.

rotación Función que mueve cada punto de una preimagen a través de un ángulo y una dirección especificados alrededor de un punto fijo.

rotational symmetry A figure can be rotated less than 360° about a point so that the image and the preimage are indistinguishable.

simetría rotacional Una figura puede girar menos de 360° alrededor de un punto para que la imagen y la preimagen sean indistinguibles.

S

sample A subset of a population.

muestra Un subconjunto de una población.

sample space The set of all possible outcomes.

espacio muestral El conjunto de todos los resultados posibles.

sampling error The variation between samples taken from the same population.

error de muestreo La variación entre muestras tomadas de la misma población.

scale The distance between tick marks on the *x*- and *y*-axes.

escala La distancia entre las marcas en los ejes *x* e *y*.

scale factor of a dilation The ratio of a length on an image to a corresponding length on the preimage.

factor de escala de una dilatación Relación de una longitud en una imagen con una longitud correspondiente en la preimagen.

scatter plot A graph of bivariate data that consists of ordered pairs on a coordinate plane.

gráfica de dispersión Una gráfica de datos bivariados que consiste en pares ordenados en un plano de coordenadas.

secant Any line or ray that intersects a circle in exactly two points; The ratio of the length of the hypotenuse to the length of the leg adjacent to the angle.

secante Cualquier línea o rayo que cruce un círculo en exactamente dos puntos; Relación entre la longitud de la hipotenusa y la longitud de la pierna adyacente al ángulo.

sector A region of a circle bounded by a central angle and its intercepted arc.

sector Una región de un círculo delimitada por un ángulo central y su arco interceptado.

segment bisector Any segment, line, plane, or point that intersects a line segment at its midpoint.

bisectriz del segmento Cualquier segmento, línea, plano o punto que interseca un segmento de línea en su punto medio.

self-selected sample Members volunteer to be included in the sample.

muestra auto-seleccionada Los miembros se ofrecen como voluntarios para ser incluidos en la muestra.

semicircle An arc that measures exactly 180°.

semicírculo Un arco que mide exactamente 180°.

semiregular tessellation A tessellation formed by two or more regular polygons.

teselado semiregular Un teselado formado por dos o más polígonos regulares.

sequence A list of numbers in a specific order.

secuencia Una lista de números en un orden específico.

series The indicated sum of the terms in a sequence.

serie La suma indicada de los términos en una secuencia.

set-builder notation Mathematical notation that describes a set by stating the properties that its members must satisfy.

notación de construción de conjuntos Notación matemática que describe un conjunto al declarar las propiedades que sus miembros deben satisfacer.

sides of an angle The rays that form an angle.

lados de un ángulo Los rayos que forman un ángulo.

sigma notation A notation that uses the Greek uppercase letter S to indicate that a sum should be found.

notación de sigma Una notación que utiliza la letra mayúscula griega S para indicar que debe encontrarse una suma.

significant figures The digits of a number that are used to express a measure to an appropriate degree of accuracy.

dígitos significantes Los dígitos de un número que se utilizan para expresar una medida con un grado apropiado de precisión.

similar polygons Two figures are similar polygons if one can be obtained from the other by a dilation or a dilation with one or more rigid motions.

polígonos similares Dos figuras son polígonos similares si uno puede ser obtenido del otro por una dilatación o una dilatación con uno o más movimientos rígidos.

similar solids Solid figures with the same shape but not necessarily the same size.

sólidos similares Figuras sólidas con la misma forma pero no necesariamente del mismo tamaño.

similar triangles Triangles in which all of the corresponding angles are congruent and all of the corresponding sides are proportional.

triángulos similares Triángulos en los cuales todos los ángulos correspondientes son congruentes y todos los lados correspondientes son proporcionales.

similarity ratio The scale factor between two similar polygons.

relación de similitud El factor de escala entre dos polígonos similares.

similarity transformation A transformation composed of a dilation or a dilation and one or more rigid motions.

transformación de similitud Una transformación compuesto por una dilatación o una dilatación y uno o más movimientos rígidos.

simple random sample Each member of the population has an equal chance of being selected as part of the sample.

muestra aleatoria simple Cada miembro de la población tiene la misma posibilidad de ser seleccionado como parte de la muestra.

simplest form An expression is in simplest form when it is replaced by an equivalent expression having no like terms or parentheses.

forma reducida Una expresión está reducida cuando se puede sustituir por una expresión equivalente que no tiene ni términos semejantes ni paréntesis.

simulation The use of a probability model to imitate a process or situation so it can be studied.

simulación El uso de un modelo de probabilidad para imitar un proceso o situación para que pueda ser estudiado.

sine The ratio of the length of the leg opposite an angle to the length of the hypotenuse.

seno La relación entre la longitud de la pierna opuesta a un ángulo y la longitud de la hipotenusa.

sinusoidal function A function that can be produced by translating, reflecting, or dilating the sine function.

función sinusoidal Función que puede producirse traduciendo, reflejando o dilatando la función sinusoidal.

skew lines Noncoplanar lines that do not intersect.

líneas alabeadas Líneas no coplanares que no se cruzan.

slant height of a pyramid or right cone The length of a segment with one endpoint on the base edge of the figure and the other at the vertex.

altura inclinada de una pirámide o cono derecho La longitud de un segmento con un punto final en el borde base de la figura y el otro en el vértice.

slope The rate of change in the *y*-coordinates (rise) to the corresponding change in the *x*-coordinates (run) for points on a line.

pendiente La tasa de cambio en las coordenadas *y* (subida) al cambio correspondiente en las coordenadas *x* (carrera) para puntos en una línea.

slope criteria Outlines a method for proving the relationship between lines based on a comparison of the slopes of the lines.

criterios de pendiente Describe un método para probar la relación entre líneas basado en una comparación de las pendientes de las líneas.

solid of revolution A solid figure obtained by rotating a shape around an axis.

sólido de revolución Una figura sólida obtenida girando una forma alrededor de un eje.

solution A value that makes an equation true.

solución Un valor que hace que una ecuación sea verdadera.

solve an equation The process of finding all values of the variable that make the equation a true statement.

resolver una ecuación El proceso en que se hallan todos los valores de la variable que hacen verdadera la ecuación.

solving a triangle When you are given measurements to find the unknown angle and side measures of a triangle.

resolver un triángulo Cuando se le dan mediciones para encontrar el ángulo desconocido y las medidas laterales de un triángulo.

space A boundless three-dimensional set of all points.

espacio Un conjunto tridimensional ilimitado de todos los puntos.

sphere A set of all points in space equidistant from a given point called the center of the sphere.

esfera Un conjunto de todos los puntos del espacio equidistantes de un punto dado llamado centro de la esfera.

square A parallelogram with all four sides and all four angles congruent.

cuadrado Un paralelogramo con los cuatro lados y los cuatro ángulos congruentes.

square root One of two equal factors of a number.

raíz cuadrada Uno de dos factores iguales de un número.

square root function A radical function that contains the square root of a variable expression.

función raíz cuadrada Función radical que contiene la raíz cuadrada de una expresión variable.

square root inequality An inequality that contains the square root of a variable expression.

square root inequality Una desigualdad que contiene la raíz cuadrada de una expresión variable.

standard deviation A measure that shows how data deviate from the mean.

desviación tipica Una medida que muestra cómo los datos se desvían de la media.

standard error of the mean The standard deviation of the distribution of sample means taken from a population.

error estandar de la media La desviación estándar de la distribución de los medios de muestra se toma de una población.

standard form of a linear equation Any linear equation can be written in this form, $Ax + By = C$, where $A \geq 0$, A and B are not both 0, and A, B, and C are integers with a greatest common factor of 1.

forma estándar de una ecuación lineal Cualquier ecuación lineal se puede escribir de esta forma, $Ax + By = C$, donde $A \geq 0$, A y B no son ambos 0, y A, B y C son enteros con el mayor factor común de 1.

standard form of a polynomial A polynomial that is written with the terms in order from greatest degree to least degree.

forma estándar de un polinomio Un polinomio que se escribe con los términos en orden del grado más grande a menos grado.

standard form of a quadratic equation A quadratic equation can be written in the form $ax^2 + bx + c = 0$, where $a \neq 0$.

forma estándar de una ecuación cuadrática Una ecuación cuadrática puede escribirse en la forma $ax^2 + bx + c = 0$, donde $a \neq 0$.

standard normal distribution A normal distribution with a mean of 0 and a standard deviation of 1.

distribución normal estándar Distribución normal con una media de 0 y una desviación estándar de 1.

standard position An angle positioned so that the vertex is at the origin and the initial side is on the positive x-axis.

posición estándar Un ángulo colocado de manera que el vértice está en el origen y el lado inicial está en el eje x positivo.

statement Any sentence that is either true or false, but not both.

enunciado Cualquier oración que sea verdadera o falsa, pero no ambas.

statistic A measure that describes a characteristic of a sample.

estadística Una medida que describe una característica de una muestra.

statistics An area of mathematics that deals with collecting, analyzing, and interpreting data.

estadísticas El proceso de recolección, análisis e interpretación de datos.

step function A type of piecewise-linear function with a graph that is a series of horizontal line segments.

función escalonada Un tipo de función lineal por piezas con un gráfico que es una serie de segmentos de línea horizontal.

straight angle An angle that measures 180°.

ángulo recto Un ángulo que mide 180°.

stratified sample The population is first divided into similar, nonoverlapping groups. Then members are randomly selected from each group.

muestra estratificada La población se divide primero en grupos similares, sin superposición. A continuación, los miembros se seleccionan aleatoriamente de cada grupo.

substitution A process of solving a system of equations in which one equation is solved for one variable in terms of the other.

sustitución Un proceso de resolución de un sistema de ecuaciones en el que una ecuación se resuelve para una variable en términos de la otra.

supplementary angles Two angles with measures that have a sum of 180°.

ángulos suplementarios Dos ángulos con medidas que tienen una suma de 180°.

surface area The sum of the areas of all faces and side surfaces of a three-dimensional figure.

área de superficie La suma de las áreas de todas las caras y superficies laterales de una figura tridimensional.

survey Data are collected from responses given by members of a group regarding their characteristics, behaviors, or opinions.

encuesta Los datos se recogen de las respuestas dadas por los miembros de un grupo con respecto a sus características, comportamientos u opiniones.

symmetric distribution A distribution in which the mean and median are approximately equal.

distribución simétrica Un distribución en la que la media y la mediana son aproximadamente iguales.

symmetry A figure has this if there exists a rigid motion—reflection, translation, rotation, or glide reflection—that maps the figure onto itself.

simetría Una figura tiene esto si existe una reflexión-reflexión, una traducción, una rotación o una reflexión de deslizamiento rígida-que mapea la figura sobre sí misma.

synthetic division An alternate method used to divide a polynomial by a binomial of degree 1.

división sintética Un método alternativo utilizado para dividir un polinomio por un binomio de grado 1.

synthetic geometry The study of geometric figures without the use of coordinates.

geometría sintética El estudio de figuras geométricas sin el uso de coordenadas.

synthetic substitution The process of using synthetic division to find a value of a polynomial function.

sustitución sintética El proceso de utilizar la división sintética para encontrar un valor de una función polynomial.

system of equations A set of two or more equations with the same variables.

sistema de ecuaciones Un conjunto de dos o más ecuaciones con las mismas variables.

system of inequalities A set of two or more inequalities with the same variables.

sistema de desigualdades Un conjunto de dos o más desigualdades con las mismas variables.

systematic sample Members are selected according to a specified interval from a random starting point.

muestra sistemática Los miembros se seleccionan de acuerdo con un intervalo especificado desde un punto de partida aleatorio.

T

tangent The ratio of the length of the leg opposite an angle to the length of the leg adjacent to the angle.

tangente La relación entre la longitud de la pata opuesta a un ángulo y la longitud de la pata adyacente al ángulo.

tangent to a circle A line or segment in the plane of a circle that intersects the circle in exactly one point and does not contain any points in the interior of the circle.

tangente a un círculo Una línea o segmento en el plano de un círculo que interseca el círculo en exactamente un punto y no contiene ningún punto en el interior del círculo.

tangent to a sphere A line that intersects the sphere in exactly one point.

tangente a una esfera Una línea que interseca la esfera exactamente en un punto.

term A number, a variable, or a product or quotient of numbers and variables.

término Un número, una variable, o un producto o cociente de números y variables.

term of a sequence A number in a sequence.

término de una sucesión Un número en una secuencia.

terminal side The part of an angle that rotates about the center.

lado terminal La parte de un ángulo que gira alrededor de un centro.

tessellation A repeating pattern of one or more figures that covers a plane with no overlapping or empty spaces.

teselado Patrón repetitivo de una o más figuras que cubre un plano sin espacios superpuestos o vacíos.

theorem A statement that can be proven true using undefined terms, definitions, and postulates.

teorema Una afirmación o conjetura que se puede probar verdad utilizando términos, definiciones y postulados indefinidos.

theoretical probability Probability based on what is expected to happen.

probabilidad teórica Probabilidad basada en lo que se espera que suceda.

transformation A function that takes points in the plane as inputs and gives other points as outputs. The movement of a graph on the coordinate plane.

transformación Función que toma puntos en el plano como entradas y da otros puntos como salidas. El movimiento de un gráfico en el plano de coordenadas.

translation A function in which all of the points of a figure move the same distance in the same direction; A transformation in which a figure is slid from one position to another without being turned.

traslación Función en la que todos los puntos de una figura se mueven en la misma dirección; El movimiento de un gráfico en el plano de coordenadas.

translation vector A directed line segment that describes both the magnitude and direction of the slide if the magnitude is the length of the vector from its initial point to its terminal point.

vector de traslación Un segmento de línea dirigido que describe tanto la magnitud como la dirección de la diapositiva si la magnitud es la longitud del vector desde su punto inicial hasta su punto terminal.

transversal A line that intersects two or more lines in a plane at different points.

transversal Una línea que interseca dos o más líneas en un plano en diferentes puntos.

trapezoid A quadrilateral with exactly one pair of parallel sides.

trapecio Un cuadrilátero con exactamente un par de lados paralelos.

trend A general pattern in the data.

tendencia Un patrón general en los datos.

trigonometric equation An equation that includes at least one trigonometric function.

ecuación trigonométrica Una ecuación que incluye al menos una función trigonométrica.

trigonometric function A function that relates the measure of one nonright angle of a right triangle to the ratios of the lengths of any two sides of the triangle.

función trigonométrica Función que relaciona la medida de un ángulo no recto de un triángulo rectángulo con las relaciones de las longitudes de cualquiera de los dos lados del triángulo.

trigonometric identity An equation involving trigonometric functions that is true for all values for which every expression in the equation is defined.

identidad trigonométrica Una ecuación que implica funciones trigonométricas que es verdadera para todos los valores para los cuales se define cada expresión en la ecuación.

trigonometric ratio A ratio of the lengths of two sides of a right triangle.

relación trigonométrica Una relación de las longitudes de dos lados de un triángulo rectángulo.

trigonometry The study of the relationships between the sides and angles of triangles.

trigonometría El estudio de las relaciones entre los lados y los ángulos de los triángulos.

trinomial The sum of three monomials.

trinomio La suma de tres monomios.

truth value The truth or falsity of a statement.

valor de verdad La verdad o la falsedad de una declaración.

two-column proof A proof that contains statements and reasons organized in a two-column format.

prueba de dos columnas Una prueba que contiene declaraciones y razones organizadas en un formato de dos columnas.

two-way frequency table A table used to show frequencies of data classified according to two categories, with the rows indicating one category and the columns indicating the other.

tabla de frecuencia bidireccional Una tabla utilizada para mostrar las frecuencias de los datos clasificados de acuerdo con dos categorías, con las filas que indican una categoría y las columnas que indican la otra.

two-way relative frequency table A table used to show frequencies of data based on a percentage of the total number of observations.

tabla de frecuencia relativa bidireccional Una tabla usada para mostrar las frecuencias de datos basadas en un porcentaje del número total de observaciones.

U

unbounded When the graph of a system of constraints is open.

no acotado Cuando la gráfica de un sistema de restricciones está abierta.

undefined terms Words that are not formally explained by means of more basic words and concepts.

términos indefinidos Palabras que no se explican formalmente mediante palabras y conceptos más básicos.

uniform motion problems Problems that use the formula $d = rt$, where d is the distance, r is the rate, and t is the time.

problemas de movimiento uniforme Problemas que utilizan la fórmula $d = rt$, donde d es la distancia, r es la velocidad y t es el tiempo.

uniform tessellation A tessellation that contains the same arrangement of shapes and angles at each vertex.

teselado uniforme Un teselado que contiene la misma disposición de formas y ángulos en cada vértice.

union The graph of a compound inequality containing *or*.

unión La gráfica de una desigualdad compuesta que contiene la palabra *o*.

union of A and B The set of all outcomes in the sample space of event A combined with all outcomes in the sample space of event B.

unión de A y B El conjunto de todos los resultados en el espacio muestral del evento A combinado con todos los resultados en el espacio muestral del evento B.

unit circle A circle with a radius of 1 unit centered at the origin on the coordinate plane.

círculo unitario Un círculo con un radio de 1 unidad centrado en el origen en el plano de coordenadas.

univariate data Measurement data in one variable.

datos univariate Datos de medición en una variable.

upper quartile The median of the upper half of a set of data.

cuartil superior La mediana de la mitad superior de un conjunto de datos.

V

valid argument An argument is valid if it is impossible for all of the premises, or supporting statements, of the argument to be true and its conclusion false.

argumento válido Un argumento es válido si es imposible que todas las premisas o argumentos de apoyo del argumento sean verdaderos y su conclusión sea falsa.

variable A letter used to represent an unspecified number or value; Any characteristic, number, or quantity that can be counted or measured.

variable Una letra utilizada para representar un número o valor no especificado; Cualquier característica, número, o cantidad que pueda ser contada o medida.

variable term A term that contains a variable.

término variable Un término que contiene una variable.

variance The square of the standard deviation.

varianza El cuadrado de la desviación estándar.

vertex Either the lowest point or the highest point of a function.

vértice El punto más bajo o el punto más alto en una función.

vertex angle of an isosceles triangle The angle between the sides that are the legs of an isosceles triangle.

ángulo del vértice de un triángulo isósceles El ángulo entre los lados que son las patas de un triángulo isósceles.

vertex form A quadratic function written in the form $f(x) = a(x - h)^2 + k$.

forma de vértice Una función cuadrática escribirse de la forma $f(x) = a(x - h)^2 + k$.

vertex of a polyhedron An intersection of the edges of a polyhedron.

vértice de un polígono Una intersección de los bordes de un poliedro.

vertex of an angle The common endpoint of the two rays that form an angle.

vértice de un ángulo El punto final común de los dos rayos que forman un ángulo.

vertical angles Two nonadjacent angles formed by two intersecting lines.

ángulos verticales Dos ángulos no adyacentes formados por dos líneas de intersección.

vertical asymptote A vertical line that a graph approaches.

asíntota vertical Una línea vertical que se aproxima a un gráfico.

vertical shift A vertical translation of the graph of a trigonometric function.

cambio vertical Una traducción vertical de la gráfica de una función trigonométrica.

volume The measure of the amount of space enclosed by a three-dimensional figure.

volumen La medida de la cantidad de espacio encerrada por una figura tridimensional.

W

work problems Problems that involve two people working at different rates who are trying to complete a single job.

problemas de trabajo Problemas que involucran a dos personas trabajando a diferentes ritmos que están tratando de completar un solo trabajo.

X

x-intercept The *x*-coordinate of a point where a graph crosses the *x*-axis.

intercepción *x* La coordenada *x* de un punto donde la gráfica corte al eje de *x*.

Y

y-intercept The *y*-coordinate of a point where a graph crosses the *y*-axis.

intercepción *y* La coordenada *y* de un punto donde la gráfica corte al eje de *y*.

Z

z-value The number of standard deviations that a given data value is from the mean.

valor *z* El número de variaciones estándar que separa un valor dado de la media.

zero An *x*-intercept of the graph of a function; a value of *x* for which $f(x) = 0$.

cero Una intercepción *x* de la gráfica de una función; un punto *x* para los que $f(x) = 0$.

Index

Q

R

S